T0155988

Chaos and Fractals

Heinz-Otto Peitgen Hartmut Jürgens Dietmar Saupe

Chaos and Fractals

New Frontiers of Science

Second Edition

With 606 illustrations, 40 in color

Heinz-Otto Peitgen
MeVis—Center for Medical Diagnostic Systems and Visualization
Universität Bremen
Universitätsallee 29
D-28359 Bremen
GERMANY
and
Department of Mathematics
Florida Atlantic University
Boca Raton, FL 33431
USA

Hartmut Jürgens
MeVis Diagnostics GmbH & Co. KG
Universität Bremen
Universitätsallee 29
D-28359 Bremen
GERMANY

Dietmar Saupe
Fachbereich Informatik und Informationswissenschaft
Universität Konstanz
D-78457 Konstanz
GERMANY

Cover design by Claus Hösselbarth.

Library of Congress Cataloging-in-Publication Data
Peitgen, Heinz-Otto, 1945–
Chaos and fractals : new frontiers of science, second edition / Heinz-Otto Peitgen, Hartmut Jürgens, Dietmar Saupe.
p. cm.
Includes bibliographical references and index.
ISBN 978-1-4684-9396-2 ISBN 978-0-387-21823-6 (eBook)
DOI 10.1007/978-0-387-21823-6
1. Chaotic behavior in systems. 2. Fractals. I. Jürgens, H. (Hartmut) II. Saupe, Dietmar, 1954– III. Title.
Q172.5C45P45 2004
003′.857—dc22 2003063341

ISBN 978-1-4684-9396-2 Printed on acid-free paper.

Softcover reprint of the hardcover 2nd edition 2004

9 8 7 6 5 4 3 2

springer.com

Preface

Almost 12 years have passed by since we wrote *Chaos and Fractals*. At the time we were hoping that our approach of writing a book which would be both accessible without mathematical sophistication and portray these exiting new fields in an authentic manner would find an audience. Now we know it did. We know from many reviews and personal letters that the book is used in a wide range of ways: researchers use it to acquaint themselves, teachers use it in college and university courses, students use it for background reading, and there is also a substantial audience of lay people who just want to know what chaos and fractals are about.

Every book that is somewhat technical in nature is likely to have a number of misprints and errors in its first edition. Some of these were caught and brought to our attention by our readers. One of them, Hermann Flaschka, deserves to be thanked in particular for his suggestions and improvements.

This second edition has several changes. We have taken out the two appendices from the first edition. At the time of the first edition Yuval Fishers contribution, which we published as an appendix was probably the first complete expository account on fractal image compression. Meanwhile, Yuvals book *Fractal Image Compression: Theory and Application* appeared and is now the publication to refer to. Moreover, we have taken out the sections at the end of each chapter, which were devoted to a focussed computer program in BASIC, which highlighted a fundamental construction in that respective chapter. Instead we direct our readers to our web-site

`http://www.cevis.uni-bremen.de/fractals/`

where we provide 10 interactive JAVA-applets.

We also like to express our sincere gratitude to the people at Springer-Verlag, New York, who made this whole project such a wonderful experience for us.

Heinz-Otto Peitgen, Hartmut Jürgens, Dietmar Saupe

Bremen and Konstanz, August 2003

Preface of the First Edition

Over the last decade, physicists, biologists, astronomers and economists have created a new way of understanding the growth of complexity in nature. This new science, called chaos, offers a way of seeing order and pattern where formerly only the random, erratic, the unpredictable — in short, the chaotic — had been observed.

James Gleick[1]

This book is written for everyone who, even without much knowledge of technical mathematics, wants to know *the details* of chaos theory and fractal geometry. This is not a textbook in the usual sense of the word, nor is it written in a 'popular scientific' style. Rather, it has been our desire to give the reader a broad view of the underlying notions behind fractals, chaos and dynamics. In addition, we have wanted to show how fractals and chaos relate to each other and to many other aspects of mathematics as well as to natural phenomena. A third motif in the book is the inherent visual and imaginative beauty in the structures and shapes of fractals and chaos.

For almost ten years now mathematics and the natural sciences have been riding a wave which, in its power, creativity and expanse, has become an interdisciplinary experience of the first order. For some time now this wave has also been touching distant shores far beyond the sciences. Never before have mathematical insights — usually seen as dry and dusty — found such rapid acceptance and generated so much excitement in the public mind. Fractals and chaos have literally captured the attention, enthusiasm and interest of a world-wide public. To the casual observer, the color of their essential structures and their beauty and geometric form captivate the visual senses as few other things they have ever experienced in mathematics. To the student, they bring mathematics out of the realm of ancient history into the twenty-first century. And to the scientist, fractals and chaos offer a rich environment for exploring and modelling the complexity of nature.

But what are the reasons for this fascination? First of all, this young area of research has created pictures of such power and singularity that a collection of them, for example, has proven to be one of the most successful world-wide series of exhibitions ever sponsored by the Goethe-Institute.[2] More important, however, is the fact that chaos theory and fractal geometry have corrected an outmoded conception of the world.

The magnificent successes in the fields of the natural sciences and technology had, for many, fed the illusion that the world on the whole functioned like a huge clockwork mechanism, whose laws were only waiting to be deciphered step by step. Once the laws were known, it was believed, the evolution or development of things could — at least in principle — be ever more accurately predicted. Captivated by the breathtaking advances in the development of computer technology and its promises of a greater command of information, many have put increasing hope in these machines.

But today it is exactly those at the active core of modern science who are proclaiming that this hope is unjustified; the ability to see ever more accurately into future developments is unattainable. One

[1] J. Gleick, *Chaos - Making a New Science*, Viking, New York, 1987.

[2] Alone at the venerable London Museum of Science, the exhibition *Frontiers of Chaos: Images of Complex Dynamical Systems* by H. Jürgens, H.-O. Peitgen, M. Prüfer, P. H. Richter and D. Saupe attracted more than 140,000 visitors. Since 1985 this exhibition has travelled to more than 100 cities in more than 30 countries on all five continents.

conclusion that can be drawn from the new theories, which are admittedly still young, is that stricter determinism and apparently accidental development are *not* mutually exclusive, but rather that their coexistence is more the rule in nature. Chaos theory and fractal geometry address this issue. When we examine the development of a process over a period of time, we speak in terms used in chaos theory. When we are more interested in the structural forms which a chaotic process leaves in its wake, then we use the terminology of fractal geometry, which is really the geometry whose structures are what give order to chaos.

In some sense, fractal geometry is first and foremost a new 'language' used to describe, model and analyze the complex forms found in nature. But while the elements of the 'traditional language' — the familiar Euclidean geometry — are basic visible forms such as lines, circles and spheres, those of the new language do not lend themselves to direct observation. They are, namely, algorithms, which can be transformed into shapes and structures only with the help of computers. In addition, the supply of these algorithmic elements is inexhaustibly large; and they are capable of providing us with a powerful descriptive tool. Once this new language has been mastered, we can describe the form of a cloud as easily and precisely as an architect can describe a house using the language of traditional geometry.

The correlation of chaos and geometry is anything but coincidental. Rather, it is a witness to their deep kinship. This kinship can best be seen in the Mandelbrot set, a mathematical object discovered by Benoit Mandelbrot in 1980. It has been described by some scientists as the most complex — and possibly the most beautiful — object ever seen in mathematics. Its most fascinating characteristic, however, has only just recently been discovered: namely, that it can be interpreted as an illustrated encyclopedia of an infinite number of algorithms. It is a fantastically efficiently organized storehouse of images, and as such it is *the* example par excellence of order in chaos.

Fractals and modern chaos theory are also linked by the fact that many of the contemporary pace-setting discoveries in their fields were only possible using computers. From the perspective of our inherited understanding of mathematics, this is a challenge which is felt by some to be a powerful renewal and liberation and by others to be a degeneration. However this dispute over the 'right' mathematics is decided, it is already clear that the history of the sciences has been enriched by an indispensable chapter. Only superficially is the issue one of beautiful pictures or of perils of deterministic laws. In essence, chaos theory and fractal geometry radically question our understanding of equilibria — and therefore of harmony and order — in nature as well as in other contexts. They offer a new holistic and integral model which can encompass a part of the true complexity of nature for the first time. It is highly probable that the new methods and terminologies will allow us, for example, a much more adequate understanding of ecology and climatic developments, and thus they could contribute to our more effectively tackling our gigantic global problems.

We have worked hard in trying to reveal the elements of fractals, chaos and dynamics in a non-threatening fashion. Each chapter can stand on its own and can be read independently from the others. Each chapter is centered around a running 'story' typeset in *Times* and printed toward the outer margins. More technical discussions, typeset in *Helvetica* and printed toward the inner margins, have been included to occasionally enrich the discussion by providing deeper analyses for those who may desire them and those who are prepared to work themselves through some mathematical notations. At the end of each chapter we offer a short BASIC program, the *Program of the Chapter*, which is designed to highlight one of the most prominent experiments of the respective chapter.

This book is a close relative of the two-volume set *Fractals for the Classroom* which was published by Springer-Verlag and the National Council of Teachers of Mathematics in 1991 and 1992. While those books were originally written for an audience which is involved with the teaching or learning of mathematics, this book is intended for a much larger readership. It combines most parts of the aforementioned books with many extensions and two important appendices.

The first appendix, written by Yuval Fisher, deals with aspects of *image compression* using fundamental ideas from fractal geometry. Such applications have been discussed for about five years and hopes of new breakthrough technologies have risen very high through the work and announcements of the group around Michael F. Barnsley. Since Barnsley has kept his work absolutely secret we still don't know what is possible and what is not. But Fisher's contribution allows us to make a fair guess. Anybody who is interested in the perspectives of image compression through fractals will appreciate this appendix.

The second appendix is written by Carl J. G. Evertsz and Benoit B. Mandelbrot and deals with *multifractal measures,* which is one of the hottest subjects in the current scientific discussion of fractal geometry. Usually we think of fractals as objects having some kind of self-similarity. The discussion of multifractal measures extends this concept to the distributions of quantities (for example, the amount of ground water found at a certain location under the surface). Furthermore, it overcomes some shortcomings of the fractal dimension when used as a tool for measurement in science.

Even with these two important contributions there remain many holes in this book. However, fortunately there are exceptional books already in print that can close these gaps. We list the following only as examples: For portraits of the personalities in the field and the genesis of the subject matter, as well as the scientific background and interrelationships, there are *Chaos – Making a New Science*,[3] by James Gleick, and *Does God Play Dice?*,[4] by Ian Stewart. For the reader who is more interested in a systematic mathematical exposition or who is ready to advance into the depths, there are the following titles: *An Introduction to Chaotic Dynamical Systems*[5] and *Chaos, Fractals, and Dynamics*,[6] both by Robert L. Devaney, and *Fractals Everywhere*,[7] by Michael F. Barnsley. An adequate technical discussion of fractal dimension can be found in the two exceptional texts, *Measure, Topology and Fractal Geometry*,[8] by Gerald A. Edgar, and *Fractal Geometry*,[9] by Kenneth Falconer. Readers more interested in fractals in physics will appreciate *Fractals*,[10] by Jens Feder, while readers who look for fractals in chemistry should not miss *The Fractal Approach to Heterogeneous Chemistry*,[11] by David Avnir. And last but not least, there is the book of books about fractal geometry written by Benoit B. Mandelbrot, *The Fractal Geometry of Nature*.[12]

We owe our gratitude to many who have assisted us during the writing of this book. Our students Torsten Cordes and Lutz Voigt have produced most of the graphics very skillfully and with unlimited patience. They were joined by two more of our students, Ehler Lange and Wayne Tvedt, during part of the preparation time. Douglas Sperry has read our text very carefully at several stages of its evolution and, in addition to helping to get our English de-Germanized, has served in the broader capacity of copy editor. Ernst Gucker, who is working on the German edition, suggested many improvements. Friedrich von Haeseler, Guentcho Skordev, Heinrich Niederhausen and Ulrich Krause have read several chapters and provided valuable suggestions. We also thank Eugen Allgower, Alexander N. Charkovsky, Mitchell J. Feigenbaum, Przemyslaw Prusinkiewicz, and Richard Voss for reading parts of the original manuscript

[3] Viking, 1987.
[4] Penguin Books, 1989.
[5] Second Edition, Addison Wesley, 1989.
[6] Addison Wesley, 1990.
[7] Academic Press, 1989.
[8] Springer-Verlag,1990
[9] John Wiley and Sons, 1990.
[10] Plenum, 1988
[11] Wiley,1989
[12] W. H. Freeman, 1982.

and giving valuable advice. Gisela Gründl has helped us with selecting and organizing third-party artwork. Claus Hösselbarth did an excellent job in designing the cover. Evan M. Maletsky, Terence H. Perciante and Lee E. Yunker read parts of our early manuscripts and gave crucial advice concerning the design of the book. Finally, we are most grateful to Yuval Fischer, Carl J. G. Evertsz and Benoit B. Mandelbrot for contributing the appendices to our book, and to Mitchell Feigenbaum for his remarkable foreword.

The entire book has been produced using the TEX and LATEX typesetting systems where all figures (except for the half-tone and color images) were integrated in the computer files. Even though it took countless hours of sometimes painful experimentation setting up the necessary macros it must be acknowledged that this approach immensely helped to streamline the writing, editing and printing.

Finally, we have been very pleased with the excellent cooperation of Springer-Verlag in New York.

Heinz-Otto Peitgen, Hartmut Jürgens, Dietmar Saupe

Bremen, May 1992

Authors

Heinz-Otto Peitgen. *1945 in Bruch (Germany). Dr. rer. nat. 1973, Habilitation 1976, both from the University of Bonn. Since 1977 Professor of Mathematics at the University of Bremen and between 1985 and 1991 also Professor of Mathematics at the University of California at Santa Cruz. Since 1991 also Professor of Mathematics at the Florida Atlantic University in Boca Raton. Visiting Professor in Belgium, Italy, Mexico and USA. Editor of several research journals on chaos and fractals. Co-author of the award-winning books *The Beauty of Fractals* (with P. H. Richter) and *The Science of Fractal Images* (with D. Saupe)

Hartmut Jürgens. *1955 in Bremen (Germany). Dr. rer. nat 1983 at the University of Bremen. Employment in the computer industry 1984–85, since 1985 Director of the Dynamical Systems Graphics Laboratory at the University of Bremen. Co-author and co-producer (with H.-O. Peitgen, D. Saupe, and C. Zahlten) of the award-winning video *Fractals: An Animated Discussion*

Dietmar Saupe. *1954 in Bremen (Germany). Dr. rer. nat. 1982 and Habilitation 1993, both from the University of Bremen. Visiting Assistant Professor of Mathematics at the University of California at Santa Cruz, 1985–87, Assistant Professor at the University of Bremen, 1987–93, Professsor of Computer Science at the University of Freiburg, 1993–1998, at the University of Leipzig, 1998–2002, and at the University of Konstanz, since 2002. Co-author of the award-winning book *The Science of Fractal Images* (with H.-O. Peitgen)

Contents

Foreword

Mitchell J. Feigenbaum[1]

The study of chaos is a part of a larger program of study of so-called 'strongly' nonlinear systems. Within the context of physics, the exemplar of such a system is a fluid in turbulent motion. If chaos is not exactly the study of fluid turbulence, nevertheless, the image of turbulent, erratic motion serves as a powerful icon to remind a physicist of the sorts of problems he would ultimately like to comprehend. As for all good icons, while a vague impression of what one wants to know is sensibly clear, a precise delineation of many of these quests is not so readily available. In a state of ignorance, the most poignantly insightful questions are not yet ripe for formulation. Of course, this comment remains true despite the fact that for technical exigencies, there are definite questions that one desperately wants the answers to.

Fluid turbulence indeed presents us with highly erratic and only partially predictable phenomena. Historically, since Laplace say, physical scientists have turned to the statistical methods when presented with problems that concern the mutual behaviors of innumerably large numbers of pieces. If for no other reason, one does so to reduce the number of details that one must measure, specify, compute, whatever. Thus, it is easier to say that 43% of the population voted for X than to offer the roster of the behavior of each of millions of voters. Just so, it is easier to specify how many gas molecules there are in an easily measurable volume than to write out the list of where and how fast each one is. This idea is altogether reasonable if not even the most desirable one. However, if one is to work out a theory of these things, so that a prediction might be rendered, then as in all matters of statistics, one must

[1]Mitchell J. Feigenbaum, Toyota Professor, The Rockefeller University, New York.

determine a so-called distribution function. This means a theoretical prediction of just how often out of uncountably many elections, etc., it is expected that each value of this average voter response occurs. For the voter question and the density of a gas question, there is just one number to determine. For the problem of fluid turbulence, even in this statistical quest, one must ask a much richer question: For example, how often do we see eddies of each size rotating at such and such a rate?

The Laws of Fluids

For the problem of voters I don't have any serious idea of how to theoretically determine this requisite distribution; nor with good frequency do the polls succeed in measuring it. After all, it might not exist in the sense that it rapidly and significantly varies from day to day. However, since physicists have long known quite reliably the laws of fluids — that is, the rules that allow you to deduce what each bit of the fluid will do later if you know what they all do now, there might be a way of doing so. Indeed, the main idea of the branch of physics called statistical mechanics is rooted in the belief that one knows in advance how to do this. The idea is, basically, that each possible detailed configuration occurs with equal likelihood. Indeed, the word 'chaos' first entered physics in Maxwell's phrase 'state of molecular chaos' in the last century to loosely mean this. Statistical mechanics — especially in its quantum-mechanical form — works very well indeed, and provides us with some of our most wonderful knowledge. However, altogether regrettably, in the context of fluid turbulence, it has persisted for the last century to roundly fail. It turns out to be a question of truly deducing from the known laws of microscopic motion of fluids what this rule of distribution must be, because the easy guess of 'everything is as random as possibly' simply doesn't work. And when that guess doesn't work, there exists as of today no methodology to provide it. Moreover, if in our present state of knowledge we should be forced to appraise the situation, then we would guess that an extraordinarily complicated distribution is required to account for the phenomena: Should it be fractal in nature, then fractal of the most perverse sort. And the worst part is that we really don't possess the mathematical power to generally say what class of object it might be sought among. Remember, we're not looking for a perfectly good quick-fix: If we are serious in seeking understanding of the analytical description of Nature, then we demand much more. When the subject of chaos and a part of that larger program called strongly nonlinear physics shall have been deemed penetrated, we shall know thoroughly how to respond to such questions, and readily image intuitively what the answers look like. To date, we can now compellingly do so for much simpler problems — and have come to possess that capability only within the last decades.

As I have said earlier, I don't necessarily care about turbulence. Rather, it serves as an icon representing a genre of problems. I was trained as a theoretical high-energy physicist, and grew deeply troubled that no methods save for that of successive improvements, so-called perturbation methods, existed. Apart from the brilliant effort of Ken Wilson, in his version of the renormalization group, that circumstance is unchanged. Knowing the microscopic laws of how things move — such schemes are called 'dynamical systems' — still leaves us almost altogether in the dark as to their larger consequences. Are the theories no good, or is it that we just can't determine what they contain? At the moment it's impossible to say. From high-energy physics to fluid physics and astrophysics our inherited ways of thinking mathematically simply fail to serve us. In a way, if perhaps modest, the questions tackled in the effort to comprehend what is now called chaos have faced these questions of methodology head on.

Nonlinearity

Let me now backtrack and discuss nonlinearity. This means first linearity. Linearity means that the rule that determines what a piece of a system is going to do next is not influenced by what it is doing now. More precisely, this is intended in a differential or incremental sense: For a linear spring, the increase of its tension is proportional to the increment whereby it is stretched, with the ratio of these increments exactly independent of how much it has already been stretched. Such a spring can be stretched arbitrarily far, and in particular will never snap or break. Accordingly, no real spring is linear.

The mathematics of linear objects is particularly felicitous. As it happens, linear objects enjoy an identical, simple geometry. The simplicity of this geometry always allows a relatively easy mental image to capture the essence of a problem, with the technicality, growing with the number of parts, basically a detail, until the parts become infinite in number, although often then too, precise answers can be readily determined.

The historical prejudice against nonlinear problems is that no so simple nor universal geometry usually exists. Until recently, the general scientific perception was that a certain nonlinear equation characterized some particular problem. If the specific problem was sufficiently interesting or demanding of resolution, then perhaps particular methods could be created for it. But it was well understood that the travail would probably be of no avail in other contexts.

Perturbation Method

Indeed, only one method was well understood and universally learned, the perturbation method. If a linear problem was viewed through distorting lenses, it qualitatively would do the same thing: If it repeated every five seconds it would persist to appear so seen through the lenses. Nevertheless, it would now no longer appear to exhibit equal tension increments for the equal elongations. After all, the tension is measurably unchanged by distorting lenses, whereas all spatial measurements are. That is, the device of distorting lenses turns a linear problem into a nonlinear one. The method of perturbation basically works only for nonlinear problems that are distorted versions of linear ones. And so, this uniquely well-learned method is of no avail in matters that aren't merely distortions of linear ones.

Geometry of Chaos

Chaos is absent in distorted linear problems. Chaos and other such phenomena that are qualitatively absent in linear problems are what we call strongly nonlinear phenomena. It is this failure to subscribe to the spectrum of configurations allowed by distorting a simple geometry that renders these problems anywhere from hard in the extreme to impenetrable. How does one ever start to intelligently describe an awkward new geometry? This question is for example intended to be loosely akin to the question of how one should describe the geometry of the surface of the Earth, not through our abstracted perceptual apparatus that allows us to visualize it immersed within a vastly larger three-dimensional setting, but rather intrinsically, forbidding this use of imagination. The solution of this question, first by Gauss and then extended to arbitrary dimensions by Riemann is, as many of you must know, at the center of the way of thinking of Einstein's General Theory of Relativity, our theory of gravity. What is to be the geometry of the object that describes the turbulent fluid's distribution function? Are there intrinsic geometries that describe various chaotic motions, that serve as a unifying way of viewing these disparate nonlinear problems, as kindred? I ask the question because I know the answer to be affirmative in certain broad circumstances. The moment this is accepted, then strongly nonlinear problems appear no longer as each one its own case, but rather coordinated and suitable for theorizing upon as their own abstract entity. This promotion from the detailed specific to the membership in a significant general class is one of the triumphs of the study of chaos in the last decade or two.

An even stronger notion than this generality of shared qualitative geometry is the notion of universality, which means no less than that this shared geometry is not only one of a qualitative similarity but also one of true quantitative identicality. After what has been, if you will, a long preamble, the fact that strongly nonlinear problems, with surprising frequency, can share a quantitatively identical geometry is what I shall pursue for the rest of this discussion, and constitutes what is termed universality in the transition to chaos.

Universality

In a qualitative way of thinking, universality can be seen to be not so surprising. There are two arguments to support this. The first part has simply to do with nonlinearity. Just as a linear object has a constant coefficient of proportionality between, for example, its tension and its expansion, a similar, but nonlinear version, has an *effective* coefficient *dependent* upon its extension. So, consider two completely different nonlinear systems. By adjusting things correctly it is not inconceivable that the effective coefficients of each part of each of the two systems could be set the same so that then their behaviors could, at least initially, be identical. That is, by setting some numerical constants (properties, so to speak, that specify the environment, mathematically called 'parameters') *and* the actual behaviors of these two systems, it is possible that they can do the identical thing. For a linear problem this is ostensibly true: For systems with the same number of parts and mutual connections, a freedom to adjust all the parameters allows one to be adjusted to be identical (truly) to the other. But, for many pieces, this is many adjustments. For a nonlinear system, adjusting a small number of parameters can be compensated, in this quest for identical behavior, by an adjustment of the momentary positions of its pieces. But then it must be that not all motions can be so duplicated between systems.

Thus, the first part of the argument is that nonlinearity confers a certain flexibility upon the adaptability of an object to desirable behavior. Nevertheless, should the precise adjustment of too many specific and subtle details be required in order to achieve a certain universal behavior, then the idea would be pedantic at best.

The Monadology of Leibniz

However, there is a second more potent argument, a paraphrasing of Leibniz in 'The Monadology' which can render this first argument potent. Let us contemplate that the motion we intend to determine to be universal over nonlinear systems has arisen by the successive imposition of more and more qualitative constraints. Should this growingly large host of impositions prove to be generally amenable to such systems (this is the hard and a priori neither obvious nor reasonable part of the discussion), then we shall ultimately discover these disparate systems to all be identically constrained by an infinite number of qualitative and, if you will, self-consistent, requirements. Now, following Leibniz, we ask, 'In how many precise, or *quantitative*, ways can this situation be tenable?' And we respond, following Leibniz, by asserting in precisely one possible uniquely determined way.

This is the best verbalization I know for explaining why such a universal behavior is possible. Both mathematics and physical experimentation confirm its rectitude perfectly. But it is perhaps difficult to have you realize how extraordinary this result appeared given the backdrop of physical and mathematical thinking in 1976 when it first appeared together with its full conceptual analysis. As anecdotal evidence, I had been directed to expound these results to one of the great mathematicians, who is renowned for his results on dynamical systems. I spoke with him at the very end of 1976. I kept trying to tell him that there was a complete *quantitative* universality to these phenomena, and he equally often understood me to have duplicated some known *qualitative* results. Finally, he said 'You mean to tell me these are metrical results?' (Metrical is a mathematical code word that means quantitative.) And I said 'Yes.' 'Well, then you're wrong!' he asserted, and turned his back on me to terminate the conversation.

The Scientific Method

Anecdote aside, what is remarkable about all this? First of all, an easy piece of methodological insight. As practitioners of a truly analytical science, physicists were trained to know that qualitative explanations are insufficient to base truth upon. Quite to the contrary, it is regarded to be at the heart of the 'scientific method' that ever more precise measurements will discriminate between rival quantitative theories to ultimately select out one as the correct encoding of the qualitative content. (Thus, think of geocentric versus heliocentric planetary theories, both qualitatively explaining the retrograde motions of the planets.) Here the method is turned on its head: Qualitatively similar phenomena, independent of any other ideational input, must ineluctably lead to the measurably *identical* quantitative result. Whence the total phenomenological support for this mighty 'scientific method?'

How Universality Works

Second, a new principle of 'economy' immediately emerges. Why put out Herculean efforts to calculate the consequences of some particular and highly difficult encoding of physical laws, when anything else — however trivial —

possessing the same qualitative properties will yield *exactly* the same predictions and results? And this is all the more satisfying because one doesn't even *know* the exact equations that describe various of these phenomena, fluid phenomena in particular. And that is because these phenomena have nothing to do, whatsoever, with the detailed, particular, microscopic laws that happen to be at play. This aspect, that is, of substituting easy problems for hard ones with no penalty, has been, as a way of thinking and performing research, the prominent fruit of the recognition of universality. When can it work? Well, in complicated interactions of scores of chemical species, in laser phenomena, in solid state phenomena, in, at least partially, biological rhythmic phenomena such as apneas and arhythmias, in fluids and, of course, in mathematics.

The Essence of Chaos

But now, as I move towards the end of this claim for virtue, let me discuss 'chaos' a bit more per se and revisit my opening 'preamble.' Much of chaos as a science is connected with the notion of 'sensitive dependence on initial conditions.' Technically, scientists term as 'chaotic' those nonrandom complicated motions that exhibit a very rapid growth of errors that, despite perfect determinism, inhibits any pragmatic ability to render accurate long-term prediction. While nomenclaturally speaking this is perforce true, I personally am not very intrigued or concerned with this facet of my subject. I've never told you what the 'transition to chaos' means, but you can readily guess from the verbiage that it's something that starts off not being chaotic, ends up being so, and hence somehow passes from one to the next. The most important fact is that there is a discernibly precise 'moment', with a corresponding behavior, which is neither chaotic nor nonchaotic, at which this transition occurs. Yes, errors do grow, but only in a marginally predictable, rather than in an unpredictable, fashion. In this state of marginal predictability inheres embryonically all the seeds of the chaotic behavior to come. That is, this transitional point, the legitimate child of universality, without full-fledged sensitive dependence upon initial conditions, knows fully how to dictate to its progeny in turn how this latter phenomenon must unfold. For a certain range of possible behaviors of strongly nonlinear systems — specifically, this range surrounding the transition to chaos — the information obtained just at the transition point fully organizes the spectrum of behaviors that these chaotic systems can exhibit.

The Geometry of Chaos

Now what is it that turns out to be universal? The answer, mostly, is a precise quantitative determination of the intrinsic geometry of the space upon which this marginal chaotic motion lives together with the full knowledge of how in the course of time this space is explored. Indeed, it was from the analysis of universality at the transition to chaos that we have come to recognize the precise mathematical object that fully furnishes the intrinsic geometry of these sort of spaces. This object, a so-called scaling function, together with the mathematically precise delineation of universality, constitutes one of the major results of the study of chaos. Granted the broad range of objects that can be termed fractal, these geometries are fractal. But not the heuristic sort of 'dragons', 'carpets', 'snowflakes', etc. Rather, these are structures which are elaborated upon at smaller and smaller scales differently at each point of the object, and so are infinitely more complicated than the above heuristic objects. There is, in more than just a way of speaking, a geometry of these dynamically created objects, and that geometry requires a scaling function to fully elucidate it. Many of you are aware of the existence of a certain object called the 'Mandelbrot set'. Virtually none of you, though, even having simulated it on your own computers, are aware that its ubiquitous existence in those sufficiently smooth contexts in which it appears, is the consequence of universality at the transition of chaos. Every one of its details is implicit in those embryonic seeds I have mentioned before.

Thus, the most elementary consequence of this deep universal geometry is that, in gross organization we notice a set of discs — the largest the main cardioid — one abutting upon the next and of rapidly diminishing radii. How rapidly do they diminish in size? In fact, each one is δ times smaller that its predecessor, with δ, a universal constant, approximately equal to $4.6692016\ldots$, the best known of the constants that characterize universality at the transition of chaos.

I have now come around full circle to my introductory comments. We have, in the last decade, succeeded in coming to know many of the correct ideas and their mathematical language in regard to the question, 'What is the nature of the objects upon which we see our statistical distributions?' 'Dimension' is a mathematical word possessing a quite broad range of technical connotations. Thus, the theory of universality is erected in a very low (that is, one- or two-) dimensional setting. However the information discussed is of an infinite-dimensional character. The physical phenomena exhibiting these behaviors can appear, for example, in the physical three-dimensional space of human experience, with the number of interacting, cooperating pieces that comprise the system investigated — also a statement of its dimension — either merely a few or an infinitude. Nevertheless, our understanding to date is of what must be admitted to be a relatively simple set of phenomena — relatively simple in comparison to the swirling and shattering complexity of fluid motions at the foot of a waterfall, phenomena that loom large and deeply impress upon us how much lies undiscovered before us.

Introduction

Causality Principle, Deterministic Laws and Chaos

Prediction is difficult, especially of the future.

Niels Bohr

For many, chaos theory already belongs to the greatest achievements in the natural sciences in this century. Indeed, it can be claimed that very few developments in natural science have awakened so much public interest. Here and there, we even hear of changing images of reality or of a revolution in the natural sciences.

Critics of chaos theory have been asking whether this popularity could perhaps only have something to do with the clever choice of catchy terms or the very human need for a theoretical explanation of chaos. Some have prophesized for it exactly the same quick and pathetic death as that of the catastrophe theory, which excited so much attention in the sciences at the end of the 1960's and then suddenly fell from grace even though its mathematical core is counted as one of the most beautiful constructions and creations. The causes of this demise were diverse and did not only have scientific roots. It can certainly be said that catastrophe theory was severely damaged by the almost messianic claims of some apologists.

Chaos theory, too, is occasionally in danger of being overtaxed by being associated with everything that can be even superficially related to the concept of chaos. Unfortunately, a sometimes extravagant popularization through the media is also contributing to this danger; but at the same time this popularization is also an important opportunity to free areas of mathematics from their intellectual ghetto and to show that mathematics is as alive and important as ever.

But what is it that makes chaos theory so fascinating? What do the supposed changes in the image of reality consist of? To these subjects we would like to pose, and to attempt to answer, some questions regarding the philosophy of nature.

Cause and Effect

The main maxim of science is its ability to relate cause and effect. On the basis of the laws of gravitation, for example, astronomical events such as eclipses and the appearances of comets can be predicted thousands of years in advance. Other natural phenomena, however, appear to be much more difficult to predict. Although the movements of the atmosphere, for example, obey the laws of physics just as much as the movements of the planets do, weather prediction is still rather problematic.

Tides Versus Weather

Ian Stewart in his article Chaos: Does God Play Dice?, Encyclopædia Britannica, 1990 Yearbook of Science and the Future, makes the following striking comparison:

"Scientists can predict the tides, so why do they have so much trouble predicting the weather? Accurate tables of the time of high or low tide can be worked out months or even years ahead. Weather forecasts often go wrong within a few days, sometimes even within a few hours. People are so accustomed to this difference that they are not in the least surprised when the promised heat wave turns out to be a blizzard. In contrast, if the tide table predicted a low tide but the beach was under water, there would probably be a riot. Of course the two systems are different. The weather is extremely complex; it involves dozens of such quantities as temperature, air pressure, humidity, wind speed, and cloud cover. Tides are much simpler. Or are they? Tides are perceived to be simpler because they can be easily predicted. In reality, the system that gives rise to tides involves just as many variables — the shape of the coastline, the temperature of the sea, its salinity, its pressure, the waves on its surface, the position of the Sun and Moon, and so on — as that which gives rise to weather. Somehow, however, those variables interact in a regular and predictable fashion. The tides are a phenomenon of order. Weather, on the other hand, is not. There the variables interact in an irregular and unpredictable way. Weather is, in a word, chaotic."

We speak of the unpredictable aspects of weather just as if we were talking about rolling dice or letting an air balloon loose to observe its erratic path as the air is ejected. Since there is no clear relation between cause and effect, such phenomena are said to have random elements. Yet there was little reason to doubt that precise predictability could, in principle, be achieved. It was assumed that it was only necessary to gather and process greater quantities of more precise information (e.g., through the use of denser networks of weather stations and more powerful computers dedicated solely to weather analysis). Some of the first conclusions of chaos theory, however, have recently altered this viewpoint. Simple deterministic systems with only a few elements can generate random behavior, and that randomness is fundamental; gathering more information does not make it disappear. This fundamental randomness has come to be called chaos.

Deterministic Chaos

An apparent paradox is that chaos is deterministic, generated by fixed rules which do not themselves involve any elements of change. We even speak of deterministic chaos. In principle, the future is completely determined by the past; but in practice small uncertainties, much like minute errors of measurement which enter into calculations, are amplified, with the effect that even though the behavior is predictable in the short term, it is unpredictable over the long term.

The discovery of such behavior is one of the important achievements of chaos theory. Another is the methodologies which have been designed for a precise scientific evaluation of the presence of chaotic behavior in mathematical models as well as in real phenomena. Using these methodologies, it is now possible, in principle, to estimate the 'predictability horizon' of a system. This is the mathematical, physical, or time parameter limit within which predictability is ideally possible and beyond which we will never be able to predict with certainty. It has been established, for example, that the predictability horizon in weather forecasting is not more than about two or three weeks. This means that no matter how many more weather stations are included in the observation, no matter how much more accurately weather data are collected and analyzed, we will never be able to predict the weather with any degree of numerical accuracy beyond this horizon of time.

But before we go into an introductory discussion of what chaos theory is trying to accomplish, let us look at some historical aspects of the field. If we look at the development of the sciences on a time-scale on which the efforts of our forbears are visible, we will observe indications of an apparent recapitulation in the present day, even if at a different level. To people during the age of early human history, natural events must have seemed largely to be pure chaos. At first very slowly, then faster and faster, the natural sciences developed (i.e., over the course of thousands of years, the area where chaos reigned seemed to become smaller and smaller). For more and more phenomena, their governing laws were wrung from Nature and their rules were recognized. Simultaneously, mathematics developed hand in hand with the natural sciences, and thus an understanding of the nature of a phenomenon soon came to also include the discovery of an appropriate mathematization of it. In this way, there was continuous nourishment for the illusion that it was only a matter of time, along with the necessary effort and means, before chaos would be completely banned from human experience.

A landmark accomplishment of tremendous, accelerating effect was made about three hundred years ago with the development of calculus by Sir Isaac Newton (1643–1727) and Gottfried Wilhelm Freiherr von Leibniz (1646–1716). Through the universal mathematical ideas of calculus, the basis was provided with which to apparently successfully model the laws of the movements of planets with as much detail as that in the development of populations, the spread of sound through gases, the conduction of heat in media, the interaction of magnetism and electricity, or even the course of weather events. Also maturing during that time was the secret belief that the terms determinism and predictability were equivalent.

The Laplace Demon

For the era of determinism, which was mathematically grounded in calculus, the 'Laplace demon' became the symbol. "If we can imagine a consciousness great enough to know the exact locations and velocities of all the objects in the universe at the present instant, as well as all forces, then there could be no secrets from this consciousness. It could calculate anything about the past or future from the laws of cause and effect."[2]

In its core, the deterministic credo means that the universe is comparable to the ordered running of a tremendously precise clock, in which the present state of things is, on the one hand, simply the consequence of its prior state, and, on the other hand, the cause of its future state. Present, past and future are bound together by causal relationships; and according to the views of the determinists, the problem of an exact prognosis is only a matter of the difficulty of recording all the relevant data. The deterministic credo was characteristic of the Newtonian era, which for the natural sciences came to an end, at the latest, through the insights of Werner Heisenberg in the 1927 proclamation of his uncertainty principle,[3] but which for other sciences is still considered valid.

Strict Causality

Heisenberg wrote: "In the strict formulation of the causality law — 'When we know the present precisely, we can calculate the future' — it is not the final clause, but rather the premise, that is false. We cannot know the present in all its determining details.

"Therefore, all perception is a selection from an abundance of possibilities and a limitation of future possibilities . . . Because all experiments are subject to the laws of quantum mechanics, and thereby also to the uncertainty principle, the invalidity of the causality law is definitively established through quantum mechanics."

Classical determinism in its fearful strictness had to be given up — a turning point of enormous importance.

How undiminished the hope in a great victory of determinism still was at the beginning of this century is impressively illustrated in the 1922 book by Lewis F. Richardson entitled *Weather Prediction by Numerical Process,*[4] in which was written: "After so much hard reasoning, may one play with a fantasy? Imagine a large hall like a theater, except that the circles and galleries go right round through the space usually occupied by the stage. The walls of this chamber are painted to form a map of the globe. The ceiling represents the north polar regions, England is the gallery, the tropics in the upper circle, Australia on the dress circle and the Antarctic in the pit. A

[2]Pièrre Simon de Laplace (1749–1829), a Parisian mathematician and astronomer.

[3]This is also called the indeterminacy principle and states that the position and velocity of an object cannot, even in theory, be exactly measured simultaneously. In fact, the very concept of a concurrence of exact position and exact velocity have no meaning in nature. Ordinary experience, however, provides no evidence of the truth of this principle. It would appear to be easy, for example, to simultaneously measure the position and the velocity of a car; but this is because for objects of ordinary size, the uncertainties implied by this principle are too small to be observable. But the principle becomes really significant for subatomic particles such as electrons.

[4]Dover Publications, New York, 1965. First published by Cambridge University Press, London, 1922. This book is still considered one of the most important works on numerical weather forecasting.

myriad of computers[5] are at work upon the weather of the part of the map where each sits, but each computer attends only to one equation or part of an equation. The work of each region is coordinated by an official of higher rank. Numerous little 'night signs' display the instantaneous values so that neighboring computers can read them. ... From the floor of the pit a tall pillar rises to half the height of the hall. It carries a large pulpit on its top. In this sits the man in charge of the whole theater; he is surrounded by several assistants and messengers. In this respect he is like the conductor of an orchestra in which the instruments are slide-rules and calculating machines. But instead of waving a baton he turns a beam of rosy light upon any region that is running ahead of the rest, and a beam of blue light upon those who are behindhand."

In his book, Richardson first laid down the basis for numerical weather forecasting and then reported on his own initial practical experience with calculation experiments. According to Richardson, the calculations were so long and complex that only by using a 'weather forecasting center' such as the one he fantasized was forecasting conceivable.

Then about the middle of the 1940's, the great John von Neumann actually began to construct the first electronic computer, ENIAC, in order to further pursue Richardson's prophetic program, among others. It was soon recognized, however, that Richardson's only mediocre practical success was not simply attributable to his equipment's lack of calculating capacity, but also to the fact that the space and time increments used in his work had not met a computational stability criterion (Courant-Friedrichs-Lewy Criterion), which was only discovered later. With the appropriate corrections, further attempts were soon under way with progressively bigger and faster computers to make Richardson's dream a reality. This development has been uninterrupted since the 1950's, and it has bestowed truly gigantic 'weather theaters' upon us.

Weak Causality

Indeed, the history of numerical weather forecasting illustrates better than anything else the undiminished belief in a deterministic (viz. predictable) world; for, in reality, Heisenberg's uncertainty principle did not at all mean the end of determinism. It only modified it, because scientists had never really taken Laplace's credo so completely seriously — as is usual with creeds. The most carefully conducted experiment is, after all, never completely isolated from the influences of the surrounding world, and the state of a system is never precisely known at any point in time. The absolute mathematical precision that Laplace presupposed is not physically realizable; minute imprecision is, as a matter of principle, always present. What scientists actually believed was this: From approximately the same causes follow approximately the same effects — in nature as well as in any good experiment. And this is indeed often the case, especially over short time spans. If this were not so, we would not be able to ascertain any natural laws, nor could we build any functioning machines.

The Butterfly Effect

But this apparently very plausible assumption is not universally true. And what is more, it does not do justice to the typical course of natural processes

[5] Richardson uses the word computer here to mean a person who computes.

over long periods of time. Around 1960, Ed Lorenz discovered this deficiency in the models used for numerical weather forecasting; and it was he who coined the term 'butterfly effect'. His description of deterministic chaos goes like this:[6] Chaos occurs when the error propagation, seen as a signal in a time process, grows to the same size or scale as the original signal.

Thus, Heisenberg's response to deterministic thinking was also incomplete. He concluded that the strong causality principle is wrong because its presumptions are erroneous. Lorenz has now shown that the conclusions are also wrong. Natural laws, and for that matter determinism, do not exclude the possibility of chaos. In other words, determinism and predictability are not equivalent. And what is an even more surprising finding of recent chaos theory has been the discovery that these effects are observable in many systems which are much simpler than the weather. In fact, they can be observed in very simple feedback systems, even as simple as the quadratic iterator $x \to ax(1-x)$.

Moreover, chaos and order (i.e., the causality principle) can be observed in juxtaposition within the same system. There may be a linear progression of errors characterizing a deterministic system which is governed by the causality principle, while (in the same system) there can also be an exponential progression of errors (i.e., the butterfly effect) indicating that the causality principle breaks down.

In other words, one of the lessons coming out of chaos theory is that the validity of the causality principle is narrowed by the uncertainty principle from one end as well as by the intrinsic instability properties of the underlying natural laws from the other end.

[6]See Peitgen, H.-O. , Jürgens, H., Saupe, D., and Zahlten, C., *Fractals — An Animated Discussion,* Video film, Freeman 1990. Also appeared in German as *Fraktale in Filmen und Gesprächen,* Spektrum der Wissenschaften Videothek, Heidelberg, 1990.

Chapter 1

The Backbone of Fractals: Feedback and the Iterator

The scientist does not study nature because it is useful; he studies it because he delights in it, and he delights in it because it is beautiful. If nature were not beautiful, it would not be worth knowing, and if nature were not worth knowing, life would not be worth living.

Henri Poincaré

When we think about fractals as images, forms or structures we usually perceive them as static objects. This is a legitimate initial standpoint in many cases, as for example if we deal with natural structures like the ones in figures 1.1 and 1.2.

But this point of view tells us little about the evolution or generation of a given structure. Often, as for example in botany, we like to discuss more than just the complexity of a ripe plant. In fact, any geometric model of a plant which does not also incorporate its dynamic growth plan for the plant will not lead very far.

The same is true for mountains, whose geometry is a result of past tectonic activity as well as erosion processes which still and will forever shape what we see as a mountain. We can also say the same for the deposit of zinc in an electrolytic experiment.

Fractals and Dynamic Processes

In other words, to talk about fractals while ignoring the dynamic processes which created them would be inadequate. But in accepting this point of view we seem to enter very difficult waters. What are these processes and what is the common mathematical thread in them? Aren't we proposing that the complexity of forms which we see in nature is a result of equally complicated processes? This is true in many cases, but at the same time the long-standing paradigm *'Complexity of structure is a result of complicated interwoven processes'* is far from being true in general. Rather, it seems — and this is one of the major surprising impacts of fractal geometry and chaos theory — that in the presence of a complex pattern there is a good chance that a very simple process is responsible for it. In other words, the simplicity of a process should not mislead us into concluding that it will be easy to understand its consequences.

California Oak Tree

California oak tree, Arastradero Preserve, Palo Alto. Photograph by Michael McGuire.

Figure 1.1

Fern

This fern is from K. Rasbach, *Die Farnpflanzen Zentraleuropas,* Verlag Gustav Fischer, Stuttgart, 1968. Reproduced with kind permission by the publisher.

Figure 1.2

1.1 The Principle of Feedback

The most important example of a simple process with very complicated behavior is the process determined by quadratic expressions such as $x^2 + c$, where c is considered to be a fixed constant, or $p + rp(1 - p)$, where r is a constant. Before we enter an initial discussion of this phenomenon — a more systematic exploration is offered in chapter 10 — let us identify and discuss one of the central icons of our presentation.

Feedback processes are fundamental in all exact sciences. In fact, they were first introduced by Sir Isaac Newton and Gottfried W. Leibniz some 300 years ago in the form of dynamic laws; and it is now standard procedure to model natural phenomena using such laws. Such laws determine, for example, the location and velocity of a particle at one time instant from its values at the preceding instant. The motion of the particle is then understood as the unfolding of that law. It is not essential whether the process is discrete (i.e., it takes place in steps) or continuous. Physicists like to think in terms of infinitesimal time steps: *natura non facit saltus*.[1] Biologists, on the other hand, often prefer to look at the changes from year to year or from generation to generation.

Iterator, Feedback and Dynamic Law

We will use the terms iterator, feedback and dynamic law synonymously. Figure 1.3 explains the idea. The same operation is carried out repeatedly, the output of one iteration being the input for the next one.

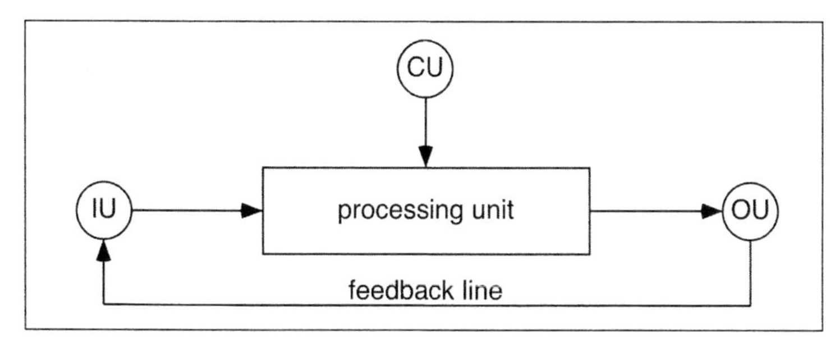

The Feedback Machine

The feedback machine with IU = input unit, OU = output unit, CU = control unit.

Figure 1.3

The Iterator: Principle of Feedback

The feedback machine has three storage units (IU = input unit, OU = output unit, CU = control unit, PU = processing unit), and one processor, all connected by four transmission lines (see figure 1.3). The whole unit is run by a clock, which monitors the action in each component and counts cycles. The control unit acts like a gear shift in an engine. That is, we can shift the iterator into a particular state and then run the unit. There are preparatory cycles and running cycles, each of which can be broken down into elementary steps:

Preparatory cycle:

Step 1: load information into IU

[1]Nature does not make radical jumps.

Step 2: load information into CU
Step 3: transmit the content of CU into PU

Running cycle:

Step 1: transmit content of IU and load into PU
Step 2: process the input from IU
Step 3: transmit the result and load into OU
Step 4: transmit the content from OU and load into IU

To initiate the operation of the machine we run one preparatory cycle. Then we start the running cycles and execute a certain number of them, the count of which may depend on observations which we make by monitoring the actual output. Execution of one running cycle is sometimes called one iteration.

What Is a Feedback Machine?

When we refer to iterations we should imagine a proper feedback machine. The dynamic behavior of such a machine can be controlled by setting certain outside parameters, similar to control levers in an engine. We will discuss the basic principles guided by the simple example of video feedback, which in fact permits real experiments. This particular feedback machine can be built using particular pieces of equipment. It is a real machine in the original sense of the word. This case is rather the exception in this book. Here the term 'feedback machine' usually refers to an abstract machine, a 'Gedankenexperiment'. Such an abstract machine may be put into operation by executing an appropriate computer program, or by using a pocket calculator or merely paper and pencil to carry out the given feedback mechanism.

Video Feedback Setup

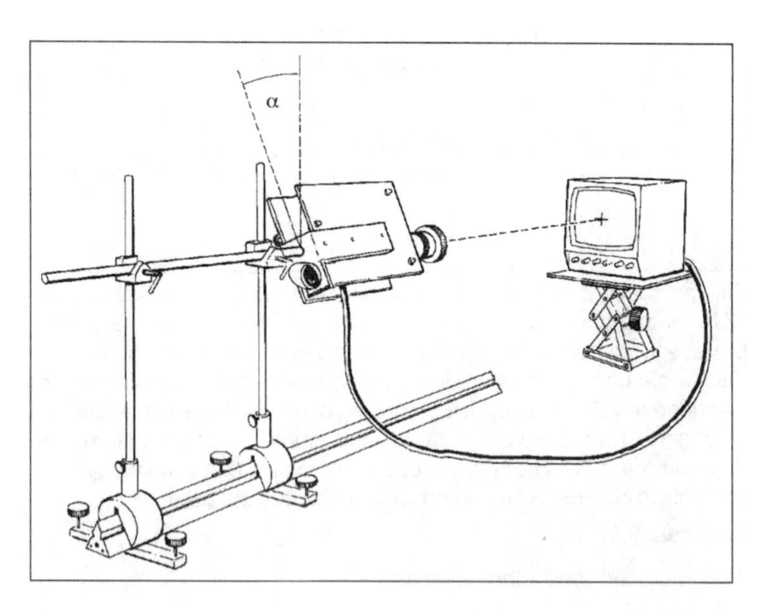

Figure 1.4

Video Feedback

Video feedback is a feedback experiment in the traditional sense of the word. Its basic configuration is probably as old as television. Nevertheless, the particular video feedback experiment which we will now present is so dramatic that its potential can excite even professionals from the television scene.[2] Figure 1.4 shows the basic setup. A video camera looks at a video monitor, and whatever it sees in its viewing zone is put onto the monitor. There are quite a few controls which have an impact on what will be seen by an outside observer, for example, the various control dials on the monitor (contrast, brightness, etc.) and video camera (focus, iris aperture, etc.), as well as the position of the camera with respect to the monitor. Below we collect some important tips which will help you to make a successful video feedback experiment yourself.

It is quite obvious how we can imbed the experiment into our logo in figure 1.3 (input unit = camera, processing unit = camera and monitor electronics, output unit = monitor screen, control unit = focus, brightness, etc.). The feedback clock runs quite fast, i.e., about 30 cycles per second, or whatever number of frames per second your TV system generates.[3]

Hints for the Video Feedback Experiment

The experiment should be set up in an almost dark room. The distance between camera and monitor should be such that the mapping ratio is approximately 1 : 1. Turn up the contrast dial on the monitor all the way and turn down the brightness dial considerably. The experiment works better if the monitor or the camera is put upside down. Moreover, the tripod should be equipped with a head that allows the camera to be turned about its long axis, while it faces the monitor. Rotate the camera some 45°(angle α) out of its vertical position. Connect the camera with the monitor. Now the basic setup is arranged. The camera should have a manual iris which is now gradually opened while the lens is focused on the monitor screen. Depending on the contrast and brightness setting you may want to light a match in front of the monitor screen in order to ignite the process.

Dramatic Impact of Controls

Each of the controls has an impact on the process, some a very dramatic one. In this regard we can think of our setup as an *analog computer* with control dials. For some kinds of controls and variables it is relatively easy to understand their mechanisms; for others it is hard; and for still others it is hard as hell. In fact, many of the phenomena which can be observed are still very poorly understood. The physicist James P. Crutchfield has probably contributed most toward a deeper and systematic understanding of the process.[4]

---

[2] It was proposed by Ralph Abraham from the University of California at Santa Cruz in the 1970's. See R. Abraham, *Simulation of cascades by video feedback*, in: "Structural Stability, the Theory of Catastrophes, and Applications in the Sciences", P. Hilton (ed.), Lecture Notes in Mathematics vol. 525, 1976, 10–14, Springer-Verlag, Berlin.

[3] NTSC is typically 30 frames per second at 480 lines per image.

[4] J. P. Crutchfield, *Space-time dynamics in video feedback*, Physica 10D (1984) 229–245.

Monitor Inside Monitor Inside ...

Effect of long distance between camera and monitor. Basic setup and mapping principle (left), real feedback — monitor inside monitor (right).

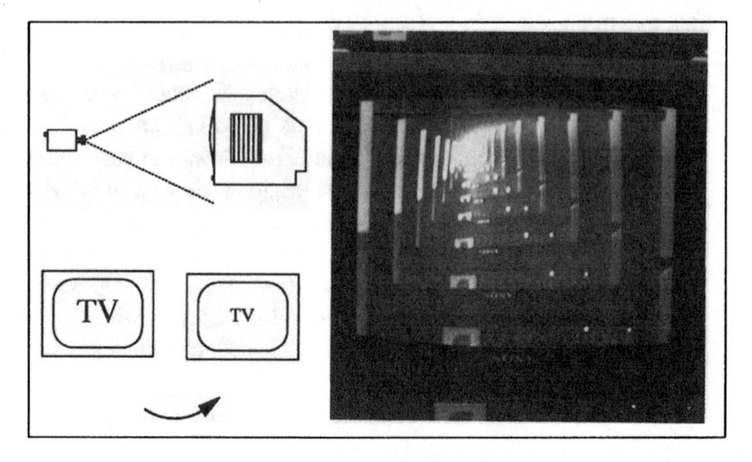

Figure 1.5

Zoom into a Zoom into ...

Effect of short distance between camera and monitor. Basic setup and mapping principle (left), real feedback — repeated magnification of the image of a pencil (right).

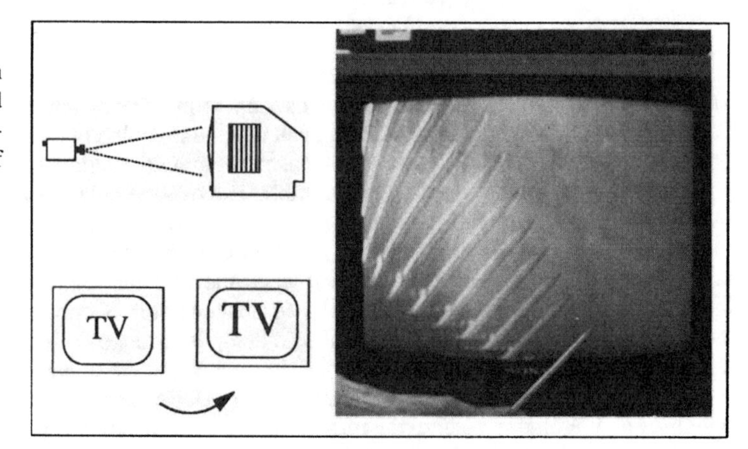

Figure 1.6

The easiest variable which has a dramatic impact on the process of image generation is the position of the camera with regard to the monitor. When the distance from the camera to the monitor is long, the monitor is just a small part of the viewing field. Consequently, the monitor will be reproduced onto a small portion of its screen, and this happens again, and again, and again, ad infinitum. In other words, we see a monitor inside a monitor inside a monitor, etc. (compare figure 1.5). The effect of the process can be described as compression, or, dynamically, as a motion to the center of the monitor. Whatever image is initially on the monitor will be squeezed and put back onto the monitor, and that image will be squeezed again, and so on. We would say that the mapping ratio is $1 : m$, where $m < 1$, i.e., something of unit length 1 on the monitor would reduce to something of length m in a single feedback cycle.

Figure 1.7 : Some examples of real video feedback. There is a more or less pronounced periodicity in these pictures which depends on the angle of the video camera. From the upper left to the lower right we can see periods $3, 5, 5, 5, 8, 8, 11, 11, >11$.

The *monitor-inside-a-monitor* effect is known by most people as video feedback. It is almost always easy to reproduce with any kind of equipment. But there is much more 'life' in this simple system than has been recognized because it is a little harder to reproduce with some equipment.

Next, let us discuss what will happen at the other extreme end of the positioning scale — when the distance between the camera and the monitor is so short that the viewing field of the camera is just a part of the monitor screen. That part is put back onto the entire screen, and again, and again, ad infinitum (compare figure 1.6). We would say that the mapping ratio is $1 : m$, where $m > 1$, i.e., something of unit length 1 on the monitor would expand to something of length m in a single feedback cycle.

Now the action in the process is best described as expansion or, dynamically, as a motion to the border of the monitor. Whatever image is initially on the monitor, a small part of it will be expanded to the full screen, and of that a small part will again be expanded, and so on. Since the TV refreshes its image about 30 times per second, it is impossible to see the individual steps in this process. The result of the close camera position can be a rather wild and almost turbulent motion on the screen.

Unchaining the Feedback

The more interesting effects occur when the position of the camera with regard to the monitor is carefully chosen to be such that the mapping ratio is nearly $1:1$. The effect is increased dramatically if the camera is turned about its axis, i.e., an image on the monitor is seen by the camera as if rotationally changed by some angle. Thus it appears on the monitor (mapping ratio $1:1$) in essentially the same size but rotated. From this point on, any simple description of the mechanisms for the wild and beautiful visual effects that can be observed breaks down. From what has been said so far, we would expect that in the rotated position we would eventually observe just a sequence of rotated images. But this prediction is far too simple. All kinds of peculiar effects occur due to many different characteristics innate to television image production. For example, the process of scanning the image on the monitor and in the camera is one of sequentially putting together a series of lines to compose the image. There is also the *memory effect* of the phosphorus on the monitor tube. In addition, there are electronic time chains and their delays in both the monitor and camera, as well as other factors.

In any event, this extremely simple feedback system demonstrates very dramatically how complicated structures can be the result of very simple feedback. In a way, this is the theme of the book. Our next set of experiments tries to bring more of a systematic light into this world of exciting phenomena. The basic principle is the same as with video feedback: An initial image is processed and then the resulting image is reprocessed by the same machine over and over again.

1.2 The Multiple Reduction Copy Machine

We now turn to a set of experiments which will provide us with a very intuitive access to the language of fractal geometry. In a sense, it is a continuation of the video feedback experiment.

First, let us consider a copy machine which is equipped with an image reduction feature. If we take an image, put it on the machine and push a button, we obtain a copy of the image. It is, however, reduced uniformly by say 50%, i.e., by a factor of 1/2. In the language of mathematics we say that the copy is *similar* to the original. The process to generate a copy is called a *similarity transformation* or *similitude*. The process just described embedded into the idea of figure 1.3 constitutes a feedback system[5] which would be very easily predictable in its long-run effect: After some ten or so cycles any initial image would be reduced to just a point. In other words, running the machine would be a waste of paper (see figure 1.8).

Single Reduction Copy Machine

Iteration by a copy machine with reduction applied to a portrait of Carl Friedrich Gauss (1777–1855).

Figure 1.8

We will now modify this principal setup. Remember, the basic action of our machine is the reduction of images. Such reductions, of course, are achieved by a lens system. As a simple modification of a stock copier, let us imagine that our custom copier has 2, or 3, or 7, or 14 532 231, or whatever number of reduction lenses. Each of them looks at the image on the copier, reduces it, and puts the result somewhere on the copy paper. One such design consists of the choice of the number of lenses, the reduction factors and the placements of the reduced images. It constitutes a particular feedback system which we can run to see what happens. We call such a machine a *Multiple Reduction Copy Machine*, abbreviated by the letters *MRCM*.

Figure 1.9 shows a first example of an MRCM which incorporates just three reduction lenses, each of them reducing by 50%, i.e., by a factor of 1/2.

What will we see emerging in the sequence of iterations as we run the feedback system? Will we see an arrangement of a smaller and smaller composite of images developing toward a point? Figure 1.10 gives the surprising

[5] Try to identify input, processing and output unit.

Multiple Reduction Copy Machine (MRCM)

The Multiple Reduction Copy Machine (MRCM): the processing unit is equipped with a three-lens system.

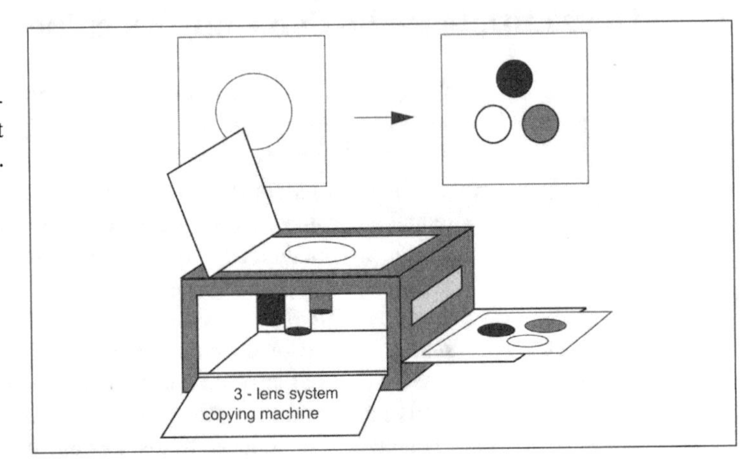

Figure 1.9

Rectangle in MRCM

Starting with a rectangle the iteration leads to the Sierpinski gasket. Shown are the first five steps and the result after some more iterations (lower right).

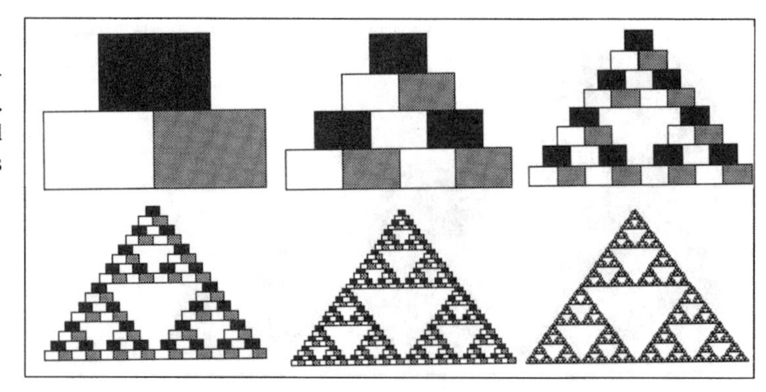

Figure 1.10

answer, the consequences of which could potentially revolutionize almost everything we have thought about images in a technical sense. Let us start with a rectangle as an initial test image. We put it onto the multiple copier, obtaining three reduced copies which we color according to the respective lens system from which each copy is produced.

A First Hieroglyph: The Sierpinski Gasket

Then, indeed, we see $3 \times 3 = 9$ smaller copies, and then $3 \times 9 = 27$ even smaller copies, then 81, 243, 729, etc., copies which rapidly decrease in size, but the resulting compound images do not reduce to a point at all. Rather, they transform into a perfect *Sierpinski gasket*, which we will use as a major example exhibiting important aspects of fractals in general. Using the imagery of a language paradigm, we have just introduced a first hieroglyph in our new fractal dialect. From what we have said so far, it is clear that this basic principle will generate an infinite variety of images. All we have to do is convert the copier into one consisting of 4, or 5, or any other number of lens systems, or with different reduction factors. We will be going into this matter in more detail in chapters 5 and 6, but there are two major surprises which are

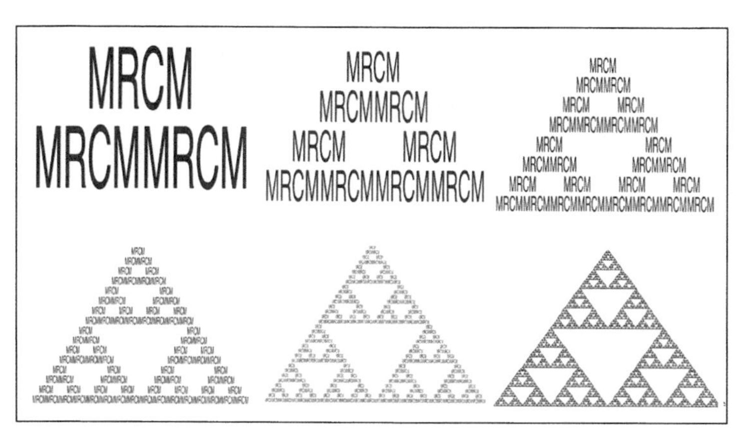

MRCM Applied to 'MRCM'

We can start with an arbitrary image — this iterator will always lead to the Sierpinski gasket.

Figure 1.11

not immediately apparent and deserve some preliminary discussion here.

Looking at figure 1.10 again, we may be led to believe that the secret to the tendency toward the formation of the Sierpinski gasket is our choice of an appropriately dimensioned rectangle as the initial image in starting the feedback process. To show that this is not the case, let us assume that instead of a rectangle as the initial image, we choose a triangle or any arbitrary image, which may be represented well enough by the letters NCTM. The question is: What will then evolve in the process? Figure 1.11 gives the answer. The same final structure is approximated as we run the machine. Each step produces a composite of images which rapidly decrease in size. It doesn't matter in the least whether these images are rectangles, triangles, or the letters NCTM; the same final composite image is approached in each case — namely, the Sierpinski gasket. In other words, the machine produces one — and only one — final image in the process, and that final image is totally independent from the image with which we start! This magnificent behavior seems to be a miracle. But in mathematical terms it just means that we have a process which produces a sequence of results tending toward *one* final object which is independent from how we start the process. This property is called *stability*.

The second surprise is that the copy machine paradigm is not just a way to recover 'mathematical monsters' like the Sierpinski gasket or its relatives (soon we will see many of them). Let us ask what the images are which we can obtain this way. What can they look like? The answer is simply incredible. For many more natural pictures there is a copy machine of the above kind which generates the desired picture. However, it is a difficult problem to design the machine for a given picture. But nevertheless, in chapter 5 and the first appendix we will introduce some of the design principles leading to the frontiers of current mathematical research.

The point here is to see some of the variety of possible images obtained by very simple feedback processes, the elements of which are easily manipulated and under our control, quite unlike the video feedback experiment.

From Similar to Affine

In our first example, each lens system behaves like a similarity transformation; i.e., a rectangle is reproduced as a rectangle, a triangle with certain angles is reproduced as a triangle with the same angles, and so on. The only thing which is changed is the scale of the image. If we pick any two points in the original image and compare their distances with that in the copy it will be scaled down by a constant factor. One principal direction for an extension will be to allow lens systems which reduce by different factors in different spatial directions. For example, the lens system may reduce by a factor of 1/2 in the horizontal direction and by a factor of 1/3 in the vertical direction. The effect of such a system is to destroy similarity: A square is reduced to a rectangle; a triangle with certain angles is reduced to a triangle with different angles. In mathematical terms we speak of *affine* transformations. Similitudes and affine transformations are, however, in one class of mathematical objects: *linear* transformations, i.e., transformations which when applied to a straight line reproduce a straight line. Only if we allow such extensions, will the metaphor of the copy machine develop its full power (see chapter 5).

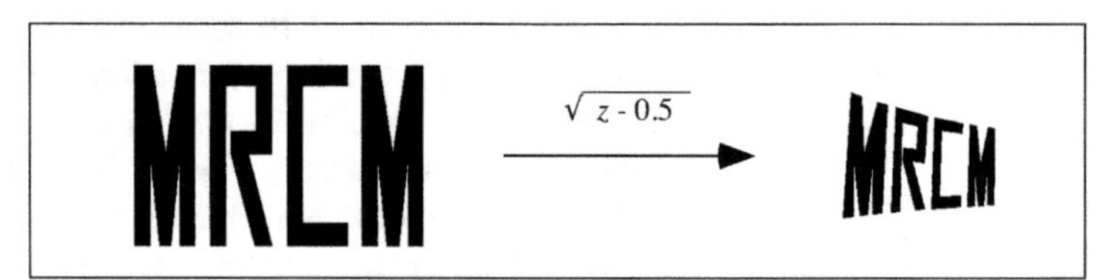

Nonlinear Transformation

The complex square root applied to the letters MRCM in the plane. Note that angles are preserved.

Figure 1.12

From Linear to Nonlinear

Real lens systems are usually not perfect similitudes. They distort an image more or less. As a radical example, a straight line seen through a fisheye lens is reproduced as a curved line. In mathematical terms we speak of nonlinear effects. Let us simulate such an effect in a simplistic model. Let us consider the numbers which are larger than 1. If we multiply such numbers by a factor of 1/3, for example, we have a perfect similitude. If we take the square root, however, we have a typical nonlinear effect: The segment between 1 and 10 is reduced to the segment between 1 and $\sqrt{10} \approx 3.16$, while the segment between 1 and 100, which is 11 times as long, is reduced to the segment between 1 and 10, which is only about 4 times as long as the segment between 1 and $\sqrt{10}$. The reduction factor changes, i.e., it depends on the location where the transformation is applied (see figure 1.12). Copy machines with nonlinear lens systems are the content of chapter 13, and will lead to the famous Julia sets as well as the Mandelbrot set. Incidentally, the systems discussed there are related to similitudes in one important sense: They preserve angles.

1.3 Basic Types of Feedback Processes

We will now turn to feedback machines which process numbers. But before we get involved in the discussion of specific examples, let us take an overview.

One-Step Machines

One-step machines are characterized by an iteration formula $x_{n+1} = f(x_n)$, where $f(x)$ can be any function of x. It requires one number as input and returns a new number — the result of the formula — as output (e.g., $f(x_n) = x_n^2 + 1$). The formula can be controlled by a fixed parameter (e.g., $x_n^2 + c$, i.e., with control parameter c), but in any case the output depends only on the input. The numbers are indexed in order to keep track of the time (cycle) in which they were obtained.

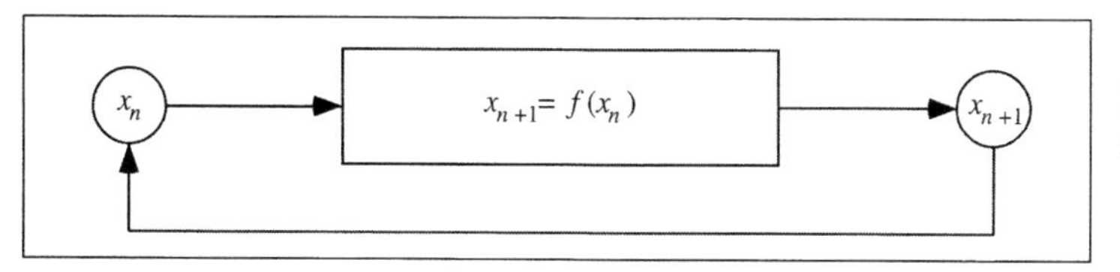

One-Step Feedback Machine

Principle of the one-step feedback machine.

Figure 1.13

One-step machines are very useful mathematical tools and have been developed in particular for the numerical solution of complex problems. They have a tradition in mathematics which goes back at least a few thousand years.

Ancient Square Root Computation

The following example of a one-step feedback machine is an algorithm which was already known to the Sumerian mathematicians some 4000 years ago. It is a beautiful example of the strength and continuity of mathematics. Mankind has seen many advances and terrible setbacks since those times, while the power and beauty of mathematical thought has remained.

Given $a > 0$. Compute a sequence $x_1, x_2, x_3, \ldots$ such that the limit is $\sqrt{a}$, i.e., x_n approaches $\sqrt{a}$ closer and closer as we proceed to larger and larger n. Here is how x_n is defined. We begin with an arbitrary guess $x_0 > 0$ and continue with

$$x_{n+1} = \frac{1}{2}\left(x_n + \frac{a}{x_n}\right), \quad n = 0, 1, 2, \ldots \tag{1.1}$$

Let us look at an example, $\sqrt{2}$. We guess $x_0 = 2$. Then

$$x_1 = \frac{1}{2}\left(x_0 + \frac{2}{x_0}\right) = \frac{1}{2}\left(2 + \frac{2}{2}\right) = 1.5$$

and

$$x_2 = \frac{1}{2}\left(x_1 + \frac{2}{x_1}\right) = \frac{1}{2}\left(1.5 + \frac{2}{1.5}\right) = \frac{17}{12} = 1.41666\ldots$$

and so on.

Let us give a brief argument why this method works in order to understand how well it works. To this end we introduce the relative error e_n of x_n, where e_n is defined by the equation

$$x_n = (1 + e_n)\sqrt{a}. \tag{1.2}$$

Replacing x_n by the equivalent $(1 + e_n)\sqrt{a}$ in eqn. (1.1) we arrive at

$$x_{n+1} = \sqrt{a}\left(1 + \frac{e_n^2}{2 + 2e_n}\right).$$

Thus, using the definition in eqn. (1.2) again, we obtain an expression for the error e_{n+1}

$$e_{n+1} = \frac{e_n^2}{2 + 2e_n}. \tag{1.3}$$

Now $x_0 > 0$ and therefore $e_0 > -1$, and thus $e_n > 0$ for $n = 1, 2, 3, \ldots$ But then $x_n > \sqrt{a}$ for all $n > 0$. Finally, we can obtain estimates out of eqn. (1.3). If we drop the '2' in the denominator we obtain

$$e_{n+1} < \frac{e_n}{2}$$

and if we drop '$2e_n$' we obtain

$$e_{n+1} < \frac{e_n^2}{2}.$$

The first inequality and the definition of e_n by eqn. (1.2) shows that

$$x_1 > x_2 > x_3 > \cdots > \sqrt{a}$$

and that the limit is $\sqrt{a}$. The second inequality shows that if $e_n < 10^{-n}$, then $e_{n+1} < 10^{-2n}/2$, i.e., in each step of the sequence the number of correct digits is nearly doubled. This algorithm for the computation of the square root is an example of a more general method for the solution of nonlinear equations, which was discovered about 4000 years later and is nowadays called Newton's method.

Two-Step Feedback Methods

One-step feedback processes represent only a particular class of a whole family of feedback methods. Another class is known as *two-step methods*. Here the output is typically computed by a formula like

$$x_{n+1} = g(x_n, x_{n-1}).$$

Take, for example, the law which generates the *Fibonacci numbers*

$$g(x_n, x_{n-1}) = x_n + x_{n-1}.$$

Fibonacci Numbers and the Rabbit Problem

Leonardo Pisano, also known as Fibonacci[6] was one of the outstanding figures in medieval Western mathematics. He traveled widely in the Mediterranean world before settling down in his native Pisa. In 1202 he published his book, Liber Abaci, which changed Europe. It acquainted Europeans with the Indian Arabic ciphers 0, 1, 2, ... His book also contained the following problem, which has inspired people ever since. There is one pair of rabbits which is born at time 0. After one month that pair is mature and a month later gives birth to a new pair of rabbits and continues to do so (i.e., every month a new pair is born to the original pair). Moreover, each new pair of rabbits matures after one month and begins producing pairs of offspring every month after that ad infinitum. One assumes that the rabbits live forever. What is the number of pairs after n months?

Let us be careful and follow the evolution of rabbits step by step. In our rabbit population, let us distinguish between adult and young pairs of rabbits. A just-born pair is young, of course, and turns adult after one time step. Moreover, an adult pair gives birth to a young pair after one time step. Now let J_n and A_n be the number of young and adult pairs after n months, respectively. Initially, at time $n = 0$, there is only one young pair ($J_0 = 1,\ A_0 = 0$). After one month the young pair has turned into an adult one ($J_1 = 0,\ A_1 = 1$). After two months the adult pair gives birth to one young pair ($J_2 = 1,\ A_2 = 1$). Then again after the next month. Moreover, the young pair turns into an adult one ($J_3 = 1,\ A_3 = 2$). The general rule, of course, is that the number of newborn pairs J_{n+1} equals the previous adult population A_n. The adult population grows by the number of immature pairs, J_n, from the previous month. Thus, the following two formulas completely describe the population dynamics

$$\begin{aligned} J_{n+1} &= A_n, \\ A_{n+1} &= A_n + J_n. \end{aligned} \tag{1.4}$$

As initial values we take $J_0 = 1$ and $A_0 = 0$. From the first of the above equations it follows that $J_n = A_{n-1}$. Inserting this into the other equation we obtain

$$A_{n+1} = A_n + A_{n-1}$$

with $A_0 = 0$ and $A_1 = 1$. This is a single equation for the total rabbit population. Using this equation, the number of pairs in successive generations is easily computed:

$$0, 1, 1, 2, 3, 5, 8, 13, 21, 34, 55, 89, 144, 233, \ldots$$

Each number in this sequence is just the sum of its two predecessors. This sequence is called the Fibonacci sequence.

We have established another feedback system, but this one is a little different from the previous systems. In all the earlier feedback loops, the state at time n was determined only by the preceding state

[6] Filius (=son) of Bonacci.

at time $n-1$. Such systems are called one-step loops. For the Fibonacci sequence the state at time $n+1$ requires information from states n and $n-1$. Such systems are called two-step loops. The simple and innocent-looking Fibonacci sequence has a variety of interesting properties. Thousands of papers have been published about them, and there is even a Fibonacci-Association with its own periodical, Fibonacci Quarterly, which reports on the never-ending stream of new results. One property has been known for a long time and has led to amazing recent research in biology, as well as having had astonishing applications in architecture and the arts for many centuries.

Apparently the Fibonacci sequence can grow beyond all limits. Our rabbits exhibit a kind of a population explosion. We can ask, however, how the population progresses from generation to generation. For that purpose we look again at the Fibonacci numbers and compute the ratios of succeeding generations (rounded to six decimals).

n	A_n	A_{n+1}/A_n	In Decimals
0	1	1/1	1.0
1	1	2/1	2.0
2	2	3/2	1.5
3	3	5/3	1.666666
4	5	8/5	1.6
5	8	13/8	1.625
6	13	21/13	1.615385
7	21	34/21	1.619048
8	34	55/34	1.617647
9	55	89/55	1.618182
10	89	144/89	1.617978
11	144	233/144	1.618056
12	233	377/233	1.618026

Apparently we are approaching steadily, if not exactly rapidly, some particular number. Have you seen that mysterious number

$$1.6180339887498948482 0\ldots$$

before? Let us open the curtain.

$$1.61803398\ldots = \frac{1+\sqrt{5}}{2},$$

which is the famous golden mean, or proportio divina,[7] as they called it in the middle ages. This number has inspired mathematicians, astronomers and philosophers like no other number in the history of mathematics.

[7] Divine proportion (Latin).

At first it seems that processes of two-step methods are not covered by the concept of a feedback machine as we have discussed it so far. Indeed, the output x_{n+1} depends not only on the last step x_n, but also on the step preceding the last, namely, x_{n-1}. Consequently, it may appear natural to extend the design of our feedback machines so that the concept incorporates a certain memory which conserves some information from the last cycles.

Feedback Machines with Memory

Machines with memory are typical for our computer age. While a machine without memory reacts on their inputs always in the same way, a machine with memory may react differently upon taking its own state or content of the memory into account. Take, for example, a soft drink machine. You will not be successful in getting a soda by just pushing a button. First you have to insert the right amount of money to make sure that the machine is in the appropriate state to accept your input.

Let us now extend the concept of a feedback machine by equipping the processing unit with an internal memory unit. Then the iteration of a two-step method $x_{n+1} = g(x_n, x_{n-1})$ can be implemented as follows. First note that to start the feedback machine two initial values x_0 and x_1 are required.

Preparation: Initialize the memory unit with x_0 and the input unit with x_1.
Iteration: Evaluate $x_{n+1} = g(x_n, x_{n-1})$, where x_n is in the input unit and x_{n-1} is in the memory unit. Then update the memory unit with x_n

Somehow it seems that feedback machines with memory should be more flexible in modeling different phenomena. But this is not at all the case. Rather, a machine with memory can be seen to be equivalent to a one-step machine which, however, works on *vectors* as input and output information. Input and output are given as pairs, or triples, or quadruples, and so on, of numbers. In other words, a pair of input variables (x_n, x_{n-1}) generates a pair of output variables (x_{n+1}, x_n).

One-Step Machines with Two Variables

Formally, we introduce a new variable, $y_n = x_{n-1}$, and extend the formula $x_{n+1} = g(x_n, x_{n-1})$ to the equivalent pair:

$$\begin{aligned} x_{n+1} &= g(x_n, y_n) \\ y_{n+1} &= x_n \end{aligned}$$

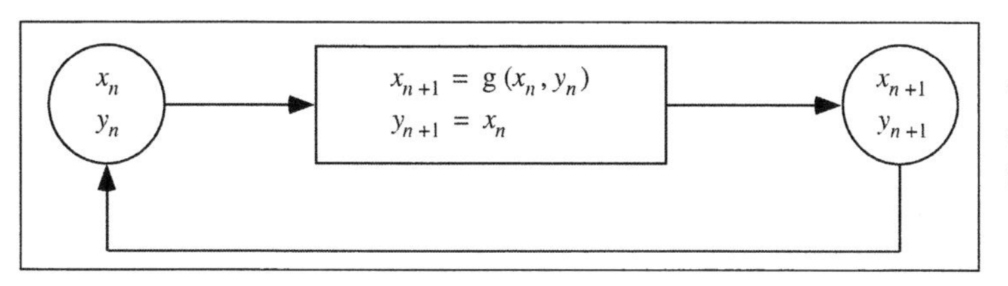

Two-Step Loop

Two-step loops are a special case of one-step feedback machines with two variables.

Figure 1.14

This simple trick can easily be generalized. For example, let us assume that the formula which determines the feedback depends on k preceding iterations. Then one can rewrite this single formula as a one-step process which is given by a set of k formulas by introducing k independent variables. Usually, the independent variables are combined into a vector of variables. The pair (x_n, y_n), for example, can be written symbolically as a single new variable Z_n. Moreover, we can then rewrite the set of formulas $x_{n+1} = g(x_n, y_n)$, $y_{n+1} = x_n$, by a single formula: $Z_{n+1} = G(Z_n)$. In other words, we do not have to go to the trouble of developing a special machine for two-step methods. They are perfectly covered by one-step machines.

The Rabbit Problem As One-Step Machine

Let us give one example, the Fibonacci numbers, defined by the two-step method

$$A_{n+1} = g(A_n, A_{n-1}) = A_n + A_{n-1}$$

with $A_0 = 0$ and $A_1 = 1$. The equivalent equations for a one-step method operating on pairs (x_n, y_n) are

$$\begin{aligned} x_{n+1} &= x_n + y_n \\ y_{n+1} &= x_n \end{aligned}$$

with initial settings $x_0 = 0$ and $y_0 = 1$. This is exactly the same as in the derivation on page 29 setting $x_n = A_n$ and $y_n = J_n$.

One-Step Machines Based on Combined Formulas

Using the compact notation $G(X_n)$ for a whole set of formulas in the processing unit considerably simplifies the description of seemingly complicated feedback processes. Here is another example, which will become important in chapters 2 and 10:

$$x_{n+1} = \begin{cases} ax_n & \text{if } x \leq 0.5 \\ a(1 - x_n) & \text{if } x > 0.5. \end{cases}$$

Here a denotes a parameter, e.g., $a = 2$ or $a = 3$. Rather than introducing a feedback machine with two formulas and an additional switch, we will rewrite the above system of two equations as a one-step process of the form $x_{n+1} = f(x_n)$, where f is the transformation, whose graph — known as the *tent transformation* — is given in figure 1.15.

The Tent Transformation

The tent transformation is given by $f(x) = ax$ if $x \leq 0.5$ and $-ax + a$ if $x > 0.5$. Here the parameter $a = 3$ has been chosen.

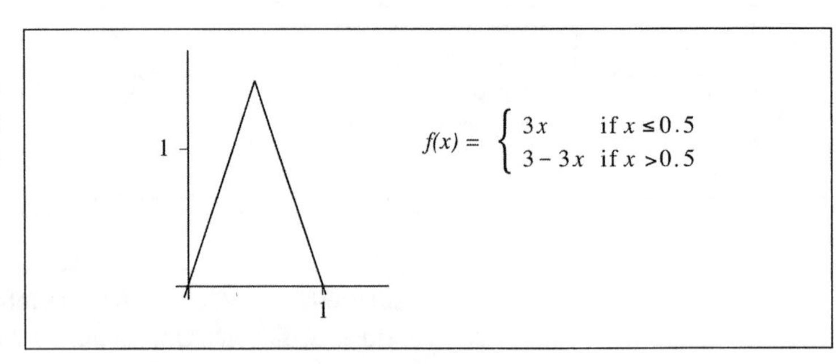

Figure 1.15

The $(3\mathrm{A}+1)$-Problem

This is an algorithm which produces sequences of integers in a most simple way, but yet its unfolding is still not completely understood. Here is the original formulation due to Lothar Collatz:

Step 1: Choose an arbitrary positive integer A.
Step 2: If $A = 1$, then STOP.
Step 3: If A is even, then replace A by $A/2$ and go to step 2.
Step 4: If A is odd, then replace A by $3A + 1$ and go to step 2.

Let us try a few choices for A:

- 3, 10, 5, 16, 8, 4, 2, 1, STOP
- 34, 17, 52, 26, 13, 40, 20, 10, 5, 16, 8, 4, 2, 1, STOP
- 75, 226, 113, 340, 170, 85, 256, 128, 64, 32, 16, 8, 4, 2, 1, STOP

The obvious conjecture is the following: the algorithm comes to a stop no matter what the initial A is. Well, it seems that the larger the initial A is the more steps we have to run until we arrive at 1. Let us try $A = 27$ to verify that guess.

- 27, 82, 41, 124, 62, 31, 94, 47, 142, 71, 214, 107, 322, 161, 484, 242, 121, 364, 182, 91, 274, 137, 412, 206, 103, 310, 155, 466, 233, 700, 350, 175, 526, 263, 790, 395, 1186, 593, 1780, 890, 445, 1336, 668, 334, 167, 502, 251, 754, 377, 1132, 566, 283, 850, 425, 1276, 638, 319, 958, 479, 1438, 719, 2158, 1079, 3238, 1619, 4858, 2429, 7288, 3644, 1822, 911, 2734, 1367, 4102, 2051, 6154, 3077, 9232, 4616, 2308, 1154, 577, 1732, 866, 433, 1300, 650, 325, 976, 488, 244, 122, 61, 184, 92, 46, 23, 70, 35, 106, 53, 160, 80, 40, 20, 10, 5, 16, 8, 4, 2, 1, STOP

Apparently our guess was not correct. Moreover, seeing this example we can really begin to wonder whether all sequences will eventually stop. As far as we know this problem is still unsolved. However, the conjecture has been verified with the aid of computers up to at least $A = 10^9$. Such a test is not as straightforward as we might think, because in the course of the calculations the sequence may exceed the largest possible number which the computer is able to accurately represent. Thus, some variable precision routines must be programmed in order to enlarge the range of numbers representable by a computer.

The algorithm can easily be extended to negative integers. Here are a few examples:

- −1, −2, −1, −2, ... CYCLE of length 2
- −3, −8, −4, −2, −1, ... runs into CYCLE of length 2
- −5, −14, −7, −20, −10, −5, −14, ... CYCLE of length 5
- −6, −3, −8, −4, −2, −1, ... runs into CYCLE of length 2
- −9, −26, −13, −38, −19, −56, ... runs into CYCLE of length 5
- −11, −32, −16, −8, −4, −2, −1, ... runs into CYCLE of length 2

Are there other cycles? Yes indeed:

- −17, −50, −25, −74, −37, −110, −55, −164, −82, −41, −122, −61, −182, −91, − 272, −136, −68, −34, −17, ... CYCLE of length 18

If we modify our algorithm by removing the STOP in Step 1 we also obtain a cycle for $A = 1$:

- 1, 4, 2, 1, ... CYCLE of length 3

and if we also allow $A = 0$:

- 0, 0, ... CYCLE of length 1.

Moreover, we may now write the algorithm as a feedback system:

$$x_{n+1} = \begin{cases} x_n/2 & \text{if } x_n \text{ is an even integer,} \\ 3x_n + 1 & \text{if } x_n \text{ is an odd integer.} \end{cases}$$

Thus, the general questions are: what are the possible cycles of the feedback system, and does any initial choice for x_0 generate a sequence which eventually runs into one of these cycles? This seems to be a moderate question which the enormous body of mathematics should have already answered — or at least be prepared to answer with no great difficulty. Unfortunately, this is not the case, which only shows that there is still a lot to do in mathematics and, moreover, simple-looking problems may be awfully hard to solve; a truly important lesson for life.

MRCM As a One-Step Machine

A more subtle and surprising case is given by our MRCM machines from the last section. They also can be interpreted as one-step machines, which are mathematically described by a single formula of the kind $X_{n+1} = F(X_n)$. Incidentally, in this case F is called the *Hutchinson operator*. We will discuss the details in chapter 5.

Wheel-of-Fortune Machines

While all previous machines are strictly deterministic our last class of machines combines determinism with randomness. Similar to the previous examples, there is a reservoir of different formulas in the processing unit. In addition, however, there is a wheel of fortune, which is used to select one of the formulas at random. The input is a single number (or a pair of numbers), and the output is a new number (or a pair of numbers), which is the result of a formula with values determined by the input. The formula is chosen randomly from a pool at each step of the feedback process. In other words, the output does not just depend on the input, much like in the case of machines with memory. Unfortunately, however, there is no standard trick to rewrite the process as a (deterministic) one-step machine. If the number of formulas is N, then the wheel of fortune has N segments, one for each formula. The size of the segment can be different for each of them in order to accommodate for different probabilities in the random selection mechanism. Random machines like this will furnish extremely efficient decoding schemes for images, which are encoded by the metaphor of a copy machine. This is the content of chapter 6.

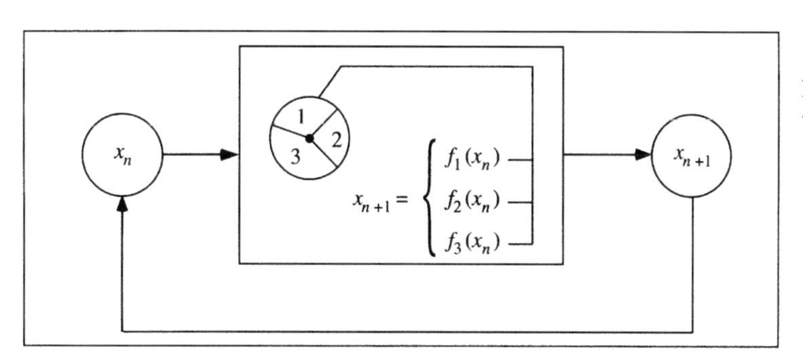

Wheel-of-Fortune Machine

Feedback machine with fortune wheel.

Figure 1.16

The Chaos Game

We want to touch upon a further exciting interpretation of what we learned in the multiple reduction copy machine, and this is another incredible relation between chaos and fractals.

The following 'game' has been termed chaos game by Michael F. Barnsley. At first glance, however, there seems to be no connection whatsoever with chaos and fractals. Let us describe the rules of the game. Well, actually, there is not just one game; there is an infinite number of them. But they all follow the same scheme. We have a die and some simple rules to choose from. Here is one of the games:

Preparations: Take a sheet of paper and a pencil and mark three points on the sheet; label them 1, 2, and 3; and call them bases. Have a die which allows you to pick the numbers 1, 2, and 3 randomly. It is obvious how to manufacture such a die. Take an ordinary die and just identify the faces 6 with 1, 5 with 2, and 4 with 3.

Rules: Start the game by picking an arbitrary point on the sheet of paper and mark it by a small dot. Call it the game point. Now roll the die. If number 2, for example, comes up, consider the line between the game point and base 2 and mark a dot exactly in the middle, i.e., halfway between the game point and base 2. This dot will be the new game point, and we have completed the first cycle of the game. Now repeat (i.e., roll the die again) to randomly get the number 1, 2, or 3; and depending on the result, mark a dot halfway between the last game point and the randomly chosen base.

The first game points shown in figure 1.17 are labeled in the order of their generation by $x_0, x_1, x_2, \ldots$ The chaos game is a very simple scheme to produce a random sequence of points; and as such it appears to be rather boring. But this first impression will immediately change when we see what is going to evolve in this feedback system.

What do you guess the outcome of the game will be after a great many cycles, i.e., what is the picture obtained by the dots $x_0, x_1, \ldots, x_{1000}$? Note that once the game point is inside the trian-

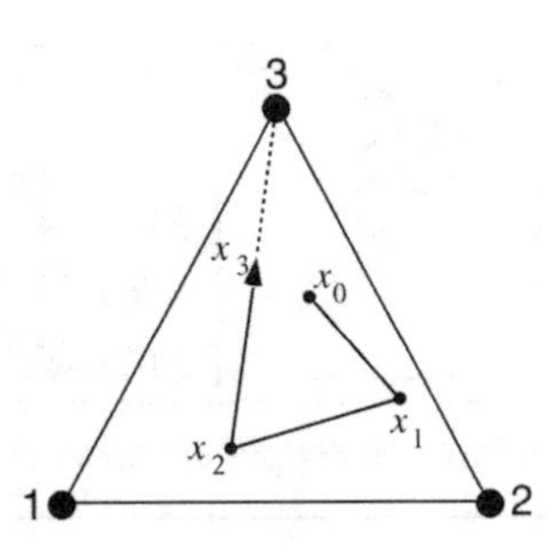

Figure 1.17 : The three base points (vertices of a triangle) and a few iterations of the game point.

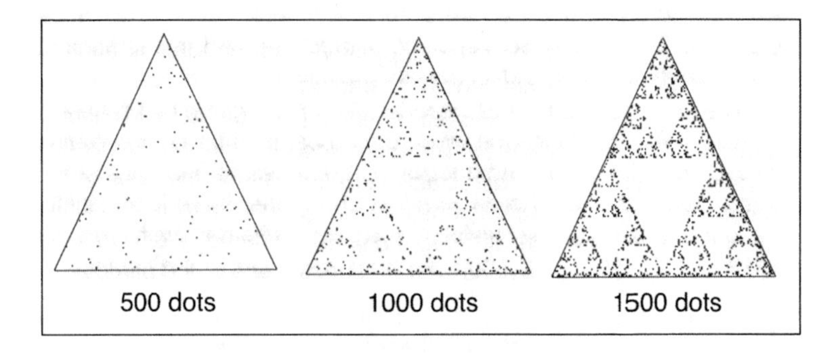

Figure 1.18 : 500, 1000 and 1500 dots of the chaos game.

gle, which is defined by the three base points, the process will remain inside forever. Moreover, it is obvious that sooner or later the game point will land inside this triangle even if we start the game outside. Therefore, intuition seems to tell us that because of the random generation we should expect a random distribution of dots somehow arranged between base 1, 2, and 3. Yes, indeed, the distribution will be random, but not so the picture or image which is generated by the dots (see figure 1.18). It isn't random at all. We see the Sierpinski gasket emerge very clearly; and this an extremely ordered structure — exactly the opposite of a random structure.

At this point this phenomenon seems to be either a small miracle or a funny coincidence, but it is not. Any picture which can be obtained by the MRCM can be obtained by an appropriately adjusted chaos game. In fact, the picture generation can generally be accelerated this way. Moreover, the chaos game is the key to extending the image coding idea which we discussed for the multiple reduction copy machine to grey scale or even color images. This will be the content of chapter 6, which will provide an elementary lesson in probability theory — though one filled with beautiful surprises.

1.4 The Parable of the Parabola — Or: Don't Trust Your Computer

Let us now turn to quadratic iterators. First, we implement the expression $x^2 + c$ in our iterator framework. Here x and c are just numbers; however, with different meanings. To iterate this expression for a fixed (control) value c means this: start with any number x, evaluate the expression, note the result and use this value as new x, evaluate the expression, and so on. Let's look at an example:

Preparation: Choose a number for c, say $c = -2$. Then choose a number x, for example $x = 0.5$.

Iteration: Evaluate the expression for x, obtaining $0.25 - 2 = -1.75$. Now repeat, i.e., evaluate the expression using the result of the first calculation as the new x, i.e., evaluate for $x = -1.75$, which yields 1.0625, and so on.

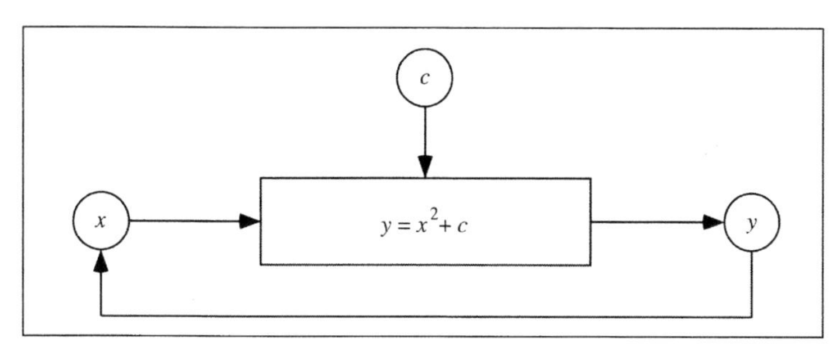

The Quadratic Iterator

The quadratic iterator interpreted as a feedback machine. The processing unit is designed to evaluate $x^2 + c$, given x and c.

Figure 1.19

The table summarizes the results for the first four iterations:

x	$x^2 + c$
0.5	−1.75
−1.75	1.0625
1.0625	−0.87109375
−0.87109375	−1.2411956787109375

Already after four cycles we are running into a problem. Because of the squaring operation the number of decimal places which are needed to represent the successive output numbers essentially doubles in each cycle. This makes it impossible to achieve exact results for more than a few iterations because computers and calculators work with only a finite number of decimal places.[8]

Do Minor Differences Matter?

This is, of course, a common problem in calculator or computer arithmetic, but we don't usually worry about it. In fact, the omnipotence of computers leads us to believe that these minor differences don't really matter. For ex-

[8]For example, the Casio *fx* 7000G has 10, and the Hewlet-Packard 28S has 12 decimal digits of accuracy.

ample, if we compute $2 * (1/3)$ we usually don't worry about the fact that the number 1/3 is not exactly representable by our calculator. We accept the answer 0.6666666667, which, of course, is different from the exact representation of 2/3. Even in a messy calculation, we are usually inclined to take the same attitude, and some put infinite confidence in the calculator or computer in the hope that these minute differences do not accumulate to a substantial error.

Scientists know (or should we say knew) very well that this assumption can be extremely dangerous. They came up with methods, which go back to ideas of Carl Friedrich Gauss (1777–1855; see figure 1.8), to estimate the error propagation in their calculations. With the advent of modern computing this practice has somehow lost ground. It seems that there are at least two reasons for this development.

The Problem of Error Propagation

Modern computing allows scientists to perform computations which are of enormous complexity and are extensive to a degree that was totally unthinkable even half a century ago. In massive computations, it is often true that a detailed and honest error propagation analysis is beyond current possibilities, and this has led to a very dangerous trend. Many scientists exhibit a growing tendency to develop an almost insane amount of confidence in the power and correctness of computers.

If we go on like this, however, we will be in great danger of neglecting some of the great heroes of science and their unbelievable struggle for accuracy in measurement and computation. Let us remember the amazing story of Johannes Kepler's model of the solar system. Kepler devised an elaborate mystical theory in which the six known planets Mercury, Venus, Earth, Mars, Jupiter, and Saturn[9] were related to the five Platonic solids (see figure 1.21).

Brahe and Kepler

Tycho Brahe, 1546–1601 (left) and Johannes Kepler, 1571–1630 (right).

Figure 1.20

[9]These planets were known in ancient times before the invention of the telescope. The seventh planet Uranus was not discovered until 1781 by the amateur astronomer Friedrich Wilhelm Herschel, and Neptune was only discovered in 1846 by Johann Gottfried Galle at the Observatory in Berlin. The ninth and most distant planet Pluto was discovered in 1930 by Clyde William Tombaugh at Lowell Observatory in Flagstaff, Arizona.

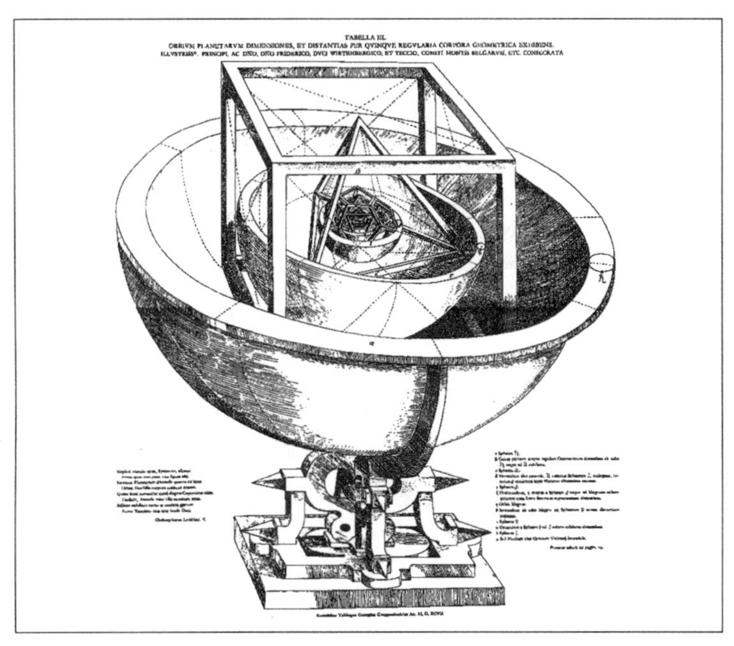

Kepler's Model of the Solar System

Each planet determines a sphere around the sun containing its orbit. Between two successive spheres Kepler inscribed a regular polyhedron such that its vertices would lie on the exterior sphere and its faces would touch the interior sphere. These are the octahedron between Mercury and Venus, the icosahedron between Venus and Earth, the dodecahedron between Earth and Mars, the tetrahedron between Mars and Jupiter, and the cube between Jupiter and Saturn.

Figure 1.21

Small Deviations with Consequences

In attempting to establish his mystical theory of celestial harmony, he had to use the astronomical data available at that time. He realized that the construction of any theory would require more precise data. That data, he knew, was in the possession of the Danish astronomer Tycho Brahe (1546–1601) who had spent 20 years making extremely accurate recordings of the planetary positions. Kepler became Brahe's mathematical assistant in February of 1600 and was assigned a specific problem: to calculate an orbit that would describe the position of Mars. He was given this particular task precisely because that orbit seemed to be the most difficult to predict. Kepler boasted that he would have the solution in eight days. Both the Copernican and the Ptolemaic theories held that the orbit should be circular, perhaps with slight modification. Thus, Kepler sought the appropriate circular orbits for Earth and Mars. In fact, the orbit for Earth, from which all observations were made, had to be determined before one could satisfactorily use the data for the positions of the planets. After years, Kepler found a solution that seemed to fit Brahe's observations. Brahe had died in the meanwhile. However, checking his orbits — by predicting the position of Mars and comparing it with more of Brahe's data — Kepler found that one of his predictions was off by at least 8 minutes of arc, which is about a quarter of the angle diameter of the moon. It would have been most natural to attribute this discrepancy to an error in Brahe's observations, especially because he had spent years in making his calculations. But having worked with Tycho Brahe, he was deeply convinced that Brahe's tables were accurate and therefore continued his attempts to find a solution.

This led him in six more years of difficult calculations filling more than 900 pages, to his revolutionary new model, according to which the orbits of the planets are elliptical rather than circular. In 1609 he published his famous *Astronomica Nova*, in which he announced two of his three remarkable laws. These are the law of elliptical paths, i.e., the orbit of each planet is an ellipse with the sun at one focus, and the law of areas, i.e., during each time interval, the line segment joining the sun and planet sweeps out an equal area anywhere on its elliptical orbit (see figure 1.22). The third law[10] was published later and helped Sir Isaac Newton formulate his law of gravity.

Kepler's First and Second Law

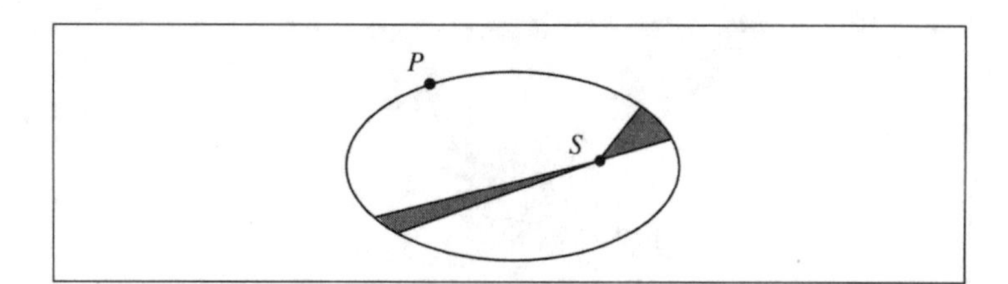

Figure 1.22

Elis Strömgren Computations for the Restricted Three-Body Problem

To demonstrate the enormous leaps which we have made through computers, we present the following instructive example. Figure 1.23 shows the result of computations, which were carried out by 56 scientists under Elis Strömgren at the Observatory of Copenhagen (Denmark) during a period of 15(!) years. The computations show particular solutions to the so-called restricted three-body problem (orbits of a moon under the influence of two planets) and were published in 1925.

Computations of this order of magnitude and complication would keep an ordinary PC busy for just a few days, if that long. This relation documents very well what some people call a scientific and technological revolution, namely, the revolution fueled by the means and power of modern scientific computation.

The Problem of Black Box Software

More and more massive computations are being performed now using black box software packages developed by sometimes very well known and distinguished centers. These packages, therefore, seem to be very trustworthy, and indeed they are. But this doesn't exclude the fact that the finest software sometimes produces total garbage, and it is an art in itself to understand and predict when and why this happens. Moreover, users often don't have a chance to carry out an error analysis simply because they have no access to the black box algorithms. More and more decisions in the development of science and technology, but also in economy and politics, are based on large-

[10]The law of times: the square of the time of revolution of a planet about the sun is proportional to the cube of its average distance from the sun.

Orbits of the Restricted Three-Body Problem

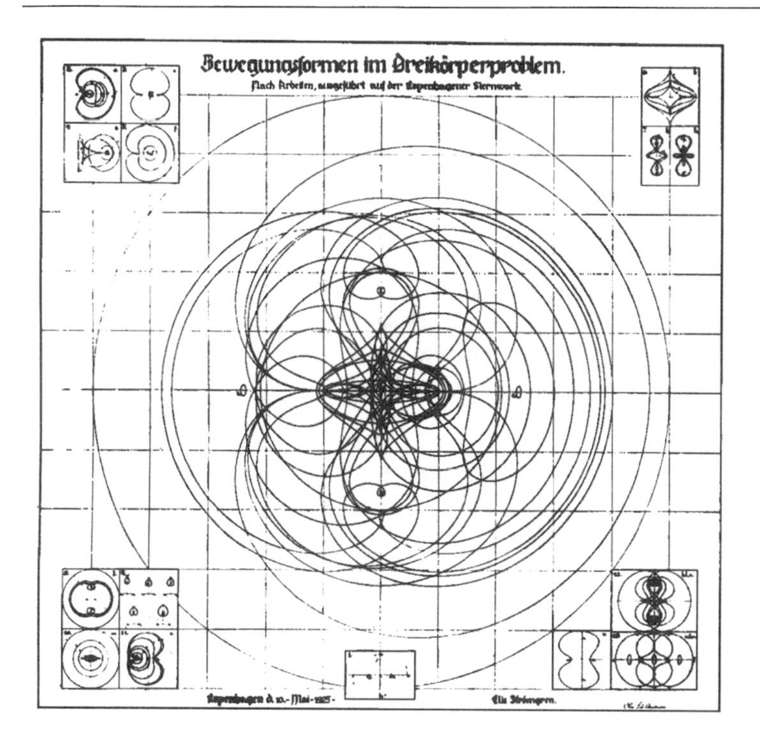

Figure 1.23

scale computations and simulations. Unfortunately, we cannot always take for granted that an honest error propagation analysis has been carried out to evaluate the results. Computer manufacturers find themselves in a race to build faster and faster machines and seem to pay comparatively little attention to the important issue of *scientific calculation quality control*.

Weather Paradigm from James Gleick

To amplify the importance of such considerations we would like to quote from James Gleick's *Chaos, Making a New Science*.[11]

"The modern weather models work with a grid of points on the order of sixty miles apart, and even so, some starting data has to be guessed, since ground stations and satellites cannot see everywhere. But suppose the earth could be covered with sensors spaced one foot apart, rising at one-foot intervals all the way to the top of the atmosphere. Suppose every sensor gives perfectly accurate readings of temperature, pressure, humidity, and any other quantity a meteorologist would want. Precisely at noon an infinitely powerful computer takes all the data and calculates what will happen at each point at 12:01, then 12:02, then 12:03, ... The computer will still be unable to predict whether Princeton, New Jersey, will have sun or rain on a day one month away. At noon the spaces between the sensors will hide fluctuations that the computer will not know about, tiny deviations from the average. By 12:01, those fluctuations will already have created small errors one foot away. Soon the errors will have

[11] James Gleick, *Chaos, Making a New Science,* Viking, New York, 1987.

multiplied to the ten-foot scale, and so on up to the size of the globe."

This phenomenon has become known as the *butterfly effect*, after the title of a paper by Edward N. Lorenz *'Predictability: Does the flap of a butterfly's wings in Brazil set off a tornado in Texas?'* Advanced calculation quality control in weather forecasting means to estimate whether the mechanisms which are at the heart of weather formation are currently in a stable or unstable state. Sooner or later the TV weather man will appear and say: 'Good evening; this is Egon Weatherbring. Because of the butterfly effect, there is no forecast this evening. The atmosphere is in an unstable state, making it impossible to take sufficiently accurate measurements for our computer models. However, we expect it to stabilize in a few days, when we will give you a prediction for the weekend.'

Logistic Feedback Iterator

Feedback machine for the logistic equation. The processing unit is designed to evaluate $p + rp(1 - p)$, given p and r.

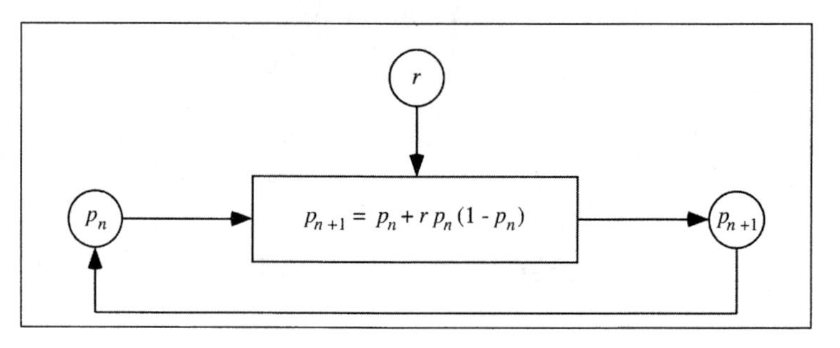

Figure 1.24

Back to the Quadratic Iteration

Let us now return to the iteration of quadratic expressions and look at the expression

$$p + rp(1 - p).$$

First, this expression can be built into an iterator as easily as we did with $x^2 + c$.

The quadratic expression $p + rp(1 - p)$ has a very interesting interpretation and history in biology. It serves as the core of a population dynamics model which in spirit goes back to the Belgian mathematician Pierre François Verhulst[12] and his work around 1845 and which led May to his famous article in *Nature*.[13]

A Population Dynamics Model

What is a population dynamics model? It is simply a law which, given some biological species, allows us to predict the population development of that species in time. Time is measured in increments $n = 0, 1, 2, \ldots$ (minutes, hours, days, years, whatever is appropriate). The size of the population is measured at time n by the actual number in the species P_n. Figure 1.25 shows a typical development.

[12] Two elaborate studies appeared in the *Mémoires de l'Académie Royale de Belgique*, 1844 and 1847.

[13] R. M. May, *Simple mathematical models with very complicated dynamics*, Nature 261 (1976) 459–467.

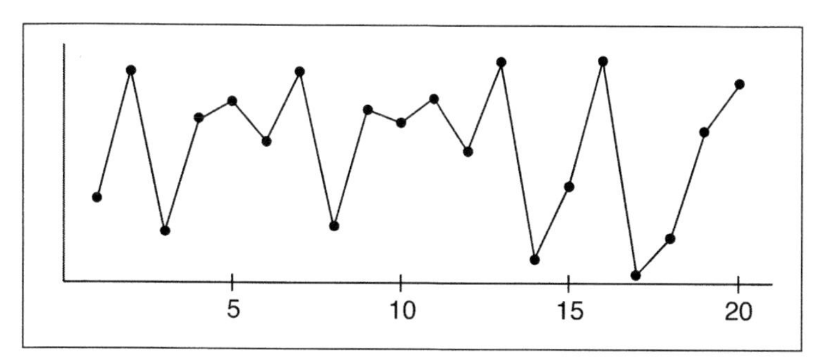

Time Series of a Population

Time series of a population — a typical development. Successive measurements are connected by line segments.

Figure 1.25

Naturally, the size of a population may depend on many parameters, such as environmental conditions (e.g., food supply, space, climate), interaction with other species (e.g., the predator/prey relationship), but also age structure, fertility, etc. The complexity of influences which determine a given population in its growth behavior is illustrated in the following medieval parable.

Of Mice and Old Maids

This year there are a lot of mice in the fields. The farmer is very concerned because he can harvest very little grain. That results in a period of very poor dowries, which leads to there being many more old maids. They all tend to love cats, which increases the cat population dramatically. That in turn is bad for the mice population. It rapidly decreases. This makes for happy farmers and very rich dowries, very few old maids, very few cats, and therefore, back come the mice. And so it goes, on and on.

Though we shouldn't take this too seriously as a model for mice and maid populations, it indicates the potential complexity of population dynamics. It also shows that populations may display cyclic behavior: up → down → up → down, and so on.

The Petri Dish Scenario

A natural modeling approach tries to freeze as many of the population parameters as possible. For example, assume that we have a species of cells which live in a constant environment, e.g., a petri dish with a constant food supply and temperature. Under such conditions we expect that there is some maximal possible population size N which is supported by the environment. If the actual population P at time n, which is P_n, is smaller than N, we expect that the population will grow. If, however, P_n is larger than N, the population must decrease.

Now we want to introduce an actual model. Just as *velocity* is one of the relevant characteristics for the motion of a body, so is *growth rate* the relevant characteristic for population dynamics. The growth rate is measured by the quantity

$$\frac{P_{n+1} - P_n}{P_n}. \tag{1.5}$$

In other words, the growth rate r at time n measures the increase of the population in one time step relative to the size of the population at time n.

Population Growth and Interest

If the population model assumes that the growth rate r is constant, then

$$\frac{P_{n+1} - P_n}{P_n} = r \tag{1.6}$$

for some number r independent of n. Solving for P_{n+1} we obtain the population growth law[14]

$$P_{n+1} = P_n + rP_n = (1 + r)P_n.$$

In such a model the population grows by a factor of $1 + r$ each time step. Indeed, the formula is equivalent to

$$P_n = (1 + r)^n P_0, \tag{1.7}$$

where P_0 is the initial population with which we start our observations at time 0. In other words, knowing r and measuring P_0 would suffice to predict the population size P_n for any point in time without even running the feedback process. In fact, eqn. (1.7) is familiar from computing the accumulation of principal and compound interest, when the rate of interest is r.

The Verhulst Population Model

The most simple population model would assume a constant growth rate, but in that situation we find unlimited growth which is not realistic. In our model we will assume that the population is restricted by a constant environment, but this premise requires a modification of the growth law. Now the growth rate depends on the actual size of the population relative to its maximal size. Verhulst postulated that the growth rate at time n should be proportional to the difference between the population count and the maximal population size, which is a convenient measure for the fraction of the environment that is not yet used up by the population at time n. This assumption leads to the Verhulst population model

$$p_{n+1} = p_n + rp_n(1 - p_n), \tag{1.8}$$

where p_n measures the relative population count $p_n = P_n/N$ and N is the maximal population size which can be supported by the environment. This is just a compact notation for our feedback process. We use integer indices to identify iterates at different time steps (p_n for input, p_{n+1} for output).

Derivation of the Verhulst Model

This population model assumes that the growth rate depends on the current size of the population. First we normalize the population count by introducing $p = P/N$. Thus we interpret $p = 0.06$, for example, as the population size being 6% of its saturation value N. Again we index p by n, i.e., we write p_n to refer to the size at time steps $n =$

[14] Note that the concept of growth rate does not depend on N, i.e., if we use a normalized count $p_n = P_n/N$, then N cancels out in $r = (p_{n+1} - p_n)/p_n$, the equivalent of eqn. (1.6).

$0, 1, 2, 3, \ldots$ Now growth rate is measured by the quantity already given corresponding to the expression (1.5),

$$\frac{p_{n+1} - p_n}{p_n}.$$

Verhulst postulated that the growth rate at time n should be proportional to $1 - p_n$ (the fraction of the environment that is not yet used up by the population at time n). Assuming that the population is restricted by a constant environment the growth should change according to the following table.

Normalized Population p	Growth Rate
small	positive, large
about 1	small
less than 1	positive
greater than 1	negative

In other words,[15]

$$\frac{p_{n+1} - p_n}{p_n} \propto 1 - p_n,$$

or, after introducing a suitable constant r,

$$\frac{p_{n+1} - p_n}{p_n} = r(1 - p_n).$$

Solving this last equation yields the population model eqn. (1.8)

$$p_{n+1} = p_n + rp_n(1 - p_n).$$

The Logistic Model

Following Verhulst this model given by eqn. (1.8) is called the *logistic* model[16] in the literature. There are several interesting remarks. First, note that it is in agreement with the table of growth rates in the technical section above. Second, it seems as if we again have a law which allows us to compute (i.e., predict) the size of the population for any point in time just as in the case of a constant growth rate. But there is a fundamental difference. For most choices of r, there is no explicit solution such as eqn. (1.7) for eqn. (1.6). That is, p_n cannot be written as a formula of r and p_0, as was previously possible. In other words, if one wants to compute p_n from p_0 one really has to run the iterator in figure 1.24 n times. We will begin our experiments with the setting $r = 3$.[17] The table below lists the first three iterates for $p_0 = 0.01$, i.e., the initial population is 1% of the maximal population size N.

[15] The $\propto$ sign means 'proportional to'. The quantity on the left side is a multiple of the expression on the right side.

[16] From *logis* (french) = house, lodging, quarter.

[17] It turns out that $r = 3$ is one of those very special choices for which there is an explicit formula of p_n in terms of r and p_0 (see chapter 10).

p	$p + rp(1-p)$
0.01	0.0397
0.0397	0.15407173
0.15407173	0.545072626044...

For the same reasons as we noted when we iterated $x^2 + c$, we observe that continued iteration requires higher and higher computational accuracy if we insist on exact results. But that appears to be unnecessary in our population dynamics model. Isn't it enough that we get some idea for how the population develops? Shouldn't we be satisfied with an answer which is reliable up to three or four digits? After all, the third decimal place controls only some tenth of a percent in our model. Thus, it seems, there is no reason not to trust that a computer or calculator will do the job. But this is definitely not true as a general rule — computed predictions in our model can be totally wrong.

The Lack of Predictability

This is at the heart of what scientists nowadays call the presence of chaos in deterministic feedback processes. One of the first ones who became aware of the significance of these effects was the MIT meteorologist Lorenz in the late fifties.[18] He discovered this effect — the lack of predictability in deterministic systems — in mathematical systems which were designed to test long-range weather predictions.

The Lorenz Experiment

As so often is the case with new discoveries, Lorenz stumbled onto the effect quite by accident. In his own words,[19] the essential part of the events were as follows.

"Well, this all started back around 1956 when some [...] methods of [weather] forecasting had been proposed as being the best methods available, and I didn't think they were. I decided to cook up a small system of equations which would simulate the atmosphere, solve them by computers which were then becoming available, and to then treat the output of this as if it were real atmospheric observational data and see whether the proposed method applied to it would work. The big task here was to get a system of equations which would produce the type of output that we could test the things on because it soon became apparent that if the solution of these equations were periodic, the proposed method would be trivial; it would work perfectly. So we had to get a system of equations which would have solutions which would not be periodic, which would not repeat themselves, but would go on irregularly and indefinitely. I finally found a system of twelve equations that would do this and found that the proposed method didn't work too well when applied to it, but in the course of doing this I wanted to examine some of the results in more detail. I had a small computer in my office then, so I typed in some of the intermediate conditions which the computer had printed out as new initial conditions to start another computation and went out for a while. When I came back I found that the solution was not the same as the one I had before;

[18] Lorenz, E. N., *Deterministic non-periodic flow,* J. Atmos. Sci. 20 (1963) 130–141.

[19] In: H.-O. Peitgen, H. Jürgens, D. Saupe, C. Zahlten, *Fractals — An Animated Discussion,* Video film, Freeman 1990. Also appeared in German as *Fraktale in Filmen und Gesprächen,* Spektrum der Wissenschaften Videothek, Heidelberg, 1990.

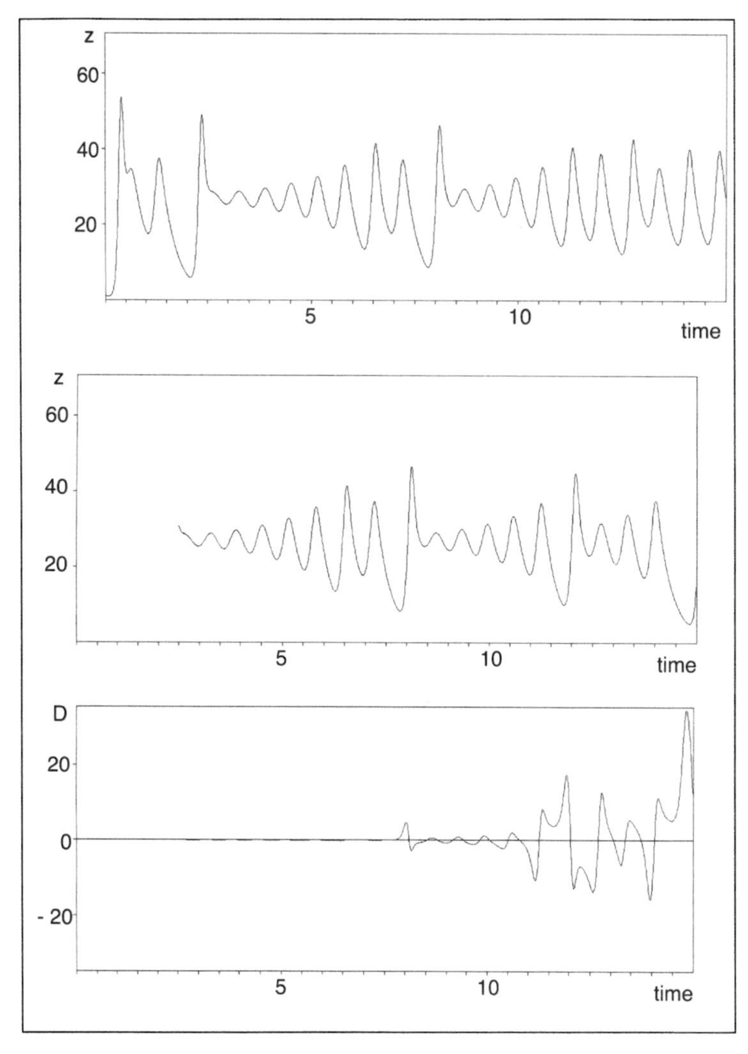

The Original Lorenz Experiment

Numerical integration of the Lorenz equation (top). It is recomputed starting at $t = 2.5$ with an initial value taken from the first integration; however, with a small error introduced (middle). The error increases in the course of the integration. The difference between the two computed results (signals) becomes as large as the signal itself (bottom).

Figure 1.26

the computer was behaving differently. I suspected computer trouble at first, but I soon found that the reason was that the numbers that I had typed in were not the same as the original ones, these [former ones] had been rounded off numbers and the small difference between something retained to six decimal places and rounded off to three had amplified in the course of two months of simulated weather until the difference was as big as the signal itself, and to me this implied that if the real atmosphere behaved as in this method, then we simply couldn't make forecasts two months ahead, these small errors in observation would amplify until they became large."

Sensitive Dependence on Initial Conditions

In other words, even if the weather models in use were absolutely correct — that is, as models for the physical development of the weather — one cannot predict with them for a long time period. This effect is nowadays called *sensitive dependence on initial conditions*. It is one of the central ingredients of what is called deterministic chaos.Our next experiment imitates Lorenz's historical one in the simplest possible way. He had used much more delicate feedback systems consisting of twelve ordinary differential equations; we simply use the logistic equation.[20] We iterate the quadratic expression $p+rp(1-p)$ for the constant $r=3$ and the initial value $p_0=0.01$ (see table 1.27). In the left column we run the iteration without interruption, while in the right column we run the iteration until the 10th iterate, stop, truncate the result 0.7229143012 after the third decimal place, which yields 0.722, and continue the iteration as if that were the last output. The experiment is carried out on a Casio *fx*–7000G pocket calculator.

The Lorenz Experiment for the Population Model

In two series of iterations (same starting point) one of the outputs of the second series is truncated to three decimal places and taken as input for the following iteration. Soon afterwards the two series of numbers lose all correlation. Underlined are those first digits which are the same on both sides.

Table 1.27

Evaluations	Without Interrupt	With Interrupt and Restart
1	0.0397	0.0397
2	0.15407173	0.15407173
3	0.5450726260	0.5450726260
4	1.288978001	1.288978001
5	0.1715191421	0.1715191421
10	0.7229143012	0.7229143012
10	0.7229143012	restart with 0.722
15	1.270261775	1.257214733
20	0.5965292447	1.309731023
25	1.315587846	1.089173907
30	0.3742092321	1.333105032
100	0.7355620299	1.327362739

Trustworthy as Rolling Dice

Now, of course, the 10th iterates of the two processes agree only in 3 decimal places and it is no surprise that there is a disagreement also in the 15th iterates. But it is a surprise — and again that indicates chaos in the system, or in the words of Lorenz, it "demonstrates lack of predictability" — that higher iterates appear totally uncorrelated. The layout of the experiment suggests that the column on the left is more trustworthy. But that is absolutely misleading, as we will see in the forthcoming experiments. Eventually the iterations become as trustworthy as if we had obtained them with a random number generator or by rolling dice. In fact, the Polish mathematician Stan Ulam discovered that remarkable property when he constructed numerical random number generators for the first electronic computer ENIAC in the late forties in connection with large-scale computations for the Manhattan Project.

[20] In fact, later on, Lorenz himself discovered that his system is strongly related to the logistic equation.

1.5 Chaos Wipes Out Every Computer

Being very skeptical, we might conclude that maybe the error — truncation after 3 decimal places — which we introduced in Lorenz's experiment was too large. Someone might conjecture that the strange behavior of the iteration would disappear if we repeated the experiment with much smaller errors in the starting values. We would not have wasted our time in calculating if that were the case. The fact is that no matter how small a deviation in the starting values we choose, the errors will accumulate so rapidly that after relatively few steps the computer prediction is worthless. To fully grasp the importance of the phenomenon, we propose a further experiment. This time we do not change the starting values for the iteration, but we use calculators produced by two different manufacturers. In other words, we conjecture that sooner or later their predictions will massively deviate from each other.

What happens if we actually carry out the iteration with two different fixed accuracy devices? What is the result after 10 iterations, or 20, or even 50? This seems to be a dull question. Doesn't one just have to evaluate 10, 20 or 50 times? Yes, of course, but the point is that the answer depends very much on the nature of the computation.

The Computer Race into Chaos

To demonstrate what we mean when we say that things depend on the computation, let us compare the results obtained by two different calculators, say a Casio and an HP. Starting with 1, let's look at 2, 3, 4, 5, 10, 15, 20, ..., 50 repeated feedback evaluations (= iterations); see table 1.28 and figure 1.29.

Evaluations	Casio	HP
1	0.0397	0.0397
2	0.15407173	0.15407173
3	0.5450726260	0.545072626044
4	1.288978001	1.28897800119
5	0.1715191421	0.171519142100
10	0.7229143012	0.722914301711
15	1.270261775	1.27026178116
20	0.5965292447	0.596528770927
25	1.315587846	1.31558435183
30	0.3742092321	0.374647695060
35	0.9233215064	0.908845072341
40	0.0021143643	0.143971503996
45	1.219763115	1.23060086551
50	0.0036616295	0.225758993390

Table 1.28 : Two different calculators at the same job do not produce the same results.

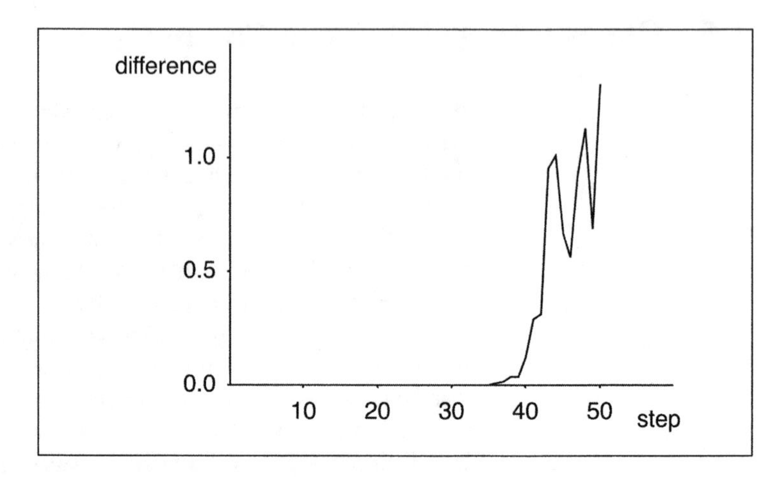

Figure 1.29 : Plot of the difference between the computed iteration values of HP and Casio.

While the first and second generation of our populations are predicted exactly the same by both calculators, they totally disagree at the 50th generation: the Casio predicts that the population is about 0.4% of the saturation value, while the HP tells us that it should be about 23%! How is that possible?

We carefully check our programs, and indeed they both are correct and use exactly the same formula $p + rp(1 - p)$. The only difference is, of course, that the Casio is restricted to 10 decimals, while the HP has 12. In other words, neither one is able to exactly represent the iterations 3 and higher. Indeed, the second iterate needs 8 decimals and therefore the third iterate would need 16, etc. Thus, there are unavoidable cut-off errors, which don't seem to matter much. At least that is suggested if we look at iterations 4 and 5. The results of the Casio and HP agree in 10 decimal places. However, for the 10th iterate we observe that the Casio and HP are in disagreement about the 10th decimal place: the Casio proposes 2 and the HP insists on 7 (see table 1.28). This suggests that we should look at the iterates between 5 and 10 in detail (table 1.30).

Indeed, while for the 5th iterate both calculators agree at the 10th decimal, they mildly disagree at the 10th decimal for the 6th iterate. The difference being 2×10^{-11}, which is so minute that one certainly finds no reason to bother with it. Looking further at the 10th iterate, however, we see how this tiny disagreement has grown to 5×10^{-10}, which is still so small that one is inclined to neglect it. But for our records let's note that the disagreement has grown by an order of magnitude (a factor of 10).

When we go back to table 1.28 and now look at 15, 20, 25, 30, 35, ... iterations we seem to observe how the tiny little infection which we noticed in the 10th decimal for the 6th iterate has migrated through all decimal places; i.e., after 40 iterations the initial tiny disagreement

Evaluations	Casio	HP
5	0.1715191421	0.171519142100
6	0.5978201201	0.597820120080
7	1.319113792	1.31911379240
8	0.05627157765	0.056271577700
9	0.2155868393	0.215586839429
10	0.7229143012	0.722914301711

Table 1.30 : The critical iterations where the two calculators begin to show signs of differing behavior.

has been amplified by a factor of 10^{10}!

But why do we say 'seem to observe'? Well, comparing the Casio and the HP we are inclined to trust the HP more because it works with higher accuracy (two extra decimal places). In other words, we tend to accept the HP answer for the 40th iterate and conclude that the Casio is totally off. But this is a little premature.

If the Casio is wrong — and of course at least one of the two must be totally wrong — we cannot assume that the error is due to a serious flaw of its design. Rather, the failure is due to a principal mathematical problem. And, of course, for that reason the HP is subject to the same disease, but with a slight delay because of its higher accuracy. In other words, all we can say for sure is that one of the two calculators is totally wrong in its predictions despite the fact that the deterministic process is very simple. But it is also very likely that both calculators are off. This dramatic effect is the unavoidable consequence of finite accuracy arithmetic and would produce the same results and dramatic effects on multimillion-dollar supercomputers.

The minute differences in the two calculators, i.e., their different accuracies, accumulate so rapidly that the predictive power of the calculators (computers) evaporates. But, believe it or not, this is still not the end of the story. Things are even wilder than we have seen so far.

We now run our example of the quadratic dynamic law, $p + rp(1-p)$, for $r = 3$ and the initial condition $p_0 = 0.01$ (as before) on one calculator (Casio) in two comparative runs. So what is the difference? If we keep all data the same and use an identical calculator, the only thing we can possibly change is the programming code in the algorithm. And there the only thing we can possibly change is the way we evaluate the quadratic expression. And even this almost ridiculously small change matters as demonstrated in table 1.31.

At first one doesn't trust one's eyes. Look at the 12th iterate. It is true. There it creeps in; the virus of unpredictability strikes again. Hereafter we are not surprised at all to see our prediction become completely unreliable.

$\mathbf{p + rp(1 - p)}$ Versus $\mathbf{(1 + r)p - rp^2}$

Two different implementations of the same quadratic law on the same calculator are not equivalent. We compare the results: there is total agreement until the 11th iterate. Then, in the 12th iterate a minute disagreement — check the last three places — 734 versus 724.

Evaluations	$p + rp(1 - p)$	$(1 + r)p - rp^2$
1	0.0397	0.0397
2	0.15407173	0.15407173
3	0.5450726260	0.5450726260
4	1.288978001	1.288978001
5	0.1715191421	0.1715191421
10	0.7229143012	0.7229143012
11	1.323841944	1.323841944
12	0.03769529734	0.03769529724
13	0.146518383	0.1465183826
14	0.5216706225	0.5216706212
15	1.270261775	1.270261774
20	0.5965292447	0.5965293261
25	1.315587846	1.315588447
30	0.3742092321	0.3741338572
35	0.9233215064	0.9257966719
40	0.0021143643	0.0144387553
45	1.219763115	0.0497855318

Table 1.31

Sooner or Later Predictability Breaks Down

If the first experiments didn't convince you that chaos is unbeatable, the last experiment should have taught you the lesson. With finite accuracy computing there is no cure for the damaging effects of chaos. Predictability sooner or later breaks down.

Now you may argue that such phenomena are very rare, or easy to detect or to foresee. Wrong! Since chaos (= breakdown of predictability) has become fashionable in the sciences, there has been literally a flood of papers demonstrating that chaos is more like the rule in nature, while order (= predictability) is more like the exception. But doesn't this contradict the phenomenal success of space missions, for example, the Voyager II mission which left our planetary system after 12 years of travel when it passed Neptune, only a few kilometers off the predicted path? No, it does not. There are strong hints that even the motion of celestial bodies is subject to the same phenomena — sooner or later... Besides, since chaos has entered upon the scientific stage — and despite its amazing historical roots in the work of Henri Poincaré at the turn of last century, this is essentially an achievement made possible by the new powers provided to science by computers — there has been remarkable progress in the deeper understanding of phenomena such as turbulence, fibrillation of the heart, laser instabilities, population dynamics, climate irregularities, brain function anomalies, etc.

Chaos Will Not Resist Deeper Understanding

Moreover, and this is truly fascinating and gives rise for a lot of hope that chaos will not resist deeper understanding forever, it has recently become clear that chaos likes to follow certain very stable patterns. This again was discovered, strangely enough, by means of computers, which otherwise seem so vulnerable to chaos. This is the main subject of chapter 11 where we will discuss the ground-breaking work of Mitchell Feigenbaum, Siegfried Großmann and Stefan Thomae, and Edward Lorenz, as well as Robert May, all of whom found order in chaos as well as routes from order into chaos.

The quadratic law $p + rp(1-p)$ which we have explored so far is just one of a universe of feedback systems which display very complicated behavior. The expression $x^2 + c$ is another example, only in a trivial sense, however. If we carried out experiments analogous to that in table 1.28 for $c = -2$, we would observe exactly the same behavior. The reason is simply that the two quadratic processes can be identified by means of a coordinate transformation, i.e., they really are the same.

Equivalence of $\mathbf{x^2 + c}$ and $\mathbf{p + rp(1+p)}$

Using indices to identify iterates at different times (index n for input, index $n+1$ for output), we can write the two quadratic laws as

$$p_{n+1} = p_n + rp_n(1-p_n), \quad n = 0, 1, 2, 3, \ldots \tag{1.9}$$

and

$$x_{n+1} = x_n^2 + c, \quad n = 0, 1, 2, 3, \ldots \tag{1.10}$$

We now verify that with the setting of

$$c = \frac{1-r^2}{4} \quad \text{and} \quad x_n = \frac{1+r}{2} - rp_n, \tag{1.11}$$

the formulas (1.9) and (1.10) are identical. More precisely, we will show that if

$$x_0 = \frac{1+r}{2} - rp_0 \tag{1.12}$$

holds, then

$$x_n = \frac{1+r}{2} - rp_n \tag{1.13}$$

holds also for $n = 1, 2, 3, \ldots$. In other words, the iteration of the population dynamics model (1.9) and the iteration of quadratic formula (1.11) using $c = (1-r^2)/4$ describe the same dynamical process. The only difference is that the x- and p-values should be interpreted using different scales (given by eqn. (1.13)). The picture is similar to a physics experiment in which the temperature is measured by two physicists, one using degrees centigrade and the other using degrees Fahrenheit. The numbers they come up with are different yet there is a very simple relation. A temperature of p degrees Fahrenheit corresponds to

$$x = \frac{5}{9}(p - 32)$$

degrees centigrade. This relation is analogous to eqn. (1.13).

We have to examine whether p_{n+1} from eqn. (1.9) can be transformed into x_{n+1} as in eqn. (1.10) when we make use of eqn. (1.11).

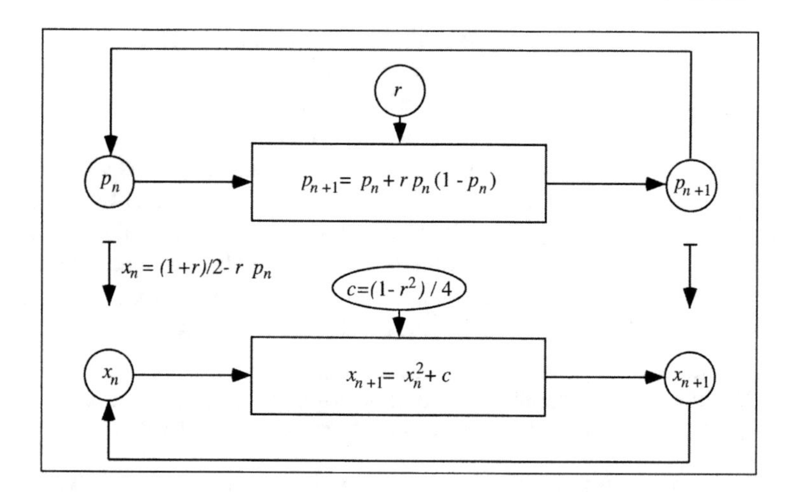

Figure 1.32 : Two quadratic iterators running in phase are tightly coupled by the transformations indicated.

Formally, this is a proof by induction. If we apply eqn. (1.11) to p_{n+1} we get

$$x_{n+1} = \frac{1+r}{2} - rp_{n+1}$$

and using eqn. (1.9) for p_{n+1}

$$x_{n+1} = \frac{1+r}{2} - rp_n - r^2 p_n(1-p_n).$$

On the other hand, eqn. (1.10) with x_n and c transformed by eqn. (1.11) yields

$$x_{n+1} = \left(\frac{1+r}{2} - rp_n\right)^2 + \frac{1-r^2}{4}.$$

Upon resolving the right-hand sides of both equations we see that they are in fact the same, namely,

$$r^2 p_n^2 - r(1+r)p_n + \frac{1+r}{2}.$$

Note that $r = 3$ corresponds to $c = -2$. This explains, indeed, that we may observe exactly the same behavior in both processes. Let us verify the equivalence of the two processes with some examples. If $r = 3$ and $p_0 = 0.01$, then $c = -2$ and $x_0 = 1.97$, according to eqn. (1.11). Computing x_n for $n = 10$ on a Casio yields $x_{10} = -0.1687429036$. After transforming x_{10} according to eqn. (1.11) we obtain $p_{10} = 0.7229143012$, which is exactly the value we can read from table 1.28 for the 10th iterate.

If, however, we repeat the same for 50 rather than 10 iterations we obtain $x_{50} = 0.2310122906$ and $p_{50} = 0.2550655142$ (according to eqn. (1.11)), which is entirely different from the 50th iteration in table 1.28. This does not disprove the validity of the equivalence of the two processes, but rather reaffirms our earlier finding that even two different ways of numerical evaluation eventually lead to disagreeing results, i.e., chaos has hit again.

Why Study Different Quadratic Iterators?

Why should we look at $x^2 + c$ when the dynamics for iterators to this formula are the same (up to some coordinate transformation) as for $p + rp(1 - p)$? There are many different problems to be solved with quadratic iterations, and indeed, in principle it does not matter which quadratic is taken because all are equivalent. However, the mathematical formulation of these problems and their solutions will be more illuminating (and perhaps less complex) depending on the particular quadratic we pick. Therefore, in each case we may choose the quadratic transformation which suits best the problem on hand.

Let us return for a moment to the question of whether there is an easy answer to why we see chaotic behavior. It seems to be obvious that whenever there is an inaccuracy in the feedback process, this error is amplified, i.e., the error propagation builds up dramatically, due to the quadratic character of the expressions. In other words, one might guess that the squaring operation is the cause of the problem. Yes, that is indeed the case; but in a much more subtle way than we might think. For the complete story, please refer to chapter 10. But let us convince ourselves that squaring alone does not explain anything! Let us look at two more simple experiments to illustrate the difficulties.

A Wild Iterator Becomes Very Tame

In our last experiment with the quadratic iterator $x_{n+1} = x_n^2 + c$ we fixed $c = -2$ and started with $x_0 = 1.97$. How about $x_0 = 1$, for example? Iteration now yields: $1, -1, -1, -1, \ldots$ And, for another example, iteration for $x_0 = 2$ yields $2, 2, 2, \ldots$ In other words, we have found initial values for x_0 with which the same wild iterator behaves perfectly tamely. We could demonstrate, however, that this is the exception, i.e., for almost all x_0 from $[-2, +2]$ one observes chaotic behavior. For example if we start with $x_0 = 1.999999999$, i.e., with a tiny deviation from $x_0 = 2$, then we will have the familiar messy behavior back again, provided we just allow sufficiently numerous iterates. This already shows that the error analysis problem is not a straightforward one, and this becomes even more apparent in our next experiment.

Let us now shift gears in our iterator by setting the control parameter to $c = -1$, rather than the previous value $c = -2$. If squaring alone were the secret to understand the lack of predictability, we should make very similar observations. Let us run an iteration where we start with $x_0 = 0.5$ (see table 1.33). Here, we observe that after a number of iterations the process settles down to a repetition of two values: 0 and -1. In fact, repeating the iteration with other initial values, for example, $x_0 = 1$, or $x_0 = 0.75$, or $x_0 = 0.25$ yields the same final answer. The feedback process is now in a perfectly stable mode.

Seventeen Iterations of $x^2 - 1$

First seventeen iterates for the starting value $x_0 = 0.5$.

Table 1.33

Evaluations	x	$x^2 - 1$
1	0.5	−0.75
2	−0.75	−0.4375
3	−0.4375	−0.80859375
4	−0.80859375	−0.3461761475
5	−0.3461761475	−0.8801620749
6	−0.8801620749	−0.2253147219
7	−0.2253147219	−0.9492332761
8	−0.9492332761	−0.0989561875
9	−0.0989561875	−0.9902076730
10	−0.9902076730	−0.0194887644
11	−0.0194887644	−0.9996201881
12	−0.9996201881	−0.0007594796
13	−0.0007594796	−0.9999994232
14	−0.9999994232	−0.0000011536
15	−0.0000011536	−1.0000000000
16	−1.0000000000	−0.0000000000
17	−0.0000000000	−1.0000000000

Stable Cycle in the Logistic Iterator

The same stability should occur in the iteration of the logistic equation if we choose the parameter r and the initial population p_0 appropriately. Solving eqn. (1.11) for r and p with the choice $c = -1$ yields

$$r = \sqrt{1-4c} = \sqrt{5},$$

$$p = \frac{1+r}{2r} - \frac{x}{r} = \frac{1-2x+\sqrt{1-4c}}{2\sqrt{1-4c}} = \frac{1-2x+\sqrt{5}}{2\sqrt{5}}.$$

Thus, for this parameter setting there is a stable cycle of two points corresponding to $x = 0$ and $x = -1$, namely,

$$p = \frac{1+\sqrt{5}}{2\sqrt{5}} = 0.723606797\ldots$$

and

$$p = \frac{3+\sqrt{5}}{2\sqrt{5}} = 1.17082039\ldots$$

We have seen this kind of behavior already in the discussion of the MRCM, where we always obtained a final image which was independent of the initial image. This property is called stability, and is very desirable in many cases. In these cases a process is predictable, and small errors along the way disappear or decay, i.e., they can be neglected. In other words, these are processes where a computer with finite precision arithmetic is a perfect tool and cannot fail.

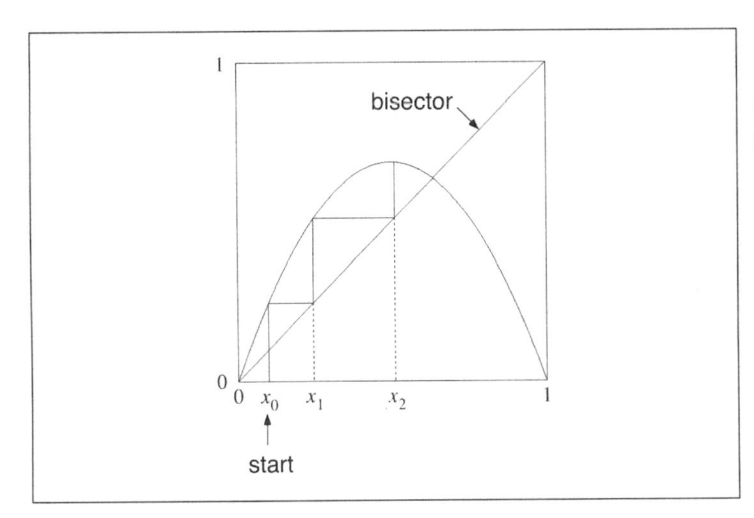

Principle of Graphical Iteration

The first steps in the graphical iteration of $x_{n+1} = ax_n(1 - x_n)$.

Figure 1.34

So far we have been able to detect the stable or unstable state of an iteration by carefully monitoring the numerical values of competing runs of the feedback process. For the particular class of quadratic processes, there is another way to detect the different kinds of behavior, which is much more visual and immediate.

Graphical Iteration of Feedback Processes

We will restrict ourselves to the iteration

$$x_{n+1} = ax_n(1 - x_n).$$

Note that if we consider the graph corresponding to the function $y = ax(1-x)$, we just obtain a parabola, which passes through the points $(0, 0)$ and $(1, 0)$ independent of the choice of the parameter a. The vertex of the parabola, which is always located at $x = 0.5$, has height $a/4$. This quadratic iteration again is equivalent to the logistic equation, or to $x_{n+1} = x_n^2 + c$. We use it here because it produces iterates which always stay in the range from 0 to 1, provided the initial value x_0 is also in this range. There is an efficient way to construct the sequence $x_0, x_1, x_2, \ldots$ by a ruler based on the graph of the parabola leading to a nice graphical visualization of the iteration, called *graphical iteration*.

To describe the iteration we plot the graph of $y = ax(1 - x)$ and draw the bisector (diagonal) (see figure 1.34). We start by marking x_0 on the x-axis. Now we draw a vertical line segment from x_0 until we hit the graph. From that point we draw a horizontal line segment until we hit the bisector. From there we continue to draw a vertical line segment until we hit the graph, and so on.

Why does this procedure work? Simply because points on the bisector have the same distance from both axes. With the aid of this method one can literally see whether the elementary iterations are in the stable or unstable state. Figure 1.35 shows the graphical iteration method for three different values of

a in the stable range of the process. For $a = 1.45$, we observe that the iteration creates a staircase which runs into the point of intersection between the graph and the bisector. For $a = 2.75$ the iteration generates a spiral which converges to the point of intersection between the graph and the bisector. For $a = 3.2$ we see how the iteration determines cyclic behavior.

Mixing

Figure 1.36 shows the iteration for $a = 4$ and one starting value x_0, however different numbers of steps of the iteration. From left to right we show the iteration after $10, 50$ and 100 steps. Apparently the process does not come to rest. Rather, it occupies the entire available space. This phenomenon, called mixing, is an indicator for the unstable state of the system. However, a rigorous analysis has to use much more subtle means to distinguish genuine instability from just a cycle of very high order. For example, what is the difference between the cobwebs in figure 1.35 ($a = 3.2$) and figure 1.36 ($a = 4$)?

The Equivalence of Graphical Iteration and the Population Model

We have already shown that the iteration process for the logistic equation is equivalent to the iteration of $x^2 + c$ (see page 53). Here we show the equivalence to the iteration based on $ax(1 - x)$ as used in the graphical method. Recall that

$$p_{n+1} = p_n + rp_n(1 - p_n). \tag{1.14}$$

We show that this is the same as

$$x_{n+1} = ax_n(1 - x_n) \tag{1.15}$$

when using the identification

$$x_n = \frac{r}{r+1} p_n \text{ and } a = r + 1. \tag{1.16}$$

We compute x_{n+1} using eqn. (1.16) and the logistic iteration and then check if the result agrees with the iteration using eqn. (1.15). We have

$$\begin{aligned} x_{n+1} &= \frac{r}{r+1} p_{n+1} = \frac{r}{r+1} (p_n + rp_n(1 - p_n)) \\ &= rp_n - \frac{r^2}{r+1} p_n^2 \end{aligned}$$

and on the other hand

$$\begin{aligned} x_{n+1} &= ax_n(1 - x_n) = (r+1) \frac{r}{r+1} p_n \left(1 - \frac{r}{r+1} p_n\right) \\ &= rp_n - \frac{r^2}{r+1} p_n^2. \end{aligned}$$

Thus we have that iterating $p_{n+1} = p_n + rp_n(1 - p_n)$ is really the same as iterating $x_{n+1} = ax_n(1 - x_n)$. In fact, the iteration of any quadratic polynomial is equivalent to the iteration of the logistic equation (with a properly chosen parameter). The proof of this assertion is similar to the above derivation.

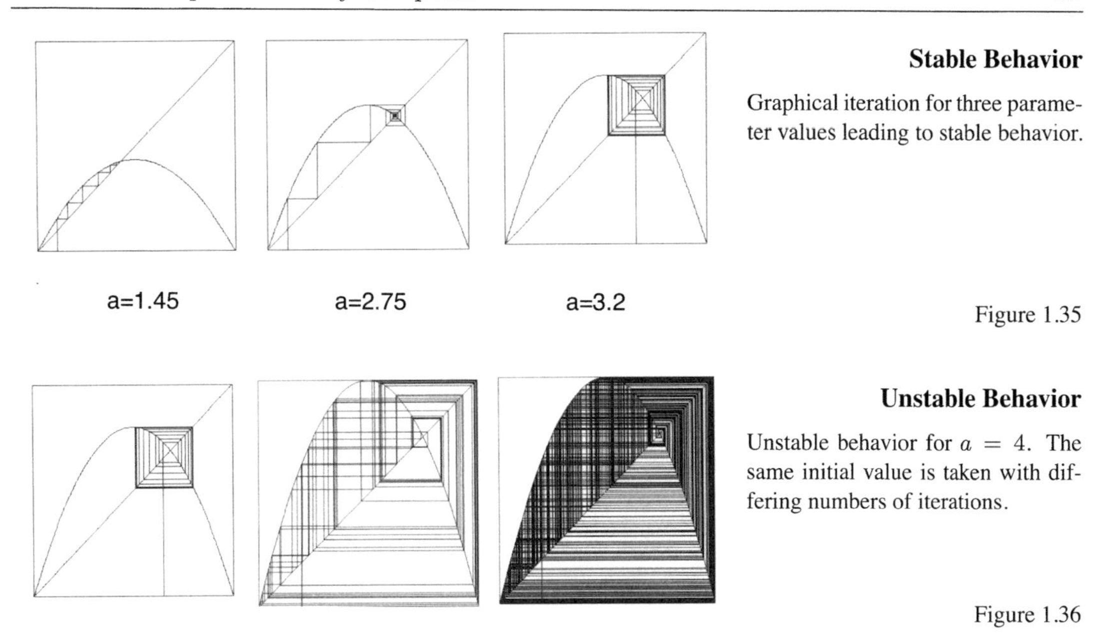

Stable Behavior

Graphical iteration for three parameter values leading to stable behavior.

Figure 1.35

Unstable Behavior

Unstable behavior for $a = 4$. The same initial value is taken with differing numbers of iterations.

Figure 1.36

Analysis of Chaos is Hard

The analysis of the quadratic feedback process is so difficult because the stable and unstable states are interwoven in an extremely complicated pattern. The feedback process can behave tamely or wildly depending solely on the setting of the control parameter.

This is much like the case with the systems used to predict weather. There are states where prediction is very reliable (like high pressure systems over the Utah deserts); and then again there are situations where any prediction breaks down, and where sophisticated multimillion dollar equipment and the brightest minds are as successful in their prediction as any Tom, Dick or Harry would be when predicting that the weather tomorrow will be the same as today. In other words, one and the same system can potentially behave both ways and there are transitions from one into the other. This is the core of the mathematics or the science of chaos. The fact that this theme is also intimately connected with fractals is the content of chapters 10 and 11. The best way to express this relation is to say that fractal geometry is the geometry of chaos.

Chapter 2

Classical Fractals and Self-Similarity

The art of asking the right questions in mathematics is more important than the art of solving them.

Georg Cantor

Mandelbrot is often characterized as the father of fractal geometry. Some people, however, remark that many of the fractals and their descriptions go back to classical mathematics and mathematicians of the past like Georg Cantor (1872), Giuseppe Peano (1890), David Hilbert (1891), Helge von Koch (1904), Waclaw Sierpinski (1916), Gaston Julia (1918), or Felix Hausdorff (1919), to name just a few. Yes, indeed, it is true that the creations of these mathematicians played a key role in Mandelbrot's concept of a new geometry. But at the same time it is true that they did not think of their creations as conceptual steps towards a new perception or a new geometry of nature. Rather, what we know so well as the Cantor set, the Koch curve, the Peano curve, the Hilbert curve and the Sierpinski gasket, were regarded as exceptional objects, as counter examples, as 'mathematical monsters'. Maybe this is a bit overemphasized. Indeed, many of the early fractals arose in the attempt to fully explore the mathematical content and limits of fundamental notions (e.g., 'continuous' or 'curve'). The Cantor set, the Sierpinski carpet and the Menger sponge stand out in particular because of their deep roots and essential role in the development of early topology.

Abnormal Monsters or Typical Nature?

But even in mathematical circles their profound meaning had been somewhat forgotten, and they were seen as shapes, intended to demonstrate the deviation from the familiar rather than to typify the normal. Then Mandelbrot demonstrated that these early mathematical fractals in fact have many features in common with shapes found in nature. Thus the title *The Fractal Geometry of Nature*[1] of his book in 1982. In other words, we could say that Mandelbrot turned the manifested mathematical interpretation and value of these fantastic

[1]Freeman, 1982.

Cauliflower Self-Similarity

The self-similarity of an ordinary cauliflower is demonstrated by dissection and two successive enlargements (bottom). The small pieces look similar to the whole cauliflower head.

Figure 2.1

inventions upside down. But in fact, he did much more. The best way to describe his contribution is to say that, indeed, some characters, such as the Cantor set, were already there. But he went on to develop the language into which the characters could be embedded. In other words, he noticed that the seemingly exceptional is more like the rule and then developed a systematic language with words and sentences and grammar. According to Mandelbrot himself, he did not follow a certain grand plan when carrying out this program; but rather summarized, in a way, his complex — one is tempted to say nomadic — scientific experiences in mathematics, linguistics, economics, physics, medical sciences and communication networks, to mention just some areas in which he was active.

Self-Similarity

Before we open our gallery of classical fractals and discuss in some detail several of these early masterpieces, let us introduce the concept of self-similarity. It will be an underlying theme in all fractals, more pronounced in some of them and in variations in others. In a way the word self-similarity needs no explanation, and at this point we merely give an example of a natural structure with that property, a cauliflower. It is not a classical mathematical fractal, but here the meaning of self-similarity is readily revealed without any math. The cauliflower head contains branches or parts, which when removed and compared with the whole are very much the same, only smaller.

These clusters again can be decomposed into smaller clusters, which again look very similar to the whole as well as to the first-generation branches. This self-similarity carries through for about three to four stages. After that the structures are too small for a further dissection. In a mathematical idealization the self-similarity property of a fractal may be continued through infinitely many stages. This leads to new concepts such as fractal dimension which are also useful for natural structures that do not have this 'infinite detail'.

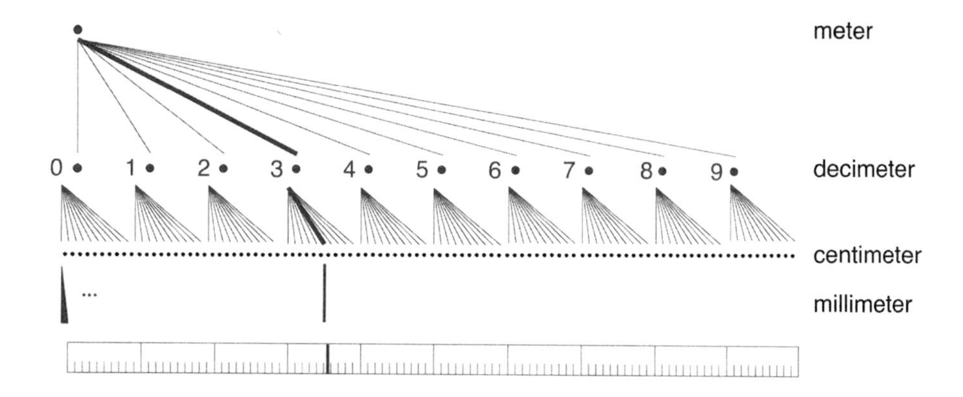

Figure 2.2 : The branches of the decimal tree leading to 357 are highlighted.

Self-Similarity in the Decimal System

Although the notion of self-similarity is only some 20 years old there are many historical constructions which make substantial use of its core idea. Probably the oldest and most important construction in that regard is our familiar decimal number system.[2] It is impossible to estimate where mathematics and the natural sciences would be without this ingenious invention. We are so used to the decimal number system that we are inclined to take it for granted. However, it evolved after a long scientific and cultural struggle and it is very closely related to the material from which fractals are made. It is also the prerequisite of the metric (measuring) system (for length, area, volume, weight, etc.). Let us look at a meter[3] stick, which carries markers for decimeters (ten make a meter), centimeters (ten make a decimeter; a hundred make a meter), and millimeters (ten make a centimeter; a thousand make a meter). In a sense a decimeter together with its markers looks like a meter with its markers, however, scaled down by a factor of 10. This is not an accident. It is

[2]Leonardo of Pisa, also known as Leonardo Fibonacci, helped introduce into mathematics the Indian-Arabic ciphers 0, 1, 2, 3, 4, 5, 6, 7, 8, and 9. His best-known work, the *Liber Abaci* (1202; 'Book of the Abacus') spends the first seven chapters explaining the place value, by which the position of a figure determines whether it is a unit, ten, hundred, and so forth, and demonstrating the use of the numerals in arithmetical operations.

[3]The metric system is now used internationally by scientists and in most nations. It was brought into being by the French National Assembly between 1791 and 1795. The spread of the system was slow but continuous, and, by the early 1970's only a few countries, notably the United States, had not adapted the metric system for general use. Since 1960 the definition of a meter has been: 1 meter = 1,650,763.73 wavelengths of the orange-red line in the spectrum of the krypton-86 atom under specified conditions. In the 1790's it was defined as 1/10,000,000 of the circumference of the quadrant of the Earth's circumference running from the North Pole through Paris to the equator.

in strict correspondence with the decimal system. When we say 357 mm, for example, we mean 3 decimeters, 5 centimeters, and 7 millimeters. In other words, the position of the figures determines their place value, exactly as in the decimal number system. One meter has a thousand millimeters to it and when we have to locate position 357 only a fool would start counting from left to right from 1 to 357. Rather, we would go to the 3 decimeter tick mark, from there to the 5 centimeter tick mark, and from there to the 7 millimeter tick mark. Most of us take this elegant procedure for granted. But somebody who has to convert miles, yards and inches can really appreciate the beauty of this system. Actually finding a position on the meter stick corresponds to a walk on the branches of a tree, the decimal number tree (see figure 2.2). The structure of the tree expresses the self-similarity of the decimal system very strongly. Similar trees reflect the self-similarity of many fractal constructions considered in this chapter.

2.1 The Cantor Set

Cantor (1845–1918) was a German mathematician at the University of Halle where he carried out his fundamental work in the foundations of mathematics, which we now call *set theory*.

Georg Cantor, 1845–1918

Figure 2.3

The Cantor set was first published[4] in 1883 and emerged as an example of certain exceptional sets.[5] It is probably fair to say that in the zoo of mathematical monsters — or early fractals — the Cantor set is by far the most important, though it is less visually appealing and more distant to an immediate natural interpretation than some of the others. It is now understood that the Cantor set plays a role in many branches of mathematics, and in fact, in a very deep sense, in chaotic dynamical systems (we will touch upon this property at least a bit), and is somehow hidden as the essential skeleton or model behind many other fractals (for example, Julia sets, as we will see in chapter 13).

The basic Cantor set is an infinite set of points in the unit interval [0,1]. That is, it can be interpreted as a set of certain numbers, as for example 0, 1, 1/3, 2/3, 1/9, 2/9, 7/9, 8/9, 1/27, 2/27, ...Plotting these and all other points (assuming we could know what they are) would not make much of a picture at all. Therefore, we use a common little trick. Rather than plotting just points we plot vertical lines all of the same length whose base points are exactly at all the different points belonging to the Cantor set. By so doing, we are able to see the distribution of these points a bit better. Figure 2.4 gives a first impression. Rather than being able to actually see the Cantor set, it is probably much more important to remember its classical construction.

[4]G. Cantor, *Über unendliche, lineare Punktmannigfaltigkeiten V*, Mathematische Annalen 21 (1883) 545–591.

[5]The Cantor set is an example of a perfect, nowhere dense subset.

The Cantor Set

The Cantor set represented by vertical lines whose base points are exactly at all the different points belonging to the set.

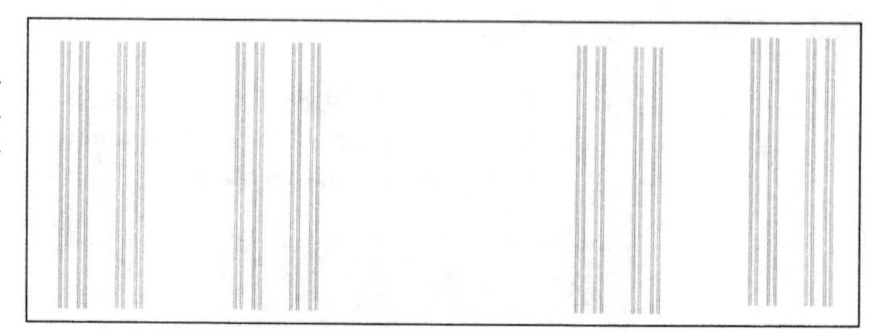

Figure 2.4

Construction of the Cantor Set

Start with the interval [0,1]. Now take away the (open) interval (1/3,2/3), i.e., remove the middle third from [0,1], but not the numbers 1/3 and 2/3. This leaves two intervals [0,1/3] and [2/3,1] of length 1/3 each and completes a basic construction step. Now we repeat, we look at the remaining intervals [0,1/3] and [2/3,1] and remove their middle thirds, which yields four intervals of length 1/9. Continue on in this way. In other words, there is a feedback process in which a sequence of (closed) intervals is generated — one after the first step, two after the second step, four after the third step, eight after the fourth step, etc. (i.e., 2^n intervals of length $1/3^n$ after the nth step). Figure 2.5 visualizes the construction.

Initial Steps of the Construction

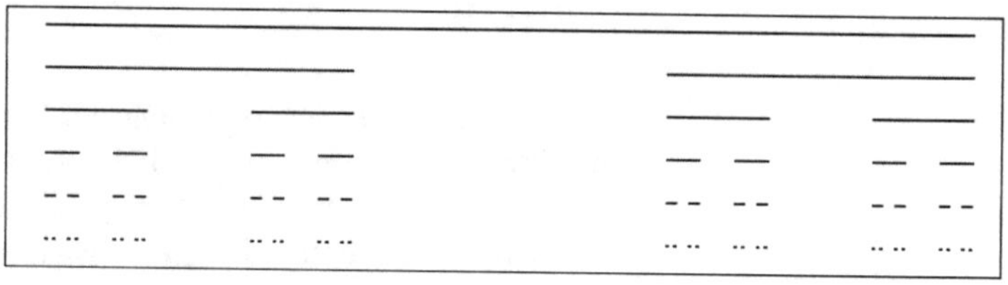

Figure 2.5

Interval End Points Are in the Cantor Set ...

What is the Cantor set? It is the set of points which remain if we carry out the *removal* steps infinitely often. How do we explain *infinitely often*? Let us try. A point, say x, is in the Cantor set if we can guarantee that no matter how often we carry out the removal process, the point x will not be taken out. Obviously $0, 1, 1/3, 2/3, 1/9, 2/9, 7/9, 8/9, 1/27, 2/27, \ldots$ are examples of such points because they are the end points of the intervals which are created in the steps; and therefore, they must remain. All these points have one thing in common. Namely, they are related to powers of 3 — or rather, to powers of 1/3. One is tempted to believe that any point in the Cantor set is of this kind, i.e., an end point of one of the small intervals generated in the process. This conclusion is categorically wrong. We will not give the complete argument but at least discuss the fact to some extent.

...But That's Not All

If the Cantor set were just the end points of the intervals of the generating process, we could easily enumerate them as shown in figure 2.6.

That means, the Cantor set would be a countable set, but it is known to be uncountable (see further below, page 73). That is, there is no way to enumerate

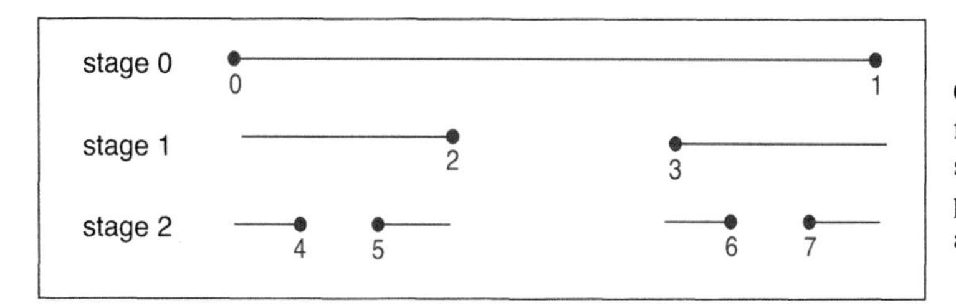

End Points of Intervals

Counting end points of intervals from the Cantor set construction. In stage $k, k > 0$ of the construction process 2^k new end points are added and enumerated as shown.

Figure 2.6

the points in the Cantor set. Thus, there must be many more points which are not end points. Can we give examples which are not end points? To name such examples, we will use a simple, but far-reaching characterization of the Cantor set, namely, by triadic numbers.

A Modification Using Decimals

But let us first see what can be done with the more familiar decimal numbers. Recall our discussion of the meter stick. Let us remove parts of the stick in several stages (see figure 2.7). Start with the meter and cut out decimeter number 5 in stage 1. This leaves 9 decimeters from each of which we take away centimeter number 5 in stage 2. Next, in stage 3, we consider the remaining 81 centimeters and remove millimeter number 5 from each one. Then we continue the process, going to tenth of millimeters in stage 4 and so on. This clearly is very similar to the basic Cantor set construction. In fact, the set of points that survive all stages in the construction, i.e., which are never taken away, are a fractal which is also called a Cantor set.

It is instructive to relate this modified Cantor set construction to the decimal number tree from figure 2.2. Removing a section of the meter stick corresponds to pruning a branch of the tree. In stage 1 the main branch with label 5 is cut off. In the following stages all branches with label 5 are pruned. In other words, only those decimals are kept which do not include the digit 5. Clearly,

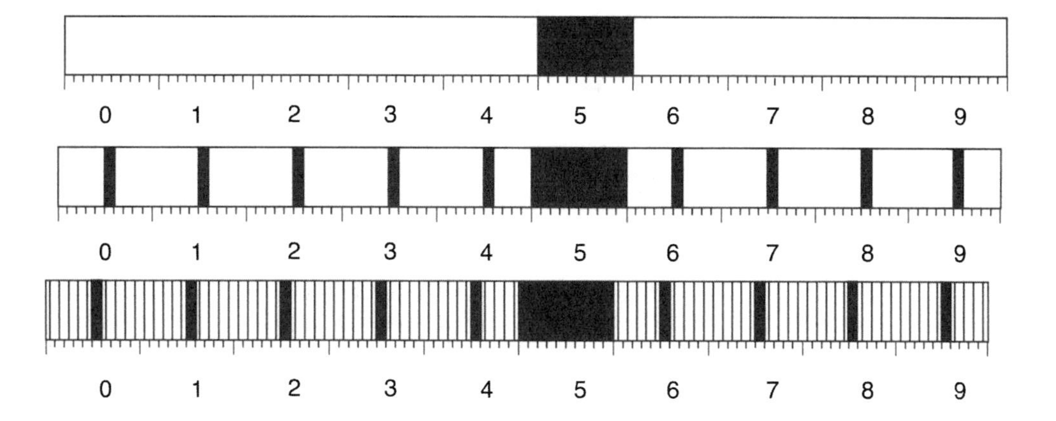

Figure 2.7 : In this meter stick the fifth decimeter (stage 1), the fifth centimeters (stage 2) and fifth millimeters (stage 3) are removed. This yields the first three stages of a modified Cantor set construction.

our choice to remove all fifth decimeters, centimeters, and so on is rather arbitrary. We could just as well have preferred to take out all numbers with a 6 in their decimal expansion, or even numbers with digits 3, 4, 5, and 6. For each choice we get another Cantor set. However, we will never obtain the classical Cantor set using this approach; this requires *triadic* numbers.

Characterization of the Cantor Set

Triadic numbers are numbers which are represented with respect to base 3. This means one only uses the digits 0, 1, and 2. We give a few examples in the following table.

Triadic Conversion

Conversion of four decimal numbers into the triadic representation.

Table 2.8

Decimal	In Powers of 3	Triadic
4	$1 \cdot 3^1 + 1 \cdot 3^0$	11
17	$1 \cdot 3^2 + 2 \cdot 3^1 + 2 \cdot 3^0$	122
0.333...	$1 \cdot 3^{-1}$	0.1
0.5	$1 \cdot 3^{-1} + 1 \cdot 3^{-2} + 1 \cdot 3^{-3} + \cdots$	0.111...

Triadic Numbers

Let us recall the essence of our familiar number system, the decimal system, and its representation. When we write 0.32573 we mean

$$3 \cdot 10^{-1} + 2 \cdot 10^{-2} + 5 \cdot 10^{-3} + 7 \cdot 10^{-4} + 3 \cdot 10^{-5}.$$

In other words, any number x in $[0, 1]$ can be written as

$$x = a_1 \cdot 10^{-1} + a_2 \cdot 10^{-2} + a_3 \cdot 10^{-3} + \ldots, \tag{2.1}$$

where the $a_1, a_2, a_3, \ldots$ are numbers from $\{0, 1, 2, \ldots, 9\}$, the decimal digits. This is called the decimal expansion of x, and may be infinite (e.g., $x = 1/3$) or finite (e.g., $x = 1/4$). When we say the expansion or representation is finite we actually mean that it ends with infinitely (redundant) consecutive zeros.

You will recall that digital computers depend on binary expansions of numbers. In computers 10 as base is replaced by 2. For example 0.11001 is

$$1 \cdot 2^{-1} + 1 \cdot 2^{-2} + 0 \cdot 2^{-3} + 0 \cdot 2^{-4} + 1 \cdot 2^{-5}.$$

There is a little bit of ambiguity in these representations. For example, we can write $2/10$ in two ways $0.1\overline{9} = 0.1999\ldots$ or $0.2 = 0.2000\ldots$, or in base 2 the number $1/4$ can be represented as $0.00\overline{1} = 0.00111\ldots$ or $0.01 = 0.01000\ldots$, where the overlining means that the respective digit (or digits) will be repeated ad infinitum.

Now we can completely describe the Cantor set by representing numbers from $[0, 1]$ in their triadic expansion, i.e., we switch to expansions of x with respect to base 3, as in eqn. (2.2);

$$x = a_1 \cdot 3^{-1} + a_2 \cdot 3^{-2} + a_3 \cdot 3^{-3} + a_4 \cdot 3^{-4} + \ldots \tag{2.2}$$

Thus, here the $a_1, a_2, a_3, \ldots$ are numbers from $\{0, 1, 2\}$.

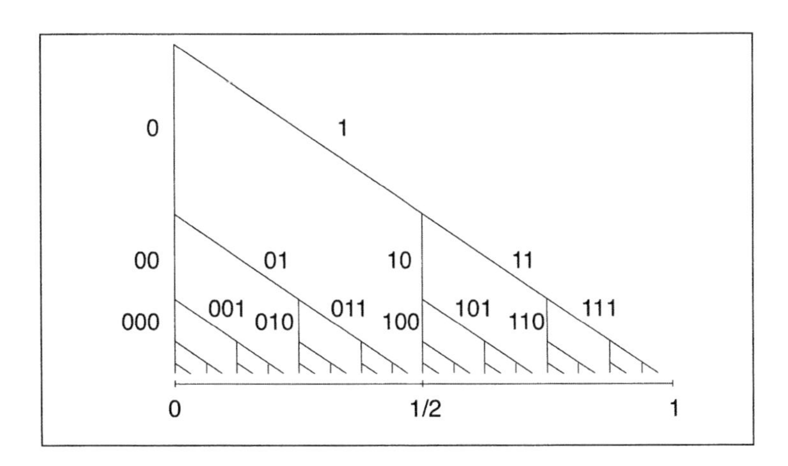

Figure 2.9 : Visualization of binary expansions by a two-branch tree. In contrast to real trees, we draw address trees with the root at the top. Any number in the interval [0,1] at the bottom can be reached from the root of the tree by following branches. Writing down the labels of these branches (0 for the left and 1 for the right branch) in a sequence will yield a binary expansion of the chosen real number. The tree has obvious self-similarity: any two branches at any node are a reduced copy of the whole tree.

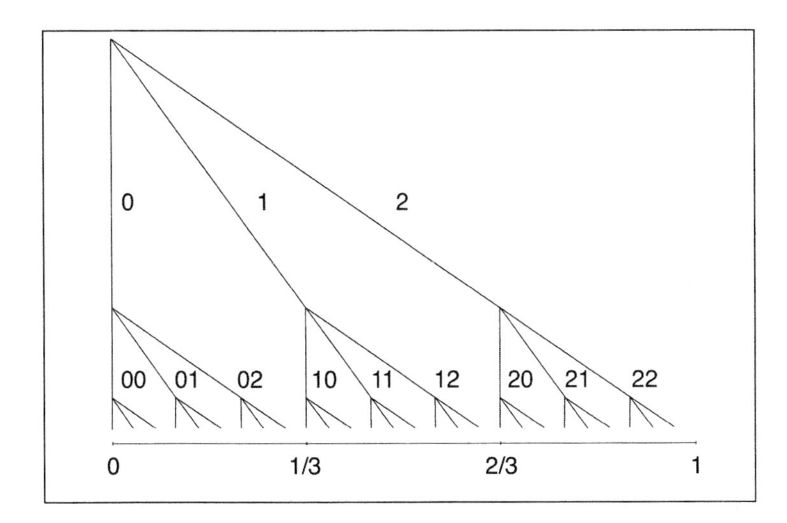

Figure 2.10 : A three-branch tree visualizes the triadic expansion of numbers from the unit interval. The first main branch covers all numbers between 0 and 1/3. Following down the branches all the way to the interval and keeping note of the labels 0, 1, and 2 for choosing the left, middle, or right branches will produce a triadic expansion of the number in the interval which is approached in this process.

Let us write some of the points of the Cantor set as triadic numbers: 1/3 is 0.1 in the triadic system, 2/3 is 0.2, 1/9 is 0.01, and 2/9 is 0.02. In general we can characterize any point of the Cantor set in the following way.

Fact. *The Cantor set C is the set of points in $[0, 1]$ for which there is a triadic expansion that does not contain the digit '1'.*

This number theoretic characterization eliminates the problem of the existence of a limit for the geometric construction of the Cantor set.

The above examples 2/3 and 2/9 are points in the Cantor set according to this statement, since their triadic expansions 0.2 and 0.02 do not contain any digits '1'. However, the other two examples 1/3 and 1/9 seem to contradict the rule, their expansions 0.1 and 0.01 clearly show a digit '1'. Yes, that is correct; but remember that we have ambiguity in our representations, and 1/3 can also be written as $0.0222\overline{2}$. Therefore, it belongs. But then, you may ask, what about $1/3 + 1/9$? This is a number from the middle third interval which is discarded in the first construction step of the Cantor set. It has a triadic expansion 0.11, and can't we also write this in different form and thus get into trouble? Yes, indeed, $1/3 + 1/9 = 0.1022\overline{2}$; but as you see, there appears a digit '1' no matter how we choose to represent that number in the triadic system. Thus, it is out and there is no problem with our description.

Distinguishing End Points from Others

Moreover, we can now distinguish points in C which are end points of some small interval occurring in the process of the feedback construction from those points which are definitely not. Indeed, end points in this sense just correspond to numbers which have triadic expansion ending with infinitely many consecutive 2's or 0's. All other possibilities, as for example

$$0.020022000222000022220000022222\ldots$$

or a number in which we pick digits 0 and 2 at random will belong to the Cantor set but are not end points, and those are, in fact, more typical for the Cantor set. In other words, if one picks a number from C at random, then with probability 1, it will not be an end point. By this characterization of the Cantor set we can understand that, in fact, any point in C can be approximated arbitrarily closely by other points from C, and yet C itself is a cloud of points. In other words, there is nothing like an interval in C (which is obvious if we recall the geometric construction, namely, the removal of intervals).

How About Decimal Expansions?

Thus, there are infinitely many points in the Cantor set that have a terminating representation in base 3. Let us add a side remark about a curious result regarding the representation of points of the Cantor set by *decimal* expansions. We may ask how many points there are in the Cantor set that have a terminating decimal expansion. The answer is quite astonishing;[6] there are exactly 14 such numbers, namely,

$$\frac{1}{4}, \frac{3}{4}, \frac{1}{10}, \frac{3}{10}, \frac{7}{10}, \frac{9}{10}, \frac{1}{40}, \frac{3}{40}, \frac{9}{40}, \frac{13}{40}, \frac{27}{40}, \frac{31}{40}, \frac{37}{40}, \frac{39}{40}.$$

[6]See C. R. Wall, *Terminating decimals in the Cantor ternary set,* Fibonacci Quart. 28, 2 (1990) 98–101.

Addresses and the Cantor Set

Let us return for a moment to the intuitive geometric construction of the Cantor set by removing middle thirds in each step from the unit interval $[0, 1]$. After the first step, we have two parts; one is left and one is right. After the second step, each of these in turn splits into two parts, a left one and a right one, and so on. Now we design an efficient labeling procedure for each part created in the steps. The two parts after the first step are labeled L and R for left and right. The four parts after the second step are labeled LL, LR, RL, RR, i.e., the L part from step one is divided into an L and an R part, which makes LL and LR, and likewise with the R part. Figure 2.11 summarizes the first three steps.

Cantor Set Addresses

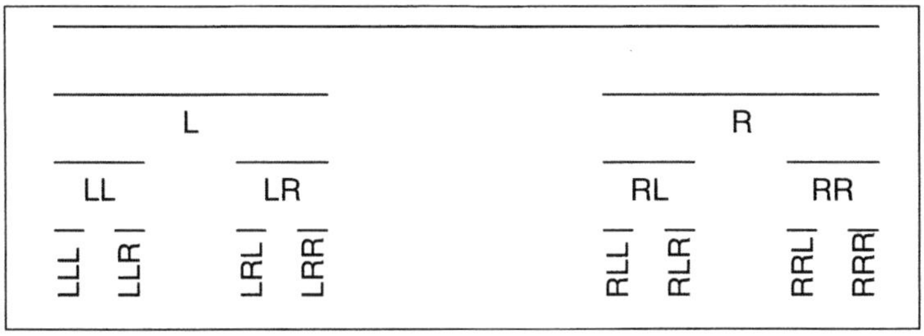

Figure 2.11

As a result, we are able to read from a label with 8 letters like $LLRLRRRL$ exactly which of the 2^8 parts of length $1/3^8$ we want to pick. It is important when reading this address, however, to remember the convention that we interpret from the left to right, i.e., letters have a place value according to their position in a word much like the numerals in the decimal system.

Addresses of Intervals Versus Addresses of Points

Finite string addresses such as $LLRLRRRL$ identify a small interval from the construction of the Cantor set. The longer the address, the higher the stage of the construction, and the smaller the interval becomes. To identify points in the Cantor set, such addresses obviously are not sufficient, as in each such interval, no matter how small it is, there are still infinitely many points from the Cantor set. Therefore we need infinitely long address strings to precisely describe the location of a Cantor set point. Let us give two examples. The first one is the point 1/3. It is in the left interval of the first stage, which has address L. Within that it is in the right interval of the second stage. This is $[2/9, 1/3]$ with address LR. Within that the point is again in the right subinterval with address LRR, and so on. Thus, to identify the position of the point exactly we write down the sequence of intervals from consecutive stages to which the point belongs: LR, LRR, $LRRR$, $LRRRR$, and so on. In other words, we can write the address of the point as the infinite string $LRRRR\ldots$, or using a bar to indicate periodic repetition $L\overline{R}$. The point 2/3 is in the right interval of the first stage. Within that and all further stages it is always in the left subinterval. Thus, the address of 2/3 is $RLLL\ldots$, or $R\overline{L}$.

Binary Tree for the Cantor Set

Another interesting way to look at the situation which is established by this systematic labeling is demonstrated in figure 2.12, where we see an infinite binary tree the branches of which repeatedly split into two branches from top

Address Tree

Addresses for points of the Cantor set form a binary tree.

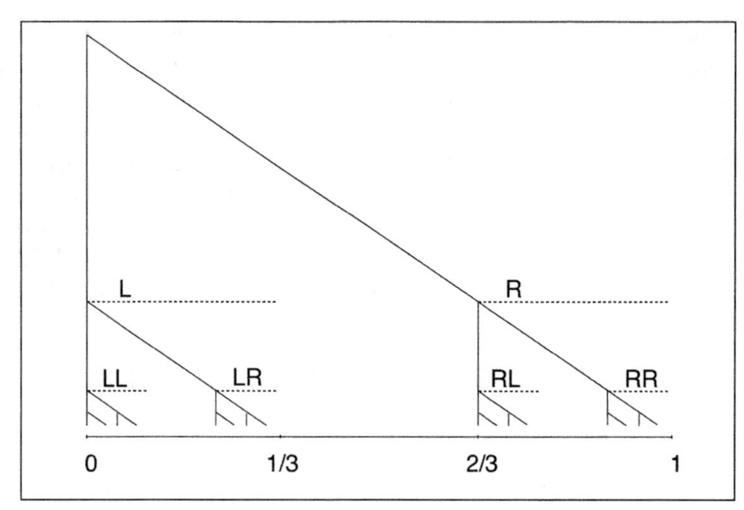

Figure 2.12

to bottom. What is the connection between the tree and the Cantor set? Well, the tree consists of nodes and branches. Each level of the tree corresponds to a certain step in the Cantor set construction; and in this way, it is actually a genealogical tree. In other words, we can compare this with a cell division process, and the tree tells us exactly from where an individual cell of some future generation is derived. This is quite nice already, but there is much more to this simple idea. For example, rather than choosing the alphabet $\{L, R\}$ we can take another two-letter alphabet and carry out a systematic replacement. Why not pick 0 and 2 as an alphabet, i.e., replace any L by a 0 and any R by a 2. Then we obtain strings of digits like 022020002 in place of $LRRLRLLLR$. You have surely guessed what we are up to. Indeed, that string of digits can be interpreted as a triadic number by just putting a decimal point in front of it, i.e., 0.022020002. Thereby, we have demonstrated the connection between the triadic representation of the Cantor set and the addressing system. In fact, this provides an argument for the validity of the triadic characterization. In other words, if we want to know where in the Cantor set a certain number is located — up to a certain degree of precision — we just have to look at its triadic expansion and then interpret each digit 0 as L and each digit 2 as R. Then, looking up the resulting address, we can find the part of the binary tree in which the number must lie.

L and R Are Not 0 and 1

The relation of L, R addresses with triadic numbers might seem to suggest going even one step further, namely, to identify L and R with 0 and 1, i.e., with binary numbers. This is, however, somewhat dangerous. Let us consider again the example point 1/3. It is represented by the address string $L\overline{R}$. This would correspond to the binary number $0.0\overline{1}$, which as a binary number, is identical to 0.1. But translated back $0.1 = 0.1000\ldots = 0.1\overline{0}$ corresponds to $R\overline{L}$, which in the Cantor set is the point 2/3! Clearly, this identification produces a contradiction. In other words, triadic numbers, but not binary numbers, are natural to the Cantor set. Or to put it another way, two-letter-based infinite strings are natural for the Cantor set, but these cannot be identified with binary numbers, despite the fact that they are strings made up of two letters/digits.

The Cardinality of the Cantor Set

We can now see that the cardinality of the Cantor set must be the same as the cardinality of the unit interval $[0,1]$. We start with the interval $[0,1]$ and show how each point in it corresponds to one point in the Cantor set.

- Each point in the interval has a binary expansion.
- Each binary expansion corresponds to a path in the binary tree for binary numbers.
- Each such path has a corresponding path in the triadic tree for the Cantor set.
- Each path in the triadic tree of the Cantor set identifies a unique point in the Cantor set by an address in triadic expansion.

Therefore, for each number in the interval, there is a corresponding point in the Cantor set. For different numbers there are different points. Thus, the cardinality of the Cantor set must be at least as large as the cardinality of the interval. On the other hand, it cannot exceed this cardinality, because the Cantor set is a subset of the interval. Therefore, both cardinalities must be the same.

Self-Similarity

The Cantor set is truly complex, but is it also self-similar? Yes, indeed, if one takes the part of C which lies in the interval [0,1/3], for example, we can regard that part as a scaled down version of the entire set. How do we see that? Let us take the definition of the Cantor set collecting all points in [0,1] which admit a triadic representation not containing digit 1. Now for every point, say

$$\xi = \alpha_1 \times 3^{-1} + \alpha_2 \times 3^{-2} + \alpha_3 \times 3^{-3} + \alpha_4 \times 3^{-4} + \ldots,$$

(with $\alpha_i \in \{0,2\}$) in the Cantor set we find a corresponding one in [0,1/3] by dividing ξ by 3, i.e.,

$$\frac{\xi}{3} = 0 \times 3^{-1} + \alpha_1 \times 3^{-2} + \alpha_2 \times 3^{-3} + \alpha_3 \times 3^{-4} + \ldots$$

Indeed, if $x = 0.200220\ldots$ and we multiply by $1/3 = 0.1$, that means that we just shift the binary digits one place to the right and obtain 0.0200220…, which is in C again. Thus, the part of the Cantor set present in [0,1/3] is an exact copy of the entire Cantor set scaled down by the factor 1/3 (see figure 2.13). For the part of C which lies in the interval [2/3,1], essentially we can do the same calculation (we only have to include the addition of 2/3 = 0.2). In the same way, any subinterval in the geometric Cantor set construction contains the entire Cantor set scaled down by an appropriate factor of $1/3^k$. In other words, the Cantor set can be seen as a collection of arbitrarily small pieces, each of which is an exact scaled down version of the entire Cantor set. This is

Self-Similarity of the Cantor Set

The Cantor set is a collection of two exact copies of the entire Cantor set scaled down by the factor 1/3.

Figure 2.13

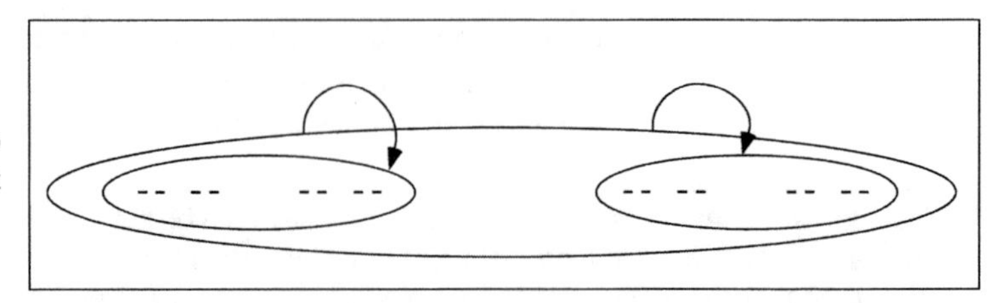

what we mean when we say the Cantor set is self-similar. Thus, taking self-similarity as an intuitive property means that self-similarity here is absolutely perfect and is true for an infinite range. Note that in our discussion of self-similarity we have carefully avoided the geometrical model of the Cantor set. Instead, we have exploited the number theoretic representation.

Note that the scaling property of the Cantor set corresponds to the following invariance property. Take a point from C and multiply it by 1/3. The result will be in C again. The same is true if we first multiply by 1/3 and then add 2/3. This is apparent from the triadic characterization, and it will be the key observation for chapter 5.

Before we continue our introduction to classical fractals with some other examples, let us touch one more property of the Cantor set which reveals an important dynamic interpretation and an amazing link with chaos.

Cantor Set as Prisoner Set

Let us look at a mathematical feedback system defined in the following way. If x is an input number, then the output number is determined by the following conditional formula (2.3).

$$x \to \begin{cases} 3x & \text{if } x \leq 0.5 \\ -3x+3 & \text{if } x > 0.5. \end{cases} \tag{2.3}$$

In other words, the output is evaluated to be $3x$ if $x \leq 0.5$ and is $-3x+3$ if $x > 0.5$.

Starting with an initial point x_0, the feedback process defines a sequence $x_0, x_1, x_2, x_3, \ldots$ The interesting question then is: what is the long-term behavior of such sequences? For many initial points x_0 the answer is very easy to derive. Take for example a number $x_0 < 0$. Then $x_1 = 3x_0$, and $x_1 < 0$. By induction it follows that all numbers x_k from this sequence are negative, and, thus $x_k = 3^k x_0$. This sequence then grows negatively without any bound, and it tends to negative infinity, $-\infty$. Let us call a sequence with such a long-term behavior an *escaping sequence* and the initial point x_0 an *escaping point*.

Let us now take $x_0 > 1$. Then $x_1 = -3x_0+3 < 0$, and again the sequence escapes to $-\infty$. But not all points are escaping points. For example, for $x_0 = 0$ we have that all succeeding numbers in the sequence are also equal to zero. We conclude, that any initial point x_0, which at some stage goes to zero will remain there forever, and thus is not an escaping point. Such points we call *prisoners*. So far we have found that all prisoner points must be in the unit interval $[0, 1]$. This leads to the interesting question: which points in the unit interval will remain and which will escape? Let us look at some examples.

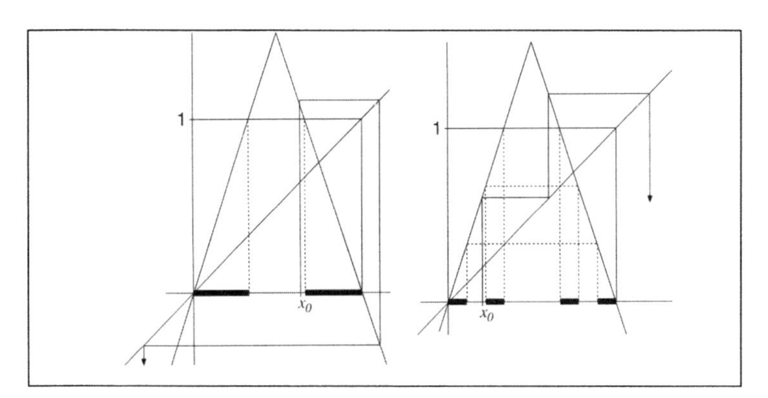

Escaping Points and Intervals

Figure 2.14

x_0	x_1	x_2	x_3	x_4	...	P/E
0	0	0	0	0		prisoner
1/3	1	0	0	0		prisoner
1/9	1/3	1	0	0		prisoner
1/2	3/2	−3/2	−9/2	−27/2		escapee
1/5	3/5	6/5	−3/5	−9/5		escapee

Clearly the entire (open) interval $(1/3, 2/3)$ escapes because when $1/3 < x_0 < 2/3$ we have $x_1 > 1$ and $x_2 < 0$. But then every point which eventually lands in that interval will also escape under iteration. Figure 2.14 illustrates these points and reveals the Cantor set construction for the points which will remain.

Fact. *The prisoner set P for the feedback system given by eqn. (2.3) is the Cantor set, while all points in $[0, 1]$ which are outside the Cantor set belong to the escape set E.*

This is a remarkable result and shows that the study of the dynamics of feedback systems can provide an interpretation for the Cantor set. This close relation between chaos and fractals will be continued in chapter 13.

2.2 The Sierpinski Gasket and Carpet

Our next classical fractal is about 40 years younger than the Cantor set. It was introduced by the great Polish mathematician Waclaw Sierpinski[7] (1882–1969) in 1916.

Waclaw Sierpinski, 1882–1969

Figure 2.15

Sierpinski was a professor at Lvov[8] and Warsaw. He was one of the most influential mathematicians of his time in Poland and had a worldwide reputation. In fact, one of the moon's craters is named after him.

The basic geometric construction of the Sierpinski gasket goes as follows. We begin with a triangle in the plane and then apply a repetitive scheme of operations to it (when we say triangle here, we mean a blackened, 'filled-in' triangle). Pick the midpoints of its three sides. Together with the old vertices of the original triangle, these midpoints define four congruent triangles of which we drop the center one. This completes the basic construction step. In other words, after the first step we have three congruent triangles whose sides have exactly half the size of the original triangle and which touch at three points which are common vertices of two contiguous triangles. Now we follow the same procedure with the three remaining triangles and repeat

[7] W. Sierpinski, *Sur une courbe dont tout point est un point de ramification,* C. R. Acad. Paris 160 (1915) 302, and W. Sierpinski, *Sur une courbe cantorienne qui contient une image biunivoquet et continue detoute courbe donnée* C. R. Acad. Paris 162 (1916) 629–632.

[8] Lvov, Ukrainian Lviv, Polish Lwów, German Lemberg, city and administrative center in the Ukrainian Republic. Founded in 1256 Lvov has always been the chief center of Galicia. Lvov was Polish between 1340 and 1772 until the first partition, when it was given to Austria. In 1919 it was restored to Poland and became a world-famous university town hosting one of the most influential mathematics schools during the 1920's and 1930's. In 1939 it was annexed by the Soviets as a result of the Hitler-Stalin Pact, and the previously flourishing Polish mathematics school collapsed. Later several of its great scientists were victims of Nazi Germany.

Sierpinski Gasket

The basic construction steps of the Sierpinski gasket.

Figure 2.16

Sierpinski Pattern

Escher's studies of Sierpinski gasket-type patterns on the twelfth-century pulpit of the Ravello cathedral, designed by Nicola di Bartolomeo of Foggia. Watercolor, ink, 278 by 201 mm. ©1923 M. C. Escher / Cordon Art – Baarn – Holland.

Figure 2.17

the basic step as often as desired. That is, we start with one triangle and then produce $3, 9, 27, 81, 243, \ldots$ triangles, each of which is an exact scaled down version of the triangles in the preceding step. Figure 2.16 shows a few steps of the process.

The Sierpinski gasket[9] is the set of points in the plane which remain if one carries out this process infinitely often. It is easy to list some points which definitely belong to the Sierpinski gasket, namely, the sides of each of the triangles in the process.

The characteristic of self-similarity is apparent, though we are not yet

[9]The Sierpinski gasket is sometimes also called the Sierpinski triangle.

LRTT

$LRTT$ denotes a subtriangle in the Sierpinski gasket which can be found following the left, right, top, top subtriangles.

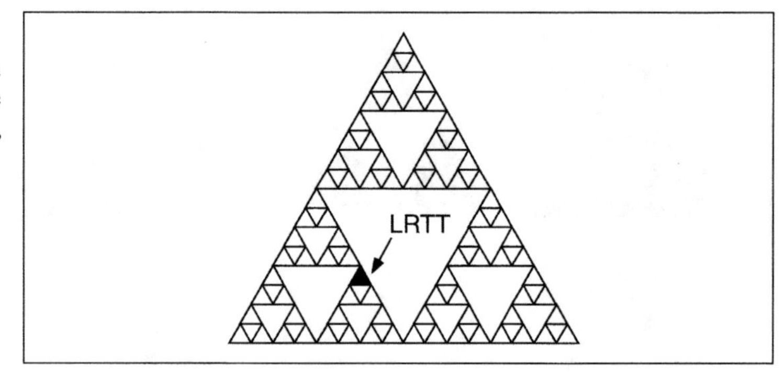

Figure 2.18

Spider-Like Tree

This tree represents not only the structure of the Sierpinski gasket but also its geometry.

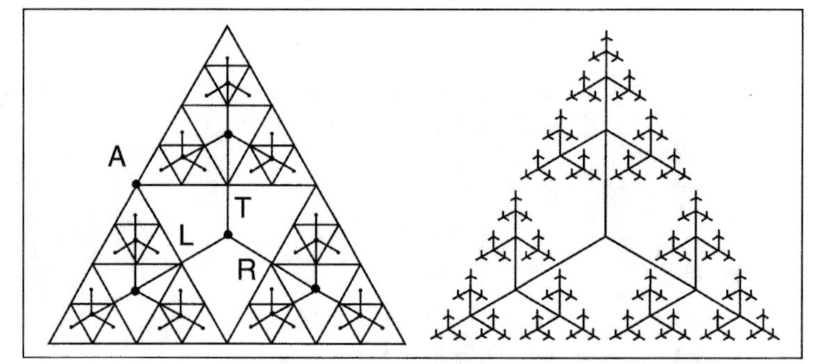

Figure 2.19

prepared to discuss it in detail. It is built into the construction process, i.e., each of the three parts in the kth step is a scaled down version — by a factor of 2 — of the entire structure in the previous step. Self-similarity, however, is a property of the limit of the geometrical construction process, and that will be available to us only in chapter 5. In chapter 8 we will explain how the Sierpinski gasket admits a number theoretic characterization from which the self-similarity follows as easily as for the Cantor set.

Addresses for the Sierpinski Gasket

Similar to our above discussion of the Cantor set we can introduce an addressing system for the subtriangles (or points) of the Sierpinski gasket. Here we must use three symbols to establish a system of addresses. If we take, for example, L (left), R (right) and T (top) we obtain sequences like $LRTT$ or $TRLLLTLR$ and read them from left to right to identify subtriangles in the respective construction stage of the Sierpinski gasket. For example, $LRTT$ refers to a triangle in the 4th generation which is obtained in the following way. Pick the left triangle in the first generation, then the right one therein, then the top one therein, and finally again the top one therein (see figure 2.18). We will discuss the importance of addresses for the Sierpinski gasket in chapter 6. They are the key to unchaining the chaos game introduced in chapter 1. We should not confuse, however, our symbolic addresses with triadic numbers.

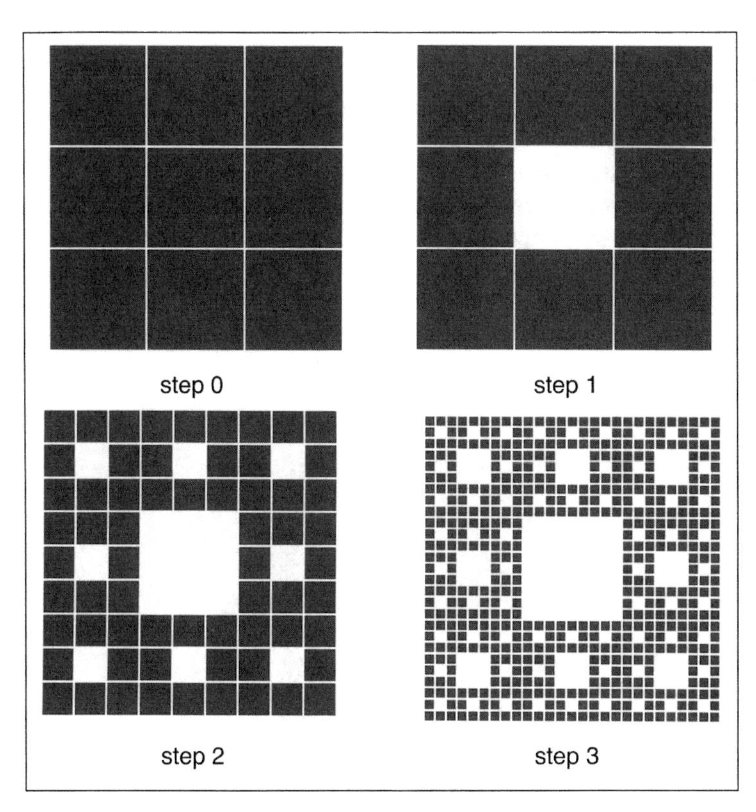

Sierpinski Carpet

The basic construction steps of the Sierpinski carpet.

Figure 2.20

There are several ways to associate trees with symbolic addresses. A particular construction is based on the triangles which are taken away in the construction process. The nodes of the tree are the centers of these triangles. The branches of the tree grow generation by generation, as shown in figure 2.19. Observe that some of the branches touch when we go to the limit. For example, the branches corresponding to $LTTT\ldots$ and $TLLL\ldots$ touch in point A.

The Sierpinski Carpet

Sierpinski has added another object to the gallery of classical fractals, the Sierpinski carpet, which at first glance just looks like a variation of the known theme (see figure 2.20). We begin with a square in the plane. Subdivide into nine little congruent squares of which we drop the center one, and so on. The resulting object which remains if one carries out this process infinitely often can be seen as a generalization of the Cantor set. Indeed, if we look at the intersection of a line which is parallel to the base of the original square and which goes through the center we observe precisely the construction of the Cantor set. We will see in section 2.7 that the complexities of the carpet and the gasket may at first look essentially the same, but there is in fact a whole world of a difference between them.

2.3 The Pascal Triangle

Blaise Pascal (1623–1662) was a great French mathematician and scientist. When only twenty years old, he built some ten mechanical machines for the addition of integers, a precursor of modern computers. What is known as the arithmetic triangle or *Pascal's triangle*, was not, however, his discovery. The first printed form of the arithmetic triangle in Europe dates back to 1527. A Chinese version of Pascal's triangle had already been published in 1303 (see figure 2.24). Pascal, however, used the arithmetic triangle to solve some problems related to chances in gambling, which he had discussed with Pierre de Fermat in 1654. This research later became the foundations of probability theory.

Blaise Pascal, 1623–1662

Figure 2.21

The Arithmetic Triangle

The arithmetic triangle is a triangular array of numbers composed of the coefficients of the expansion of the polynomial $(1+x)^n$. Here n denotes the row starting from $n=0$. Row n has $n+1$ entries. For example, for $n=3$ the polynomial is

$$(1+x)^3 = 1 + 3x + 3x^2 + x^3.$$

Thus, row number 3 reads $1, 3, 3, 1$ (see figure 2.22).

There are two ways to compute the coefficients. The first one inductively computes one row based on the entries of the previous row. Assume that the coefficients $a_0, \ldots, a_n$ in row n are given:

$$(1+x)^n = a_0 + a_1 x + \cdots + a_n x^n,$$

and the coefficients $b_0, \ldots, b_{n+1}$ of the following row are required:

$$(1+x)^{n+1} = b_0 + b_1 x + \cdots + b_{n+1} x^{n+1}.$$

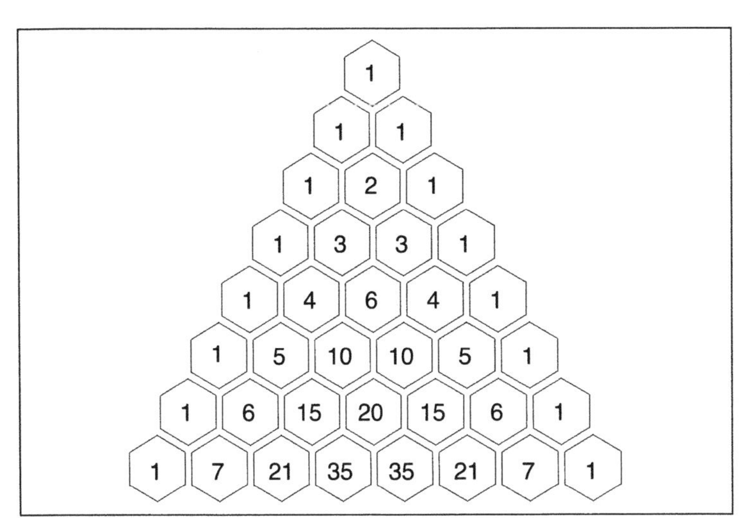

Pascal's Triangle

The first eight rows of Pascal's triangle in a hexagonal web.

Figure 2.22

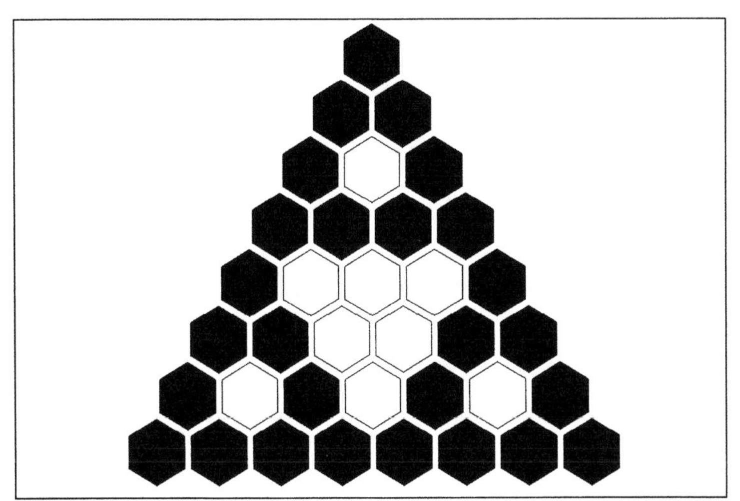

Color Coding

Color coding of even (white) and odd (black) entries in the Pascal triangle with eight rows.

Figure 2.23

These are directly related to the known coefficients $a_0, \ldots, a_n$:

$$\begin{aligned}(1+x)^{n+1} &= (1+x)^n(1+x) \\ &= (a_0 + a_1x + \cdots + a_nx^n)(1+x) \\ &= a_0 + a_1x + \cdots + a_nx^n \\ &\quad + a_0x + a_1x^2 + \cdots + a_nx^{n+1} \\ &= a_0 + (a_0+a_1)x + \cdots + (a_{n-1}+a_n)x^n + a_nx^{n+1}.\end{aligned}$$

Chinese Arithmetic Triangle

Already in 1303 an arithmetic triangle had appeared in China at the front of Chu Shih-Chieh's *Ssu Yuan Yii Chien* which tabulates the binominal coefficients up to the eighth power.

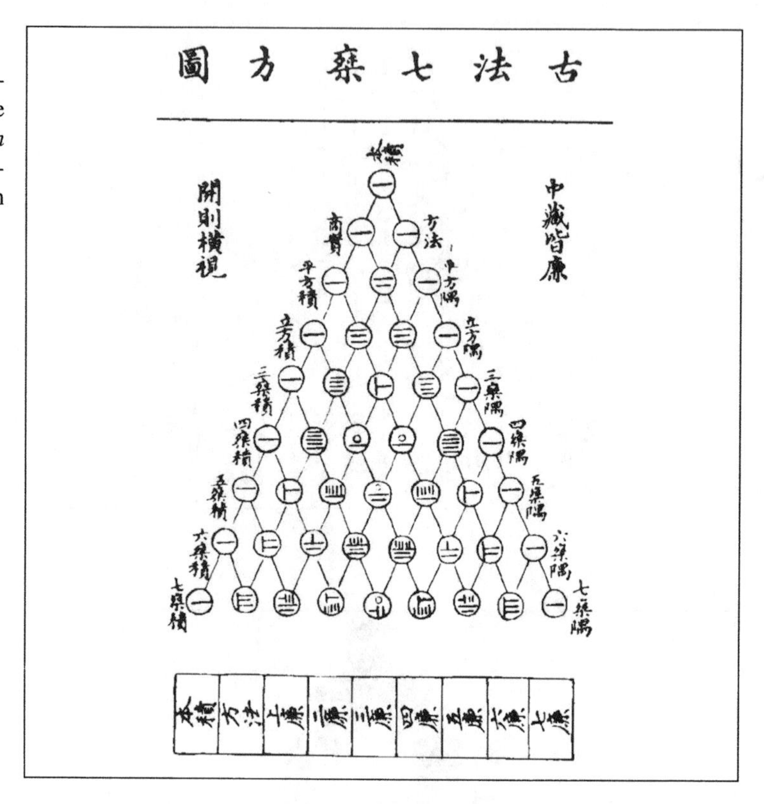

Figure 2.24

Comparing coefficients we obtain the result

$$\begin{aligned} b_0 &= a_0, \\ b_k &= a_{k-1} + a_k \quad \text{for } k = 1, \ldots, n, \\ b_{n+1} &= a_n. \end{aligned}$$

The recipe to compute the coefficients of a row is thus very simple. The first and last numbers are copied from the line above. These will always be equal to 1. The other coefficients are just the sum of the two coefficients in the row above. In this scheme it is most convenient to write Pascal's triangle in the form with the top vertex centered on a line above it as shown in figure 2.22.

For computing small numbers of rows from Pascal's triangle the inductive method as outlined above is quite satisfactory. However, sometimes it is of advantage to have a direct approach available. It is based on the binomial theorem, which states

$$(x+y)^n = \sum_{k=0}^{n} \frac{n!}{k!(n-k)!} x^{n-k} y^k$$

where the notation $n!$ is 'factorial n' and defined as

$$n! = 1 \cdot 2 \cdots (n-1) \cdot n$$

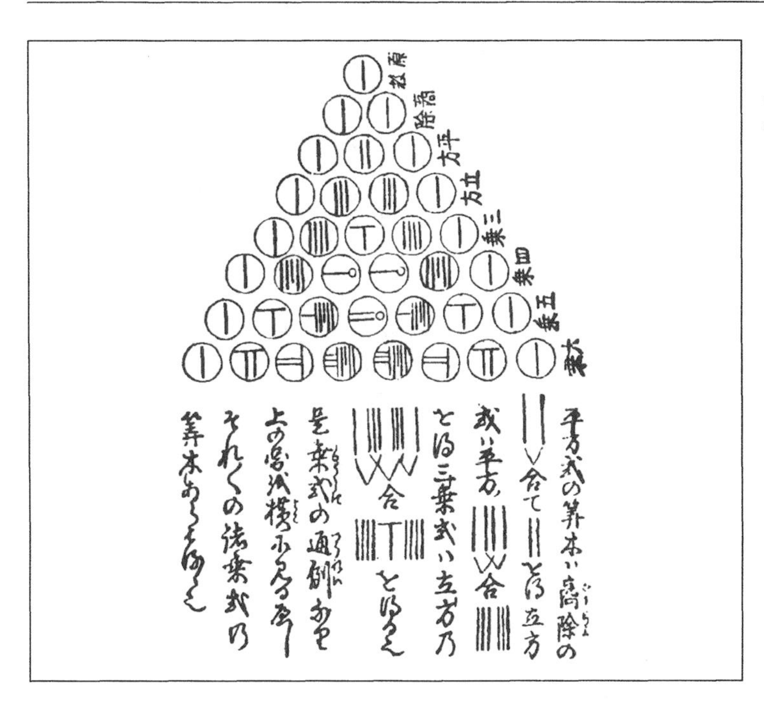

Pascal's Triangle in Japan

Appeared 1781 in Murai Chūzen's *Sampō Dōshi-mon*.

Figure 2.25

for positive integers n and $0! = 1$. The coefficients in the polynomial are called binomial coefficients.[10]

Applying the formula to $(1+x)^n$ we immediately obtain the kth coefficient b_k (k runs from 0 to n) of row number n of Pascal's triangle via

$$b_k = \frac{n!}{k!(n-k)!} = \frac{n(n-1)\cdots(n-k+1)}{1\cdot 2\cdots k}.$$

For example, the coefficient for $k=3$ in row $n=7$ is

$$b_3 = \frac{7\cdot 6\cdot 5}{2\cdot 3} = 35;$$

see the fourth entry in the last row in figure 2.22.

Another identity is easy to derive: the sum of all coefficients in row number n of Pascal's triangle is equal to 2^n, which is seen by setting $x = y = 1$ in the binomial formula.

[10]The notion of binomial coefficient was introduced in 1544 by Michael Stifel, who also showed how to calculate $(1+x)^{n+1}$ from $(1+x)^n$. The notation $n!$ (factorial n) for the product $1\cdot 2\cdot 3\cdots n$ was introduced by Christian Kramp in an algebra book around 1808. Euler wrote $[n]$; Gauss used the notation $\pi(n)$.

Figure 2.26 : Color coding of even and odd entries in the Pascal triangle with 16, 32, and 64 rows.

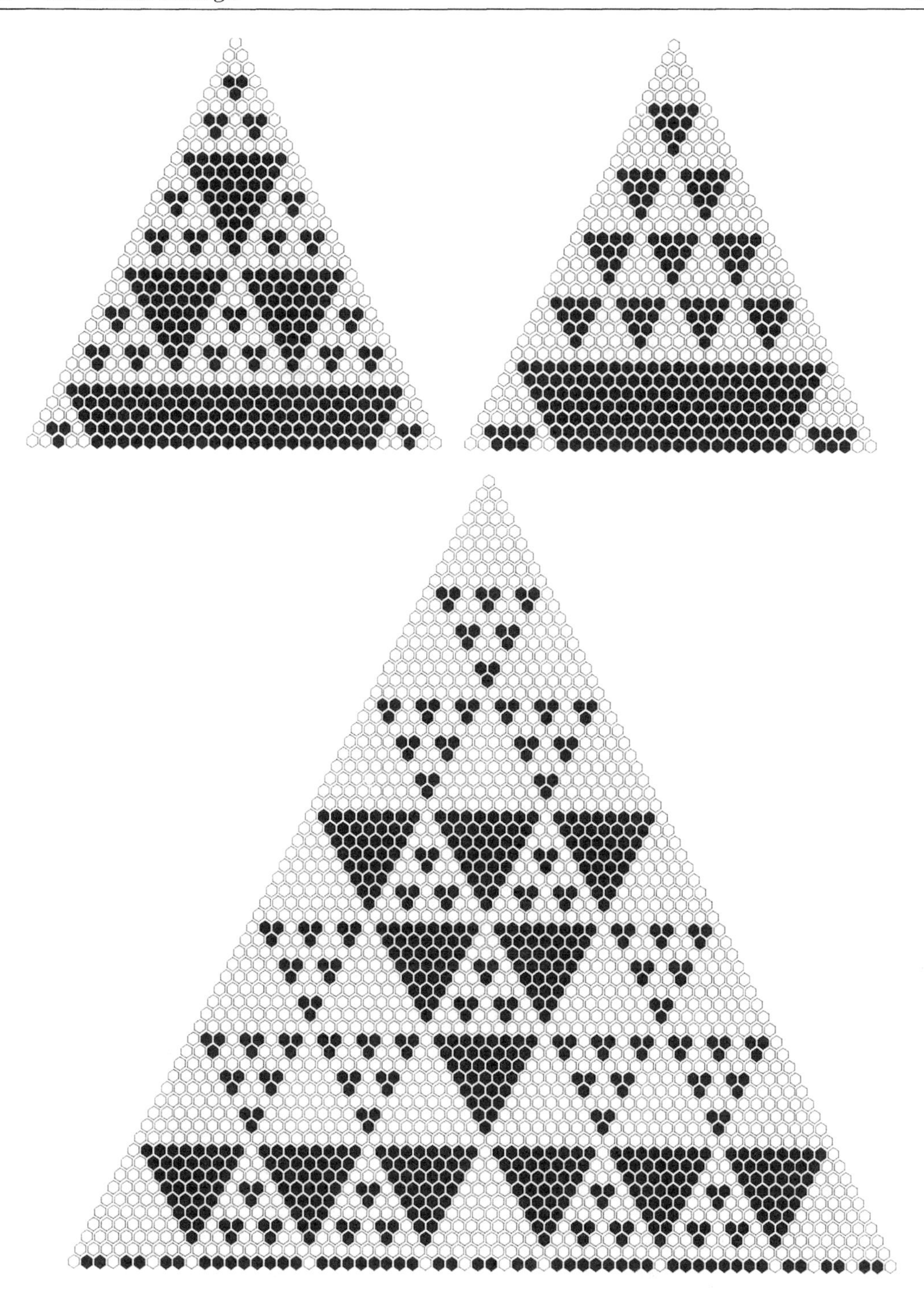

Figure 2.27 : Color coding the Pascal triangle. Black cells denote divisibility by 3 (top left), by 5 (top right) and by 9 (bottom).

To keep order we have put the first eight rows of a Pascal triangle into a hexagonal web (see figure 2.22). Now let us color code properties of the numbers in the triangle. For example, let us mark each hexagonal cell which is occupied by an odd number with black ink, i.e., the even ones are left white. Figure 2.23 shows the result.

It is worthwhile to repeat the experiment with more and more rows (see figure 2.26). The last image of that series already looks very similar to a Sierpinski gasket. Is it one? We have to be very careful about this question and will give a first answer in chapter 3. These number theoretic patterns are just one of an infinite variety of related ones. You will recall that even/odd just means divisible by 2 or not. Now, how do the patterns look when we color code divisibility by $3, 5, 7, 9$, etc., by black cells and nondivisibility by white cells? Figure 2.27 gives a first impression.

Each of these patterns has beautiful regularities and self-similarities which describe elementary number theoretic properties of the Pascal triangle. Many of these properties have been observed and studied for several centuries. The book by B. Bondarenko,[11] lists some 406 papers by professional and amateur mathematicians over the last three hundred years.[12]

[11] B. Bondarenko, *Generalized Triangles and Pyramids of Pascal; Their Fractals, Graphs and Applications*, Tashkent, Fan, 1990, in Russian.

[12] In chapter 8 we will demonstrate how the fractal patterns and self-similarity features can be deciphered by the tools which are the theme of chapter 5.

2.4 The Koch Curve

Helge von Koch was a Swedish mathematician who, in 1904, introduced what is now called the *Koch curve*.[13] Fitting together three suitably rotated copies of the Koch curve produces a figure, which for obvious reasons is called the *snowflake curve* or the *Koch island* (see figures 2.29 and 2.30).

Koch's Original Construction

Excerpt from von Koch's original 1906 article.

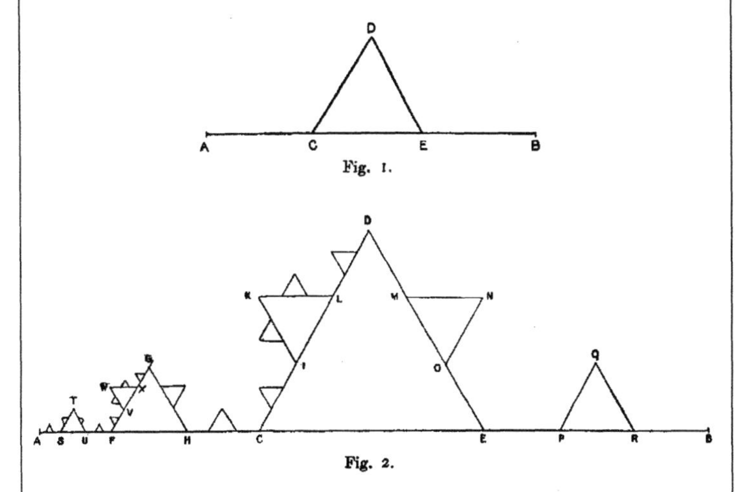

sidérer comme positif le côté laissé à gauche quand on parcourt le segment dans le sens positif. Pour abréger, nous désignons par Ω cette opération au moyen de laquelle on passe d'un segment rectiligne *AB* à la ligne polygonale *ACDEB* déviant de *AB* vers le côté positif.

Fig. 1.

Fig. 2.

2. Partons maintenant d'une ligne droite déterminée *AB*, le sens de *A* vers *B* étant considéré comme positif (fig. 2). Par l'opération Ω, *AB* est remplacée par la ligne brisée *ACDEB*, les segments *AC*, *CD*, *DE*, *EB* étant égaux entre eux et leur sens positif étant respectivement celui de *A* vers *C*, de *C* vers *D*, de *D* vers *E*, de *E* vers *B*.

Effectuons l'opération Ω sur chacun de ces segments; la ligne *ACDEB* sera remplacée par la ligne brisée *AFGHCIKLDMNOEPQRB* composée de 16 segments égaux *AF*, *FG* etc.

Figure 2.28

Little is known about von Koch, whose mathematical contributions were certainly not in the same category as those of the stars like Cantor, Peano, Hilbert, Sierpinski or Hausdorff. But in this chapter on classical fractals, Koch's construction must have its place simply because it leads to many interesting generalizations and must have inspired Mandelbrot immensely. The Koch curve is as difficult to understand as the Cantor set or the Sierpinski gasket. However, the problems with it are of a different nature. First of all — as the name already expresses — it is a curve, but this is not immediately

[13] H. von Koch, *Sur une courbe continue sans tangente, obtenue par une construction géometrique élémentaire*, Arkiv för Matematik 1 (1904) 681–704. Another article is H. von Koch, *Une méthode géométrique élémentaire pour l'étude de certaines questions de la théorie des courbes planes*, Acta Mathematica 30 (1906) 145-174.

The Koch Snowflake

The outline of the Koch snowflake (also called Koch island) is composed of three congruent parts, each of which is a Koch curve as shown in figures 2.31 and 2.33.

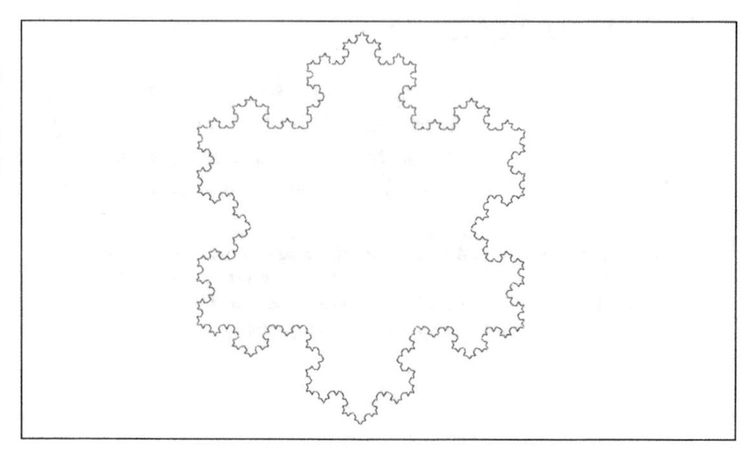

Figure 2.29

Some Natural Flakes

The Koch snowflake obviously has some similarities with real flakes, some of which are pictured here.

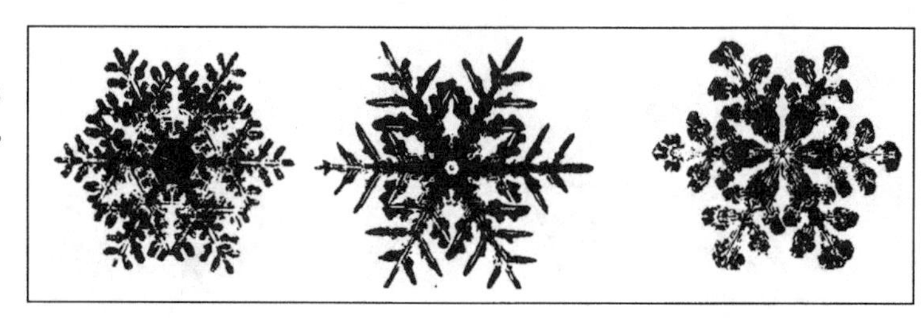

Figure 2.30

clear from the construction. Secondly, this curve contains no straight lines or segments which are smooth in the sense that we could see them as a carefully bent line. Rather, this curve has much of the complexity which we would see in a natural coastline, folds within folds within folds, and so on.

Geometric Construction

Here is the simple geometric construction of the Koch curve. Begin with a straight line. This initial object is also called the *initiator*. Partition it into three equal parts. Then replace the middle third by an equilateral triangle and take away its base. This completes the basic construction step. A reduction of this figure, made of four parts, will be reused in the following stages. It is called the *generator*. Thus, we now repeat, taking each of the resulting line segments, partitioning them into three equal parts, and so on. Figure 2.31 illustrates the first steps. Self-similarity is built into the construction process, i.e., each of the four parts in the kth step is again a scaled down version — by a factor of 3 — of the entire curve in the previous $(k-1)^{\text{st}}$ step.

Actually, Koch wanted to provide another example for a discovery first made by the German mathematician Karl Weierstraß, who in 1872 had precipitated a minor crisis in mathematics. He had described a curve that could not be differentiated, i.e., a curve which does not admit a tangent at any of its points. The ability to differentiate (i.e., to calculate the slope of a curve from point to point) is a central feature of calculus, which was invented indepen-

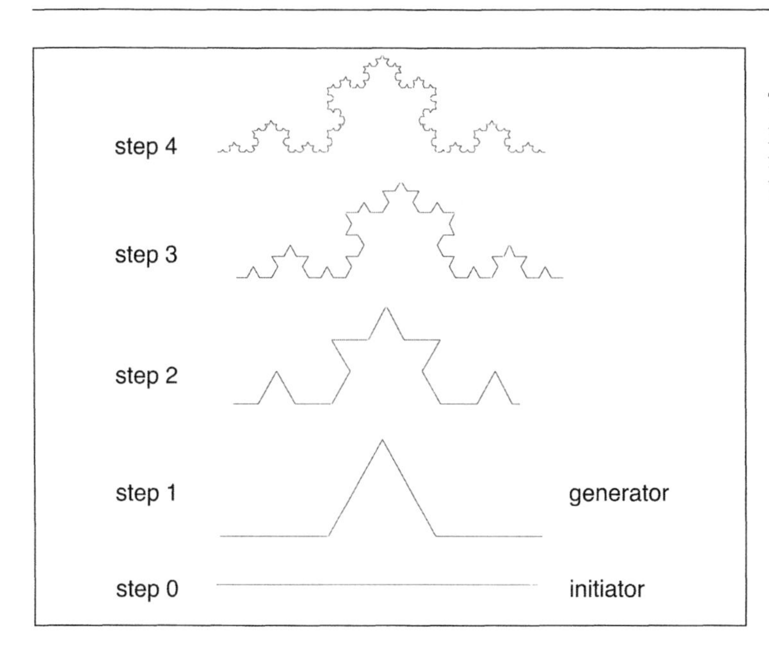

Koch Curve Construction

The construction of the Koch curve proceeds in stages. In each stage the number of line segments increases by a factor of 4.

Figure 2.31

dently by Newton and Leibniz some 200 years before Weierstraß. The idea of slope is a fairly intuitive one and goes hand in hand with the notion of a tangent (see figure 2.32).

If a curve has a corner, then there is a problem. There is no way to fit a unique tangent. The Koch curve is an example of a curve which in a sense is made out of corners everywhere, i.e. there is no way to fit a tangent to any of its points.

Generalized Koch Constructions

It is almost obvious how one can generalize the construction to obtain a whole universe of self-similar structures. Such a Koch construction is defined by an initiator, which may be a collection of line segments, and a generator, which is a polygonal line, composed of a number of connected line segments. Beginning with the initiator, one replaces each line segment by a properly scaled down copy of the generator curve. Here it is necessary to carefully match end points of the line segment and the generator. This procedure is

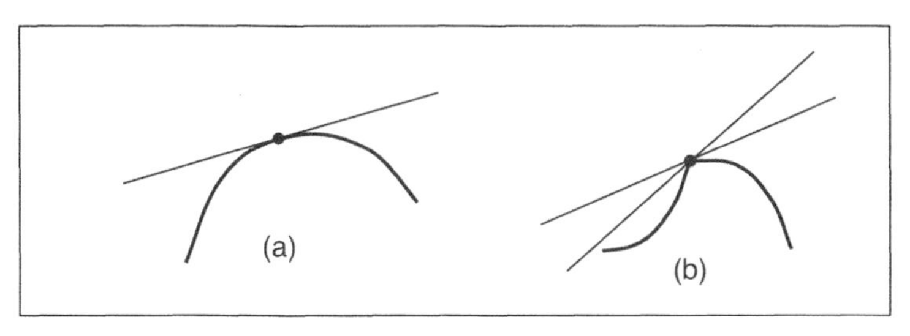

Tangents of Curves

At corners the tangent of a curve is not uniquely defined.

Figure 2.32

Comparing Koch Curve Construction Steps

Construction process of the Koch curve, step 5 (top) and step 20 (bottom).

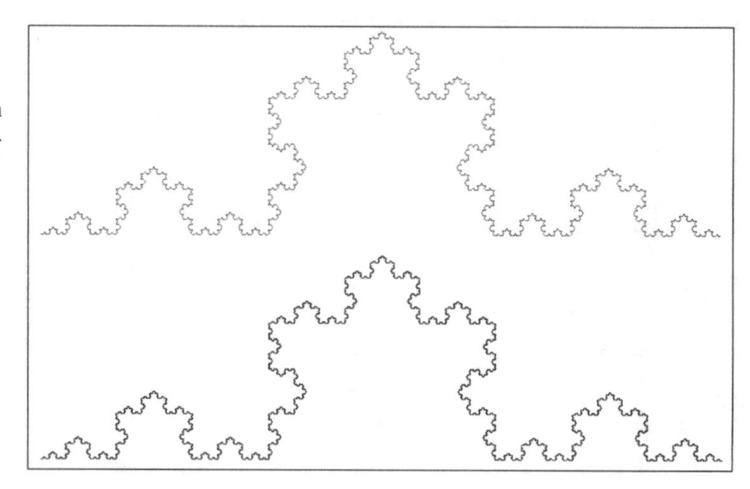

Figure 2.33

repeated ad infinitum. In practice, of course, one stops as soon as the length of the longest line segment in the graph is below the resolution of the graphics device. Whether or not the Koch construction yields a converging sequence of images or even curves depends on the choice of initiator and generator. Figure 2.34 shows an example.

The Length of the Koch Curve

Let us return to the original Koch curve and discuss its length. In each stage we obtain a curve. After the first, we are left with a curve which is made up of four line segments of the same length, after the second step we have 4×4, and then $4 \times 4 \times 4$ line segments after the third step, and so on. If the original line had length L, then after the first step a line segment has length $L \times 1/3$, after the second step we have $L \times 1/3^2$, then $L \times 1/3^3$, and so on. Since each of the steps produces a curve of line segments, there is no problem in measuring their respective lengths. After the first step it is $4 \times L \times 1/3$, then $4^2 \times L \times 1/3^2$, and so on. After the kth step, it is $L \times 4^k/3^k$. We observe that from step to step the length of the curves grows by a factor of 4/3.

Now there are several problems. First of all, the Koch curve is the object which one obtains if one repeats the construction steps infinitely often. But what does that mean? Next, even if we could answer this question, why is it a curve which comes out? Or, why is it that the curves in each step do not intersect themselves?

In figure 2.33 we see two curves which we can hardly distinguish. But they are different. The top one shows the result of the construction after 5 steps, while the other curve shows the result after 20 steps. In other words, since the length of the individual line segments is $1/3^k$, where k is the number of steps, it is clear that any of the changes in the construction are soon below visibility unless one works under a microscope. Thus, for practical purposes, one is tempted to be satisfied with a display of something like the 10th step, or whatever is appropriate to fool the eye. But such an object is not the Koch curve. It would have finite length and would still show its straight line con-

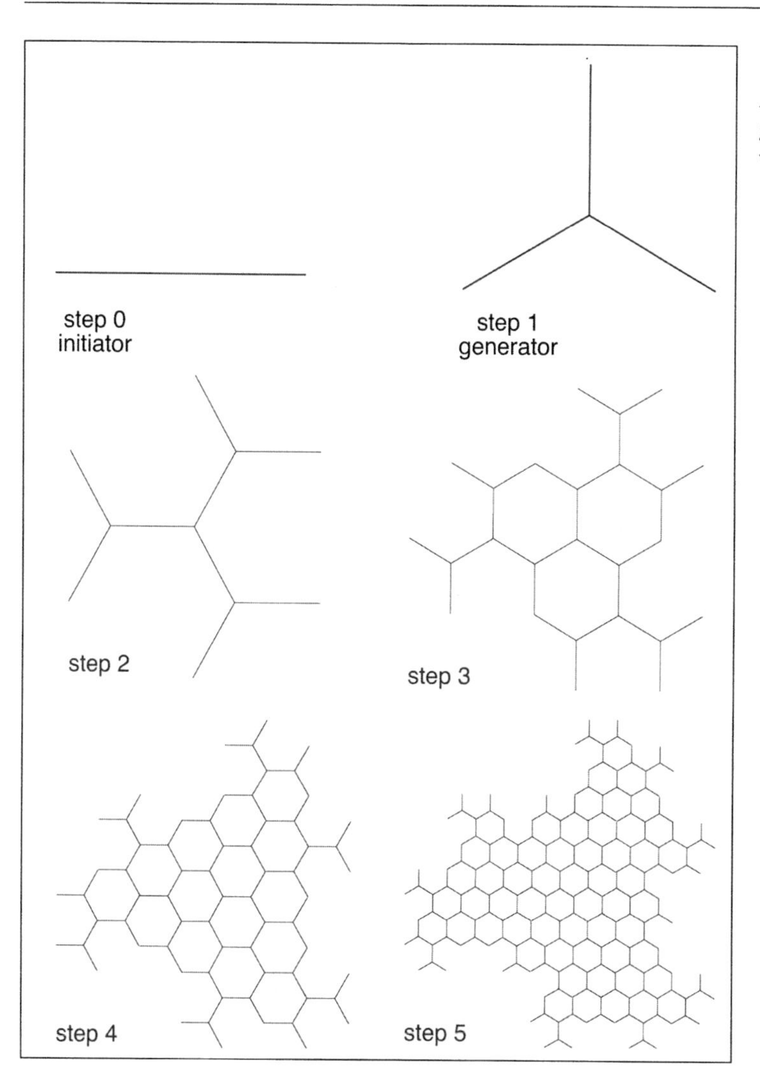

Another Koch Construction

A different choice of initiator and generator produces another fractal with self-similarities.

Figure 2.34

struction segments under sufficient magnification. It is of crucial importance to distinguish between the objects which we obtain at any (single) step in the construction and the final object. We will pick up this difficulty, which of course is also present in the previous classical fractals, in the following chapters.

2.5 Space-Filling Curves

Talking about dimension in an intuitive way, we perceive lines to be typical for one-dimensional objects and planes as typical for two-dimensional objects. In 1890 Giuseppe Peano[14] (1858–1932) and immediately after that in 1891, David Hilbert[15] (1862–1943), discussed curves which live in a plane and which dramatically demonstrate that our naive idea about curves is very limited.[16] They discussed curves which 'fill' a plane, i.e., given some patch of the plane, there is a curve which meets every point in that patch. Figure 2.35 indicates the first steps of the iterative construction of Peano's original curve.

Space-Filling Structures in Nature

In Nature the organization of space-filling structures is one of the fundamental building blocks of living beings. An organ must be supplied with the necessary supporting substances such as water and oxygen. In many cases these substrates will be transported through vessel systems that must reach every point in the volume of the organ. For example, the kidney houses three interwoven tree-like vessel systems, the arterial, the venous, and the urinary systems (see the color plates). Each one of them has access to every part of the kidney. Fractals solve the problem of how to organize such a complicated structure in an efficient way. Of course, this was not what Peano and Hilbert were interested in almost 100 years ago. It is only now, after Mandelbrot's work, that the omnipresence of fractals becomes apparent.

Construction with Initiator and Generator

The Peano curve is obtained by another version of the Koch construction. We begin with a single line segment, the initiator, and then substitute the segment by the generator curve as shown in figure 2.35. Apparently the generator has two points of self-intersection. More precisely, the curve touches itself at two points. Observe that this generator curve fits nicely into a square, which is shown in dotted lines. It is this square whose points will be reached by the Peano curve.

Let us carefully describe the next step. Take each straight line piece of the curve in the first stage and replace it by the properly scaled down generator. Obviously, the scaling factor is 3. This constitutes stage 2. There are a total of 32 self-intersection points in the curve. Now we repeat, i.e., in each step, line segments are scaled down by a factor of 3. Thus, in the kth step, a line segment has length $1/3^k$, which is a very rapidly declining number. Since each line segment is replaced by nine line segments of one-third the length of the previous line segments, we can easily calculate the length of the curves in each step. Assume that the length of the original line segment constituting the initiator was 1, then we obtain in stage 1: $9 \times 1/3 = 3$, and stage 2: $9 \times 9 \times 1/3^2 = 9$. Expressed as a general rule, in each step of the construction, the resulting curve increases in length by a factor of 3. In stage k, the length thus is 3^k.

[14] G. Peano, *Sur une courbe, qui remplit toute une aire plane*, Math. Ann. 36 (1890) 157–160.

[15] D. Hilbert, *Über die stetige Abbildung einer Linie auf ein Flächenstück*, Math. Ann. 38 (1891) 459–460.

[16] Hilbert introduced his example in Bremen, Germany, during the annual meeting of the *Deutsche Gesellschaft für Naturforscher und Ärzte*, which was the meeting at which he and Cantor were instrumental in founding the *Deutsche Mathematiker Vereinigung*, the German Mathematical Society.

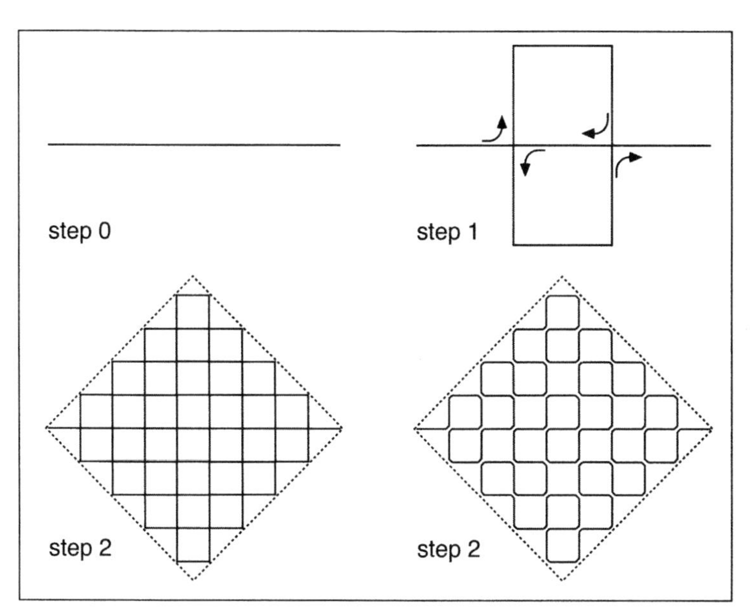

Peano Curve Construction

Construction of a plane-filling curve with initiator and generator. In each step, one line segment is replaced by nine line segments scaled down by a factor of 3. For reasons of clarity the corners in these polygonal lines, where the curve may intersect itself, have been slightly rounded.

Figure 2.35

Self-Similarity

The Peano curve construction, though as easy, or as difficult, as the construction of the Koch curve, bears within it several difficulties which did not occur or were hidden in the latter (construction). For example, take the intuitive concept of self-similarity. For the construction of the Koch curve, it seemed that we could say that the final curve (i.e., the curve which you see on a graphics terminal after many steps) has similarity with each of the preceding steps. If you look at the Peano curve in the same intuitive way, each of the steps has similarity with the preceding steps; but if you look at the final curve (i.e., the curve which results after many steps on a graphics terminal), essentially we see a 'filled out' square which doesn't look at all similar to the early steps of the construction. In other words, either the Peano curve is not self-similar, or we have to be much more careful in describing what self-similarity means. In fact, we will see in chapter 7 that the Peano curve is perfectly self-similar. The problem is to 'see' the final object as a curve because, in any graphical representation, it 'looks' much more like a piece of the plane.

Parametrization of a Square by the Peano Curve

Let us explore the space-filling property a bit further. When you look at the displayed stages of the development of the curve, you notice that approximately the first $1/9^{th}$ of the curve stays within the left subsquare, and in fact, seems to fill just that area. Corresponding statements hold for the other subsquares. You will also notice that each subsquare can also be tiled into nine sub-subsquares, each one reduced by 1/9 when compared to the whole one. The curve first traces out all tiles of a

Hilbert's Paper — Page 1

The first page of Hilbert's original 1890, 2-page paper with the first visualization of a fractal, his space-filling curve.

Figure 2.36

Ueber die stetige Abbildung einer Linie auf ein Flächenstück.*)

Von

DAVID HILBERT in Königsberg i. Pr.

Peano hat kürzlich in den Mathematischen Annalen**) durch eine arithmetische Betrachtung gezeigt, wie die Punkte einer Linie stetig auf die Punkte eines Flächenstückes abgebildet werden können. Die für eine solche Abbildung erforderlichen Functionen lassen sich in übersichtlicherer Weise herstellen, wenn man sich der folgenden geometrischen Anschauung bedient. Die abzubildende Linie — etwa eine Gerade von der Länge 1 — theilen wir zunächst in 4 gleiche Theile 1, 2, 3, 4 und das Flächenstück, welches wir in der Gestalt eines Quadrates von der Seitenlänge 1 annehmen, theilen wir durch zwei zu einander senkrechte Gerade in 4 gleiche Quadrate 1, 2, 3, 4 (Fig. 1). Zweitens theilen wir jede der Theilstrecken 1, 2, 3, 4 wiederum in 4 gleiche Theile, so dass wir auf der Geraden die 16 Theilstrecken 1, 2, 3, . . . , 16 erhalten; gleichzeitig werde jedes der 4 Quadrate 1, 2, 3, 4 in 4 gleiche Quadrate getheilt und den so entstehenden 16 Quadraten

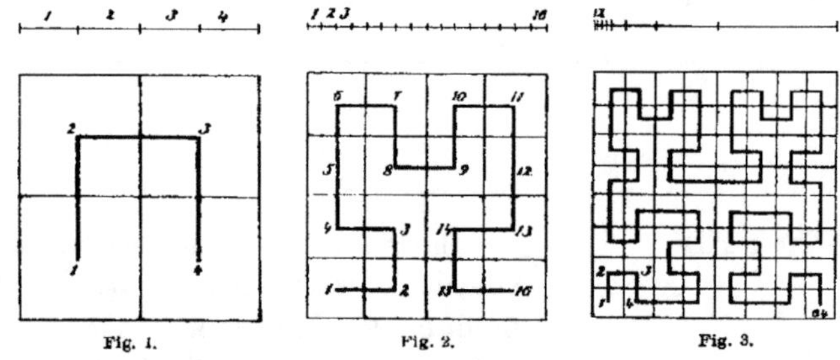

Fig. 1. Fig. 2. Fig. 3.

werden dann die Zahlen 1, 2 . . . 16 eingeschrieben, wobei jedoch die Reihenfolge der Quadrate so zu wählen ist, dass jedes folgende Quadrat sich mit einer Seite an das vorhergehende anlehnt (Fig. 2). Denken wir uns dieses Verfahren fortgesetzt — Fig. 3 veranschaulicht den

*) Vergl. eine Mittheilung über denselben Gegenstand in den Verhandlungen der Gesellschaft deutscher Naturforscher und Aerzte. Bremen 1890.

**) Bd. 36, S. 157.

30*

subsquare before it enters the next subsquare. This process goes on and on through all stages of the curves.

The implication of this is as follows: if we trace out a stage of the Peano curve up to a certain percentage, let us say to 10/27 of its total length, i.e., about 37%, then we end up at a certain point in the square. Now we go to the next stage and again trace out 37% of the new, longer curve (see figure 2.38). Again we end up at a point in the square, and this point is not far from the first one. When repeating this procedure for the following stages, we obtain a sequence of points. These points will converge to a unique point in the square. This point may be called the point with address 10/27. In this manner, we can define for all percentages — for all numbers between 0.0 and 1.0 — a point in the

Hilbert's Paper — Page 2

nächsten Schritt —, so ist leicht ersichtlich, wie man einem jeden gegebenen Punkte der Geraden einen einzigen bestimmten Punkt des Quadrates zuordnen kann. Man hat nur nöthig, diejenigen Theilstrecken der Geraden zu bestimmen, auf welche der gegebene Punkt fällt. Die mit den nämlichen Zahlen bezeichneten Quadrate liegen nothwendig in einander und schliessen in der Grenze einen bestimmten Punkt des Flächenstückes ein. Dies sei der dem gegebenen Punkte zugeordnete Punkt. *Die so gefundene Abbildung ist eindeutig und stetig und umgekehrt einem jeden Punkte des Quadrates entsprechen ein, zwei oder vier Punkte der Linie.* Es erscheint überdies bemerkenswerth, dass durch geeignete Abänderung der Theillinien in dem Quadrate sich leicht *eine eindeutige und stetige Abbildung finden lässt, deren Umkehrung eine nirgends mehr als dreideutige ist.*

Die oben gefundenen abbildenden Functionen sind zugleich einfache Beispiele für überall stetige und nirgends differentiirbare Functionen.

Die mechanische Bedeutung der erörterten Abbildung ist folgende: *Es kann sich ein Punkt stetig derart bewegen, dass er während einer endlichen Zeit sämmtliche Punkte eines Flächenstückes trifft.* Auch kann man — ebenfalls durch geeignete Abänderung der Theillinien im Quadrate — zugleich bewirken, *dass in unendlich vielen überall dichtvertheilten Punkten des Quadrates eine bestimmte Bewegungsrichtung sowohl nach vorwärts wie nach rückwärts existirt.*

Was die analytische Darstellung der abbildenden Functionen anbetrifft, so folgt aus ihrer Stetigkeit nach einem allgemeinen von K. Weierstrass bewiesenen Satze*) sofort, dass diese Functionen sich in unendliche nach ganzen rationalen Functionen fortschreitende Reihen entwickeln lassen, welche im ganzen Intervall absolut und gleichmässig convergiren.

Königsberg i. Pr., 4. März 1891.

*) Vergl. Sitzungsberichte der Akademie der Wissenschaften zu Berlin, 9. Juli 1885.

Figure 2.37

square. These points will form a curve that passes through every point in the square! Using mathematical terms, we say that "the square can be parametrized by the unit interval". Thus, a curve, which by nature is something one-dimensional, can fill something two-dimensional. It seems that the use of the intuitive notion of dimension here is rather slippery.

To make the argument precise, one would have to introduce an addressing system, which for the case of the Peano curve would be based on strings composed of nine symbols, or digits. For each point in the square there is an address, which is an infinite string. This string also identifies points from each stage of the Peano curve construction.

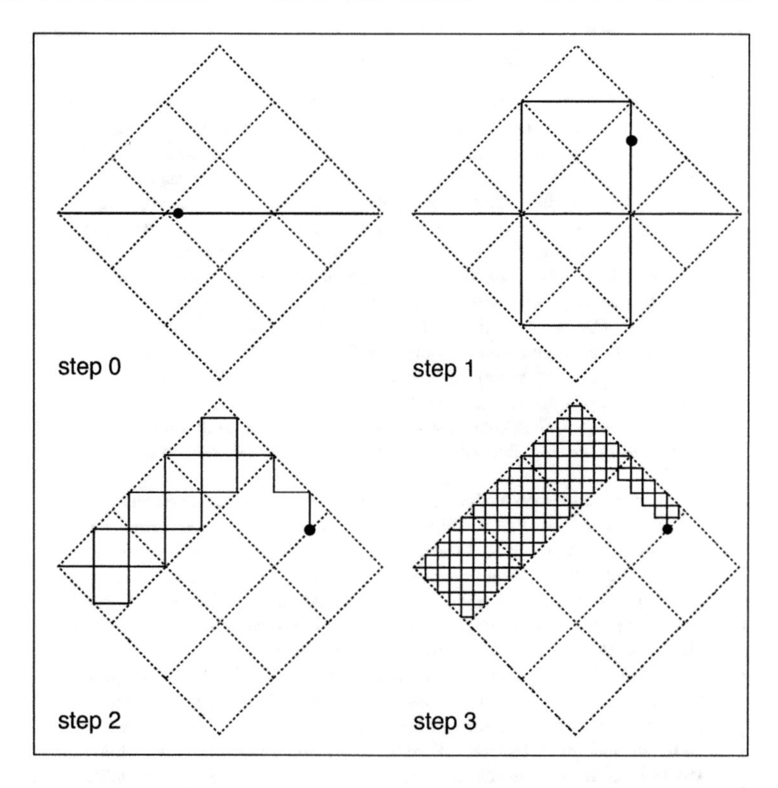

Figure 2.38 : The Peano curves of four different stages are traced out to $1/3 + 1/27 = 10/27$ of the total length. The rest of the curve is not shown in the bottom figures. The parameter $10/27$ identifies a point marked in each graph. These points converge to a unique point in the square as the number of steps increases.

That sequence of points (one from each stage) will converge to the original point in the square.

Is There a Better Way to Fill a Square by a Curve?

The space-filling Peano curve, or rather any finite stage of the construction, certainly is very awkward to draw by hand or even by a plotter under computer control. The number of small line segments that must be drawn to fill the square is just enormous. Moreover, there is a 90 degree turn after every segment. Therefore it is fair to ask whether there is a much simpler way to fill a square with a line. Think of how you would approach that problem with a pencil in your hand. It seems the easiest way would be to just draw a zigzagging line from one side of the square to the other, making sure that the turns are narrow enough in order to avoid any white spaces on the paper.

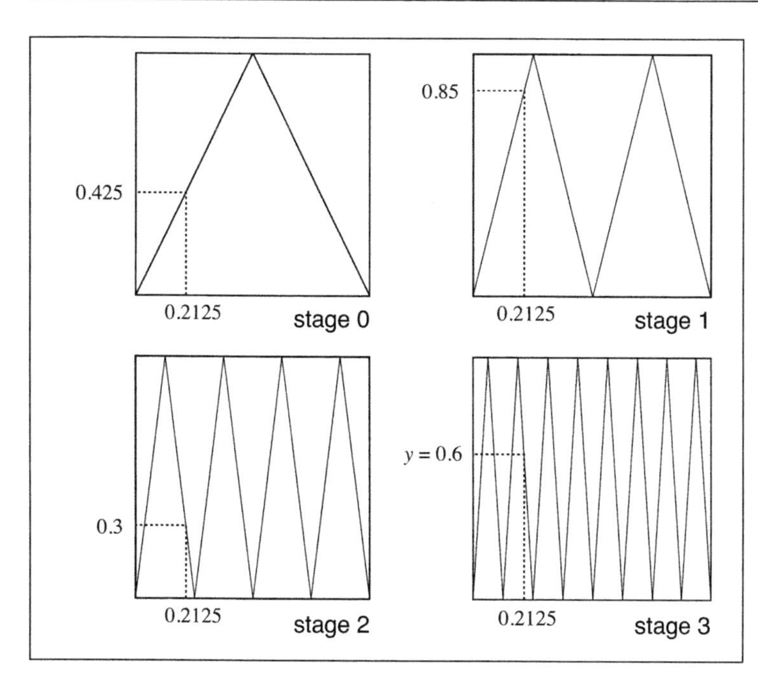

Filling a Square the Naive Way

The first four stages in an attempt to construct another space-filling curve using zigzag curves.

Figure 2.39

The Naive Construction ...

Let us put this procedure into a more mathematical description analogous to the stages in the Peano curve construction. Stage 0 is a simple line from the lower left corner to the middle of the top side of the square and back to the base line ending at the lower right corner. For the next stage let us double the resolution in the sense that a horizontal line somewhere in the middle of the square will intersect the next curve at twice as many points. This is easy to achieve just by doing two zigzags at half the distance (see figure 2.39).

It is obvious how to continue the construction. For each stage we just double the number of zigzags. For any given resolution $\varepsilon > 0$ we can certainly find a stage at which the generated curve passes by every point in the square with a distance less than ε, and we would say that the job is done. Moreover, we could even claim to have invented a space-filling curve, which is in some sense self-similar, because in each stage the curve is composed of two copies of the curve of the previous stage, properly scaled in the horizontal direction.[17] Any child will accomplish something like this at an early age. Certainly, the brilliant minds of Peano and Hilbert must have been aware of this. Then, what was it that drove them to invent such complicated constructions, which then were even accepted for publication in most renowned mathematical journals?

[17]For cases like this one, where the scaling factor is different in different directions, the term *self-affine* is more appropriate. Affine transformations are discussed further in chapters 5 and 6.

...Does Not Lead Anywhere

The answer seems contra-intuitive, but logical after a little bit of analysis. What comes out of the Peano curve construction in the limit is a curve, as pointed out in the technical section starting on page 93. This curve has infinite length, self-similarities and it reaches every point in the square. By contrast, the naive construction from above will *not* lead to a curve, although every stage of it is a curve! Let us explore this astonishing fact. We label the horizontal and vertical axes of the square by x and y, which both range from 0 to 1. A curve from some stage, say the nth stage, of the construction then is given by the graph of a zigzag function, which we call y_n. Let us now fix some coordinate x between 0 and 1 and look at the corresponding values $y_n(x)$ as the stage n increases. If the construction really leads to a well-defined limit curve, then one must expect that the sequence of points $y_1(x), y_2(x), \ldots$ also converges, namely, to the y-value of the limit curve at position x. Clearly this is true for all points x which allow a finite binary representation such as 1/4 or 139/256 because by construction at such points all curves will have a y-value of 0, provided the stage is large enough. But then there are other points which clearly violate the crucial convergence property. For example, at $x = 1/7$ the y-values of the curves are 2/7, 4/7, 6/7, 2/7, 4/7, 6/7, ... and so on in a periodic fashion. Therefore, there is no limit object, no space-filling curve, and no new insight. This naive way of filling the square is essentially the same as just filling a finite array of pixels in an image by assigning 'black' to every pixel. After a certain number of steps we are done, and continuing for higher resolutions would not make sense. There is no self-similarity and certainly not a fractal behind the picture. So this is the real ingenuity of Peano and Hilbert — they created a 'monster' with unforeseen properties which were never thought possible before.

Analysis of the Naive Approach to Space-Filling

It is not hard to analyze the sequence of curves from the naive space-filling construction. For this purpose let us introduce the periodically extended tent transformation

$$h(x) = \begin{cases} 2\,\text{frac}(x) & \text{if } \text{frac}(x) \leq 0.5 \\ 2(1 - \text{frac}(x)) & \text{if } \text{frac}(x) > 0.5, \end{cases}$$

where $\text{frac}(x)$ denotes the fractional part of x, i.e.,

$$\text{frac}(x) = x - \max\{k \mid k \leq x, k \text{ integer}\}.$$

With this notation we can write the curves in the construction simply as

$$\begin{aligned} y_0(x) &= h(x), \\ y_1(x) &= h(2x), \\ y_2(x) &= h(4x), \\ &\vdots \\ y_n(x) &= h(2^n x), \\ &\vdots \end{aligned}$$

where in all cases x ranges from 0 to 1. We see that only the fractional parts of $x, 2x, 4x, \ldots, 2^n x, \ldots$ determine the y-values of the curves at position x. Considering the example $x = 1/7$ from the text, we now see that

$$\begin{aligned} \text{frac}\,(1/7) &= 1/7 \\ \text{frac}\,(2/7) &= 2/7 \\ \text{frac}\,(4/7) &= 4/7 \\ \text{frac}\,(8/7) &= 1/7 \end{aligned}$$

Thus, the fractional part of $16/7$ again is $2/7$ and so on in a periodic fashion. Applying the tent transformation to these fractional parts $1/7, 2/7, 4/7, 1/7, \ldots$ finally yields the sequence of values $2/7, 4/7, 6/7, 2/7, 4/7, 6/7, \ldots$ as claimed in the text. Therefore the limit

$$\lim_{n\to\infty} y_n(1/7)$$

does not exist, and the sequence of curves $y_0, y_1, \ldots$ cannot have a limit curve.

To conclude we ask whether our choice of $x = 1/7$ for the convergence test is a rather artificial and rare case. The answer here is no. In fact, it is true for almost all positions x that the sequence $y_0(x), y_1(x), \ldots$ has no limit. Let us briefly elaborate. The fractional parts of $2^n x$ are most easily found, when the position x is given in a binary representation. The example from the text, $x = 1/7$, has a binary expansion $0.001001\ldots$:

$$x = \frac{1}{8} + \frac{1}{64} + \frac{1}{512} + \cdots = \frac{\frac{1}{8}}{1 - \frac{1}{8}} = \frac{1}{7}.$$

Multiplying a binary number by 2 is equivalent to just shifting all binary digits one position to the left. Taking the fractional part of the result means deleting any leading digits before the binary point. For example, the repeated binary shifts of $1/7$ are $0.010010\ldots$, $0.100100\ldots$, and $0.001001\ldots$, which is equal to $1/7$ again. The complete operation is also known as the binary shift, and we remark in passing that it is central to the analysis of chaos in chapter 10. Applying the binary shift repeatedly to a number is thus the same as placing that number at the corresponding position on an infinitely precise ruler, and looking at it through a microscope, increasing the magnifying power continuously by a factor of two. If we take a random number between 0 and 1, with random binary digits, then the binary shift will produce a sequence of random numbers, which certainly will never settle down to a limiting value. Because 'most' numbers have random digits, the lack of convergence with regard to our naive curve construction is typical and not the exception.

Two Dithering Methods

Dithering with the Hilbert curve (right) versus traditional dithering is shown in the upper row. The squares are continuously shaded from white (lower left corner) to black (upper right corner). The bottom row shows the test image "Lenna" (left) and a dithering using the Hilbert curve (right).

Figure 2.40

An Application of Space-Filling Curves

It may seem that space-filling curves are mostly an academic curiosity — regarded as 'monsters' initially. However, they became important roots in Mandelbrot's development of fractals as models of nature. Moreover — and this may come as a real surprise — those early monsters are also good for down-to-earth technical applications 100 years after their discovery. Let us briefly describe an image processing application published at the prestigious SIGGRAPH[18] convention in 1991. It introduces a novel digital halftoning technique useful to render a grey-scale image on a bilevel graphic device such as a laser printer.[19] The problem lies in the fact that a printer renders a bitmap, an array of black and white pixels, while shades of grey cannot be represented at the pixel level. To cope with this difficulty, several so-called dithering techniques have been used. They are based on scanning a given grey-scale image line by line or in small square blocks. A black and white approximation of the image is produced with the objective to minimize an overall error. Usually, there are artefacts in the result which make it obvious that a dithering process was involved. How can space-filling curves help? Imagine a Hilbert curve that passes through all pixels of the given grey-scale image. The curve offers an alternative to scanning the image line by line,

[18]SIGGRAPH is the Special Interest Group Graphics of the Association for Computing Machinery (ACM). Their yearly conventions draw about 30,000 professionals from the field of computer graphics.

[19]L. Velho, J. de Miranda Gomes, *Digital Halftoning with Space-Filling Curves*, Computer Graphics 25,4 (1991) 81–90.

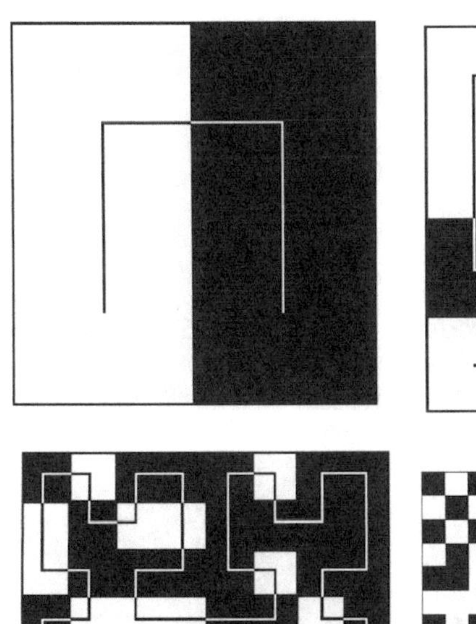

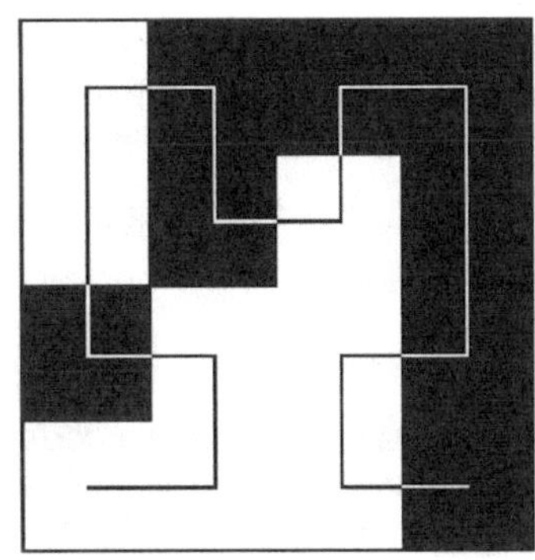

Dithering with the Hilbert Curve

The principle of the dithering algorithm based on four successive stages of the Hilbert scan of an image. The same shaded square is used as in figure 2.40.

Figure 2.41

namely, to sample the image pixel by pixel along the Hilbert curve. Now a sequence of consecutive pixels along this convoluted path can be replaced by a black and white approximation. The advantage of the image scan along the Hilbert curve is that it is free of any directional features present in the traditional methods. It produces aperiodic patterns of clustered dots which are perceptually pleasant with characteristics similar to photographic grain structures. Figure 2.40 compares the traditional approach, called clustered-dot ordered dither, with the new method. Besides this dithering algorithm, there are other earlier applications of space-filling curves in image processing.[20]

Hilbert Curve Dithering Algorithm

Let us describe the details of a simplified version of the dithering algorithm with the Hilbert curve. We consider a square image with continuously varying grey shades which must be approximated by an image which may contain only black and white pixels. The resolution of the output image must be a power of 2. For example, we consider images with 2, 4, 8, and 16 pixels per row and per column in figure 2.41. As shown for the first few of these cases, we can fit a corresponding Hilbert curve exactly to such a tiling of the image. This introduces an ordering

[20] R. J. Stevens, A. F. Lehar, F. H. Preston, *Manipulation and Presentation of Multidimensional Image Data Using the Peano Scan,* IEEE Transactions on Pattern Analysis and Machine Intelligence 5 (1983) 520–526.

of the pixels. For the 4 by 4 pixel example, where we label columns by letters A, B, C, and D, and rows by 1, 2, 3, and 4, the pixels of the image are ordered as follows

$$A1, B1, B2, A2, \dots, D2, C2, C1, D1.$$

Let us denote by

$$I_1, I_2, \dots, I_n$$

the intensity values of the corresponding pixels of the shaded input image (ranging from 0 for black to 1 for white). Here n is a power of 2, the total number of pixels in the image. For the definition of the output image we have to compute corresponding intensity values

$$O_1, O_2, \dots, O_n \in \{0, 1\}.$$

To begin we set

$$O_1 = \begin{cases} 0 & \text{if } I_1 \leq 0.5 \\ 1 & \text{if } I_1 > 0.5. \end{cases}$$

This approximation carries an error

$$E_1 = I_1 - O_1.$$

Instead of ignoring this error we can pass it along to the next pixel in the sequence. More precisely, we set

$$\begin{aligned} O_k &= \begin{cases} 0 & \text{if } I_k + E_{k-1} \leq 0.5 \\ 1 & \text{if } I_k + E_{k-1} > 0.5 \end{cases} \\ E_k &= I_k + E_{k-1} - O_k. \end{aligned}$$

In other words, the error diffuses along the sequence of pixels. The goal of this error diffusion is to minimize the overall error in the intensities averaged over blocks of various sizes of the image. For example, we have that the errors, summed up over the complete image, are equal to

$$\sum_{k=1}^{n} E_k = E_n$$

which is expected to be relatively small. The crucial point of the algorithm is that the error diffuses along the Hilbert curve which traces out the image in a way that is conceived as very irregular to our sensory system. If we replace the Hilbert curve for example by a curve which scans the image row by row, the regular error diffusion will produce disturbing artefacts. The algorithm proposed by Velho and de Miranda Gomes at SIGGRAPH is a generalization of this method. It considers blocks of consecutive pixels from the Hilbert scan at a time, instead of individual pixels.[21]

[21] The simplified version presented here was first published in I. H. Witten and M. Neal, *Using Peano curves for bilevel display of continuous tone images,* IEEE CG&A, May 1982, 47–52.

Conclusion

In conclusion, we have shown that the notion of self-similarity in a strict sense requires a discussion of the object which finally results from the constructions of the underlying feedback systems; and it can be dangerous to use the notion without care. One must carefully distinguish between a finite construction stage and the fractal itself. But if that is so, then how can we discuss the forms and patterns which we see in nature, as for example the cauliflower, from that point of view?

The cauliflower shows the same forms — clusters are composed of smaller clusters of essentially the same form — over a range of several, say five or six, magnification scales. This suggests that the cauliflower should be discussed in the framework of fractal geometry very much like our planets are suitably discussed for many purposes as perfect spheres within the framework of Euclidean geometry. But a planet is not a perfect sphere and the cauliflower is not perfectly self-similar. First, there are imperfections in self-similarity: a little cluster is not an exact scaled down version of a larger cluster. But more importantly, the range of magnification within which we see similar forms is finite. Therefore, fractals can only be used as models for natural shapes, and one must always be aware of the limitations.

2.6 Fractals and the Problem of Dimension

The invention of space-filling curves was a major event in the development of the concept of dimension. They questioned the intuitive perception of curves as one-dimensional objects, because they filled the plane (i.e., an object which is intuitively perceived as two-dimensional). This contradiction was part of a discussion which lasted several decades at the beginning of this century. Talking about fractals we usually think of the fractal dimension, Hausdorff dimension or boxcounting dimension (we will discuss this in detail in chapter 4), but the original concepts reside in the early development of topology.

A World Behaving Like Rubber

Topology is a branch of mathematics which has essentially been developed in this century. It deals with questions of form and shape from a qualitative point of view. Two of its basic notions are 'dimension' and 'homeomorphism'. Topology deals with the way shapes can be pulled and distorted in a space that behaves like rubber.

Circle, Square and Koch Island

A circle can be continuously deformed into a triangle. A triangle can be deformed into a Koch Island. Topologically they are all equivalent.

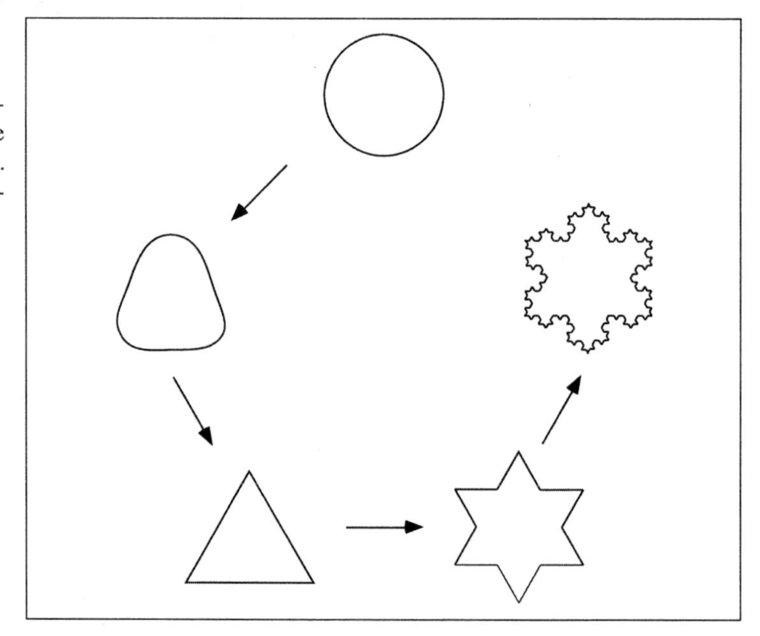

Figure 2.42

In topology straight lines can be bent into curves and circles can be pinched into triangles or pulled out as squares. For example, from the point of view of topology a straight line and the Koch curve cannot be distinguished. Or the coast of a Koch island is the same as a circle. Or a plain sheet of paper is equivalent to one which is infinitely crumpled. However, not everything is topologically changeable. Intersections of lines, for example remain intersections. In mathematicians' language an intersection is invariant; it cannot be destroyed nor can new ones be born, no matter how much the lines are stretched and twisted. The number of holes in an object is also topologically invariant, meaning that a sphere may be transformed into the surface of a horseshoe, but never into a doughnut. The transformations which are allowed are called homeomorphisms,[22] and when applied, they must not change the invariant properties of the objects. Thus, a sphere and the surface of a cube are homeomorphic, but the sphere and a doughnut are not.

Topological Equivalence

We have mentioned already that a straight line and the Koch curve are topologically the same. Moreover, a straight line is the prototype of an object which is of dimension one. Thus, if the concept of dimension is a topological notion, we would expect that the Koch curve also has topological dimension one. This is, however, a rather delicate matter and it troubled mathematicians around the turn of the century.

The history of the various notions of dimension involves the greatest mathematicians of that time: men like H. Poincaré, H. Lebesgue, L. E. J. Brouwer, G. Cantor, K. Menger, W. Hurewicz, P. Alexandroff, L. Pontrjagin, G. Peano, P. Urysohn, E. Čech, and D. Hilbert. That history is very closely related to the creation of the early fractals. Hausdorff remarked that the problem of creating the right notion of dimension is very complicated. People had an intuitive idea about dimension: the dimension of an object, say X, is the number of independent parameters (coordinates), which are required for the unique description of its points.

Poincaré's idea was inductive in nature and started with a point. A point has dimension 0. Then a line has dimension 1, because it can be split into two parts by a point (which has dimension 0). And a square has dimension 2, because it can be split into two parts by a line (which has dimension 1). A cube has dimension 3, because it can be split into two parts by a square (which has dimension 2).

Topological Dimension

In the development of topology, mathematicians looked for qualitative features which would not change when the objects were transformed properly (technically by a homeomorphism). The (topological) dimension of an object certainly should be preserved. But it turned out that there were severe difficulties in arriving at a proper and detailed notion of dimension which would behave that way. For example, in 1878 Cantor found a transformation f from the unit interval $[0, 1]$ to the unit square $[0, 1] \times [0, 1]$ which was one-to-one and onto.[23] Thus it seemed that we need only one parameter for the description of the points in a square. But Cantor's transformation is not a homeomorphism. It is not continuous, i.e., it does not yield a space-filling *curve*!

But then the plane-filling constructions of Peano and later Hilbert yielded transformations g from the unit interval $[0, 1]$ to the unit square $[0, 1] \times [0, 1]$ which were even continuous. But they were not one-to-one (i.e., there are points, say x_1 and x_2, $x_1 \neq x_2$, in the unit interval which are mapped to the same point of the square $y = g(x_1) = g(x_2)$).

[22] Two objects X and Y (topological spaces) are homeomorphic if there is a homeomorphism $h : X \to Y$ (i.e., a continuous one-to-one and onto mapping that has a continuous inverse h^{-1}).

[23] The notion 'onto' means here that for every point z of the unit square there is one point x in the unit interval that is mapped to $z = f(x)$.

Construction of the Menger Sponge

An object which is closely related to the Sierpinski carpet is the Menger sponge, after Karl Menger (1926). Take a cube, subdivide its faces into nine congruent squares and drill holes as shown from each central square to the opposite central square (the cross-section of the hole must be a square). Then subdivide the remaining eight little squares on each face into nine smaller squares and drill holes again from each of the central little squares to their opposite ones, and so on.

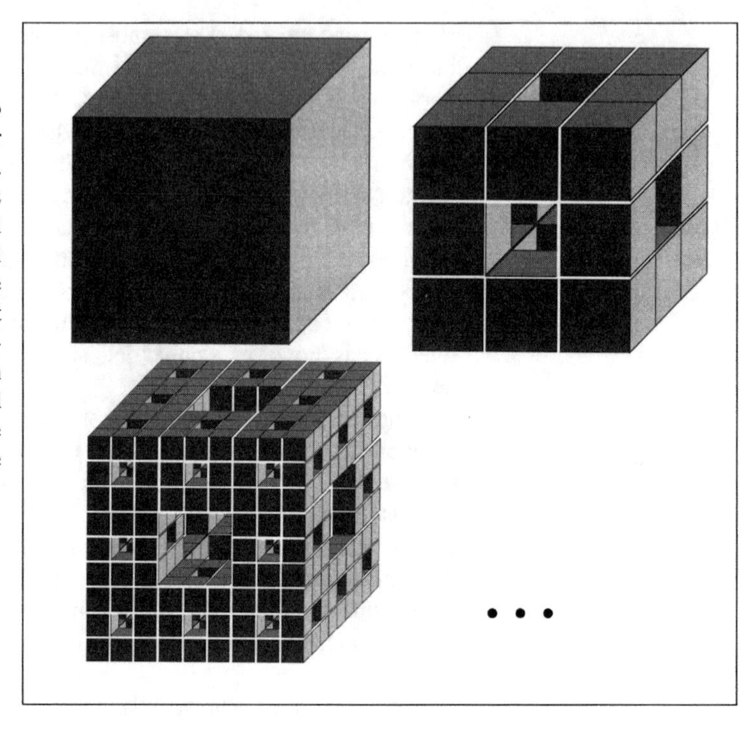

Figure 2.43

This raised the question — which so far seemed to have an obvious answer — whether or not there is a one-to-one and onto transformation between $I = [0,1]$ and $I^2 = [0,1] \times [0,1]$ which is continuous in both directions. Or more generally, is the n-dimensional unit cube $I^n = [0,1]^n$ homeomorphic to the m-dimensional one, $I^m = [0,1]^m, n \neq m$? If there were such a transformation, mathematicians felt that they were in trouble: a one-dimensional object would be homeomorphic to a two-dimensional one. Thus, the idea of topological invariance would be wrong.

Line and Square Are Not Equivalent

Between 1890 and 1910 several 'proofs' appeared showing that I^n and I^m are not homeomorphic when $n \neq m$, but the arguments were not complete. It was the Dutch mathematician Brouwer who ended that crisis in 1911 by an elegant proof, which enriched the development of topology enormously. But the struggle for a suitable notion of dimension and a proof that obvious objects — like I^n — had obvious dimensions went on for two more decades. The work of the German mathematician Hausdorff (which led eventually to the fractal dimension) also falls in this time span.

During this century mathematicians came up with many different notions of dimension (small inductive dimension, large inductive dimension, covering dimension, homological dimension).[24] Several of them are topological in nature; their value is always a natural number (or 0 for points) and does not

[24]C. Kuratowski, *Topologie II*, PWN, 1961. R. Engelking, *Dimension Theory*, North Holland, 1978.

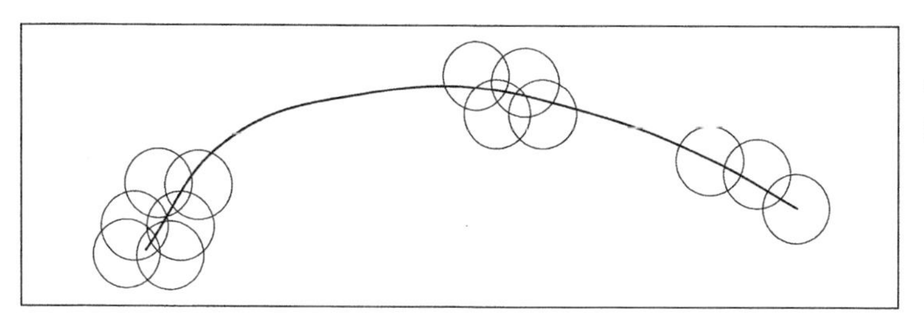

Covering a Curve

Three different coverings of a curve by circles.

Figure 2.44

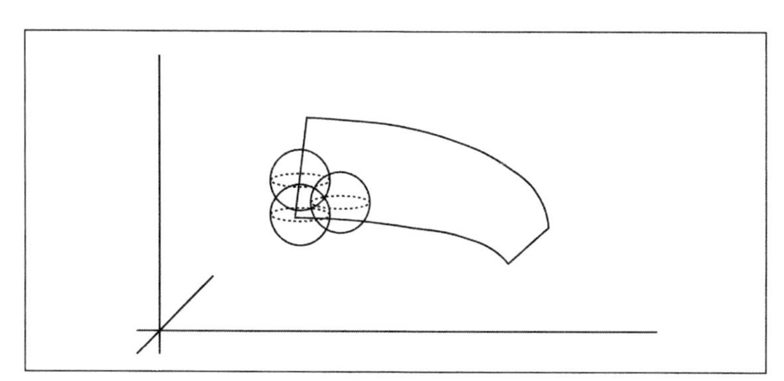

Covering a Surface

The covering of a surface by balls.

Figure 2.45

change for topologically equivalent objects. As an example of these notions we will discuss the covering dimension. Other notions of dimension capture properties which are not at all topologically invariant. The most prominent one is the Hausdorff dimension. The Hausdorff dimension for the straight line is 1, while for the Koch curve it is $\log 4/\log 3$. In other words, the Hausdorff dimension has changed, though from a topological point of view the Koch curve is just a straight line. Moreover, $\log 4/\log 3 = 1.2619\ldots$ is not an integer. Rather it is a fraction, which is typical for fractal objects. Other examples which are of (covering) dimension 1 are the coast of the Koch island, the Sierpinski gasket and also the Sierpinski carpet. Even the Menger sponge, whose basic construction steps are indicated in figure 2.43, is of (covering) dimension 1. Roughly, the gasket, carpet and sponge have (covering) dimension 1 because they contain line elements but no area, or volume elements. The Cantor set is of dimension 0 because it consists of disconnected points and it does not contain any line segment.

The Covering Dimension

Let us now discuss the topologically invariant cover dimension. The idea behind its concept — which is attributable to Lebesgue — is the following observation: let us take a curve in the plane (see figure 2.44), and try to cover it with disks of a small radius. The arrangement of disks on the left part of the curve is very different from the one in the middle, which in turn is very different from the one on the right part of the curve. What is the difference? In the right part we can only find pairs of disks which have nonempty intersection, while in the center part we can find triplets and in the left part even a quadruplet of disks which have nonempty intersection.

This is the crucial observation, which leads to a definition. We say that the curve has covering dimension 1 because we can arrange coverings of the curve with disks of small radius so that there are no triplets and quadruplets, but only pairs of disks with nonempty intersection, and moreover, there is no way to cover the curve with sufficiently small disks so that there are no pairs with nonempty intersection.

This observation generalizes to objects in space (in fact also to objects in higher dimensions). For example, a surface in space (see figure 2.45) has covering dimension 2, because we can arrange coverings of the surface with balls of small radius so that there are no quadruplets but only triplets of balls with nonempty intersection, and there is no way to cover the surface with sufficiently small balls so that there are only pairs with nonempty intersection.

Refinement of Covers and Covering Dimension

We are accustomed to associating dimension 1 with a curve, or dimension 2 with a square, or dimension 3 with a cube. The notion of covering dimension is one way to make this intuition more precise. It is one of several notions in the domain of topological dimensions. Let us first discuss the covering dimension for two examples, a curve in the plane and a piece of a surface in space, in figure 2.44 and figure 2.45, respectively.

We see a curve covered with little disks and focus our attention on the maximal number of disks in the cover which have nonempty intersection. This number is called the order of the cover. Thus, at the left end of figure 2.44 the order is 4, while in the center it is 3, and at the right end it is 2. In figure 2.45 we see a piece of a surface in space covered with balls and the order of the cover is 3.

We have almost arrived at the definition. For that, we introduce the notion of a refinement of a cover. For us, covers of a set X in the plane (or in space) are just collections of finitely many open disks (or balls) of some radius,[25] say $A = \{D_1, \ldots, D_r\}$, such that their union covers X. More precisely we assume that we have a compact metric space X. A finite cover, then, is a finite collection of open sets, such that X is contained in the union of these open sets. An open cover $B = \{E_1, \ldots, E_l\}, l \geq r$ is called a refinement of $A = \{D_1, \ldots, D_r\}$ provided for each E_i there is D_k such that $E_i \subset D_k$. The order of an open cover A is the maximal integer k, such that there are disjoint indices $i_1, \ldots, i_k$ with $D_{i_1} \cap \cdots \cap D_{i_k} \neq \emptyset$. If the intersection of all pairs of sets from a cover is empty, then the order is 1. If a cover has order n then any $n+1$ sets from the cover have empty intersection.

We will now define the covering dimension of X,[26] $dim X$. Let $n \geq 0$ be an integer. Then we define $dim X \leq n$, provided any finite open cover of X has a finite open refinement of order $\leq n+1$. Finally, $dim X = n$, provided $dim X \leq n$, but not $dim X \leq n-1$. In other words, the latter condition means that there is a finite open cover of X

[25] 'Open' means that we consider a disk (resp. ball) without the bounding circle (resp. sphere) or, more generally, unions of such disks (resp. balls).

[26] We assume that X is a compact metric space.

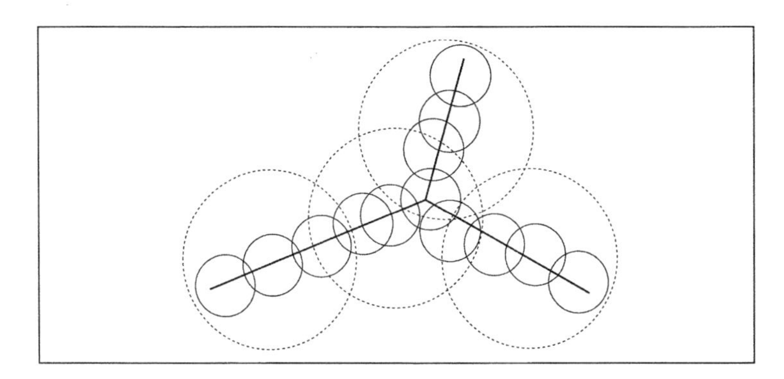

Figure 2.46 : The covering at a branching point and a refinement.

such that all finite open refinements have order $\geq n + 1$.

Figure 2.46 illustrates that the notion of refinement is really crucial. The large cover (dotted circles) covers the y-shaped object so that in one case three disks intersect. But there is a refinement with smaller disks (solid circles, each smaller open disk is contained in a large open disk), such that at most two disks intersect.

Now it is intuitively clear that a finite number of points can be covered so that there is no intersection. Curves can be covered so that the order of the cover is 2 and there is no cover of sufficiently small disks or balls with order 1. Surfaces can be covered so that the order of the cover is 3 and there is no cover of sufficiently small disks or balls with order 2. Thus the covering dimension of points is 0, that of curves is 1, and that of surfaces is 2.

The same ideas generalize to higher dimensions. Moreover, it does not matter whether we consider a curve imbedded in the plane or in space and use disks or balls to cover it.[27]

---

[27] For more details about dimensions we refer to Gerald E. Edgar, *Measure, Topology and Fractal Geometry*, Springer-Verlag, New York, 1990.

2.7 The Universality of the Sierpinski Carpet

We have tried to obtain a feeling for what the topological notion of dimension is and we have learned that from this point of view not only a straight line, but also, for example, the Koch curve is a one-dimensional object. Indeed, from the topological point of view the collection of one-dimensional objects is extremely rich and large, going far beyond objects like the one in figure 2.47, which come to mind at first.

A Tame One-Dimensional Object

This wild-looking curve is far from a really complex one-dimensional object.

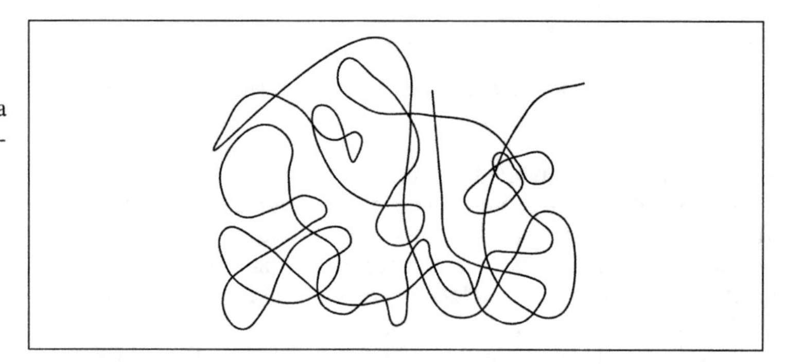

Figure 2.47

The House of One-Dimensional Objects

We are now prepared to get an idea of what Sierpinski was trying to accomplish when he developed the carpet. We want to build a house or hotel for all one-dimensional objects. This house will be a kind of *super-object* which contains all possible one-dimensional objects in a topological sense. This means that a given object may be hidden in the super-object not exactly as it appears independently, but rather as one of its topologically equivalent mutants. Just imagine that the object is made out of rubber and can adjust its form to fit into the super-object. For example, the spider with five arms in figure 2.48 may appear in any one of the equivalent forms in the super-object.

In which particular form a spider with five arms will be hidden in the super-object is irrelevant from a topological point of view. In other words, if one of the arms were as wild as a Koch curve, that would be acceptable too.

Topologically Equivalent Spiders

All these spiders with five arms are topologically equivalent.

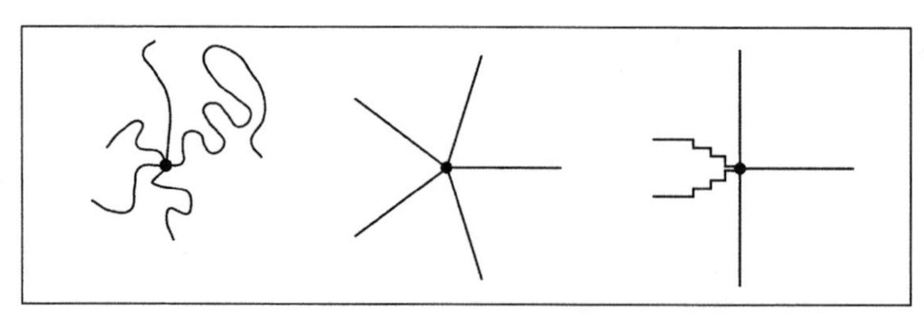

Figure 2.48

Sierpinski's marvelous result[28] in 1916 says that the carpet is such a super-object. In it we can hide any one-dimensional object whatsoever. In other words, any degree of (topological) complexity a one-dimensional object may have must also be present in the Sierpinski carpet. Sierpinski's exact result is:

Fact. *The Sierpinski carpet is universal for all compact*[29] *one-dimensional objects in the plane.*

Let us get some initial idea about the meaning of the above statements. Take a piece of paper and draw a curve (i.e., a typical one-dimensional object) which fits (this makes it compact) on the paper. Try to draw a really complicated one, as complicated as you can think, with as many self-intersections as you may wish. Even draw several curves on top of each other. Whatever complication you may invent, the Sierpinski carpet is always ahead of you. That is, any complication which is in your curve is also in some subset (piece) of the Sierpinski carpet. More precisely, within the Sierpinski carpet we may find a subset which is topologically the same as the object which you have drawn on your paper. The Sierpinski carpet is really a super-object. It looks orderly and tame, but its true nature goes far beyond what can be seen. In other words, what we can see with our eyes and what we can see with our mind are as disparate as they can be. We might say the Sierpinski carpet is a hotel which accommodates all possible (one-dimensional, compact) species living in flatland. But not everything can live in flatland.

Planar and Non-planar Curves

We can draw curves in a plane or in space. But can we draw all curves that we can draw in space also in a plane? At first glance yes, but there is a problem. Take the figure "8" (a) in the plane in figure 2.49, for example.

Is it a real figure "8" (with one self-intersection) or does it just look like a figure "8" because it is a projection of the twisted circle which lies in space as in (b)? Without further explanation it could be both. However, from a qualitative point of view a figure "8" is very different from a circle because it has a self-intersection and it separates the plane into three regions rather than only two as for a circle. Thus, topologically we have to distinguish them from each other. The curve in (b) is a circle from a topological point of view and can be embedded easily into a plane without self-intersections once we untwist it.

This raises the question whether any curve in space can be embedded into a plane without changing its topological character. The answer is no. The WG&E example in figure 2.50 is a simple illustration. Imagine that we have three houses A, B, and C which have to be supplied with water, gas and electricity from W, G, and E so that the supply lines do not cross (if drawn in a plane). There is no way to

[28] W. Sierpinski, *Sur une courbe cantorienne qui contient une image biunivoquet et continue detoute courbe donnée* C. R. Acad. Paris 162 (1916) 629–632.

[29] Compactness is a technical requirement which can be assumed to be true for any drawing on a sheet of paper. For instance, a disk in the plane without its boundary would not be compact, or a line going to infinity would also not be compact. Technically, compactness for a set X in the plane (or in space) means that it is bounded, i.e., it lies entirely within some sufficiently large disk in the plane (or ball in space) and that every convergent sequence of points from the set converges to a point from the set.

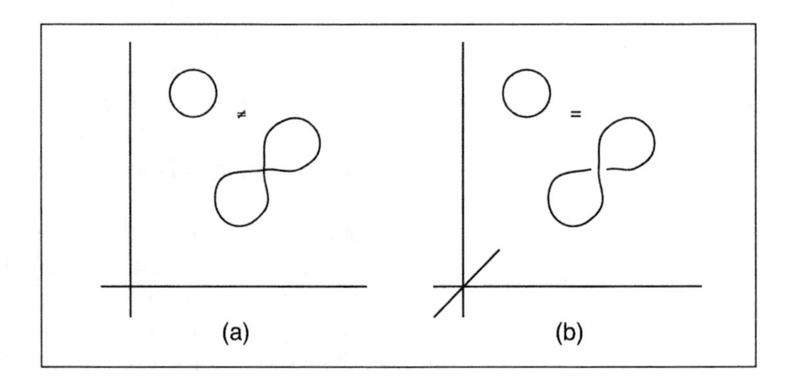

Figure 2.49 : The figure "8" is not equivalent to the circle (a). The twisted circle is equivalent to a circle (b).

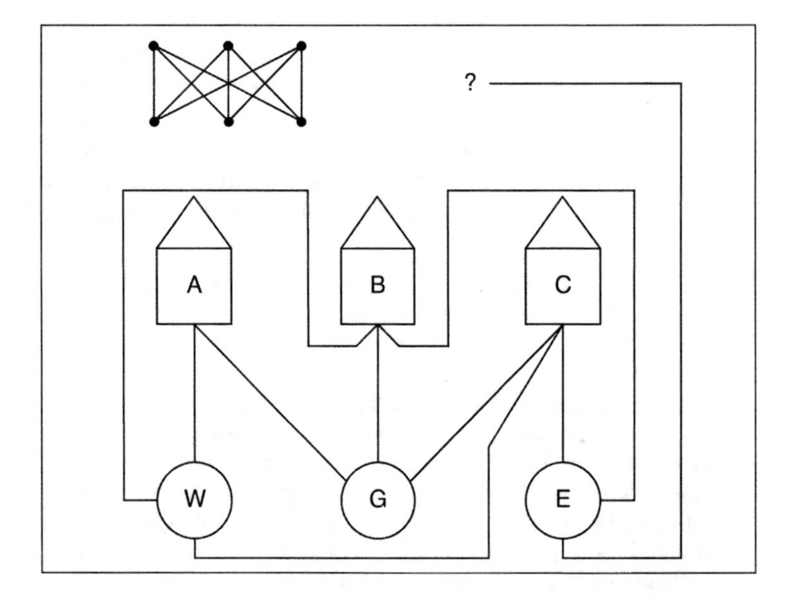

Figure 2.50 : A tough problem: can one get water, gas and electricity to all three houses without any intersection? The complete graph (with intersections!) is shown in the upper left corner of the figure.

bring a line from E to A without a crossing. The only way to escape a crossing is to go into space (i.e., run the supply lines at different levels).

Thus, if we are interested in maintaining the topological character of one-dimensional objects, we may have to go into space. In fact, every one-dimensional object can be embedded into three-dimensional space. Generalizations of this question are at the heart of topology. It is this branch which goes beyond the intuitive understanding of why the skeleton in figure 2.50 cannot be embedded into the plane by providing very deep methods which generalize to higher dimensions.

For instance, any two-dimensional object can be embedded into a five-dimensional space, and the five dimensions are actually needed in order to avoid obtaining effects similar to self-intersections, which would change the topological character.

Note that the graph in figure 2.50 cannot be drawn in the plane without self-intersections. Thus this graph cannot be represented in the Sierpinski carpet. This leads to the question, what is the universal object for one-dimensional objects in general (i.e., both in the plane and in space)?

The Universality of the Menger Sponge

About ten years after Sierpinski had found his result, the Austrian mathematician Karl Menger solved this problem and found a hotel for all one-dimensional objects. He proved around 1926 the following.[30]

Fact. *The Menger sponge is universal for all compact one-dimensional objects.*

Roughly, this means that for any admissible object (compact, one-dimensional) there is a part in the Menger sponge which is topologically the same as the given object.[31] That is, imagine that the given object again is made of rubber. Then some deformed version of it will fit exactly into the Menger sponge.

We cannot demonstrate the proofs of Menger or Sierpinski's amazing results; they are beyond the scope of this book. But we want to give an idea of the complexity of the one-dimensional objects we can find here. Let us discuss just one of many methods to measure this complexity. In particular this will allow us to distinguish between the Sierpinski gasket and the Sierpinski carpet. Since their basic construction steps are so similar (see section 2.2), we may ask whether the gasket is also universal. In other words, how complicated is the Sierpinski gasket? Is it as complicated as the carpet, or less? And if less, how much less complicated is it? Would you bet on your guess or visual intuition?

The answer is really striking: the Sierpinski gasket is absolutely tame when compared with the carpet, though visually there seems to be not much of a differencc. The Sierpinski gasket is a hotel which can accommodate only a few (one-dimensional, compact) very simple species from flatland. Thus, there is, in fact, a whole world of a difference between these two fractals. Let us look at objects like the ones in figure 2.51.

What we see are line segments with crossings. Or, we could say we see a central point to which there are different numbers of arms attached. We like to count the number of arms by a quantity which we call the branching order of a point. This will be a topological invariant. That is, this number will not change when we pass from one object to a topologically equivalent one. We can easily come up with one-dimensional objects of any prescribed branching order.

[30] K. Menger, *Allgemeine Räume und charakteristische Räume, Zweite Mitteilung: „Über umfassendste n-dimensionale Mengen"*, Proc. Acad. Amsterdam 29, (1926) 1125–1128. See also K. Menger, *Dimensionstheorie,* Leipzig 1928.

[31] Formally, for any compact one-dimensional set A there is a compact subset B of the Menger sponge which is homeomorphic to A.

Order of Spiders

Spiders with increasing branching.

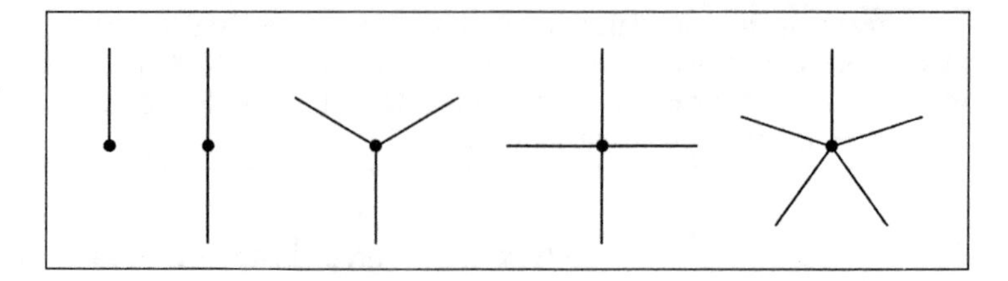

Figure 2.51

Branching Order

There is a very instructive way to distinguish one aspect of many different (topological) complexity features for one-dimensional objects. This concerns their branching and is measured by the branching order,[32] which we introduce next. Figure 2.52 shows some different types of branched structures.

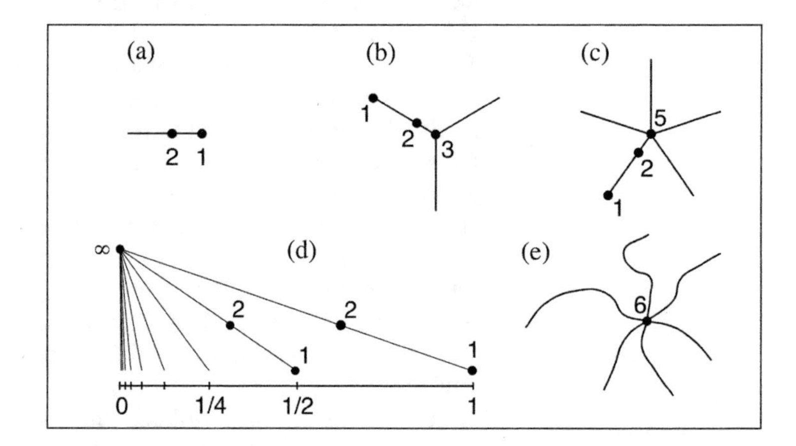

Figure 2.52 : Several examples of finite and (countably) infinite branching order. The numbers indicate the branching order of the corresponding points.

The branching order is a local concept. It measures the number of branches which meet in a point. Thus, for a point on a line we count two branches, while for an endpoint we count one branch. In example (d) of figure 2.52 we have one point — labelled ∞ — from which there are infinitely (countably) many line segments. Thus, this point would have branching order ∞, while points on the generating line segments (disregard the limit line) again would have branching order 2. Let us call the objects in figure 2.52 spiders. Thus, the spiders shown have 2, 3, 5, ∞, and 6 arms.

[32] See A. S. Parchomenko, *Was ist eine Kurve,* VEB Verlag, 1957.

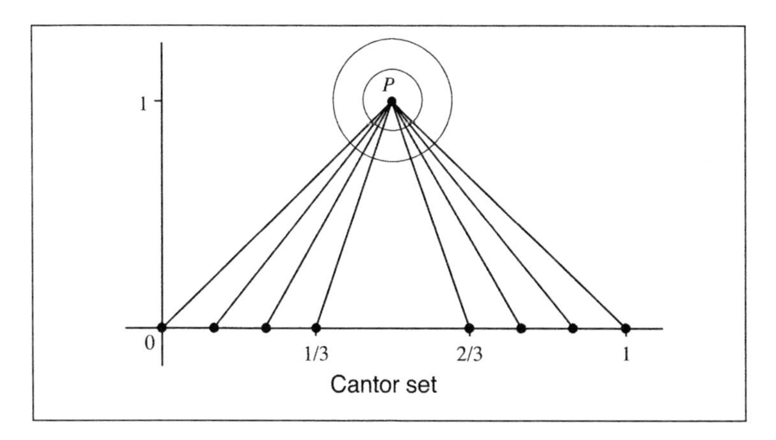

Figure 2.53 : An example of uncountably infinite branching order; the Cantor brush.

Let X be a set[33] and let $p \in X$ a point. Then we define the branching order of X at p to be[34]

$$\text{ord}_X(p) = \text{the number of branches of } X \text{ at } p.$$

One way to count these branches would be to take a sufficiently small disk around the point of interest and count the number of intersection points of the boundary of the disks with X.

Let us now construct a monster spider whose branching order has the cardinality of the continuum, i.e., there are as many branches as there are numbers in the unit interval $[0, 1]$.

We begin by taking a single point, say P, in the plane at $(1/2, 1)$ (see figure 2.53), together with the Cantor set C in $[0, 1]$. Now from each point in C we draw a line segment to P. You will recall that the cardinality of the Cantor set is the same as the cardinality of [0,1]. Therefore the cardinality of the points in the boundary of a small disk around P will again be the same. We call this set a Cantor brush.

Any pictorial representation of the Cantor brush can be a bit misleading. It might suggest that there is a countable number of bristles, while in fact they are uncountable. It can be shown that the Cantor brush is, however, a set of (covering) dimension 1, as is, of course any spider with a finite number of arms.

Now let us look at the Sierpinski gasket in terms of the branching order (see figure 2.54). Which spiders can be found in the gasket? It

[33]Formally, we need that X is a compact metric space.

[34]A formal definition goes like this. Let α be a cardinal number. Then one defines $\text{ord}_X(p) \leq \alpha$, provided for any $\varepsilon > 0$ there is a neighborhood U of p with a diameter $\text{diam}(U) < \varepsilon$ and such that the cardinality of the boundary ∂U of U in X is not greater than α, $\text{card}(\partial U) \leq \alpha$. Moreover, one defines $\text{ord}_X(p) = \alpha$, provided $\text{ord}_X(p) \leq \alpha$ and additionally there is $\varepsilon_0 > 0$, such that for all neighborhoods U of x with diameter less than ε_0 the cardinality of the boundary of U is greater or equal to α, $\text{card}(\partial U) \geq \alpha$.

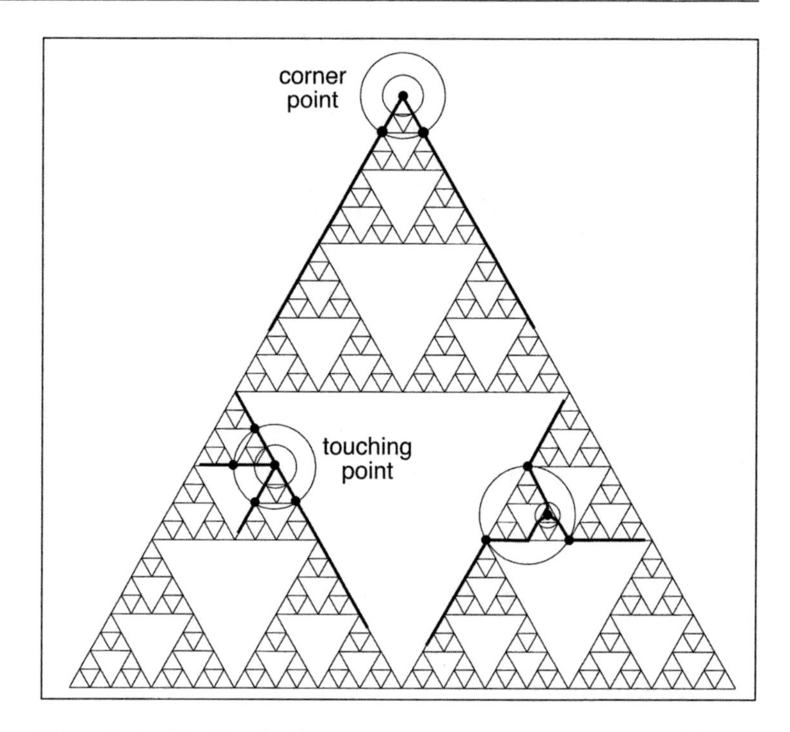

Figure 2.54 : The Sierpinski gasket allows only the branching orders 2, 3, and 4.

can be shown that if p is any point in the Sierpinski gasket S then

$$\text{ord}_S(p) = \begin{cases} 2, & \text{if } p \text{ is a corner of the initial triangle} \\ 4, & \text{if } p \text{ is a touching point} \\ 3, & \text{if } p \text{ is any other point.} \end{cases}$$

If p is a corner point, exactly two arms lead to this point. Observe how the circles drawn around the corner point (they need not to be centered at p) intersect the Sierpinski gasket at just two points. If p is a touching point, we can trace four arms to this point. In this case one can see circles around p that intersect the gasket at exactly four points. Now if p is any other point it must be right within an infinite sequence of subtriangles.[35] These subtriangles are connected to the rest of the Sierpinski gasket at just three points. Thus we can find smaller and smaller circles around p which intersect the Sierpinski gasket at exactly three points and we can construct three arms which pass through these points leading to p.

[35]Compare page 291.

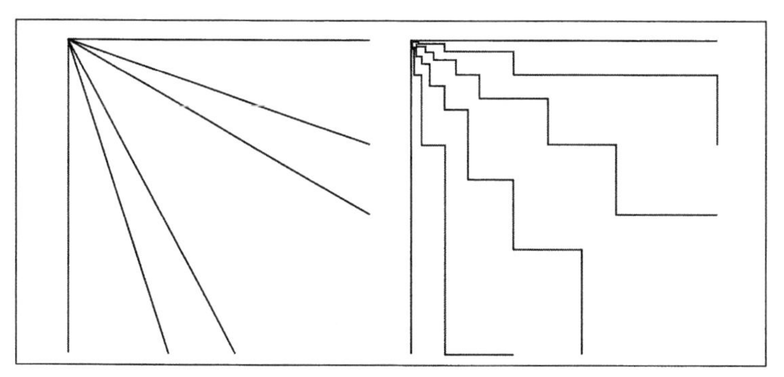

Six-Armed Spiders

These two spiders are topologically equivalent.

Figure 2.55

The Universality of the Sierpinski Carpet

We can observe that the Sierpinski gasket has points with branching order 2, 3 and 4 (see figure 2.54). These are the only possibilities. In other words, a spider with five (or more) arms *cannot* be found in the Sierpinski gasket![36]

On the other hand, the Sierpinski carpet is universal. Therefore it must accommodate spiders with any branching order, and, in particular, it must even contain a (topological) version of the Sierpinski gasket. Let us try to construct, as a very instructive example, a spider with five or six arms in the Sierpinski carpet. This is demonstrated in figure 2.56 and figure 2.57. Figure 2.55 shows the actual spider which we have found in the carpet (right) and a topologically equivalent spider (left).

Let us summarize. Our discussion of the universality of the Sierpinski carpet shows that fractals in fact have a very firm and deep root in a beautiful area of mathematics, and in varying an old Chinese saying[37] we may say fractals are more than pretty images.

[36] This is rather remarkable and it is therefore very instructive to try to construct a spider with five arms and understand the obstruction!

[37] A picture is worth a thousand words.

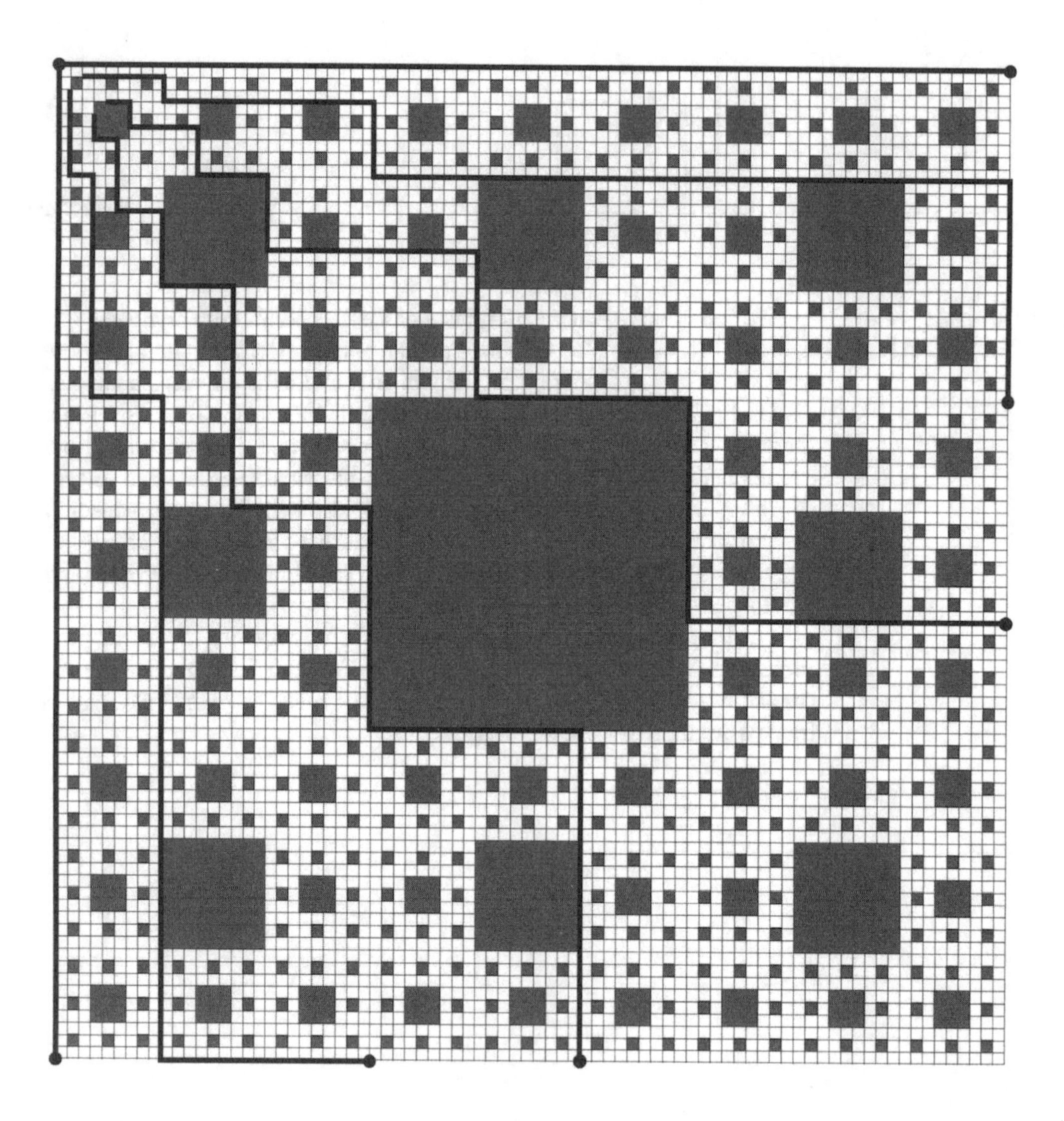

Figure 2.56 : Construction of a spider with six arms using symmetry and a recursive construction.

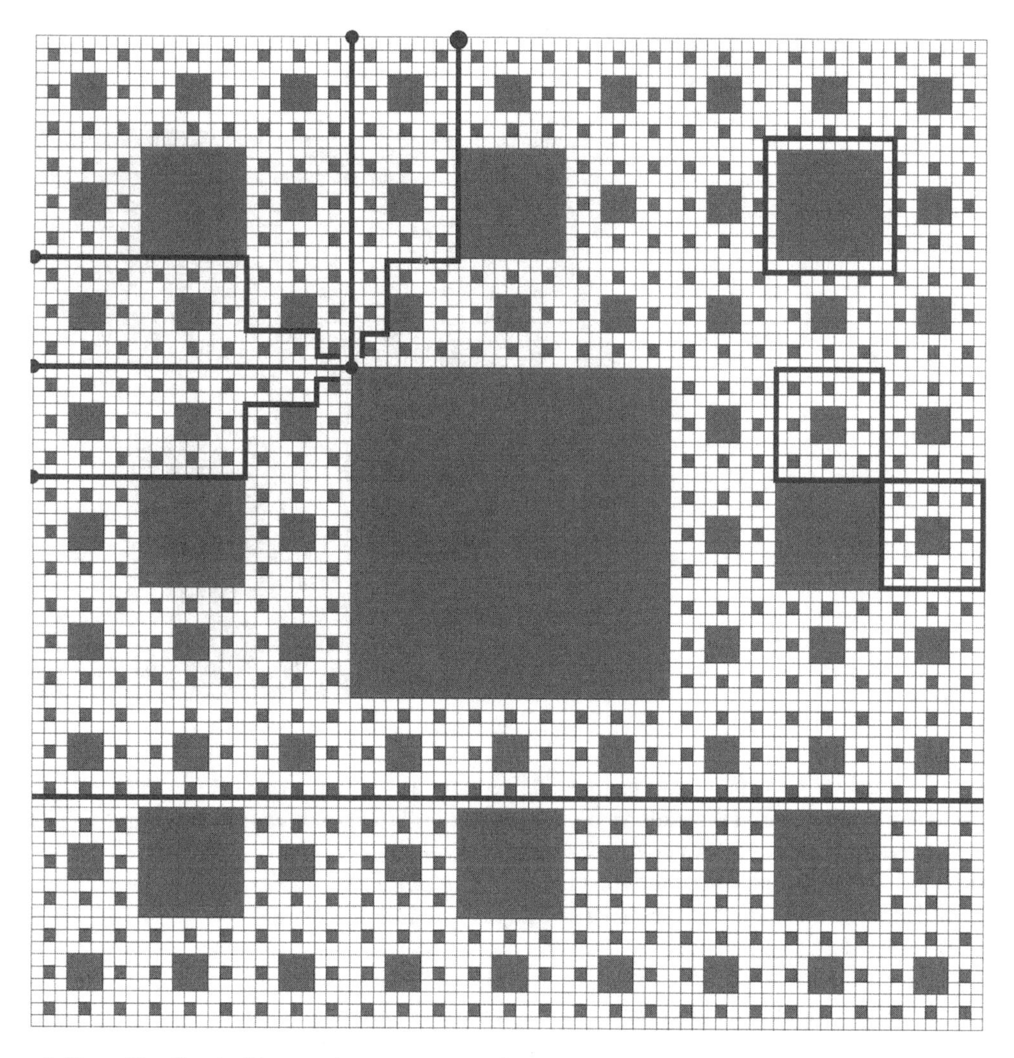

Figure 2.57 : The Sierpinski carpet houses any one-dimensional object: lines, squares, figure "8" like shapes, five-arm spiders or even deformed versions of the Sierpinski gasket (this is not shown — can you construct it?).

2.8 Julia Sets

Gaston Julia (1893–1978) was only 25 when he published his 199-page masterpiece[38] in 1918, which made him famous in the mathematics centers of his day. As a French soldier in the First World War, Julia had been severely wounded, as a result of which he lost his nose. Between several painful operations, he carried on his mathematical research in a hospital. Later he became a distinguished professor at the École Polytechnique in Paris.

Gaston Julia

Gaston Julia, 1893–1978, one of the forefathers of modern dynamical systems theory.

Figure 2.58

Julia's Work Was Famous in the 1920's

Although Julia was a world-famous mathematician in the 1920's, his work was essentially forgotten until Mandelbrot brought it back to light at the end of the seventies through his fundamental experiments. Mandelbrot was introduced to Julia's work through his uncle Szolem Mandelbrojt, who was a mathematics professor in Paris and the successor of Jacques Salomon Hadamard at the prestigious Collège de France.

Mandelbrot was born in Poland in 1924, and after his family had emigrated to France in 1936, his uncle felt responsible for his education. Around 1945, his uncle recommended Julia's paper to him as a masterpiece and source of good problems, but Mandelbrot didn't like it. Somehow he could not relate to the style and kind of mathematics which he found in Julia's paper and chose his own very different course, which, however, brought him back to Julia's work around 1977 after a path through many sciences, which some characterize as highly individualistic or nomadic. With the aid of computer graphics, Mandelbrot showed us that Julia's work is a source of some of

[38] G. Julia, *Mémoire sur l'iteration des fonctions rationnelles,* Journal de Math. Pure et Appl. 8 (1918) 47–245.

the most beautiful fractals known today. In this sense, we could say that this masterpiece is full of classical fractals which had to wait to be kissed awake by computers. In the first half of this century Julia was indeed world famous. To learn about his results, Hubert Cremer organized a seminar at the University of Berlin in 1925 under the auspices of Erhard Schmidt and Ludwig Bieberbach. The list of participants reads almost like an excerpt of a 'Who's Who' in mathematics at that time. Among them were Richard D. Brauer, Heinrich Hopf, and Kurt Reidemeister. Cremer also produced an essay on the topic which contains the first visualization of a Julia set (see figure 2.59).[39]

First Visualization

First drawing by Cremer in 1925 visualizing a Julia set.

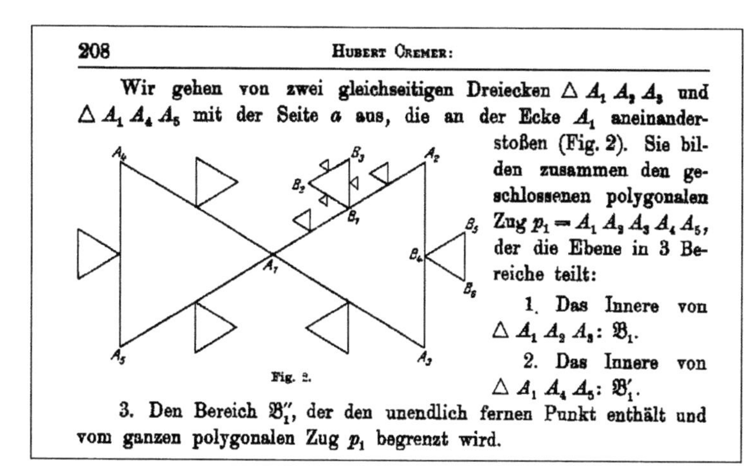

208 HUBERT CREMER:

Wir gehen von zwei gleichseitigen Dreiecken $\triangle A_1 A_2 A_3$ und $\triangle A_1 A_4 A_5$ mit der Seite a aus, die an der Ecke A_1 aneinanderstoßen (Fig. 2). Sie bilden zusammen den geschlossenen polygonalen Zug $p_1 = A_1 A_2 A_3 A_4 A_5$, der die Ebene in 3 Bereiche teilt:

1. Das Innere von $\triangle A_1 A_2 A_3$: $\mathfrak{B}_1$.
2. Das Innere von $\triangle A_1 A_4 A_5$: $\mathfrak{B}'_1$.
3. Den Bereich $\mathfrak{B}''_1$, der den unendlich fernen Punkt enthält und vom ganzen polygonalen Zug p_1 begrenzt wird.

Figure 2.59

The Quadratic Feedback System

Julia sets live in the complex plane. They are crucial for the understanding of iterations of polynomials like $x^2 + c$, or $x^3 + c$, etc. A detailed introduction will be given in chapter 13, but here we assume that you are familiar with the concept of complex numbers. If you aren't, we propose that for now you simply think of real numbers. Let us look at $x^2 + c$ as an example. Iterating means that we fix c and choose some value for x and obtain $x^2 + c$. Now we substitute this value for x and evaluate $x^2 + c$ again, and so on. In other words, for an arbitrary but fixed value of c we generate a sequence of complex numbers

$$x \to x^2 + c \to (x^2 + c)^2 + c \to ((x^2 + c)^2 + c)^2 + c \to \ldots$$

This sequence must have one of two following properties:

The Julia Set Dichotomy

- either the sequence becomes unbounded: the elements of the sequence leave any circle around the origin;
- or the sequence remains bounded: there is a circle around the origin which is never left by the sequence.

The collection of points which lead to the first kind of behavior is called the *escape set* for c, while the collection of points which lead to the second kind

[39] H. Cremer, *Über die Iteration rationaler Funktionen*, Jber. d. Dt. Math.Verein. 33 (1925) 185–210.

Some Samples of Julia Sets

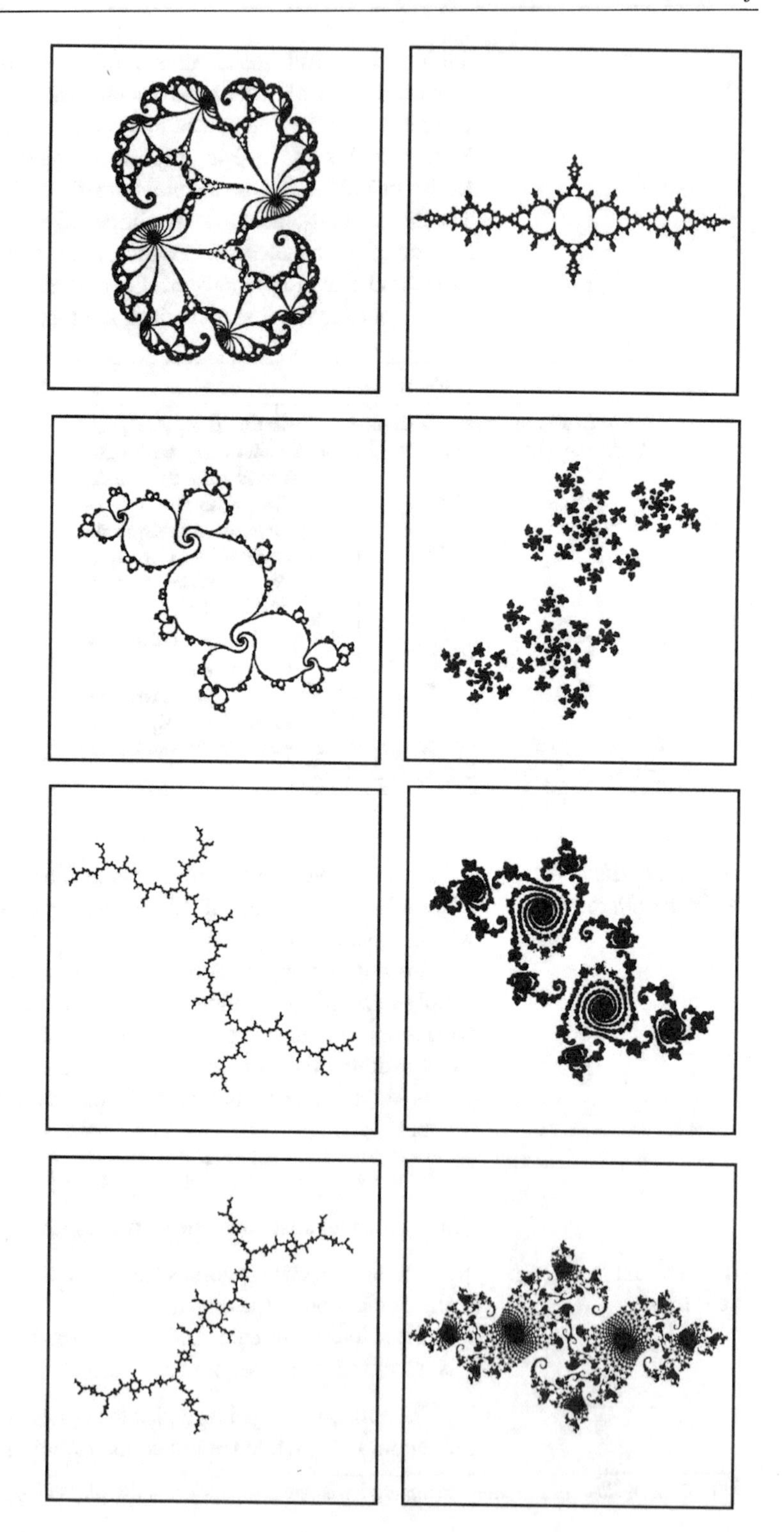

Figure 2.60

of behavior is called the *prisoner set* for c. This terminology has already been used in the section on the Cantor set. Both sets are nonempty. For example, given c, then for x sufficiently large, $x^2 + c$ is even larger than x. Thus, the escape set contains all points x which are very large. On the other hand, if we choose x so that $x = x^2 + c$, then iteration remains stationary. Starting with such an x the sequence produced by the iteration will be constant $x, x, x, \ldots$. In other words, neither can the prisoner set be empty.

The Julia Set

Both sets cover some part of the complex plane and complement each other. Thus, the boundary of the prisoner set is simultaneously the boundary of the escape set, and that is the Julia set for c (or rather $x^2 + c$). Figure 2.60 shows some Julia sets obtained through computer experiments.

Is there self-similarity in the Julia sets? Already from our first crude figure it seems obvious that there are structures that repeat at different scales. In fact, any Julia set may be covered by copies of itself, but these copies are obtained by a *nonlinear* transformation. Thus, the self-similarity of Julia sets is of a very different nature as compared to the Sierpinski gasket, which is composed of reduced but otherwise congruent copies of itself.

2.9 Pythagorean Trees

Pythagoras, who died at the beginning of the fifth century B.C., was known to his contemporaries, and later even to Aristotle, as the founder of a religious brotherhood in southern Italy, where Pythagoreans played a political role in the sixth century B.C. The linking of his name to the Pythagorean theorem is, however, rather recent and spurious. In fact, the theorem was known long before the life-time of Pythagoras. An important discovery ascribed to Pythagoras, or in any case to his school, is that of the *incommensurability* of side and diagonal of the square; that is, the ratio of diagonal and side of the square is not equal to the ratio of two integers.

The Pythagorean Theorem
$\mathbf{a^2 + b^2 = c^2}$

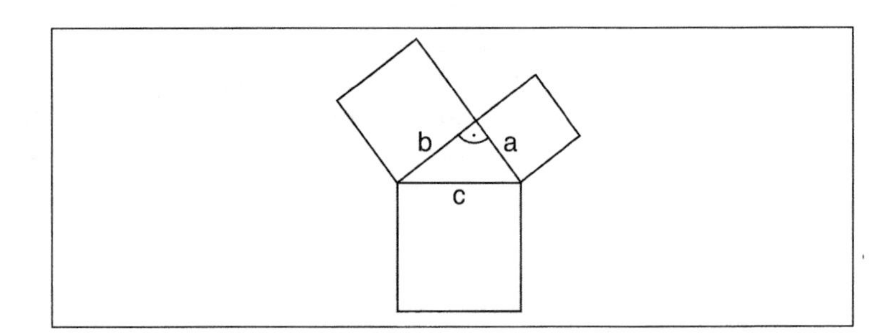

Figure 2.61

The Incommensurability of Side and Diagonal of the Square

The discovery that the ratio of diagonal and side of the square is not equal to the ratio of two integers produced the necessity to extend the number system to irrational numbers. The length of the diagonal in the unit square, $\sqrt{2}$, is irrational. Let us give the argument. Assume that p and q are positive integers with $\sqrt{2} = p/q$. We may also assume that p and q have no common divisor. Then $p^2 = 2q^2$, i.e., p^2 must be an even number. But this implies that p itself must be even, because the square of an odd number is odd. Thus $p = 2r$. But then $p^2 = 2q^2$ means that $4r^2 = 2q^2$ or $2r^2 = q^2$, which means that q must be even as well. But this contradicts the assumption that p and q have no common divisor. Thus, $\sqrt{2}$ is irrational. This proof is found in the tenth book of Euclid around 300 B.C.

The computation of square roots is a related problem and has inspired mathematicians to discover some wonderful geometric constructions. One of them allows us to construct $\sqrt{n}$ for any integer n. It could be called the square root spiral, and it is a geometric feedback loop. Figure 2.62 explains the idea.

The Construction of Pythagorean Trees

The construction which yields the family of Pythagorean trees and their relatives is very much related to the construction of the square root spiral. The construction proceeds along the following steps and is shown in figure 2.63.

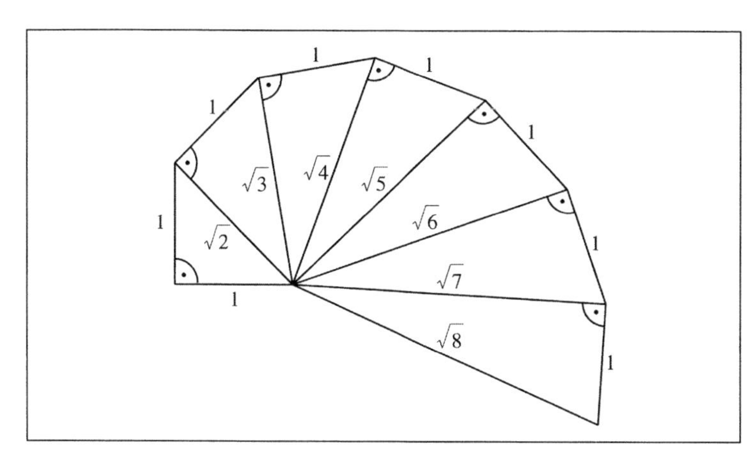

The Square Root Spiral

Construction of a square root spiral. We begin with a right-angled triangle so that the sides forming the right angle are of length 1. Then the hypotenuse is of length $\sqrt{2}$. Now we continue by constructing another right triangle so that the sides adjacent to the right angle have length 1 and $\sqrt{2}$. The hypotenuse of that triangle has length $\sqrt{3}$, and so on.

Figure 2.62

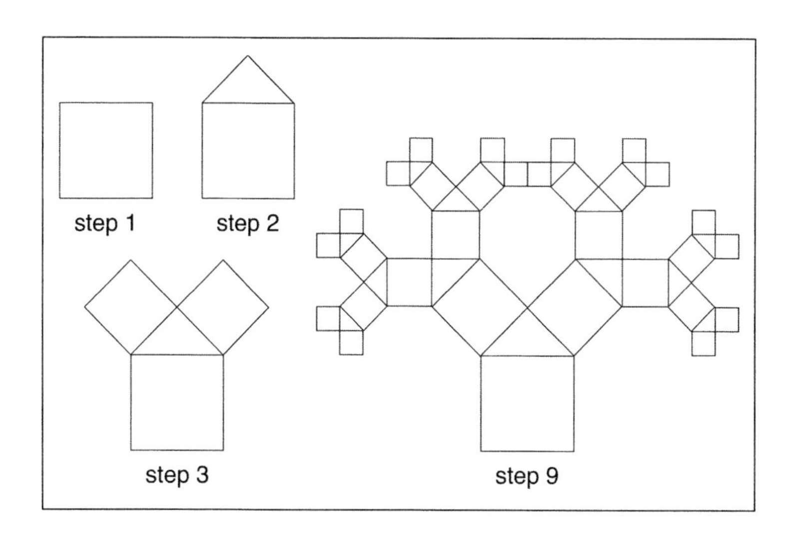

Pythagorean Tree Construction

Figure 2.63

Step 1: Draw a square.
Step 2: Attach a right triangle to one of its sides along its hypotenuse (here with two equal sides).
Step 3: Attach two squares along the free sides of the triangle.
Step 4: Attach two right triangles.
Step 5: Attach four squares.
Step 6: Attach four right triangles.
Step 7: Attach eight squares.

Once we have understood this basic construction we can modify it in various ways. For example, the right triangles which we attach in the process need not be isosceles triangles. They can be any right triangle. But once we allow such variations we have, in fact, an additional degree of freedom. The

Two Constructions with Non-Isosceles Triangles

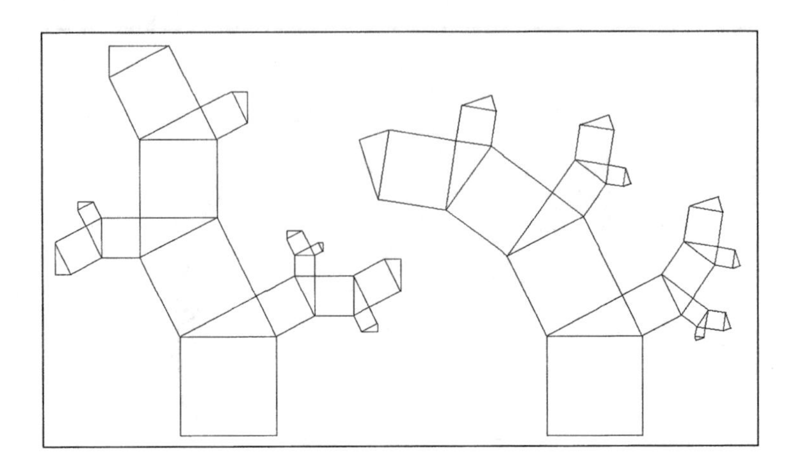

Figure 2.64

right triangles can always be attached in the same orientation, or we may flip their orientation after each step. Figure 2.64 shows the two possibilities.

Figure 2.65 shows the results of these constructions after some 50 steps. It is most remarkable that the only thing which we have changed is the orientation of the triangles, not their size. The results, however, could not look more different. In the first case we see some kind of spiraling leaf, while the second reminds us of a fern or pine tree. Note that in the bottom construction in figure 2.65 we see a major stem from which branches radiate off in a right, left, right, left, ... pattern. This seems to be quite different in the other construction. There we see a major stem which curls and from which we have only a branching away to one side. Would you have guessed that both 'plants' derive from the same feedback principle? Didn't they look at first as if they belonged to totally different families? They are, however, very close relatives, and this becomes apparent when analyzing the corresponding construction processes. This is one way that fractals may help to introduce new tools into botany. The biologist Aristid Lindenmayer (1925–1989) introduced the concept of *L-systems* along these lines, and we will discuss that approach in some detail in chapter 7.

Let us continue to look into our primitive, but amazing, constructions using some other modifications. Why not take just any kind of triangle? To keep some regularity we should take similar ones. This opens the door to a large variety of fascinating forms which range from plant-like ones to tilings to who knows what. In figure 2.66 we have attached equilateral triangles, and notice that the construction becomes periodic.

Passing from equilateral triangles to isosceles triangles with angles greater than 90° yields another surprise — a form which is broccoli-like (see figure 2.67). These constructions raise a number of interesting questions. When does the construction lead to an overlap? By what law do the lengths of the sides of the triangles or squares decrease in the process? Moreover, we have beautiful examples of structures which are self-similar, i.e., each structure subdivides

Figure 2.65 : The two constructions carried out some 50 times. Note that the size of the triangles is the same in both.

Periodic Tiling

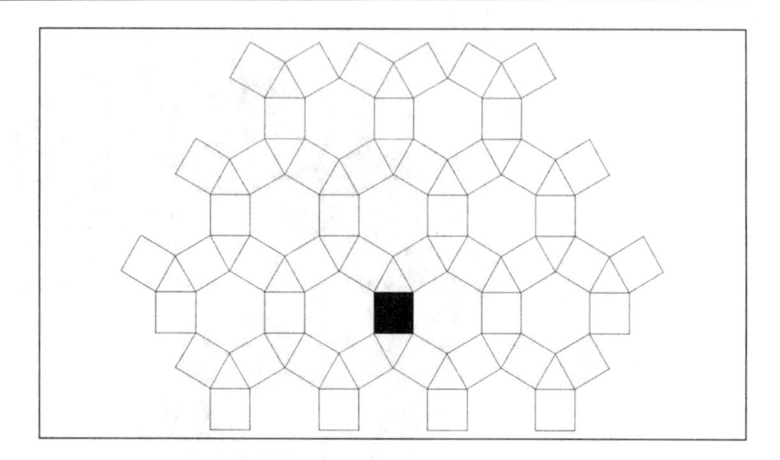

Figure 2.66

Broccoli-Like Pythagoras Tree

Construction with isosceles triangles which have angle greater than 90°.

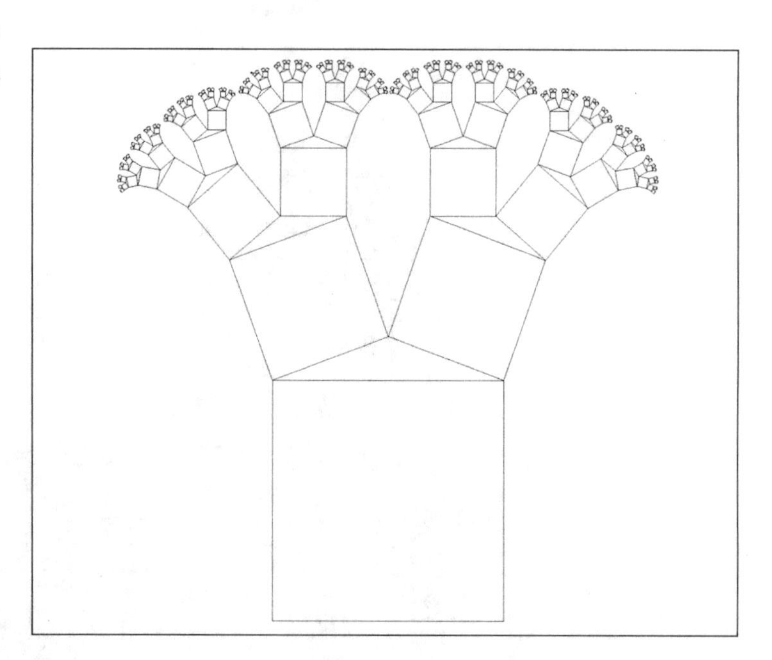

Figure 2.67

during construction into two major branches, and these again into two major branches, and so on; and each of these branches is a scaled down version of the entire structure.

Our gallery of historical fractals ends here, though we have not discussed the contributions of Henry Poincaré, Karl Weierstraß, Felix Klein, L. F. Richardson, or A. S. Besicovitch. They all would deserve more space than we could give them here, but we refer the interested reader to Mandelbrot's book.[40]

[40]B. Mandelbrot, *The Fractal Geometry of Nature*, Freeman, New York, 1982.

Chapter 3

Lim and Self-Similarity

Now, as Mandelbrot points out [...] nature has played a joke on the mathematicians. The 19th-century mathematicians may have been lacking in imagination, but nature was not. The same pathological structures that the mathematicians invented to break loose from 19th-century naturalism turn out to be inherent in familiar objects all around us in nature.

Freeman Dyson[1]

Dyson is referring to mathematicians, like G. Cantor, D. Hilbert, and W. Sierpinski, who have been justly credited with having helped to lead mathematics out of its crisis at the turn of the century by building marvelous abstract foundations on which modern mathematics can now flourish safely. Without question, mathematics has changed during this century. What we see is an ever-increasing dominance of the algebraic approach over the geometric. In their striving for absolute truth, mathematicians have developed new standards for determining the validity of mathematical arguments. In the process, many of the previously accepted methods have been abandoned as inappropriate. Geometric or visual arguments were increasingly forced out. While Newton's *Principia Mathematica*, laying the fundamentals of modern mathematics, still made use of the strength of visual arguments, the *new objectivity* seems to require a dismissal of this approach. From this point of view, it is ironic that some of the constructions which Cantor, Hilbert, Sierpinski and others created to perfect their extremely abstract foundations simultaneously hold the clues to understanding the patterns of nature in a visual sense. The Cantor set, Hilbert curve, and Sierpinski gasket all give testimony to the delicacy and problems of modern set theory and at the same time, as Mandelbrot has taught us, are perfect models for the complexity of nature.

Finding the right abstract formulation for the old concept of a *limit* was part of the struggle to build a safer foundation for modern mathematics. As we know, the concept of limits is one of the most powerful and fundamental ideas

[1]Freeman Dyson, *Characterizing irregularity,* Science 200 (1978) 677–678.

Romanesco

The new bread romanesco, a crossing between cauliflower and broccoli, exhibits striking self-similarity.

Figure 3.1

in mathematics and the sciences. At the same time, it is one which troubles many nonmathematicians. This is very unfortunate, especially because of the fact that contemporary mathematics seems to tell us that the concept of limits is trivial. The truth is, of course, that building the right mathematical framework for the understanding of limits took the best mathematicians thousands of years. It is therefore very inappropriate for us to ignore the problems of nonmathematicians today, for they are sometimes of the same quality and depth as those which puzzled the great mathematicians in the past.

Self-similarity, by contrast, seems to be a concept which can be understood without any trouble. The term self-similarity hardly needs an explanation. One would guess that the term has been around for centuries, but it has not. It is only some 25 years old. The new bread romanesco, a crossing between

Plate 1: Stilben (used in some detergents) dendrites in polarized light, © Manfred Kage, Institut für wissenschaftliche Fotografie.

Plate 2: Cast of a child's kidney, venous and arterial system, © Manfred Kage, Institut für wissenschaftliche Fotografie.

Plate 3: Broccoli Romanesco.

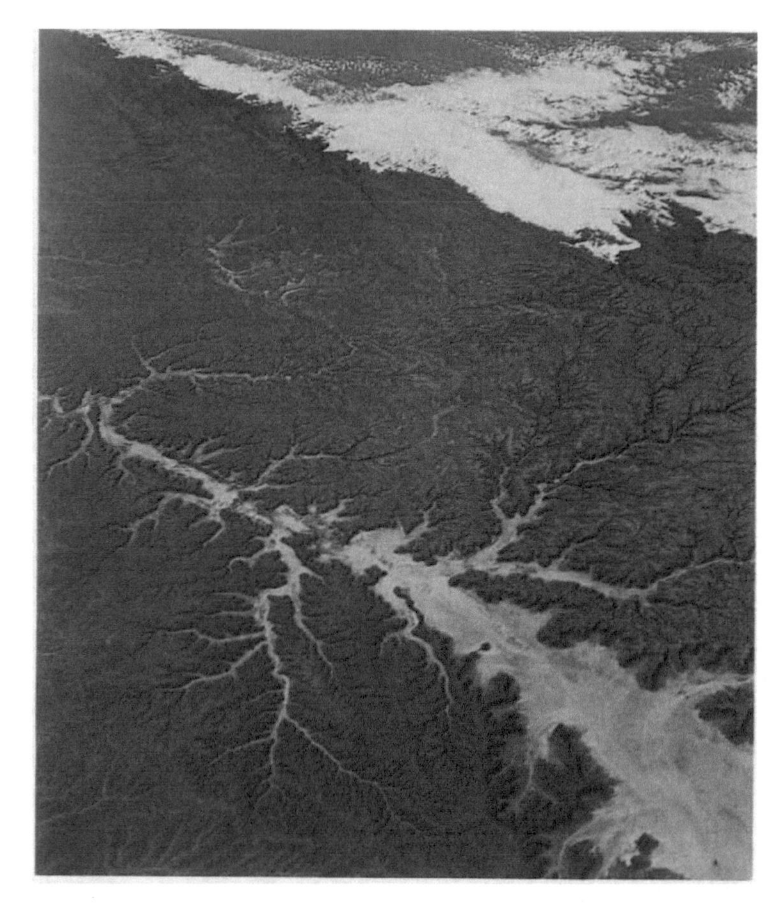

Plate 4: Wadi Hadramaut, Gemini IV image, © Dr. Vehrenberg KG.

Plate 5: Broccoli Romanesco, detail.

Plate 7: Fractal forgery of mountain range (top left), inverted mountain range, showing valleys as mountains and mountains as valleys (bottom left), inverted mountain range rendered as cloud pattern (top right) used in Plate 6, © R.F. Voss.

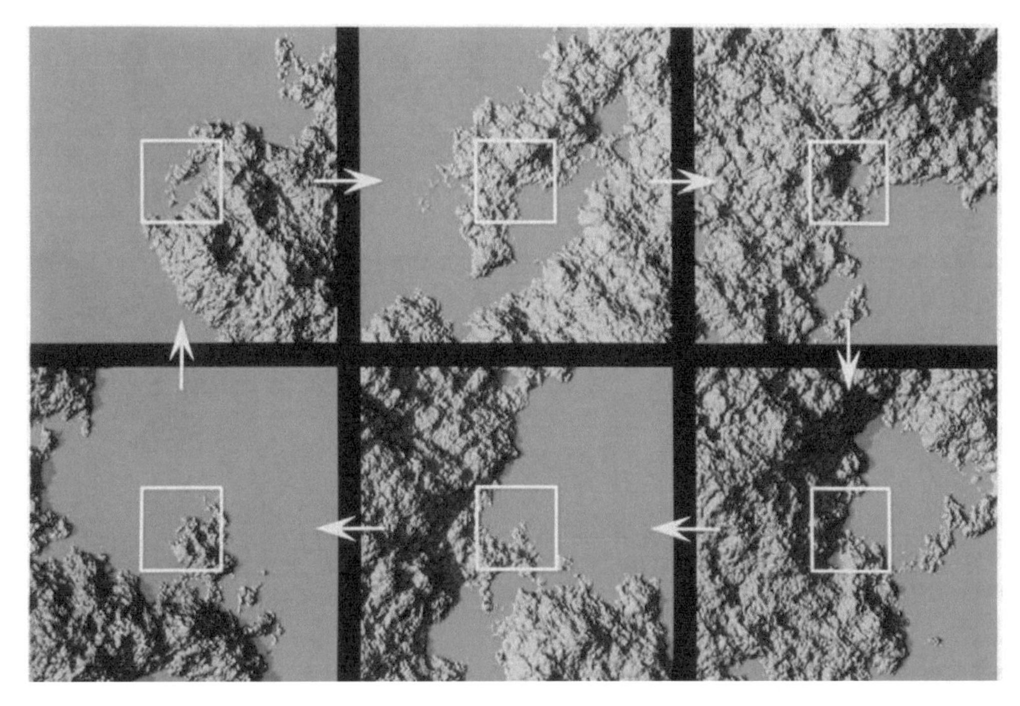

Plate 8: Fractal coast, repeating after 6 magnifications, © R.F. Voss.

Plate 9: Fractal Moon Craters, © R.F. Voss.

Plate 10: "Zabriski Point", fractal forgery of a mirage, © K. Musgrave, C. Kolb, B.B. Mandelbrot.

Plate 11: "Carolina", fractal forgery, © K. Musgrave.

Plate 12: Fractal forgery of planet rise, © K. Musgrave.

Plate 13: "Ein kleines Nachtlicht", fractal forgery, stereoscopic image. View the left image with your right eye and the right image with your left eye. © K. Musgrave, C. Kolb, B.B. Mandelbrot.

Plate 14: Dawn over the Himalayas, Gemini IV image, © Dr. Vehrenberg KG.

cauliflower and broccoli, illustrates the concept (see figure 3.1 and the color plates 3 and 5). Macroscopically we see a form which is best described as a cluster. That cluster is composed of smaller clusters, which look almost identical to the entire cluster, however scaled down by some factor. Each of these smaller clusters again is composed of smaller ones, and those again of even smaller ones. Without difficulty we can identify three generations of clusters on clusters. The second, third, and all the following generations are essentially scaled down versions of the previous ones. In a rough sense, this is what we call *self-similarity*.

We will see that a rigorous discussion of the concept of self-similarity is intimately related to the concept of limit, and therefore it will require some care. The visual observation in nature, however, is simple and immediate. Once one has been introduced to this basic phenomenon, it is hard to walk through the fields and woods without constantly examining plants and other structures.

Fractals add a new dimension to the problems of dealing with limits; but also — and this is our point here — a refreshingly new perspective from which to understand the concept of limits. On one hand fractals may visualize the limit object in a feedback process; on the other hand some fractals demonstrate self-similarity in its purest form. In fact, many fractals can be completely characterized and defined by their self-similarity properties.

3.1 Similarity and Scaling

What Is Similarity?

Self-similarity extends one of the most fruitful notions of elementary geometry: similarity. Two objects are similar if they have the same shape, regardless of their size. Corresponding angles, however, must be equal, and corresponding line segments must all have the same factor of proportionality. For example, when a photo is enlarged, it is enlarged by the same factor in both horizontal and vertical directions. Even an oblique, i.e., nonhorizontal, nonvertical, line segment between two points on the original will be enlarged by the same factor. We call this enlargement factor the *scaling factor*. The transformation between the objects is called similarity transformation.

Similarity Transformations

Similarity transformations are compositions involving a scaling, a rotation and a translation. A reflection may additionally be included, but we skip the details of that. Let us be more specific for similarity transformations in the plane. Here we denote points P by their coordinate pairs $P = (x, y)$. Let us apply scaling, rotation and translation to one point $P = (x, y)$ of a figure. First, a scaling operation, denoted by S, takes place, yielding a new point $P' = (x', y')$. In formulas,

$$\begin{aligned} x' &= sx, \\ y' &= sy, \end{aligned}$$

where $s > 0$ is the scale factor. A scale reduction occurs, if $s < 1$, and an enlargement of the object will be produced when $s > 1$. Next, a rotation R is applied to $P' = (x', y')$, yielding $P'' = (x'', y'')$:

$$\begin{aligned} x'' &= \cos\theta \cdot x' - \sin\theta \cdot y', \\ y'' &= \sin\theta \cdot x' + \cos\theta \cdot y'. \end{aligned}$$

This describes a counterclockwise (mathematically positive) rotation of P' about the origin of the coordinate system by an angle of θ. Finally, a translation T of P'' by a displacement (T_x, T_y) is given by

$$\begin{aligned} x''' &= x'' + T_x, \\ y''' &= y'' + T_y \end{aligned}$$

which yields the point $P''' = (x''', y''')$. Summarizing, we may write

$$P''' = T(P'') = T(R(P')) = T(R(S(P)))$$

or, using the notation

$$W(P) = T(R(S(P))),$$

we have $P''' = W(P)$. W is the similarity transformation. In formulas,

$$\begin{aligned} x''' &= s\cos\theta \cdot x - s\sin\theta \cdot y + T_x, \\ y''' &= s\sin\theta \cdot x + s\cos\theta \cdot y + T_y. \end{aligned}$$

Applying W to all points of an object in the plane produces a figure which is similar to the original.

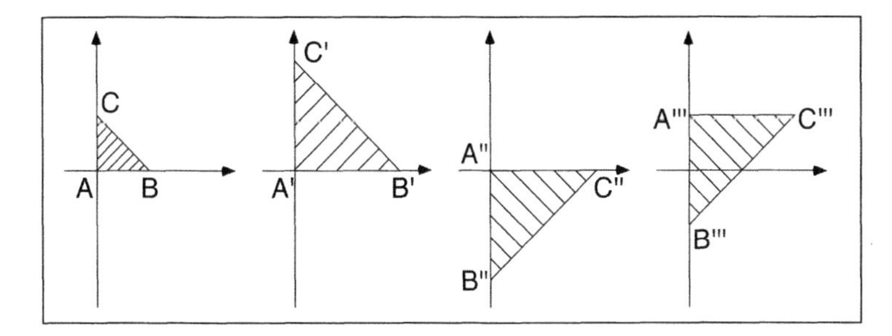

Figure 3.2 : A similarity transformation is applied to the triangle ABC. The scaling factor is $s = 2$, the rotation is by 270°, and the translation is given by $T_x = 0$ and $T_y = 1$.

The similarity transformations can also be formulated mathematically for objects in other dimensions, for example, for shapes in three or only one dimension. In the latter case we have points x on the real line, and the similarity transformation can simply be written as $W(x) = sx + t$, $s \neq 0$.

Consider a photo which is enlarged by a factor of 3. Note that the area of the resulting image is $3 \cdot 3 = 3^2 = 9$ times the area of the original. More generally, if we have an object with area A and scaling factor s, then the resulting object will have an area which is $s \cdot s = s^2$ times the area A of the original. In other words, the area of the scaled up object increases as the square of the scaling factor.

Scaling Three-Dimensional Objects

What about scaling three-dimensional objects? If we take a cube and enlarge it by a scaling factor of three, it becomes three times as long, three times as deep, and three times as high as the original. We observe that the area of each face of the enlarged cube is $3^2 = 9$ times as large as the face of the original cube. Since this is true for all six faces of the cube, the total surface area of the enlargement is nine times as much as the original. More generally, for objects of any shape whatever, the total surface area of a scaled-up object increases as the square of the scaling factor.

What about volume? The enlarged cube has three layers, each with $3 \cdot 3 = 9$ little cubes. Thus the total volume is $3 \cdot 3 \cdot 3 = 3^3 = 27$ times as much as the original cube. In general, the volume of a scaled up object increases as the cube of the scaling factor.

These elementary observations have remarkable consequences, which were the object of discussion by Galileo (1564–1642) in his 1638 publication *Dialogues Concerning Two New Sciences*. In fact Galileo[2] suggested 300

[2]We quote D'Arcy Thompson's account from his famous 1917 *On Growth and Form* (New Edition, Cambridge University Press, 1942, page 27): "[Galileo] said that if we tried building ships, palaces or temples of enormous size, yards, beams and bolts would cease to hold together; nor can Nature grow a tree nor construct an animal beyond a certain size, while retaining the proportions and employing the material which suffice in the case of a smaller structure. The thing will fall to pieces of its own weight unless we either change its relative proportions, which will at length cause it to become clumsy, monstrous and inefficient,

Galileo Galilei's Dialogues Concerning Two New Sciences, 1638

DISCORSI
E
DIMOSTRAZIONI
MATEMATICHE,
intorno à due nuoue ſcienze
Attenenti alla
MECANICA & i MOVIMENTI LOCALI;
del Signor
GALILEO GALILEI LINCEO,
Filoſofo e Matematico primario del Sereniſſimo
Grand Duca di Toſcana.
Con una Appendice del centro di grauità d'alcuni Solidi.

IN LEIDA,
Appreſſo gli Elſevirii. M. D. C. XXXVIII.

Figure 3.3

feet as the limiting height for a tree. Giant sequoias, which live only in the Western United States and hence were unknown to Galileo, grow as high as 360 feet. However, Galileo's reasoning was correct; the tallest giant sequoias adapt their form in ways that evade the limits of his model.

What was his reasoning? The weight of a tree is proportional to its volume. Scaling up a tree by a factor s means that its weight will be scaled by s^3. At the same time the cross-section of its stem will only be scaled by s^2. Thus

or else we must find new material, harder and stronger than was used before. Both processes are familiar to us in Nature and in art, and practical applications, undreamed of by Galileo, meet us at every turn in this modern age of cement and steel."

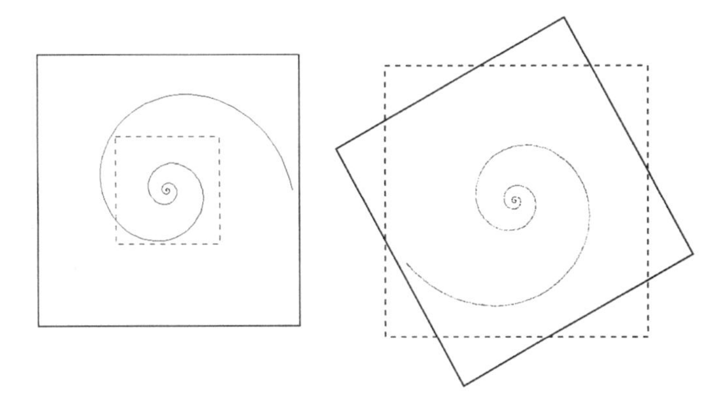

Magnifying a Logarithmic Spiral

The magnifying of a logarithmic spiral by a factor b shows the same spiral, however, rotated by an angle θ (about 210°).

Figure 3.4

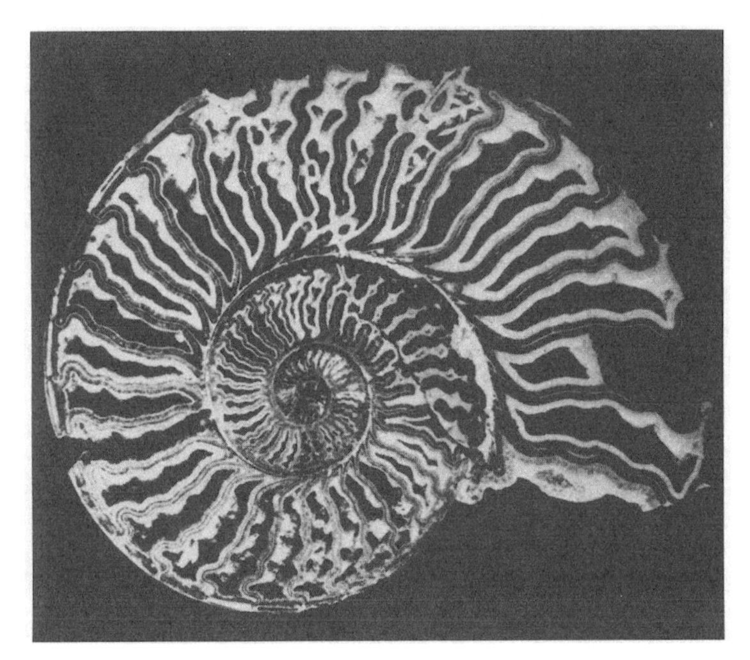

Ammonite

The growth pattern of an ammonite follows a logarithmic spiral.

Figure 3.5

the pressure inside the stem would scale by $s^3/s^2 = s$. That means that if s increases beyond a certain limit, then the strength of wood will not be sufficient to bear the corresponding pressure.[3]

This tension between volume and area also explains why mountains do not exceed a height of 7 miles, or why different creatures respond differently to falling.[4] For example, a mouse may be unharmed by a ten-story fall, but

[3]Here is a related problem. Suppose a nail in a wall supports some maximum weight w; how much weight would a nail which is twice as big support?

[4]See J. B. S. Haldane, *On Being the Right Size*, 1928, for a classic essay on the problem of scale.

a man may well be injured by just falling from his own height. Indeed, the energy which has to be absorbed is proportional to the weight, i.e., proportional to the volume of the falling object. This energy can only be absorbed over the surface area of the object. With scaling up, volume, hence weight, hence falling energy, goes up much faster than area. As volume increases the hazards of falling from the same height increase.

Similarity and Growth of Ammonites

In chapter 4 we will continue to discuss scaling properties. In particular we will look at spirals, as for example the logarithmic spiral. We all have seen how a spiral drawn on a disk seems to grow continuously as it is turned around its center. In fact, the logarithmic spiral is special in that magnifying it is the same as rotating the spiral. Figure 3.4 illustrates this remarkable phenomenon, which as such is another example of a self-similar structure. Figure 3.5 shows an ammonite which is a good example of a logarithmic spiral in nature. In other words, an ammonite grows according to a law of similarity. It grows in such a way that its shape is preserved.

Babies Are Not Similar to Their Parents

Most living things, however, grow by a different law. An adult is not simply a baby scaled up. In other words, when we wonder about the similarity between a baby and its parents we are not talking about (the mathematical term of) geometric similarity! In the growth from baby to adult, different parts of the body scale up, each with a different scale factor. Two examples are:

- Relative to the size of the body, a baby's head is much larger than an adult's. Even the proportions of facial features are different: in a baby, the tip of the nose is about halfway down the face; in an adult, the nose is about two thirds above the chin. Figure 3.6 illustrates the deformation of a square grid necessary to measure the changes in shape of a human head from infancy to adulthood.
- If we measure the arm length or head size for humans of different ages and compare it with body height, we observe that humans do not grow in a way that maintains geometric similarity. The arm, which at birth is one-third as long as the body, is by adulthood closer to two-fifths as long. Figure 3.7 shows the changes in shape when we normalize the height.

Isometric and Allometric Growth

In summary, the growth law is far from being a similarity law. A way to get insight into the growth law of, for example, the head size versus the body height, is by plotting the ratio of these two quantities versus age. In table 3.8, we list these data for a particular person.[5] Entering the ratio and the age in a diagram we obtain figure 3.9. If the growth were proportional, that is, according to similarity, the ratio would be constant throughout the lifetime of the person, and we would have gotten all points on a straight horizontal line. Graphing therefore provides a way to test for proportional growth. In our example we do not have an overall proportional growth. We can discern two different phases: one that fits early development, up to the age of about three years, and another that fits development after that time. In the first period

[5] The data in this table is taken from D'Arcy Thompson, *On Growth and Form*, New Edition, Cambridge University Press, Cambridge, 1942, page 190.

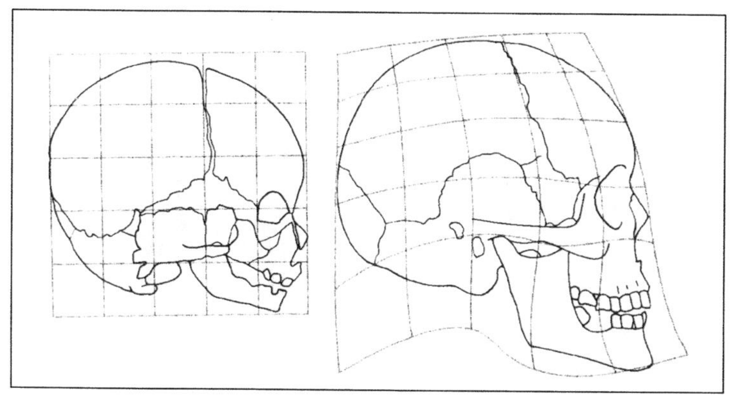

Nonlinear Growth

The head of a baby and an adult are not similar, i.e., they do not transform by a simple scaling. Figure adapted from *For All Practical Purposes,* W. H. Freeman, New York, 1988.

Figure 3.6

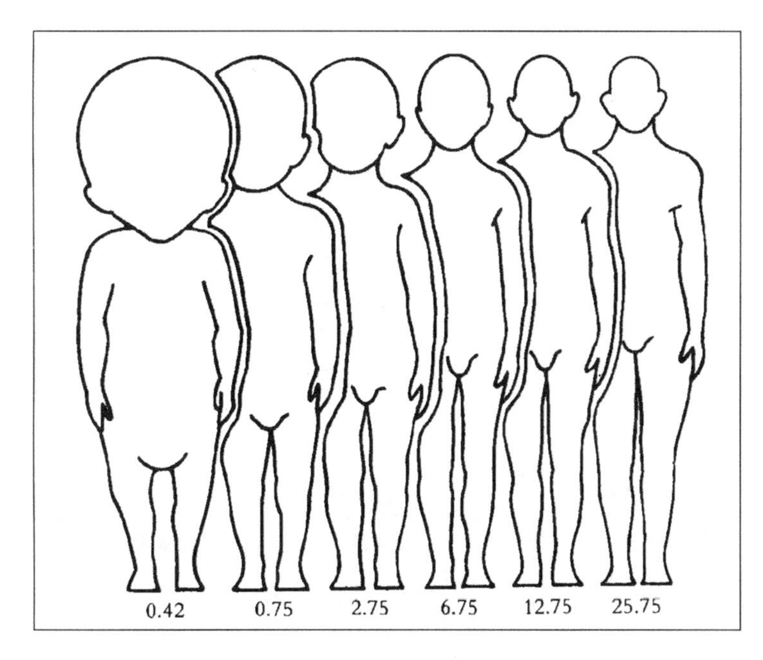

Changing Proportions

Changes in shape between 0.5 and 25 years. Height is normalized to 1. Figure adapted from *For All Practical Purposes,* W. H. Freeman, New York, 1988.

Figure 3.7

we have proportional growth, sometimes called *isometric growth*. After the age of three years, however, the ratio drops significantly, indicating that body height is growing relatively faster than head size. This is called *allometric growth*. At about the age of 30 the growth process is completed and the ratio is constant again. A more sophisticated analysis of this data yielding mathematical growth laws will be presented in the next chapter. In fact, the well-known phenomenon of nonproportional growth above is at the heart of fractal geometry, as we will see shortly. Having discussed similarities and ways of scaling, let us now return to the central theme of this chapter: what is self-similarity?

Head Size Versus Body Height Data

Body height and head size of a person. The last column lists the ratio of head size to body height. The first few years this ratio is about constant, while later it drops, indicating a change from isometric to allometric growth.

Table 3.8

Age (years)	Body Height (cm)	Head Size (cm)	Ratio
0	50	11	0.22
1	70	15	0.21
2	79	17	0.22
3	86	18	0.21
5	99	19	0.19
10	127	21	0.17
20	151	22	0.15
25	167	23	0.14
30	169	23	0.14
40	169	23	0.14

Graphing Growth

Growth of head relative to height for the data from table 3.8. On the horizontal axis age is marked off, while the vertical axis specifies the head to body height ratio.

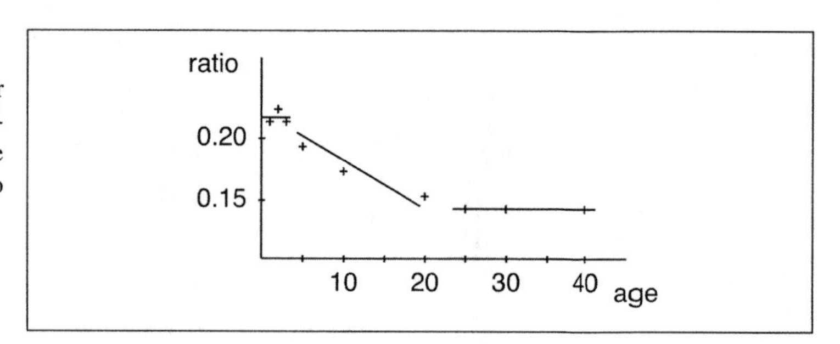

Figure 3.9

What Is Self-Similarity?

Intuitively, this seems clear; the word self-similar hardly needs a definition — it is self-explanatory. However, talking in precise mathematical terms about self-similarity really is a much more difficult undertaking. For example, in the romanesco, or for that matter, in any physically existing object, the self-similarity may hold only for a few orders of magnitude. Below a certain scale, matter decomposes into a collection of molecules, atoms, and, going a bit further, elementary particles. Having reached that stage, of course, it becomes ridiculous to consider miniature scaled-down copies of the complete object. Also, in a structure like a cauliflower the part can never be exactly equal to the whole. Some variation must be accounted for. Thus, it is already clear at this point that there are several variants of mathematical definitions of self-similarity. In any case, we like to think of mathematical fractals as objects which possess recognizable details at all microscopic scales — unlike real physical objects. When considering cases of fractals where the small copies, while looking like the whole, have variations, we have so-called *statistical self-similarity*, a topic which we will get back to in chapter 9. Moreover, the miniature copies may be distorted in other ways, for example, somewhat skewed. For this case there is the notion of *self-affinity*.

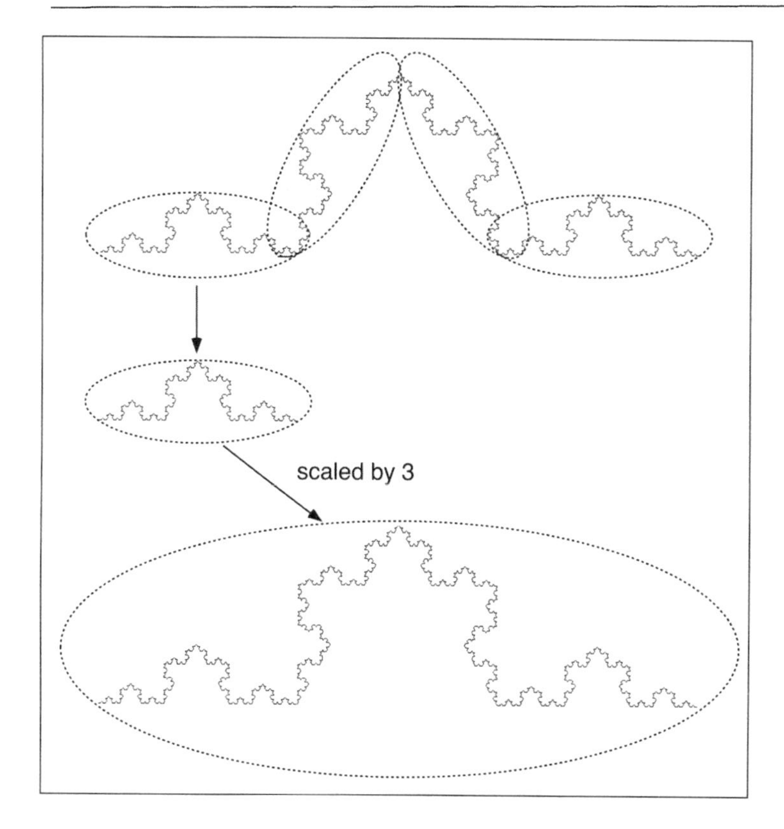

Blowup of Koch Curve

One-quarter of the Koch curve (top) is magnified by a factor of 3. Due to the self-similarity of the Koch curve the result is a copy of the whole curve.

Figure 3.10

Self-Similarity of the Koch Curve

To exemplify the concept, we choose the Koch curve which is already familiar from the second chapter. Can we find similitudes (similarity transformations) in the Koch curve? The Koch curve looks like it is made up of four identical parts. Let us look at one of these, say the one at the extreme left. We take a variable zoom lens and observe that at exactly $\times 3$ magnifying power the little piece seems to be identical to the entire curve. Each one of the little pieces breaks into four identical pieces again, and each of them seems to be identical to the entire Koch curve when we apply a magnifying lens of $\times 9$, and so on ad infinitum. This is the self-similarity property in its mathematically purest form.

Different Degrees of Self-Similarity

But even in this case, where copies of the whole appear at all stages and are exact and not distorted in any way, there are still various degrees of self-similarity possible. Consider, for example, a cover of a book that contains on it a picture of a hand holding that very book. Surprisingly, this somewhat innocent-sounding description leads to a cover with a rather complex design. As we look deeper and deeper into the design, we see more and more of the rectangular covers. Contrast that with an idealized structure of a two-branch tree as shown in figure 3.11. Also pictured is the self-similar Sierpinski gasket. All three examples are self-similar structures: they contain small replicas of

Three Different Self-Similar Structures

The Sierpinski gasket (left) is self-similar at all of its points, while the two-branch tree (middle) is self-similar only at the leaves. The structure on the right is self-similar only at the center point.

Figure 3.11

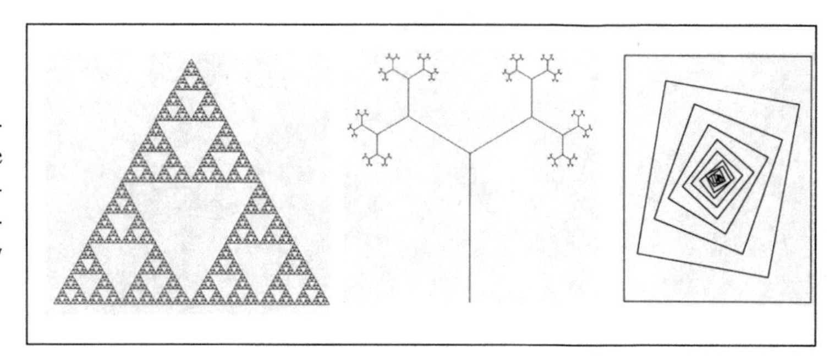

the whole. However, there is a significant difference. Let us try to find points which have the property that we can identify small replicas of the whole in their neighborhoods at any degree of magnification.

Self-Similarity at a Point

In the case of the book design, the copies are arranged in one nested sequence, and clearly the self-similarity property can be found only at *one particular point*. This is the limit point at which the size of the copies tends to zero. The book cover is self-similar at this point.

Self-Affinity

The situation is much different in the two-branch tree. The complete tree is made up of the stem and *two* reduced copies of the whole. Thus, smaller and smaller copies accumulate near the leaves of the tree. In other words, the property of self-similarity condenses in the set of leaves. The whole tree is not strictly self-similar, but *self-affine*. The stem is not similar to the whole tree but we can interpret it as an affine copy which is compressed to a line.

Strict Self-Similarity

Finally, in the Sierpinski gasket, similar to the Koch curve above, we can find copies of the whole near *every* point of it, which we have already discussed. The gasket is composed from small but exact copies of itself. Considering these differences, we call all three objects self-similar, while only the Sierpinski gasket and the Koch curve are in addition called *strictly self-similar*. Also the set of leaves without the stem and all the branches is strictly self-similar. Now what would a cauliflower be in these categories? It would be a physical approximation of a self-similar, but not *strictly* self-similar, object akin to the two-branch tree.

3.2 Geometric Series and the Koch Curve

Fractals such as the Koch curve, the Sierpinski gasket and many others are obtained by a construction process. Ideally, however, this process should never terminate. Any finite stage of it produces an object, which may have a lot of fine structure, depending on how far the process has been allowed to proceed; but essentially it still is far from a true fractal. Thus, the fractal only exists as an idealization. It is what we would get if we let the process run 'indefinitely'. In other words, fractals really are limit objects, and their existence is not as natural as it may seem. This is important, and the mathematical foundation of such limits is one of the goals of this chapter and some others.

Limits often lead to new quantities, objects or qualities; this is true particularly for fractals (we will come back to that later on). However, given a sequence of objects, there are cases where it is not immediately obvious whether a limit exists at all. As for example, the first sum in

$$\sum_{k=1}^{\infty} \frac{1}{k} = \frac{1}{1} + \frac{1}{2} + \frac{1}{3} + \cdots, \quad \sum_{k=1}^{\infty} \frac{1}{k^2} = \frac{1}{1} + \frac{1}{4} + \frac{1}{9} + \cdots$$

is divergent[6] (i.e., the sum is infinite) whereas the second one converges to $\pi^2/6$, as shown by Euler.

Let us recall for a moment the discussion of geometric series. For a given number $-1 < q < 1$ the question is, does

$$\sum_{k=0}^{\infty} q^k = 1 + q + q^2 + q^3 + \cdots$$

have a limit, and what is the limit? To this end one defines

$$S_n = 1 + q + q^2 + q^3 + \cdots + q^n.$$

Then on the one hand we have $S_n - qS_n = 1 - q^{n+1}$, and on the other hand $S_n - qS_n = S_n(1-q)$. Putting these two identities together we obtain

$$S_n = \frac{1 - q^{n+1}}{1 - q}. \tag{3.1}$$

In other words, as n becomes larger, q^{n+1} becomes smaller, which means that S_n gets closer and closer to $1/(1-q)$. In short, we have justified

$$\sum_{k=0}^{\infty} q^k = \frac{1}{1-q}. \tag{3.2}$$

[6]The sum $1+1/2+1/3+1/4+\cdots$ is infinite. An argument for this fact goes as follows. Assume that the sum has a finite value, say S. Then, clearly $1/2+1/4+1/6+\cdots = S/2$. It follows that $1+1/3+1/5+\cdots = S-(1/2+1/4+1/6+\cdots) = S/2$. But also since $1 > 1/2, 1/3 > 1/4, 1/5 > 1/6, \ldots$ we must have that $1 + 1/3 + 1/5 + \cdots > 1/2 + 1/4 + 1/6 + \cdots$. This is a contradiction, as both sums should equal $S/2$. Therefore, our assumption, namely, that the sum $1 + 1/2 + 1/3 + \cdots = S$, must be wrong. A finite limit of this sum cannot exist.

Koch Island Construction

The Koch island is the limit object of the construction and has area $\frac{2}{5}\sqrt{3}a^2$.

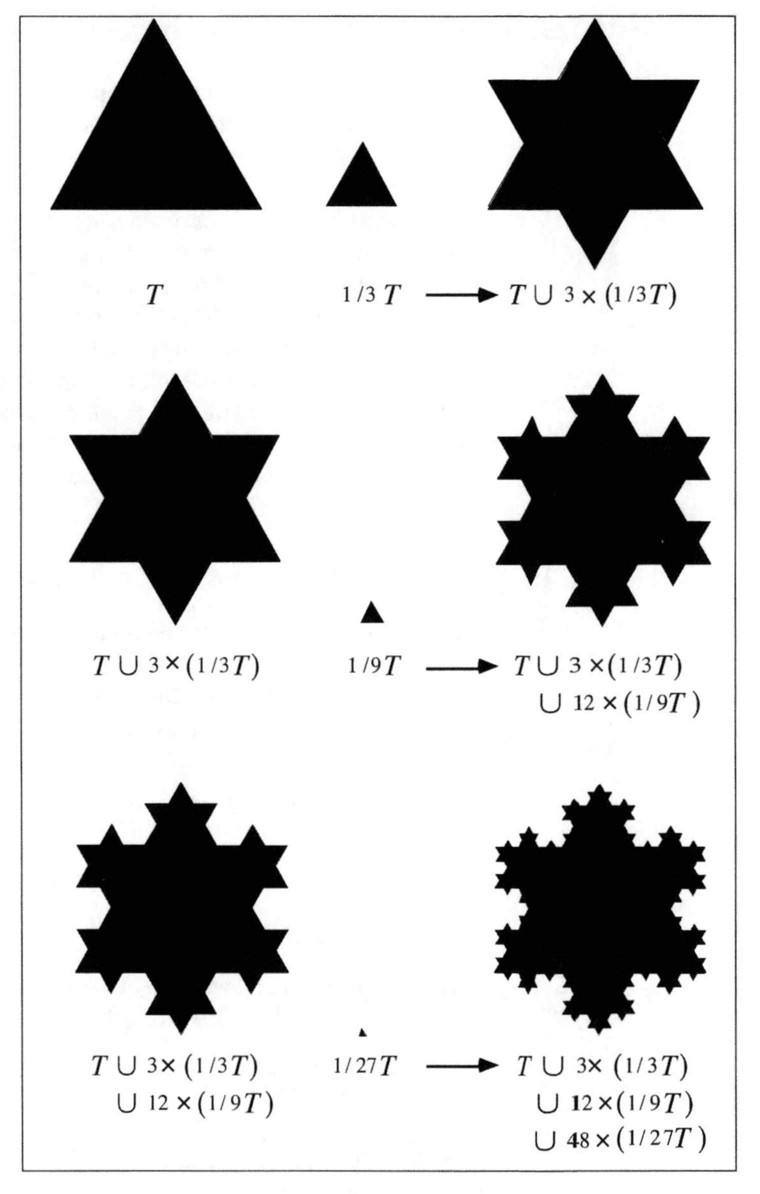

Figure 3.12

This is one of the elementary limit considerations which is useful,[7] even though the limit $1/(1-q)$ is not at all enlightening. Nevertheless, it will help us to understand a particular point about fractal constructions. In theory S_n

[7]Remember, for example, the problem of understanding infinite decimal expansions of the form $0.1543999\ldots$ We know that it is just 0.1544, but why? Well, first $0.1543999\ldots = 0.1543\ldots + 9\cdot 10^{-5}(1+10^{-1}+10^{-2}+\cdots)$. Then one can apply eqn. (3.2) with $q = 10^{-1}$ and obtain $1+10^{-1}+10^{-2}+\cdots = 10/9$. Thus, $9\cdot 10^{-5}(1+10^{-1}+10^{-2}+\cdots) = 9\cdot 10^{-5}\cdot 10/9$, which is 10^{-4}. Finally, $0.1543999\ldots = 0.1543 + 10^{-4} = 0.1544$.

will be different from the limit $1/(1-q)$, no matter how large we choose n. But practically — for example, in a finite accuracy computer — both will be indistinguishable provided n is large enough.

The Construction Process of Geometric Series

The geometric series has an analogy in the construction of basic fractals. There is an initial object, here the number 1, and a scaling factor, here q. The important property of the scaling factor is that it be less than 1 in magnitude. Then there is a construction process.

Step 1: Start with 1.
Step 2: Scale down 1 by the scaling factor q and add.
Step 3: Scale down 1 by the scaling factor $q \cdot q$ and add.
Step 4: ...

The point is that this infinite construction leads to a new number, representing that process — the limit of the geometric series.

The Construction Process of the Koch Island

The Koch island, which we see in its basic construction in figure 3.12, is obtained in an analogous manner, except that rather than adding up numbers, we 'add up' geometrical objects. 'Addition', of course, is here interpreted as a union of sets; and the important point is that in each step we add a certain number of scaled down versions of the initial set.

Step 1: We choose an equilateral triangle T with sides of length a.
Step 2: We scale down T by a factor of 1/3 and paste on three copies of the resulting little triangle as shown. Now the island is bounded by $3 \cdot 4$ line segments, each of length $a/3$.
Step 3: We scale down T by a factor of $1/3 \cdot 1/3$ and paste on $3 \cdot 4$ copies of the resulting little triangle as shown. The resulting island is bounded by $3 \cdot 4 \cdot 4$ straight line segments, each of length $1/3 \cdot 1/3 \cdot a$.
Step 4: ...

The point here is that this infinite construction leads to a *new* geometric object, the Koch island. In fact, the analogy between the geometric process and geometric series goes much further. Let us get a first impression. What is the area of the Koch island, the geometric object which we see as a limit of the above process?

The Area of the Koch Island

Well, let us try to figure out how much area we add in each step. At the beginning we have the area A_1 for the initial triangle T, and calculate $A_1 = \sqrt{3}/4 \cdot a^2$. In each step k, we have to add the area of n_k little equilateral triangles with sides s_k. Convince yourself that $n_1 = 3, n_2 = 3 \cdot 4, n_3 = 3 \cdot 4 \cdot 4$, ... Thus, $n_k = 3 \cdot 4^{k-1}$. The sides s_k of the little triangles are obtained by successively scaling down the side of the original triangle by a factor 1/3. Therefore, $s_k = (1/3)^k a$. Combining these results we get

$$\begin{aligned} A_{k+1} &= A_k + n_k \cdot \frac{\sqrt{3}}{4} \cdot s_k^2 = A_k + 3 \cdot 4^{k-1} \cdot \frac{\sqrt{3}}{4} \cdot \frac{1}{3^{2k}} a^2 \\ &= A_k + \frac{\sqrt{3}}{12} \cdot \left(\frac{4^{k-1}}{9^{k-1}} \right) a^2 . \end{aligned}$$

In other words, if we develop the terms step by step we have the series

$$A_{k+1} = A_1 + \frac{\sqrt{3}}{12}\left(1 + \frac{4}{9} + \frac{4^2}{9^2} + \cdots + \frac{4^{k-1}}{9^{k-1}}\right) a^2.$$

The expression in the parentheses is a partial sum of the geometric series $1 + \frac{4}{9} + \frac{4^2}{9^2} + \frac{4^3}{9^3} + \cdots$ which has the limit $\frac{1}{1-4/9} = \frac{9}{5}$. That means that the Koch island, the geometric object in the limit, has area

$$A = A_1 + \frac{\sqrt{3}}{12} \cdot \frac{9}{5} a^2$$

and since $A_1 = \sqrt{3}/4 \cdot a^2$, we finally obtain

$$A = \frac{2}{5}\sqrt{3}a^2.$$

This is quite a convincing argument that there is indeed a new geometric object resulting from the infinite process. But a rigorous argument would need much more.

It would need a language which would allow us to talk about the process of adding new and smaller shapes in the above construction exactly in the same way as is used to discuss adding smaller and smaller numbers in a series. In fact, this language already exists. One of the great achievements of what is called *point set topology* was to extend the idea of limits as known when dealing with numbers to far-reaching abstractness. This, together with a notion called *Hausdorff distance*, which is a generalization of the usual distance between points to the distance between two *point sets*, provides the right framework in which we can indeed find a perfect analogy between the infinite process of adding numbers in a geometric series and its limit behavior on the one hand, and the infinite adding of smaller and smaller triangles in the Koch island construction and its limit behavior on the other. In some sense, nothing new and exciting happens or has to be understood. Everything is just an appropriate translation of how we are used to thinking about geometric series. In that sense the Koch island is a visualization of the limit of a geometric series.

Limits Lead to New Qualities

Let us now look at properties of the limit, which are not shared by any of its finite stage approximations. The most important property is that of self-similarity. For example, the self-similarity of the Koch curve is reflected by the fact that the curve is made up of four identical parts. Can we actually verify the self-similarity with our images on paper? Of course not. There are two reasons: a technical one and a mathematical one.

The Technical Problem with Demonstrating Self-Similarity

The technical reason is obvious. Black ink on white paper comes in little dots which under a sufficiently high powered microscope look more or less like random specks and certainly not like a Koch curve. This effect could be called limited resolution and is very similar to the problem of representing numbers in a computer. Recall that $\sqrt{2}$ in a computer representation is never really $\sqrt{2}$, but rather some approximation such as 1.414214. Magnifying an image can be compared with multiplying a number by some factor greater than 1. For example, if we multiply $\sqrt{2}$ by $\sqrt{2}$ again and again we will get $2, 2\sqrt{2}, 4, 4\sqrt{2}, 8, 8\sqrt{2}, \ldots$ In other words, we will get powers of $\sqrt{2}$. If we compute increasing powers of an approximation of $\sqrt{2}$, then for a while we obtain results which are close to the true powers of $\sqrt{2}$. But sooner or later our numerical results will deviate more and more, and they will eventually totally disagree with our theoretical expectations.

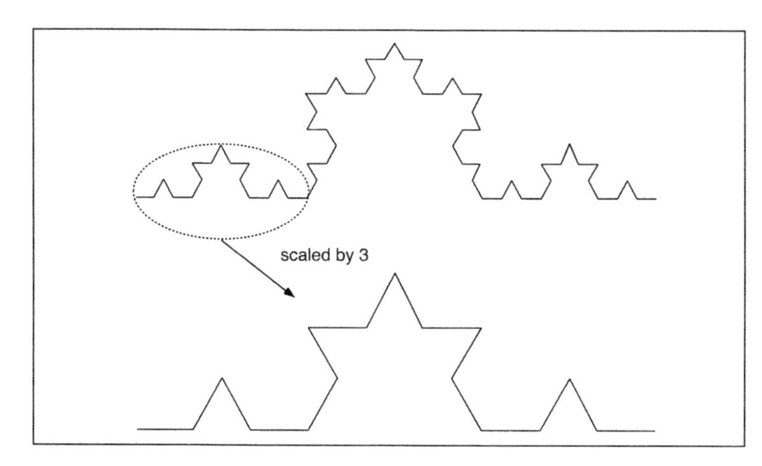

No Self-Similarity at Finite Stage

Each single construction stage of the Koch curve does not generate a self-similar curve. For example, scaling a part of the stage 3 approximation (top) by a factor of 3 (bottom) does not yield a curve equal to the stage 3 curve .

Figure 3.13

The Mathematical Problem with Demonstrating Self-Similarity

The mathematical reason for the impossibility of running these experiments on paper is similar. Only the limit structure, but none of the intermediate construction stages, has the property of perfect self-similarity; and the limit structure cannot be obtained by any computer whatsoever. This is very much like the true and precise numerical value of $\sqrt{2}$ not being representable by any computer. It would need infinitely many digits. The only pictures of the Koch curve which are possible are approximating images. For example, if we draw an image of the 5th stage and compare it with an image of the 10th stage, we do not see a difference. But there is, of course, a dramatic difference. The change, however, is below the resolution of the device (printer or screen). No matter which stage we may choose to represent our Koch curve, it will be indistinguishable from the true image of the Koch curve if the stage is sufficiently advanced. But theoretically the two objects (i.e., some stage in the process here and the Koch curve there) are dramatically different. For example, for advanced stages, the boundary of the respective object is made by tiny little straight line segments. Thus under sufficient magnification those will become macroscopically visible. In other words, if we look at one of the pieces in, say, the 10th stage under the microscope, we will see a piece which is familiar, say, from the 2nd stage, while magnifications (with the correct magnifying factor) of the boundary in the limit structure will look exactly like the Koch curve. Also, an approximation of the Koch curve by any of its finite stage constructions cannot be self-similar, no matter how accurate the approximation is (see figure 3.13). The fact is, however, the Koch curve contains no straight line segment of any length.[8]

A Second New Quality of the Limit Object

Another property of the Koch curve, which is not shared by any of its finite stage approximations, is that its length is infinite (see section 2.4). As the Koch curve is one-third of the boundary of the Koch island, we have that the boundary of the Koch island is also infinitely long. In contrast to this, the area of the Koch island is a finite and well-defined number, as seen above. That, in fact, is the metaphoric message of Mandelbrot's 1967 article in the magazine *Science*, entitled *How long is the coast of Britain?* We will discuss this in more detail in chapter 4.

Self-Similarity in Geometric Series

Looking back at the geometric series one may see a remarkable correspondence to the self-similarity of the Koch curve. If we formally scale the series

$$\sum_{k=0}^{\infty} q^k = 1 + q + q^2 + q^3 + \cdots$$

with the factor q, we obtain $q\sum_{k=0}^{\infty} q^k = q + q^2 + q^3 + q^4 + \cdots$. Therefore,

$$\sum_{k=0}^{\infty} q^k = 1 + q\sum_{k=0}^{\infty} q^k. \qquad (3.3)$$

This is the 'self-similarity' of the geometric series. The value of the sum is 1 plus the scaled down version of the whole series. As in the case of the Koch curve, the self-similarity only holds for the limit but not for any finite stage. For example, denote $S_2 = 1 + q + q^2$, then $1 + qS_2 = 1 + q + q^2 + q^3 \neq S_2$.

In summary, we have linked the Koch curve and island to the geometric series, providing strong evidence for the existence of these fractals. Let us see in the next two sections how we can approach these objects from another direction, namely, as solutions of appropriate equations.

[8]Mathematically it is a continuous curve which is nowhere differentiable. It was invented by Helge von Koch just to provide an example for such a curve; see H. v. Koch, *Une méthode géometrique élémentaire par l'étude de certaines questions de la théorie des courbes planes,* Acta. Mat. 30 (1906) 145–174.

3.3 Corner the New from Several Sides: Pi and the Square Root of Two

Limits have always had something mysterious about them, and it would be a great loss not to communicate that. Therefore, let us make an excursion and see how limits can reach out into the unknown. Limits create and characterize new quantities and new objects. The study of these unknowns was the pacemaker in early mathematics and has led to the creation of some of the most beautiful mathematical inventions. When Archimedes computed π by his approximation of the circle by a sequence of polygons, or when the Sumerians approximated $\sqrt{2}$ by an incredible numerical scheme, which was much later rediscovered by no one less than Newton, they were well aware of the fact that π and $\sqrt{2}$ were unusual numbers. The beautiful relation between the Fibonacci sequence $1, 1, 2, 3, 5, 8, 13, 21, 34, 55, \ldots$ and the golden mean $\frac{1}{2}(1+\sqrt{5})$ has, over several centuries, inspired scientists and artists alike to wonderful speculations. It is almost ironic that mathematics and physics at the most advanced levels have recently taught us that some of these speculations, which motivated Kepler, among others, to speculate about the harmony of our cosmos, have an amazing parallel in modern science: it has been understood that in scenarios, which describe the breakdown of order and the transition to chaos, the golden mean number characterizes something like the last barrier of order before chaos sets in. Moreover, the Fibonacci numbers occur in a most natural way in the geometric patterns which can occur along those routes.

In this section we focus on two numbers, $\pi = 3.14\ldots$ and $\sqrt{2} = 1.41\ldots$, and their approximations from various directions. While the story of π is in some sense a diversion from fractals, the central theme of the book, the other example will be developed along lines which parallel the definition and approximation of fractals as worked out in the following sections.

Archimedes' Method for π

The method used by Archimedes for the calculation of π is based on inscribed and circumscribed regular polygons. In our presentation we use modern mathematical tools such as the sine and tangent functions which were not known to Archimedes, of course. We start with an inscribed hexagon to a circle of radius r. It has $n = 6$ sides. The angle covered by one half side is $\theta = \pi/6$ (see figure 3.14).

The length of the inscribed side is $2r\sin\theta$. The length of a side of the circumscribed hexagon is $2r\tan\theta$. Thus, for the length of the circle $U = 2\pi r$ we have

$$2rn\sin\theta < U < 2rn\tan\theta.$$

Dividing by $2r$ we obtain a lower and an upper estimate for π,

$$n\sin\theta < \pi < n\tan\theta.$$

In numbers this is $3 < \pi < 3.464$, not a very accurate result. But we can easily improve the result simply by doubling the number n of sides

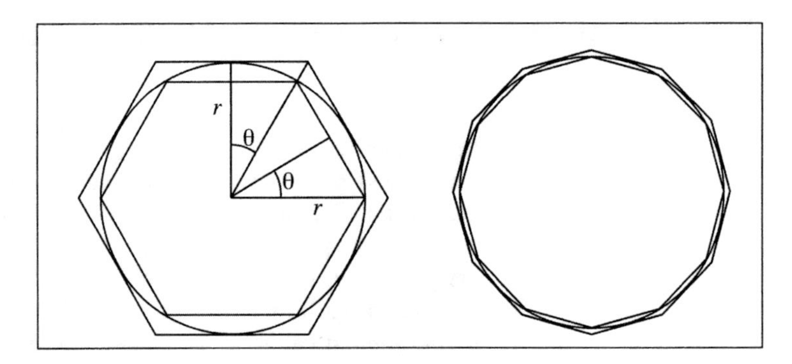

Figure 3.14 : Inscribed and circumscribed regular polygons.

and replacing θ by $\theta/2$, which yields

$$2n\sin\frac{\theta}{2} < \pi < 2n\tan\frac{\theta}{2}.$$

This is $3.106 < \pi < 3.215$. Further doubling, i.e., going from a regular polygon of 12 sides to one of 24 sides, and then to 48, 96, and so on, we can obtain an estimate that can be as sharp as we want. After k such doubling steps the formula is

$$2^k n\sin\frac{\theta}{2^k} < \pi < 2^k n\tan\frac{\theta}{2^k}.$$

It is not clear, exactly what method Archimedes used to compute the sines and tangents. Probably he used an iteration method based on formulas similar to

$$\begin{aligned}\sin\frac{\theta}{2} &= \sqrt{\frac{1-\cos\theta}{2}},\\ \tan\frac{\theta}{2} &= \frac{\sin\theta}{1+\cos\theta}.\end{aligned}$$

π and the Length of a Circle

The computation of the length of a circle, i.e., the computation of π, is a problem which challenged ancient mathematicians to a great extent. This problem has a history which is more than 4000 years old. The *Old Testament* uses $\pi = 3$ (see 1. Kings 7:23). The Babylonians used $\pi = 3.125$ and the Egyptians[9] (around 1700 B. C.) proposed $\pi = 3.1604\ldots$ Also in China philosophers and astronomers were very active in deriving approximations of π. One of the best goes back to Zu Chong-Zhi (430–501), who used the value 355/113, which has seven correct digits. At that time Chinese silk was sold as far as Rome. But it is not clear whether the fundamental work of Archimedes

[9] In fact they proposed an algorithm for the computation of the area of a circle: take away $1/9$ of the diameter and square the remaining $8/9$ of the result.

was also known to the Chinese. Archimedes was the first (around 260 B.C.) to provide a definite solution to the problem. He considered the circle with radius 1 and approximated half of its circumference by a sequence of regular polygons. In fact, he considered a sequence of approximating regular polygons which were inscribed and another sequence of regular polygons which were circumscribed. He carried out the approximation a few steps and obtained the numerical value 3.141031951 which has already four leading correct digits. He could have gone to even higher accuracy because his method was absolutely correct.

A more elegant method was discovered by the medieval scholar and philosopher Nicolaus Cusanus around 1450. It is another example for a feedback system and a forerunner of the very sophisticated methods used nowadays to compute π on mainframe computers up to millions of digits.

Cusanus' Method of Computation of π

Archimedes considered a fixed circle and approximated its circumference by a sequence of polygons. In a way Cusanus turned this approach around and employed a sequence of regular polygons with fixed circumference. More precisely, the regular polygons have 2^n, $n = 2, 3, 4, \ldots$ vertices such that the circumference always has length 2! He then computed the length of the corresponding circles which were inscribed and circumscribed (see figure 3.15).

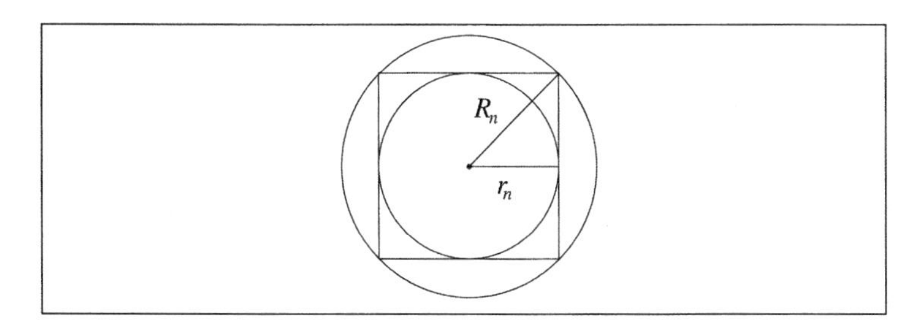

Figure 3.15 : Initial square and circles in Cusanus' method. For a given regular polygon with 2^n sides which sum up to a circumference of two units, the inscribed and circumscribed circles are considered.

Let R_n and r_n denote the radius of the circumscribed and inscribed circle of the nth polygon. Then we have

$$2\pi r_n < 2 < 2\pi R_n,$$

or equivalently

$$\frac{1}{R_n} < \pi < \frac{1}{r_n}. \tag{3.4}$$

For $n = 2$ we have a square with circumference 2 (see figure 3.15), and thus we compute using the Pythagorean theorem $R_2 = \sqrt{2}/4$ and $r_2 = 1/4$. Then Cusanus continued to extract the following useful

n	r_n	R_n	p_n	Error
2	0.250000	0.353553	3.313708	0.172116
3	0.301777	0.326641	3.182598	0.041005
4	0.314209	0.320364	3.151725	0.010132
5	0.317287	0.318822	3.144118	0.002526
6	0.318054	0.318438	3.142224	0.000631
7	0.318246	0.318342	3.141750	0.000158
8	0.318294	0.318318	3.141632	0.000039
9	0.318306	0.318312	3.141603	0.000010
10	0.318309	0.318310	3.141595	0.000002
11	0.318310	0.318310	3.141593	0.000001

Table 3.16 : The first few steps of Cusanus' method for the iterative calculation of π. The approximation p_n in the fourth column is computed by $p_n = 2/(r_n + R_n)$. The error $p_n - \pi$ decreases by about a factor of four in each step.

relations from geometric considerations which were already known to Archimedes, Pythagoras, and others.

$$\begin{aligned} r_{n+1} &= \frac{R_n + r_n}{2} \\ R_{n+1} &= \sqrt{R_n r_{n+1}} \end{aligned}$$

for $n = 2, 3, \ldots$ It turns out that $r_n < R_n$ for all n and that r_n increases while R_n decreases as n grows. Thus, both sequences have limits, and these limits must be the same.[10] But then eqn. (3.4) implies that the limit must be $\frac{1}{\pi}$. It turns out that Cusanus' method yields π up to 10 correct decimal places if one computes the feedback system for up to $n = 18$. Table 3.16 lists the first eleven steps, the corresponding approximations of π, and their errors.

Other Approaches to π

F. Vieta (1540–1603):

$$\frac{2}{\pi} = \frac{\sqrt{2}}{2} \cdot \frac{\sqrt{2+\sqrt{2}}}{2} \cdot \frac{\sqrt{2+\sqrt{2+\sqrt{2}}}}{2} \cdots$$

J. Wallis (1616–1703):

$$\frac{\pi}{2} = \frac{2 \cdot 2}{1 \cdot 3} \cdot \frac{4 \cdot 4}{3 \cdot 5} \cdot \frac{6 \cdot 6}{5 \cdot 7} \cdot \frac{8 \cdot 8}{7 \cdot 9} \cdots$$

J. Gregory (1638–1675) and G. W. Leibniz (1646–1716):

$$\frac{\pi}{4} = 1 - \frac{1}{3} + \frac{1}{5} - \frac{1}{7} + \frac{1}{9} - \frac{1}{11} + \frac{1}{13} - \cdots \tag{3.5}$$

[10] If they were not the same, say $R_n \to R$ and $r_n \to r$ with $r \neq R$, then $(R + r)/2 \neq r$, as it should.

L. Euler (1707–1783):

$$\frac{\pi^2}{6} = \frac{1}{1^2} + \frac{1}{2^2} + \frac{1}{3^2} + \frac{1}{4^2} + \frac{1}{5^2} + \cdots$$
$$\frac{\pi^4}{90} = \frac{1}{1^4} + \frac{1}{2^4} + \frac{1}{3^4} + \frac{1}{3^4} + \frac{1}{4^4} + \cdots$$

C. F. Gauss (1777–1855):

$$\pi = 48 \arctan \frac{1}{48} + 32 \arctan \frac{1}{57} - 20 \arctan \frac{1}{239}. \qquad (3.6)$$

S. Ramanujan (1887–1920):

$$\frac{1}{\pi} = \frac{\sqrt{8}}{9801} \sum_{n=0}^{\infty} \frac{(4n!)(1103 + 26\,390n)}{(n!)^4 396^{4n}}.$$

J. M. Borwein and P. M. Borwein (1984):

$$x_{n+1} = \frac{1 - \sqrt{1 - x_n^2}}{1 + \sqrt{1 - x_n^2}} \qquad x_0 = \frac{1}{\sqrt{2}}$$
$$y_{n+1} = (1 + x_{n+1})^2 y_n - 2^{n+1} x_{n+1} \qquad y_0 = \frac{1}{2}.$$

With these settings y_n converges quadratically to $1/\pi$.

The following is another interesting characterization.[11] An integer is called square free provided it is not divisible by the square of a prime number. For example, 15 is square free ($15 = 3 \cdot 5$), but 50 is not ($50 = 2 \cdot 5^2$). Now let $h(n)$ be the number of, and $q(n) = h(n)/n$ be the fraction of, square free numbers between 1 and n. Then

$$\lim_{n \to \infty} q(n) = \frac{6}{\pi^2}.$$

World Records in π

Like no other irrational number, π has fascinated the giants of science as well as amateurs around the world. For hundreds and even thousands of years more and more digits of π have been worked out using sometimes extremely tedious methods. This enormous effort stands in absolutely no relation to its use. It would be hard to find applications in scientific computing, where more than some 20 digits of π are necessary. Nonetheless, people have been pushing the number of known digits of π higher and higher as if it were a sport like the high jump, where athletes are driven to equal and surpass the standing world record. When asking mountaineers about their motivation to painfully climb a particularly high peak, they very well might answer, that they do it 'because it is there'. In this sense the number π is even better than Mount Everest because the number of digits in π is unlimited. Once arrived at a world record, there

[11] C. R. Wall, *Selected Topics in Elementary Number Theory*, University of South Carolina Press, Columbia, 1974, page 153.

is already the challenge to also conquer the next ten, or hundred, or million digits.

Ludolph's Number

Let us give some examples of the craze that went on in the previous centuries and that is still continuing today with the help of computers. The Dutch mathematician Ludolph van Ceulen (1540–1610) dedicated a large portion of his work to the computation of π. In 1596 he reported 20 digits of π, and just before his death he succeeded in computing 32 and even 35 digits pushing the method of Archimedes to its extreme: he used inscribed and circumscribed polygons with $2^{62} \approx 10^{18}$ vertices. The last three digits are inscribed on his tombstone, and henceforth the number π was also known as *Ludolph's number*.

Approximation of π Using Series and Hand Calculation

Partial list of the world records of the computation of π from 1700 until computing machines became available. Only 527 of the 707 digits computed by Shanks were correct.

Year	Name	Digits
1700	Sharp	72
1706	Machin	100
1717	Delaney	127
1794	Vega	140
1824	Rutherford	208
1844	Strassnitzky and Dase	200
1847	Clausen	248
1853	Rutherford	440
1855	Richter	500
1873	Shanks	707
1945	Ferguson	620

Table 3.17

Machin's Formula for π

In 1706 John Machin discovered an elegant and computable way to represent π as a limit. Before, in 1671, Gregory had discovered that the area under the curve $1/(1+x^2)$ from 0 to x was $\arctan x$. The arctangent series

$$\arctan x = x - \frac{x^3}{3} + \frac{x^5}{5} - \frac{x^7}{7} + \cdots \tag{3.7}$$

was a direct conclusion of this. Substituting $x = 1$ then gives an easy formula for $\pi/4$; see eqn. (3.5). However, this series is very slowly convergent and, thus, not useful for actual computations. Machin devised a neat like trick to modify the Gregory series and improve its convergence dramatically. The derivation is easy using the trigonometric identities

$$\tan(\alpha \pm \beta) = \frac{\tan\alpha \pm \tan\beta}{1 \mp \tan\alpha \tan\beta}.$$

Let β be the unique angle less than $\pi/4$ such that

$$\tan\beta = \frac{1}{5}.$$

Using the above trigonometric formulas, we compute

$$\tan 2\beta = \frac{2\tan\beta}{1-\tan^2\beta} = \frac{\frac{2}{5}}{1-\frac{1}{25}} = \frac{5}{12}$$

and

$$\tan 4\beta = \frac{2\tan 2\beta}{1-\tan^2 2\beta} = \frac{\frac{5}{6}}{1-\frac{25}{144}} = \frac{120}{119}.$$

From the last result we see that $\tan 4\beta \approx 1$, and therefore $4\beta \approx \pi/4$. Now the tangent of the difference between these two angles can again be computed

$$\tan(4\beta - \frac{\pi}{4}) = \frac{\tan 4\beta - 1}{1+\tan 4\beta} = \frac{\frac{1}{119}}{\frac{239}{119}} = \frac{1}{239}.$$

In other terms,

$$4\beta - \frac{\pi}{4} = \arctan\frac{1}{239},$$

or, solved for $\pi/4$, we obtain the final result

$$\frac{\pi}{4} = 4\arctan\frac{1}{5} - \arctan\frac{1}{239}.$$

In contrast to Gregory's formula there are two series to be computed here, but this drawback is more than compensated for by the fact that these series converge much more rapidly, especially the second one. Following Machin's idea many more similar formulas expressing π as a sum of arctangents have been developed, among them one from Gauss; see eqn. (3.6).

The Number Cruncher's Pain ...

After the discovery of differential calculus in the 17th century, new and better methods were devised for the computation of π. These methods used the series expansions of the arcsine and arctangent. The most convenient one for calculation with paper and pencil was provided by John Machin (1680–1752). Table 3.17 lists the progress made along these lines.[12] Computations typically took several months. Of course, some mistakes in such immense work were unavoidable. So when Vega computed his 140 digits in 1794, he discovered an error in the 113th place of Delaney's result. The 200 digits of Strassnitzky and Dase also were not in agreement with Rutherford's. Clausen then showed that the error was in Rutherford's calculation. Also Shanks' result was wrong from digit 527 on. Of all these, Strassnitzky deserves special mention. The actual calculations were carried out by Johann Martin Zacharias Dase (1824–1861), who was a calculating prodigy. His extraordinary calculating powers were

[12] Our exposition here is based in part on the book *A History of Pi* by Petr Beckmann, Second Edition, The Golem Press, Boulder, 1971.

verified by renowned mathematicians. He multiplied two 8-digit numbers in 54 seconds, two 20-digit numbers in 6 minutes, and two 100-digit numbers in under 9 hours, all of it in his head! There are at least two abilities that such prodigies must have: rapid execution of arithmetical operations, and something like a photographic memory to store the vast amount of information. On the other hand it seems that extraordinary intelligence is not necessary; on the contrary, this would be counter-productive. Dase, for example, was no exception. All who knew him agree, that except for calculating and numbers, he was quite dull. At the age of 20 Strassnitzky taught him an arctangent formula for π similar to Machin's formula, and in two months time Dase produced 200 correct digits. But that was not all. In three years he computed natural logarithms of the first million integers, each to seven decimal places, and continued to work on a table of hyperbolic functions. He was brought to the attention of Gauss, and upon Gauss' recommendation, he started to make a listing of the factors of all numbers from 7,000,000 to 10,000,000, a work sponsored by the Hamburg Academy of Sciences. However, Dase died in 1861, after finishing about half of them.

...and the Circle Squarer's Ease

In 1885 F. Lindemann succeeded in proving a fundamental theorem on transcendental numbers which also solved an age-old problem: π is a transcendental number,[13] which implies that squaring the circle is an impossible task. Nonetheless, people continued to find 'solutions' to the circle squaring problem, some more obscure than others. Here is just one example. In 1897, the house of representatives of Indiana, USA, passed a bill "for an act introducing a new mathematical truth", which defined two(!) values of π, namely 3.2 and 4. Fortunately, the senate of Indiana postponed further consideration of the law indefinitely.

Approaching π with Technology

In the 20th century it became more and more difficult to break the record in the computation of π — until computers came on the scene. It was a relatively simple task to program a computer to evaluate, for example, Machin's formula up to a thousand digits. Of course, as soon as it became possible, it was done. Table 3.18 lists the records in this second phase.

The computations up until the seventies were all based on arctangent series that had already been used by the pre-computer age pioneers. A complete listing of the first 100,000 digits of π was published by Shanks and Wrench in 1962.[14] In the last section of the paper the authors speculate about the possibility of computing a million digits, concluding that "One would really want a computer 100 times as fast, 100 times as reliable, and with memory 10 times as large. No such machine now exists. [...] In 5 to 7 years such a computer [...] will, no doubt become a reality. At that time a computation of π to 1,000,000 digits will not be difficult." The authors were too optimistic; it took 12 more years until Jean Guilloud and Martine Bouyer checked off that millionth digit.

[13] A number x is called algebraic provided that it is the root of a polynomial equation with rational coefficients. A transcendental number is one that is not algebraic.

[14] D. Shanks and J. W. Wrench, Jr., *Calculation of π to 100,000 Decimals,* Mathematics of Computation 16, 77 (1962) 76–99.

Year	Name	Computer	Digits
1949	Reitwiesner	ENIAC	2,037
1954	Nicholson et al.	NORC	3,089
1958	Felton	Pegasus	10,000
1958	Genuys	IBM704	10,000
1959	Unpublished	IBM704	16,167
1961	Shanks, Wrench	IBM7090	100,000
1973	Guilloud, Bouyer	CDC7600	1,000,000
1983	Kanada et al.	Hitachi S-810	16,000,000
1985	Gosper	Symbolics	17,000,000
1986	Bailey	Cray2	29,300,000
1987	Kanada	SX 2	134,000,000
1989	Kanada, Tamura	HITAC S-820/80	1,073,741,799
1994	Borwein, Borwein, and Kanada	HITACS-3900/480	4,294,967,286
1999	Takahashi, Kanada	HITACHI SR8000	206,158,430,000
1999	Kanada et al	HIT. SR8000/MPP	1,241,100,000,000

Approximation of π by Computer

World records of the computation of π in the computer age. The computing times are mostly on the order of 5 to 30 hours, the shortest one being 13 minutes (1945) and a long one (400 hours) gave the 2002 record.

Table 3.18

The simplicity of the method using, for example, the Gauss formula (3.6) in connection with the arctan series (3.7) is a temptation for any ambitious programmer. It provides an excellent exercise for a programming course. We have tried it out and succeeded in computing the first 200,000 digits.[15] However, the undertaking turned out to be not quite as easy as initially assumed. In the first run only the first 60,000 digits were correct. The mistake was due to insufficient treatment of overflow errors.

How Far Can We Go?

The question arises, how many digits one can possibly hope to be able to compute. The algorithms based on arctangent expansions have the property that doubling the number of digits in the result requires a computation which is *four* times as long. The 1973 computation of a million decimals took 23 hours. For example, to get from one million to 128 million digits, one must double the number of digits seven times ($128 = 2^7$). On the same computer, the time of 23 hours, quadrupled seven times, would have yielded a computing time requirement of about 43 years ... Even though computers were becoming faster and faster, it was clear that there would be an end to that development sooner or later. Thus, a couple more millions of digits seemed possible, but certainly not hundreds of millions of digits. So the record of a million decimals stood for 10 years. But the grounds had already been prepared for yet another escalation.

Another Breakthrough: New Algorithms

A major breakthrough occurred in 1976 when algorithms which yield a quadratically convergent iteration procedure were discovered independently by Brent and Salamin.[16] This means that in each iteration step of these methods the number of correct digits is doubled. More recently the Borwein brothers

[15] The program ran about 15 hours on a Macintosh IIfx.

[16] R. P. Brent, *Fast multiple-precision evaluation of elementary functions,* Journal Assoc. Comput. Mach. 23 (1976) 242–251. E. Salamin, *Computation of π using arithmetic-geometric mean,* Math. Comp. 30, 135 (1976) 565–570.

have worked out a family of even more efficient methods.[17] All the new algorithms are more efficient than the good old arctangent series, however, only because of another breakthrough in a different area — arithmetic. Addition of two n-digit numbers costs about n operations (add all corresponding pairs of digits and add up). However, the direct, naive multiplication of two n-digit numbers would essentially have to be carried out in n^2 operations (multiply each digit with all other digits and add up). Thus, when the number of digits n is of the order of a million or more, the difference between addition and multiplication is dramatic. Thus, the discovery, that the complexity of the multiplication of numbers, is effectively not significantly larger than that of the addition of numbers is almost unbelievable: a multiplication can be made almost as fast as an addition.[18] Practical implementations make use of a form of Fast Fourier Transformation techniques. The combination of the new feedback methods for π and the fast multiplication algorithms for very long numbers facilitated computations of π with millions of digits of precision.[19] The record at the end of the 20th century stood at about 206 billion digits. Two different algorithms (Gauss-Legendre and Borwein's 4-th order convergent algorithm) running each about 40 to 50 hours provided coinciding results. Then, in 2002, Yasumasa Kanada reported haviong calculated 1.24 trillion digits. The prospects are good to get even one more trillion digits in the near future. Of course, for practical computations almost all of these digits are completely useless.

Two Reasons to Compute π

However, there are two new reasons for this excessive digit hunting. The first one is related to a longstanding conjecture which states that the digits in π as well as the pairs of digits, the triplets of digits, and so on are uniformly distributed. In mathematical terms, π is believed to be a normal number. By extensive computer study, one may be able to find signs about the truth or falsity of this conjecture. At least up to the digits computed so far, statistical tests indicate that π is, in fact, normal.[20] Of course, this is far from a proof. The other reason why programs for the calculation of π should be written is that they can be used to effectively test the reliability of computer hardware. It is claimed that some computer manufacturers indeed perform such tests.[21] Even the smallest error at any operation in the calculation will invariably produce wrong digits from some place on, and these errors are very obviously detectable.

[17] See the book, J. M. Borwein, P. B. Borwein, *Pi and the AGM — A Study in Analytic Number Theory,* Wiley, New York, 1987.

[18] More precisely, the way the computing requirements grow as the number of digits in the factors of the multiplication is increased is not much worse than the corresponding (linear) growth of computing time for the addition of long numbers. The interested reader is referred to the survey in D. Knuth, *The Art of Computer Programming, Volume Two, Seminumerical Algorithms,* Addison Wesley, Reading, 1981, pages 278–299.

[19] For a description of techniques and algorithms see J. M. Borwein, P. B. Borwein, and D. H. Bailey: *Ramanujan, modular equations, and approximations to pi, or how to compute one billion digits of pi,* Am. Math. Monthly 96 (1989) 201–219.

[20] In the first 200 billion digits computed in 1999 by Takahashi and Kanada, the digit zero appeared 20,000,030,841 times, while the digit one came up 19,999,914,711 times and so on.

[21] In fact, in the 1962 paper by Shanks and Wrench, one instance of such hardware failure was reported, and an auxiliary run of the program was made to correct for the error. Thus, at least in the time about 30 years ago, reliability of the arithmetic was an important practical issue even for the 'end user'.

Is There a Message in π?

The advanced and more recent efforts to compute π may have inspired Carl Sagan to a part of his novel *Contact*[22] where he presents speculation about a hidden pattern or message God may have provided in the digits of π. In the story a super computer makes a discovery after countless hours of number crunching: the sequence of digits of π, located very far from the beginning, interpreted bitwise and displayed as a rectangular picture, shows a well-known figure — a circle. The novel concludes:

"In whatever galaxy you happen to find yourself, you take the circumference of a circle, divide by its diameter, measure closely enough, and uncover a miracle — another circle, drawn kilometers downstream of the decimal point. There would be richer messages further in. It doesn't matter what you look like, or what you're made of, or where you come from. As long as you live in this universe, and have a modest talent for mathematics, sooner or later you'll find it. It's already here. It's inside everything. You don't have to leave your planet to find it. In the fabric of space and in the nature of matter, as in a great work of art, there is, written in small, the artist's signature. Standing over humans, gods, and demons, [...] there is an intelligence that antedates the universe."

We now return to more worldly issues of numbers. Although limits are very useful for numerical computation of irrational numbers such as π, e or square roots, it is more satisfying from a theoretical point of view to have a more direct definition of the numbers. This could be an implicit definition in the form of a suitable equation that simultaneously prescribes an approximation by a feedback process, namely, just by iterating the equation. Let us look at this issue in the remainder of this section.

$\sqrt{2}$ and Incommensurability

We recall the problem of the *incommensurability* of the side and the diagonal of a square: the ratio of the diagonal and side of a square is not equal to the ratio of two integers.[23] In other words, $\sqrt{2} \neq p/q$ for any integer p and q. No doubt the diagonal is real, but does that mean that $\sqrt{2}$ exists as a number in some sense? This was a great question; and though it sounds naive from today's point of view, it was not and still is not. Just ask yourself how you would *convince* somebody (of the existence of such a number). Certainly you could not expect much aid from the decimal expansion, which goes on and on in a seemingly totally disorganized fashion: the first 100 digits in the decimal expansion are

$$\begin{aligned}\sqrt{2} = 1.&41421\ 35623\ 73095\ 04880\ 16887\\ &24209\ 69807\ 85696\ 71875\ 37694\\ &80731\ 76679\ 73799\ 07324\ 78462\\ &10703\ 88503\ 87534\ 32764\ 15727\ldots\end{aligned}$$

But there is a different way to expand $\sqrt{2}$. Namely, to represent it as a special kind of limit, and then $\sqrt{2}$ looks almost as 'natural' as an integer does. This

[22] Carl Sagan, *Contact,* Pocket Books, Simon & Schuster, New York, 1985.

[23] Compare chapter 2, page 124.

and some of the other most beautiful and mysterious limits are related to continued fraction expansions.

Continued Fractions

Let us begin with a seemingly strange way of writing rational numbers. Here is an example:

$$\frac{57}{17} = 3 + \cfrac{1}{2 + \cfrac{1}{1 + \cfrac{1}{5}}}.$$

Let us see how this representation is obtained step by step:

$$\begin{aligned}\frac{57}{17} &= 3 + \frac{6}{17} = 3 + \frac{1}{\frac{17}{6}} = 3 + \frac{1}{2 + \frac{5}{6}} \\ &= 3 + \frac{1}{2 + \frac{1}{\frac{6}{5}}} = 3 + \frac{1}{2 + \frac{1}{1 + \frac{1}{5}}}.\end{aligned}$$

In this way any rational number can be written as a *continued fraction expansion*. The point is that a rational number has a finite expansion (i.e., the process terminates after some definite number of steps). In our example we write for short

$$\frac{57}{17} = 3 + [2, 1, 5].$$

The same algorithm applies to irrational numbers. However, in this case the algorithm never stops. It produces an infinite continued fraction representation.

Continued Fraction Expansion of $\sqrt{2}$

Let us look into a slightly more general situation which brings us back to $\sqrt{2}$. We begin with the equation:

$$x^2 + 2x - 1 = 0.$$

The positive root of this equation is $x = \sqrt{2} - 1 < 1$. Note that $x^2 + 2x - 1 = 0$ can be rewritten as $x^2 + 2x = 1$, or $x(2 + x) = 1$, or

$$x = \frac{1}{2 + x}.$$

Moreover, after replacing x by $\frac{1}{2+x}$ on the right side,

$$x = \cfrac{1}{2 + \cfrac{1}{2 + x}}$$

and then, doing it again,

$$x = \cfrac{1}{2 + \cfrac{1}{2 + \cfrac{1}{2 + x}}}$$

etc. In other words, there will be an infinite repetition of 2's in the continued fraction expansion of $\sqrt{2}-1$. Naturally, this implies that $\sqrt{2}$ has the expansion

$$x = 1 + \cfrac{1}{2 + \cfrac{1}{2 + \cfrac{1}{2 + \cfrac{1}{2 + \cdots}}}}.$$

This remarkable identity relates $\sqrt{2}$ with the sequence of numbers $1 + [2, 2, 2, 2, \ldots]$, the digits of the continued fraction expansion of $\sqrt{2}$. We write $\sqrt{2} = 1 + [2, 2, 2, 2, \ldots]$ and mean that $1, 2, 2, 2, \ldots$ are placed into the fractions as above. In other words, $\sqrt{2}$ is the limit of the sequence 1, $1 + [2] = 1.5$, $1 + [2, 2] = 1.4$, $1 + [2, 2, 2] = 1.416\ldots$, and so on. Thus, $\sqrt{2}$ has a perfectly regular and periodic continued fraction expansion, while in an expansion with respect to some base like 10 the expansion looks like a big mess. It will never be periodic, because otherwise $\sqrt{2}$ would be a rational number.

Continued Fraction Expansion of the Golden Mean

The process which we discussed in detail for the equation $x^2 + 2x = 1$ works the same in a slightly different case,

$$x^2 = ax + 1$$

where a is an integer. After dividing by x and substituting for x twice we obtain

$$x = a + \frac{1}{x} = a + \cfrac{1}{a + \cfrac{1}{x}} = a + \cfrac{1}{a + \cfrac{1}{a + \cfrac{1}{x}}}$$

and so on. Thus, the continued fraction expansion will be

$$x = a + [a, a, a, \ldots].$$

Specifically, if $a = 1$, then the positive root of $x^2 - x - 1 = 0$ is the golden mean $x = (1 + \sqrt{5})/2$ and we obtain

$$x = \frac{1+\sqrt{5}}{2} = 1 + [1, 1, 1, \ldots] = 1 + \cfrac{1}{1 + \cfrac{1}{1 + \cdots}}.$$

Therefore the golden mean has the simplest possible continued fraction expansion. All roots of quadratic equations with integer coefficients have continued fraction expansions, which are eventually periodic, like $2 + [2, 2, 3, 2, 3, 2, 3, \ldots]$ or $2 + [1, 1, 4, 1, 1, 4, 1, 1, 4, \ldots]$. Rational numbers are characterized by a finite continued fraction expansion.

Characterization by Equations

Let us summarize what our main point is about irrational numbers so far. If we only had a limit representation such as the decimal expansion of $\sqrt{2}$, we would feel quite uncomfortable. Comfort comes from some other characterization:

1. $\sqrt{2}$ has an elementary continued fraction expansion, $1 + [2, 2, 2, \ldots]$.
2. $\sqrt{2}$ solves an equation, $x^2 - 2 = 0$.

But we can do even better. Consider the function

$$N(x) = \frac{1}{2}\left(x + \frac{2}{x}\right)$$

and its fixed points $x = N(x)$. Compute

$$x = \frac{1}{2}\left(x + \frac{2}{x}\right) = \frac{x}{2} + \frac{1}{x}, \quad \frac{x}{2} = \frac{1}{x}, \quad x^2 = 2.$$

Thus, the fixed points of the function $N(x)$ are just the square roots of two, and we may replace $x^2 - 2 = 0$ in our list above by

$$x = N(x) = \frac{1}{2}\left(x + \frac{2}{x}\right).$$

There is an important reason for favoring this fixed point formulation over $x^2 - 2 = 0$: we can use $N(x)$ as the governing of the feedback process,

$$x_{n+1} = N(x_n), \quad n = 0, 1, 2, \ldots \tag{3.8}$$

This iteration process will surely converge to the positive root of two, provided we start with a positive initial number $x_0 > 0$. We have discussed this already in chapter 1, page 27, and just give an example here, choosing $x_0 = 100$; see table 3.19.

Approximation of the Square Root of 2

Approximation of square root of 2 using the iteration $x_{n+1} = (x_n + 2/x_n)/2$. The initial guess is $x_0 = 100$. Once the method is about the same magnitude as the true value $1.4142135623730950\ldots$, the iterates converge very rapidly, and the number of correct digits doubles in each step.

n	x_n	Correct Digits
0	100.0000000000000000	0
1	50.0100000000000000	0
2	25.0249960007998400	0
3	12.5524580467459030	0
4	6.3558946949311400	0
5	3.3352816092804338	0
6	1.9674655622311490	1
7	1.4920008896897231	2
8	1.4162413320389438	3
9	1.4142150140500532	6
10	1.4142135623738401	13
11	1.4142135623730950	all

Table 3.19

We see that the iteration process converges to $\sqrt{2}$ very rapidly after some initial iterations have brought the number x_n into a region close to the root.

The number of correct leading digits roughly doubles in each step. Of course, this is no coincidence, but, in fact, the predominantly used method for the calculation of square roots, called *Newton's method*. Let us summarize our findings:

1. There is a well-defined approximation procedure for $\sqrt{2}$, the feedback process

$$x_{n+1} = \frac{1}{2}\left(x_n + \frac{2}{x_n}\right), \quad x_0 > 0$$

with a rapid convergence.

2. There is a corresponding fixed point equation

$$x = \frac{1}{2}\left(x + \frac{2}{x}\right)$$

which characterizes the limit, $\sqrt{2}$.

The fixed-point equation should be seen in connection with symmetries, e.g., a regular hexagon is rotationally symmetric by a rotation of 60°, and it also has a reflectional symmetry. In other words, one has an object, applies some operation (transformation) to it and obtains the same object. Our goal will be to corner fractals in the same way as one does irrational numbers, i.e., by an elementary limit process stemming from a fixed-point equation, which characterizes the fractal by an invariance property.

3.4 Fractals as Solutions of Equations

Let us return to fractals and find out how we can carry over the concepts we have learned from dealing with the square root of 2. Summarizing the main point about the Koch curve we have that the curve is a limit of a process, a limit which has special properties, and which we can characterize in a similar way as $\sqrt{2}$ is characterized by its beautiful continued fraction expansion. But does the Koch curve really exist? Well, this question is very much of the same nature as the question of the existence of irrational numbers. Recall that in that case we take comfort from the fact that we believe in the validity of some closely related and characterizing concept. For example, for $\sqrt{2}$ we argue that this is the number which solves the equation $x^2 - 2 = 0$ or $x = (x + 2/x)/2$. Or for 2π we argue that this is the number which gives a length to the unit circle. Observe that here neither number is characterized as a limit of a sequence, and this really helps us to accept these numbers! The hypothesis that π might still not be known in mathematics if it did not relate to a circle so beautifully is speculative. Nevertheless, would Euler have discovered that $1 + 1/2^2 + 1/3^2 + 1/4^2 + \cdots$ is some very special number ($\pi^2/6$) worth being investigated even if π was not somehow a reality?

In other words, we need some further reasons to accept the existence of the Koch curve, as well as characterizations which relate it to different ideas and concepts or principles. This is a major desire in mathematics. If an object or result suddenly becomes interpretable from a new point of view, mathematicians usually feel that they have made progress and are satisfied.

Is There an Invariance Property for the Koch Curve?

We may ask: is there an invariance property for the Koch curve? Can we find a characterization which is similar to that of $\sqrt{2}$? One type of invariance transformation is apparent. The Koch curve has an obvious reflectional symmetry. But this is not characterizing in the sense that it singles out the Koch curve. Ideally, we would like to find a transformation or a set of transformations which leave the Koch curve invariant. Such a transformation then could be viewed as some kind of symmetry. Recall the discussion of the self-similarity of the Koch curve at the end of section 3.2. Let us now be a little bit more formal and precise. Figure 3.22 illustrates the similarity transformation of the Koch curve. First, we reduce the Koch curve by a factor of 1/3. We put it onto a copier with reduction features and produce four copies. Then we paste the four identical copies as shown in the bottom part of figure 3.22 and obtain a curve which looks like the original one. The Koch curve is a collage of the four copies.

The Similarity Transformations of the Koch Curve

The following table lists the details of the similarity transformations w_1 to w_4 of the Koch curve as shown in figure 3.22. When we take into account that

$$\begin{aligned} \cos 60^\circ &= \cos(-60^\circ) = \tfrac{1}{2}, \\ \sin 60^\circ &= -\sin(-60^\circ) = \tfrac{\sqrt{3}}{2}, \end{aligned}$$

Number	Scale	Rotation	Translation	
k	s	θ	T_x	T_y
1	1/3	0°	0	0
2	1/3	60°	1/3	0
3	1/3	−60°	1/2	$\sqrt{3}/6$
4	1/3	0°	2/3	0

Table 3.20 : Similarity transformations of the Koch curve collage. The transformations are carried out first by applying the scaling, then the rotation, and finally the translation (see section 3.1).

Transformation	x-Part	y-Part
$w_1(x,y)$	$\frac{1}{3}x$	$\frac{1}{3}y$
$w_2(x,y)$	$\frac{1}{6}x - \frac{\sqrt{3}}{6}y + \frac{1}{3}$	$\frac{\sqrt{3}}{6}x + \frac{1}{6}y$
$w_3(x,y)$	$\frac{1}{6}x + \frac{\sqrt{3}}{6}y + \frac{1}{2}$	$-\frac{\sqrt{3}}{6}x + \frac{1}{6}y + \frac{\sqrt{3}}{6}$
$w_4(x,y)$	$\frac{1}{3}x + \frac{2}{3}$	$\frac{1}{3}y$

Table 3.21 : Explicit formulas for the similarity transformations of the Koch curve collage.

we obtain explicit formulas for the transformations as given in table 3.21.

Characterization by an Equation for the Self-Similarity

This collage-like operation can be described by a single mathematical transformation. We let $w_1, \ldots, w_4$ be the four similarity transformations given by a reduction with factor 1/3 composed with a positioning (rotation and translation) at piece k along the polygon as shown in figure 3.22 (bottom). Then, if A is any image, let $W(A)$ denote the collection (union) of all four transformed copies

$$W(A) = w_1(A) \cup w_2(A) \cup w_3(A) \cup w_4(A). \tag{3.9}$$

This is a transformation of images, or more precisely, subsets of the plane. Figure 3.23 shows the result of this transformation when applied to an arbitrary image, for example, when A is the word 'KOCH'. When comparing the results in figure 3.22 and figure 3.23, we make a fundamental observation. In the case where we apply the transformation W from eqn. (3.9) to the image of the Koch curve, we obtain the Koch curve back again. That is, if we formally introduce a symbol K for the Koch curve, we have the important identity

$$W(K) = K,$$

which is the desired invariance (or fixed-point) property. In other words, if we pose the problem of finding a solution X to the equation $W(X) = X$, then

The Koch Collage

The Koch curve is invariant under the transformations w_1 to w_4.

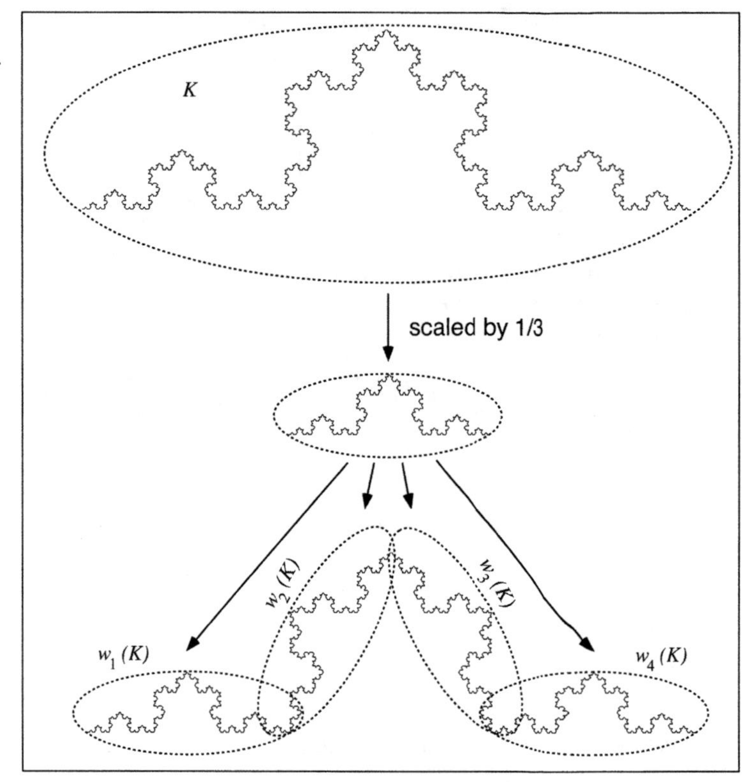

Figure 3.22

The KOCH Collage

The word 'KOCH' is not invariant under W.

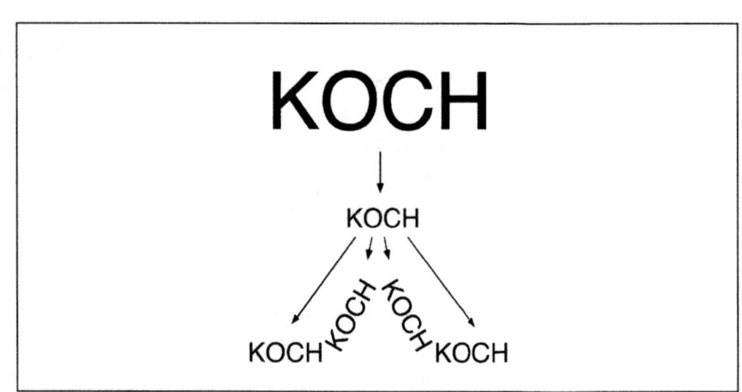

Figure 3.23

the Koch curve K solves the problem. Moreover, this equation also shows the self-similarity of K since

$$K = w_1(K) \cup w_2(K) \cup w_3(K) \cup w_4(K)$$

states that K is composed of four similar copies of itself. In other words, we have characterized K by its self-similarity. If we further substitute for K the

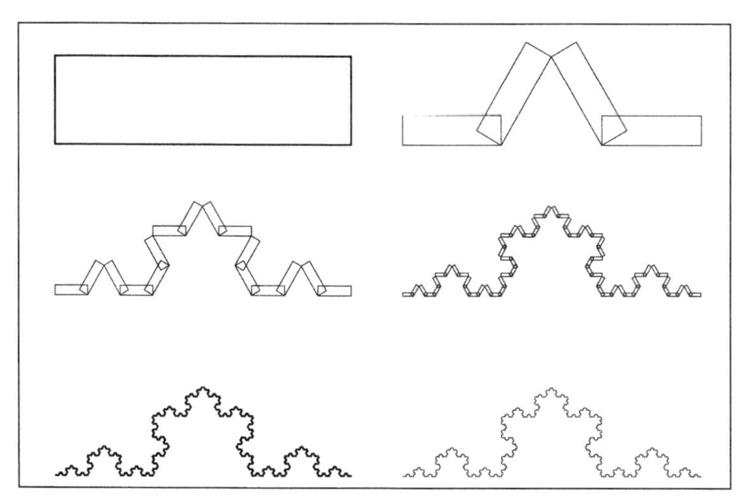

Limit Object Koch Curve

Starting with an arbitrary shape, a rectangle, iteration of the Hutchinson operator produces a sequence of images, which converge to the Koch curve.

Figure 3.24

collection of the four copies on the right-hand side of the equation, then it becomes clear that K is made of 16 similar copies of itself, and so on. We will come back to this interpretation of self-similarity later in this section.

When we apply the same transformation W to the name KOCH (i.e., X is the image 'KOCH'), we do not get back the name KOCH at all. Rather, we see some strange collage.

Only the Koch Curve Is Invariant Under W

We are led to conjecture that maybe the only image which is left invariant under the collage transformation W is the Koch curve. Indeed, that is a theorem which has far-reaching consequences which will be discussed in chapter 5. A collage transformation like W above is called a *Hutchinson operator*, after J. Hutchinson, who was the first to discuss its properties.[24]

The Koch Curve As a Limit Object

Having characterized the Koch curve as a fixed point of the Hutchinson operator, we now conclude the analogy to the computation of $\sqrt{2}$ (see eqn. (3.8)). It remains to show that mere iteration of the operator W applied to a starting configuration A_0 yields a sequence

$$A_{n+1} = W(A_n), \quad n = 0, 1, 2, \ldots,$$

which converges to the limit object, the Koch curve. This is indeed the case, and figure 3.24 visualizes the limit process, providing pictorial evidence that there is such a self-similar object. Let us summarize.

1. There is a well-defined approximation procedure for the Koch curve, the feedback process

   $$A_{n+1} = W(A_n), \quad n = 0, 1, 2, \ldots$$

   where A_0 can be any initial image and W denotes the Hutchinson operator

   $$W(A) = w_1(A) \cup w_2(A) \cup w_3(A) \cup w_4(A)$$

[24] J. Hutchinson, *Fractals and self-similarity*, Indiana University Math. J. 30 (1981) 713–747.

for the Koch curve.

2. There is a corresponding fixed-point equation

$$A = W(A)$$

which uniquely characterizes the limit, the Koch curve.

How can we make sure that what we see — W applied to the Koch curve yields the Koch curve again — is actually true? Can we really trust an image, or better, a graphic experiment? The answer is that we should take it as some supporting evidence, but not more than that. After all, it might be that in some invisibly small detail there is a difference between $W(K)$ and K itself. In other words, we must go on and convince ourselves that this remarkable self-similarity property is actually a fact and not just an experimental artefact. This will be our next goal. However, we will first discuss this property in two simpler examples, the *Cantor set* and the *Sierpinski gasket*.[25]

The Cantor Set Construction

The geometric feedback construction of the Cantor set.

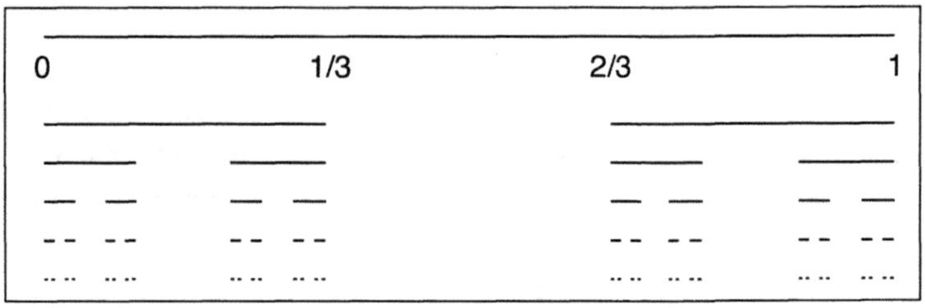

Figure 3.25

Equation for the Cantor Set

In chapter 2 the Cantor set was introduced as a limit in a geometric feedback process (begin with the unit interval, remove the open interval of length 1/3 centered at 1/2, then remove the middle thirds of the remaining intervals, and so on). Moreover, it has been described as the set of numbers between 0 and 1, for which there exists a triadic expansion that does not contain the digit 1. This last characterization allows us to verify that the Cantor set is the fixed point of the appropriate Hutchinson operator W given by the two transformations

$$\begin{aligned} w_1(x) &= \frac{x}{3}, \\ w_2(x) &= \frac{x}{3} + \frac{2}{3}. \end{aligned}$$

Thus, for a given set A, $W(A) = w_1(A) \cup w_2(A)$. Figure 3.26 shows how the transformations act when applied to the unit interval.

Our claim is that the Cantor set is a solution to the equation

$$W(X) = X$$

i.e., the Cantor set C is invariant under W, and $W(C) = C$.

[25]The mathematical discussion must be postponed to chapter 5 where we will look at the convergence of images and the characterization of fractals by Hutchinson operators in detail.

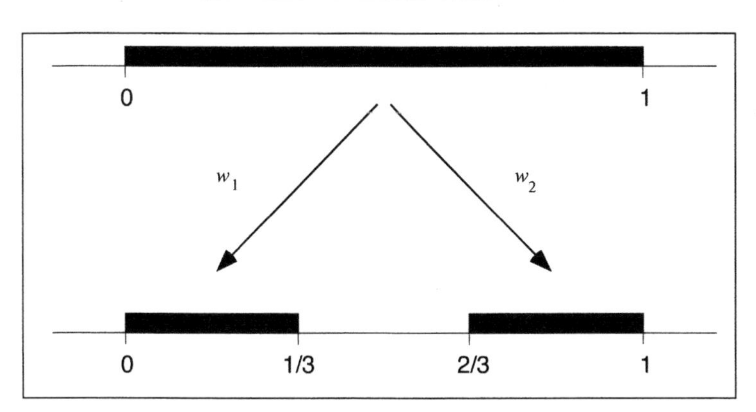

The Cantor Set Transformations

The similarity transformations w_1 and w_2 for the Cantor set.

Figure 3.26

The Invariance of C under W

Cantor himself gave a characterization of the set named after him in terms of numbers expanded with respect to base 3, triadic numbers. Recall that any number x, $0 \leq x \leq 1$, can be expanded in

$$x = a_1 \cdot 3^{-1} + a_2 \cdot 3^{-2} + a_3 \cdot 3^{-3} + a_4 \cdot 3^{-4} + \ldots,$$

where the digits a_k are from $\{0, 1, 2\}$. Then x is written in the form $x = 0.a_1a_2a_3\ldots$, i.e., we take the coefficients $a_1, a_2, a_3, \ldots$ as triadic digits. Then the Cantor set is determined by

$$C = \{x \mid x = 0.a_1a_2a_3\ldots, a_i \in \{0, 2\}\},$$

i.e., by all numbers which admit a triadic expansion that misses the triadic digit 1. Using this characterization we can, in fact, convince ourselves that the invariance property, which characterizes self-similarity, is true: first, we have to understand how w_1 and w_2 act on triadic numbers, but that is really easy to explain: if $x = 0.a_1a_2a_3\ldots$, then $w_1(x) = 0.0a_1a_2a_3\ldots$, and $w_2(x) = 0.2a_1a_2\ldots$ Thus, if $a_k \in \{0, 2\}$, then the triadic digits of $w_1(x)$ and $w_2(x)$ will also have that property, i.e., $w_k(C)$, $k = 1, 2$, is contained in C again. But can we get all points of C in this way? Indeed, if $y \in C$, i.e., $y = 0.a_1a_2a_3\ldots$ and $a_i \in \{0, 2\}$, then there is x in C and exactly one of the two transformations w_k, $k = 1, 2$, will have the property $w_k(x) = y$. Simply take $x = 0.a_2a_3\ldots$ Then if $a_1 = 0$, choose w_1; otherwise choose w_2. This establishes that $W(C) = C$ holds.

The Invariance Property and Self-Similarity

The invariance property explains self-similarity. We start with

$$C = w_1(C) \cup w_2(C)$$

i.e., C is a collage of two similar copies of itself — scaled down by a factor of 1/3. Then we obtain

$$C = w_1(w_1(C) \cup w_2(C)) \cup w_2(w_1(C) \cup w_2(C)),$$

The Sierpinski Gasket Revisited

Construction of the Sierpinski gasket as a limit. Stages 0 to 3 are shown.

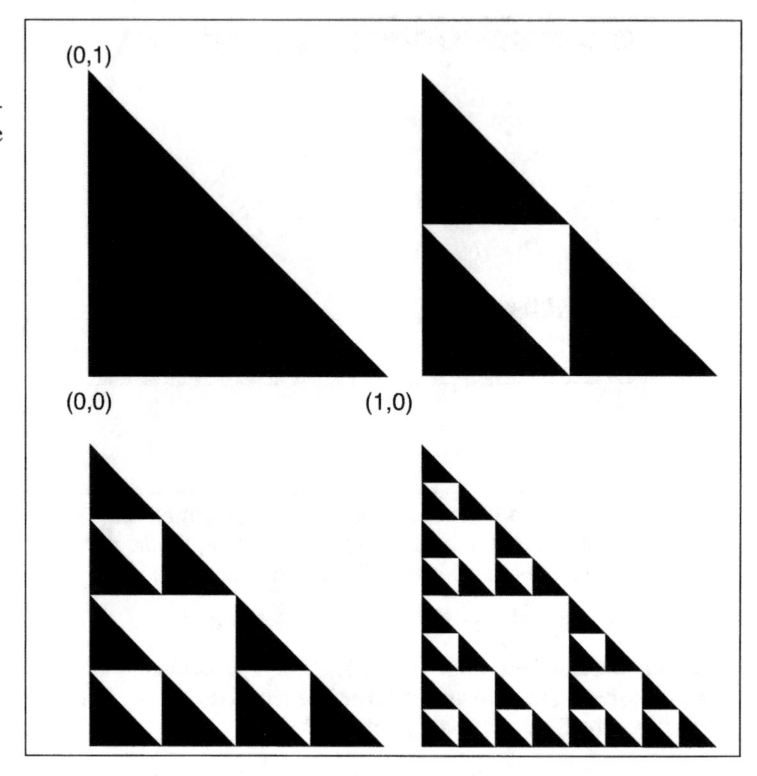

Figure 3.27

which leads to

$$C = w_1(w_1(C)) \cup w_1(w_2(C)) \cup w_2(w_1(C)) \cup w_2(w_2(C)),$$

i.e., C is a collage of four similar copies of itself — scaled down by a factor of 1/9, and so on. That is to say, we can identify smaller and smaller pieces in C which are just scaled down versions of C.

Let us discuss in a similar fashion the Sierpinski gasket. Again we begin by a limit characterization which is actually the one given by Sierpinski in 1916.

The Sierpinski Gasket as a Limit

Start with a triangle. It can be any kind, but for reasons which will soon become apparent, we will let T be a right triangle with two of its sides having length one. Now pick the midpoints of the sides. These define a center triangle, which we remove. We are left with three similar triangles, and for each we pick the midpoints of their sides, take away the center triangles, and are left with nine smaller triangles, and so on (see figure 3.27).

Also, the Sierpinski gasket is self-similar. To discuss this feature, we think of it in a plane, so that the vertices are at coordinates (0,0), (1,0), and (0,1). Then we introduce three similarity transformations w_1, w_2, and w_3. Each of these transformations can be interpreted as a scaling by a factor of 1/2, together with a positioning such that

$$\begin{aligned} w_1(0,0) &= (0,0) \\ w_2(1,0) &= (1,0) \\ w_3(0,1) &= (0,1). \end{aligned}$$

We claim that if S denotes the Sierpinski gasket, then

$$S = w_1(S) \cup w_2(S) \cup w_3(S). \tag{3.10}$$

The Invariance of the Sierpinski Gasket

In other words, if we introduce the Hutchinson operator

$$W(A) = w_1(A) \cup w_2(A) \cup w_3(A)$$

where A is any image in the plane, then

$$W(S) = S,$$

i.e., the Sierpinski gasket is invariant under W, or it solves the equation $W(X) = X$.

This means that the Sierpinski gasket can be broken down into 3, or 9, or 27 (abstractly 3^k) triangles which are scaled down versions of the entire Sierpinski gasket S by a factor of 1/2, or $(1/2)^2$, or $(1/2)^3$ (abstractly $(1/2)^k$). In other words, once we have given an argument for eqn. (3.10), we have completely understood the self-similarity of the Sierpinski gasket.

The Binary Characterization of the Sierpinski Gasket and the Invariance of S under W

Though it seems to be obvious from the geometric construction that S should satisfy $S = w_1(S) \cup w_2(S) \cup w_3(S)$, we prefer to give a solid argument. The geometric removal process in figure 3.27 is equivalent to looking at certain points in the plane and taking away a certain subset in a systematic fashion. If (x, y) is a point in the plane with nonnegative coordinates and $x + y \leq 1$, then (x, y) is in the triangle with vertices $(0,0), (1,0), (0,1)$. Given any point (x, y) from this triangle, it can be tested for membership in the Sierpinski gasket in the following way.

Write down a binary expansion of both coordinates

$$\begin{aligned} x &= 0.a_1a_2a_3\ldots, \text{ where } a_k \in \{0,1\}, \\ y &= 0.b_1b_2b_3\ldots, \text{ where } b_k \in \{0,1\}. \end{aligned}$$

The point (x, y) belongs to the Sierpinski gasket if and only if corresponding digits a_k and b_k are never both equal to 1. In other words, $a_k = 1$ must imply $b_k = 0$ and $b_k = 1$ must imply $a_k = 0$, and this holds for all $k = 1, 2, 3\ldots$ We will derive this characterization in chapter 5, section 5.4.

Thus, a point z is disregarded if a binary expansion of its coordinates x and y have a pair of coefficients $a_k = 1$, $b_k = 1$, respectively. At first there seems to be a problem with some points like $(x, y) = (0.5, 0.5)$ for example. This is clearly a point in the Sierpinski gasket, although it seems obvious from the equality of x and y that one can always find corresponding binary digits a_k and b_k of x and y which are both equal to 1. But note that 0.5 has two binary representations:

Transformation	x-Part	y-Part
$w_1(x, y)$	$0.0a_1a_2a_3\ldots$	$0.0b_1b_2b_3\ldots$
$w_2(x, y)$	$0.1a_1a_2a_3\ldots$	$0.0b_1b_2b_3\ldots$
$w_3(x, y)$	$0.0a_1a_2a_3\ldots$	$0.1b_1b_2b_3\ldots$

Table 3.28 : Explicit formulas in binary expansions for the similarity transformations of the Sierpinski gasket. The point $z = (x, y)$ is defined by $x = 0.a_1a_2a_3\ldots$ and $y = 0.b_1b_2b_3\ldots$

one is $0.5 = 0.1000\ldots$ and the other one is $0.5 = 0.0111\ldots$ Choosing the first for x and the second for y we see that the point belongs to S also according to the binary characterization of the Sierpinski gasket.

Using the binary characterization of S we can now argue that Hutchinson's formula $S = w_1(S) \cup w_2(S) \cup w_3(S)$ is correct. All we have to do is understand how w_k acts upon a point (x, y) in S. The details are a bit tedious, but they are of the same nature as with the ternary characterization of the Cantor set from chapter 2. Table 3.28 lists the three points to which (x, y) is transformed under w_1, w_2 and w_3. Note that the points of the Sierpinski gasket can be grouped into three sets depending on the first binary digits of x and y. The first set collects points with $a_1 = b_1 = 0$, the second points with $a_1 = 1$ and $b_1 = 0$, and in the third set we find all points with $a_1 = 0$ and $b_1 = 1$. There are just three points which are contained in two of the above categories simultaneously, namely, $(0.5, 0)$, $(0, 0.5)$, and $(0.5, 0.5)$. But this does not pose any problem for the following conclusion. Using the above table it becomes clear that $w_1(S)$ is equal to the first subset, $w_2(S)$ is the second and $w_3(S)$ is the third. Thus, indeed, we have that

$$W(S) = w_1(S) \cup w_2(S) \cup w_3(S) = S.$$

In the discussion of the Koch curve, Cantor set, and Sierpinski gasket we have learned that each of these basic fractals can be obtained by a limit process. But simultaneously there is a fixed-point characterization by a Hutchinson operator which is a composition of appropriate similarity transformations. This is a very far reaching insight. For one thing, it explains the meaning of self-similarity. But in fact, the Hutchinson operator gives us much more. It also provides us an alternative way to talk about the existence of the Koch curve, or Cantor set, or Sierpinski gasket.

A Unique Identification of Objects

It can be shown that each of the Hutchinson operators which we introduced earlier identifies a unique object in a plane (for the Koch curve and Sierpinski gasket) and on a line (for the Cantor set), which it leaves fixed. Or in other words, if W is the appropriate Hutchinson operator, then the solution to the equation $W(X) = X$ will automatically be either the Koch curve, the Cantor set, or the Sierpinski gasket. Thus we have a characteristic equation for each of these fractals. Naturally, these equations are not unique. Also, for $\sqrt{2}$ there are several possible characterizations by equations, and the same is true here. This is a topic with very interesting variations, which we will pick up again in chapter 7. There are also characterizations of traditional geometric objects in terms of similarity invariance properties. Take, for example, a square or simple triangle. The breakdown in figure 3.29 shows how these objects can be split up in a self-similar way. Thus, we can see fractals like the Sierpinski gasket in the same family as traditional geometrical objects. In fact, they solve the same kind of equations. Or in other words, from this point of view, fractals can be seen as extensions of traditional geometry, very much like irrational numbers can be seen as extensions of rational numbers by solving appropriate equations.

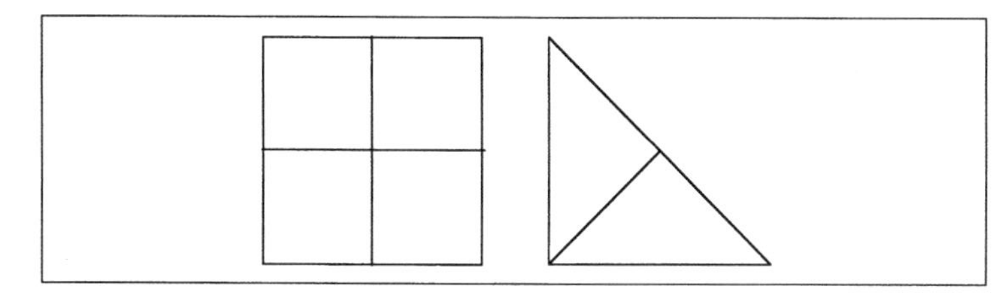

Tilings of Square and Triangle

Breakdown of square into four scaled down squares, and of triangle into two scaled down and similar triangles.

Figure 3.29

Self-Similarity in the Series of Hutchinson Operators

Using the Hutchinson operator W we can complete the analogy to the geometric series. Let us start out with a triangle T of the coast of the Koch island, see figure 3.30. We now apply the Hutchinson operator to T and add the result. Correspondingly, in a geometric series we would start with the number 1, and the first step would consist of a multiplication of the number with a factor q with succeeding addition. Here, after the first application, we have

$$T \cup W(T) = T \cup w_1(T) \cup w_2(T) \cup w_3(T) \cup w_4(T).$$

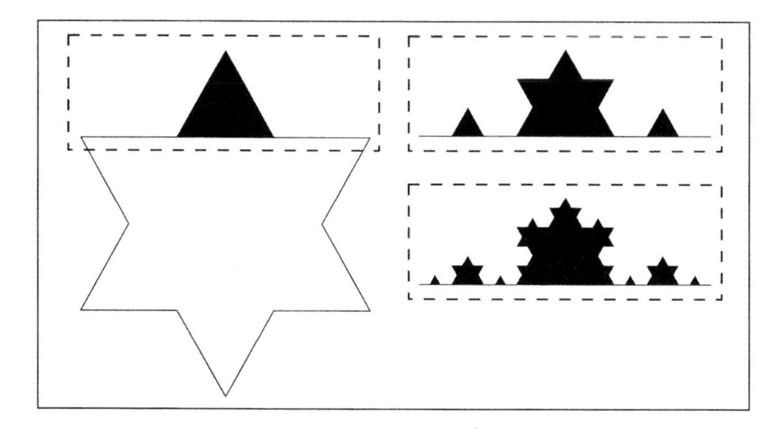

Figure 3.30 : The starting configuration (left) and the first two steps in the construction of a part of the Koch island in analogy to the geometric series.

Thus, we have added four triangles. In the next step, we again apply the Hutchinson operator W to our current configuration $T \cup W(T)$ and add the result:

$$T \cup W(T) \cup W^2(T).$$

Here $W^2(T)$ denotes the repeated application of W, i.e., $W(W(T))$, and this is the collection of 16 triangles given by

$$w_1(w_1(T)), w_1(w_2(T)), \ldots, w_4(w_3(T)), w_4(w_4(T)).$$

The next step yields

$$T \cup W(T) \cup W^2(T) \cup W^3(T).$$

In analogy to the geometric series we may even write down the limit object of this construction as

$$\bigcup_{k=0}^{\infty} W^k(T)$$

where we imply the convention $W^0(T) = T$.

Chapter 4

Length, Area and Dimension: Measuring Complexity and Scaling Properties

Nature exhibits not simply a higher degree but an altogether different level of complexity. The number of distinct scales of length of natural patterns is for all practical purposes infinite.

Benoit B. Mandelbrot[1]

Geometry has always had two sides, and both together have played very important roles. There has been the analysis of patterns and forms on the one hand; and on the other, the measurement of patterns and forms. The incommensurability of the diagonal of a square was initially a problem of measuring length but soon moved to the very theoretical level of introducing irrational numbers. Attempts to compute the length of the circumference of the circle led to the discovery of the mysterious number π. Measuring the area enclosed between curves has, to a great extent, inspired the development of calculus.

Today measuring length, area and volume appears to be no problem. If at all, it is a technical one. In principle, we usually think these problems have long since been solved. We are used to thinking that what we see can be measured if we really want to do so. Or we look up an appropriate table. Mandelbrot tells the story that the length of the border between Spain and Portugal has two very different measurements: an encyclopedia in Spain claims 616 miles, while a Portuguese encyclopedia quotes 758 miles. Who is right? If you look up the length of the coast of Britain in various sources, again you will find

[1] Benoit B. Mandelbrot, *The Fractal Geometry of Nature,* Freeman, 1982.

that the results vary between 4500 and 5000 miles.[2] What is happening here? There seems to be a problem. That is the theme of Mandelbrot's 1967 article[3] *How long is the coast of Britain?* For a moment we are led to believe that somebody has done a sloppy job. We have all seen those people surveying in the field with their high-precision optical gear. Is it possible that they made a mistake? And who made it; who is right and who is wrong? How do we find out?[4] And today with satellite surveying and laser precision, do we have more reliable results? The answer is no. And the fact is, we never will.

We will demonstrate that for all practical purposes, typical coastlines do not have a meaningful length! This statement seems to be ridiculous or at least counter-intuitive. An object like an island with some definitive area should also have some definitive length to its boundary.

We know that if we measure the circumference of a circular object, we will not obtain πd, d the diameter, but rather something close to it. We know we are inaccurate, but we don't worry because if we need a more accurate result, we just increase the level of precision in our measurement. Measurement requires units such as miles, yards, inches, etc.: all idealized straight-line segments. If we have a curved object such as a circle, then there is no doubt that the object has a definitive length and that it can be measured as accurately as necessary. Somehow our experience is that objects which fit on a piece of paper have finite length. But this is a misleading intuition. We usually measure the length only of objects, for which the result in fact does make sense and is of some practical value. But coastlines (and fractals) are not the only exceptions.

[2]The *Encyclopedia Americana*, New York, 1958, states "Britain has coasts totaling 4650 miles = 7440 km". *Collier's Encyclopedia*, London, 1986 states "The total mileage of the coastline is slightly under 5000 miles = 8000 km."

[3]B. B. Mandelbrot, *How long is the coast of Britain? Statistical self-similarity and fractional dimension,* Science 155 (1967) 636–638.

[4]Here are several methods of getting an answer: (1) Ask all the people in Britain and take the average of their answers. (2) Check encyclopedias. (3) Take a very detailed map of Britain and measure the coast using compasses. (4) Take a very detailed map of Britain and a thin thread, fit it on the coast, and then measure the length of the thread. (5) Walk the coast of Britain.

4.1 Finite and Infinite Length of Spirals

One possible class of objects which defies length measurement are spirals, it seems. Spirals fit on a piece of paper, and obviously do have infinite length. Well, do spirals really have infinite length? This is a very delicate question. Some have, and others don't.

Spirals have fascinated mathematicians throughout the ages. Archimedes (287–212 B.C.) wrote a treatise on spirals, and one of them is even named after him. The Archimedean spiral is a good model for the grooves on a record, or the windings of a rolled carpet. The characteristic feature of an Archimedean spiral is that the distance between its windings is constant. The mathematical model for such a spiral is very easy to obtain once we introduce polar coordinates: a point P in the plane is described by a pair (r, ϕ), where r is the distance to the origin of a coordinate system (the radius) and ϕ is the angle of the radius to the positive x-axis, measured in radians, i.e., $0 \leq \phi \leq 2\pi$ (see figure 4.1).

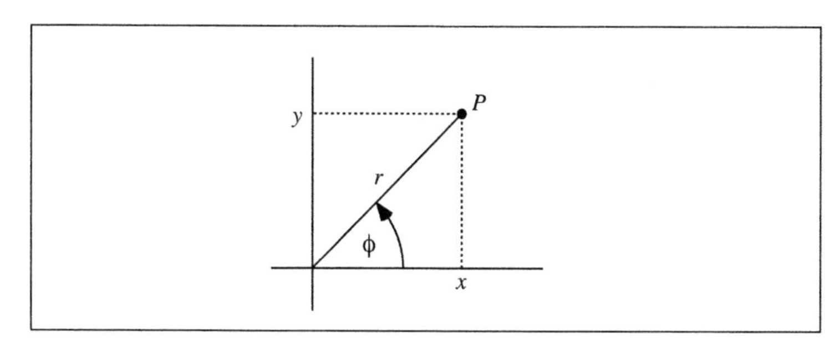

Polar Coordinates

The polar coordinates of the point P with Cartesian coordinates (x, y) are (r, ϕ), where $r = \sqrt{x^2 + y^2}$ is the distance to the origin and ϕ is the angle with the positive x-axis. Thus, $x = r \cos \phi$, and $y = r \sin \phi$.

Figure 4.1

In this frame of reference, an Archimedean spiral (seen from its center) can be modelled by the equation

$$r(\phi) = q\phi$$

where we now allow ϕ to be any nonnegative number, i.e., $\phi = 2\pi$ is one turn, $\phi = 4\pi$ corresponds to two turns, and so on. When drawing this spiral we start in the center. As we make one complete turn, the angle ϕ increases by 2π and r increases by $2\pi q$, the constant distance between two successive windings.

If we replace $r(\phi)$ by the natural logarithm $\log r(\phi)$, we obtain a formula for the logarithmic spiral: $\log r(\phi) = q\phi$, or, equivalently,

$$r(\phi) = e^{q\phi}.$$

When $q > 0$ and ϕ grows beyond all bounds, the spiral goes to infinity. When $q = 0$, we obtain a circle. And when $q < 0$, we obtain a spiral which winds into the center of the coordinate system as ϕ goes to infinity. This spiral is related to geometric sequences and has a remarkable property which is related to

Archimedean Spiral

Stepping along the Archimedean spiral in steps of a constant angle α yields an arithmetic sequence of radii $r_1, r_2, \ldots$

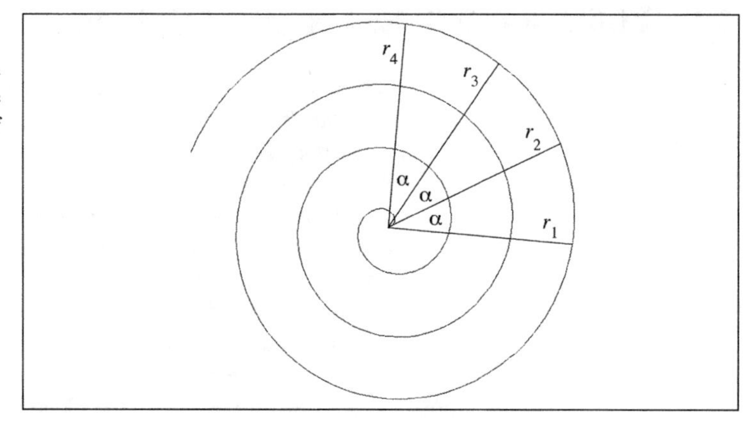

Figure 4.2

Logarithmic Spiral

Stepping along the logarithmic spiral in steps of a constant angle α yields a geometric sequence of radii $r_1, r_2, \ldots$

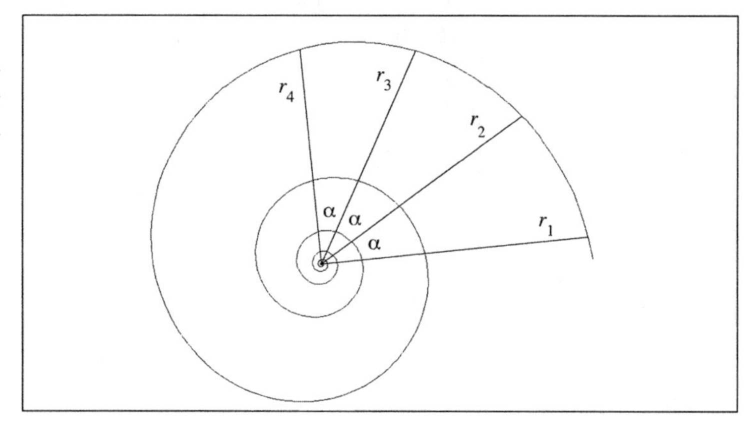

Figure 4.3

fractals. It is self-similar in a way which has equally inspired mathematicians, scientists and artists.

The great Swiss mathematician Jacob Bernoulli (1654–1705) devoted a treatise, entitled *Spira Mirabilis* (Wonderful Spiral), to the logarithmic spiral. He was so impressed by its self-similarity that he chose the inscription *Eadem Mutata Resurgo* (In spite of changes — resurrection of the same) for his tombstone in the Cathedral of Basel.

Spirals and the Arithmetic and Geometric Means

An Archimedean spiral is related to arithmetic sequences in the following way: we choose an arbitrary angle, say α, and points on the spiral, whose radii $r_1, r_2, r_3, \ldots$ have angle α to each other (see figure 4.2). Then the numbers r_i constitute an arithmetic sequence, i.e., the differences between consecutive numbers are the same. Thus, $r_3 - r_2 = r_2 - r_1$, and so on. Indeed, if $r_1 = a\phi_1$, then $r_2 = a(\phi_1 + \alpha)$, and $r_3 = a(\phi_1 + 2\alpha)$. In other words, $r_2 = (r_1 + r_3)/2$. That means any radius is the arithmetic mean of its two neighboring radii.

If we replace the arithmetic mean $(r_1 + r_3)/2$ by the geometric mean $\sqrt{r_1 r_3}$, we obtain the other classical spiral, the famous logarithmic spiral (see figure 4.3). Let us see why. Squaring the equation for the geometric mean gives $r_2^2 = r_1 r_3$, or, equivalently,

$$\frac{r_1}{r_2} = \frac{r_2}{r_3}.$$

Taking logarithms this identity reads

$$\log r_3 - \log r_2 = \log r_2 - \log r_1.$$

That means, the logarithms of successive radii form an arithmetic sequence. Thus, we obtain $\log r = q\phi$, the formula for the logarithmic spiral.

The radii r_i of the logarithmic spiral form a geometric sequence. We have

$$\frac{r_1}{r_2} = \frac{r_2}{r_3} = \frac{r_3}{r_4} = \frac{r_4}{r_5} = \cdots.$$

Thus, there is a constant, say a, such that for any index n

$$\frac{r_n}{r_{n+1}} = a$$

and

$$r_{n+1} = \frac{r_n}{a} = \frac{r_{n-1}}{a^2} = \cdots = \frac{r_1}{a^n}.$$

Self-Similarity of the Logarithmic Spiral

What is the amazing property which Bernoulli admired so much? He observed that a scaling of the spiral with respect to its center has the same effect as simply rotating the spiral by some angle. Indeed, if we rotate the logarithmic spiral $r(\phi) = e^{q\phi}$ by an angle of ψ clockwise, then the new spiral will be

$$r(\phi) = e^{q(\phi+\psi)}.$$

Since

$$e^{q(\phi+\psi)} = e^{q\psi} \cdot e^{q\phi},$$

rotating by ψ is the same as scaling by $s = e^{q\psi}$.

Now what is the length of the spiral? Let us look at an example of a spiral where the construction process makes calculation easy. It is, by the way, just another example of a geometric feedback system.

The Construction of Polygonal Spirals

We generate an infinite polygon. First, choose a decreasing sequence $a_1, a_2, a_3, \ldots$ of positive numbers. Now a_1 is the length of our initial piece. We construct the polygon in the following way: draw a_1 vertically from bottom to top. At the end make a right turn and draw a_1 again (from left to right). At the end of that line start to draw a_2 (first, continue in the same direction, from

Spiral or Not Spiral?

A 'spiral' by Nicholas Wade. Reproduced with kind permission by the artist. From: Nicholas Wade, *The Art and Science of Visual Illusions,* Routledge & Kegan Paul, London, 1982.

Figure 4.4

left to right). At the end make another right turn and draw a_2 again (now from top to bottom). At the end of that line take a_3. Continue on using the same principles. Figure 4.5 shows the first steps of this construction.

How long is this polygonal spiral? Well, each segment a_k appears twice in the construction, and thus the length is twice the sum of all a_k, i.e., $2(a_1 + a_2 + a_3 + \cdots)$. Let us now choose particular values of a_k. Let q be a positive number. If we take $a_k = q^{k-1}$, we obtain as total length $2\sum_{k=0}^{\infty} q^k$, which is a geometric series. Provided that $q < 1$, the limiting length[5] is equal to $2/(1-q)$. Thus, this polygonal spiral has finite length.

An Infinitely Long Spiral In a Finite Area

If we take, however, $a_k = 1/k$, $k = 1, 2, \ldots$, we obtain a series which is known not to have a limit.[6] In other words, the associated spiral is infinitely long, although it fits onto a finite area! Figure 4.6 shows both cases. Can you *see* which of the two spirals is finite and which is infinite?

The above polygonal spiral constructions can easily be used to support a smooth spiral construction. Observe that the polygons are composed of right angles with equal sides a_k. Each of them encompasses a segment of a circle — in fact, exactly a quarter of a circle with radius a_k. Putting together these segments appropriately produces a smooth spiral. Figure 4.7 shows the first

[5]Recall that the limit of the geometric series $1 + q + q^2 + q^3 + q^4 + \cdots$ is $1/(1-q)$, when $|q| < 1$.

[6]The sum $1 + 1/2 + 1/3 + 1/4 + \cdots$ is infinite (see the footnote on page 141).

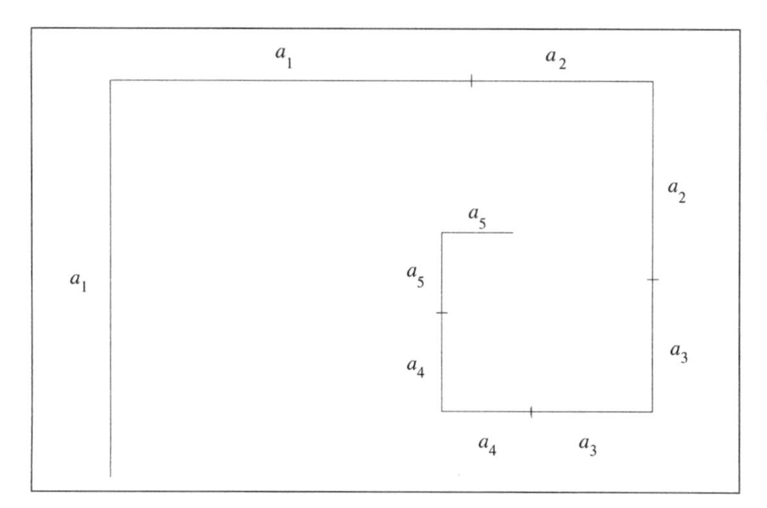

Polygonal Spiral

The first construction steps of a polygonal spiral.

Figure 4.5

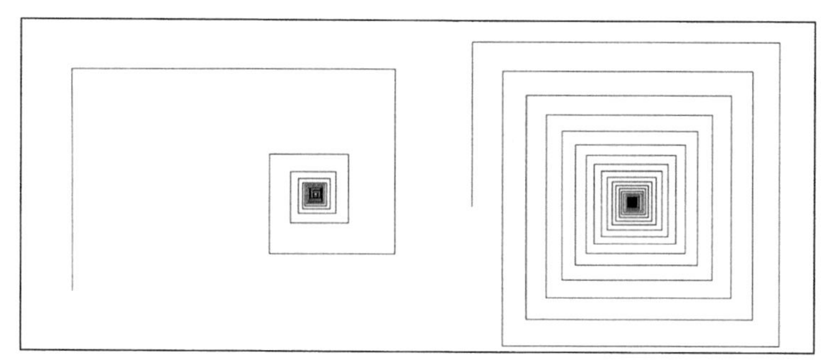

Infinite and Finite Polygonal Spirals

The spiral on the left is the one for $a_k = 1/k$ (i.e., the length is infinite). The spiral on the right is the one for $a_k = q^{k-1}$ with $q = 0.95$, a value slightly below 1 (i.e., it has a finite length).

Figure 4.6

two steps of this construction.

What is the length of these smooth spirals? Observe that the radii of the circle segments are of length a_k, while the circle segments then are of length $s_k = 2\pi a_k/4 = (\pi/2)a_k$. In other words, we have the total length

$$\sum_{k=1}^{\infty} s_k = \frac{\pi}{2}\sum_{k=1}^{\infty} a_k,$$

which is finite for $a_k = q^{k-1}$ (where $q < 1$) but infinite for $a_k = 1/k$. Figure 4.8 shows both spirals.

Again, it is amazing how little our visual intuition helps us to 'see' finite or infinite length. In other words, the fact that a curve fits on a piece of paper does not tell us whether its length is finite or not. Fractals add a new dimension to that problem.

Smooth Polygonal Spiral

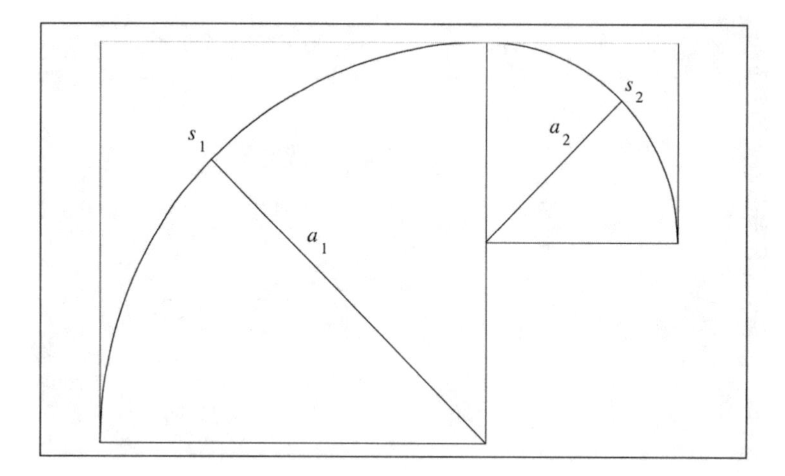

Figure 4.7

Infinite and Finite Smooth Spirals

The smooth spiral construction from figure 4.7 is carried out for the polygonal spirals from figure 4.6: $a_k = 1/k$ (left) and $a_k = q^{k-1}$ with $q = 0.95$ (right). Again the left spiral has infinite length while the right hand one has a finite length.

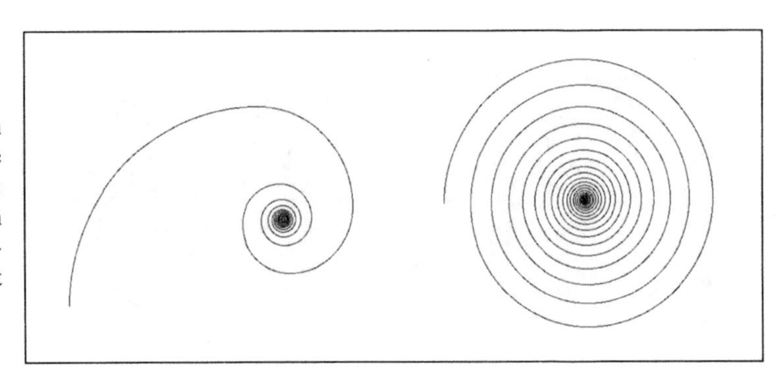

Figure 4.8

The Golden Spiral

If we take for our polygonal spiral construction $a_k = 1/g^{k-1}$, $k = 1, 2, \ldots$, where $g = (1 + \sqrt{5})/2$ is the golden mean, we obtain the famous golden spiral. For the length of this spiral we compute

$$\frac{2}{1 - \frac{1}{g}} = \frac{2g}{g - 1} = 2g^2 = 3 + \sqrt{5}.$$

Here we have used that g satisfies $g^2 - g - 1 = 0$ (i.e., $g - 1 = 1/g$).

The golden spiral can also be obtained in another beautiful construction: start with a rectangle with sides a_1 and $a_1 + a_2$, where $a_1 = 1$ and $a_2 = 1/g$ (i.e., $a_1/a_2 = g$). The rectangle breaks down into a square with sides a_1 and a smaller rectangle with sides a_2 and a_1. This smaller rectangle again breaks down into a square with sides a_2 and an even smaller rectangle with sides a_3 and a_2, and so on (see figure 4.9). Note that

$$\frac{a_2}{a_3} = \frac{\frac{1}{g}}{a_1 - a_2} = \frac{\frac{1}{g}}{1 - \frac{1}{g}} = \frac{1}{g - 1} = \frac{1}{\frac{1}{g}} = g.$$

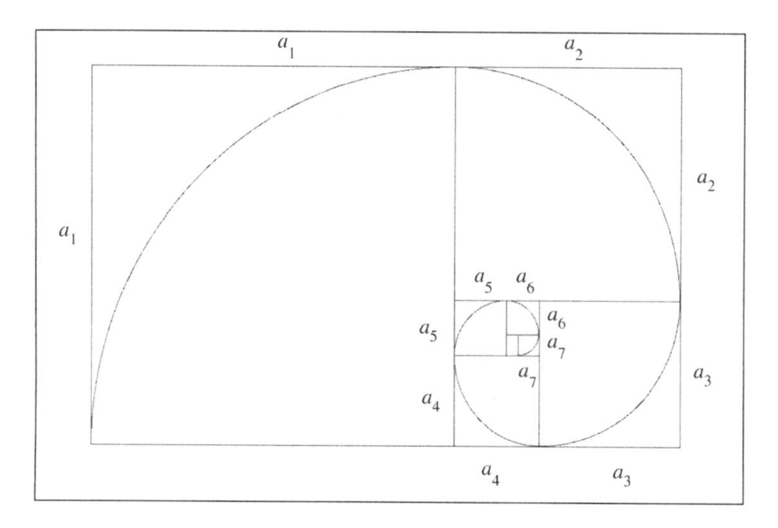

Figure 4.9 : The golden spiral.

In the same way we obtain that $a_k/a_{k+1} = g$. With that we compute $a_k = a_{k-1}/g = \cdots = a_1/g^{k-1} = 1/g^{k-1}$. The length of the inscribed smooth spiral is equal to $\frac{\pi}{2}g^2 = \frac{\pi}{4}(3+\sqrt{5})$.

4.2 Measuring Fractal Curves and Power Laws

The computation of the length of the various spirals — finite or infinite — is based on the corresponding mathematical formulas. The result on the infinite length of the Koch curve and the coast of the Koch island in chapter 2 is derived from the precise construction process of these fractals. Both of these methods for length computation of course fail when we consider fractals in nature such as real coastlines. There is no formula for the coastline of Great Britain, and also there is no defined construction process. The shape of the island is the result of countless years of tectonic activities of the earth on the one hand and the never-stopping erosion and sedimentation processes on the other hand. The only way to get a handle on the length of the coastline is to measure. In practice we measure the coast on a geographical map of Britain rather than the real coast. We take compasses set at a certain distance. For example, if the scale of the map is 1:1,000,000 and the compass width is 5 cm, then the corresponding true distance is 5,000,000 cm or 50 km (approximately 30 miles). Now we carefully walk the compasses along the coast counting the number of steps. Figure 4.10 shows a polygonal representation of the coast of Britain. The vertices of the polygons are assumed to be on the coast. The straight-line segments have constant length and represent the setting of the compasses. We have carried out this measurement using four different compass settings.[7]

Compass Setting	Length
500 km	2600 km
100 km	3800 km
54 km	5770 km
17 km	8640 km

Smaller Scales Give Longer Results

This elaborate experiment reveals a surprise. The smaller the setting of the compasses, the more detailed the polygon and — the surprising result — the longer the resulting measurement will be. In particular, up in Scotland the coast has a very large number of bays of many different scales. With one compass setting many of the smaller bays are still not accounted for, while in the next smaller one they are, while still smaller bays are still ignored at that setting, and so on.

Measuring a Circle

Let us compare this phenomenon with an experimental measurement of the perimeter of a circle. We use a circle of diameter 1000 km, so that the perimeter is of the same order of magnitude as the measured length of the coast of Britain. We do not have to go through the process of actually walking compasses around the circle. Rather, we make use of the classical approach of Archimedes who had worked out a procedure to calculate what these measurements would

[7]In: H.-O. Peitgen, H. Jürgens, D. Saupe, C. Zahlten, *Fractals — An Animated Discussion,* Video film, Freeman, New York, 1990. Also appeared in German as *Fraktale in Filmen und Gesprächen,* Spektrum der Wissenschaften Videothek, Heidelberg, 1990.

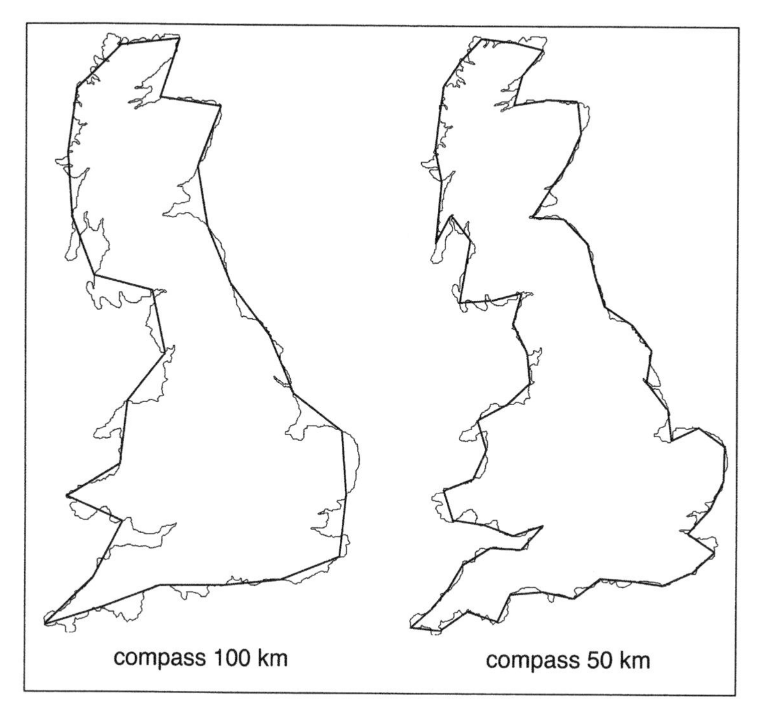

Approximations of Britain

Polygonal approximation of the coast of Britain.

Figure 4.10

Number of Sides	Compass Setting	Length
6	500.00 km	3000 km
12	258.82 km	3106 km
24	130.53 km	3133 km
48	65.40 km	3139 km
96	32.72 km	3141 km
192	16.36 km	3141 km

Measuring the Circle

Length of a circle of diameter 1000 km approximated using inscribed regular polygons. The entries are computed from the formula of Archimedes; see page 147.

Table 4.11

be (see page 147 and table 4.11). In order to compare the results we enter the measurements in a graph. However, because the size of our compass setting varies over a broad spectrum from a few kilometers to several hundred, a length-versus-setting diagram is difficult to draw. In such a situation one usually passes to a log/log diagram. On the horizontal axis the logarithm of the inverse compass setting (1/setting) is marked. This quantity can be interpreted as the precision of the measurement. The smaller the compass setting is, the more precise is the measurement. The vertical axis is for the logarithms of the length. We take logarithms with respect to base 10, but that doesn't really matter. Moreover, we like to interpret $1/s$ as a *measure of precision*, i.e., when s is small then the precision $1/s$ is large. Our log/log plots will always show how the total length ($\log(u)$) changes with an increase in precision ($\log(1/s)$). Figure 4.12 shows the results for the coastline of Britain and the circle.

Log/Log Diagram for Coast of Britain and Circle

Log/log diagram for measurements of coast of Britain and circle of diameter 1000 km (table 4.11). u = length in km, s = setting of compasses in km. Rather than looking at $\log(s)$, we prefer to consider $\log(1/s)$ as a measure of the precision of the length.

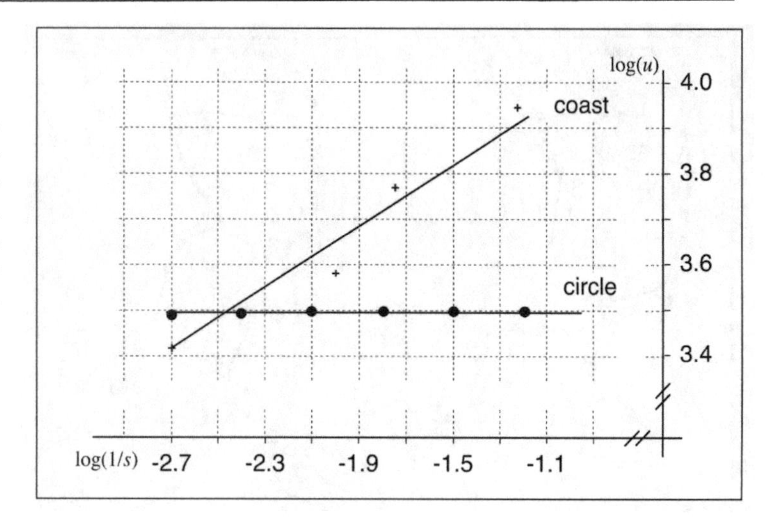

Figure 4.12

Fitting a Straight Line to a Series of Points

We make a remarkable observation. Our points in the diagram roughly fall on straight lines. It is a topic of mathematical statistics how to define a line that approximates the points in such a diagram. Obviously, we cannot expect that the points fall exactly on a line, because of the nature of the measurements. However, a measure of the deviation of the line from the collection of points can be minimized. This leads to the widely used *method of least squares*. In our case we obtain a horizontal line for the circle and a line with some slope $d \approx 0.3$ for the coast of Britain.

Assume that we take these data and use them to make a forecast of the changes when passing to more precise measurements, i.e., when we use a smaller compass setting s. For this purpose we would simply extrapolate the lines to the right. This would yield about the same result for the circle since the line is approximately horizontal. In other words, the circle has a finite length. However, the measured length of the coast would increase at smaller scales of measurement.

Let us denote by b the intercept of the fitting line with the vertical axis. Thus, b corresponds to the logarithm of the measured length at scale $s = 1$ corresponding to 1 km. The relationship between the length u and the scale or size s covered by the compasses can be expressed[8] by

$$\log u = d \cdot \log \frac{1}{s} + b. \tag{4.1}$$

Equation 4.1 expresses how the length changes when the setting of the compasses is changed, assuming that in a log/log plot the measurements fall on a straight line. In that case the two constants, d and b, characterize the growth

[8]Recall that a straight line in a x-y diagram can be written as $y = dx + b$, where b is the y-intercept and d is the slope of the line (i.e., $d = (y_2 - y_1)/(x_2 - x_1)$), for any pair of points (x_1, y_1) and (x_2, y_2) on the line.

law. The slope d of the fitting line is the key to the fractal dimension of the underlying object. We will discuss this in the next section.

Power Laws

We would not like to take it for granted that the reader is familiar with log/log diagrams. To explain the main idea, let us take some data from an experiment in physics. To investigate the laws governing free fall we may drop an object from different levels of a tall tower or building (of course with proper precautions taken). With a stopwatch we measure the time necessary for the object to reach the bottom. With height differences between levels being 4 meters, we get the following table of data (table 4.13).

Height h (in m)	Drop Time t (in sec)	$\log h$	$\log t$
4	0.89	0.60	−0.051
8	1.26	0.90	0.100
12	1.55	1.08	0.190
16	1.79	1.20	0.253
20	2.00	1.30	0.301
24	2.19	1.38	0.340
28	2.37	1.45	0.375
32	2.53	1.51	0.403

Table 4.13 : Drop time versus height of free fall. The last two columns list the logarithms (base 10) for the data. The original and the logarithmic data is plotted in figure 4.14.

Figure 4.14 shows the data graphically. Clearly, the plotted points are not on a straight line (top curve). Thus, the relation between height and drop time is not linear. The corresponding plot on double logarithmic paper at the bottom, however, reveals that there is a law describing the relationship between height and drop time. This law is a *power law* of the form

$$t = c \cdot h^d. \tag{4.2}$$

Such a law is called a power law because t changes as if it were a power of h. The problem, then, is to verify the conjecture and determine c and d. To begin, let us assume that in fact eqn. (4.2) holds. Now we take the base 10 logarithm (of course, any base will work) on both sides and obtain

$$\log t = d \cdot \log h + \log c.$$

In other words, if one plots $\log t$ versus $\log h$, rather than t versus h, one should see a straight line with slope d and y-intercept $b = \log c$ ($c = 10^b$). This is done in figure 4.14 on the bottom.

Thus, when measurements in a log/log plot essentially fall onto a straight line, then it is reasonable to work with a power law which governs the relationship between the variables; and moreover, the log/log plot allows us to read off the exponent d in that power law as the slope

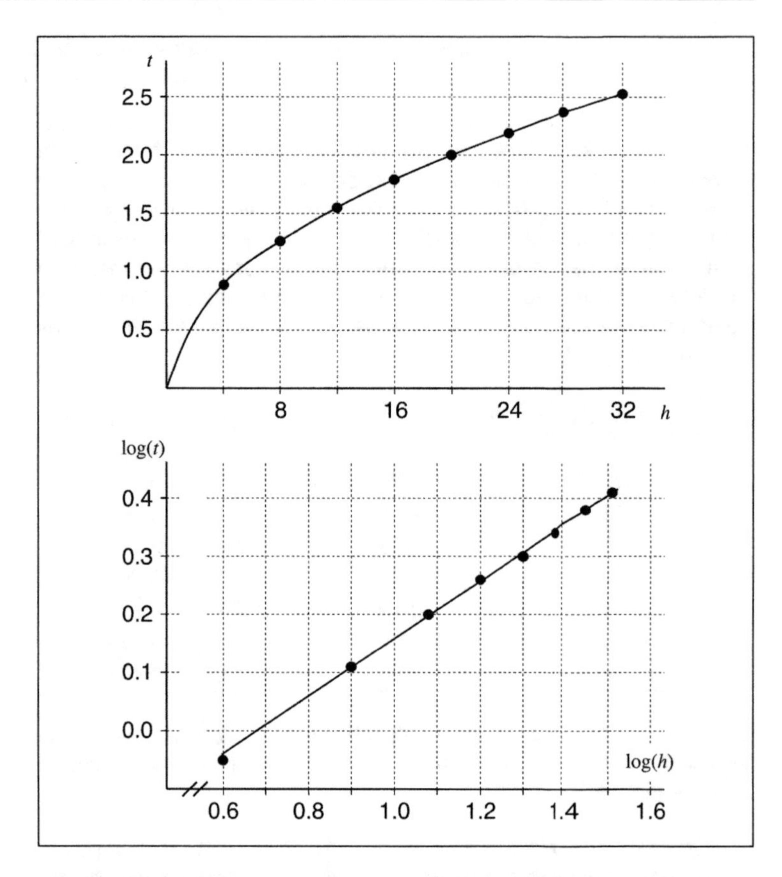

Figure 4.14 : The data from table 4.13 shows graphically the dependence of the drop time on the height of the fall. The data is displayed on the top using linear scales resulting in a parabola-like curve. On the bottom a double logarithmic representation of the same data is given. The data points appear to lie on a line.

of the straight line. In our example we can draw the fitting line in the double logarithmic plot and read off the slope d and the y-intercept $\log c$:

$$d = 0.50 \text{ and } \log c = -0.34.$$

Thus, $c = 10^{-0.34}$, and the power law determined from the measurements is

$$t = 0.45\, h^{0.50}. \tag{4.3}$$

By the way, this is in good agreement with the Newtonian law of motion, which implies that the distance fallen is proportional to the square of the drop time. Formally,

$$h = \frac{g}{2}\, t^2,$$

where $g \approx 9.81\text{m/sec}^2$ is the gravitational acceleration. Solving this equation for t yields

$$t = \sqrt{\frac{2h}{g}} \approx 0.452 \cdot h^{0.5},$$

which is to be compared with our empirical result in eqn. (4.3).

Power Law for Allometric Growth

When we discussed allometric growth in chapter 3, we saw an interesting example of a power law. Let us remember that we compared measured head sizes with the body height as a baby developed into a child and then grew to adulthood. We learned that there were two phases — one up until the age of three, and the second after that until the growth process terminates. Using the approach of power laws with the tools of double logarithmic graphs we now try to model the allometric phase of growth by an appropriate power law. To this end we reconsider the original data from table 3.8 and extend it by corresponding logarithms (see table 4.15).

Age (years)	Height (cm)	Head Size (cm)	Height (logarithm)	Head Size (logarithm)
0	50	11	1.70	1.04
1	70	15	1.85	1.18
2	79	17	1.90	1.23
3	86	18	1.93	1.26
5	99	19	2.00	1.28
10	127	21	2.10	1.32
20	151	22	2.18	1.34
25	167	23	2.22	1.36
30	169	23	2.23	1.36
40	169	23	2.23	1.36

Table 4.15 : Body height and head size of a person with logarithms of the same data.

The plot in figure 4.16 on log/log scales reconfirms the two-stage growth of the measured person. We can fit two lines to the data, the first one reaching until age three and the second one for the rest of the data. The first line has a slope of about one. This corresponds to an equal growth rate (see page 43) of head size and body height; the two quantities are proportional and the growth is called isometric. The second line has a much lower slope, about $1/3$. This yields a power law stating that the head size should be proportional to the cube root of the body size. Or — turned around — we have that the body height is proportional to the cube of the head size,

$$\text{body height} \propto (\text{head size})^3.$$

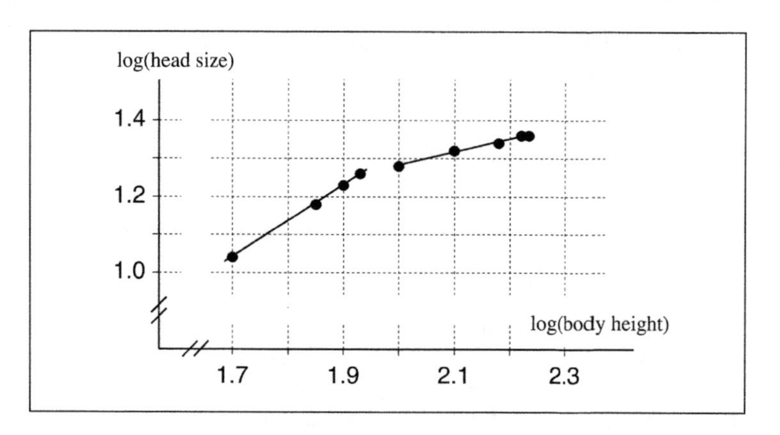

Figure 4.16 : Double logarithmic plot of head size versus body height data.

The body grows much faster than the head; here we speak of allometric growth. Of course our little analysis should not be mistaken for a serious research result. The measurements were taken from only one person and only at large time intervals. Moreover, the test person was born in the 19th century. Thus, the cubic growth law above is probably neither exact nor representative.

Let us summarize. If the x and y data of an experiment range over very large numerical scales, then it is possible that there is a power law which expresses y in terms of x. To test the power law conjecture we plot the data in a log/log plot. If, then, the measurements fit a straight line, we can read off the exponent of the law as the slope of that line.

Figure 4.12 supports that there is a power law (i.e., that eqn. (4.1) is true). Or equivalently, we may then conclude that ($u =$ length, $s =$ compass setting)

$$u = c \cdot \left(\frac{1}{s}\right)^d . \tag{4.4}$$

For the coast of Britain, we would then find that $d \approx 0.36$. The result of this graphical analysis is, thus, that the measured length u of the coast grows in proportion to the precision $1/s$ raised to the power 0.36,

$$u \propto \frac{1}{s^{0.36}} .$$

Maps with More and More Detail

At this point we have to discuss several aspects of relation (4.4). One immediate consequence is that the length goes to infinity like $1/s^d$ as $s \to 0$. But can we really let the compass setting s go to zero? Of course we can, but there is some danger. If we let the size of the compass setting go to zero on some particular map of Britain, then the law (4.4) would be invalid due to the finite resolution of the map. In fact, in this case the measured length would tend to a limit. The power law and its consequences are only valid in a measured range of compass settings based on simultaneously picking maps with more and more detail. In other words, the power law characterizes the complexity of the coast of Britain over some range of scales by expressing how quickly the length increases if we measure with ever finer accuracy. Eventually, such measurements do not make much sense anymore because we would run out of maps and would have to begin measuring the coast in reality and face all the problems of identifying where a coast begins and ends, when to measure (at low or high tide), how to deal with river deltas and so on. In other words, the problem becomes somewhat ridiculous. But nevertheless, we can say that in any practical terms the coast of Britain has no length. The only meaningful thing we can say about its length is that it behaves like the above power law over a range of scales to be specified and that this behavior will be characteristic.

Characteristic Power Laws

What do we mean when we say 'characteristic'? Well, we mean that the exponents in the power laws are likely to be different when we compare the coast of Britain with those of Norway or California. The same will be true if we carry out an analogous experiment for the length of borders, e.g., the border of Portugal and Spain. Now we understand why the Portuguese encyclopedia came out with a larger value than the one in Spain. Since Portugal is very small in comparison with Spain, it is very likely that the map used in Portugal for the measurement of the common border had much more detail — was of much smaller scale — than the one in Spain. The same reasoning explains the differences for the measurements of the coast of Britain.[9]

Measuring Utah

Let us look at the border of the state of Utah, one of the 50 states in the U.S.A. Figure 4.17 shows a map and collects a few measurements of the border of Utah. Obviously (if you did not already know it) the border of Utah is very straight.[10] If we represent the measurements in a log/log diagram, we obtain insight into the power law behavior. Apparently, the best way to fit a straight line to the points is by using a practically horizontal line. That is to say, the border of Utah has a power law with exponent $d \approx 0$ comparable with that of a circle, and that means that the border has, for all practical purposes, a finite length.

Measuring the Koch Curve

Let us now try to understand the importance and meaning of the power law behavior in a pure mathematical situation. Recall the Koch island from chapter 3. The Koch island has a coast which is formed by three identical Koch curves. Now remember that each Koch curve can be divided into four self-similar parts, which are similar to the entire curve via a similarity transformation which reduces by a factor of 3.

Therefore, it is natural to choose compass settings covering sizes of the form $1/3$, $1/3^2$, $1/3^3$, $\ldots, 1/3^k$. Of course there are two ways to work with these compass settings: an impossible one and the obvious one. It would be technically impossible to set compasses precisely to say $1/3^4 =$

[9]The first measurements of this kind go back to the British scientist R. L. Richardson and his paper *The problem of contiguity: an appendix of statistics of deadly quarrels,* General Systems Yearbook 6 (1961) 139–187.

[10]We like Utah for many reasons. One of them is that we were introduced to fractals during a sabbatical in Salt Lake City during the 1982/83 academic year. And it was there where we did our first computer graphical experiments on fractals in the Mathematics and Computer Science Departments of the University of Utah.

The Western United States

The table collects a few measurements of the border of Utah based on maps of various scales.

Setting	Length
500 km	1450 km
100 km	1780 km
50 km	1860 km
20 km	1890 km

Figure 4.17

Log/Log Diagram

Log/log representation of measurements of the border of Utah, where u = length measured in units of 1 km; s = compass setting measured in units of 1 km.

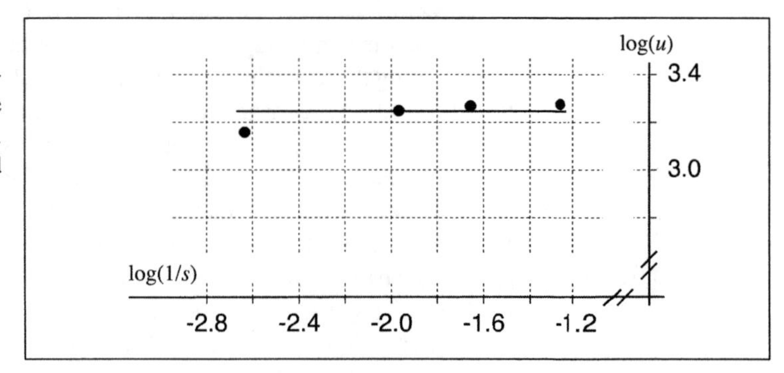

Figure 4.18

0.012345679012... The thing to do would be to keep the compass setting constant and look at magnifications by a factor of $1, 3, 3^2, 3^3, \ldots$ Even that would be a waste of time because, from the construction of the Koch curve, we know exactly what the measurements would be, namely, 4/3 for compass setting $s = 1/3$, 16/9 for $s = 1/9, \ldots, (4/3)^k$ for $s = (1/3)^k$.

Let us now represent these measurements in a log/log diagram (figure 4.20). Since we are free to choose a logarithm with respect to a convenient base, we take $\log_3$ so that for compass setting $s = 1/3^k$ and length $u = (4/3)^k$ we obtain

$$\log_3 \frac{1}{s} = k \text{ and } \log_3 u = k \log_3 \frac{4}{3}.$$

Combining the two equations we obtain for the desired growth law

$$\log_3 u = d \log_3 \frac{1}{s},$$

with

$$d = \log_3 \frac{4}{3} \approx 0.2619.$$

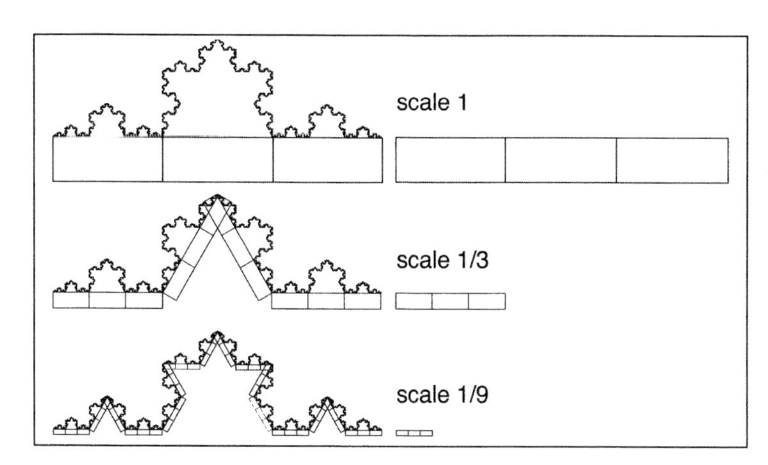

Measuring the Koch Curve

Measuring the length of the Koch curve with different compass settings (scales).

Figure 4.19

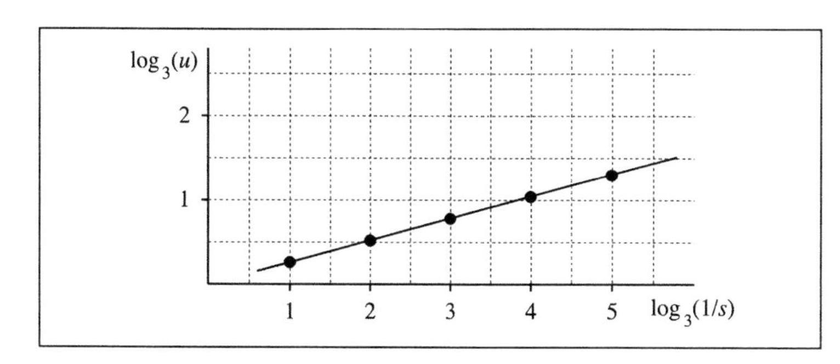

Log/Log Plot for the Koch Curve

Diagram of $\log_3(u)$ versus $\log_3(1/s)$.

Figure 4.20

This number is smaller than the value $d \approx 0.36$ which we found for the coastline of Britain. In other words, from this point of view, the coastline is even more convoluted and rugged than the Koch snowflake curve.

4.3 Fractal Dimension

In our attempts to measure the length of the coast of Britain, we learned that the question of length — and likewise in other cases, of area or volume — can be ill-posed. Curves, surfaces, and volumes can be so complex that these ordinary measurements become meaningless. However, there is a way to measure the degree of complexity by evaluating how fast length, or surface, or volume increases if we measure with respect to smaller and smaller scales. The fundamental idea is to assume that the two quantities — length or surface, or volume and scale — do not vary arbitrarily but rather are related by a law, which allows us to compute one quantity from the other. The kind of law which seems to be relevant, as we explained previously, is a power law of the form $y \propto x^d$.

The Notion of Dimension

Such a law also turns out to be very useful for the discussion of *dimension*. Dimension is not easy to understand. At the turn of the century it was one of the major problems in mathematics to determine what dimension means and which properties it has (see chapter 2). And since then the situation has become somewhat worse because mathematicians have come up with some ten different notions of dimension: topological dimension, Hausdorff dimension, fractal dimension, self-similarity dimension, box-counting dimension, capacity dimension, information dimension, Euclidean dimension, and more. They are all related. Some of them, however, make sense in certain situations, but not at all in others, where alternative definitions are more helpful. Sometimes they all make sense and are the same. Sometimes several make sense but do not agree. The details can be confusing even for a research mathematician.[11] Thus, we will restrict ourselves to an elementary discussion of three of these dimensions:

- self-similarity dimension
- compass dimension (also called divider dimension)
- box-counting dimension

All are special forms of Mandelbrot's *fractal dimension*[12] which in turn was motivated by Hausdorff's[13] fundamental work from 1919. Of these three notions of dimension the box-counting dimension has the most applications in science. It is treated in the next section.

Self-Similar Structures

We discussed the concept of self-similarity in the last chapter. Let us recall the essential points. A structure is said to be (strictly) self-similar if it can be broken down into arbitrarily small pieces, each of which is a small replica of the entire structure. Here it is important that the small pieces can in fact be obtained from the entire structure by a *similarity transformation*. The best

[11] Two good sources for those who want to pursue the subject are: K. Falconer, *Fractal Geometry, Mathematical Foundations and Applications*, Wiley, New York, 1990, and J. D. Farmer, E. Ott, J. A. Yorke, *The dimension of chaotic attractors*, Physica 7D (1983) 153–180.

[12] Fractal is derived from the Latin word *frangere*, which means 'to break'.

[13] Felix Hausdorff (1868–1942) was a mathematician at the University of Bonn. He was a Jew, and he and his wife committed suicide in 1942, after he had learned that his deportation to a concentration camp was only one week away.

Felix Hausdorff, 1868–1942

Figure 4.21

way to think of such a transformation is what we obtain from a photocopier with a reduction feature. For example, if we take a Koch curve and put it on a copying machine, set the reduction to 1/3 and produce four copies, then the four copies can be pasted together to give back the Koch curve. It then follows that if we copy each of the four reduced copies by a reduction factor of 1/3 four times (i.e., produce 16 copies which are reduced by a factor of 1/9 compared to the original), then these 16 copies can also be pasted together to reproduce the original. With an ideal copier, this process could be repeated infinitely often. Again, it is important that the reductions are similarities.

It would be a mistake to believe that if a structure is self-similar, then it is also fractal. Take, for example, a line segment, or a square, or a cube. Each can be broken into small copies which are obtained by similarity transformations (see figure 4.22). These structures, however, are not fractals.

Scaling Factors Can Be Characteristic

Here we see that the reduction factor is 1/3, which is, of course, arbitrary. We could as well have chosen 1/2, or 1/7 or 1/356. But precisely in this fact lies the difference between these figures and fractal structures. In the latter the reduction factors — if they exist — are characteristic. For example, the Koch curve only admits 1/3, 1/9, 1/27, etc. The point, however, which is common to all strictly self-similar structures — fractal or not — is that there is a relation between the reduction factor (scaling factor) and the number of scaled down pieces into which the structure is divided.

Self-Similarity of Line, Square, Cube

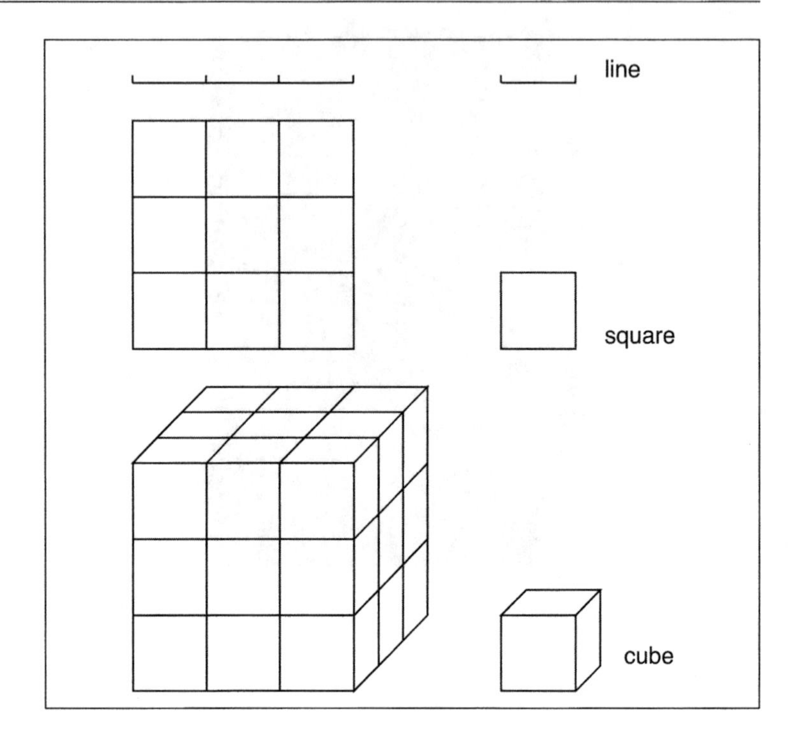

Figure 4.22

Object	Number of Pieces	Reduction Factor
line	3	1/3
line	6	1/6
line	173	1/173
square	$9 = 3^2$	1/3
square	$36 = 6^2$	1/6
square	$29\,929 = 173^2$	1/173
cube	$27 = 3^3$	1/3
cube	$216 = 6^3$	1/6
cube	$5\,177\,717 = 173^3$	1/173
Koch curve	4	1/3
Koch curve	16	1/9
Koch curve	4^k	$1/3^k$

Apparently, for the line, square, and cube there is a nice power law relation between the number of pieces a and the reduction factor s. This is the law

$$a = \frac{1}{s^D}, \tag{4.5}$$

where $D = 1$ for the line, $D = 2$ for the square, and $D = 3$ for the cube. In other words, the exponent in the power law agrees exactly with those numbers which are familiar as (topological) dimensions of the line, square, and cube.

If we look at the Koch curve, however, the relationship of $a = 4$ to $s = 1/3$ and $a = 16$ to $s = 1/9$ is not so obvious.

But being guided by the relation for the line, square, and cube, we try a little bit harder. We postulate that eqn. (4.5) holds anyway. In other words, $4 = 3^D$. Taking logarithms on both sides, we get

$$\log 4 = D \cdot \log 3,$$

or equivalently

$$D = \frac{\log 4}{\log 3} \approx 1.2619.$$

But do we get the same if we take smaller pieces, as with a reduction factor of 1/9? To check this out, we would postulate that $16 = 9^D$, or $\log 16 = D \cdot \log 9$, or $D = \log 16/\log 9$, from which we compute

$$D = \frac{\log 4^2}{\log 3^2} = \frac{2\log 4}{2\log 3} = \frac{\log 4}{\log 3} \approx 1.2619.$$

And as a general rule,

$$D = \frac{\log 4^k}{\log 3^k}$$

implies that $D = \log 4/\log 3$. Hence the power law relation between the number of pieces and the reduction factor gives the same number D, regardless of the scale we use for the evaluation. It is this number D, a number between 1 and 2, that we call the self-similarity dimension of the Koch curve.

Self-Similarity Dimension

More generally, given a self-similar structure, there is a relation between the reduction factor s and the number of pieces a into which the structure can be divided; and that is

$$a = \frac{1}{s^D}$$

or equivalently

$$D = \frac{\log a}{\log 1/s},$$

where D is called the *self-similarity dimension*. In cases where it is important to be precise, we use the symbol D_s for the self-similarity dimension in order to avoid confusion with the other versions of fractal dimension. For the line, the square and the cube we obtain the expected self-similarity dimensions $1, 2$, and 3, respectively. For the Koch curve we get $D \approx 1.2619$, a number whose fractional part is familiar from measuring the length of the Koch curve in the last section. The fractional part 0.2619... is exactly equal to the exponent of the power law describing the measured length in terms of the compass setting used! Before we discuss this in more detail let us try a few more self-similar objects and compute their self-similarity dimensions. Figure 4.24 shows the Sierpinski gasket, Sierpinski carpet, and the Cantor set. Table 4.23 compares the number of self-similar parts with the corresponding scaling factors.

Dimensions of Some Fractals

Self-similarity dimensions for other fractal objects.

Table 4.23

Object	Scale s	Pieces a	Dimension D_s
Cantor set	$1/3^k$	2^k	$\log 2 \,/\, \log 3 \approx 0.6309$
Sierpinski gasket	$1/2^k$	3^k	$\log 3 \,/\, \log 2 \approx 1.5850$
Sierpinski carpet	$1/3^k$	8^k	$\log 8 \,/\, \log 3 \approx 1.8928$

Self-Similarity Dimension and Length Measurement

What is the relation between the power law of the length measurement using different compass settings and the self-similarity dimension of a fractal curve? It turns out that the answer is very simple, namely,

$$D_s = 1 + d$$

where d, as before, denotes the slope in the log/log diagram of length u versus precision $1/s$, i.e., $u = c/s^d$. Let us see why. First, we simplify by choosing appropriate units of length measurements such that the factor c in the power

Three More Fractals

The Sierpinski gasket, Sierpinski carpet, and Cantor set are shown with their building blocks, scaled down copies of the whole.

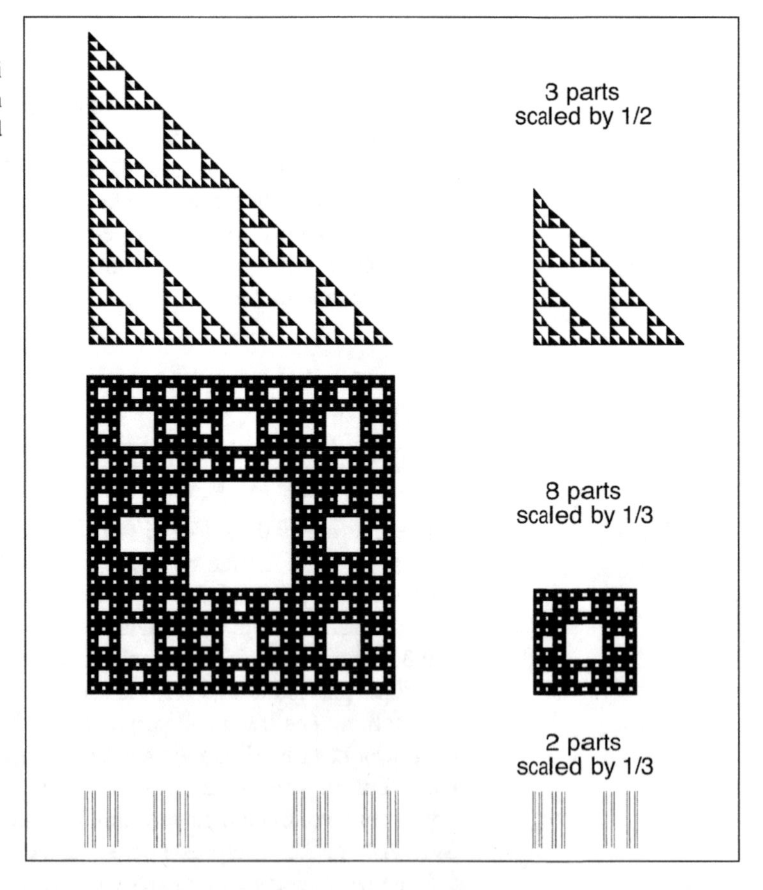

Figure 4.24

law becomes unity

$$u = \frac{1}{s^d}. \tag{4.6}$$

Taking logarithms, we obtain

$$\log u = d \cdot \log \frac{1}{s}, \tag{4.7}$$

where u is the length with respect to compass setting s. On the other hand we have the power law $a = 1/s^D$, where a denotes the number of pieces in a replacement step of the self-similar fractal with scaling factor s. In logarithmic form, this is

$$\log a = D_s \cdot \log \frac{1}{s}. \tag{4.8}$$

Now we can note the connection between length u and number of pieces a. At scaling factor $s = 1$ we measure a length $u = 1$. This is true by construction: above in equation (4.6) we have set up units so that $u = 1$ when $s = 1$. Thus, when measuring at some other scale s, where the whole object is composed of a small copies each of size s, then we measure a total length of a times s,

$$u = a \cdot s.$$

This is the key to the following conclusion. Taking logarithms again we get

$$\log u = \log a + \log s.$$

In this equation we can substitute the logarithms $\log u$ and $\log a$ from equations (4.7) and (4.8). This yields

$$d \cdot \log \frac{1}{s} = D_s \cdot \log \frac{1}{s} + \log s.$$

Since $\log 1/s = -\log s$ we get

$$-d \cdot \log s = -D_s \cdot \log s + \log s$$

and dividing by $\log s$ and sorting terms we finally arrive at

$$D_s = 1 + d.$$

The result is that the self-similarity dimension can be computed in two equivalent ways:

- Based on the self-similarity of geometric forms find the power law describing the number of pieces a versus $1/s$, where s is the scale factor which characterizes the parts as copies of the whole. The exponent D_s in this law is the self-similarity dimension.
- Using the compass-type length measurement, find the power law relating the length with $1/s$, where s is the compass setting. The exponent d in this law, incremented by 1, is the self-similarity dimension, $D_s = 1 + d$.

Motivated by this result we may now also generalize the dimension found in the alternative to shapes that are not self-similar curves such as coastlines and the like. Thus, we define the *compass dimension* (sometimes also called divider or ruler dimension) by

$$D_c = 1 + d$$

where d is the slope in the log/log diagram of the measured length u versus precision $1/s$. Thus, since $d \approx 0.36$ for the coast of Britain, we can say that the coast has a fractal (compass) dimension of about 1.36. The fractal dimension of the state border of Utah of course is equal to 1.0, the fractal dimension of the straight line.

3/2-Curve: Two Steps

The first two replacement steps in the construction of the 3/2-curve.

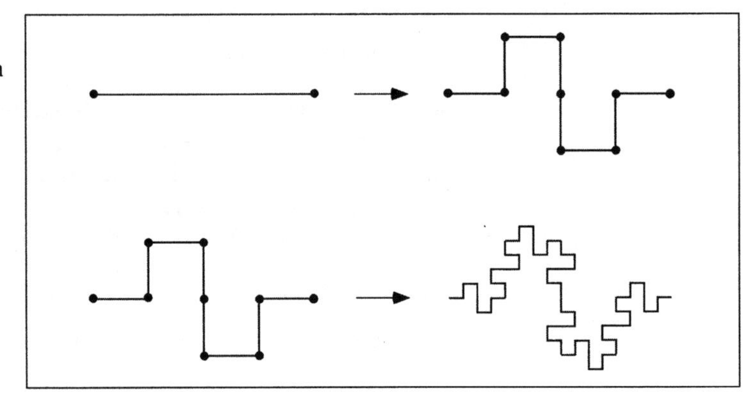

Figure 4.25

Measuring the 3/2-curve

We continue with another basic example of a self-similar curve, the 3/2-curve. The construction process starts from a line segment of length 1. In the first step we replace the line segment by the *generator* curve, a polygonal line of 8 segments, each of length 1/4 (see figure 4.25). That is to say, the polygonal line has length 8/4, the length has doubled. In the next step, we scale down the polygonal line by a factor of 1/4 and replace each line segment of length 1/4 in step 1 by that scaled down polygonal line.

After the second step, we have 8^2 line segments, each of length $1/4^2$, so that the total length is now $8^2/4^2 = 2^2$. In the next step, we scale down the generator by a factor of $1/4^2$ and replace each line segment of length $1/4^2$ in step 2 by that scaled down generator, and so on. Apparently, the length of the resulting curve is doubled in each step (i.e., after the step k, the length is 2^k). The number of line segments grows by a factor of 8 in each step (i.e., after the k-th step we have 8^k line segments of length $1/4^k$). Entering these data in a log/log diagram (preferably working with $\log_4$), we obtain figure 4.26.

Measuring the slope of the fitted line, we obtain $d = 0.5$. More directly, the length computed with line segments of length $1/4^k$ is 2^k, and this is reflected in the power law

$$u = \sqrt{\frac{1}{s}} = \frac{1}{s^{0.5}}$$

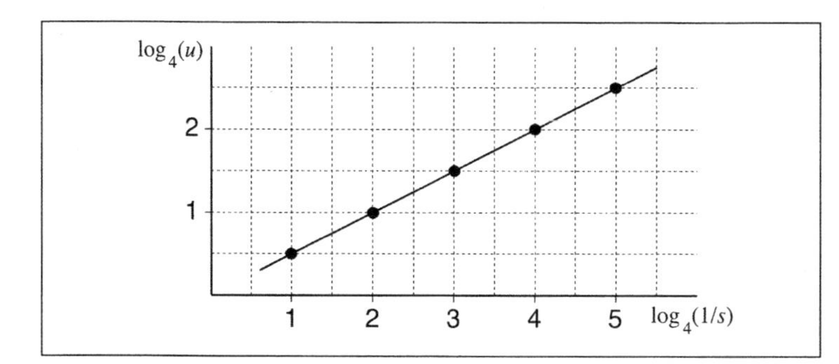

Log/Log Plot for the 3/2-Curve

Length versus 1/scale in the 3/2-curve. The result is a line with slope 1/2.

Figure 4.26

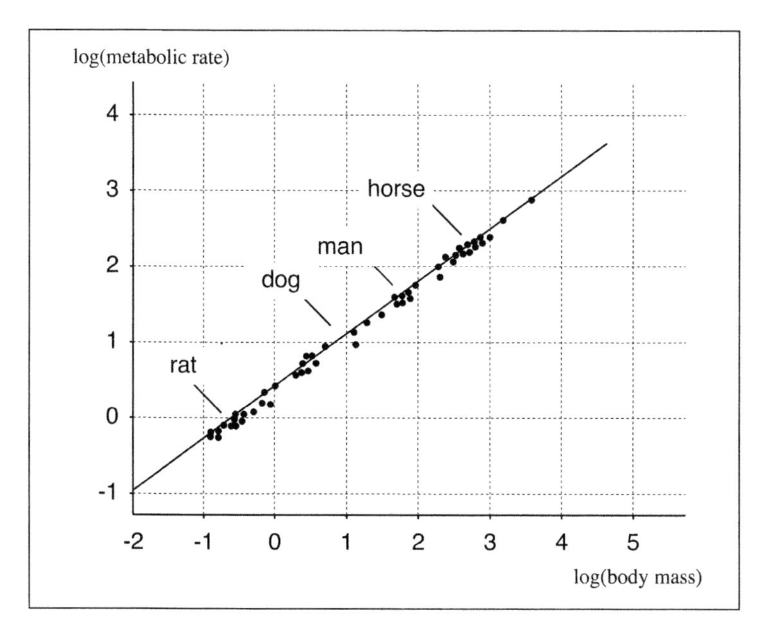

Metabolic Rate As Power Law

The reduction law of metabolism, demonstrated in logarithmic coordinates, showing basal metabolic rate as a power function of body mass.

Figure 4.27

with exponent $d = 1/2$. Thus, the compass dimension and the self-similarity dimension are equal to $D = 1 + d = 1.5$, which justifies the name *3/2-curve*.

The Fractal Nature of Organisms

We conclude this section with some fascinating speculations, which go back to a 1985 paper by M. Sernetz and others,[14] concerning the fractal nature of organs. This paper discusses the metabolic rates of various animals (e.g., rats, dogs and horses) and relates them to their respective body masses. The metabolic rate is measured in Joules per second and the mass in kilogram. Since body mass is proportional to volume and since volume scales as r^3, when r is the scaling factor, a first guess would be that the metabolic rate should be proportional to the body mass (i.e., proportional to r^3). Figure 4.27 reveals, however, that the exponent in the power law is significantly different

[14]From M. Sernetz, B. Gelléri, F. Hofman, *The organism as a bioreactor: interpretation of the reduction law of metabolism in terms of heterogeneous catalysis and fractal structure*, Journal Theoretical Biology 117 (1985) 209–230.

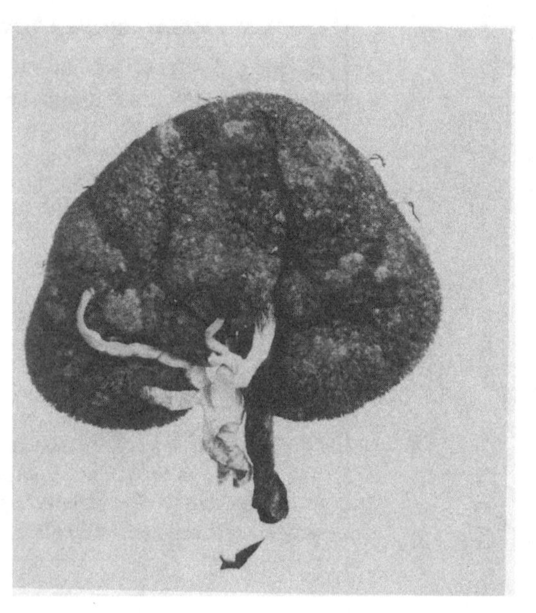

Figure 4.28 : Arterial and venous casts of a kidney of a horse as an example of fractal structures in organisms. Both systems in the natural situation fit entirely into each other and yet represent only the negative of the kidney. The remaining interspace between the vessels corresponds to the actual tissue of the organ (see also the color plate 2). Pictures courtesy of Manfred Sernetz.

from this expected value 1.

The slope α for the fitted line is approximately 0.75. In other words, if m denotes the metabolic rate and w the body mass, then

$$\log m = \alpha \log w + \log c,$$

where $\log c$ is the m-intercept. Thus, $m = cw^{\alpha}$. Using $w \propto r^3$, this is equivalent to $m \propto r^{3\alpha}$, where $3\alpha \approx 2.25$.

This means that our guess, according to which the metabolic rate should be proportional to the mass or volume, is wrong. It merely scales according to a fractal surface of dimension 2.25. How can that be explained? One of the speculations is that the above power law for the metabolic rate in organisms is a reflection of the fact that an organism is, in some sense, more like a highly convoluted surface than a solid body. In carrying this idea a little further — maybe too far — we could say that animals, including humans, look like three-dimensional objects, but they are much more like fractal surfaces. Indeed, if we look beneath the skin, we find all kinds of systems (e.g., the arterial and venous systems of a kidney) which are good examples of fractal surfaces in their incredible vascular branching (see color plate 2). From a physiological point of view, it is almost self-evident that the exchange functions of a kidney are intimately related to the size of the surfaces of its urinary and blood vessel systems. It is obvious that the volume of such a system is finite; it fits into the kidney! At the same time, the surface is in all practical terms infinite! And the relevant measuring task, quite like the ones for coastlines, would be to determine how the measured surface area grows as we use higher and higher accuracy. This leads to the fractal dimension, which characterizes some aspects of the complexity of the bifurcation structure in such a system. This numerical evaluation of characteristic features of vessel systems can potentially become an important new tool in physiology. For example, questions like the following have been asked: What are the differences between systems of various animals? Or, is there a significant change in the fractal dimension when measured for systems with certain malfunctions?

4.4 The Box-Counting Dimension

In this section we discuss our third and final version of Mandelbrot's fractal dimension: the *box-counting dimension*. This concept is related to the self-similarity dimension: it gives the same numbers in many cases, but also different numbers in some others.

Non-Self-Similar Structures

So far, we have seen that we can characterize structures which have some very special properties such as self-similarity, or structures like coastlines, where we can work with compasses of various settings. But what can be done if a structure is not at all self-similar and as wild as figure 4.29, for example?

A Wild Fractal

A wild structure with some scaling properties.

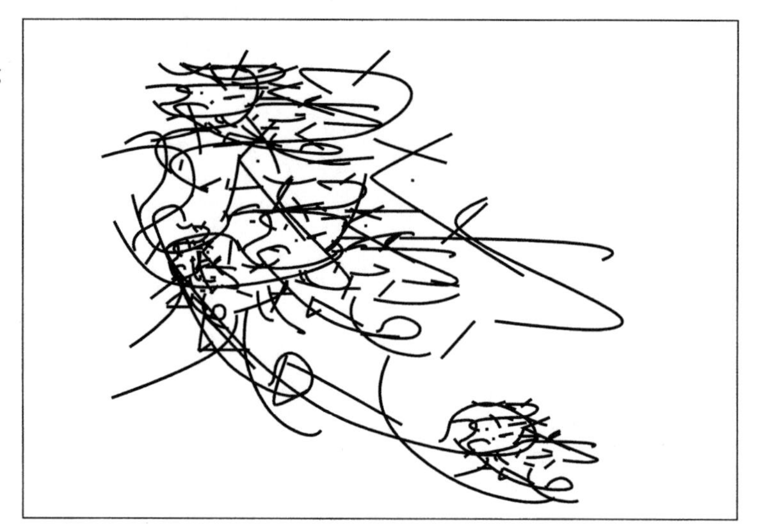

Figure 4.29

In such a case, there is no curve which can be measured with compasses; and there is no self-similarity, though there are some scaling properties. For example, the 'cloud' in the lower right corner looks somewhat similar to the large 'cloud' in the upper portion. The box-counting dimension proposes a systematic measurement, which applies to any structure in the plane and can be readily adapted for structures in space. The idea is very much related to the coastline measurements.

We put the structure onto a grid with mesh size s, and count the number of grid boxes which contain some of the structure. This gives a number, say N. Of course, this number will depend on the size s. Therefore we write $N(s)$. Now we change s to progressively smaller sizes and count the corresponding numbers $N(s)$. Next we make a log/log diagram; we plot the logarithms $\log(N(s))$ versus $\log(1/s)$.

The Box-Counting Dimension

We then try to fit a straight line to the plotted points of the diagram and measure its slope D_b. This number is the *box-counting dimension*, another special form of Mandelbrot's fractal dimension. Figure 4.30 illustrates this procedure using only two measurements. We find a slope of about $D_b = 1.45$.

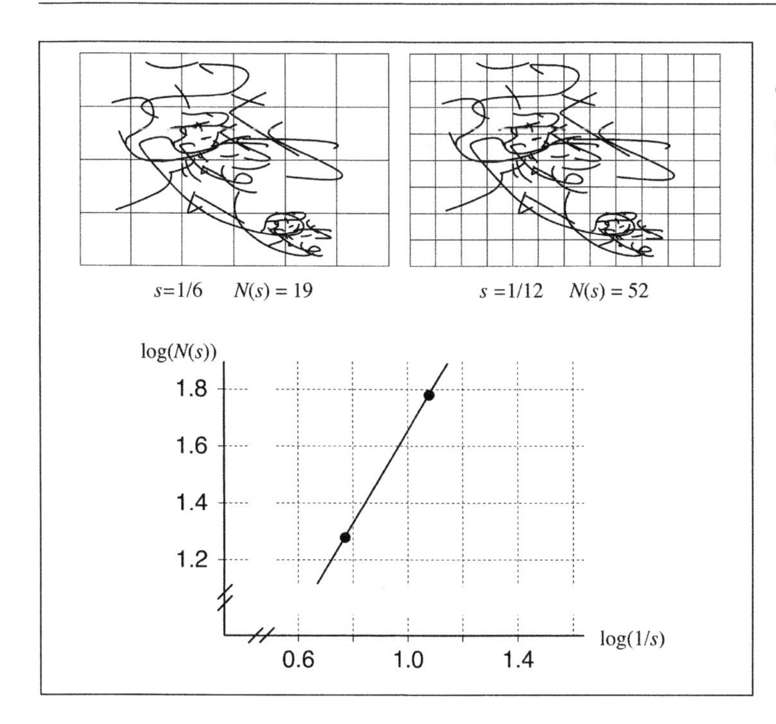

Box-Count

The wild structure is box-counted using two grids. The slope of the line is $\log(52/19)/\log 2 \approx 1.45$.

Figure 4.30

For practical purposes it is often convenient to consider a sequence of grids where the mesh size is reduced by a factor of 1/2 from one grid to the next. In this approach each box from a grid is subdivided into four boxes each of half the size in the next grid. When box-counting a fractal using such grids we arrive at a sequence of counts $N(2^{-k}), k = 0, 1, 2, \ldots$ Here we have adopted the convention to set $s = 2^0 = 1$ for the coarsest grid. The slope of the line from one data to the next in the corresponding log/log diagram is

$$\frac{\log N(2^{-(k+1)}) - \log N(2^{-k})}{\log 2^{k+1} - \log 2^k} = \log_2 \frac{N(2^{-(k+1)})}{N(2^{-k})}$$

where in the term on the right we have used logarithms with base 2 while the term on the left holds for any base. The result thus is the base 2 logarithm of the factor by which the box-count increases from one grid to the next. This slope is an estimate for the box-counting dimension of the fractal. In other words, if the number of boxes counted increases by a factor of 2^D when the box size is halved, then the fractal dimension is equal to D.

Self-Similarity and Box-Counting Dimension are Different

It is a nice exercise to experimentally verify the fact that the box-counting dimensions D_b of the Koch curve and the 3/2-curve are the same as the respective self-similarity and compass dimensions. Note, however, that in the plane a box-counting dimension D_b will never exceed 2. The self-similarity dimension D_s, however, can easily exceed 2 for a curve in the plane. To convince ourselves, we need only construct an example where the reduction

Self-Intersection

First steps of a curve generation with self-intersections.

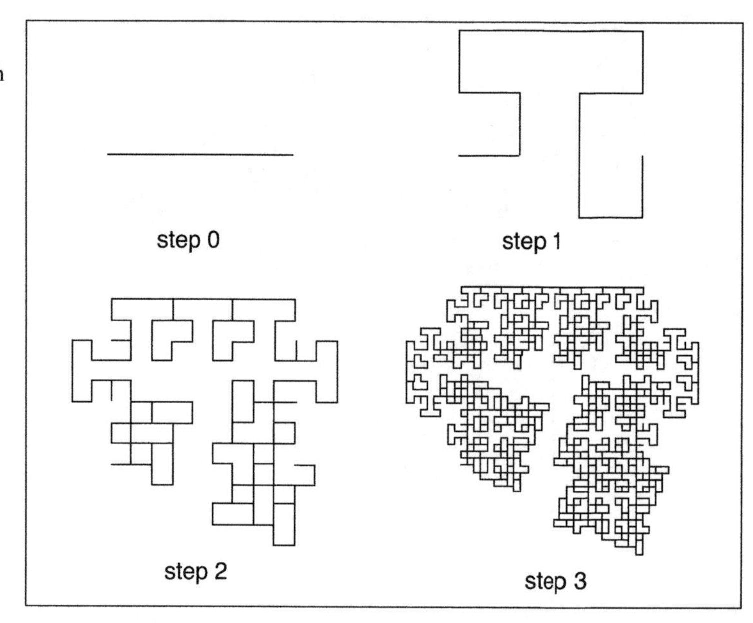

Figure 4.31

factor is $s = 1/3$ and the number of pieces in a replacement step is $a > 9$ (see figure 4.31). Then

$$D_s = \frac{\log a}{\log 1/s} > 2.$$

The reason for this discrepancy is that the curve generated in figure 4.31 has overlapping parts, which, by principle, are counted only once in the box-counting method, but with corresponding multiplicities in the computation of the self-similarity dimension. For this curve we have $s = 1/3$ and $a = 13$ and, thus, the self-similarity dimension is

$$D_s = \frac{\log 13}{\log 3} \approx 2.335.$$

Advantages of Box-Counting Dimension

The box-counting dimension is the one most used in measurements in all the sciences. The reason for its dominance lies in the easy and automatic computability by machine. It is straightforward to count boxes and to maintain statistics allowing dimension calculation. The program can be carried out for shapes with and without self-similarity. Moreover, the objects may be embedded in higher dimensional spaces. For example, when considering objects in common three-dimensional space, the boxes are not flat but real three-dimensional boxes with height, width, and depth. But the concept also applies to fractals such as the Cantor set which is a subset of the unit interval, in which case the boxes are small intervals.

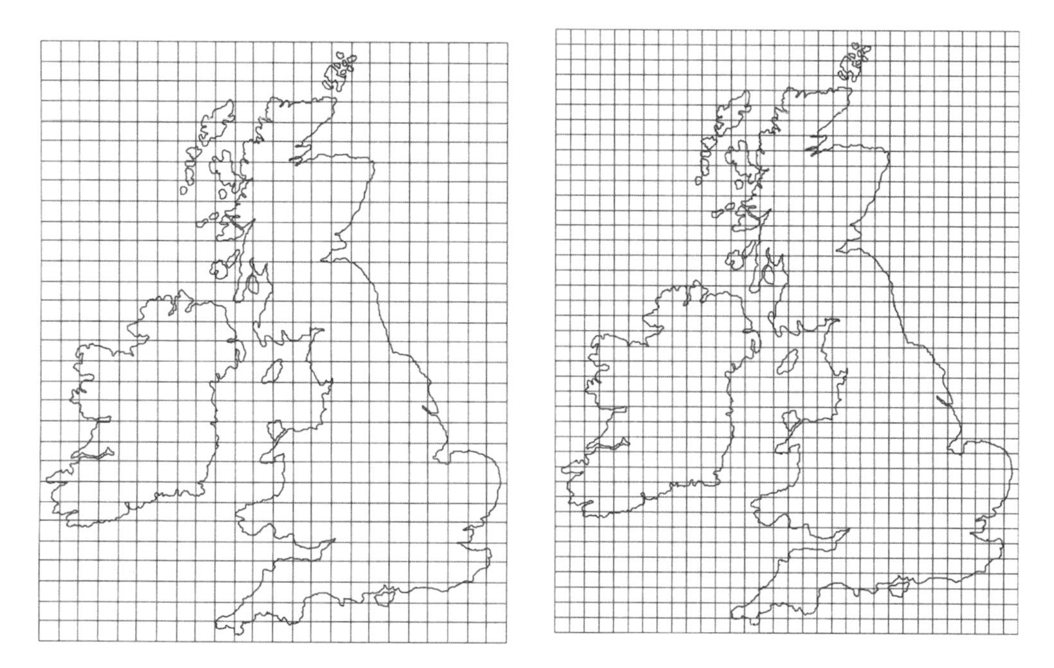

Figure 4.32 : Count all boxes that intersect (or even touch) the coastline of Great Britain, including Ireland.

Box-Counting Dimension of the Coast of Great Britain

As an example let us reconsider the classic example, the coastline of Great Britain. Figure 4.32 shows an outline of the coast with two underlying grids. Having normalized the width of the entire grid to 1 unit, the mesh sizes are 1/24 and 1/32. The box-count yields 194 and 283 boxes that intersect the coastline in the corresponding grids (check this carefully, if you have the time). From these data it is now easy to derive the box-counting dimension. When entering the data into a log/log diagram, the slope of the line that connects the two points is

$$d = \frac{\log 283 - \log 194}{\log 32 - \log 24} \approx \frac{2.45 - 2.29}{1.51 - 1.38} \approx 1.31.$$

This is in nice agreement with our previous result from the compass dimension.

Fractal Dimensions and Their Limitations

The concept of fractal dimension has inspired scientists to a host of interesting new work and fascinating speculations. Indeed, for a while it seemed as if the fractal dimensions would allow us to discover a new order in the world of complex phenomena and structures. This hope, however, has been dampened by some severe limitations. For one thing, there are several different dimensions which give different answers. We can also imagine that a structure is a mixture of different fractals, each one with a different value of box-counting dimension. In such a case, the conglomerate will have a dimension which is simply the dimension of the component(s) with the largest dimension. That means the resulting number cannot be characteristic for the mixture. What we would really like to have is something more like a spectrum of numbers which gives information about the distribution of fractal dimensions in a structure. This program has, in fact, been carried out and runs under the theme *multifractals*.[15]

The historical roots of fractal dimensions are in Hausdorff's work from 1918[16] although his definition of what became later known as Hausdorff dimension is not practical in the sense that it is very difficult to compute even in elementary examples and nearly impossible to estimate in practical applications. Nevertheless, it is very important in theory and we will see a glimpse of that in the appendix dealing with multifractal measures. For an account of the various notions of dimensions related to fractal dimensions as well as their mutual relation we refer to the excellent books of Gerald A. Edgar[17] and Kenneth Falconer.[18] We conclude this section with both the definition of the Hausdorff dimension, which is quite technical, and its relation to the box-counting dimension.

Definition of Hausdorff Dimension

We will restrict ourselves to a definition of the Hausdorff dimension for sets A which are imbedded in Euclidean space

$$\mathbf{R}^n = \{x \mid x = (x_1, \ldots, x_n), x_i \in \mathbf{R}\}$$

for some natural number n. We need some mathematical notation to arrive at a definition. Firstly, there is a distance function $d(x,y)$, the Euclidean distance of x and y in $\mathbf{R}^n$,

$$d(x,y) = \sqrt{\sum_{i=0}^{n}(x_i - y_i)^2}.$$

Secondly, there is the infimum and supremum of a subset X of real numbers,

$$\begin{aligned}\inf\{x \in X\} &= \text{largest lower bound of } X,\\ \sup\{x \in X\} &= \text{smallest upper bound of } X.\end{aligned}$$

This means that $a = \inf\{x \in X\}$ provided $a \leq x$ for all $x \in X$ and for any $\varepsilon > 0$ there is $x \in X$ such that $x - a < \varepsilon$. Similarly, $b = \sup\{x \in X\}$ provided $b \geq x$ for all $x \in X$ and for any $\varepsilon > 0$ there is $x \in X$ such that $b - x < \varepsilon$. Using these notions we can now define the diameter of a subset U of $\mathbf{R}^n$.

$$\text{diam}(U) = \sup\{d(x,y) \mid x, y \in U\}.$$

The last notion we need is that of an open cover of a subset A of $\mathbf{R}^n$. A subset U of $\mathbf{R}^n$ is called open provided for any $x \in U$ there is a

[15] See B. B. Mandelbrot, *An introduction to multifractal distribution functions,* in: Fluctuations and Pattern Formation, H. E. Stanley and N. Ostrowsky (eds.), Kluwer Academic, Dordrecht, 1988; J. Feder, *Fractals,* Plenum Press, New York, 1988; K. Falconer, *Fractal Geometry, Mathematical Foundations and Applications*, John Wiley & Sons, Chichester, 1990.

[16] F. Hausdorff, *Dimension und äußeres Maß,* Math. Ann. 79 (1918) 157–179.

[17] G. A. Edgar, *Measure, Topology and Fractal Geometry,* Springer-Verlag, New York, 1990.

[18] K. Falconer, *Fractal Geometry, Mathematical Foundations and Applications,* John Wiley & Sons, Chichester, 1990.

small ball $B_\varepsilon(x) = \{y \in \mathbf{R}^n \mid d(x,y) < \varepsilon\}$ of radius $\varepsilon > 0$ centered at x which is entirely in U. A family of open subsets $\{U_1, U_2, U_3, \ldots\}$ is called an open cover (countable) of A provided

$$A \subset \bigcup_{i=1}^{\infty} U_i.$$

Now we are ready to define the Hausdorff dimension of A: Let s and ε be positive real numbers. Then define

$$h_\varepsilon^s(A) = \inf \left\{ \sum_{i=0}^{\infty} \text{diam}(U_i)^s \;\middle|\; \begin{array}{l} \{U_1, U_2, \ldots\} \\ \text{open cover of } A \text{ with} \\ \text{diam}(U_i) < \varepsilon \end{array} \right\}.$$

Thus, the infimum is extended over all open covers of A for which the covering sets U_i have diameter less than ε. For each such cover we take the diameters of the open sets of the cover, raise them to the sth power, and take the sum. This sum may be finite or infinite. As we decrease ε the class of permissible covers of A is reduced. Therefore, the infimum increases and so approaches a limit as $\varepsilon \to 0$ whiuch can be infinite or a real number. We write

$$h^s(A) = \lim_{\varepsilon \to 0} h_\varepsilon^s(A).$$

The limit $h^s(A)$ is called the s-dimensional Hausdorff measure of A. In particular, it follows that s-dimensional Hausdorff measure of the empty set is 0 and $h^s(A) \le h^s(B)$ if $A \subset B$. Moreover, $h^1(A)$ is the length of a smooth curve A; $h^2(A)$ is the area of a smooth surface A up to a factor of $\pi/4$; $h^3(A)$ is the volume of a smooth three-dimensional manifold A up to a factor of $4\pi/3$. Another important property is this: If $f : A \to \mathbf{R}^n$ satisfies a Hölder condition for all pairs $x, y \in A$, i.e.,

$$d(f(x), f(y)) \le c(d(x,y))^\alpha$$

for some constants $c > 0$ and $\alpha > 0$, then

$$h^{s/\alpha}(f(A)) \le c^{s/\alpha} h^s(A).$$

For example, if f is a similarity transformation with contraction factor $0 \le c < 1$, then f satisfies a Hölder condition with $\alpha = 1$, and $h^s(f(A)) \le c^s h^s(A)$. Moreover, Hausdorff proved that for any set A the following holds true. There is a number $D_H(A)$ such that

$$h^s(A) = \begin{cases} \infty & \text{for } s < D_H(A) \\ 0 & \text{for } s > D_H(A). \end{cases}$$

This number $D_H(A)$ is defined as the Hausdorff dimension

$$D_H(A) = \inf\{s \mid h^s(A) = 0\} = \sup\{s \mid h^s(A) = \infty\}.$$

If $s = D_H(A)$, then $h^s(A)$ may be zero, infinite, or some positive real number. We finally collect some fundamental properties of the Hausdorff dimension:

(1) If $A \subset \mathbf{R}^n$, then $D_H(A) \leq n$.
(2) If $A \subset B$, then $D_H(A) \leq D_H(B)$.
(3) If A is a countable set, then $D_H(A) = 0$.
(4) If $A \subset \mathbf{R}^n$ and $D_H(A) < 1$, then A is totally disconnected.
(5) Let C_∞ be the Cantor set. Then $D_H(C\infty) = \log 2/\log 3$.

Let us give a heuristic argument for property (5), assuming that $0 < h^s(C_\infty) < \infty$ for $s = D_H(C_\infty)$. Note that C_∞ splits into two parts, $C_L = C_\infty \cap [0, 1/3]$, and $C_R = C_\infty \cap [2/3, 1]$ which are both similar to C_∞ but scaled by a factor of $c = 1/3$. Thus

$$h^s(C_\infty) = h^s(C_L) + h^s(C_R) = c^s h^s(C_\infty) + c^s h^s(C_\infty).$$

Now we divide by $h^s(C_\infty) \neq 0$ and obtain $1 = 2c^s$, or $s = \log 2/\log 3$.

There are several difficulties in evaluating the Hausdorff dimension in a concrete case. The box-counting dimension in some sense is motivated by avoiding these difficulties.

Hausdorff Dimension Versus Box-Counting Dimension

The central difficulty in evaluating the Hausdorff dimension is the one given by the terms $\sum_{i=0}^{\infty} \text{diam}(U_i)^s$. The box-counting dimension simplifies this problem by replacing the terms $\text{diam}(U_i)^s$ by the terms δ^s. A formal definition of the box-counting dimension D_b of any bounded subset A of $\mathbf{R}^n$ proceeds as follows. Let $N_\delta(A)$ be the smallest number of sets of diameter at most δ which cover A.[19] Then

$$D_b(A) = \lim_{\delta \to 0} \frac{\log N_\delta(A)}{\log 1/\delta},$$

provided that limit exists.

There are several equivalent definitions of $D_b(A)$. For example, consider a subdivision of $\mathbf{R}^n$ into a lattice of grid size δ. That is a tessellation of $\mathbf{R}^n$ by cubes of side length δ. Now let $N'_\delta(A)$ be the number of cubes that intersect A. It is a fact that

$$D_b(A) = \lim_{\delta \to 0} \frac{\log N'_\delta(A)}{\log 1/\delta},$$

provided the limit exists. Roughly speaking the definition says that $N_\delta(A) \propto \delta^{-s}$ for small δ, where $s = D_b(A)$. More precisely it says that

$$N_\delta(A)\delta^s \to \begin{cases} \infty & \text{for } s < D_b(A) \\ 0 & \text{for } s > D_b(A). \end{cases}$$

But

$$N_\delta(A)\delta^s = \inf \left\{ \sum_i \delta^s \;\middle|\; \begin{array}{l} \{U_1, U_2, \ldots\} \text{ finite cover of} \\ A \text{ with } \text{diam}(U_i) \leq \delta \end{array} \right\}.$$

[19] Since A is bounded we can always assume that the cover is finite.

This should be compared with the definition of the Hausdorff dimension to see that the only difference is in the terms diam$(U_i)^s$ versus the term δ^s.

Unfortunately it is not true that the Hausdorff dimension and the box-counting dimension always are the same.[20] For example, it can be shown, that $D_b(A) = n$ for any dense subset of $\mathbf{R}^n$. In other words, the box-counting dimension of the set of rational numbers in $[0, 1]$ is 1, the Hausdorff dimension of the same (countable) set is 0. Another striking example is the set $A = \{0, 1/2, 1/3, 1/4, \ldots\}$. This set has a fractional box-counting dimension. In fact $D_b(A) = 1/2$. In other words, if $D_b(A)$ is not an integer we may not blindly conclude that A has fractal properties. But it is true that the Hausdorff dimension and the box-counting dimension do agree for a large class of sets which includes the classical fractals like the Cantor set, the Sierpinski gasket, Sierpinski carpet and many others, as we will report at the end of chapter 5.

[20] For details see K. Falconer, *Fractal Geometry, Mathematical Foundations and Applications*, John Wiley & Sons, Chichester, 1990.

4.5 Borderline Fractals: Devil's Staircase and Peano Curve

The fractals discussed in this chapter so far have a noninteger fractal dimension, but not all fractals are of this type. Thus, we want to expand our knowledge with two examples of fascinating fractals which represent very extreme cases: the first is the so-called devil's staircase, which implies a fractal curve of dimension 1.0. The second is a Peano curve of dimension equal to 2.0.

Devil's Staircase: Construction

The column construction of the devil's staircase.

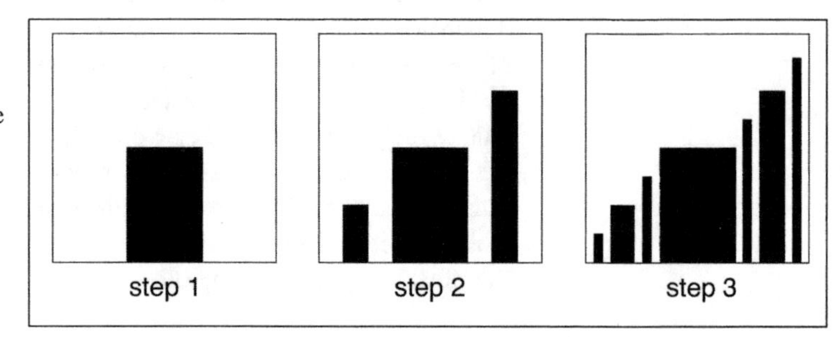

Figure 4.33

The Complete Devil's Staircase

The devil's staircase is the boundary line between the black and the white part of the square.

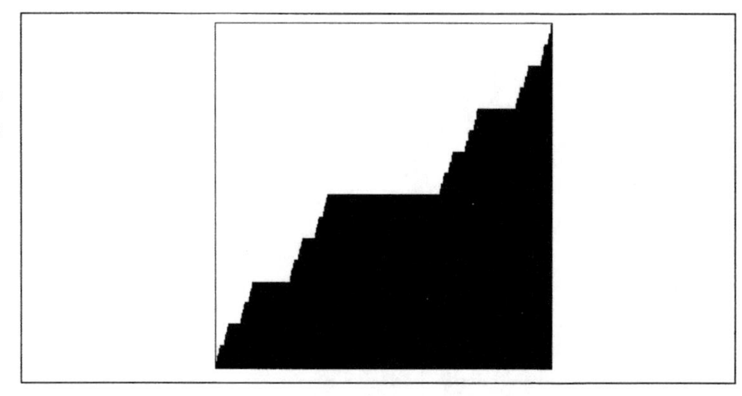

Figure 4.34

Devil's Staircase

The first one of these objects, the devil's staircase, is intimately related to the Cantor set and its construction. We take a square with sides of length 1. Then we start to construct the Cantor set on the base side (i.e., we take away successively middle thirds in the usual way). For each middle third of length $1/3^k$ which we take away, we paste in a rectangular column with width $1/3^k$ and a certain height. Let us see how this is done in figure 4.33. In the first step, a column is erected over the middle third of the base side — the interval $[1/3, 2/3]$ — of the square with height $1/2$. In the second step, we erect two columns, one of height $1/4$ over the interval $[1/9, 2/9]$ and the other of height $3/4$ over the interval $[7/9, 8/9]$. In the third step, we erect four columns of heights $1/8, 3/8, 5/8, 7/8$, and in the kth step, we erect 2^{k-1} columns of heights $1/2^k, 3/2^k, \ldots, (2^k - 1)/2^k$. In the limit, we obtain an area, the upper border of which is called the *devil's staircase*. Figure

4.34 shows an image obtained in a computer rendering. We see a staircase ascending from left to right, a staircase with infinitely many steps whose step heights become infinitely small. As the process continues, the square in figure 4.33 gets an upper white and a lower black part. In the limit, there will be a perfect symmetry. The white part will be an exact copy of the black part. Put another way, the white part is obtained from the black by a rotation of 180°. In this sense the devil's staircase divides the square in two halves fractally. One immediate consequence of this argument is that in the limit the area beneath the staircase is exactly half of the initial square.

Area under the Devil's Staircase

We will look again at the columns in figure 4.33 and observe that the two narrow columns of width 1/9 in step 2 make one column of height 1; likewise, the four columns of width 1/27 in step 3 make two columns of height 1; and so on. In other words, if we move the columns from the right side over to the left and vertically cut the center column of width 1/3 into two equal parts, which we put on top of each other, we obtain a figure which eventually fills half the square (see figure 4.35).

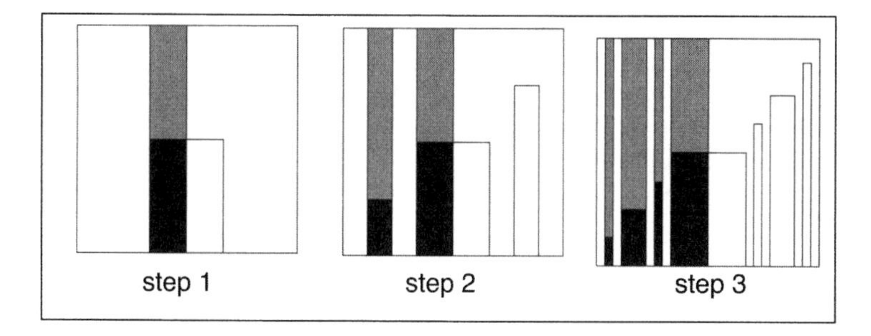

Figure 4.35 : The area under the devil's staircase is 1/2.

With the devil's staircase, we can also check an explicit argument. If we group the areas of the columns according to figure 4.35 we obtain the total area A under the staircase using a geometric series as follows:

$$\begin{aligned} A &= \frac{1}{3}\cdot\frac{1}{2}+\frac{1}{9}\left(\frac{1}{4}+\frac{3}{4}\right)+\frac{1}{27}\left(\frac{1}{8}+\frac{3}{8}+\frac{5}{8}+\frac{7}{8}\right)+\cdots \\ &= \frac{1}{6}+\frac{1}{9}\cdot\left(1+\frac{2}{3}+\frac{4}{9}+\cdots+\frac{2^k}{3^k}+\cdots\right). \end{aligned}$$

The sum of the geometric series in the bracket is 3. Thus, the result is $A=\frac{1}{6}+\frac{3}{9}=\frac{1}{2}$.

Length of the Devil's Staircase

Now, to move on to our next question: how long is the devil's staircase? A polygonal approximation of the staircase makes it obvious that

- the staircase is a curve, which has no gaps, and
- the length of that curve is exactly 2!

Thus, we have constructed a curve which is fractal, yet it has a finite length. In other words, the slope d in the log/log diagram of length versus 1/scale is $d = 0$, and the fractal dimension would be $D = 1 + d = 1$! This result is important because it teaches us that there are curves of finite length which we would like to call fractal nevertheless. Moreover, the devil's staircase looks self-similar at first glance, but is not. One may ask, of course, why those curves are called fractal in the first place? An argument in support of spending the characterization 'fractal' in this case is the fact that the devil's staircase is the graph of a very strange function, a function, that is constant everywhere except in those points that are in the Cantor set.

Polygon Construction of Devil's Staircase

It will be helpful in following the construction if you compare figure 4.33 with the following figure 4.36. We construct a polygonal line for each step in figure 4.33 by walking in horizontal and vertical directions only. We always start in the lower left corner and walk horizontally until we hit a column. At this point, we play fly and walk up the column until we reach the top. There we again walk horizontally until we hit the next column which we again surmount. At the top, we again walk on horizontally and then vertically, continuing on in the same pattern as often as necessary until we reach the right upper corner. In each step, the polygonal lines constructed in this way have length 2 because, when summing up all horizontal lines the result is 1 and the sum of all vertical lines is also 1.

Polygonal Construction

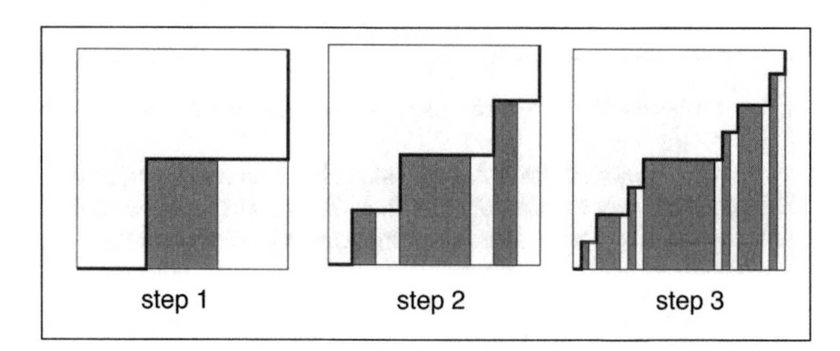

Figure 4.36

Self-Affinity

The area under the devil's staircase is not self-similar. Let us explain (see figure 4.37). The area can be broken down into six identical building blocks. Block 1 is obtained from the entire area by contracting the image horizontally by a factor of 1/3 and vertically by a factor of 1/2 (i.e., two different factors). This is why the object is not self-similar. For a self-similarity transformation, the two factors would have to be identical. Block 6 is the same as block 1. Moreover, a rectangle with sides of length 1/3 and 1/2 can house exactly one copy of block 1 together with a copy obtained by rotation by 180 degrees. This explains blocks 2 and 3, or 4 and 5. A contraction which reduces an image

Self-Affinity

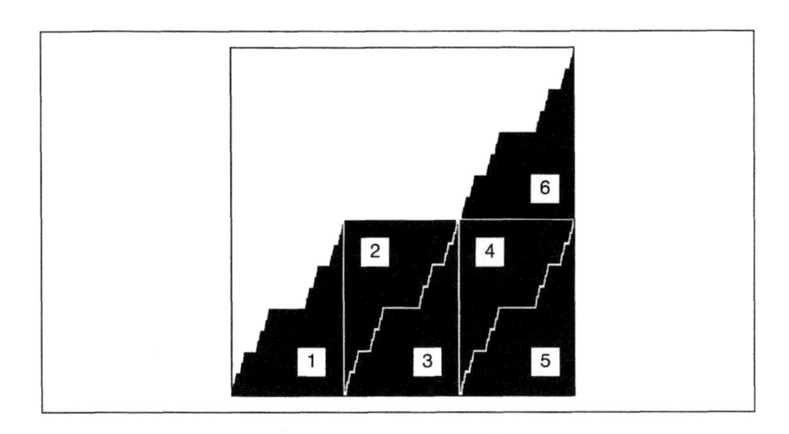

Figure 4.37

by different factors horizontally and vertically is a special case of a so-called *affine transformation*. Objects that are composed of affine copies of the whole are called *self-affine*. The area under the devil's staircase is an example.

Devil's Staircase by Curdling

The devil's staircase may look like an odd mathematical invention. It is, indeed, a mathematical invention; but it isn't really so odd, for it has great importance in physics.[21] We now discuss a problem — not really from physics, though it points in that direction — where the staircase comes out naturally.

Let us modify the Cantor set (see figure 4.38). Our initial object is no longer a line segment but rather a bar with density $r_0 = 1$. We suppose that we can compress and stretch the bar arbitrarily. The initial bar has length $l_0 = 1$ and therefore mass $m_0 = 1$. Now we cut the bar in the middle, obtaining two identical pieces of equal mass $m_1 = 1/2$. Next we hammer them so that the length of each reduces to $l_1 = 1/3$ without changing the cross-section. Since mass is conserved, the density in each piece must increase to $r_1 = m_1/l_1 = 3/2$. Repeating this process, we find that in the nth generation we have $N = 2^n$ bars, each with a length $l_n = 1/3^n$ and a mass $m_n = 1/2^n$. Mandelbrot calls this process *curdling* since an originally uniform mass distribution by this process clumps together into many small regions with a high density. The density of each of the small pieces is $r_n = m_n/l_n$. Figure 4.38 shows the density as height of the bars in each generation.

Assume now that the curdling process has been applied infinitely often, and we think of the resulting structure as put on the unit interval. Then we can ask: what is the mass $M(x)$ of the structure in the segment from 0 to x?[22] The mass does not change in the gaps, but it increases by infinitesimal jumps at the points of the Cantor set. The graph of the function $M(x)$ turns out to be none other than the devil's staircase.

The Peano Curve

Having the fractal dimension $D = 1$, and yet not being an ordinary curve, the devil's staircase is one extreme case. Let us now look at extreme cases of

[21] P. Bak, *The devil's staircase,* Phys. Today 39 (1986) 38–45.

[22] This can be written formally as $M(x) = \int_0^x dm(t)$.

Curdling

Density shown as height in the successive generations of Cantor bars.

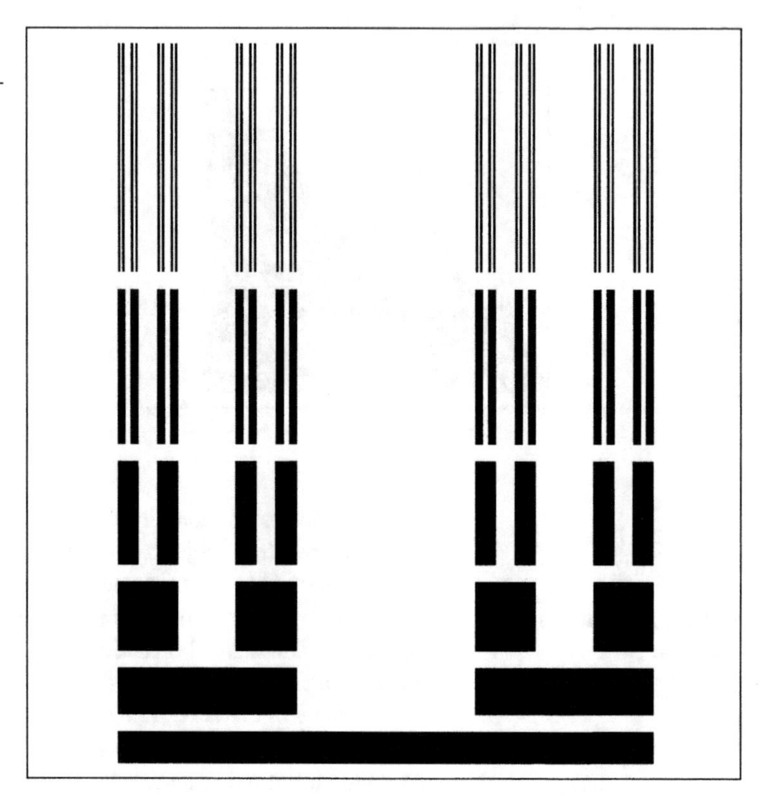

Figure 4.38

the opposite kind, curves which have fractal dimension $D = 2$. The first curve of this kind was discovered by G. Peano in 1890. His example created quite a bit of uncertainty about possible or impossible notions of curves, and for that reason also for dimension. We have already introduced the Peano curve in chapter 2 (see figure 2.35). Recall that in its construction, line segments are replaced by a generator curve consisting of nine segments, each one being one-third as long.

Based on the scaling factor 1/3 we measure the curve with $s = 1/3^k$, $k = 0, 1, 2, \ldots$ as the size s of the compass setting. This yields total lengths of $u = (9/3)^k = 3^k$. Assuming the power law $u = c \cdot 1/s^d$, we first note that $c = 1$ because for $s = 1$ we have $u = 1$. Moreover, we conclude from the equation $\log u = d \cdot \log 1/s$ that

$$d = \frac{\log u}{\log 1/s} = \frac{\log 3^k}{\log 3^k} = \frac{k}{k} = 1.$$

In other words, $D = 1 + d = 2$ (i.e., the Peano curve has fractal dimension 2). This reflects on the area-filling property of the Peano curve. The discussion of the self-similarity and area-filling properties of the Peano curve is continued in chapter 7.

Chapter 5

Encoding Images by Simple Transformations

Fractal geometry will make you see everything differently. There is danger in reading further. You risk the loss of your childhood vision of clouds, forests, flowers, galaxies, leaves, feathers, rocks, mountains, torrents of water, carpets, bricks, and much else besides. Never again will your interpretation of these things be quite the same.

Michael F. Barnsley[1]

So far, we have discussed two extreme ends of fractal geometry. We have explored fractal monsters, such as the Cantor set, the Koch curve, and the Sierpinski gasket; and we have argued that there are many fractals in natural structures and patterns, such as coastlines, blood vessel systems, and cauliflowers. We have discussed features, such as self-similarity, scaling properties, and fractal dimensions shared by those natural structures and the monsters; but we have not yet seen that they are close relatives in the sense that maybe a cauliflower is just a 'mutant' of a Sierpinski gasket, and a fern is just a Koch curve 'let loose'. Or phrased as a question, is there a framework in which a natural structure, such as a cauliflower, and an artificial structure, such as a Sierpinski gasket, are just examples of one unifying approach; and if so, what is it? Believe it or not, there is such a theory, and this chapter is devoted to it. It goes back to Mandelbrot's book, *The Fractal Geometry of Nature,* and a beautiful paper by the Australian mathematician Hutchinson.[2] Barnsley and

---

[1] Michael F. Barnsley, *Fractals Everywhere*, Academic Press, 1988.

[2] J. Hutchinson, *Fractals and self-similarity,* Indiana Journal of Mathematics 30 (1981) 713–747. Some of the ideas can already be found in R. F. Williams, *Compositions of contractions,* Bol. Soc. Brasil. Mat. 2 (1971) 55–59.

Berger have extended these ideas and advocated the point of view that they are very promising for the encoding of images.[3] In fact, this will be the focus of the appendix on image compression.

Fractal Geometry As a Language

We may regard fractal geometry as a new language in mathematics. As the English language can be broken down into letters and the Chinese language into characters, fractal geometry promises to provide a means to break down the patterns and forms of nature into primitive elements, which then can be composed into 'words' and 'sentences' describing these forms efficiently.

The word 'fern' has four letters and communicates a meaning in very compact form. Imagine two people talking over the telephone. One reports about a walk through a botanical garden admiring beautiful ferns. The person on the other end understands perfectly. As the word fern passes through the lines, a very complex amount of information is transmitted in very compact form. Note that 'fern' stands for an abstract idea of a fern and not exactly the one which was admired in the garden. To describe the individual plant adequately enough that the admiration can be shared on the other end, one word is not sufficient. We should be constantly aware of the problem that language is extremely abstract. Moreover, there are different hierarchical levels of abstractness, for example, in the sequence: tree, oak tree, California oak tree, ...

Here we will discuss one of the major dialects of fractal geometry as if it were a language. Its elements are primitive transformations, and its words are primitive algorithms. For these transformations together with the algorithms, in section 1.2 we introduced the metaphor of the Multiple Reduction Copy Machine (MRCM),[4] which will be the center of interest in this chapter.

[3] M. F. Barnsley, V. Ervin, D. Hardin, and J. Lancaster, *Solution of an inverse problem for fractals and other sets,* Proceedings of the National Academy of Sciences 83 (1986) 1975–1977; M. Berger, *Encoding images through transition probabilities,* Math. Comp. Modelling 11 (1988) 575–577. A survey article is: E. R. Vrscay, *Iterated function systems: Theory, applications and the inverse problem,* in: Bélair, J. and Dubuc, S., (eds.), *Fractal Geometry and Analysis,* Kluwer Academic, Dordrecht, 1991. A very promising approach seems to be presented in the recent paper A. E. Jacquin, *Image coding based on a fractal theory of iterated contractive image transformations,* to appear in: IEEE Transactions on Signal Processing. See also the chapter *Fractal Image Compression* by Y. Fisher, R. D. Boss, and E. W. Jacobs, to appear in *Data Compression*, J. Storer (ed.), Kluwer Academic, Norwell, MA.

[4] A similar metaphor has been used by Barnsley in his popularizations of iterated function systems (IFS), which is the mathematical notation for MRCMs.

5.1 The Multiple Reduction Copy Machine Metaphor

MRCM = IFS

Let us briefly review the main idea of the MRCM, the multiple reduction copy machine. This machine provides a good metaphor for what is known as *deterministic iterated function systems* (IFS) in mathematics. From here on we use both terminologies interchangeably; sometimes it is more convenient to work with the machine metaphor, while in more mathematical discussions we tend to prefer the IFS notion. The reader may wish to skip back to the first chapter to take a look at figures 1.8 and 1.9. The copy machine takes an image as input. It has several independent lens systems, each of which reduces the input image and places it somewhere in the output image. The assembly of all reduced copies in some pattern is finally produced as output. Here are the dials of the machine

Dial 1: number of lens systems,

Dial 2: setting of reduction factor for each lens system individually,

Dial 3: configuration of lens systems for the assembly of copies.

The crucial idea is that the machine runs in a feedback loop; its own output is fed back as its new input again and again. While the result of this process is rather silly when there is only one reduction lens in the machine (only one point remains as shown in figure 1.8), this banal experiment turns into something extremely powerful and exciting when several lens systems are used. Moreover, we allow other transformations besides ordinary reductions (i.e., transformations more general than similarity transformations).

Imagine that such a machine has been built and someone wants to steal its secret — its construction plan. How much time and effort is necessary to get all the necessary information? Not very much at all. Our spy just has to run the machine once on an arbitrary image.[5] One copy reveals all the geometric features of the machine which we now start to operate in feedback mode.

An MRCM for Sierpinski Gaskets

Consider an MRCM with three lens systems, each of which is set to reduce by a factor of 1/2. The resulting copies are assembled in the configuration of an equilateral triangle. Figure 5.1 shows the effect of the machine run three times beginning with different initial images. In (a) we take a disk and use different shadings to keep track of the effect of the individual lens systems. In (b) we try a truly 'arbitrary' image. In just a few iterations the machine, or abstractly speaking the process, throws out images which look more and more like a Sierpinski gasket. In (c) we start with a Sierpinski gasket and observe that the machine has no effect on the image. The assembled reduced copies are the same as the initial image. That is, of course, because of the self-similarity property of the Sierpinski gasket.

[5] Almost any image can be used for this purpose. Images with certain symmetries provide some exceptions. We will study these in detail further below.

MRCM for the Sierpinski Gasket

Three iterations of an MRCM with three different initial images.

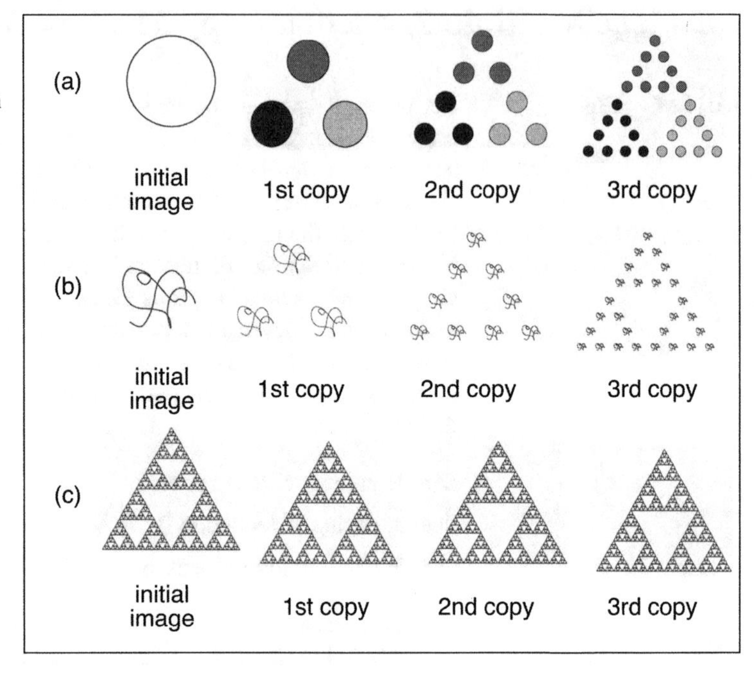

Figure 5.1

The Attractor of the MRCM

Let us summarize this first experiment. No matter which initial image we take and run the MRCM with, we obtain a sequence of images which always tends towards one and the same final image. We call it the *attractor* of the machine or process. Moreover, when we start the machine with the attractor, then nothing happens, one says the attractor is left *invariant* or *fixed*. Perhaps it will help to explain this result if we compare our experiment with a physical one in which we have a bowl (figure 5.2, left) and observe how a little iron ball put into different initial positions and then let loose always comes to rest at the bottom, the rest point. But if we put the ball right at the bottom to begin with, nothing happens.

The bowl here corresponds to our machine. Initial positions of the ball here correspond to initial images in the machine. Observing the path of the ball in time corresponds to running the machine repeatedly, and the rest point of the ball corresponds to the final image. The fact that the ball moves continuously

Bowls

Bowls with one and two dishes (attractors).

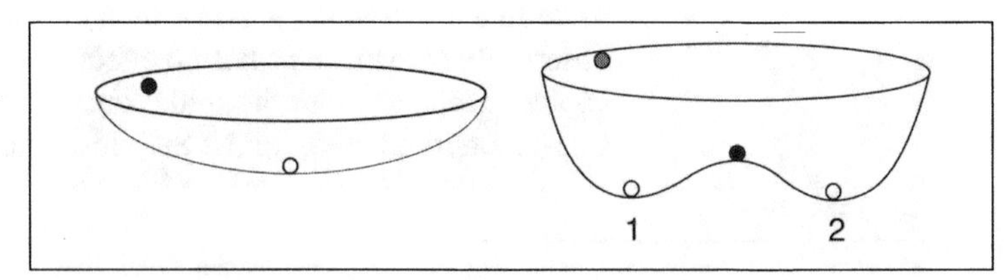

Figure 5.2

with time, while our machine operates in discrete steps, is not an essential difference. The point is that the ball in the bowl provides a metaphor for a *dynamical system* with only one attractor. The right-hand image in figure 5.2 shows a situation with two different attractors. There the final development depends on where we start.

Is the MRCM more like a bowl with one dish or like a bowl with two or more dishes? And, how does the answer depend on the setting of the control dials? In other words, can it be that with one setting of the dials, the MRCM has one attractor, while there are several attractors with another setting? These are typical questions for modern mathematics, questions typical for a field called *dynamical systems theory*, which provides the framework for discussing chaos as well as the generation of fractals.

There are two ways to answer such a question. If we are lucky, we will be able to find a general principle in mathematics which is applicable and gives an answer. If that is not the case, we can either try to find a new theory, or if that turns out to be too hard at the moment, we can try to gain insight into the situation by carefully controlled experiments. It is quite clear that experiments alone will not be satisfactory in many cases. Often we do not know how the bowl is shaped. Then, if we find, for example, that for all tested initial positions, we always arrive at the same rest point, what does that tell us? Not much. We still could be in a situation with several rest points. That is to say, that quite by accident, the tested initial positions were not taken sufficiently arbitrarily.

Experiments Need Theoretical Support

In other words, finding that our MRCM seems to always run towards the same final image is a wonderful experimental discovery, but it needs theoretical support. It turns out that using some general mathematical principles and results developed by Felix Hausdorff and Stefan Banach, we can in fact show that any MRCM always has a unique final image as an attractor, and that this final image is invariant under the iteration of the MRCM. This is Hutchinson's beautiful and fundamental contribution to fractal theory. When we say 'any MRCM' here, we mean that the number and design of the lens systems may change in the MRCM. The only property which must be satisfied to have Hutchinson's result is that each lens system contracts images.

5.2 Composing Simple Transformations

The Multiple Reduction Copy Machine is based on a collection of contractions. The term contraction means, roughly speaking, that points are moved closer together when one contraction is applied. Of course, similarity transformations (compare section 3.1) describing reduction by lenses are contractions. But we may also use transformations which reduce by different factors in different directions. For example, a transformation which reduces by one factor, say 1/3, horizontally and by a different factor, say 1/2, vertically is also allowed (see, for example, the devil's staircase in section 4.5). Note that a *similarity transformation* maintains angles unchanged, while more general contractions may not.

Transformations of the MRCM

We may also take transformations of the latter kind combined with a shearing and/or rotation, and/or reflection. Figure 5.3 illustrates some admissible 'lens systems' for our MRCM. Mathematically, these are described as *affine linear transformations* of the plane.

Admissible Transformations

Transformations with scaling, shearing, reflection, rotation and translation (not shown) are admissible in an MRCM.

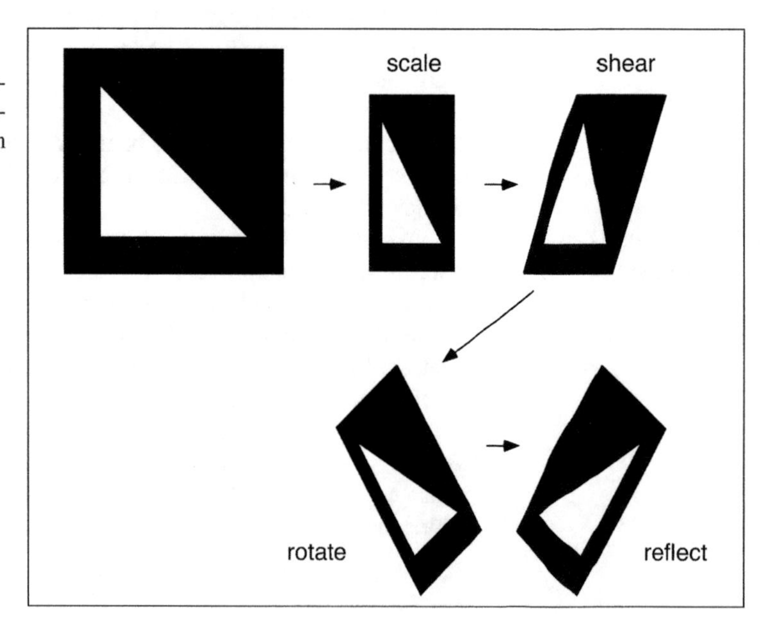

Figure 5.3

Affine Linear Transformations

The lens systems of our MRCMs can be described by affine linear transformations of the plane. Talking about a plane means that we fix a coordinate system, an x-axis, and a y-axis. Relative to that coordinate system every point P in the plane can be written as a pair (x, y). Sometimes we write $P = (x, y)$. In this way, points can be added together and can be multiplied by real numbers: if $P_1 = (x_1, y_1)$ and $P_2 = (x_2, y_2)$, then

$$P_1 + P_2 = (x_1 + x_2, y_1 + y_2)$$

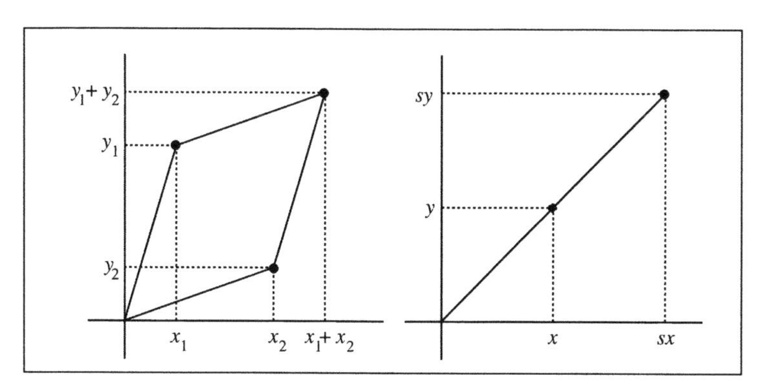

Sum and Multiplication with Scalar

(Left) Two points (x_1, y_1) and (x_2, y_2) are added: $(x_1, y_1) + (x_2, y_2) = (x_1 + x_2, y_1 + y_2)$. (Right) A point is multiplied by a number: $s \cdot (x, y) = (sx, sy)$.

Figure 5.4

and

$$sP = (sx, sy).$$

A linear mapping F is a transformation which associates with every point P in the plane a point $F(P)$ such that

$$F(P_1 + P_2) = F(P_1) + F(P_2)$$

for all points P_1 and P_2 and

$$F(sP) = sF(P)$$

for any real number s and all points P. A linear transformation F can be represented with respect to the given coordinate frame by a matrix

$$\begin{pmatrix} a & b \\ c & d \end{pmatrix}$$

where, if $P = (x, y)$ and $F(P) = (u, v)$, then

$$\begin{aligned} u &= ax + by, \\ v &= cx + dy. \end{aligned}$$

In other words, a linear transformation is determined by four coefficients a, b, c, and d. There are special representations which help us to discuss contractions more conveniently. To this end we write the four coefficients in our matrix as

$$\begin{pmatrix} r\cos\phi & -s\sin\psi \\ r\sin\phi & s\cos\psi \end{pmatrix}.$$

Such a representation is always possible. Just set

$$r = \sqrt{a^2 + c^2}$$

and

$$\phi = \arccos \frac{a}{\sqrt{a^2 + c^2}}$$

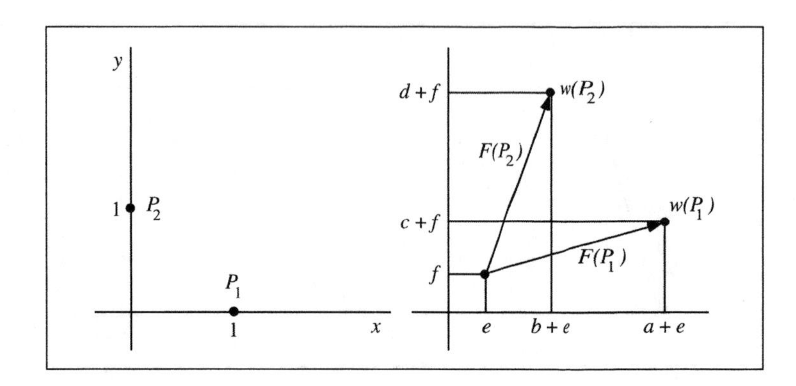

Figure 5.5 : The affine transformation described by six numbers a,b,c,d,e,f is applied to two points $P_1 = (1, 0)$ and $P_2 = (0, 1)$.

to obtain r and ϕ. Similar formulas hold for s and ψ. In this way it is easier to discuss reductions, rotations and reflections. For example:

- $s = r$, $0 \leq r < 1$, and $\psi = \phi$ fixes a mapping which reduces by a factor of r and simultaneously rotates by the angle ϕ counterclockwise (the mapping is just a reduction, if $\phi = 0$).
- $s = r$, $0 \leq r < 1$, $\phi = \pi$ and $\psi = 0$ fixes a mapping which reduces by a factor of r and simultaneously reflects with respect to the y-axis.
- $r = a$, and $s = b$, $0 \leq a < 1$, $0 \leq b < 1$, and $\phi = \psi = 0$ fixes a mapping which reduces by a factor of a in x-direction and by a factor of b in y-direction.

Affine linear mappings are simply the composition of a linear mapping together with a translation. In other words, if F is linear and Q is a point, then the new mapping $w(P) = F(P) + Q$, where P is any point in the plane, is said to be affine linear. Affine linear mappings allow us to describe contractions which involve positioning in the plane (i.e., the translation by Q). Since F is given by a matrix and Q is given by a pair of coordinates, say (e, f), an affine linear mapping w is given by six numbers,

$$\left(\begin{array}{cc|c} a & b & e \\ c & d & f \end{array}\right)$$

and if $P = (x, y)$ and $w(P) = (u, v)$ then

$$\begin{aligned} u &= ax + by + e, \\ v &= cx + dy + f. \end{aligned}$$

Another notation for the same equations is also sometimes used in this text,

$$w(x, y) = (ax + by + e, cx + dy + f).$$

In the discussion of iterated function systems, it is crucial to study the objects which are left invariant under iteration of an IFS. Now, given an affine linear mapping w, one can ask which points are left invariant under w? This is an exercise with a system of linear equations. Indeed, $w(P) = P$ means

$$\begin{aligned} x &= ax + by + e, \\ y &= cx + dy + f. \end{aligned}$$

Solving that system of equations yields exactly one solution, as long as the determinant $(a-1)(d-1) - bc \neq 0$. This point $P = (x, y)$ is called the fixed point of w. Its coordinates are

$$\begin{aligned} x &= \frac{-e(d-1) + bf}{(a-1)(d-1) - bc}, \\ y &= \frac{-f(a-1) + ce}{(a-1)(d-1) - bc}. \end{aligned}$$

The First Step: Blueprint of MRCM

Already the first application of the MRCM to a given image will usually reveal its internal affine linear contractions. This could be called the *blueprint* of the machine. Note, that it is necessary to select an initial image with sufficient structure in order to uniquely identify the transformations. Otherwise

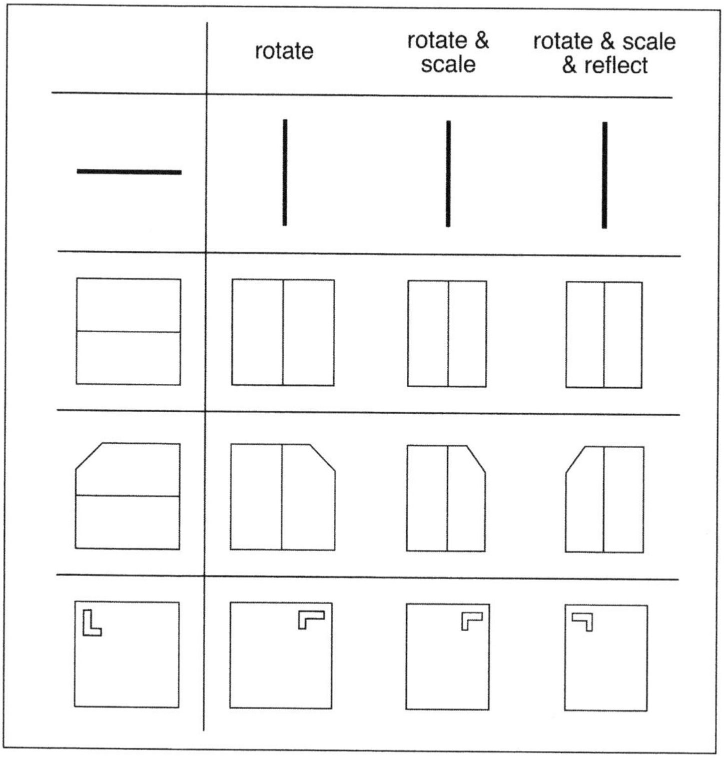

rotate | rotate & scale | rotate & scale & reflect

Unfolding the Blueprint

We consider three transformations (see column headings) and four initial images (left column). The first two images obviously are not suitable to fully unfold the blueprint of the machine. They cannot detect the reflection in the last transformation.

Figure 5.6

one cannot safely detect some of the possible rotations and reflections. Figure 5.6 illustrates this problem with three transformations. In the following images in this chapter we typically use a unit square $[0,1] \times [0,1]$ with an inscribed letter 'L' in the top left corner as an initial image to unfold the blueprint.

The lens systems of an MRCM are described by a set of affine transformations $w_1, w_2, \ldots, w_N$. For a given initial image A, small affine copies $w_1(A), w_2(A), \ldots, w_N(A)$ are produced. Finally, the machine overlays all these copies into one new image, the output $W(A)$ of the machine:

$$W(A) = w_1(A) \cup w_2(A) \cup \cdots \cup w_N(A).$$

Iterated Function System

W is called the *Hutchinson operator*. Running the MRCM in feedback mode thus corresponds to iterating the operator W. This is the essence of a deterministic iterated function system (IFS). Starting with some initial image A_0, we obtain $A_1 = W(A_0)$, $A_2 = W(A_1)$, and so on. Figures 5.7 and 5.8 show the MRCM feedback system and its blueprint for the Sierpinski gasket (3 transformations).

MRCM Feedback System

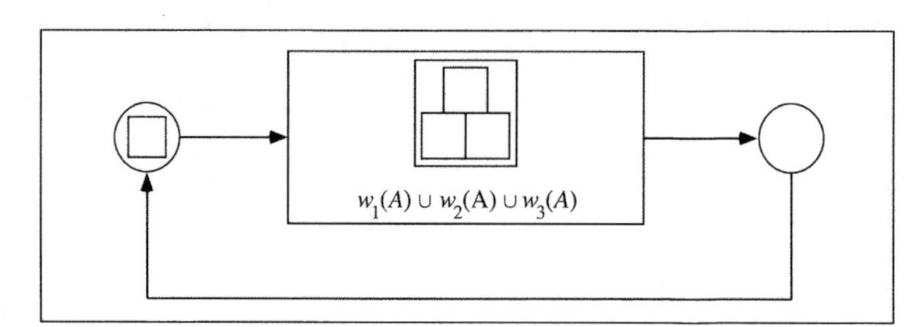

Figure 5.7

First MRCM Blueprint

Blueprint of an MRCM using a unit square with an inscribed letter 'L' in the top left corner as an initial image. The purpose of the outline of the initial image on the output on the right is to allow the identification of the relative positioning of the images.

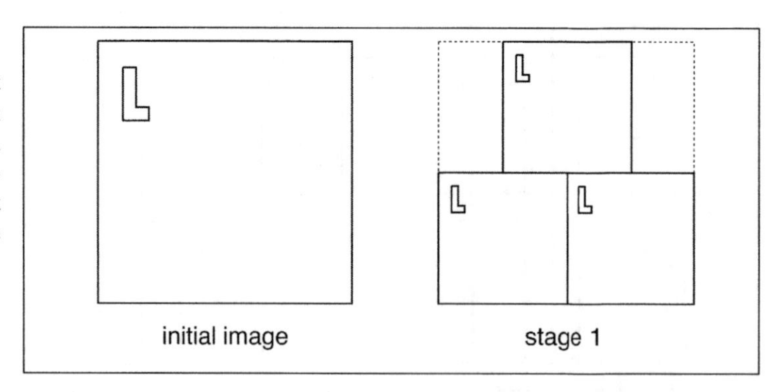

Figure 5.8

IFS and the Hutchinson Operator

Let $w_1, w_2, \ldots, w_N$ be N contractions of the plane (we will carefully discuss this term a little bit later). Now we define a new mapping — the Hutchinson operator — as follows: let A be any subset of the plane.[6] Here we think of A as an image. Then the collage obtained by applying the N contractions to A and assembling the results can be expressed by the collage mapping:

$$W(A) = w_1(A) \cup w_2(A) \cup \cdots \cup w_N(A). \quad (5.1)$$

The Hutchinson operator turns the repeated application of the metaphoric MRCM into a dynamical system: an IFS. Let A_0 be an initial set (image). Then we obtain

$$A_{k+1} = W(A_k), k = 0, 1, 2, \ldots,$$

a sequence of sets (images), by repeatedly applying W. An IFS generates a sequence which tends towards a final image A_∞, which we call the attractor of the IFS (or MRCM), and which is left invariant by the IFS. In terms of W this means that

$$W(A_\infty) = A_\infty.$$

We say that A_∞ is a fixed point of W. Now how do we express that A_n tends towards A_∞? How can we make the term contraction precise? Is A_∞ a unique attractor? We will find answers to these questions in this chapter.

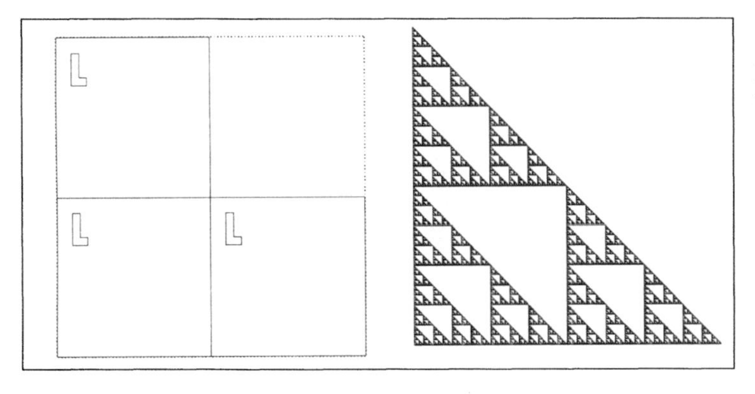

Sierpinski Gasket Variation

IFS with three similarity transformations with scaling factor 1/2.

Figure 5.9

What happens if we change the transformations or, in other words, if we play with the dials of the machine (i.e., if one changes the number of lenses, or changes their contraction properties, or assembles the individually contracted images in a different configuration)? In the following figures we show the

[6] Being more mathematically technical, we allow A to be any compact set in the plane. Compactness means that A is bounded and that A contains all its limit points, i.e., for any sequence of points from A with a cluster point, we have that the cluster point also belongs to A. The open unit disk of all points in the plane with a distance less than 1 from the origin is not a compact set, but the closed unit disk of all points with a distance not exceeding 1 is compact.

The Twin Christmas Tree

Another IFS with three similarity transformations with scaling factor 1/2.

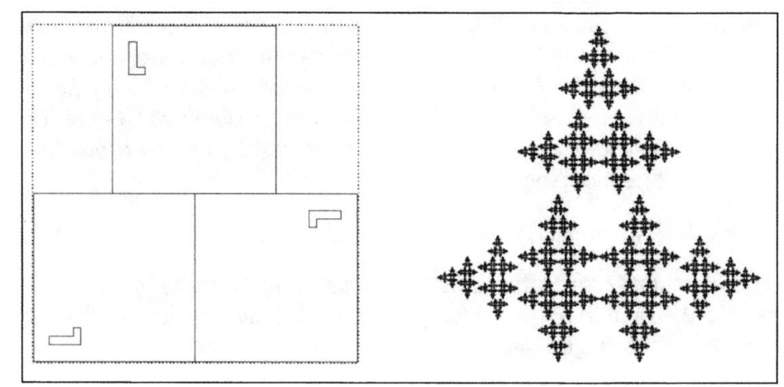

Figure 5.10

A Dragon With Threefold Symmetry

The white lines are inserted only to show that the figure can be made up from three parts similar to the whole.

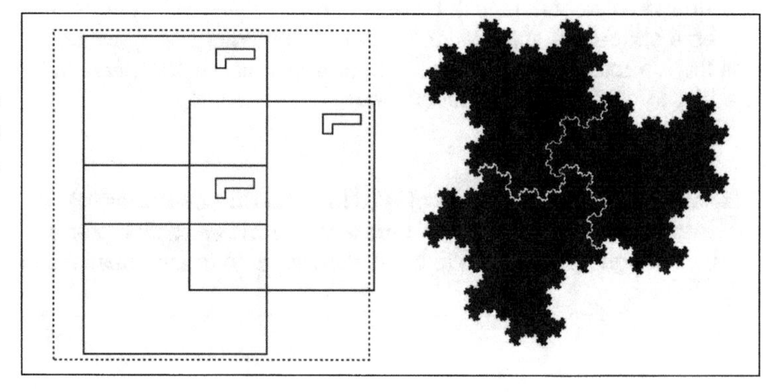

Figure 5.11

results of some IFSs with different settings: the blueprint and the attractor. The blueprint is represented in a single drawing: the dotted square is for the initial image, and the solid-line polygons represent the contractions.

Our first example is a small modification of the IFS which generates the Sierpinski gasket (see figure 5.9). It consists of three transformations, each of which scales by a factor of 1/2 and translates as shown in the blueprint.

We are tempted to conjecture that all IFSs of three transformations that scale by 1/2 produce something very similar to the Sierpinski gasket. But this is far from the truth. In figure 5.10 we try another such IFS which differs from the original one for the Sierpinski gasket only by the addition of rotations. The lower right transformation rotates 90 degrees clockwise, while the lower left rotates by 90 degrees counter-clockwise. The result, called the *twin Christmas tree*, is clearly different from the Sierpinski gasket.

Now we start to also change the scaling factors of the transformations. In figure 5.11 we have chosen the factor of $s = 1/\sqrt{3}$ for all three transformations. Moreover, a clockwise rotation by 90 degrees is also included in each transformation. The result is a two-dimensional object with a fractal boundary: a type of dragon with threefold symmetry. It is invariant under a rotation of 120 degrees. It would be a good exercise at this point to compute

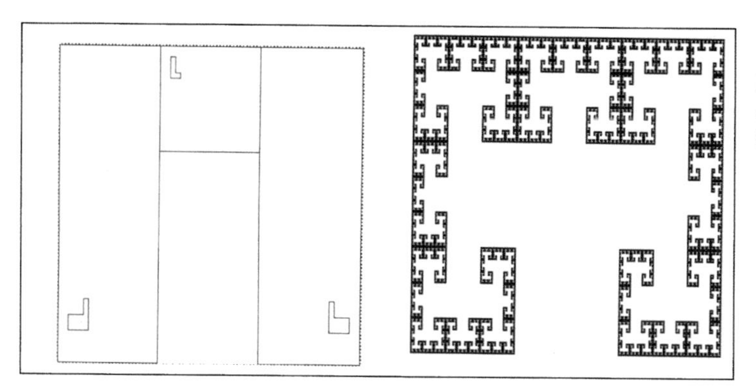

The Cantor Maze

IFS with three transformations, one of which is a similarity. The attractor is related to the Cantor set.

Figure 5.12

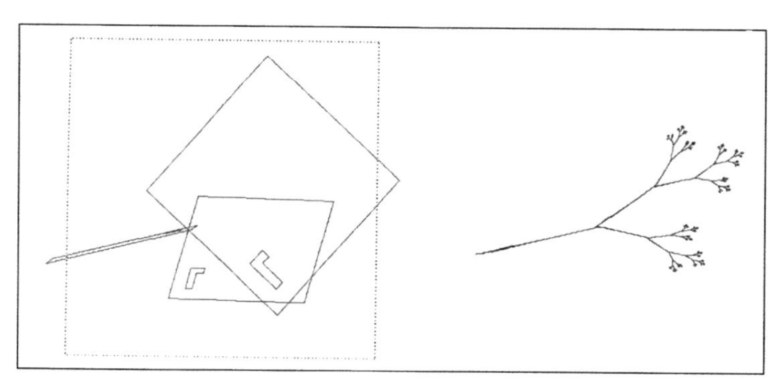

IFS for a Twig

IFS with three affine transformations (only one similarity).

Figure 5.13

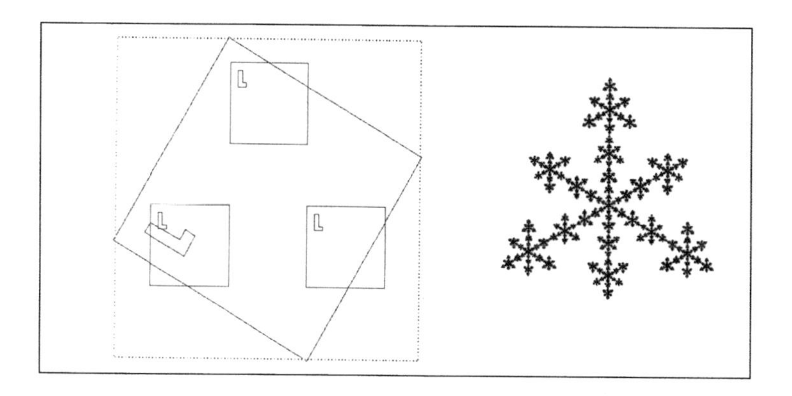

Crystal with Four Similarity Transformations

Figure 5.14

the self-similarity dimension of the attractor using the techniques from the last chapter.

So far we have made use of only similarity transformations. In figure 5.12 there is only one similarity (which scales by 1/3) and two other transformations, which are rotations followed by a horizontal scaling by 1/3 and a reflection in one of the two cases. The result is a sort of maze, for which we have

Crystal with Five Transformations

IFS with five similarity transformations. Can you see Koch curves in the attractor?

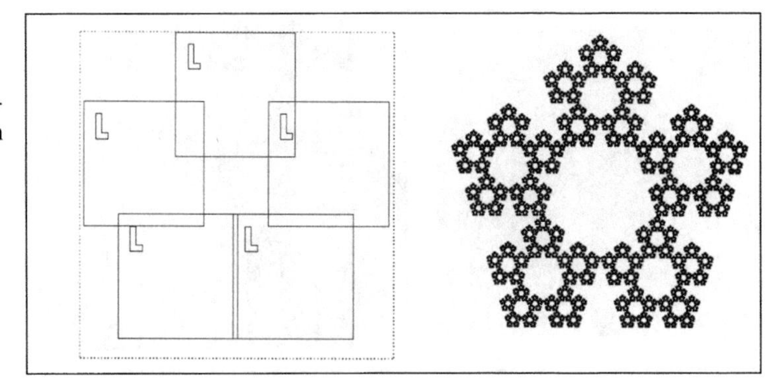

Figure 5.15

A Tree

The attractor of an MRCM with five transformations can even resemble the image of a tree (the attractor is shown twice as large as the blueprint indicates).

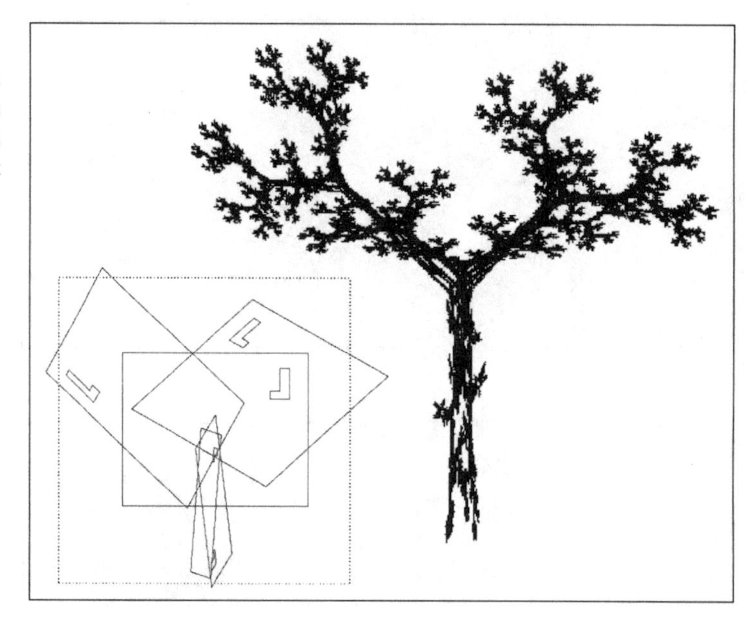

Figure 5.16

reason to introduce the name *Cantor maze*. The Cantor set is woven into the construction in all its details; all points of the cross product of two Cantor sets are connected in a systematic fashion.

Here is our last example of an MRCM with only three transformations (see figure 5.13). The transformations involve rotations; some have different horizontal and vertical scaling factors; and one involves even a shear. What we get is very familiar: a nice twig.

We continue with two examples with more than three transformations (figures 5.14 and 5.15). All transformations are similarities with only scaling and translation. Only one transformation in figure 5.14 includes an additional rotation. These amazingly simple constructions already reveal quite complex

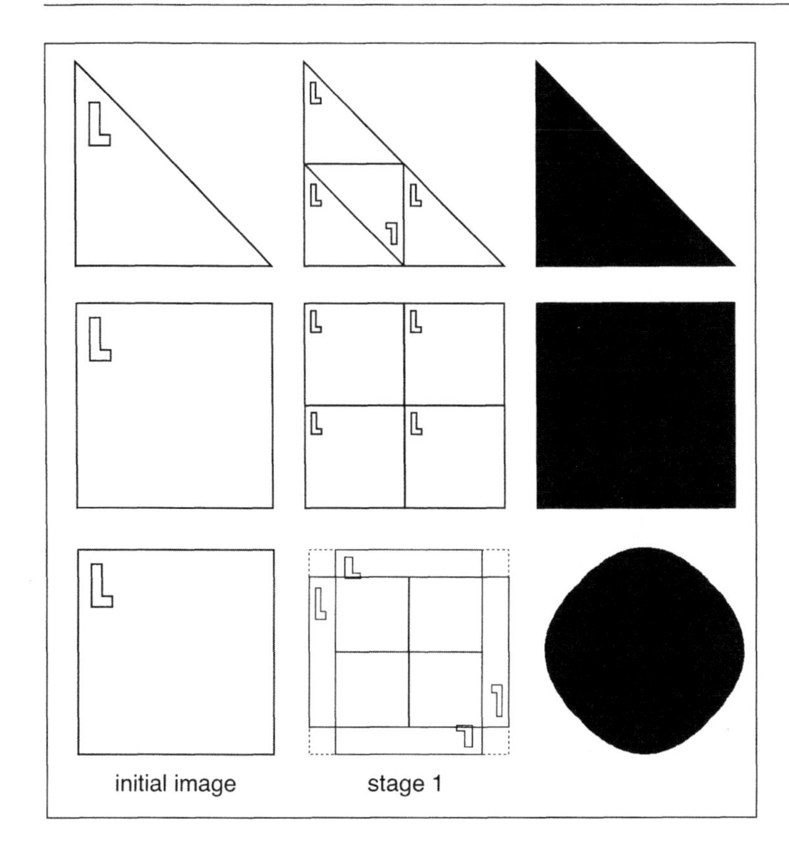

IFS Encodings of Triangle, Square, and Circle

Figure 5.17

and beautiful structures reminiscent of ice crystals.

Finally, let us close our little gallery by a surprisingly realistic drawing of a tree. Can you believe that even this image is a simple IFS attractor? In fact, it is encoded by just five affine transformations (see figure 5.16). This example convincingly demonstrates the capabilities of IFSs in drawing fractal images.

Fractal Geometry Extends Classical Geometry

Given an arbitrarily designed MRCM, what is the final image (its attractor) which it will generate? Will it always be a fractal? Certainly not. Many objects of classical geometry can be obtained as attractors of IFSs as well. But often this way of representation is neither more enlightening nor simpler than the classical description. We illustrate in figure 5.17 how the areas of a square and a triangle can be obtained as IFS attractors. Representations of a plain circle, however, remain somewhat unsatisfactory using IFSs; only approximations are possible.

5.3 Relatives of the Sierpinski Gasket

We have seen already quite impressively how rich and varied the patterns and structures are that can be obtained by MRCMs. In this section we want to explore some close relatives of the Sierpinski gasket or rather of the skewed variation of the gasket shown in figure 5.9. What do we mean by *relatives*? The blueprint of the Sierpinski gasket was given by three contractions reducing an initial square as laid out in figure 5.18.

Blueprint for Relatives

The blueprint determines an MRCM only up to the eight symmetry transformations of a square.

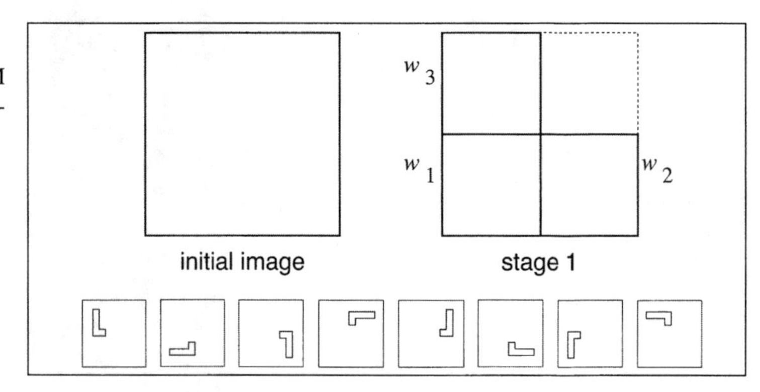

Figure 5.18

There are several possibilities to transform a square into a square by a linear transformation which involve rotations and reflections. Our blueprint is not specific in this respect. In other words, it describes a whole family of MRCMs. Each choice determines an MRCM of this family. So far we have only seen the one producing the Sierpinski gasket. Before we introduce the other members let us define a kind of alphabet which enables us to give names to the different family members. First we set

$$\begin{aligned} v_1(x,y) &= (x/2, y/2), \\ v_2(x,y) &= ((x+1)/2, y/2), \\ v_3(x,y) &= (x/2, (y+1)/2). \end{aligned}$$

These are three contractions which reduce an initial square by a factor of $1/2$ and position the resulting square appropriately. Note, that the choice $w_1 = v_1$, $w_2 = v_2$, and $w_3 = v_3$ provides the MRCM we already have seen. Next we specify the eight symmetry transformations of a square, i.e., the four rotations $d_0, \ldots, d_3$ and the four reflections $d_4, \ldots, d_7$. For example, d_1 is the counterclockwise rotation by 90 degrees, $d_2 = d_1^2$ is the rotation by 180 degrees, d_4 is the horizontal reflection, and d_6 is the reflection about the diagonal. Figure 5.19 provides the definitions.

The Symmetries of a Square

The eight symmetries of a square $d_0, \ldots, d_7$ form an elementary example of a finite group G. Being a group means that there is a composition '$\circ$' of its elements such $d_k \circ d_l \in G$ (d_k follows d_l) for all pairs d_k, d_l which satisfies:

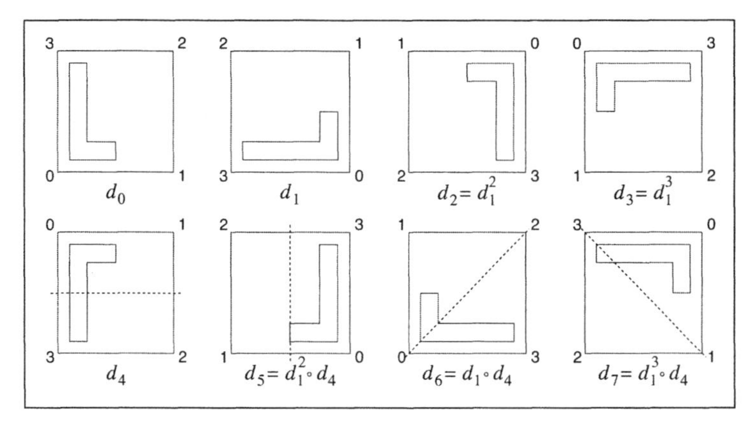

Symmetry Transformations of the Square

The result of the transformations d_0 through d_7 applied to a square with labeled vertices and inscribed 'L'.

Figure 5.19

(1) There is a neutral element $e \in G$ such that $d_k \circ e = d_k$, and $e \circ d_k = d_k$, for any $d_k \in G$.

(2) For any $d_k \in G$ there is an inverse element $d_l \in G$ such that $d_k d_l = e$.

In our example the neutral element is $e = d_0$. The composition is the usual composition of transformations. The composition table establishes the group structure.

		k							
		0	1	2	3	4	5	6	7
	0	0	1	2	3	4	5	6	7
	1	1	2	3	0	7	6	4	5
	2	2	3	0	1	5	4	7	6
l	3	3	0	1	2	6	7	5	4
	4	4	6	5	7	0	2	1	3
	5	5	7	4	6	2	0	3	1
	6	6	5	7	4	3	1	0	2
	7	7	4	6	5	1	3	2	0

Table 5.20 : The results of the composition $d_k \circ d_l$, e.g., $d_4 \circ d_3 = d_6$.

There is another useful way of looking at the transformation which is revealed when labelling the vertices of the square (counterclockwise) from 0 to 3. Then a symmetry transformation is given by a permutation of the four elements. The group G contains a subgroup given by the four rotations. Since d_2 and d_3 can be expressed as compositions of d_1 this subgroup is called cyclic. Note that the elements d_5, d_6, and d_7 can be expressed by the compositions of d_1 and d_4 alone (see figure 5.19).

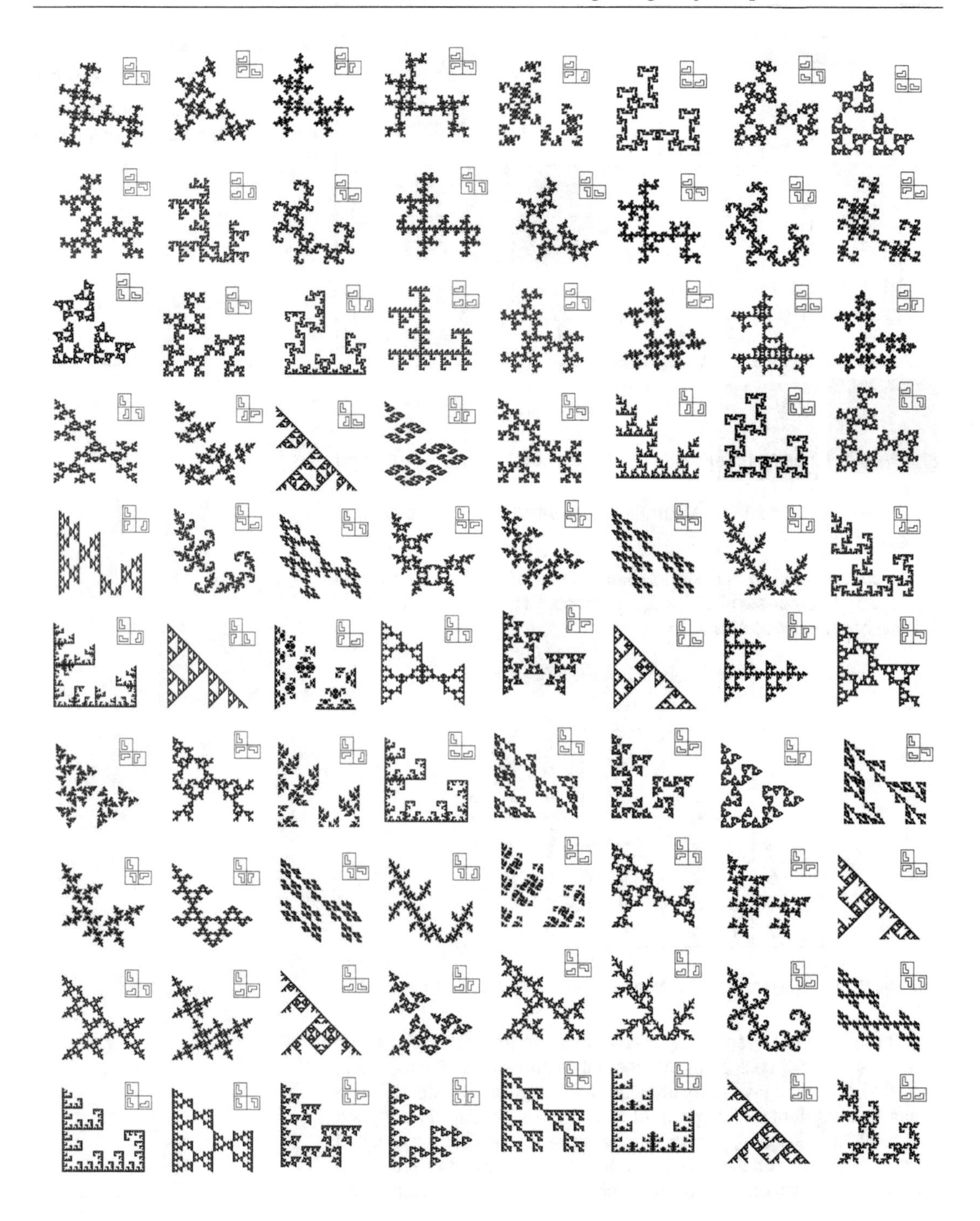

Figure 5.21 : The first 80 variations for MRCMs with blueprint 5.18.

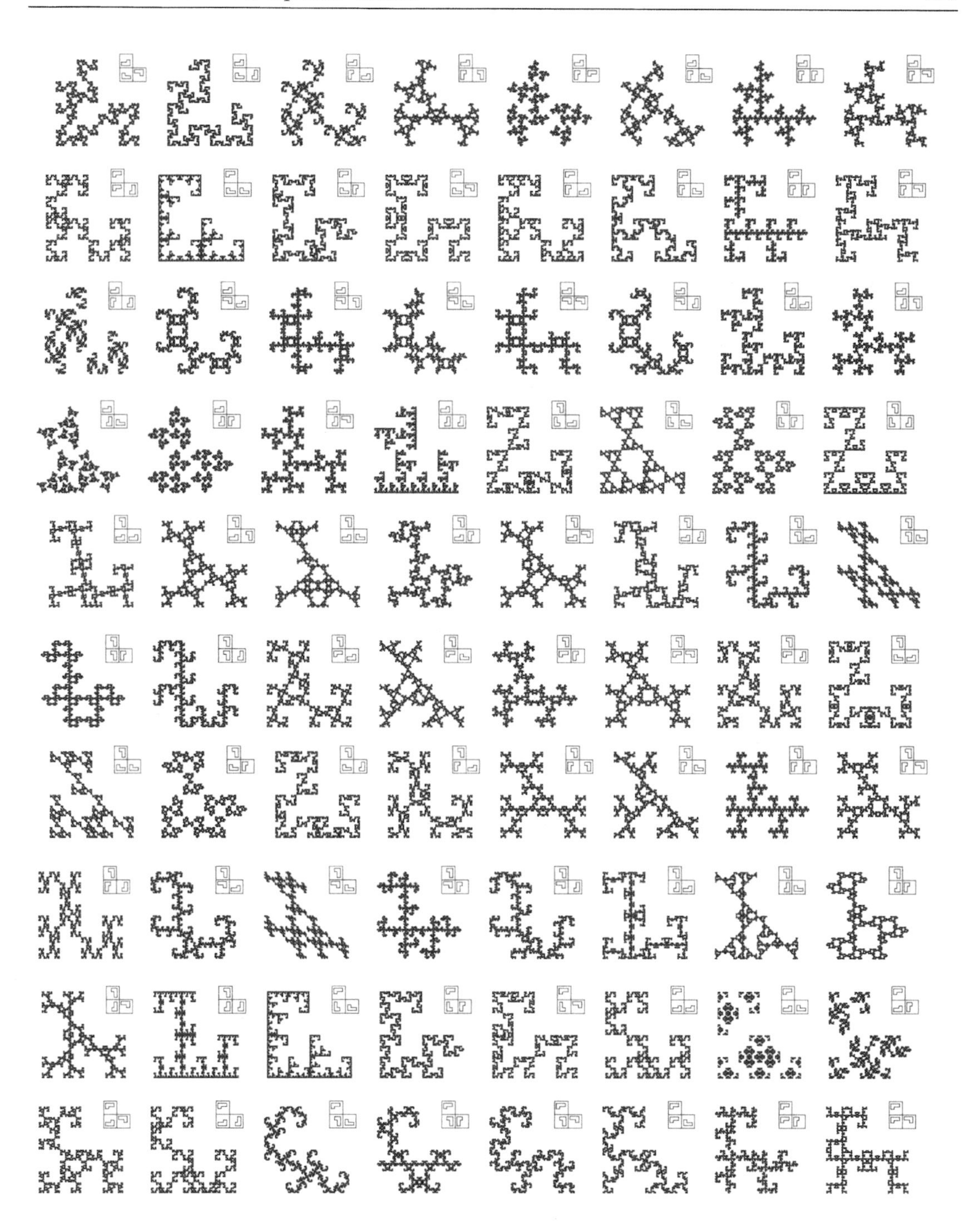

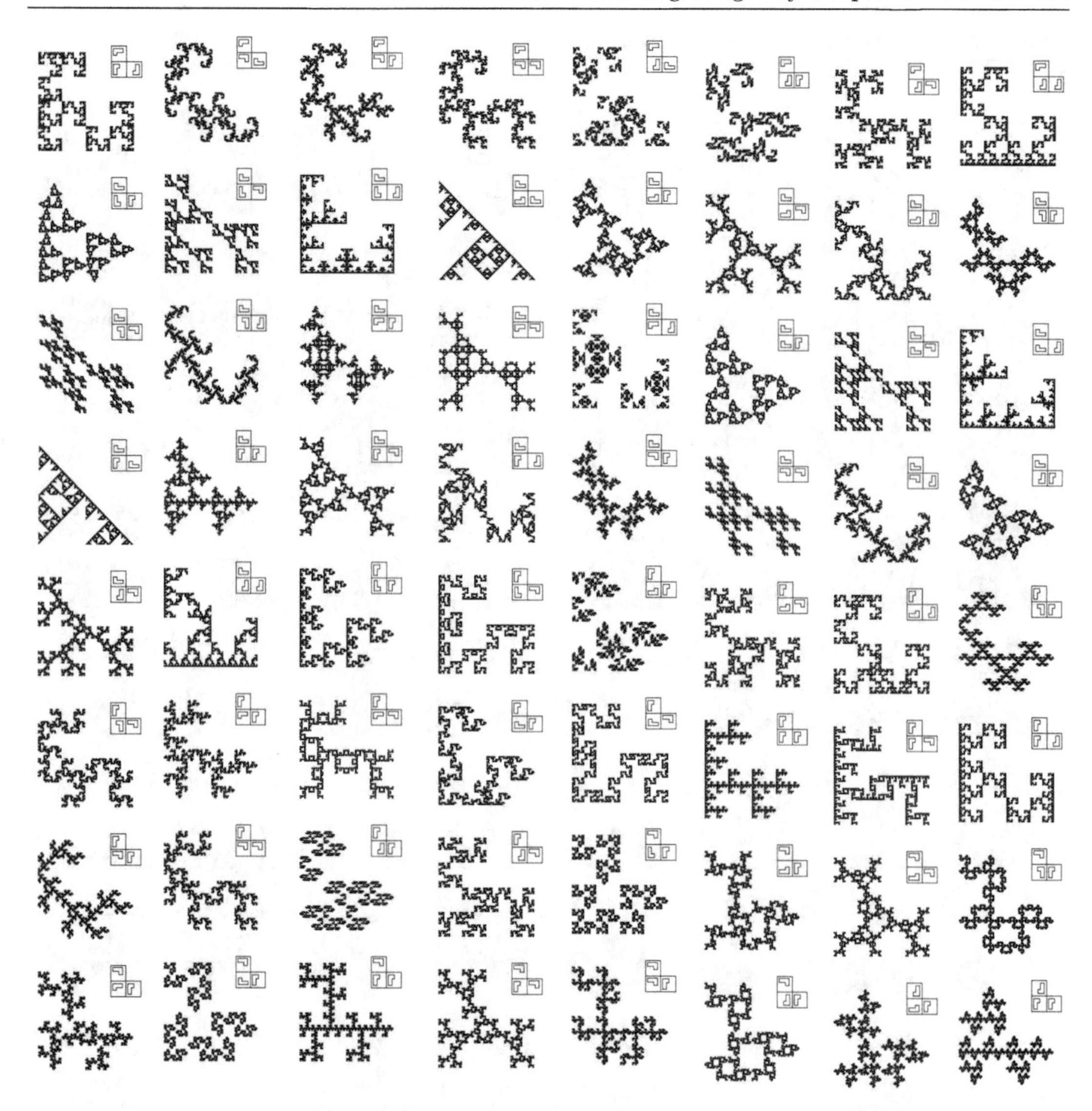

Figure 5.23 : The last 64 variations for MRCMs with blueprint 5.18.

Now we can describe the alphabet for our family. A family member is specified by a triplet w_1, w_2, w_3, where each w_i is given by

$$w_i = v_i d_k,$$

where $k = 0, 1, \ldots, 7$ and $i = 1, 2, 3$. In other words, there are eight choices of d_k's for each w_i, which makes altogether $8^3 = 512$ different triplets w_1, w_2, w_3, each one describing a specific MRCM.

Let us now look at the family picture of all 512 MRCMs. Figures 5.21–5.23 show 224 of these close relatives of the Sierpinski gasket. Where are the remaining images? First note that none of the configurations are symmetric with respect to the diagonal. Therefore, if A_∞ represents one of the images; then the image which is obtained by reflection at the diagonal, i.e., $d_6(A_\infty)$, should be another member of the family of the 512. Indeed, if the triplet $v_1 d_r, v_2 d_s, v_3 d_t$ generates A_∞, then $v_1 d'_r, v_2 d'_t, v_3 d'_s$ (where $d'_k = d_6 d_k d_6$) is the corresponding triplet of contractions that generates $d_6(A_\infty)$.[7] Figure 5.24 shows an example of such a pair of twins. This makes $2 \times 224 = 448$ nonsymmetric images.

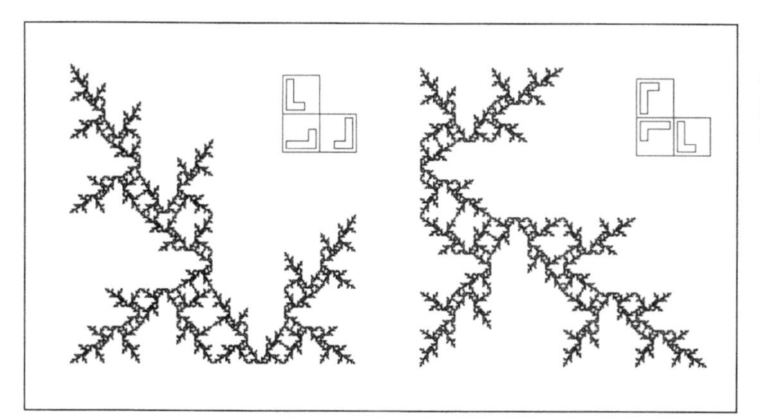

Symmetric Counterparts

One example chosen from the 224 images of figures 5.21–5.23 and its symmetric counterpart.

Figure 5.24

Where are the remaining 64 images? It turns out that there are eight more images which come in multiplicities of eight and which are symmetric with respect to the diagonal. Figure 5.25 shows these eight images together with their equivalent MRCMs.

Let us explain why there are exactly eight symmetric images which occur with multiplicity eight. Our first observation is that each of the images under consideration has to be symmetric with respect to the diagonal. That is, if A_∞ is one of them then $d_6(A_\infty) = A_\infty$. Figure 5.26 shows an image (top) that has this symmetry. The key of understanding the multiplicity is in monitoring the black subsquare of the upper right corner.

First note, that for w_1 we can only choose $\{v_1 d_0, v_1 d_6\}$ and $\{v_1 d_2, v_1 d_7\}$. Other transformations $v_1 d_k$ would turn the black square off the diagonal and thus break the symmetry. Finally we need to discuss which choices are admissible for w_2 and w_3. Clearly, once we choose w_2, then we must choose w_3 suitably in order to preserve the symmetry (see figure 5.27). There are four positions of the black square which are specified by the four pairs of symmetry transformations:

$$\{d_0, d_6\},\ \{d_1, d_5\},\ \{d_2, d_7\},\ \{d_3, d_4\}.$$

[7]Note that $A_\infty = v_1 d_r(A_\infty) \cup v_2 d_s(A_\infty) \cup v_3 d_t(A_\infty)$ implies $d_6(A_\infty) = v_1 d_6 d_r(A_\infty) \cup v_2 d_6 d_t(A_\infty) \cup v_3 d_6 d_s(A_\infty) = v_1 d_6 d_r d_6(d_6(A_\infty)) \cup v_2 d_6 d_t d_6(d_6(A_\infty)) \cup v_3 d_6 d_s d_6(d_6(A_\infty))$.

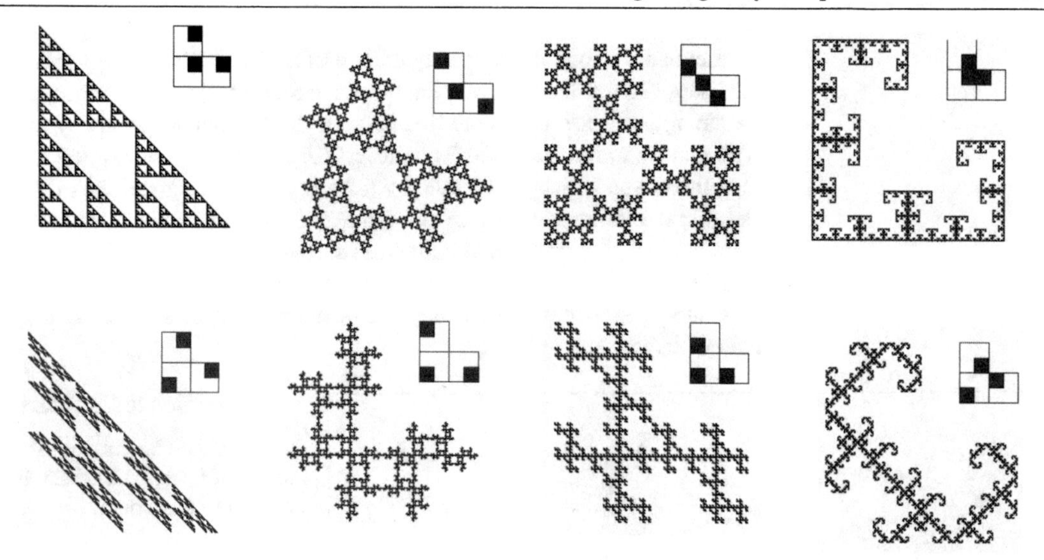

Figure 5.25 : There are eight different symmetric attractors. Each one can be encoded by eight different sets of transformations.

Admissible Transformations

A symmetric image (top) is transformed by w_1. The two left configurations yield two pairs of choices for w_1, namely, $d_0 v_1, d_6 v_1$ and $d_2 v_1, d_7 v_1$. The right two choices yield configurations which are not symmetric with respect to the diagonal.

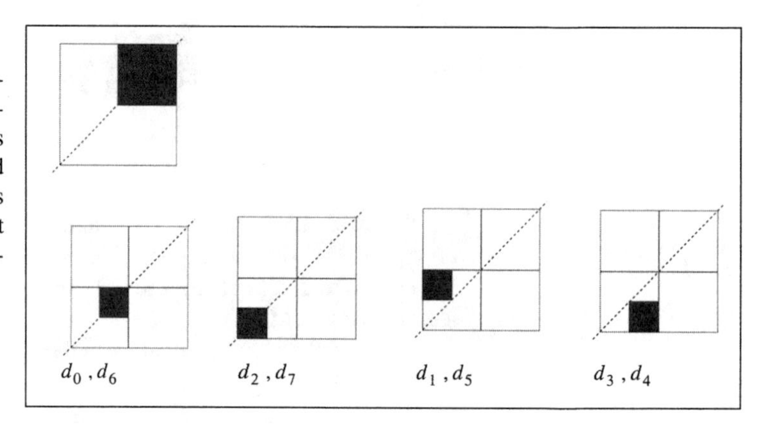

Figure 5.26

Thus if we pick $w_2 = v_2 \circ d_k$ with d_k from one of these pairs, say $d_k \in \{d_r, d_s\}$, then we have to choose $w_3 = v_3 d_6 d_r$ or $w_3 = v_3 d_6 d_s$. Using the composition table we find the admissible pairs for w_2 and w_3, which are illustrated in figure 5.27:

$$\begin{aligned}
w_2 &= v_2 \circ d_k, \ d_k \in \{d_0, d_6\}, \ w_3 = v_3 \circ d_l, \ d_l \in \{d_0, d_6\}, \\
w_2 &= v_2 \circ d_k, \ d_k \in \{d_1, d_5\}, \ w_3 = v_3 \circ d_l, \ d_l \in \{d_3, d_4\}, \\
w_2 &= v_2 \circ d_k, \ d_k \in \{d_2, d_7\}, \ w_3 = v_3 \circ d_l, \ d_l \in \{d_2, d_7\}, \\
w_2 &= v_2 \circ d_k, \ d_k \in \{d_3, d_4\}, \ w_3 = v_3 \circ d_l, \ d_l \in \{d_1, d_5\}.
\end{aligned}$$

In summary, we have $2 \times 2 \times 2$ choices for each configuration in figure 5.27 and that makes $8 \times 8 = 64$ different MRCMs with multiplicity eight.

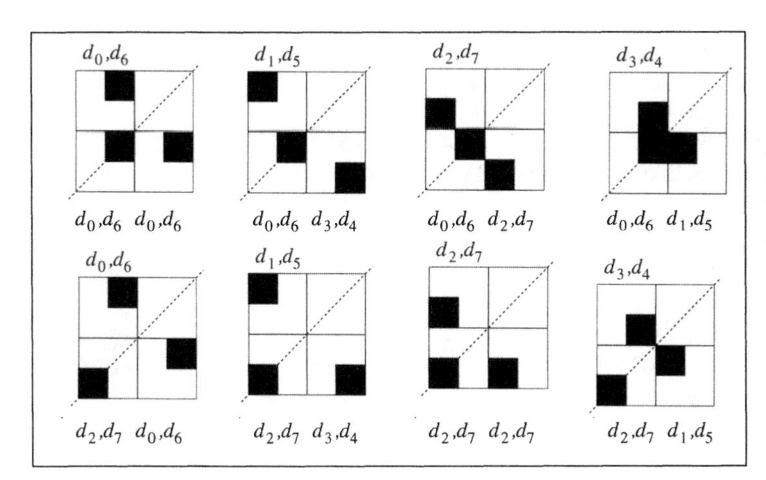

Transformations for Symmetric Attractors

These 64 MRCMs (each icon represents eight MRCMs) generate the eight symmetric images of our family. The order of the icons corresponds to the images in figure 5.25.

Figure 5.27

Analyzing the $224+8=232$ different images we find an amazing variety of patterns and forms, all of which are close relatives of the Sierpinski gasket. Some are closer than others. For some of them it is hard to believe that they actually are the result of such a simple MRCM or an MRCM at all. The mathematical properties in the family are quite interesting as well. Some of the images are connected (one piece); others are not. Those which are not connected are in fact totally disconnected (something like a Cantor set). Those which are connected again split into two classes. One class is the type of patterns which are simply connected (no holes) and the other consists of patterns with infinitely many holes like the Sierpinski gasket itself. Figure 5.28 shows examples of these three cases.

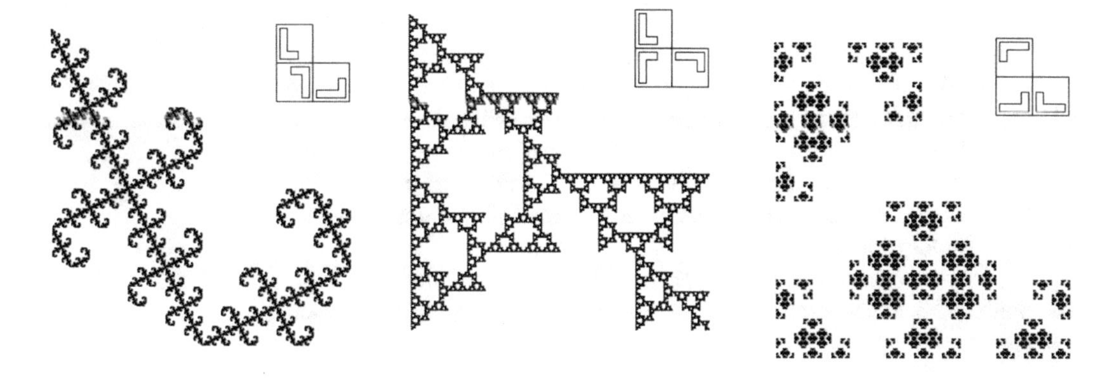

Figure 5.28 : Our family can be divided into three cases: simply connected, connected (but not simply connected), totally disconnected.

5.4 Classical Fractals by IFSs

The concept of Iterated Function Systems allows us to make the construction of classical fractals much more transparent. They can be obtained as attractors of appropriate IFSs. In other words, the question of their existence as discussed in chapter 3 (we discussed the problem in detail for the Koch curve) can finally be settled by showing that for a given IFS there is a unique attractor. This will be done in the course of this chapter. But IFSs also allow us to better understand the number theoretical characterizations of some classical fractals like the Cantor set or the Sierpinski gasket.

Cantor Set

You will recall the characterization of the Cantor set by ternaries: it is the set of points of the unit interval which have a triadic expansion that does not contain the digit 1 (see chapter 2). Now we look at an IFS with

$$w_0(x) = \frac{1}{3}x, \quad w_1(x) = \frac{1}{3}x + \frac{1}{3}, \quad w_2(x) = \frac{1}{3}x + \frac{2}{3}.$$

Note, that this machine operates only on one variable (i.e., not in the plane).

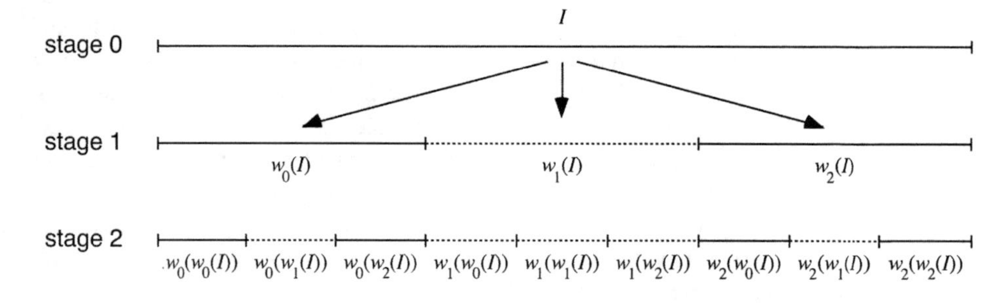

Figure 5.29 : First iteration stages of the triadic IFS. If w_1 is left out, the Cantor set is generated as the attractor.

Figure 5.29 shows the first stages of its iteration (using the unit interval as initial image). The attractor of this machine is clearly the unit interval (again and again, the unit interval is simply transformed into the unit interval). But what would happen if we used only the two transformations w_0 and w_2? In this case we obviously would obtain the Cantor set as the attractor (the iteration would correspond to the classical construction steps of the Cantor set: again and again middle thirds would be left out).

Now observe that w_1 transforms the unit interval to the interval $[\frac{1}{3}, \frac{2}{3}]$, i.e., points with triadic expansion from 0.1 to $0.1222\ldots = 0.1\overline{2}$. In fact, whenever w_1 is involved in the iteration of the IFS this leads to points with an expansion that contains the digit 1. In other words, leaving out everything which comes from w_1 just amounts to the ternary description of the Cantor set.

Sierpinski Gasket

Let us now turn to the Sierpinski gasket (or to its variation as already shown in figure 5.9). Now we look at the IFS that is given by four similarity transformations transforming the unit square Q into its four congruent subsquares (see figure 5.30).

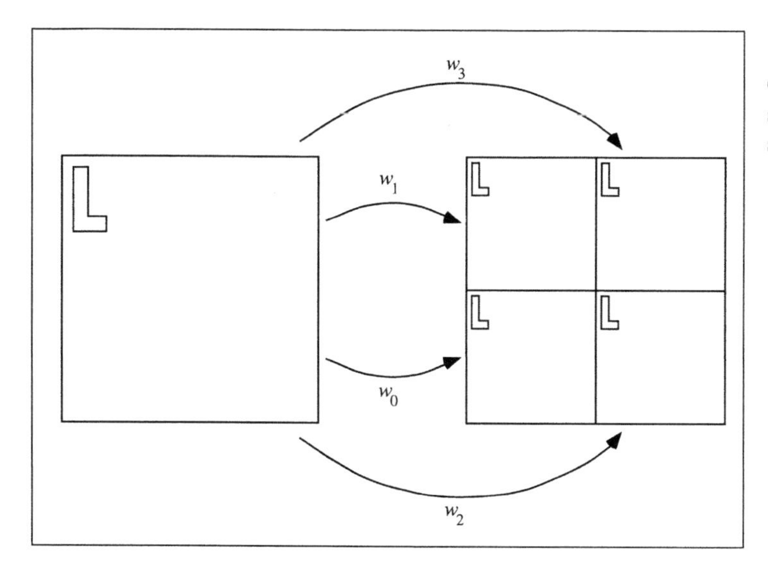

Four Contractions

Contractions transforming the unit square into its four congruent subsquares.

Figure 5.30

	0	1
1	w_{01}	w_{11}
0	w_{00}	w_{10}

	00	01	10	11
11	$w_{01}(w_{01})$	$w_{01}(w_{11})$	$w_{11}(w_{01})$	$w_{11}(w_{11})$
10	$w_{01}(w_{00})$	$w_{01}(w_{10})$	$w_{11}(w_{00})$	$w_{11}(w_{10})$
01	$w_{00}(w_{01})$	$w_{00}(w_{11})$	$w_{10}(w_{01})$	$w_{10}(w_{11})$
00	$w_{00}(w_{00})$	$w_{00}(w_{10})$	$w_{10}(w_{00})$	$w_{10}(w_{10})$

First Stages

The first two stages of the IFS. Observe that the generated subsquares can be identified by a binary coordinate system.

Figure 5.31

It is convenient to label these transformations in binary form (i.e., 00, 01, 10 and 11 instead of 0, 1, 2, 3):

$$\begin{aligned} w_{00}(x,y) &= (\tfrac{1}{2}x, \tfrac{1}{2}y), & w_{01}(x,y) &= (\tfrac{1}{2}x, \tfrac{1}{2}y+\tfrac{1}{2}), \\ w_{10}(x,y) &= (\tfrac{1}{2}x+\tfrac{1}{2}, \tfrac{1}{2}y), & w_{11}(x,y) &= (\tfrac{1}{2}x+\tfrac{1}{2}, \tfrac{1}{2}y+\tfrac{1}{2}). \end{aligned}$$

Using all the four similarity transformations in an IFS will generate the unit square as an attractor. Figure 5.31 shows the first stages of the iteration of this machine. Note that we use a binary coordinate system to identify the subsquares which are generated in each step. The binary coordinate system provides a very convenient way to do bookkeeping.

For example, in the first stage, w_{01} has transformed the unit square Q into the subsquare $w_{01}(Q)$ at (0,1), w_{11} into the subsquare at (1,1), and so on. In the second stage we find, for example, the square at $(10, 11)$ is $w_{11}(w_{01}(Q))$ (i.e., first apply w_{01} to Q and then w_{11} to the result). Here is another example: $w_{10}(w_{00}(w_{11}(Q)))$ would produce the square in the third stage at (101,001). Do you see the labelling system? In the composition $w_{10}(w_{00}(w_{11}(Q)))$, take the first digits from left to right, i.e., 101. This gives the binary x-coordinate of the subsquare. Then take the second digits in the composition from left to right, i.e., 001; this gives the y-coordinate.

We know already that the attractor of the IFS given by w_{00}, w_{01} and w_{10} will be the Sierpinski gasket. In other words, if we leave out everything in the unit square IFS which comes from w_{11}, we will also get the Sierpinski gasket. Now the binary bookkeeping pays off. Given any stage k the 4^k little subsquares are identified by pairs of binary coordinates (with k digits). How can we recover whether w_{11} was involved in the production of a subsquare by the IFS? We just take the two binary coordinates which identify the little square and write them on top of each other, for example, $(100111, 010000)$ and $(100111, 001100)$:

```
100111 100111
010000 001100
  NO     YES
```

If we find the digit 1 simultaneously in corresponding places, then w_{11} was involved, otherwise not. Thus, omitting all these squares step by step will generate the Sierpinski gasket from the unit square.[8] This is in the same spirit as in the ternary description of the Cantor set. Moreover, we note that we have just built the interface to the geometrical patterns in Pascal's triangle, because our omission criterion here is exactly Kummer's number theoretical criterion for even binomial coefficients. We will explore this marvellous relation more in chapter 8.

Sierpinski Carpet

The Sierpinski carpet (see figure 2.56) has a very similar number theoretical description. Just start with the unit square and subdivide into nine congruent squares. For the appropriate IFS we use the transformations which transform the unit square into these subsquares; see figure 5.32 (again no rotations or reflections are allowed).

This time we label the transformations using ternary numbers like $w_{00}, w_{01}, w_{02}, w_{10}, \ldots, w_{22}$. Accordingly, each square in the kth stage is identified by a pair of ternary coordinates (with k digits). In the limit, each point in the unit square is described by a pair of infinite ternary digit strings like $(011201\ldots,\ 210201\ldots)$. Now the Sierpinski carpet is obtained by omitting everything which comes from the transformation w_{11}. This means that we keep only those points in the unit square which admit a description by a pair of ternary numbers without the digit 1, or if the digit 1 appears in one of the coordinates, it must not appear at the same place in the other coordinate. For example, we keep $(11\overline{0}, 00\overline{1})$. Also $(201\overline{2}, 101\overline{0})$ belongs to the carpet,

[8] This explains the binary characterization of the Sierpinski gasket which we have used in chapter 3, page 169, for the discussion of self-similarity.

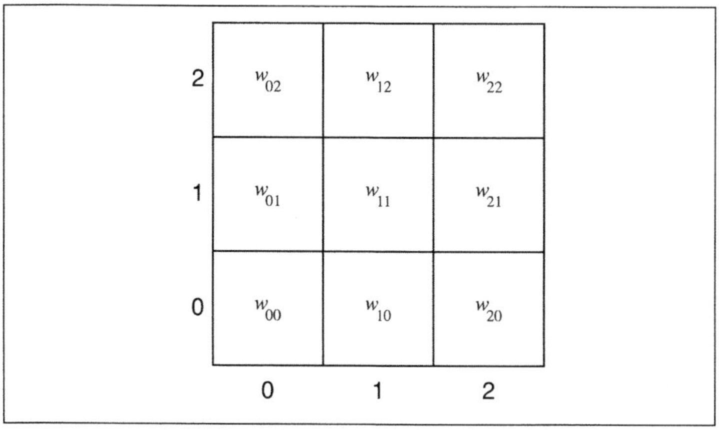

2	w_{02}	w_{12}	w_{22}
1	w_{01}	w_{11}	w_{21}
0	w_{00}	w_{10}	w_{20}
	0	1	2

Nine Contractions

The contractions transform the unit square into nine congruent sub squares that can be conveniently identified by a triadic coordinate system.

Figure 5.32

because it is equal to $(202\bar{0}, 101\bar{0})$. But we omit $(201\bar{0}, 101\bar{0})$, and so on. We remark that in this precise sense the Sierpinski carpet is the logical extension of the Cantor set into the plane.

In this book we have presented a gallery of classical fractals. This gallery has had no essential addition until very recently. B. Mandelbrot opened the doors wide to many new rooms in the gallery and added some potentially eternal masterpieces — like the Mandelbrot set — to it. But there are also two other creations or discoveries which have given current research significant momentum. One is the first *strange attractor* discovered by E. Lorenz at MIT in 1962, and the second is what we would like to call *Barnsley's fern*. The Mandelbrot set, Lorenz attractor, and Barnsley's fern have each opened a new and separate division in the gallery of mathematical monsters. Of all these, Barnsley's fern belongs to the subject of this chapter.

Barnsley was able to encode the image in figure 5.34 with only four lens systems. Figure 5.35 shows the design of his MRCM by means of its application to an initial rectangular image. Note that contraction number 3 involves a reflection. Also, contraction number 4 is obviously not a similarity transformation; it contracts the rectangle to a mere line segment. The attractor which is generated by the IFS will not be self-similar in the precise mathematical

	Translations		Rotations		Scalings	
	e	f	ϕ	ψ	r	s
1	0.0	1.6	-2.5	-2.5	0.85	0.85
2	0.0	1.6	49	49	0.3	0.34
3	0.0	0.44	120	-50	0.3	0.37
4	0.0	0.0	0	0	0.0	0.16

Barnsley Fern Transformations

The angles are given in degrees.

Table 5.33

Barnsley's Fern

Barnsley's fern generated by an MRCM with only four lens systems.

Figure 5.34

meaning of the word. The original transformations are given in table 5.33.[9]

The importance of Barnsley's fern to the development of the subject is that his image looks like a natural fern, but lies in the same mathematical category of constructions as the Sierpinski gasket, the Koch curve, and the Cantor set. In other words, that category not only contains extreme mathematical monsters which seem very distant from nature, but it also includes structures which are related to natural formations and which are obtained by only slight modifications of the monsters. In a sense, the fern is obtained by shaking an

[9]M. F. Barnsley, *Fractal Modelling of Real World Images,* in: *The Science of Fractal Images,* H.-O. Peitgen and D. Saupe (eds.), Springer-Verlag, New York, 1988, page 241.

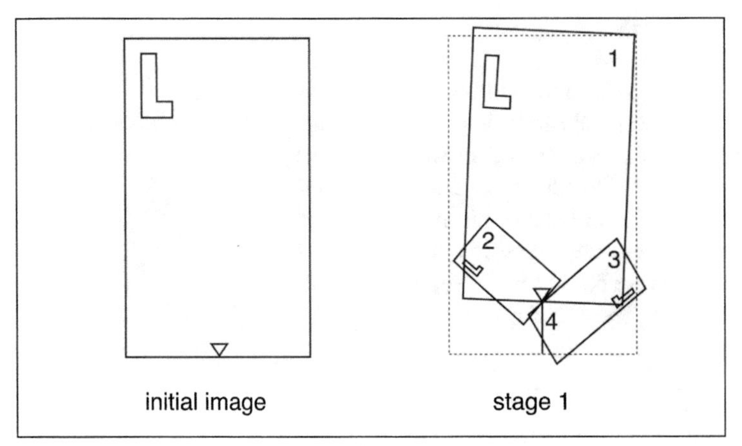

Blueprint of Barnsley's Fern

The small triangle in the initial image and its first copy on the right indicate where the 'stem' of the fern is attached to the rest of the leaf.

Figure 5.35

MRCM which generates the Koch curve so that the lens systems alter their positions and contraction factors (see figure 5.36).

Let us now turn to another aspect of our concept of MRCM. The message which is expressed by the image of the fern is very impressive. Something as complicated and structured as a fern seems to have a lot of information content. But as figure 5.35 demonstrates, the information content from the point of view of IFSs is extremely small. This observation suggests viewing the IFS as a tool for coding and compressing images. In the following section we will discuss some basic ideas. A detailed technical discussion can be found in Fisher's appendix on image compression.

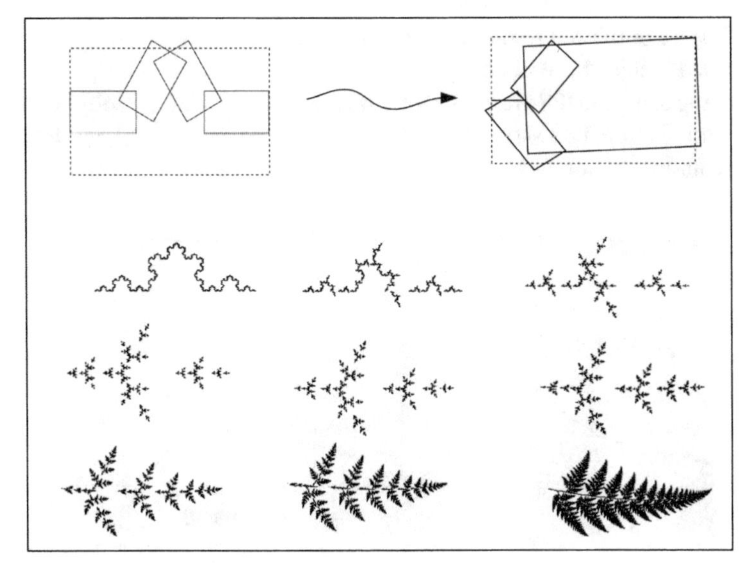

Koch Curve Transformed into the Fern

By changing the parameters of the transformations for the Koch curve continuously to those of the fern, the generated image smoothly transforms from one fractal into the other. The lower nine images of the figure show some intermediate stages of this metamorphosis.

Figure 5.36

5.5 Image Encoding by IFSs

Each of the images in our gallery is obtained by a very simple machine, the blueprint of which is revealed by stage 1 in each experiment. How many images are there which can be generated this way? The answer is obvious — infinitely many. Any number and particular choice of lenses and their position define a new image. In other words, we can think of the blueprint of the MRCM (i.e., the set of transformations which describe the IFS) as the blueprint (or encoding) of an image. In figure 5.37 we have summarized this interpretation using the twig-like structure. The transformations are:

	a	b	c	d	e	f
1	−0.467	0.02	−0.113	0.015	0.4	0.4
2	0.387	0.43	0.43	−0.387	0.256	0.522
3	0.441	−0.091	−0.009	−0.322	0.421	0.505

Let us summarize what we have learned so far. We have introduced a machine, called an MRCM, which is essentially an arrangement of lens systems which contract images. The MRCM generates a dynamical system, an IFS. That is, running the machine in a feedback environment leads to a sequence of images $A_0, A_1, A_2, \ldots$, where A_0 is an arbitrary initial image. The sequence of images will lead to a final image, A_∞, which is independent of the initial image A_0. If we choose A_∞ as the initial image, then nothing happens (i.e., the IFS leaves A_∞ invariant). We say that A_∞ is a *fixed point* of the IFS, or that A_∞ is an *attractor* for the dynamical system. In this sense we can identify the resulting attractor with the IFS. The mathematical description of the lens systems of the machine is given by a set of affine linear transformations, each one specified by six real numbers. We may interpret these data as a coding of the final image A_∞. For the decoding we only need to run the machine with any initial image. Eventually, the coded image A_∞ will emerge.

The Problem of Decoding

However, in some cases the decoding using the IFS presents a serious problem. For example, take Barnsley's fern. Figure 5.38 shows the first stages of the IFS. Obviously, even after 10 iterations we still have a long way to go to

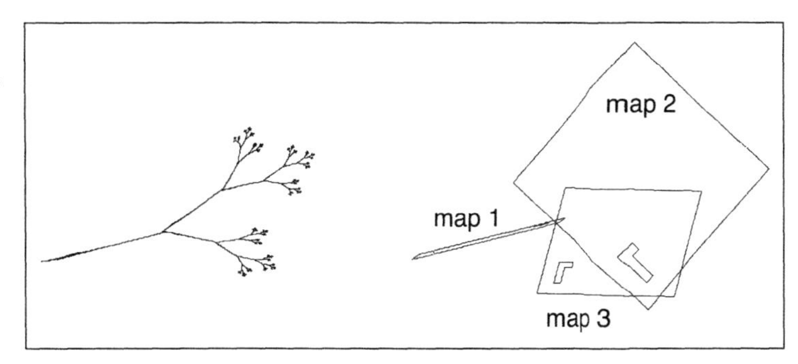

Twig Blueprint

Encoding of a twig by three transformations.

Figure 5.37

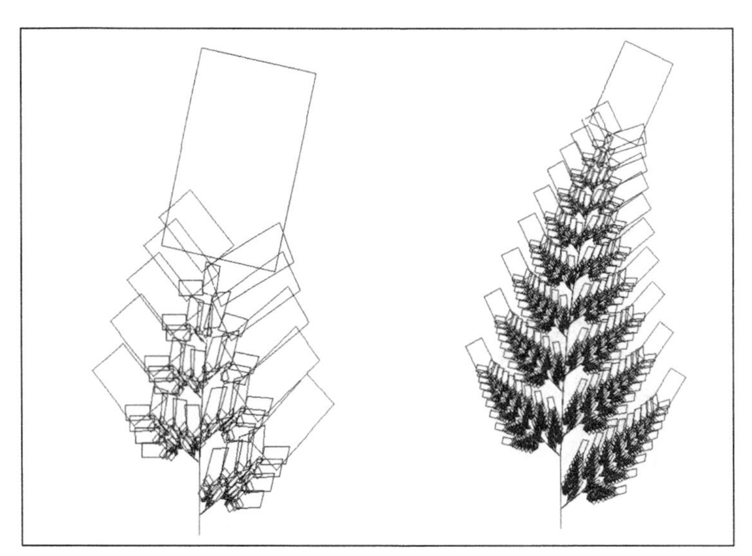

The First Iterates

Stage 5 and stage 10 of the fern copy machine.

Figure 5.38

reach the complete fern. Thus, we are led to the general question: after how many steps can one assume that the final image has been approximated sufficiently well? To answer this we need to clarify what we mean by *sufficiently well*. There are two criteria which seem to be reasonable.

The first would require that two successive iterations change so little that the change is below graphic resolution. This compares very well to computational problems. A solution to a square root calculation, for example, is accepted when the first 10 digits no longer change. The second criterion is more practical and allows an a priori estimate of the number of necessary iterations. This estimate derives from the following worst-case scenario. Recall that the initial image may be completely arbitrary. At this point, however, let us require that it covers the attractor. For example, it could be a sufficiently large square. Since the final image is independent of the initial image, we will not accept a given iteration as an approximation for the final image as long as we still see contracted versions of the initial image in that iteration. This is the case in figure 5.38. It is apparent that even after 10 iterations the dynamical system is still far from the final image, the attractor. The reason is that contraction number 1 (see figure 5.35) reduces by only a factor of 85%. Therefore, in order to reduce the initial rectangle to a size below pixel size — to the point at which the rectangular structure becomes unrecognizable — we have to carry out at least N iterations, where N is estimated in the following way. Assume that the initial rectangle is drawn on a 1000×1000 pixel screen and covers 500×200 pixels. Then N approximately solves the equation

$$500 \cdot 0.85^N = 1.$$

Thus, $N \approx 39$. In a straightforward implementation of the IFS one has to calculate and draw

$$M = 1 + 4 + 4^2 + 4^3 + \cdots + 4^N = \frac{4^{N+1} - 1}{3}$$

rectangles for N iterations. With $N = 39$ we compute the incredibly large number $M \approx 4^{39} \approx 3 \times 10^{23}$. Even if we assume that our computer calculates and draws a whopping million rectangles per second, then to see the final image we would have to wait 3×10^{17} seconds, which is about 10^{10} years, which is a time span of the order of the estimated age of the universe. This gives some flavor of the decoding problem. In chapter 6, however, we will learn a very elementary and powerful decoding method which generates a good approximation of the final image on a computer screen within seconds! We will also modify the above inefficient algorithm to the point where it will produce the fern (and other attractors) with the same precision and in a reasonable time.

Encoding: The Inverse Problem

In order to make use of IFSs for image coding, one first has to solve another crucial problem, namely, to construct a suitable MRCM for a given image. This is the inverse problem; encoding is inverse to decoding. Of course, we cannot expect to be universally able to build an MRCM which produces exactly the given image. However, approximations should be possible. We can make these as close to the original as we desire, as explained next.

Assume we are given a black and white picture, digitized at a resolution of $n \times m$ pixels. This image can be *exactly* reproduced by an MRCM simply by requiring that for every black pixel of the image, there exists a lens which contracts the whole image to that particular pixel. Running the machine just once starting out with any image will produce the prescribed black and white pixel image. Naturally, this is not an efficient way to code an image because for every black pixel we need to store one affine transformation. However, the argument demonstrates that in principle it is possible to achieve approximations of any desired accuracy. Thus, the problem is to find ways to construct a better MRCM which does not need as many transformations but still produces a good approximation. Several difficult questions are raised in this context:

(1) How can the quality of an approximation be assessed? How do we quantify differences between images?
(2) How can we identify suitable transformations?
(3) How can we minimize the necessary number of affine transformations?
(4) What is the appropriate class of images suitable for this approach?

Most of these questions have been intensively studied. The first fully automated fractal image compression algorithm was given in Arnaud Jacquin's Ph.D. thesis in 1989 and later published in a journal.[10] Given an image, the encoder finds a contractive affine image transformation (fractal transform) T such that the fixed point of T is close to the given image. The clou is that

[10] A. E. Jacquin, *Image coding based on a fractal theory of iterated contractive image transformations*, IEEE Trans. Image Processing, 1:18-30, 1992.

the transform does not operate on the entire image. Instead it copies only scaled pieces of it to other locations. The decoding is as usual by iteration of the fractal transform starting from an arbitrary image. Due to the contraction mapping principle, that is to be discussed in the next section, the sequence of iterates converges to the fixed point of T. Jacquin's original scheme showed promising results. Since then, several researchers have improved the original algorithm. In just 14 years after Jacquin's development over 600 research papers have appeared, which we have collected in a depository in the world wide web.[11] In spite of the huge effort, fractal image encoding has not reached the point where it can clearly outperform state-of-the-art algorithms based on wavelets such as those that are in the recent JPEG2000 still image coding standard.

[11] See the links on http://www.inf.uni-konstanz.de/cgip/.

5.6 Foundation of IFS: The Contraction Mapping Principle

The image coding problem has led us to one of the central questions: how images can be compared or what the distance between two images is. In fact, this is crucial for the understanding of iterated function systems. Without an answer to these questions we will not be able to precisely verify the conditions under which the machine will produce a limiting image. Felix Hausdorff, whom we have already mentioned as the man behind the mathematical foundations of the concept of fractal dimension, proposed a method of determining this distance which is now named after him — the Hausdorff distance. Introducing the Hausdorff distance $h(A, B)$ has two marvelous consequences. First, we can now talk about the sequence of images A_k having the limit A_∞ in a very precise sense: A_∞ is the limit of the sequence $A_0, A_1, A_2, \ldots$ provided that the Hausdorff distance $h(A_\infty, A_k)$ goes to 0 as k goes to ∞. But even more importantly, Hutchinson showed that the operator W, which describes the collage

$$W(A) = w_1(A) \cup w_2(A) \cup \cdots \cup w_N(A)$$

is a contraction with respect to the Hausdorff distance. That is, there is a constant c, with $0 \leq c < 1$, such that

$$h(W(A), W(B)) \leq c \cdot h(A, B)$$

for all (compact) sets A and B in the plane. In establishing this fundamental property, Hutchinson was able to inject into consideration one of the most powerful and beautiful principles in mathematics — the contraction mapping principle, which has a long history and owes its final formulation to the great Polish mathematician Stefan Banach (1892–1945).

If the works and achievements of mathematicians could be patented, then the contraction mapping principle would probably be among those with the highest earnings up to now and for the future. Once he allowed himself a certain degree of abstraction, Banach understood that many individual and special cases which floated in the work of earlier mathematicians can be subsumed under one very brilliant principle. The result is nowadays a theorem in *metric topology*, a branch of mathematics which is basic for a great part of modern mathematics and is usually a topic reserved for students of an advanced university-level mathematics courses. We will explain the core of Banach's ideas in a nonrigorous style.

Measuring Distance: The Metric Space

The Hausdorff distance determines the distance of images. It is based on the concept of distance of points to be explained here. Expressed generally, the distance between points of a space X can be measured by a function $d : X \times X \to \mathbf{R}$. Here $\mathbf{R}$ denotes the real numbers and the function d must have the properties that

(1) $d(x, y) \geq 0$
(2) $d(x, y) = 0$ if and only if $x = y$

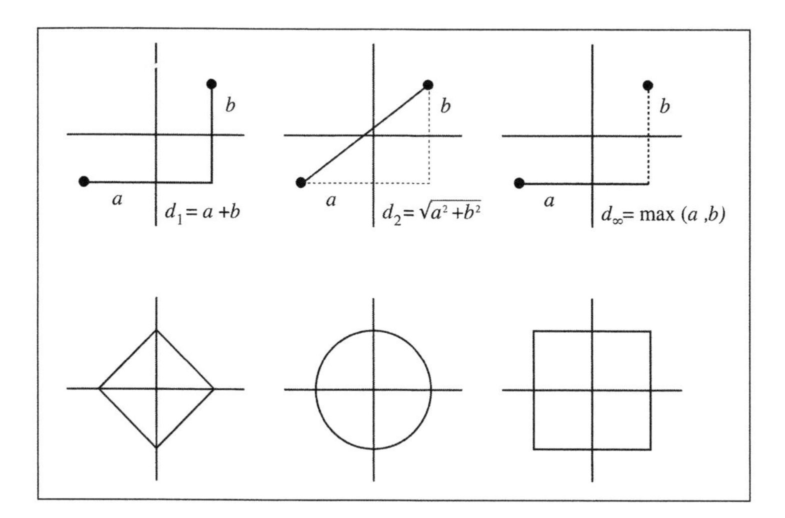

Figure 5.39 : Three methods of measuring distance in the plane (the lattice distance, the Euclidean distance, the maximum norm distance) and the corresponding unit sets (the set of points which have the distance 1 to the origin of the coordinate system).

(3) $d(x,y) = d(y,x)$
(4) $d(x,y) \leq d(x,z) + d(z,y)$ (triangle inequality),

hold for all $x, y, z \in X$. We call such a mapping d a metric. A space together with a metric is called a metric space. Some examples are (see figure 5.39):

(1) For real numbers x and y we can set

$$d(x,y) = |x - y|.$$

(2) For points $P = (x,y)$, $Q = (u,v)$ in the plane we can define

$$d_2(P,Q) = \sqrt{(x-u)^2 + (y-v)^2}.$$

This is the Euclidean metric.
(3) Another metric in the plane would be

$$d_\infty(P,Q) = \max\{|x-u|, |y-v|\}.$$

This is the maximum metric.
(4) A further metric illustrated in figure 5.39, the lattice metric, is given by

$$d_1(P,Q) = |x-u| + |y-v|.$$

The last metric d_1 on the list is sometimes also referred to as the Manhattan metric, because it is the distance a cab driver in Manhattan would have to drive to get from P to Q.

Once we have a metric for a space X we can talk about limits of sequences. Let $x_0, x_1, x_2, \ldots$ be a sequence of points from X and a an element from X. Then a is the limit of the sequence provided

$$\lim_{k\to\infty} d(x_k, a) = 0.$$

In other words, for any $\varepsilon > 0$ we can find a point x_n in the sequence so that any point later in the sequence has distance to a less than ε:

$$d(x_k, a) < \epsilon, \quad k > n.$$

In this case we say that the sequence converges to a. Often it is very desirable to test the convergence of a sequence without knowledge of the limit. This, however, works only if the underlying space X has a special nature (i.e., it is a complete metric space). Then one may discuss limits by monitoring the distance of consecutive points in the sequence.

The space X is called a complete metric space if any Cauchy sequence has a limit which belongs to X. More precisely, this means the following: Let $x_0, x_1, x_2, \ldots$ be a given sequence of points in X. It is a Cauchy sequence if for any given number $\varepsilon > 0$ we can find a point x_m in the sequence so that any two points later in the sequence have a distance less than ε:

$$d(x_i, x_j) < \varepsilon, \quad i, j \geq m.$$

Then the limit of the sequence exists and is a point of X. Two examples are:

(1) The set of rational numbers is not complete. There are Cauchy sequences of rational numbers whose limits exist but are not rational numbers. An example of such a sequence is given by

$$x_n = \sum_{k=1}^{n} \frac{1}{k^2}.$$

This sequence of rational numbers converges to the irrational limit $\pi^2/6$.

(2) The plane $\mathbf{R}^2$ is complete with respect to any of the metrics, d_1, d_2, or d_∞.

The Environment of the Contraction Mapping Principle

In chapter 1 we learned that a large variety of dynamic processes and phenomena can be seen from the point of view of a feedback system. A sequence of events $a_0, a_1, a_2, \ldots$ is generated starting with an initial event a_0, which can be chosen from a pool of admissible choices. As time elapses (as n grows), the sequence can show all kinds of behavior. The central problem of dynamical systems theory is to forecast the long-term behavior. Often that behavior will not depend very much on the initial choice a_0. That is exactly the environment for the contraction mapping principle. It provides everything which we can hope for to make a forecast. But having in mind the variety of both wild and tame behavior which feedback systems can produce, it is clear that the principle will select some subclass of feedback systems for which it can be applied. Let us collect the two features which characterize this class:

(1) **The Space.** The objects — numbers, images, transformations, etc., which we call a_n — must belong to a set in which we can measure the distance between any two of its elements, for example, the distance between x and y is $d(x, y)$. Furthermore, the set must be saturated in some sense. That means, if an arbitrary sequence satisfies a special test which examines the possible existence of a limit, then a limit exists and belongs to the set (technically, the space is a *complete metric* space).
(2) **The Mapping.** The sequence of objects is obtained by a mapping, say f. That means that for any initial object a_0, a sequence $a_0, a_1, a_2, \ldots$ is generated by $a_{n+1} = f(a_n), n = 0, 1, 2, \ldots$ Furthermore, f is a contraction. That means that for any two elements of the space, say x and y, the distance between $f(x)$ and $f(y)$ is always strictly less than the distance between x and y.[12]

The Result of the Contraction Mapping Principle

For this class of feedback systems the contraction mapping principle gives the following remarkable result:

(1) **The Attractor.** For any initial object a_0 the feedback system $a_{n+1} = f(a_n)$ will always have a predictable long-term behavior. There is an object a_∞ (the limit of the feedback system) to which the system will go. That limit object is the same no matter what the initial object a_0 is. We call a_∞ the unique *attractor* of the feedback system.
(2) **The Invariance.** The feedback system leaves a_∞ invariant. In other words, if we start with a_∞, then a_∞ is returned. a_∞ is a fixed point of f, i.e., $f(a_\infty) = a_\infty$.
(3) **The Estimate.** We can predict how fast the feedback system will arrive close to a_∞ when it is started at a_0. We only have to test the feedback loop once on the initial object. That means, if we measure the distance between a_0 and $a_1 = f(a_0)$, we can already safely predict how often we have to run the system to arrive near a_∞ within a prescribed accuracy. Moreover, we can estimate the distance between a_0 and a_∞.

The Attractor of a Contractive Mapping

A mapping f is a contraction of the metric space X, provided that there is a constant c, $0 \leq c < 1$, such that for all x, y in X one has that

$$d(f(x), f(y)) \leq c d(x, y).$$

The constant c is called the contraction factor for f. Let $a_0, a_1, a_2, \ldots$ be a sequence of elements from a complete metric space X defined by $a_{n+1} = f(a_n)$. The following holds true:

(1) There is a unique attractor $a_\infty = \lim_{n \to \infty} a_n$.

[12] Technically, $d(f(x), f(y)) \leq c \cdot d(x, y)$ with a constant $0 \leq c < 1$.

(2) a_∞ is invariant, $f(a_\infty) = a_\infty$.
(3) There is an a priori estimate for the distance from a_n to the attractor, $d(a_n, a_\infty) \leq c^n d(a_0, a_1)/(1-c)$.

Let us explain the estimate in property (3). From the contraction property of f we derive

$$d(f(a_0), a_\infty) = d(f(a_0), f(a_\infty)) \leq cd(a_0, a_\infty).$$

Applying the triangle inequality, we further obtain

$$\begin{aligned} d(a_0, a_\infty) &\leq d(a_0, f(a_0)) + d(f(a_0), a_\infty) \\ &\leq d(a_0, f(a_0)) + cd(a_0, a_\infty) \end{aligned}$$

thus,

$$d(a_0, a_\infty) \leq \frac{d(a_0, f(a_0))}{1-c}$$

and likewise

$$d(a_n, a_\infty) \leq \frac{d(a_n, a_{n+1})}{1-c}$$

for all $n = 0, 1, 2, \ldots$ Finally, with

$$\begin{aligned} d(a_n, a_{n+1}) &\leq cd(a_{n-1}, a_n) \\ &\leq c^2 d(a_{n-2}, a_{n-1}) \\ &\leq \cdots \\ &\leq c^n d(a_0, a_1) \end{aligned}$$

we arrive at the result

$$d(a_n, a_\infty) \leq \frac{c^n}{1-c} d(a_0, a_1).$$

This allows us to predict n so that a_n is within a prescribed distance to the limit.

We now examine the operation of an IFS and how it can be described by means of the contraction mapping principle. To start we need to define the distance between two images. For simplicity let us consider only black and white images. Mathematically speaking an image is a compact set[13] in the plane.

The Hausdorff Distance

Given an image A, we introduce the ε-collar of A, written A_ε, which is the set A together with all points in the plane which have a distance from A of not more than ε (see figure 5.40). Hausdorff measured the distance between two (compact) sets A and B in the plane using ε-collars. Formally, we write $h(A, B)$ for that distance. To determine its value we try to fit A into an ε-collar of B, and B into an ε-collar of A. If we take ε large enough, this will be possible. The Hausdorff distance $h(A, B)$ is just the smallest ε such that the ε-collar A_ε absorbs B and the ε-collar B_ε absorbs A.

[13] Technically, compactness for a set X in the plane means that it is bounded, i.e., it lies entirely within some sufficiently large disk in the plane and that every convergent sequence of points from the set converges to a point from the set.

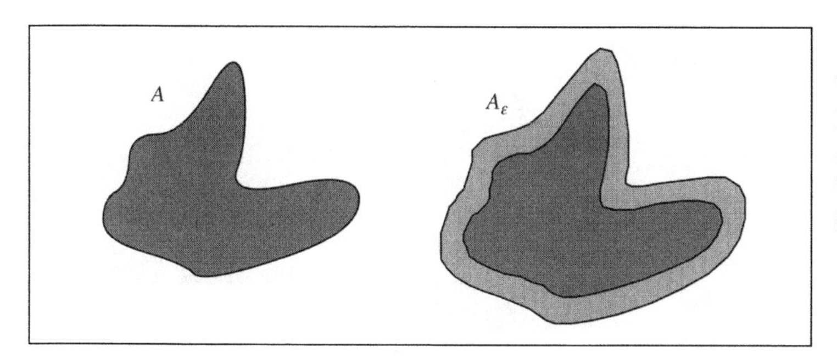

The ε-Collar

The ε-collar of a set A in the plane. Note that the ε-collar of A includes A and is not just a set of points close to A, as the term 'collar' might suggest.

Figure 5.40

Definition of the Hausdorff Distance

In precise mathematical terms the definition of the Hausdorff distance is as follows. Let X be a complete metric space with metric d. For any compact subset A of X and $\varepsilon > 0$, define the ε-collar of A by

$$A_\varepsilon = \{x \in X \mid d(x, y) \leq \varepsilon \text{ for some } y \in A\}.$$

For any compact subsets A and B of X the Hausdorff distance is

$$h(A, B) = \inf\{\varepsilon \mid A \subset B_\varepsilon \text{ and } B \subset A_\varepsilon\}.$$

According to Hausdorff the space of all compact subsets of X, equipped with the Hausdorff distance, is another complete metric space. This implies that the space of all compact subsets of X is a suitable environment for the contraction mapping principle.

With this definition it follows that $h(A, B) = 0$ when A is equal to B. Also, if A is just a point and B is just a point, then $h(A, B)$ is the distance between A and B in the ordinary sense. Figure 5.41 illustrates that fact and gives a few more examples useful for getting acquainted with the notion of Hausdorff distance.

The Hutchinson Operator

Let us now return to the state of affairs which Hutchinson obtained when analyzing the operator W

$$W(A) = w_1(A) \cup w_2(A) \cup \cdots \cup w_N(A),$$

where the transformations $w_i, i = 1, \ldots, N$, are contractions with contraction factors c_i. Hutchinson was able to show that W is also a contraction, however, with respect to the Hausdorff distance. Thus, the contraction mapping principle can be applied to the iteration of the Hutchinson operator W. Consequently, whatever initial image is chosen to start the iteration of the IFS, for example, A_0, the generated sequence

$$A_{k+1} = W(A_k), \quad k = 0, 1, 2, 3, \ldots$$

will tend towards a distinguished image, the attractor A_∞ of the IFS. Moreover, this image is invariant:

$$W(A_\infty) = A_\infty.$$

Four Examples of Hausdorff Distance

To obtain the Hausdorff distance between two planar sets A and B we compute $a_\varepsilon = \inf\{\varepsilon \mid B \subset A_\varepsilon\}$ (left figures) and $b_\varepsilon = \inf\{\varepsilon \mid A \subset B_\varepsilon\}$ (right figures). B barely fits into the a_ε-collar of A, and A barely fits into the b_ε-collar of B. The Hausdorff distance is the maximum of both values, $h(A, B) = \max\{a_\varepsilon, b_\varepsilon\}$. The sets A and B are two points (top row), a disk and a line segment (second row), a disk and a large square (third row, here $b_\varepsilon = 0$), and two intersecting disks (bottom row).

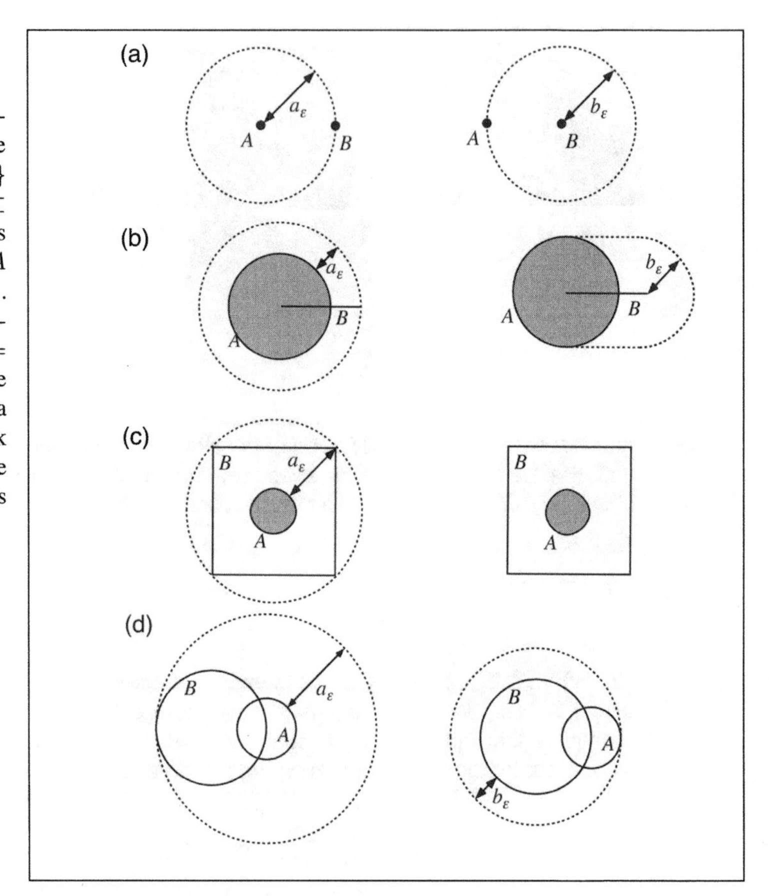

Figure 5.41

This solves a central problem raised in chapter 3. The Koch curve, the Sierpinski gasket, etc., all seem to be objects in the plane, and there are convergent processes for them, namely, the iteration of the corresponding Hutchinson operators. But we could not prove that these fractals really exist and are not just some impossible artifact of a self-referential scheme such as the assumption of a barber who shaves all men who do not shave themselves — obviously a falsehood. However, now, with Hutchinson and Hausdorff's results in hand, we are sure that the limit object with the self-similarity property truly exists.

The contraction mapping principle even gives us something in addition for free. Knowing the contraction factor c of the Hutchinson operator W, we can estimate how fast the IFS will produce the final image from just applying the Hutchinson operator one time to A_0. Since the contraction factor c of W is determined by the contraction w_i with the worst contraction factor c_i, i.e., $c = \max\{c_i\}$, the efficiency of the IFS is determined by this individual contraction. This is the theoretical background of our experiments in figure 5.1 and the encoding of images by IFSs.

The Contractivity of the Hutchinson Operator

Hutchinson applied the contraction mapping principle to the operator W. The principle requires that the space in which W operates is complete. The completeness of this space of compact subsets of a space X, which itself is complete (e.g., the Euclidean plane), was already known to Hausdorff. So it remained to show that the Hutchinson operator W is a contraction. Let us briefly illustrate the idea of the argument with the example of two contractions w_1 and w_2 with contraction factors $c_1, c_2 < 1$. We take any two compact sets A and B, and show that the Hausdorff distance $h(W(A), W(B))$ between

$$W(A) = w_1(A) \cup w_2(A)$$

and

$$W(B) = w_1(B) \cup w_2(B)$$

is strictly less than the distance $h(A, B)$ between A and B.

Compare figure 5.42 for the following. Let ε be the Hausdorff distance between A and B, $h(A, B) = \varepsilon$. Then B is in the ε-collar of A, $B \subset A_\varepsilon$. Applying the transformations w_1 and w_2 yields

$$w_1(B) \subset w_1(A_\varepsilon) \text{ and } w_2(B) \subset w_2(A_\varepsilon).$$

From the contraction property of the two transformations it follows that

$$\begin{aligned} w_1(A_\varepsilon) &\subset (c_1 \cdot \varepsilon)\text{-collar of } w_1(A), \\ w_2(A_\varepsilon) &\subset (c_2 \cdot \varepsilon)\text{-collar of } w_2(A). \end{aligned}$$

Setting $c = \max(c_1, c_2)$ we obtain that both $w_1(B)$ and $w_2(B)$ are contained in the $(c \cdot \varepsilon)$-collar of $w_1(A) \cup w_2(A)$. The same argument applied to A and ε-collar B_ε also yields that both $w_1(A)$ and $w_2(A)$

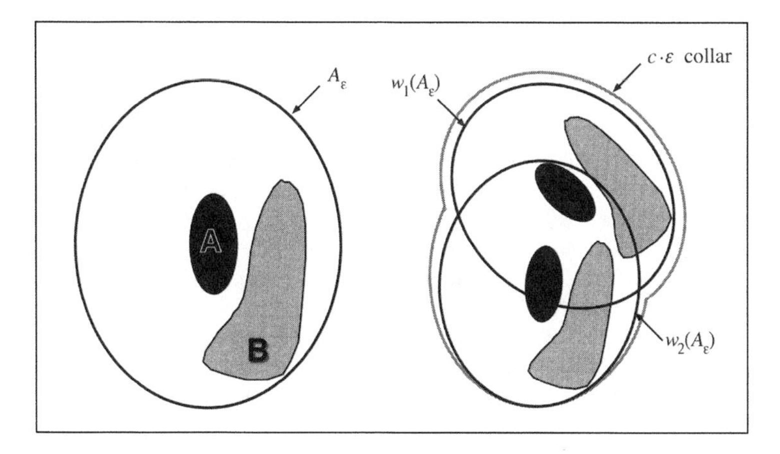

Figure 5.42 : The Hausdorff distance between the sets A and B, ε, shrinks at least by the factor $c = \max(c_1, c_2) < 1$ when the Hutchinson operator is applied.

are contained in the $(c \cdot \varepsilon)$-collar of $w_1(B) \cup w_2(B)$. With that it is clear from the definition that the Hausdorff distance $h(W(A), W(B))$ is less than $c \cdot \varepsilon$. Thus, the Hutchinson operator W is a contraction with contraction factor $c < 1$. Therefore the worst contraction of the transformations in the IFS determines the overall contraction factor of the machine.

In summary, our experiments are built on very firm ground and are not just the results of some lucky or accidental choices. Hutchinson's work lays the ground for a whole new discussion of images and their encoding. But as we have seen, there are still some open and very serious problems, for example, the problem of decoding. We have seen that the fern can be encoded by an IFS, but we have not yet given away the secret of how the image has been obtained (i.e., how the fern has been decoded). In a sense this means that we can lock up images into very tiny little boxes, which makes them invisible; but we don't yet know the keys needed to get them out again into the visible world. What we need is some artist who unchains our encodings. But this is the subject of the next chapter. On the other hand, there is the inverse problem, the problem to find the encoding of a given image.

Fractal Dimension for IFS Attractors

We have seen that an attractor A_∞ generated by a simple IFS whose contractions are similarities is self-similar. In this case, we can compute the self-similarity dimension, provided the N contractions $w_1, \ldots, w_N$ have the property that $w_i(A_\infty) \cap w_k(A_\infty) = \emptyset$, for all i, k with $i \neq k$, and the w_i are one-to-one. This type of attractor is said to be totally disconnected. There is no overlapping of the small copies of the attractor. If in addition, the contractions are reductions with the same factor c, $0 \leq c < 1$, then the self-similarity dimension $D_s = d$ of the attractor A_∞ can be computed from the equation $Nc^d = 1$. This is the same as

$$D_s = \frac{\log N}{\log 1/c}.$$

Moreover, we can show that the self-similarity dimension is the same as the box-counting dimension. Note, that the formula can be useless when there is substantial overlap of the contractions of the attractor. To see this, consider the example of a square covered by four reduced copies of it, each one reduced by a contraction factor of, say, 3/4 (implying substantial overlap). Then the formula gives $D_s = \log 4 / \log 4/3 > 2$!

If we have N similarities with reduction factors $c_1, \ldots, c_N$, then Hutchinson showed that we can still compute the fractal dimension $D_s = d$, by solving an equation which includes the special case where $c_1 = c_2 = \cdots = c_N$. He showed that

$$c_1^d + c_2^d + \cdots + c_N^d = 1.$$

Of course, in most cases one cannot solve this equation by hand for the dimension d. Rather, a numerical procedure must be employed.

The condition that the attractor must be totally disconnected for the formula to hold can be relaxed somewhat.[14] It is still true if the attractor is just touching. An example of such a situation can be constructed in a straightforward manner as follows. Consider the unit square $[0,1] \times [0,1]$ and its regular square subdivision in $n \times n$ cells (see figure 5.43 for $n = 5$). We select k of the n^2 subsquares and imagine an MRCM with k contractions each one of which contracts the entire unit square to one of the k subsquares. Thus, such a contraction involves a scaling by the factor of $1/n$ and a translation by a vector of the form $(i/n, j/n)$, where $i, j \in \{0, 1, \ldots, n-1\}$. But note, that there may be in addition a rotation by $0, 90, 180$, or 270 degrees and also a reflection involved (giving rise to eight variations). Thus, we may choose a contraction from a pool of a total of $8n^2$ possibilities. We have already seen some examples of this form: the Cantor set, the Sierpinski gasket variation (figure 5.9), the Sierpinski carpet (figure 2.56), and the square (figures 5.30 and 5.32). There will be many more in chapter 7.

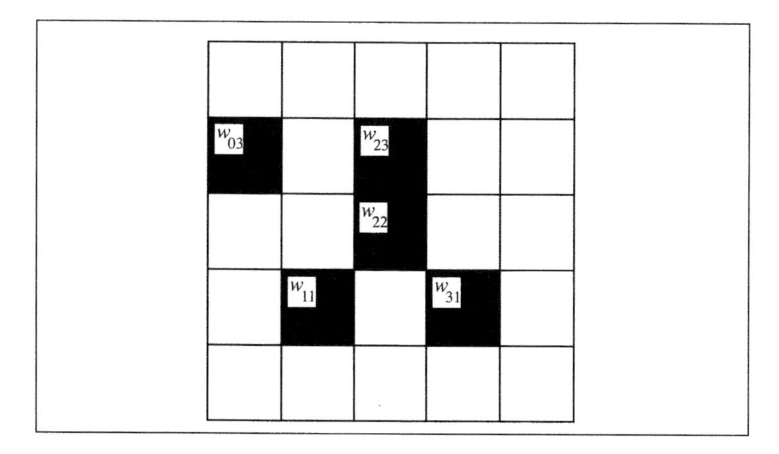

Figure 5.43 : Schematic diagram of an IFS of five transformations with contraction factor $1/n = 1/5$. The fractal dimension of the corresponding attractor is $\log 5/\log 5 = 1$.

For each IFS of this form with k contractions which transform the unit square to one of its n^2 subsquares, the self-similarity or box-counting dimension is given by $\log k/\log n$. Indeed, to verify the formula for the box-counting dimension we just have to choose grids of mesh size $s = 1/n$, $s = 1/n^2$, $s = 1/n^3$, ...Then the number $N(1/p^r)$ of boxes which contain some portion of the attractor will be exactly k^r. In other words $D = \lim_{r \to \infty} \log(k^r)/\log(n^r) = \log k/\log n$.

[14] See J. Hutchinson, *Fractals and self-similarity,* Indiana University Journal of Mathematics 30 (1981) 713–747, and G. Edgar, *Measures, Topology and Fractal Geometry,* Springer-Verlag, New York, 1990.

5.7 Choosing the Right Metric

In the last section we mentioned several possible definitions of a distance of points in the plane. The Hausdorff distance between images is also affected by the choice of that distance. So it is no surprise, and in fact important, to note that the contraction mapping principle also depends on the choice of distance.

Dependence on the Distance Notion

Let us recall the methods of measuring distance in the plane discussed in the last section. For example, if P and Q are two points we can measure the Euclidean distance d_2 (this is the length of a straight-line segment between P and Q), the lattice distance d_1 (this is the sum of the length of two horizontal and vertical line segments which connect P and Q), or the maximum norm distance d_∞ (see figure 5.39). These are only three of a great many possible definitions. It is interesting to note the various geometrical shapes that are given by the sets of points that have a distance less than or equal to 1 from the origin. Naturally these shapes depend on the metric. For the Euclidean metric we obtain the unit disk, and for the maximum metric we get the unit square. But even more important for our purposes is the fact that it also depends on the metric whether or not a given transformation is a contraction. It seems counter-intuitive that a transformation may be a contraction in one case but not with respect to another metric.

The Metric Determines Contractiveness: An Example

It is important to note that everything depends on the choice of the metric. A given transformation may be a contraction with respect to one metric, but not a contraction with respect to another one. For example, consider the map w which is given by the matrix

$$\left(\begin{array}{cc|c} 0.55 & -0.55 & 0 \\ 0.55 & 0.55 & 0 \end{array}\right)$$

which defines a rotation by 45 degrees, a scaling by $0.55\sqrt{2} \approx 0.778$, and no translation (see figure 5.44). The transformation w is a contraction for the metric d_2 but not with respect to d_1 or d_∞.

To see the argument let us fix the point $P = (0,0)$ and consider points Q for each metric. Note that the transformation w leaves the origin P invariant ($w(P) = P$).

For the d_1-metric we choose $Q = (1,0)$. Then Q is transformed into $w(Q) = (0.55, 0.55)$, and we have

$$d_1(w(P), w(Q)) = 0.55 + 0.55 = 1.1 > 1.0 = d_1(P,Q).$$

Thus, in terms of the metric d_1, the transformation w does not shrink the distance between P and Q; w is not a contraction.

For the d_∞-metric we look at $Q = (1,1)$. It is mapped to $w(Q) = (0, 1.1)$, thus

$$d_\infty(w(P), w(Q)) = \max(0, 1.1) = 1.1 > 1.0 = d_\infty(P,Q).$$

Thus, w is not a contraction with respect to d_∞ either.

Finally, let us examine the situation for the Euclidean metric. To show that w is a contraction, we need to consider arbitrary points $P = (x, y)$ and $Q = (u, v)$. Recall, that

$$d_2(P, Q) = \sqrt{(x-u)^2 + (y-v)^2}.$$

We compute the transformed points

$$\begin{aligned} w(P) &= (0.55x - 0.55y, 0.55x + 0.55y) \\ &= 0.55(x - y, x + y) \\ w(Q) &= (0.55u - 0.55v, 0.55u + 0.55v) \\ &= 0.55(u - v, u + v) \end{aligned}$$

and their distance:

$$\begin{aligned} & d_2(w(P), w(Q)) \\ &= 0.55\sqrt{((x-y)-(u-v))^2 + ((x+y)-(u+v))^2} \\ &= 0.55\,(\,(x-y)^2 + (x+y)^2 - 2(x-y)(u-v) \\ &\quad - 2(x+y)(u+v) + (u-v)^2 + (u+v)^2\,)^{1/2} \\ &= 0.55\sqrt{2(x^2+y^2) - 2(2xu + 2yv) + 2(u^2+v^2)} \\ &= 0.55\sqrt{2((x-u)^2 + (y-v)^2)} \\ &= 0.55\sqrt{2}d_2(P, Q) \\ &< d_2(P, Q). \end{aligned}$$

Since the contraction factor is $c = 0.55\sqrt{2} \approx 0.778 < 1$, we have that w is a contraction with regard to the Euclidean metric d_2.

Let us take as an example a similarity transformation which is a composition of a rotation of 45°and a scaling by a factor of about 0.778. Figure 5.44 shows how this transformation acts upon the different unit sets.[15] In each case the transformed image is reduced in size, but only the transformed image of the Euclidean unit disk is contained in the disk. In all other cases there is some overlap, indicating that the transformation is not a contraction with respect to the underlying metric.

The Euclidean Metric Is Not Always the Choice

Based on the above observation one might conjecture that the Euclidean metric is special in the sense that it captures the contractivity of a transformation when other metrics do not. However, this is not the case. Take, for example, a transformation which first rotates by 90°and then scales the x-component of the result by 0.5, i.e.,

$$(x, y) \rightarrow (-0.5y, x).$$

[15] The unit sets are defined to be the sets of points with a distance not greater than 1 from the origin. Thus, they depend on the metric used. For example, the unit set for the Euclidean metric is a disk, while it is a square for the maximum metric (see figures 5.44 and 5.46).

Contraction and Metric

A transformation which rotates by 45° and scales by 0.778 is a contraction with respect to the Euclidean metric (center) but not with respect to the lattice metric (left) or the maximum metric (right).

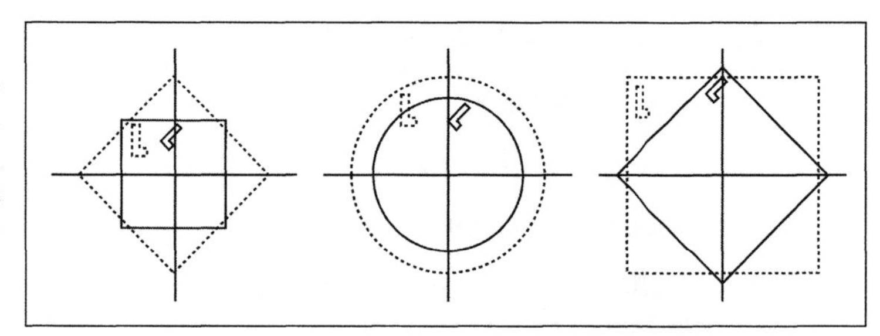

Figure 5.44

Square Code

Coding of a square with only two transformations. The rotation of 90° is crucial; without it the transformations would not be contractions.

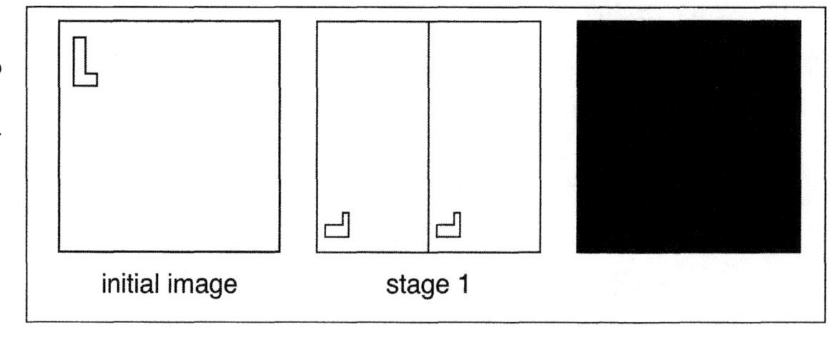

Figure 5.45

Using two such transformations with appropriate translations added, we have coded a square (see figure 5.45). It is easy to check that the square is in fact the fixed point of the corresponding Hutchinson operator. But the transformations are not contractions with respect to the Euclidean metric d_2 (the point (1,0) is rotated to (0,1), and the subsequent scaling does not have an effect here). Moreover, they are not contractions with respect to the lattice metric d_1 or the maximum metric d_∞ either. Therefore it seems an open question, whether the corresponding IFS in fact does have the square as an attractor.

The question can be settled since there are metrics which make the transformations contractive, see figure 5.46. The trick is to design the metric so that it measures differently in the x- and y-directions. In this way the unit set of all points with distance one or less from the origin becomes a rectangle, which contains its transformed image.

Contraction Mapping Principle and IFS

Thus, we see that it may be important to find a suitable metric for an application of the contraction mapping principle. In particular the third part of the principle, which predicts how fast the iteration of the IFS will approach the attractor, is effected by the quality of the metric. The smaller the contraction ratio the better is the estimate for the speed of convergence of the IFS, and the contraction ratio of course depends heavily on the choice of the metric. The power to make a good prediction will be important in the context of the *inverse problem* mentioned in section 5.5.

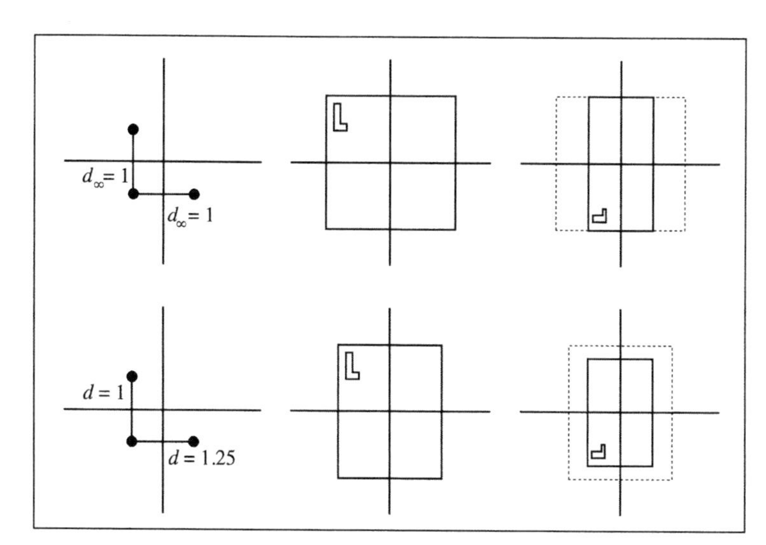

A Special Metric

The transformation, which rotates by 90° and scales in x-direction with factor 0.5, is neither a contraction relative to the maximum metric (top) nor to the Euclidean or lattice metric. But it is a contraction with respect to a metric which measures using different weights in x- and y-direction (bottom). An example is given by the metric $d(P,Q) = \max\{1.25|x-u|, |y-v|\}$. We show the unit sets (center) and their images under $w(x,y) = (-0.5y, x)$ (right).

Figure 5.46

5.8 Composing Self-Similar Images

Several methods have been proposed for the automatic solution of the inverse problem, i.e., the encoding of images, and it is still open what will be the right choice. Therefore, we should discuss a few ideas, some of which go back to Barnsley in the early 1980's. These ideas, however, do not (yet) lead to automatic algorithms, they are more suitable for interactive computer programs requiring an intelligent human operator. Some automatic strategies will be discussed in Fisher's appendix on image compression.

IFS Attractor and MRCM Blueprint

Assume that we already have approximated a given original image by an MRCM. Recall that the blueprint of an MRCM is already determined by the first copy it produces. The copy is a collage of transformed images. Applying the MRCM to the original image, called *target image*, one also determines the quality of the approximation. When the copy is identical to the original, then the corresponding IFS codes the target image perfectly. When the distance of the copy to the target is small, then we know from the contraction mapping principle, that the attractor of the IFS is not far from the initial image, which is equal to the target image in this case. Figure 5.47 illustrates this principle for the Sierpinski gasket.

Encoding Self-Similar Images

These properties enable us to find the code for a given target image, in particular for target images which contain apparent self-similarities such as the fern. With a little practice it is easy to identify portions of the picture which are affine copies of the whole. For example, in the fern in figure 5.48 the part $R^{(1)}$ is a slightly smaller and rotated copy of the whole fern. This observation leads to the numerical computation of the first affine transformation w_1. The same procedure applies to the copies $R^{(2)}$ and $R^{(3)}$ in the figure. Even the bottom part of the stem (part $R^{(4)}$) is a copy of the whole. However, this copy is degenerate in the sense that the corresponding transformation contains a scaling in one direction by a factor of 0.0, i.e., the fern transformed by w_4 is reduced to a line. The resulting four transformations already comprise the complete system since the portions $R^{(1)}$ to $R^{(4)}$ completely cover the fern.

Interactive Encoding: The Collage Game

In general we need a procedure to generate a set of transformations such that the union of the transformed target images cover the target image as closely as possible. Taking the example of a leaf we illustrate how this can be done with an interactive computer program. In the beginning the leaf image must be entered in the computer using an image scanner. Then the leaf boundary can be extracted from the image using standard tools in image processing. The result in this case is a closed polygon which can be rapidly displayed on the computer screen. Moreover, affine transformations of the polygon can also be computed instantly and displayed. Using interactive input devices such as the mouse, knobs or even just the keyboard, the user of the program can easily manipulate the six parameters that determine one affine transformation. Simultaneously the computer displays the transformed copy of the initial polygon of the leaf. The goal is to find a transformation such that the copy fits snugly onto a part of the original leaf. Then the procedure is repeated, and the user next tries to fit another affine copy onto another part of the leaf that is not yet covered by the first. Continuing in this way the complete leaf will be covered by small and possibly distorted copies of itself. Figure 5.49 shows some of the intermediate stages that might occur in the design of the leaf transformations.

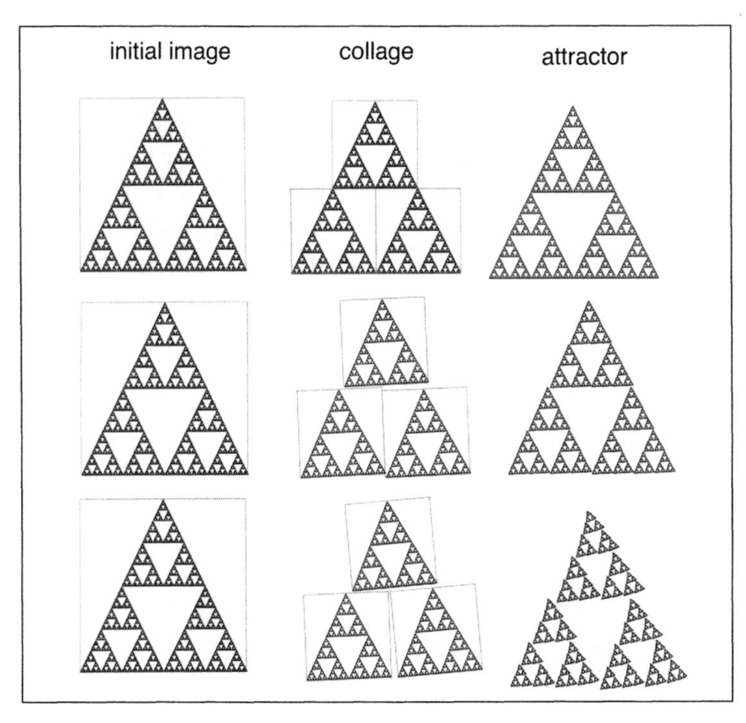

Testing Collages

Application of three MRCMs to a Sierpinski gasket. Top: the correct MRCM leaves the image invariant; middle: a reasonable approximation; bottom: a bad approximation

Figure 5.47

Contraction Mapping Principle and Collages

Let us exploit the contraction mapping principle from page 251 to analyze the results of figure 5.47. The a priori estimate for a sequence $a_0, a_1, a_2, \ldots$ which is generated by a contraction f in a metric space with attractor a_∞ yields

$$d(a_n, a_\infty) \leq \frac{c^n}{1-c} d(a_0, a_1).$$

Here c is the contraction factor of f and $a_{k+1} = f(a_k)$ for $k = 0, 1, 2, \ldots$. In particular, this means that

$$d(a_0, a_\infty) \leq \frac{1}{1-c} d(a_0, f(a_0)). \tag{5.2}$$

Thus, a single iteration starting from the initial a_0 gives us an estimate for how far a_0 is from the attractor a_∞ with respect to the metric d. Now let us interpret this result for the Hutchinson operator W with respect to the Hausdorff distance h. Let c be the contraction factor of W and let P be an arbitrary image (formally a compact subset of the plane). We would like to test how good a given Hutchinson operator will encode

Fern Collage

This fern is a slight modification of the original Barnsley fern, allowing an easier identification of its partition into self-similar components $R^{(1)}$ to $R^{(4)}$.

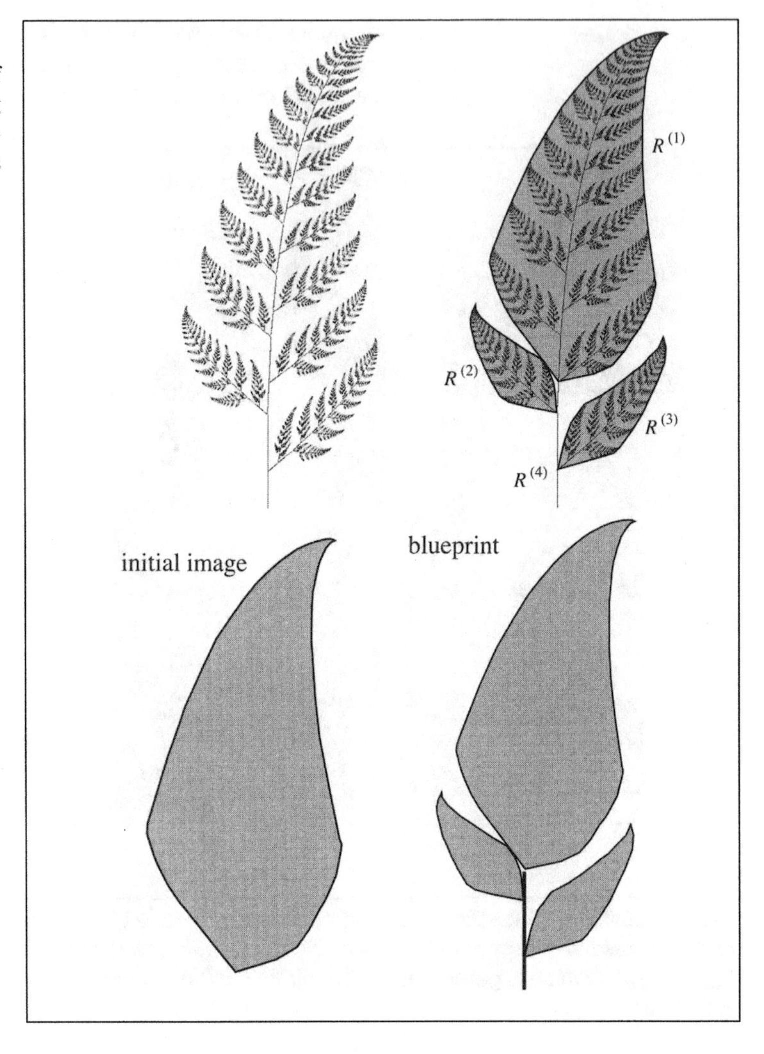

Figure 5.48

the given image P. This can be obtained from eqn. (5.2). Indeed, in this setting (5.2) now reads

$$h(P, A_\infty) \leq \frac{1}{1-c} h(P, W(P)), \tag{5.3}$$

where A_∞ is the attractor of the IFS given by W. In other words, the quality of the encoding, measured by the Hausdorff distance between P and A_∞ is controlled by applying the Hutchinson operator just once to P and quantified by $h(P, W(P))$. Barnsley calls eqn. (5.3) the 'collage theorem for iterated function systems'.

Leaf Collage

Design stages for a leaf: scanned image of a real leaf and a polygon capturing its outline (top); collage by seven transformed images of the polygon and the attractor of the corresponding IFS (bottom).

Figure 5.49

Again it is the contraction mapping principle which says that the attractor of the IFS will be close to the target image, the leaf, when the design of the collage is also close to the leaf. In the attempt to produce as accurate a collage as possible, there is a second goal that hinders exactness, namely, the coding should also be efficient in the sense that as few transformations as possible are used. The definition of an optimal solution to the problem must thus find a compromise between quality of the collage and efficiency. The automatic generation of collages for given target images is a challenging topic of current research (see Fisher's appendix on image compression).

The collage game is just one example of an entire class of mathematical problems which goes under the name optimization problems. Such problems are typically very easily stated but are often very difficult to solve even with high-powered, supercomputer technology and sophisticated mathematical algorithms.

Optimization Problem for Collages

The a priori estimate of the contraction mapping principle

$$h(P, A_\infty) \leq \frac{1}{1-c} h(P, W(P))$$

gives rise to an optimization problem. Assume we are given a picture P which we want to encode by an IFS. We decide to limit ourselves to N contractions in the IFS, which have to be determined. Any N-tuple $w_1, \ldots, w_N$ defines a Hutchinson operator W. We may further assume that the contraction factors of the transformations we want to consider are all less than or equal to some $c < 1$. Following the above estimate we have to minimize the Hausdorff distance[16] $h(P, W(P))$ among all admissible choices of W.

The Curse of Computational Complexity

A well-known example of this class of problems is the traveling salesman problem, which goes as follows. Choose some number of towns (for example, all U.S. towns with more than 10,000 inhabitants) and find the shortest route which a salesman must travel to reach all these towns. We would really think that a problem as simple as this should be no trouble for computers. But the truth is that computers become totally useless as soon as the number of towns chosen is larger than a few hundred. Problems of this kind are said to be *computationally complex* and it is understood by now that they are invariably resistant to quick solutions and always will be. The message from such examples is that simple problems may not have simple answers, and we can say that the sea of mathematics is filled with such animals. Unfortunately, it is not yet clear whether the collage game can be mathematically formulated in a way which avoids extreme computational complexity.[17] In any case, it is very likely that the computational complexity will be terrible for some images and very manageable for others. The guess is that images which are dominated by self-similar structures might be very manageable. That alone would be reason enough to continue exploring the field simply because we see such characteristics in so many of nature's formations and patterns.

There are some other problems which lead directly into current research problems which we want to at least mention.

[16] The computational problem evaluating the Hausdorff distance for digitized images is addressed in R. Shonkwiller, *An image algorithm for computing the Hausdorff distance efficiently in linear time,* Info. Proc. Lett. 30 (1989) 87–89.

[17] The algorithms discussed in the appendix on image compression try to circumvent this problem.

5.9 Breaking Self-Similarity and Self-Affinity: Networking with MRCMs

Creating an image with an MRCM quite naturally leads to a structure which has repetition in smaller and smaller scales. In the cases where each of the contractions involved in the corresponding IFS is a similarity with the same reduction factor (for example, the Sierpinski gasket), we call the resulting attractor *strictly self-similar*. Also when different reduction factors occur, the resulting attractor is said to be *self-similar*. When the contractions are not similarities, but affine linear transformations (for example, the devil's staircase), we call the resulting attractor *self-affine*.

Two Ferns

Two ferns different from Barnsley's fern. Observe that in both cases the placement of the major leaves on the stem differ from that of the small leaves on the major ones. The two ferns look the same at this scale, but the blowups in the next figure reveal important differences.

Figure 5.50

In any case, an IFS produces self-similar, or self-affine, images. As we have pointed out, IFSs can also be used to approximate images that are not self-similar or self-affine. The approximation can be made as accurate as desired. However, the very small features of the corresponding attractor will still reveal the self-similar structure. In this section of this chapter we will generalize the concept of IFSs so that this restriction is removed.[18]

Non-Self-Similar Ferns

Figure 5.50 shows two ferns which almost look like the familiar Barnsley fern, but they are different. Upon close examination of the two ferns, we observe that the phyllotaxis has changed. The placement of the major leaves on the stem is different from that of the small leaves on the major ones. That

[18]Similar concepts are in M. F. Barnsley, J. H. Elton, and D. P. Hardin, *Recurrent iterated function systems,* Constructive Approximation 5 (1989) 3–31. M. Berger, *Encoding images through transition probabilities,* Math. Comp. Modelling 11 (1988) 575–577. R. D. Mauldin and S. C. Williams, *Hausdorff dimension in graph directed constructions,* Trans. Amer. Math. Soc. 309 (1988) 811–829. G. Edgar, *Measures, Topology and Fractal Geometry,* Springer-Verlag, New York, 1990. The first ideas in this regard seem to be in T. Bedford, *Dynamics and dimension for fractal recurrent sets,* J. London Math. Soc. 33 (1986) 89–100.

Blowups of the Major Lower Right Leaf

Left: blowups of the left fern of figure 5.50 reveal the hierarchy (a): all subleaves are placed opposing each other. Right: blowups of the right fern reveal the hierarchy (b): the subleaves of the major leaf again show a placement with offset.

Figure 5.51

means that the major leaves are no longer scaled down copies of the entire fern. In other words, these ferns are neither self-similar nor self-affine in a strict sense. Nevertheless, we would say that they have some features of self-similarity. But what are these features and how are these particular ferns encoded? The answers to these questions will lead us to *networked* MRCMs, or, in other words, *hierarchical* IFSs.

To see some of the hierarchical structure we now look at a blowup of one of the major leaves from each of the ferns (see figure 5.51). This reveals the different hierarchies in their encoding. The placement of the sub-subleaves is different. On the left, the subleaves of all stages are always placed opposing each other, while on the right, this placement alternates from stage to stage: in one stage subleaves are placed opposing each other and in the next stage subleaves are placed with an offset. For ease of reference let us call these hierarchies type (a) and type (b).

We begin to see that the encoding by IFSs goes much beyond the problem of image encoding. Understanding the self-similarity hierarchies of plants,

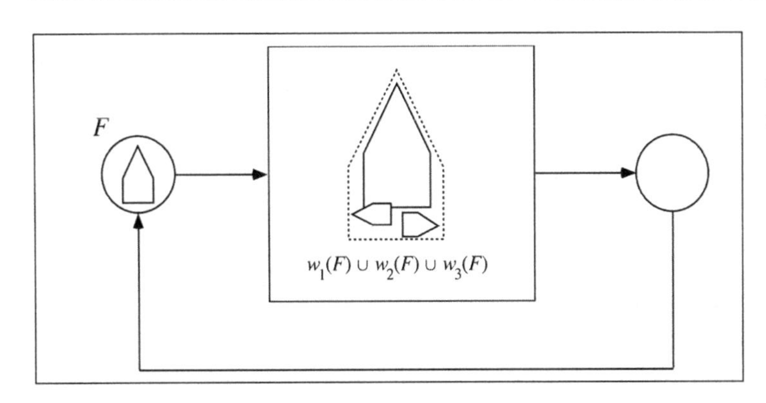

Basic Machine for Fern

The feedback system of Barnsley's fern (without stem).

Figure 5.52

for example, in terms of IFSs opens a new door to a formal mathematical description of phyllotaxis in botany. We will see that self-similarity structures can even be mixed.

Networking MRCMs

We expand the concept of the MRCM to include several MRCMs operating in a network. We will illustrate, how a non-self-similar fern can be obtained by two networked MRCMs. To keep things as simple as possible, we disregard the stems. Figure 5.52 displays the basic machine for a fern without stem.

Let us first consider the fern with hierarchy of type (a) from figure 5.51. We can identify two basic structures: the entire fern $D^{(1)}$, and one of its major leaves, say the one at the lower right, $D^{(2)}$ (see figure 5.53).

The leaf in this case is a self-similar, or more precisely, a self-affine structure. All subleaves are copies of the whole leaf and vice versa. The complete

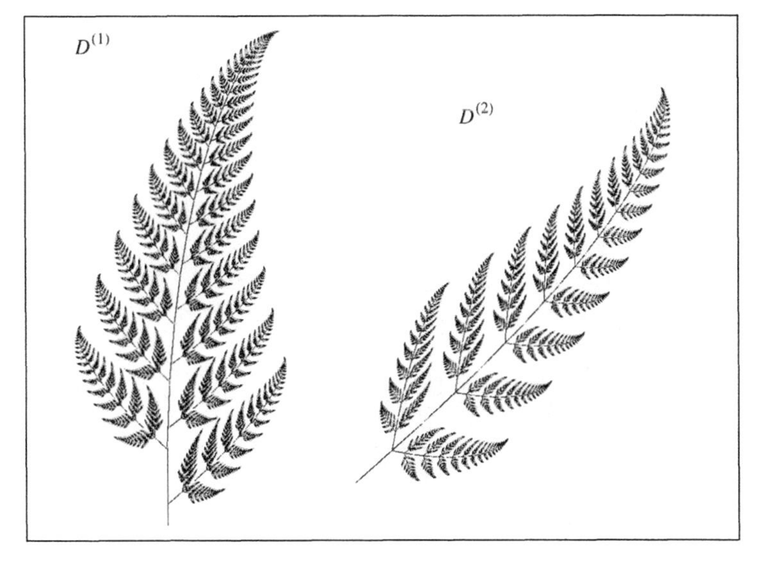

Basic Structure

Division of a fern of hierarchy type (a) into its basic structures: the whole fern and one of its major leaves.

Figure 5.53

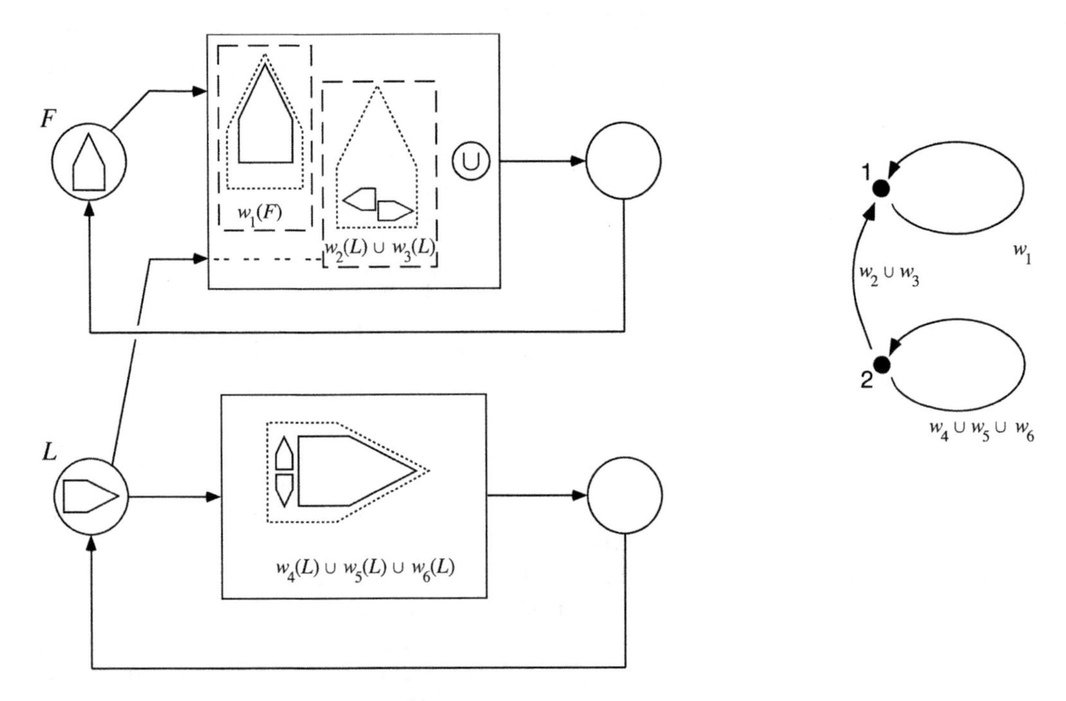

Figure 5.54 : This network of two MRCMs generates the fern with the leaf placement given by the hierarchy of type (a). The graph of the corresponding IFS is shown on the right.

fern is made up of copies of this leaf, but it is *not* simply a copy of the leaf. This is due to the different placement of the leaves and subleaves. This is the crucial difference between Barnsley's self-affine fern and this one, where the self-affinity is broken. Due to this breaking of self-similarity the fern cannot be generated with an ordinary MRCM. However, we may join two different machines to form a networked MRCM as shown in figure 5.54 which will accomplish the task.

One of the machines (bottom) is used to produce the main leaf alone. This machine works like the one for Barnsley's fern (disregarding the stem for simplicity). Thus, it has three transformations: one transformation maps the entire leaf to its lower left subleaf, the second maps to the corresponding upper left subleaf, and finally, the third transformation maps the leaf to all subleaves except for the bottom subleaves, which are already covered by the other two transformations.

The other machine (top) produces the whole fern. It has two inputs and one output. One input is served by its own output. The other input is served by the bottom MRCM. There are also three transformations in this machine. However, each transformation is applied to only one particular input image. Two transformations (w_2 and w_3 in the figure 5.54) operate on the results produced by the bottom MRCM. These produce the left and right bottom leaves at the proper places on the fern. The other transformation (w_1 in the figure)

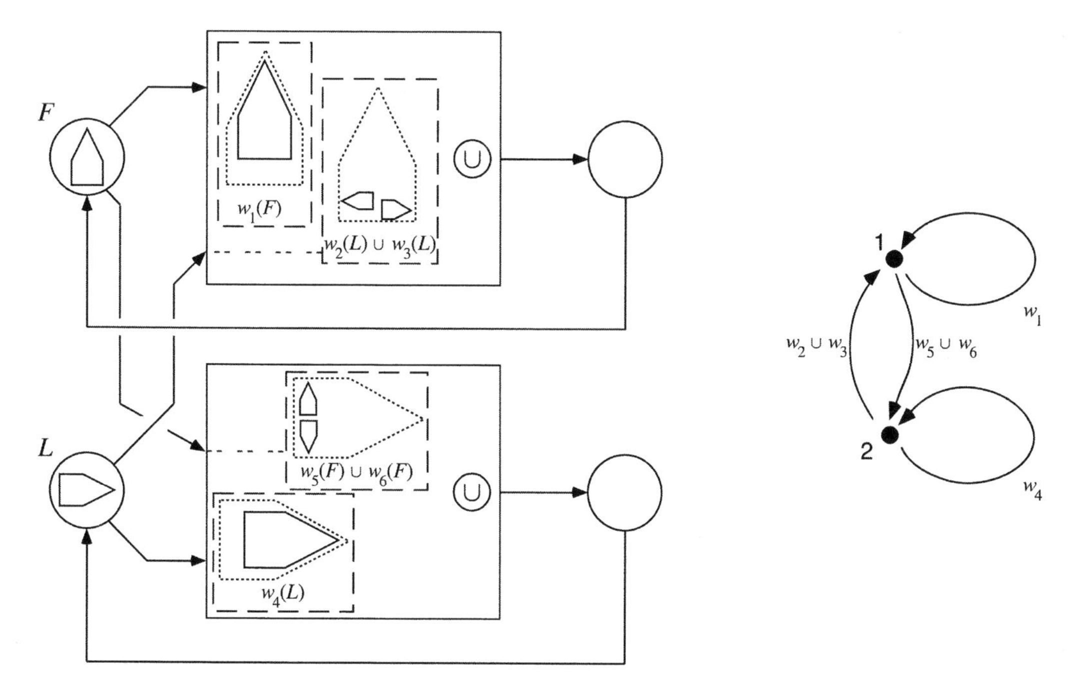

Figure 5.55 : This network of two MRCMs generates the other fern with the leaf placement given by hierarchy of type (b).

operates on the results from the top MRCM. The results of all transformations are merged when they are transferred to the output of the top machine. This is indicated by the '∪' sign. Transformation w_1 maps the entire fern to its upper part (i.e., the part without the two bottom leaves). This was also the case in the plain MRCM for Barnsley's fern. In this way the fern with the prescribed pattern for the leaf placement from hierarchy of type (a) will be generated.

Rearranging Input Connections

In order to produce the fern as in hierarchy of type (b), we need to go just a small step further interconnecting the two MRCMs both ways. This fern is characterized by the fact that the entire fern reappears in the main leaves as subleaves, while the main leaves themselves are not copies of the entire fern. This is easy to do as shown in figure 5.55. The only change relative to the network for the hierarchy (a) fern is given by the extra input in the bottom MRCM. This input image (in the limit it is the entire fern) will be transformed to make the two lowest subleaves of the leaf.

But how do we run these networks? Well, we just take any initial image, like a rectangle, and put it on the two copy machines. The machines take these input images following to the connections of the input lines and produce two outputs, one for the main leaf and one for the fern. These outputs are now used as new inputs as indicated by the feedback connections. When we iterate this process we can observe how the leaf-MRCM creates the major lower right hand leaf and how the fern-MRCM generates the complete fern.

The Contraction Mapping Principle Does It Again

Is the successful operation of this machinery just a pure accident? Not at all! Above, we discussed the contraction mapping principle. It turns out that we can also subsume the network idea under that principle, which shows the value of that rather abstract but very powerful mathematical tool. In conclusion, the networking machine has exactly one limit image, its attractor, and this attractor is independent of the initial images. To put it another way, the networking machines are encodings of non-self-similar ferns, and their hierarchies decipher the self-similarity features of these ferns. In fact, the hierarchy of the network deciphers the self-similarity of an entire class of attractors. Imagine that we change the contraction properties and positioning of the individual lens systems. As a result, we will obtain an entire cosmos of structures. However, each of them has exactly the same self-similarity features. We have thus reached the beginning of a new and very auspicious theory which promises to systematically decipher all possible self-similarity properties. The mathematical description of networked MCRMs is the topic of the remainder of this section.

Formalism of Hierarchical IFSs

There is an extension of the concept of a Hutchinson operator for a network of MRCMs. It requires working with matrices. Let

$$\mathbf{A} = \begin{pmatrix} a_{11} & a_{12} & \dots & a_{1m} \\ \vdots & \vdots & & \vdots \\ a_{m1} & a_{m2} & \dots & a_{mm} \end{pmatrix}$$

be an $(m \times m)$-matrix with elements a_{ij} and let

$$\mathbf{b} = \begin{pmatrix} b_1 \\ \vdots \\ b_m \end{pmatrix}$$

be an m-vector. Then $\mathbf{Ab}$ is the m-vector $\mathbf{c} = \mathbf{Ab}$ with components c_i , where

$$c_i = \sum_{j=1}^{m} a_{ij} b_j .$$

In analogy to this concept of ordinary matrices, a hierarchical IFS (corresponding to a network of M MRCMs) is given by an $(M \times M)$-matrix

$$\mathbf{W} = \begin{pmatrix} W_{11} & \dots & W_{1M} \\ \vdots & & \vdots \\ W_{M1} & \dots & W_{MM} \end{pmatrix},$$

where each W_{ij} is a Hutchinson operator (i.e., W_{ij} is given by a finite number of contractions). This is the matrix Hutchinson operator $\mathbf{W}$, which acts on an M-vector $\mathbf{B}$ of images

$$\mathbf{B} = \begin{pmatrix} B_1 \\ \vdots \\ B_M \end{pmatrix},$$

where each B_i is a compact subset of the plane $\mathbf{R}^2$. The result of $\mathbf{W}(\mathbf{B})$ is an M-vector $\mathbf{C}$ with components C_i, where

$$C_i = \bigcup_{j=1}^{M} W_{ij}(B_j).$$

It is convenient to allow that some of the Hutchinson operators are 'empty', $W_{ij} = \emptyset$. Here the symbol $\emptyset$ plays a similar role as 0 in ordinary arithmetic: the $\emptyset$ operator transforms any set into the empty set (i.e., for any set B we have $\emptyset(B) = \emptyset$).

Next we make a natural identification. The network of MRCMs corresponds to a graph with nodes and directed edges. For the output of each MRCM there is exactly one node, and for each output-input connection in the network there is a corresponding directed edge. These graphs, displayed next to our MRCM networks, are a compact representation of the hierarchy of the IFSs (see, for example, the non-self-similar ferns and the Sierpinski fern).

Note that a directed edge from node j to node i means that the output of j is transformed according to a specific Hutchinson operator (i.e., the one that operates on the corresponding input of the MRCM) and then fed into node i. The output of this node is the union of all the transformed images which are fed in. Now we define W_{ij}. If there is a directed edge from node j to node i, then W_{ij} denotes the corresponding Hutchinson operator. In the other case we set $W_{ij} = \emptyset$. For our examples we thus obtain

$$\mathbf{W} = \begin{pmatrix} w_1 & w_2 \cup w_3 \\ \emptyset & w_4 \cup w_5 \cup w_6 \end{pmatrix}$$

for the fern of type (a),

$$\mathbf{W} = \begin{pmatrix} w_1 & w_2 \cup w_3 \\ w_5 \cup w_6 & w_4 \end{pmatrix}$$

for the fern of type (b), and

$$\mathbf{W} = \begin{pmatrix} w_1 & w_2 \cup w_3 & \emptyset \\ \emptyset & w_4 & w_5 \cup w_6 \\ \emptyset & \emptyset & w_7 \cup w_8 \cup w_9 \end{pmatrix}$$

for the Sierpinski fern. Observe that here we have used a short form for writing Hutchinson operators. For example, when transforming any set B by $w_2 \cup w_3$ we write

$$w_2 \cup w_3(B) = w_2(B) \cup w_3(B).$$

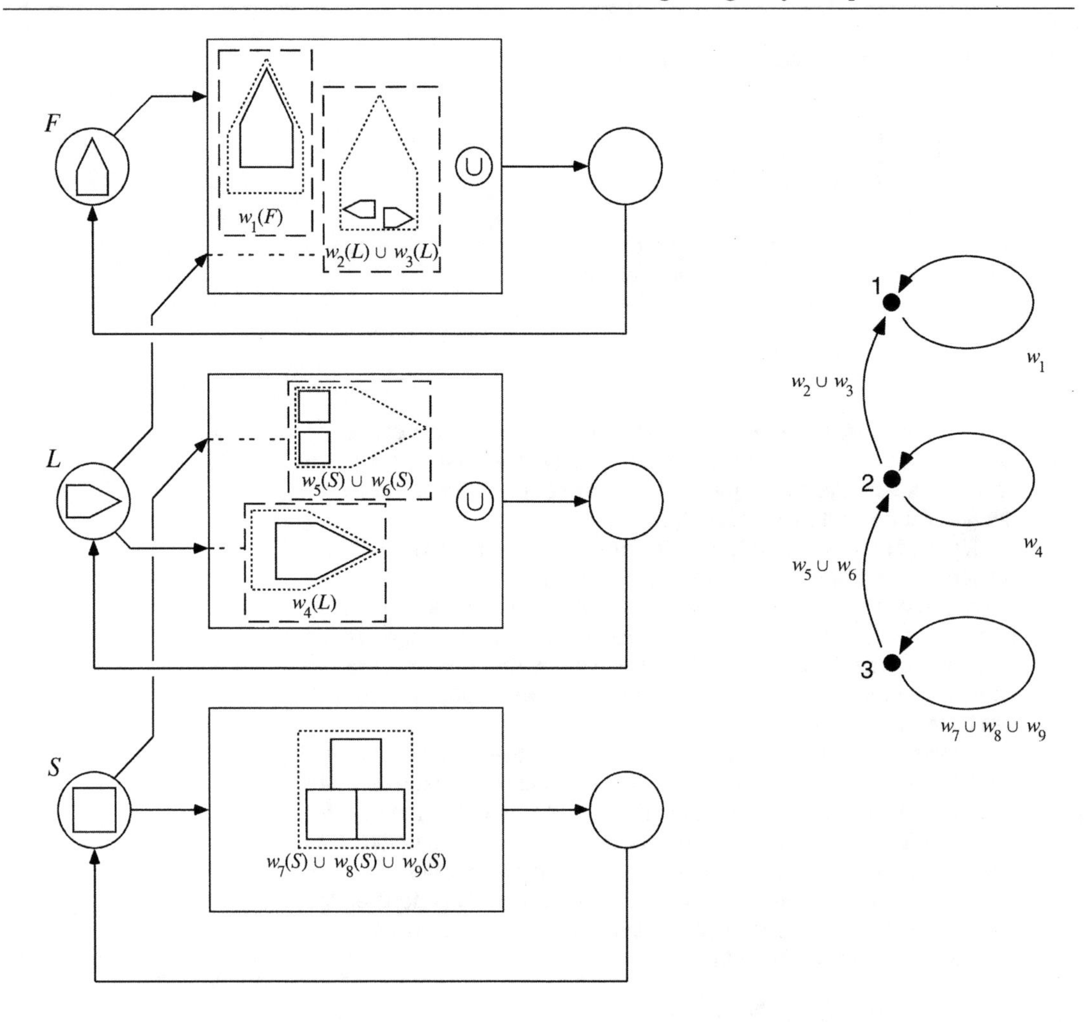

Figure 5.56 : A network of three MRCMs to generate a fern which is made up of Sierpinski gaskets.

With these definitions we can now describe the iteration of a hierarchical IFS formally. Let $\mathbf{A}_0$ be an initial M-vector of images. The iteration defines the sequence of M-vectors

$$\mathbf{A}_{k+1} = \mathbf{W}(\mathbf{A}_k), \ \ k = 0, 1, 2, \ldots$$

It turns out that this sequence again has a limit $\mathbf{A}_\infty$, which we call the attractor of the hierarchical IFS.

The proof is again by the contraction mapping principle. We start with the plane equipped with a metric such that the plane is a complete metric space. Then the space of all compact subsets of the plane with the Hausdorff distance as a metric is also a complete metric space. Now we take the M-fold Cartesian product of this space and call it H. On H there is a natural metric $d_{\max}$ which comes from the Hausdorff

The Sierpinski Fern and One of Its Main Leaves

Figure 5.57

distance: Let $\mathbf{A}$ and $\mathbf{B}$ be in H , then

$$d_{\max}(\mathbf{A},\mathbf{B}) = \max\{h(A_i, B_i) \mid i = 1,\dots,M\},$$

where A_i and B_i denote the components of $\mathbf{A}$ and $\mathbf{B}$ and $h(A_i, B_i)$ denotes their Hausdorff distance. It follows almost from the definitions that

- H is again a complete metric space, and
- $\mathbf{W} : H \to H$ is a contraction.

For completeness we must add the requirement that the iterates $\mathbf{W}^n$ of the matrix Hutchinson operator do not consist entirely of $\emptyset$-operators. Thus, the contraction mapping principle applies with the same consequences as for the ordinary Hutchinson operator.

The Sierpinski Fern

To finish this section we will use the networked MRCMs for a rather strange looking fern, which we may call the *Sierpinski fern* (see figure 5.57). It is the fern of hierarchy (a) with subleaves replaced by small Sierpinski gaskets. The network incorporates three MRCMs as shown in figure 5.56. The first two are responsible for the overall structure of the fern as before, while the third is busy with producing a Sierpinski gasket which is fed to one of the other machines.

The experiment generating a Sierpinski fern demonstrates that networked MRCMs are suitable discussing and encoding hierarchies of self-similarity features, and, moreover, is the appropriate concept mixing several fractals together.

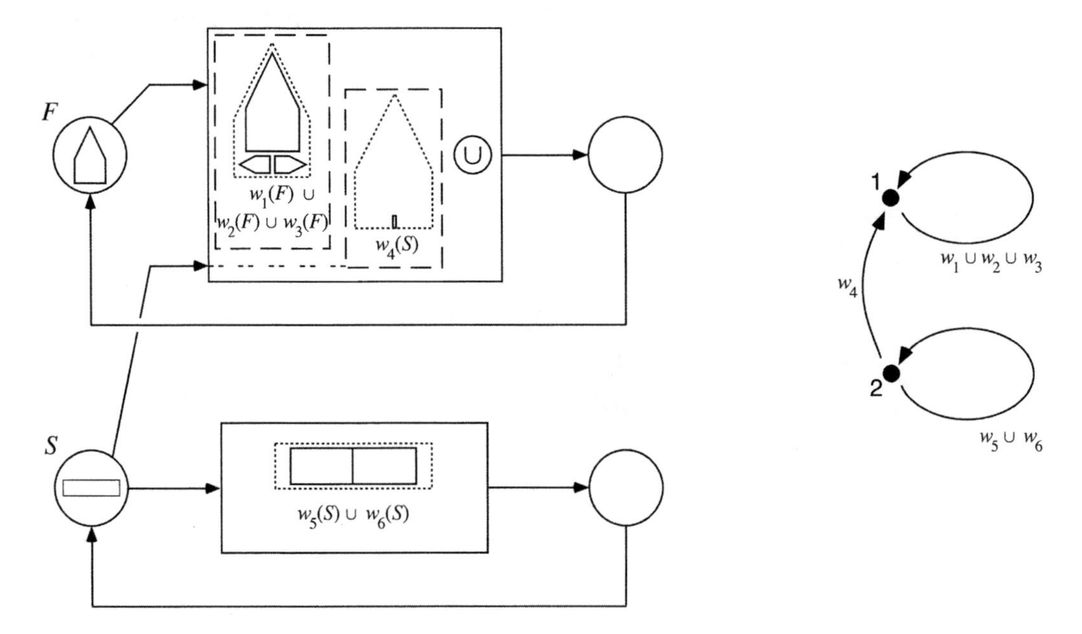

Figure 5.58 : The lower MRCM generates a line which is fed into the upper MRCM to build the stem of the fern.

The Stem in Barnsley's Fern

When we introduced Barnsley's fern by an MRCM we observed that it was not strictly self-similar, the problem being first of all in the stem. There we obtained the stem from a degenerate affine linear copy of the whole fern (i.e., collapsed to a line). From the point of view of networked MRCMs this aspect becomes much clearer. The design in figure 5.58 is a network with two MRCMs. The top machine produces the leaves and the bottom machine the stems. From that point of view Barnsley's fern is essentially a mix of two (strictly) self-similar structures.[19]

The variety of structures which can be obtained by networked MRCMs is unimaginable. As an application of networked MRCMs we present in chapter 8 an elegant solution to some long-standing open problems: the deciphering of the global geometric patterns in Pascal's triangle, which are obtained from divisibility properties of binomial coefficients.

Networked MRCMs also bring us a step closer to a solution of the problem of automatic image encoding. The encoding by iterated function systems only leads to self-similar approximations of the target image. With networked MRCMs we break the target image into pieces which can be encoded more or less independently. This results in an approximation with mixed self-similarity structures. This concept can be formalized by the so-called *partitioned IFS* (see the appendix on image compression).

[19]More precisely, the fern without the stem is self-affine, not self-similar, because the transformations which produce the leaves are only approximate similitudes.

Chapter 6

The Chaos Game: How Randomness Creates Deterministic Shapes

Nothing in Nature is random... A thing appears random only through the incompleteness of our knowledge.

Spinoza

Our idea of randomness, especially with regard to images, is that structures or patterns which are created randomly look more or less arbitrary. Maybe there is some characteristic structure, but if so, it is probably not very interesting — like a box of nails poured out onto a table.

Brownian Motion

Or look at the following example. Small particles of solid matter suspended in a liquid can be seen under a microscope moving about in an irregular and erratic way. This is the so called *Brownian motion*,[1] which is due to the random molecular impacts of the surrounding particles. It is a good example of what we expect from a randomly steered motion. Let us describe such a particle motion step by step. Begin at a point in the plane. Choose a random direction, walk some distance and stop. Choose another random direction, walk some distance and stop, and so on. Do we have to carry out the experiment to be able to get a sense of what the evolving pattern will be? How would the pattern look after 100, or 1000, or even more steps? There seems to be no problem forecasting the essential features: we would say that more or less the same patterns will evolve, however, just a bit more dense.

In any event, there doesn't seem to be much to expect from randomness in conjunction with image generation. But let us try a variant which, at first glance, could well belong to that category. Actually, following Barnsley[2] we

[1] The discovery was made by the botanist Robert Brown around 1827.

[2] M. F. Barnsley, *Fractal modelling of real world images*, in: The Science of Fractal Images, H.-O. Peitgen and D. Saupe (eds.), Springer-Verlag, New York, 1988.

The Chaos Game Board and the First Steps ...

The first six steps of the game. Game points are connected by line segments.

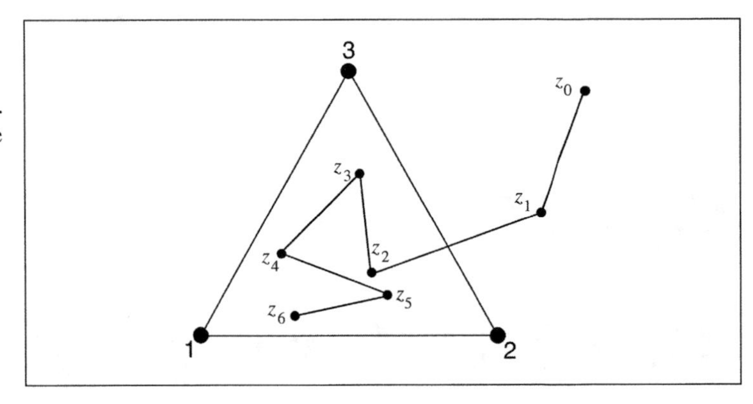

Figure 6.1

are going to introduce a family of games which can potentially change our intuitive idea of randomness quite dramatically.

The Chaos Game

Here is the first game of this sort. We need a die whose six faces are labeled with the numbers $1, 2$, and 3. An ordinary die, of course, uses numbers from 1 to 6; but that does not matter. All we have to do is, for example, identify 6 with $1, 5$ with $2, 4$ with 3 on an ordinary die. Such a die will be our generator of random numbers from the reservoir 1, 2, and 3. The random numbers which appear as we play the game, for example, 2, 3, 2, 2, 1, 2, 3, 2, 3, 1, ..., will drive a process. The process is characterized by three simple rules. To describe the rules we have to prepare the game board. Figure 6.1shows the setup: three markers, labeled $1, 2$, and 3, which form a triangle.

Now we are ready to play. Let us introduce the rules as we play. Initially we pick an arbitrary point on the board and mark it by a tiny dot. This is our current *game point*. For future reference we denote it by z_0. Now we throw the die. Assume the result is 2. Now we generate the new game point z_1, which is located at the midpoint between the current game point z_0 and the marker with label 2. This is the first step of the game. Now you can probably guess what the other two rules are. Assume we have played the game for k steps. We have thus generated $z_1, \ldots, z_k$. Roll the die. When the result is n generate a new game point z_{k+1}, which is placed exactly at the midpoint between z_k and the marker labeled n. Figure 6.1 illustrates the game. To help identify the succession of points, we connect the game points by line segments as they evolve. A pattern seems to emerge which is just as boring and arbitrary as the structure of a random walk. But that observation is a far cry from the reality. In figure 6.2 we have dropped the connecting line segments and have only shown the collected game points. In (a) we have run the game up to $k = 100$, in (b) up to $k = 500$, in (c) up to $k = 1000$, and in (d) up to $k = 10,000$ steps.

Randomness Creates Deterministic Shapes

The impression which figure 6.2 leaves behind is such that we are inclined, at first, not to believe our eyes. We have just seen the generation of the Sierpinski gasket by a random process, which is amazing because the Sierpinski gasket has become a paragon of structure and order for us. In other words, we

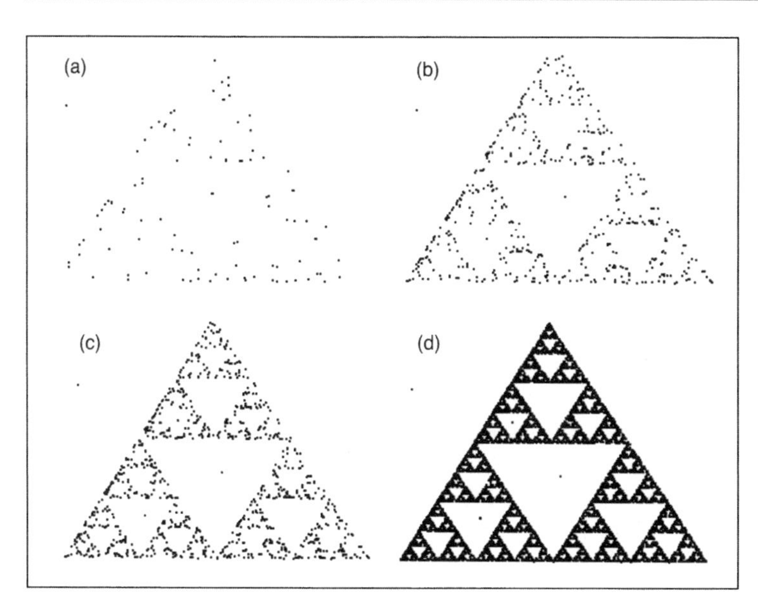

...and the Next Game Points

The chaos game after 100 steps (a), 500 steps (b), 1000 steps (c), and 10,000 steps (d). Only the game points are drawn without connecting lines. (Note that there are a few spurious dots that are clearly not in the Sierpinski gasket.)

Figure 6.2

have seen how randomness can create a perfectly deterministic shape. To put it still another way, if we follow the time process step by step, we cannot predict where the next game point will land because it is determined by throwing a die. But nevertheless, the pattern which all the game points together leave behind is absolutely predictable. This demonstrates an interesting interplay between randomness and deterministic fractals.

But there are a few — if not many — questions about this interaction. For example, how can we explain the small specks which we observe upon close examination of the images in figure 6.2 and which definitely do not belong to the Sierpinski gasket? Or what happens if we use another die, maybe one which is slightly or severely biased? In other words, does the random process itself leave some imprint or not? Or is this creation the result of a special property of the Sierpinski gasket? In other words, are there chaos games which produce some other, or even any other, fractal as well as the Sierpinski gasket?

6.1 The Fortune Wheel Reduction Copy Machine

As you may have guessed, there are many variations of the chaos game, which produce many different fractals. In particular, all images that can be generated by means of a Multiple Reduction Copy Machine of the last chapter are also accessible using the chaos game played with appropriate rules. This is the topic of this section.

Random Affine Transformations

The basic rule of the above chaos game is: generate a new game point z_{k+1} by picking the midpoint between the last game point z_k and the randomly chosen marker, which is represented by a number from the set $\{1, 2, 3\}$. The three possible new game points can be described by three transformations, say w_1, w_2, and w_3 applied to the last game point. What kind of transformations are these? It is crucial to observe that they are the (affine linear) transformations which we discussed for the Sierpinski gasket in chapter 5. There we interpreted them as mathematical descriptions of lens systems in an MRCM. In fact, here each w_n is just a similarity transformation which reduces by a factor of 1/2 and is centered at the marker point n. That implies, that w_n leaves the marker point with label n invariant. In the language of the above rules: if a game point is at a marker point with label n and one draws number n by rolling the die, then the succeeding game point will stay at the marker point. As we will see, it is a good idea to start the chaos game with one of these fixed points.

Chaos Game and IFS Transformations for the Sierpinski Gasket

Our first chaos game generates a Sierpinski gasket. Let us try to derive a formal description of the transformations which are used in this game. To that end we introduce a coordinate system with x- and y-axes. Now suppose that the marker points have coordinates

$$P_1 = (a_1, b_1), \quad P_2 = (a_2, b_2), \quad P_3 = (a_3, b_3).$$

The current game point is $z_k = (x_k, y_k)$, and the random event is the number n (1, 2, or 3). Then the next game point is

$$z_{k+1} = w_n(z_k) = (x_{k+1}, y_{k+1}),$$

where

$$\begin{aligned} x_{k+1} &= \tfrac{1}{2}x_k + \tfrac{1}{2}a_n, \\ y_{k+1} &= \tfrac{1}{2}y_k + \tfrac{1}{2}b_n. \end{aligned}$$

In terms of a matrix (as introduced in the last chapter) the affine linear transformation w_n is given by

$$\left(\begin{array}{cc|c} \frac{1}{2} & 0 & \frac{1}{2}a_n \\ 0 & \frac{1}{2} & \frac{1}{2}b_n \end{array} \right).$$

Note that with $w_n(P_n) = P_n$, the marker points are fixed. Now we can play the chaos game following this algorithm:

Preparation: Pick z_0 arbitrarily in the plane.

Iteration: For $k = 0, 1, 2, \ldots$ set $z_{k+1} = w_{s_k}(z_k)$, where s_k is chosen randomly (with equal probability) from the set $\{1, 2, 3\}$ and plot z_{k+1}.

In other words, in step k, s_k keeps track of the random choice, i.e., the result of throwing the die. The sequence $s_0, s_1, s_2, \ldots$, together with the initial point z_0, is a complete description of a round of the chaos game. We abbreviate the sequence with (s_k). More formally, we would say that (s_k) is a random sequence with elements from the 'alphabet' $\{1, 2, 3\}$.

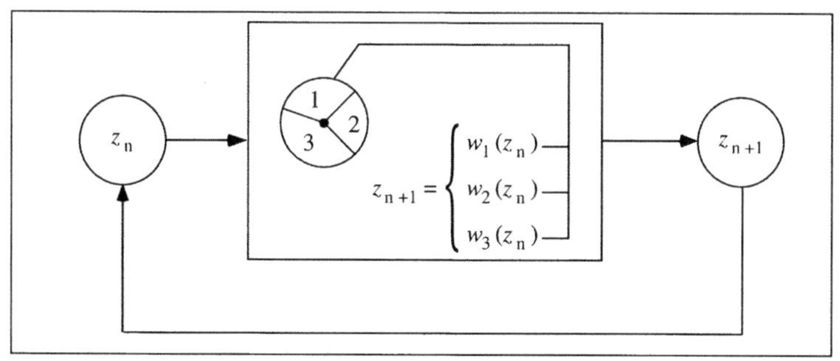

The Fortune Wheel

Feedback machine with fortune wheel (FRCM).

Figure 6.3

MRCM and FRCM

Note that our concept of an MRCM (or IFS) is strictly deterministic. We describe now a modification of our machine which corresponds to the chaos game: rather than applying the copy machine to entire images, we apply it to single points. Moreover, we do not apply all lens systems simultaneously. Rather, in each step we pick one at random (with a certain probability) and apply it to the previous result. And finally the machine does not draw just a single point; it accumulates the generated points. All these accumulated points form the final image of the machine. This would be a random MRCM. Correspondingly, we call it a *Fortune Wheel Reduction Copy Machine* (FRCM). Running this machine is the same as playing a particular chaos game.

What is the relation of an MRCM and its random counterpart? For the Sierpinski gasket we have just seen the answer. The corresponding FRCM also generates a Sierpinski gasket. And indeed this is a case of a general rule: the final image of an MRCM (its IFS attractor) can be generated by a corresponding FRCM, which is the same as playing the chaos game according to a specific set of rules.

Random Iterated Function Systems: Formal Description of FRCM

We have shown that an MRCM is determined by N affine-linear contractions

$$w_1, w_2, \ldots, w_N.$$

One copying step in the operation of the machine is described by the Hutchinson operator

$$W(A) = w_1(A) \cup \cdots \cup w_N(A).$$

Starting with any initial image A_0, the sequence of generated images $A_1 = W(A_0)$, $A_2 = W(A_1)$,... converges to a unique attractor A_∞, the final image of the machine. A corresponding FRCM is given by the same contractions

$$w_1, w_2, \ldots, w_N$$

and some (positive) probabilities

$$p_1, p_2, \ldots, p_N > 0,$$

where

$$\sum_{i=1}^{N} p_i = 1.$$

This setup is called a random iterated function system, while the corresponding MRCM is called a deterministic iterated function system. Let $s_1, s_2, s_3, \ldots$ be a sequence of random numbers which are chosen from the set $\{1, 2, \ldots, N\}$ independently and with probability p_k for the event $s_i = k$. Assume z_0 is a fixed point of one of the transformations (e.g., $w_1(z_0) = z_0$); then

(1) all points of the sequence $z_0, z_1 = w_{s_1}(z_0), z_2 = w_{s_2}(z_1), \ldots$ lie in the attractor A_∞;
(2) the sequence $z_0, z_1, z_2, \ldots$ almost surely fills out the attractor A_∞ densely.

The first fact is immediate from the invariance property of the attractor. The second one will be investigated in the next section. In summary, an MRCM and a corresponding FRCM encode the same image A_∞; we can produce the attractor by playing the chaos game with this machine. The restriction 'almost surely' in the second property is only a fine technical point. Theoretically, it may happen for example, that even though the sequence $s_1, s_2, \ldots$ is random, all events are identical. This is like having a die that forever rolls the number '1', even though it is a perfect and fair die. In this case the chaos game would certainly fail to fill out the attractor. However, the chance of such an abnormal outcome is zero.

The Fern

100,000 game points of the chaos game. Left: FRCM with equal probability for all contractions. Right: Tuned FRCM. Here the probabilities for choosing the different transformations are not the same.

Figure 6.4

A New Approach to the Decoding Problem

In other words, the chaos game provides a new approach to the problem of decoding images from a set of transformations. Let us recall the problem of computational complexity which occurred when we tried to obtain Barnsley's fern by straightforward IFS iteration. We estimated in chapter 5 that we would need about 10^{10} years of computer time for a computer which calculates and draws about a million rectangles per second. If we switch to a chaos game interpretation of the fern, the situation becomes rather different. Now we only have to keep track of one single point. This can be done easily by a computer even if we perform millions of iterations. So let us play the chaos game with the FRCM which is determined by the four transformations $w_1, \ldots, w_4$ which generate the fern. We assume equal probability for all transformations (as we did in our first chaos game). We start with a point z_0, choose at random a transformation — say w_2 — and apply it to z_0. Then we continue with our new game point $z_1 = w_2(z_0)$ and choose another transformation at random, etc. The left part of figure 6.4 shows the disappointing result after more than 100,000 iterations. Indeed, the incompleteness of this image corresponds to the difficulty in obtaining the fern image by running the MRCM. Playing this chaos game for even millions of iterations, we would not obtain a satisfying result.

Now you certainly will wonder how we obtained the right-hand image of figure 6.4. It is also produced by playing the chaos game, and it shows only about 100,000 iterations. What is the difference? Well, with respect to this image we could say we used a 'tuned' fortune wheel where we did not use equal probabilities for all transformations but made an appropriate choice for the probability to use a certain transformation.[3] The satisfactory quality of the right image is a very convincing proof of the potential power of the chaos game as a decoding scheme for IFS encoded images. But how does one select the probabilities, and why does a careful choice of the probabilities reduce the required time of the decoding process from 10^{10} years to a few seconds? And why does the chaos game work at all?

Chaos Game for Networked IFSs

We can play the chaos game also for networked MRCMs, i.e., hierarchical IFSs. From a formal point of view a hierarchical IFS is given by a matrix Hutchinson operator

$$\mathbf{W} = \begin{pmatrix} W_{11} & \dots & W_{1M} \\ \vdots & & \vdots \\ W_{M1} & \dots & W_{MM} \end{pmatrix},$$

which operates on M planes, where each W_{ik} is a Hutchinson operator mapping subsets from the kth to the ith plane.[4] It is important to allow that some of the W_{ik} are the $\emptyset$-operator, i.e., the operator which maps any set into the empty set $\emptyset$. Let us recall, how the chaos game works for an ordinary Hutchinson operator given by N contractions $w_1, \dots, w_N$. We need probabilities $p_1, \dots, p_N$ and an initial point, say x_0. Then we generate a sequence $x_0, x_1, x_2, \dots$ by computing

$$x_{n+1} = w_{i_n}(x_n), \quad n = 0, 1, 2, \dots$$

where $i_n = m \in \{1, \dots, N\}$ is chosen randomly with probability p_m. The chaos game for a matrix Hutchinson operator generates a sequence of vectors $\mathbf{X}_0, \mathbf{X}_1, \mathbf{X}_2, \dots$, the components of which are subsets of the plane. Here $\mathbf{X}_{n+1}$ is obtained from $\mathbf{X}_n$ by applying randomly selected contractions from $\mathbf{W}$ to the components of $\mathbf{X}_n$. To avoid writing many indices we describe one single generation step using the notation $\mathbf{X} = \mathbf{X}_n$ and $\mathbf{Y} = \mathbf{X}_{n+1}$. The components of these two vectors are denoted by $x_1, \dots, x_M$ and $y_1, \dots, y_M$. The random selection of contractions from $\mathbf{W}$ is best described in two steps. For each row i in $\mathbf{W}$ we make two random choices:

Step 1: Choose a Hutchinson operator in the ith row of $\mathbf{W}$ randomly, W_{ik} (it must not be the $\emptyset$-operator.) Assume that this operator is given by N contractions $w_1, \dots, w_N$.

[3] We will present details of tuned fortune wheels in section 6.3.

[4] See the technical section on page 272.

Step 2: Choose a contraction from these $w_1, \dots, w_N$ at random, say w_m.

Then to determine the ith component y_i of $\mathbf{Y}$ we apply the random choice w_m to the kth component x_k of $\mathbf{X}$ (w_m is part of the Hutchinson operator W_{ik}). That is, we compute $y_i = w_m(x_k)$. Again, to obtain the components of $\mathbf{X}_{n+1}$ the contractions are applied to the components of $\mathbf{X}_n$ according to the column index of their Hutchinson operator in $\mathbf{W}$.

The random choices are governed by probabilities. The way we have set up the random iteration it is natural to associate probabilities for both of the above steps. For step 1 we pick probabilities P_{ik} for each Hutchinson operator in $\mathbf{W}$ so that the sum of each row i is 1,

$$P_{i1} + \cdots + P_{iM} = 1, \ \ i = 1, \dots, M$$

where

$$P_{ik} = 0 \text{ if } W_{ik} = \emptyset.$$

This ensures that the $\emptyset$-operators are never chosen. Now assume that W_{ik} is given by the contractions $w_1, \dots, w_N$. We pick probabilities $p_1, \dots, p_N$ for each w_j such that $p_j > 0$ and $p_1 + \cdots + p_N = 1$. Now the probability of choosing w_m is p_m, provided the Hutchinson operator W_{ik} has already been selected in step 1.

Why Does the Chaos Game Work?

But before looking at the issue of efficiency, we need to discuss why the chaos game fills out the attractor of an IFS in the first place. From chapter 5 it is clear that starting the IFS iteration with any initial image A_0 we obtain a sequence $A_1, A_2, \dots$ of images that converge to the attractor image A_∞. Without any loss we may pick just a single point as the initial image, say $A_0 = \{z_0\}$. Assume that the IFS is given by N affine transformations. Then the result after the first iteration is an image consisting of N points, namely

$$A_1 = \{w_1(z_0), w_2(z_0), \dots, w_N(z_0)\}.$$

After the second iteration we get N^2 points, and so on. Of course, these points will get arbitrarily close to the attractor and eventually provide an accurate approximation of the whole attractor.

Playing the chaos game with the same initial point z_0 is very similar to this procedure. It produces a sequence of points $z_1, z_2, \dots$ where the kth point z_k belongs to the kth image A_k obtained from the IFS. Thus, the points z_k get closer and closer to the attractor in the process. If the initial point z_0 is already a point of the final image, then so are all the generated points as shown in figure 6.5. It is very easy to name some points which must be in the attractor, namely the fixed points of the affine transformations involved. Note that z_0 is such a fixed point if $z_0 = w_k(z_0)$ for some $k = 1, \dots, N$.[5] This explains the spurious dots in the Sierpinski gasket back in figure 6.2. There the initial point

[5]Compare the section on affine transformations on page 220.

FRCM Versus MRCM

The first five iterations of the MRCM for the Sierpinski gasket starting out from a single point (the top vertex of the triangle). The points generated by the chaos game starting out from the same initial point are shown in solid black.

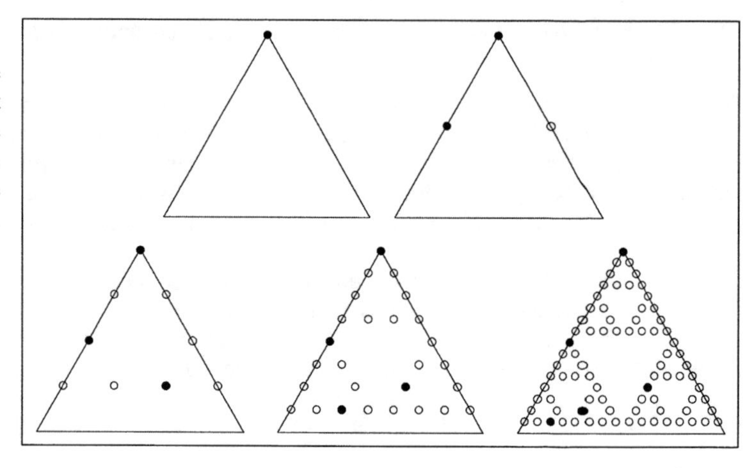

Figure 6.5

was not chosen to be already a point in the gasket. Thus, the first iterations of the chaos game produce points which come closer to the gasket but still are visibly different. The visible difference vanishes after only a few iterations, of course.

To fully understand the success of the chaos game, it remains to be shown in the next section that the sequence of generated points comes arbitrarily close to *any* point in the attractor.

6.2 Addresses: Analysis of the Chaos Game

To analyze the chaos game we need a suitable formal framework which allows us to precisely specify the points of the attractor of an IFS and also the positions of the moving game point. This framework consists of a particular addressing scheme which we will develop using the example of the Sierpinski gasket.

The Metric System As an IFS

The basic idea of such an addressing system is in fact some thousand years old. The decimal number system with the concept of place values explains well the way we look at addresses and the idea behind the chaos game. Let us look at the decimal system in a more materialistic form: a meter stick subdivided into decimeters, centimeters, and millimeters. When specifying a three-digit number like 357 we refer to the 357th out of 1,000 mm. Reading the digits from left to right amounts to following a decimal tree (see figure 6.6) and arriving at location 357 in three steps.

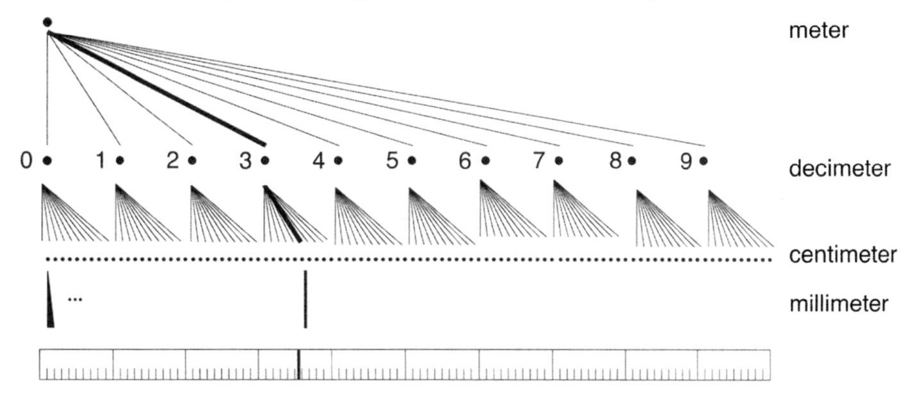

Figure 6.6 : Locating 357 by the decimal tree on a meter stick.

It is crucial for our discussion of the chaos game that there is another way to arrive at location 357 reading the digits *from right to left*. This makes us familiar with the *decimal MRCM*. The decimal MRCM is an IFS consisting of ten contractions (similarity transformations) $w_0, w_1, \ldots, w_9$ given explicitly by

$$w_k(x) = \frac{x}{10} + \frac{k}{10}, \quad k = 0, 1, \ldots, 9.$$

In other words w_k reduces the meter stick to the kth decimeter. Running the decimal MRCM establishes the familiar metric system on the meter stick.[6]

Start with the 1 meter unit. The first step of the decimal MRCM generates all the decimeter units of the meter stick. The second step generates all the centimeters, and so on. In this sense the decimal — together with its ancient relatives like the hexagesimal — system is probably the oldest MRCM.

Let us now read 357 from right to left by interpreting digits as contractions. Thus, starting with a 1 meter unit, we first apply the transformation w_7, which

[6]Here is an exercise: can one also construct an MRCM for the British/American system relating miles to feet and inches?

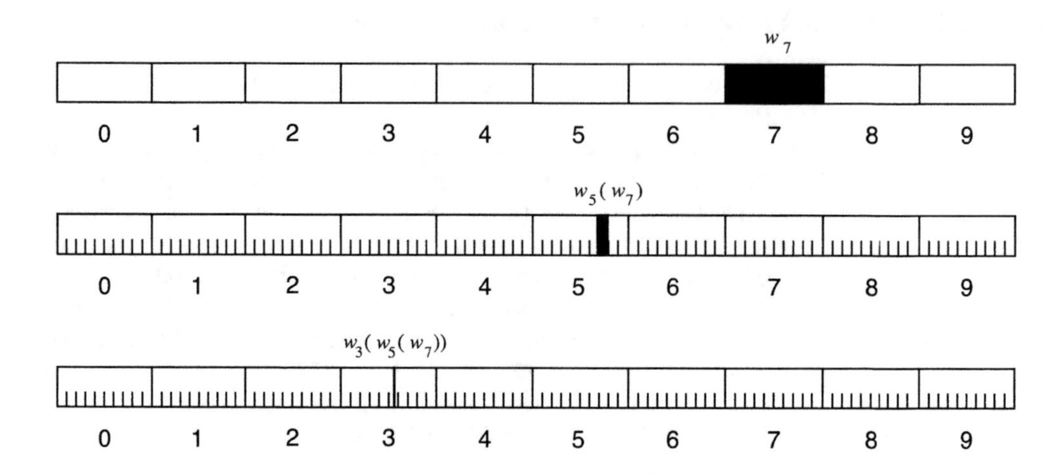

Figure 6.7 : Locating 357 by applying contractions of the decimal MRCM.

leads to the decimeter unit starting at 7 (see figure 6.7). Next we apply w_5 and arrive at the 57th centimeter. Finally w_3 brings us to the 357th millimeter location again. Thus, reading from left to right and interpreting in terms of place values, or reading from right to left and interpreting in terms of decimal contractions is the same.

The Chaos Game on the Meter Stick

We now play the chaos game on the meter stick. We generate a random sequence of digits from $\{0, \ldots, 9\}$, start with an arbitrary game point (= a millimeter location) and move to a new location according to the random sequence. We would consider the chaos game to be successful if it eventually visits all millimeter locations. Let us look at a random sequence like

$$\ldots 765016357,$$

where we write from right to left for convenience. After the third step in the game we arrive at the millimeter location 357. The next random number is 6. Which millimeter location is visited next? Clearly 635! The initial number 7 has therefore lost its relevance; no matter what this number is we visit the millimeter site 635 in the fourth step. For the same reason we continue to visit 163, then 016, and so on. In other words, running the chaos game amounts to moving a slider with a window three digits wide from right to left over the random sequence (see figure 6.8).

When will we have visited all millimeter locations? This will obviously be the case, when the slider window will have shown us all possible three-digit combinations. Is that likely to happen, if we produce the digits by a random number generator? The answer is 'yes', because that is one of the fundamental features which are designed into random number generators on computers. It is just a lazy way to generate all possible three-digit addresses. Even a random number generator which is miserable with respect to the usual statistical tests

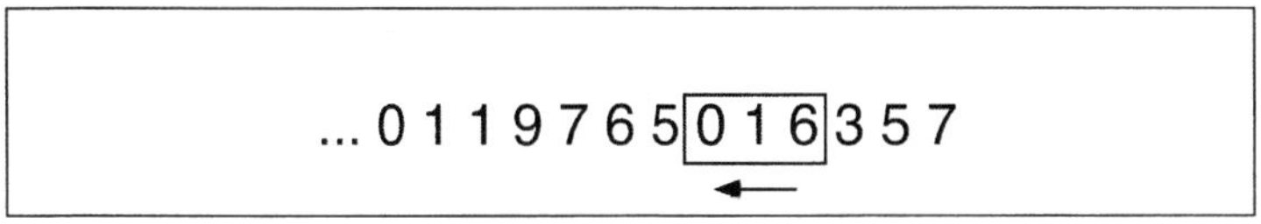

...0 1 1 9 7 6 5 |0 1 6| 3 5 7

A three-digit window sliding over the sequence ...0119765016357 allows to address millimeter locations.

Figure 6.8

would do the job provided it generates all three-digit combinations.[7]

Let us now see how the same idea works for the Sierpinski gasket, the fern, and in general. We know that there is a very definite hierarchy in the Sierpinski gasket. At the highest level (level 0) there is one triangle. At the next level (level 1) there are three. At the level 2 there are nine. Then there are $27, 81, 243$, and so on. Altogether, there are 3^k triangles at the kth level. Each of them is a scaled down version of the entire Sierpinski gasket, where the scaling factor is $1/2^k$ (refer to figure 2.16).

Triangle Addresses

We need a labeling or addressing scheme for all these small triangles in all generations. The concept for this is similar to the construction of names in some Germanic languages as, for example, Helga and Helgason, John and Johnson, or Nils and Nilsen. We will use numbers as labels instead of names:

Level 1	Level 2	Level 3
	11	111 112 113
1	12	121 122 123
	13	131 132 133
	21	211 212 213
2	22	221 222 223
	23	231 232 233
	31	311 312 313
3	32	321 322 323
	33	331 332 333

Unfortunately, we run out of space very rapidly when we try to list the labels at more than a very few levels. The system, however, should be apparent. It is a labeling system using lexicographic order much like that in a telephone book, or like the place value in a number system. The labels $1, 2$, and 3 can be interpreted in terms of the hierarchy of triangles or in terms of the hierarchy of a tree (see figure 6.9). For triangles:

- 1 means lower left triangle;
- 2 means lower right triangle;
- 3 means upper triangle.

[7] Barnsley explains the success of the chaos game by referring to results in ergodic theory (M. F. Barnsley, *Fractals Everywhere*, Academic Press, San Diego, 1988). This is mathematically correct but practically useless. There are two questions: One is, why does the properly tuned chaos game produce an image on a computer screen so efficiently? The other is, why does the chaos game generate sequences which fill out the IFS attractor densely? These are not the same questions! The ergodic theory explains only the latter, while it cannot rule out that it may take some 10^{10} years for the image to appear. In fact, this could actually happen, if computers lasted that long.

Address Trees

Sierpinski tree (left), symbolic tree (right).

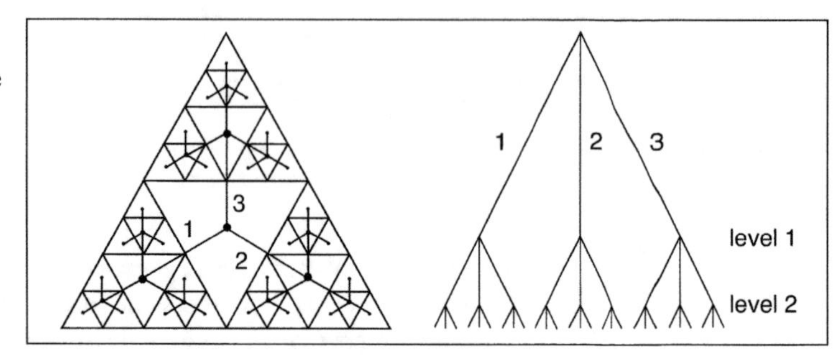

Figure 6.9

Locating Addresses

Locating the subtriangle with address 132 in the Sierpinski gasket by a sequence of nested subtriangles.

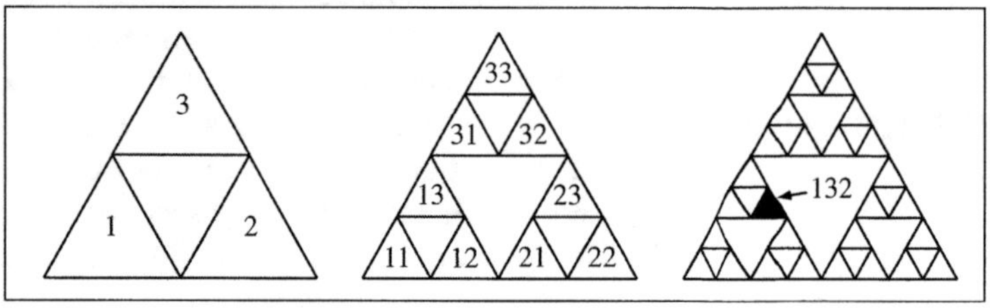

Figure 6.10

The Triangle With Address 13213

Thus, the address 13213 means that the triangle we are looking for is in the 5th level. The address 13213 tells us exactly where to find it. Let us now read the address. We read it from left to right much like a decimal number. That is, the places in a decimal number correspond here to the levels of the construction process. Start at the lower left triangle of the first level. Within that find the upper triangle from the second level. Therein locate the lower right triangle from the third level. We are now in the subtriangle with address 132 (see figure 6.10). Within that take the lower left triangle from the fourth level, and finally therein come to rest at the upper triangle. In other words, we just follow the branches of the Sierpinski tree in figure 6.9 five levels down.

Let us summarize and formalize. An address of a subtriangle is a string of digits $s_1 s_2 \ldots s_k$ where each s_i is an integer from the set $\{1, 2, 3\}$. The index k can be as large as we like. It identifies the level in the construction of the Sierpinski gasket. The size of a triangle decreases by 1/2 from level to level, at the kth level it is $1/2^k$.

Address of a Point

Let us now pick a *point* in the Sierpinski gasket. How can we specify an address in this case? The answer is that we have to carry on the addressing scheme for subtriangles ad infinitum specifying smaller and smaller subtriangles all of which contain the given point. Thus, we can identify any given point z by a sequence of triangles, $D_0, D_1, D_2, \ldots$ There is one triangle from each level such that D_{k+1} is a subtriangle of D_k and z is in D_k for all $k = 0, 1, 2, \ldots$ That sequence of triangles determines a sequence of integers $s_1, s_2, \ldots$ such

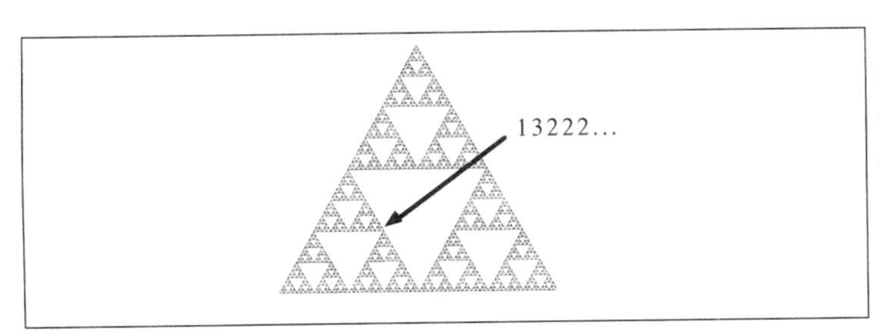

Address of a Point

The point address 13222... Note that this point could also be identified by address 12333...

Figure 6.11

that

$$\text{address}(D_1) = s_1$$
$$\text{address}(D_2) = s_1 s_2$$
$$\text{address}(D_3) = s_1 s_2 s_3$$

and so on. Selecting more and more terms in that sequence means locating z in smaller and smaller triangles (i.e., with more and more precision). This is just like fixing a location on a meter stick in higher and higher precision. Therefore, taking infinitely many terms identifies z exactly:

$$\text{address}(z) = s_1 s_2 s_3 \ldots \tag{6.1}$$

Reading Left to Right

It is important to remember how to read the address. The address is read from left to right and is thus interpreted as a sequence of nested triangles. The position of the digit in the sequence determines the level in the construction.

Touching Points in the Sierpinski Gasket

We must point out, though, that our addressing system for the points of the Sierpinski gasket does not always yield unique addresses. This means that there are points with two different possible addresses much like $0.4\overline{9}$ and 0.5 in the decimal system. Let us explore this fact. When constructing a Sierpinski gasket, we observe that in the first stage there are three triangles, and any two of these meet at a point. In the next stage there are nine triangles, and any adjacent pair of these nine triangles meet at a point. What are the addresses of the points were subtriangles meet? Let us check one example (see figure 6.11). The point where the triangles with label 1 and 3 meet has the addresses: 1333... and also 3111... Likewise, the point where the triangles with labels 13 and 12 meet has two different addresses: namely, 13222... and 12333... As a general rule all points z where two triangles meet must have addresses of the form

$$\text{address}(z) = s_1 \ldots s_k r_1 r_2 r_2 r_2 \ldots$$

and

$$\text{address}(z) = s_1 \ldots s_k r_2 r_1 r_1 r_1 \ldots$$

where s_i, r_1, r_2 are from the set $\{1, 2, 3\}$ and r_1 and r_2 are different. Points which have that nature are called *touching points*. They are characterized by twin addresses (compare figure 6.12).

Touching Points

The touching point with twin addresses 1333... and 3111...

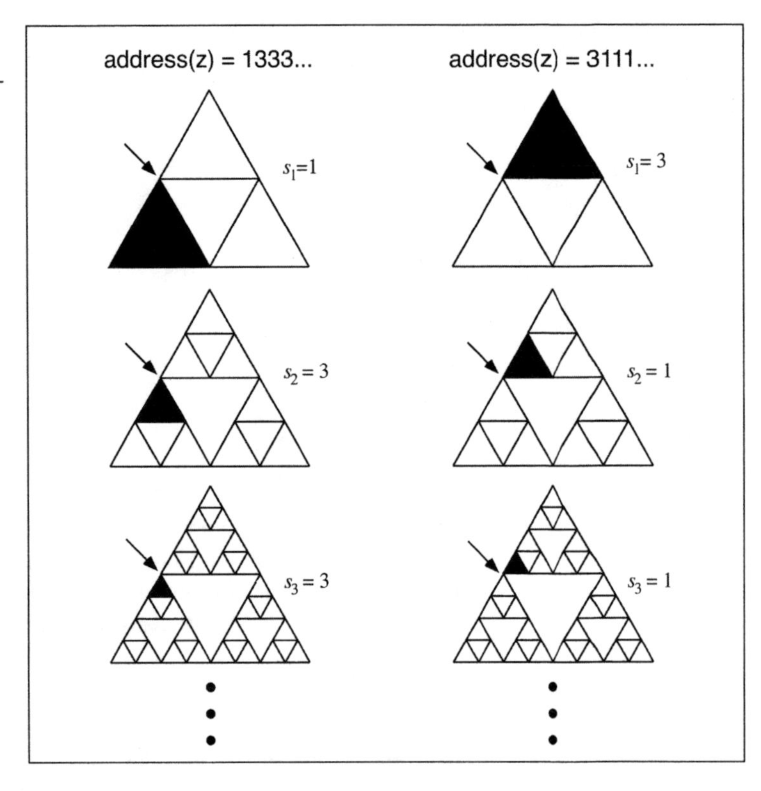

Figure 6.12

From the construction process of the Sierpinski gasket one might be misled to conjecture that, except for the three outside corner points, all its points are touching points. This conjecture is wrong; and here, using the language of addresses, is a nice argument which makes the issue clear. If all points of the Sierpinski gasket were touching points they could be characterized by twin addresses of the above form. But obviously most addresses we can imagine are not of this particular form (e.g., address$(z) = s_1 s_2 s_3 \ldots$ where each s_i is randomly chosen). In other words most points are not touching points.

Space of Addresses

Let us now develop the formalism of addresses a little more. To this end we introduce a new object, $\sum_3$, the space of addresses. An element σ from that space is an infinite sequence $\sigma = s_1 s_2 \ldots$, where each s_i is from the set $\{1, 2, 3\}$. Each element σ from that space identifies a point z in the Sierpinski gasket. However, different elements in $\sum_3$ may correspond to the same point. This is the case for all touching points.

Addresses for the Cantor Set

At this point let us see how the concept of addresses works in another example of fractals, the Cantor set C. Here we would address with only two labels, 1 and 2. All infinite strings of 1's and 2's together are the address space $\sum_2$. There is a significant difference when we compare the Cantor set with the Sierpinski gasket. Points on the Cantor set have only one address. We say that addresses for the Cantor set are *unique*. Thus, for each address there

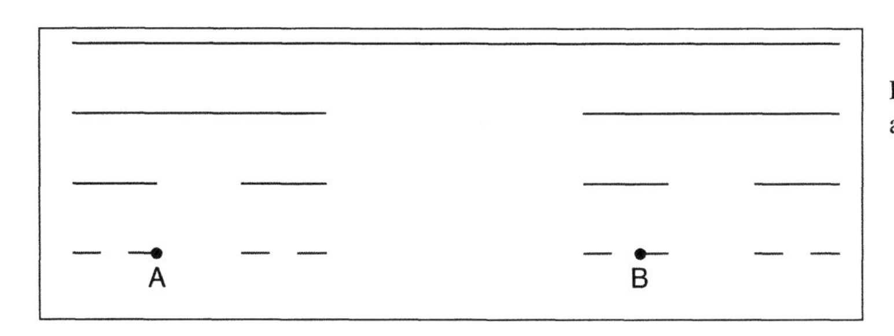

Cantor Set Addresses

Point A has address 11222... and the address of point B is 212111...

Figure 6.13

is exactly one point, and vice versa. In other words, $\sum_2$ is in a one-to-one correspondence with C: we can identify $\sum_2$ and C. In the case of $\sum_3$ and the Sierpinski gasket, this is not possible because there exist points with two different addresses.

Addresses for IFS Attractors

Any fractal which is the attractor of an IFS has a space of addresses attached to it. More precisely, if the IFS is given by N contractions $w_1, \ldots, w_N$, then any point z from the attractor A_∞ has an address in $\sum_N$, the space of all infinite sequences $s_1 s_2 s_3 \ldots$, where each number s_i is from the set $\{1, 2, \ldots, N\}$.

To be specific, let us take any point z in the attractor A_∞ of the IFS, which is given by the set of N contractions $w_1, \ldots, w_N$. These contractions applied to A_∞ yield a covering of the attractor as explained in chapter 5,

$$A_\infty = w_1(A_\infty) \cup \cdots \cup w_N(A_\infty).$$

Our given point z surely resides in at least one of these sets, say in $w_k(A_\infty)$. This determines the first part of the address of z, namely $s_1 = k$. The set $w_k(A_\infty)$ is further subdivided into N (not necessarily disjoint) subsets

$$\begin{aligned} w_k(A_\infty) &= w_k(w_1(A_\infty) \cup \cdots \cup w_N(A_\infty)) \\ &= w_k(w_1(A_\infty)) \cup \cdots \cup w_k(w_N(A_\infty)). \end{aligned}$$

Again our given point surely is in at least one of these subsets, say $w_k(w_l(A_\infty))$, and this determines the second part of the address of z, namely $s_2 = l$. Note that there may be several choices for s_2, in which case we would have several different addresses for one point. This procedure can be carried on indefinitely. Computing more and more components of the address specifies the point z more and more precisely because the subsets considered get smaller and smaller due to the contraction property of the affine transformations of the IFS.

As in the case of the Sierpinski gasket, we get a sequence of nested subsets D_k of increasing level of the attractor, all of which contain the given point z. If $\sigma = s_1 s_2 \ldots$ denotes an address of z these subsets are

$$D_k = w_{s_1}(w_{s_2}(\cdots w_{s_k}(A_\infty))).$$

For brevity of notation we often omit the brackets. So

$$w_k w_l(A_\infty) = w_k(w_l(A_\infty))$$

Address Interpretation

Reading the address backwards when applying the contractions.

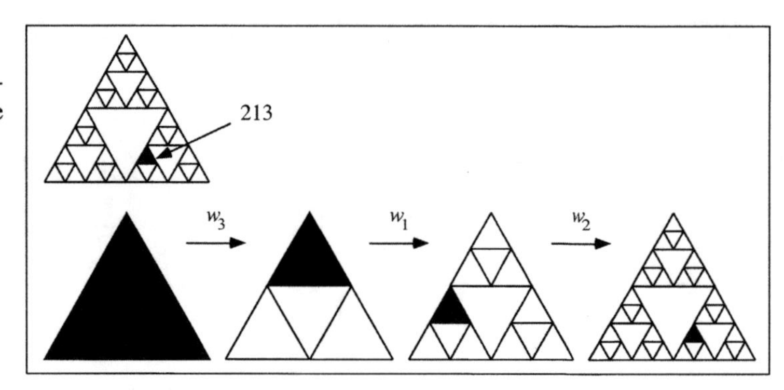

Figure 6.14

and

$$w_{s_1} w_{s_2} \cdots w_{s_k}(A_\infty) = w_{s_1}(w_{s_2}(\cdots w_{s_k}(A_\infty))).$$

Reading Right to Left

It is important to note here that in a sense we have now read the sequence $s_1 s_2 \ldots s_k$ from right to left because we first apply w_{s_k} to A_∞. Then to the result of that, we apply $w_{s_{k-1}}$, and so on, until we finally apply w_{s_1}. Let us look at an example. In figure 6.14 we show how the subtriangle with address 213 is obtained by $w_2(w_1(w_3(S)))$. In words, this is 'the top (3) of the left (1) of the right (2) subtriangle' (note the correspondence in the order). This simple observation that addresses may be interpreted from left to right or from right to left in different operational ways leading however to the same location is very crucial for understanding why the chaos game works.

One-to-One or Many-to-One

Assume there is a one-to-one correspondence between points in A_∞ and $\sum_N$. In other words, there is a unique address for each point z in the attractor A_∞. Then we call the attractor *totally disconnected*. Figure 6.15 shows the attractors of three IFSs, one totally disconnected, one just touching, and the third with overlap. In the overlapping case it is hard to see an address for a given point by visual inspection. However, it is always easy to compute the corresponding point for a given address.

So far we have developed a labelling scheme for points of the IFS attractor. Using this technique we now discuss why the chaos game works and how it can generate the attractor. We will once again discuss the basic ideas for the Sierpinski gasket and then see how we can extend to the general case.

Chaos Game with Equal Probabilities

Let us begin with some straightforward observations. In the ideal case our die will be perfect. Any of the numbers 1, 2, or 3 will appear with the same statistical frequency. If p_k denotes the probability of the event of our throwing number k, $k = 1, 2, 3$, then $p_1 = p_2 = p_3 = 1/3$.

Let us now play the chaos game with such a perfect die. We assume that the actual game point z_n is in the Sierpinski gasket, but we do not know where. The Sierpinski gasket can be broken down into three sets in the first level, nine in the second, and 3^k in the kth level. Let us pick one of these sets, for example, one from the first level. What is the probability that we will see

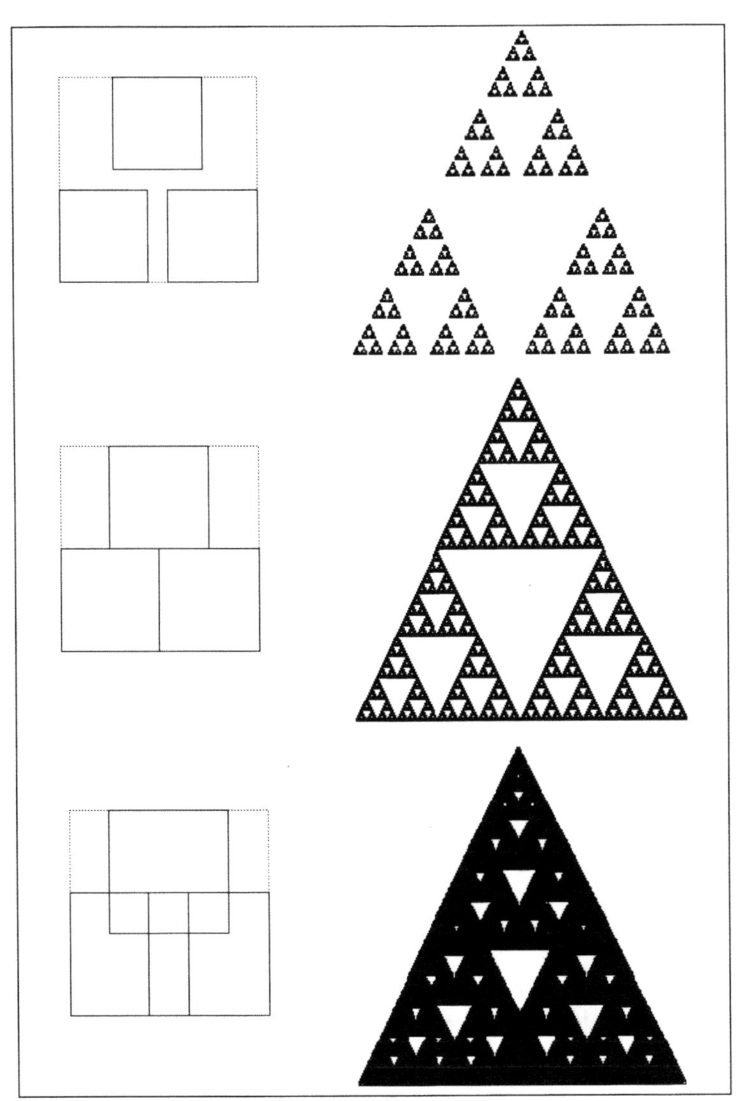

Three Cases

Three IFS attractors, totally disconnected, just touching, and overlapping (MRCM blueprints appear reduced).

Figure 6.15

the next game point z_{n+1} in this subtriangle? Obviously, the probability will be 1/3, no matter where z_n, and for that same reason $z_{n-1}, z_{n-2}, \ldots$ are (see figure 6.16).

Now let us pick a set from the second level. Again, we assume no information about the location of z_n other than its being in the Sierpinski gasket. Therefore, if we want to see a succeeding game point on a selected set D of the second level, we should generate two new game points z_{n+1}, and z_{n+2}. What is the probability that we will see z_{n+2} in D? Obviously it is 1/9. Thus, if we pick a set D from the kth level, then the probability that the chaos game

Probabilities

The probability that a game point lands in a selected set of the first level after one iteration is 1/3. For a set of the second level after two iterations the probability is 1/9.

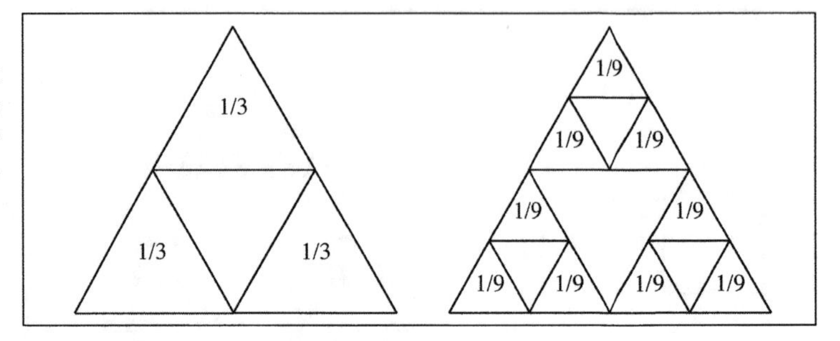

Figure 6.16

will produce a game point z_{n+k} which lands in D after k iterations is $1/3^k$.

Let us repeat in terms of the contractions w_1, w_2, and w_3. Each of the contractions w_1, w_2, and w_3 is drawn with probability 1/3. Consequently, each pair $w_i w_k$ is drawn with probability 1/9, and in general each of the 3^k possible compositions $w_{s_1} w_{s_2} \cdots w_{s_k}$ with s_i from the set $\{1, 2, 3\}$, is selected with probability $1/3^k$.

We can now explain why the chaos game produces a sequence of game points which will eventually fill out the entire Sierpinski gasket at any resolution. Mathematically, the chaos game produces a sequence

$$\begin{aligned} z_0 &= \text{starting point (initial game point)} \\ z_1 &= w_{s_1}(z_0) \\ z_2 &= w_{s_2} w_{s_1}(z_0) \\ &\vdots \\ z_k &= w_{s_k} \cdots w_{s_2} w_{s_1}(z_0), \end{aligned}$$

where the sequence of events $s_1, s_2, \ldots, s_{k-1}, s_k$ is randomly generated. The last point z_k is in a kth generation subtriangle D which has the address $s_k s_{k-1} \ldots s_2 s_1$.

The Game Points Get Close to Any Point of the Gasket

Pick a test point P on the Sierpinski gasket. We need an argument which establishes that if we play the chaos game sufficiently long, we will produce points which come as close to P as we wish. To this end we want to see a game point which has, at most, a small distance ε from P. Let us assume that the diameter of the Sierpinski gasket is d. We know that the triangles in the mth level of the Sierpinski gasket have the diameter $d/2^m$. In other words, if we choose m so large that $d/2^m < \varepsilon$ and pick a triangle D in the mth generation which contains P, then every point in D has, at most, distance ε from P. This set D has an address

$$\text{address}(D) = t_1 t_2 \ldots t_m, \quad t_i \in \{1, 2, 3\}.$$

Now let us look at a run of the chaos game with many events. We like to write the sequence in reverse order, $\ldots, s_k, s_{k-1}, \ldots, s_2, s_1$. As soon as we detect a block of length m within the sequence $\ldots, s_k, s_{k-1}, \ldots, s_2, s_1$ which is identical to $t_1, t_2, \ldots, t_m$, we are finished. For example, let us take

$$\ldots, s_k, \ldots, s_{j+m+1}, t_1, t_2, \ldots, t_m, s_j, \ldots, s_2, s_1.$$

Then z_j is some point in the Sierpinski gasket, and therefore $w_{t_1} \cdots w_{t_m}(z_j)$ will be in D. Thus, everything is settled if we can trust to eventually seeing block $t_1 t_2 \ldots t_m$ as we play. But the probability that any sequence of length m matches up with $t_1 t_2 \ldots t_m$ is equal to $1/3^m$. Therefore, playing the chaos game with a perfect die will sooner or later produce such a sequence, and thus a point which is in the subtriangle D and therefore as close to the test point P as we required.

Chaos Game Generates IFS Attractor

The chaos game will produce points which densely cover the Sierpinski gasket. We can generalize this fact for the attractor of an arbitrary IFS. Let us briefly sketch the arguments for this general situation. Let the IFS be given by N contractions $w_1, \ldots, w_N$, and let A_∞ be its attractor. The attractor is invariant under the Hutchinson operator $H(X) = w_1(X) \cup \cdots \cup w_N(X)$. A corresponding random IFS is given by these contractions w_i and associated probabilities p_i (with $p_i > 0$ and $p_1 + \cdots + p_N = 1$). We need to show that we can get arbitrarily close to any point P of the attractor A_∞ by playing the chaos game with this setup. Now let us try to generate a point which lies within a distance ε from P. Let the address of P be given by

$$\text{address}(P) = t_1 t_2 \ldots$$

where $t_i \in \{1, \ldots, N\}$. Then the point P is contained in all sets A_m,

$$A_m = w_{t_1} w_{t_2} \cdots w_{t_m}(A_\infty), \quad m = 1, 2, \ldots$$

We have

$$A_1 \supset A_2 \supset A_3 \supset \cdots$$

and the diameter[8] of A_m decreases to zero as m increases. Knowing the contraction ratios $c_1, \ldots, c_N$ of the transformations $w_1, \ldots, w_N$, these diameters can be estimated. By definition of the contraction ratios we have for any set B with diameter $\text{diam}(B)$ that after transformation by w_i the diameter is reduced by the contraction ratio $c_i < 1$:

$$\text{diam}(w_i(B)) \leq c_i \text{diam}(B).$$

Therefore, the diameter of A_m can be bounded:

$$\begin{aligned} \text{diam}(A_m) &= \text{diam}(w_{t_1} \cdots w_{t_m}(A_\infty)) \\ &\leq c_{t_1} c_{t_2} \cdots c_{t_{m-1}} c_{t_m} \text{diam}(A_\infty). \end{aligned}$$

Since the contraction ratios are all less than 1, we can make this diameter as small as ε just by considering a sufficiently large number m of transformations in this sequence. Thus, all points with addresses starting with $t_1 \ldots t_m$ have a distance of at most ε from the given point P. In other words, we need to see this sequence $t_1, \ldots, t_m$ sometime during the chaos game. The chance that any given block of

[8]Recall from chapter 4 that the diameter of a bounded set B is $\text{diam}(B) = \sup\{d(x, y) \mid x, y \in B\}$ where $d(x, y)$ denotes the distance between x and y.

length m exhibits this sequence is equal to the product of probabilities $p_{t_1} p_{t_2} \cdots p_{t_m}$, a positive number. In other words, we can get from any point of A_∞ arbitrarily close to P when playing the chaos game sufficiently long.

Addresses for Rectangular Array

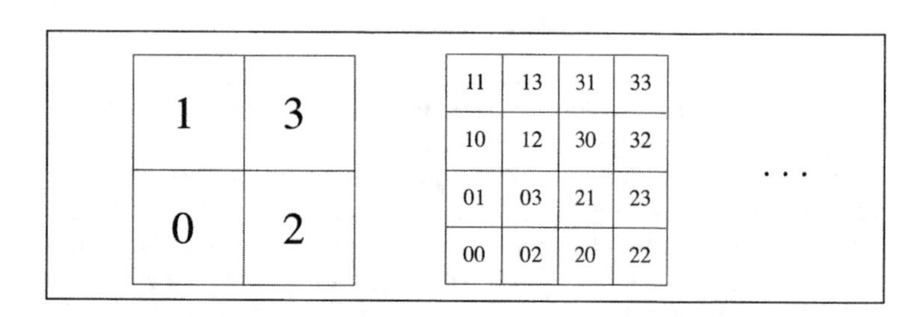

Figure 6.17

Chaos Game on a Computer Screen

So far we have discussed the chaos game from a mathematical point of view. Let us try to materialize the setup a bit (i.e., discuss the chaos game in a form that is closer to the situation encountered on a computer screen). The pixels of a computer screen form a rectangular array. Usually they are identified by coordinates (e.g., pixel number 5 in row number 12). But we can also use an address system as discussed in this chapter.

First we divide the screen into four equal squares and assign the addresses 0 to 3 as shown in figure 6.17. Next we divide each square into four equal parts, each of which is then identified by two-digit addresses. In this way we can describe the pixels of an 8×8 pixel screen by three-digit addresses and a 16×16 pixel screen by four-digit addresses (in general a $2^n \times 2^n$ pixel screen by n-digit addresses).

Assume we work with an 8×8 pixel screen. How do we find the pixel with address 301? We read the address from left to right and follow a sequence of nested squares which finally identifies the pixel at screen coordinates (4,5) — see figure 6.18. Now we set up four contractions w_0, w_1, w_2 and w_3 as in section 5.4 (i.e., the transformation w_i transforms the whole screen into the square i) in correspondence to our address system.

For example, if the square Q represents the whole screen, the sequence

$$w_3(Q) \supset w_3(w_0(Q)) \supset w_3(w_0(w_1(Q)))$$

is just the sequence of nested rectangles enclosing our pixel P in figure 6.18.

In section 5.4 we have demonstrated what happens if we drop the contraction w_3 which transforms the whole screen into the upper right square. As a result we obtain only those pixels that have an address which does not contain the digit 3. We have shown that the IFS associated with w_0, w_1 and w_2 generates a Sierpinski gasket as in figure 5.9 (or in this case its 8×8 pixel approximation). Let us now see how the chaos game played with the contractions w_0, w_1 and w_2 also generates exactly these pixels.

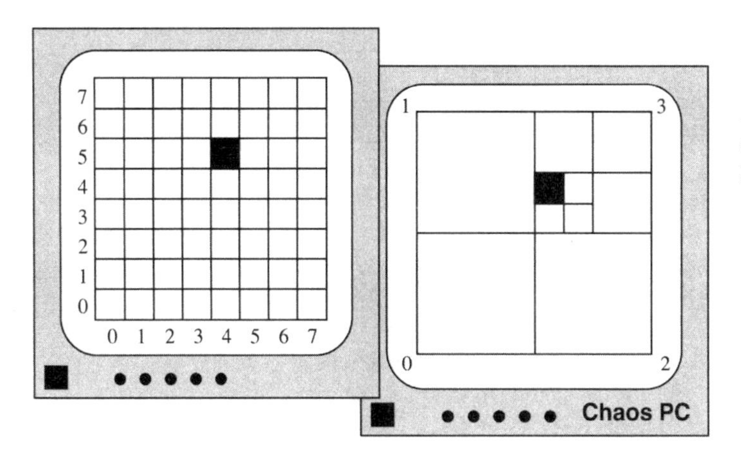

Screen Coordinates and Addresses

Addressing pixels: pixel at screen coordinates (4,5) has the address 301.

Figure 6.18

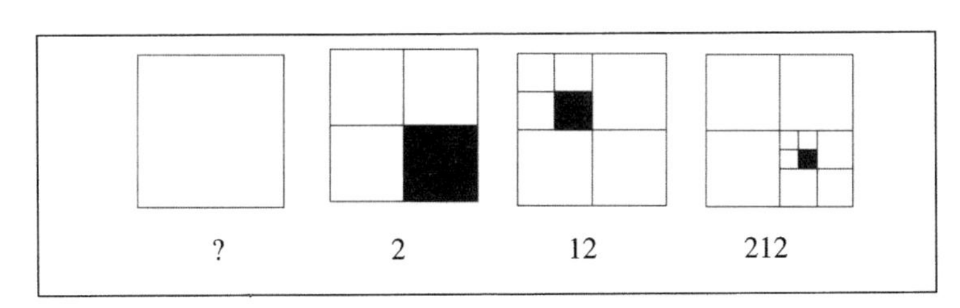

Route to Pixel 212

First steps of the chaos game leading to pixel 212.

Figure 6.19

Again let us look at a concrete example of random numbers, say,

$$\ldots 01211210010212.$$

Starting with a game point anywhere on the screen, the first move brings us to square 2, the second one into the square with address 12 and the third to pixel 212 (see figure 6.19). The next step would lead to address 0212 (a subsquare of pixel 021) provided we had four-digit addresses. But since we work with an 8×8 resolution and corresponding three-digit addresses we can forget about the fourth digit in the address, the trailing 2. Next we would visit pixel 102, then 010, and so on. In other words, we slide a three-digit window over our random sequence and switch on all pixels whose addresses appear.

In this setup the chaos game will be successful if the sequence of numbers which drives the game contains all possible three-digit combinations of the numerals 0, 1 and 2. In fact, the drawing would not even have to be random. The only property which we need is that all possible addresses would appear. Moreover, the efficiency of the chaos game would be governed by how fast all possible combinations would be exhausted. That is the true secret of the chaos game. It has nothing to do with deep mathematical results like ergodic theory, as is sometimes argued in some research literature.[9]

[9] This was first observed by Gerald S. Goodman; see G. S. Goodman, *A probabilist looks at the chaos game,* in: *Fractals in the Fundamental and Applied Sciences,* H.-O. Peitgen, J. M. Henriques, L. F. Peneda (eds.), North-Holland, Amsterdam, 1991.

6.3 Tuning the Fortune Wheel

Our discussion of the Fortune Wheel Reduction Copy Machine has been based on the assumption that the die used for the chaos game is perfect. The probabilities for the transformations are the same. But what effect would a change of these probabilities have on the outcome of the chaos game?

How Many Hits Are in a Subtriangle?

Let us discuss this in a little more formal way for the triangles D in the mth level of the Sierpinski gasket. If we play n times and z_1 to z_n are the game points, the question is, how many points among $z_1, \ldots, z_n$ fall into D? Let us denote this count by $h(z_1, \ldots, z_n; D)$. When the die is perfect we correctly expect that in the long run, each of the small triangles in the mth level will be hit the same number of times. More precisely, the statement is that the fraction of points from $z_1, \ldots, z_n$ which are in D tends to $1/3^m$ as we consider more and more points, i.e., as the total number of points n increases. Note, that there are 3^m subtriangles of the mth generation, all of which should be equally probable. Expressed in a formula,

$$\lim_{n \to \infty} \frac{h(z_1, \ldots, z_n; D)}{n} = \frac{1}{3^m}. \tag{6.2}$$

In other words, counting the events falling in D generates a measure $\mu(D)$, which is nothing but the probability which we have attached to D in our earlier discussion.

The following table lists the counts of points falling in each subtriangle of the second generation when 1000 points are generated as in figure 6.20. In the long run each subtriangle should collect 11.1% of all points.

Address	Count	In %
11	103	10.3
12	122	12.2
13	105	10.5
21	107	10.7
22	112	11.2
23	117	11.7
31	108	10.8
32	108	10.8
33	118	11.8

Now let us change the situation slightly. Let us assume that our die is biased. This means that the probabilities p_1 for number 1 to come up, or p_2 for number 2 to appear, or p_3 for number 3 to be shown are no longer the same. In other words, we have that $p_1 + p_2 + p_3 = 1$, and for example, $p_1 = 0.5$, $p_2 = 0.3$, and $p_3 = 0.2$. Before we discuss what will happen to the chaos game under these circumstances, we want to explain how we can produce a die with exactly this bias.

Simulating Loaded Dice

Naturally we will simulate the rolling of a die with a computer by calling random numbers from a *random number generator*. Usually, whatever

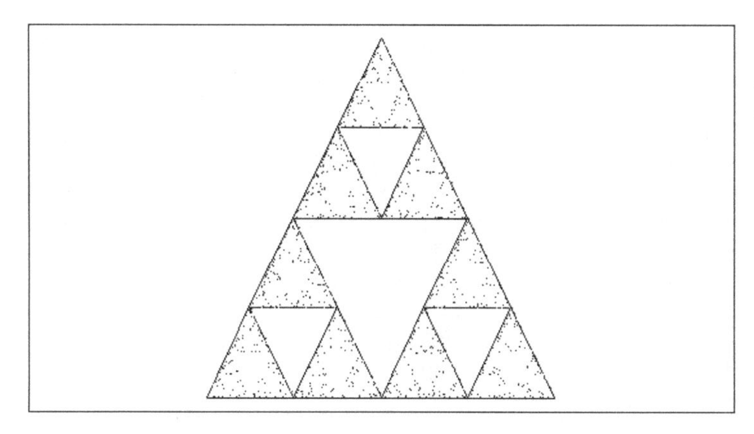

Game Points with Perfect Die

One-thousand game points, when all transformations are chosen with equal probability.

Figure 6.20

the actual algorithm may be, random numbers which are supplied in a programming environment are normalized (i.e., take values between 0 and 1) and are uniformly distributed. Uniform distribution means that the probability of producing a random number in a small interval $[a, b]$ with $0 \leq a < b \leq 1$ is equal to $b - a$. Thus, if the interval $[0, 1]$ is broken down into say 100 subintervals $[0.00, 0.01], [0.01, 0.02]$, etc. we can expect in the long run that in each subinterval we will collect about 1% of all generated random numbers.

Tuning of Random Number Generators

The random number generator returns a sequence of numbers $r_1, r_2, \ldots$ from the interval $[0, 1]$ which we divide into N subintervals

$$[0, 1] = [x_0, x_1) \cup [x_1, x_2) \cup \cdots \cup [x_{N-1}, x_N]$$

where

$$0 = x_0 < x_1 < \cdots < x_{N-1} < x_N = 1.$$

Let us denote by I_k the kth such interval. After n calls for random numbers, we have obtained $r_1, \ldots, r_n$ and count the number of results r_i falling in the subinterval I_k. Let us call the result of this count $h(r_1, \ldots, r_n; I_k)$. For a good random number generator, we expect that $h(r_1, \ldots, r_n; I_k)$ will only depend on the length of I_k and is in fact equal to this length. In terms of a formula, we expect

$$\lim_{n \to \infty} \frac{h(r_1, \ldots, r_n; I_k)}{n} = \text{length } (I_k) = x_k - x_{k-1}.$$

It is interesting to note that we can turn this relation around to compute the length of the interval by counting random numbers! Let us remark in passing that a whole class of methods for the numerical computation for many different types of problems have been based on a similar use of random numbers. For apparent reasons these methods are called Monte Carlo methods.

For example, in the year 1777 Georges L. L. Comte de Buffon (1707–1788) suggested computing the probability that a needle, dropped on a page of ruled paper, will intersect one of the lines. He solved the problem, and the answer turned out to reveal a relation to the number $\pi = 3.141592\ldots$ Assuming that the distance d of the parallel lines is greater than the length l of the needle, it is not hard to show that the probability p the needle will hit one of the lines is equal to $2l/d\pi$. Later Pierre Simon de Laplace (1749–1827) interpreted this relation as an entirely new way of computing π. Just throw the needle many times and count the intersections. This count divided by the total number of throws approximates this probability p, and thus, facilitates the computation of $\pi = 2l/dp$.[10]

But let us return to the tuning of the chaos game. We obtain an arbitrarily tuned random number generator (a die with N faces and N prescribed corresponding probabilities) in the following way: if we want probabilities $p_k, k = 1, \ldots, N$, we define

$$\begin{aligned} I_1 &= [0, p_1) \\ I_2 &= [p_1, p_1 + p_2) \\ &\vdots \\ I_k &= [p_1 + p_2 + \cdots + p_{k-1}, p_1 + p_2 + \cdots + p_k) \\ &\vdots \\ I_N &= [p_1 + p_2 + \cdots + p_{N-1}, 1] \end{aligned}$$

and choose event k provided the random number r_i is in the interval I_k.

With such a random number generator on hand it is easy to simulate a biased die. Given probabilities p_1, p_2, and p_3, one defines three intervals

$$I_1 = [0, p_1), \quad I_2 = [p_1, p_1 + p_2), \quad \text{and } I_3 = [p_1 + p_2, 1].$$

Note that the length of I_k is equal to p_k. Therefore, if we choose number k whenever the random number falls into I_k, then event k will be drawn with probability p_k. For example, when $p_1 = 0.5, p_2 = 0.3$, and $p_3 = 0.2$, then

$$I_1 = [0, 0.5), \quad I_2 = [0.5, 0.8), \quad \text{and } I_3 = [0.8, 1].$$

Rerunning the Chaos Game with a Biased Die

Let us now see how the chosen probabilities p_k effect the generation of game points. Figure 6.21 shows the result of plotting 1000 and 10,000 points. In the long run we again obtain a Sierpinski gasket. But we observe an obvious additional pattern in the distribution of game points. The density of points varies in different subtriangles, however, in a very systematic way. For example, look at the pattern in the lower left subtriangle. Here the distribution of points looks the same as the distribution in the whole triangle, although this subtriangle contains only about 50% of the points. Indeed, it turns out that also the distribution of points is self-similar.

[10] Of course, this approach to the computation of π is rather inefficient. It can be shown that, for example, the probability of obtaining π correct to five decimal places in 3400 throws is less than 1.5%.

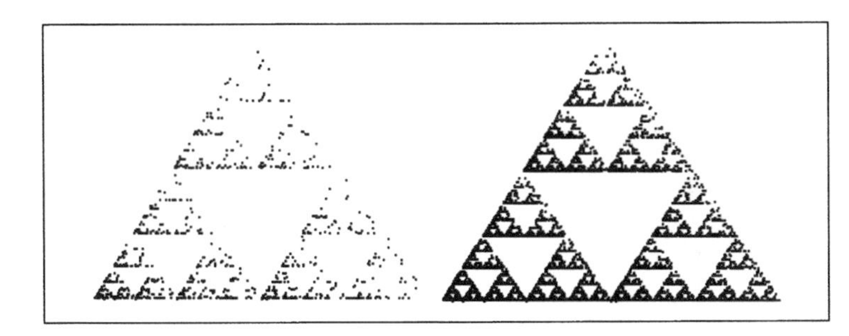

Game Points with Loaded Die

Here are 1000 (left) and 10,000 (right) game points, when w_1 is chosen with 50%, w_2 with 30%, and w_3 with 20% probability.

Figure 6.21

Let us try to estimate the probabilities by which we will see events falling within the triangles of the different levels of the Sierpinski gasket. The answer is simple for the three triangles which make the first level. Figure 6.22 illustrates the result.

But already for the nine triangles of the second level, it is crucial that we recall how each of these triangles is obtained from the entire triangle by compositions of transformations of the form $w_{s_2}w_{s_1}$ with s_1, s_2 from the set $\{1, 2, 3\}$. This is exactly what the addresses tell us. Thus, if D is one of these triangles with address $s_2 s_1$, then the probability that we will see an event in D after two iterations is $p_{s_2}p_{s_1}$ (see figure 6.23).

The probabilities vary between 0.04 and 0.25. The lower left triangle is hit about six times as often as the topmost triangle. We verify these estimates in the left image which we obtained in figure 6.21 by counting points in the subtriangles as before. The last column in the following table lists the expected outcomes in percent.

Address	Count	In %	Expected
11	238	23.8%	25%
12	139	13.9%	15%
13	108	10.8%	10%
21	146	14.6%	15%
22	91	9.1%	9%
23	64	6.4%	6%
31	101	10.1%	10%
32	72	7.2%	6%
33	41	4.1%	4%

Expressed as a general rule, we can say that when we pick a triangle D in the kth level which has

$$\text{address}(D) = s_1 s_2 \dots s_k,$$

then the probability that this triangle will be hit after k iterations is the product $p_{s_1} \cdots p_{s_k}$. This implies that this construction also yields *self-similarity of the probability distribution* of the game points in the Sierpinski gasket. The

Probabilities Level 1

Probabilities attached to triangles in first level.

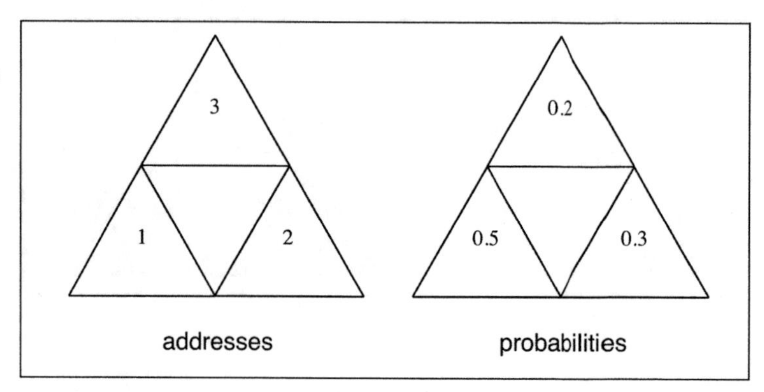

Figure 6.22

Probabilities Level 2

Addresses and corresponding probabilities attached to triangles of second level.

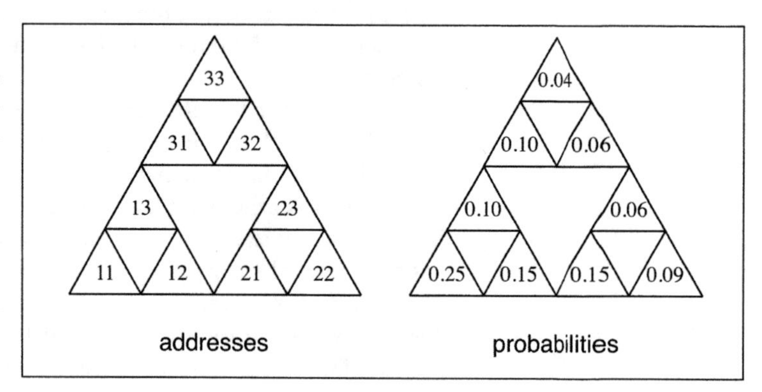

Figure 6.23

geometric self-similarity means that subtriangles of level $k+1$ are reduced copies of the subtriangles of level k (e.g., w_1 copies subtriangle $s_1 s_2 \dots s_k$ to subtriangle $1 s_1 s_2 \dots s_k$). Likewise we obtain the probabilities for subtriangles of level $k+1$ from the probabilities of the corresponding subtriangles of level k (e.g., the probability to hit triangle $1 s_1 s_2 \dots s_k$ is p_1 times the probability to hit triangle $s_1 s_2 \dots s_k$).

We can test the probability to hit a triangle D of the kth generation by sampling the relative counts, and expect to get for all such subtriangles

$$\lim_{n \to \infty} \frac{h(z_1, \dots, z_n; D)}{n} = p_{s_1} \cdots p_{s_k}. \tag{6.3}$$

As before, $h(z_1, \dots, z_n; D)$ denotes the number of hits among the first n game points $z_1, \dots, z_n$ in D. In fact, this number is visualized in our chaos game images in the form of the density of the displayed points. This result leads to two major consequences:

(1) A strategy to design efficient decoding schemes for IFS codes.
(2) An extension of the concept of IFSs from an encoding of black-and-white images to an encoding of color images. This point will be discussed further below (see page 308).

Unbiased Dice Are Best for the Sierpinski Gasket ...

We have seen that even with a biased die, we will eventually generate the Sierpinski gasket. Depending on the chosen probabilities it may, however, take a very, very long time to see its final shape. According to eqn. (6.3), the relative number of hits in some parts of the Sierpinski gasket can be extremely small, though always greater than zero, while it will be very large in other parts. In other words, for reasons of efficiency, we should keep all probabilities the same for the generation of the Sierpinski gasket. But is that a general rule of thumb for all IFS attractors?

...But Not for the Fern

Let us recall the problems we had when we tried to generate the Barnsley fern by the chaos game. Namely, we were not able to see the final image when playing the game with equal probabilities for all transformations. To analyze the situation, we pick one of the tiny primary leaves T at the top of the fern (see figure 6.24). We can describe this leaf in terms of the contractions w_1 to w_4:

$$T = w_1 w_1 \cdots w_1 w_3(F), \tag{6.4}$$

where F denotes the entire fern, and the total number of transformations is k. Therefore, leaf T represents one of the sets of level k, with address

$$\text{address}(T) = 11\ldots13,$$

where 1 occurs $k-1$ times. In other words, the probability q of seeing a game point in T after k iterations is just $q = p_1^{k-1} p_3$.

Now let us take k in the order of 15. Thus, T is the 15th leaf on the right side of the fern. Breaking down the fern F to the 15th level means that we have $4^{15} \approx 10^9$ sets, and T is just one of them. If we take uniform probabilities, $p_k = 0.25, k = 1,2,3,4$, then the probability to see a game point up there is $q = 0.25^{15} \approx 0.931 \cdot 10^{-9}$! Thus, for all practical purposes the probability is zero; and therefore the fern on the left of figure 6.4 is as incomplete as it is. In other words, this way of generating one or two hundred thousand game points to picture 10^9 sets is doomed to failure.

Changing the Probabilities

If we take, however, a relatively large probability for w_1 and small probabilities for w_2, w_3, and w_4, then we can arrange $q = p_1^{14} p_3$ as more appropriate. For example, with $p_1 = 0.85$ and $p_3 = 0.05$, we estimate $q \approx 0.00514$. In only, say, 100,000 iterations we expect about 500 points in the tiny leaf T. Thus, by modifying the probabilities we can push the likelihood of seeing a game point in T after k iterations from practically zero to a very reasonable likelihood. In other words, the strongly biased die here creates a distribution of the 10^5 game points over the 10^9 sets, which is sufficiently efficient for the decoding of the fern image as shown on the right of figure 6.4.

Recipe for Choosing the Probabilities

It is a difficult and still unsolved mathematical problem to determine the best choice for the probabilities p_i. The problem can be stated in the following way. Let ε be a prescribed precision of approximation in the sense that for every point of the attractor there is at least one point generated by the chaos game close by, namely at a distance of

A Tiny Leaf on the Fern

Description of one of the tiny leaves of the fern.

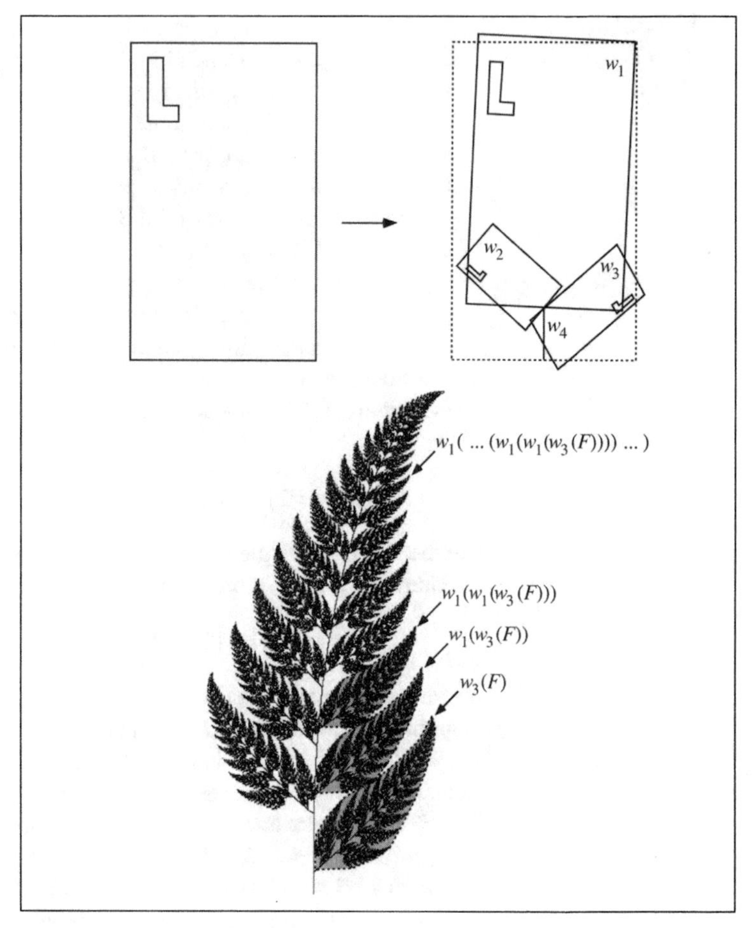

Figure 6.24

not greater than ε. In other words, the Hausdorff distance between the attractor and its approximation is at most ε. Now the optimization problem consists in finding the probabilities p_1 to p_N so that the expected number of iterations in the chaos game needed to reach this required approximation is minimal.

Although the problem is unsolved, there are some heuristic methods for choosing 'good' probabilities. Below we present one of them which has been popularized by Barnsley.[11] In the last section of this chapter we will discuss improved methods.

We consider an IFS with N transformations $w_1, \ldots, w_N$ and assume that its attractor is totally disconnected. So, if A denotes the attractor,[12] then the transformed images $w_1(A), \ldots, w_N(A)$ form a disjoint covering of the attractor. These small affine copies of the attractor are called attractorlets. If in the course of running the corre-

[11] M. F. Barnsley, *Fractals Everywhere,* Academic Press, San Diego, 1988.

[12] For ease of notation we drop the ∞-index in the symbol for the attractor.

sponding chaos game we generate a total of n points, then we may ask how these points are distributed among the N attractorlets. Allotting the same number of points in each attractorlet, i.e., n/N points, will yield a point set which is uniformly distributed over the attractor in the case of the Sierpinski gasket, but not in the case of the Barnsley fern. Let us now consider an ε-collar A_ε of the attractor,[13] the set of all points which have a distance not greater than ε from the attractor. Then

$$A \subset w_1(A_\varepsilon) \cup \cdots \cup w_N(A_\varepsilon) \subset A_\varepsilon .$$

For small values $\varepsilon > 0$ the sets $w_i(A_\varepsilon)$ are close approximations of the attractorlets. We now assign the number of points to fall into the ith attractorlet according to the percentage that the area of the corresponding set $w_i(A_\varepsilon)$ contributes to the area of A_ε.[14] Thus, in order to achieve a uniform distribution of points the number of points in each attractorlet should be proportional to the corresponding area, where we assume that there is no significant overlap of the attractorlets. It is a well-known fact from linear algebra that the factor by which an area changes after undergoing an affine transformation w_i is the absolute value of the determinant of the corresponding matrix of coefficients C_i. Thus, the area of the ε-collar of the ith attractorlet is roughly p_i times the area of the ε-collar of the entire attractor, where

$$p_i = \frac{|\det C_i|}{|\det C_1| + \cdots + |\det C_N|}, \quad i = 1, \ldots, N .$$

Therefore, we aim at collecting $n \cdot p_i$ points of the chaos game in the ith attractorlet. This is easily achieved just by choosing the probabilities $p_1, \ldots, p_N$ according to the above formula.

This recipe for choosing the probabilities also usually works fine in cases when there are some small overlapping parts of the attractor. However, special consideration is necessary in cases when there is large overlap or one of the transformations has a zero determinant. In the latter case the recipe from above would just prescribe a zero probability; and consequently the transformation would never be used. The transformation which yields the stem of Barnsley's fern is an example. Here we arbitrarily assign a small probability, say $\delta = 0.01$. The whole procedure may be summarized by the formula

$$p_i = \frac{\max(\delta, |\det C_i|)}{\sum_{k=1}^{N} \max(\delta, |\det C_k|)}, \quad i = 1, \ldots, N$$

where $\delta > 0$ is a small constant.

---

[13] See section 5.6 for a definition of an ε-collar.

[14] We need the above construction using the ε-collar of the attractor because the area of the attractor itself may not be meaningful. For example, the area of the Sierpinski gasket is zero.

Half-Tone and Color

So far we have discussed only black-and-white images and their encoding through IFSs together with their decoding through the chaos game. We have seen that the probabilities p_i give us very explicit control over the distribution of the game points falling within the sets into which an attractor can be broken down. In other words, counting the relative frequency of points falling into subsets of the attractor establishes a measure on the attractor. This so-called *invariant measure* can be interpreted as a half-tone image.

Assume we were looking at a raster of screen pixels, say m rows and n columns. Each of the pixels, $P_{ij}, i = 1, \ldots, m, j = 1, \ldots, n$, carries a unit of half-tone information, a value which we interpret as a number Q_{ij} somewhere between 0 and 1. The value 1 corresponds to black and 0 to white. Now, let us pick an IFS with contractions $w_1, \ldots, w_N$ and probabilities $p_1, \ldots, p_N$. Then we can look at the statistics of the chaos game in pixel P_{ij}:

$$\lim_{k\to\infty} \frac{h(z_1, \ldots, z_k; P_{ij})}{k} = R_{ij}.$$

That means we run the chaos game with our probabilities and count the relative number of hits R_{ij} in pixel P_{ij}. Now, we can interpret the relative pixel count in terms of half-tone information. We simply select the intensity of a pixel proportional to the count (with some factor of proportionality α)

$$Q_{ij} = \alpha R_{ij}, \quad i = 1, .., m, \; j = 1, \ldots, n.$$

In other words, we encoded an image up to an overall brightness setting, which can be adjusted afterwards.[15] The code for the image simply consists of the necessary transformations and the corresponding probabilities

$$\{w_1, \ldots, w_N\}, \; \{p_1, \ldots, p_N\}.$$

Running the chaos game and evaluating pixel hits turns this information into a half-tone image. Now the question arises, can we use this approach to encode a given image with pixel intensity Q_{ij} by an IFS? This leads to the following inverse problem: find a code $\{w_1, \ldots, w_N\}, \{p_1, \ldots, p_N\}$, so that

$$\lim_{k\to\infty} \frac{h(z_1, \ldots, z_n; P_{ij})}{k} \propto Q_{ij} \tag{6.5}$$

where '$\propto$' means proportional? This means that the half-tone images would be encoded into $7N$ real numbers. From a solution to that problem to an understanding of *color images* is only a small step. Any color image can be regarded as an image which is composed of three components: a red, a green and a blue image. This is just the RGB technique of producing a color image on a TV screen. Each of the components can, of course, be interpreted as a half-tone image combined with the respective color information red, green, or blue.

As we have seen it is possible to use the chaos game and pixel hit statistic to visualize the invariant measure. However, there are even more promising methods for half-tone image encoding and decoding by IFS. They will be discussed in Fisher's appendix on image compression.

[15] Thus, a picture, which is uniformly white has the same encoding as a picture which is uniformly grey or black.

The Invariant Measure

The contractions $w_1, w_2, \ldots, w_N$ and the probabilities $p_1, p_2, \ldots, p_N$ determine how frequently a certain pixel P_{ij} will be hit by the chaos game. The average fraction of hits

$$\lim_{k\to\infty} \frac{h(z_1,\ldots,z_k;P_{ij})}{k} = R_{ij}.$$

is the result of a particular measure μ which has the attractor A_∞ of the IFS as its support (i.e., $\mu(A_\infty) = 1$).[16] In other words,

$$\mu(P_{ij}) = R_{ij}.$$

This measure μ is a Borel measure and is invariant under the Markov operator $M(\nu)$, which is defined in the following way. Let X be a large square in the plane which contains A_∞, the attractor of the IFS, and ν a (Borel) measure on X. Then this operator is defined by

$$M(\nu) = p_1\nu w_1^{-1} + p_2\nu w_2^{-1} + \cdots + p_N\nu w_N^{-1}.$$

In other words, $M(\nu)$ defines a new normalized Borel measure on X. We evaluate this measure for a given subset B in the following way: first we take the preimages $w_i^{-1}(B)$ with respect to X, then evaluate ν on that, and finally we multiply with the probabilities p_i and add up the results.

Here is an example. Let

$$\begin{aligned} w_1(x) &= 1/2\,x, & p_1 = 1/3, \\ w_2(x) &= 1/2\,x + 1/2, & p_2 = 2/3. \end{aligned}$$

This is an IFS which has the unit interval as its attractor $A_\infty = [0,1]$. Now assume that we start with a measure given by the density

$$h_0(x) = \begin{cases} 1 & \text{if } x \in [0,1], \\ 0 & \text{otherwise,} \end{cases}$$

i.e., the initial measure is $\nu_0(A) = \int_A h_0(x)dx$. For a subset $A \subset [0,1/2]$ of the left half of the unit interval we have $w_2^{-1}(A) \subset [-1,0]$ and $\nu_0(w_2^{-1}(A)) = 0$. Thus, $\nu_1(A) = \nu_0(w_1^{-1}(A))$. A corresponding argument holds for the right half interval $[1/2,1]$. Intuitively speaking, w_1 draws the measure ν_0 multiplied by p_1 into the left half-interval $[0,1/2]$ while w_2 does the same with the right half-interval, multiplying the result by p_2. Thus, after the first step we obtain the density

$$h_1(x) = \begin{cases} p_1 & \text{if } x \in [0,1/2), \\ p_2 & \text{if } x \in [1/2,1), \\ 0 & \text{otherwise,} \end{cases}$$

and $\nu_1(A) = \int_A h_1(x)dx$. We construct the measure $\nu_2 = M(\nu_1)$ along the same lines and obtain the density function h_2 as shown in

[16] This is of a more mathematical nature requiring notions from measure theory. Readers without corresponding mathematical background may wish to skip this section.

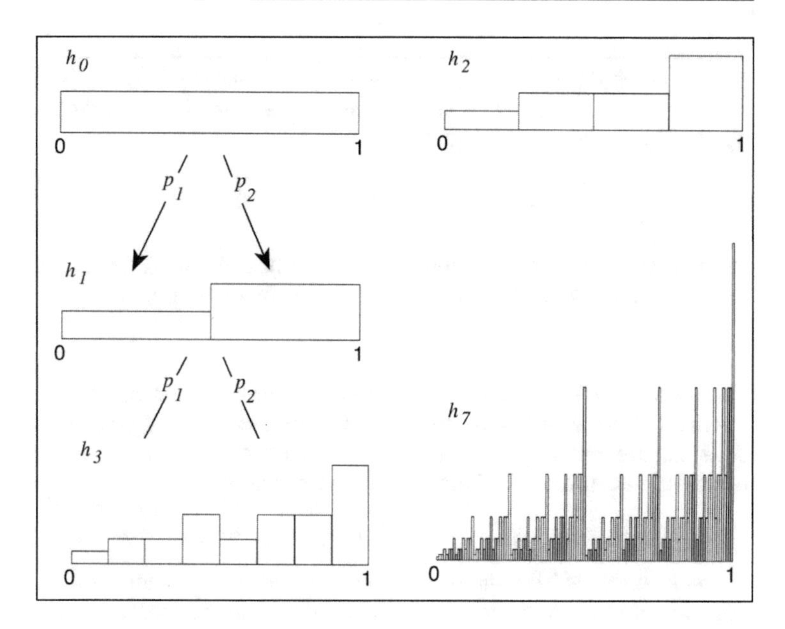

Figure 6.25 : The probabilities p_1 and p_2 generate a binomial measure on the unit interval. The figure shows the corresponding densities developing under iteration of the Markov operator M.

figure 6.25. In the limit this process generates a binomial measure, that is a self-similar multifractal measure. Details will be discussed in the appendix by Evertsz and Mandelbrot.

The Markov operator M turns out to be a contraction in the space of normalized Borel measures of X, equipped with the Hutchinson distance[17]

$$d_H(\nu_1, \nu_2) = \sup \left| \int f d\nu_1 - \int f d\nu_2 \right|$$

where the supremum is taken over all functions $f : X \to \mathbf{R}$ with the property that $|f(x) - f(y)| \leq d(x, y)$. $d(x, y)$ denotes the distance in the plane. The contraction mapping principle can be applied because the space of normalized (Borel) measures is complete with this distance. Thus, there exists a unique fixed point μ of the Markov operator M, $M(\mu) = \mu$. This is exactly the measure which we are looking for when we try to find a solution to the inverse problem for half-tone images.

[17] J. Hutchinson, *Fractals and self-similarity*, Indiana University Journal of Mathematics 30 (1981) 713–747.

6.4 Random Number Generator Pitfall

Anyone who considers arithmetical methods of producing random digits is, of course, in a state of sin.

John von Neumann (1951)

The chaos game played on a computer inherently needs a random number generator. So far we have not examined this topic other than noting how to obtain random numbers with a prescribed distribution provided that the generator on the computer supplies numbers with a uniform distribution. On a computer, random numbers are not really random; they are obtained using deterministic rules actually stemming from a feedback system. Thus, the produced numbers only appear to be random, while, in fact, they are even completely reproducible in another run of the same program. For this reason the random numbers produced by computers are also called *pseudo-random*. There are quite a lot of methods in use for random number generation, which often are not apparent to the programmer. Thus, the statistical properties of the numbers coming out of the machine are typically unknown except for a claim of uniform distribution. In this section we demonstrate that when playing the chaos game a lot more than the simple uniform distribution of the random numbers is required. These requirements are naturally fulfilled by using a perfect die.

Random Numbers from the Logistic Equation

In the first chapter we studied the chaos presented by iteration of the simple quadratic functions. It would seem possible to make use of the logistic equation

$$x_{k+1} = 4x_k(1 - x_k) \tag{6.6}$$

as a method for generating random numbers. In fact, this approach was suggested by Stanislaw M. Ulam and John von Neumann, who had been interested in the design of algorithms for random numbers to be implemented on the first electronic computer ENIAC. The iteration of eqn. (6.6) produces numbers in the range from 0 to 1. Let us divide this range into three equal intervals [0,1/3), [1/3,2/3) and [2/3,1]. Each number generated will be in one of the three intervals. Now we drive the chaos game for the Sierpinski gasket by this 'random number generator'. Figure 6.26 shows the result after 1000 iterations.

Failure to Produce the Sierpinski Gasket

The result is rather strange looking because only some very limited details of the Sierpinski gasket show through.[18] All of the points are exactly in the triangle. However, most parts of the Sierpinski gasket seem to be missing, even when iterating for a much longer time. Recalling the last section, one is tempted to believe that the probabilities are perhaps not adjusted correctly. To perform a test we compute a histogram[19] of a total of 10,000 computed random numbers.

[18] This procedure was suggested among other pseudo-random number generators by Ian Stewart, *Order within the chaos game?* Dynamics Newsletter 3, Nos. 2 & 3, May 1989, 4–9. Stewart ends his article: 'I have no idea why these results are occurring [...] Can these phenomena be explained? [...]' Our arguments will give some first insight. They were worked out by our students E. Lange and B. Sucker in a semester project of an introductory course on fractal geometry.

[19] It is important to do the histogram computation using double precision calculations. Otherwise it is very likely, that the

Sierpinski Gasket via Logistic Equation I

The first attempt to generate the Sierpinski gasket using a random number generator based on the logistic equation.

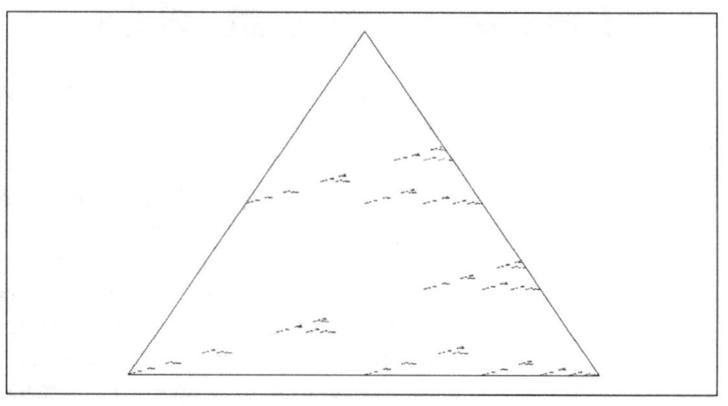

Figure 6.26

Sierpinski Gasket via Logistic Equation II

Another attempt to generate the Sierpinski gasket. The 'improved' random number generator based on the logistic equation is used.

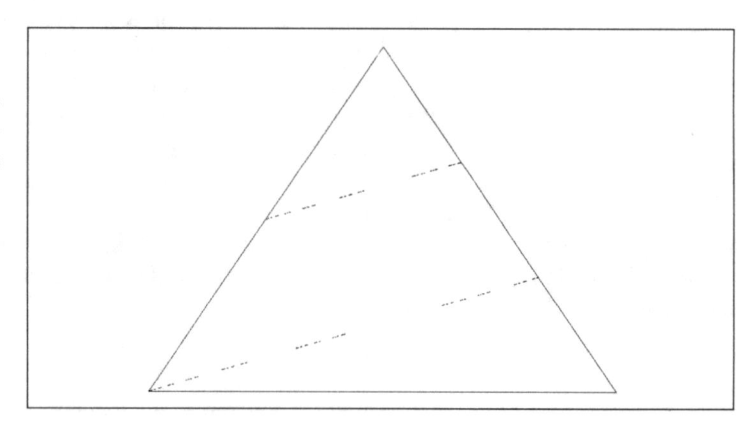

Figure 6.27

Interval	Count	Frequency
[0,1/3)	3910	39%
[1/3,2/3)	2229	22%
[2/3,1)	3861	39%

The result demonstrates a substantial deviation from the optimal frequencies of 1/3 per interval (33.3%). In order to revise the way our choices of the affine transformations are derived from the random numbers we must perform a more detailed empirical analysis. Let us subdivide the unit interval into 20 small intervals of length 0.05 each, and count the corresponding numbers for 100,000 iterates (see table 6.28).[20]

iteration for the logistic equation will run into a periodic cycle of a low period (perhaps even less than 1000), and, as a consequence a histogram based on such an orbit would be a numerical artifact. This effect and the topic of histograms will be continued in chapter 10.

[20]This experiment approximates the so-called natural measure of the quadratic iterator. See chapter 10 for more details.

Interval	Count	Interval	Count
[0.00, 0.05)	14403		
[0.05, 0.10)	6145		
[0.10, 0.15)	4812		
[0.15, 0.20)	4256		
[0.20, 0.25)	3809		
		[0.00, 0.25)	33425
[0.25, 0.30)	3487		
[0.30, 0.35)	3389		
[0.35, 0.40)	3303		
[0.40, 0.45)	3244		
[0.45, 0.50)	3097		
[0.50, 0.55)	3240		
[0.55, 0.60)	3251		
[0.60, 0.65)	3196		
[0.65, 0.70)	3459		
[0.70, 0.75)	3621		
		[0.25, 0.75)	33287
[0.75, 0.80)	3882		
[0.80, 0.85)	4164		
[0.85, 0.90)	4821		
[0.90, 0.95)	6012		
[0.95, 1.00)	14409		
		[0.75, 1.00]	33288

Statistics for the Quadratic Iterator

100,000 iterations of the quadratic iterator are produced and for each interval in the table we count the corresponding number of iterations. Clearly, the intervals near the endpoints of $[0, 1]$ receive the highest counts.

Table 6.28

Adjusting the Probabilities Correctly

Based on the results in the table, we divide the unit interval into the three subintervals [0,1/4), [1/4,3/4) and [3/4,1]. Now iteration of the logistic equation seems to produce about the same number of iterates in each subinterval. Thus, this setting will produce a random number generator for the three outcomes $1, 2$ and 3 with about equal probability of 1/3 each. Using this scheme we again play the chaos game hopeful that now we will produce the complete Sierpinski gasket fairly rapidly. But figure 6.27 reveals a great disappointment; the result is even worse than before.

The conjecture that the problem is due to badly chosen probabilities has obviously proven false. To get to the core of the matter we need to take a look at other properties we need to have guaranteed by the random number generator to make the chaos game work. Recall that the addressing system was the key to understanding the operation of the game. For each point of the attractor there was an address consisting of an infinite string of digits from $\{1, 2, 3\}$. The chaos game will produce a point close by any point of the attractor provided it is set up to generate all possible finite addresses with some appropriate probability. Looking back at the poor results of the last two experiments, we realize that we were not able to produce most of the

Rerun with Addresses

Rerun of the last experiment with addresses inscribed and fat points.

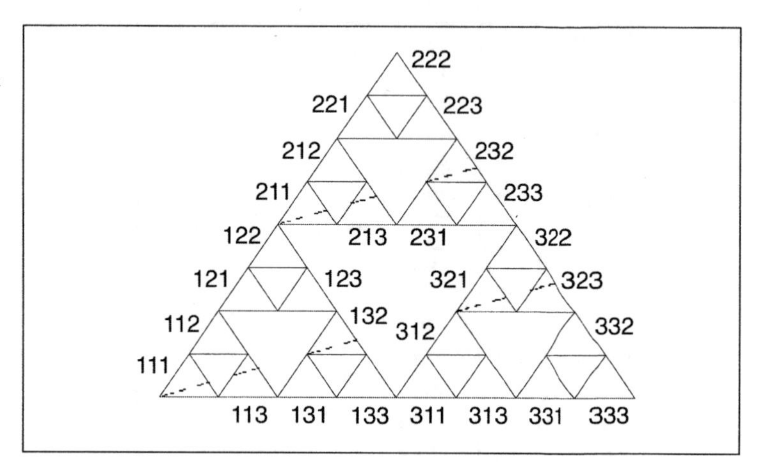

Figure 6.29

addresses. In our adjustment of the iteration (6.6) to the intervals $[0, 1/4)$, $[1/4, 3/4)$, and $[3/4, 1]$ we made sure that addresses starting with $1, 2$ and 3 would occur with the same frequency. But how about the addresses starting with $11, 12, 13$ and so on? Let us try to find out which addresses do not occur by rerunning the last experiment and plotting the points on a grid inscribed with 3-digit addresses.

Where Are the No-Shows?

We discover that certain combinations of three digits in the addresses never show up, namely,

$$222, 221, 223, 212, 231, 233, 122, 121, 123, 322, \ldots$$

In other words, only the following eight 3-digit addresses actually appear:

$$111, 113, 132, 211, 213, 232, 321, 323.$$

Having reached this far we should also be able to deduce these artifacts directly from the construction of the random numbers by means of iteration of the logistic equation. Figure 6.30 shows the graphical iteration for $4x(1-x)$. The three intervals [0,1/4), [1/4,3/4) and [3/4,1] are also marked on both axes. Looking at the diagram it becomes clear that in the iteration certain combinations of random number outcomes are simply impossible. If we start with a point in the first interval [0,1/4), then the next point must necessarily be either again in the first interval [0,1/4) or in the second interval [1/4,3/4). Therefore, the combination 13 can never appear in the scheme. Continuing, we realize, that a number in the second interval [1/4,3/4) will be transformed into a number of the third interval [3/4,1], and all numbers of the third will be in the first or second interval after one iteration.[21] Does this mean that the combinations 13, 21, 22, and 33 are not possible? Careful! Yes, our digital computer die, after rolling a 1, cannot roll a 3 next. But in terms of addresses,

[21]There is only one exception to this rule, namely the point 3/4. This point stays fixed, i.e., $4 \cdot 3/4 \cdot (1 - 3/4) = 3/4$. This is irrelevant for our discussion.

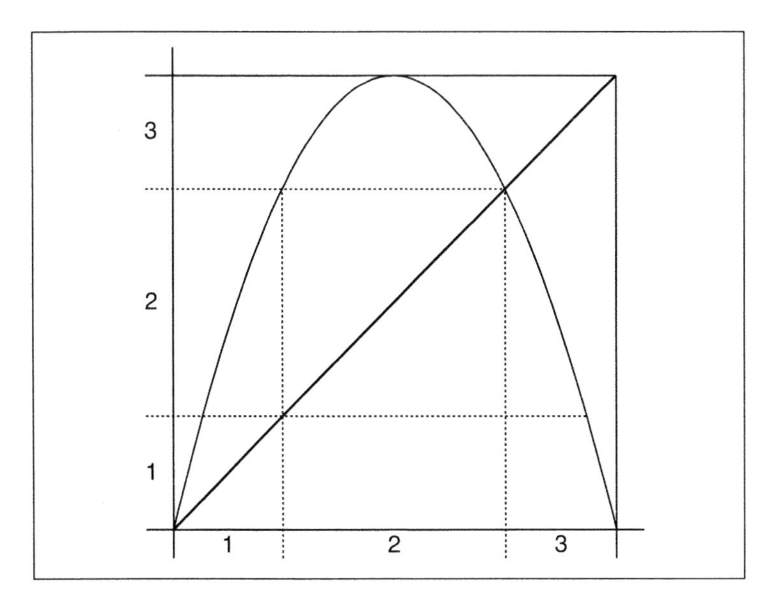

The Logistic Parabola

Graphical iteration for $4x(1 - x)$ with indicated regions used by the random number generator for the chaos game.

Figure 6.30

this translates into the reverse sequence 31! Recall, that addresses are read from left to right, and rolls in the chaos game right to left. Thus, we have verified that the chaos game played with our random number generator will not be able to produce points which have one of the following 2-digit strings in their address: 31, 12, 22, and 33. This is exactly, what we observed above in the experiment.

Of course, we can now extend our analysis to the case of our first attempt with the logistic equation (with intervals [0,1/3), [1/3,2/3), and [2/3,1]). The possible addresses are somewhat different, but in principle the failure to render the Sierpinski gasket stems from the same source: not every finite length sequence of interval indices can be produced. In other words, the events (the individual indices 1, 2, and 3) are not independent.

Modeling the Attractor of a Hierarchical IFS Driven by the Quadratic Iterator

There is an astonishing relation between driving the chaos game with the quadratic iterator $x_{k+1} = 4x_k(1 - x_k)$, eqn. (6.6), and hierarchical IFSs. This stresses once again the significance of the concept of hierarchical IFSs as a new mathematical tool.

Driving the chaos game with the quadratic iterator and appropriately adjusted probabilities as used in figure 6.27 means that transformation w_1 cannot be followed by w_3, w_2 not by w_2, w_2 not by w_1, and w_3 not by w_3. Or, turned the other way around, we can see the

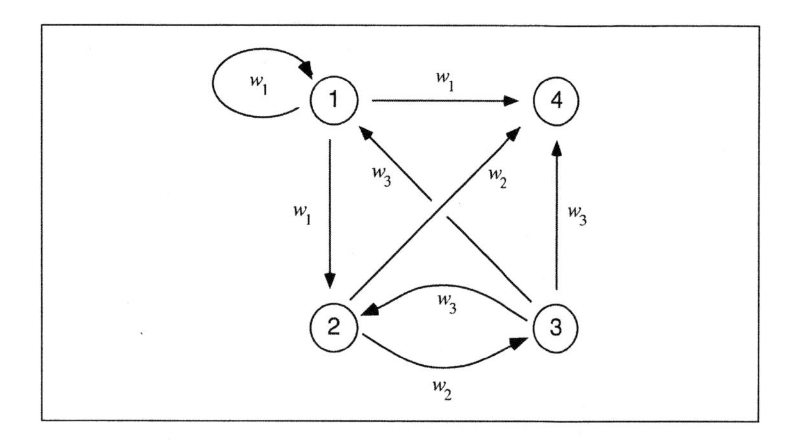

Figure 6.31 : Graph of hierarchical IFS corresponding to the chaos game based on the 'random numbers' produced by the logistic equation.

following admissible sequences of transformations

w_1 followed by w_1,
w_1 then w_2,
w_2 then w_3,
w_3 then w_1,
w_3 then w_2.

Building an IFS as indicated by the graph in figure 6.31 leads to exactly the same result. First, consider the nodes 1, 2, and 3 and their connection by directed edges. Informally speaking, these nodes prescribe the 'next admissible transformation'. You can check that the transformation w_i can only be applied in the order we just discussed. Now consider node 4. It collects all admissible compositions. The corresponding matrix Hutchinson operator would be

$$\mathbf{W} = \begin{pmatrix} w_1 & \emptyset & w_3 & \emptyset \\ w_1 & \emptyset & w_3 & \emptyset \\ \emptyset & w_2 & \emptyset & \emptyset \\ w_1 & w_2 & w_3 & \emptyset \end{pmatrix}.$$

The relevant attractor would appear in the 4th component (shown in figure 6.31).

Independent Rolls

This leads us to an important requirement for random number generators, which we have implicitly assumed in the chaos game but have not yet formulated explicitly. The rolls of the die, or the outcomes of the digital computer die, must be independent from each other. Without that, it is possible that even though the three outcomes 1, 2, and 3 appear with the same frequency, we can have a sequence of events which is rather restricted. The chance of rolling a '3' may be 100% when the previous result had been '2', and it may be 0% when the last roll was a '1' or a '3'. The right way to play the chaos game, however, requires a die that produces a '3' with a fixed probability no matter what the previous roll (or any of the previous rolls for that matter) has been. A real unbiased die with six faces naturally has this property. To roll a '1', and then a '2' occurs with probability 1/36 independent of all the previous results.

The Linear Congruential Generator

The most widely used random number generators on current computers use variants of the linear congruential method. With a modulus m and a starting value r_0, $0 \leq r_0 < m$, the numbers r_k are computed with the formula

$$r_{k+1} = (ar_k + c) \bmod m$$

where the multiplier a and the increment c are nonnegative integers less than the modulus m. The modulus usually is conveniently chosen to be a power of 2 matching the word length of the particular machine. This method produces integers in the range from 0 to $m - 1$. Each number is completely determined by its predecessor according to the above formula. In conclusion, all sequences of pseudo random numbers generated in this way must be periodic. The multiplier a and the increment c can be chosen so that the period has the maximal value m. Because of the periodicity, which is inherent in the construction, the sequences of random numbers thus generated cannot be truly random. The randomness comes in various flavors, and there exist a large number of statistical tests, e.g., frequency test, run test, collision test and spectral test.[22]

We conclude that it is important for the chaos game played on a computer to rely on a random number generator which guarantees the independence of all numbers produced. Only in this way is it possible to generate points for all necessary addresses. Most random number generators supplied with computers nowadays seem to fulfill this property sufficiently well. But it is by no means true for all generators in use. We illustrate this with two examples which were considered in the 1950's: the middle-square generator and the Fibonacci generator.

The First Random Number Generator

Before computers existed people who needed random numbers used to roll dice, draw cards from a deck or, later, use mechanical devices. Tables of random digits had also been published. For example, in 1927 L. H. C. Tippet produced a table of over 40,000 digits 'taken at random from census reports'. In 1946 John von Neumann was the first to suggest that random numbers could be computed on a machine using a deterministic algorithm, the *middle-square generator*. In this method a decimal number r_0 with n digits is given as a seed. This seed value is squared and the middle n digits of the

[22]For an introduction into the topic of random number generation see D. E. Knuth, *The Art of Computer Programming, Volume 2, Seminumerical Algorithms*, Second Edition, Addison-Wesley, Reading, Massachusetts, 1981.

The Middle-Square Generator

The middle-square method for random number generation fails to produce the Sierpinski gasket (points fattened).

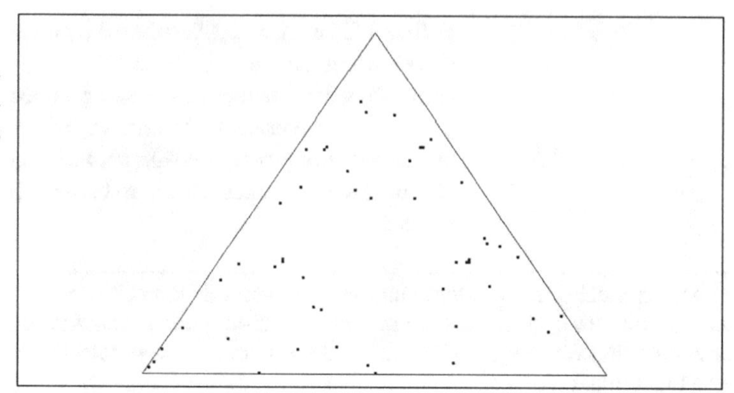

Figure 6.32

The Fibonacci Generator

The Fibonacci generator for random number generation also fails to produce the Sierpinski gasket.

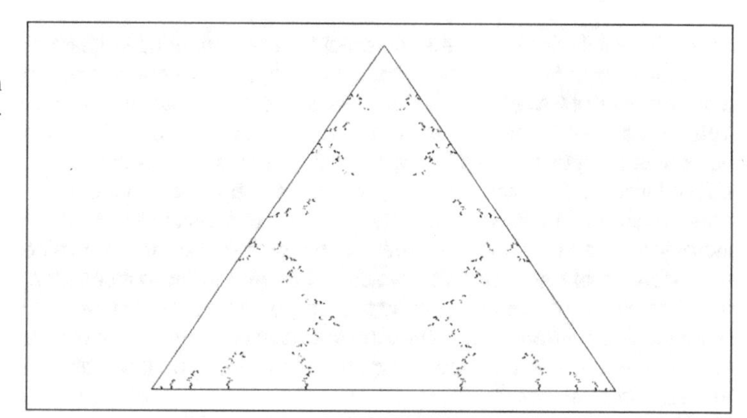

Figure 6.33

result are extracted, yielding the next number r_1. Then square r_1, extract the middle n digits to obtain r_2, and so on. The range of numbers produced in this way extends from 0 to $10^n - 1$, and by dividing the results by 10^n we get numbers distributed over the unit interval $[0, 1]$ as required for the normalized random numbers. The middle-square generator, however, has been shown to be a rather poor source of random numbers, although when set up with certain numbers of digits and initial seeds, it may produce a long sequence of numbers that passes all practical tests for randomness. Figure 6.32 shows our attempt for the Sierpinski gasket.

A Second-Order Formula

The Fibonacci generator is probably the simplest method of second order for production of random numbers. Each number is computed not only from its predecessor, but also from the second predecessor. The formula is

$$r_i = (r_{i-1} + r_{i-2}) \bmod m.$$

In Figure 6.33 we have chosen $m = 2^{18}$. The result is a rather surprising fractal — but it is far from the complete Sierpinski gasket it should be.

6.5 Adaptive Cut Methods

As shown in this chapter and the last one, a wide range of fractals can be encoded as attractors of an IFS. We have discussed two alternatives for the rendering of this attractor: a deterministic algorithm (i.e., the iteration of the MRCM), and a probabilistic method, the chaos game. However, both approaches have limitations. The deterministic algorithm performs poorly when the contraction ratios of the affine transformations vary significantly, as in the case of Barnsley's fern. On the other hand, the performance of the chaos game depends strongly on the choice of probabilities; and so far we have only seen a rule of thumb for the determination of probabilities (see page 307). But now we will discuss a method of computing improved probabilities from a deterministic approximation of the attractor.

In fact, the algorithm we are going to discuss can be used to implement a deterministic rendering of the attractor which avoids the drawbacks of simply iterating an MRCM.[23] In many cases this algorithm is a better choice than the probabilistic rendering. It can especially render the attractor up to a prescribed precision, whereas with the probabilistic method there is no criterion that ensures that the chaos game has been played sufficiently long to achieve the desired approximation. On the other hand, a well-tuned chaos game is extremely efficient in rapidly rendering a first impression of the global appearance of the attractor.

Covering Sets for the Attractor

First, let us discuss the problem of approximating the attractor A_∞ of an IFS given by the contractions $w_1, \ldots, w_N$ to a prescribed precision $\varepsilon > 0$. In other words, we are looking for a covering of the attractor by sets measuring less than ε in diameter. The iteration of the Hutchinson operator can provide such a covering. Starting with any set A which includes the attractor ($A_\infty \subset A$), the first iteration gives a covering by N sets,

$$A_\infty \subset w_1(A) \cup \cdots \cup w_N(A),$$

the second a covering by N^2 sets,

$$A_\infty \subset w_1 w_1(A) \cup w_2 w_1(A) \cup \cdots \cup w_N w_N(A),$$

and so on. All transformations w_k, $k = 1, \ldots, N$ are contractions. Thus, after a certain number of iterations, say m, all N^m covering sets of the form

$$w_{s_1} w_{s_2} \cdots w_{s_m}(A), \quad s_i \in \{1, \ldots, N\}$$

have a diameter less than ε. However, from the example of the fern we know that the number N^m of these sets can be astronomically high, ruling out any practical computation by machine.

A Good Idea

We note that most of the final N^m covering sets are much smaller than necessary. Thus, it would be a great improvement if we could adaptively

[23]Details have appeared in the paper *Rendering methods for iterated function systems* by D. Hepting, P. Prusinkiewicz and D. Saupe, in: *Fractals in the Fundamental and Applied Sciences*, H.-O. Peitgen, J. M. Henriques, L. F. Peneda (eds.), North-Holland, Amsterdam, 1991.

Adaptive MRCM Iteration

Restricting the iteration of the Hutchinson operator to those sets which are larger than a prescribed tolerance finally provides only sets of the desired size. However, in general they do not cover the whole attractor.

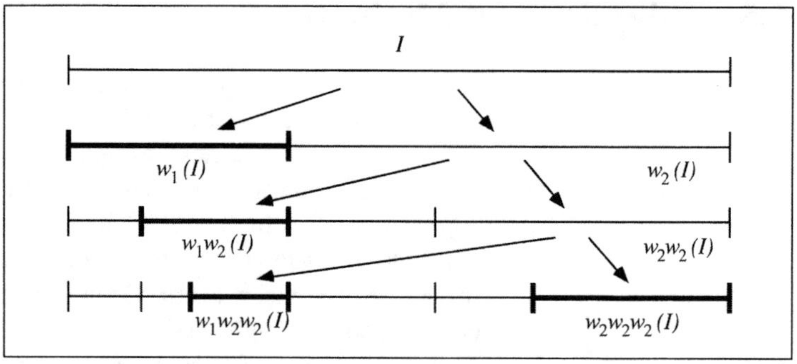

Figure 6.34

stop the iteration depending on the size of the sets $w_{s_1}\cdots w_{s_k}(A)$ at the intermediate steps $k = 1, \ldots, m$.

Setting Up a Simple Test Example

Let us discuss a simple example to make the issue clear. Consider the following system of only two transformations

$$\begin{aligned} w_1(x) &= \frac{x}{3} \\ w_2(x) &= \frac{2x}{3} + \frac{1}{3} \end{aligned}$$

which operate on the set of real numbers. Note that the contraction factor for w_1 is $1/3$ but for w_2 it is $2/3$. There is a strong connection between these transformations and those corresponding to the Cantor set.[24] However, the attractor of this set of transformations is not a Cantor set; it is not even a fractal but simply the unit interval $I = [0, 1]$. This can be seen by noting that the interval is invariant under the associated Hutchinson operator:

$$\begin{aligned} w_1(I) &= [0, 1/3] \\ w_2(I) &= [1/3, 1] \end{aligned}$$

thus

$$H(I) = w_1(I) \cup w_2(I) = [0, 1] = I.$$

From the characterization of the attractor of an IFS by its invariance property we conclude that the unit interval is indeed the attractor of our simple system above.

First Try of an Adaptive Method

Let us try to cover the attractor by sets not exceeding the size $\varepsilon = 1/3$. If we start with $I = [0, 1]$ the first iteration step provides the sets $w_1(I) = [0, 1/3]$ and $w_2(I) = [1/3, 1]$ which — both together — should be used as input for the next step of the iteration. But since the first one of these, $w_1(I)$, is already of the desired size we continue the iteration only with the other one, $w_2(I)$ (see figure 6.34). This yields

$$\begin{aligned} w_1w_2(I) &= w_1[1/3, 1] = [1/9, 1/3] \\ w_2w_2(I) &= w_2[1/3, 1] = [5/9, 1]. \end{aligned}$$

[24] Recall that those are $w_1(x) = x/3$ and $w_2(x) = x/3 + 2/3$ (see page 166).

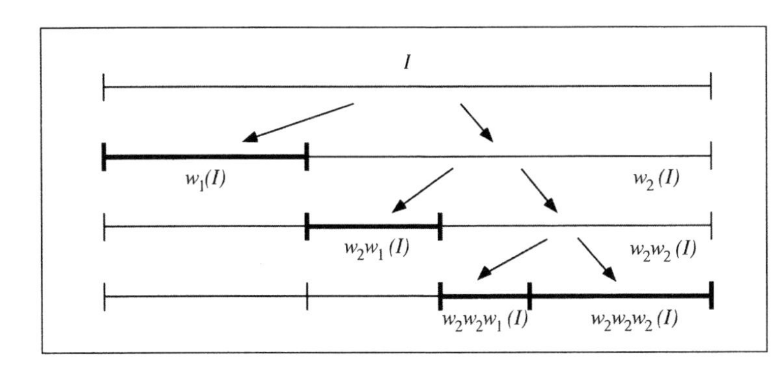

Adaptive Cut Hierarchy

The hierarchical subdivision introduced in the discussion of addresses provides the appropriate framework for the adaptive cut algorithm.

Figure 6.35

The first of these sets is less than ε in size, but the other one is not yet of that size. Thus, we repeat the procedure one more time for $[5/9, 1]$ obtaining

$$\begin{aligned} w_1w_2w_2(I) &= w_1[5/9,1] = [5/27,1/3] \\ w_2w_2w_2(I) &= w_2[5/9,1] = [19/27,1]. \end{aligned}$$

The sizes of these intervals are $4/27$ and $8/27$, respectively, both less than $\varepsilon = 1/3$. We are finished and expect that the collection of small intervals obtained in this procedure covers the attractor. But our check

$$\begin{aligned} &[0,1/3] \cup [1/9,1/3] \cup [5/27,1/3] \cup [19/27,1] \\ &\qquad = [0,1/3] \cup [19/27,1] \neq [0,1] \end{aligned}$$

reveals that some portions of the unit interval are still missing (see also figure 6.34). In other words, simply cutting off some parts of the IFS iteration does not work.

The Correct Adaptive Method

We need a different kind of hierarchical refinement. Figure 6.35 shows the ideal strategy of subdivision for our example. It ends up with the sets $w_1(I)$, $w_2w_1(I)$, $w_2w_2w_1(I)$ and $w_2w_2w_2(I)$, each of which has the desired size and all together cover the unit interval I. But how can we obtain this kind of subdivision? Upon closer examination of figure 6.35 we see that it shows the hierarchical subdivision used for the addressing mechanism for the attractor. Figure 6.36 shows a corresponding address tree. The branches in the tree have different lengths; the nodes of one level are not at the same height. Rather, the height coordinate represents the size to which the unit interval is reduced when the corresponding composed contraction is applied to it. Thus, the idea of the method consists in pruning precisely those branches of the address tree which pass the height 1/3.

Expressed more formally, we subdivide in the first step

$$A_\infty = w_1(A_\infty) \cup \cdots \cup w_N(A_\infty). \tag{6.7}$$

In the next step we subdivide each of the sets $w_k(A_\infty)$ according to

$$w_k(A_\infty) = w_kw_1(A_\infty) \cup \cdots \cup w_kw_N(A_\infty).$$

Adaptive Address Tree

The left address tree corresponding to figure 6.35 is pruned at branches where the corresponding composed contractions reach the contractivity 1/3. The right tree has branches pruned at contraction factor 1/6 yielding a covering of the unit interval at a higher resolution. (The widths of the columns in the figure have no meaning; they do not match the size of the corresponding attractorlets.)

Figure 6.36

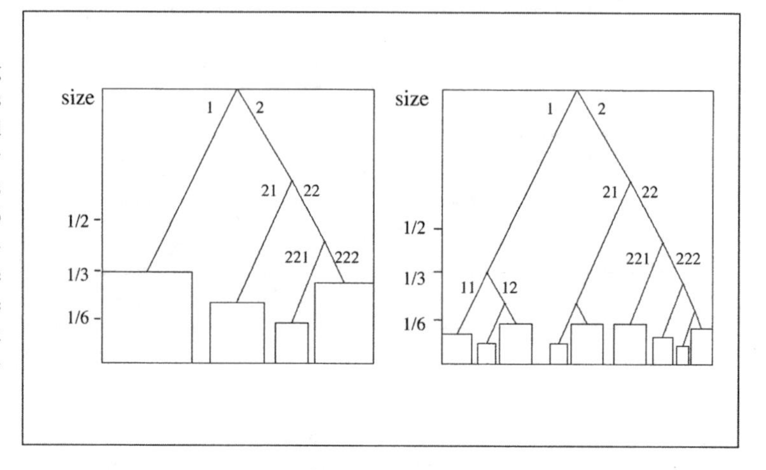

In the third step we subdivide each of the sets $w_k w_l(A_\infty)$ following

$$w_k w_l(A_\infty) = w_k w_l w_1(A_\infty) \cup \cdots \cup w_k w_l w_N(A_\infty).$$

In other words, in stage n each subset of the attractor with address $s_1 \ldots s_{n-1}$ is subdivided into those with addresses

$$s_1 \ldots s_{n-1}1, \; s_1 \ldots s_{n-1}2, \; \ldots, \; s_1 \ldots s_{n-1}N.$$

These subsets of A_∞ are the *attractorlets* already described on page 306. Whenever we have reached an attractorlet which is smaller than ε in diameter we keep it. All other attractorlets must be subdivided further. In our example this process generated the attractorlets with the addresses $1, 21, 221$ and 222.

On the basis of these ideas we can efficiently compute approximations of the attractor A_∞. We simply select one point from each of the final attractorlets. In this way we obtain a representative point set. Accordingly, in this construction we have that for each point in the attractor there will be an approximation in our point set that is not further away than ε.

Let us demonstrate this for our example. We know that $y_0 = 0$ is a point in A_∞ (0 is the fixed point of w_1). This allows us to compute points of the attractorlets:

$$\begin{aligned}
y_1 &= w_1(0) = 0,\\
y_2 &= w_2 w_1(0) = w_2(0) = 1/3,\\
y_3 &= w_2 w_2 w_1(0) = w_2(1/3) = 5/9,\\
y_4 &= w_2 w_2 w_2(0) = w_2 w_2(1/3) = w_2(5/9) = 19/27.
\end{aligned}$$

Since the size of the corresponding attractorlets is not larger than ε we have the approximation

$$A_\varepsilon = \left\{0, \frac{1}{3}, \frac{5}{9}, \frac{19}{27}\right\} \subset A_\infty$$

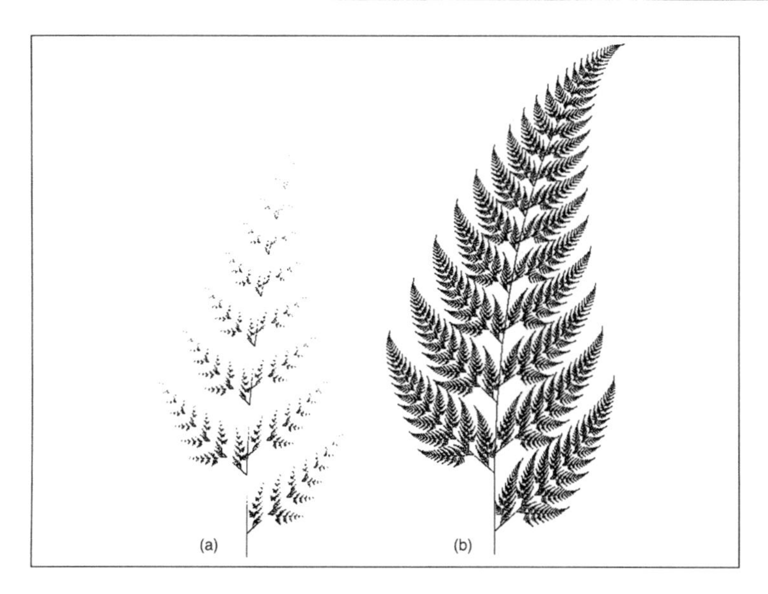

IFS Iteration Versus Adaptive Cut Method

A comparison of two methods for the rendering of the attractor of an IFS. (a) Iteration of the Hutchinson operator (starting with a point) for a total of $m = 9$ iterations, resulting in $N_1 = 4^9 = 262,144$ points. (b) The adaptive cut algorithm using $N_2 = 198,541$ points.

Figure 6.37

and surely no point in A_∞ has a distance greater than $\varepsilon = 1/3$ from all points in A_ε.

Let us compare this with the simple IFS iteration starting with the fixed point 0. Achieving the desired precision would require three complete steps. Thus we would generate eight points in contrast to the four points of A_ε, doubling the necessary workload. We note that actually a much worse factor of inefficiency is typical.

Comparing Algorithms: The Fern Example

In summary, the adaptive algorithm subdivides an attractorlet recursively until the diameter is guaranteed to be less than or equal to the tolerance ε. For the final collection of attractorlets we pick one point of each set as a representative. These points cover the attractor with precision ε. Figure 6.37 compares the different methods when applied to rendering Barnsley's fern. Since the contraction ratios of the transformations involved in constructing the fern are significantly different from each other, running the MRCM some number of cycles yields a highly uneven distribution of points in the attractor. The adaptive cut algorithm removes this drawback as can be seen impressively in the figure.

The Adaptive Cut Method

The adaptive cut method allows us to compute approximations A_ε of A_∞ to a prescribed precision ε so that the Hausdorff distance[25] $h(A_\varepsilon, A_\infty)$ is less than or equal to ε. Thus, all points of A_ε lie within the distance ε to points of A_∞ and vice versa. Now we look at w_1, w_2, $\ldots$, w_N and compute the contraction ratios of these transformations. Let us introduce the symbol $\rho(w_k)$ for these ratios. All transformations w_k which satisfy $\rho(w_k) \leq \varepsilon/\text{diam}(A_\infty)$ can be eliminated from further

[25] See chapter 5 for the definition of Hausdorff distance.

subdivision because the corresponding attractorlets $w_k(A_\infty)$ are less than ε in diameter. For all other transformations w_k we continue with the second level and compute the contraction factors of the composite transformations

$$\rho(w_k w_1), \rho(w_k w_2), \ldots, \rho(w_k w_N).$$

The procedure is repeated, i.e., we eliminate those compositions whose contraction ratio is less than or equal to $\varepsilon/\text{diam}(A_\infty)$ and continue with the others, considering composite transformations with three elements, and so on.

Thus, the general procedure is the following. Having reached a composition with sufficiently small contraction factor,

$$\rho(w_{s_1} w_{s_2} \cdots w_{s_m}) \leq \frac{\varepsilon}{\text{diam}(A_\infty)}$$

we have found an attractorlet with address $s_1 s_2 \cdots s_m$ and size

$$\text{diam}(w_{s_1} w_{s_2} \cdots w_{s_m}(A_\infty)) \leq \varepsilon,$$

which is what we are looking for. When a composition $w_{s_1} \cdots w_{s_m}$ has a contraction factor which is still too large, i.e., greater than $\varepsilon/\text{diam}(A_\infty)$, then we continue to consider the N compositions of the next level,

$$\begin{gathered} w_{s_1} w_{s_2} \cdots w_{s_m} w_1 \\ w_{s_1} w_{s_2} \cdots w_{s_m} w_2 \\ \vdots \\ w_{s_1} w_{s_2} \cdots w_{s_m} w_N \,. \end{gathered}$$

In this way we construct all compositions which satisfy

$$\rho(w_{s_1} \cdots w_{s_m}) \leq \frac{\varepsilon}{\text{diam}(A_\infty)} \leq \rho(w_{s_1} \cdots w_{s_{m-1}}).$$

Let S be the set of the corresponding attractorlet addresses $s_1 \ldots s_m$. For any point $x_0 \in A_\infty$ (e.g., take the fixed point of w_1) the Hausdorff distance between the attractor and the set

$$A_\varepsilon = \{x \,:\, x = w_{s_1} \cdots w_{s_m}(x_0),\ (s_1, \ldots, s_m) \in S\}$$

is bounded by ε, i.e., $h(A_\varepsilon, A_\infty) \leq \varepsilon$. This completes the basic description of the adaptive cut method.

For practical purposes the method can be even more accelerated using the fact that the attractor can be displayed only with a finite screen resolution. The idea is to eliminate from further consideration images of more than one point falling in the same pixel.[26] This is of particular importance in cases of overlapping attractors, since the adaptive algorithm ignores such overlapping.

[26] See S. Dubuc and A. Elqortobi, *Approximations of fractal sets*, Journal of Computational and Applied Mathematics 29 (1990) 79–89.

The remaining problem is to compute or estimate the contraction factors $\rho(w_{s_1} \cdots w_{s_m})$. Here we propose three methods which vary in ease of computation and in the quality of the estimates. In all cases distances are measured using the Euclidean metric.

The first and simplest method relies on the property $\rho(w_1 w_2) \leq \rho(w_1)\rho(w_2)$ and estimates the contraction factor of a composite affine transformation $w_{s_1} \cdots w_{s_n}$ as the product of the individual contraction factors. Thus,

$$\rho(w_{s_1} \cdots w_{s_m}) \leq \rho(w_{s_1}) \cdots \rho(w_{s_m}).$$

Unfortunately this formula may grossly overestimate the actual value of the contraction ratio of the composite transformation. For example, consider the following two affine mappings:

$$\begin{aligned} w_1(x,y) &= (0.01x, 0.99y), \\ w_2(x,y) &= (0.99x, 0.01y). \end{aligned}$$

Since $\rho(w_1) = \rho(w_2) = 0.99$, we obtain $\rho(w_1)\rho(w_2) = 99^2/10,000$. On the other hand,

$$w_1 w_2(x,y) = (0.0099x, 0.0099y)$$

and $\rho(w_1 w_2) = 0.0099$. Thus, the use of the product $\rho(w_1)\rho(w_2)$ overestimates the actual value of $\rho(w_1 w_2)$ by a factor of 99.

We can use an alternative method for estimating the contraction ratio using the following property.[27] The contraction ratio $\rho(w)$ of an affine transformation

$$w(x,y) = (ax + by + e, cx + dy + f)$$

satisfies the inequality

$$\rho(w) \leq 2 \max\{|a|, |b|, |c|, |d|\}.$$

In the above example this result provides a bound 0.0198 which is much improved over $99^2/10,000$ but still off by a factor of 2.

The third method is computationally more expensive than the previous two, but provides exact values. It uses the fact that the contraction ratio $\rho(w)$ of the affine transformation $w(z) = Az + B$ (where A denotes a matrix and B a column vector) can be expressed as the square root of the maximal eigenvalue of $A^T A$ (A^T denotes the transpose of the matrix A)

$$\rho(w) = \sqrt{\max\{|\lambda_i| \ : \ \lambda_i \text{ eigenvalue of } A^T A\}}.$$

This formula is valid for the affine transformations in spaces of arbitrary dimension n. In the two-dimensional case ($n = 2$) with

$$A = \begin{pmatrix} a & b \\ c & d \end{pmatrix}$$

[27] See G. H. Golub and C. F. van Loan, *Matrix Computations*, Second Edition, Johns Hopkins, Baltimore, 1989, page 57.

the evaluation of $\rho(w)$ involves the computation of two square roots. Explicitly, the result is

$$\rho(w) = \sqrt{\frac{p + \sqrt{p^2 - 4q}}{2}},$$

where

$$p = a^2 + b^2 + c^2 + d^2,$$
$$q = (ad - bc)^2.$$

As we are not interested in the exact contraction ratio but only in a tight bound of it, we may replace the square root computation by a properly organized table look-up procedure, which will speed up the method considerably.

As we stated at the start, the distances in this discussion are measured using the Euclidean metric. Alternatively we could switch to a different metric. For example, for the maximum metric d_∞ (see page 249) the contraction ratio can be computed efficiently by the formula

$$\rho_\infty(w) = \max\{|a| + |b|, |c| + |d|\},$$

where the coefficients $a, \ldots, d$ are the elements of the matrix A as above.

Covering Sets for the Fern

The adaptive cut algorithm may produce renderings of different resolution depending on the choice of the tolerance ε for the Hausdorff distance. Three different values are used in the above plots $\varepsilon = 0.5, 0.1, 0.015$. Each point was drawn as a small disk with appropriate radius such that the attractor is guaranteed to be covered by the image.

Figure 6.38

The adaptive cut method provides a list of points approximating the attractor A_∞ with prescribed accuracy. When these points are taken as centers of small disks, then they provide a covering of A_∞. Let

$$D_\varepsilon(y) = \{x \in \mathbf{R}^2 : |x - y| \leq \varepsilon\}$$

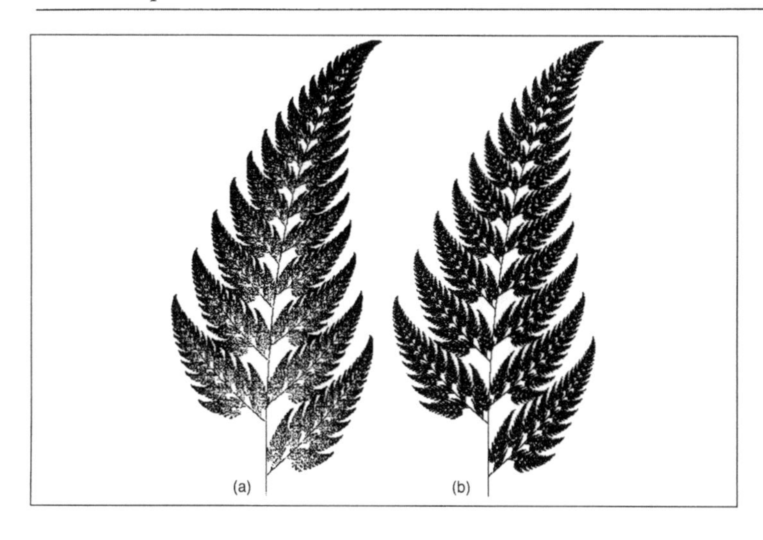

Chaos Game — Two Sets of Probabilities

The chaos game using 198,541 points. (a) The probabilities used are 0.85, 0.07, 0.07 and 0.01. (b) The improved probabilities are 0.73, 0.13, 0.11, and 0.03.

Figure 6.39

the set of points in the plane which lie within the distance ε from the point y. Then for our simple one-dimensional example the set

$$C_\varepsilon = D_\varepsilon(0) \cup D_\varepsilon(1/3) \cup D_\varepsilon(5/9) \cup D_\varepsilon(19/27)$$

would cover the attractor (i.e, $A_\infty \subset C_\varepsilon$) and all points of C_ε would have a distance to A_∞ which is at most $\varepsilon = 1/3$. Figure 6.38 shows such covering sets for the fern.

Estimating Probabilities for the Chaos Game

The chaos game requires a set of probabilities p_k, $k = 1, \ldots, N$ that determine which of the transformations $w_1, \ldots, w_N$ should be taken in each step of the algorithm. As demonstrated, the choice of these probabilities is not obvious. The adaptive cut method may provide another way to assign values for the probabilities. We subdivide the points plotted by the adaptive method into N subsets, each one collecting points drawn in the corresponding attractorlet $w_k(A_\infty)$.[28] The relative number of points in each subset determines the corresponding probability. For example, for the fern we obtain the numbers $0.73, 0.13, 0.11, 0.03$.[29] When used as probabilities in the chaos game, these values result in an image with more evenly distributed points as compared to images based on the chaos game using probabilities suggested by the above formula from page 307 (see figure 6.39).

[28]This specification is not precise because the attractorlets $w_k(A_\infty)$ of the first stage may be overlapping. The algorithm counts a point representing the attractorlet $w_{s_1} w_{s_2} \cdots w_{s_m}(A_\infty)$ as a point belonging to $w_{s_1}(A_\infty)$.

[29]These probabilities should not be taken as absolute because their values depend to some extent on the resolution of the image. Other weight factors may be better for other resolutions.

Chapter 7

Recursive Structures: Growing Fractals and Plants

The development of an organism may [...] be considered as the execution of a 'developmental program' present in the fertilized egg. The cellularity of higher organisms and their common DNA components force us to consider developing organisms as dynamic collections of appropriately programmed finite automata. A central task of developmental biology is to discover the underlying algorithm from the course of development.

Aristid Lindenmayer and Grzegorz Rozenberg[1]

The historical constructions of fractals by Cantor, Sierpinski, von Koch, Peano, etc., have been labeled as 'mathematical monsters'. Their purpose has been mainly to provide certain counterexamples, for example, showing that there are curves that go through all points in a square. Today a different point of view has emerged due to the ground-breaking achievements of Mandelbrot. Those strange creations from the turn of the century are anything but exceptional counterexamples; their features are in fact typical of nature. Consequently, fractals are becoming essential components in the modeling and simulation of nature. Certainly, there is a great difference between the basic fractals shown in this book and their counterparts in nature: mountains, rivers, trees, etc. Surely, the artificial fractal mountains produced today in computer graphics already look stunningly real. But on the other hand they still lack something we would certainly feel while actually camping in the real mountains. Maybe it is the (intentional) disregarding of all developmental processes in the fractal models which is one of the factors responsible for this shortcoming.

---

[1] In: *Automata, Languages, Development*, A. Lindenmayer, G. Rozenberg (eds.), North-Holland, Amsterdam, 1975.

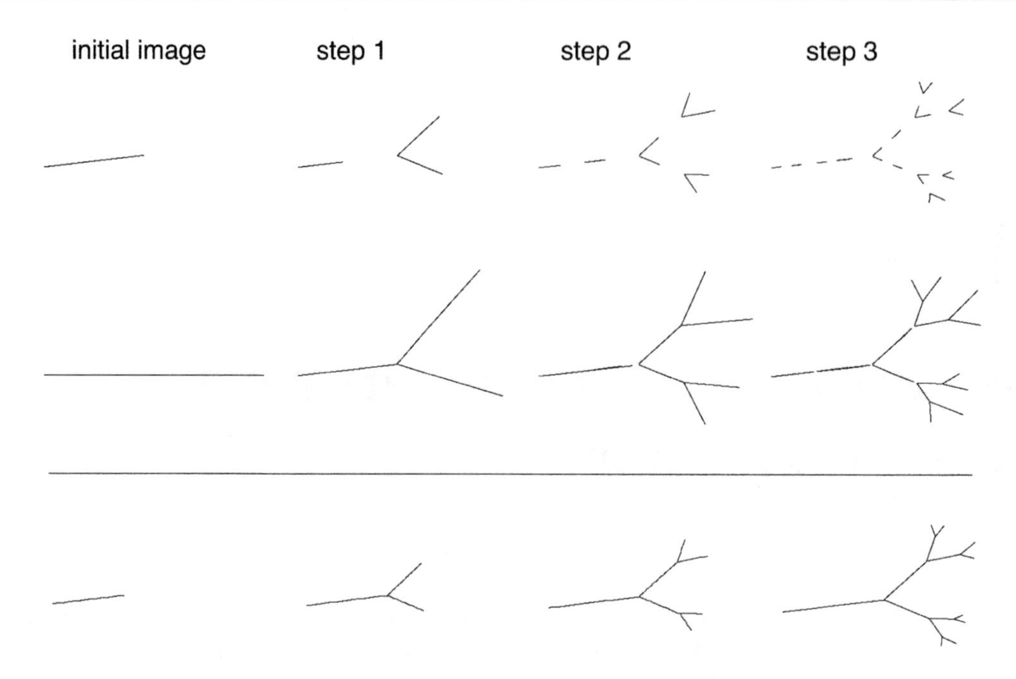

Figure 7.1 : An attempt to simulate growth by an MRCM. The two top rows show the first steps of the iteration of a single MRCM with two different initial images. In the top row a small 'stem', and in the middle row a tall 'stem' is taken as the initial image. Both cases reveal deficiencies. The small stem seems to grow, but unnatural gaps develop between the parts of the structure. The iteration of the tall stem does not produce gaps; however, the branches seem to split and do not grow. In contrast, the bottom row shows the development we would like to simulate. Small new twigs are growing from the stem and the other branches. Note that the bottom figures are not just scaled down copies from the middle row.

Fractals in nature are always a result of some growth process. In contrast, we have regarded fractals as *static*. Indeed, it was a goal to characterize fractals as solutions to equations. And nothing can be more eternal than the solution of an equation. You may argue on the other hand that these fractals have usually been obtained from dynamical processes such as the one in the Multiple Reduction Copy Machine introduced in chapter 1. However, our attention has always been focused on the end result, the attractor of the machine. The process leading to this end product was important merely when it revealed properties of the attractor. In this chapter we pay more attention to the intermediate stages in the production of a fractal. We begin the discussion with another dialect of the language of fractals which has been created specifically for the description of natural growth processes. This dialect is called *L-systems*.

In figure 7.1 we illustrate the contrast between the way an MRCM may generate a twig from a bush or tree and how one would actually expect a real twig to grow with time.

Aristid Lindenmayer, 1925–1989

Figure 7.2

Development in Time

If we want to include the development in time in our fractal models we have to consider, for example, erosion models for the generation of fractal mountains or growth and evolution models for plants. While erosion models in the context of computer graphics[2] have just started to play a role in current research and will not be discussed in this presentation, growth models for plants have been around for a while; and some of them are also applicable to imaging. The view that growth and form are interrelated actually has a long tradition in biology. In his monumental work *On Growth and Form* D'Arcy Thompson traces its origins back to the late seventeenth century and comments:[3]

> *The rate of growth deserves to be studied as a necessary preliminary to the theoretical study of form, and organic form itself is found, mathematically speaking, to be a function of time. [...] We might call the form of an organism an event in space-time, and not merely a configuration in space.*

In 1968 the biologist Aristid Lindenmayer invented a formalization of the description of plant growth that is also very suitable in computer implementations. These formal descriptions are now known as *parallel rewriting systems* or *L-systems*.[4] A recent account of the state of the art of L-systems in the context of pattern formation in botany is presented in the beautiful book

[2] K. Musgrave, C. Kolb and R. Mace, *The synthesis and the rendering of eroded fractal terrain*, SIGGRAPH '89, Computer Graphics 24 (1988).

[3] D'Arcy Thompson, *On Growth and Form*, New Edition, Cambridge University Press, Cambridge, 1942.

[4] A. Lindenmayer, *Mathematical models for cellular interaction in development, Parts I and II,* Journal of Theoretical Biology 18 (1968) 280–315.

The Algorithmic Beauty of Plants by Przemyslaw Prusinkiewicz and Aristid Lindenmayer.[5]

But the modeling of growth for plants by L-systems is only one theme of this chapter. We also reexamine many of the fractals we have seen in the preceding chapters and demonstrate how we can describe 'natural' developmental processes which finally create these fractals. We will use two approaches towards this issue: our familiar concept of MRCMs (carefully set up for this problem) and the language of L-systems. In other words, we will see the growth of fractals.

[5] P. Prusinkiewicz and A. Lindenmayer, *The Algorithmic Beauty of Plants*, Springer-Verlag, New York, 1990.

7.1 L-Systems: A Language for Modeling Growth

To begin, let us consider a plant, namely, a blue-green alga, a species called *anabaena catenula*. The alga grows in filaments of strings of cells. There are two types of cells: specialized cells, which do not divide and which are called *heterocysts*, and unspecialized cells, which divide and are responsible for the entire growth of the alga. In a laboratory experiment[6] cells divided about every 14 hours. The cell division was asymmetric in the sense that one of the offspring was generally smaller than the other at the time of division. Moreover, division was governed by a simple rule: if a given cell arose as the left offspring of a division, then its left progeny would be the smaller one in the next generation. Correspondingly, for a right offspring, the right progeny would be the smaller one in the next generation. In a figure illustrating the procedure (figure 7.3) we indicate the reproductive history of a cell by an arrow. The arrow points to the left, when the cell has emerged as a left daughter cell, and it points to the right for a right daughter cell. The smaller cells need about 20% more time to mature. This difference, however, is ignored in our first model as pictured in the figure. Therefore, in each following stage there is no difference between the children of large and small cells. A second model that takes these differences into account will be derived further on.

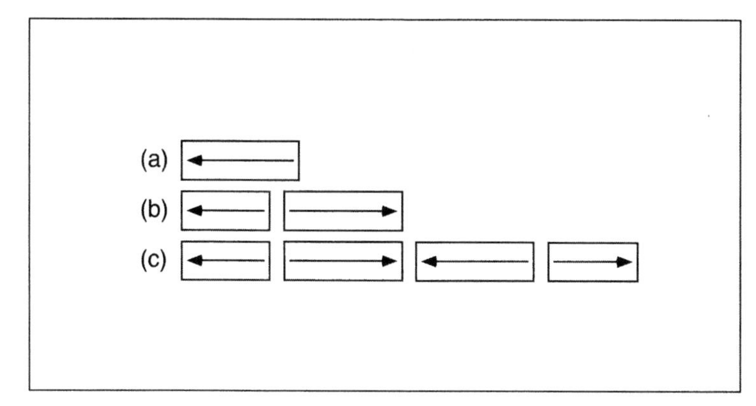

First Model of Cell Division

In (a) one large cell is shown with an arrow to the left, indicating that the cell is a left daughter cell. In (b) the cell has subdivided into two cells, a small one on the left and a big one on the right. The arrows again indicate whether a cell has been created on the left or on the right. In the next division, (c), there are a total of four cells.

Figure 7.3

Reproduction Rules

Let us now write down the reproduction rules in a more formal manner. For that purpose we denote a string of cells by a string of symbols, each symbol denoting an individual cell. Let $\overleftarrow{A}$ stand for a small cell which was created as a left daughter cell (i.e., in the diagram it would appear with an arrow pointing to the left). Likewise, $\overrightarrow{A}$ denotes a small right daughter cell, and $\overleftarrow{B}$ and $\overrightarrow{B}$ are the large left and right cells, respectively. Now the observed rule for $\overleftarrow{A}$ states that this cell produces a small cell on the left and a large cell on the right. Thus, we can signify the reproduction rule for a small left cell with

$$\overleftarrow{A} \rightarrow \overleftarrow{A}\overrightarrow{B}. \tag{7.1}$$

[6]G. J. Mitchison and M. Wilcox, *Rule governing cell division in anabaena*, Nature 239 (1972) 110–111.

Development of Anabaena Catenula

Sequence of photographs of a growing blue-green alga (anabaena catenula) filament, demonstrating the asymmetric division rule and the appearance of a new heterocyst (specialized) cell at arrow. Photographs (a) to (e) were taken at 0, 6, 10, 15 and 21 hours, respectively. The division of the cell bracketed in (a) can be followed by referring to the bracketed cells in (b), (c) and (d). The picture is from G. J. Mitchison and M. Wilcox, *Rule governing cell division in anabaena*, Nature 239 (1972) 110–111.

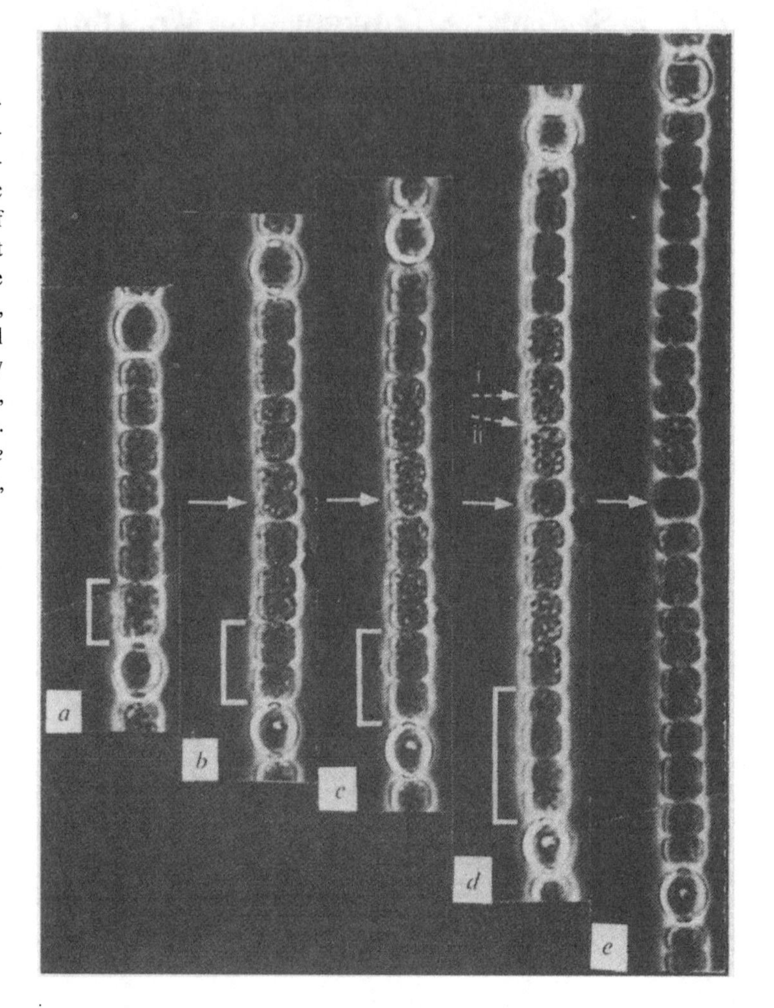

Figure 7.4

Using the same principles of notation, we write the reproduction rules for the other cells as

$$\begin{aligned} \overrightarrow{A} &\to \overleftarrow{B}\,\overrightarrow{A}, \\ \overleftarrow{B} &\to \overleftarrow{A}\,\overrightarrow{B}, \\ \overrightarrow{B} &\to \overleftarrow{B}\,\overrightarrow{A}. \end{aligned} \tag{7.2}$$

With this notation it is now very easy to predict the sequence of cells as they appear. For example, if we start with $\overleftarrow{A}$, we look up its reproduction in the above table, and we find $\overleftarrow{A}\,\overrightarrow{B}$. Now we apply the same procedure to $\overleftarrow{A}$ and $\overrightarrow{B}$ simultaneously to obtain four cells

$$\overleftarrow{A}\,\overrightarrow{B}\,\overleftarrow{B}\,\overrightarrow{A}.$$

In the next cell division we have 8 cells, namely,

$$\overleftarrow{A}\overrightarrow{B}\overleftarrow{B}\overrightarrow{A}\overleftarrow{A}\overrightarrow{B}\overleftarrow{B}\overrightarrow{A},$$

then 16 cells

$$\overleftarrow{A}\overrightarrow{B}\overleftarrow{B}\overrightarrow{A}\overleftarrow{A}\overrightarrow{B}\overleftarrow{B}\overrightarrow{A}\overleftarrow{A}\overrightarrow{B}\overleftarrow{B}\overrightarrow{A}\overleftarrow{A}\overrightarrow{B}\overleftarrow{B}\overrightarrow{A},$$

and so forth. Obviously, there is a repeating behavior in this sequence. The cell sequence $\overleftarrow{A}\overrightarrow{B}\overleftarrow{B}\overrightarrow{A}$ is a periodic cycle. This is the case because $\overleftarrow{A}\overrightarrow{B}$, its first half, produces the same sequence as the second half $\overleftarrow{B}\overrightarrow{A}$.

Modeling Anabaena Catenula with Ages

As a first step to improve the model for cell division for anabaena catenula, we may consider the different time spans needed by small and large cells to mature and to divide. The laboratory observation has been that small cells take about 20% more time on the average between successive divisions. Thus, whereas an initially large cell takes five time steps to divide, an initially small cell takes a total of six to grow and divide. A large cell B might develop in four stages into states denoted by C, D, E and F. Then, a division takes place producing either two cells AB or BA as before, depending on whether the initial cell B was a left daughter or a right one. A small cell A, on the other hand, needs one additional time step for the whole process. This can be simply modeled by introducing a transition from cell type A to type B. Thus, for small cells A the development is in six stages either

$$A \rightarrow B \rightarrow C \rightarrow D \rightarrow E \rightarrow F \rightarrow AB$$

or

$$A \rightarrow B \rightarrow C \rightarrow D \rightarrow E \rightarrow F \rightarrow BA$$

again depending on whether the initial cell was a left or right daughter. As before, let us use arrows to denote left and right daughter cells and summarize the production rules of the model:

$$\overleftarrow{A} \rightarrow \overleftarrow{B},\ \overleftarrow{B} \rightarrow \overleftarrow{C},\ \overleftarrow{C} \rightarrow \overleftarrow{D},\ \overleftarrow{D} \rightarrow \overleftarrow{E},\ \overleftarrow{E} \rightarrow \overleftarrow{F},$$
$$\overrightarrow{A} \rightarrow \overrightarrow{B},\ \overrightarrow{B} \rightarrow \overrightarrow{C},\ \overrightarrow{C} \rightarrow \overrightarrow{D},\ \overrightarrow{D} \rightarrow \overrightarrow{E},\ \overrightarrow{E} \rightarrow \overrightarrow{F},$$

and

$$\overleftarrow{F} \rightarrow \overleftarrow{A}\overrightarrow{B},\ \overrightarrow{F} \rightarrow \overleftarrow{B}\overrightarrow{A}.$$

Starting with $\overleftarrow{A}$ the following transitions will take place:

$$\begin{aligned}\overleftarrow{A} &\rightarrow \overleftarrow{B} \rightarrow \overleftarrow{C} \rightarrow \overleftarrow{D} \rightarrow \overleftarrow{E} \rightarrow \overleftarrow{F}\\ &\rightarrow \overleftarrow{A}\overrightarrow{B} \rightarrow \overleftarrow{B}\overrightarrow{C} \rightarrow \overleftarrow{C}\overrightarrow{D} \rightarrow \overleftarrow{D}\overrightarrow{E} \rightarrow \overleftarrow{E}\overrightarrow{F}\\ &\rightarrow \overleftarrow{F}\overleftarrow{B}\overrightarrow{A} \rightarrow \overleftarrow{A}\overrightarrow{B}\overleftarrow{C}\overrightarrow{B} \rightarrow \overleftarrow{B}\overrightarrow{C}\overleftarrow{D}\overrightarrow{C} \rightarrow \overleftarrow{C}\overrightarrow{D}\overleftarrow{E}\overrightarrow{D}\\ &\rightarrow \overleftarrow{D}\overrightarrow{E}\overleftarrow{F}\overrightarrow{E} \rightarrow \overleftarrow{E}\overrightarrow{F}\overleftarrow{A}\overrightarrow{B}\overrightarrow{F} \rightarrow \overleftarrow{F}\overleftarrow{B}\overrightarrow{A}\overleftarrow{B}\overrightarrow{C}\overleftarrow{B}\overrightarrow{A}.\end{aligned}$$

Obviously, the behavior of the cells is more complicated than in the first crude model. Cells in different cytological states, of different sizes,

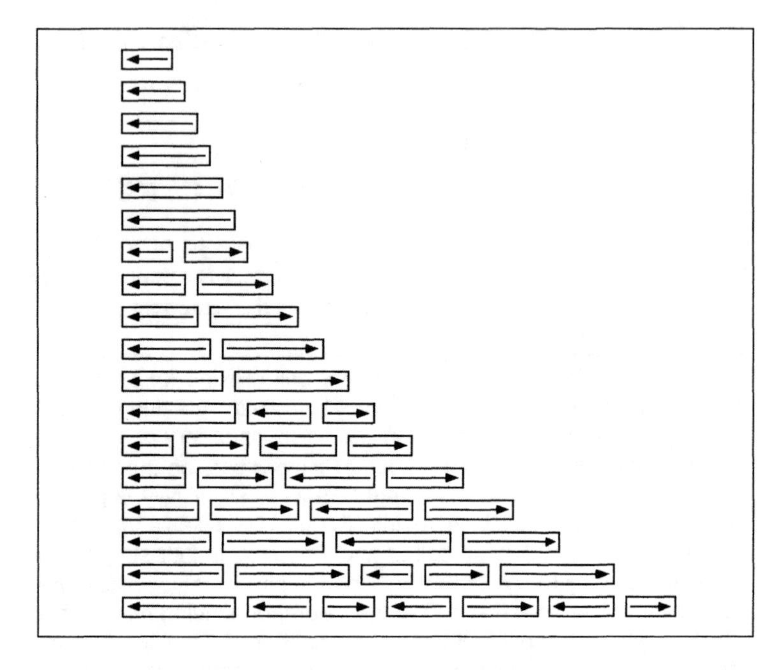

Figure 7.5 : Action of the modified L-system for cell division of blue-green alga anabaena catenula as given in the long transition formula for $\overleftarrow{A}$. Note that cells in different cytological states and of different sizes coexist.

and even from different generations, coexist (see figure 7.5). The photograph from the original laboratory experiment confirms that this model captures the actual process more accurately (see figure 7.4).

L-Systems: Another Example of an Iterator

What we have described here in (7.1) and (7.2) is a first very simple L-system for the vegetative part of the filament. It consists of the four (re)*production rules* and the definition of an initial state, which we have chosen to be $\overleftarrow{A}$. This initial state in L-systems is called the *axiom*. The concept is just another concrete form of an iterator as pictured in figure 7.6. It is a feedback machine which operates on strings. The production (or *rewriting*) rules determine how a given input string is transformed into an output string.

In fact, string rewriting feedback machines lie at the heart of what is known as formal languages and formal grammars in computer science. On these topics an enormous amount of work has been done. The concept has been explored in many directions and several classes (or types) of feedback machines have been identified. L-systems are string rewriting machines which are characterized by the fact that the production rules are applied simultaneously to all symbols of the input string[7] This property reflects the biological origin of L-systems. Lindenmayer intended to capture, for example, cell divi-

[7]This is in contrast to sequential application, as is typical for Chomsky grammars.

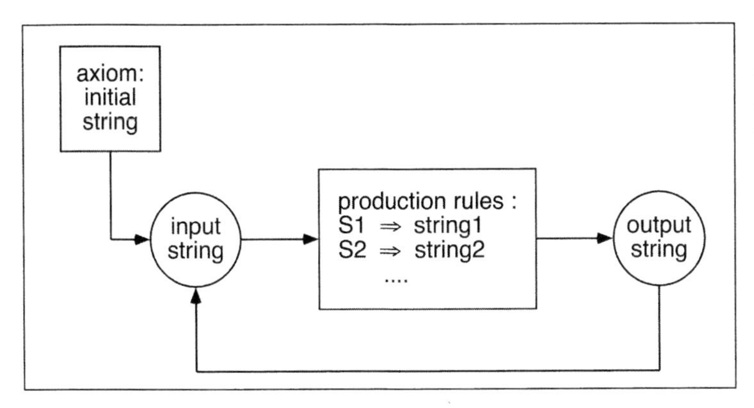

The L-System Machine

The internal operation of the machine is given by the production rules. Input and output consist of strings of symbols. The initial input string is called the axiom.

Figure 7.6

sions in multicellular organisms, where many divisions may occur at the same time.

In the following we will use so called deterministic context-free L-systems. This means that the rewriting rules depend only on single, isolated symbols (i.e., the substitution of a particular symbol of an input string depends only on this symbol and not on its neighbors). Moreover, for each symbol the machine can work on, there is exactly one rewriting rule.[8]

Deterministic L-Systems

In this chapter we only use context-free L-systems. Such systems are formally defined by an *alphabet*

$$V = \{a_1, a_2, \ldots, a_n\},$$

the *production map*

$$\begin{aligned} P : V &\rightarrow V^* \\ a &\rightarrow P(a), \end{aligned}$$

where V^* is the set of all strings formed by symbols from V, and an *axiom*

$$\alpha^{(0)} \in V^*,$$

the initial string.

Note that for all symbols of the alphabet $a \in V$ there is exactly one production (or rewriting) rule $P(a)$. Starting with the axiom $\alpha^{(0)}$, the L-system generates a sequence of strings: $\alpha^{(0)}, \alpha^{(1)}, \alpha^{(2)}, \ldots$ where the string $\alpha^{(i+1)}$ is obtained from the preceding string $\alpha^{(i)}$ by applying the production rules to all (e.g., m) symbols $\alpha_1^{(i)}, \ldots, \alpha_m^{(i)}$ of the string simultaneously:

$$\alpha^{(i+1)} = P(\alpha_1^{(i)})P(\alpha_2^{(i)})P(\alpha_3^{(i)}) \cdots P(\alpha_m^{(i)}).$$

[8]In this special case it does not matter whether the symbols in a string are sequentially rewritten one by one or simultaneously all at once.

Formally, we write an L-system using the following notation (exemplified by the basic anabaena model):

L-system: Anabaena (basic model)
Axiom: $\overleftarrow{A}$
Production rules: $\overleftarrow{A} \rightarrow \overleftarrow{A}\,\overrightarrow{B}$
$\overrightarrow{A} \rightarrow \overleftarrow{B}\,\overrightarrow{A}$
$\overleftarrow{B} \rightarrow \overleftarrow{A}\,\overrightarrow{B}$
$\overrightarrow{B} \rightarrow \overleftarrow{B}\,\overrightarrow{A}$

It seems apparent that the approach of L-systems can be very useful in describing growth phenomena in a short and precise manner. The axiom and the production rules constitute the system. These rules are typically derived from research, in our example from controlled laboratory experiments with a strain of the alga organism. Once the rules have been set, they can easily be implemented on a computer in order to check in a simulation if they really capture the essential phases of the development of the plant. This verification would hardly be possible without the tool of a modern computer, at least for L-systems of such complexity as can be expected for real living plants that are somewhat more complicated than the simple string of alga cells.

There is, however, one point we have not yet discussed. Namely, it would be hopeless to carry out the verification process if we had to look at long, confusing strings of symbols. Each symbol might have a precise and understandable interpretation, but the whole string would be too long and too complicated to be understood. This is already very obvious from looking at the results of our above calculations in our very simple model.

Visualization of L-Systems

Again, the computer can help. A visual translation of the generated strings is required, and here it is literally true that a picture is worth a thousand words. On the other hand, the graphical interpretation should reflect the meaning of symbols. In this sense it should be closely coupled to the underlying problem. In the above case of alga, we may picture the cells $\overleftarrow{A}$ and $\overleftarrow{B}$ (and also $\overleftarrow{C}$, $\overleftarrow{D}$, $\overleftarrow{E}$ and $\overleftarrow{F}$) as cylinders or rectangles of varying lengths enclosing an arrow to the left or to the right as in figure 7.5. In this way a long string is interpreted graphically, and a visual inspection of the result will readily reveal important qualities of the underlying L-system model.

The graphical interpretation of strings of symbols is not predetermined in any way. This is one important strong point of the L-system approach to developmental models. The quality of the pictorial representation is completely independent of the generation process of the symbol string and can be adjusted to the available graphics environment and to the intentions of the experiment. For example, the interpretation may be as simple as a straightforward line drawing, or it may be the outcome of an elaborate ray-tracing computation yielding highly realistic three-dimensional images.

Let us summarize. To develop an L-system for a particular biological species we can proceed in the following steps:

- analyze the object in nature (observation) and/or in a laboratory;

- set up the rules in an informal way;
- formulate rules and the initial state as an L-system;
- run a simulation on a computer generating a long string of symbols;
- translate the result into a graphical output;
- compare the picture (or several pictures for different developmental stages) with the behavior of the real object.

Finally, corrections can be made in the model, if necessary, and the steps may be repeated.

In 1968, Lindenmayer was interested in the development of filamentous plants on the cellular level. He considered two classes of filaments, simple (consisting of sequences of cells) and branching. The L-system symbols that he used corresponded to individual cells, thus only organisms consisting of relatively small numbers of cells (of the order of hundreds) could be handled in practice. He chose two species of algae, which are multicellular organisms with a small enough number of cells, to illustrate his concepts. Later on (in the seventies) Lindenmayer extended the interpretation of L-systems, so that L-system symbols were applied to represent entire plant modules, such as an internode, a leaf, or a bud. Using this interpretation, he analyzed the structure of compound inflorescences. This work provided a basis for the realistic visualization of the models of herbaceous (non-woody) plants. This was the first realistic visualization of plant models generated using L-systems.[9] Towards the end of this chapter we will sketch a few of the more elementary of these recent results. But before we turn to these applications of L-systems, let us start the second theme of this chapter: the growth of fractal patterns.

[9] See P. Prusinkiewicz, A. Lindenmayer, J. Hanan, *Developmental models of herbaceous plants for computer imagery purposes,* Computer Graphics 22, 4 (1988) 141–150. Previous application of L-systems to image synthesis, pioneered by Alvy Ray Smith in 1978, used L-systems to generate abstract branching structures that did not correspond to the existing species. See A. R. Smith, *Plants, fractals, and formal languages,* Computer Graphics 18, 3 (1984) 1–10.

7.2 Growing Classical Fractals with MRCMs

Classical fractals like the Cantor set, the Koch curve, the Sierpinski gasket, the Peano curve, etc., were very strange objects at the time they were introduced. Their creators were extremely careful in their definitions and exactly described the construction processes. We will demonstrate that L-systems provide a new language to efficiently and precisely define such constructions. On the other hand, this approach is in some respects rather formalistic and it is a good idea to start in a more visual and more familiar manner. Thus, before we let fractals grow using L-systems, let us find out what we can do with Multiple Reduction Copy Machines and iterated function systems.[10]

Encoding Images ...

We introduced the concept of the MRCM as a tool to encode images. We described several ways of decoding; the basic method is the iteration of the feedback machine, the MRCM. Starting with an arbitrary image one application of the machine produces a new image, a collage of contracted copies of the first image. Applying the machine to this new image and iteration of the process leads to a final image, the attractor of the machine, which is the decoded image.

... Versus Pattern Generation

The fact that the outcome of the process, the attractor of the machine, is completely independent of the starting image has been stressed in the preceding chapters on MRCMs. However, the intermediate stages in the iteration very much depend on the initial image. In this sense, the iteration of an MRCM is an example of recursive pattern generation. In this section we take a closer look at these patterns and observe that by choosing an appropriate initial pattern, we may very well reproduce the historical construction process given for many classical fractals.

The Cantor Set

As a first example, we compare the MRCM encoding of the Cantor set with the original construction process (divide a line segment into three equal parts, remove the middle part, repeat this process ad infinitum). Figure 7.7 shows the blueprint of the MRCM. We use the word 'Cantor' as initial image. Figure 7.8 shows the iteration (left) leading to the Cantor set as final image and compares this with the classical construction (right). Observe that the classical construction seems to somehow capture the Cantor set C more accurately in every stage: the Cantor set is a subset of all the intermediate stages C_0, C_1, C_2, ... This is not true for the MRCM representation when started with an arbitrary pattern as shown in the left part of the figure.

In some sense the MRCM for the Cantor set seems to grasp only the structure of the final image, namely, its self-similarity. But we can do more. In the case of the Cantor set this is not a difficult observation. We simply have to do two things:

[10]The relation between L-systems and IFSs was first studied in T. Bedford, *Dynamics and dimension for fractal recurrent sets,* J. London Math. Soc. 33 (1986) 89–100. Another discussion more oriented towards formal languages is presented in P. Prusinkiewicz and M. Hammel, *Automata, languages, and iterated function systems,* in: *Fractals Modeling in 3-D Computer Graphics and Imaging,* ACM SIGGRAPH '91 Course Notes C14 (J. C. Hart, K. Musgrave, eds.), 1991.

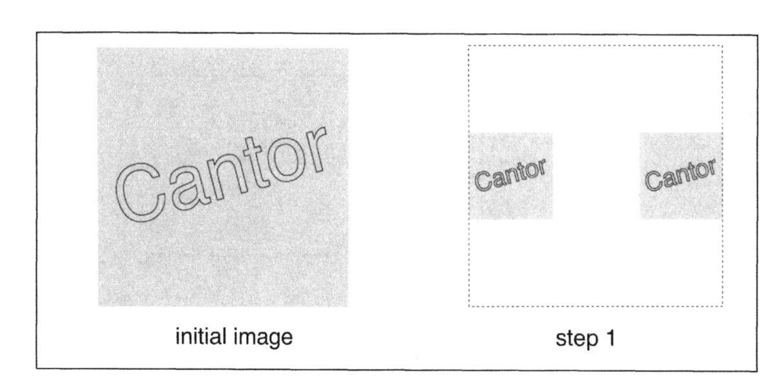

MRCM Blueprint of Cantor Set

MRCM encoding of the Cantor set by two affine transformations, which reduce by a factor of 1/3 and position as indicated by the squares.

Figure 7.7

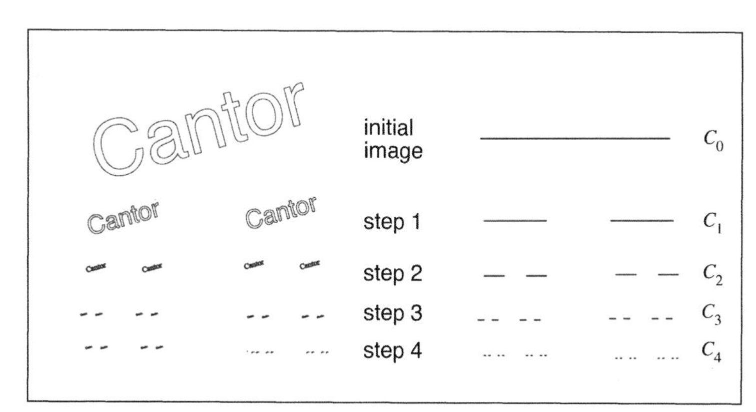

Cantor Set

MRCM iteration (left) is compared with the original construction process (right). In the course of the iteration of the MRCM, the initial pattern vanishes and the shape of the Cantor set becomes visible. The classical construction of the Cantor set proceeds by repeatedly removing the middle thirds of all line segments.

Figure 7.8

- use the line segment C_0 as initial image;
- set up the two transformations w_1, w_2 of the MRCM such that C_0 is mapped onto the two line segments of C_1, i.e., the left and right thirds of the initial segment C_0.

Then the iteration of the machine exactly generates the sets of the classical construction as shown on the right in figure 7.8.

The Koch Curve

Next let us examine some further examples. A close relative of the Cantor set is the Koch curve. Figure 7.9 shows the classical construction of the Koch curve, which is just a variation of the Cantor set construction. Here in each line segment the middle third is not omitted but replaced by two segments forming a tentlike angle. In chapter 2 we noted that self-similarity is built into the construction process: the curve K_{i+1} is composed from four contracted similar copies of K_i. This observation leads directly to the correct setup of an MRCM:

- use K_0, a line segment, as initial image;
- set up four contractions $w_1, \ldots, w_4$ such that the transformed copies $w_i(K_0)$ are exactly mapped onto the line segments of K_1.

The Koch Curve

The first stages of Koch's classical construction. All end points of the generated line segments are part of the final curve.

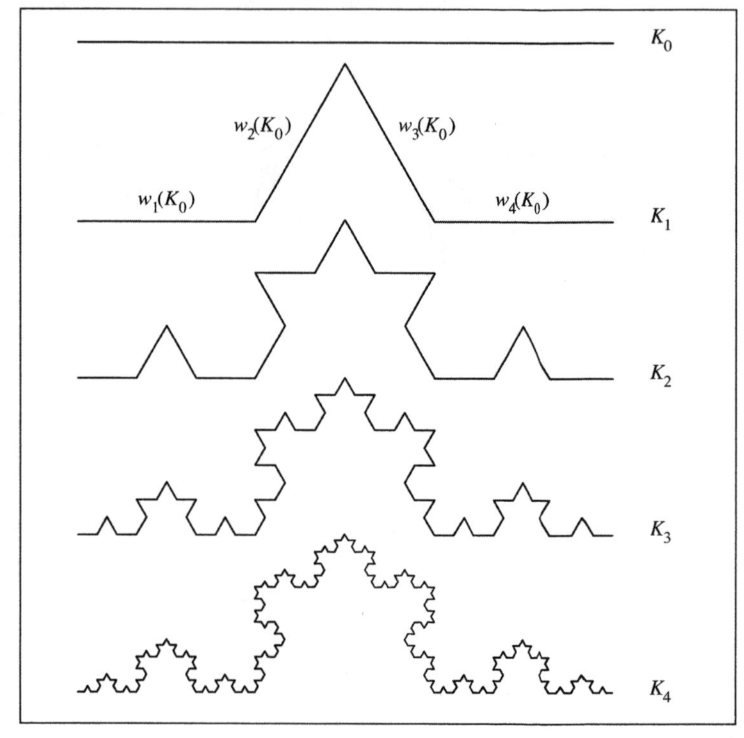

Figure 7.9

The Sierpinski Gasket

The first stages of the classical construction of the Sierpinski gasket. Starting with a shaded triangle S_0 we repeatedly remove the middle parts of the remaining triangles. Note that in this construction the final set is a subset of all the preceding stages.

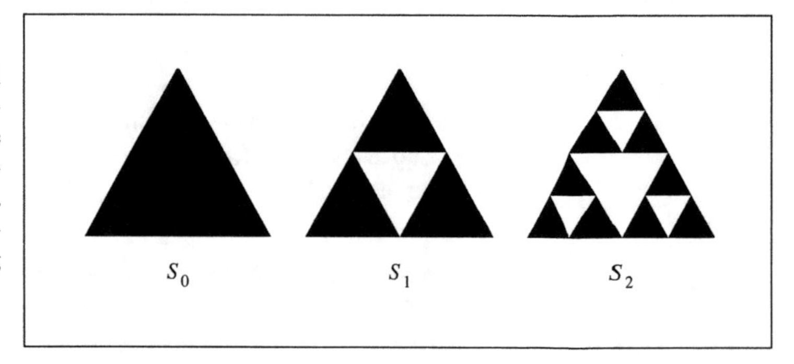

Figure 7.10

The iteration of this machine exactly generates the sequence K_1, K_2, K_3, ...of the classical construction.

The Sierpinski Gasket

Let us now turn to our favorite example, the Sierpinski gasket, and illuminate some further aspects. Figure 7.10 shows its classical construction. The setup of an appropriate MRCM is straightforward:

- use S_0, a shaded equilateral triangle, as the initial image;
- choose three transformations w_1, w_2, and w_3 such that the copies $w_1(S_0), \ldots, w_3(S_0)$ are aligned with the three triangles in S_1.

Nonunique Transformations ...

You may have noted that the selection of the transformations w_i is not completely determined by S_1. Indeed, even if we assume that $w_1(S_0)$ is the left, $w_2(S_0)$ is the top and $w_3(S_0)$ is the right triangle of S_1, there are still $6^3 = 216$ possible choices because each of the transformations may include reflection and rotations of 120 degrees, 240 degrees or 0 degrees (which may be the only case you may have thought of). But for all possible choices the iteration of the machine generates the desired sequence $S_0, S_1, S_2, \ldots$ of the classical construction.

The same kind of ambiguity is true for the selection of transformations for the MRCMs which generate the Cantor set and the Koch curve. But in these cases they do not matter either. So why is it interesting to discuss this at all? Let us demonstrate this point for the MRCMs which generate the Sierpinski construction.

Sierpinski's Rose Garden

Rose gardens obtained by the iteration of another MRCM which encodes the Sierpinski gasket.

Figure 7.11

...Uncovered by Choice of Initial Pattern

Figure 7.11 shows the iteration of a particular MRCM which encodes the Sierpinski gasket. Taking the triangle S_0 as the initial image would plainly generate the usual classical sequence $S_0, S_1, S_2, \ldots$ as shown in figure 7.10. However, choosing a somewhat different initial pattern in figure 7.11, a 'rose', we uncover the details of the MRCM. From the top two images (the blueprint of the machine) we can conclude that the lower left transformation involves a 120 degree rotation and a horizontal reflection, the top transformation is just a reduction and the lower right transformation is a horizontal reflection followed by a 120 degree rotation. Because of the unusual initial image it is nearly impossible to see from the blueprint or the first iteration that this

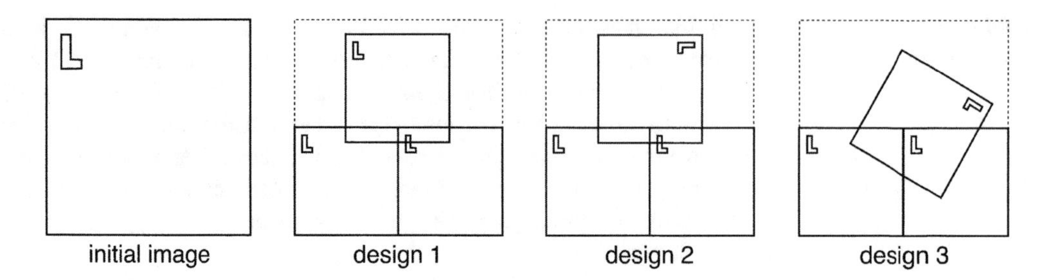

Figure 7.12 : The blueprint for three choices of transformations, the first one of which is the standard configuration for the Sierpinski gasket. Note, that the overlap is required to obtain an equilateral Sierpinski gasket.

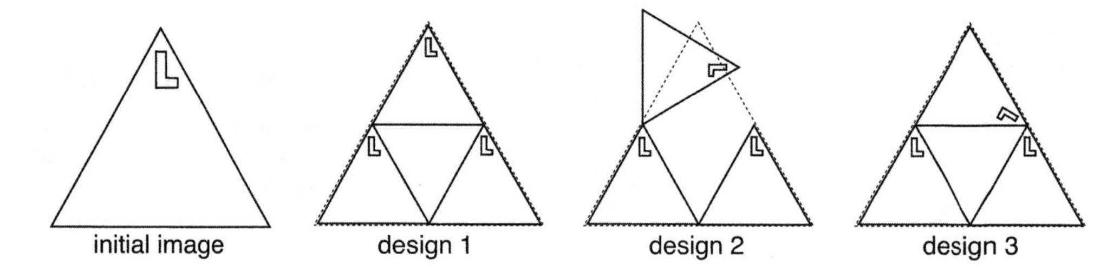

Figure 7.13 : The same choices of transformations as in 7.12, however, using an equilateral triangle as initial image.

machine encodes a Sierpinski gasket. Rather, it may remind you of some computer artistic attempt. But there is much more hidden, something which makes this a very special rose garden.

Figure 7.15 shows the iteration of precisely the same MRCM yielding some kind of interesting maze. In contrast to the rose garden, each generated pattern is a subset of the Sierpinski gasket. However, is it possible to find an initial image such that in each stage all line segments join up to form a single path that visits all subtriangles of that stage? The answer is yes. The path always starts at the left corner and ends at the right corner. In the limit it visits all points of the Sierpinski gasket. Figure 7.16 reveals this very surprising construction called the *Sierpinski arrowhead*.

In the figure we choose an arrow pointing from left to right as the initial image. It is positioned at the bottom of S_0. For reference we show S_0 in a dotted pattern to provide a better orientation. You will observe that in the first step the MRCM generates a curve which visits all subtriangles of S_1 from this initial image. In the next step we obtain a curve which visits all subtriangles of S_2, and so on. In the limit this defines a curve which visits all points of the Sierpinski gasket.

MRCM Blueprint Pitfalls

There are 216 possible choices of transformations to encode the equilateral Sierpinski gasket with the left vertex at $(0,0)$, the right vertex at $(1,0)$, and the top vertex at $(1/2, \sqrt{3}/2)$. Our standard method for

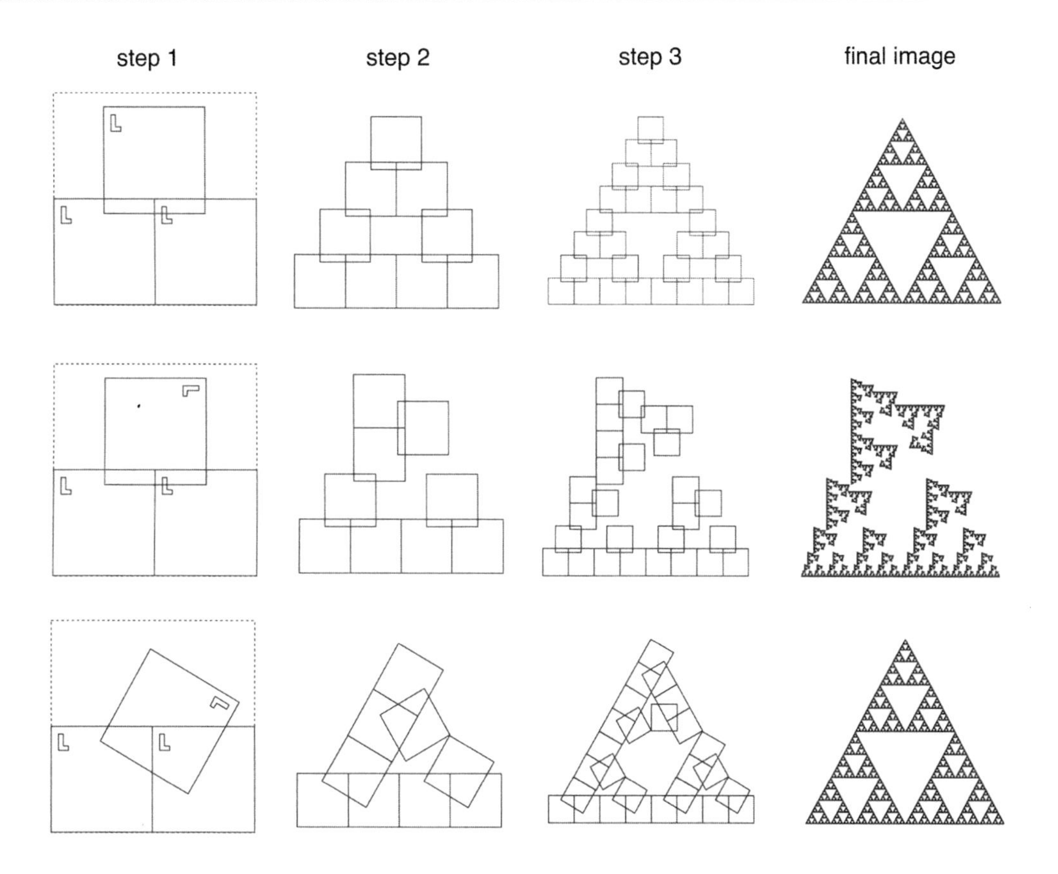

Figure 7.14 : Running the MRCM with the transformations specified in figure 7.12 reveals that only one of the modifications of the standard design produces the Sierpinski gasket.

visualizing the MRCM blueprint uses the unit square with the inscribed letter 'L' as an indication of the orientation as the initial image. The 216 different possible choices of transformations are not at all obvious. In fact, one can easily be mislead.

Figure 7.12 shows three different design choices. Guess which of these three settings will generate the Sierpinski gasket! Without much experience we would be inclined to bet on the designs 1 and 2, but certainly not on design 3. Figure 7.14 reveals the surprising answers. Designs 3 yields the Sierpinski gasket while design 2 is as far from the gasket as it can be. What's going on here? Observe that in design 2 the upper transformation involves a 90 degree (clockwise) rotation which does not appear to be dramatically different from design 1 when we just look at the first iteration. In contrast, design 3 involves a 120 degree (clockwise) rotation and the first iteration certainly does not immediately suggest that a Sierpinski gasket will result. But as figure 7.14 shows, it does, and the reason is obvious after the experiment. The Sierpinski gasket decomposes into three pieces, each of which is

The Sierpinski Maze

The iteration of the same MRCM as in figure 7.11 started with a different initial image.

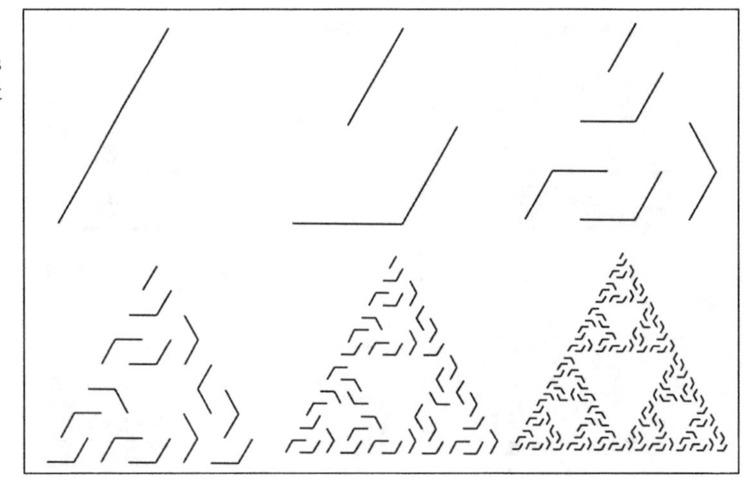

Figure 7.15

The Sierpinski Arrowhead

The iteration of the same MRCM as in the previous two figures (started with an arrow as initial image) generates a sequence of curves which visit all subtriangles of the classical construction. The limit curve touches all points of the Sierpinski gasket. The resulting structure is called the *Sierpinski arrowhead* by Mandelbrot.

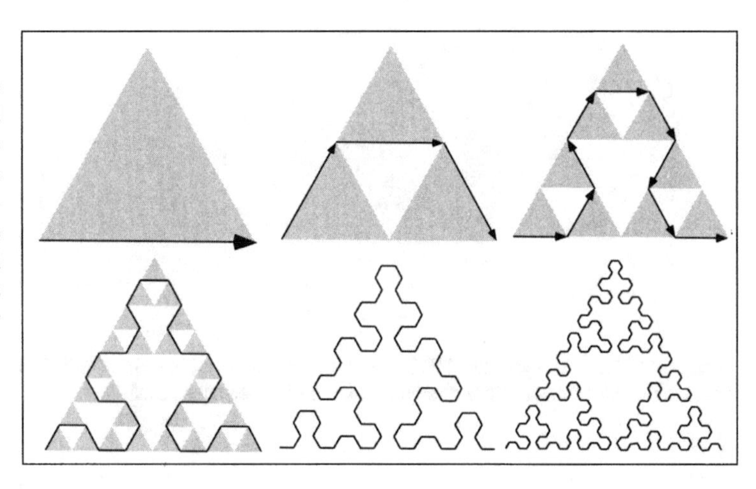

Figure 7.16

a reduced copy of the whole. Moreover, each piece is invariant under multiples of 120 degree rotations and reflections, as we remarked above. Thus, in other words, symmetries which may or may not be suggested by the first iteration and the arbitrary choice of an initial image — like the square in design 2 — of an MRCM are irrelevant. In fact, as we have seen they can be quite misleading. The only symmetries which provide us with choices for the transformations of an MRCM are those which are valid for the limit image. For the design of the Sierpinski gasket an equilateral triangle is appropriate since it has suitable symmetries (see figure 7.13). Note, that such a triangle is used to obtain an MRCM which generates the classical construction as shown in figure 7.10.

Sierpinski Gasket As a Curve

The Sierpinski gasket can be interpreted as a parametrized curve, i.e., there exists a continuous transformation of an interval to the gasket which covers all of its points. This fact can be understood using an addressing scheme which is induced by the transformations w_0, w_1 and w_2 of the MRCM which generates the path sequence of the Sierpinski arrowhead from figure 7.16 (to make things more obvious we use the indices 0, 1 and 2 instead of 1, 2 and 3). Figure 7.17 shows the first two stages of the hierarchy of addresses, which we obtain in this way.

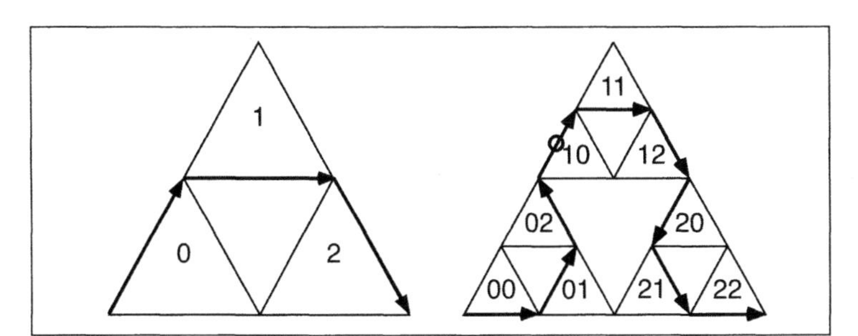

Figure 7.17 : A modified address hierarchy for the Sierpinski gasket.

In the first stage we label the subtriangles of S_0 according to $w_0(S_0)$ (label 0), $w_1(S_0)$ (label 1), $w_2(S_0)$ (label 2). In the next stage we label according to $w_0w_0(S_0)$, $w_0w_1(S_0)$, $w_0w_2(S_0)$, $w_1w_0(S_0)$, ..., $w_2w_2(S_0)$. The corresponding labels are $00, 01, 02, 10, \ldots, 22$. This labeling method is then continued to all further stages. Observe that the arrangement of the transformations implies that in all stages the arrows form a directed curve. The starting and end points of curves obtained from a preceding stage are matched to build up a new curve. This has an important consequence for our addressing scheme, namely, that the neighborhood relations in the address space carry over to geometric neighborhood relations of corresponding subtriangles. For example, in the sequence $(0, 1, 2)$ of subtriangles of the first stage each subtriangle shares a vertex with the subtriangle whose address follows in the list. This is not different from the traditional addressing scheme. But the same property also holds for the next stage and all following ones. For example, neighbors in the sorted lists $(00, 01, 02, 10, 11, 12, 20, 21, 22)$ and $(000, 001, 002, 100, \ldots, 222)$ are also neighbors geometrically, and this is new. Recall from chapters 2 and 6 that the points of the Sierpinski gasket can be identified by infinitely long addresses of three symbols. These addresses were not to be confused with triadic numbers. But it is a good idea to use this type of identification in the case we are discussing here. Identifying the triadic numbers of the unit interval with point addresses as derived in this section defines a continuous transformation of the unit interval onto the Sierpinski gasket. For example the triadic number 0.101 is identified with the address $101000\ldots$ which represents the marked point in figure 7.17 (the circle within triangle 10). Also, note that

0.101 is equal to 0.100222... in the triadic system, and that the corresponding address 100222... identifies the same point in the Sierpinski gasket.

Peano Curve Construction

The first stages of the original space-filling curve construction of Peano. The grey background indicates the square which is finally filled by the curve. Furthermore it shows the resulting tiling associated with the first construction step (including rotations) which is used for the design of the MRCM.

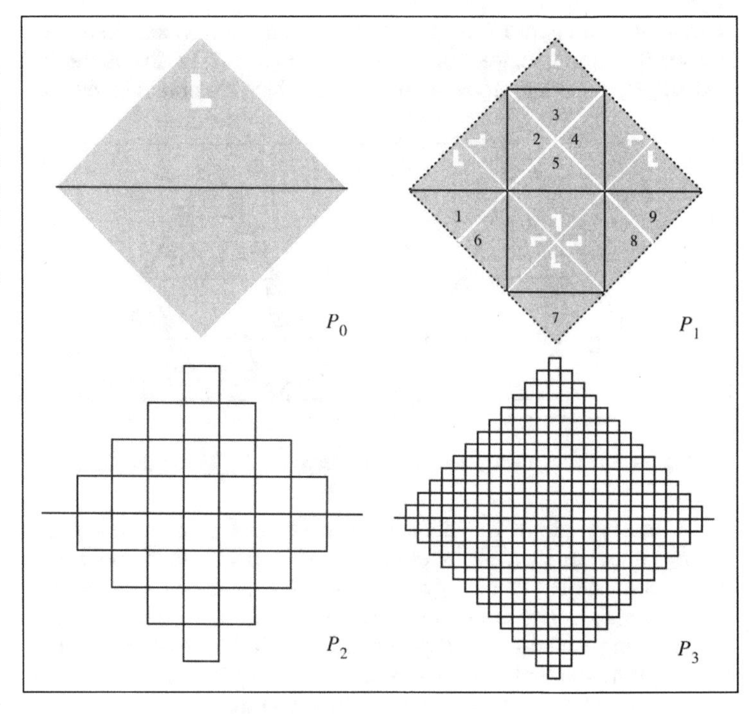

Figure 7.18

The Peano Curve

Our next example is another classical and important curve construction: the original Peano curve. It was proposed by Giuseppe Peano in 1890 as a curve which fills a square and therefore had a strong impact on the discussion of the concept of dimension. Although Peano did not provide any illustrations of his construction, let us describe the approach in terms of collections of lines. He started with a straight line P_0, the diagonal of the square which the curve is going to fill (see figure 7.18). For the next step he reduced P_0 by the factor 1/3 and fitted 9 copies into the square in the order shown. The second and eighth copies are rotated by 90 degrees counter-clockwise, the fourth and sixth copies are rotated clockwise and the fifth copy is rotated 180 degrees. All segments together form the new curve P_1. In the following steps this process is repeated: the curve P_{n-1} is scaled down by the factor 1/3 and 9 copies are fitted as in the first step to form the new curve P_n. You will observe that already the curve P_1 has two points of self-intersection (or touching points). P_2 has 32 such points and in the limit the number increases by the factor 9 from step to step. In this respect Peano's construction is a bit unsatisfying. It seems that the great mathematician David Hilbert shared this feeling: one year later he

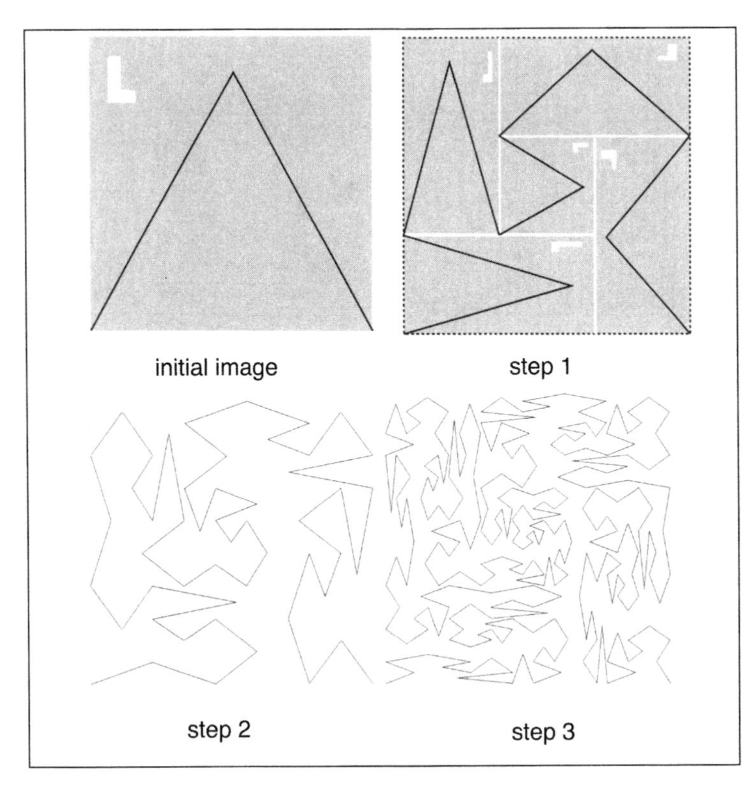

Self-Avoiding Space-Filling Curve

Blueprint (top) of an MRCM which generates a self-avoiding and space-filling curve. Looking at stage three (bottom right) it is hard to believe that this construction is really self-avoiding.

Figure 7.19

presented a space-filling curve construction which is self-avoiding. We will discuss his ideas a bit later. First let us return to the original Peano curve and set up an appropriate MRCM. This is a simple task:

- use the curve P_0 as the initial image;
- choose transformations $w_1, \ldots, w_9$ such that the square is mapped onto the subsquares as indicated in figure 7.18. Thereby we map P_0 onto the line segments of P_1.

Iterating this machine produces exactly the sequence of curves $P_0, P_1, P_2, \ldots$ which lead to the Peano curve in the limit. This limit object actually is a curve as already discussed in chapter 2 (see page 93). The fact that the limit curve fills a complete square follows from the MRCM interpretation. Indeed, just observe that the shaded square in figure 7.18 remains invariant under the 9 contractions of the MRCM. Thus, it is the attractor.

Curves That Fill a Cube

In passing, let us add that there even exist curves that cover a full cube of three dimensions. The discovery that such seemingly impossible objects really exist shocked the world of mathematics around the turn of the century. Such excitement can be experienced today not only by first-rate mathematicians but by anybody with access to a computer. It takes only a matter of seconds to produce recursive geometric patterns and, with a suitable program on hand,

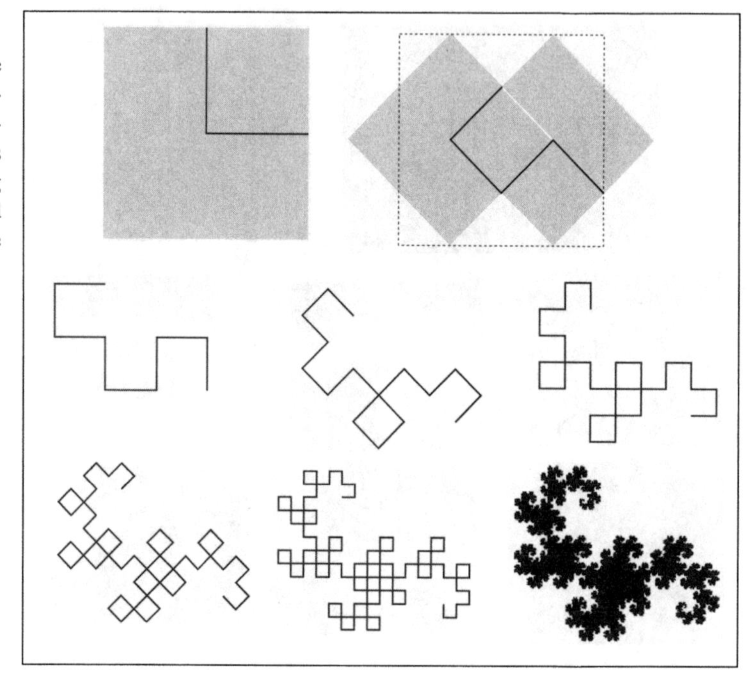

The Dragon Curve

The dragon curve is another example of a space-filling curve. Its boundary is a fractal. In the bottom right-hand corner the limit curve of this construction is shown. The starting curve, shown on a grey square, and the first stage (top) are used as the blueprint of an appropriate MRCM.

Figure 7.20

one may join the early pioneers in their hunt for new curve constructions. An implementation of our MRCM should be a first step in that direction. However, with a plain MRCM it is not possible to create a self-avoiding and space-filling curve that is aesthetically pleasing. Figure 7.19 shows one attempt. There is a 'chaotic' ensemble of lines which makes it hard to believe that the construction is in fact space-filling and self-avoiding.[11]

The Dragon Curve

Let us conclude this section with another more recent example, the Harter-Heighway dragon.[12] This is a construction of a space-filling curve which finally fills an area with a fractal boundary. We start with a right angle as shown in figure 7.20. The first stage of the construction is obtained by fitting two copies of this curve (each one reduced by the factor $1/\sqrt{2}$ and rotated clockwise by 45 or 135 degrees respectively) to form a new curve. Again this procedure is repeated stage by stage ad infinitum and can be described by a simple MRCM,[13] for which we use the right angle curve as initial image and choose transformations w_1, w_2 as indicated by the blueprint in figure 7.20. Several relatives of this curve are known. This one (and you will see this in the next section) is especially well suitable for an L-system approach.

---

[11] It is an interesting and open question whether one can construct such a curve with fewer transformations than five as used in figure 7.19. Note that the space-filling property again follows from the invariance of the initial square under the MRCM.

[12] See C. Davis and D. E. Knuth, *Number Representations and Dragon Curves,* Journal of Recreational Mathematics 3 (1970) 66–81 and 133–149. Also page 66 in B. B. Mandelbrot, *The Fractal Geometry of Nature,* Freeman, New York, 1982.

[13] Note that the curves in the first two stages are self-avoiding while those of all subsequent stages are not. The self-similarity dimension is $\log 2/\log\sqrt{2} = 2$.

7.3 Turtle Graphics: Graphical Interpretation of L-Systems

In order to grow the classical fractals from the previous section by using L-systems the resulting symbol strings must be interpreted graphically. This interpretation is independent of the string generation. We have already pointed out that this feature is of special value. Data generation and data visualization can be separated into independent modules.

In this section we present a very simple graphical interface for symbol strings.[14] It is based on Seymour Papert's concept of *turtle graphics*[15] and is especially suited for curves in the plane. With such an interpretation we can formulate the original constructions of the classical fractals in a very concise and compact form, namely, as L-systems with just a few production rules. Of course, these production rules must be very carefully set up in order to correctly interface with the graphical interpretation. In other words, all aspects of the graphical interpretation have to be firmly defined.

Let us imagine a turtle sitting on a sheet of paper facing in a certain direction. The tail of the turtle is a bit dirty. Thus, it leaves a trace on the paper as soon as the turtle starts to move. The turtle is thoroughly trained. It understands several commands, which we transmit by remote control. The commands are given in the form of symbols (from the list of symbols used in the L-system). These symbols will be just ordinary letters from the alphabet and some special symbols such as $+$ or $-$. Here is a first set of instructions to the turtle:

- F move forward by a certain fixed step length l and draw a line from the old to the new position
- f move forward as for F but do not draw the line (raise the tail)
- $+$ turn left (counterclockwise) by a fixed angle δ
- $-$ turn right (clockwise) by the angle δ

Figure 7.21 shows a short string of symbols and the corresponding graphical interpretation. In this case the angle δ for left and right turns is 90 degrees. Thus the turtle starts to move one step to the right (thereby drawing a line of length l). Then it turns left and continues to draw a line upward. At this point it raises its tail and makes another step upward (without drawing). Then it turns right and draws a line, and so on. With such few and simple directives we can already let the turtle draw amazingly complex images.

State Changes Must Be Defined Firmly

Instead of drawing a line the turtle could draw an arrow (as shown in 7.21), a dashed line or even a thin cylinder. This is some of the freedom we have in the graphical interpretation. On the other hand, after each command the turtle may have a new position and direction. This change must be defined firmly.

Especially the size of the step length l and the angle δ must be specified before the interpretation can be started. In fact, the size of the angle has an

[14] It was introduced in P. Prusinkiewicz, *Graphical applications of L-systems*, Proceedings of Graphics Interface '86, Kaufmann, 1986, 247–253. See also P. Prusinkiewicz and J. Hanan, *Lindenmayer Systems, Fractals and Plants*, Vol. 79 of Lecture Notes on Biomathematics, Springer-Verlag, New York, 1989.

[15] S. Papert, *Mindstorms: Children, Computers, and Powerful Ideas*, Basic Books, New York, 1980.

Turtle Interpretation

Initially the turtle is headed left to right. When the interpretation of all 12 symbols is completed the turtle will be headed up.

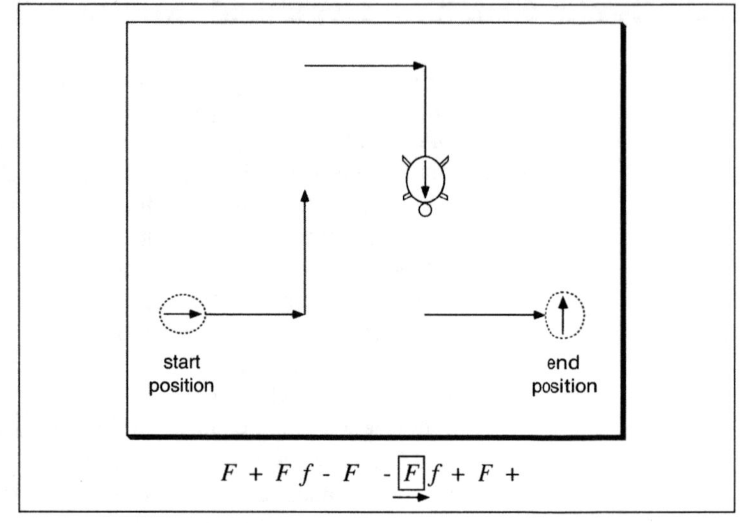

Figure 7.21

Two Strings — Three Angles

Graphical interpretation of two strings, $F+F+F-F-F$ (upper images) and $FF+FF+F+F-F-F+F+FF+F$ (lower images), and for three different angles δ: 60 degrees (left), 90 degrees (middle) and 120 degrees (right).

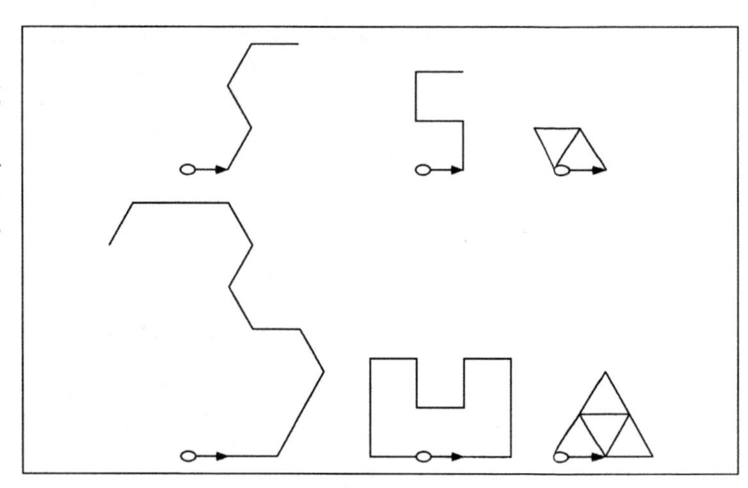

Figure 7.22

important impact on the shape of the resulting graphics, while the step length only influences the overall size of the image. Let us demonstrate this fact. Figure 7.22 shows the graphical interpretation of two strings:

Top: $F+F+F-F-F$
Bottom: $FF+FF+F+F-F-F+F+FF+F$.

We show the interpretation for three different angles (60, 90 and 120 degrees). The results are quite different. Some are even surprising. In particular this is true for the bottom right-hand curve which shows only 9 lines although there are 12 symbols F in the string. But if you follow the path of the turtle you will observe that some of the lines are drawn twice.

State of the Turtle

In mathematical terms (and more useful for programming) we say that the turtle has a state consisting of a current position, given by two coordinates x and y, and a current heading, specified by an angle α. This is written as a triplet (x, y, α). The state of the turtle is changed every time a command is interpreted. Using elementary trigonometry the following table can be deduced:

command	state (x, y, α) is changed to
F	$(x + l\cos\alpha, y + l\sin\alpha, \alpha)$
f	$(x + l\cos\alpha, y + l\sin\alpha, \alpha)$
$+$	$(x, y, \alpha - \delta)$
$-$	$(x, y, \alpha + \delta)$

Here l denotes the step length (by which the turtle moves forward) and δ the angle (by which the turtle turns left or right). We always start with the state $(0, 0, 0)$, i.e., the turtle is headed to the right.

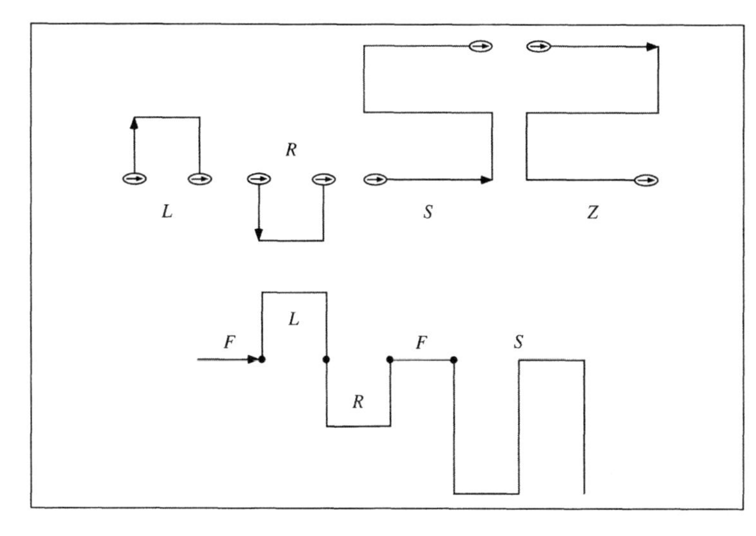

More Symbols

The symbols L, R, S and Z represent curves which are composed from the commands F, $+$ and $-$. The bottom curve is the result of the string $FLRF - S$.

Figure 7.23

Symbols with Complex Interpretation

It is very convenient to assume that our turtle can also accept commands which represent composite movements. In other words, it should be able to interpret a single symbol as a certain defined sequence of commands like F, f, $+$ and $-$. In the following we will use the symbols L, R, S and Z. They all represent small curves and you will find the choice of these symbols and their graphical interpretation quite obvious. The turtle will interpret these symbols in the following way:

symbol	interpretation
L	$+F-F-F+$
R	$-F+F+F-$
S	$FF+F+FF-F-FF$
Z	$FF-F-FF+F+FF$

Figure 7.23 shows the graphical interpretation of these symbols for the angle $\delta = 90$ degrees. In all cases the start and end positions of the turtle are marked by an arrow which indicates the direction the turtle is headed to. The symbol L represents a kind of small detour from a direct step forward (it starts with a left turn). The symbol R represents a small detour in the other direction (it starts with a right turn). The S-shaped curve is represented (you would have guessed this) by the symbol S and the reflected S-shaped curve is represented by the symbol Z.

7.4 Growing Classical Fractals with L-Systems

Let us now explore how we can use L-systems for constructing fractals. As a first example we will use the Koch curve. Its description by an L-system is pleasingly simple. In the classical construction we repeatedly replace a straight line segment by a sequence of four lines, as shown in figure 7.24 (right). This sequence can be described by the string $F+F--F+F$ (where we choose the angle $\delta = 60$ degrees). This is a straight line segment forward (F), a left turn by 60 degrees ($+$), another line forward (F), then two turns to the right by 60 degrees each ($--$) (i.e., a total turn of 120 degrees), a line forward (F), a left turn ($+$) and another line forward (F).

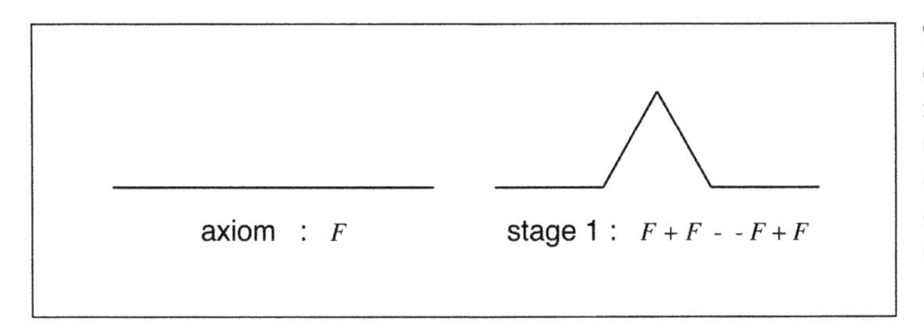

The elements of the classical Koch construction for the curve can be represented by the symbol F (left) and the string $F + F - -F + F$ (right). But note that F represents different step lengths l in the two stages shown (see page 356).

Figure 7.24

Thus, the process of the Koch construction can easily be described by the following L-system:

L-system: Koch curve
Axiom: F
Production rules: $F \rightarrow F + F - -F + F$
$+ \rightarrow +$
$- \rightarrow -$
Parameter: $\delta = 60$ degrees

The second and third production rules simply state that the symbols $+$ and $-$ should be replaced by the same symbols, respectively (i.e., they should not change in the substitution process). For now we can assume that the step length l is reduced to one-third of the previous step length in each stage of the derivation (we will return to this issue in a moment). Elaborating the L-system, we obtain the following strings of symbols for the first two stages:

Axiom: F
Stage 1: $F + F - -F + F$
Stage 2: $F + F - -F + F + F + F - -F + F - -F + F - -F + F +$
$F + F - -F + F$

The L-system prescribes replacing an F (a straight line segment) by a sequence of lines $F + F - -F + F$. If we interpret this geometrically, it is exactly the classical replacement. Thus, it is not surprising that the L-system works in exactly the same fashion as the original Koch construction. The advantage

Cantor Set by L-System — First Try

The first stages of L-system $F \rightarrow FfF$ for the Cantor set.

axiom :	F	
stage 1 :	FfF	
stage 2 :	$FfFfFfF$	
stage 3 :	$FfFfFfFf$ $FfFfFfF$	

Figure 7.25

of the L-system is that it is directly applicable as a method for (graphical) computer generation of the curve. The graphical interpretation according to the defined rules is the same as in figure 7.9.

Determining the Step Length

In the case of the Koch snowflake curve, it is quite obvious how to choose the step lengths for the turtle for any given stage. For each stage we have to reduce the length by a factor of three. Then the size of the total curve is always the same since a line segment will be replaced by four segments, the first and last of which come to lie on the two ends of the original segment. The situation may be much more complex in many other cases. There is no general theory that permits us to compute the scale of the generated curves.[16] Thus, the only practical way to proceed is to first assume an arbitrary step length, for example, equal to 1.0, to compute the whole graphical interpretation of the turtle command string, and then to rescale the result so that it will fit conveniently on the monitor for viewing or on the page for printing.

Our next example is the generation of the Cantor set. Again we start with a straight line. The line is divided into three equal parts and the middle segment is removed. This process is repeated ad infinitum. For an L-system we can formalize this by the production rule $F \rightarrow FfF$ (i.e., we replace a line segment by three line segments, of which the middle one is not drawn). This suggests the following L-system:

L-system: Cantor set (first try)
Axiom: F
Production rules: $F \rightarrow FfF$
$f \rightarrow f$

Figure 7.25 shows the first three stages of this system and its graphical interpretation. Obviously this is not the desired result. But what is wrong? Where is the mistake in our L-system? Let us look at the second stage. We observe that the size of the middle gap is too small. It should be three times as large as it is. The gap is represented by a single symbol f. This particular symbol derived from the f which was already generated in stage one. Obviously it should have been replaced by a sequence of three symbols fff (which represents a gap three times as long as a single f). Thus the L-system for the

[16]For a solution to this problem for a restricted class of L-systems see F. M. Dekking, *Recurrent sets*, Advances in Mathematics 44, 1 (1982) 78–104.

axiom : F

stage 1 : FfF

stage 2 : $FfFfffFfF$

stage 3 : $FfFfffFfF$ $fffffffff$ $FfFfffFfF$

Cantor Set by L-System — Second Try

The first stages of the correct L-system for the Cantor set.

Figure 7.26

Cantor set should be:

L-system:	Cantor set (second try)
Axiom:	F
Production rules:	$F \to FfF$
	$f \to fff$

Graphics Uncover Hidden Pitfalls

Indeed, figure 7.26 shows the desired result. This simple example already demonstrates that there are many hidden pitfalls in the derivation of an appropriate L-system for a concrete example. But the graphical interpretation of the generated strings will uncover most mistakes immediately.

Let us now return to the Sierpinski arrowhead and try to find out how we can describe the curve construction (which we introduced using an MRCM) in terms of an L-system. Figure 7.27 shows the first two stages. First, we obviously have to choose the angle $\delta = 60$ degrees. Then we could describe the curve on the left by $+F - F - F+$. The first and the last symbol $+$ are needed to have the turtle facing in the appropriate direction for its first or next movement. But wait, let us recall the symbol L. By definition the turtle interprets this symbol just as the string $+F - F - F+$. Thus, the left curve is simply described by the symbol L. Likewise the right curve is $+R - L - R+$ where R is interpreted as $-F + F + F-$.

Now imagine flipping the curves of figure 7.27 upside down. Then the left curve would be represented by the symbol R and the right curve would

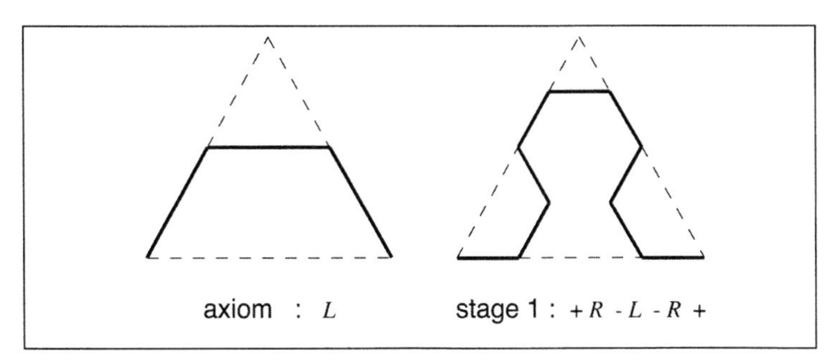

The Sierpinski Arrowhead Generator

The first two stages of an L-system Sierpinski curve.

Figure 7.27

The Peano Curve

A space-filling construction — the Peano curve: (top) axiom and production; (bottom) the second and third stages.

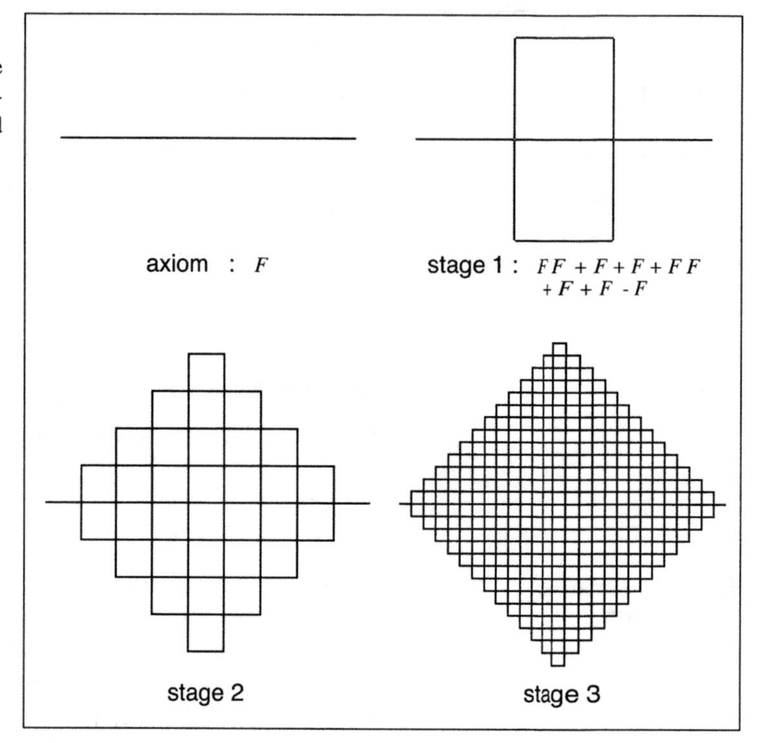

Figure 7.28

be described by $-L+R+L-$. Combining both observations suggests the following L-system:

L-system: Sierpinski arrowhead
Axiom: L
Production rules: $L \to +R-L-R+$
$R \to -L+R+L-$
$+ \to +$
$- \to -$
Parameter: $\delta = 60$ degrees

For the first three stages this L-system gives:

Axiom: L
Stage 1: $+R-L-R+$
Stage 2: $+-L+R+L--+R-L-R+--L+R+L-+$
Stage 3: $+-+R-L-R++-L+R+L-++R-L-R+-$
$-+-L+R+L--+R-L-R+--L+R+L-+$
$--+R-L-R++-L+R+L-++R-L-R+-+.$

These strings exactly describe the curves which we have already seen in figure 7.16.

Multiple Choices for Correct Production Rules

Our next example demonstrates a minor complication, which is rather typical for L-systems. Figure 7.28 once again shows the original space-filling Peano curve. Its construction starts with just one line segment. We certainly will not hesitate to encode this by a simple F command. But the curve in the upper right part might cause some doubt. This curve (which we are going to substitute for plain line segments) can be encoded in several ways. After a forward step, the turtle can be told to make a left or a right turn, or to go straight ahead. For example, after a right turn there must be two left turns with line segments in between all three turns. Then, the turtle has again arrived at a point where the curve meets itself, and there is a choice, namely, to go straight ahead or to make a left turn. In either case the turtle will trace the top loop of the generator counterclockwise or clockwise, and then it just has to finish up the last line segment to the end of the curve. In terms of turtle commands, these two alternatives can be described as follows (in any case we choose $\delta = 90$ degrees):

$$F - F + F + F + F - F - F - F + F \text{ and}$$
$$F - F + F + FF + F + F + FF.$$

The options for an initial left turn at the first bifurcation of the curve are:

$$F + F - F - F - F + F + F + F - F \text{ and}$$
$$F + F - F - FF - F - F - FF.$$

If the turtle goes straight at the first decision point, we have

$$FF - F - F - FF - F - F + F \text{ or}$$
$$FF + F + F + FF + F + F - F.$$

The complete L-system may work with any of the above choices, for example, we can set:

L-system:	Peano curve
Axiom:	F
Production rules:	$F \rightarrow FF + F + F + FF + F + F - F$
	$+ \rightarrow +$
	$- \rightarrow -$
Parameter:	$\delta = 90$ degrees

Our last example, which we already discussed from the viewpoint of MR-CMs, is the dragon curve. The construction starts with an L-shaped curve. Let us call this K_0. The next stage K_1 is obtained by fitting two copies of K_0 rotated clockwise — one rotated by 45 degrees, the second rotated by 135 degrees (see figure 7.29).

First we note that we obviously should use an angle $\delta = 45$ degrees. Then K_0 can be encoded by $--F++F$. It is convenient to define for this curve a new symbol: D (i.e., the turtle will interpret D as $--F++F$). Let us now analyze K_1. The first part of the curve can be encoded directly (i.e., by the symbol D), but we have to be careful with the second part. Loosely speaking

The Dragon Construction

L-system construction of the Dragon curve. The new symbols D and E (top) are the main building blocks.

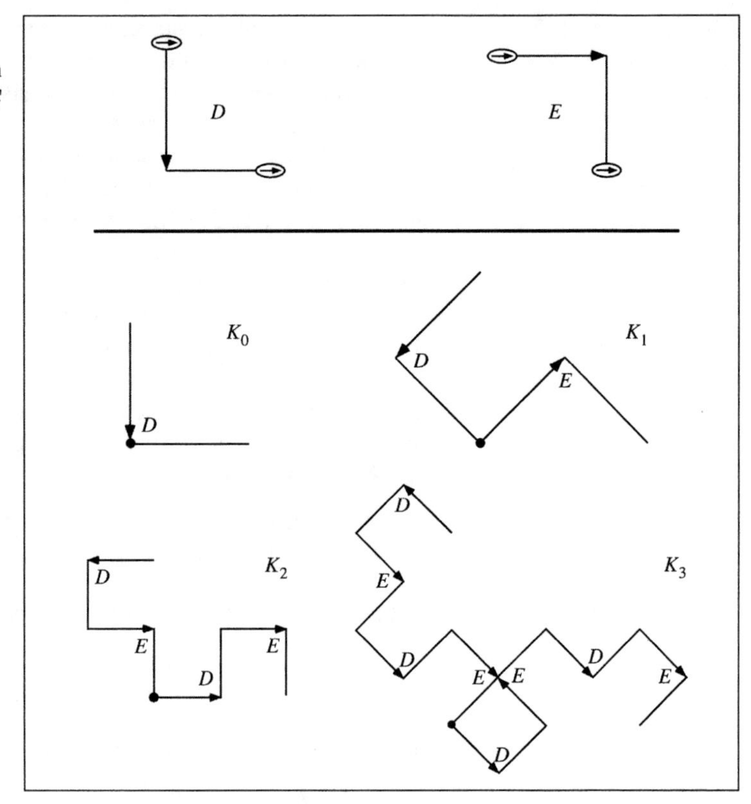

Figure 7.29

the end points of the two copies of K_0 are fitted together to make up the new curve. Therefore, we have to trace the second copy of K_0 in reversed order: $F - -F + +$. For this we introduce the new symbol E. We arrive at the encoding $-D + +E$ for the curve K_1 (the minus sign rotates everything by 45 degrees) which also gives us the first rewriting rule: $D \rightarrow -D + +E$. The rewriting rule for E is symmetric and reflects the reversed order we found in the second part of the curve. Thus we obtain the L-system:

L-system:	Dragon curve
Axiom:	D
Production rules:	$D \rightarrow -D + +E$
	$E \rightarrow D - -E+$
	$+ \rightarrow +$
	$- \rightarrow -$
Parameter:	$\delta = 45$ degrees

We check the first three stages:

Axiom: D
Stage 1: $-D + +E$

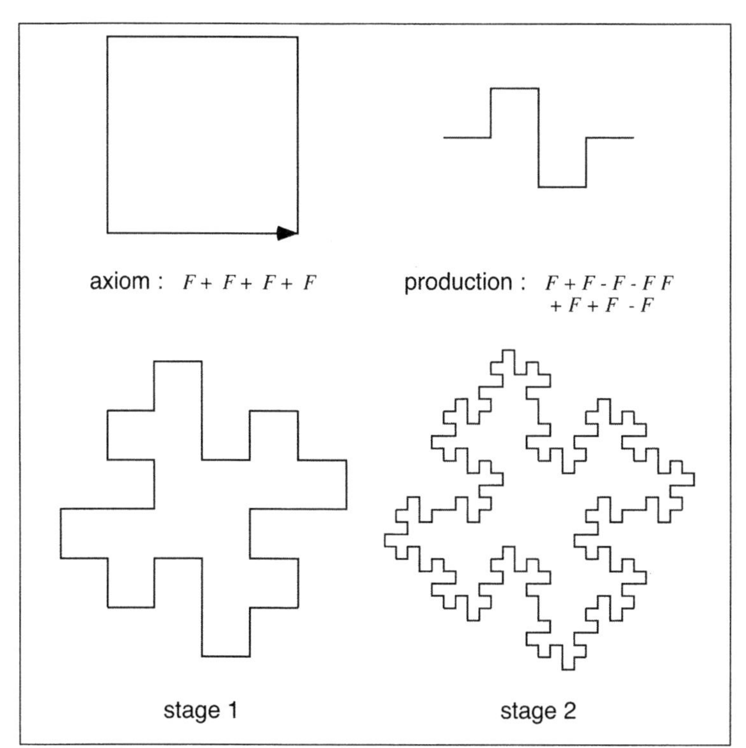

The Koch Island

The quadratic Koch island: (top) axiom and production of the L system; (bottom) first and second stages.

Figure 7.30

Stage 2: $--D++E++D--E+$
Stage 3: $---D++E++D--E+$
$++-D++E--D--E++.$

The graphical interpretation of these strings indeed generates the curves K_0, K_1, K_2 and K_3 (see figure 7.29).

Not Suitable for MRCMs

Let us now turn to fractal curves which we have not seen in this chapter so far because we cannot obtain them using an MRCM.

Our first example is the quadratic Koch island shown in figure 7.30. The construction starts with a closed polygon of four sides (a square). If we choose an angle $\delta = 90$ degrees this square can be described by the string $F+F+F+F$. In the following stages all line segments are replaced by the zig-zag curve which is shown in the upper right-hand corner of figure 7.30. We can encode this curve by $F+F-F-FF+F+F-F$. We get:

L-system:	Quadratic Koch island
Axiom:	$F+F+F+F$
Production rules:	$F \rightarrow F+F-F-FF+F+F-F$
	$+ \rightarrow +$
	$- \rightarrow -$
Parameter:	$\delta = 90$ degrees

At the bottom of figure 7.30 we show the first stages of this curve construction. Now, why is it not possible to describe this construction by an MRCM?

The answer is simple. This construction does not produce a self-similar image in the sense of an MRCM. On the other hand, an MRCM always generates a self-similar final image, and this property is tightly built into the feedback operation of the machine. In other words, the curve produced in a certain stage is made up of complete (although transformed) copies of the curve which was produced at the preceding stage. The construction process of the quadratic Koch island violates this principle. Thus, it cannot be described by an MRCM.

But restricting the construction to just one side of the Koch island (e.g., the bottom segment, which stems from the first F of the axiom) changes the situation completely. Each of the four sides is made of a curve construction which is self-similar (in the limit) and which can be described by an MRCM. In other words, the only reason why the property of self-similarity is violated is the axiom $F + F + F + F$. Because of this axiom the limit curve is not simply self-similar but a composition of four self-similar curves.

Let us now discuss two curve constructions which are really not feasible for a simple MRCM. Both examples are space-filling curves like the original Peano curve. But whereas the Peano construction creates curves which have self-intersections the following constructions are self-avoiding. This means that each curve (at all stages) has neither self-intersections nor touching points. The first of the two examples is attributed to David Hilbert (1862–1943) and was published in 1891 just one year later than Peano's.[17]

Hilbert's Construction

Hilbert's construction is extraordinarily elegant, especially when we take into consideration that it was created decades before tools like L-systems, MRCMs or recursive computer programming were invented. Let us try a description which follows Hilbert's ideas.[18] In figure 7.31 the dotted square shows the area which we are going to fill by the curve. We divide this square into four quarters. The construction starts with a curve H_0 which connects the centers of the quadrants by three line segments. Assume the size of the segments to be 1. In the next step we produce four copies (reduced by 1/2) of this initial stage and place the copies into the quarters as shown. Thereby we rotate the first copy clockwise and the last one counterclockwise by 90 degrees. Then we connect the start and end points of these four curves using three line segments (of size 1/2) as shown and call the resulting curve H_1. In the second step we scale H_1 by 1/2 and place four copies into the quadrants of the square as in step one. Again we connect using three line segments (now of size 1/4) and obtain H_2. This curve contains 16 copies of H_0, each of size 1/4. As a general rule, in step n we obtain H_n from four copies of H_{n-1} which are connected by three line segments of length $1/2^n$ and this curve contains $4n$ copies of H_0 (scaled down by $1/2^n$).

[17] D. Hilbert, *Über die stetige Abbildung einer Linie auf ein Flächenstück*, Mathematische Annalen 38 (1891) 459–460.

[18] The original paper of Hilbert is reproduced in the figures 2.36 and 2.37 on pages 94–95 in chapter 2.

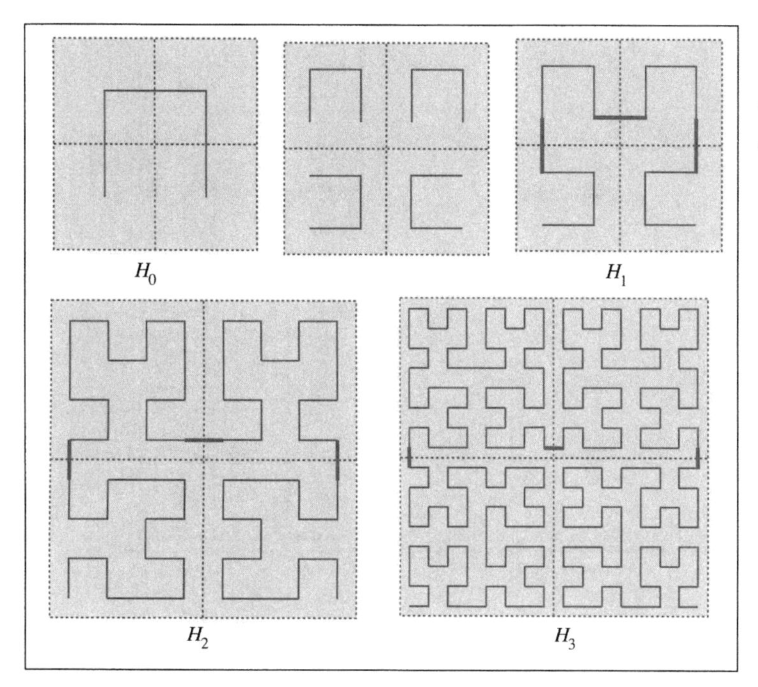

Four Stages of the Hilbert Curve

The first stages of the construction of a Hilbert curve.

Figure 7.31

Inner Versus Outer Replacements

What we have just demonstrated is a geometric construction which is similar to the curve generation using an MRCM. We generate the curve at a certain stage from copies of the curve in the previous stage. This could be called a macroscopic (or outer) replacement. In this respect the L-system construction is different. It is restricted to the replacements of the smallest elements (symbols), and we could call this a microscopic (or inner) replacement. Macroscopic replacements are easy to visualize and therefore easy to understand whereas microscopic replacements seem to be much harder to comprehend. Let us explore this point of view. How should we define the second stage H_2 of the Hilbert curve suitable for the L-system approach? Let us look at H_1 again. There are four copies of H_0 in H_1. Loosely speaking each of these copies has to be replaced by a copy of H_1 to obtain H_2. However, this is not quite what we want because the connecting line segments must also be modified. L-systems provide a solution to this problem. With L-systems we only use formal substitutions, and the appropriate size of the line segments is adjusted automatically. Let us demonstrate this in detail.

First we have to choose the angle $\delta = 90$ degrees. Then H_0 can be encoded by the symbol L (which is interpreted by the turtle as $+F-F-F+$) and H_1 can be encoded by $+RF-LFL-FR+$ (see figure 7.32). Please note that in both cases the turtle is headed right at the start and end positions. Since we obtain H_1 from H_0 this determines the first production rule of the L-system. For the next step, the generation of H_2, we also need a rewriting rule for

L-System for Hilbert Curve

Geometric interpretation of the encoding of the curves H_0 and H_1 using the symbols L and R (left). When starting with axiom R in place of L, we obtain mirror images of the corresponding stages of the Hilbert curve (right).

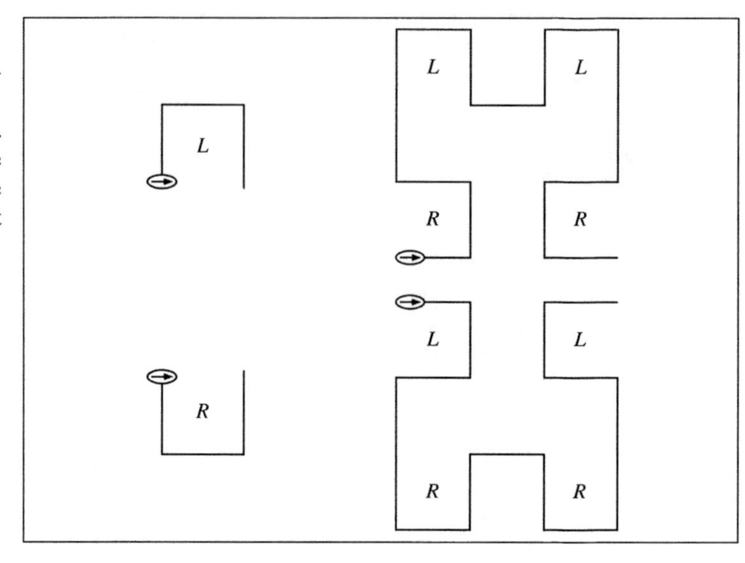

Figure 7.32

R. But the only difference between the mirrored symbols L and R is their orientation. The turtle traces the L-curve in clockwise direction — starting with a left turn — whereas the R-curve is traced counterclockwise, starting with a right turn. Therefore, the rewriting rule for R is a mirror of the rule for L. Instead of right turns, make left turns and vice versa. Moreover, we must exchange the symbols R and L. Thus we arrive at:

L-system: Hilbert curve
Axiom: L
Production rules: $L \to +RF - LFL - FR+$
$R \to -LF + RFR + FL-$
$F \to F$
$+ \to +$
$- \to -$
Parameter: $\delta = 90$ degrees

For the first three stages this L-system gives:

Axiom: L
Stage 1: $+RF - LFL - FR+$
Stage 2: $+ - LF + RFR + FL - F - +RF - LFL - FR + F + RF - LFL - FR + -F - LF + RFR + FL - +$
Stage 3: $+ - +RF - LFL - FR + F + -LF + RFR + FL - F - LF + RFR + FL - +F + RF - LFL - FR + -F - + - LF + RFR + FL - F - +RF - LFL - FR + F + RF - LFL - FR + -F - LF + RFR + FL - +F + -LF + RFR + FL - F - +RF - LFL - FR + F + RF - LFL - FR + -F - LF + RFR + FL - + - F - +RF - LFL - FR + F + -LF + RFR + FL - F -$

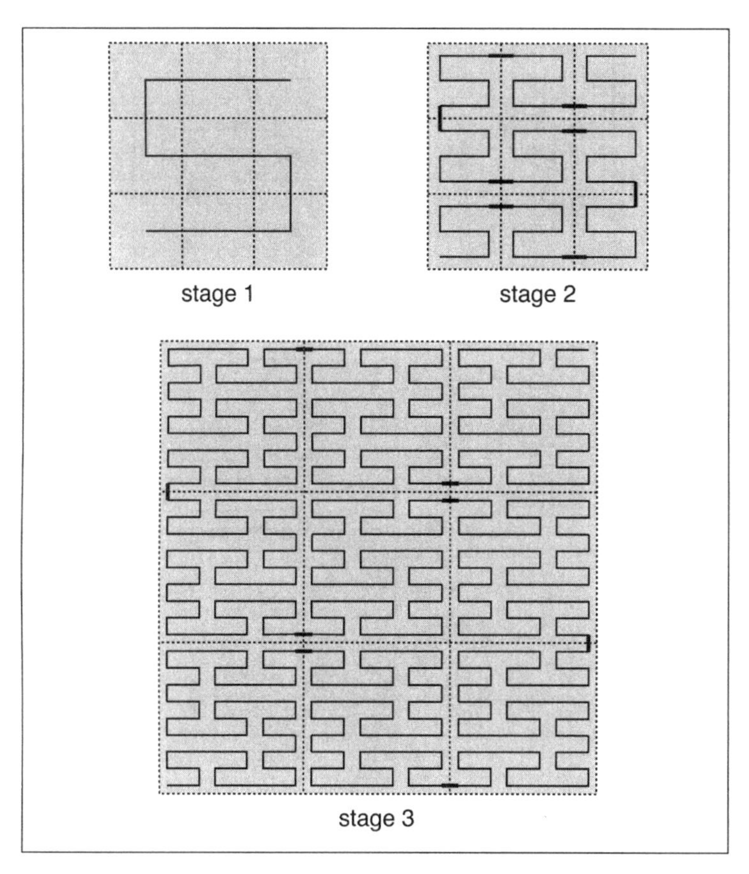

Another Space-Filling Peano Curve

The first stages of the space-filling S-curve. It is a collage of S- and Z-shaped curves.

Figure 7.33

$$LF + RFR + FL - +F + RF - LFL - FR + -+.$$

which is a valid description of H_0, H_1, H_2 and H_3 (see figure 7.31).

The SZ-Curve

Let us complete this section with a close relative of the Peano curve. The first stages of this space-filling and self-avoiding construction are shown in figure 7.33. It starts with an S-shaped curve and continues by joining small S- and Z-shaped curves. Setting the angle $\delta = 90$ degrees we can use our symbols S and Z to encode these curves. The initial curve is simply encoded by the symbol S. The curve of the second stage is given by $SFZFS + F + ZFSFZ - F - SFZFS$ (see figure 7.34). You should observe that the turtle traces the middle part ($ZFSFZ$) of the curve from right to left; thus, this part is a vertical reflection of the first part (we have tried to indicate this by the placement of the letters S and Z in the graphics). The complete L-system is given by:

L-system: S-shaped Peano curve
Axiom: S
Production rules: $S \to SFZFS + F + ZFSFZ - F - SFZFS$

L-System for SZ-Curve

The axiom of the L-system 'S-shaped Peano curve' is given by the symbol S. The production rules can be interpreted graphically as the arrangement of S- and Z-shaped curves shown.

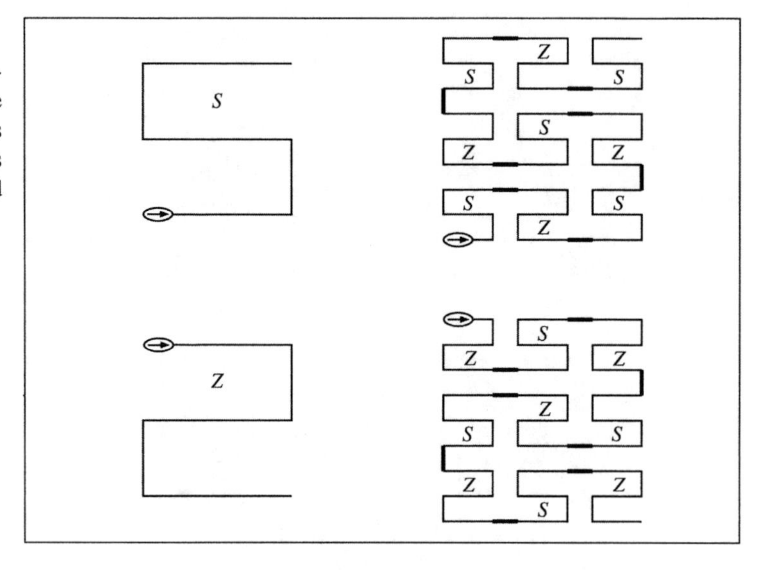

Figure 7.34

$$Z \to FSFZ - F - SFZFS + F + ZFSFZ$$
$$F \to F$$
$$+ \to +$$
$$- \to -$$

Parameter: $\delta = 90$ degrees

7.5 Growing Fractals with Networked MRCMs

From the discussion in the last section it is clear that it requires some (maybe even a bit more) experience to derive appropriate L-systems for fractals that are not strictly self-similar. Therefore it would be a pity if the more geometric approach using MRCMs really could not be extended to such constructions. But you already know the solution to this problem from chapter 5: networked MRCMs or hierarchical iterated function systems. Let us find out how we can use this concept in the present context.

Composing Images by Networked MRCMs

You will recall that all machines of a network of MRCMs work in parallel and all machines produce (in each step) their own image. The transformations of a particular machine may operate not only on its own image, but also on all images of the other machines in the network. In other words, each of the machines may compose its image by transforming the images of all the other machines. We will see immediately that this is exactly what we need.

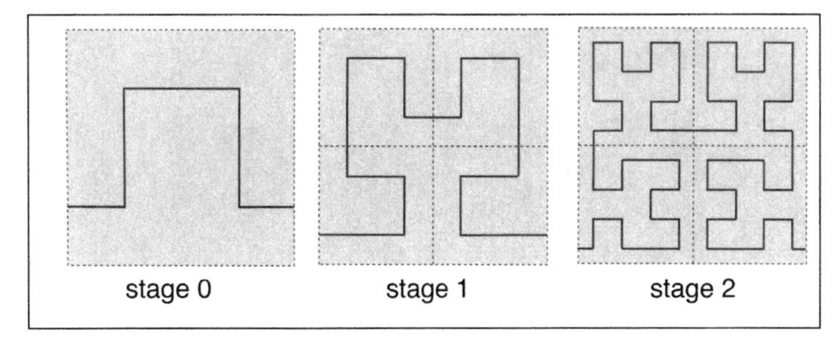

First Stages of Hilbert Curve — Review

The first stages of a Hilbert curve construction: from stage to stage the size of the line segments is reduced by the factor 1/2.

Figure 7.35

Let us start with the Hilbert curve construction. Figure 7.35 shows the first stages. However, note that the curves are extended slightly. In contrast to the original construction the curve reaches from the right to the left boundary of the grey square. The purpose of this extension is to simplify the discussion. Now consider the curve of the first stage, which is subdivided into four parts — one in each quadrant (see the center curve in figure 7.35). These curve segments are obviously not small copies of the initial image (the stage 0 curve). Rather, they look a bit like four copies of a P-shaped curve. Some of these are rotated or reflected.

In figure 7.36 we show a close-up of this P-shaped curve segment (from the upper left quadrant) and compare it with the corresponding close-up of the second stage. We have also divided the close-up of the second stage curve into four parts as indicated. This time we find three copies of the P-shaped curve and one copy of the stage 0 curve rotated 90 degrees in the lower left-hand quadrant. Based on these observations we are able to describe two networked MRCMs. Let us call them the H- and the P-machines. The H-machine will produce the images which show the curves of the Hilbert construction (as in figure 7.35). The P-machine will start with the P-shaped curve and continue as indicated in figure 7.36. It will operate on its own images and on the product

Close-Up of One Quadrant

Close-up of the upper left quadrant of the first and second stage curves as indicated in figure 7.35. The close-up of the second stage curve (right) is divided into four parts again.

Figure 7.36

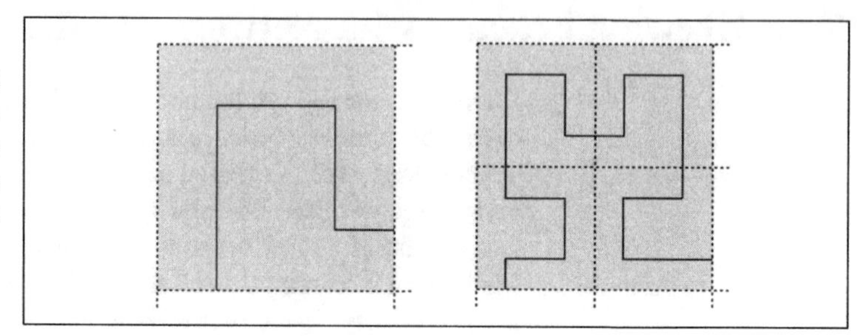

Blueprint of Networked Hilbert MRCMs

Blueprint of the two MRCMs which are required to generate the Hilbert construction. The H-machine is shown above, the P-machine is at the bottom.

Figure 7.37

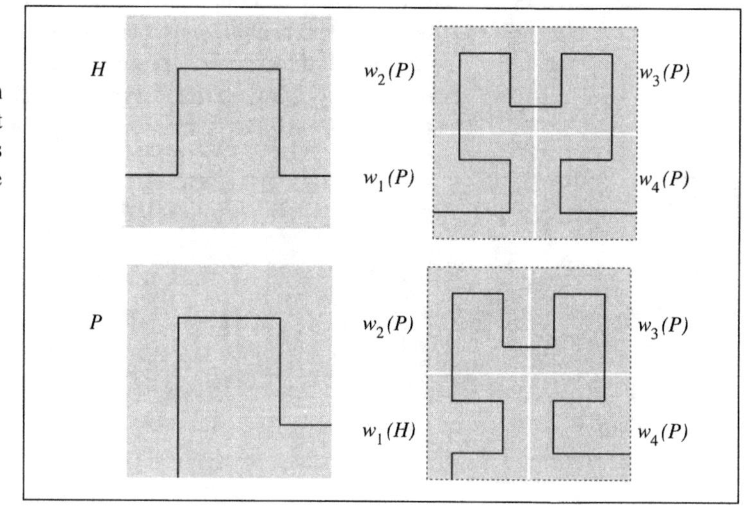

of the H-machine. The H-machine will only operate on the images of the P-machine. Figure 7.37 shows the blueprint of the two MRCMs. Note that both machines use the same transformations.

Operation of the Hilbert MRCMs

In symbolic notation we can describe the operation of the machines by the equations:

$$\begin{aligned} H_{n+1} &= w_1(P_n) \cup w_2(P_n) \cup w_3(P_n) \cup w_4(P_n) \\ P_{n+1} &= w_1(H_n) \cup w_2(P_n) \cup w_3(P_n) \cup w_4(P_n). \end{aligned}$$

The transformations $w_1, \ldots, w_4$ are completely determined by the blueprint. They all reduce by the factor 1/2. The design is set up such that $H_1, H_2, \ldots$ is the sequence of Hilbert curves provided that H_0 is chosen to be equal to the first curve of figure 7.35. P_0, P_1, P_2, …is the sequence of supplementary curves which starts with the P-shaped curve P_0 as indicated on the left of figure 7.36. In the definition of H_{n+1} all transformations are applied to P_n. This reflects that all inputs for the H-machine are connected to the output of the P-machine. On the other hand, in the definition of P_{n+1} observe that only w_1 is applied to H_n. So only one input line of the P-machine is connected to

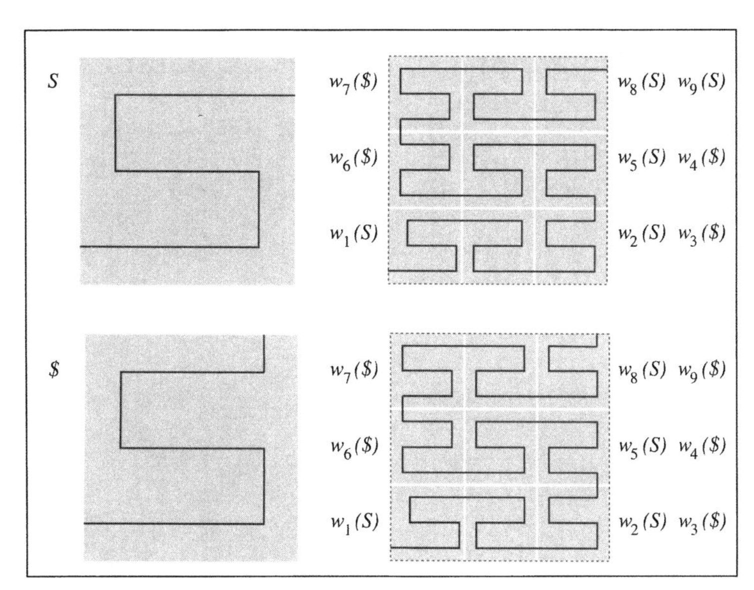

Blueprint of Networked Peano MRCMs

Blueprint of the MRCMs which are required to generate the Peano curve from figure 7.33.

Figure 7.38

the output of the H-machine. All other input is feedback from the output of the P-machine.

A Small Difference

Obviously there is a minor difference between the classical construction and the one just presented, namely, those small line segments which extend to the boundary. In fact, those line segments indicate the neighborhood relation of quadrants in the construction (i.e., how the curve leaves and enters the adjacent quadrants). In the original construction this is hidden by the process of adding the three short line segments in each stage. Finally, let us remark that the MRCM construction can be extended to precisely generate the classical construction but this would require two more MRCMs. These again use the same transformations but start with different initial images. We omit these technical details.

Let us briefly discuss the other example of a self-avoiding Peano curve that we have generated using an L-system. The same kind of analysis that we have demonstrated for the Hilbert curve leads in this case to the blueprint of two networked MRCMs as shown in figure 7.38.

Operation of the Network for Peano's Curve

Let us call the first of these two machines the S-machine. Because of the short vertical line at the top of the S-shaped curve in the blueprint of the other machine, we call that one the \$-machine. The S-machine has four input lines which are connected to the \$-machine (the transformation w_3, w_4, w_6 and w_7 are applied to the product of this machine). The \$-machine also has four input lines which are connected to the other machine (this involves the transformation w_1, w_2, w_5 and w_8). Thus, the operation of this network of

Blueprint of a Network for a Twig

The two networked MRCMs geometrically simulate the growing of a twig (bottom).

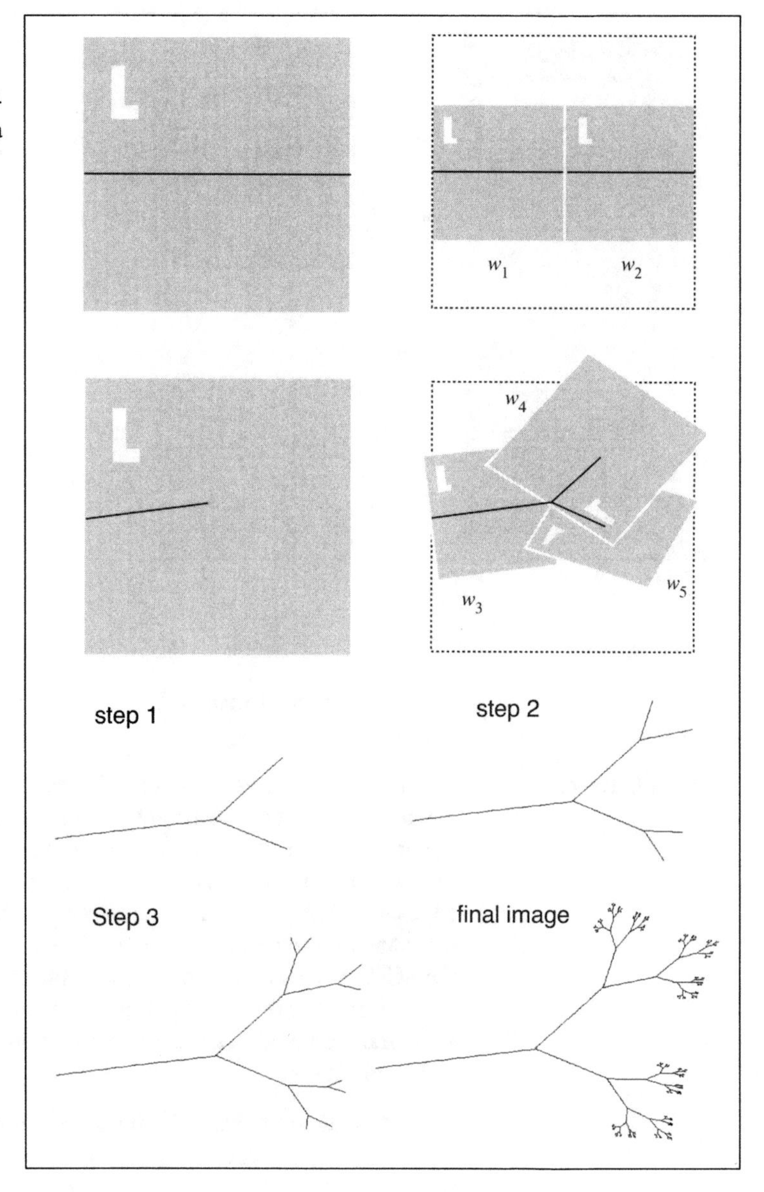

Figure 7.39

MRCMs can be described by the equations:

$$\begin{aligned} S_{n+1} &= w_1(S_n) \cup w_2(S_n) \cup w_3(\$_n) \cup w_4(\$_n) \cup w_5(S_n) \cup \\ &\quad w_6(\$_n) \cup w_7(\$_n) \cup w_8(S_n) \cup w_9(S_n) \\ \$_{n+1} &= w_1(S_n) \cup w_2(S_n) \cup w_3(\$_n) \cup w_4(\$_n) \cup w_5(S_n) \cup \\ &\quad w_6(\$_n) \cup w_7(\$_n) \cup w_8(S_n) \cup w_9(\$_n). \end{aligned}$$

The transformations $w_1, \ldots, w_9$ are chosen according to the blueprint. They all reduce by the factor 1/3. The initial images S_0 and $\$_0$ have to be selected exactly as indicated by the left side images for the S- and $-curves in figure 7.38.

Growing of a Twig

Let us conclude this discussion with a network of MRCMs which can grow a twig. Figure 7.39 shows the blueprint of two networked MRCMs which can be used as an example. The attractor of the first machine (L) is a straight line. Here we use two transformations which reduce by a factor of 1/2. Since in this set-up we take just this line as the initial image the machine simply provides this image as a constant output which we use for the stem of the twig. The other MRCM (T) is set up to let the twig grow. Its initial image represents the stem of the twig. The transformation w_3 is applied to the output of the first MRCM. It superimposes the line exactly onto the stem of the twig, thereby fixing the stem once and for all. The other two transformations w_4 and w_5 are chosen in such a way that they fit small copies of the initial image to the end of the stem.

The iteration of the two MRCMs can be described by the equations

$$\begin{aligned} L_{n+1} &= w_1(L_n) \cup w_2(L_n) \\ T_{n+1} &= w_3(L_n) \cup w_4(T_n) \cup w_5(T_n). \end{aligned}$$

Using the attractor L as the initial image L_0 we obtain $L_n = L$. If we define $S = w_3(L)$ (which we also use as initial image T_0) we obtain a single equation for the production of the twig machine:

$$T_{n+1} = S \cup w_4(T_n) \cup w_5(T_n).$$

L-Systems Are More Appropriate

Figure 7.39 (bottom) shows the result of the iteration of this network (T_2, T_3, T_4). But it is obvious from the construction that this machinery does not really simulate growth. It rather creates a geometric illusion of this process. In this sense L-systems are much more appropriate. They are designed to capture realistic growth processes. This is an important aspect which we have already mentioned in the introduction to this chapter. So let us now extend this concept and see how we can describe the growing of more complex structures than just strings of algae.

7.6 L-System Trees and Bushes

The representation of a string of cells by a string of symbols is straightforward. But how can we represent a branching structure (as is typical for trees) by a string of symbols which is just a linear sequential list of characters? For a moment this seems to be a problem. But there is a simple solution: we introduce a new symbol which indicates a branch point (we will denote this by a left square bracket '['). This left bracket will be matched by a corresponding closing bracket ']' indicating that at that point the definition of the branch is completed. Take as an example the string

$$ABA[BBAA][CCBB]ABA[AABB]ABA.$$

Removing the bracketed portions of the string we obtain

$$ABAABAABA$$

which represents the segments of the stem of a plant from which three branches spring off (at one point of the stem the branches $BBAA$ and $CCBB$ and a bit further up one branch $AABB$).

Turtle Interpretation of Branch Commands

Now what should our turtle do when it encounters these new symbols? First, upon receipt of the [-command it should remember its current position and direction. Technically speaking the turtle state has to be stored and saved. Then the branch can be drawn by the usual interpretation of the subsequent commands. Termination of the branch is triggered by the]-command. The turtle must then return to the location of the branch point, which it remembers.

Stacking of Turtle States

A computer implementation is most conveniently carried out by stacking the turtle states. This allows us to keep track of even complex branching hierarchies. When encountering a [-command, the current state of the turtle is saved at the top of the stack. The]-command, on the other hand, pops the top state from the stack and puts the turtle into this state. Technically the state S of the turtle is described by three numbers, its position (x, y) and its direction, given by the angle α. Thus, $S = \{x, y, \alpha\}$. Initially the stack is empty. Whenever the turtle encounters a [-command we save the current state $S_{n+1} = S$ and increment by 1 the counter n for the number of states on the stack. On the other hand when it receives a]-command we restore its state ($S = S_n$) and decrease the stack counter by 1.

A Weedlike Plant

Now let us examine some simple examples. The first one looks like some weed-like plant. There is a stem with three segments and two main branches. Each segment and branch looks the same (i.e., it also has three smaller segments and two branches). Here is the L-system, again using the symbols which we introduced for turtle graphics. For compactness we omit those production rules from the list which do not change any symbols such as $F \to F$. We do this also in all the remaining L-systems.

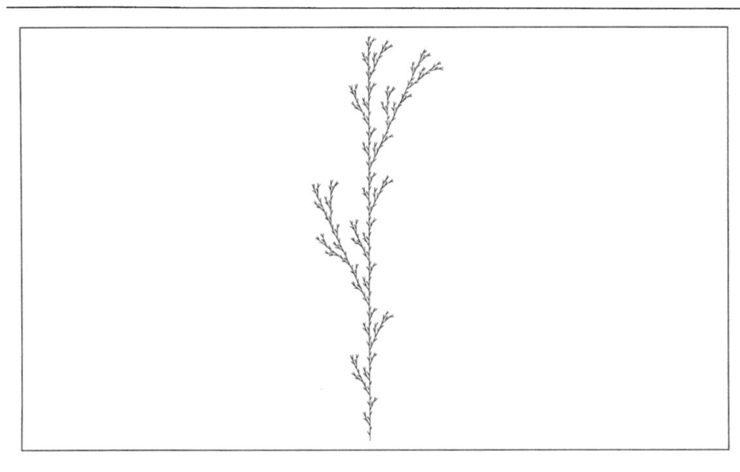

A Regular Weed

The first stages of a weedlike plant.

Figure 7.40

L-system:	Weedlike plant I
Axiom:	F
Production rules:	$F \rightarrow F[+F]F[-F]F$
Parameter:	$\delta = 25.7$ degrees

Figure 7.40 shows the first five stages. The first two of them are:

Axiom: F
Stage 1: $F[+F]F[-F]F$
Stage 2: $F[+F]F[-F]F[+F[+F]F[-F]F]F$
$[+F]F[-F]F[-F[+F]F[-F]F]F[+F]F[-F]F$

Figure 7.41 shows one more example, a two-dimensional bush. The complexity of this figure increases dramatically from step to step. Indeed the image of stage 4 is represented by 160,000 symbols.

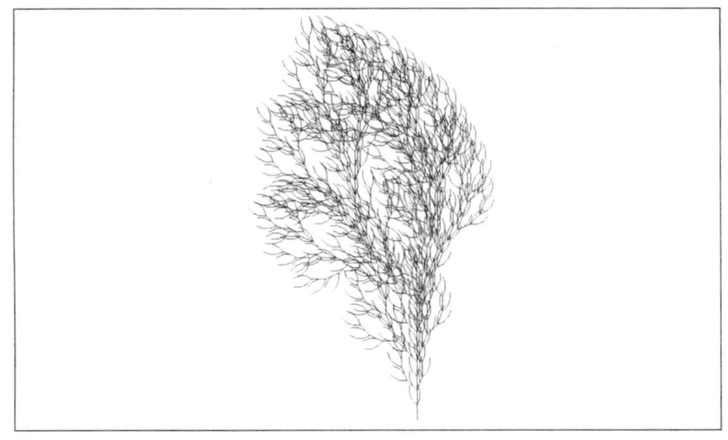

A Simple Bush

This simple bush is produced by an L-system with axiom F, production $F \rightarrow FF+[+F-F-F]-[-F+F+F]$ and angle $\delta = 25$ degrees.

Figure 7.41

A Different Weed

Note the two different types of branching.

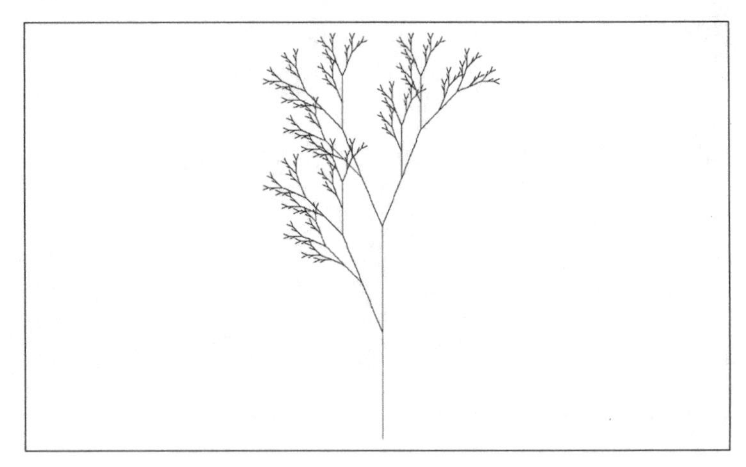

Figure 7.42

Grass

The next example involves two important production rules, one for F and another one for B, where B is an auxiliary symbol which we introduce to model this kind of branching. The turtle simply ignores this symbol. Here is the L-system:

L-system: Weedlike plant II
Axiom: B
Production rules: $F \rightarrow FF$
$B \rightarrow F[+B]F[-B]+B$
Parameter: $\delta = 20$ degrees

Figure 7.42 shows the first stages. You should be able to easily identify two kinds of branching in this plant. They correspond to the strings $[+B]$ and $[-B]+B$ in the production rule for B.

Adding Randomness to L-Systems

So far we only have presented L-systems for the generation of images of single plants. But when making pictures of fields of plants, we need to include variations for the same plant so that we are not merely copying one plant image many times. This might lead us from deterministic rewriting systems, which we have considered so far, to non-deterministic L-systems. These systems are characterized by multiple production rules for the same symbol (e.g., there may be several production rules for the symbol F). The decision as to which one of the rules to apply in a particular instance is determined by a random process. Most conveniently, we can attach certain probabilities to the individual rules and then choose the rule according to these probabilities. For example, if there are six production rules for one symbol, each one with the same probability (1/6), then the choice could be made rolling a die. Such L-systems are said to be *stochastic*. Figure 7.43 shows an example for the first weedlike structure from above.

L-system: Stochastic weedlike plant III
Axiom: F

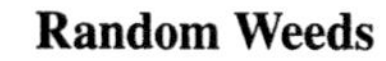

Random Weeds

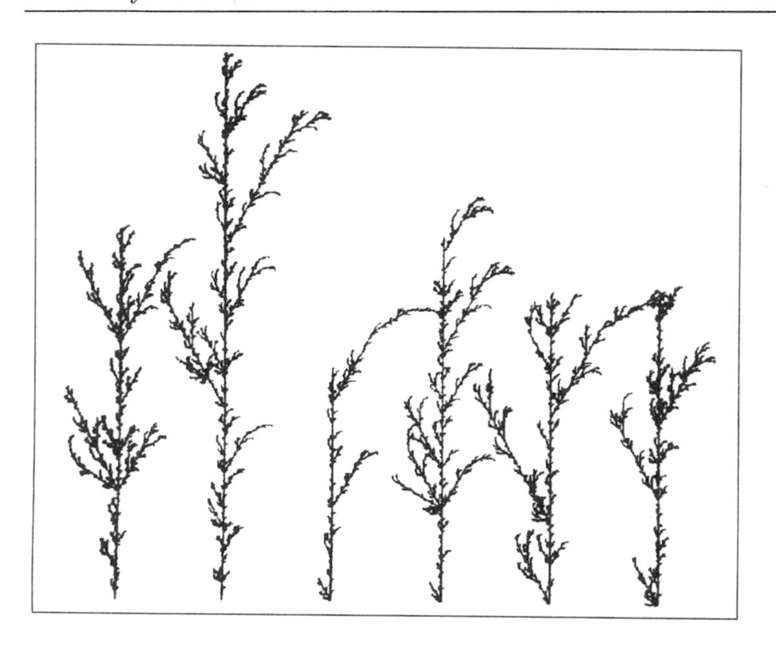

Figure 7.43

Production rules: $F \rightarrow F[+F]F[-F]F$ (probability 1/3)
$F \rightarrow F[+F]F$ (probability 1/3)
$F \rightarrow F[-F]F$ (probability 1/3)
Parameter: $\delta = 25.7$ degrees

The random Koch island from chapter 9 also may be described by a stochastic L-system.

L-system: Random Koch curve
Axiom: F
Production rules: $F \rightarrow F - F + +F - F$ with probability 0.5
$f \rightarrow F + F - -F + F$ with probability 0.5
Parameters: $\delta = 60$ degrees

For the random Koch island, just replace the axiom by $F - -F - -F$.

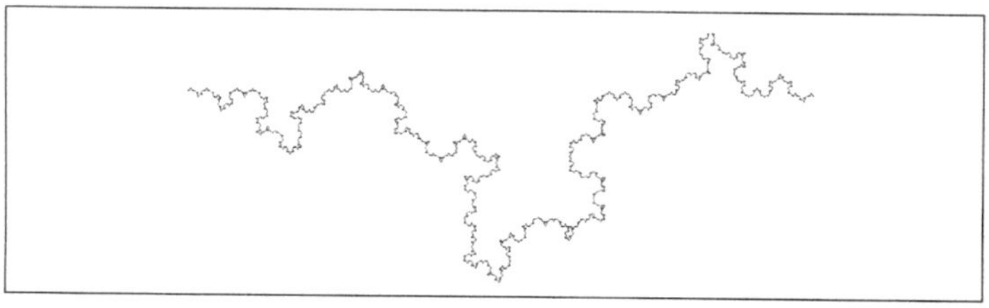

Random Koch Curve

Figure 7.44

The method of stochastic L-systems has in fact been used to produce stunningly realistic images of plants.[19] Of course, the full developmental models for such structures are much more complicated than the simple weeds presented here.

Further Extensions

There are many possible ways to extend the concept of L-systems. For example, the strategy for the expansion of characters can be modified so that one or another rule may apply according to preceding and succeeding symbols. In addition, the turtle which interprets the expanded strings may be allowed to learn a wider vocabulary of commands. For example, parentheses can be used to group drawing commands which define the boundary of a polygon to be filled. Naturally, line-style and color are also parameters of interest, and the turtle may be instructed to move and draw in three dimensions. Curved surfaces may be considered in addition to polygons. Another restriction is the fixed step length associated for example with the symbol F. One can create parameters attached to symbols that specify their interpretation also with precise numerical values such as step length or angle increment. This feature, present in so-called parametric L-systems, is very powerful and helps enormously to create more realistic biological models. These extensions are beyond the scope of this chapter, but they can be found along with further references and applications in the book by Prusinkiewicz and Lindenmayer, from which we drew many of the examples given in our presentation.

[19] See P. Prusinkiewicz and A. Lindenmayer, *The Algorithmic Beauty of Plants*, Springer-Verlag, New York, 1990. The figure 7.43 is also from this book.

Chapter 8

Pascal's Triangle: Cellular Automata and Attractors

Mathematics is often defined as the science of space and number [...] It was not until the recent resonance of computers and mathematics that a more apt definition became fully evident: mathematics is the science of patterns.

Lynn Arthur Steen, 1988

Being introduced to the Pascal triangle for the first time, one might think that this mathematical object was a rather innocent one. Surprisingly it has attracted the attention of innumerable scientists and amateur scientists over many centuries. One of the earliest mentions (long before Pascal's name became associated with it) is in a Chinese document from around 1303.[1] Boris A. Bondarenko,[2] in his beautiful recently published book, counts several hundred publications which have been devoted to the Pascal triangle and related problems just over the last two hundred years. Prominent mathematicians as well as popular science writers such as Ian Stewart,[3] Evgeni B. Dynkin and Wladimir A. Uspenski,[4] and Stephen Wolfram[5] have devoted articles to the marvelous relationship between elementary number theory and the geometrical patterns found in the Pascal triangle. In chapter 2 we introduced one example: the relation between the Pascal triangle and the Sierpinski gasket.

One Theme — Many Faces

This relationship is indeed a wonderful marvel, and we want to take this opportunity to demonstrate how approaching one mathematical question from totally different angles can beautifully lead to a thorough understanding of

[1]See figure 2.24 in chapter 2.

[2]B. Bondarenko, *Generalized Pascal Triangles and Pyramids: Their Fractals, Graphs and Applications,* Tashkent, Fan, 1990, in Russian.

[3]I. Stewart, *Game, Set, and Math*, Basil Blackwell, Oxford, 1989.

[4]E. B. Dynkin and W. Uspenski: *Mathematische Unterhaltungen II*, VEB Deutscher Verlag der Wissenschaften, Berlin, 1968.

[5]S. Wolfram, *Geometry of binomial coefficients*, Amer. Math. Month. 91 (1984) 566–571.

Capturing Pascal's Triangle

Three approaches to the patterns in Pascal's triangle.

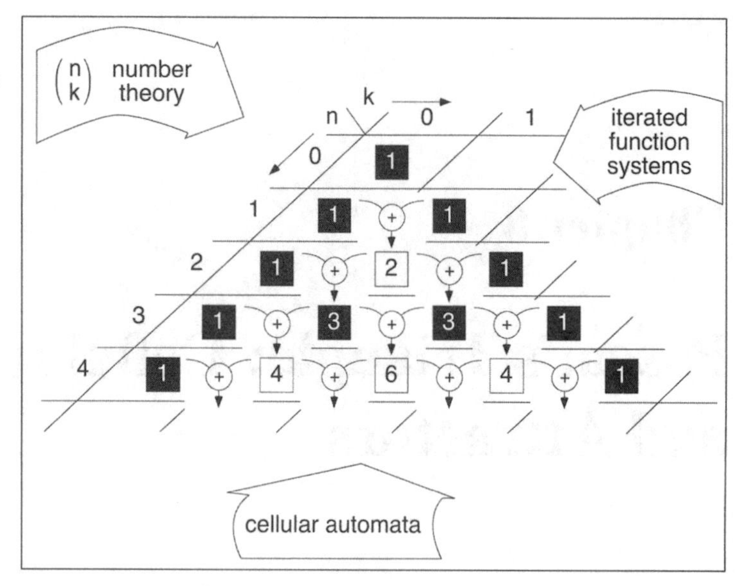

Figure 8.1

that matter. Let us restate the problem.[6] Look at the Pascal's triangle in figure 8.1. It has long been observed that by coloring

- all odd entries black, and
- all even entries white,

we obtain a geometrical pattern which is very closely related to the Sierpinski gasket. Figure 8.1 shows the first 5 rows and the beginning of this pattern formation (first, the black cells only outline a triangle) and figure 8.2 shows the first 128 rows.[7] In fact, the more rows we take into account (e.g., 256, 512, etc.), the more details of the Sierpinski gasket become visible in the geometric pattern.

But we can also use different coloring rules. This leads to all kinds of amazing fractal structures in the triangle. Thus, it is a very interesting question whether there is a way to describe these global pattern formations and how we can find their mathematical foundations for them.

The most important mathematical interpretation of Pascal's triangle is through binomial coefficients, i.e., the coefficients of the polynomials:

$$\begin{aligned}(1+x)^0 &= 1\\ (1+x)^1 &= 1+1x\\ (1+x)^2 &= 1+2x+1x^2\\ &\vdots\\ (1+x)^n &= a_0+a_1x+\cdots+a_nx^n.\end{aligned}$$

[6]For a more complete discussion see also M. Sved, *Divisibility – With Visibility*, Mathematical Intelligencer 10, 2 (1988) 56–64.

[7]See also figure 2.26 in chapter 2.

Modulo 2 Pattern

The first 128 rows in Pascal's triangle (each cell carries one entry) colored according to divisibility by 2.

Figure 8.2

These coefficients[8] are explicitly given by

$$a_k = \binom{n}{k} = \frac{n!}{(n-k)!k!}, \quad 0 \leq k \leq n \tag{8.1}$$

where, as usual, factorial n is defined as

$$n! = 1 \cdot 2 \cdot 3 \cdots n$$

for $n \geq 1$, and $0! = 1$.[9] Here are some particular cases, which directly follow from these definitions.

$$\binom{n}{0} = 1, \quad \binom{n}{1} = n, \quad \binom{n}{n-1} = n, \quad \binom{n}{n} = 1.$$

Moreover,

$$\binom{n}{k} = \binom{n}{n-k}.$$

In other words, introducing a coordinate system for the cells in the triangular array as in in figure 8.1, where $n = 0, 1, 2, \ldots$ is the row index and $k = 0, 1, 2, \ldots$ is the column index, then the entry in cell with coordinates (n, k) is $\binom{n}{k}$.

[8]The notation $\binom{n}{k}$ was introduced by Andreas von Ettingshausen in his book *Die kombinatorische Analysis*, Vienna, 1826.

[9]For later consideration in the context of cellular automata, we also adopt the convention $\binom{n}{k} = 0$ for $k < 0$ and $k > n$.

Divisibility of Binomial Coefficients

Thus, one approach to the patterns in Pascal's triangle would be to understand the divisibility properties of binomial coefficients. However, computing the entries a_k according to eqn. (8.1) for figures like 8.2 does not lead very far. The reason is that factorials grow extremely rapidly.

n	$n!$	n	$n!$
1	1	6	720
2	2	7	5,040
3	6	8	40,320
4	24	9	363,880
5	120	10	3,628,800

The number 100! has 158 digits,

$$\begin{aligned} 100! = {} & 9332621544\ 3944152681\ 6992388562\ 6670049071 \\ & 5968264381\ 6214685929\ 6389521759\ 9993229915 \\ & 6089414639\ 7615651828\ 6253697920\ 8272237582 \\ & 5118521091\ 6864000000\ 0000000000\ 00000000, \end{aligned}$$

and 1000! about 2568 digits, which surely is beyond the range of common computer arithmetic.[10]

As a first step to overcome these difficulties we use the recursive definition of Pascal's triangle (as indicated in figure 8.1) which is obtained from the addition rule[11]

$$\binom{n+1}{k} = \binom{n}{k-1} + \binom{n}{k}. \tag{8.2}$$

This fundamental relation avoids the computation of large factorials. However, the binomial coefficients themselves also grow rapidly as the row index n increases. Already in row $n = 34$ we find an entry

$$\binom{34}{17} = 2{,}333{,}606{,}220 > 2^{31} - 1 = 2{,}147{,}483{,}647$$

which cannot be represented exactly in normal computer arithmetic. Fortunately, we do not need the actual numerical values of the binomial coefficients when testing for divisibility. For example, the decision whether a binomial coefficient is odd or even follows directly from the addition rule. Observe that $\binom{n+1}{k}$ is odd provided $\binom{n}{k-1}$ is odd and $\binom{n}{k}$ is even, or vice versa. Systematically we have:

[10] The estimate of 2568 digits is obtained by a famous formula developed by James Stirling in 1730 which approximates $n! \approx \sqrt{2\pi n}(n/e)^n$, where $e = 2.71828\ldots$ denotes Euler's number.

[11] See section 2.3.

$\binom{n}{k-1}$	$\binom{n}{k}$	$\binom{n+1}{k}$
even	even	even
odd	even	odd
even	odd	odd
odd	odd	even

Cellular Automata ...

This elementary observation is not only of computational importance; it also provides the link to cellular automata. This is another approach to Pascal's triangle which will be explored in the following. We will see that there is a whole class of cellular automata which are closely related to the evolution of divisibility patterns in Pascal's triangle.

...and IFS

However, running a cellular automaton and testing divisibility properties of binominal coefficients have a common property, namely, that they are local (or microscopic) procedures. They allow the generation of a geometric pattern but do not at all explain the global (or macroscopic) appearance of the pattern. For example, why do we begin to see the Sierpinski gasket when coloring the odd entries in Pascal's triangle?

To address this problem we once again will bring iterated function systems (IFS) into play. If you recall section 5.4, this does not come as a total surprise and you might have a initial vague idea how this approach to Pascal's triangle could look. We will guide you to this point and explore its relation to divisibility properties and cellular automata, and you will watch the pieces of the puzzles falling into place.

8.1 Cellular Automata

Cellular automata have starting points far back in the sciences. In some sense we might say that Pascal's triangle is the first cellular automaton. Their recent development is rooted in the work of Konrad Zuse, Stanislaw Ulam and John von Neumann and is closely related to the first computing machines. During the 1970's and 1980's cellular automata had a strong revival through the work of Stephen Wolfram, who published an interesting survey.[12] Today cellular automata have become a very important modeling and simulation tool in science and technology, from physics, chemistry, and biology, to computational fluid dynamics in airplane and ship design, and to philosophy and sociology.

One-Dimensional Two-State Automaton

The first steps of a one-dimensional cellular automaton with two states (black and white cells).

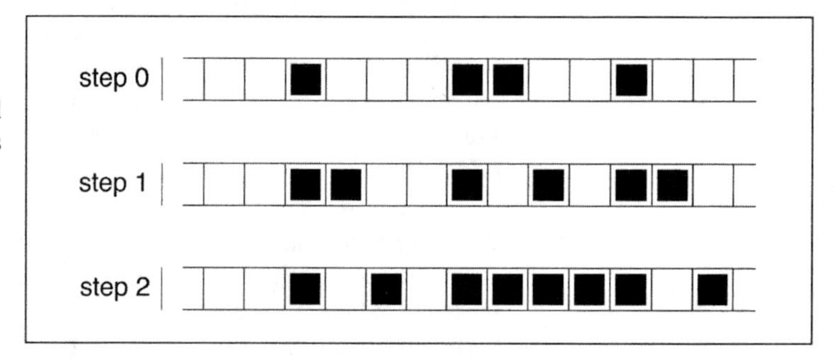

Figure 8.3

Feedback Machines

Cellular automata are perfect feedback machines. More precisely, they are mathematical finite state machines which change the state of their cells step by step. Each cell has one out of p possible states represented by the numbers $0, 1, \ldots, p-1$. Sometimes we speak of a *p-state cellular automaton*. The automaton can be one-dimensional where its cells are simply lined up like a chain or two-dimensional where cells are arranged in an array covering the plane.[13] Figure 8.3 shows the first steps of a one-dimensional two-state automaton. Sometimes we like to draw the succeeding steps of a one-dimensional cellular automaton one below the other and call the steps *layers*. When running the machine it grows layer by layer as shown in figure 8.4.

To run a cellular automaton we need two entities of information: an initial state of its cells (i.e., an initial layer) and a set of rules or laws. These rules describe how the state of a cell in a new layer (in the next step) is determined from the states of a group of cells from the preceding layer. The rules should *not depend* on the position of the group within the layer. Thus, it can be specified by a look-up table or if possible by a formula. Figure 8.6 shows look-up tables for two-state cellular automata which are given by configurations as in (a) and (b) of figure 8.5. These are just two examples of rules for one-dimensional cellular automata. The look-up table (a) was used in figures 8.3 and 8.4.

[12] S. Wolfram (ed.), *Theory and Application of Cellular Automata*, World Scientific, Singapore, 1986.

[13] In fact, the automaton can have any dimension m, where m is a natural number.

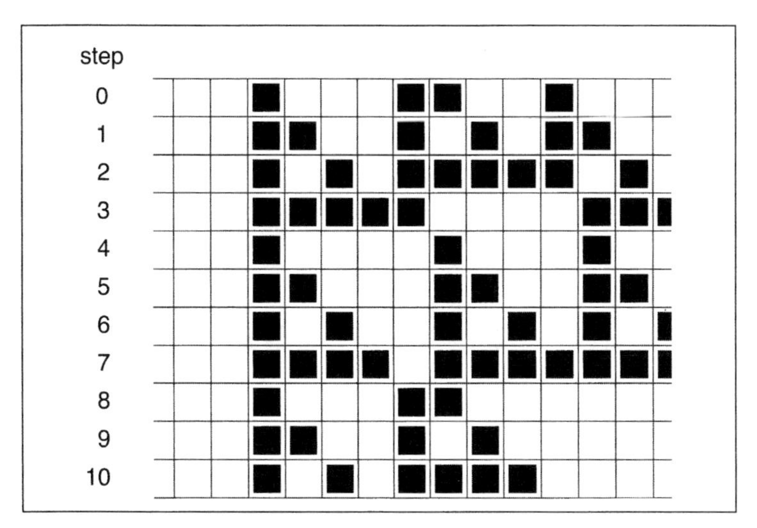

Growing of Layers of a One-Dimensional Automaton

The iterations of the one-dimensional cellular automaton from figure 8.3 are continued. The first 10 steps of are drawn from top to bottom.

Figure 8.4

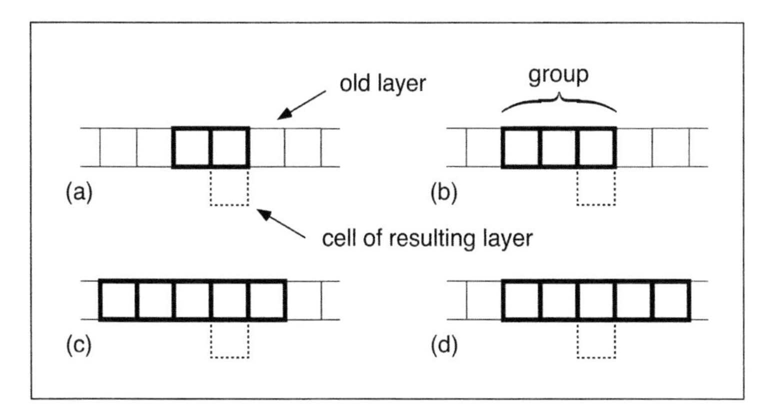

Automata Rules

There are several ways a rule may determine the state of a cell in the succeeding layers. In (a) the state of a new cell is determined by the states of two cells, in (b) by the state of three cells. In (c) and (d) the states of five cells determine the state of a new cell, but note that the position of the new cell with respect to the group is different in (c) and (d).

Figure 8.5

Game of Life

Particular two-dimensional cellular automata became very popular as the *Game of Life* through the work of John Horton Conway in the 1970's. In the Game of Life each cell is either dead (0) or alive (1) and changes its state according to the states in its immediate neighborhood, including its own state. More precisely, a cell that is alive (symbolized as a black cell) at one time step will stay alive in the next step when precisely two or three cells among its eight neighbors (see figure 8.7) in a square lattice are alive. If more than three neighbors are alive, the cell will die from overcrowdedness. If fewer than two neighbors are alive, the cell will die from loneliness. A dead cell will come to life when surrounded by exactly three live neighbors. Figure 8.8 shows the evolution of the Game of Life in some steps. One of the challenges of the game is to design cell clusters which exhibit a particularly interesting behavior. For example, there are clusters, called *blinkers*, which reproduce

Automata Look-Up Tables

Two examples of look-up tables. (a) four rules for a configuration based on two cells and two states. (b) eight rules for a configuration based on three cells and two states.

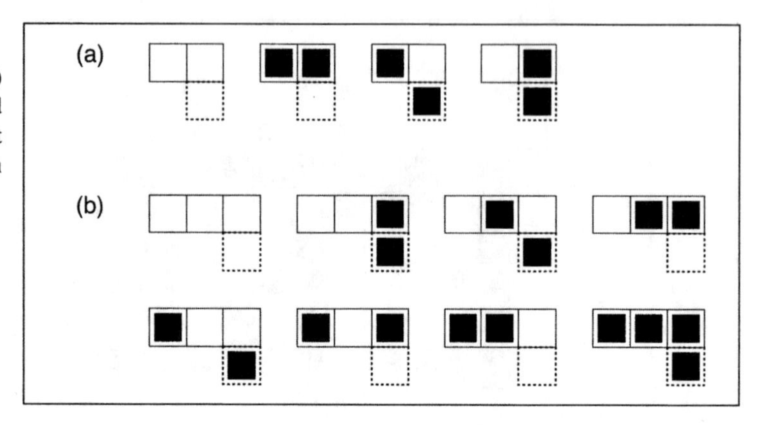

Figure 8.6

Neighborhood in Two-Dimensional Automata

In the Game of Life the neighborhood of a cell decides over life or death. Cell (a) and (b) will stay alive but cell (c) and (d) will die. At (f) a cell will come into life but not at (e).

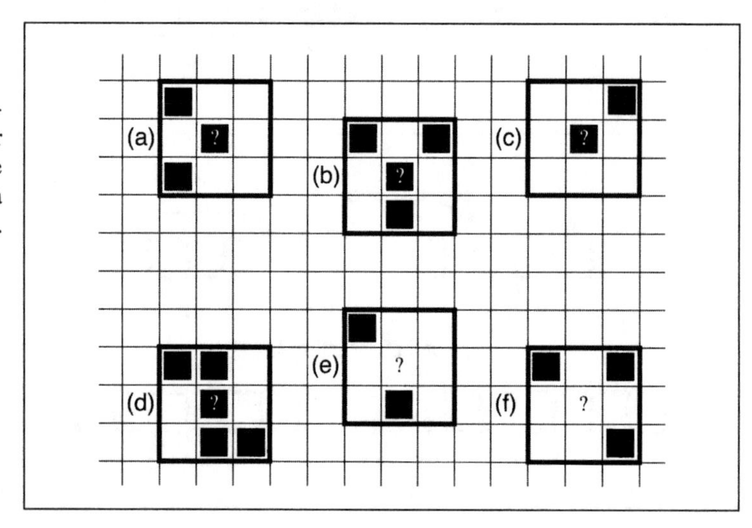

Figure 8.7

themselves after some steps, *gliders* move in a certain direction, *star ships* leave a trace of blinkers, and *guns* periodically eject gliders.

The Number of Games

The rules of the Game of Life are only one choice out of many imaginable sets of rules. For the two possible states and a neighborhood of eight cells generating a new center cell there are $2^{2^9} \approx 10^{154}$ different possible sets.

Let us briefly touch some variants of the Game of Life. The *one-out-of-eight rule* is given by the following set of rules: a cell becomes alive if exactly one of its neighbors is alive; otherwise it remains unchanged. Figure 8.9 shows the resulting pattern which evolves after 29 steps starting with just one living cell. Apparently some self-similarity is built into the formation of this pattern.

Another example, the *majority rule*, is obtained by these conventions: if five or more of the neighborhood of nine cells (including the cell itself) are

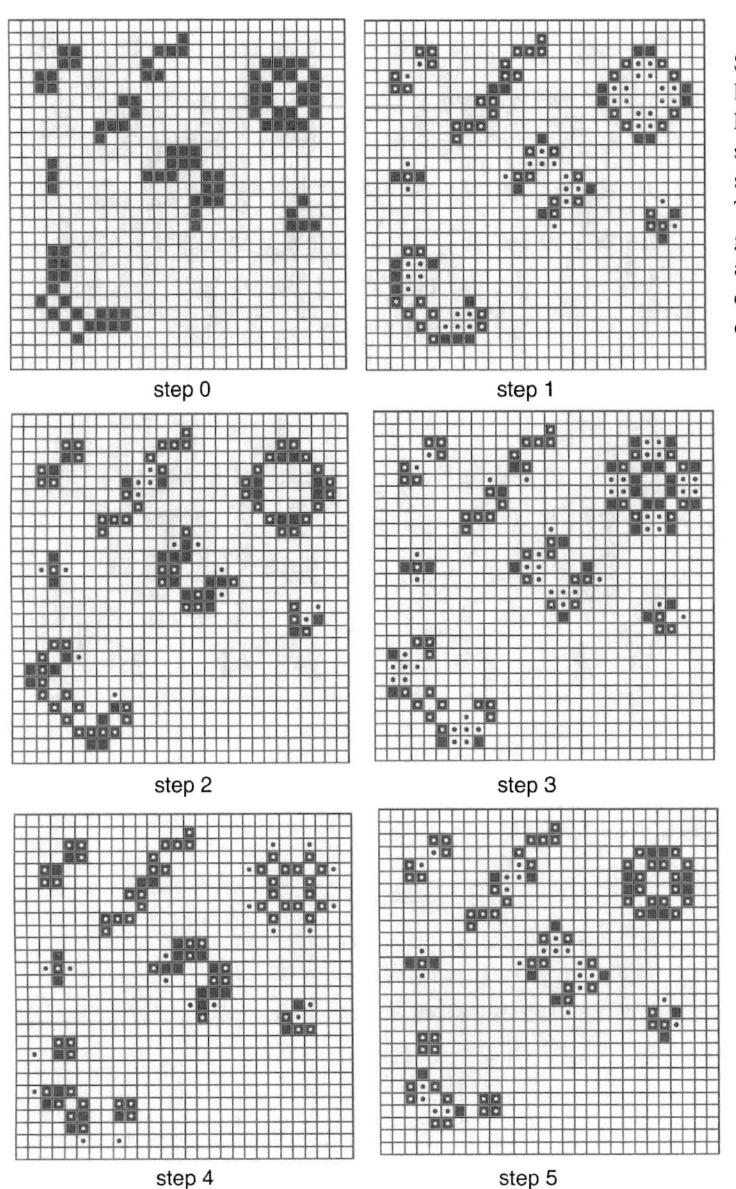

The Game of Life

Six successive steps of the Game of Life. Dots indicate the position of living cells of the previous step. Observe that some of the cell clusters shown exhibit a periodic behavior. The center right one is a so-called glider which slowly moves to the left as long as it does not hit another cell cluster. It takes 4 steps to move one cell to the left.

Figure 8.8

alive, then this cell will also become or remain alive. Otherwise it will die or remain dead. In other words, the center cell adjusts to the majority in the neighborhood. The resulting patterns resemble some phenomena in statistical physics such as percolation[14] or Ising spin systems. Figure 8.11 shows some

[14]See section 9.2 in chapter 9.

One-Out-of-Eight Rule

Starting with just one cell this self-similar pattern evolves after 29 steps.

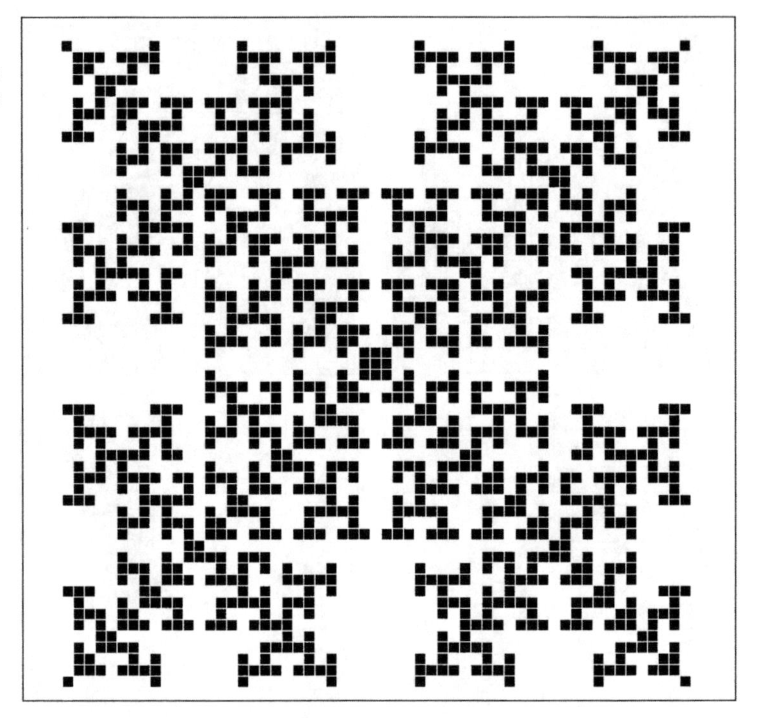

Figure 8.9

experiments which evolve as stable pattern after some 30 steps starting in each case with a random initial distribution of living cells.

NWSE Neighbors

The north, west, south, and east neighbors.

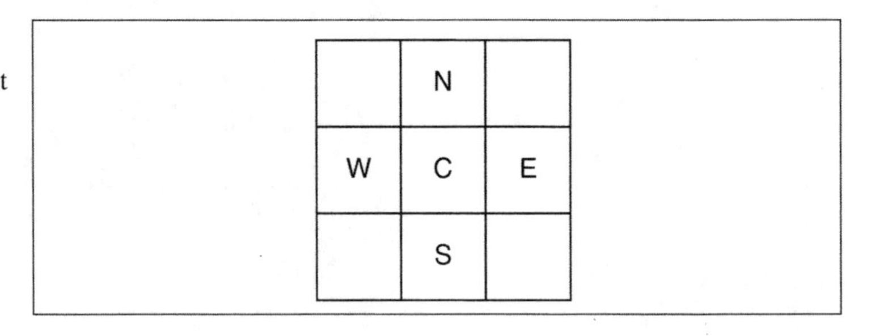

Figure 8.10

Finally, we consider rules which only take four neighbors (again in a two-dimensional square lattice) into account (see figure 8.10). Following Tommaso Toffoli and Norman Margolus[15] we label the center cell by C = center, and the four neighbors are labelled E = east, W = west, S = south, and N = north. If we allow two states for each cell of this configuration of five cells ($CSWNE$) then the state of $CSWNE$ will be given by five binary digits. For example,

[15]T. Toffoli and N. Margolus, *Cellular Automata Machines: A New Environment For Modelling*, MIT Press, Cambridge, Mass., 1987.

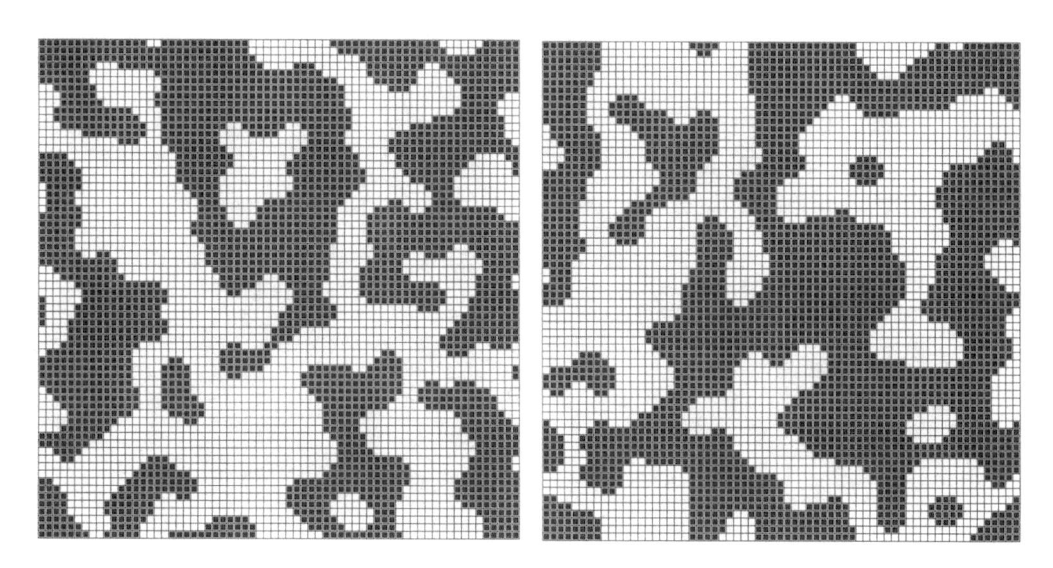

Figure 8.11 : Two examples for a game with the majority rule. Starting from a random distribution of black cells the game settles down (i.e., further iterations do not change the state of cells) to the patterns shown. For the two images, two different initial distributions were used.

$CSWNE = 11010$ indicates that cells C, S, and N are alive, while the other two are dead. A complete set of rules can be given by a table of the 32 possible states of $CSWNE$ and the subsequent values of the center cell C. Note that for such a configuration there are $2^{32} \approx 4 \cdot 10^9$, i.e., 4 billion possible different tables.

CSWNE	C	CSWNE	C	CSWNE	C	CSWNE	C
00000	0	01000	1	10000	1	11000	1
00001	0	01001	1	10001	1	11001	1
00010	0	01010	1	10010	1	11010	1
00011	0	01011	1	10011	1	11011	1
00100	1	01100	0	10100	1	11100	1
00101	1	01101	0	10101	1	11101	1
00110	1	01110	0	10110	1	11110	1
00111	1	01111	0	10111	1	11111	1

Applying the rules from the above table we obtain a familiar pattern. Figure 8.12 shows the 5th and 15th step starting with just one live cell near the lower left corner. Studying the table you can find a rather simple rule for producing the entries. Note that if the center cell C is dead (0) then the new value depends only on cells W and S. On the other hand, if the cell is alive (1) then it will remain alive in the next step. In fact, we have just seen an example of how we can construct a two-dimensional automaton with the behavior of

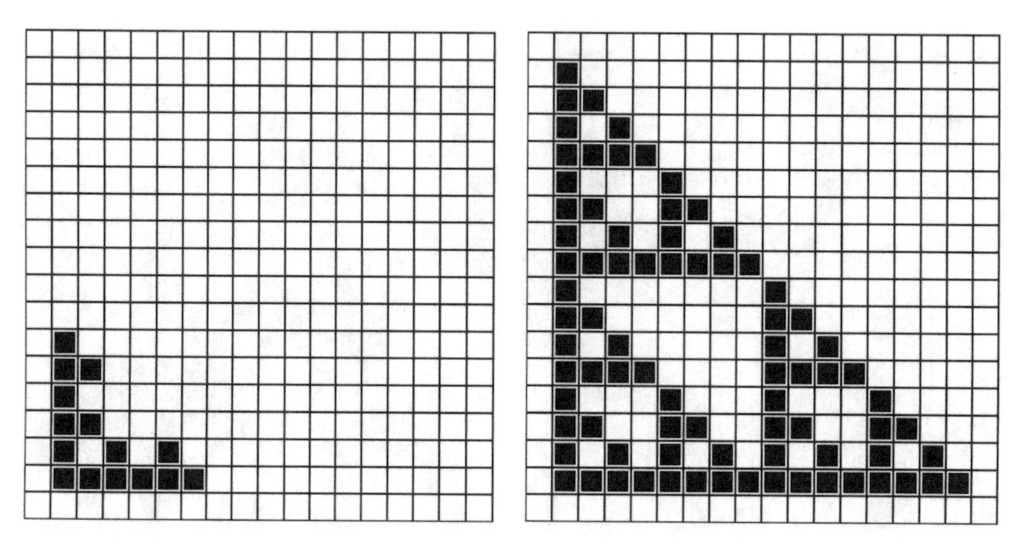

Figure 8.12 : Starting with just one cell this two-dimensional automaton behaves like a one-dimensional one, but the layers grow diagonally from the bottom left corner.

Figure 8.13 : Starting with an 8×8 cell cluster these patterns evolve after 13 (left) and 27 (right) steps using the parity rule (8.3).

a one-dimensional one.[16] In other words, cells grow layer by layer like the layers of a one-dimensional cellular automaton (although in this example the layers are diagonals and the pattern grows from bottom left to top right).

[16] In fact, the same basic idea works for any given one-dimensional automaton.

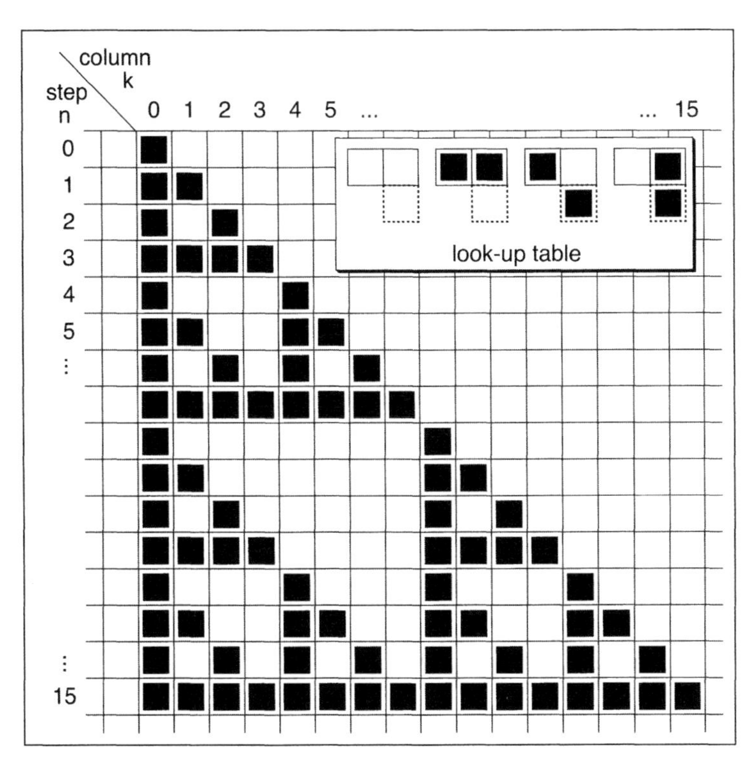

Sierpinski Automaton

The first 16 layers of a cellular automaton with look-up table displayed in the upper right.

Figure 8.14

Many interesting rules can be expressed by a simple formula. For example, the *parity rule* is given simply by

$$C_{\text{new}} = C_{\text{old}} + S_{\text{old}} + W_{\text{old}} + N_{\text{old}} + E_{\text{old}} \bmod 2. \tag{8.3}$$

which means that C_{new} is 0 or 1 if the sum on the right-hand side is even or odd, respectively. Here E, W, S, N and C represent the old and new cell states as indicated by the indices 'old' and 'new'. Thus for $CSWNE = 11010$ we obtain $C_{\text{new}} = 1$, for $CSWNE = 11011$ $C_{\text{new}} = 0$, and so on. Figure 8.13 shows the evolution of this cellular automaton after 13 and 27 steps starting with a square block of 8×8 black cells.

Pascal's Triangle ...

Let us return to one-dimensional automata. The look-up table used in figure 8.14 reflects the addition of even and odd binomial coefficients. An odd entry is colored black. That is, the evolution of the corresponding cellular automaton will produce the pattern which is obtained from the Pascal triangle when we color cells with odd entries black and cells with even entries white and start with an appropriate initial layer. This is seen in the figure, where we follow the evolution of the first 16 layers starting with the initial layer, which has one black cell.

...and Polynomials

Let us explore the connection between cellular automata and coefficients of polynomials a bit further. First we look at an example involving the powers of the polynomial $r(x) = 1 + x$:

$$\begin{aligned}
(r(x))^0 &= 1 \\
(r(x))^1 &= 1 + x \\
(r(x))^2 &= 1 + 2x + x^2 \\
(r(x))^3 &= 1 + 3x + 3x^2 + x^3 \\
(r(x))^4 &= 1 + 4x + 6x^2 + 4x^3 + x^4 \\
&\vdots \\
(r(x))^n &= a_0(n) + a_1(n)x + a_2(n)x^2 + \cdots + a_n(n)x^n.
\end{aligned}$$

Now let $a_k(n)$ for all integers k and $n \geq 0$ denote the state of cell number k of the n-th layer of a one-dimensional automaton.[17] Starting with $a_0(0) = 1$ and $a_k(0) = 0$ for $k \neq 0$ the rule

$$a_k(n) = a_{k-1}(n-1) + a_k(n-1) \tag{8.4}$$

generates the coefficients of $(r(x))^n$. Equation (8.4) is nothing else but the addition rule in eqn. (8.2) for binomial coefficients.

Now we want to look at the divisibility properties of $a_k(n)$ with respect to an integer p. We write

$$a \equiv b \pmod{p}$$

provided $a - b$ is a multiple of p.[18] Using this language, our test for odd and even binominal coefficients $\binom{n}{k}$ or cells $a_k(n)$ is simply to check whether

$$a_k(n) \equiv 0 \pmod{2} \text{ or } a_k(n) \equiv 1 \pmod{2}.$$

Moreover, our addition rules (8.2) and (8.4) imply in mod 2 arithmetic

$a_{k-1}(n)$	$a_k(n)$	$a_k(n+1)$
0	0	0
1	0	1
0	1	1
1	1	0

This is just the look-up table of the 2-state automaton shown in figure 8.14 where black corresponds to 1 and white to 0. Thus, this figure shows the coefficients of the powers of $r(x) = 1 + x$ modulo 2.

[17] Strictly speaking, this is not a finite state automaton because the numbers $a_k(n)$ grow beyond all bounds. However, we will arrive at a finite state machine when restricting our attention to the divisibility properties.

[18] In other words, $a \equiv b \pmod{p}$ provided a and b differ by a multiple of p. For $p = 2$ this means that $a-b$ is even. Furthermore, $a \equiv 0 \pmod{2}$ means that a is even and $a \equiv 1 \pmod{2}$ means that a is odd. The notation was introduced by Carl Friedrich Gauss.

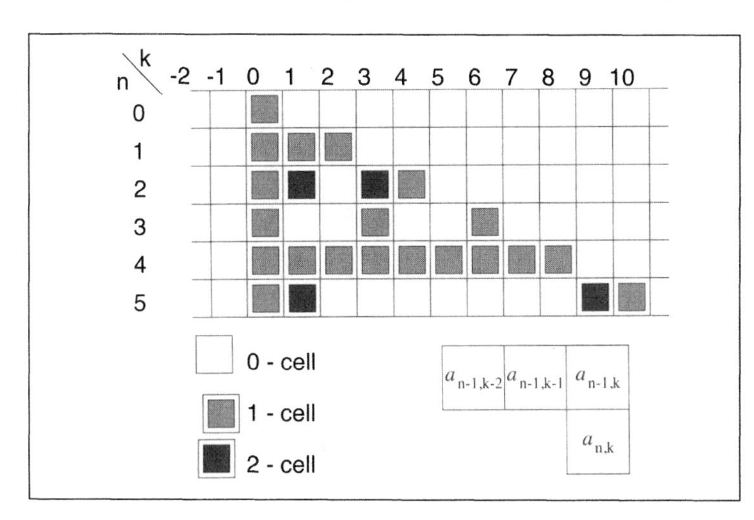

Mod 3 Automaton

Cellular automaton generated by the polynomial $r(x) = 1 + x + x^2$ and $p = 3$. There are two types of indices, a row index n which runs $n = 0, 1, 2, \ldots$, and a column index k which runs in the integers $\ldots, -2, -1, 0, 1, 2, \ldots$ The initial layer, $n = 0$, consists of cells of state 0 except for the cell at $k = 0$, which is a cell with state 1. The rule for the cells of the new layer is: $a_{n+1,k} = a_{n,k-2} + a_{n,k-1} + a_{n,k} \bmod 3$

Figure 8.15

Generalizations

Now we can generalize in two ways: we can look at coefficients modulo integers other than 2 and we can look at arbitrary polynomials. Let us take the example: $r(x) = 1 + x + x^2$:

$$\begin{aligned}
(r(x))^0 &= 1 \\
(r(x))^1 &= 1 + x + x^2 \\
(r(x))^2 &= 1 + 2x + 3x^2 + 2x^3 + x^4 \\
(r(x))^3 &= 1 + 3x + 6x^2 + 7x^3 + 6x^4 + 3x^5 + x^6 \\
&\vdots \\
(r(x))^n &= a_0(n) + a_1(n)x + a_2(n)x^2 + \cdots + a_{2n}(n)x^{2n}.
\end{aligned}$$

Do you see an extension of the addition rule for binomial coefficients, eqn. (8.4)? You can check in the first few lines that the law

$$a_k(n) = a_{k-2}(n-1) + a_{k-1}(n-1) + a_k(n-1) \tag{8.5}$$

holds. A proof of this relation would proceed by induction. When looking at the divisibility properties with respect to $p = 3$ we obtain the coefficients

$$\begin{aligned}
(r(x))^0 &\to 1 \\
(r(x))^1 &\to 1\,1\,1 \\
(r(x))^2 &\to 1\,2\,0\,2\,1 \\
(r(x))^3 &\to 1\,0\,0\,1\,0\,0\,1
\end{aligned}$$

and so on. Figure 8.15 shows the evolution of the corresponding 3-state automaton.

Linear Cellular Automata

In a similar way we could start with any polynomial $r(x) = a_0 + a_1 x + \cdots + a_d x^d$ of degree d and integer coefficients a_i, and then look at the coefficients of $(r(x))^n$ modulo some positive integer p for $n = 0, 1, 2, \ldots$ and the result would be that the kth coefficient of $(r(x))^{n+1}$ is obtained by an addition formula involving $d+1$ coefficients from $(r(x))^n$. In other words, given a polynomial with integer coefficients and a positive integer p there is an associated cellular automaton which generates the coefficients modulo p of the powers $(r(x))^n, n = 0, 1, 2, \ldots$ Since the look-up table is generated by an addition formula these automata are called *linear cellular automata* (LCA).

States and Colors

The choice of the positive integer p determines the number of states of the automaton. If $p = 2$, i.e., we are considering arithmetic modulo 2, then we have an automaton which can be graphically represented in black and white. For $p > 2$ we would need colors to adequately represent the evolution of an automaton. We can often simplify a p-state to a 2-state automaton by the following modification:

- Cells representing a nonzero coefficient are colored black.
- Cells representing a zero coefficient are colored white.

With this background of linear cellular automata, we can state a number of very interesting problems:

- *Pattern Formation*. Given a polynomial with integer coefficients and a positive integer p, discuss the global pattern formation which evolves when the automaton has produced for a long time.
- *Colors*. What is the relationship between the global patterns which are obtained for different choices of p and a fixed, given polynomial?
- *Fractal Dimension*. What is the fractal dimension of the global pattern?
- *Higher Dimensions*. Polynomials in one variable generate one-dimensional linear cellular automata. A polynomial in m variables determines a linear cellular automaton in m dimensions. How can we generalize the results to m-dimensional automata?
- *Factorization*. If a polynomial $r(x)$ is the product of two polynomials $s(x)$ and $t(x)$, how is the pattern determined by $r(x)$ related to the patterns determined by $s(x)$ and $t(x)$, and how are the dimensions related?

Actually, the last problem critically depends on the choice of p, the number of states, because what actually counts is whether

$$r(x) \equiv s(x)t(x) \pmod p.$$

For example, the polynomial $r(x) = 1 + x$ is irreducible with respect to the integers, i.e., if $r(x) = s(x)t(x)$, then the factorization must be trivial, i.e., $s(x) = 1$ and $t(x) = 1 + x$. If we use arithmetic modulo p, and p is not a prime number, however, then $r(x) = 1 + x$ admits nontrivial factorizations like, for example,

$$1 + x \equiv (1 + 3x)(1 + 4x) \pmod 6.$$

Several of these problems are still wide open while others have been understood only recently through new tools provided by fractal geometry (namely, hierarchical iterated function systems) which stresses again that fractals are more than pretty images.

8.2 Binomial Coefficients and Divisibility

In the remaining part of this chapter we will discuss some of the problems listed at the end of the last section for the particular choice $r(x) = 1 + x$ and positive integers p. Thus, in the following we will only look at the divisibility properties of binomial coefficients,[19] although a similar discussion can be done for general polynomials.[20]

In our discussion we will primarily address the question of whether a binominal coefficient is divisible by p or not. In other words, we consider the black and white coloring of the Pascal triangle interpreted modulo p (see figure 8.16). The question of divisibility can be approached with the aid of prime number factorization. Below we will see that in order to understand the patterns in Pascal's triangle formed by the coefficients divisible by an integer p, we should build on the patterns generated by the prime factors of p.

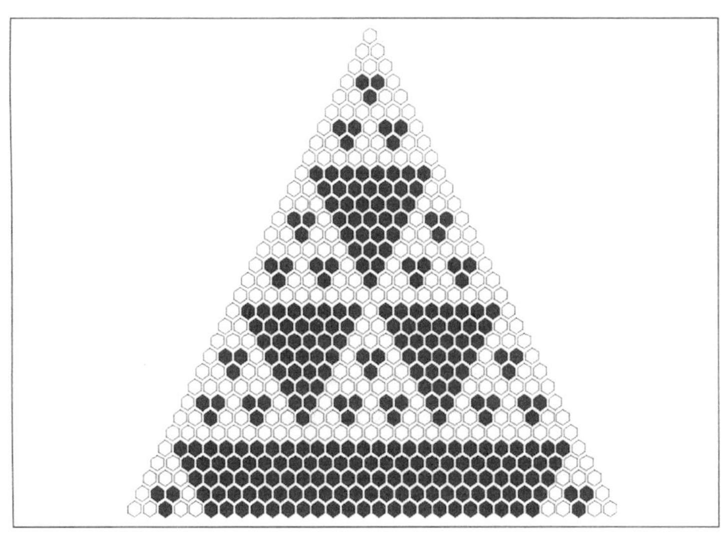

Pattern in Pascal's Triangle Mod 3

Coefficients in the Pascal triangle which are divisible by 3 are shown in black.

Figure 8.16

We have seen that we can answer the question of divisibility by recursively computing binominal coefficients using an addition rule like eqn. (8.2) modulo p with a subsequent test as to whether the result is 0 or not. On the other hand, we know very well how to describe the coefficients without a recursion, namely, by

$$\binom{n}{k} = \frac{n!}{(n-k)!k!}.$$

[19] F. v. Haeseler, H.-O. Peitgen, G. Skordev, *Pascal's triangle, dynamical systems and attractors,* Ergod. Th. & Dynam. Sys. 12 (1992) 479–486.

[20] F. v. Haeseler, H.-O. Peitgen, G. Skordev, *Linear cellular automata, substitutions, hierarchical iterated function systems and attractors,* in: *Fractal Geometry and Computer Graphics,* J. L. Encarnacao, H.-O. Peitgen, G. Sakas, G. Englert (eds.), Springer-Verlag, Heidelberg, 1992.

Coordinate Systems

Two coordinate systems for the presentation of binomial coefficients. The new modified system is on the right.

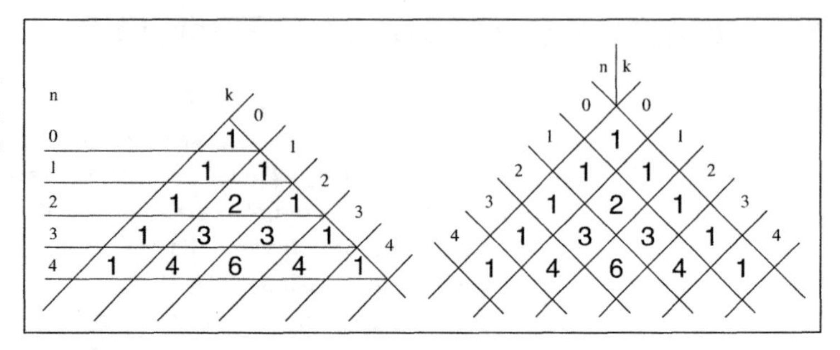

Figure 8.17

The major question for now will be to understand whether or not these coefficients are divisible by p also by means of a direct, nonrecursive computation. It turns out that this problem was solved in a most elegant manner some 150 years ago by the German mathematician Ernst Eduard Kummer.[21] The careful development of Kummer's criterion, which is local in nature, will build a solid foundation for the next step towards understanding the global pattern formation in Pascal's triangle.

It turns out that for the following it will be more convenient to work in a new (n, k)-coordinate system (see figure 8.17). The connection between the old and new representation is easy. In the old system we find at position (n, k) the coefficient $\binom{n}{k}$, while in the new system we have at position (n, k) the binomial coefficient

$$\binom{n+k}{k} = \frac{(n+k)!}{n!k!}$$

Figure 8.18 shows the array in the new system, however, rotated and right angled, together with the usual coloration corresponding to even and odd entries.

Divisibility Sets P(r)

We will now describe our problem more formally. We define the following set:

$$P(r) = \left\{ (n,k) \;\middle|\; \binom{n+k}{k} \text{ is not divisible by } r \right\},$$

where r is some integer. Thus, figure 8.18 is a graphical representation of a part of $P(2)$.

Observe that if p and q are two different prime numbers and a given integer is not divisible by $p \cdot q$, then it is also not divisible by p or q alone. Thus,

$$P(pq) = P(p) \cup P(q), \quad \text{if } p \neq q, \quad p, q \text{ prime}.$$

For example, $P(6) = P(2) \cup P(3)$ (see figure 8.19). This observation can be generalized to the factorization in prime powers. If we consider the prime factorization of an integer r,

$$r = p_1^{r_1} \cdots p_s^{r_s},$$

[21] E. E. Kummer, *Über Ergänzungssätze zu den allgemeinen Reziprozitätsgesetzen*, Journal für die reine und angewandte Mathematik 44 (1852) 93–146. For the result relevant to our discussion see pages 115–116.

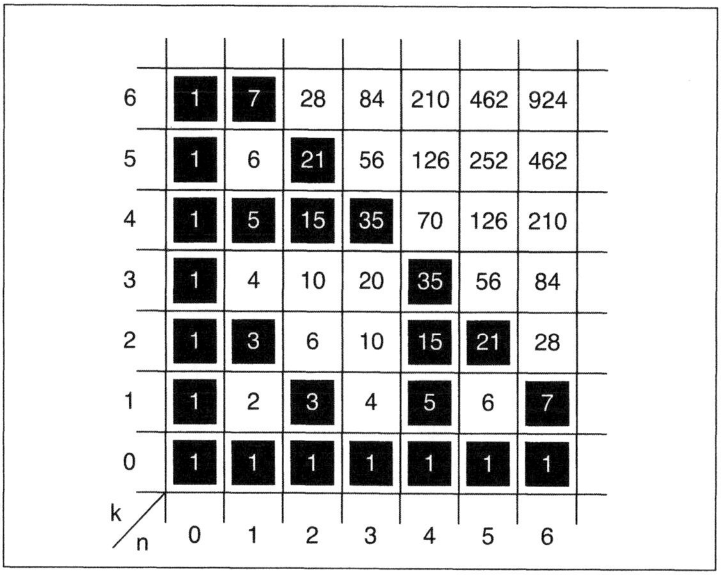

k \ n	0	1	2	3	4	5	6
6	1	7	28	84	210	462	924
5	1	6	21	56	126	252	462
4	1	5	15	35	70	126	210
3	1	4	10	20	35	56	84
2	1	3	6	10	15	21	28
1	1	2	3	4	5	6	7
0	1	1	1	1	1	1	1

Divisibility by 2

Presentation of Pascal's triangle in the new (n, k)-coordinate system. Position (n, k) shows $\binom{n+k}{k}$. Odd numbers are colored black.

Figure 8.18

where the prime numbers p_k are different and the exponents τ_k are natural numbers, then

$$P(r) = P({p_1}^{\tau_1}) \cup \cdots \cup P({p_s}^{\tau_s}).$$

Thus, to understand the pattern formation for $P(r)$ it suffices to understand $P(p^\tau)$,

$$P(p^\tau) = \left\{ (n, k) \;\middle|\; \binom{n+k}{k} \text{ is not divisible by } p^\tau \right\},$$

for a prime number p and some positive integer τ.

The p-adic Approach

Now let us discuss some gems from elementary number theory attributable to Adrien Marie Legendre (1808), Ernst Eduard Kummer (1852), and Edouard Lucas (1877). These results, together with hierarchical iterated function systems, will completely decipher the patterns in $P(p^\tau)$ for any prime number p and positive integer τ.

We would like to have a direct method to check whether $\binom{n+k}{k}$ is divisible by p^τ. Kummer observed that this information is encoded in the p-adic representation of n and k. You are familiar with decimal expansions like

$$n = d_0 + d_1 \cdot 10 + d_2 \cdot 10^2 + \cdots + d_m \cdot 10^m,$$

where the numbers $d_i \in \{0, \ldots, 9\}$ are the decimal digits. Thus n can be represented as the decimal number

$$n = d_m d_{m-1} \ldots d_1 d_0.$$

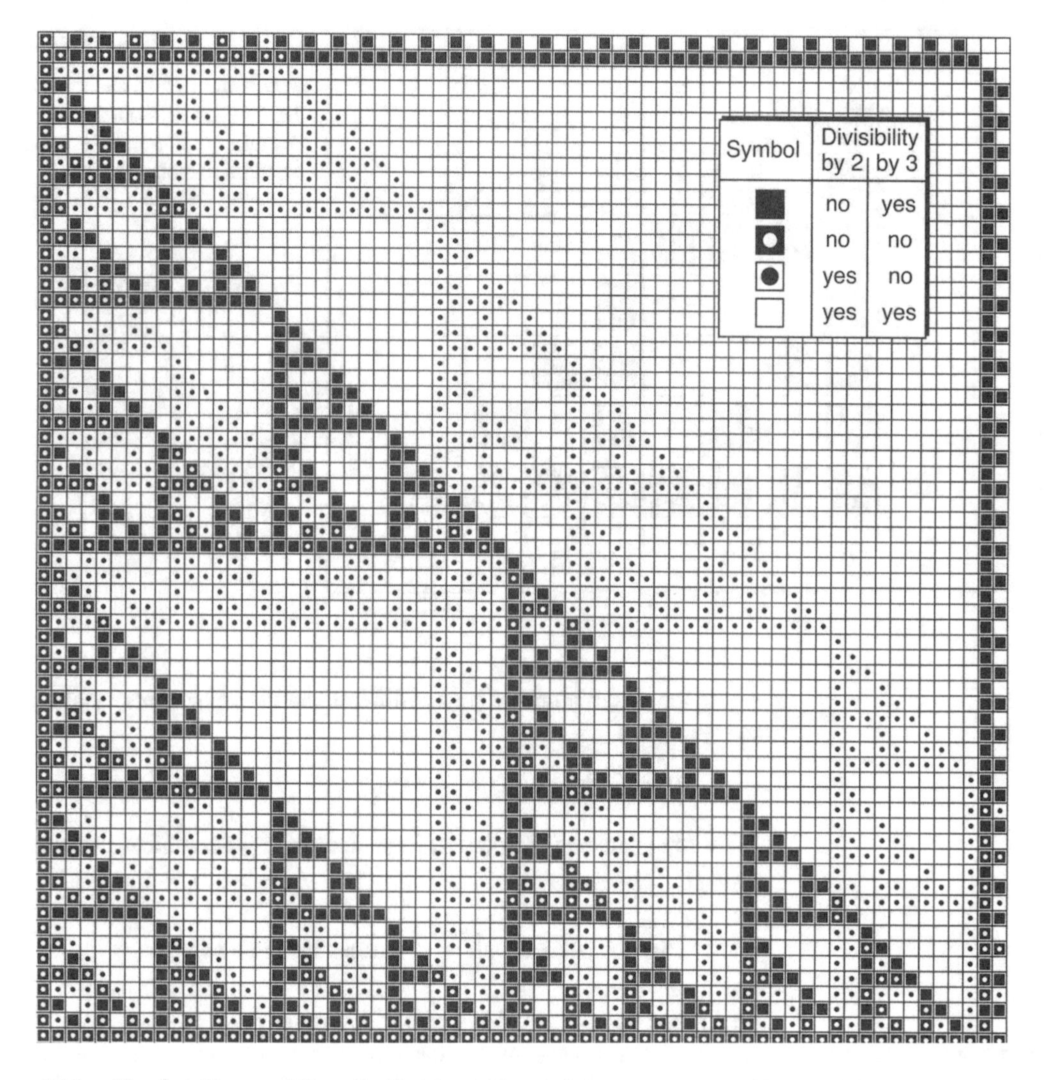

Figure 8.19 : The first 66 rows of Pascal's triangle and its mod-6 pattern generated by a cellular automaton using the rule: $a_{n,k} = a_{n-1,k} + a_{n,k-1} \bmod 6$.

Now the p-adic expansion of an integer n is given analogously by

$$n = a_0 + a_1 p + a_2 p^2 + \cdots + a_m p^m$$

where the digits are now $a_i \in \{0, \ldots, p-1\}$ and m may be different from the m used above in the decimal number. Corresponding to decimal numbers, we introduce the p-adic representation

$$n = (a_m a_{m-1} \ldots a_1 a_0)_p.$$

For example, $n = 17$ and $k = 8$ have the expansions

$$\begin{aligned} n &= (17)_{10} = (10001)_2 = (122)_3 = (32)_5 = (23)_7 = (16)_{11} \\ k &= (08)_{10} = (01000)_2 = (022)_3 = (13)_5 = (11)_7 = (08)_{11}. \end{aligned}$$

Kummer's observation concerns the number of carries which occur when we add n and k in the p-adic representation when p is a prime. For example, let us add the triadic representations:

$$\begin{array}{r} 1\;2\;2 \\ +0_1 2_1 2 \\ \hline 2\;2\;1 \end{array}$$

Observe that when adding the rightmost digits we obtain a carry to the next digit. The same is true when adding the second digits of the two numbers. Thus, we obtain two carries. On the other hand, if we add the corresponding binary representations:

$$\begin{array}{r} 10001 \\ +01000 \\ \hline 11001 \end{array}$$

we obtain no carry at all. In other words, if we define

$$c_p(n, k) = \text{number of carries in the } p\text{-adic addition of } n \text{ and } k,$$

we have demonstrated that $c_2(17, 8) = 0$ and $c_3(17, 8) = 2$. Now we can state Kummer's result.

Kummer's Result *Let $\tau = c_p(n, k)$. $\binom{n+k}{k}$ is divisible by the prime power p^τ but not by $p^{\tau+1}$.*

In other words, the prime factorization of $\binom{n+k}{k}$ contains exactly $c_p(n, k)$ factors of p. Applied to our example, $n = 17$ and $k = 8$, we should have that $\binom{n+k}{k}$ has no factors of 2, because $c_2(17, 8) = 0$, and exactly two factors of 3, because $c_3(17, 8) = 2$. Thus, it is an odd number and divisible by 9, but not by 27. In fact, we compute that

$$\binom{n+k}{k} = \binom{25}{8} = 3^2 \cdot 5^2 \cdot 11 \cdot 19 \cdot 23. \tag{8.6}$$

which confirms our conclusions.

Interpreting Kummer's result the other way around, we conclude from the factorization in eqn. (8.6), that

$$c_p(17, 8) = \begin{cases} 2, & \text{for } p = 3, 5 \\ 1, & \text{for } p = 11, 19, 23 \\ 0, & \text{otherwise} \end{cases}$$

In fact, we check, for example, $c_{11}(17,8) = 1$ by adding 17 and 8 in modulo 11 arithmetic, obtaining one carry — as expected,

$$\begin{array}{r} 1\ 6 \\ +\ {}_1 8 \\ \hline 2\ 3. \end{array}$$

Lucas' Criterion

To determine whether $\binom{n}{k}$ is odd or even we can use Lucas' criterion as follows.[22] We compute the binary form of n and k, say, $n = 23 = (10111)_2$ and $k = 17 = (10001)_2$. Then we write them one over the other.

$$\begin{array}{l} 10111 \\ 10001 \end{array}$$

Now $\binom{n}{k}$ is odd, if and only if every digit of the bottom number k is less than or equal to the digit of n above. This is the case for our example, $n = 23$ and $k = 17$. In fact, $\binom{23}{17} = 100947$ is odd.

Let us see how Lucas' criterion follows directly from Kummer's result. Let the binary expansions of n and k be

$$\begin{aligned} n &= a_m 2^m + a_{m-1}2^{m-1} + \cdots + a_0 \\ k &= b_m 2^m + b_{m-1}2^{m-1} + \cdots + b_0 \end{aligned}$$

with binary digits $a_i, b_i \in \{0, 1\}$. Since $k \leq n$ some of the leading binary digits of k may be 0. First observe that we know from Kummer's result that $\binom{n}{k} = \binom{n-k+k}{k}$ is not divisible by 2 if and only if we have for the 2-adic expansion

$$n - k = c_m 2^m + c_{m-1}2^{m-1} + \cdots + c_0$$

the property

$$c_i + b_i \leq 1, \quad i = 0, \ldots, m.$$

In other words, in the p-adic addition of $n - k$ and k there will be no carry.

To complete the argument, we have to show two implications. First, if $\binom{n}{k}$ is odd, then Lucas' criterion follows, i.e., $a_i \geq b_i$, for all i. Second, if Lucas' criterion is satisfied, then it follows that $\binom{n}{k}$ is odd. Let i denote an arbitrary index $i \in \{0, \ldots, m\}$. Now we start with the first part, assuming that $\binom{n}{k}$ is odd. Kummer's result above states that $c_i + b_i \leq 1$. And this implies that $a_i = b_i + c_i$ and, in conclusion, also $a_i \geq b_i$, which is what was to be shown. Now we do the second part. We assume $a_i \geq b_i$. Then $c_i = a_i - b_i$, which implies that $c_i + b_i = a_i \leq 1$. Thus, according to Kummer's result $\binom{n}{k}$ is odd. This finishes the second part and completes the proof of Lucas' criterion.

[22] It is related to several published criteria, like the one in I. Stewart, *Game, Set, and Math*, Basil Blackwell, Oxford, 1989, which Stewart attributes to Edouard Lucas following Gregory J. Chaitin's book *Algorithmic Information Theory*, Cambridge University Press, 1987.

Mod-p Condition

As a particular case we obtain from Kummer's criterion that $\binom{n+k}{k}$ is not divisible by the prime number p provided $c_p(n,k) = 0$. In other words, we have

$$P(p) = \{(n,k) \mid c_p(n,k) = 0\}.$$

Moreover, the number of carries $c_p(n,k)$ is 0 if and only if

$$a_i + b_i \leq p - 1, \quad i = 0, \ldots, m$$

where a_i and b_i denote the p-adic digits of n and k, i.e.,

$$\begin{aligned} n &= a_0 + a_1 p + a_2 p^2 + \cdots + a_m p^m, \\ k &= b_0 + b_1 p + b_2 p^2 + \cdots + b_m p^m. \end{aligned}$$

This we will call the *mod-p condition*.

For prime powers, Kummer's result implies that

$$P(p^\tau) = \{(n,k) \mid c_p(n,k) < \tau\}$$

The proof of Kummer's observation can be based on a formula by Legendre dating from 1808 which determines the largest exponent μ of the prime power p^μ which divides $n!$.

Kummer's Result and Legendre's Identity

We recall Kummer's theorem of 1852.

- Let $c_p(n,k)$ be the number of carries in the p-adic addition of n and k and $\tau = c_p(n,k)$. Then $\binom{n+k}{k}$ is divisible by the prime power p^τ but not by $p^{\tau+1}$.

In order to derive this beautiful result we will use a formula by Legendre from 1808 which deals with the divisibility of $n!$ by a prime power. The formula is as follows.

- Let $\mu(n)$ be the largest integer exponent of the prime power $p^{\mu(n)}$ which divides $n! = 1 \cdot 2 \cdots n$. Thus, $n!$ is divisible by $p^{\mu(n)}$ but not by $p^{\mu(n)+1}$. Then

$$\mu(n) = \frac{n - \sigma}{p - 1} \tag{8.7}$$

where σ is the sum of the p-adic coefficients $a_i \in \{0, 1, \ldots, p-1\}$ of n,

$$\begin{aligned} n &= a_0 + a_1 p + a_2 p^2 + \cdots + a_m p^m, \\ \sigma &= a_0 + a_1 + \cdots + a_m. \end{aligned} \tag{8.8}$$

To show Legendre's formula we first establish the useful identity

$$\mu(n) = \sum_{i=1}^{\infty} \left\lfloor \frac{n}{p^i} \right\rfloor, \tag{8.9}$$

where the brackets $\lfloor\ \rfloor$ denote the greatest integer less than or equal to the enclosed quantity. Thus, e.g., $\lfloor 6.1 \rfloor = 6$, $\lfloor 5.9 \rfloor = 5$, $\lfloor \pi \rfloor = 3$.

Note that $\lfloor n/p^i \rfloor = 0$ for large i. Thus, the sum in eqn. (8.9) is a finite sum. Let us first prove this identity (8.9). We observe that the term $\lfloor n/p^i \rfloor$ is the number of elements from $\{1, 2, \ldots, n\}$ which are divisible by p^i. For example, if $p = 2$, $i = 3$ and $n = 17$, then $\lfloor n/p^i \rfloor = 2$. In other words, there are two integers less than or equal to 17 which are divisible by 2^3, namely, 8 and 16. Next we observe that the sum in eqn. (8.9) counts any factor in the product

$$1 \cdot 2 \cdot 3 \cdots (n-1) \cdot n$$

which is divisible by p^i but not by p^{i+1} exactly i times, namely, once in $\lfloor n/p \rfloor$, once in $\lfloor n/p^2 \rfloor, \ldots$, and once in $\lfloor n/p^i \rfloor$. This accounts for all occurrences of p as a factor of $n!$, i.e., identity (8.9) is established.

Here is an example: $n = 10$ and $p = 2$. Thus, we consider factors of 2 in the product $10! = 2 \cdot 3 \cdot 4 \cdot 5 \cdot 6 \cdot 7 \cdot 8 \cdot 9 \cdot 10$. Indeed, 2 is divisible by 2^1, 4 by 2^2, 6 by 2^1, 8 by 2^3 and 10 by 2^1. Thus, $10!$ is divisible by 2^8. This is shown in the following representation,

$$\begin{array}{ccccccccccc}
1 \cdot & 2 & \cdot 3 \cdot & 4 & \cdot 5 \cdot & 6 & \cdot 7 \cdot & 8 & \cdot 9 \cdot & 10 & \\
 & 1 & & 1 & & 1 & & 1 & & 1 & \lfloor 10/2 \rfloor = 5 \\
 & & & 1 & & & & 1 & & & \lfloor 10/4 \rfloor = 2 \\
 & & & & & & & 1 & & & \lfloor 10/8 \rfloor = 1 \\
 & p^1 & & p^2 & & p^1 & & p^3 & & p^1 &
\end{array}$$

showing the 8 occurrences of the factor $p = 2$. Thus, $\mu(10) = 8$. Observe that the sum

$$\sum_{i=1}^{\infty} \left\lfloor \frac{10}{2^i} \right\rfloor = \left\lfloor \frac{10}{2} \right\rfloor + \left\lfloor \frac{10}{4} \right\rfloor + \left\lfloor \frac{10}{8} \right\rfloor = 5 + 2 + 1 = 8$$

is just the number of these occurrences. Here is another example: $n = 1000$ and $p = 3$. $1000!$ is divisible by 3^{498} but not by 3^{499}, since

$$\begin{aligned}
\mu(1000) &= \left\lfloor \frac{1000}{3} \right\rfloor + \left\lfloor \frac{1000}{9} \right\rfloor + \left\lfloor \frac{1000}{27} \right\rfloor \\
&\quad + \left\lfloor \frac{1000}{81} \right\rfloor + \left\lfloor \frac{1000}{243} \right\rfloor + \left\lfloor \frac{1000}{729} \right\rfloor \\
&= 333 + 111 + 37 + 12 + 4 + 1 = 498.
\end{aligned}$$

Let us now establish Legendre's identity. By means of eqn. (8.9) Legendre's formula (8.7) is equivalent to

$$\sum_{i=1}^{\infty} \left\lfloor \frac{n}{p^i} \right\rfloor (p-1) = n - \sigma. \tag{8.10}$$

We proceed by showing this relation. Using the p-adic representation of n in (8.8) and the definition of the brackets $\lfloor\ \rfloor$ we get

$$\left\lfloor \frac{n}{p^i} \right\rfloor = a_i + a_{i+1}p + \cdots + a_m p^{m-i}, \quad i \leq m.$$

With that we compute the two sums

$$\begin{aligned}\sum_{i=1}^{\infty}\left\lfloor\frac{n}{p^i}\right\rfloor p &= \sum_{i=1}^{m}\left(a_i p + a_{i+1}p^2 + \cdots + a_m p^{m-i+1}\right)\\ &= a_1 p + a_2 p^2 + a_3 p^3 + \cdots + a_m p^m\\ &\quad + a_2 p + a_3 p^2 + \cdots + a_m p^{m-1}\\ &\quad + a_3 p + \cdots + a_m p^{m-2}\\ &\quad\vdots\\ &\quad + a_m p\end{aligned}$$

$$\begin{aligned}\sum_{i=1}^{\infty}\left\lfloor\frac{n}{p^i}\right\rfloor &= \sum_{i=1}^{m}\left(a_i + a_{i+1}p + \cdots + a_m p^{m-i}\right)\\ &= a_1 + a_2 p + a_3 p^2 + \cdots + a_m p^{m-1}\\ &\quad + a_2 + a_3 p + \cdots + a_m p^{m-2}\\ &\quad + a_3 + \cdots + a_m p^{m-3}\\ &\quad\vdots\\ &\quad + a_m.\end{aligned}$$

The difference between the two sums is to be computed,

$$\begin{aligned}\sum_{i=1}^{\infty}\left\lfloor\frac{n}{p^i}\right\rfloor (p-1) &= (a_1 p + a_2 p^2 + \cdots + a_m p^m)\\ &\quad - (a_1 + a_2 + \cdots + a_m)\\ &= (n - a_0) - (\sigma - a_0)\\ &= n - \sigma.\end{aligned}$$

This establishes Legendre's identity (8.7).

Now we derive Kummer's criterion from Legendre's identity. Thus, let the p-adic expansions of n and k be

$$\begin{aligned}n &= a_0 + a_1 p + a_2 p^2 + \cdots + a_m p^m,\\ k &= b_0 + b_1 p + b_2 p^2 + \cdots + b_m p^m.\end{aligned}$$

where $a_i, b_i \in \{0, 1, \ldots, p-1\}$. Now, if p^ν is the largest prime power of p which divides $\binom{n+k}{k}$ then

$$\nu = \mu(n+k) - \mu(n) - \mu(k),$$

since

$$\binom{n+k}{k} = \frac{(n+k)!}{n!k!}.$$

In other words, we have to show that

$$c_p(n,k) = \mu(n+k) - \mu(n) - \mu(k), \tag{8.11}$$

where $c_p(n,k)$ is the number of carries in the p-adic addition of

$$n = (a_m a_{m-1} \ldots a_1 a_0)_p \text{ and}$$
$$k = (b_m b_{m-1} \ldots b_1 b_0)_p.$$

Carrying out the addition of these two numbers in the p-adic representation produces carries $\varepsilon_0, \varepsilon_1, \varepsilon_2, \ldots$ which are either 0 or 1. Formally, they are obtained from

$$\varepsilon_0 = \left\lfloor \frac{a_0 + b_0}{p} \right\rfloor \text{ and } \varepsilon_i = \left\lfloor \frac{a_i + b_i + \varepsilon_{i-1}}{p} \right\rfloor, \quad i = 1, 2, \ldots$$

The sum of carries in Kummer's theorem is

$$c_p(n,k) = \sum_{i=0}^{\infty} \varepsilon_i.$$

Now we consider the sum $n + k$ in p-adic representation, i.e.,

$$n + k = \sum_{i=0}^{\infty} c_i p^i$$

where $c_i \in \{0, 1, \ldots, p-1\}$. If we define for convenience of notation $\varepsilon_{-1} = 0$ then we can express the p-adic digits c_i of $n + k$ in terms of those of n and k and the carries ε_i as follows:

$$c_i = a_i + b_i + \varepsilon_{i-1} - \varepsilon_i p \text{ for } i = 0, 1, 2, \ldots$$

Finally, we use Legendre's identity to show (8.11):

$$\begin{aligned}
\nu &= \mu(n+k) - \mu(n) - \mu(k) \\
&= \frac{n + k - \sum_{i=0}^{\infty} c_i}{p-1} - \frac{n - \sum_{i=0}^{\infty} a_i}{p-1} - \frac{k - \sum_{i=0}^{\infty} b_i}{p-1} \\
&= \frac{1}{p-1} \left(\sum_{i=0}^{\infty} a_i + \sum_{i=0}^{\infty} b_i - \sum_{i=0}^{\infty} (a_i + b_i + \varepsilon_{i-1} - \varepsilon_i p) \right) \\
&= \frac{1}{p-1} \sum_{i=0}^{\infty} \varepsilon_i p - \varepsilon_{i-1} \\
&= \frac{1}{p-1} \left(\sum_{i=0}^{\infty} \varepsilon_i p - \sum_{i=0}^{\infty} \varepsilon_{i-1} \right) \\
&= \frac{1}{p-1} \left(\sum_{i=0}^{\infty} \varepsilon_i p - \sum_{i=0}^{\infty} \varepsilon_i \right)
\end{aligned}$$

finishing with

$$\nu = \frac{1}{p-1} \sum_{i=0}^{\infty} \varepsilon_i (p-1) = \sum_{i=0}^{\infty} \varepsilon_i = c_p(n,k).$$

With this computation, Kummer's result is established.

Coloring Pascal's Triangle

So far we have only discussed whether a binomial coefficient is divisible by a prime number p or not, which we used for black and white coloring. However, if a coefficient is not divisible by p we could color the respective entry in Pascal's triangle with one of $p-1$ colors (depending on the modulus). Let us close this section with a short remark on the computation of the color. The crucial result which determines the color without having to run the associated linear cellular automaton is attributable to Lucas (1877). Let

$$\begin{aligned} n+k &= a_0 + a_1 p + a_2 p^2 + \cdots + a_m p^m, \\ k &= b_0 + b_1 p + b_2 p^2 + \cdots + b_m p^m, \end{aligned}$$

where $a_i, b_i \in \{0, \ldots, p-1\}$ are the p-adic digits. Then

$$\binom{n+k}{k} \equiv \binom{a_0}{b_0} \cdot \binom{a_1}{b_1} \cdots \binom{a_h}{b_h} \pmod{p} \tag{8.12}$$

Let us look at an example. According to the above criterion

$$\binom{7}{4} \equiv \binom{1}{1} \cdot \binom{2}{1} \pmod{3}$$

because $7 = (21)_3$ and $4 = (11)_3$. In fact,

$$\binom{7}{4} = 35 \equiv 2 \pmod{3}$$

and also

$$\binom{1}{1} \cdot \binom{2}{1} = 1 \cdot 2 \equiv 2 \pmod{3}.$$

Again, the criterion follows from Legendre's identity, as did Kummer's result.[23] We will, however, skip these details and turn now to the description of the global pattern formation in Pascal's triangle.

In summary, we use the mod-p condition to test a binomial coefficient for the divisibility by a prime number, and we resort to Lucas' factorization eqn. (8.12), when we want to know in addition what the value of the coefficient is in the modulo p sense.

[23]For a proof not using Legendre's identity, see also N. J. Fine: *Binomial coefficients modulo a prime number*, Amer. Math. Monthly 54 (1947) 589. Lucas' identity can also be used to analyze the global structure of the colored Pascal triangle, i.e., the color patterns which are obtained if one uses p colors, one for each modulus ($\equiv 0 \pmod{p}$, $\equiv 1 \pmod{p}$, ..., $\equiv p-1 \pmod{p}$). In fact Sved derived Lucas' result from the geometrical patterns of Pascal's triangle mod p. In other words, the fractal patterns in Pascal's triangle are equivalent to number theoretical properties of binomial coefficients, and understanding more about the fractal properties will lead to a wider understanding of these number theoretical properties.

8.3 IFS: From Local Divisibility to Global Geometry

You will recall the surprisingly short program 'Skewed Sierpinski gasket' from chapter 2. Its secret was hidden in just one BASIC statement:

```
IF (x AND y) = 0 THEN PSET (x+30, y+30)
```

The logical expression of this statement determined whether a point was drawn or not. The expression '`x AND y`' in the if-clause stands for the bitwise logical `AND` operation. For example, `101 AND 010` is 0 while `101 AND 110` is 1. In other words, the expression is equal to 0 (false) only if no two matching binary digits of `x` and `y` are both 1. Thus, this expression allows us to test for the occurrence of a carry in the binary addition of the coordinates `x` and `y`. In other words this program uses the Kummer criterion for $p = 2$ and $c_2(n,k) = 0$ setting $n =$ `x` and $k =$ `y`. Figure 8.20 gives an impression of the resulting pattern showing more and more details of the Sierpinski gasket. Let us now try to explain why this criterion really is able to generate the Sierpinski gasket.

Mod-2 Pattern and Binary Addresses

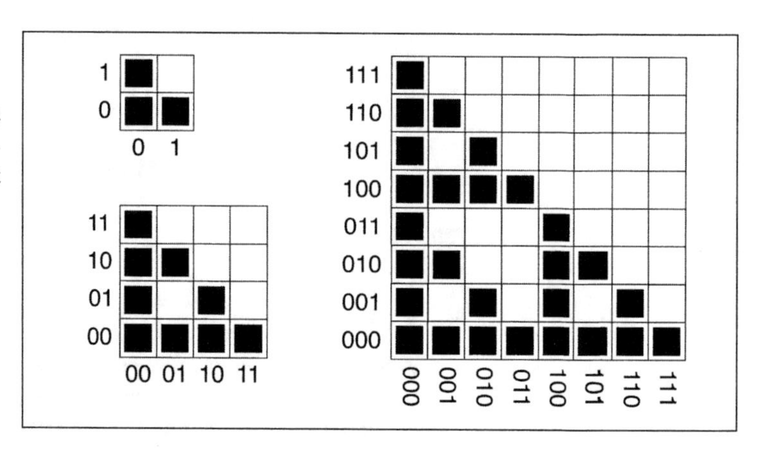

Address testing on a 2 × 2, 4 × 4 and 8 × 8 grid. More and more details of the familiar Sierpinski gasket are revealed.

Figure 8.20

Sierpinski Gasket and Base 2

Let us consider the unit square in the plane,

$$Q = \{(x,y) \mid 0 \le x \le 1, 0 \le y \le 1\}.$$

Now we expand x and y in base 2, i.e.,

$$x = \sum_{i=1}^{\infty} a_i 2^{-i}, \quad a_i \in \{0,1\},$$

$$y = \sum_{i=1}^{\infty} b_i 2^{-i}, \quad b_i \in \{0,1\}.$$

With this notation we can provide a number theoretical description of the Sierpinski gasket:

$$S = \{(x,y) \in Q \mid \text{there are binary expansions of } x \text{ and } y \text{ with } a_i + b_i \le 1, i = 1,2,\ldots\} \tag{8.13}$$

Let us look at some examples:

(x, y)	(0,0)	(1,0)	(1/2,1/2)	(3/4,3/4)
x	0.000...	0.111...	0.1000...	0.11000...
y	0.000...	0.000...	0.0111...	0.10111...
	$\in S$	$\in S$	$\in S$	$\notin S$

The first three points, (0,0), (1,0), and (1/2,1/2) are in S. In the last example we also see why we have to say 'there is an expansion...' in the characterization of S. Otherwise we would not have that $(1/2, 1/2)$ is in S. On the other hand $(3/4, 3/4)$ is not in S, no matter how we expand, because a_1 and b_1 must both be 1, i.e., $a_1 + b_1 = 2$.

Note that there is a direct relation to Kummer's carry-condition above. Indeed, saying that there is a base 2 expansion for x and y with $a_i + b_i \leq 1$, is the same as saying that in adding x and y in the binary number system there is no carry.

In section 5.4 we used iterated function systems (IFS) to convince ourselves that (8.13) indeed characterizes the Sierpinski gasket. We introduced four contractions w_{00}, w_{01}, w_{10}, and w_{11}, which contract the unit square Q as in figure 8.21 by a factor of 2:

$$\begin{aligned} w_{00}(x, y) &= (x/2, y/2) \\ w_{01}(x, y) &= (x/2, y/2 + 1/2) \\ w_{10}(x, y) &= (x/2 + 1/2, y/2) \\ w_{11}(x, y) &= (x/2 + 1/2, y/2 + 1/2). \end{aligned}$$

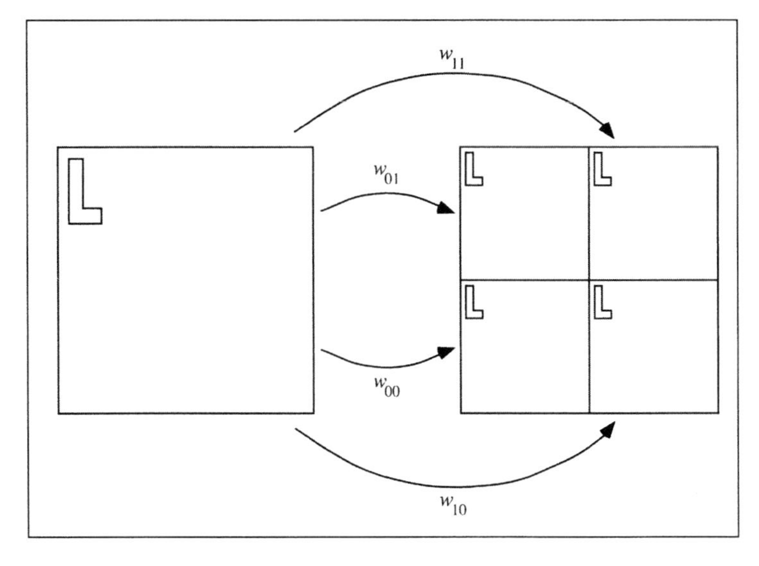

Four Similarity Transformations

Transformations of the square Q.

Figure 8.21

Three of these transformations w_{00}, w_{01}, w_{10} provide a Hutchinson equation for the Sierpinski gasket S:

$$S = w_{00}(S) \cup w_{01}(S) \cup w_{10}(S). \tag{8.14}$$

The Number Theoretical Description

We base the proof of the number theoretical description in eqn. (8.13) on the definition that the Sierpinski gasket is given by the contractions w_{00}, w_{01}, w_{10} and the corresponding Hutchinson equation (8.14). Any object (compact, nonempty set) which satisfies this equation must be the Sierpinski gasket because there is only one solution. Thus, to verify that S from (8.13) is the Sierpinski gasket, we must prove that S as in (8.13) satisfies (8.14). We proceed by showing the two relations

$$w_{00}(S) \cup w_{01}(S) \cup w_{10}(S) \subset S$$

and

$$w_{00}(S) \cup w_{01}(S) \cup w_{10}(S) \supset S.$$

For the first, take any point $(x,y) \in S$ and its binary expansion

$$(x,y) = (0.a_1a_2\ldots, 0.b_1b_2\ldots)$$

Following eqn. (8.13), $a_i + b_i \leq 1$ holds for all indices $i = 1, 2, \ldots$ Now we apply the three transformations, w_{00}, w_{01}, and w_{10},

$$\begin{aligned} w_{00}(0.a_1a_2\ldots, 0.b_1b_2\ldots) &= (0.0a_1a_2\ldots, 0.0b_1b_2\ldots) \\ w_{01}(0.a_1a_2\ldots, 0.b_1b_2\ldots) &= (0.0a_1a_2\ldots, 0.1b_1b_2\ldots) \\ w_{10}(0.a_1a_2\ldots, 0.b_1b_2\ldots) &= (0.1a_1a_2\ldots, 0.0b_1b_2\ldots) \end{aligned}$$

Clearly, all three resulting points are also in S, because the first digits of the x- and y-components of the results are never both equal to 1, and for the remaining pairs of digits the same holds because $a_i + b_i \leq 1$ for $i = 1, 2, \ldots$

To show the second relation, we again take any point $(x,y) \in S$ with binary expansion as above and have to provide another point in $(x', y') \in S$ such that one of the images $w_{00}(x', y'), w_{01}(x', y')$, or $w_{10}(x', y')$ is equal to the given point (x, y). We may choose

$$(x', y') = (0.a_2a_3\ldots, 0.b_2b_3\ldots)$$

Note that a_1 and b_1 cannot both be equal to 1. Therefore, we immediately obtain

$$(x,y) = \begin{cases} w_{00}(x', y') \text{ if } a_1 = 0 \text{ and } b_1 = 0 \text{ or} \\ w_{01}(x', y') \text{ if } a_1 = 0 \text{ and } b_1 = 1 \text{ or} \\ w_{10}(x', y') \text{ if } a_1 = 1 \text{ and } b_1 = 0 \end{cases}$$

and there are no other cases. This concludes our proof, and, thus, (8.13) characterizes the Sierpinski gasket.

The binary representation also allows us to see how the iteration of the Hutchinson operator, applied to an arbitrary point in the square Q, yields a sequence of points that get closer and closer to the Sierpinski gasket. Observe that if $(x,y) = (0.a_1a_2\ldots, 0.b_1b_2\ldots)$, with arbitrary a_i, and b_i, then applying the maps w_{00}, w_{01}, and w_{10} again and again as in an IFS, yields points with coordinates for which more and more of the leading binary decimals satisfy $a_i + b_i \leq 1$. In other words, starting with

$$A_0 = Q$$

and then running the IFS, generates the sequence

$$A_n = w_{00}(A_{n-1}) \cup w_{01}(A_{n-1}) \cup w_{10}(A_{n-1}), n = 1, 2, \ldots,$$

where the coordinates of the points of A_n satisfy $a_i + b_i \leq 1$ in the leading n binary decimals. Furthermore, the sequence will lead towards the Sierpinski gasket as an attractor, i.e.,

$$A_\infty = S.$$

The first steps are shown in figure 8.22. Now observe that this would be exactly the result of figure 8.20 if the coordinates used in that figure had been preceded by a decimal point. In this case the patterns found on the 2 × 2, 4 × 4 or 8 × 8 grid would exactly match the steps A_n of our iterated function system. But introducing a decimal point in figure 8.20 simply means that we look at rescaled versions of Pascal's triangle (i.e., scaled by 1/2, 1/4 or $1/2^n$ in general). In other words, the mod-2 pattern which we see in Pascal's triangle is exactly the pattern which we obtain when iterating the IFS which encodes the Sierpinski gasket.

Divisibility by Primes

Now we are prepared to look at the patterns obtained from the divisibility of binominal coefficients primes. Or more formally, we want to describe the global pattern formations in

$$P(p) = \left\{ (n,k) \;\middle|\; \binom{n+k}{k} \text{ not divisible by } p \right\},$$

First we construct an appropriate iterated function system. We consider the unit square Q and subdivide it into p^2 congruent squares $Q_{a,b}$ with $a, b \in \{0, \ldots, p-1\}$. Then we introduce corresponding contraction mappings

$$w_{a,b}(x,y) = \left(\frac{x+a}{p}, \frac{y+b}{p} \right),$$

where

$$w_{a,b}(Q) = Q_{a,b},\ a, b \in \{0, \ldots, p-1\}.$$

This is the generalization of what we have already done for the case $p = 2$ in figure 8.21. Now we define a set of admissible transformations by imposing the restriction

$$a + b \leq p - 1.$$

Patterns of A_n

The first three steps of the iterated function system coding the Sierpinski gasket.

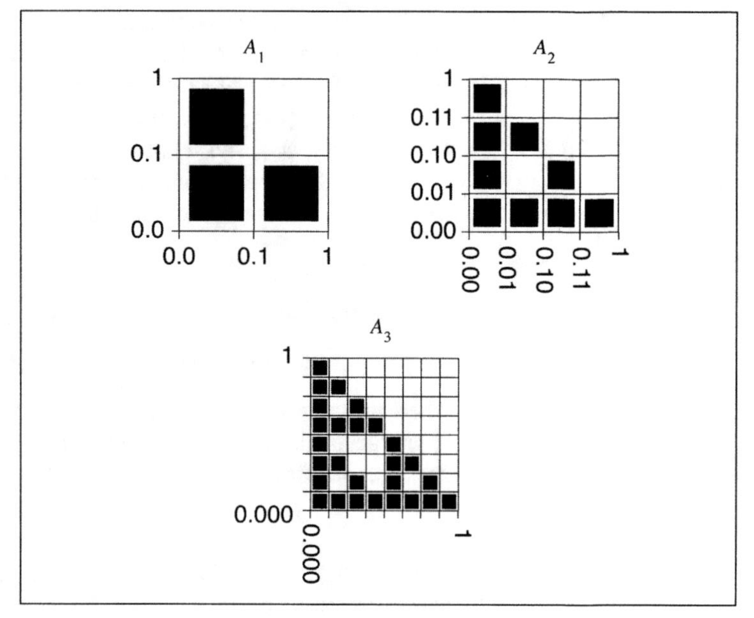

Figure 8.22

This yields a total number of $N = p(p+1)/2$ contractions, each with contraction factor $1/p$. We now introduce the Hutchinson operator W_p corresponding to these N contractions,

$$W_p(A) = \bigcup_{a+b \leq p-1} w_{a,b}(A),$$

where A is any subset of the plane. With the initial set $A_0 = Q$ we can start the iteration

$$A_m = W_p(A_{m-1}), \; m = 1, 2, \ldots$$

and figure 8.23 shows the first two steps for the choice $p = 3$.

In order to keep track of the iteration, we subsequently subdivide each of the p^2 subsquares of Q into p^2 even smaller ones, and so on repeatedly. Having indexed the first subdivision of Q by $Q_{a,b}$, we continue to label the subsquares of the second subdivision by $Q_{ac,bd}$ and so on. For the example $p = 3$ shown in figure 8.23, the square $Q_{10,12}$ are identified in the following way: the pair (1,1), made from the leading digits in the index of $Q_{10,12}$, determines the center square in the first subdivision, and the pair (0,2) determines the upper left corner square therein. In other words, the square

$$Q_{a_{m-1} \ldots a_0, b_{m-1} \ldots b_0}$$

is a square of the mth generation. We find it by reading the double p-adic addresses given by the pair $(a_{m-1} \ldots a_0, b_{m-1} \ldots b_0)$. This natural address-

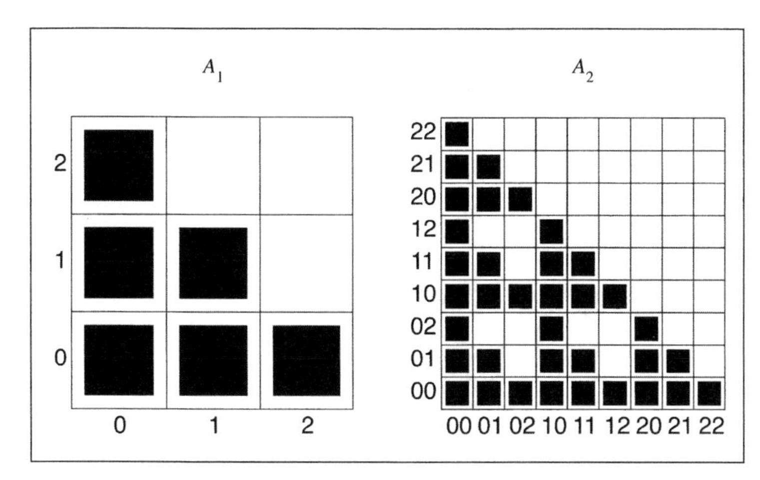

Mod-3 Machine

First two steps of the iteration of the function system W_p for $p = 3$.

Figure 8.23

ing system helps us to keep track of all the iterations of W_p, the Hutchinson operator. For example, we obtain

$$Q_{a_{m-1}\dots a_0, b_{m-1}\dots b_0} = w_{a_{m-1},b_{m-1}}(\cdots w_{a_0,b_0}(Q)),$$

where $a_i + b_i \le p - 1$.

In other words, we now can say that A_m is the collection of all those squares of the mth subdivision of Q into p^{2m} little squares, whose addresses $(a_{m-1}\dots a_0, b_{m-1}\dots b_0)$ satisfy the condition $a_i + b_i \le p - 1$, i.e.,

$$A_m = \bigcup_{a_i+b_i\le p-1} Q_{a_{m-1}\dots a_0, b_{m-1}\dots b_0}.$$

Rescaling the Pascal Triangle

Let us now relate the subsquares $Q_{a_{m-1}\dots a_0, b_{m-1}\dots b_0}$ to the entries of the Pascal triangle. First we generate a geometric model of the divisibility pattern in the Pascal triangle. To this end we equip the first quadrant of the plane with a square lattice so that each square has side length 1. Thus, each square is indexed by an integer pair (n, k) and we call it $R_{n,k}$.

$$R_{n,k} = \{(x, y) \mid n \le x \le n+1, k \le y \le k+1\}.$$

The geometrical model of $P(p)$ will now be obtained by selecting all squares $R_{n,k}$ for which p does not divide $\binom{n+k}{k}$:

$$\mathcal{P}(p) = \left\{ R_{n,k} \;\middle|\; \binom{n+k}{k} \text{ is not divisible by } p \right\}.$$

We will now relate this infinite pattern to the evolution of the Hutchinson operator, i.e., to the sequence of patterns A_m. Note that all A_m are within Q and that A_m is a union of a finite number of squares of side length $1/p^m$. To see the relation between A_m and $\mathcal{P}(p)$ we will look at $\mathcal{P}(p)$ through a sequence of square 'windows' $[0, p^m] \times [0, p^m]$ of side length p^m. Now for

The $\mathcal{P}_m(3)$ Subsquares

The squares $\mathcal{P}_1(3)$ (the six black squares with grey underlay in the lower left-hand group) and $\mathcal{P}_2(3)$ (all black squares). Compare with figure 8.23.

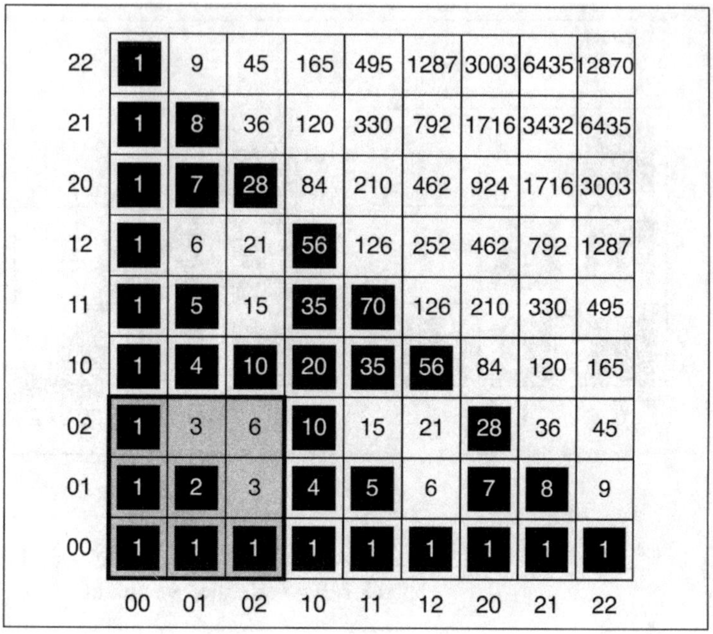

22	1	9	45	165	495	1287	3003	6435	12870
21	1	8	36	120	330	792	1716	3432	6435
20	1	7	28	84	210	462	924	1716	3003
12	1	6	21	56	126	252	462	792	1287
11	1	5	15	35	70	126	210	330	495
10	1	4	10	20	35	56	84	120	165
02	1	3	6	10	15	21	28	36	45
01	1	2	3	4	5	6	7	8	9
00	1	1	1	1	1	1	1	1	1
	00	01	02	10	11	12	20	21	22

Figure 8.24

$m = 1, 2, \ldots$ we pick that part from the geometrical model $\mathcal{P}(p)$ which falls in the corresponding window:

$$\mathcal{P}_m(p) = \mathcal{P}(p) \cap [0, p^m] \times [0, p^m].$$

Figure 8.24 displays $\mathcal{P}_1(p)$ and $\mathcal{P}_2(p)$ for $p = 3$. Comparing $\mathcal{P}_1(p)$ and $\mathcal{P}_2(p)$ with the pattern of A_1 and A_2 in figure 8.23 we observe that they are identical, though A_1 and A_2 are in the unit square and $\mathcal{P}_1(p)$ (resp. $\mathcal{P}_2(p)$) fit into a square of side length p (resp. p^2). In other words, if we rescale the patterns $\mathcal{P}_m(p)$ by a factor of $1/p^m$ we obtain an object which we want to show is identical with A_m. To this end we introduce

$$\mathcal{S}_m(p) = \frac{1}{p^m} \cdot \mathcal{P}_m(p)$$

or more explicitly

$$\mathcal{S}_m(p) = \left\{ \frac{z}{p^m} \,\middle|\, z \in \mathcal{P}_m(p) \right\}.$$

Indeed, each subsquare in $\mathcal{S}_m(p)$ is indexed by an integer pair (n, k) such that p does not divide $\binom{n+k}{k}$. In other words each such subsquare is identical with a $Q_{a_{m-1}\ldots a_0, b_{m-1}\ldots b_0}$, where $n = (a_{m-1}\ldots a_0)_p$ and $k = (b_{m-1}\ldots b_0)_p$ according to Kummer's mod-p condition. Summarizing we have that

$$A_m = \mathcal{S}_m(p), \quad m = 1, 2, \ldots$$

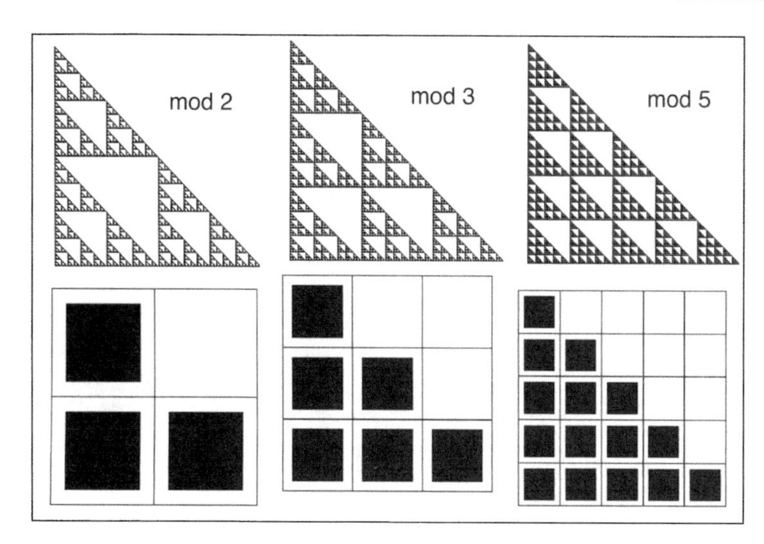

The First Primes

The limit sets of the rescaled geometric models of $P(2), P(3), P(5)$, and their associated IFSs (i.e., a graphical representation of the transformations w_{ab}).

Figure 8.25

As we let m go to infinity, we know that A_m will converge towards the attractor of the IFS, and consequently the rescaled geometric models $\mathcal{S}_m(p)$ will also converge to the attractor of the IFS.[24] We denote the limit set by $\mathcal{S}(p)$. In this manner we have just seen that the rescaled geometric models have a limit set. It represents a rescaled geometric model of the (infinite) Pascal triangle modulo p which we denoted by $P(p)$.[25]

Fractal Dimension

Figure 8.25 shows the resulting geometric models $\mathcal{S}(p)$ when running the IFSs corresponding to $P(p)$ for $p = 2, 3, 5$. The approach by iterated function systems allows us to compute the fractal dimensions of these sets:

	Dimension
$\mathcal{S}(2)$	$\log 3 / \log 2 \approx 1.585$
$\mathcal{S}(3)$	$\log 6 / \log 3 \approx 1.631$
$\mathcal{S}(5)$	$\log 15 / \log 5 \approx 1.683$
$\mathcal{S}(7)$	$\log 28 / \log 7 \approx 1.712$

If p is prime then the formula for the self-similarity dimension of $\mathcal{S}(p)$ is

$$D_s = \frac{\log p(p+1)/2}{\log p}.$$

Can you explain why? Just note, that these sets are strictly self-similar and recall the definition of the self-similarity dimension from chapter 4.

[24]Convergence is with respect to the Hausdorff metric.

[25]In this regard we also refer to S. J. Willson, *Cellular automata can generate fractals*, Discrete Applied Math. 8 (1984) 91–99 who studied limit sets of linear cellular automata via rescaling techniques.

8.4 HIFS and Divisibility by Prime Powers

We have seen that iterated function systems are in some sense the natural framework in which to decipher the global pattern formation obtained by the divisibility properties in Pascal's triangle (or of pattern formations in linear cellular automata). We have, however, only taken the first step, namely, with respect to the divisibility by a prime p. Our next step, considering divisibility by prime powers p^τ, is rather long compared with the first one. We will describe how in this case the global patterns arising in Pascal's triangle[26] can be completely understood through hierarchical IFSs.[27] The fractal patterns in

$$P(p^\tau) = \left\{(n,k) \;\middle|\; \binom{n+k}{k} \text{ is not divisible by } p^\tau\right\}$$

can be deciphered by a hierarchical IFS whose design and properties are in very close correspondence with Kummer's criterion for finding the largest prime power which divides $\binom{n+k}{k}$. Figure 8.26 shows two examples where we observe that the straightforward self-similarity properties of the sets $P(p)$ have been replaced by hierarchies of self-similarity features.

First Prime Powers

Rescaled geometric models of $P(4)$ and $P(8)$.

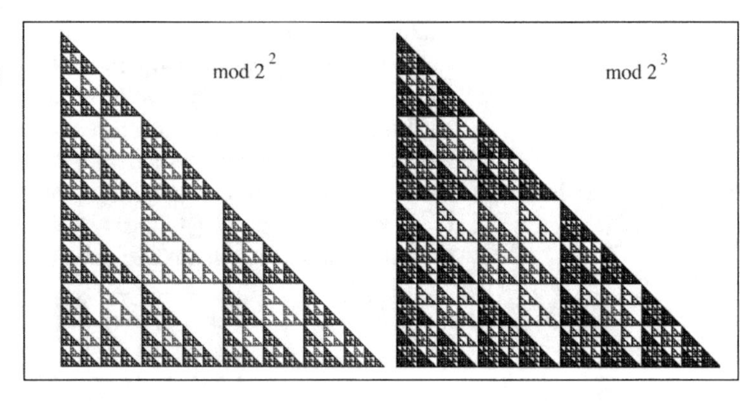

Figure 8.26

Let us start with some observations concerning the example $P(4)$. In this case Kummer's criterion implies:

$$P(4) = \{(n,k) \mid c_2(n,k) = 0 \text{ or } c_2(n,k) = 1\}.$$

If $c_2(n,k) = 0$, then $\binom{n+k}{k}$ is not divisible by 2. If $c_2(n,k) = 1$, then $\binom{n+k}{k}$ is divisible by 2, but not by 4. Thus, if either one of the conditions is satisfied, then $\binom{n+k}{k}$ is not divisible by 4. Therefore, we also have to take into account those coordinates (n,k) whose p-adic addition have exactly one carry (i.e., where $\binom{n+k}{k}$ is divisible by 2 but not by 4). How can we reflect this property in an iterated function system?

[26] Sketching some recent work from F. v. Haeseler, H.-O. Peitgen, G. Skordev, *Pascal's triangle, dynamical systems and attractors*, Ergod. Th. & Dynam. Sys. 12 (1992) 479–486.
[27] See section 5.9.

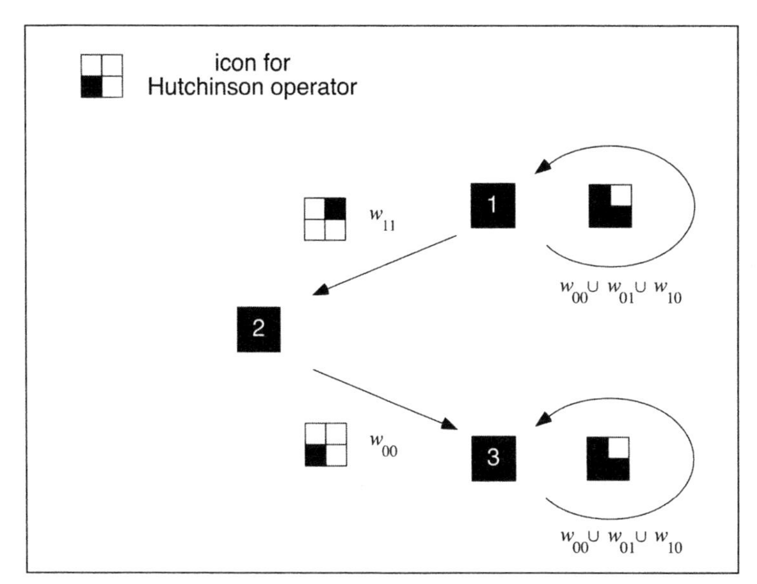

HIFS for Mod-4

The graph of a hierarchical iterated function system for the mod-4 example. Iterating this system generates three images, one for each node (the nodes representing three networked MRCMs). The final image of node 1 is a Sierpinski gasket. The final image of node 3 is the desired mod-4 pattern.

Figure 8.27

Again we look at the unit square Q and the contractions $w_{00}, \ldots, w_{11}$ of figure 8.21.

We have seen that m iterations of these transformations, first applying $w_{a_0 b_0}$ then $w_{a_1 b_1}$, etc. lead to the subsquare

$$w_{a_{m-1}b_{m-1}}(\;\cdots\; w_{a_0 b_0}(Q)) \;=\; Q_{a_{m-1}\ldots a_0, b_{m-1}\ldots b_0}.$$

So far we have only considered the case where the binary addition of $(a_{m-1}\ldots a_0)_2$ and $(b_{m-1}\ldots b_0)_2$ had no carry (i.e., $a_i + b_i \le 1$). Now, how can we achieve exactly one carry? At first glance it appears that all we have to use is the transformation w_{11}. For example, applying the sequence $w_{01}, w_{10}, w_{11}, w_{01}, w_{00}$ would lead to

$$Q_{00110,01101} = w_{00}(w_{01}(w_{11}(w_{10}(w_{01}(Q))))).$$

And indeed in this example the addition would provide a carry in the third binary decimal. But wait; there is also a carry in the fourth binary decimal counted from the right! What is wrong? Obviously we have to be a bit more careful. The transformation w_{11} provides a carry — so far okay — but the transformation which follows has to be w_{00}. Otherwise we would obtain another carry. Therefore, having w_{11}, followed by w_{01}, as in our example, is not allowed.

Figure 8.27 shows the graph of a hierarchical iterated function system which reflects our observations. The nodes 1, 2 and 3 represent three networked MRCMs. The first one operates in a feedback loop applying the Hutchinson operator $w_{00} \cup w_{01} \cup w_{10}$. The second one transforms the output of the first one using w_{11}. The third machine again operates in a feedback loop with $w_{00} \cup w_{01} \cup w_{10}$, but additionally it merges the output of the second machine, transformed by w_{00}.

How do we iterate this network? We start the iteration with three copies of the unit square Q, one for each node, $A_0(1) = Q$, $A_0(2) = Q$, and $A_0(3) = Q$. The first step provides for these nodes:

$$\begin{aligned}
A_1(1) &= w_{00}(Q) \cup w_{01}(Q) \cup w_{10}(Q) \\
&= Q_{0,0} \cup Q_{0,1} \cup Q_{1,0} \\
A_1(2) &= w_{11}(Q) \\
&= Q_{1,1} \\
A_1(3) &= w_{00}(Q) \cup w_{00}(Q) \cup w_{01}(Q) \cup w_{10}(Q) \\
&= Q_{0,0} \cup Q_{0,1} \cup Q_{1,0}
\end{aligned}$$

and for the second step we obtain

$$\begin{aligned}
A_2(1) &= \bigcup_{a+b\leq 1} w_{a,b}(Q_{0,0} \cup Q_{0,1} \cup Q_{1,0}) \\
&= Q_{00,00} \cup Q_{00,01} \cup Q_{01,00} \\
&\quad \cup Q_{00,10} \cup Q_{00,11} \cup Q_{01,10} \\
&\quad \cup Q_{10,00} \cup Q_{10,01} \cup Q_{11,00} \\
A_2(2) &= w_{1,1}(Q_{0,0} \cup Q_{0,1} \cup Q_{1,0}) \\
&= Q_{10,10} \cup Q_{10,11} \cup Q_{11,10} \\
A_2(3) &= \bigcup_{a+b\leq 1} w_{a,b}(Q_{0,0} \cup Q_{0,1} \cup Q_{1,0}) \cup w_{0,0}(Q_{1,1}) \\
&= Q_{00,00} \cup Q_{00,01} \cup Q_{01,00} \\
&\quad \cup Q_{00,10} \cup Q_{00,11} \cup Q_{01,10} \\
&\quad \cup Q_{10,00} \cup Q_{10,01} \cup Q_{11,00} \cup Q_{01,01}.
\end{aligned}$$

These steps and the next are visualized in figure 8.28. In step m we obtain in node 1 all subsquares $Q_{a_{m-1}\cdots a_0, b_{m-1}\cdots b_0}$ whose indices produce no carry, which yields the rescaled geometric model $S_m(2)$. For node 2 we obtain subsquares with the carry produced by the leading binary decimal of its indices, and for node 3 we obtain subsquares whose indices provide no carry or just one. In other words, $A_m(3)$ provides the desired geometric model of $P(4)$ after rescaling by $1/2^m$ (see also figure 8.30).

Construction of a Hierarchical IFS

Let us now build a hierarchical iterated function system for general prime powers p^τ. As before, we consider contractions with contraction factor $1/p$

$$w_{a,b} : Q \to Q, \text{ where } w_{a,b}(Q) = Q_{a,b} \tag{8.15}$$

and $w_{a,b}(x, y) = ((x + a)/p, (y + b)/p)$ for $a, b \in \{0, \ldots, p-1\}$. Again the unit square Q is subdivided into p^2 congruent squares $Q_{a,b}$, which are indexed by the p-adic pair (a, b).

Our hierarchical IFS will have $2\tau - 1$ nodes (i.e., there will be $2\tau - 1$ individual IFSs which are networked with each other). Each IFS will use mappings from definition 8.15, which operate on a unit square Q. In order to distinguish them, we designate them $Q^1, \ldots, Q^{2\tau-1}$.

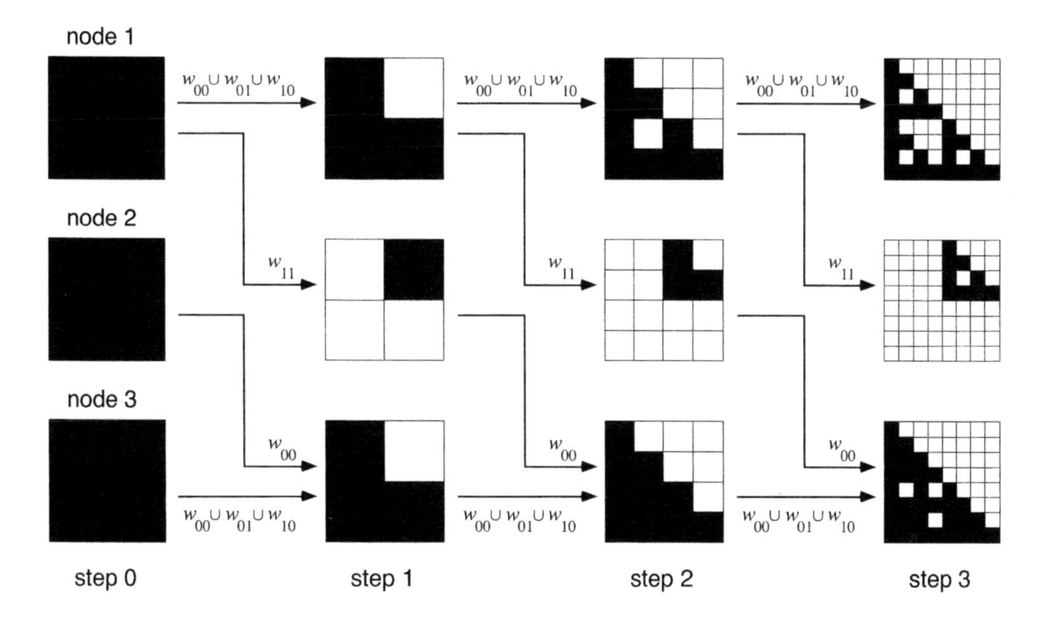

Figure 8.28 : The first three steps of the networked MRCM for $P(4)$ shown in figure 8.27.

In principle, a hierarchical IFS can have mappings from any Q^i to any other Q^j. However, our particular hierarchical IFS will only have particular connections. They are the systematical extension of what we already have seen for $P(4)$. Figure 8.29 shows the resulting network. The black squares are numerated from 1 to $2\tau - 1$ and represent the individual nodes of our HIFS. The arrows between squares specify Hutchinson mappings. More precisely, we have four types of such mappings. They are distinguished by an index to the letter W, which determines a selection of the contraction mappings $w_{a,b}$. More precisely, let B be any (compact) subset of the plane and $a, b \in \{0, \dots, p-1\}$. Then:

$$\begin{aligned} W_{\leq p-1}(B) &= \cup_{a+b \leq p-1} w_{a,b}(B), \\ W_{\leq p-2}(B) &= \cup_{a+b \leq p-2} w_{a,b}(B), \\ W_{\geq p-1}(B) &= \cup_{a+b \geq p-1} w_{a,b}(B), \\ W_{\geq p}(B) &= \cup_{a+b \geq p} w_{a,b}(B). \end{aligned}$$

In our hierarchical IFS we have

$$\begin{aligned} W_{\leq p-1} &: Q^{2i-1} \to Q^{2i-1}, i = 1, \dots, \tau, \\ W_{\geq p} &: Q^{2i-1} \to Q^{2i}, \quad i = 1, \dots, \tau - 1, \\ W_{\geq p-1} &: Q^{2i} \quad \to Q^{2i+2}, i = 1, \dots, \tau - 2, \\ W_{\leq p-2} &: Q^{2i} \quad \to Q^{2i+1}, i = 1, \dots, \tau - 1. \end{aligned} \tag{8.16}$$

Now we run this IFS. Starting with a unit square for each node, we see an evolution of patterns in each node. In fact, as we run the machine sufficiently long, the evolution of these patterns will begin to stagnate, i.e., run into a

Tower Machine

Tower of IFSs and their network channels.

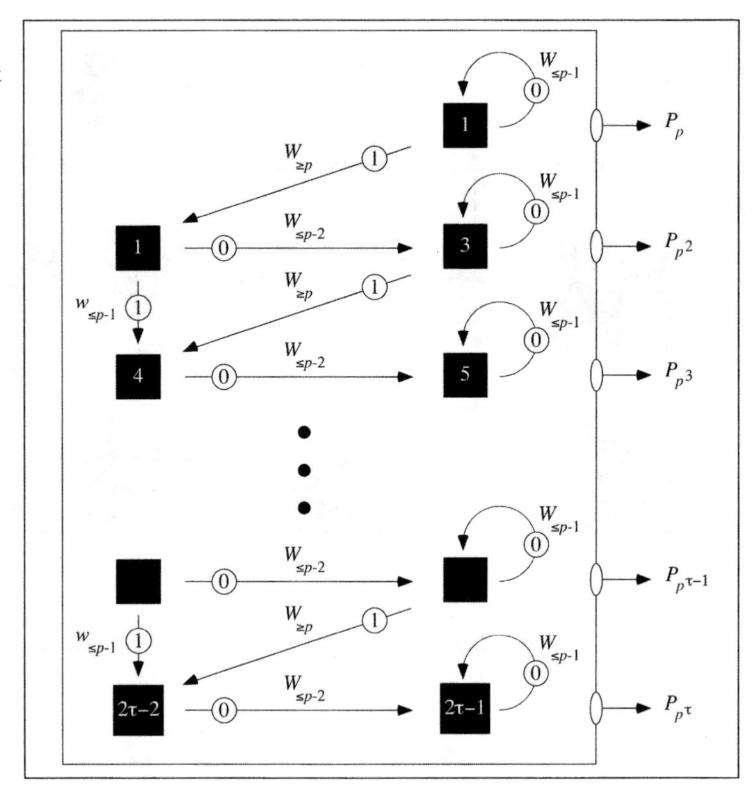

Figure 8.29

limit. Imagine that what we have termed the tower machine in figure 8.29 has viewing windows through which we can watch this evolution. Our particular interest would be to monitor the evolution in the nodes shown on the right-hand side (i.e., the nodes corresponding to $Q^{2k-1}, k = 1, \ldots, \tau$). The result is that we see exactly the patterns which are given by $P(p^k)$: the pattern in node $2k - 1$ corresponds to the pattern in $P(p^k)$.

Deciphering the Hierarchy of Self-Similarity Features

In particular the global pattern of $P(p^\tau)$ can be found in $Q^{2\tau-1}$. But also note that the network in figure 8.29 shows how we have to 'mix together' all the $P(p^k), 1 \le k \le \tau - 1$ to obtain $P(p^\tau)$. In other words, the hierarchical IFS not only allows the generation of the pattern of $P(p^\tau)$. Even more importantly, it exactly deciphers the hierarchy of the self-similarity features in $P(p^\tau)$.

Our result once again shows very strongly that hierarchical IFSs are not just there to make pretty pictures. They are deeply rooted in pure mathematics. They appear here as entirely natural for the explanation of the discussed geometrical patterns in the Pascal triangle.

Running the Tower IFS

Let us spend a little more effort to explain how the tower machine in figure 8.29 actually runs. We construct a $(2\tau - 1) \times (2\tau - 1)$ matrix F_{p^τ}. The entries in this matrix are two kinds of symbols, the empty symbol $\emptyset$, or one of the Hutchinson mappings from 8.16. Let us denote the general element of F_{p^τ} by f_{ij}, i.e., we think of this element as a mapping from Q^j to Q^i. Thus, if according to figure 8.29 there is no mapping, we put the empty symbol:

$$F_{p^\tau} = (f_{i,j}),\ i,j = 1,\ldots,2\tau - 1.$$

Then

$$\begin{array}{rcll} f_{ll} &=& W_{\leq p-1}, & \text{whenever } l \text{ is odd,} \\ f_{l+1,l} &=& W_{\geq p}, & \text{whenever } l \text{ is odd,} \\ f_{l+1,l} &=& W_{\leq p-2}, & \text{whenever } l \text{ is even,} \\ f_{l+2,l} &=& W_{\geq p-1}, & \text{whenever } l \text{ is even,} \\ f_{l+1,l} &=& W_{\geq p}, & \text{whenever } l \text{ is odd,} \\ f_{kl} &=& \emptyset, & \text{otherwise.} \end{array} \tag{8.17}$$

Let us look at the example of $\tau = 3$. Here we have a 5×5 matrix:

$$\begin{pmatrix} W_{\leq p-1} & \emptyset & \emptyset & \emptyset & \emptyset \\ W_{\leq p} & \emptyset & \emptyset & \emptyset & \emptyset \\ \emptyset & W_{\geq p-2} & W_{\leq p-1} & \emptyset & \emptyset \\ \emptyset & W_{\geq p-1} & W_{\geq p} & \emptyset & \emptyset \\ \emptyset & \emptyset & \emptyset & W_{\leq p-2} & W_{\leq p-1} \end{pmatrix}$$

Now we let this matrix F_{p^τ} operate on a stack of subsets of the plane, say $B = (B_1, \ldots, B_{2\tau-1})$, and let $C = F_{p^\tau}(B)$. i.e., if $C = (C_1, \ldots, C_{2\tau-1})$, then

$$C_k = \bigcup_{l=1,..,2\tau-1} f_{kl}(B_l)$$

where $f_{kl}(B_l) = \emptyset$, provided $f_{kl} = \emptyset$.

Now let $A_0 = (Q^1, \ldots, Q^{2\tau-1})$ and $A_{m+1} = F_{p^\tau}(A_m), m = 0, 1, 2, \ldots$ We can analyze the content of component $2s - 1, s = 1, \ldots, \tau$, in the stack of objects A_m for each m and compare it with a rescaled geometrical model of a colored part of the Pascal triangle, which we call $\mathcal{S}_m(p^s)$, as in the analysis of divisibility by p. More precisely we introduce unit squares at a lattice point (n, k) in $\mathbf{R}^2$

$$R_{n,k} = \left\{(x, y) \in \mathbf{R}^2 | x = n + u, y = k + v, (u, v) \in Q\right\}$$

representing a black cell in the colored version of $\mathcal{P}_m(p^s)$, i.e., an entry (n, k) in $\mathcal{P}_m(p^s)$ for which $\binom{n+k}{k}$ is not divisible by p^s. If we rescale $R_{n,k}$ by $1/p^m$ we obtain a square (in the unit square Q) which has width $1/p^m$, i.e., a square which can be identified with one of the squares $Q_{a_{m-1}\ldots a_0, b_{m-1}\ldots b_0}$, where $n = (a_{m-1} \ldots a_0)_p$ and $k =$

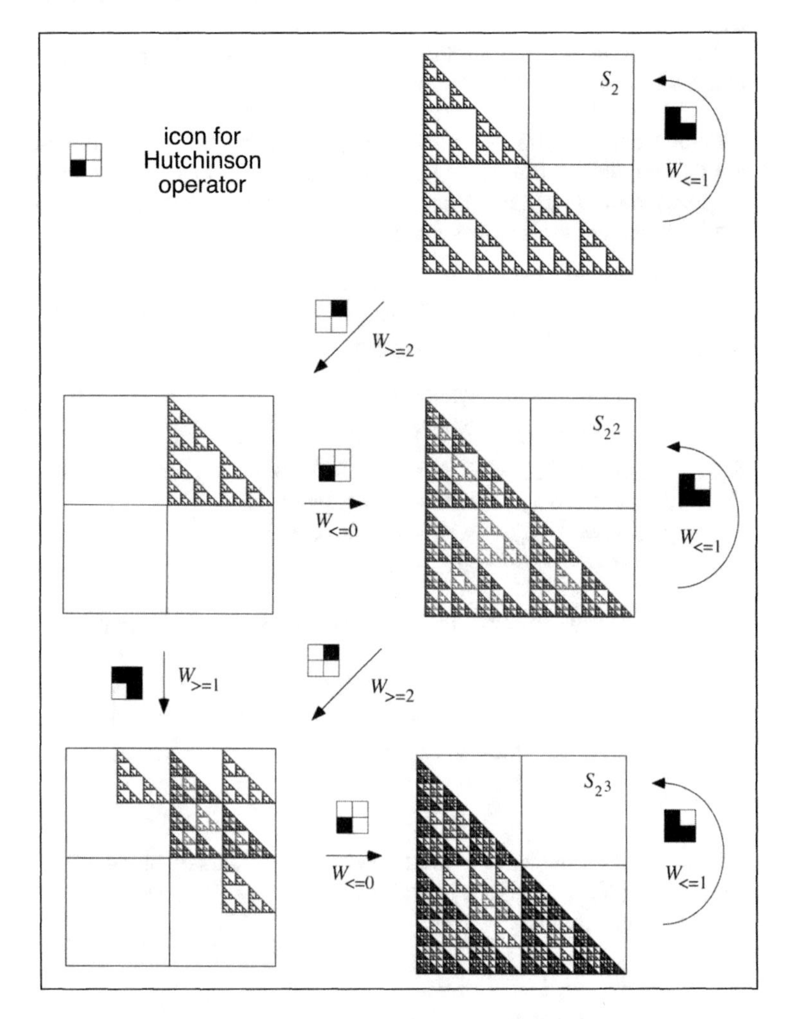

Figure 8.30 : The top node of this IFS network generates the Sierpinski gasket S_2. The upper three nodes form the part for S_4 (compare with figures 8.27 and 8.28). The whole network generates S_8.

$(b_{m-1} \ldots b_0)_p$. We will write alternatively $Q_{n,k}$. Then we introduce the rescaled geometrical model for the colored version of $\mathcal{P}_m(p^s)$:

$$\mathcal{S}_m(p^s) = \bigcup_{n,k} Q_{n,k}, \text{ where } 0 \le n+k \le p^m - 1 \text{ and } \binom{n+k}{k} \text{ not divisible by } p^s.$$

The collection of little squares in $\mathcal{S}_m(p^s)$ will be identical with the $(2s-1)^{\text{th}}$ component of A_m. Figure 8.30 shows the results in the hierarchical IFS for $p = 2$ and $\tau = 3$. The layout is adopted from that in figure 8.29.

To complete the argument we use Kummer's criterion, according to which $\binom{n+k}{k}$ is not divisible by p^s if $s > c_p(n,k)$, where $c_p(n,k)$ is

Plate 15: 3-dimensional cross section of a Julia set in 4-dimensional quarternion space, © R. Lichtenberger.

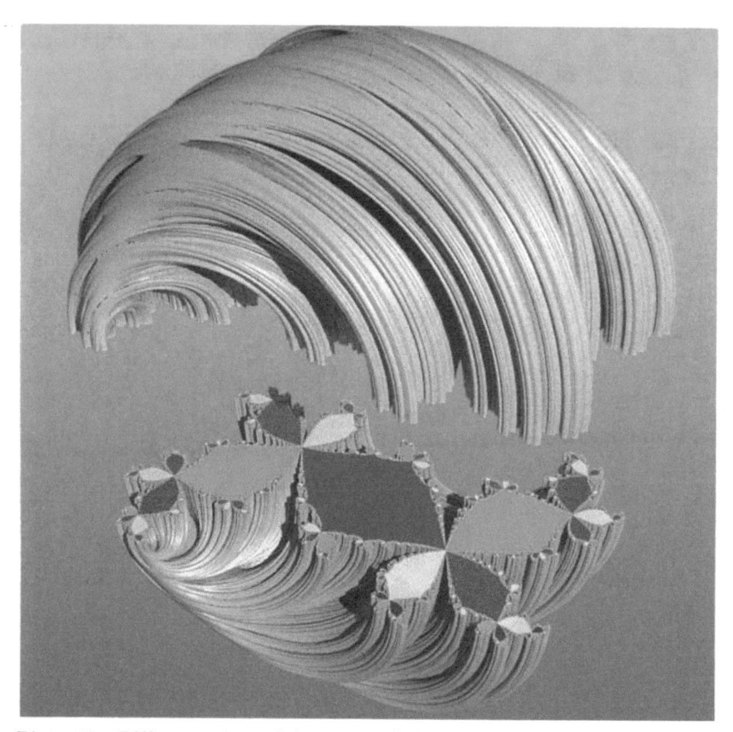

Plate 16: Different view of the same Julia set with cut open 2-dimensional cross section revealing the corresponding Julia set in the complex plane, © R. Lichtenberger.

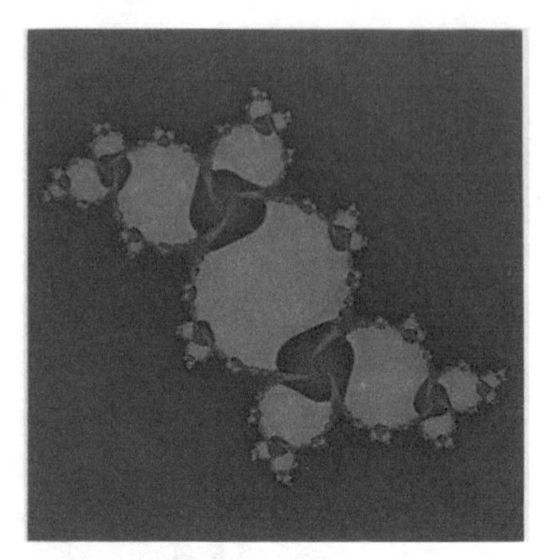

Plate 17: Julia set of the quadratic family for $c=-11+0.67i$. This is close to a parabolic situation.

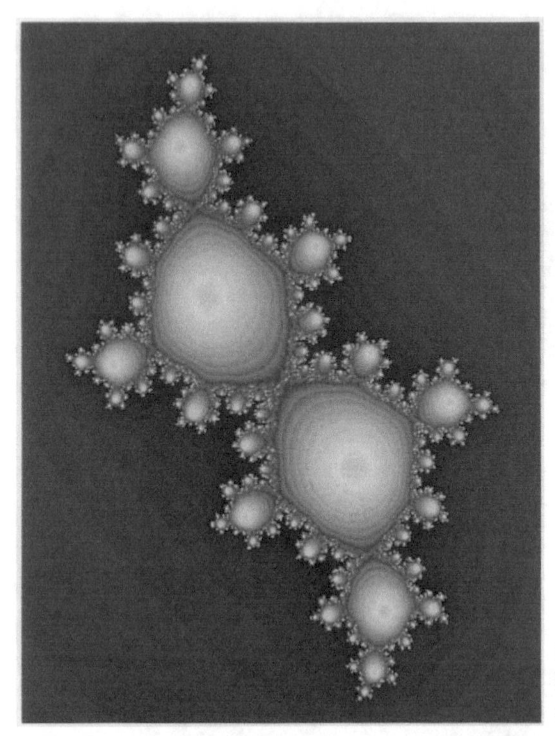

Plate 18: Julia set of the quadratic family x^2+c. For $c=-0.39054-0.58679i$ a Siegel disk is obtained.

Plate 19: 3D-rendering of the potential of a connected Julia set.

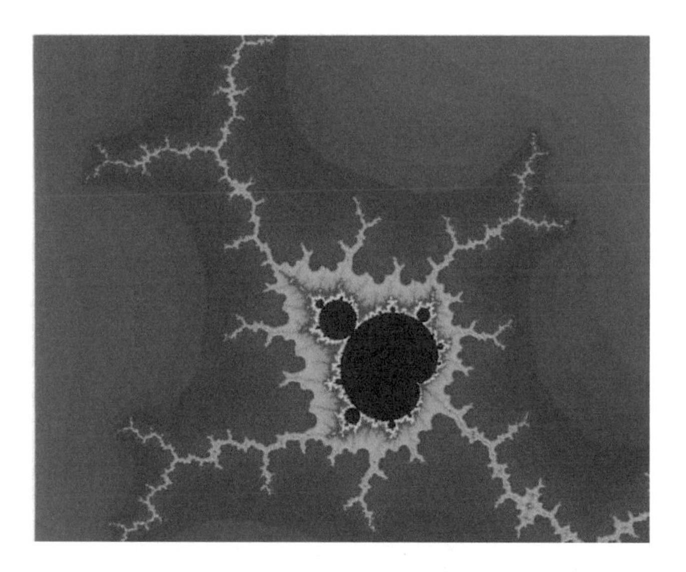

Plate 20: Four different renderings of one detail of the Mandelbrot set. The coloring of the first image (top) is computed by the escape-time-method and corresponds to equipotential lines. The 3D-rendering (middle) shows the potential of the Mandelbrot set. This image is the cover of the book *The Beauty of Fractals.* The distance-estimator rendering (bottom left) uses colors to represent the distance to the Mandelbrot set while the 3D-rendering (bottom right) shows height corresponding to distance.

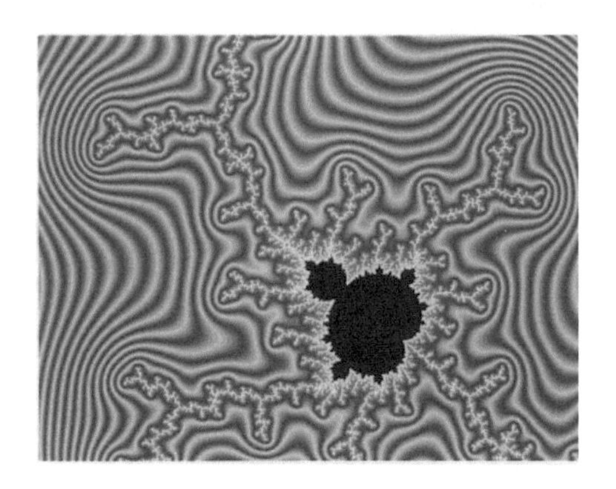

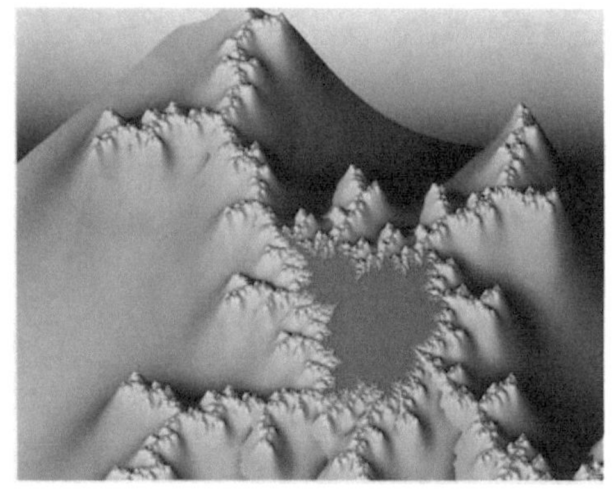

Plate 21: High resolution image of the potential of a piece of the Mandelbrot.

Plate 22: Natural ice formation on Mount Kilimanjaro, © John Reader.

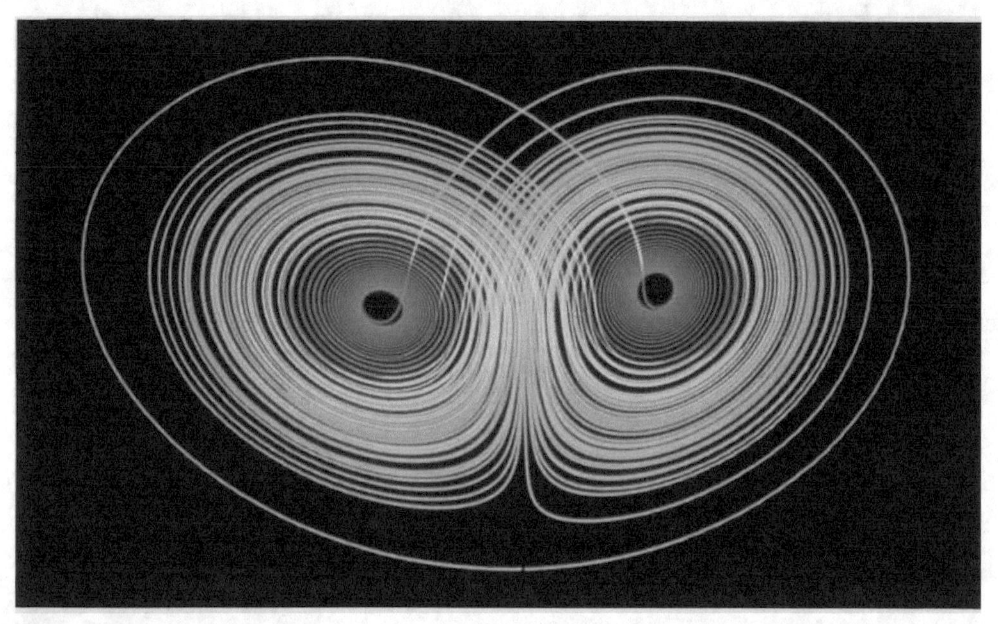

Plate 23: Two trajectories on the Lorenz attractor with color indicating distance to unstable steady states.

Plate 24: 3D-rendering of a piece of the potential of the Mandelbrot set with random fractal clouds, cover image of *The Science of Fractal Images,* Springer-Verlag, New York, 1988.

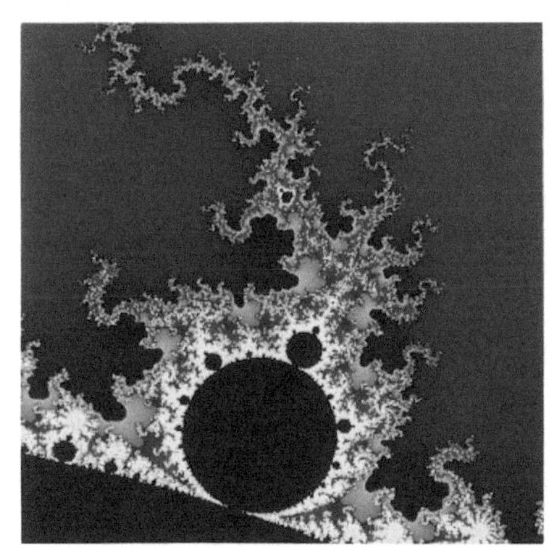

Plate 25: Original enlargement of the Mandelbrot set used for the rendering in plate 24, cover of *Scientific American,* August 1985.

Plate 26: Variation of the rendering in plate 24.

Plate 27: The pendulum experiment from section 12.8. The basins of attraction of the three magnets are colored red, blue, and yellow.

Plate 28: Detail of plate 27 showing the intertwined structure of the three basins.

the number of carries in the p-adic addition of n and k. Observe that each arrow in figure 8.29 is marked by a 0 or 1 which represents the carry.

Now we can complete the comparison of the $(2s-1)^{\text{th}}$ component of A_m in the iteration of the hierarchical IFS and the rescaled part of the Pascal triangle $\mathcal{S}_m(p^s)$. Any of the little squares in A_m is of the form $Q_{a_{m-1}\ldots a_0, b_{m-1}\ldots b_0}$, where the pair $(a_{m-1}\ldots a_0, b_{m-1}\ldots b_0)$ satisfies $c_p(a_{m-1}\ldots a_0, b_{m-1}\ldots b_0) \leq s-1$ by the construction of the hierarchical IFS (see figure 8.29). Indeed, the mappings of type $W_{\leq p-1}$ and $W_{\leq p-2}$ will produce no carry, while a mapping of type $W_{\geq p}$ will obviously produce a carry. But also a mapping of type $W_{\geq p-1}$ in figure 8.29 will produce a carry because it will always be preceded by a mapping of type $W_{\geq p}$. Finally, we note that

$$\begin{aligned} n &= a_0 + a_1 p + \cdots + a_{m-1}p^{m-1}, \text{ and} \\ k &= b_0 + b_1 p + \cdots + b_{m-1}p^{m-1}. \end{aligned}$$

This shows that the two patterns in the $(2s-1)^{\text{th}}$ component of A_m and the rescaled model of $\mathcal{S}_m(p^s)$ agree. As a consequence of the above observations, we also obtain that the rescaled geometric models have a limit set $\mathcal{S}(p^s)$ as $m \to \infty$.

Let us finally remark that there is an illuminating formula for the fractal dimension of these objects appearing in the components of a limit set of a hierarchical IFS.[28] Applying this formula to our example we obtain that the Hausdorff dimension of $\mathcal{S}(p^\tau)$ is equal to the Hausdorff dimension of $\mathcal{S}(p)$, i.e., it is independent of τ. This result has been obtained in a different way by John M. Holte by exploiting Kummer's result.[29] Intuitively the independence of the Hausdorff dimension of $\mathcal{S}(p^\tau)$ from τ is suggested from the images of $\mathcal{S}(p^\tau)$, where we can observe that the patterns in $P(p^\tau)$ are in some hierarchical fashion just mixtures of the patterns for $P(p)$.

[28] R. Mauldin and S. C. Williams, *Hausdorff dimension in graph directed constructions*, Trans. Amer. Math. Soc. 309 (1988) 811–829.

[29] See J. Holte, *A recurrence-relation approach to fractal dimension in Pascal's triangle*, International Congress of Mathematicians, Kyoto, Aug. 1990.

8.5 Catalytic Converters, or How Many Cells Are Black?

Pascal's triangle has been in existence for many centuries and has inspired beautiful investigations. We have seen the first step of how it has laid the foundation for the understanding of pattern formation for linear cellular automata in one dimension.[30] But it has also recently sparked the investigation of a problem which at first glance seems to have no relation to the triangle at all.

Assume we were to play darts with a large Pascal triangle as a target. What are the probabilities that we would hit a black cell or a white one, or more precisely, an odd or an even number, or a number which is divisible by 3 or one which is not, or a number which is divisible by p^τ or one which is not? Our discussion of the global patterns in the Pascal triangle makes it possible to answer such questions. We just have to evaluate the corresponding areas in the structures corresponding to the rescaled geometric models of $P(p^\tau)$. Depending on the parameters, this may turn out to be a rather technical computation.

Number of Black Cells in a Row

Let us look at a related question which allows a more immediate answer. We again use the original coordinate system for the Pascal triangle (see the left option of figure 8.17). How many black cells are there in the rth row? In other words, how many of the numbers which appear in the rth row are not divisible by 2, or 3, or 5, or any other prime number p? There is a remarkably direct procedure to arrive at the result. First, take r and expand it with respect to base p,

$$r = c_0 + c_1 p + c_2 p^2 + \cdots + c_m p^m, \quad c_i \in \{0, \ldots, p-1\}.$$

Now let $h_p(r)$ be defined by

$$h_p(r) = \prod_{i=0}^{m} (c_i + 1) = (c_0 + 1) \cdot (c_1 + 1) \cdots (c_m + 1). \tag{8.18}$$

Then $h_p(r)$ is the number of entries in the rth row of the Pascal triangle which are not divisible by p.

Determining the Count h_p

Let us give an argument using again the modified coordinate system as in figure 8.17 (right). Here the rth row is characterized by $n+k=r$. Thus, we are asking for the cardinality of

$$\left\{ (n,k) \;\middle|\; n+k = r \text{ and } \binom{n+k}{k} \text{ is not divisible by } p \right\}.$$

Consider the p-adic representations of n, k, and r,

$$\begin{aligned} n &= a_m p^m + \cdots + a_1 p + a_0, & a_i &\in \{0, \ldots, p-1\}, \\ k &= b_m p^m + \cdots + b_1 p + b_0, & b_i &\in \{0, \ldots, p-1\}, \\ r &= c_m p^m + \cdots + c_1 p + c_0, & c_i &\in \{0, \ldots, p-1\}. \end{aligned}$$

[30] For a discussion of higher dimensions see F. v. Haeseler, H.-O. Peitgen, G. Skordev, *On the hierarchical and global structure of cellular automata and attractors of dynamical systems*, to appear.

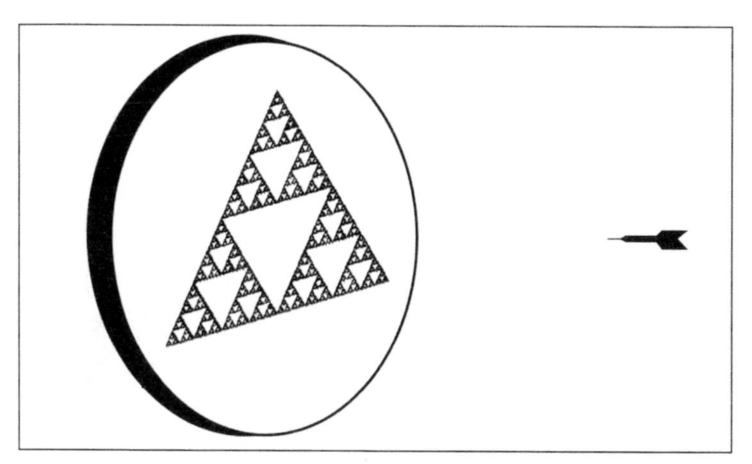

Pascal Dart

What is the probability of hitting a point of the mod-2 pattern in the Pascal triangle?

Figure 8.31

According to Kummer's mod-p criterion, $\binom{n+k}{k}$ is not divisible by p if and only if $a_i + b_i \le p-1$ for all $i = 0, \ldots, m$. In this case there is no carry in the p-adic addition of n and k. Thus, the coefficients of the sum $r = n + k$ must satisfy

$$c_i = a_i + b_i, \text{ where } i = 0, \ldots, m.$$

How many choices of n are there such that this condition is satisfied? For the ith p-adic digit there are $c_i + 1$ such choices, namely, $a_1 = 0, \ldots, c_i$. Thus, the total number of possible choices is the product

$$h_p(n) = \prod_{i=0}^{m} (c_i + 1).$$

Exactly that many entries (n, k) in row r have the property that $\binom{n+k}{k}$ is not divisible by p.

Connection to the Invariant Measure

Figure 8.32 shows $h_p(r)$ for $p = 2$ as a function of r. You might recognize this graph. It seems that we already obtained the same function in our discussion of the invariant measure for the chaos game (see figure 6.25). Let us give a short argument for this coincidence. Observe that for $p = 2$ the count of black cells in row $r + 2^m$ (with $r < 2^m$) is $h_p(r + 2^m) = 2h_p(r)$ (compare equation 8.18, where $(c_m + 1) = 2$). Thus, it is twice as large as the count in row r. In other words, if we look at the first 2^{m+1} rows of Pascal's triangle with mod-2 coloring, 1/3 of the black cells fall into the first 2^m rows and 2/3 fall into the second 2^m rows, and this is true for all $m = 1, 2, \ldots$ This observation links the count $h_p(r)$ to the construction of the invariant measure in figure 6.25: the density plot h_k corresponds exactly to the count $h_p(r)$, for $0 < r < 2^k$. A detailed discussion of this measure (and the various connections to the chaos game and Pascal's triangle) are carried out in the appendix on multifractal measures.

Number of Entries

The number of entries in the rth row of the Pascal triangle.

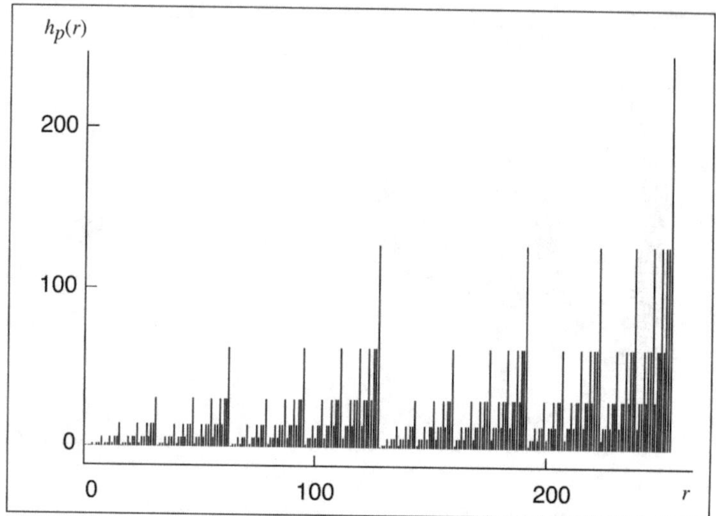

Figure 8.32

Reaction Rate Measurement

Chemical reaction rate in a catalytic oxidation process.

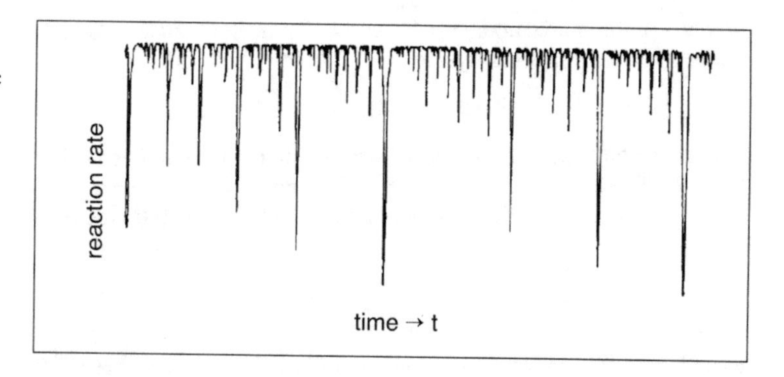

Figure 8.33

Towards Catalytic Converters

Now imagine the graph of figure 8.32 flipped over and compare with figure 8.33, which shows the measurement of the chemical reaction rate as a function of time in a catalytic oxidation process. The remarkable resemblance of the flipped-over graph and this kind of measurements provided the motivation to model a catalytic converter by one-dimensional cellular automata.[31] Thus, we are back to the relation of cellular automata and polynomials which we discussed at the beginning of this chapter. This relation allows us to interpret $h_p(r)$ as a count of 'oxidized' cells in an appropriate cellular automaton. In this sense our discussion has provided a first glimpse of an idea of why modeling a catalytic converter by cellular automata could be a successful approach and exhibits the qualitative behavior found in real chemical experiments.

[31] A. W. M. Dress, M. Gerhardt, N. I. Jaeger, P. J. Plath, H. Schuster, *Some proposals concerning the mathematical modelling of oscillating heterogeneous catalytic reactions on metal surfaces*. In L. Rensing and N. I. Jaeger (eds.), *Temporal Order*, Springer-Verlag, Berlin, 1984.

Chapter 9

Irregular Shapes: Randomness in Fractal Constructions

Why is geometry often described as 'cold' and 'dry'? One reason lies in its inability to describe the shape of a cloud, a mountain, a coastline, or a tree. Clouds are not spheres, coastlines are not circles, and bark is not smooth, nor does lightning travel in a straight line. [...] The existence of these patterns challenges us to study those forms that Euclid leaves aside as being 'formless', to investigate the morphology of the 'amorphous'.

Benoit B. Mandelbrot[1]

Self-similarity seems to be one of the fundamental geometrical construction principles in nature. For millions of years evolution has shaped organisms based on the survival of the fittest. In many plants and also organs of animals, this has led to fractal branching structures. For example, in a tree the branching structure allows the capture of a maximum amount of sun light by the leaves; the blood vessel system in a lung is similarly branched so that a maximum amount of oxygen can be assimilated. Although the self-similarity in these objects is not strict, we can identify the building blocks of the structure — the branches at different levels.

In many cases the 'dead' world also carries some fractal characteristics. An individual mountain, for example, may look like the whole mountain range in which it is located. The distribution of craters on the moon obeys some scaling power laws, like a fractal. Rivers, coastlines, and clouds are other examples. However, it is generally impossible to find hierarchical building blocks for these objects as in the case of organic living matter. There is no apparent self-similarity, but still the objects look the same in a statistical sense — which will be specified — when magnified.

[1] Benoit B. Mandelbrot, *The Fractal Geometry of Nature,* Freeman, New York, 1982, p. 1.

In summary, many natural shapes possess the property that they are irregular but still obey some scaling power law. One of the consequences — as discussed in chapter 4 — is that it is impossible to assign quantities such as length or surface area to these natural shapes. There cannot be a simple numerical answer to the question 'How long is the coastline of Great Britain?'. If 5000 miles would be measured by someone as the length of the coastline, someone else with a better (finer) measuring technique would come up with a result longer than 5000 miles. The more appropriate question to ask would be: how irregular, how convoluted is a coastline, or what is its fractal dimension? In this chapter this question is turned around. Methods for generating models of coastlines (and other shapes) with *prescribed* fractal dimension are given. Well, you may propose that the Koch snowflake curve, for example, may already serve as a good model for the coastline of an island. However, even though such exact self-similar curves have the desired scaling invariance and fractal dimension, they still are not perceived as realistic models of a coastline. The reason lies in their lack of randomness. To model coastlines, we need curves that look different when magnified but still invoke the same characteristic impression. In other words, looking at the magnified version of the coastline one should not be able to tell that it is indeed a magnification of the original. Rather, it ought to be regarded just as well as a different part of the coastline drawn at the same scale.

We begin our discussion at just that point — introducing some element of randomness into the otherwise rigorously organized classical fractals.[2] This leads to physical, so-called percolation models with applications ranging from the fragmentation of atomic nuclei to the formation of clusters of galaxies. An experiment which yields random fractal dendritic (tree-like) structures at an intermediate scale — useful for practical demonstration in a classroom — is an electrochemical aggregation process discussed in section 9.3. A mathematical model of this aggregation process is based on Brownian motion of particles. This can be implemented on a computer without much trouble. The underlying scaling laws of Brownian motion and one important generalization (fractional Brownian motion) are the topic of the fourth section. With these tools in hand, fractal landscapes and coastlines can be simulated on a computer, as shown in the last section and on the color pages.

[2]The randomization of branching structures obtained from MRCMs is more conveniently discussed in the context of chapter 7.

9.1 Randomizing Deterministic Fractals

Randomizing a deterministic classical fractal is the first and simplest approach generating a realistic 'natural' shape. We consider the Koch curve, the Koch snowflake, and the Sierpinski gasket.

Introducing Randomness in the Koch Snowflake Curve

The method for including randomness in the Koch snowflake construction requires only a very small modification of the classical construction. A straight line segment will be replaced as before by a broken line of four segments, each one one-third as long as the original segment. Also the shape of the generator is the same. However, there are two possible orientations in the replacement step: the small angle may go either to the left or to the right (see figure 9.1).

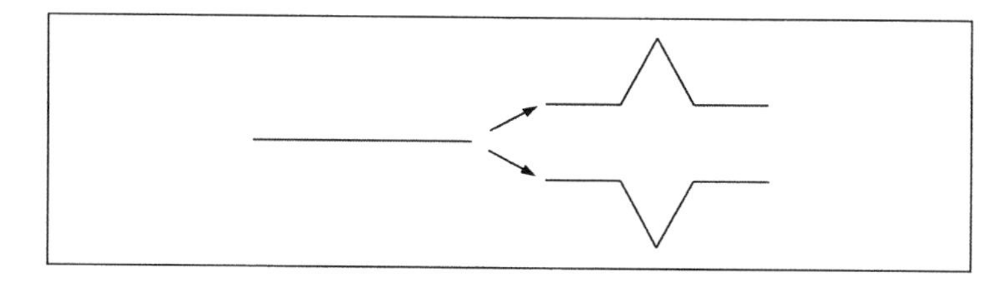

Two Possible Replacement Steps in the Koch Construction

Figure 9.1

Let us now choose one of these orientations at random in each replacement step. Let us call the result the *random Koch curve*. Composing three different versions of the random Koch curve, placed so that the end points meet, yields the *random Koch snowflake*. In this process some characteristics of the Koch snowflake will be retained. For example, the fractal dimension of the curve will be the same (about 1.26). But the visual appearance is drastically different; it looks much more like the outline of an island than the original snowflake curve (see figures 9.2 and 9.3). A different island can be constructed using the same ideas applied to the 3/2-curve introduced in chapter 4 (see figures 9.4 and 9.5). Here the dimension of the curve is higher, exactly 1.5.

Two Ways to Randomize the Sierpinski Gasket

In these first two above examples of random fractals, random decisions had to be made in the construction process. Each decision was based on a random choice of one out of two possibilities. Let us now give an example where a random number from an entire interval is used, the randomized Sierpinski gasket. The construction process is identical to the original one. Thus, in each step a triangle is subdivided into four subtriangles, the central one of which is removed. However, in the subdivision we can now allow for subtriangles which are not equilateral. On each side of a triangle to be subdivided we pick a point at random, then we connect the three points and obtain the four subtriangles. The center subtriangle is removed and the procedure repeats (see figure 9.6).

Let us next discuss a modification of the Sierpinski gasket. This will lead us directly to the topic of the next section, a physical phenomenon with many applications: *percolation*. We again use the standard subdivision into equilateral triangles. One easy modification is then to simply choose one of the four subtriangles in a replacement step at random and remove it. Thus, the

Random Koch Curve

One realization of the random Koch curve. The replacement steps are the same as in the original Koch curve, with the exception that the orientation of the generator is chosen randomly in each step.

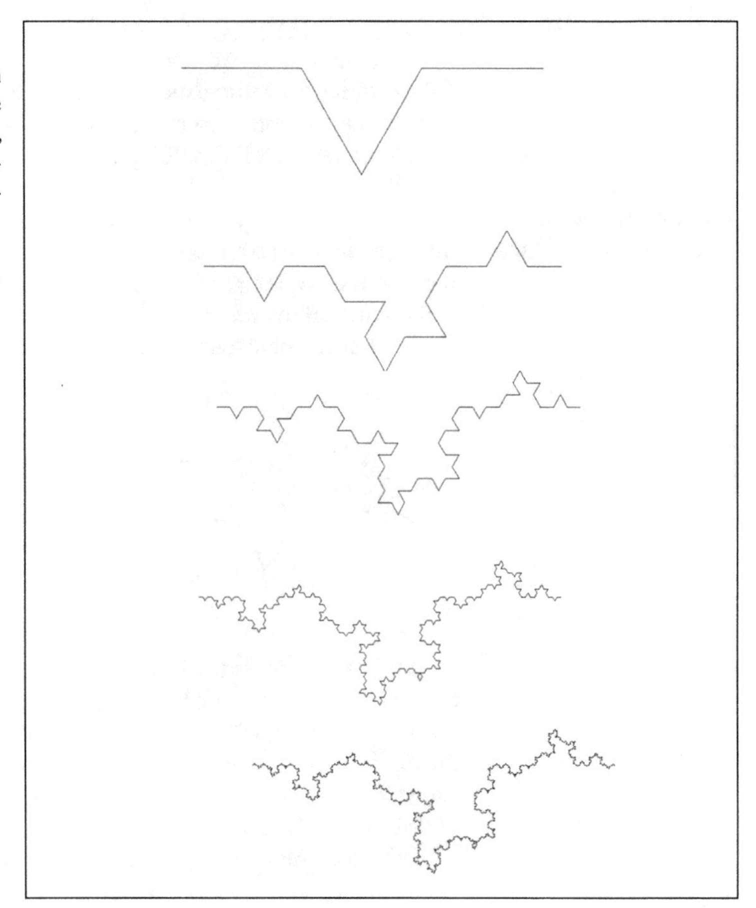

Figure 9.2

Random Koch Island

The random Koch island is composed of three different versions of the random Koch curve, placed such that the end points meet.

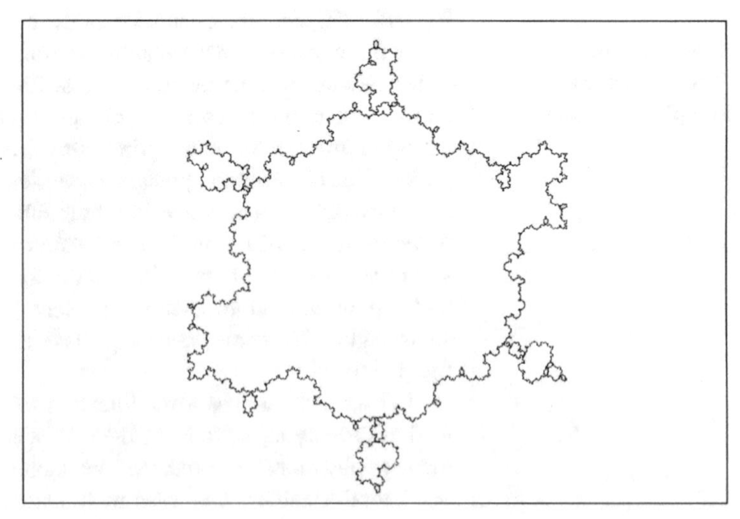

Figure 9.3

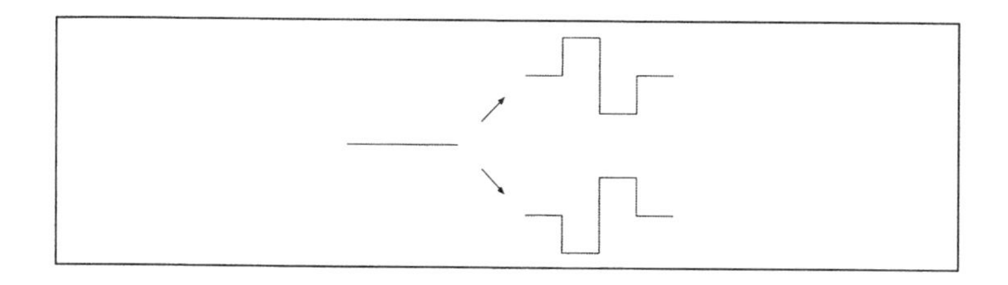

Initiator and Generator for the Randomized 3/2-Curve

Figure 9.4

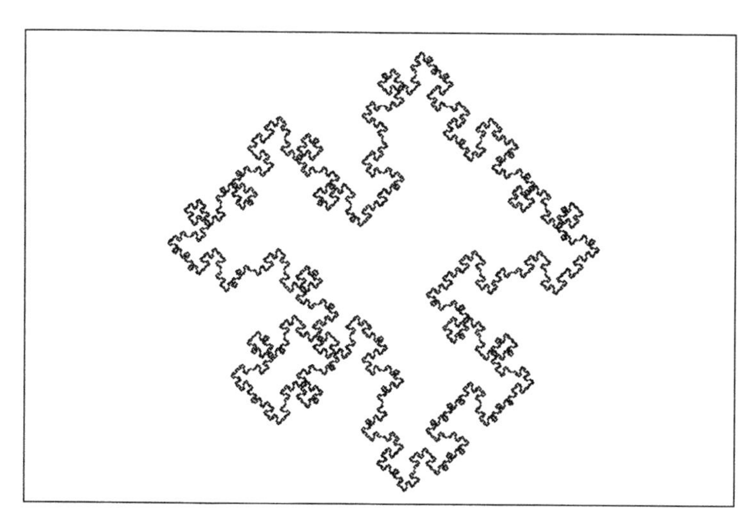

Random 3/2 Island

Composing four different versions of the random 3/2-curve produces an island with a coastline of dimension 1.5.

Figure 9.5

Modified Sierpinski Gasket 1

The points subdividing the sides are picked at random. Stage 4 of the construction process is shown.

Figure 9.6

center subtriangle may be removed, but a different one might just as well be selected and removed (see figure 9.7).

In the figure we can see small and large clusters of connected triangles. A cluster is defined as a collection of black triangles, which are connected across their sides and which are completely surrounded by white triangles.[3] Is there

[3]Two triangles touching each other only at a vertex are not considered a cluster.

Modified Sierpinski Gasket 2

In each replacement step the subtriangle to be deleted is picked at random. The black triangles are those from stage 5 of the construction process.

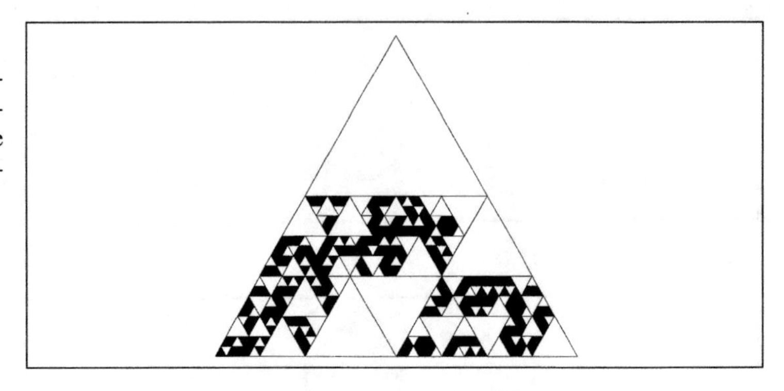

Figure 9.7

a cluster that connects all three sides of the underlying large triangle? What is the probability for that? What is the expected size of the largest cluster? Questions of this type are relevant in the percolation theory, which is discussed in the following section.

9.2 Percolation: Fractals and Fires in Random Forests

Let us carry the idea one step further. We consider a triangular lattice of some resolution and deal with each subtriangle independently. Such a subtriangle is colored black or not according to a random event which occurs with a prescribed probability $0.0 \leq p \leq 1.0$. The overall shape of the result depends dramatically on the probability p chosen. Obviously, for $p = 0$ we get nothing, while for $p = 1$ we obtain a solid triangle of full size. For intermediate values of p the object has a density that increases with p (see figure 9.8).[4] Initially, for small values of p we get only a few specs here and there. For higher values of the probability the specs grow larger, until at some critical value of $p = p_c$ the shape seems to become glued together into one big irregular lump. Further increases in the probability, of course, thicken the cluster even more.

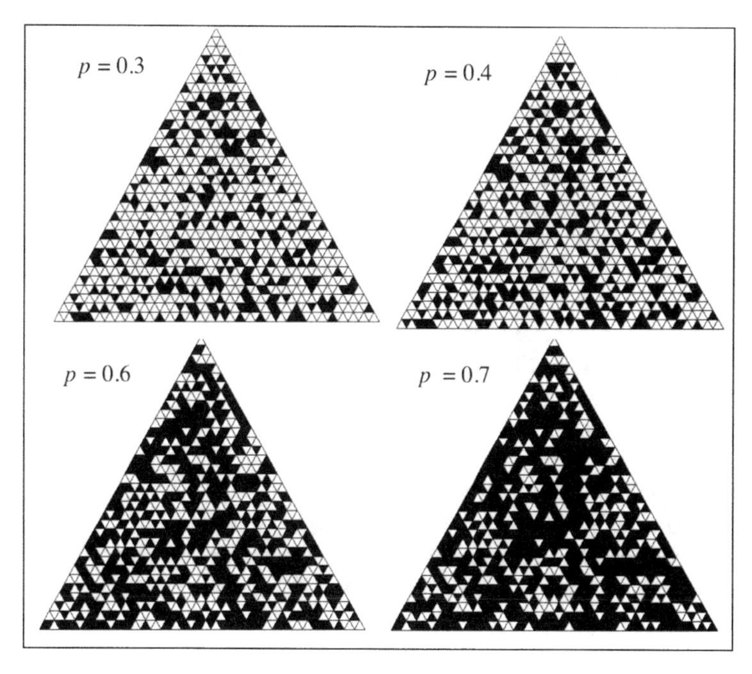

Triangular Lattice with Randomly Occupied Sites

Subtriangles of stage 5 are chosen with probability p. The four values of p used in this figure are (from top left to bottom right): 0.3, 0.4, 0.6, 0.7. In the first two graphs many small clusters coexist ($p = 0.3, 0.4$). In the graph for $p = 0.7$ one major big cluster exists and allows only very few further small clusters.

Figure 9.8

Percolation — When Things Start to Flow

When a structure changes from a collection of many disconnected parts into basically one big conglomerate, we say that *percolation*[5] occurs. The name stems from an interpretation of the solid parts of the structure as open pores. Assume that the whole two-dimensional plane is partitioned into a regular array of such pores which are either open (with probability p) or closed (with probability $1 - p$). Let us pick one of the open pores at random and try to inject a fluid at that point. What happens? If the formation is 'below the percolation threshold', i.e., if the probability p is less than p_c, we

[4] Formally, the average density is equal to p.

[5] 'Percolation' originates from the Latin words 'per' (through) and 'colare' (to flow).

expect that the pore is a part of a relatively small cluster of open pores. By a cluster we mean a collection of connected open pores, which is completely surrounded by closed ones. In other words, below the threshold we will be able to inject only some finite amount of fluid until the cluster is filled, but no more. If the probability is above the threshold value, then chances are good that the corresponding cluster is infinitely large. We can inject as much of the fluid as we like. In a practical example, the water percolates through the coffee grains and drips into the pot as coffee. The most interesting phenomena happen while increasing the probability from below the percolation threshold to a value above p_c. For example, the probability that a pore picked at random indeed belongs to the cluster of maximal size changes at $p = p_c$ from zero to a positive value. Moreover, right at the percolation threshold p_c, this maximal cluster is a fractal! It has a dimension that can be determined experimentally and, in some cases, also analytically.[6]

Dangerous Forest Fires Beyond Percolation

A paradigm often used for percolation is given by forest fires. Points in the clusters correspond to trees in the forest, and the fire cannot spread across gaps between trees. So the question whether the forest is below or above the percolation threshold is a vital one. In the first case trees are relatively sparse and only a small portion of all trees will burn down, while the other case is devastating: almost the complete forest will be destroyed. Let us elaborate the model a bit further. For simplicity we assume that the forest is not a natural one. Rather, the trees are planted in rows and columns in a square lattice. When all sites from this array are occupied by trees, the situation is clear — a fire ignited anywhere will spread over the whole forest (unless disturbed by strong winds or fire fighters not accounted for in this model). Thus, let us assume the more interesting case where each site in the lattice is occupied by a tree with a fixed probability $p < 1$. A burning tree may ignite its immediate neighboring trees. In the square lattice these are the trees at the four locations beside, above and below the burning one. In physicists' jargon these sites are called 'nearest neighbors'. Given a square with L^2 sites we distribute trees according to the chosen probability p and start a fire. Let us set all those trees on fire which are located along the left side of the square. In this simple model we can now simulate how the fire spreads. We proceed in discrete time steps. In each step a burning tree ignites all those neighbor trees that are not already on fire. After the tree has burned out, it leaves a stump, which from then on is no longer relevant. See figure 9.9 for a small simulation of this kind. It goes without saying that this type of percolation model will be of little or no help in fighting or analyzing actual forest fires. The point is that this paradigm is very suitable for an explanation and introduction of the topic.

Forest Fires at Percolation Burn Longest

How long will such a fire last? If trees are very sparse — due to a low probability p — then the fire has not much material to burn, and it dies out very quickly, leaving most of the forest unharmed. On the other hand, if the trees are very dense (p is large), the forest does not have much of a chance

[6] A very nice and enjoyable introduction to the topic for nonspecialists is given in Dietrich Stauffer, *Introduction to Percolation Theory,* Taylor & Francis, London, 1985. There is a new expanded edition by D. Stauffer and A. Aharony, 1992.

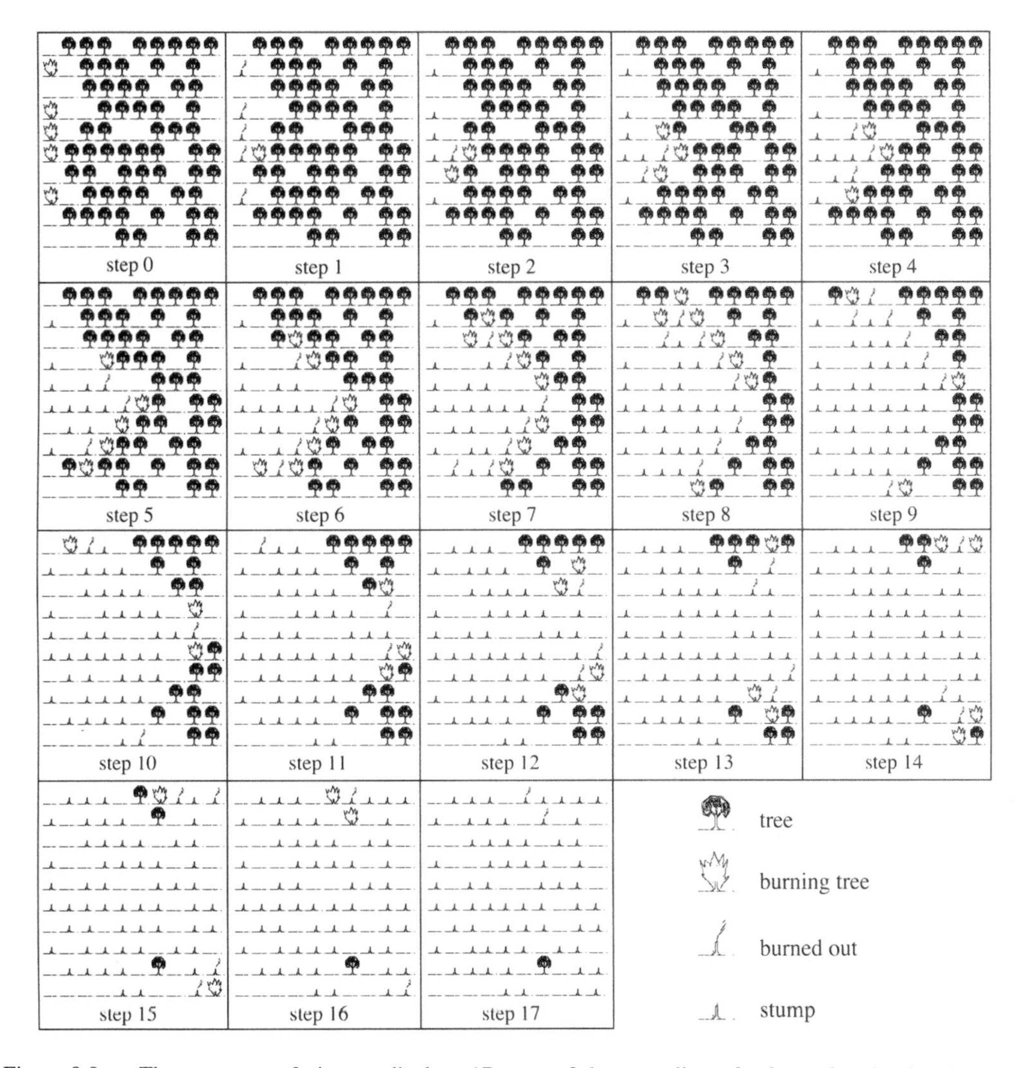

Figure 9.9 : The sequence of pictures displays 17 steps of the spreading of a forest fire simulated on a square 10×10 lattice. Initially, trees are placed at the lattice sites with probability 0.6 and ignited along one side of the lattice (step 0). After 17 steps the fire is dead; only a single tree survives.

for survival. Virtually the whole forest will be destroyed. Moreover, this will happen rather quickly; in not many more than L time steps the fire will have swept across the whole square leaving almost nothing but blackened stumps. There must be an intermediate probability which leads to a maximal duration of the forest fire (see figure 9.10).

In the diagram there is a sharp peak near the probability 0.6: this is the percolation threshold. The peak is indeed very sharp, if we increase the size of the forest, i.e., the number L of columns and rows, then the amplitude of

Forest Fire Duration

Average duration of forest fires versus probability p, simulated on square lattices with 20 rows (bottom), 100 rows (middle), and 500 rows (top). For each point in the plot 1000 runs were performed and averaged. The larger the lattice chosen, the more pronounced the peak of the forest fire duration near the percolation threshold at about $p = 0.60$.

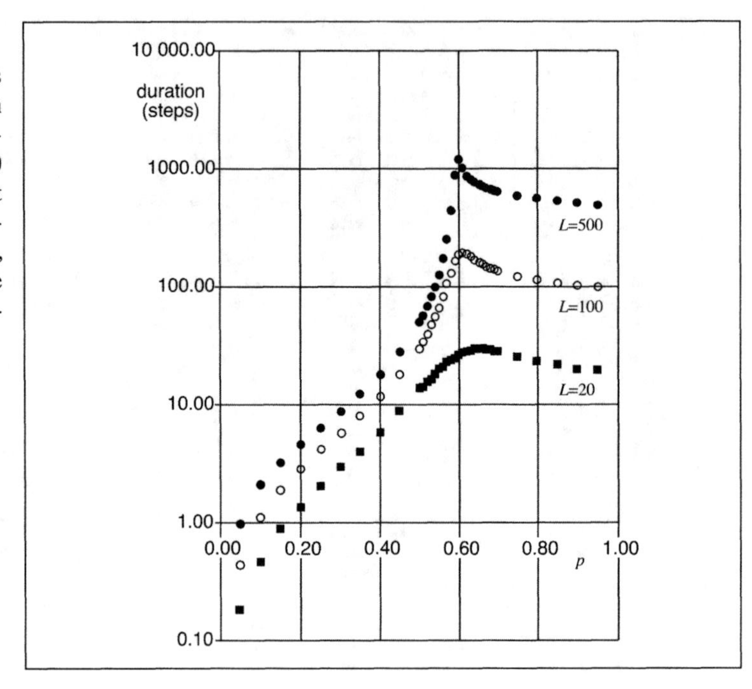

Figure 9.10

the peak grows without bound.[7] In mathematical terms, there is a *singularity* at the percolation. The probability corresponding to the percolation has been measured experimentally very carefully; the accepted value is $p_c \approx 0.5928$.

A logical activity to pursue at this point would be to analyze the scaling laws of the forest fire duration. How long does the fire burn as the size of the underlying lattice grows without bound? There are three very different cases corresponding to $p < p_c$, $p = p_c$, and $p > p_c$, where the special case right at the percolation reveals a power law with a noninteger exponent — evidence for a fractal structure.

The Maximal Tree Cluster Size

A quantity that has been more thoroughly studied is the maximal cluster size, which is closely related to the fire duration. Let us denote the number of trees in the largest cluster in a lattice of size L by $M(L)$. As indicated, the cluster size will vary with L. A normalized measure of the maximal cluster size may be more convenient. It is given by the probability that a lattice site, picked at random, is a member of the maximal cluster. The notation is $P_L(p)$. It depends on the probability p and (to a lesser degree) on the lattice size L. In order to estimate $P_L(p)$ we can average the relative cluster size $M(L)/L^2$ over many samples of the random forest. With larger and larger lattices, the dependence on L diminishes. In other words, we arrive at a limit

$$P_\infty(p) = \lim_{L \to \infty} P_L(p).$$

[7] The amplitude will increase faster than the width L of the forest, but not as fast as the area L^2.

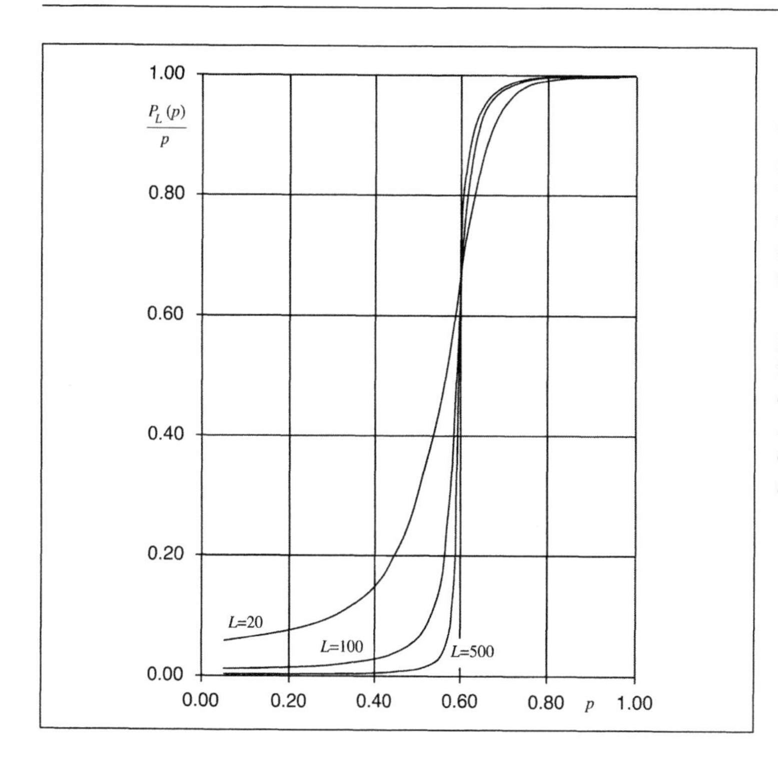

Phase Transition at Percolation

The graph shows the probability that a site with a tree picked at random will be reached by the forest fire in the simulation. It corresponds to the data shown in figure 9.10. The lattice sizes (numbers of rows) are 20, 100, and 500. The probability that a given tree will be burned decreases to zero below the percolation threshold when the lattice size is increased. Above the percolation the portion of trees burned grows asymptotically according to a power law.

Figure 9.11

For low values of p, the probabilities $P_L(p)$ are negligible, and in the limit $L \to \infty$ they tend to zero. But there is a critical value — namely the percolation threshold p_c — beyond which $P_\infty(p)$ grows rapidly. In other words, if p is above percolation, the maximal cluster is infinitely large and comes close to all lattice sites, while for $p < p_c$, the probability that a site picked at random belongs to the maximal cluster is negligible.

The Phase Transition at the Percolation Threshold

At the percolation probability this likelihood increases abruptly. In fact, for $p > p_c$ and p close to p_c the probability $P_\infty(p)$ is given by a power law[8]

$$P_\infty(p) \propto (p - p_c)^\beta$$

with an exponent $\beta = 5/36$. In terms of our forest fire simulation, we may equivalently consider the fraction of trees burned down after termination of the fire (see figure 9.11). There is a sharp increase near the critical value p_c, which becomes more drastic for lattices with more rows. This effect is also called a *phase transition* like similar phenomena in physics. For example, when heating water there is a phase transition from liquid to gas at 100 degrees Celsius.[9]

[8] The symbol '$\propto$' means 'proportional'.
[9] Under standard pressure.

A Fractal Called the Incipient Percolation Cluster

From this observation we can make some conclusions about the size of the maximal cluster: if $p > p_c$, then the probability $P_\infty(p) > 0$ implies that the cluster size scales as L^2. On the other hand, for $p \leq p_c$ we may conjecture that a power law holds so that the size is proportional to L^D with $D < 2$. This would indicate a fractal structure of the maximal cluster. This is true, however, only for one special value of p, namely exactly at the percolation $p = p_c$. The fractal percolation cluster at the threshold is often called the *incipient* percolation cluster. Its dimension has been measured; it is $D \approx 1.89$. For values below p_c the maximal cluster size scales only as $\log(L)$.

Other Aspects of Percolation

This analysis, of course, is not the whole truth about percolation. For example, there are other lattices. We can consider three-dimensional or even higher-dimensional lattices. We can also include other neighborhood relations. There are many other quantities besides p_c, D, and $P_\infty(p)$ of interest. For example, the *correlation length* ξ is an important characteristic number. It is defined as an average distance between two sites belonging to the same cluster. As p approaches p_c from below, the correlation length ξ grows beyond all bounds. This growth is again described by a power law

$$\xi \propto |p - p_c|^{-\nu}$$

with an exponent ν, which is equal to 4/3 for two-dimensional lattices. The correlation length is of relevance for numerical simulations. When the lattice size L is smaller than the correlation length ξ, all clusters look fractal with the same dimension. Only when the resolution of the lattice is sufficiently high ($L \gg \xi$) is it possible to determine that clusters are in fact finite and have dimension 0 for $p < p_c$.

Some Constants Are Universal

It is important to note that the percolation threshold p_c depends on the choices in the various models, e.g., on the type of lattice and the neighborhood relations of a site. However, the way quantities such as the correlation length scale near the percolation do not depend on these choices. Thus, numbers characterizing this behavior, such as the exponents β and ν and the fractal dimension of the percolation cluster, are called *universal*. The values of many constants, for example, $p_c \approx 0.5928$ and $D \approx 1.89$, are, however, only approximations obtained by elaborate computer studies. It is a current challenge to derive methods for the exact computation of these constants, but we cannot go into any more details here — a lot of problems are still open and are areas of active research.

From Square Back to Triangular Lattices

To conclude this section let us return to the triangular lattice with which we started and which was motivated by the Sierpinski gasket. The quantities $M(L)$, $P_L(p)$, and $P_\infty(p)$ can also be analogously defined for this case. The first numerical estimates in 1960 indicated that the percolation threshold is about 0.5. Then it took about 20 years from the first nonrigorous arguments to a full mathematical proof to show that $p_c = 0.5$ exactly. Moreover, it was shown that the fractal dimension of the incipient percolation cluster is

$$D = \frac{91}{48} \approx 1.896$$

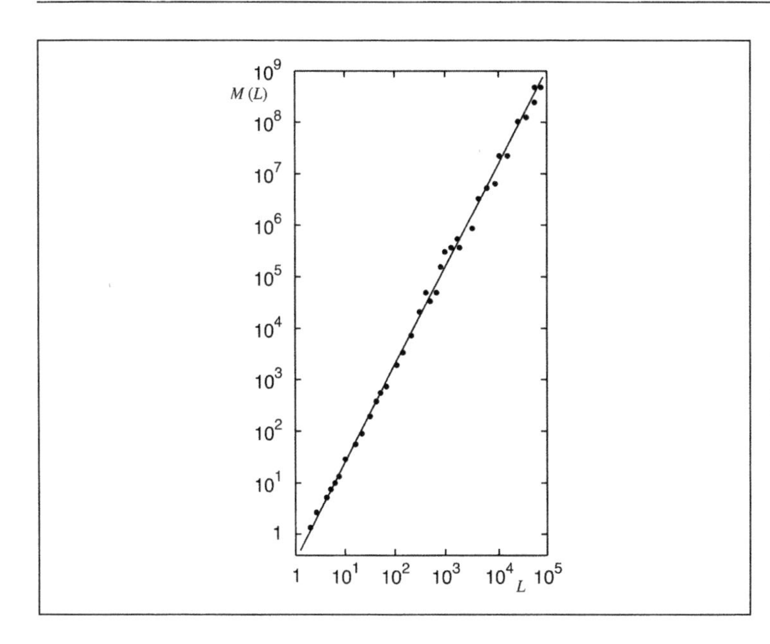

Fractal Dimension of the Incipient Percolation Cluster

The fractal dimension D of the incipient percolation cluster in a triangular lattice is determined here in a log-log diagram of the cluster size $M(L)$ versus the grid size L. The percolation threshold is $p_c = 0.5$. The slope of the approximating line confirms the theoretical value $D = 91/48$. (Figure adapted from D. Stauffer, *Introduction to Percolation Theory,* Taylor & Francis, London, 1985.)

Figure 9.12

(compare figure 9.12). This is about the same value as the one determined numerically for the square lattice. Therefore it has been conjectured that it is the correct dimension of the incipient percolation cluster in all two-dimensional lattices.

The Renormalization Technique

Instead of looking at the proof for the result $p_c = 0.5$, we may provide a different interesting argument, which opens the door to another method for the analysis of fractals which we have not discussed so far: *renormalization*. One of the keys to understanding fractals is their self-similarity, which reveals itself when applying an appropriate scaling of the object in question. Is there a similar way to understand the incipient fractal percolation cluster? The answer is yes; and it is not hard to investigate, at least for the triangular lattice. The claim is that a reduced copy of the cluster looks the same as the original from a statistical point of view. But how can we compare the two? For this purpose we systematically replace collections of lattice sites by corresponding so-called super-sites. In a triangular lattice it is natural to join three sites to form one super-site. This super-site inherits information from its three predecessors which determines whether it is occupied or not. The most natural rule for this process is the majority rule; if two or more of the three original sites are occupied, then — and only in this case — the super-site is also occupied.

Figure 9.13 shows the procedure and also the geometrical placement of the sites.[10] The super-sites themselves form a new triangular lattice which can now be reduced in size to match the original lattice, allowing a comparison. The concentration of occupied sites — let us call it p' — in the renormalized

[10]Note that this is not the same as the triangular lattice obtained in figure 9.8.

Renormalization of Sites in Triangular Lattice

Three neighboring sites are joined to form a super-site. The super-site is occupied if two or all three of the sites are occupied. The super-sites form another triangular grid, however, rotated by 30 or 90 degrees. Scaling down the size of this grid of super-sites completes one cycle of the renormalization scheme. See figure 9.15 for examples.

Figure 9.13

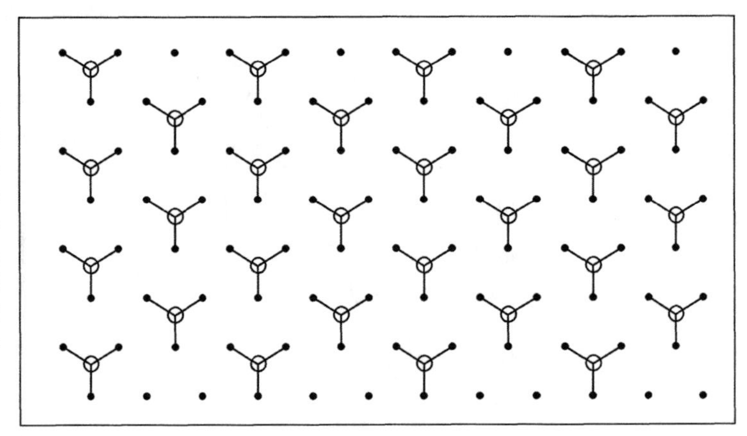

lattice will not generally be the same as in the old lattice. For example, if p is low, then there are only a few isolated occupied sites, most of which will have vanished in the process of renormalization, thus $p' < p$. At the other end of the scale, when p is large, many more super-sites will be formed, which close up the gaps left in the original lattice, resulting in $p' > p$. Only at the percolation threshold can we expect similarity. There the renormalized super-cluster should be the same as before, in other words,[11]

$$p' = p.$$

In this case we are lucky; we can compute at which probability p the above equation holds! A super-site will be occupied if all three of the original sites are occupied, or if exactly one site is not occupied. The probability for an occupied site is p. Thus, the first case occurs with probability p^3. In the other case, we have a probability of $p^2(1-p)$ that any particular site is not occupied while the other two are. There are three such possibilities. Thus, summing up we arrive at

$$p' = p^3 + 3p^2(1-p)$$

as the probability for the super-site being occupied. Now we are almost there. What is p such that $p' = p$? For the answer we have to solve the equation

$$p = p^3 + 3p^2(1-p)$$

or, equivalently,

$$p^3 + 3p^2(1-p) - p = 0.$$

It is easy to check that

$$p^3 + 3p^2(1-p) - p = -2p(p-0.5)(p-1).$$

[11] Here we see a remarkable interpretation of self-similarity in terms of a fixed point of the renormalization procedure. This relation of ideas from renormalization theory has turned out to be extremely fruitful in the theory of critical phenomena in statistical physics.

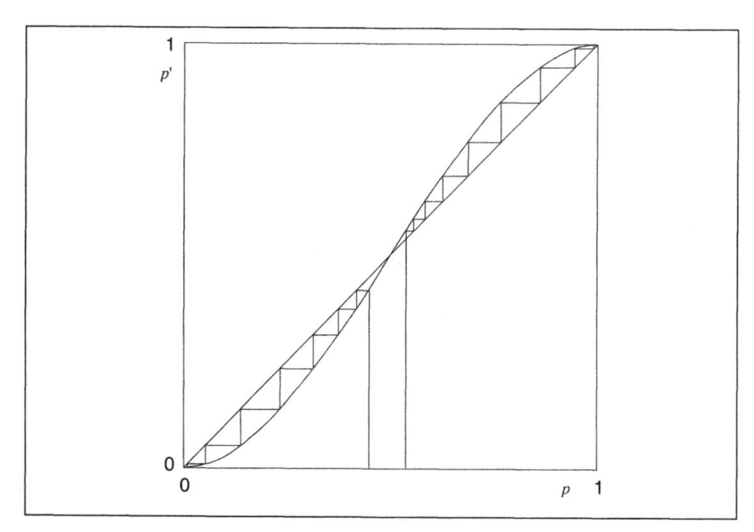

The Renormalization Transformation

Graphical iteration for the renormalization transformation $p \to p^3 + 3p^2(1-p)$ of the triangular lattice.

Figure 9.14

Thus, there are three solutions, $p = 0, p = 0.5$, and $p = 1$. Of these three solutions, two are not of interest, namely $p = 0$ and $p = 1$. A forest without trees ($p = 0$) renormalizes to another forest without trees, which is not surprising. Likewise, a saturated forest ($p = 1$) also does not change when renormalized. But the third solution, $p = 0.5$, is the one we are after. It corresponds to a nontrivial configuration, i.e., the forest does have a structure, which after renormalization is still statistically the same. The super-sites have the same probability 0.5 of being occupied as the sites in the original lattice. This is the statistical self-similarity expected at the percolation threshold. Thus, an elementary renormalization argument tells us that $p_c = 0.5$ in accordance to the actual result.

At the percolation threshold, renormalization does not change anything — even when applied many times over. This is not the case for all other probabilities $0 < p < 1, p \neq 0$. To study the effect of repeated renormalization we need to consider something very familiar, a feedback system, which relates the probabilities of a site being occupied before and after a renormalization. Thus, we have to consider the iteration of the cubic polynomial

$$p \to p^3 + 3p^2(1-p).$$

This study is best presented in the corresponding diagram of the graphical iteration (see figure 9.14). The situation is very clear. Starting with an initial probability $p_0 < 0.5$, the iterations converge to 0, while an initial $p_0 > 0.5$ leads to the limit 1. Only right at the percolation threshold $p_c = 0.5$ do we obtain a dynamical behavior different from the above, namely a fixed point.

Using Renormalization as an Investigative Tool

As an application of this renormalization transformation we could check whether a given lattice is above or below percolation. We would carry out the renormalization procedure a number of times. If the picture converges to an

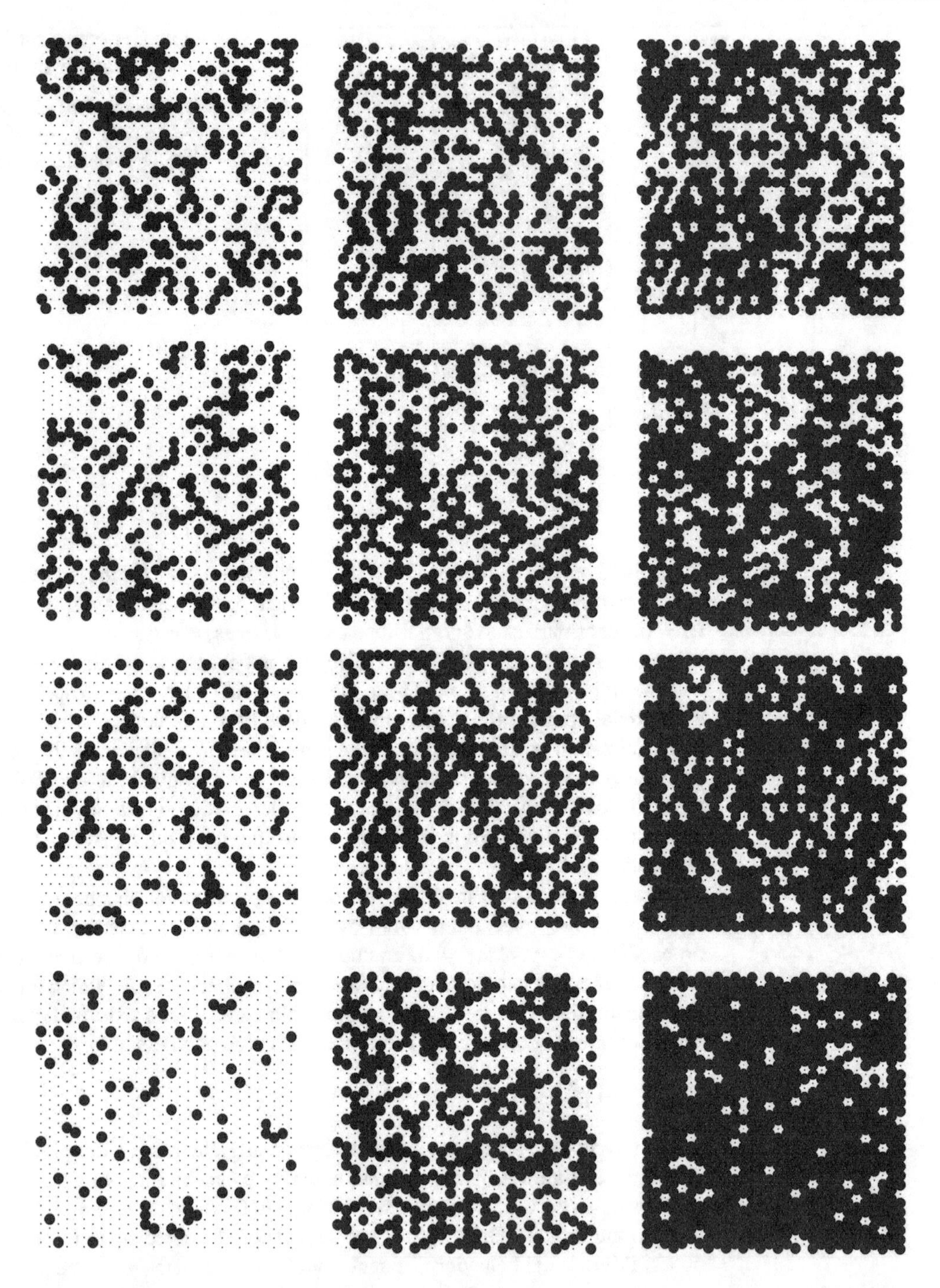

Figure 9.15 : Three renormalization stages for three given configurations (top) corresponding to the cases $p = 0.35 < p_c, p = 0.5 = p_c$ and $p = 0.65 > p_c$ (from left to right).

empty configuration (no sites occupied), then the parameter p belonging to the lattice in question is below percolation; and if in the long run all sites tend to be occupied, the original configuration is above the percolation threshold. This program is carried out in figure 9.15. From the three original configurations in the top row it is not quite clear by visual inspection whether they are above or below percolation, but the renormalization reveals this information after already three steps.

The triangular lattice is a rather special case. When applying the technique to other lattices, only approximations of p_c can be expected. But it is remarkable how this new idea has created a method to approach the very difficult problem of determining percolation parameters. The basic idea of renormalization came from the physicist Leo P. Kadanoff in 1966, in connection with critical phenomena in a different area of theoretical physics. The idea eventually led to quantitative results and explained the physics of phase transitions in a satisfactory way. After all, the way from the idea of renormalization to its concrete, final form was so elusive that Kadanoff did not find it. Rather, Ken G. Wilson at Cornell in 1970 surmounted the difficulties and developed the method of renormalization into a technical instrument that has proven its worth in innumerable applications. About a decade later he was honored with the Nobel prize for his work.

The situation at the percolation threshold, or more generally at the renormalization fixed point, has an analogy in fractal constructions. Recall, for example, the Koch curve construction where the object must be scaled by a factor of $s = 1/3$ in each step of the construction. If we scale by a factor $s < 1/3$, then in the limit we arrive at just a point. On the other hand, scaling by $s > 1/3$ in each step lets the construction grow beyond any bounds. Only if we scale exactly by 1/3 from step to step will we get an interesting limit with self-similarity. In most other cases besides the Koch construction it is not at all obvious how to choose the 'right' scaling.[12]

Percolation is a widely used model and applies to many phenomena observed in nature and the engineering sciences. An example is the formation of thin gold films on an amorphous substrate, where the parameter in question corresponds to the amount of gold provided.[13] At the percolation threshold the metal provides electrical conductivity. On the other hand, percolation is also relevant at scales as large as in the formation of galaxies and clusters of galaxies.[14]

---

[12] See F. M. Dekking, *Recurrent Sets,* Advances in Mathematics 44, 1 (1982) 78–104.

[13] See R. Voss, *Fractals in Nature,* in: *The Science of Fractal Images,* H.-O. Peitgen and D. Saupe (eds.), Springer-Verlag, New York, 1988, pages 36–37.

[14] For a survey of the state-of-the-art of percolation theory see A. Bunde and S. Havlin (eds.), *Fractals and Disordered Systems,* Springer-Verlag, Heidelberg, 1991.

9.3 Random Fractals in a Laboratory Experiment

There exists a wealth of fractal structures observed in nature and laboratory experiments.[15] In this section we concentrate on one particularly interesting example: aggregation.

Cluster by Aggregation of Small Particles

The research on the aggregation of small particles to form large clusters in polymer science, material science and immunology, among other fields, has been going on for a long time. However, their study has recently been revitalized tremendously by concepts from fractal geometry.[16] In this section we describe only one particular experiment from this volume, reported on by Mitsugu Matsushita, dealing with electrochemical deposition leading to dendritic structures. It has the advantage that its setup is small and easy to build, and the necessary chemical substances are easily obtained and not dangerous.[17] The complete experiment takes only about 20 minutes. Thus, it may be conveniently conducted right in the classroom. The setup may be filmed and projected by video equipment or even put directly on an overhead projector[18] for viewing by a larger audience.

Dendritic Electrochemical Deposition

Let us quote the description of the experiment directly from Matsushita's article:[19]

"Electrochemical deposition has for a long time been one of the most familiar aggregation phenomena in chemistry. Only very recently has it received attention from the entirely new viewpoint of fractal geometry. In practice, electrodeposition processes may be complex, and the resulting deposits may exhibit a variety of complex structures. However, if the metal deposition is controlled mainly by a single process, e.g., diffusion, then the deposits usually exhibit statistically simple, self-similar, i.e., fractal, structures.

"In this experiment metallic zinc in the form known as zinc metal-leaves was grown two-dimensionally. The experimental procedures used to grow zinc metal-leaves are as follows. A Petri-dish of diameter approx. 20 cm and depth approx. 10 cm is filled with 2 M $ZnSO_4$ aqueous solution (depth approx. 4 mm), and a layer of n-butyl acetate [$CH_3COO(CH_2)_3CH_3$] is added to make an interface (Fig. 9.16). A tip of a carbon cathode (pencil core of diameter approx. 0.5 mm) is polished carefully so as to make it flat perpendicularly to the axis. The cathode is then set at the center of the Petri-dish so that the flat tip is placed just on the interface (Fig. 9.16). The electrodeposition is

[15] E. Guyon and H. E. Stanley (eds.), *Fractal Forms*, Elsevier/North-Holland and Palais de la Découverte, 1991.

[16] See *The Fractal Approach to Heterogeneous Chemistry: Surfaces, Colloids, Polymers*, D. Avnir(ed.), Wiley, Chichester, 1989, and *Aggregation and Gelation*, F. Family and D. P. Landau (eds.), North-Holland, Amsterdam, 1984.

[17] Of course, after the experiment, the used liquids must be disposed of properly (not in the sink). Moreover, a good ventilation of the room is recommended.

[18] However, the heat of the lamp in the projector disturbs the experiment, which runs best at constant temperature (large solid zinc leaves form). It is therefore advisable to leave the overhead projector off most of the time. It is best to video film the experiment for immediate viewing on monitors of a projection unit.

[19] M. Matsushita, *Experimental observation of aggregations*, in: *The Fractal Approach to Heterogeneous Chemistry: Surfaces, Colloids, Polymers*, D. Avnir (ed.), Wiley, Chichester 1989.

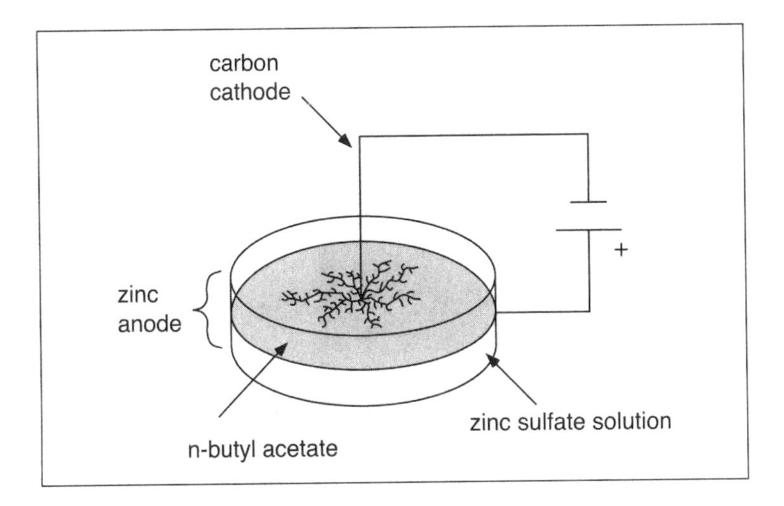

Figure 9.16 : In the Petri-dish a solution of zinc sulfate is covered by a thin layer of n-butyl-acetate.

initiated by applying a d.c. voltage between the carbon cathode and a zinc ring-plate anode of diameter approx. 17 cm, width approx. 2.5 cm and thickness approx. 3 mm placed in the Petri-dish. A zinc metal-leaf grows two-dimensionally at the interface between the two liquids from the edge of the flat tip of the cathode towards the outside anode with an intricately branched random pattern (Fig. 9.17). If the cathode tip is rounded or is immersed in the $ZnSO_4$ solution the deposit grows three-dimensionally into the solution. Usually, the zinc metal-leaves grow to a size of about 10 cm in about 10 min by applying a constant d.c. voltage of about 5 V. The temperature of the system was kept fixed, e.g., at about room temperature.

"The investigation of fractal structures of electrodeposits and their morphological changes is also of practical importance. The electrodeposition experiments presented here are clearly relevant to processes such as metal migration on ceramic or glass substrati and to zinc deposits on cathodes in various batteries. In both cases the growth of deposits is the main factor limiting the lifetime of many electronic parts and batteries."

The mathematical modeling of the electrochemical deposition of zinc metal-leaves is based on the fundamental concept of Brownian motion. Brownian motion refers to the erratic movements of small particles of solid matter suspended in a liquid. These movements can only be seen under a microscope. After the discovery of such movement of pollen it was believed that the cause of the motion was biological in nature. However, about 1828 the botanist Robert Brown realized that a physical explanation, rather than the biological one, was correct. The effect is due to the influence of very light collisions

Sample Result

This dendritic growth pattern was produced in only about 15 minutes by Peter Plath, University of Bremen. The reproduction here is in about the original size. The real zinc dendrite looks very attractive due to its metallic shiny character.

Figure 9.17

with the surrounding molecules. In the electrochemical experiment zinc ions randomly wander around in the solution until they are caught by the attractive pull of the carbon cathode. The aggregation of a zinc ion is most likely where the density of field lines is greatest. This is the case at the interface between the solution and the acetate, in particular at the tips of the dendrite. We derive a simple method for the computer simulation of such Brownian motion, which will enable us to also simulate the results of the electrochemical experiment.

Simulation of Diffusion Limited Aggregation (DLA)

To simulate diffusion limited aggregation (= DLA) based on the Brownian motion of particles is not hard.[20] We fix a single 'sticky' particle somewhere, say at the origin of a two-dimensional coordinate system. This particle is not allowed to move. Next we select a region of interest centered around the initial sticky particle, say a circular area of some radius, which could be chosen as 100 or perhaps 500 particle diameters. We inject a free particle at the boundary of the region and let it move about randomly. Two things may happen in the course of this motion. Either the particle leaves the region of interest, in which case we forget about it and start a new free particle at a random position at the boundary of the region, or it stays in this region until it gets close to the

[20]The model presented here originates from T. A. Witten and L. M. Sander, Phys. Rev. Lett. 47 (1981) 1400–1403 and Phys. Rev. B27 (1983) 5686–5697.

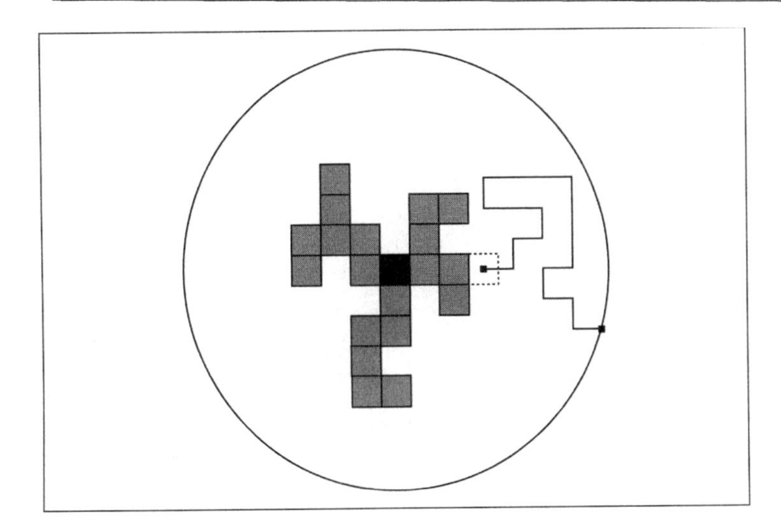

Simulation of the Electrochemical Aggregation Experiment

Simulation of Brownian motion in two dimensions is used for the paths of the zinc ions in the liquid. Particles move from pixel to pixel until they 'attach' to the existing dendrite.

Figure 9.18

sticky particle. In that latter case it attaches and also becomes a sticky particle (with some probability). Now the procedure is repeated, in effect growing a cluster of connected sticky particles which very much resembles the dendrites resulting from DLA in electrochemical deposition (see figure 9.19).

The practical computation is usually based on a square lattice of pixels (see figure 9.18) and the free particle may move to one of its four neighboring pixels in one time step. For a large cluster the process may take *very* long, and several tricks should be used to accelerate the process. For example, at each step the particle may be allowed to move a larger distance than just one pixel. This is possible when the current particle is relatively far from the cluster. More precisely, the distance it may jump in one step is limited by the distance of the particle to the cluster.

Problems

Based on time records of both the real electrochemical experiment and the computer simulation, several questions are of interest.

1. What is the fractal dimension of the aggregate?
2. Clearly, the density of the particles decreases the greater the distance to the center of the dendrite is. Is there a mathematical (power law) relation between density and distance?
3. Does the voltage between the ring anode and the carbon cathode in the experiment have an effect on the value of the fractal dimension of the aggregate? If so, how can we modify the simulation to take account of that?
4. How is the electrical current related to the size of the aggregate?

Answers to some of these questions have been found, but the research on aggregation is far from complete.[21] The fractal dimension, for example, has

[21] See the review article by H. Eugene Stanley and Paul Meakin, *Multifractal phenomena in physics and chemistry,* Nature 335

Simulation Results

Result of the numerical simulation of DLA based on Brownian motion of single particles.

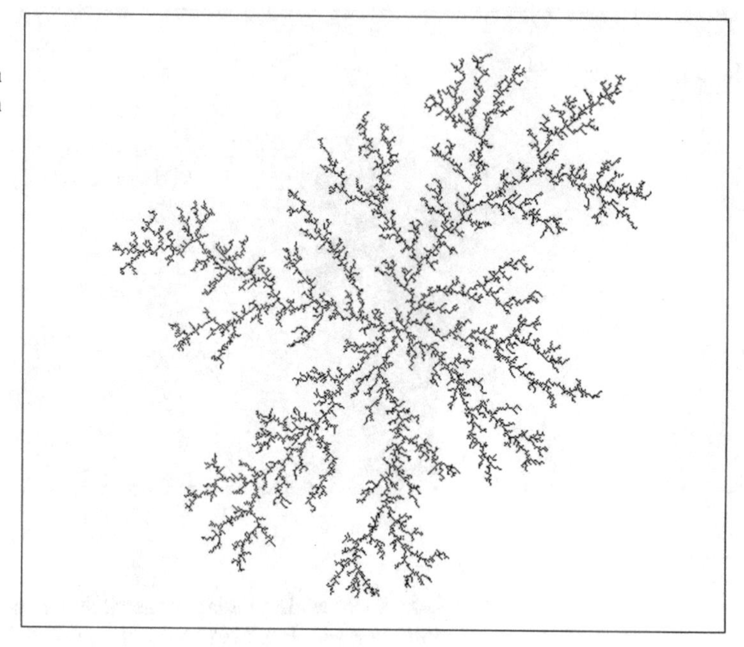

Figure 9.19

been measured extensively in experiments and simulations, both yielding the same value 1.7. When the dendrites grow in three dimensions instead of two, the dimension is about 2.4 to 2.5. The dependence of the dimension on the voltage has been studied also (see figure 9.20).

The mathematical model for diffusion limited aggregation can be extended and improved. For example, the sticking probability mentioned further above, which determines whether an ion close to the dendrite attaches to the structure or continues to wander around, is a parameter of interest. It allows variations of the small-scale structures. The smaller the sticking probability, the farther particles may reach in the fjords of the dendrite, thickening the dendrite to form a moss-like structure.[22]

Extensions of the DLA Model

Some interesting extensions of the simple model for DLA have been studied. Instead of tracing a single particle, many particles can be considered simultaneously.[23] Moreover, alternatively, the dendritic cluster may be allowed to move about, picking up particles that are close by. There is another seemingly unrelated very different model for DLA. Instead of tracing particles, an equation is solved which reflects that in reality there are infinitely many particles moving about simultaneously. Thus, in place of individual particles

(1988) 405–409 and the survey by A. Aharony, *Fractal growth,* in: *Fractals and Disordered Systems,* A. Bunde and S. Havlin (eds.), Springer-Verlag, Heidelberg, 1991.

[22]At large scales, however, the dendritic structure obtained using a small sticking probability does not look 'thick'. The fractal dimension, measured at large scales, is independent of the sticking probability.

[23]See R. F. Voss and M. Tomkiewicz, *Computer simulation of dendritic electrodeposition,* Journal of Electrochemical Society 132, 2 (1985) 371–375.

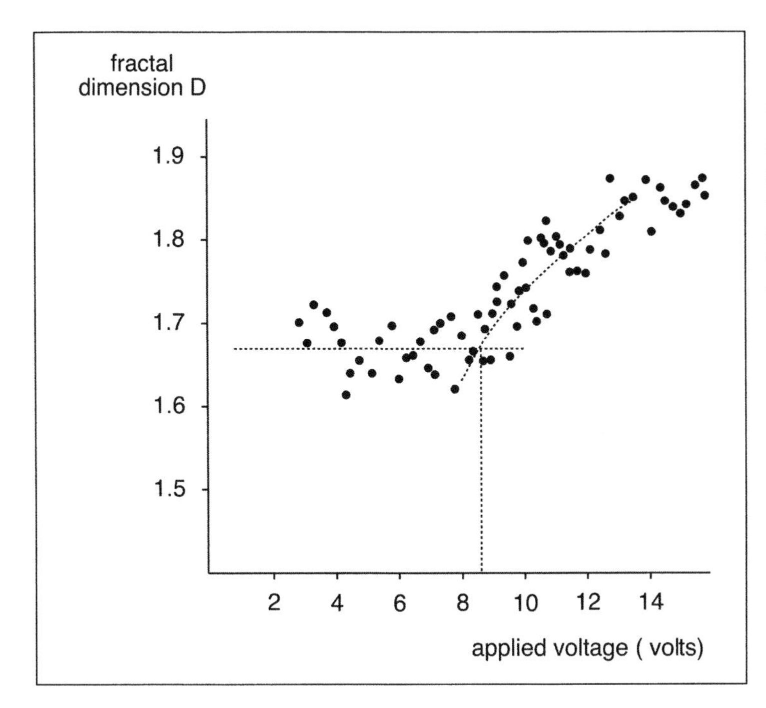

Fractal Dimension Versus Voltage

This graph shows the experimental results relating the fractal dimension of the DLA aggregate to the applied voltage. For low voltages the dimension seems to be about constant. Then there is a critical voltage after which the dimension grows abruptly.

Figure 9.20

some continuous density function is considered. The equation governing the electrostatic potential is a partial differential equation, known as the *Laplace equation*. Aggregation occurs along the boundary of the dendrite where the gradient of the potential is greatest. Sometimes fractals such as DLA clusters are therefore also called *Laplacian fractals*. It is easy to introduce a parameter in the Laplacian model which controls the dimension.[24] For visualizations of simulated DLA clusters and their electrostatic potential see color plates 33–37 and 39.

Phenomena similar to the aggregation discussed here occur at all scales of measurement, in distribution of galaxies as well as in the microcosm. In addition to diffusion limited aggregation and percolation, which we have already mentioned, a partial list of phenomena would range from molecular fractal surfaces to viscous fingering in porous media and clouds and rainfall areas.[25]

[24]For details see L. Pietronero, C. Evertz, A. P. Siebesma, *Fractal and multifractal structures in kinetic critical phenomena*, in: *Stochastic Processes in Physics and Engineering*, S. Albeverio, P. Blanchard, M. Hazewinkel, L. Streit (eds.), D. Reidel Publishing Company (1988) 253–278.

[25]For a discussion of these and other phenomena from a physical point of view see the book *Fractals* by J. Feder, Plenum Press, New York,1988.

9.4 Simulation of Brownian Motion

Brownian motion is not only important part of the model for diffusion limited aggregation, but it also serves as the basis of many other models for natural fractal shapes such as landscapes. In order to study these models it is necessary to better understand Brownian motion and its generalizations. In this and the following section we take a closer look at Brownian motion and methods for its simulation.

Before stating the results and the extensions, we simplify and consider Brownian motion in only one space variable. Thus, the motion of particles is restricted to a line. The tiny molecular impacts affect the particle only from the left or the right causing a displacement of a certain length l in either direction. Can we make any prediction about the total displacement after a number of such time steps, say n steps? If so, we could also simulate Brownian motion for larger time intervals, thus reducing the cost for the simulation.

The Mean Square Displacement

Let us solve this problem; it is not hard. First of all, we realize that it is not very sensible to ask for the total expected displacement, i.e., the displacement of a particle averaged over many samples. This would be *zero* because all individual displacements are $+l$ or $-l$, both with equal probability 0.5. Instead of the average overall displacement, let us consider the *square* of the displacement, a nonnegative number. The average of the square displacements, called the *mean square displacement*, tells us how much the particles spread in a given number of time steps on the average. The result of its computation is nl^2, the number of steps times the square of the step length l. Thus, the more steps we allow, the farther the particles spread out. Moreover, we have quantified this relation: *on the average* the square of the displacement is equal to the number of steps n, multiplied by l^2.

Computing the Mean Square Displacement

To compute the mean square displacement denote by $d_1, d_2, \ldots, d_n$ the n displacements of length l. We consider the value of the square

$$(d_1 + d_2 + \cdots + d_n)^2 = \sum_{i=1}^{n} \sum_{j=1}^{n} d_i d_j .$$

The individual terms $d_i d_j$ in the sum on the right-hand side are easy to analyze. Each factor is either $+l$ or $-l$ with the same probability 0.5, and, moreover, the factors are independent from each other in the case $i \neq j$. From this it follows that there are four cases for the product, which are all equally likely, as given in the table.

d_i	d_j	$d_i d_j$	Probability
l	l	l^2	0.25
l	$-l$	$-l^2$	0.25
$-l$	l	$-l^2$	0.25
$-l$	$-l$	l^2	0.25

Thus, the product is equal to $+l^2$ or $-l^2$ with the same probability 0.5, and the expected value of such a product for $i \neq j$ is zero. Of course, the value of the terms $d_i d_i$ are always equal to $+l^2$ for all $i = 1, \ldots, n$. Therefore, the result is clear: the expected value of the squared total displacement is equal to the number of steps, n, multiplied by the square l^2.

The number n of steps corresponds to the number of impacts on a particle and cannot be directly measured in an experiment. To arrive at more useful representation of Brownian motion we consider the time duration t. Assuming an average number of n impulses during a time span t, the particle travels a total length of nl. Denoting by v the average speed of the particle we get the relation $vt = nl$. Using this formula, we obtain for the mean square displacement nl^2 the expression $nl^2 = vlt$. In other words, the mean square displacement is *proportional* to the time span t. The factor of proportionality depends on the average speed v of the particle and the step length l. This is the fundamental property of Brownian motion, verified in 1908 in a series of seminal experiments by the French physicist Jean Perrin. It is also true in spaces having two or more dimensions.

Up to this point we know that the total displacement after some time t is zero on the average, and that the expected square of the displacement is proportional to t. What more can we say about the distribution of the displacement after time t? In other words, if we sample Brownian motion (or Brownian motion simulated on a computer) at regular time intervals of length t, what is the distribution of the resulting measured displacements? In table 9.22 we list the outcome of such an experiment, which is graphed in figure 9.21. For the sake of simplicity unit length displacements have been introduced here, i.e., the step length l was set equal to 1. In each time interval of length t, 100 unit length displacements have been carried out and added up. The sum is listed as the total displacement Δ during a time period of length t corresponding to the $n = 100$ steps.[26] The mean square displacement is 99.82, very close to the theoretically expected number 100.[27]

The shape of the curve in the graph is very familiar to most of us. It is a graph belonging to a distribution which is commonly known as the *Gaussian* or *bell-shaped* distribution. For example, consider the deviation in the body heights of a large group of people, or the variation of several measurements of the length of some (nonfractal) object. Sometimes the Gauss distribution is taken as a model for a statistically healthy sample — which may not always have desirable practical consequences. For example, grades in a class are often given so that the fluctuations of the grades around the average match the prescribed bell-shaped form. Taken to the extreme, this implies that in any class — no matter how brilliant the students may be — there must be a couple of students who flunk the course because Gauss' distribution demands it.

---

[26] Note that these sums must be even numbers, because $\Delta = a - b$, where $a + b = 100$ and a and b denote the number of times a positive or negative unit length displacement occurred. Thus, $b = 100 - a$, and $\Delta = 2a - 100$, an even number.

[27] With $l = 1$ and arbitrarily choosing $t = 1$, we obtain an average speed of $v = 100$. The expected mean square displacement thus is $nl^2 = vlt = 100$.

Statistics for Simulated One-Dimensional Brownian Motion

Graph corresponding to the data from table 9.22.

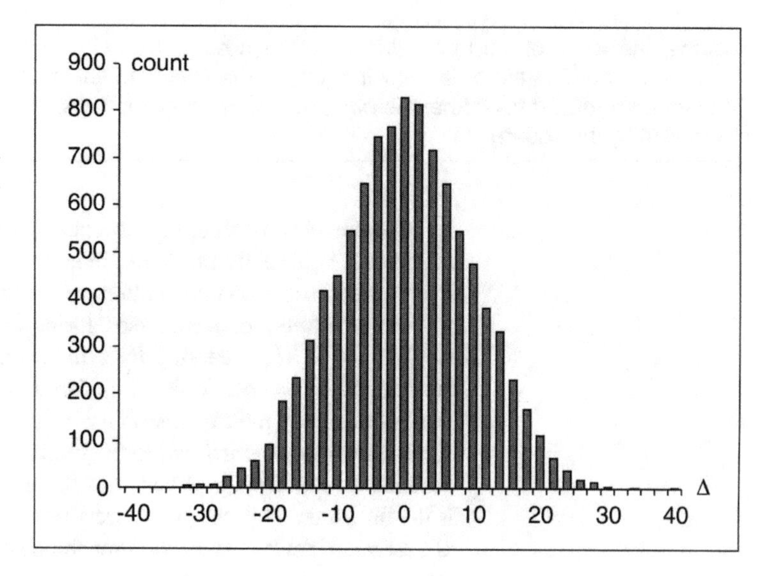

Figure 9.21

Brownian Motion in One Dimension

Displacements Δ of Brownian motion sampled $10,000$ times at regular time intervals. In each time interval 100 unit length displacements have been carried out and added up. For example, the displacement $\Delta = 14$ occurred 335 times out of the total of $10,000$ trials.

Δ	Count	Δ	Count	Δ	Count	Δ	Count
0	828			26	21	−26	28
2	815	−2	767	28	17	−28	9
4	718	−4	746	30	6	−30	10
6	648	−6	648	32	1	−32	7
8	547	−8	547	34	2	−34	1
10	478	−10	453	36	0	−36	0
12	383	−12	421	38	0	−38	0
14	335	−14	315	40	2	−40	1
16	233	−16	234	42	1	−42	0
18	171	−18	185	44	0	−44	0
20	116	−20	94	46	0	−46	0
22	66	−22	60	48	0	−48	0
24	42	−24	44	50	0	−50	0

Table 9.22

Returning to the results of the above experiment on Brownian motion in one dimension, we note that in this case they are not an accident, nor are they due to an arbitrary decision of some statistically minded individual that the outcome should match the Gaussian distribution so well. In fact, the Gaussian distribution arises in all cases where independent and similar (i.e., identically distributed) random events are summed up or averaged. This is the content of an important mathematical theorem called the *central limit theorem*.[28] Thus, the characterization of Brownian motion in one variable is now complete. The displacement after time t is a so-called random variable

[28]See any textbook on probability theory or statistics.

Pnts	Count	Pnts	Count	Pnts	Count	Pnts	Count
1	0	10	249	19	8503	28	2449
2	0	11	538	20	8961	29	1608
3	0	12	1033	21	9268	30	960
4	0	13	1573	22	9127	31	549
5	0	14	2541	23	8238	32	255
6	4	15	3574	24	7314	33	110
7	15	16	4836	25	5985	34	39
8	48	17	6051	26	4894	35	17
9	110	18	7527	27	3621	36	3

Throwing Six Dice 100,000 Times

Six dice are thrown 100,000 times. The points of all six dice are totalled up and a statistic of these sums is shown in the table.

Table 9.23

with a Gaussian distribution, which is specified by its mean zero and the mean square displacement proportional to the time difference t. Samples from such a Gaussian distribution, where the mean square is normalized to 1, are called (normalized) Gaussian random numbers.

Gaussian Random Numbers

From these observations it is clear that a simulation of Brownian motion can be based on such Gaussian random numbers. They are equivalent to the displacements corresponding to some time interval. If a displacement for a different time interval is desired, for example, a time twice as long, then simply multiply the Gaussian random number by the appropriate factor — here $\sqrt{2}$. There are efficient and accurate methods available for producing Gaussian random numbers.[29] For our purposes it is sufficient, however, to consider only a simple method based on the above-mentioned central limit theorem. We can even construct a Gaussian random number using the rolls of a die. This would initially produce random numbers from the list 1, 2, 3, 4, 5, 6, where each number in this set carries the same probability, 1/6, of being chosen. This is called a *uniform distribution* of a random variable. On most computers such random numbers are available with a much wider range of outcomes, usually $0, 1, 2, \ldots, A$ with $A = 2^{15} - 1$ or even $A = 2^{31} - 1$. If we divide the result by A then we obtain a number in the interval from 0 to 1; and the probability that the result of such an evaluation is, for example, between 0.25 and 0.75 is 50% or 0.50.[30] More generally, the probability, that the random number lies between a and b is $b - a$, when a and b are chosen with $0 \leq a \leq b \leq 1$. To simulate Gaussian random numbers, simply take any number of dice, e.g., 6, and roll them. Here the result will be defined as the sum of all dice values, which is a number from 6 to 36. Let us repeat the throw many times and keep a record of how many times we come up with each number between 6 and 36 (see table 9.23 and figure 9.24).

The distribution has the characteristic bell shape. In fact, the central limit theorem again ensures that the Gaussian distribution is approximated by the

[29] For example, the Box-Muller method, see W. H. Press, B. P. Flannery, S. A. Teukolski, W. T. Vetterling, *Numerical Recipes*, Cambridge University Press, Cambridge, 1986, p. 202.

[30] In many programming environments this division is internally carried out, and those random numbers are already uniformly distributed in the unit interval.

Approximate Gaussian Distribution by Throwing Dice

The data from table 9.23 of throwing six dice many times is drawn as a graph. The distribution is approximately Gaussian.

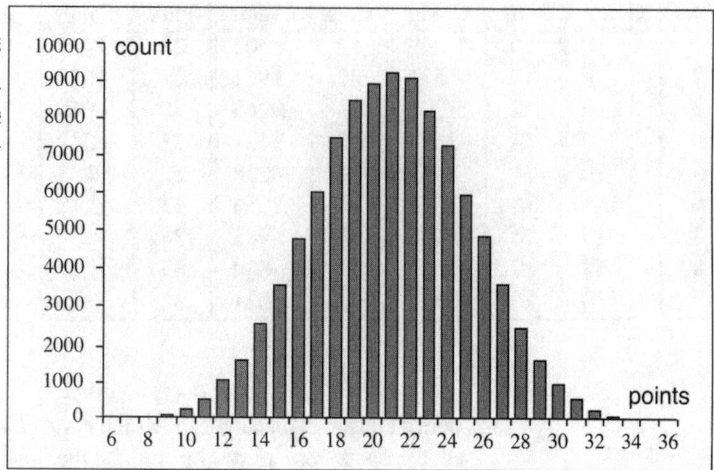

Figure 9.24

above experiment, and, moreover, the quality of the approximation is improved by raising the number of throws and the number of dice.

For practical purposes it is advisable to normalize the results before actually making use of them in a fractal construction. One reason is that the results do not belong to a Gaussian distribution centered around 0. For example, they are always positive, and the expected value, which is the average of all numbers depends on the number of dice used. The recipe for the normalization can easily be derived using elementary probability theory, but here we only state the final formulae. Let us define the notation

A	the upper limit of our random number generator, which returns numbers $0, 1, \ldots, A$ (as above)
n	number of 'dice' used
$Y_1, \ldots, Y_n$	results of one throw of all n 'dice'.

Then an approximate Gaussian random variable is given by

$$D = \frac{1}{A}\sqrt{\frac{12}{n}}(Y_1 + Y_2 + \cdots + Y_n) - \sqrt{3n}$$

when A and n are large. It is normalized so that the expected value is zero and the variance[31] is one. This formula is easily implemented on a computer. For our purposes it suffices to use a small number for n, e.g., $n = 3$. Then the formula simplifies to

$$D = \frac{2}{A}(Y_1 + Y_2 + Y_3) - 3.$$

[31] The variance is the mean square deviation from the expectation. In our case a variance equal to 1 implies that about 68.27% of all outcomes D are less than 1, 95.45% are less than 2, and 99.73% are less than 3 in magnitude.

In the special case above with six real dice, we have to take into account that the dice values range from 1 (and not 0) to a rather small maximum, 6. Using the exact variance we obtain

$$D = \sqrt{\frac{2}{35}}\,(Y_1 + \cdots + Y_6 - 21)\,.$$

The following table 9.25 is based on this formula.

Pnts		Pnts		Pnts		Pnts	
1		10	−2.63	19	−0.48	28	1.67
2		11	−2.39	20	−0.24	29	1.91
3		12	−2.15	21	0.00	30	2.15
4		13	−1.91	22	0.24	31	2.39
5		14	−1.67	23	0.48	32	2.63
6	−3.59	15	−1.43	24	0.72	33	2.87
7	−3.35	16	−1.20	25	0.96	34	3.11
8	−3.11	17	−0.96	26	1.20	35	3.35
9	−2.87	18	−0.72	27	1.43	36	3.59

Normalizing the Throw of Six Dice

The table lists the conversion of the dice sum to an approximate normalized Gaussian random number.

Table 9.25

The Next Step: Summing Up Independent Gaussian Random Numbers

The Gaussian random numbers above can be used in a simulation of Brownian motion in one dimension. Let us proceed in the time direction t in equal steps δt. Within each time slot of length δt we accumulate the impacts of all molecules that bump into our particle resulting in a total displacement which is correctly modelled as a Gaussian random number. We set the position of the particle at the starting time to 0, written in shorthand as $X(0) = 0$. After a time step of length δt we evaluate our (normalized) Gaussian random number, call the output D_1, and the position thus is changed to $X(\delta t) = D_1$. After two time steps we get another displacement, a number D_2 returned from our second call to the random number generator. The position is now the sum

$$X(2\delta t) = X(\delta t) + D_2 = D_1 + D_2.$$

Continuing in this way we sum up our Gaussian random numbers, in formula,

$$X(k\delta t) = D_1 + D_2 + \cdots + D_k,\quad k = 1, 2, 3, \ldots$$

The outcome is displayed in figure 9.26.

If an approximation is desired only at every other time, we can shorten the computation because we know that the mean *square* displacement for twice the time differences is also twice as large. Thus, a multiplication of the Gaussian random numbers by $\sqrt{2}$ suffices. In other words,

$$X(2k\delta t) = \sqrt{2}(D_1 + D_2 + \cdots + D_k),\quad k = 1, 2, 3, \ldots$$

Brownian Motion by Summing Up Gaussian Random Variables

Independent Gaussian random numbers (upper curve) ares summed up yielding a crude model of Brownian motion in one variable (bottom curve). The particle's position $X(t)$ is plotted in the vertical direction, horizontally time t is varied. The particle moves up and down without any correlation, i.e., if the particle gains height in one time step, the chance for a continuation and the chance for a change of this trend are exactly the same (50 : 50).

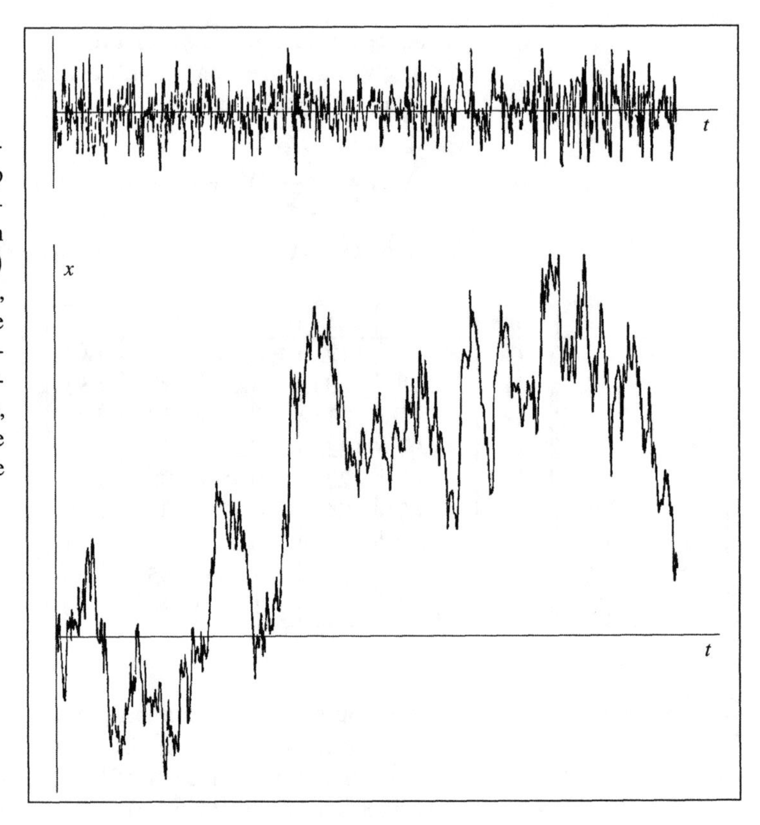

Figure 9.26

An Alternative: The Random Midpoint Displacement Method

Another straightforward and the most popular way to produce Brownian motion is called *random midpoint displacement*.[32] It has several advantages over the method of summing up white noise, the most important one being that it can be generalized to several dimensions useful, for example, for modelling height fields of landscapes.[33]

If the process $X(t)$ is to be computed for times t between 0 and 1, then we start out by setting $X(0) = 0$ and by selecting $X(1)$ as a sample of a Gaussian random number. Next $X(\frac{1}{2})$ is constructed as the average of $X(0)$ and $X(1)$, i.e., $\frac{1}{2}(X(0)+X(1))$ plus an offset D_1. Compare the visualization of this step and the next one in figure 9.27. The offset D_1 is a Gaussian random number, which should be multiplied by a scaling factor $\frac{1}{2}$. Then we reduce the scaling factor by $\sqrt{2}$, i.e., it is now $1/\sqrt{8}$, and the two intervals from 0 to $\frac{1}{2}$ and from $\frac{1}{2}$ to 1 are divided again. $X(\frac{1}{4})$ is set as the average $\frac{1}{2}(X(0) + X(\frac{1}{2}))$ plus an offset $D_{2,1}$, which is a Gaussian random number multiplied by the current

[32]The method was introduced in the paper by A. Fournier, D. Fussell and L. Carpenter, *Computer rendering of stochastic models*, Comm. of the ACM 25 (1982) 371–384.

[33]Another advantage is that we can prescribe the values of $X(t)$ for various times t and have the random midpoint displacement compute intermediate values. In this sense, the method could be interpreted as fractal interpolation.

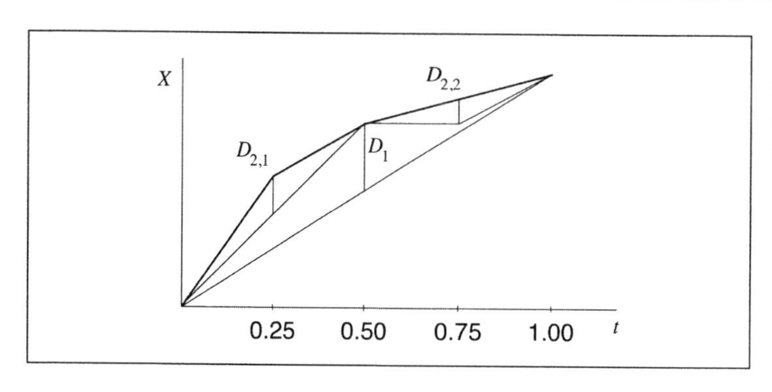

Displacing Midpoints

The first two stages of the midpoint displacement technique as explained in the text.

Figure 9.27

scaling factor $1/\sqrt{8}$. The corresponding formula holds for $X(\frac{3}{4})$, i.e.,

$$X(\tfrac{3}{4}) = \frac{X(\frac{1}{2}) + X(1)}{2} + D_{2,2}$$

where $D_{2,2}$ is a random offset computed as before.

The third stage proceeds in the same manner: reduce the scaling factor by $\sqrt{2}$, i.e., it is $1/\sqrt{16}$. Then set

$$\begin{aligned} X(\tfrac{1}{8}) &= \tfrac{1}{2}(X(0) + X(\tfrac{1}{4})) + D_{3,1} \\ X(\tfrac{3}{8}) &= \tfrac{1}{2}(X(\tfrac{1}{4}) + X(\tfrac{1}{2})) + D_{3,2} \\ X(\tfrac{5}{8}) &= \tfrac{1}{2}(X(\tfrac{1}{2}) + X(\tfrac{3}{4})) + D_{3,3} \\ X(\tfrac{7}{8}) &= \tfrac{1}{2}(X(\tfrac{3}{4}) + X(1)) + D_{3,4}. \end{aligned}$$

In each formula, D_3 is computed as a (different) Gaussian random number multiplied by the current scaling factor $1/\sqrt{16}$. The following step computes $X(t)$ at $t = 1/16, 3/16, \ldots, 15/16$ using a scaling factor again reduced by $\sqrt{2}$, and continues as indicated above and illustrated in figure 9.28.

Analysis of the Random Midpoint Displacement Method

If the Brownian motion is to be computed for times t between 0 and 1, then one starts by setting $X(0) = 0$ and selecting $X(1)$ as a sample of a Gaussian random variable with mean 0 and variance (mean square) var $(X(1)) = \sigma^2$. Then var $(X(1) - X(0)) = \sigma^2$ also, and we expect

$$\text{var}\ (X(t_2) - X(t_1)) = |t_2 - t_1|\sigma^2 \tag{9.1}$$

for $0 \le t_1 \le t_2 \le 1$. We set $X(\frac{1}{2})$ to be the average of $X(0)$ and $X(1)$ plus some Gaussian random offset D_1 with mean 0 and variance Δ_1^2. Then

$$X(\tfrac{1}{2}) - X(0) = \frac{1}{2}(X(1) - X(0)) + D_1$$

and thus $X(\frac{1}{2}) - X(0)$ has mean value 0 and the same holds for $X(1) - X(\frac{1}{2})$. Further, for (9.1) to be true we must require that

$$\text{var}\ (X(\tfrac{1}{2}) - X(0)) = \frac{1}{4}\text{var}\ (X(1) - X(0)) + \Delta_1^2 = \frac{1}{2}\sigma^2.$$

Eight Stages of Midpoint Displacement

Brownian motion via midpoint displacement. Eight stages are shown depicting approximations of Brownian motion using $3, 5, 9, \ldots, 257$ points.

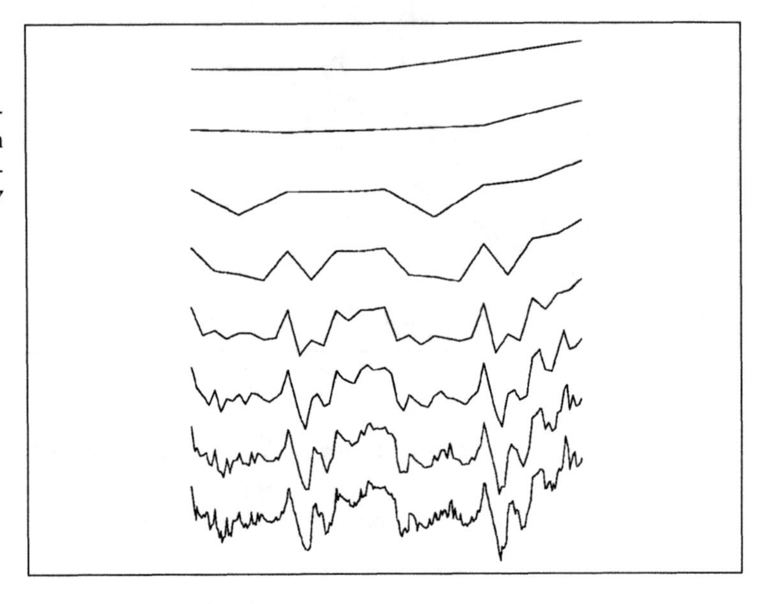

Figure 9.28

Therefore

$$\Delta_1^2 = \frac{1}{4}\sigma^2.$$

In the next step we proceed in the same fashion setting

$$X(\tfrac{1}{4}) - X(0) = \frac{1}{2}(X(0) + X(\tfrac{1}{2})) + D_2$$

and observe that again the increments in X, here $X(\frac{1}{2}) - X(\frac{1}{4})$ and $X(\frac{1}{4}) - X(0)$ are Gaussian and have mean 0. So we must choose the variance Δ_2^2 of D_2 such that

$$\text{var}\,(X(\tfrac{1}{4}) - X(0)) = \frac{1}{4}\text{var}\,(X(\tfrac{1}{2}) - X(0)) + \Delta_2^2 = \frac{1}{4}\sigma^2$$

holds, i.e.,

$$\Delta_2^2 = \frac{1}{8}\sigma^2.$$

We apply the same idea to $X(\frac{3}{4})$ and continue to finer resolutions yielding

$$\Delta_n^2 = \frac{1}{2^{n+1}}\sigma^2$$

as the variance of the displacement D_n. Thus, corresponding to time differences $\Delta t = 2^{-n}$, we add a random element of variance $2^{-(n+1)}\sigma^2$ which is proportional to Δt as expected.

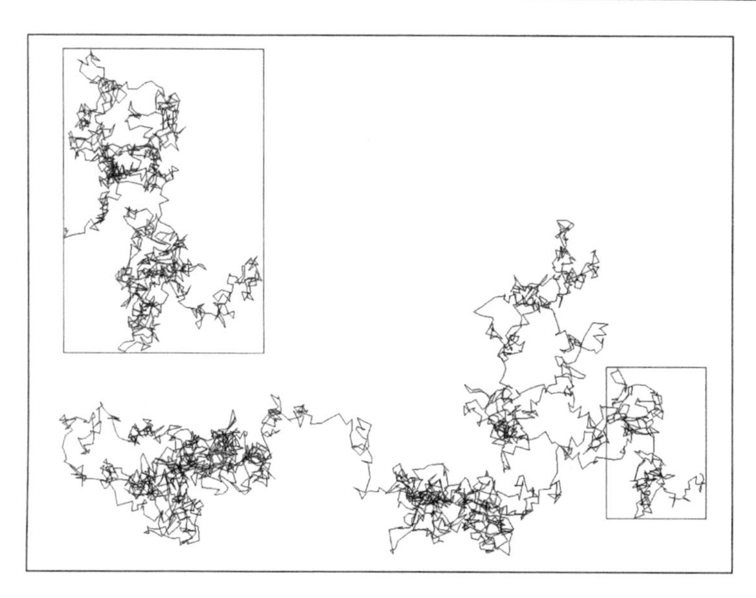

Trace of Brownian Motion in the Plane

Shown is the trace of the Brownian motion of a particle. The boxed detail of the trace (magnified in the upper left portion of the figure) suggests an invariance of scale or self-similarity: the detail looks like the whole.

Figure 9.29

Moving Up to the Next Degree of Freedom

Having produced Brownian motion in one dimension it is now an easy task to generalize to the two-dimensional case. The small impacts on a particle are no longer restricted to only two possible directions, a bump from the left or a bump from the right. Rather the direction may be chosen arbitrarily from a range of angles between zero and 180 degrees, in radians from 0 to π.[34] All angles are equally likely; thus, in a simulation, a random variable with a uniform distribution will suffice. In summary, the displacement of the particle is computed by choosing the direction in this specified manner and the amount of the displacement as before by means of a normalized Gaussian random variable.[35]

A graphical record of the Brownian motion of a particle looks as expected, a very erratic trace (see figure 9.29). The particle wanders around without any pattern. Some regions of the plane are filled densely by the trace. In fact, the fractal dimension of such a trace is equal to two. The enlargement of a section of the path reveals the self-similarity of the motion, it looks very much the same as the whole curve. This resemblance is true, of course, only in a statistical sense and not exactly.

[34] It is not necessary to consider larger angles because the displacement may be positive or negative.

[35] Note that this is not the generalization of Brownian motion which yields height field models of landscapes mentioned earlier on page 452 (see also section 9.6).

9.5 Scaling Laws and Fractional Brownian Motion

What Is the Scaling Invariance in the Graph of One-Dimensional Brownian Motion?

Let us now return to the one-dimensional Brownian motion and discuss its self-similarities that define it as a fractal. By construction and also just by looking at the graph in figure 9.26 it is clear that we cannot expect a similarity of the usual type in which we can take the graph of the Brownian motion and scale it up or down in the time direction and in the amplitude (with possibly different scaling factors) to obtain the original graph. Such an exact affine self-similarity is obviously not possible due to the randomness in the generation mechanism. However, in figure 9.30 we have tried the construction of the scaled copies of the original anyway. Here we have used the enlargement factor of two for the horizontal direction, while the amplitudes were kept unchanged. We note that the curves do not look very similar; there is much less variation in the bottom curves where we have stretched time with factors $2, 4, 8, \ldots$, and 64.

In the next figure we repeat the experiment with the same factor of two in the horizontal as well as in the vertical direction. As we scale by two horizontally we also multiply amplitudes by two. This changes the curves dramatically as displayed in figure 9.31. Now the bottom curves have greatly increased variation in amplitude; the graphs look much more erratic.

From these observations we can conclude that between the two scaling factors 1 (figure 9.30) and 2 (figure 9.31), there should be a scaling factor r that yields curves that are visually the same, i.e., when scaling Brownian motion in time by a factor of 2 and in amplitude by a factor of r we see no striking general differences, even when we repeat the scaling procedure many times. To find this number r we may continue by trial and error. However,

Brownian Motion Rescaled Wrongly

In this experiment we scale up a sample of Brownian motion in one variable by factors of two in the horizontal direction, while maintaining amplitudes. The result over six such steps is given here with the original curve. Note how the peaks of the 'mountains' are shifted to the right when going down to the next curves. In each plot half of the data from the previous curve disappears due to clipping at the right boundary of the plot.

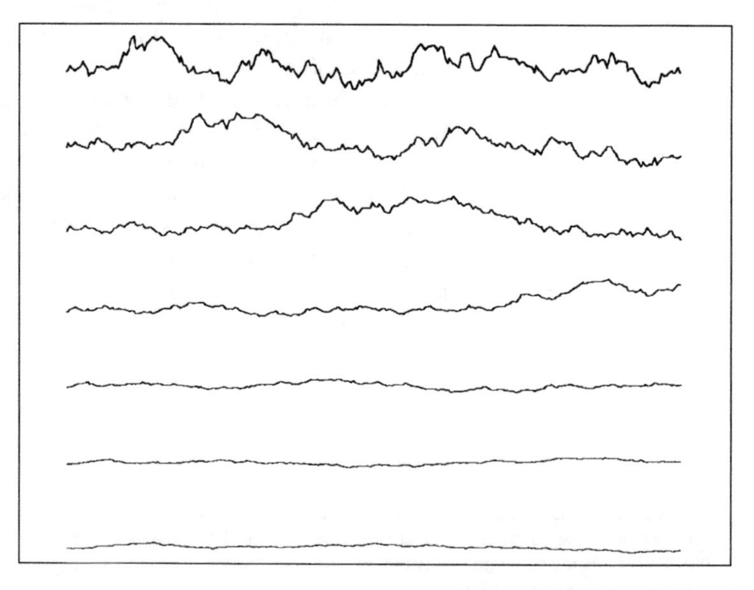

Figure 9.30

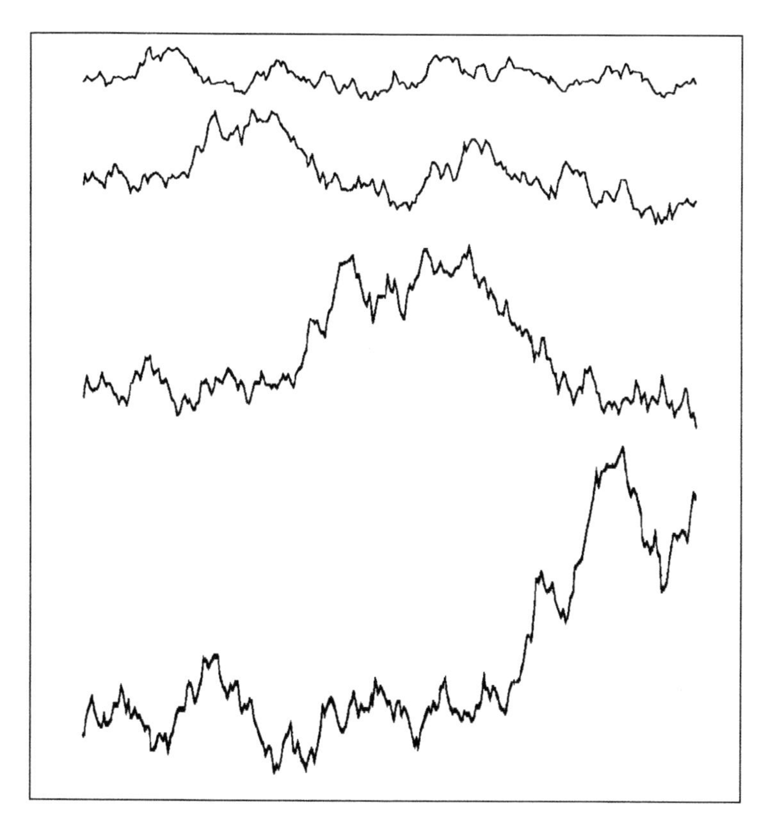

Brownian Motion Rescaled Wrongly Again

The same experiment as in figure 9.30, but with the same horizontal and vertical scaling factors 2.

Figure 9.31

the result is known to be $r = \sqrt{2}$. This follows directly from our analysis of the mean squared displacements Δ^2 of the Brownian motion $X(t)$, which showed proportionality to the time differences t,

$$\Delta^2 \propto t.$$

Consider the rescaled random function

$$Y(t) = rX\left(\frac{t}{a}\right),$$

i.e., the graph of X is stretched in the time direction by a factor of a and in the amplitude by r. The displacements in Y for time differences t are the same as those in X multiplied by r for corresponding time differences t/a. Thus, the squared displacements are proportional to r^2t/a. In order to ensure the same constant of proportionality as in the original Brownian motion, we simply have to require $r^2/a = 1$, or, equivalently, $r = \sqrt{a}$. When replacing t by $t/2$, i.e., stretching the graph by a factor of 2 as in the figures 9.30 and 9.31, we have $a = 2$, and thus, $r = \sqrt{2}$ as stated.

The last figure in this sequence, figure 9.32, demonstrates the result. Indeed, the curves look about the same. In fact, they are the same, statistically

Properly Rescaled Brownian Motion

The same experiment as in figure 9.30, but with horizontal scaling factor 2 and the proper vertical scaling factor $r = \sqrt{2}$. The curves are statistically equivalent, revealing the scaling law for Brownian motion. The shaded regions shows the same shape properly rescaled for the different stages.

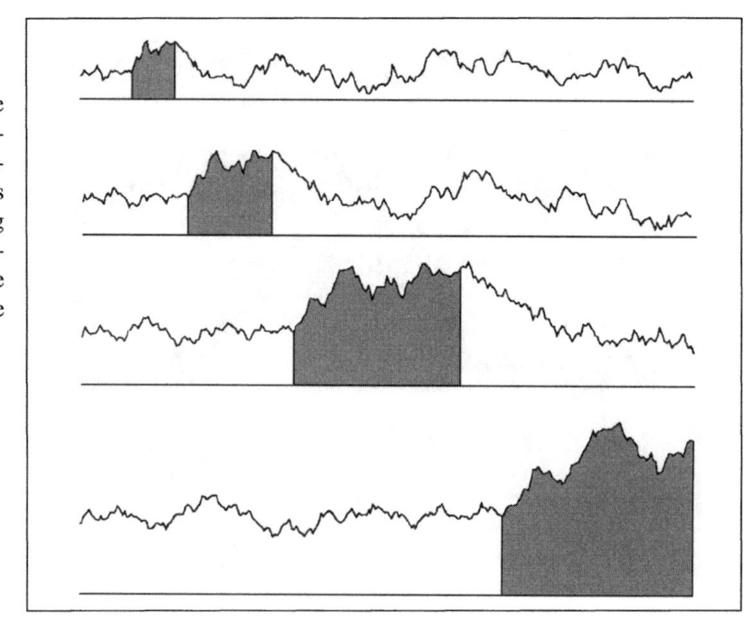

Figure 9.32

speaking. An analysis of mean values, variances, moments and so forth would give the same statistical properties of the rescaled curves. This is the scaling invariance of the graph of Brownian motion.

Are Other Scaling Factors Possible?

In the discussion of scaling invariance we have shown that for ordinary Brownian motion, we need to scale amplitudes by $\sqrt{2}$ when time (the horizontal direction) is scaled by a factor of 2. Scaling amplitudes by other factors, such as 1 or 2, changes the statistical properties of the graphs as the figures 9.30 and 9.31 show. Now we may ask the next logical question: for an arbitrary vertical scaling factor between 1 and 2 what would a curve look like if it *did* exhibit scaling invariance? Such curves in fact do exist and what they describe is called *fractional Brownian motion*. The figures 9.33 and 9.34 show examples for scaling factors $2^{0.2} = 1.148\ldots$ and $2^{0.8} = 1.741\ldots$ In general, fractional Brownian motion is characterized by the exponent that occurs in the scaling factor (0.2 or 0.8 in the figures above mentioned, 0.5 for ordinary Brownian motion). This exponent is usually written as H and sometimes called the *Hurst exponent,* after Harold Edwin Hurst, a hydrologist who did some early work, together with Mandelbrot, on scaling properties of river fluctuations. The proper range for the exponent is from 0, corresponding to very rough random fractal curves, to 1 corresponding to rather smooth looking random fractals. In fact, there is a direct relation between H and the fractal dimension of the graph of a random fractal. This relation is explained in a paragraph further below.

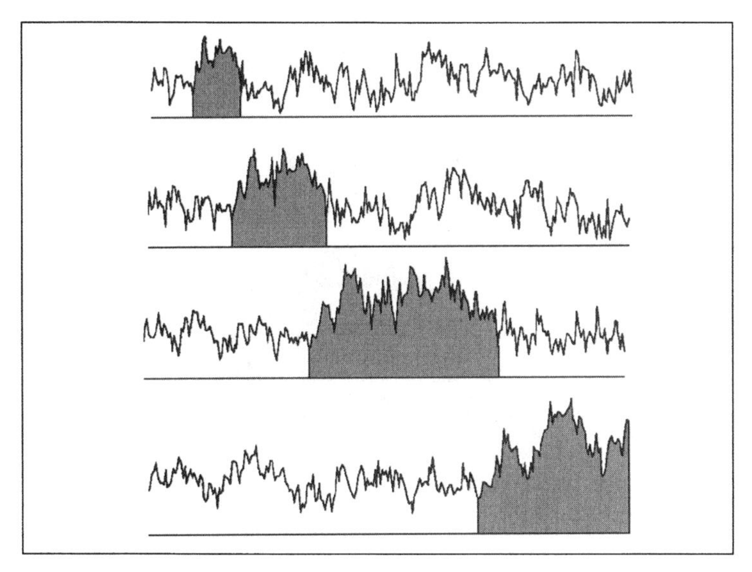

Fractional Brownian Motion 1

Properly rescaled fractional Brownian motion with vertical scaling factor $2^{0.2} = 1.148\ldots$ The curves are much rougher as compared to usual Brownian motion (see figure 9.32).

Figure 9.33

Fractional Brownian Motion and Statistical Self-Similarity

Ordinary Brownian motion is a random process $X(t)$ with Gaussian increments and

$$\text{var}\,(X(t_2) - X(t_1)) \propto |t_2 - t_1|^{2H},$$

where $H = \frac{1}{2}$. The generalization to parameters $0 < H < 1$ is called fractional Brownian motion. We say that the increments of X are statistically self-similar with parameter H, in other words

$$X(t) - X(t_0) \text{ and } \frac{X(rt) - X(t_0)}{r^H}$$

are statistically indistinguishable, i.e., they have the same finite dimensional joint distribution functions for any t_0 and $r > 0$. For convenience let us set $t_0 = 0$ and $X(t_0) = 0$. Then the two random functions

$$X(t) \text{ and } \frac{X(rt)}{r^H}$$

can be clearly seen as statistically indistinguishable. Thus 'accelerated' fractional Brownian motion $X(rt)$ is properly rescaled by dividing amplitudes by r^H.

Although possible, it is hard to obtain curves of different fractal dimensions by modifying the method of summing up white noise, the method described first on page 451 and in figure 9.26. But a small change in the random midpoint displacement method yields approximations of fractional Brownian motion. For a random fractal with a prescribed Hurst exponent $0 \leq H \leq 1$, we only

Fractional Brownian Motion 2

Properly rescaled fractional Brownian motion with vertical scaling factor $2^{0.8} = 1.741\ldots$ The curves are much smoother as compared to usual Brownian motion (see figure 9.32).

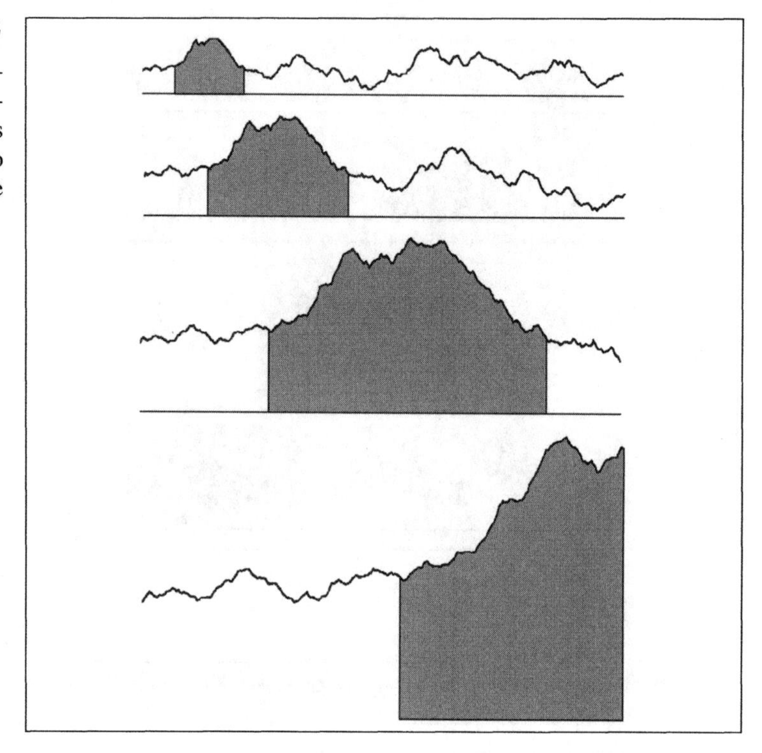

Figure 9.34

have to set the initial scaling factor for the random offsets to $\sqrt{1 - 2^{2H-2}/2^H}$; and in further steps the factor must undergo reductions by $1/2^H$.

The Relation Between H and Dimension D

In this section we give a simple formula for the fractal dimension of the graph of a random fractal. The graph is a line drawn in two dimensions. Thus its dimension should be at least 1 but must not exceed 2. In fact, the exact formula for the fractal dimension of the graph of a random fractal with Hurst exponent H is

$$D = 2 - H.$$

Thus, we obtain the whole possible range of fractal dimensions when we let the exponent H vary from 0 to 1. This yields dimensions D decreasing from 2 to 1.

Box-Counting Graphs of Fractional Brownian Motion

Let us employ the box-counting method for the estimation of fractal dimension of the graph of a random fractal $X(t)$. Recall, that all statistical properties of the graph remain unchanged when we replace $X(t)$ by $X(2t)/2^H$. Suppose we have covered the graph of $X(t)$ for t between 0 and 1 by N small boxes of size r. Now consider boxes of half the size $r/2$. From the scaling invariance of the fractal we see that the range of $X(t)$ in the first half interval from 0 to 1/2 is expected to be $1/2^H$ times the range of $X(t)$ over the whole interval. Of course the same holds for the second half interval from 1/2 to 1. For each half interval we would expect to need $2N/2^H$ boxes of the smaller size $r/2$. For both half intervals combined we therefore would need $2^{2-H}N$ smaller boxes. When we carry out the same idea again for each quarter interval, we will find again that the number of boxes must be multiplied by 2^{2-H}, i.e., we need $(2^{2-H})^2 N$ boxes of size $r/4$. Thus, in general, we get

$$(2^{2-H})^k N \text{ boxes of size } \frac{r}{2^k}.$$

Using the limit formula for the box-counting dimension and a little bit of calculus we compute

$$D = \lim_{k \to \infty} \frac{\log[(2^{2-H})^k N]}{\log \frac{2^k}{r}} = 2 - H.$$

This result is in accordance with chapter 4, page 203, where it is shown that the fractal dimension is equal to D if the number of boxes increases by a factor of 2^D when the box size is halved.

At this point, however, a word of caution must be given. It is important to realize that the above derivation implicitly fixes a scaling between the amplitudes and the time variable, which really have no natural relation. Therefore the result of the computation, the fractal dimension, may depend on the choice of this association of scales. This is particularly visible when one tries to estimate the dimension based on measurements of length.[36]

Fractional Brownian motion can be divided into three quite distinct categories: $H < \frac{1}{2}$, $H = \frac{1}{2}$ and $H > \frac{1}{2}$. The case $H = \frac{1}{2}$ is the ordinary Brownian motion, which has independent increments, i.e., $X(t_2) - X(t_1)$ and $X(t_3) - X(t_2)$ with $t_1 < t_2 < t_3$ being independent in the sense of probability theory; their correlation is 0. For $H > \frac{1}{2}$ there is a positive correlation between these increments, i.e., if the graph of X increases for some t_0, then it tends to continue to increase for $t > t_0$. For $H < \frac{1}{2}$ the opposite is true. There is a negative correlation between the increments, and the curves seem to oscillate more erratically.

[36] For details we refer to R. Voss, *Fractals in nature*, in: *The Science of Fractal Images*, H.-O. Peitgen, D. Saupe (eds.), Springer-Verlag, New York, 1988, pages 63–64 and B. B. Mandelbrot, *Self-affine fractals and fractal dimension,* Physica Scripta 32 (1985) 257–260.

9.6 Fractal Landscapes

The next big step is to leave the one-dimensional setting and to generate graphs that are not lines but surfaces. One of the first ways to accomplish this uses a triangular construction. In the end the surface is given by heights above node points in a triangular mesh such as shown in figure 9.35.

Triangular Mesh

The fractal surface is built over the mesh with surface heights specified at each of the node points.

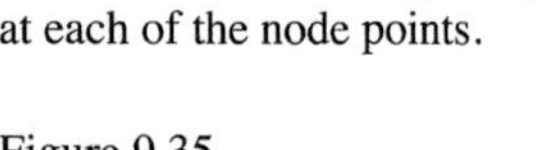

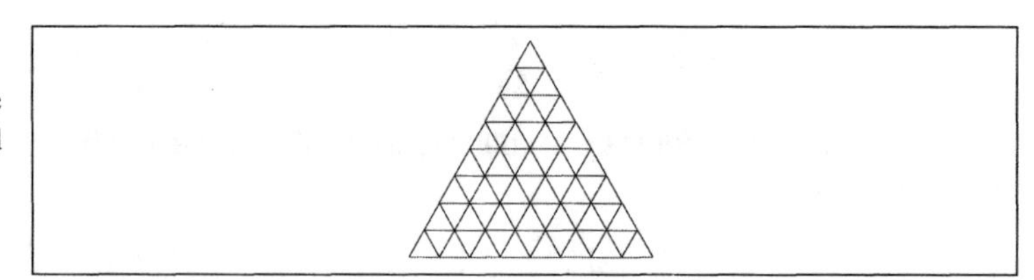

Figure 9.35

Extension to Two Dimensions Using Triangles

The algorithm proceeds very much in the spirit of the midpoint displacement method in one dimension. We start out with a big base triangle and heights chosen at random at the three vertices. The triangle is subdivided into four subtriangles. Doing this introduces three new node points, at which the height of the surface is first interpolated from the heights of its two neighbor points (two vertices of the original big triangle) and then displaced in the usual fashion. In the next stage we obtain a total of 16 smaller triangles, and heights for nine new points must be determined by interpolation and offsetting. The random displacements necessary in each stage must be performed using the same recipe as used in the usual midpoint displacement algorithm, i.e., in each stage we have to reduce the scaling factor for the Gaussian random number by $1/2^H$. Figure 9.36 shows the procedure and a perspective view of the first few approximations of the resulting surface.

The Method Using Squares

The actual programming of the fractal surface construction is made a little easier when triangles are replaced by squares. Going from one square grid to the next with half the grid size proceeds in two steps (see figure 9.37). First the surface heights over the midpoints of all squares are computed by interpolation from the heights of their four neighbor points and an appropriate random offset. In the second step the elevation of the remaining intermediate points are computed. Note that these points also have four neighbor points (except at the border of the square) whose heights are already known, provided the first step has been carried out. Again interpolation from the heights of these four neighbors is used and the result offset by a random displacement. Care has to be taken at the boundary of the base square, where the interpolation can only incorporate three neighbor points. The reduction of the scaling factors must also be modified slightly when using squares. Since there are two steps necessary to reduce the grid size by a factor of two, we should reduce the scaling factor in each step not by $1/2^H$ but rather by the square root $\sqrt{1/2^H}$. A sample of a result of the algorithm is presented in figure 9.38.

Let us note that the fractal dimension of the graphs of our functions are determined, just as in the case of curves, by the parameter H. The graphs are

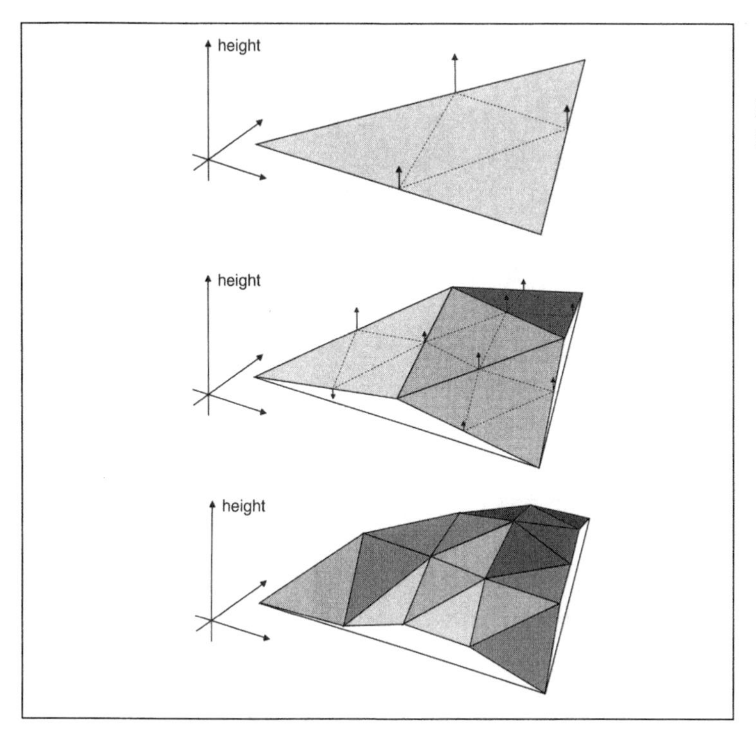

Fractal Surface on Triangle Base

Fractal construction of a surface using the tessellation of a triangle.

Figure 9.36

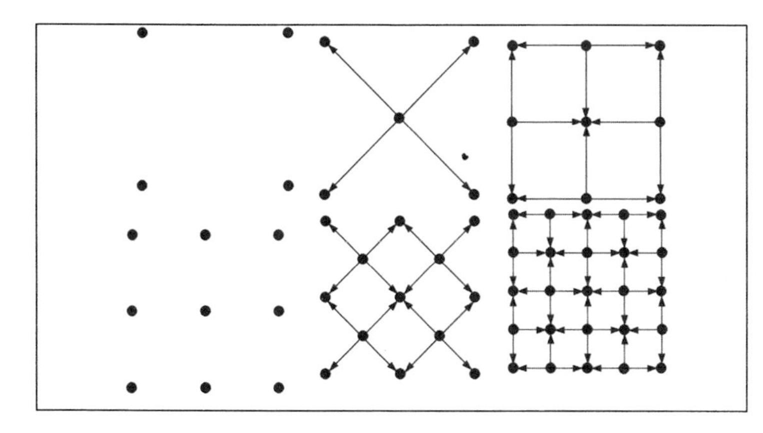

Square Mesh

The two refinement steps for the first two stages of the algorithm for generating fractal surfaces on a square by midpoint displacement are shown.

Figure 9.37

surfaces residing in a three-dimensional space, thus, the fractal dimension is at least 2 but not larger than 3: it is $D = 3 - H$.

Refinements and Extensions

Many refinements of the algorithm exist. The approximation of a true, so-called Brownian surface may be improved by adding additional 'noise', not only at the new nodes created in each step but to all nodes of the current mesh. This has been termed *random successive additions*. Another algorithm is based on the spectral characterization of the fractal. Here one breaks down

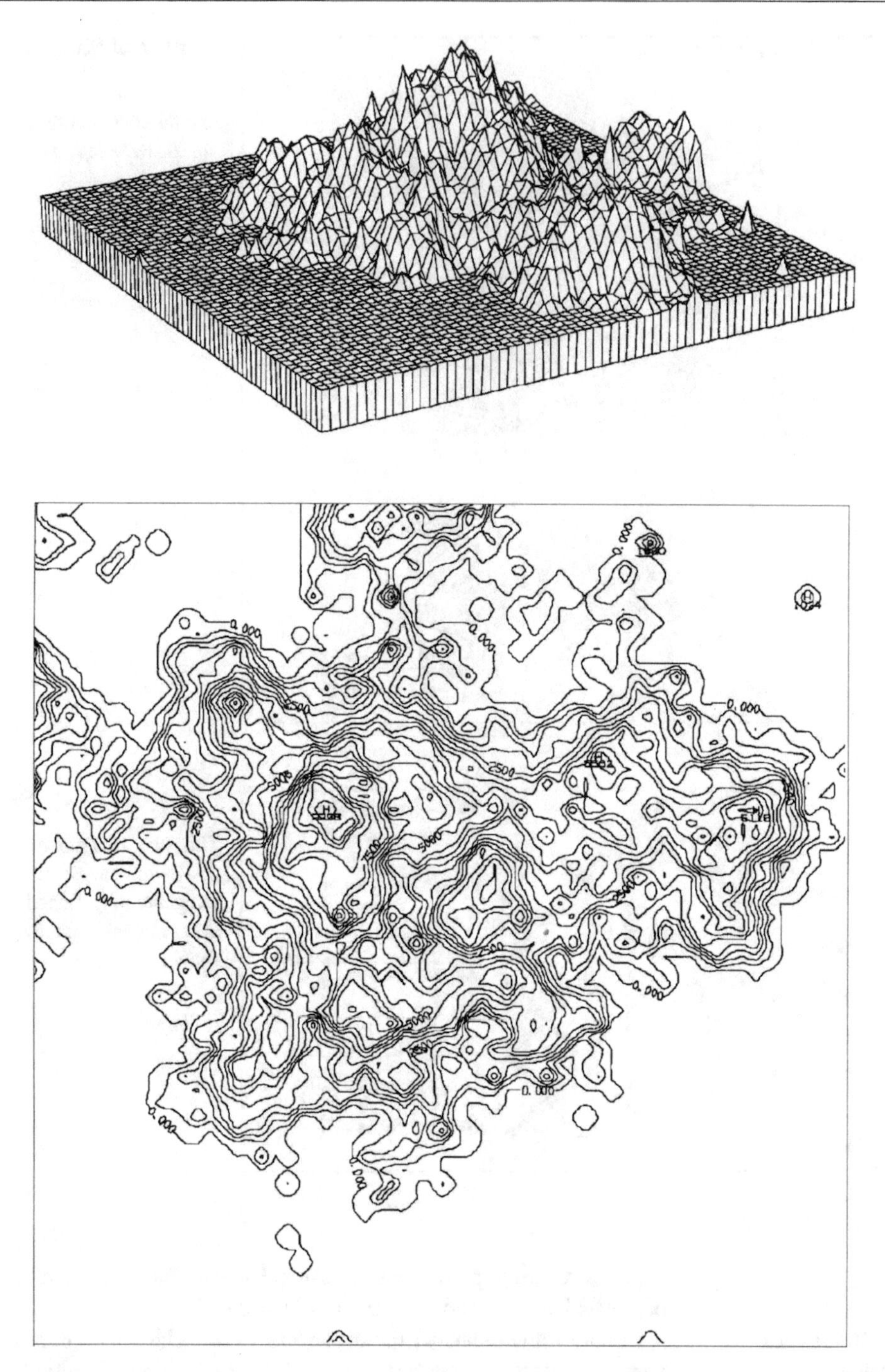

Figure 9.38 : A fractal landscape with corresponding topographical map. The midpoint displacement technique is applied for a mesh of 64 × 64 squares. Height values less than zero are ignored so that the resulting landscape looks like a rugged island with mountains.

the function into many sine and cosine waves of increasing frequencies and decreasing amplitudes.[37] Current research has focussed on local control of the fractal. For example, it is desirable to let the fractal dimension of the surface depend on the location. The 'valleys' of a fractal landscape, for example, should be smoother than the high mountain peaks. Of course the computer graphical representation of the resulting landscapes including the removal of hidden surfaces can be very elaborate; and proper lighting and shading models may provide topics for another whole book.[38]

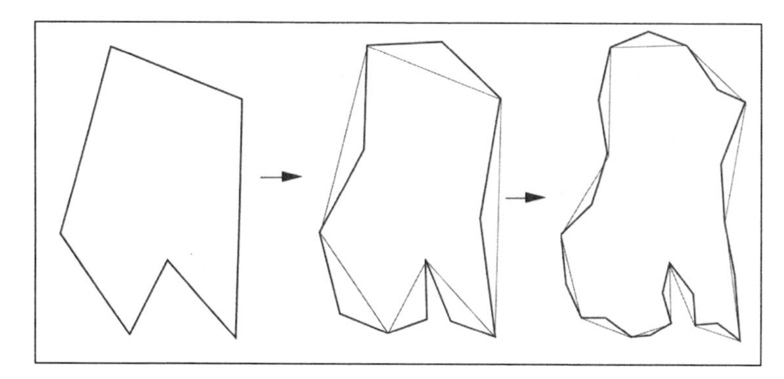

Simple Generation of a Fractal Coast

Generating a fractal coastline by means of successive random midpoint displacements.

Figure 9.39

Extracting Fractal Coastlines from Fractal Landscapes

In this final section we return to one of the leading questions, namely how to create imitations of coastlines. There are several ways. One cheap version is given first. It is a direct generalization of the midpoint displacement technique in one dimension (compare figure 9.39). We start out with a coarse approximation of the coastline of an island. The approximation could, for example, be done by hand, just plotting a polygon with a few vertices. Each side of the polygon is then subdivided simply by displacing the center point of the side along the direction perpendicular to the side. The amount of the displacement is determined by the Gaussian random number generator multiplied by a scaling factor as in the usual midpoint displacement algorithm. Thus, in this step the number of edges of the polygon is doubled. We may then repeat the step with the new sides of the refined polygon using a scaling factor for the random numbers which should be reduced by $1/2^H$. The parameter H between 0 and 1 again determines the roughness, i.e., the fractal dimension D, of the resulting fractal curve; the larger H the smoother is the curve. There are three shortcomings in this procedure.

1. The limit curve may have self-intersections.
2. There are no islands possible.
3. The statistical properties of the algorithm do not specify mathematically 'pure' random fractals, i.e., statistically the curves are not the same everywhere.

[37]Several more algorithms, including pseudo code, are discussed in the first two chapters of *The Science of Fractal Images,* H.-O. Peitgen and D. Saupe (eds.), Springer-Verlag, New York, 1988.

[38]See, for example, *Illumination and Color in Computer Generated Imagery,* R. Hall, Springer-Verlag, New York, 1988.

At least the first two problems of the above method are overcome by a more elaborate approach. The basis of it is a *complete* fractal landscape that may be computed by any method, e.g., by the method using squares, described in detail further above. One chooses an intermediate height value as a 'sea level' as in figure 9.38. The task is then to extract the corresponding coastline of the given fractal. The easiest way of accomplishing this is to push the subdivision of the underlying triangle or square so far that there are as many points computed as one wishes to plot in a picture, e.g., 513×513 for display on a computer graphics screen. Note that 513 is a good number, because $513 = 2^9 + 1$ and, thus, it occurs naturally in the subdivision process of the square. All height values, about a quarter-million in this case, are scanned and a black dot is produced at the appropriate pixel, provided the corresponding height value exceeds the selected sea level. The fractal dimension of the coastline is controlled by the parameter H that is used in the generation of the landscape. It is $D = 2 - H$, the same formula as for the fractal dimension of the graphs of fractional Brownian motion.

Two-Dimensional Fake Clouds

With a color computer display, convincing clouds can be generated very fast using the fractal landscapes. Consider such a landscape generated at some resolution of the order of about 513×513 mesh points as above. For each pixel there is an associated height value, which we now interpret rather as a color. The very high peaks of the mountains in the landscape correspond to white, intermediate height values to some bluish white, and the lowlands to plain blue. This is very easy to adjust using a so-called color map, which in most computer graphics hardware is a built-in resource. The display of a top view of this data with a one-to-one correspondence between the mesh points in the fractal and the pixels of the screen will show a very nice cloud. The parameter H in the fractal, which controls the fractal dimension, can be adjusted to the preferences of the viewer. The only drawback of such a rendering is that the model of the cloud is a two-dimensional one. There is no thickness to the cloud; a side view of the same object is impossible.

Animation of True Three-Dimensional Clouds

But the concept of fractals can be extended. We can produce random fractal functions based not only on a line or a square, but based on a cube. The function then specifies a numerical value for all points inside a cube. This value may be interpreted as a physical quantity such as temperature, pressure or water vapor density. The volume that contains all points of the cube with water vapor density exceeding a given threshold may be seen as a cloud. One can go even a step further. Clouds are fractal not only in their geometry but also in time. That is, we may introduce a fourth dimension and interpret random fractals in four variables as clouds that change in time, allowing animation of clouds and similar shapes.[39]

[39] This method has been used in the opening scene of the video *Fractals: An Animated Discussion,* H.-O. Peitgen, H. Jürgens, D. Saupe, C. Zahlten, Freeman, 1990.

Chapter 10

Deterministic Chaos: Sensitivity, Mixing, and Periodic Points

A dictionary definition of chaos is a 'disordered state of collection; a confused mixture'. This is an accurate description of dynamical systems theory today — or of any other lively field of research.

Morris W. Hirsch[1]

Mathematical research in chaos can be traced back at least to 1890, when Henri Poincaré studied the stability of the solar system. He asked if the planets would continue on indefinitely in roughly their present orbits, or might one of them wander off into eternal darkness or crash into the sun. He did not find an answer to his question, but he did create a new analytical method, the geometry of dynamics. Today his ideas have grown into the subject called topology, which is the geometry of continuous deformation. Poincaré made the first discovery of chaos in the orbital motion of three bodies which mutually exert gravitational forces on each other.

Others followed Poincaré's pioneering trail. In the former Soviet Union, for example, the mathematician Andrey Kolmogorow made basic advances in the irregular features of dynamics. By the 1960's, the American mathematician Stephen Smale had formulated a plan to classify all the typical kinds of dynamic behavior. Within Smale's world view, chaos found a place as a natural phenomenon completely on a par with such regular behavior as periodic cycles.

Practical applications of the concept of chaos as a natural phenomenon followed. For example, sometimes a fluid flows smoothly, but sometimes it becomes turbulent and irregular for no apparent reason. In attempting to explain why, two European mathematicians, David Ruelle and Floris Tak-

[1] In: *Chaos, Fractals, and Dynamics*, P. Fischer, W. R. Smith (eds.), Marcel Dekker, Inc., New York, 1985.

ens, suggested in 1970 that turbulent flow might be an example of dynamic chaos. At about the same time, chaos began to get attention in the sciences. Experimental scientists, notably the American physicists Harry Swinney and Jerry Golub and the French physicist Albert Libchaber, showed that Ruelle and Takens were partly right. Chaos does occur in turbulent flow but not in precisely the way they suggested.

This work raised important questions. How can chaotic models be tested experimentally, and how can different types of chaos be distinguished? The usual technique for testing a theory is to make a long series of observations and compare the results with the theoretical predictions. With chaos, however, the butterfly effect invalidates the results; they will vary widely because of even the slightest errors in the observations.

Following up on these questions, here are some which we would like to settle in this chapter. How can we make the notion of chaos more precise? How can we be sure that what looks like chaos is really chaotic and not just very complicated but perfectly predictable? For example, when we see a seemingly chaotic time series, how can we be sure that it is not periodic, just with an extremely long period? In other words, what are the signs of chaos? How can they be measured? What is the value of numerical calculations in the presence of chaos? How can we build examples of chaos which can be intuitively understood?

There are many dynamical systems that can produce chaos. However, in this chapter and the next the focus of our presentation is the iteration of only one particular transformation. It is the quadratic transformation which comes in different forms, for example, $x \to ax(1-x)$. This may seem like a rather artificial choice which may not bear much significance for the many other chaotic systems studied in mathematics and observed in physical experiments. However, on the contrary, it has turned out that the qualitative phenomena of the quadratic transformation are in fact the paradigm of chaos in dynamical systems. Moreover, for the quadratic transformation the properties of chaos can be observed and completely analyzed mathematically. Thus, their study is time well spent and later, in chapter 12, we will see how the quadratic transformation reappears as the basic chaos generator in other systems.

10.1 The Signs of Chaos: Sensitivity

Sensitivity Versus Stability — The Main Issue

In Chapter 1 we experienced a big surprise. The computer, paradigm of reliability and precision was knocked out by a simple feedback process, the quadratic iterator. The greatest problem that computers are confronted with when dealing with chaos is the extreme sensitivity of an iterator. It is reflected in a host of unexpected numerical phenomena which lead us to the conclusion that we must be very careful with the interpretation of the computer output. We will now reveal some of the difficulties that the machine cannot cope with in the presence of chaos.

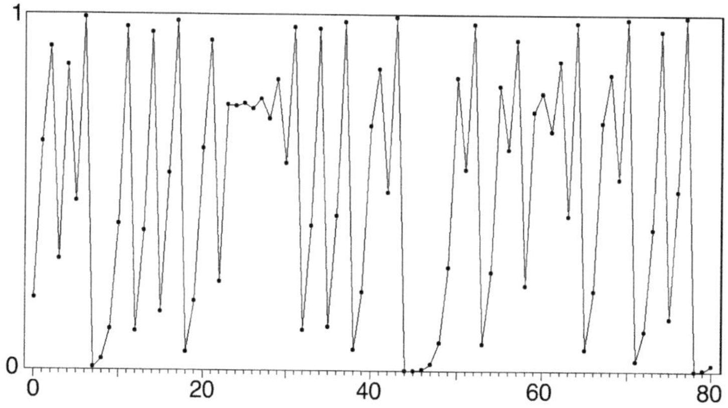

First Time Series

Time series for the quadratic iterator starting with $x_0 = 0.2027$ and parameter $a = 4$.

Figure 10.1

We have seen three different versions of the quadratic iterator: $x \to x^2 + c$ (for $c = -2$), $p \to p + rp(1-p)$ (for $r = 3$) and $x \to ax(1-x)$ (for $a = 4$). In this chapter we will base our discussion mainly on the quadratic iterator of the form $x \to ax(1-x)$. We start our tour with a couple of very simple time series experiments. Figure 10.1 shows the computed time series of x-values starting at $x_0 = 0.2027$ with the parameter set at $a = 4$. This is called the *orbit* of x_0. On the horizontal axis the number of iterations ('time') is marked, while on the vertical axis the amplitudes (ranging from 0 to 1) are given for each iteration. The points are connected by line segments. The picture shows an irregular pattern much like stock market indices, which are difficult to predict. The only thing that one can be sure of seems to be the fact that the graph cannot escape the bounds 0 and 1. This is clear from the formula $ax(1-x)$: when x is between 0 and 1, then so is $ax(1-x)$ when $0 \leq a \leq 4$.

Now let us look at the second time series in figure 10.2, which is based on the same formula and the same initial value. The only difference lies in the choice of the parameter a. At first glance the graph looks just as chaotic as the previous one. But this is categorically wrong; the two could not be any more different from a qualitative point of view. If you consider the last quarter of the second graph you will notice a periodicity. In fact, the graphics suggest that the orbit settles in on a cycle with an apparent period 5 or 10.

Second Time Series

Time series for the quadratic iterator starting with $x_0 = 0.2027$ and parameter $a = 3.742718$.

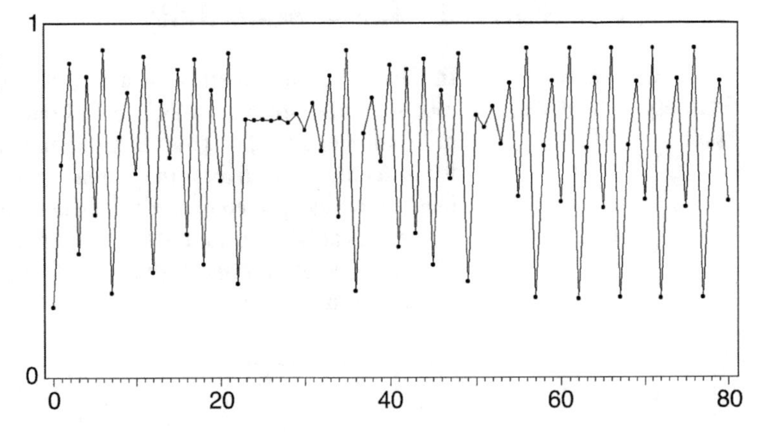

Figure 10.2

Stability

Nonsensitive, stable behavior: all initial values between 0 and 1 lead to the same final state. $a = 2.75$.

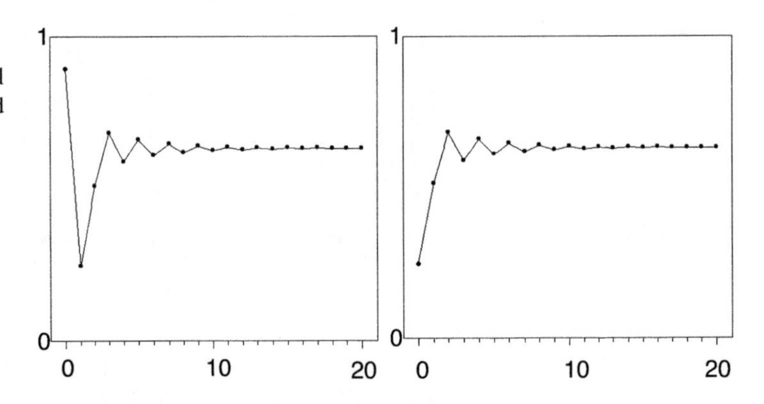

Figure 10.3

Every fifth or tenth iteration seems to give the same amplitude, and the cycle repeats.[2] Although this sort of behavior is not totally trivial, computers will not be seriously troubled by such phenomena. We will discuss the scenario of periodic cycles much more in chapter 11, but at this point let us already make the difference clear.

Figure 10.3 shows a very similar, but simpler, case. Here the two orbits shown converge to a single value x_∞ rather than some periodic orbit. It does not matter where we start the iteration, we will always end up at this same *final state*. The underlying mechanism is well observable by means of *graphical iteration*, which was introduced in chapter 1.[3] We briefly summarize: the left part of figure 10.4 shows the graphical iteration starting at the initial point $x_0 = 0.22$. The parabola is the graph of the iteration function $ax(1-x)$ and is the locus of points (x_n, x_{n+1}), because by definition of the iteration one

[2] Looking more closely at the numerical values of the iterates we see that the period is really 20.
[3] See page 57.

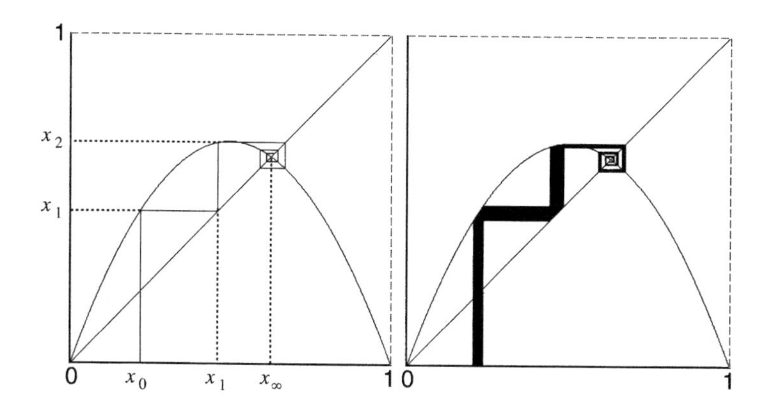

Attraction to a Point

Graphical iteration of an initial point leads to an attracting final state, the intersection of the parabola and the diagonal (left). On the right the iteration of an entire interval of initial values is contracted into the final stable state.

Figure 10.4

must have

$$x_{n+1} = ax_n(1 - x_n).$$

Thus, above x_0 the parabola has height x_1. On the diagonal at the same height we find the point with coordinates (x_1, x_1). Above that point we find the parabola with height x_2, and so on. Continuing the graphical iteration leads us to the intersection of the parabola and the diagonal, this is the point (x_∞, x_∞).

In the right half of figure 10.4 we repeat the experiment with a variation. Not only do we iterate the same initial point $x_0 = 0.22$, but also all values between $x_0 - 0.015$ and $x_0 + 0.015$, in effect iterating an entire interval. We observe that all values in the interval are attracted by the same final state, and this is an example of stability, which definitely cannot survive in the presence of chaos.

Sensitivity Amplifies Even the Smallest Error

The phenomenon of sensitivity, however, magnifies even the smallest error. This is demonstrated dramatically by setting the parameter a to the value 4. Repeating the experiment from figure 10.4 we obtain the plot shown in figure 10.5. The initial small interval has already grown considerably after just a few iterations. Allowing some more iterations would show us that every number from the whole interval [0,1] will be covered. To convince ourselves that this is not an artifact resulting from the width of the initial interval being too large, we repeat the same experiment with an even smaller initial interval. In figure 10.5 (right) the width is only 0.0005, less than in our imitation of the Lorenz experiment; see page 48 in chapter 1. What was shown there in a table as plain numbers can be seen here in graphical iteration: even the smallest deviation will escalate in the course of the iteration.

Let us summarize. Sensitivity implies that any arbitrarily small interval of initial values will be enlarged significantly by iteration. More precisely, this behavior is called *sensitive dependence on initial conditions*. For the quadratic iterator discussed here it is even true that any interval will expand under iteration to the full interval from 0 to 1.

Sensitivity

Sensitivity demonstrated by graphical iteration: in the course of the iteration even a small deviation increases substantially (left). The experiment is repeated with an even smaller interval of initial points (right).

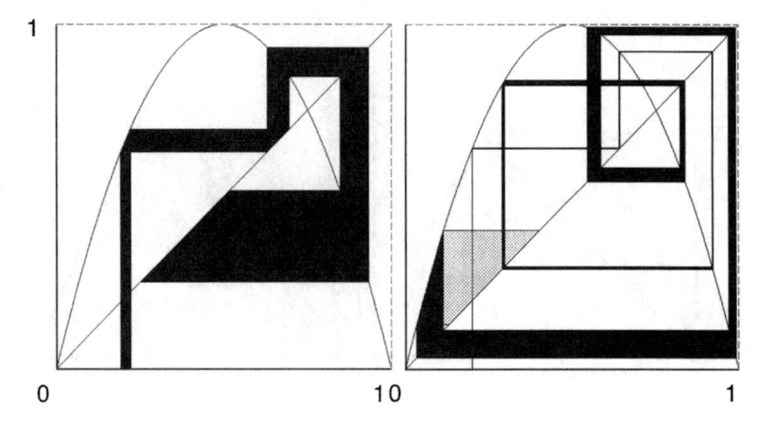

Figure 10.5

Example of a Linear Transformation

The property of sensitivity is central to chaos. Sensitivity, however, does not *automatically* lead to chaos. Indeed, there are sensitive systems which certainly do not behave chaotically. This is demonstrated with the simple example $x \to cx$, where the parameter c is greater than 1. This is a linear transformation, and it exhibits sensitive dependence on initial conditions: any deviation is magnified during the course of the iteration (see figure 10.6). Given an initial point x_0, then after n iterations we have $x_n = c^n x_0$. The deviations or errors of an orbit started nearby behave in the same fashion. Given an initial error $E_0 = \varepsilon$, i.e., we start with $u_0 = x_0 + E_0$, then after n iterations we have $u_n = c^n(x_0 + E_0)$ and the error has developed to

$$E_n = u_n - x_n = c^n(x_0 + \varepsilon) - c^n x_0 = c^n \varepsilon.$$

Therefore, any deviation ε is magnified by the factor $c > 1$ in each iteration. The system is sensitive, but it is certainly not chaotic.

The Characterization by Lorenz

The meteorologist Lorenz described the sensitivity of chaotic systems in a way that differentiates between the quadratic iterator and the tame linear transformation. Given an initial deviation, then in a chaotic system, it will become as large as the true 'signal' itself. In other words, after some iterations the error will be in the same order of magnitude as the correct values. Consider the quadratic iterator ($a = 4$) and the initial value $x_0 = 0.202$ again. Introducing a very small deviation of $\varepsilon = 10^{-6}$, we have another nearby initial value $u_0 = 0.202001$. Figure 10.7 shows the corresponding time series $x_0, x_1, x_2, x_3, \ldots$ (top), $u_0, u_1, u_2, u_3, \ldots$ (center), and the development of the absolute value of deviation, i.e., the error $|u_n - x_n|$ (bottom). In other words, in the top the original signal is shown along with the approximating signal, while in the bottom the difference or error signal is displayed.

In the beginning the error remains very small. For about the first 15 iterations it is not distinguishable from zero in our figure. Although the error is growing during these iterations, it is still too small to be seen. But after it has attained some critical magnitude it seems to explode. For the following

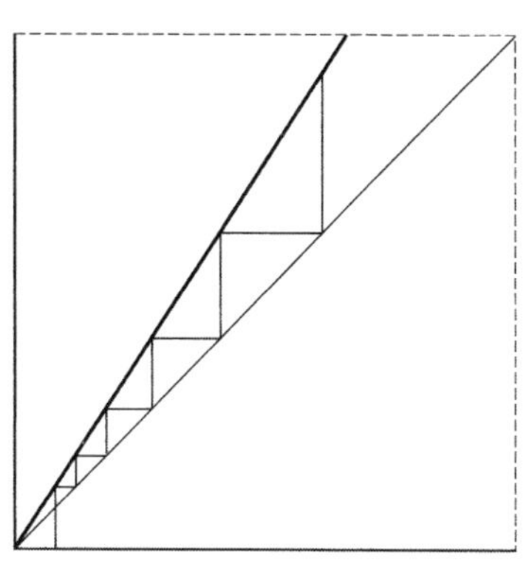
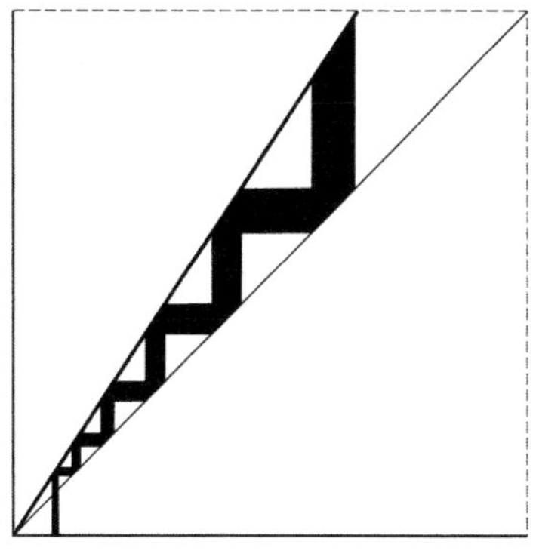

Sensitivity in Linear Systems

Iteration of the linear transformation $x \to 1.5x$. Left: single initial value; right: interval of initial values.

Figure 10.6

iterations the error signal looks just as erratic as the original time series. Moreover, the amplitudes of the error are of the same order as the amplitudes of the time series on the left, namely, 1. This behavior is typical for chaotic systems.

We can contrast this to the error development in the linear system $x \to cx$. Here the error and the true values grow in harmony. The relation between the signal, the values of the iteration, and the error remain the same at all times. The *relative* error computes as

$$\frac{E_n}{x_n} = \frac{c^n \varepsilon}{c^n x_0} = \frac{\varepsilon}{x_0} = \text{const.}$$

Thus, the iteration is quite harmless and may be safely carried out on a computer until the limit of the allowable range of numbers of the machine has been reached.

Analysis of Error Propagation

The phenomenon of sensitive dependence on initial conditions as described above is a *quality* that all chaotic systems definitely have. Other systems may or may not exhibit sensitivity. This is already a big step forward in the direction of understanding chaos. However, one of the major driving forces of science has always been to also *quantify* the qualities it has discovered. How would that be possible in the case of error propagation as discussed here?

Let us begin with the error propagation in the simple linear system $x \to cx$. Per iteration the error grows by a factor of c. Therefore

$$\left| \frac{E_n}{E_0} \right| = c^n. \tag{10.1}$$

The experiment in figure 10.7 clearly shows that such a simple law cannot be expected for the error propagation of the quadratic iterator $x \to 4x(1-x)$. On the other hand, during the first 15 iterations or so, the error must have grown more or less uniformly. This motivates another experiment. We count the number of iterations necessary for the error to exceed a certain threshold for the first time, say 0.1. We can do this for various initial values and also for different initial errors. We choose three initial values, 0.202, 0.347, and

Error Development

The quadratic iterator $x \to 4x(1-x)$ applied to two initial values differing by 10^{-6} (top and center) and the (absolute) difference of the two signals (bottom).

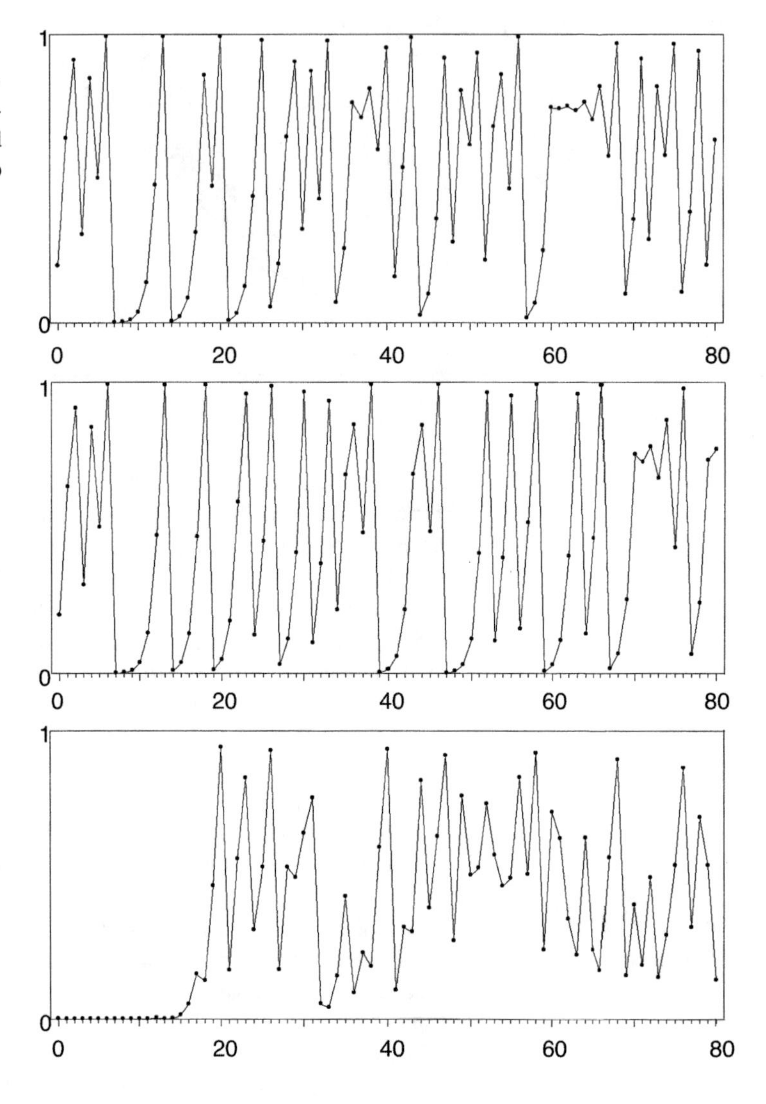

Figure 10.7

0.869, and four possible initial errors, to arrive at a total of 12 cases. Table 10.8 documents the results.

For example, when starting the iteration with the correct initial value 0.202 and comparing the iteration with that of a perturbed initial value 0.203, then the error of 0.001 at the beginning builds up in $n = 9$ steps to $E_n = 0.25622$. To derive a measure of how fast the error grows per iteration we take another look at eqn. (10.1). It relates the ratio E_n/E_0 to the factor c which characterizes the error growth. In our experiment above, we have E_n and E_0 conveniently at our disposal and can now pretend that these data come from a linear system. This allows us to derive the corresponding factor c, which numerically characterizes

x_0	Error E_0	Steps n	Error E_n	Exponent
0.202	0.001000	9	0.25622	0.61623
0.202	0.000100	11	−0.12355	0.64720
0.202	0.000010	15	0.25730	0.67703
0.202	0.000001	17	0.15866	0.70438
0.347	0.001000	7	−0.12331	0.68781
0.347	0.000100	11	−0.18555	0.68417
0.347	0.000010	15	0.31390	0.69028
0.347	0.000001	18	0.19490	0.67668
0.869	0.001000	8	−0.25072	0.69054
0.869	0.000100	10	0.14068	0.72491
0.869	0.000010	13	0.11428	0.71876
0.869	0.000001	18	0.32095	0.70439

Error Propagation

For three initial values orbits are started nearby using the quadratic iterator $x \to 4x(1-x)$. Four different initial errors are tried. After n steps the error has accumulated to a magnitude exceeding the threshold 0.1 for the first time; see the columns 'Steps n' and 'Error E_n'. The last column lists a measure of how much the error increases per iteration; see the text for an explanation of the exponent.

Table 10.8

the growth process of the error. To find the appropriate formula we take natural logarithms on both sides of eqn. (10.1)

$$\ln \left| \frac{E_n}{E_0} \right| = n \ln c.$$

Dividing by n yields

$$\ln c = \frac{1}{n} \ln \left| \frac{E_n}{E_0} \right| . \tag{10.2}$$

The quantity on the right side thus gives us the logarithm of the error growth constant c. This number is listed in the last column of table 10.8, headed by 'exponent'. It is interesting to note that the results are all about 0.7. Thus, the factor c is

$$c \approx e^{0.7} \approx 2.$$

The result of the experiment is thus the following:

- *In the quadratic iterator $x \to 4x(1-x)$, small errors will roughly double in each iteration.*

Of course, the errors can double only if they are sufficiently small. Moreover, the error doubling occurs only on the average. There are points x in the unit interval where small errors do not magnify. For example, this is the case near $x = 0.5$, where the graph of the parabola is rather flat, and errors are compressed. On the other hand, near the end points of the unit interval, errors are enlarged by factors up to 4.

The Ljapunov Exponent

The reasoning above directly leads to the concept of *Ljapunov exponents* $\lambda(x_0)$. It quantifies the average growth of infinitesimally small errors in the initial point x_0. Indeed, in eqn. (10.2) and table 10.8 we have derived a method to approximate the exponent. It is interesting that it seems to be independent

Alexander Michailowitsch Ljapunov, 1857–1918

Figure 10.9

of the initial value x_0. How can we make the computation more precise? In order to compute the average growth of an error it would be of advantage to consider many more iterations than just a dozen or so as in table 10.8. It seems that this necessarily implies that we start with a *very* small initial error E_0, because it will be roughly doubled in each iteration until it exceeds the threshold 0.1 for the first time. It is clear that this approach does not reach very far. Even if we reduce the initial error to the smallest possible given by the machine precision (which is of the order of 10^{-12} or so) we can expect to be able to perform only a few more iterations, but not hundreds and thousands of them. Thus, a different procedure is asked for. Let us assume that we can work with arbitrarily small initial errors E_0. We rewrite the total error amplification factor $|E_n/E_0|$ as

$$\left|\frac{E_n}{E_0}\right| = \left|\frac{E_n}{E_{n-1}}\right| \cdot \left|\frac{E_{n-1}}{E_{n-2}}\right| \cdots \left|\frac{E_1}{E_0}\right|$$

and attempt to estimate each factor separately! Assume for now that there is a workable method for this task. Now, however, when computing the total error amplification $|E_n/E_0|$ we must multiply all the individual factors which surely will result in a number much too large to be represented by an ordinary computer number (causing an overflow error). Fortunately we can avoid this because what we really are interested in is the geometric mean of the factors or rather the logarithm of the mean (see eqn. (10.2)). In formula,

$$\begin{aligned}\frac{1}{n}\ln\left|\frac{E_n}{E_0}\right| &= \frac{1}{n}\ln\left|\frac{E_n}{E_{n-1}} \cdot \frac{E_{n-1}}{E_{n-2}} \cdots \frac{E_1}{E_0}\right| \\ &= \frac{1}{n}\sum_{k=1}^{n}\ln\left|\frac{E_k}{E_{k-1}}\right| .\end{aligned}$$

Summing up the logarithms of the amplification factors surely avoids the overflow problems.

Now how can we estimate the error amplification factors in each iteration? Let us consider

$$\left| \frac{E_{k+1}}{E_k} \right| .$$

This describes by how much a small error E_k in x_k, the kth iterate, is enlarged (or reduced) in the following iteration. This error amplification factor is in essence independent of the size of the (small) error E_k. For example, if we consider an error in x_k being only half as large, i.e., $E_k/2$, then we can also expect the error in x_{k+1} to be half as large, i.e., $E_{k+1}/2$. Thus, it does not really matter whether we use the precise number E_k or just some arbitrary small error ε, say $\varepsilon = 0.001$.[4] This provides a solution to the problem; we fix an arbitrary small error ε and estimate the error amplification factors $|E_{k+1}/E_k|$ for very small initial errors E_0 by $|\tilde{E}_{k+1}|/\varepsilon$ where

$$\tilde{E}_{k+1} = f(x_k + \varepsilon) - f(x_k)$$

and $f(x) = 4x(1 - x)$ is the quadratic iterator. We do not have to worry about errors growing too large and are in the position to compute as many error amplification factors as we desire.

In summary, we have arrived at an improved and feasible method to compute the Ljapunov exponent more reliably by averaging over many iterations. The formula for this procedure is

$$\frac{1}{n} \ln \left| \frac{E_n}{E_0} \right| \approx \frac{1}{n} \sum_{k=1}^{n} \ln \left| \frac{\tilde{E}_k}{\varepsilon} \right| .$$

The results for the initial values used in table 10.8 are given in table 10.10. An error of 0.001 has been used in each iteration. As the number of iterations grows the exponent converges to $\lambda(x_0) = \ln 2 = 0.69314\ldots$

Measuring these exponents in this more careful analysis results in the number $\lambda(x_0) = 0.693$. With that we have succeeded in quantifying the sensitive dependence on initial conditions for the quadratic iterator. Now we are in a position to compare the sensitivity found here with that in other chaotic systems.

The Ljapunov exponent $\lambda(x_0)$ is a powerful experimental device to separate unstable, chaotic behavior from that which is stable and predictable and to measure these properties. Especially where $\lambda(x_0) > 0$ is large, sensitivity

[4] To justify this argument consider an error $E_k = \varepsilon$ in x_k, i.e., $\tilde{x}_k = x_k + \varepsilon$. Let $f(x) = 4x(1-x)$. Then the error E_{k+1} in x_{k+1} is

$$E_{k+1} = f(\tilde{x}_k) - f(x_k) = \tilde{x}_{k+1} - x_{k+1} = 4(x_k + \varepsilon)(1 - x_k - \varepsilon) - 4x_k(1 - x_k) = 4\varepsilon(1 - 2x_k) - 4\varepsilon^2.$$

Dividing the error by E_k yields $E_{k+1}/E_k = 4(1 - 2x_k) - 4\varepsilon$, a quantity dominated by $4(1 - 2x_k)$. The term -4ε is comparatively small. Thus, we see that the error amplification factor E_{k+1}/E_k is independent of the error $E_k = \varepsilon$ as long as ε is small. It is given by $4(1 - 2x_k)$.

For three initial values x_0 (compare table 10.8) Ljapunov exponents are calculated from averaging error amplification factors over varying numbers of iterations.

Table 10.10

Iterations	$x_0 = 0.202$	$x_0 = 0.347$	$x_0 = 0.869$
10	0.62475	0.68873	0.72803
100	0.69435	0.69360	0.69708
1000	0.69368	0.69199	0.69004
10000	0.69322	0.69290	0.69337
100000	0.69307	0.69306	0.69327

with respect to small changes in initial conditions is large. It is important to note that the concept of Ljapunov exponents has been generalized so that it applies to many interesting dynamical systems in mathematics and the sciences. It has become one of the keys to measuring, evaluating and detecting chaotic behavior.[5]

The Ljapunov Exponent for Smooth Transformations

The characterization of the Ljapunov exponent needs a bit of calculus. To this end we assume that we consider the iteration of a smooth transformation f, which certainly is the case when f is a polynomial such as the quadratic $f(x) = ax(1-x)$. In this iteration we have $x_{n+1} = f(x_n)$ for $n = 0, 1, 2, \ldots$ The equation (10.2) will guide us to an appropriate definition. First we rewrite the relative growth of the error after n steps as a product

$$\frac{E_n}{E_0} = \frac{E_n}{E_{n-1}} \cdot \frac{E_{n-1}}{E_{n-2}} \cdots \frac{E_1}{E_0}$$

and taking absolute values and natural logarithms we see that

$$\frac{1}{n} \ln \left| \frac{E_n}{E_0} \right| = \frac{1}{n} \sum_{k=1}^{n} \ln \left| \frac{E_k}{E_{k-1}} \right| .$$

By definition of the error terms we have

$$\frac{E_k}{E_{k-1}} = \frac{f(x_{k-1} + E_{k-1}) - f(x_{k-1})}{E_{k-1}}$$

and from calculus we obtain

$$\lim_{E_0 \to 0} \frac{E_k}{E_{k-1}} = f'(x_{k-1}).$$

Thus,

$$\lim_{E_0 \to 0} \frac{1}{n} \sum_{k=1}^{n} \ln \left| \frac{E_k}{E_{k-1}} \right| = \frac{1}{n} \sum_{k=1}^{n} \ln |f'(x_{k-1})|.$$

Now letting $n \to \infty$ we obtain the Ljapunov exponent

$$\lambda(x_0) = \lim_{n \to \infty} \frac{1}{n} \sum_{k=1}^{n} \ln |f'(x_{k-1})|. \tag{10.3}$$

[5] See for example H. G. Schuster, *Deterministic Chaos*, Physik-Verlag, Weinheim and VCH Publishers, New York, 1984.

Let us present an explicit formula for the Ljapunov exponent $\lambda(x_0)$ for the special case that the orbit of x_0 is periodic with a period $m > 0$. Thus,

$$x_m = f^m(x_0) = x_0.$$

Because of this periodicity we have

$$\lambda(x_0) = \frac{1}{m}\sum_{k=1}^{m} \ln|f'(x_{k-1})|.$$

In other words, the average of the logarithmic amplification factors, taken over one periodic cycle, is the same as the average taken over two, three, or more periods, and, thus, equal to the limit in eqn. (10.3). Thus, if x_0 is a fixed point, i.e., $m = 1$, we get

$$\lambda(x_0) = \ln|f'(x_0)|.$$

For a 2-cycle the Ljapunov exponent is

$$\lambda(x_0) = \lambda(x_1) = \frac{1}{2}(\ln|f'(x_0)| + \ln|f'(x_1)|).$$

In the case of the quadratic iterator, where $f(x) = ax(1-x)$ we get

$$\begin{aligned} \lambda(x_0) &= \lim_{n\to\infty} \frac{1}{n}\sum_{k=1}^{n} \ln|a - 2ax_{k-1}| \\ &= \ln a + \lim_{n\to\infty} \frac{1}{n}\sum_{k=0}^{n-1} \ln|1 - 2x_k|. \end{aligned}$$

Note that since $x = 0$ is a fixed point of the quadratic iterator, we have $\lambda(0) = \ln a$ (and also $\lambda(1) = \ln a$). For $a = 4$ this is approximately 1.39, which indicates the instability of this fixed point. For most other points[6] x between 0 and 1 we compute $\lambda(x) \approx 0.693$.

[6] To be technical and more precise, we note in passing that for $a = 4$ the result $\lambda(x) = \ln 4$ holds for all points x whose orbits eventually end in the fixed point 0. This is a dense subset of the interval $[0, 1]$. On the other hand, almost all points in the interval will have a Ljapunov exponent equal to $\ln 2 \approx 0.693$. This means that if an initial point is picked at random, then the orbit (almost surely) fills the interval densely and the exponent is $\ln 2$.

10.2 The Signs of Chaos: Mixing and Periodic Points

Let us now turn to what is called the *mixing behavior* of the quadratic iterator. In figure 10.5 we saw how a small error is amplified in the course of the iteration. But we can also interpret this figure from a slightly different point of view. The points of the small interval of initial values finally become spread over the whole unit interval (the numbers from 0 to 1). In fact, we can start with an arbitrary interval of initial values. When iterated they become spread over the whole interval.

Mixing

An intuitive way to interpret mixing is to subdivide the unit interval into subintervals and require that by iteration we can get from any starting subinterval to any other target subinterval. If this requirement is fulfilled for all finite subdivisions, we say that the system exhibits mixing.

Let us check an example for $f(x) = 4x(1-x)$ using the subdivision of $I = [0,1]$ into 10 subintervals of length 1/10,

$$I_k = \left[\frac{k-1}{10}, \frac{k}{10}\right], \quad k = 1, \ldots, 10.$$

Consider the starting subinterval $I_2 = [0.1, 0.2]$. It is transformed by iteration of the quadratic function giving the sequence of intervals listed in table 10.11.

Intervals Reached by Subinterval I_2

The subinterval $I_2 = [0.1, 0.2]$ is iterated four times. The subintervals which are reached at each iteration are marked with a bullet.

Table 10.11

Iteration	Interval	Intervals Reached									
n		I_1	I_2	I_3	I_4	I_5	I_6	I_7	I_8	I_9	I_{10}
0	$[0.100, 0.200]$		•								
1	$[0.360, 0.640]$				•	•	•	•			
2	$[0.922, 1.000]$										•
3	$[0.000, 0.289]$	•	•	•							
4	$[0.000, 0.822]$	•	•	•	•	•	•	•	•	•	

After one iteration subintervals I_4 to I_7 are reached. The next iterate produces only points from I_{10}. However, the third and fourth iteration yield points from the remaining intervals I_1, I_3, I_8, and I_9. We would have obtained a similar result if we had taken a starting subinterval different from I_2, or if we had subdivided the unit interval into 100, 1000, or even one million subintervals. This is the essence of mixing. Expressed informally, 'mixing is if we can get everywhere from anywhere'.

The Definition

A bit more formally we describe this mixing property in the following way:

- *For any two open intervals I and J (which can be arbitrarily small, but must have a nonzero length) one can find initial values in I which, when iterated, will eventually lead to points in J.*

Figure 10.12 demonstrates mixing for the quadratic iterator for the value $a = 4$. We show two very small intervals I and J and you can follow the iteration of one initial value taken from I which leads (after 11 steps) to a point in J. You might compare this with stirring a drop of milk into your coffee or mixing a small pocket of spice into dough by kneading it thoroughly. Indeed, this will be discussed in more detail in section 10.4.

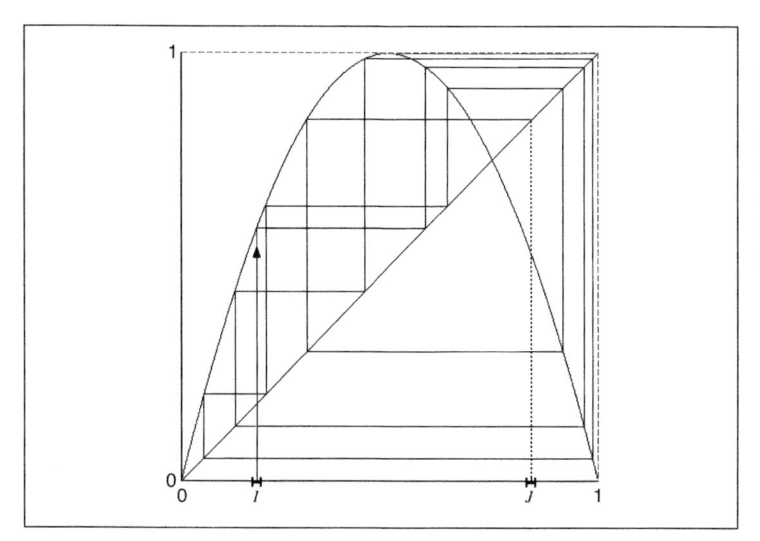

From I to J

Mixing: the iteration of an initial value from the interval I is spread all over the unit interval. For any interval J there is an initial point in I with an orbit that reaches J. In this example 11 iterations suffice. The parameter is $a = 4$.

Figure 10.12

Measuring the Mixing Property

Given a small interval I for initial points and a small target interval J we may choose points in I, compute their orbits and check whether they ever enter the target interval. Some orbits will accomplish this sooner than others. Some initial points may have orbits that never reach the target interval. Let us do a straightforward numerical experiment that collects some statistics about the behavior of these orbits and which leads to an interesting exponential law.[7] We select a family of $10{,}000$ initial points equally spaced in the interval I and follow their orbits until they have hit the target interval and are discarded. We call the orbits which remain after some number of iterations the survivors. How many survivors will there be after, say 100 iterations? After 1000 iterations? Is there some power law by which the number of survivors decays? The answers, of course, depend on the choice of the intervals. For example, we could arrange them so that all orbits fall into the target already in the first iteration. That would be rather uninteresting. Thus, let us make the interval I very small and choose J in such a way so that at least for a few dozen iterations no orbits land in the target interval J. For example, we may choose

$$I = [0.2, 0.2 + 10^{-11}], \quad J = [0.68, 0.69].$$

During the first 36 iterations none of the initial $10{,}000$ orbits reach the target. In the following iterations some of the orbits finally succeed; in

[7] Such an experiment has been carried out by James Yorke and also Robert Shaw. See R. Shaw, *Strange attractors, chaotic behavior, and information flow,* Z. Naturforsch. 36a (1981) 80–112.

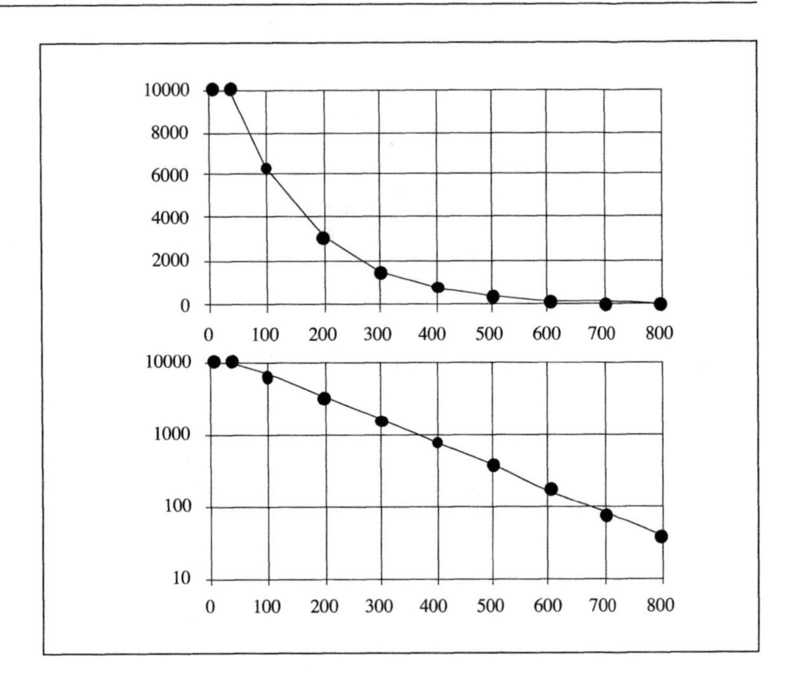

Figure 10.13 : Survivors $S(n)$ are plotted versus number n of iterations. Below the same data are shown logarithmically.

the 37th iteration 63 of them end up in the target set, then again 63, then 62, 63, and so on. When the number of survivors has decreased to about half the initial number of $10{,}000$, we expect that the decay has dropped to about one-half of 63 per iteration. In fact, we obtain a decrease of 31 orbits from 5037 survivors down to 5006 after 133 iterations. This leads us to the hypothesis that the number $S(n)$ of survivors decays exponentially with the number n of iterations according to

$$S(n) \propto e^{-\frac{n}{\tau}}. \tag{10.4}$$

The number τ can be interpreted as an estimate for the number of iterations necessary to reduce the survivors by a factor of $1/e \approx 0.368$ and is also called the average lifetime of the orbits. In order to check this formula we use a semi-logarithmic plot of $\log S(n)$ versus n (see figure 10.13).

After an initial transient period of 36 iterations the logarithm $\log S(n)$ decreases linearly which confirms the conjecture of exponential decay. Based on the results $S(100) = 6346$ and $S(600) = 171$ we estimate that

$$\tau \approx 138 \text{ iterations.}$$

These facts can be summarized as follows. The information contained in the small interval persists for a certain number of iterations depending on the size and location of the interval I. After that it is no longer possible to see that the orbits initially had started very close to each other. Moreover, once this state is achieved the number of the survivors decays exponentially, in this case with the rate of $1/e$ per 138 iterations. This number depends on the choice of the target interval J. Below we derive histograms and invariant distributions which will reconfirm the numerical result presented here.

Small Intervals Iterated Expand to $[0, 1]$

The quadratic iterator exhibits not only mixing as specified in the above definition, but an even stronger type of spreading. Take an arbitrary small subinterval and consider its iteration. In each step the result will be an interval again, however, of different size and location. Small intervals will grow in size on average and after a certain finite number of iterations the whole unit interval is covered. For example, carrying out one more iteration of $I_2 = [0.1, 0.2]$ in table 10.11 would yield the entire unit interval,

$$f^5([0.1, 0.2]) = f([0.000, 0.822]) = [0, 1].$$

In other words, any given point of the unit interval has preimages in the small initial subinterval. Or, any subinterval will expand to the full unit interval in the course of the iteration. The following question now comes naturally. On the average, how many iterations are necessary for this to happen when considering the collection of subintervals of a given width? Table 10.14 provides the answer. We start with the two subintervals of width 1/2. Already the first iteration produces the unit interval. Then we double the number of subintervals over and over again and compute the desired average number of iterations: $2.5, 4.0, 5.12, \ldots$

N	k	N	k	N	k	N	k
2	1.00	32	6.19	512	10.27	8192	14.28
4	2.50	64	7.25	1024	11.28	16384	15.28
8	4.00	128	8.25	2048	12.28	32768	16.28
16	5.12	256	9.27	4096	13.28	65536	17.28

N is the number of subintervals of $[0, 1]$, k is the average number of iterations of $4x(1-x)$ necessary to expand subintervals to $[0, 1]$.

Table 10.14

The result is surprisingly clear. When doubling the number of subintervals, i.e., reducing their size by a factor of 1/2, the average number of iterations seems to grow by exactly 1. This is not a coincidence. We recall that the numerical computation of the Ljapunov exponent for this iterator yielded $\lambda = \ln 2$, which states that small errors are amplified by a factor of 2, which is another way expressing the observation above.

Periodic Points Are Central to Chaos ...

You already know that there are also points which are of an entirely different nature: *periodic points*. If we start our iteration with a periodic point, this leads only to some few intervals which are visited again and again. Theoretically we can find infinitely many points of this type in each interval. Besides sensitivity and mixing, the existence of periodic points in any given subinterval is regarded as one of the necessary conditions for chaos and will be discussed next. At the end of this chapter we will even learn how the periodic points can be found by a simple formula. This formula applied for $a = 4$ yields that, for example,

$$\sin^2 \frac{\pi}{7} \rightarrow \sin^2 \frac{2\pi}{7} \rightarrow \sin^2 \frac{4\pi}{7}$$

is a cycle of period three for the quadratic iterator. One of the three points, namely, the first, is in the interval I from figure 10.12. Thus, even though the interval I is very small it contains a point from a 3-cycle besides initial points whose orbits reach everywhere. So what kind of behavior will we see if we choose an initial value at random?

An Unstable 3-Cycle

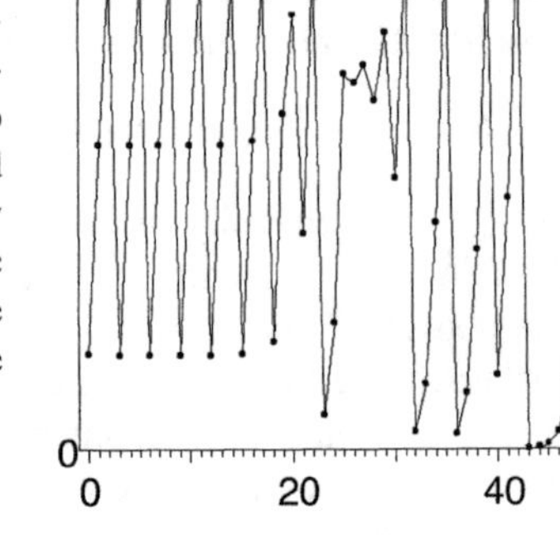

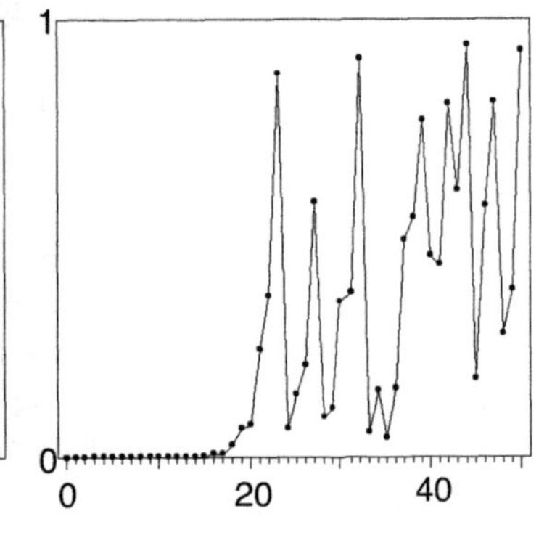

Sensitivity strikes again: the machine computed iteration of a periodic point with period 3. Due to roundoff errors in the computer and the sensitivity the orbit eventually moves away from the exact periodic cycle (at about the middle of the time series on the left). On the right the development of the error is plotted.

Figure 10.15

...But Cannot Be Detected

Well, assume we have chosen $\sin^2(\pi/7) = 0.1882550\ldots$ as an initial point. This is certainly a number which cannot be represented precisely in the computer. The error in representing this number by a single-precision floating point number is approximately 0.00000005.[8] Based on the sensitive dependence on initial conditions, we can predict what will happen. In each step of the iteration the error will become about twice as large. After 20 steps it will have multiplied by a factor of a million: about 1/20. In other words, for at most 20 iterations we will be close to the 3-cycle, and then sensitivity strikes again. This is demonstrated convincingly in figure 10.15. The same problem arises for all the other unstable periodic points which we can compute. And even when a periodic point can be represented exactly in computer arithmetic, after one iteration a minute round-off error will throw the orbit off the periodic one and open the door to letting sensitivity play its destructive part again.[9] Summarizing, we conclude that in a system with sensitivity there is no possibility of detecting a periodic orbit by running time series on a computer.

[8] The number depends on the particular machine.

[9] There are some exceptions to this, namely, when the complete periodic orbit is computed exactly by the machine. For the quadratic iterator, this will be the case when starting with $x_0 = 0$: all the following points are also zero, and the machine has no problems with that because there is no round-off error when multiplying by zero.

10.3 Ergodic Orbits and Histograms

Ergodic Orbits

For the discussion of mixing we use small intervals and their iteration, but why don't we simply take single initial points? This question seems to be justified by figure 10.16. Here we have continued the iteration of the initial value which took us from the starting interval I to the target interval J in figure 10.12. It seems that the iteration eventually fills out the whole unit interval similar to the iteration of a small subinterval. Indeed, we will see that in each interval there are infinitely many points of this type.[10] When iterated they are mixed throughout the whole unit interval. The iteration of such a point $x_0, x_1, x_2, \ldots$ gets arbitrarily close to any other point of the unit interval. Such orbits are called *ergodic*. But we must be careful: not all initial points of an interval like I produce ergodic orbits. For example, periodic points and their preimages yield orbits of only finitely many different points which, thus, cannot reach all subintervals of arbitrary small size.

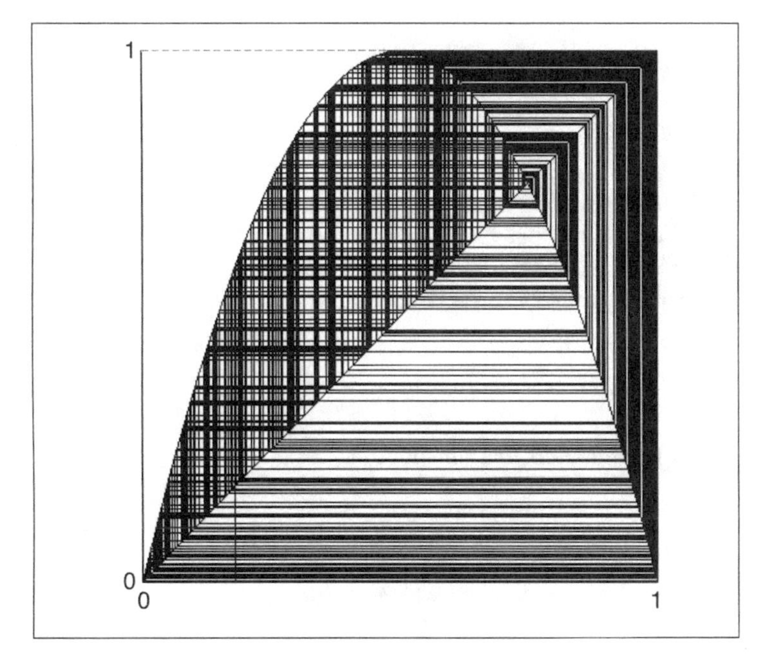

An Ergodic Orbit

The orbit from figure 10.12 is continued here for a few hundred iterations. It densely covers the unit interval.

Figure 10.16

Computing a Histogram

So let us look for numerical detection of ergodicity. Here is a simple experiment. We pick an initial point x_0 and iterate our quadratic iterator, say $m = 10^6$ times. Now we would like to see which parts of the unit interval are visited by the orbit $x_0, \ldots, x_m$ and how often. To this end we divide the unit

[10] This is only a theoretical result. On a computer there is only a finite collection of numbers representable. Thus, it is impossible to get arbitrarily close to all of the numbers in the unit interval.

Distribution of an Orbit

Distribution of the orbit $x_0, \ldots, x_m$. Hits are counted in 600 different subintervals of $[0, 1]$. The number of hits in an interval is proportional to the area of the columns drawn on this interval. The total area is 1.

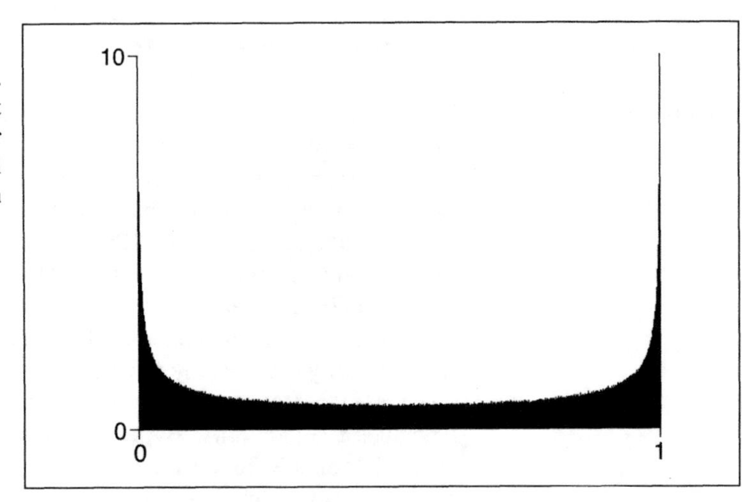

Figure 10.17

interval again into a large number N of small subintervals given by

$$I_k = \left[\frac{k-1}{N}, \frac{k}{N}\right), \quad k = 1, \ldots, N.$$

Then we note the number of events in I_k: we count how many of the iterates $x_0, \ldots, x_m$ fall into each interval I_k. Let this number be n_k. Naturally, however, we want to produce a count which is somewhat independent from the length of the orbit m. Thus, noting that the orbit has $m + 1$ elements, we define

$$\mu_k = \frac{n_k N}{m+1}.$$

These numbers μ_k vary between 0 and N by definition, and sum up to N,

$$\mu_1 + \mu_2 + \cdots \mu_N = N.$$

In other words, μ_k/N can be interpreted as a probability: it is the probability that we guess correctly (without calculations) the interval I_k into which a point falls, randomly chosen from the first $m + 1$ points of the orbit of x_0. Now we can plot columns of height μ_k and width $1/N$ into a histogram as in figure 10.17, where we have chosen $m = 10^6$, and $N = 600$. The columns cover an area which is equal to 1.

A Histogram for Mixing

We see a distribution for the μ_k which is symmetric with respect to 1/2, and which is rather flat in the center while having steep boundary layers at 0 and 1. This means that during the course of the quadratic iteration $x \to 4x(1-x)$ the probability that we see a point of the orbit near 0 or 1 is comparatively much higher than that of seeing it in the center of the unit interval. Running the same experiment again for different initial values x_0 results in histograms which are indistinguishable from the one above. As we would increase the number N of subintervals and the length m of the underlying orbit, the effect would be a smoothing of the shape seen in figure 10.17. In the limit we would approximate a well-known curve:

$$\nu(x) = \frac{1}{\pi\sqrt{x(1-x)}}. \tag{10.5}$$

Derivation of the Invariant Distribution

If X is a random variable with the density $\nu(x) = 1/(\pi\sqrt{x(1-x)})$, it is easy to check that $Y = f(X) = 4X(1-X)$ has the same distribution.

The key for this property is that the probability density must satisfy a functional equation which we derive from conservation of probability. Consider the probability contained in some small interval J of length Δy centered around some point $y = 4x(1-x)$ in the unit interval. This is

$$P(Y \in J) \approx \nu(y)\Delta y.$$

For an orbit to land in that small interval about y it must first be in some small interval I_1 near x or another small interval I_2 around $1-x$. Just how small these intervals must be, depends on how small Δy was chosen and on the slope of the generic parabola $4x(1-x)$ at x and $1-x$ (see figure 10.18). The larger the slope, the closer the orbit. More precisely, if Δx_1 and Δx_2 denote the sizes of the two intervals we have

$$\Delta y \approx |f'(x)|\Delta x_1, \quad \Delta y \approx |f'(1-x)|\Delta x_2. \tag{10.6}$$

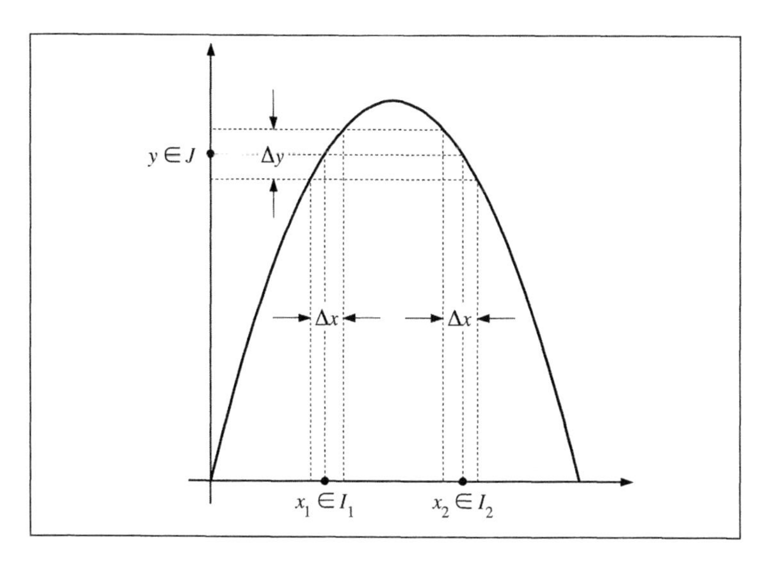

Figure 10.18 : The point y has two preimages, $x_1 = x$ and $x_2 = 1-x$. To hit the interval J of width Δy, an orbit must first pass through either I_1 or I_2.

In other words, the probability to see a point in the interval J near y must be equal to the sum of the probabilities to have points in the intervals I_1 and I_2 near x and $1-x$. Thus,

$$\begin{aligned} P(Y \in J) &= P(X \in I_1) + P(X \in I_2) \\ &\approx \nu(x)\Delta x_1 + \nu(1-x)\Delta x_2. \end{aligned} \tag{10.7}$$

Let us put the three equations together. Substitute $\nu(y)\Delta y$ for $P(Y \in J)$ in eqn. (10.7), solve eqn. (10.6) for Δx_1 and Δx_2, and substitute the results in eqn. (10.7). Finally, divide by Δy to obtain the functional equation

$$\nu(y) = \frac{\nu(x)}{|f'(x)|} + \frac{\nu(1-x)}{|f'(1-x)|}. \tag{10.8}$$

This is called a Frobenius-Perron equation. Now we check that this equation holds true when the density function

$$\nu(x) = \frac{1}{(\pi\sqrt{x(1-x)})}$$

is inserted. Using the identity

$$1 - 4x(1-x) = (2x-1)^2$$

we compute the left side of eqn. (10.8).

$$\begin{aligned} \nu(y) &= \nu(4x(1-x)) \\ &= \frac{1}{\pi\sqrt{4x(1-x)(1-4x(1-x))}} \\ &= \frac{1}{2\pi|2x-1|\cdot\sqrt{x(1-x)}}. \end{aligned}$$

Observe that $\nu(1-x) = \nu(x)$ and $f'(1-x) = -f'(x)$ implying $|f'(1-x)| = |f'(x)|$. With that we compute the right side of eqn. (10.8).

$$\begin{aligned} \frac{\nu(x)}{|f'(x)|} + \frac{\nu(1-x)}{|f'(1-x)|} &= \frac{2\nu(x)}{|f'(x)|} \\ &= \frac{2}{\pi\sqrt{x(1-x)}\cdot 4|1-2x|}. \end{aligned}$$

This establishes the equality of both sides for the given distribution ν. Moreover, it can be shown that $\nu(x) = 1/(\pi\sqrt{x(1-x)})$ is the only possible density function that can satisfy the functional equation.[11]

Furthermore, we can produce the invariant distribution by iteration of the corresponding Frobenius-Perron operator.[12] Namely, the right-

[11] See P. Collet, J.-P. Eckmann, *Iterated Maps on the Interval as Dynamical Systems*, Birkhäuser, Boston, 1980.

[12] See, for example, R. Shaw, *Strange attractors, chaotic behavior, and information flow*, Z. Naturforsch. 36a (1981) 80–112.

hand side of the functional equation can be interpreted as an operator which assigns a new probability density $\Phi(\nu)$ to a given one, ν,

$$\Phi(\nu)(y) = \sum_{x \in f^{-1}(y)} \frac{\nu(x)}{|f'(x)|}, \quad y \in [0,1].$$

Starting with an arbitrary initial probability density ν_0 we can iterate this Frobenius-Perron operator Φ,

$$\nu_{k+1} = \Phi(\nu_k), \quad k = 0, 1, 2, \ldots$$

to arrive at a sequence of probability distributions that rapidly converges to the invariant one, $\nu(x) = 1/(\pi\sqrt{x(1-x)})$.

The Exponential Decay Rate

Let us apply the invariant density function $\nu(x)$ to estimate the persistence parameter τ in the exponential law in eqn. (10.4) for the decay of the survivors. Consider the collection of initial points and their orbits disregarding the target interval for the moment. After some small number of iterations (here about 36) the points of the orbits are scattered over the entire interval with a distribution which is approximately given by eqn. (10.5). Now we ask the question how many of these points fall into the target interval $J = [0.68, 0.69]$. This can be estimated by the value of the invariant distribution at the center of the interval, multiplied by the width, i.e.,

$$\frac{\nu(0.685)}{100} = \frac{1}{100\pi\sqrt{0.685(1-0.685)}} \approx 0.0068525$$

times the total number of points.[13]

Thus, based on a total of $10{,}000$ points, approximately 68.525 points are removed per iteration which roughly agrees with our numerical findings of 63 points. To estimate τ we need to compute the number of iterations necessary to reduce the number of remaining points to a fraction of $1/e$ based on the removal rate of a fraction of 0.0068525 of all points per iteration. Thus, we solve

$$(1 - 0.0068525)^{\tau} = \frac{1}{e}$$

for τ. The result is $\tau \approx 145$ iterations which again is in accordance with our numerical findings.

[13] Precisely the expectation of this number is

$$\int_{0.68}^{0.69} \frac{1}{\pi\sqrt{x(1-x)}}\,dx = \left[\frac{2}{\pi}\arctan\sqrt{\frac{x}{1-x}}\right]_{0.68}^{0.69} = \frac{2}{\pi}\left(\arctan\sqrt{\frac{0.69}{0.31}} - \arctan\sqrt{\frac{0.68}{0.32}}\right) \approx 0.0068527$$

times the total number of points.

Histogram Detail

Histogram for the interval $[0.4, 0.5]$ and the interval $[0.45, 0.46]$, corresponding to an increase in resolution of 10 times and 100 times respectively.

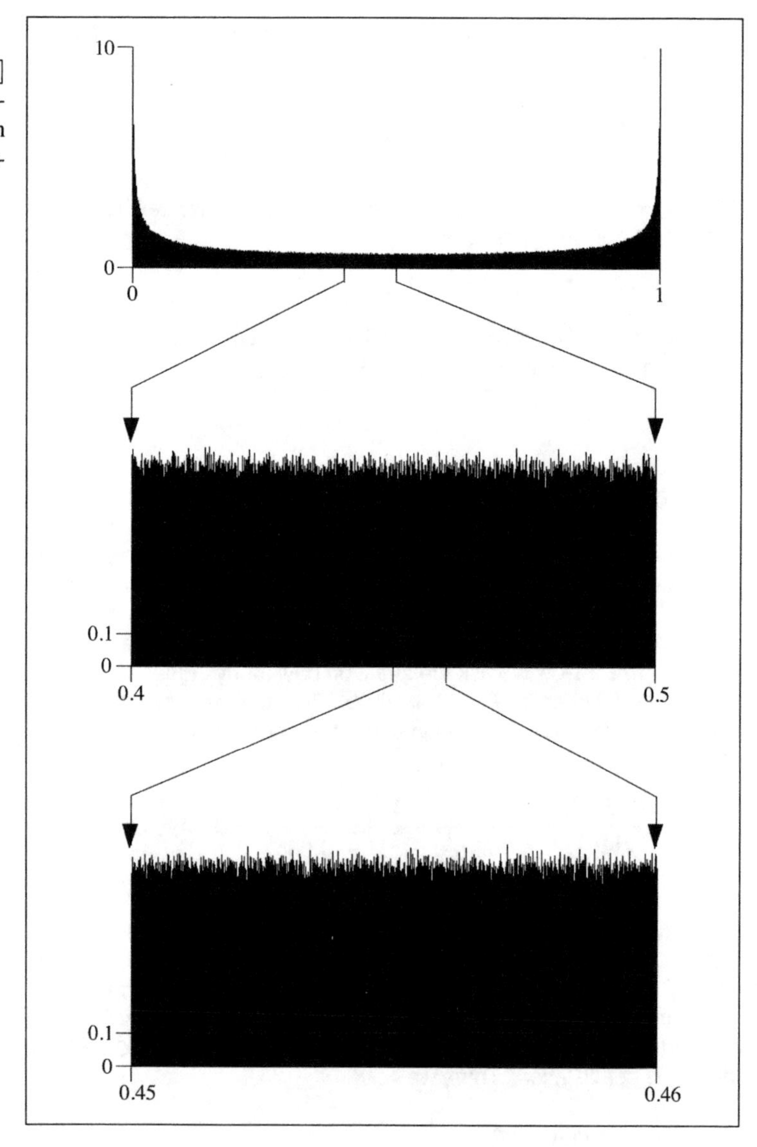

Figure 10.19

Testing for Ergodicity

If we want to be sure, at least from an experimental point of view, that a given orbit is in fact ergodic, then we have to look at a whole sequence of experiments refining that of figure 10.17. One way to do this would be to take an arbitrary subinterval, say $[0.4, 0.5]$, and subdivide this interval into N, say again $N = 600$, subintervals and repeat the counting experiment. In other words, we increase the resolution by a factor of 10. Of course, the length of the orbit must be increased correspondingly to provide enough data to support the statistical evaluation we have in mind. If we still see that this

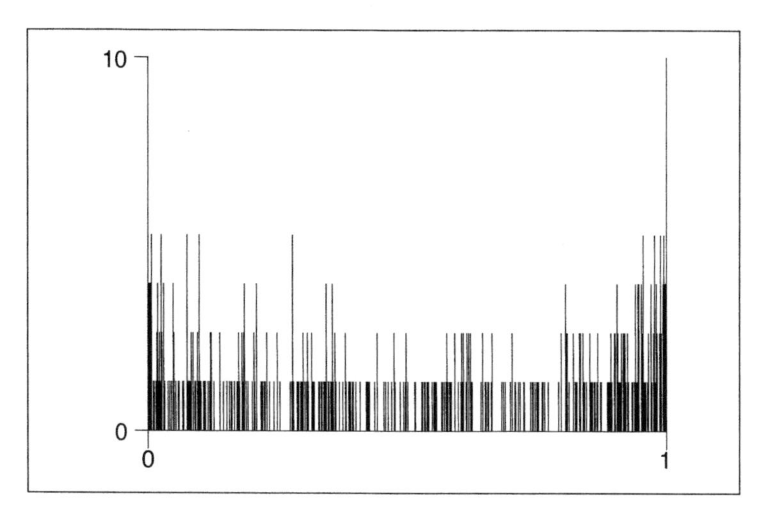

Single-Precision Histogram

Histogram of 5000 iterations for $a = 4$ with intial point $x_0 = \sin(2)$ using single-precision arithmetic.

Figure 10.20

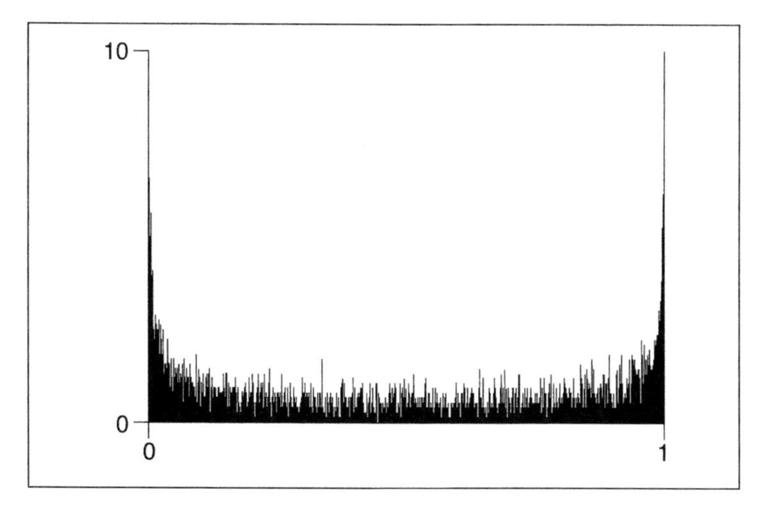

Another Single-Precision Histogram

Histogram of 5000 iterations for $a = 4$ with intial point $x_0 = 0.55$ using single-precision arithmetic.

Figure 10.21

subinterval $[0.4, 0.5]$ has positive values μ_k everywhere, then we have gone one step further in supporting the ergodicity hypothesis. Figure 10.19 shows the result.

Let us summarize what we have seen in our experiments so far. Starting the iteration numerically with any number different from $0, 1,$ and $1/2$ the behavior will always be ergodic. In particular, we do not see the periodic points which are theoretically present. But now with the following experiments let us turn everything upside down. And this really should show you how delicate it is to reason about chaos based on computer simulations.

A Bad Histogram

When we did the first quick tests to compute histograms, as in figure 10.17, we were not careful enough and stumbled into a big surprise. Our

Periodicity After All

(a) Implementation of $4x(1-x)$: Periods $1, 436, 836, 4344$;
(b) Implementation of $4x-4x^2$: Periods $1, 1135$.

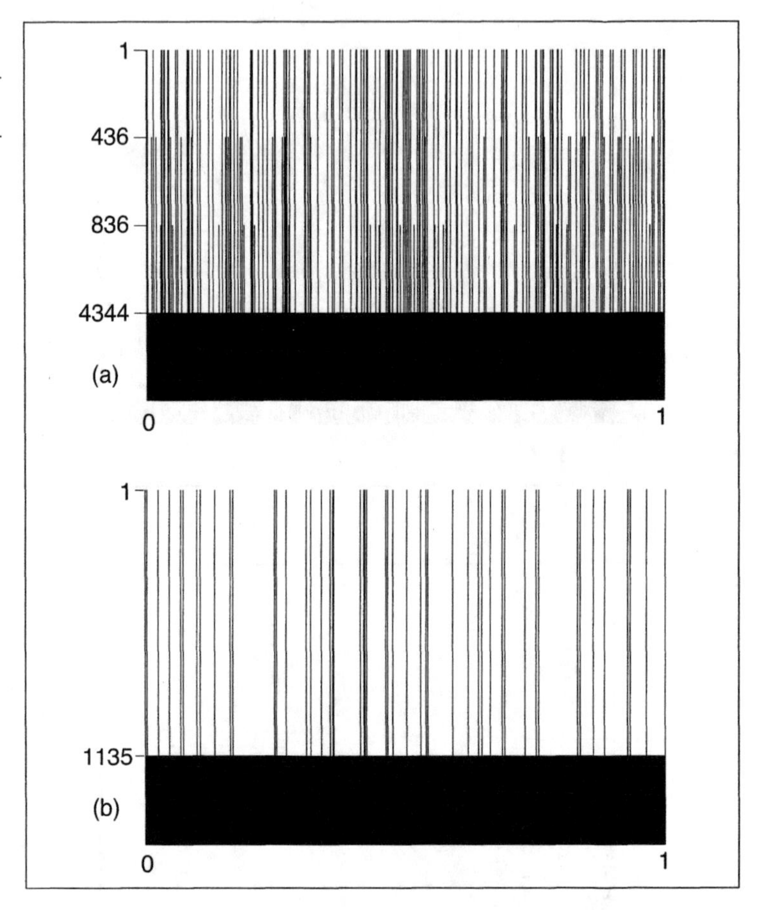

Figure 10.22

computations were done in single-precision arithmetic BASIC, and we picked an initial value which should reflect some arbitrariness, namely, $x_0 = \sin(2)$. When we looked at the resulting histogram reproduced in figure 10.20 it did not at all show the distribution which we expected.

Very distinct gaps in the histogram are visible. They correspond to small subintervals which have not been visited by the orbit. We repeated the experiment with another initial value, this time $x_0 = 0.55$, and obtained something more satisfactory (see figure 10.21). But shouldn't the histogram be independent from the initial value?

Maybe we did not iterate enough times? We increased m from $m = 10^5$ to $m = 10^6$ and obtained exactly the same images. Now it was clear that something was really wrong. But what? The bad histogram and its stability versus an increase in the number m of iterations suggested that for $x_0 = \sin(2)$ the iteration had run into a periodic cycle. But didn't that contradict what we have discussed so far? Shouldn't small unavoidable errors in the computation guarantee that we never run into a periodic cycle? Unfortunately, not at all!

Our next experiment confirmed that. We simply asked the following questions: Given initial sampling values spread over the entire unit interval, $x_0 = i/600, i = 1, \ldots, 599$, is there a periodic cycle into which the iteration will run if started with x_0? And if so, what is the corresponding period? The setup for the experiment was very simple. Starting at x_i, 5000 iterations were carried out, and we assumed that the last point x_{5000} was already a periodic point. To compute the corresponding period, we performed another 10,000 iterations and checked at each iteration whether x_n was equal to x_{5000}. The first time that such an x_n was found, the period was determined as $n - 5000$.

Periodic Cycles

Indeed, it turned out that any of these sampling initial values eventually led to a periodic cycle of period $1, 436, 836$, or 4344. Figure 10.22 (top) shows which periodic cycle was eventually reached for the different initial values. The majority of initial values led to the cycle of length 4344. By the way, $x_0 = \sin(2)$ led to a cycle of length 436, which was short enough to remain visible in the histogram reproduced in figure 10.20. Once this became clear we changed the implementation of the quadratic function in the code. Instead of $4x(1-x)$ we then used $4x - 4x^2$, which is the same mathematically, of course, but makes a big difference in the experiment, as can be seen in the bottom part of the figure. Then almost all initial values led to a periodic cycle of period 1135, and a few to the fixed point 0, which corresponds to the number 1 in the figure.

So what is the conclusion? Once again we have experienced extreme sensitivity in the iteration which renders the computer results worthless and even misleading. These periodic orbits are artifacts of the computation, the result of systematic rounding-off or truncation errors. They seem to vanish as soon as the precision of the arithmetic is increased, as for example in figure 10.17, where double-precision arithmetic is used.

Very Low Precision Arithmetic

Let us investigate the mechanism, which produces the periodic orbits, a bit further. To make things clearer we decrease the precision dramatically. In figure 10.23 we show a situation where we distinguish only 25 different values $v_0, \ldots, v_{24}$ in the range from $v_0 = 0$ to $v_{24} = 1$. Our purpose is to demonstrate the effect of truncation.[14] In this example all real numbers between v_n and v_{n+1} are truncated to the value v_n. Since the corresponding truncation is carried out for the evaluation of $4x(1-x)$, we obtain a staircase function as an approximation of the parabola as shown in our figure 10.23. The table 10.24 captures the essence of the staircase function approximating the parabola in figure 10.23. For the 25 points $v_0, \ldots, v_{24}$ the indices of the transformed points are listed. Thus, $v_0 \to v_0, v_1 \to v_3, v_2 \to v_7$, and so on.

[14]Truncating a number is the process of keeping the first significant digits, discarding or chopping off all others. Most computers nowadays use rounding instead which may modify the last possible significant digit depending on the following digits. For example, in a machine working with a precision of five decimal digits, the number 3.14159265 would be stored as 3.1415 when truncation is used, while the result would be 3.1416 with rounding.

Grid Arithmetic

Graphical iteration on a coarse grid. The parabola $4x(1-x)$ becomes a staircase function. This table below captures the essence of this function approximating the parabola. For the 25 points $v_0, \ldots, v_{24}$ the indices of the transformed points are listed. For example, $v_{18} \to v_{23}$.

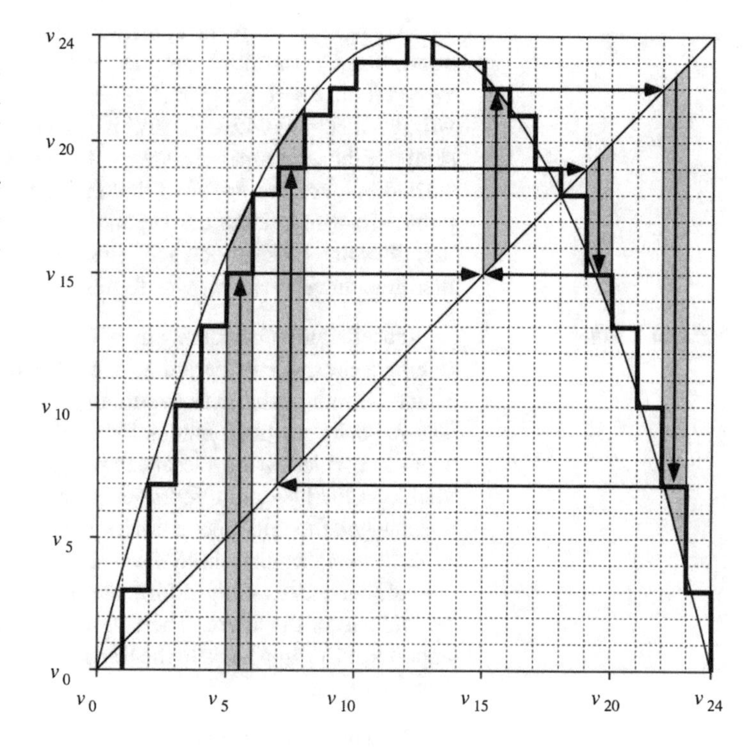

Figure 10.23

Staircase Function

Only the indices of points are listed. Read $v_0 \to v_0$, $v_1 \to v_3$, $v_2 \to v_7$, and so on.

Table 10.24

0	0	5	15	10	23	15	22	20	13
1	3	6	18	11	23	16	21	21	10
2	7	7	19	12	24	17	19	22	7
3	10	8	21	13	23	18	18	23	3
4	13	9	22	14	23	19	15	24	0

Let us examine the iteration of v_5 which represents all real numbers from $5/24 \approx 0.2083$ to $6/24 = 0.25$ (not including the last number).[15] The quadratic iterator transforms this range of numbers to the interval from $95/144 \approx 0.6697$ to $3/4 = 0.75$. Thus, the size of the interval is expanded by the factor $13/6 \approx 2.17$, which agrees with our expectation from the analysis of the sensitivity to initial conditions. But what we observe using our low precision arithmetic is drastically different: v_5 is simply mapped onto v_{15}. We can interpret this as the introduction of an additional contraction in the iteration process: the range of real numbers represented by v_5 is contracted to v_{15}. The next step of the iteration leads us to v_{22}, then to v_7, v_{19} and again to v_{15}. Thus we have run into a periodic 4-cycle. We can also find a 3-cycle (v_3, v_{10}, v_{23}) and two fixed points (v_0 and v_{18}). There are no other periodic

[15]In general, v_k represents all numbers $k/24 + r$, where $r \in [0, 1/24)$.

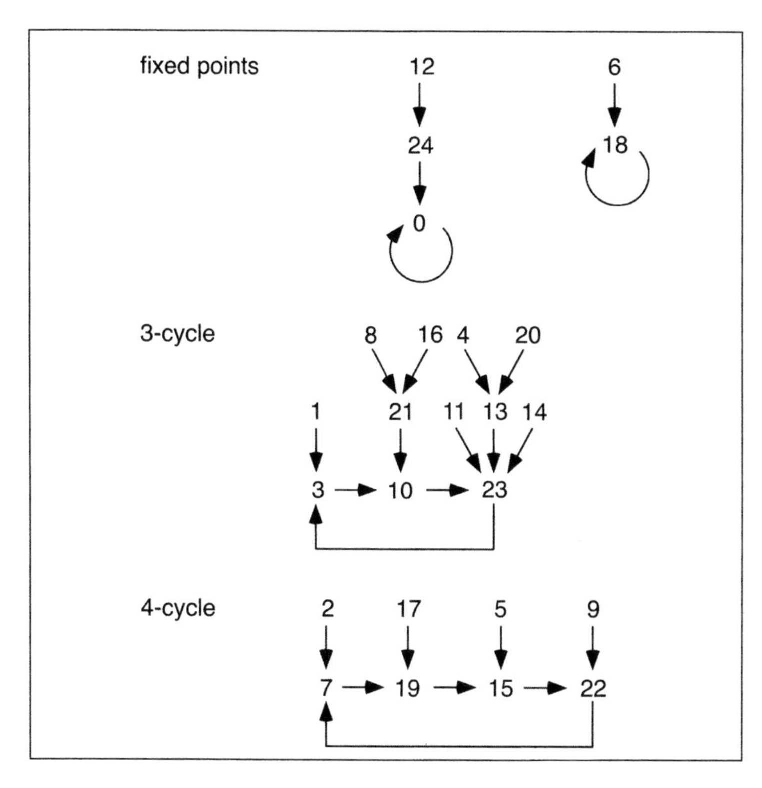

The Complete Picture

The very low precision arithmetic allows only four different long-term behaviors of the quadratic iteration, two fixed points and two periodic cycles of period three and four. The figure shows the complete list of all transformations as listed in table 10.24 so that orbits can be seen by following the arrows. For the cycles and the initial values only the corresponding indices are given. For example, 15 corresponds to v_{15}.

Figure 10.25

cycles, especially no periodic 2-cycles. All initial values of the iteration lead to one of these periodic orbits or fixed points. These facts are summarized in figure 10.25.

Periodic Orbits Are Artifacts

If we increase the precision of our arithmetic (perhaps to 100 or 1000 different values) everything changes: the periodic orbits vanish and new ones arise. In other words, the periodic orbits are artifacts of the computation with limited precision. Indeed, essentially the same mechanisms can be found if we use floating point arithmetic. When iterating $4x(1-x)$ more and more orbits fall onto each other and only a few periodic cycles survive. This result explains figure 10.22.

Do you recall how we obtained the 'good' histogram in figure 10.17? We used double-precision arithmetic rather than single precision. Indeed, we still have that any iteration will eventually run into a periodic cycle — there is no way out simply because there are only finitely many, though very many, machine numbers — but these cycles are different from those found for single-precision arithmetic; and the length of the cycles will be so large that it doesn't destroy the experiment immediately. In fact, we have tested for periodic cycles for some samples of initial values and have not found that they become periodic for the first billion iterations.

10.4 Metaphor of Chaos: The Kneading of Dough

Generally speaking, the analysis of chaos is extremely difficult. However, in the specific model case of the quadratic iterator there is a beautiful and illuminating way to really understand chaotic behavior through and through. First we prepare the ground for one of the most important metaphors in chaos theory: the kneading of dough. The kneading process guarantees that a pocket of spices inserted into the dough will be mixed thoroughly throughout the mass.

A Definition of Chaos

While a general definition for chaos applicable to most cases of interest is still lacking, mathematicians agree that for the special case of iteration of transformations there are three common characteristics of chaos:[16]

1. sensitive dependence on initial conditions,
2. mixing, and
3. dense periodic points.

The kneading of dough provides an intuitive access to all of these mathematical properties of chaos. Moreover, we will see that the kneading process is closely related to the quadratic iterator! In this way the scenario of chaos that we have been discussing so far can be completely understood without having to resort to some higher mathematics.

Kneading with a Rolling Pin

Kneading as a feedback process: stretch, fold, stretch, and so on.

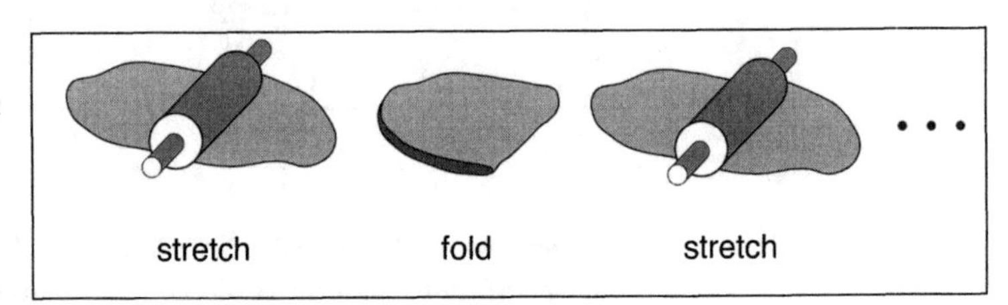

Figure 10.26

Kneading, a Deterministic Process

There is nothing random about the kneading process itself. Rather, a baker applies a certain action over and over again. We imagine kneading as the process of stretching the dough and folding it over, repeated many times. But in spite of this deterministic definition, the results have many features in common with randomness. Let us see why.

Stretch-and-Fold Kneading

We look at an idealized situation, which certainly only a highly trained baker can achieve. The dough is homogeneously stretched to twice the length. Then it is bent at the center and folded over. Figure 10.27 shows a side view of this operation. Let us see how this kneading works on different parts of the dough by dividing the dough into 12 blocks which then are put through two stretch-and-fold operations (see figure 10.28). The resulting layers of transformed blocks of dough already look a bit mixed.

We idealize the situation even one step further. Imagine we work with infinitely thin layers of dough. Folding these layers does not change the thick-

[16]See R. L. Devaney, *An Introduction to Chaotic Dynamical Systems*, Addison-Wesley, Redwood City, Second Edition, 1989. Devaney uses the notion of *transitivity* for mixing.

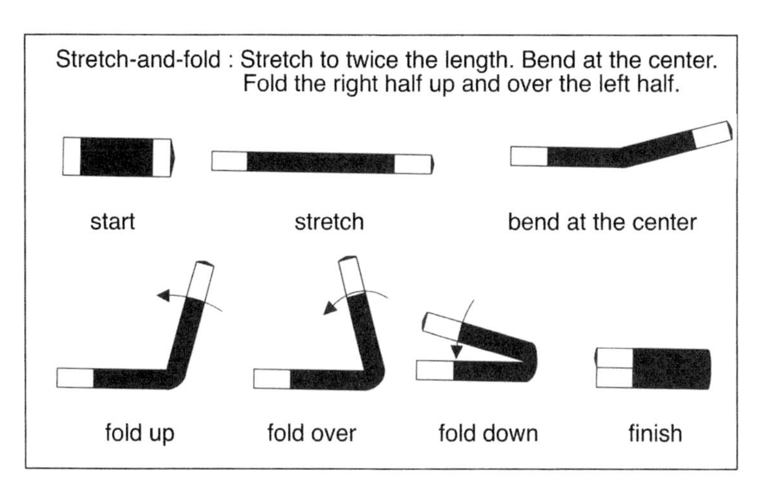

Stretch-and-Fold

Uniform kneading by stretch-and-fold.

Figure 10.27

ness and we can represent the dough by a line segment. Figure 10.29 shows some stages of the corresponding representation of the kneading process. We mark two grains of spice and follow their paths. The two grains are rather close together initially. But where will they be after a dozen kneadings? It is very likely that we will find them in very different places in the dough. In fact, that would be a consequence of the mixing properties of kneading. In other words, kneading destroys *neighborhood relations*. Grains which are very close initially will likely not be close neighbors after a while. This is

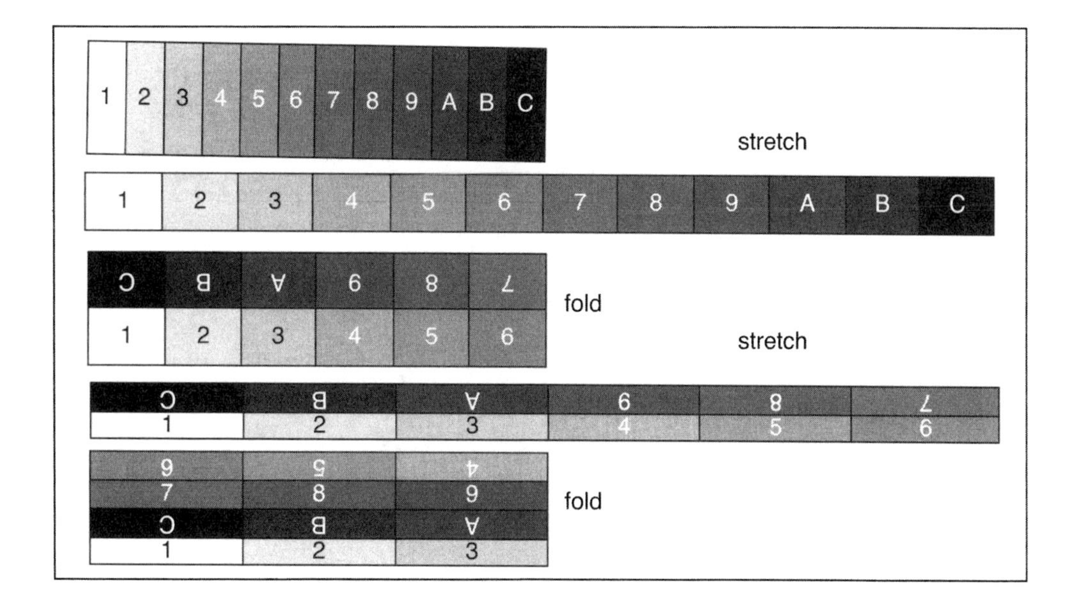

Figure 10.28 : Two operations of stretch-and-fold kneading applied to 12 textured blocks of dough.

Kneading

Two grains, symbolized by a dot and a square, are subjected to four stages of the stretch-and-fold. They are mixed throughout the dough.

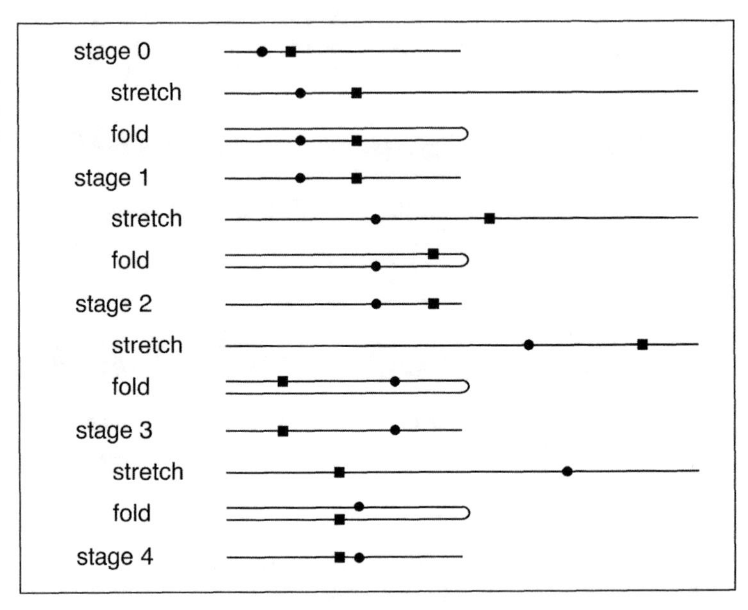

Figure 10.29

Stretch-Cut-and-Paste

Uniform kneading by stretch-cut-and-paste.

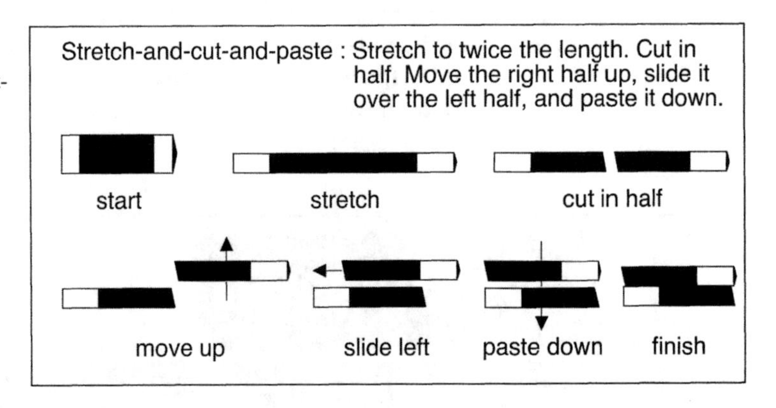

Figure 10.30

the effect of *sensitive dependence on initial conditions*. Small deviations in initial positions lead to large deviations in the course of the process.

Stretch-Cut-and-Paste Kneading

Let us discuss a second kneading operation. Here we again stretch the dough uniformly to twice its length. But then we cut the dough at the center into two parts and paste them on top of each other (see figure 10.30).

When comparing the stretch-and-fold operation with the stretch-cut-and-paste operation, our intuition would be that both kneading operations apparently mix particles around, but in a very different manner, generating quite distinct iteration behaviors. The surprise is, however, that both kneadings are essentially the same! A first idea of this fact can be obtained from figure 10.31. Again we divide the dough into 12 blocks. Then we apply the stretch-

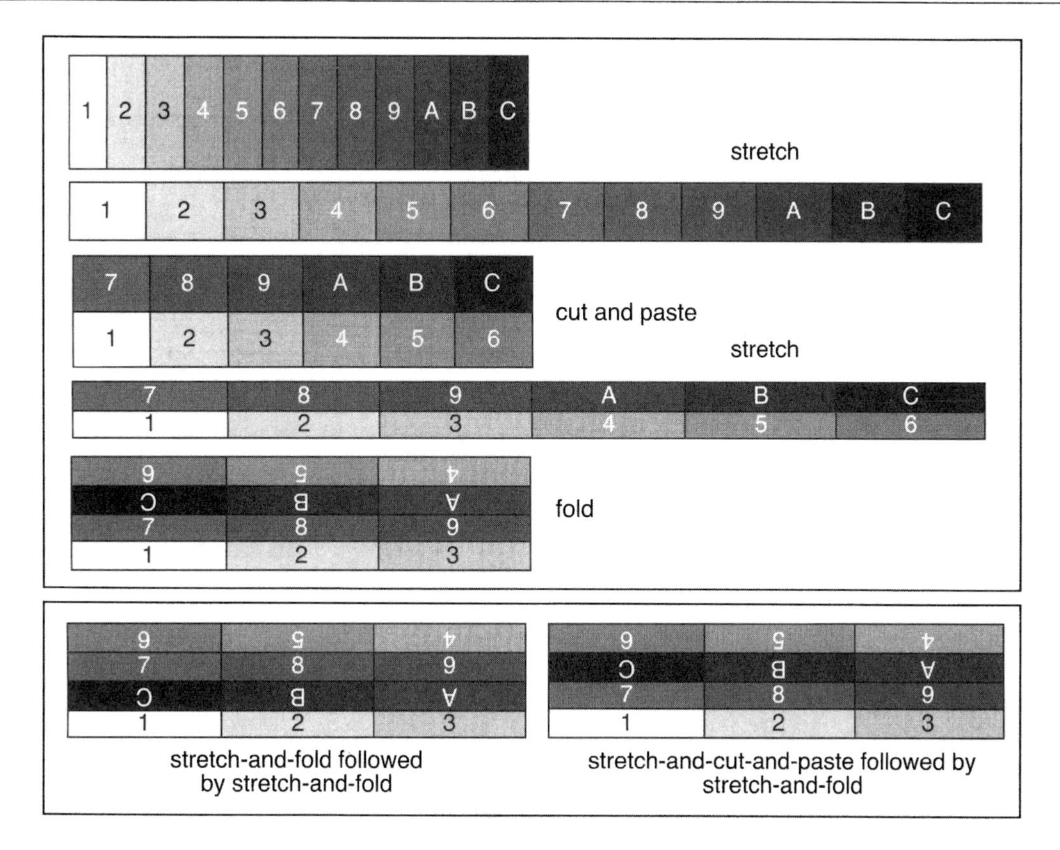

Figure 10.31 : Stretch-cut-and-paste followed by stretch-and-fold applied to 12 textured blocks of dough. The resulting horizontal order of the blocks is identical to the one obtained from the application of the two succeeding stretch-and-fold operations in figure 10.28.

cut-and-fold operation followed by one stretch-and-fold operation. The result is compared with the one obtained for two succeeding stretch-and-fold operations in the bottom part of the figure. We observe that they are identical when ignoring the vertical order of the pieces!

This again suggests to neglect any thickness of the dough. Thus, from now on we think of the dough being represented by a line segment. And after a kneading operation — for example stretch-cut-and-paste — the result will be represented again by an immaterial straight line. This is the first step towards a mathematical model of the kneading operations. Taking the interval $[0, 1]$ as the original line segment modeling the dough, we can now check how the two different kneading operations act. Let us use the symbol T for the stretch-and-fold operation and the symbol S for stretch-cut-and-paste.

As an example, let us follow a particle through several kneading steps once applying the stretch-and-fold operation T three times (see figure 10.32, left) and another time using two stretch-cut-and-paste transformations S followed by one stretch-and-fold operation T (see figure 10.32, right). We observe in

Tracing a Particle

The interval $[0, 1]$ at the top is the model for the dough. We follow the path of a particle at $x = 9/10$ when the following kneading operations are applied: left T, T, T; right S, S, T.

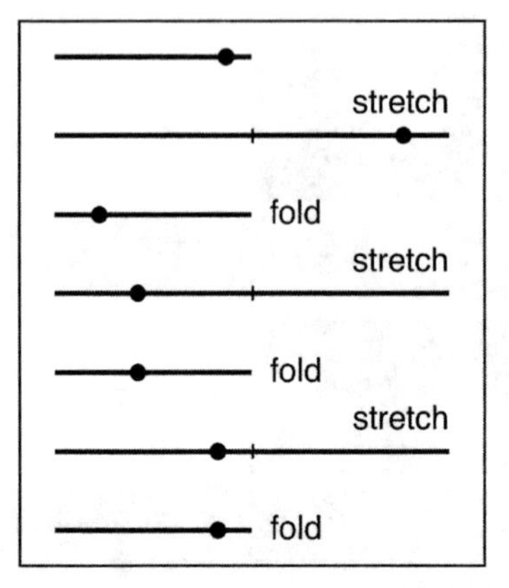

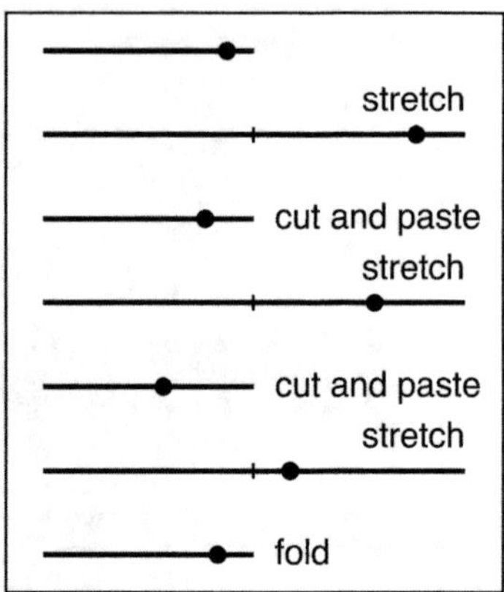

Figure 10.32

both experiments that the particle arrives exactly at the same position, though the route in between is different. Thus, if the particle is initially at position x, we have seen that

$$T(T(T(x))) = T(S(S(x))).$$

This experience along with the result in figure 10.31 motivates us to conjecture an substitution property of the two kneading operations. In fact, we will prove this conjecture below starting on page 502.

Substitution Property

Fact. *N kneading steps using the stretch-and-fold operation T, i.e.,*[17]

$$T^N = \underbrace{TT \cdots T}_{N \text{ times}}$$

yield the same material in each vertical column as $N - 1$ *kneading steps by the stretch-cut-and-paste operation S followed by one kneading step of T, i.e.,*

$$TS^{N-1} = T \underbrace{SS \cdots S}_{N-1 \text{ times}} .$$

Formula for Stretch-and-Fold Kneading

The mathematical model for kneading of the one-dimensional ideal of dough is a function. The stretch-and-fold kneading operation is represented by the following transformation, for which we use the symbol T again:

$$T(x) = \begin{cases} 2x & \text{if } x \le 0.5 \\ -2x + 2 & \text{if } x > 0.5. \end{cases} \tag{10.9}$$

Figure 10.33 shows the graph of this transformation, which is called the *tent transformation* because of its shape. This graph looks like a simplification of the parabola.

[17] It is common mathematical notation to interpret a composition of operators such as $TSSS$ from right to left. In this case S is applied three times followed by one application of T.

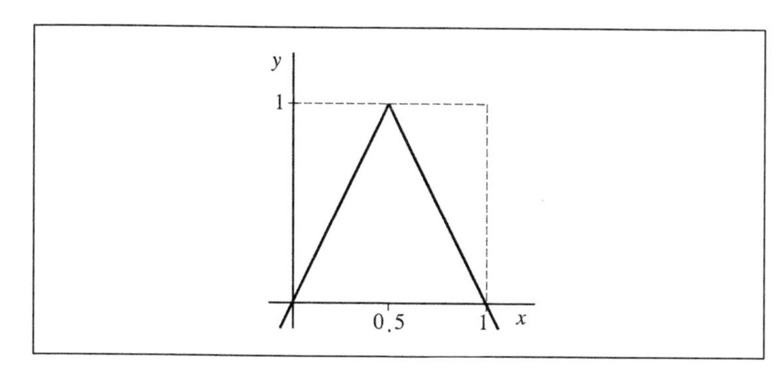

Tent Transformation

Graph of piecewise-linear tent transformation eqn. (10.9) corresponding to stretch-and-fold transformation.

Figure 10.33

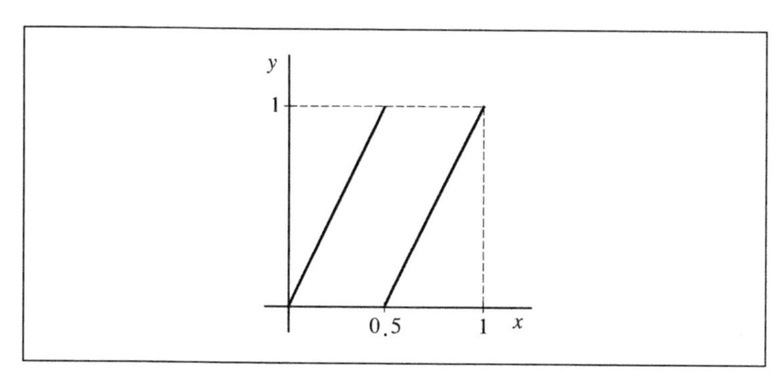

Saw-Tooth Transformation

Graph of saw-tooth transformation eqn. (10.10) corresponding to the stretch-cut-and-paste transformation.

Figure 10.34

The justification of this model is almost self-evident. The dough is modeled by the unit interval $[0, 1]$. The stretching operation is taken care of by the factor 2 in front of x. The first half interval of $[0, 1]$ is only stretched and not folded. Thus, the first part of the definition of T is in order, $T(x) = 2x$ for $x \leq 1/2$. The second half interval becomes $[1, 2]$ after the stretching, and must be folded over its left end point. This is equivalent to folding at $x = 0$, i.e., multiplying by -1 and shifting to the right by two units. Thus, $T(x) = -2x + 2$ for $x \geq 1/2$.

Formula for Stretch-Cut-and-Paste Kneading

The model for the second procedure, the stretch-cut-and-paste kneading operation, is another elementary mathematical transformation, the *saw-tooth transformation* S, defined for numbers x from the unit interval $[0, 1]$:

$$S(x) = \begin{cases} 2x & \text{if } x < 0.5 \\ 2x - 1 & \text{if } x \geq 0.5 \end{cases} . \tag{10.10}$$

Its graph (see figure 10.34) justifies the name. Again the verification of the model is evident, and we omit the details.[18]

[18]Only for the point $x = 1/2$ is it not self-evident what the corresponding value, $S(1/2)$, of the saw-tooth function should be. We cut the dough precisely at $x = 1/2$. Now the right end point of the left portion of the dough, which is 1/2, moves to the point 1. However, the left end point of the right dough segment also corresponds to 1/2, but it is moved to 0. Thus, we may choose either $S(1/2) = 0$ or $S(1/2) = 1$ in the definition of S. However, all of the following arguments can be carried through with either definition using only very minor modifications. For the presentation in this book we have chosen the first option which leads to a

Substitution Property of the Kneading Operations

As announced we can now show the substitution property of the kneading operations using their appropriate mathematical descriptions. Let T denote the piecewise-linear tent transformation (10.9) and S the saw-tooth transformation (10.10). Then we can verify by straightforward calculation that for any x in $[0, 1]$

$$TT(x) = TS(x). \tag{10.11}$$

This is visualized in figure 10.35. Let us look at an example: $x = 0.9$. Then

$$\begin{aligned} T(0.9) &= -2 \cdot 0.9 + 2 = 0.2, \\ T(0.2) &= 2 \cdot 0.2 = 0.4. \end{aligned}$$

On the other hand,

$$\begin{aligned} S(0.9) &= 2 \cdot 0.9 - 1 = 0.8, \\ T(0.8) &= -2 \cdot 0.8 + 2 = 0.4. \end{aligned}$$

Now let us check eqn. (10.11) formally. On the left side we compute the result in four cases as follows.

$$\begin{array}{rcccll} T(T(x)) &=& T(2x) &=& 4x & \text{for } 0 \le x \le 0.25, \\ T(T(x)) &=& T(2x) &=& -4x+2 & \text{for } 0.25 < x \le 0.5, \\ T(T(x)) &=& T(-2x+2) &=& 4x-2 & \text{for } 0.5 < x \le 0.75, \\ T(T(x)) &=& T(-2x+2) &=& -4x+4 & \text{for } 0.75 < x \le 1, \end{array}$$

and the corresponding right-hand sides yield

$$\begin{array}{rcccll} T(S(x)) &=& T(2x) &=& 4x & \text{for } 0 \le x \le 0.25, \\ T(S(x)) &=& T(2x) &=& -4x+2 & \text{for } 0.25 < x < 0.5, \\ T(S(x)) &=& T(2x-1) &=& 4x-2 & \text{for } 0.5 \le x \le 0.75, \\ T(S(x)) &=& T(2x-1) &=& -4x+4 & \text{for } 0.75 < x \le 1, \end{array}$$

which completes the proof. (Note that $-4x + 2 = 4x - 2$ for $x = 0.5$.)

The identity eqn. (10.11) is the key to obtaining identities for higher iterations very elegantly: for example

$$TTT(x) = TSS(x),$$

or

$$TTTT(x) = TSSS(x),$$

and so on. Indeed, when we apply T to both sides of

$$TT(x) = TS(x),$$

we obtain

$$TTT(x) = TTS(x).$$

slightly simpler discussion.

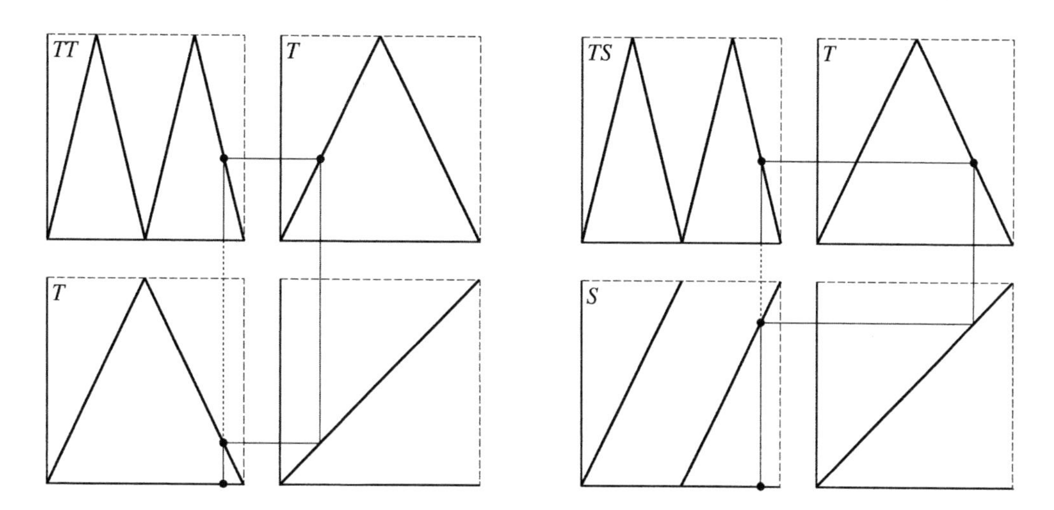

Figure 10.35 : The composition of two tent functions (left) and the saw-tooth function followed by the tent function (right). Both approaches lead to the same result, a double tent. The lower left graph in each part provides for a given x the value $T(x)$ (left) or $S(x)$ (right) which then is fed to the upper right graph leading to the value $T(T(x))$ (left) and $T(S(x))$ (right). The upper left graph is the graph of the composition TT (left part) and TS (right part). Both are the same demonstrating the substitution property.

Now we can substitute TS for TT on the right side and obtain

$$TTT(x) = TSS(x),$$

and so on. This actually means that N iterations of T — stretch-and-fold — lead to the same result as $N-1$ iterations of S — stretch-cut-and-paste — followed by one application of T. This is the substitution property.

Kneading and the Quadratic Iterator

Let us introduce an argument which makes a connection between the feedback system $x \to 4x(1-x)$ and the kneading of dough. When we graph the transformation $y = 4x(1-x)$ in an xy-coordinate system, we obtain the generic parabola shown in figure 10.36.

Here we are interested only in x-values ranging from 0 to 1. Note that corresponding y-values also range from 0 to 1. We have that y monotonically increases for $x < 1/2$ and monotonically decreases for $x > 1/2$. We observe that the interval $[0, 1/2]$ on the x-axis is stretched out to the interval $[0, 1]$ on the y-axis, and the same is true for the interval $[1/2, 1]$. In other words, the transformation $4x(1-x)$ stretches both intervals to twice their length.

Nonuniform Stretching

The stretching, however, is nonuniform. In fact, small intervals which are close to 0 or 1 are stretched a great deal, while intervals, which are close to the midpoint $1/2$ are compressed. To show this we have put markers on the x-axis which are equally spaced and observe how their corresponding y-values are not uniformly spaced in figure 10.36.

The Generic Parabola

Graph of the generic parabola $y = 4x(1-x)$.

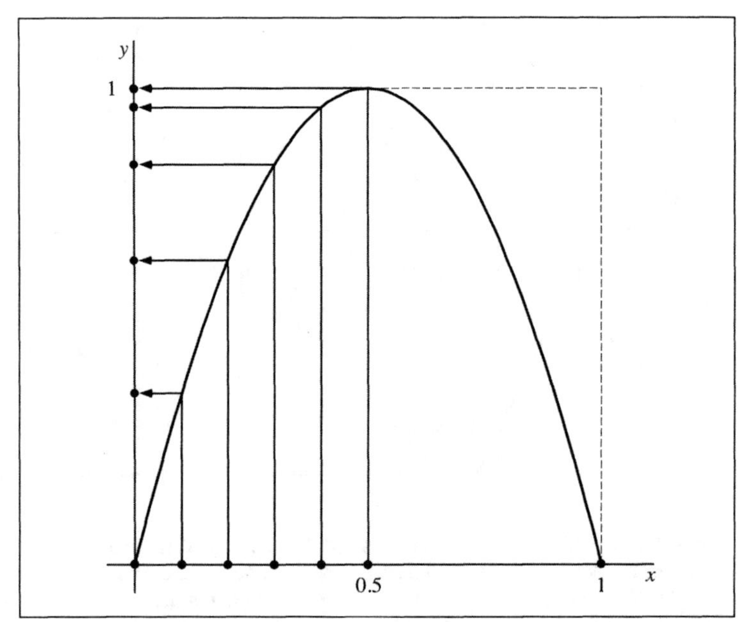

Figure 10.36

Now we can get to the point of making a connection with kneading. What happens if we apply the transformation $4x(1-x)$ to the interval $[0,1]$? We already know that each half of the interval is stretched to twice its length. Moreover, checking the end points of the intervals, we find: $0 \to 0, 1/2 \to 1$, $1 \to 0$. This means that the result of one application of the transformation $4x(1-x)$ to the interval $[0,1]$ can be interpreted as a combination of a stretching and folding operation (see figure 10.37). In other words, the iteration of $x \to 4x(1-x)$ is a relative of the uniform stretch-and-fold kneading operation. Actually we will see that it is a very close relative.

To understand the properties of chaos for the quadratic iterator we will study those of the iteration of the tent function and show that both are equivalent. The goal of this section is to derive the central tool for this purpose, a formula that allows the direct computation of any iterate for the tent transformation without having to carry out the iteration process over and over again. In other words, we will obtain an explicit expression for the result $T^k(x_0)$ for all initial points x_0 and all iteration stages k. The first piece of the solution of this problem is the substitution property of the tent and the saw-tooth transformations explained above. This yields $T^k(x_0) = TS^{k-1}(x_0)$. The remaining piece is an explicit formula for the iteration of the saw-tooth function needed for the evaluation of $S^{k-1}(x_0)$.

A New Notation for the Saw-Tooth Function

There is an elegant formula for the saw-tooth function which is different from the original definition in eqn. (10.10). It uses a function which computes

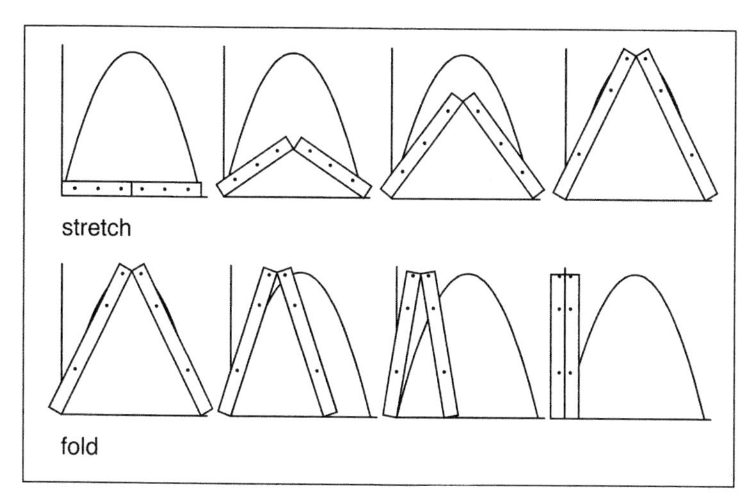

Nonuniform Stretching With Parabola

Interpretation of the quadratic transformation as a stretch-and-fold operation using elastic bars.

Figure 10.37

the fractional part[19] Frac (x) of a number x.

$$\text{Frac}\,(x) = x - k \text{ if } k \le x < k+1, \text{ } k \text{ integer.} \tag{10.12}$$

Some examples for the evaluation of this function are

$$\begin{aligned}\text{Frac}\,(0.4) &= 0.4,\\ \text{Frac}\,(5.123) &= 0.123,\\ \text{Frac}\,(18) &= 0,\\ \text{Frac}\,(24/7) &= 3/7.\end{aligned}$$

With this notation the saw-tooth transformation can be written as

$$S(x) = \text{Frac}\,(2x) \text{ for } 0 \le x < 1. \tag{10.13}$$

This is straightforward to check. If $0 \le x < 1/2$, then $0 \le S(x) = 2x < 1$ and Frac $(2x) = 2x$, yielding the same result. On the other hand, if $1/2 \le x < 1$, then $0 \le S(x) = 2x - 1 < 1$, thus, $1 \le 2x < 2$ and Frac $(2x) = 2x - 1$, giving the same result again. Only for the point $x = 1$ the formula in eqn. (10.13) does not work. But this is not significant, because $x = 1$ is a fixed point of the operator S and, moreover, there are no other points in the unit interval which are transformed to this fixed point. In other words, the dynamics of the iteration of the saw-tooth transformation S in the unit interval can be split into two independent parts, the interval $[0, 1)$, which does not include the point $x = 1$, and the singleton $\{1\}$. Of course, all of the interesting dynamics happens in $[0, 1)$. Thus, it is no loss to neglect the fixed point $x = 1$ in the following discussion, where we will use mostly the representation of eqn. (10.13) for the saw-tooth function.

[19]In the mathematical literature the fractional part Frac (x) of a number x is usually expressed using the modulo function, written as $x \bmod 1$.

The first advantage of the Frac-version of the saw-tooth transformation is the fact that iterating this transformation is possible in a closed form. What does that mean? Assume we start with $0 \leq x_0 < 1$ and compute

$$x_1 = \text{Frac}\,(2x_0)\,,$$

and then

$$x_2 = \text{Frac}\,(2x_1)\,,$$

and so on. In other words

$$x_{k+1} = \text{Frac}\,(2x_k)\,, \quad k = 0, 1, 2, 3, \ldots$$

Assume we want to know what x_k will be for some very large value of k. Do we have to carry out the iteration process k times? Not at all. We can express x_k explicitly in terms of x_0 alone. The result of the following straightforward derivation is the simple closed form

$$x_k = \text{Frac}\left(2^k x_0\right). \tag{10.14}$$

Let us look at an example. We iterate $x_0 = 8/25$ ten times using the saw-tooth function and check the closed form.

k	x_k	$2^k x_0$	Verify $x_k = \text{Frac}\left(2^k x_0\right)$
0	8/25	8/25	$8 = 0 \cdot 25 + 8$
1	16/25	16/25	$16 = 0 \cdot 25 + 16$
2	7/25	32/25	$32 = 1 \cdot 25 + 7$
3	14/25	64/25	$64 = 2 \cdot 25 + 14$
4	3/25	128/25	$128 = 5 \cdot 25 + 3$
5	6/25	256/25	$256 = 10 \cdot 25 + 6$
6	12/25	512/25	$512 = 20 \cdot 25 + 12$
7	24/25	1024/25	$1024 = 40 \cdot 25 + 24$
8	23/25	2048/25	$2048 = 81 \cdot 25 + 23$
9	21/25	4096/25	$4096 = 163 \cdot 25 + 21$
10	17/25	8192/25	$8192 = 327 \cdot 25 + 17$

Derivation of the Closed Form Iterate

The closed form eqn. (10.14) of the iteration of the saw-tooth function eqn. (10.13) follows from two basic properties of the operator Frac. Let m be any integer. Then

$$\begin{aligned} \text{Frac}\,(x + m) &= \text{Frac}\,(x)\,, \\ \text{Frac}\,(mx) &= \text{Frac}\,(m\,\text{Frac}\,(x))\,. \end{aligned}$$

The first formula follows directly from the definition in eqn. (10.12). To verify the other we let k be an integer such that $k \leq x < k + 1$. Thus, $\text{Frac}\,(x) = x - k$. Then

$$\begin{aligned} \text{Frac}\,(m\text{Frac}\,(x)) &= \text{Frac}\,(m(x - k)) = \text{Frac}\,(mx - mk) \\ &= \text{Frac}\,(mx) \end{aligned}$$

where the last equality follows from the first property. With these results we show the validity of the closed form iterate eqn. (10.14) by induction. The first iterate

$$x_1 = \text{Frac}\,(2x_0)\,,$$

is already in the closed form as claimed. For the induction step from k to $k+1$ let us assume the hypothesis

$$x_k = \text{Frac}\left(2^k x_0\right).$$

Then we compute

$$\begin{aligned} x_{k+1} &= \text{Frac}\,(2x_k) = \text{Frac}\left(2\,\text{Frac}\left(2^k x_0\right)\right) \\ &= \text{Frac}\left(2^{k+1} x_0\right) \end{aligned}$$

using the second property of the Frac-operator in the last equality. This concludes the proof by induction.

We do not recommend the closed form of the saw-tooth transformation for the numerical computation of the iterations because the required powers of 2 rapidly grow large and become untractable. However, the closed form is of great value for the theoretical discussion of chaos in the iteration as we will see in the next section.

Direct Computation of Iterates of the Tent Transformation

We now have collected the two necessary tools for the final result of this section, namely, the substitution property of the kneading operations and the closed form of the iteration of the saw-tooth transformation. Assume we have x_0 and would like to know the result x_k after k stretch-and-fold operations T. Without the substitution property there would be no other way to do this than to compute iteration after iteration. But with the aid of the substitution property, we would first compute $k-1$ iterations of the stretch-cut-and-paste operation S, based on the shortcut according to the explicit formula eqn. (10.14), which gives

$$y = S^{k-1}(x_0) = \text{Frac}\left(2^{k-1} x_0\right)$$

and then apply the stretch-and-fold operation T once,

$$x_k = T(y) = \begin{cases} 2y & \text{if } y \le 0.5 \\ -2y+2 & \text{if } y > 0.5. \end{cases}$$

In other words, rather then iterating k times, we just have to compute a power of 2, and carry out two multiplications and one addition! Extracting the fractional part, i.e., evaluation the function Frac costs almost nothing because it just means neglecting the integer part of the result.

Let us present an example. The initial point $x_0 = 8/25$ is a periodic point of the tent transformation with period 10. We can now check this fact without iterating ten times! Instead we compute x_{10} directly, using the above method for $k = 10$. We obtain

$$\begin{aligned} y &= S^9(x_0) = \text{Frac}\left(2^9 \cdot 8/25\right) = \text{Frac}\,(4096/25) \\ &= \text{Frac}\,(163 + 21/25) = 21/25. \end{aligned}$$

The result $y = 21/25$ is greater than 1/2. Thus,

$$x_{10} = T(y) = -2y + 2 = -42/25 + 2 = 8/25.$$

Voilà! $x_{10} = x_0$.

Summary of Results

Let us summarize the mathematical results of this section. They will be used in the following. The saw-tooth function

$$S(x) = \begin{cases} 2x & \text{if } x < 0.5 \\ 2x - 1 & \text{if } x \geq 0.5 \end{cases}$$

is the mathematical model for the stretch-cut-and-paste kneading. Using the notation of the fractional part

$$\text{Frac}\,(x) = x - k \text{ if } k \leq x < k + 1, \ \ k \text{ integer},$$

we can put the saw-tooth function into the simplified form

$$S(x) = \text{Frac}\,(2x) \ \text{ for } 0 \leq x < 1.$$

There is an explicit formula for the kth iterate,

$$S^k(x_0) = \text{Frac}\left(2^k x_0\right).$$

The tent transformation

$$T(x) = \begin{cases} 2x & \text{if } x \leq 0.5 \\ -2x + 2 & \text{if } x > 0.5 \end{cases}$$

is the mathematical model for the stretch-and-fold kneading. The substitution property relates the two kneading transformations in the sense that $k - 1$ applications of S followed by one transformation T yield the same result as the tent transformation T, applied k times,

$$T^k(x) = TS^{k-1}(x).$$

This relation allows the direct computation of iterates also for the tent transformation. Let

$$y = S^{k-1}(x_0) = \text{Frac}\left(2^{k-1} x_0\right)$$

and compute

$$x_k = T^k(x_0) = T(y) = \begin{cases} 2y & \text{if } y \leq 0.5 \\ -2y + 2 & \text{if } y > 0.5. \end{cases}$$

10.5 Analysis of Chaos: Sensitivity, Mixing, and Periodic Points

We are now prepared to carry out the next step in the plan of attack for unraveling the chaos for the quadratic iterator. We start with the iteration of the saw-tooth transformation using its closed form description to derive the central properties of chaos: sensitivity, mixing, and dense periodic points. The substitution property allows us to carry over these features to the iteration of the tent transformation. In the following section we conclude the analysis of chaos by exploiting another equivalence, namely, between the tent transformation and the quadratic iterator.

We begin with the saw-tooth transformation

$$S(x) = \text{Frac}\,(2x) \quad \text{for} \quad 0 \leq x < 1$$

and reveal a new interpretation by passing to binary representations of the real number x between 0 and 1. Recall that any real number x from the unit interval can be written as $x = 0.a_1a_2a_3\ldots$, where the a_k are *binary digits*, i.e., each a_k is either 0 or 1 and

$$x = a_1 2^{-1} + a_2 2^{-2} + a_3 2^{-3} + \cdots.$$

For example $1/2 = 0.100\ldots$, $3/4 = 0.1100\ldots$, $1/3 = 0.\overline{01}$ (overlining means periodic repetition), $1/7 = 0.\overline{001}$. One useful observation related to binary expansions is the following. Let x and y be numbers in the unit interval having binary expansions $x = 0.a_1a_2a_3\ldots$ and $y = 0.b_1b_2b_3\ldots$ Then

$$|x - y| \leq 2^{-k} \tag{10.15}$$

provided that x and y agree in the first k binary digits, i.e., $a_i = b_i$ for $i = 1,\ldots,k$.[20]

The Shift Operator

Now what does the iteration of the saw-tooth function mean in terms of binary expansions? Very simply, multiplication by 2 means passing from $0.a_1a_2a_3\ldots$ to $a_1.a_2a_3\ldots$ Therefore one application of the transformation is accomplished by first shifting all binary digits one place to the left and then erasing the digit that is moved in front of the point,

$$x = 0.a_1a_2a_3\ldots \rightarrow S(x) = 0.a_2a_3a_4\ldots \tag{10.16}$$

Because of the type of this almost mechanical procedure the transformation is also called the *shift operator* when interpreted in the context of binary representations. For the examples above, we list the results in the following table 10.38.

Resolving the Ambiguity

There is a technicality which we must address here, namely, the ambiguity of the binary representations. For example, the decimal number 0.5 has two possible binary versions, 0.1 and $0.0111\ldots = 0.0\overline{1}$. The application of the

[20] To derive this property, we assume $x > y$ and compute

$$|x - y| = |0.a_1a_2a_3\ldots - 0.b_1b_2b_3\ldots| \leq |0.a_1\ldots a_k\overline{1} - 0.b_1\ldots b_k\overline{0}| = |0.0\ldots0\overline{1}| = 2^{-k}.$$

Shift Operator

Four examples for the transformation of the shift operator.

Table 10.38

x_0	Binary	Transformed Binary	x_1
1/2	0.1	0.0	0
3/4	0.11	0.1	1/2
1/3	$0.\overline{01}$	$0.\overline{10}$	2/3
1/7	$0.\overline{001}$	$0.\overline{010}$	2/7

shift operator will have different results for the two numbers, 0 and $0.\overline{1} = 1$. To arrive at the correct result, we require that binary numbers must not end with repeating digits '1'. Thus, we represent $1/2 = 0.1$ and $1/4 = 0.01$, but not as $0.0\overline{1}$ or $0.00\overline{1}$.[21]

Binary Encoding of a Number

How can one obtain the binary representation of a decimal or rational number? There are several small algorithms for this purpose. It may come as a surprise that the shift operator provides a direct method to do the transformation. Interpreted this way it is a binary encoder. Here are the details.

Assume the binary representation of a number x with $0 \leq x < 1$ is desired. Set $x_0 = x$ and start the iteration of the shift operator. Thus, we compute

$$x_{n+1} = \text{Frac}\,(2x_n)\,, \quad n = 0, 1, \ldots$$

Then the binary representation of x is given by

$$0.a_0 a_1 a_2 \ldots$$

where the binary digits a_k are related to the kth iterate x_k:

$$a_k = \begin{cases} 0 & \text{if } x_k < 1/2 \\ 1 & \text{otherwise.} \end{cases}$$

Let us take two examples, $3/4$ and $1/7$. In the first case the iteration gives $3/4,\ 1/2,\ 0,\ 0,\ 0,\ \ldots$ Thus, the binary encoding of $3/4$ is $0.11000\ldots$ For $1/7$ we obtain a periodic cycle $1/7,\ 2/7,\ 4/7,\ 1/7,\ \ldots$ Thus, the binary encoding is also periodic, namely, $0.001001\ldots = 0.\overline{001}$.

The motion of a spice particle in the dough (when kneading by the stretch-cut-and-paste operation) can now be studied by investigating eqn. (10.16), which is also called the *shift on two symbols*. As simple as it may look, the dynamics which it generates are extremely complicated. We now turn to the three characteristics of chaos in iterated transformations: sensitivity, mixing,

[21] Two remarks are in order here. Firstly, our convention to outlaw all binaries ending with repeating digits '1' implies that we cannot represent the number 1.0 in the form $0.111\ldots$ But we have already argued that the point $x = 1$ is uninteresting and irrelevant for the dynamics of the iteration of the saw-tooth transformation. Secondly, referring to the footnote on page 501, note that in order to comply with the alternative definition of the saw-tooth function, where $S(1/2) = 1$, we would have to forbid all finite binaries, i.e., numbers $0.a_1a_2a_3\ldots$ that end with repeating digits '0'.

and periodic points merged in everywhere. All three of them can be verified for the stretch-cut-and-paste kneading.

Sensitivity to Initial Conditions

Imagine that we pick an initial number $x_0 = 0.a_1a_2a_3\ldots$ but only specify it up to N digits, say $N = 100$. Then the true number will differ from the specified one by at most 2^{-100}, a difference so small that we would say it should not matter at all. In any event, we can consider this difference to be something like an error of measurement. Since we don't know the digits $a_{101}a_{102}a_{103}\ldots$, we can assume that in each step of our calculation somebody flips a coin and thereby determines those digits a_{101} and so on (head = 0, tail = 1). Thus, we could say that our initial number is known to us only up to some degree of uncertainty — sometimes called noise — in the data which only affects the digits at position 101 and higher. Now let us run the iteration in eqn. (10.16).

At the beginning everything is tame. But as we continue iterating the noise creeps closer and closer to the decimal point, and after precisely 100 iterations, the results will become perfectly random. That is the phenomenon of sensitive dependence on initial conditions, but it is simultaneously an accurate and solid argument for the properties of the corresponding kneading operation on the dough.

Moreover, we can now provide an argument for the uniform distribution of spices in the dough after kneading. If the spice originally comes in a clump, the coordinates of its particles should be given as $0.a_1a_2\ldots a_ka_{k+1}\ldots$, where the first k digits are the same for all particles, because they are clustered. The remaining digits are uniformly distributed, modeling the random mixing of the spice in the cluster. After k applications of the shift, the common coordinates are gone, and only the random digits are left, which yields a uniform distribution of spice in the entire dough.

Sensitivity — A Closer Look

A more precise definition of sensitivity is the following: given any point x_0 between 0 and 1, there exists a point y_0 arbitrarily close to x_0 such that the outcome of the iteration started at points x_0 and y_0 will eventually differ by a certain threshold. This threshold must be the same for all points x_0 in the interval and is called the sensitivity constant. Note that it is not required that all orbits started close to x_0 will develop this deviation exceeding the threshold.

Let us argue that this definition of sensitivity holds for the iteration of the shift operator. We claim that the threshold in this case can be as large as $1/2$. Let us pick an arbitrary sample starting point in binary representation by rolling a die, writing 0 for an even roll and 1 for an odd one. The result might be

$$x_0 = 0.01011010111000011001\ldots$$

Now we try to find a starting point y_0 close by which should develop a difference to the orbit of x_0 which reaches the threshold $1/2$ at some point. For y_0 we may pick the same number as x_0 except for one of the binary digits, which we change. For example, if we require that y_0

has a distance to x_0 of at most 2^{-5}, then it suffices to flip the sixth digit of x_0,

$$y_0 = 0.0101111011100011001\ldots$$

After five iterations we have

$$x_5 = 0.01011100011001\ldots \le \frac{1}{2}$$

and

$$y_5 = 0.11011100011001\ldots = x_5 + \frac{1}{2}$$

Thus,

$$|y_5 - x_5| = \frac{1}{2}$$

as required. Clearly, we can find starting points y_0 arbitrarily close to x_0 with the same property. All we need to do is to just flip one of the binary digits which must be of sufficiently high order. At one point in the iteration this digit will be the most significant one and the difference with the orbit of x_0 will be $1/2$ again. Note that all further iterates in both orbits are identical. A difference also in those iterations is not required by the definition. Of course one may devise other strategies for the choice of y_0 that produce an orbit which is different from that of x_0 in all iterations.

A Cycle of Period 4

The point $0.\overline{0111}$ is a periodic point. The binary expansion allows to immediately read off the iterative behavior (here visualized as graphical iteration).

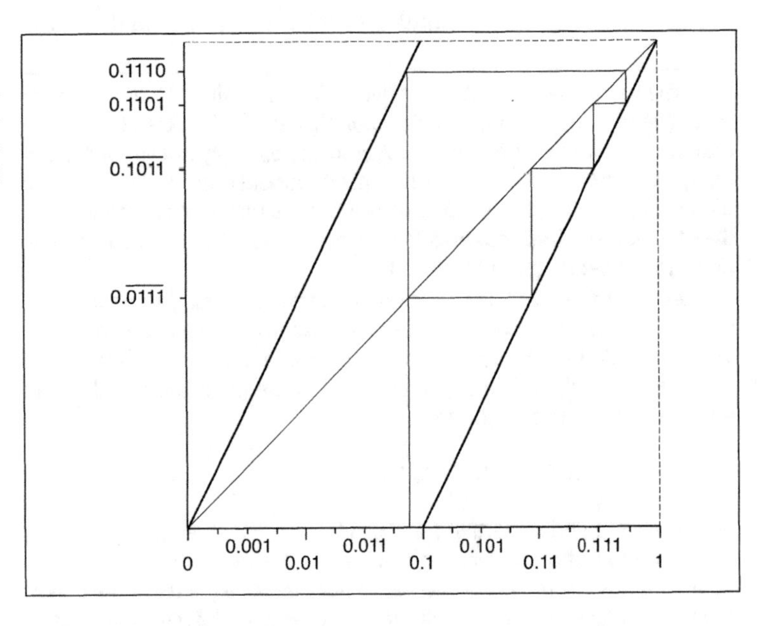

Figure 10.39

Periodic Points ...

Let us proceed to an understanding of another phenomenon which goes with chaos. What happens if we specify

$$x_0 = 0.\overline{a_1 a_2 a_3 \dots a_k}.$$

In other words, we have an infinite string of binary digits which repeats after k digits. Now running the iteration means that after k steps we will see x_0 again, and then after another k steps again, and again, and so forth. In other words, we see a cycle of length k. We call x_0 periodic with respect to the binary shift. Clearly, we can produce cycles of any length. But more importantly, for any given number x_0, we can find a number w_0 arbitrarily close to x_0, which is periodic. Let us see how this works.

... Are Dense

Well, if $x_0 = 0.a_1 a_2 a_3 \dots$ then choose

$$w_0 = 0.\overline{a_1 a_2 a_3 \dots a_k}.$$

for some k. Then x_0 and w_0 differ only by (at most) 2^{-k}, and w_0 is periodic. This means that *periodic points are dense*. An illustrative example for a periodic point is given in figure 10.39. Let us discuss another example in more detail. We want to approximate the irrational number

$$\frac{1}{\pi} = 0.3183098861\dots$$

better and better by periodic points. In binary representation $1/\pi$ is

$$\frac{1}{\pi} = 0.010100010111110011000001101101\dots$$

Allowing longer and longer periods we can approach the initial point as closely as we like. In the table below we use the first 5, 10, 15, 20, and 25 binary digits and repeat them periodically to generate the approximations, which we list also as decimal fractions.

Initial Point		Difference	Period
Binary	Fraction	$\lvert x_0 - 1/\pi \rvert$	
$0.0101000101111100110000011\dots$	$1/\pi$	$0.000 \cdot 2^{0}$	aperiodic
$0.\overline{01010}$	$10/31$	$0.137 \cdot 2^{-5}$	5
$0.\overline{0101000101}$	$325/1023$	$0.632 \cdot 2^{-10}$	10
$0.\overline{010100010111110}$	$10430/32767$	$0.060 \cdot 2^{-15}$	15
$0.\overline{01010001011111001100}$	$333772/1048575$	$0.211 \cdot 2^{-20}$	20
$0.\overline{0101000101111100110000011}$	$10680707/334554431$	$0.113 \cdot 2^{-25}$	25

Mixing

The final property is *mixing*; see page 480. Choose any two arbitrarily small intervals I and J. For mixing, one requires that one can find a starting point x_0 in I, whose orbit will enter the other interval at some iteration (see figure 10.40). It is straightforward to check this property for the shift operator.

Derivation of Mixing

Choose any two intervals I and J and let $n > 0$ such that interval I has a length greater than $1/2^{n-1}$. Further, let

$$0.a_1a_2a_3\ldots$$

be the binary representation of the midpoint of the first interval I. Moreover, let

$$y = 0.b_1b_2b_3\ldots$$

be the binary representation of a point y of the second interval J. Now we construct an initial point x_0 in I which, after exactly n iterations of the shift operator, will be equal to y, thus, providing the required point in the target interval J. To define x_0 we copy the first n digits of the center of I and then append all digits of the target point y:

$$x_0 = 0.a_1\ldots a_nb_1b_2b_3\ldots$$

Now we check: x_0 differs from the center of I by at most 2^{-n} which is at most half the width of interval I. Thus, it must be contained in the interval I. Secondly, after n iterations we have

$$x_n = 0.b_1b_2b_3\ldots = y.$$

So we have even over-fulfilled the requirement. In the case of the shift operator (saw-tooth transformation) we can hit any target point in the interval J.

Ergodic Behavior

So far we have verified the three properties of chaos for the saw-tooth transformation: sensitivity, dense periodic points, and mixing. Closely related to mixing is *ergodic* behavior. Ergodicity means that if we pick a number x_0 in the unit interval at random, then almost surely[22] the results of the shift operation will produce numbers which will get arbitrarily close to any number in the unit interval. Numbers x_0 with a periodic pattern in their binary expansion do not show such behavior and in some way they are extremely scarcely populated in the unit interval.

[22] The technical term 'almost surely' means that the probability for the following assertion is equal to one. For example, a number picked at random from the unit interval almost surely is irrational.

Example for Ergodicity

Typical numbers are more like this number:

$$x_0 = 0.0100011011000001010011100101110111 0000\ldots$$

Can you identify the pattern? Here is the rule: first write all numbers which need one binary digit, i.e., 0 and 1, then all strings of two binary digits, i.e., 00, 01, 10, and 11. Continue in this fashion with strings of three digits (000, 001, 010, 011, 100, 101, 110, 111) and so forth. Clearly by construction the resulting number will get close, and in fact arbitrarily close, to any given number under iteration of the shift operator. Indeed, let us take any number y in the interval and expand it in binary form

$$y = 0.a_1a_2a_3\ldots$$

Cutting off the expansion after k digits results in a number

$$z = 0.a_1a_2\ldots a_k$$

which is very close to y:

$$|z - y| \leq 2^{-k}.$$

Now we observe that the string of digits $a_1a_2\ldots a_k$ must appear in x_0 by construction at some place, and, therefore, sufficiently many shifts will bring this string to the leading digits. This provides a number that agrees with y and z in the first k leading digits; thus, it is as close to y as z.

It may seem that our example of the initial point x_0 is rather artificial. However, selecting the binary digits for x_0 at random has the same effect. The resulting orbit almost surely is ergodic. To see this just note that any block of binary digits must appear at some point in the binary representation of x_0 and the same reasoning as above applies.

The Next Step: Chaos for the Tent Transformation

We have seen that the iteration of the saw-tooth transformation S (or the shift operator, or the stretch-cut-and-paste kneading) exhibits the three properties of chaos. Now we proceed to the next stage and unfold the chaos for the tent transformation T (or the stretch-and-fold kneading). We recall that by means of the substitution property the iteration of T can be reduced to the iteration given by the saw-tooth transformation S. The kth iterate x_k is obtained by $k-1$ binary shifts followed by a single stretch-and-fold operation T. Since the first part is just a shift by $k-1$ binary digits, we now can easily carry all the complicated dynamic behavior — sensitive dependence, denseness of periodic points, and mixing — of the shift transformation over to the stretch-and-fold transformation. As a first example, let us see how a periodic point for the shift transformation generates a periodic point for the stretch-and-fold transformation. Or to phrase this differently, let us see how a seemingly impossible question turns into a very simple one.

Periodic Points for Stretch-and-Fold

Assume that we ask: what are periodic points for the iteration of the tent transformation T? Or, more precisely, find x_0, so that $x_n = x_0$ for a given integer n where $x_i = T(x_{i-1})$ for $i = 1, \ldots, n$. We claim that all we have to do is to take a point w_0 which is periodic for the shift transformation with period n, and to apply the tent transformation to obtain a periodic point $x_0 = T(w_0)$ of T. Indeed, let $w_0 = S^n(w_0)$ be a periodic point of S. Then we check whether $x_n = x_0$ using our definition of x_0 and the substitution property of the two kneading transformations.

$$\begin{aligned} x_n &= T^n(x_0) = T^n(T(w_0)) = T^{n+1}(w_0) \\ &= T(S^n(w_0)) = T(w_0) = x_0. \end{aligned}$$

Hence, it is true: if w_0 is a periodic point for the binary shift, then $x_0 = T(w_0)$ is a periodic point for the stretch-and-fold transformation T with the same period.[23]

Using this result it is not hard to reason that periodic points of T are dense in the unit interval. Likewise it is a bit technical but not difficult to derive sensitivity and mixing for the tent transformation. The details are given in the following technical section.

Chaos for the Tent Transformation

The goal of this technical section is to derive the three central properties of chaos for the iteration of the tent transformation T: dense periodic points, sensitivity, and mixing.

Binary Representation of the Tent Transformation

As for the analysis of chaos for the saw-tooth function, the representation of the transformation in terms of binary expansions is essential for the discussion of the three properties of chaos. We recall that the tent transformation is given by

$$T(x) = \begin{cases} 2x & \text{if } x \le 0.5 \\ -2x + 2 & \text{if } x > 0.5. \end{cases}$$

Let

$$x = 0.a_1 a_2 a_3 \ldots$$

be a binary expansion of $x \in [0, 1]$. Clearly, if $x < 1/2$ the tent transformation is identical to the saw-tooth transformation, thus

$$T(x) = 0.a_2 a_3 a_4 \ldots \quad \text{if } x < 1/2.$$

If $x \ge 1/2$, then $S(x) = 2x - 1$, and

$$\begin{aligned} T(x) &= -2x + 2 = 1 - (2x - 1) \\ &= 1 - S(x) = 1 - 0.a_2 a_3 a_4 \ldots \end{aligned}$$

Introducing the dual binary digit

$$a^* = \begin{cases} 1 & \text{if } a = 0 \\ 0 & \text{if } a = 1 \end{cases}$$

[23] To ensure that n is also the *minimal* period, i.e., $x_m \ne x_0$ for $m = 1, \ldots, n-1$, certain restrictions on the choice of the binary digits in $w_0 = 0.\overline{a_1 \ldots a_n}$ must be applied.

we have in that case, $x \geq 1/2$, $T(x) = 0.a_2^* a_3^* a_4^* \ldots$ because

$$0.a_2 a_3 a_4 \ldots + 0.a_2^* a_3^* a_4^* \ldots = 0.111 \ldots = 1.$$

The binary representation of the tent transformation therefore is

$$T(0.a_1 a_2 a_3 \ldots) = \begin{cases} 0.a_2 a_3 a_4 \ldots & \text{if } a_1 = 0 \\ 0.a_2^* a_3^* a_4^* \ldots & \text{if } a_1 = 1. \end{cases}$$

To deal with the ambiguous binary representations of rational numbers we only have to require to use 0.1 for $1/2$. The above binary version of T works also for $x = 1$ when representing 1 as $0.111\ldots$

Dense Periodic Points

Let us start the discussion of chaos with periodic points. We have already shown that a periodic point $w \in [0, 1)$ of the saw-tooth function S with $S^n(w) = w$ induces a periodic point $x = T(w)$ of T with $T^n(x) = x$. To show the denseness of these periodic points of T we demonstrate that we can find periodic points whose binary expansions start with an arbitrary sequence $a_1 \ldots a_n$. The point $w = 0.\overline{0a_1 \ldots a_n} < 1/2$ is periodic under S with period $n+1$. Thus, recalling that T is the shift transformation when the first digit of the argument is 0, we obtain $x = T(w) = 0.\overline{a_1 \ldots a_n 0}$, and x is periodic under T with period $n+1$.

For an example we take the interval $I = [27/32, 28/32]$ consisting of all binary numbers whose expansion starts with 0.11011 and look for a periodic point in I. The midpoint of the interval is $55/64 = 0.110111$. Then, according to the above we may choose

$$x = T(0.\overline{0110111}) = 0.\overline{1101110} = \frac{110}{127} \in I.$$

Indeed, x is in I and periodic under T with period 7.

Sensitive Dependence on Initial Conditions

We continue with the sensitive dependence on initial conditions. Thus, let $x_0 = 0.a_1 a_2 a_3 \ldots$ be an arbitrary initial point from the unit interval and its binary expansion. For $n > 0$ given we search for a point z_0 near x_0, such that $|z_0 - x_0| \leq 2^{-n}$ and such that the orbits of x_0 and z_0 differ at some iterate by $\varepsilon = 1/2$. We choose

$$z_0 = 0.a_1 \ldots a_n a_{n+1}^* a_{n+2} a_{n+3} a_{n+4} \ldots$$

i.e., we flip bit number $n+1$. Using the estimate (10.15), we obtain $|z_0 - x_0| \leq 2^{-n}$ and claim that $|z_n - x_n| = 1/2$. For the proof we consider the two cases $a_n = 0$ and $a_n = 1$ separately.

Case $a_n = 0$. Then

$$\begin{aligned} x_n &= T^n(x_0) = T(S^{n-1}(x_0)) = T(0.a_n a_{n+1} a_{n+2} \ldots) \\ &= T(0.0 a_{n+1} a_{n+2} a_{n+3} \ldots) = 0.a_{n+1} a_{n+2} a_{n+3} \ldots \\ z_n &= T^n(z_0) = T(S^{n-1}(z_0)) = T(0.a_n a_{n+1}^* a_{n+2} a_{n+3} \ldots) \\ &= T(0.0 a_{n+1}^* a_{n+2} a_{n+3} \ldots) = 0.a_{n+1}^* a_{n+2} a_{n+3} a_{n+4} \ldots \end{aligned}$$

and the claim follows.

Case $a_n = 1$. Then

$$\begin{aligned} x_n &= T(0.1 a_{n+1} a_{n+2} a_{n+3} \ldots) = 0.a_{n+1}^* a_{n+2}^* a_{n+3}^* \ldots \\ z_n &= T(0.1 a_{n+1}^* a_{n+2} a_{n+3} \ldots) = 0.a_{n+1} a_{n+2}^* a_{n+3}^* \ldots \end{aligned}$$

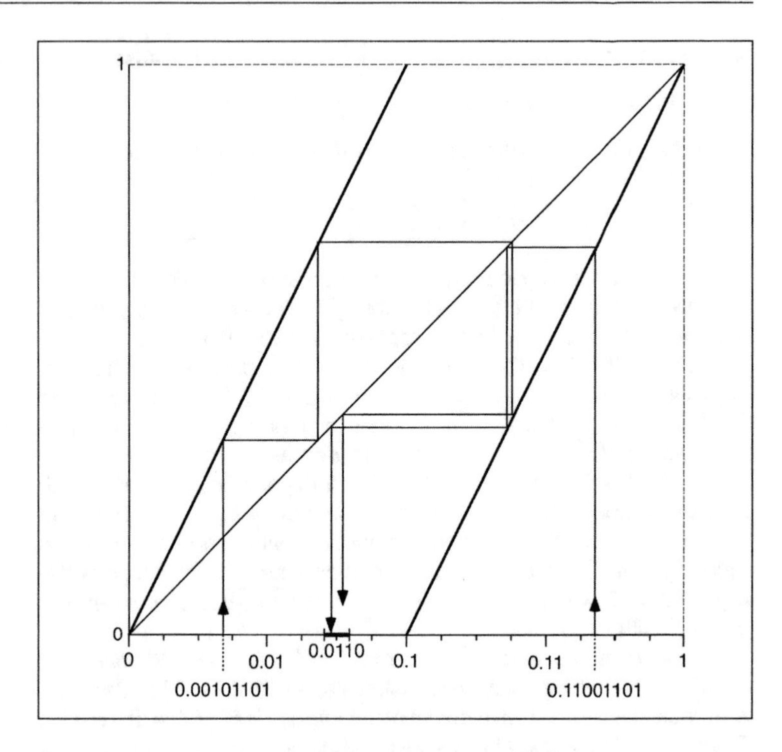

Figure 10.40 : Mixing requires that any given interval J can be reached from any other interval. Here two examples are shown how we can reach a small interval at 0.0110.

and again, the claim follows also for this case.

For an example let us reconsider the interval $I = [27/32, 28/32]$ and its midpoint $x_0 = 0.1101110\overline{0}$. We choose $n = 6$ so that we are looking for an initial point z_0 in I which will drift away from the orbit of x_0 during the course of iteration of the tent transformation. We flip the seventh bit of x_0 yielding $z_0 = 0.1101111$ and compute

$$\begin{aligned} x_6 &= T^6(x_0) = T(S^5(0.110111\overline{0})) = T(0.1\overline{0}) = 0.\overline{1} = 1 \\ z_6 &= T^6(z_0) = T(S^5(0.1101111\overline{0})) = T(0.11\overline{0}) \\ &= 0.0\overline{1} = 0.1\overline{0} = 1/2 \end{aligned}$$

Thus, $|z_6 - x_6| = 1/2$, as required.

Mixing

We conclude with the derivation of the mixing property. Given are two arbitrary open intervals I and J in the unit interval. It is always possible to choose n large enough and bits $a_1 \ldots a_n$ and $b_1 \ldots b_n$ so that all binary numbers in $[0, 1]$ whose binary expansion begins with $a_1 \ldots a_n$ are in the interval I and all binaries starting with $b_1 \ldots b_n$ are in J. Now we specify an initial point $x_0 \in I$ such that the nth iterate is in J, $x_n = T^n(x_0) \in J$. Again we treat the two cases $a_n = 0$ and $a_n = 1$ separately.

Case $a_n = 0$. Then we choose $x_0 = 0.a_1 \dots a_n b_1 \dots b_n$ and verify

$$\begin{aligned} x_n &= T^n(x_0) = T(S^{n-1}(x_0)) = T(0.a_n b_1 \dots b_n) \\ &= T(0.0b_1 \dots b_n) = 0.b_1 \dots b_n \in J. \end{aligned}$$

Case $a_n = 1$. Then we choose $x_0 = 0.a_1 \dots a_n b_1^* \dots b_n^*$ and verify

$$\begin{aligned} x_n &= T^n(x_0) = T(S^{n-1}(x_0)) = T(0.a_n b_1^* \dots b_n^*) \\ &= T(0.1b_1^* \dots b_n^*) = 0.b_1 \dots b_n 111 \dots \in J. \end{aligned}$$

For an example we use $I = [27/32, 28/32]$, as before, and $J = [14/32, 15/32]$. Here $n = 5$ and all binaries of the form $0.11011\dots$ are in I and all binaries $0.01110\dots$ are in J. Since $a_5 = 1$, we choose

$$x_0 = 0.110110^*1^*1^*1^*0^* = 0.1101110001.$$

Then

$$\begin{aligned} x_5 &= T^5(x_0) = T(S^4(x_0)) = T(0.110001) \\ &= T(0.01110\overline{1}) \in J. \end{aligned}$$

Thus, the knowledge of the shift operator and its relation to the tent transformation indeed provides us with the key to derive the three properties of chaos for the tent transformation.

This concludes the theoretical discussion of chaos for the iteration of the kneading operations and the binary shift operator.

10.6 Chaos for the Quadratic Iterator

The Last Step

What we have learned in the previous sections may seem to be a rather special case, but in fact the contrary is true. This section is devoted to the *equivalence* of the iteration of the uniform kneading operator given by the tent transformation and the quadratic iteration $x \to 4x(1-x)$ (the non-uniform kneading). In other words, all the complex behavior which we were able to show first for the shift operator and then for the tent transformation can also be found in the quadratic iterator. And in some respects this gives a theoretical background for what we have learned in our experiments in the first part of this chapter: there is sensitivity on initial conditions; there is mixing; we can determine ergodic and also periodic points. Thus, we have made a full circle in our story about chaos and kneading.

Coordinate Transformation

The function $h(x) = \sin^2(\pi x/2)$ is a coordinate transformation. For each x there is a corresponding value $x' = h(x)$ and vice versa. As indicated along the axes, intervals do not retain their lengths when subjected to the transformation.

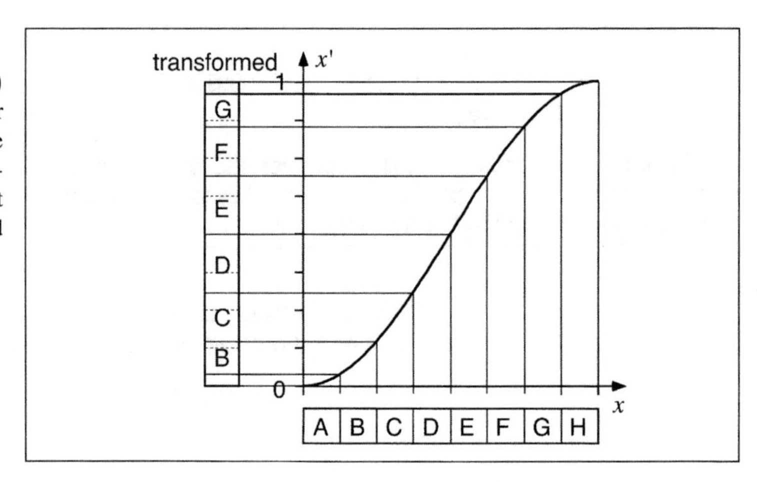

Figure 10.41

Tent Function Versus Parabola

The equivalence of the iteration of the tent transformation $T(x)$ and the quadratic parabola $4x(1-x)$ is established by a nonlinear change of coordinates given by

$$x' = h(x) = \sin^2\left(\frac{\pi x}{2}\right).$$

Before we show why this is true let us first explain what it means. The S-shaped graph of the function h is shown in figure 10.41. Note that h transforms the unit interval $[0,1]$ to itself in a one-to-one fashion, i.e., for all $x' \in [0,1]$ there is exactly one $x \in [0,1]$ with $x' = h(x)$. How does this function translate the dynamics of the tent transformation to that of the quadratic iterator? Assume we are looking at an initial point x_0 and its orbit $x_0, x_1, x_2, \ldots$ under the tent transformation. Thus,

$$x_1 = T(x_0), x_2 = T^2(x_0), \ldots, x_k = T^k(x_0), \ldots \qquad (10.17)$$

The transformed initial point is

$$x'_0 = h(x_0).$$

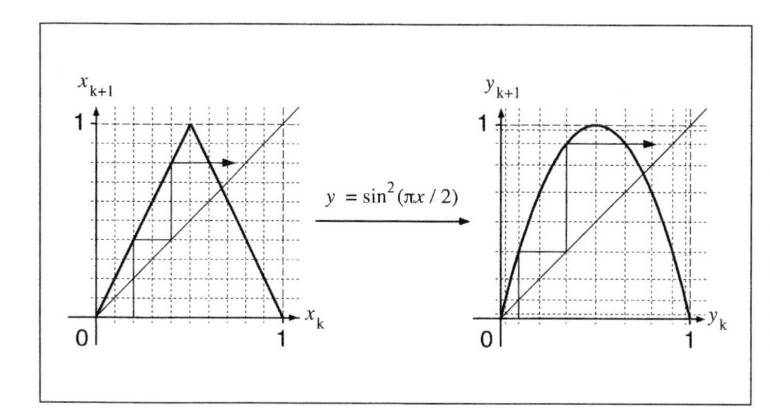

Coordinate Transformation of Graphical Iteration

Changing the coordinates according to the function $h(x)$ transforms the graph of the tent transformation to that of the quadratic function $f(x) = 4x(1-x)$. Furthermore, graphical iteration for T (left) is transformed into graphical iteration for f (right) also using h. The two orbits are equivalent.

Figure 10.42

This is — so to speak — the initial point x_0 in new coordinates, namely, those belonging to the iteration of the parabola. Now we compute the iteration of $y_0 = x'_0$, using the quadratic transformation $f(x) = 4x(1-x)$, obtaining

$$y_1 = f(y_0), y_2 = f^2(y_0), \ldots, y_k = f^k(y_0), \ldots$$

The claim is that this is really the same orbit as the above in eqn. (10.17), however, given in the modified coordinates (see figure 10.42). In other words, not only is $y_0 = h(x_0)$, but also

$$y_1 = h(x_1), y_2 = h(x_2), \ldots, y_k = h(x_k), \ldots$$

Thus, iterating x_0 under T produces an orbit which is — after change of coordinates — the same as that of $y_0 = x'_0 = h(x_0)$ under the quadratic f. In terms of the functions f and T this equivalence can be put in the form of the *functional equation*

$$f^k(h(x)) = h(T^k(x)), \quad k = 1, 2, \ldots \tag{10.18}$$

for all $x \in [0,1]$.

Example

Let us present an example, the iteration of $x_0 = 8/25$. The data for the first nine iterations is collected in table 10.43 and the equivalence is visualized in figure 10.44. Indeed we find that the transformed coordinates of the orbit for the tent transformation perfectly matches the orbit of the quadratic iterator started at $y_0 = x'_0$.

Sensitivity May Destroy the Equivalence

However, this result must be interpreted with caution! Although the mathematics seem to provide the definite and assuring claim that the two rightmost columns in table 10.43 are identical for as many iterations as we wish, we must not conclude that this equivalence holds in practice when computing more and more entries with a calculator or computer program. The reason, of course, lies in the sensitive dependence on initial conditions. Although $x_0 = 8/25$ is exact, x'_0 cannot be exactly represented in the machine, and, moreover, there

Checking the Equivalence

The table lists the first nine iterates of $x_0 = 8/25$ under T, the corresponding iterates of $y_0 = x'_0$ under f, and makes the check, computing the transformed numbers $x'_k = h(x_k)$.

k	$x_k = T^k(x_0)$	$x'_k = \sin^2(\pi x_k/2)$	$y_k = f^k(y_0)$
0	8/25	0.232	0.232
1	16/25	0.713	0.713
2	18/25	0.819	0.819
3	14/25	0.594	0.594
4	22/25	0.965	0.965
5	6/25	0.136	0.136
6	12/25	0.469	0.469
7	24/25	0.996	0.996
8	2/25	0.016	0.016
9	4/25	0.062	0.062

Table 10.43

will be additional small errors introduced at every step of the iteration. Thus, after a finite number of iterations, depending on the precision of the arithmetic, we end up producing a numerical orbit that has no resemblance to the true orbit belonging to x'_0, and, thus, the numerical observation of the equivalence is in fact destroyed by chaos.

The Equivalence of the Tent Transformation and the Quadratic Iterator

In this section we present the mathematics behind the equivalence of the iteration of the tent transformation and the quadratic iterator. To see that they are really the same, all the tools we need are two familiar trigonometric identities,

$$\sin^2\alpha + \cos^2\alpha = 1 \quad \text{and} \quad \sin 2\alpha = 2\sin\alpha\cos\alpha.$$

Iterating an initial point x_0 under the tent function and iterating the transformed point $x'_0 = \sin^2(x_0\pi/2)$ under the parabola $f(x) = 4x(1-x)$ produces iterations that correspond to each other by means of the transformation $x' = h(x) = \sin^2(x\pi/2)$.

To establish this algebraically, we start with x_0 for the tent function and use $y_0 = x'_0$ for the parabola. Thus, $x_0, x_1, \ldots$ is the iteration under the tent function and $y_0, y_1, \ldots$ is the corresponding iteration under the parabola. We can show by induction that, in fact, $y_k = x'_k = h(x_k)$ for all numbers $k = 0, 1, 2, \ldots$, proving the equivalence.

We start with the transformation $y_0 = \sin^2(x_0\pi/2)$, where $0 \leq x_0 \leq 1$. We substitute for y_0 in the formula for the quadratic iteration.

$$y_1 = 4y_0(1-y_0) = 4\sin^2\left(\frac{\pi x_0}{2}\right)\left(1 - \sin^2\left(\frac{\pi x_0}{2}\right)\right).$$

We substitute using the trigonometric identity $\cos^2\alpha = 1 - \sin^2\alpha$.

$$y_1 = 4\sin^2\left(\frac{\pi x_0}{2}\right)\cos^2\left(\frac{\pi x_0}{2}\right).$$

Simplify using the double-angle identity $\sin 2\alpha = 2\sin\alpha\cos\alpha$.

$$y_1 = \sin^2(x_0\pi).$$

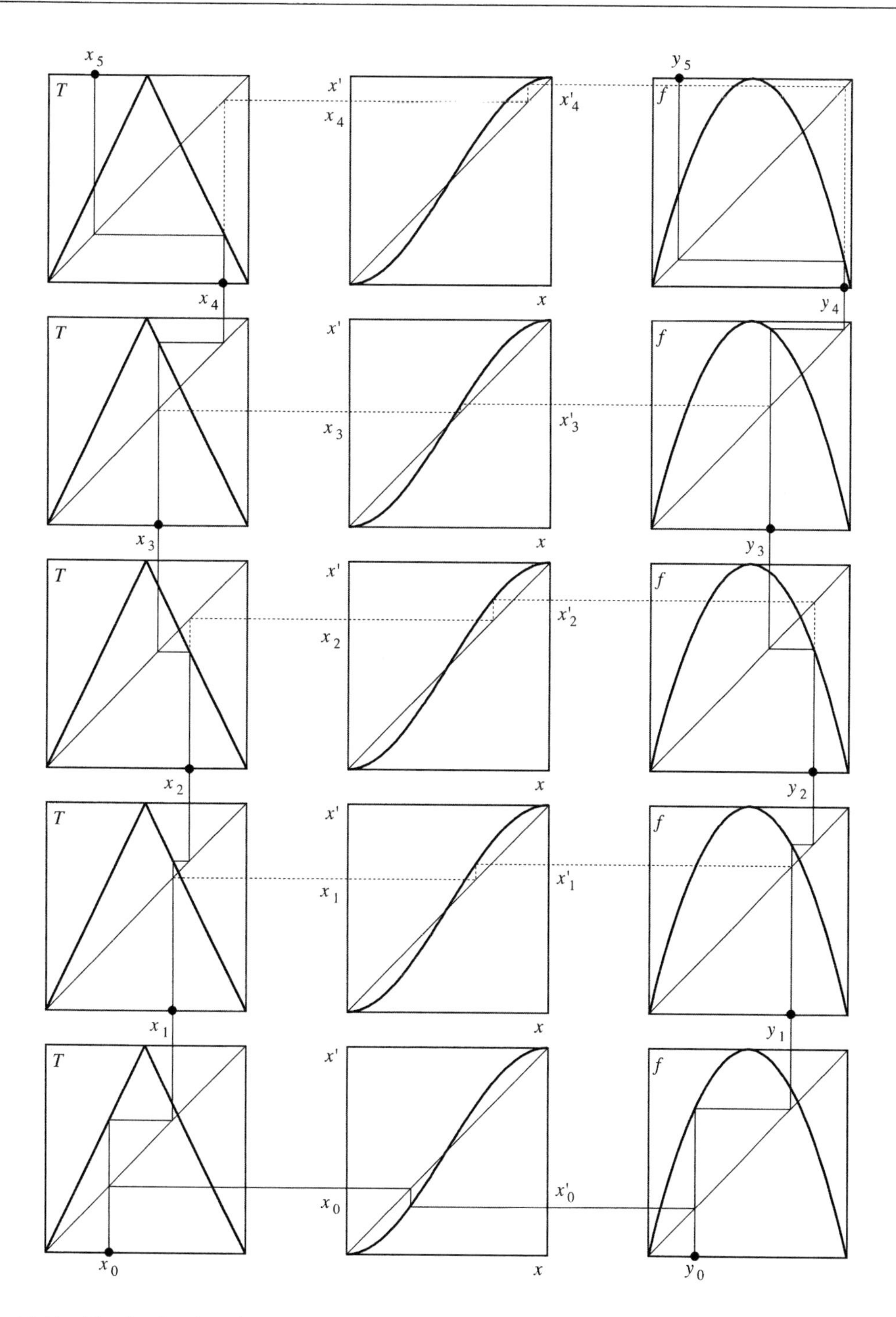

Figure 10.44 : The first few iterations of the tent transformation in table 10.43 and their counterparts for the parabola are visualized here together with the coordinate transformation establishing the equivalence.

The first iterate of x_0 under the tent function is $x_1 = T(x_0)$. We now show that y_1 above is in fact identical to x_1 after change of coordinates, i.e.,

$$x'_1 = h(x_1) = y_1.$$

We begin with the case $0 \leq x_0 \leq 1/2$. Thus, $x_1 = T(x_0) = 2x_0$ and

$$x'_1 = \sin^2\left(\frac{\pi x_1}{2}\right) = \sin^2(x_0\pi) = y_1.$$

Now we do the other case, $1/2 < x_0 \leq 1$. First we substitute $x_1 = T(x_0) = 2 - 2x_0$.

$$x'_1 = \sin^2\left(\frac{\pi x_1}{2}\right) = \sin^2(\pi - x_0\pi).$$

Then we simplify, first using $\sin^2(\alpha + \pi) = \sin^2(\alpha)$, and then $\sin^2(-\alpha) = \sin^2\alpha$.

$$x'_1 = \sin^2(-x_0\pi) = \sin^2(x_0\pi) = y_1.$$

The result shows $x'_1 = y_1$ and the conclusion $x'_k = y_k$ for all k now follows by induction. Thus, since $x'_k = h(T^k(x_0))$ and $y_k = f^k(h(x_0))$, we have shown the functional equation (10.18).

By the way it is also possible to define a transformation that relates the iteration of the saw-tooth transformation S to that of the quadratic iterator. For this purpose use $h(x) = \sin^2(x_0\pi)$. The corresponding functional equation again is $f^k(h(x)) = h(S^k(x))$. Note, however, that h does not induce an equivalence relation, because h is not one-to-one. This means that we only can conclude from properties of S to properties of f, but not vice versa. We omit the details.

Exploiting the Equivalence: Explicit Periodic Points

As an application of the above formulas for the change of coordinates we can easily name periodic points for the quadratic iterator. All we need is a periodic point for the tent transformation, say x_0. Then we apply the equivalence transformation to obtain $\sin^2(\pi x_0/2)$ which is guaranteed to be periodic in the quadratic iterator. For example, $x_0 = 2/7$ is periodic with period three for the tent transformation,

$$2/7 \to 4/7 \to 6/7 \to 2/7.$$

Therefore, the initial value $\sin^2(\pi/7) = 0.188255099\ldots$ is also a point from a periodic orbit of period three in the quadratic iterator. This is what we have used in the first section of this chapter on page 484.

From Chaos for T to Chaos for f

Thus, the iteration of the tent transformation and the parabola are totally equivalent. All the signs of chaos are found when iterating the quadratic function $f(x) = 4x(1-x)$.

- Points that are periodic for the tent transformation correspond to points that are periodic for the parabola.

- Points that show mixing by leading from one given interval to another for the tent transformation correspond to points that have the same behavior for the parabola.
- Points that exhibit sensitivity for the tent transformation correspond to points that show sensitivity for the parabola.

However, we remark that these conclusions are not self-evident. In the following technical section we present the proof for the first two properties.

Dense Periodic Points and Mixing

Let T be the tent transformation, $f(x) = 4x(1-x)$ the quadratic transformation, and $h(x) = \sin^2(\pi x/2)$ the transformation for the change of coordinates. The functional equation $f^k(h(x)) = h(T^k(x))$ for $k = 1, 2, \ldots$ and $x \in [0,1]$ has been shown above. Furthermore, we know that periodic points of T are dense in $[0,1]$ and T is mixing. **(a)** We claim that periodic points of f are dense in $[0,1]$. Let $y \in [0,1]$. We will show that there is a sequence of periodic points of f with limit y. We may choose x as a preimage of y under h, i.e., $h(x) = y$, because h is onto. Since periodic points of T are dense in $[0,1]$ we find a sequence of points $x_1, x_2, \ldots$ with limit x and such that each point x_k is a periodic point of T of some period, say p_k.[24] Thus, $T^{p_k}(x_k) = x_k$ for $k = 1, 2, \ldots$ We claim that the sequence $y_1, y_2, \ldots$ with $y_k = h(x_k)$ has limit y and is a sequence of periodic points of f. The first claim is true because h is continuous.[25] The second claim follows from the functional equation $f^k h = hT^k$. Indeed,

$$f^{p_k}(y_k) = f^{p_k}(h(x_k)) = h(T^{p_k}(x_k)) = h(x_k) = y_k.$$

(b) We claim that the transformation f is mixing. Let U and V be two open intervals in $[0,1]$. We have to find a point $y \in U$ and a natural number k so that $f^k(y) \in V$. We start with taking the preimages $A = h^{-1}(U) = \{x \in [0,1] \mid h(x) \in U\}$ and $B = h^{-1}(V)$. Note that A and B are open, because h is continuous. Thus there exists a natural number k and $x \in A$ such that $T^k(x) \in B$, since T is mixing. Set $y = h(x)$. Now, using the functional equation we obtain $f^k(y) = f^k(h(x)) = h(T^k(x))$. And since $T^k(x) \in B$ we conclude that $h(T^k(x)) \in h(B) = V$. Thus, f is mixing.

These proofs are rather straightforward and require only to properly use the appropriate definitions together with the functional equation $f^k h(x) = hT^k(x)$. Dense periodic points for the tent transformation and the equivalence of T and f yield that also f has dense periodic points. Mixing for T and the equivalence yield that also f is mixing. Now, this approach does *not* work

[24] Note that — in contrast to our standard use of notation — the sequence $x_1, x_2, \ldots$ is *not* an orbit of T.

[25] One way of defining what continuity for a function $f : X \to X$ (where X is, for example, a subset of the real line) means is the following. The function f is said to be continuous provided that for any $x \in X$ and any sequence $x_1, x_2, \ldots$ with limit x we have that the sequence $f(x_1), f(x_2), \ldots$ has also a limit which is $f(x)$. An alternative and equivalent definition is the following. The transformation f is continuous provided for any open set U in X the preimage $f^{-1}(U) = \{x \in X \mid f(x) \in U\}$ is open in X. A subset U of the real line $\mathbf{R}$ is said to be open provided for any $x \in U$ there is an open interval I containing x which is entirely in U.

for the third property of chaos, sensitivity. This can be demonstrated by the following rather simple counterexample.

Sensitivity: Counterexample

The iterations of the functions

$$f(x) = 2x, \quad x \in [1, \infty)$$

and

$$g(y) = y + \log 2, \quad y \in [0, \infty)$$

are equivalent by means of the coordinate transformation

$$h(x) = \log x.$$

Indeed, $h : [1, \infty) \to [0, \infty)$ is a continuous, one-to-one and onto transformation and the inverse of h, $h^{-1}(y) = e^y$, is continuous as well. Moreover,

$$h(f(x)) = g(h(x)) \text{ for all } x \in [1, \infty),$$

since

$$\log 2x = \log x + \log 2.$$

From this we get the functional equation $h(f^k(x)) = g^k(h(x))$ for $k = 1, 2 \ldots$ Note that f has sensitive dependence on initial conditions but g has not, because g is just a translation. Initial errors are magnified as powers of 2 in the course of the iteration of f, while initial errors remain constant using g. As a consequence of this observation, sensitive dependence on initial conditions is not generally inherited from one dynamical system to another which has iterations that are equivalent by change of coordinates. In contrast, the properties of mixing and dense periodic points are passed over to the equivalent system.

Therefore, the derivation of sensitivity for the quadratic transformation requires more than just the sensitivity of T and the equivalence of f and T. In this case we must exploit the fact that the underlying space is just a (compact) interval. This is the crucial difference to the counterexample presented above.

Derivation of Sensitivity

We claim that f has sensitive dependence on initial conditions. Let y be an arbitrary point in $[0, 1]$. We will show that there is a sequence of initial points $y_1, y_2, \ldots$ with limit y such that the corresponding orbits will drift away from that of y by at least a distance of some certain $\delta_f > 0$. We may choose x as the preimage of y under h, i.e., $h(x) = y$. Now T is sensitive, and, thus, there is a constant $\delta_T > 0$ and a sequence of initial points $x_1, x_2, \ldots$ for T with limit x, such that the corresponding orbits under T will drift away from that of x by at least a distance of δ_T. Precisely, this means that for each initial point x_k there is an iteration count n_k such that the n_k^{th} iterate of x_k differs from the n_k^{th} iterate of x by at least δ_T,

$$|T^{n_k}(x_k) - T^{n_k}(x)| \geq \delta_T. \tag{10.19}$$

We define

$$\delta_f = \inf\{|h(x) - h(y)| \mid |x - y| \geq \delta_T,\ x, y \in [0, 1]\}.$$

Since $h(x) = \sin^2(\pi x/2)$ is a strictly monotonically increasing function we conclude that δ_f is the minimum of the continuous function $h(x) - h(x - \delta_T)$ defined for $\delta_T \leq x \leq 1$. In our case this minimum is $\delta_f = h(\delta_T) > 0$.[26]

Now we consider the sequence $y_1, y_2, \ldots$ with $y_k = h(x_k)$, i.e., we apply the change of coordinates to the sequence of initial points for T. Firstly, since the transformation h is continuous, we have that this sequence has a limit, namely,

$$\lim_{k\to\infty} y_k = \lim_{k\to\infty} h(x_k) = h(x) = y.$$

Secondly, we apply the functional equation $f^{n_k}h = hT^{n_k}$ to obtain

$$\begin{aligned} |f^{n_k}(y_k) - f^{n_k}(y)| &= |f^{n_k}(h(x_k)) - f^{n_k}(h(x))| \\ &= |h(T^{n_k}(x_k)) - h(T^{n_k}(x))|. \end{aligned}$$

Because of the inequality (10.19) and the definition of δ_f we get

$$|f^{n_k}(y_k) - f^{n_k}(y)| \geq \delta_f.$$

Thus, the orbit with initial point y_k achieves a distance greater than or equal to δ_f after n_k iterations. Since $y_k \to y$, we have found initial points arbitrarily close to y that have this property. Thus, f has sensitivity at the point y.

Mixing and Dense Periodic Points Imply Sensitivity

There is an alternative and elegant solution of the problem of deducing sensitivity which works not only for the quadratic transformation but for all similar cases. The result has recently been worked out by a group of five Australian mathematicians.[27] They showed in a theorem that the properties of mixing and dense periodic points already suffice to show the third property of chaos, sensitivity. In other words, if f is chaotic and f and g are equivalent via a change of coordinates h, i.e., $f(h(x)) = h(g(x))$, then also g is chaotic. Thus, we do not need to undertake the task of deriving sensitivity of f from that of T. The proof of this theorem is technical but not difficult and will be discussed in the following section.

Chaos for the Gauss Map

We close this section with an example of a chaotic transformation which is similar to the quadratic transformation in the sense that it allows a complete analysis of its qualitative and quantitative properties of chaos. The transformation is

$$G(x) = \begin{cases} 0 & \text{if } x = 0 \\ \text{Frac}\left(\frac{1}{x}\right) = \frac{1}{x} \bmod 1 & \text{for } x \in (0, 1) \end{cases}$$

[26] Note that this definition of δ does not work in the counterexample presented further above where the change of coordinates is given by $h(x) = \log x$. In that case $\inf\{|h(x) - h(y)| \mid |x - y| \geq \delta_T,\ x, y > 1\} = 0$, which is the point where the proof would collapse if applied to the counterexample.

[27] J. Banks, J. Brooks, G. Cairns, G. Davis, P. Stacey, *On Devaney's definition of chaos*, American Math. Monthly 99.4 (1992) 332–334.

which is called the *Gauss map*. The iteration of G is related to the continued fraction expansion of numbers (see chapter 3). Consider $x_0 \in [0, 1)$ and define

$$x_{k+1} = G(x_k), \quad k = 0, 1, 2, \ldots$$

Then there is a corresponding sequence of integers $n_1, n_2, \ldots$ with

$$n_{k+1} = \frac{1}{x_k} - x_{k+1}.$$

In other words, $x_{k+1} = G(x_k)$ is the fractional part of $1/x_k$, while n_{k+1} is the corresponding integer part of $1/x_k$. It is an easy exercise to verify that the continued fraction expansion of x_0 is simply given by

$$x = [n_1, n_2, n_3, \ldots] = \cfrac{1}{n_1 + \cfrac{1}{n_2 + \cfrac{1}{n_3 + \cfrac{1}{n_4 + \cdots}}}}.$$

For this purpose make use of the relation

$$[n_1, \ldots, n_k] = \frac{1}{n_1 + [n_2, \ldots, n_k]}.$$

We note that the iteration of the Gauss map terminates, i.e., lands in the fixed point 0, provided that the initial point x_0 is a rational number. For an irrational number x_0 the sequence $x_k, k = 0, 1, 2, \ldots$ never reaches the fixed point zero, and

$$x_0 = \lim_{k \to \infty} [n_1, n_2, \ldots, n_k].$$

In terms of continued fraction expansions the Gauss map can be written as

$$G([n_1, n_2, n_3, \ldots]) = [n_2, n_3, n_4, \ldots]$$

and this provides an analogy to the shift transformation corresponding to the quadratic iterator. In fact, the three properties of chaos — dense periodic points, mixing, and sensitivity — can be derived in a similar way as for the shift transformation.[28] Furthermore, one knows that the Ljapunov exponent for the Gauss map is equal to $\pi^2/6 \log 2 \approx 2.373 > 0$ for almost all initial points in the unit interval. The invariant distribution of G was already known to Gauss. Starting with a uniform distribution of initial points x_0, Gauss considered the resulting distributions F_1 of x_1, F_2, of x_2, and so on. The invariant distribution, F_∞, is obtained in the limit of F_k as $k \to \infty$,

$$F_\infty(x) = \log_2(1 + x).$$

The problem that troubled Gauss — and that he was not able to solve to his satisfaction – was the analysis of the speed of convergence with which the limit is attained. Only many years later, this question could be settled.[29]

[28] See R. M. Corless, *Continued fractions and chaos,* American Math. Monthly 99, 3 (1992) 203–215.

[29] For an account of this story see D. E. Knuth, *The Art of Computer Programming, Volume 2, Seminumerical Algorithms,* Addison-Wesley, Reading, Massachusetts, 1981, pages 345–349.

10.7 Mixing and Dense Periodic Points Imply Sensitivity

How Is Chaos Inherited?

We have followed Devaney's book[30] for a definition of chaos by the three fundamental properties: mixing,[31] dense periodic points, and sensitive dependence on initial conditions. We have seen above that we were able to analyze the chaos properties of the saw-tooth transformation quite easily. Then we used the functional relation $TS = TT$ to establish chaos also for the tent transformation T. It is quite natural to ask whether the three chaos properties are in fact independent. That is, whether one or two of these conditions could imply the other(s) or not. Another natural question is that of inheritance. Given that a mapping f is chaotic and that g is related to f, can we conclude that g is chaotic as well? We will answer these basic questions following an elementary discussion in a short note published recently.[32] We have seen several examples of mappings in the previous sections which are chaotic and which are related to each other. Let us first sharpen the notion of being related. The proper notions are those of *topological conjugacy* and *topological semi-conjugacy*. The discussion could be carried out in a very general situation but we prefer to stay with transformations of the real line.

Let X and Y be two subsets of the real line and let f and g be transformations, $f : X \to X$ and $g : Y \to Y$. Then f and g are said to be *topologically conjugate* provided f and g are continuous and there is a homeomorphism $h : X \to Y$ such that the functional equation

$$h(f(x)) = g(h(x))$$

holds for all $x \in X$.

Definitions

The transformation $f : X \to Y$ is said to be continuous provided for any $x \in X$ and any sequence $x_1, x_2, \ldots$ with limit x we have that the sequence $f(x_1), f(x_2), \ldots$ has also a limit which is $f(x)$. An alternative and equivalent definition is the following. The transformation f is continuous provided for any open set U in Y the preimage $f^{-1}(U) = \{x \in X \mid f(x) \in U\}$ is open in X. A subset V of the real line $\mathbf{R}$ is said to be open provided for any $x \in V$ there is an open interval I containing x which is entirely in V. If X is any subset of the real line we can also introduce open subsets of X, as we have used it in the above definition of continuity. A subset U of X is said to be open (in X) provided there is an open subset of the real line $\mathbf{R}$, say V, such that $U = X \cap V$.

A mapping $h : X \to Y$ is said to be a homeomorphism provided h is continuous, one-to-one and onto, and the inverse mapping h^{-1} is also continuous.

Note that topological conjugacy is an equivalence relation. In other words the following three properties are true:

[30] R. Devaney, *An Introduction to Chaotic Dynamical Systems, Second Edition,* Addison-Wesley, Redwood City, CA, 1989.

[31] Devaney uses the notion of transitivity for mixing.

[32] J. Banks, J. Brooks, G. Cairns, G. Davis, P. Stacey, *On Devaney's definition of chaos,* American Math. Monthly 99.4 (1992) 332–334.

a) f is topologically conjugate to f.
b) If f is topologically conjugate to g then g is topologically conjugate to f.
c) If f is topologically conjugate to g and g is topologically conjugate to h then f is topologically conjugate to h.

Indeed, in order to show property a) choose as a homeomorphism the identity transformation. For b) choose as a homeomorphism h^{-1}. Then $hf = gh$ implies $h^{-1}g = fh^{-1}$. For c) let h_1 be the homeomorphism satisfying $h_1 f = gh_1$, and let h_2 be the homeomorphism satisfying $h_2 g = hh_2$. Then $h_2 h_1$ satisfies the functional equation $h_2 h_1 f = hh_2 h_1$.

Let us look at a class of examples. In several sections above we have related the dynamics of $x \to ax(1-x)$ to that of $z \to z^2 + c$. We now show that for any polynomial of second degree like $f(x) = \alpha x^2 + \beta x + \gamma$ there is a homeomorphism h such that

$$h(f(x)) = g(h(x)) \text{ for all } x \in \mathbf{R},$$

where $g(x) = x^2 + c$. In fact, h can be chosen as an affine linear mapping $h(x) = mx + n$. It is easy to verify that

$$m = \alpha, \ n = \frac{\beta}{2}, \ c = \alpha\gamma + \frac{\beta}{2}\left(1 - \frac{\beta}{2}\right).$$

Equivalence of Quadratic Polynomials

Let us derive the coefficients for h from the assumption that $h(x) = mx + n$ and that h solves the functional equation $h(f(x)) = g(h(x))$ where $f(x) = \alpha x^2 + \beta x + \gamma$ and $g(x) = x^2 + c$. Using the explicit forms of f, g, and h this yields

$$\begin{aligned} m\alpha x^2 + m\beta x + m\gamma + n &= (mx+n)^2 + c \\ &= m^2x^2 + 2mnx + n^2 + c. \end{aligned}$$

Comparing coefficients this gives

$$m = \alpha, \ n = \frac{\beta}{2}, \ c = \alpha\gamma + \frac{\beta}{2}\left(1 - \frac{\beta}{2}\right).$$

For example, taking the logistic population model

$$f(x) = x + rx(1-x) = -rx^2 + (r+1)x,$$

we have $\alpha = -r$, $\beta = r + 1$, and $\gamma = 0$. This yields the topologically conjugate function $g(x) = x^2 + c$ with

$$c = \alpha\gamma + \frac{\beta}{2}\left(1 - \frac{\beta}{2}\right) = \frac{r+1}{2}\left(1 - \frac{r+1}{2}\right) = \frac{1-r^2}{4}.$$

which reconfirms our result from page 53 of chapter 1.

The consequences of having a topological conjugacy between f and g are strong. For example, when f is mixing then g is mixing, and likewise when g is mixing then f is mixing. Similarly, when f has dense periodic points then so has g, and when g has dense periodic points then so has f. The reason for this equivalence is that mixing and dense periodic points are topological properties. Recall, however, that the property of sensitive dependence on initial conditions is not in general inherited under topological conjugacy.

Saw-Tooth and Tent Transformation Revisited

In our discussion of the saw-tooth and tent transformation we established the crucial relation $TT = TS$. In other words, if we let $h = T$ we have a functional relation of the form $hS = Th$. But there is a problem in using this for a topological conjugacy between T and S. There are in fact two problems. The transformation T is continuous and S is not. If h were a homeomorphism then also S would have to be continuous because of the functional equation $hS = Th$ (which would be equivalent to $S = h^{-1}Th$), but it is not. The reason is that h is not a homeomorphism. It is continuous and onto, but not one-to-one (each $y \neq 1$ in $[0, 1]$ has two preimages $x_1 \neq x_2$ such that $h(x_1) = y = h(x_2)$), and therefore there is no inverse transformation for h. This situation leads to a very useful modification of the notion of topological conjugacy.

Topological Semi-Conjugacy

Let X and Y be two subsets of the real line and let f and g be transformations, $f : X \to X$ and $g : Y \to Y$. Then g is said to be *topologically semi-conjugate* to f provided there is a continuous and onto transformation $h : X \to Y$ such that the functional equation

$$h(f(x)) = g(h(x))$$

holds for all $x \in X$.

Note that this is not an equivalence relation, because when g is semi-conjugate to f we may still have that f is not semi-conjugate to g. Moreover, note that we did not require that f or g is continuous.

This is exactly the situation which we found for T and S in the previous sections. In other words T is semi-conjugate to S. Let us now draw two important consequences.

Fact. *The functional equation $hf = gh$ implies $hf^n = g^n h$ for any natural number n.*

Indeed, $hf = gh$ implies $ghf = g^2h$ and using $gh = hf$ this implies $hf^2 = g^2h$. Likewise, we obtain the general case by induction.

The next fact shows that our work in section 10.5, where we established the chaos property for T from that of S has to be seen on a very general background.

Fact. *Assume that g is semi-conjugate to f via a continuous and onto transformation h. Moreover, assume that f has dense periodic points and is mixing. Then also g has dense periodic points and is mixing.*

Dense Periodic Points and Mixing Are Inherited

Let us prove the above fact. Let $f : X \to X$ and $g : Y \to Y$.

(a) Periodic points of g are dense in Y. Let $y \in Y$. We will show that there is a sequence of periodic points of g with limit y. We may choose x as a preimage of y under h, i.e., $h(x) = y$, because h is onto. Since periodic points of f are dense in X we find a sequence $x_1, x_2, \ldots$ with limit x and such that $f^{n_k}(x_n) = x_n$. In other words, x_n is a periodic point of period n_k. We claim that the sequence $\{y_n\}$ with $y_n = h(x_n)$ has limit y and is a sequence of periodic points of g. The first claim is true because h is continuous. The second claim follows from the functional equation $hf^n = g^n h$. Indeed,

$$g^{n_k}(y_n) = g^{n_k}(h(x_n)) = h(f^{n_k}(x_n)) = h(x_n) = y_n.$$

(b) The transformation g is mixing. Let U and V be two open sets in Y. We have to find $y \in U$ and a natural number n so that $g^n(y) \in V$. We start with taking the preimages $A = h^{-1}(U)$ and $B = h^{-1}(V)$. Note that A and B are open, because h is continuous. Thus there exists a natural number n and $x \in A$ such that $f^n(x) \in B$, since f is mixing. Set $y = h(x)$. Now, using the functional equation we obtain $g^n(y) = g^n(h(x)) = h(f^n(x))$. And since $f^n(x) \in B$ we conclude that $h(f^n(x)) \in h(B) = V$.

Are Properties of Chaos Independent?

Let us finally come to a discussion of whether the three properties of chaos are independent from each other. Here we follow the exposition of a recent paper.[33]

Fact. *Let X be an arbitrary subset of the real line and $f : X \to X$ be a continuous transformation which has the properties of mixing and has dense periodic points. Then f has also sensitive dependence on initial conditions.*

In other words, if f is chaotic and g is topologically semi-conjugate to f, then also g is chaotic.[34]

Dense Periodic Points and Mixing Imply Sensitivity

Let us sketch the proof of this fact. We have to find $\delta > 0$, so that for any $x \in X$ and any open subset J in X containing x we have to find a point z in J and and a natural number n so that $|f^n(x) - f^n(z)| > \delta$. The argument has two steps.

Firstly, there is a $\delta_0 > 0$, such that for any $x \in X$ there is a periodic point $p \in X$ with the property that $|f^k(p) - x| \geq \delta_0/2$, for all $k = 0, 1, 2, \ldots$ We prove this assertion by contradiction. We choose two arbitrary periodic points r and s with disjoint orbits, i.e., such that $f^k(r) \neq f^l(s)$ for all k, l. We define δ_0 to be the distance between these orbits, i.e.;

$$\delta_0 = \min\{|f^k(r) - f^l(s)| \mid k, l \in \{0, 1, 2, \ldots\}\}.$$

Now let $x \in X$. We show that x has a distance of at least $\delta_0/2$ to at least one of the two periodic orbits. Let us assume the contrary, i.e.,

[33] J. Banks, J. Brooks, G. Cairns, G. Davis, P. Stacey, *On Devaney's definition of chaos,* American Math. Monthly 99,4 (1992) 332–334.

[34] This fact can be generalized to arbitrary metric spaces (see the paper by Banks et al).

that x has a distance less than $\delta_0/2$ to the orbit of r and also to the orbit of s. In other words, $|x - f^k(r)| < \delta_0/2$ and $|x - f^l(s)| < \delta_0/2$ for some k and l. Then by the triangle inequality

$$\begin{aligned} |f^k(r) - f^l(s)| &= |f^k(r) - x + x - f^l(s)| \\ &\leq |f^k(r) - x| + |x - f^l(s)| \\ &< \tfrac{\delta_0}{2} + \tfrac{\delta_0}{2} = \delta_0. \end{aligned}$$

But this is a contradiction to our defintion of δ_0, which proves our first claim.

Secondly, we will show that f has sensitive dependence on initial conditions with sensitivity constant $\delta = \delta_0/8$. To this end let x be any point in X and let J be any open subset of X containing x. Since periodic points are dense in X we find a periodic point p in $U = J \cap (x - \delta, x + \delta)$. Let n denote the period of p. Now there exists a periodic point $q \in X$ whose orbit is of distance at least $4\delta = \delta_0/2$ from x. Set

$$\begin{aligned} W_i &= (f^i(q) - \delta, f^i(q) + \delta) \cap X, \;\; i = 1, \ldots, n, \\ V &= f^{-1}(W_1) \cap f^{-2}(W_2) \cap \cdots \cap f^{-n}(W_n), \end{aligned}$$

where $f^{-i}(A) = \{z \in X \mid f^i(z) \in A\}$ denotes the ith preimage of the set A under f. Note that $q \in V$, so that V is not empty. Moreover, V is an open set, because f is continuous.[35] Since f is also mixing we find a natural number k and $y \in U$ such that $f^k(y) \in V$. Now choose j to be the integer part of $1 + k/n$. Thus, $1 \leq nj - k \leq n$. By construction we have that the point

$$f^{nj}(y) = f^{nj-k}(f^k(y)) \in f^{nj-k}(V)$$

is contained in the open set W_{nj-k}. On the other hand $f^{nj}(p) = p$, so that by the triangle inequality,[36]

$$\begin{aligned} |f^{nj}(p) - f^{nj}(y)| &= |p - f^{nj}(y)| \\ &= |x - f^{nj-k}(q) + f^{nj-k}(q) - f^{nj}(y) + p - x| \\ &\geq |x - f^{nj-k}(q)| - |f^{nj-k}(q) - f^{nj}(y)| - |p - x|. \end{aligned}$$

Therefore, since $p \in (x - \delta, x + \delta)$ and

$$f^{nj}(y) \in (f^{nj-k}(q) - \delta, f^{nj-k}(q) + \delta) \cap X,$$

we obtain

$$|f^{nj}(p) - f^{nj}(y)| \geq 4\delta - \delta - \delta = 2\delta.$$

Thus, we have that either $|f^{nj}(x) - f^{nj}(y)| \geq \delta$ or that $|f^{nj}(x) - f^{nj}(p)| \geq \delta$. Indeed, if both distances would be strictly less than δ, then by the triangle inequality

$$\begin{aligned} |f^{nj}(p) - f^{nj}(y)| &= |f^{nj}(p) - f^{nj}(x) + f^{nj}(x) - f^{nj}(y)| \\ &\leq |f^{nj}(p) - f^{nj}(x)| + |f^{nj}(x) - f^{nj}(y)| \\ &< 2\delta, \end{aligned}$$

[35] That means that for any x in V there is an open interval I containing x such that $I \cap X$ is entirely in V. To see that V is open we first note that W_i is open in X. Then $f^{-i}(W_i)$ is open as a preimage of an open set under a continuous transformation. Finally, the finite intersection of open sets is open.

[36] $|a + b| \geq |a| - |b|$.

which is a contradiction. Since $p, y \in U \subset J$ we have that the nj^{th} iterate of f at p or y is more than a distance δ from the nj^{th} iterate of f at x.

The reader may have wondered why we have presented the inheritance of chaos under topological semi-conjugation in the setting of an arbitrary subset X on the real line and not just for intervals of the real line. Here is the reason. Let us conclude this section with a brief discussion of the transformation

$$T(x) = \begin{cases} 3x & \text{if } x \leq 0.5 \\ 3 - 3x & \text{if } x > 0.5. \end{cases}$$

This transformation is familiar from chapter 2 (page 74) where we showed that T leaves the Cantor set invariant. In other words, if $x \in C$, then $T(x) \in C$. Recall that the Cantor set C consists of all number $x \in [0, 1]$ that admit a ternary representation

$$x = \sum_{i=1}^{\infty} a_i 3^{-i} \text{ with } a_i \in \{0, 2\}.$$

Let us argue that T is chaotic on the Cantor set C. To this end we introduce a saw-tooth function $S : [0, 1] \to [0, 1]$ with three teeth by setting

$$S(x) = \begin{cases} 3x & \text{if } 0 \leq x < 1/3 \\ 3x - 1 & \text{if } 1/3 \leq x < 2/3 \\ 3x - 2 & \text{if } 2/3 \leq x \leq 1. \end{cases}$$

Observe that in terms of ternary representations of numbers the transformation S is just the shift operator, i.e.,

$$S(0.a_1 a_2 a_3 \ldots) = 0.a_2 a_3 a_4 \ldots$$

Moreover, S leaves the Cantor set C invariant, $S(C) = C$, and S is chaotic on C by exactly the same arguments as the old saw-tooth function was seen to be chaotic on $[0, 1]$ in section 10.5. Finally, we observe that there is a semi-conjugation $h : C \to C$ between T and S. More, precisely, with $h = T$ we have $Th = hS$. Thus, we can argue that also the tent transformation T is chaotic on the Cantor set.

10.8 Numerics of Chaos: Worth the Trouble or Not?

In chapter 1 (and also here) we demonstrated that the theoretical and computational behavior of the iteration of chaotic systems will almost always differ significantly. This is a consequence of two central aspects of these systems. The first one is the extreme sensitivity which is inherent to chaotic systems. The other reason is due to the limited precision when doing floating point arithmetic on a computer. It is only natural to question whether it is worthwhile at all to compute any orbits for chaotic systems. Strictly speaking, all computed values clearly will be wrong. The first numbers will have only small errors, but soon the errors become as large in magnitude as the points in the true orbit. So what is the use of the computed orbit? The answer is stunning and unbelievable at first sight. Although the computed orbit is fundamentally wrong, it is also equally correct in that it approximates a true orbit of the same chaotic system very well. Moreover, this approximation holds not only for the first few iterations but for all of them! This means that the computed orbit always stays close to a true one, like the hiker's shadow on the ground stays close to the hiker wherever he chooses to go. How can that be true in the presence of sensitivity? There seems to be some contradiction. But if the statement of the shadowing of an orbit is really true — and we will in fact show that further below — then the question raised in the title of this section can be answered affirmatively.

Before we go into details of this *shadowing lemma*, as it is called in the initiated circles of chaos researchers, we would like to raise your attention to another problem illustrating again the fact that computed orbits must be carefully interpreted. In the last section we clearly made the point that the shift operator is the central chaos generating mechanism. Perhaps some readers have already felt inspired to try a quick and straightforward implementation of the shift operator on a computer or calculator. But to their dismay, the orbits will certainly have been anything but chaotic. What is your guess as to what we will see if we iterate the shift operator on a computer?

Shift Operator on the Calculator

Most people who know a little bit (or more) about the internal works of computers regard this as a rather boring question. They reckon that independent of the initial value, after only a few iterations we end in the fixed point 0. The reasoning is simple. Computers encode real numbers in the binary system, but of course they can only use a finite number of digits. Therefore they can only represent binary numbers like:

$$x_0 = 0.a_1a_2a_3 \ldots a_m000 \ldots$$

where the digits a_k are 0 or 1 and m is a constant which depends on the chosen precision and the type of the computer. Thus, after m iterations of the shift operator we reach $x_m = 0$. In other words, on the computer we do not expect to see ergodic or periodic behavior. But let us be safe and try to confirm our argument by an experiment on the computer.

Most people take it for granted that a calculator is a computer, just smaller. We enter (for example) 0.6 into a pocket calculator. With this initial value x_0

we start some iterations of the shift operator. First, we multiply by 2, subtract 1 and obtain $x_1 = 0.2$. For the next step we obtain $x_2 = 0.4$, then $x_3 = 0.8$, $x_4 = 0.6$ and again $x_5 = 0.2$, etc. Yes, we see a perfect periodic behavior. If you want to check this, go ahead. We tried several pocket calculators and always found the same result. If you should not have a pocket calculator at hand but rather a personal computer on your desk, you might want to check it on that machine. Perhaps you can write a small BASIC program which can be done in a couple of minutes. The outcome of the experiment will depend on the BASIC interpreter on the machine. Some interpreters really compute in binary arithmetic and after 22 iterations you have reached 0 but some others work more like a pocket calculator and show the periodic behavior. So, what is the trick?

The Trick of Pocket Calculators ...

Most pocket calculators use binary coded decimal arithmetic. This means that the individual digits of decimal numbers are binary coded in four bits per digit. In that way up to a certain number of digits it is possible to encode decimals exactly. For example the number 0.6 is encoded by

$$0.0110.$$

Note that 110 is simply the binary encoding of 6. Let us compare this with the ordinary binary encoding of 0.6 which is the periodic extension

$$0.100110011001\ldots$$

The exact binary encoding on a computer would require infinitely many binary digits which of course are not available. Therefore 0.6 cannot be encoded exactly in binary form whereas it can be encoded exactly as a binary coded decimal. And this carries over to the iteration of the shift operator. Using (plain) binary encoding we can represent only those decimals exactly which have a finite encoding (e.g., 0.375 is binary encoded 0.011), thus the iteration will always lead to 0. Using binary coded decimal arithmetic we can also represent decimals exactly which have an infinite binary encoding (like 0.6), and therefore we can see some periodic behavior.

...Knocked Out by Chaos

But be warned. This trick does not really change the problems. Sensitivity is present, and its problems cannot be circumvented by any machine with finite precision. Just recall the experiments from chapter 1. Somehow pocket calculators only fake precision; they do not really provide it. And even a simple choice like 1/3 taken as initial value will knock out the tricky pocket calculator.

The Shadowing Lemma

Let us now discuss why any computed orbit of a chaotic system, although exposed to sensitivity, manages to be a close approximation of a truthful orbit of the same system. Since the quadratic iterator is equivalent to the shift transformation, we concentrate on the latter.

Given an initial point x_0 and the corresponding exact orbit under the shift operator

$$x_{k+1} = \text{Frac}\,(2x_k)\,, \quad k = 0, 1, \ldots$$

we allow errors to be made in each step of the computation. First of all, the initial point x_0 may not be represented exactly. A machine can only use a number y_0 close to x_0. Let us call the error made in this approximation ε_0,

$$y_0 = x_0 + \varepsilon_0.$$

Based on this value y_0 an orbit is calculated, for which we use the notation $y_0, y_1, y_2, \ldots$ However, this is not the exact orbit for y_0 because in each iteration step there will be an error. Let us call the error in the kth step ε_k. More precisely, we define

$$y_k = \text{Frac}\,(2y_{k-1} + \varepsilon_k)\,.$$

From our discussion of sensitivity it is clear that any error introduced anywhere doubles in each iteration. After only a few steps, there is absolutely no correlation between what is computed and either of the true orbits started at x_0 or y_0. Still we can show that there is some exact orbit started at some initial point near x_0 and y_0, say z_0, which is approximated closely for all of the computed iterations! The situation is sketched in figure 10.45. The only assumption we must require to prove that fact and to derive the initial point z_0 is that the errors are bounded by some constant $\varepsilon > 0$,

$$|\varepsilon_k| \leq \varepsilon, \quad k = 0, 1, 2, \ldots$$

The conclusion is that for any iteration the exact orbit generated out of z_0 will fall within ε-distance of the computed orbit. In the kth iteration we have

$$|z_k - y_k| \leq \varepsilon. \tag{10.20}$$

Derivation of the Shadowing Lemma

The proof for the shadowing lemma as stated in the text requires a fair amount of formula writing. This should not distract the reader, because everything is elementary, and the underlying idea is simple. To begin with the derivation let us consider a finite sequence of n computed points for the shift operator as already stated above. For a given initial point x_0 we get the computed orbit which is listed explicitly here for completeness:

$$\begin{aligned} y_0 &= x_0 + \varepsilon_0 \\ y_1 &= \text{Frac}\,(2y_0 + \varepsilon_1) \\ &\vdots \\ y_n &= \text{Frac}\,(2y_{n-1} + \varepsilon_n)\,. \end{aligned}$$

The errors ε_k are bounded by $\varepsilon > 0$

$$|\varepsilon_k| \leq \varepsilon, \quad k = 0, 1, \ldots, n.$$

The Shadow of the Computed Orbit

Starting at x_0 the exact orbit would be $x_0, x_1, \ldots, x_n$. The computed orbit starts within ε-distance at y_0 and soon will deviate from the exact orbit. However, in the ε-shadow of the computed orbit $y_0, y_1, \ldots, y_n$ there will be an exact orbit which starts at z_0.

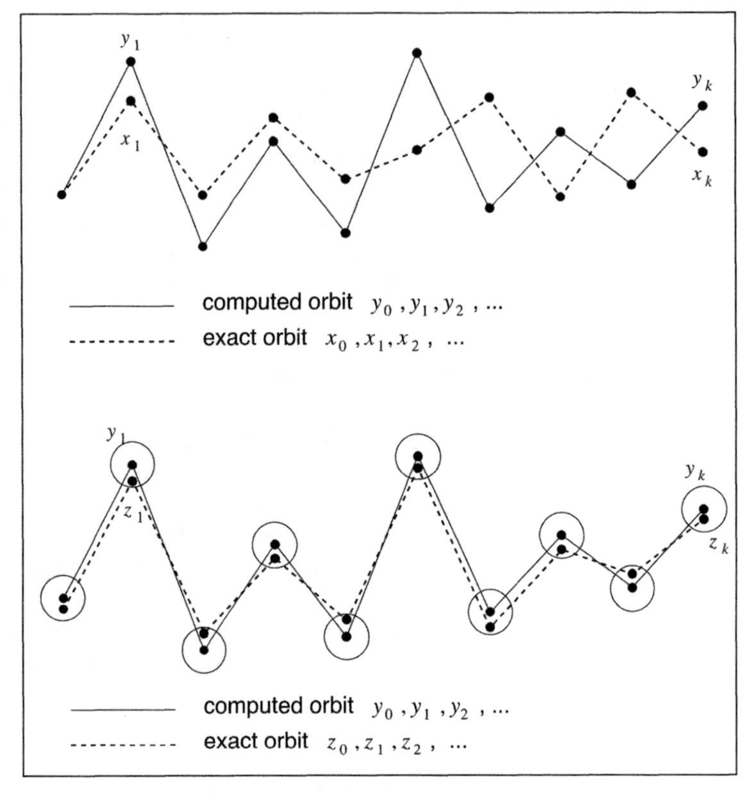

Figure 10.45

We can express the points y_k in terms of x_0 and errors alone using the properties of Frac on page 506:

$$\begin{aligned} y_0 &= x_0 + \varepsilon_0 \\ y_1 &= \text{Frac}\,(2(x_0+\varepsilon_0)+\varepsilon_1) \\ &= \text{Frac}\,(2x_0 + 2\varepsilon_0 + \varepsilon_1) \\ y_2 &= \text{Frac}\,(2(2x_0+2\varepsilon_0+\varepsilon_1)+\varepsilon_2) \\ &= \text{Frac}\,(4x_0 + 4\varepsilon_0 + 2\varepsilon_1 + \varepsilon_2)\,. \end{aligned}$$

For each iteration we must multiply the previous result by two and add the corresponding error term. Thus, in the kth step we obtain

$$y_k = \text{Frac}\left(2^k x_0 + 2^k \varepsilon_0 + 2^{k-1}\varepsilon_1 + \cdots + 2\varepsilon_{k-1} + \varepsilon_k\right).$$

More conveniently this is expressed in mathematical shorthand as

$$y_k = \text{Frac}\left(2^k x_0 + \sum_{i=0}^{k} 2^{k-i}\varepsilon_i\right). \tag{10.21}$$

In particular, in the last iteration ($k = n$) we obtain

$$y_n = \text{Frac}\left(2^n x_0 + \sum_{i=0}^{n} 2^{n-i}\varepsilon_i\right).$$

Now we can specify z_0, the initial point of the exact orbit whose shadow we have computed as the orbit $y_0, \ldots, y_n$. The idea is simply to define z_0 as that point, which when multiplied with 2^n will yield exactly the argument of the Frac-function in the last formula. This is

$$\begin{aligned} z_0 &= \text{Frac}\left(x_0 + \sum_{i=0}^{n} 2^{-i}\varepsilon_i\right) \\ &= \text{Frac}\left(x_0 + \varepsilon_0 + \frac{\varepsilon_1}{2} + \frac{\varepsilon_2}{4} + \cdots + \frac{\varepsilon_n}{2^n}\right). \end{aligned} \tag{10.22}$$

With that choice we have made sure that $z_n = y_n$. Thus, the error between the computed and the exact orbit $z_0, \ldots, z_n$ is zero in the last iterate. To analyze the deviation between the orbit of z_0 and the computed orbit $y_0, \ldots, y_n$ in the other iterates we first derive an explicit formula for the exact k^{th} iterate z_k.

$$\begin{aligned} z_k &= \text{Frac}\left(2^k z_0\right) = \text{Frac}\left(2^k\left(x_0 + \sum_{i=0}^{n} 2^{-i}\varepsilon_i\right)\right) \\ &= \text{Frac}\left(2^k x_0 + \sum_{i=0}^{n} 2^{k-i}\varepsilon_i\right). \end{aligned} \tag{10.23}$$

To discuss the difference in the kth iterate between the exact orbit of z_0 and the computed one of y_0 we consider first the difference Δ_k in the arguments of the Frac-function in eqn. (10.21) and eqn. (10.23),

$$\begin{aligned} \Delta_k &= \left(2^k x_0 + \sum_{i=0}^{n} 2^{k-i}\varepsilon_i\right) - \left(2^k x_0 + \sum_{i=0}^{k} 2^{k-i}\varepsilon_i\right) \\ &= \sum_{i=k+1}^{n} 2^{k-i}\varepsilon_i. \end{aligned}$$

To obtain an error bound for this expression we use the bound ε on the errors ε_i and compute

$$\begin{aligned} |\Delta_k| &\le \sum_{i=k+1}^{n} 2^{k-i}|\varepsilon_i| \le \sum_{i=k+1}^{n} 2^{k-i}\varepsilon \\ &= \varepsilon\left(\frac{1}{2} + \frac{1}{4} + \cdots + \frac{1}{2^{n-k}}\right) < \varepsilon. \end{aligned}$$

Here we must include a word of caution because from these inequalities we cannot conclude that $|z_k - y_k| \le \varepsilon$ holds for $k = 0, \ldots, n$, although in most cases this will be true. The reason lies in the discontinuities of the Frac-function at integer points. Thus, for example, in the case that the argument in eqn. (10.21) is slightly below an integer, and that in eqn. (10.23) is slightly above an integer we arrive at $|z_k - y_k| \approx 1 > \varepsilon$. Formally, in order to resolve this problem correctly, we would have to define a new metric for the unit interval which identifies the points 0 and 1. In other words, we think of the unit interval as a circle with circumference equal to 1. Then the closeness of the arguments in equations (10.21) and (10.23) carries over to that of z_k and y_k, as claimed. Speaking rigorously, the statement in eqn. (10.20) must be modified in this sense to be true.

Moreover, a straightforward modification is in order to show that the claim holds for all integers $k = 0, 1, 2, \ldots$ We just have to consider the limit as $n \to \infty$ in equations (10.22). Then

$$\begin{aligned} |\Delta_k| &= \left| \sum_{i=k+1}^{\infty} 2^{k-i} \varepsilon_i \right| \leq \sum_{i=k+1}^{\infty} 2^{k-i} |\varepsilon_i| \\ &\leq \varepsilon \sum_{i=k+1}^{\infty} 2^{k-i} |\varepsilon_i| = \varepsilon \end{aligned}$$

and the rest follows as above.

Let us summarize and interpret this result. We have learned that when we compare computed orbits with exact orbits then the deviation due to accumulated error propagation will soon amplify so rapidly in the course of the computation that typically any correlation between exact orbits and computed orbits will vaporize. Nevertheless, the iteration behavior of our system is so enormously rich that within the shadow of the computed orbit there will be some exact orbit traveling along. This is a truly amazing fact, in particular if we remind ourselves that the errors in the computed orbit may be chosen randomly, as long as they remain bounded overall, i.e., the individual errors ε_k in each step of the computation could be thought of as being even chosen by a random number generator, as long as $|\varepsilon_k| < \varepsilon$ for some appropriate choice of ε.[37]

On the other hand, the shadowing lemma should not mislead us to think that it provides us with a way to escape the consequences of sensitivity to initial condition and some value for the long-term *prediction* of orbits in our chaotic system. However, the shadowing lemma does ensure us that statistical properties measured by computer experiments are in fact significant.

The assertions of the shadowing lemma hold true in many chaotic systems. They can be interpreted in a rather dramatic way: Under the circumstances of shadowing a deterministic model supports almost any prediction.

[37] A complete discussion of the shadowing property for tent transformations has been presented by E. Coven, I. Kan, J. A. Yorke in *Pseudo-orbit shadowing in the family of tent maps*, Trans. Amer. Math. Soc. 308,1 (1988) 227–241.

Chapter 11

Order and Chaos: Period-Doubling and Its Chaotic Mirror

...there is a God precisely because Nature itself, even in chaos, cannot proceed except in an orderly and regular manner.

Immanuel Kant

Routes to Chaos

Chaos theory began at the end of the last century with some great initial ideas, concepts and results of the monumental French mathematician Henri Poincaré. Also the more recent path of the theory has many fascinating success stories. Probably the most beautiful and important one is the theme of this chapter. It is known as the *route from order into chaos*, or *Feigenbaum's universality*.

Chaos and order have long been viewed as antagonistic in the sciences. Special methods of investigation and theory have been designed for both. Natural laws like Newton's law or Kepler's law represent the domain of order. Chaos was understood to belong to a different face of nature where simple — or even complicated — laws would not be valid. In other words, chaos was seen not just as a higher degree of complexity or as a more complex form of order, but as a condition in which nature fails to obey laws. Even more challenging was the observation that natural systems seem to have no difficulty switching from one state into the other, from laminar flow into turbulent flow, from a regular heart beat into a fibrillating heart beat, from predictability into unpredictability.

One of the great surprises revealed through the studies of the quadratic iterator

$$x_{n+1} = ax_n(1 - x_n), \quad n = 0, 1, 2, \dots \tag{11.1}$$

is that both antagonistic states can be ruled by a *single* law. An even bigger surprise was the discovery that there is a very well defined 'route' which leads

Time Series and Final State

The long-term behavior of the quadratic iterator for $a = 2$: plot of time series (left) and final state marked in a final-state diagram (right).

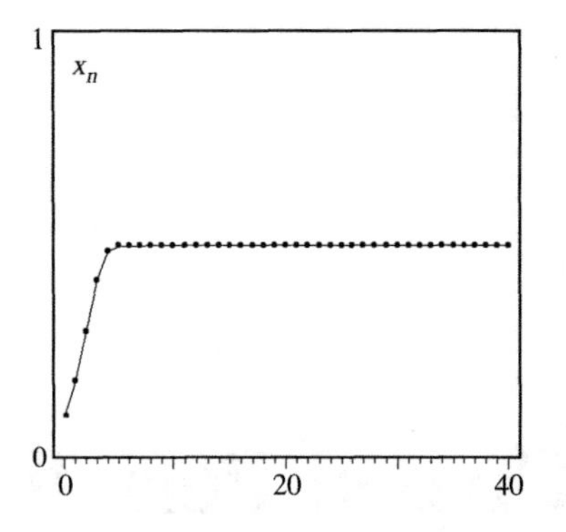

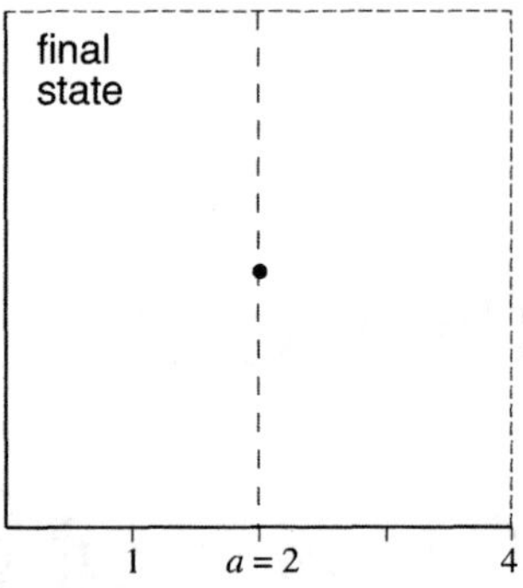

Figure 11.1

from one state — order — into the other state — chaos. Furthermore, it was recognized that this route is *universal*. 'Route' means that there are abrupt qualitative changes — called bifurcations — which mark the transition from order into chaos like a schedule, and 'universal' means that these bifurcations can be found in many natural systems both qualitatively and quantitatively.

Long-Term Behavior

The following computer experiment turns out to be loaded with marvellous scientific discoveries. We will soon see that it raises more questions than we can answer. Some of them are still open today and present tough research problems. Here is the experiment. We want to explore the behavior of the quadratic iterator (11.1) for all values of the parameter a between 1 and 4. To be more precise, we are only interested in the long-term behavior, i.e., we would like to know what happens to the iterates x_n when the dependence on the initial choice x_0 is diluted to almost zero.

Clearly, the iteration produces values x_n which remain in the interval $[0, 1]$ as long as the initial value x_0 is from that interval. For example, what do we obtain for $a = 2$? Figure 11.1 (left) shows the time series for this parameter and a randomly chosen initial value x_0. After a transient phase of a few iterations the orbit settles down to a fixed point, which we call the *final state*. If we repeat this experiment for different initial values between 0 and 1 we always reach the same final state, the value 0.5. Let us enter this point into a *final-state diagram:* we draw the final state versus the value of the chosen parameter a. This is done in figure 11.1 (right) where we have marked the point ($a = 2$, final state $= 0.5$).

The Feigenbaum Diagram

Now let's complete the diagram. But how can we compute the final state of the iteration for a particular parameter a? Here is a procedure which works fine to obtain a first draft:

1. We choose an initial value x_0 at random from the interval $[0, 1]$ and carry out, say 200, iterations computing $x_1, x_2, \ldots, x_{200}$.
2. We drop the first 100 iterations $x_1, x_2, \ldots, x_{100}$.
3. We plot the remaining iterations $x_{101}, \ldots, x_{200}$ in the diagram.

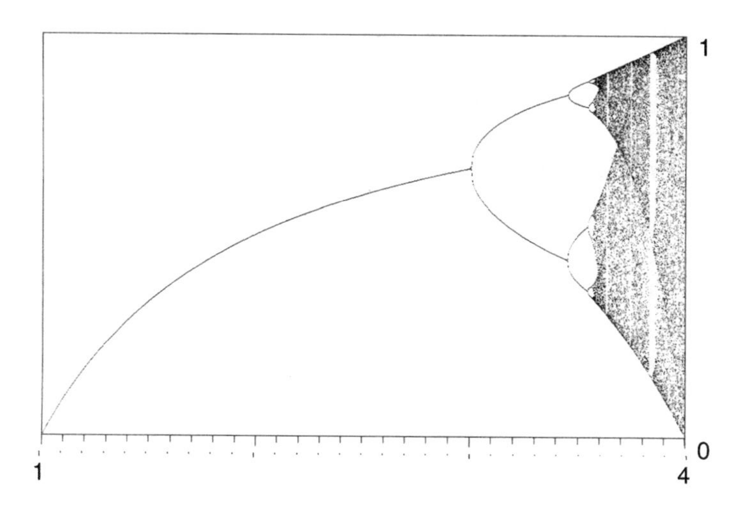

Final-State Diagram

Final-state diagram for the quadratic iterator (11.1) and parameter a between 1 and 4.

Figure 11.2

Applying this procedure only for the parameter $a = 2$ we obtain the diagram shown in figure 11.1 (right). Thus, all plotted points x_{101} to x_{200} fall onto just one dot. Figure 11.2 shows the result for all parameters between 1 and 4. We note that for parameters $a > 3$ the final state is not a mere point but a collection of $2, 4$, or more points. For $a = 4$, of course we have the chaos discussed in the previous chapter, and the points of the final state densely fill up the complete interval. Sometimes this image is also called the *Feigenbaum diagram* because it is intimately connected with the ground-breaking work of the physicist Mitchell Feigenbaum. Indeed this diagram is a remarkable fractal, and later we will even see that it is closely related to the famous Mandelbrot set.

The Feigenbaum diagram has become the most important icon of chaos theory. It will most likely be an image which will remain as a landmark of the scientific progress of this century. The image is a computer-generated image and is necessarily so. That is to say that the details of it could have never been obtained without the aid of a computer. Consequently, the beautiful mathematical properties attached to the structure would definitely be still in the dark and would probably have remained there if the computer had not been developed. The success of modern chaos theory would be unimaginable without the computer.

One essential structure seen in the Feigenbaum diagram 11.2 is that of a branching tree which portrays the qualitative changes in the dynamical behavior of the iterator $x \to ax(1-x)$. Out of a major stem we see two branches bifurcating, and out of these branches we see two branches bifurcating again, and then two branches bifurcating out of each of these again, and so on. This is the *period-doubling regime* of the scenario.

The Period-Doubling

Let us explain very crudely what period-doubling means. Where we see just one branch the long-term behavior of the system tends towards a fixed final state, which, however, depends on the parameter a. This final state will be reached no matter where — at which initial state x_0 — we start. When we see two branches this just means that the long-term behavior of the system is now alternating between two different states, a lower one and an upper one. This is called *periodic* behavior. Since there are two states now, we say that the *period* is two. Now, when we see four branches all that has happened is that the period of the final-state behavior has increased from two to four. That is period-doubling: $1 \to 2 \to 4 \to 8 \to 16 \to \cdots$ Beyond this period-doubling cascade at the right end of the figure we see a structure with a lot of detailed and remarkable designs. Chaos has set in, and eventually, at $a = 4$, chaos governs the whole interval from 0 to 1.

The Feigenbaum diagram has features that are both of a qualitative nature and a quantitative one. The qualitative features are best analyzed through the methodology of fractal geometry. The structure in figure 11.2 has self-similarity properties, which, we will now show, means that the route from order to chaos is one with infinite detail and complexity.

Figure 11.3 shows a sequence of close-ups. We start this sequence with a reproduction of figure 11.2 and magnify the rectangular window indicated in the initial diagram, but showing it upside-down. This is our first close-up image, which indeed looks like the whole diagram. Again we make a magnification of the rectangle indicated and show the result upside-down obtaining the second close-up. The third close-up is the last one in our demonstration. Theoretically, we could go on infinitely often, which we have symbolized by drawing the next two succeeding close-up windows into the bottom image. In other words, the final-state diagram is a self-similar structure.

To get the flavor of some quantitative features let us note that the branches in the period-doubling regime become shorter and shorter as we look from left to right. It is therefore a tempting thought to imagine that the lengths of the branches (in the direction of the a-axis) relative to each other might decrease according to some law, perhaps a geometric law. This idea leads to several consequences.

The Feigenbaum Point

First of all, if true, it would constitute a threshold, i.e., a parameter a beyond which the branches of the tree could never grow. This would mark the end of the period-doubling regime. Indeed, there is such a threshold, which has become known as the *Feigenbaum point* $a = s_\infty = 3.5699456\ldots$ It is precisely the a-value at which the sequence of rectangles shown in figure 11.3 converges. The Feigenbaum point splits the final-state diagram into two very distinct parts, the period-doubling tree on the left and the area governed by chaos on the right. However, the right part is not simply a region of utter chaos; it hides a variety of beautiful structures which we will reveal in this chapter.

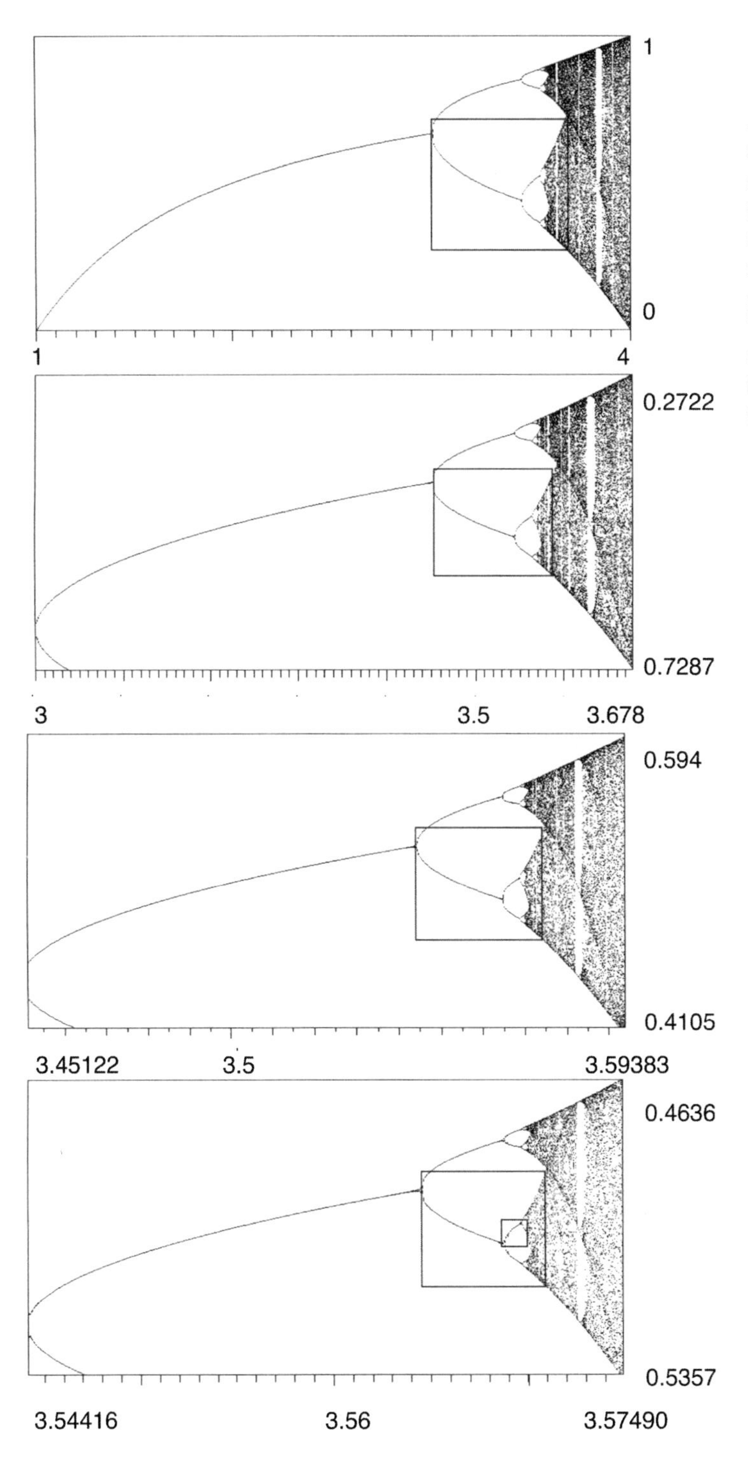

Self-Similarity in the Feigenbaum Diagram

A close-up sequence of the final-state diagram of the quadratic iterator reveals its self-similarity. Note that the vertical values in the first and third magnifications have been reversed to reflect the fact that the previous diagram has been inverted. The second magnification is, of course, also a vertical inversion of the first; the values, however, are in their 'normal' relationship.

Figure 11.3

Feigenbaum's δ

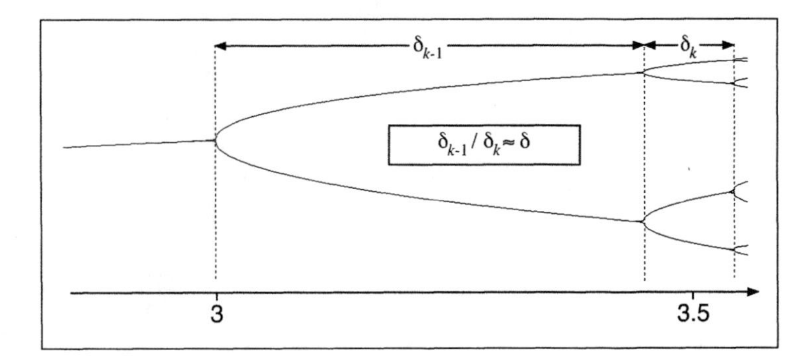

Figure 11.4

The Feigenbaum Constant and Universality

Secondly, if there is a rule that quantifies the way the period-doubling tree approaches the Feigenbaum point, one could try to compare it with the laws which one might observe in related iterators. In fact, experiments regarding these fascinating ideas were carried out by Feigenbaum around 1975; and to the great surprise of the scientific community, he found that the law could be isolated from the branching behavior and that this law, in fact, was exactly the same for many different iterators. Naturally, his discoveries stimulated a whole new mathematical research direction. Actually, in a very precise sense the law can be captured in just one number which Feigenbaum measured at first by numerical experiments to be $\delta = 4.6692\ldots$ and then found that this number was the same for related iterators. This number δ became known as the *Feigenbaum constant* and its appearance in many different systems was called *universality*.

Roughly the meaning of the constant δ is this: if we measure the length of two succeeding branches (in the a-axis direction) then their ratio turns out to be approximately δ (see figure 11.4). In fact, this number is also reflected in the sequence of magnifications in figure 11.3; δ is the magnification factor from one enlargement to the next.

The number $\delta = 4.6692\ldots$ is a constant of chaos comparable only to the fundamental importance of numbers like π. Feigenbaum's discovery was the first of many footprints by which the tracks of chaos are now recognized. The number δ has been observed in systems as varied as dripping faucets, the oscillation of liquid helium, and the fluctuation of gypsy moth populations. It is a predictable constant in the world of chaos.

One possible and very useful interpretation of the universality of δ would be by using it for predictions. For example, by just measuring two successive bifurcations we would be able to predict the bifurcations thereafter and also predict where the threshold would be. This interpretation became, in fact, an achievement of incredible consequences some years after Feigenbaum's work, when experimental physicists discovered that the scenario of period-doubling and the value of δ manifest themselves in real physical experiments like the dynamic behavior in certain fluid flows. In other words, the meaning of universality was suddenly covering not just very primitive mathematical

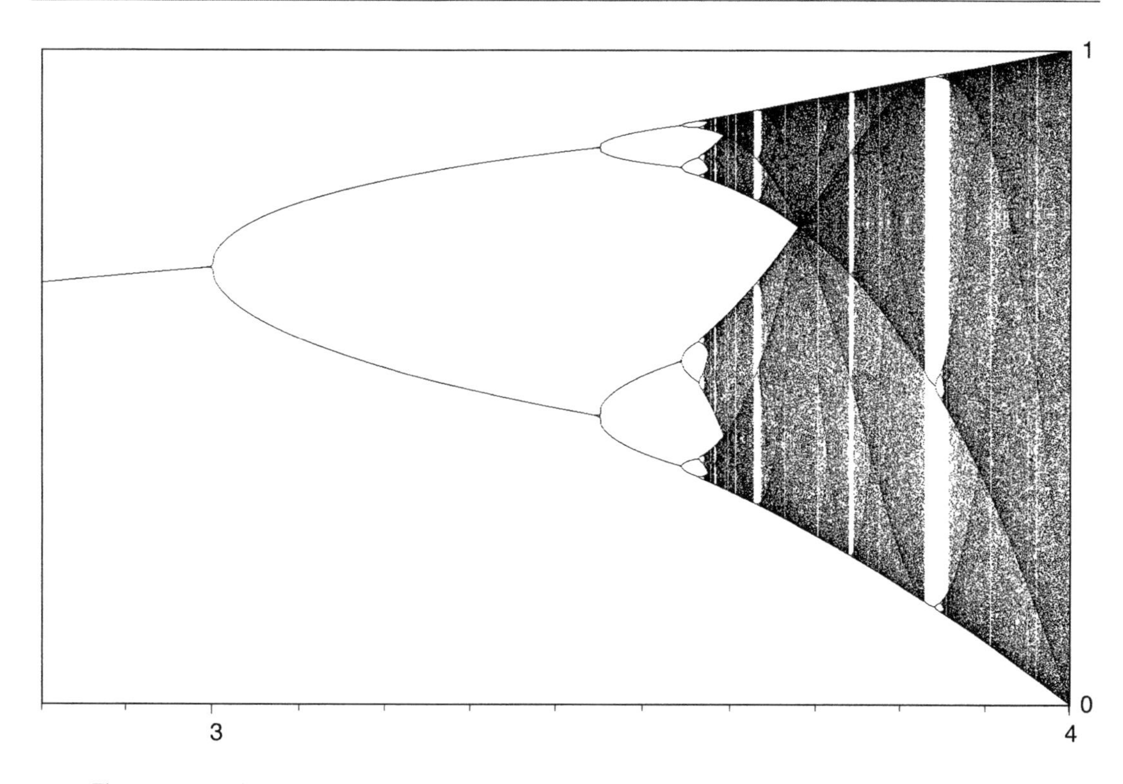

Figure 11.5 : Final-state diagram for the quadratic iterator (11.1) and parameter a between 2.8 and 4.

models but also real physical phenomena. Soon a gold rush in the experimental sciences set in and the region of validity for the universality became larger and larger. Simultaneously, it was understood that there was not just one universal aspect in the structure of figure 11.2 but many other quantitative and qualitative ones. In essence that means whenever a system (even a very complex real system) behaves in a period-doubling fashion then it is very likely that one will see the full structure of the Feigenbaum diagram in it. In other words, although the quadratic iterator in some sense is certainly much too simple to carry any information about real systems, in a very striking and general sense it, in fact, does carry the essential information about how systems may develop chaotic behavior.

This discovery is one of first-rate importance and gives testimony to the beauty and adventure of mathematical thinking.

11.1 The First Step from Order to Chaos: Stable Fixed Points

The portion of the final-state diagram to the left of the Feigenbaum point s_∞ is a self-similar fractal tree (see figure 11.6). It describes the *period-doubling scenario* of the quadratic iterator, which leads from a very simple and orderly behavior of the dynamics right to the beginning of the chaotic region. Let us try to understand the mechanism lying at the base of its generation and leading to the self-similarity of the tree.

The Period-Doubling Tree

The first portion of the final-state diagram — the period-doubling tree.

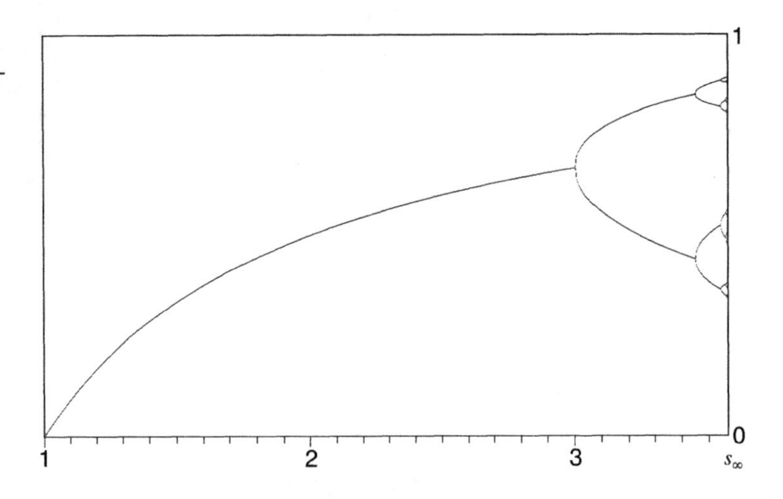

Figure 11.6

At the Stem ...

We start our discussion at the 'stem' of the tree, the part which lies between $a = 1$ and $a = 3$. This part represents rather simple dynamics, a stable situation where the iteration is always led to a rest point. In figure 11.7 we show two typical examples. The left plot shows the time series of the initial point 0.1 for the parameter $a = 1.75$. The iteration settles down on a final state which we denote by $A(1.75)$. In this case the final state is simply the value $3/7$. Indeed, starting with any initial value between 0 and 1, we approach this limiting value. The right graph shows the situation for $a = 2.75$. In this case the final state $A(2.75)$ is $7/11$, but the course of the iteration does not approach this value directly. Rather, it oscillates around the final state while settling down.

In both cases we have a situation which, from the point of view of making predictions, could not be better. No matter where we choose an initial value x_0, we can predict that in the long run we will be near or at the attractor $A(a)$. It seems as if that point is equipped with a magic force which attracts iterations independent of x_0.

For ease of notation in this chapter let us introduce the symbol f_a for the quadratic function

$$f_a(x) = ax(1 - x).$$

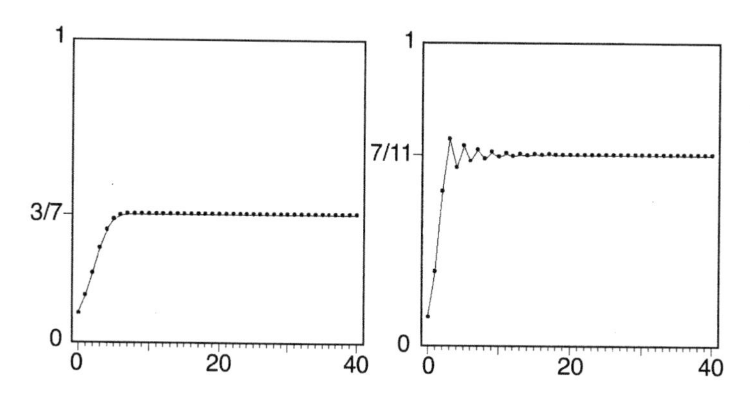

Stability of Final State

Stable behavior of time series for initial value 0.1 and $a = 1.75$ (left) and $a = 2.75$ (right).

Figure 11.7

...a Stable Fixed Point

What would we observe when starting the iteration of the quadratic iterator f_a exactly with the value of the attractor, i.e., $x_0 = A(a)$? Then we would have for all iterates $x_0 = x_1 = x_2 = \ldots$ In other words, x_0 would be a fixed point of f_a (sometimes also called a rest point). Thus x_0 would solve the fixed point equation $f_a(x) = x$,

$$ax(1 - x) = x. \tag{11.2}$$

There are two solutions, which we call p_0 and p_a,

$$p_0 = 0 \text{ and } p_a = \frac{a-1}{a}.$$

Moreover, we note that if $x_0 = 1$, then $x_1 = 0$ and $x_1 = x_2 = \cdots = 0$. We say that 1 is a *preimage* of the fixed point $p_0 = 0$. But $x_0 = 0$ and $x_0 = 1$ are the only initial values which lead to 0; all other values are attracted by $p_a = (a-1)/a$. Indeed, if we take any x_0 between 0 and $(a-1)/a$ then $x_1 > x_0$ for all parameters $a > 1$.[1] In other words, the iteration is pushed away from the rest point $p_0 = 0$. We say that p_0 is a *repeller* or an *unstable fixed point*. On the other hand $p_a = (a-1)/a$ is a *stable fixed point* (or *attractive* fixed point) for all parameters a between 1 and 3.[2] Let us verify these facts using graphical iteration as shown in figure 11.8. You will recall that here the iteration is represented by a path with horizontal and vertical steps, which for convenience, we will now call a poly-line (see chapter 1, page 57). The left image shows the situation for $a = 1.75$. The parabola $ax(1-x)$ intersects the bisector at the fixed points $p_0 = 0$ and $p_a = (a-1)/a$. Between these values the parabola lies above the bisector, but its vertex lies beyond the intersection. Thus the iteration must be repelled away from 0. On the other hand the poly-line representing the iteration is trapped between the parabola and the bisector and thus is led directly to the second intersection point at $(a-1)/a$.

[1]Assume $0 < x_0 < p_a = (a-1)/a$. We compute $x_1 - x_0 = ax_0(1-x_0) - x_0 = x_0(a - ax_0 - 1)$. From the assumption it follows that $a - 1 > ax_0$, thus, $a - ax_0 - 1 > 0$ and $x_1 - x_0 > 0$. In other words, $x_1 > x_0$.

[2]However for $a > 3$ this point also looses its stability, and this will be discussed later.

Graphical Iteration Near a Stable Fixed Point

Graphical iteration for $a = 1.75$ (left) and $a = 2.75$ (right). In both cases we start with $x_0 = 0.1$. The iteration settles down at $p_a = (a-1)/a$ which is $3/7$ (left) and $7/11$ (right).

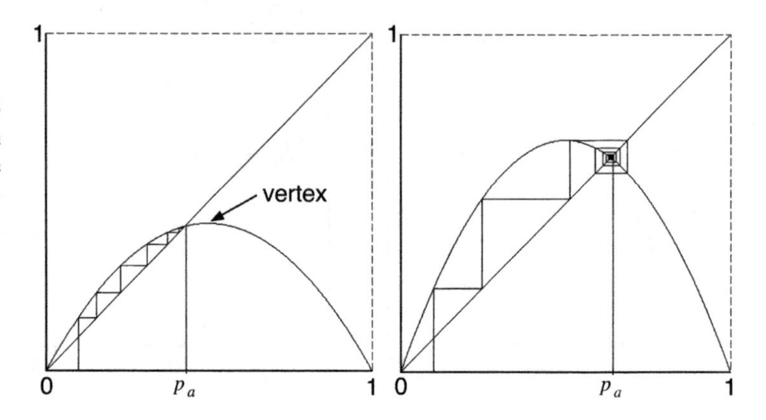

Figure 11.8

The right graph in figure 11.8 shows the situation for $a = 2.75$. In this case the bisector intersects the parabola beyond its vertex. Thus the poly-line, which represents the iteration, begins to spiral around the point of intersection. This spiraling is directed inwards; the process again settles down at $(a-1)/a$. In other words, the fixed point is still attractive although the local behavior (i.e., the way orbits are attracted) has changed. The spiraling of the poly-line explains the oscillations which we observed in figure 11.7.

Four Types of Local Dynamics

There are four basic types of iteration behavior near a fixed point. As an example we consider the linear feedback process

$$x_{n+1} = sx_n$$

i.e., the iteration function is sx, and its graph is a straight line intersecting the bisector at the origin, which is the only fixed point if $s \neq 1$. The slope of the straight line determines the way it intersects the bisector (see figure 11.9). The slope is given by the parameter s. Accordingly we classify the different iteration behavior into four cases:

S1: $0 < s < 1$	iteration as staircase inwards to stable fixed point 0
S2: $-1 < s < 0$	iteration as spiral inwards to stable fixed point 0
U1: $s > 1$	iteration as staircase outwards from unstable fixed point 0
U2: $s < -1$	iteration as spiral outwards from unstable fixed point 0.

The cases $s = \pm 1$ are special. If $s = 1$, every point is a fixed point, while for $s = -1$, all orbits have the form $x, -x, x, -x, \ldots$

We proceed by considering nonlinear systems. The type of the fixed point x^* of a nonlinear feedback process $x_{n+1} = f(x_n)$ can be characterized by the derivative $f'(x^*)$, which can be interpreted as the slope of the graph of f at x^*. In the case of a linear transformation $x \rightarrow sx$, the type of dynamics near the fixed point are characterized by the parameter s. For a nonlinear transformation the same characterization

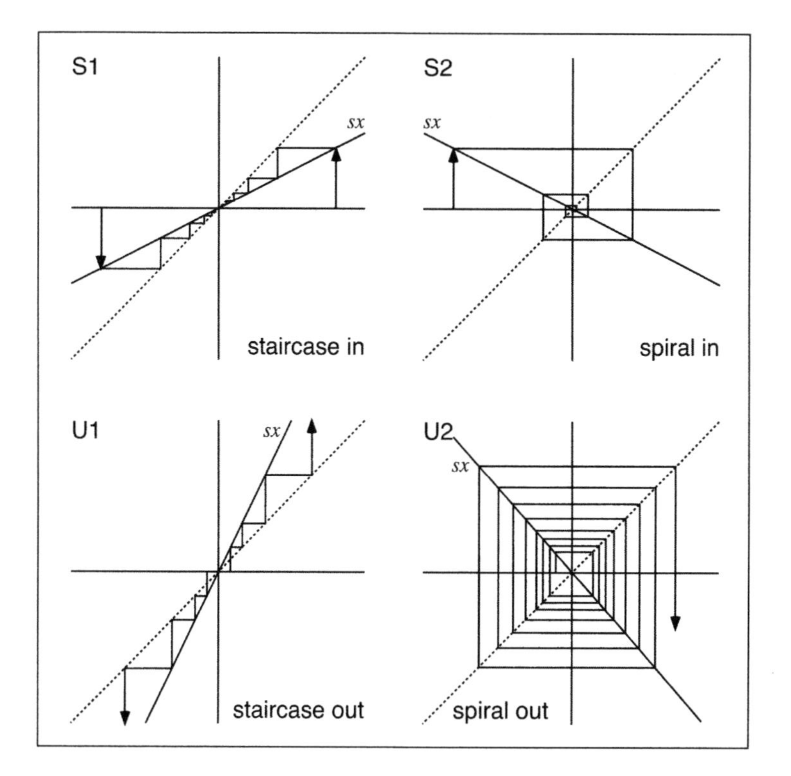

Figure 11.9 : The fixed point at the origin is stable (top) or unstable (bottom) depending on the slope of the function at the fixed point.

is possible by the derivative $f'(x^*)$. In particular, the fixed point is attractive if

$$|f'(x^*)| < 1$$

and repelling if

$$|f'(x^*)| > 1.$$

Let us apply this fact to the two fixed points $p_0 = 0$ and $p_a = (a-1)/a$ of the quadratic iterator $x \to f_a(x) = ax(1-x)$. The derivative of the transformation f_a is given by

$$f_a'(x) = a(1-2x).$$

Consider the first fixed point $p_0 = 0$. The derivative is $f_a'(0) = a$. Thus,

$$\begin{aligned} &|f'(p_0)| < 1 \text{ for } 0 \le a < 1 \\ &|f'(p_0)| = 1 \text{ for } a = 1 \\ &|f'(p_0)| > 1 \text{ for } a > 1. \end{aligned}$$

Thus, as soon as the parameter a passes from $a < 1$ to $a > 1$ the fixed point p_0 loses its stability and becomes a repelling fixed point (type U1, staircase out).

To discuss the other fixed point $p_a = (a-1)/a$ we compute the derivative

$$f_a'(p_a) = 2 - a.$$

Thus, for this fixed point we obtain:

Parameter	Derivative	Type of Fixed Point
$0 < a < 1$	$1 < f'(p_a) < 2$	repelling
$1 < a < 2$	$0 < f'(p_a) < 1$	stable (staircase in)
$2 < a < 3$	$-1 < f'(p_a) < 0$	stable (spiral in)
$3 < a \leq 4$	$-2 \leq f'(p_a) < -1$	repelling (spiral out)

Iteration to Super Attractive Fixed Point

For $a = 2$ we observe a super attractive situation. The graphical iteration demonstrates how the orbit rushes into the fixed point.

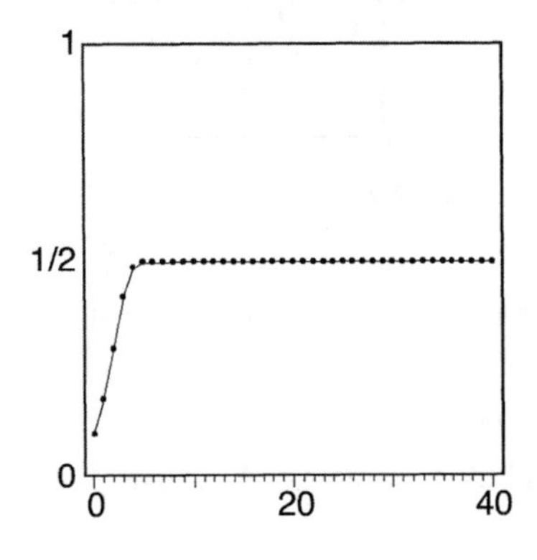

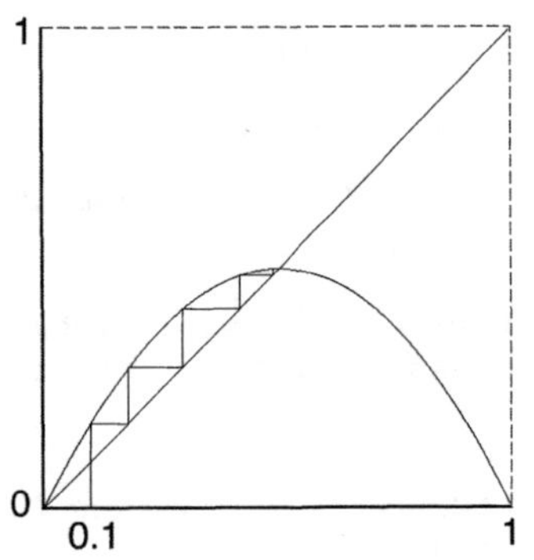

Figure 11.10

Now we can ask: is there a parameter a for which the behavior at the rest point changes to spiraling (or oscillation)? Yes, the spiraling sets in as soon as the vertex of the parabola surpasses the right intersection point of the parabola and the bisector. In other words, we are looking for the case where the intersection point and the vertex of the parabola fall together. Since the parabola has its maximum at $x_{\max} = 0.5$, we must solve the equation $0.5 = 0.5a(1 - 0.5)$.

The Super Attractive Case

The solution is $a = 2$. Figure 11.10 shows this situation. The left graph is the plot of the time series for the initial value 0.1. Comparing it with the time series for $a = 1.75$ and $a = 2.75$ (figure 11.7), we find that for a = 2 the attractive fixed point is approached much faster. This impression is also supported by the graphical iteration; when the poly-line representing the iteration approaches the vertex of the parabola it rushes to the fixed point very quickly. Indeed, this point is called *super attractive*. This case is very special; in the interval of parameters from 1 to 3 it occurs only at one point, $a = s_1 = 2$.

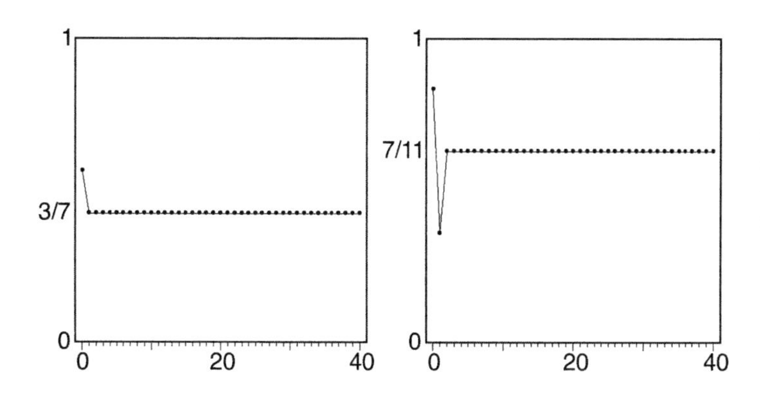

A Direct Approach to the Fixed Point

Examples of initial values which directly lead to the fixed points: $x_0 = 0.5714\ldots$ and $a = 1.75$ (left), $x_0 = 0.8432\ldots$ and $a = 2.75$ (right).

Figure 11.11

It seems reasonable to conjecture that for a given starting point x_0 the iteration reaches the attractive fixed point fastest when the parameter a is of the super attractive type. In fact, this is true for most initial values x_0, but not for all; there are a number of exceptions. Figure 11.11 shows just two of them (compare figure 11.7). On the left (for $a = 1.75$) we start with $x_0 = 4/7$. This leads within one step directly to the fixed point.[3] On the right ($a = 2.75$) we set $x_0 = 0.8432\ldots$, which leads within two steps exactly to the fixed point. Somehow it seems to be a paradox, but in the super attractive case $a = 2$ this behavior is not possible; there are no starting points x_0 which end up right in the fixed point after a finite number of iterations.

Super Attractive Fixed Points

We have characterized a fixed point x^* of a nonlinear feedback process $x_{n+1} = f(x_n)$ by its derivative $f'(x^*)$. It is stable provided $|f'(x^*)| < 1$. It is said to be super attractive if $f'(x^*) = 0$. Let us see what happens if the iteration approaches such a rest point for the example of the quadratic iterator $f_a(x) = ax(1-x)$. The fixed point in question is $x^* = p_a = (a-1)/a$. Let us start the iteration close to p_a; we set

$$x_0 = \frac{a-1}{a} + \varepsilon$$

equal to the fixed point plus a small perturbation ε. Then we obtain $x_1 = ax_0(1-x_0)$ and calculate

$$x_1 = \frac{a-1}{a} + 2\varepsilon - a\varepsilon - a\varepsilon^2.$$

Now the fixed point is attractive provided

$$|2\varepsilon - a\varepsilon - a\varepsilon^2| < |\varepsilon|.$$

Then x_1 will be closer to x_* than x_0, and — using the same line of arguments — x_2 will be even closer, and so on. This is always the

[3] We check that for $a = 7/4$ the fixed point is $p_a = (a-1)/a = 3/7$. With the starting point $x_0 = 4/7$ we get $x_1 = 7/4 \cdot 4/7 \cdot (1 - 4/7) = 3/7 = p_a$.

Time Profile for Super Attractive Case

Time profile for the super attractive case $a = s_1 = 2$ and the graphical iteration of a typical orbit ($x_0 = 0.1$).

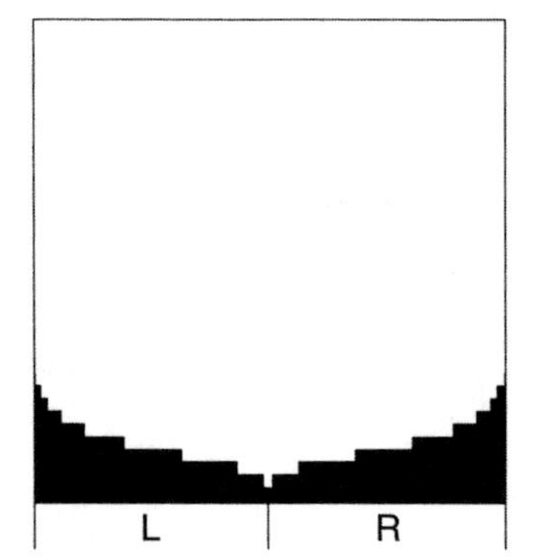

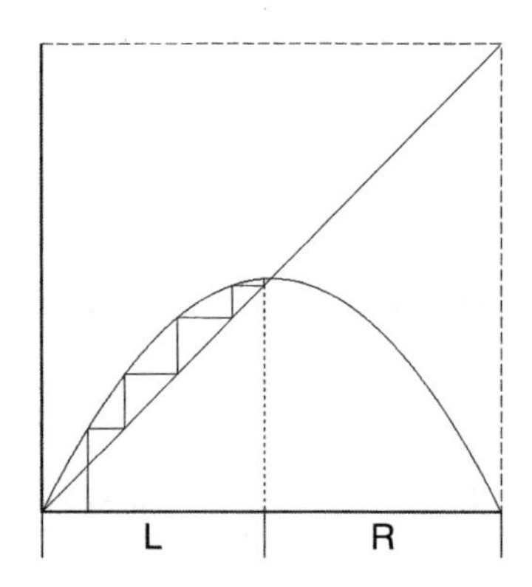

Figure 11.12

case when $1 < a < 3$ and ε is sufficiently small. In the special case when $a = 2$ the term on the left becomes $2\varepsilon^2$, which means that the initial perturbation ε is reduced quadratically. This implies that the number of digits which agree when we compare p_a with x_k will double in each step. However, for $a \neq 2$ the linear term $2\varepsilon - a\varepsilon$ remains and therefore the reduction is less powerful. Thus, it becomes apparent why we call the fixed point in the case $a = 2$ super attractive.

Time Profiles

These observations raise the question how long it will take the iteration of an arbitrary initial value to settle down at the fixed point. Let us set up an experiment to investigate this question. From the unit interval, we choose equally spaced initial values

$$x_0 = \frac{k}{600}, \quad k = 1, 2, \ldots, 599.$$

For each of these starting values we begin the iteration and count how many steps it takes to get to the fixed point $(a-1)/a$ within a short distance, say $1/6000$. The result of each sequence of iterations is a count $N_a(x_0)$, which depends on the initial value x_0 and the parameter a. We present the result of the experiment in a diagram which we call the *time profile*. On the horizontal axis the initial values are marked. For each starting value x_0, we plot a column whose height represents the computed iteration count $N_a(x_0)$. Figure 11.12 shows the time profile for the super attractive case $a = 2$.

The diagram shows a valley shape which agrees with a first intuitive guess. The attractor $A(2) = 0.5$ lies in the center of the interval and exactly where the valley has its deepest point. In other words, the number $N_a(x_0)$ increases as x_0 moves from the center towards either one of the endpoints of the unit interval. Note that the parabola of the quadratic iterator and its time profile exhibit the same symmetry. We can verify this property if we compare the iteration of an initial value from the left part L of the unit interval $x_0 = c$ and its symmetric counterpart $x_0 = 1 - c$ from the right side R. In both cases the

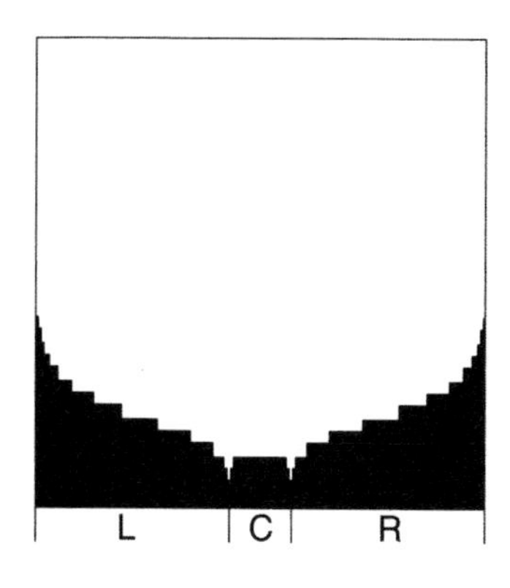

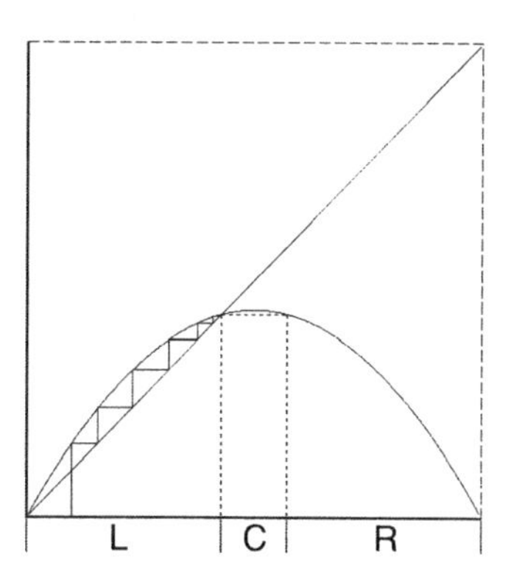

Time Profile $a = 1.75$

Time profile for $a = 1.75$ and graphical iteration of a typical orbit ($x_0 = 0.1$).

Figure 11.13

first step of the iteration gives $x_1 = ac(1-c)$ (which lies on the left side). Thus $N_a(c)$ and $N_a(1-c)$ are the same; the valley must be symmetric.

Let us now examine the time profile for a parameter a below the super attractive case, say $a = 1.75$ (see figure 11.13). Surprisingly, we observe two small sub-valleys within the large valley and one intermediate flat-topped mountain. Further experiments for other values of a between 1 and 2 provide the same result: two sharp sub-valleys within one large valley. Again the reason for this behavior can be studied with graphical iteration, also shown in figure 11.13.

Every x_0 chosen in the left interval L generates a poly-line which creeps up towards the fixed point. In other words, the iteration monotonically approaches the fixed point $p_a = (a-1)/a = 3/7$. The number of steps of the staircase which it takes to get close to the attractor increases as we move x_0 closer to 0. This explains the steep angle of the slope on the extreme left of the time profile. What, however, about initial points x_0 in the remaining part of the unit interval? The graphical iteration makes it clear that there is exactly one new choice $x_0 > 1/2$ such that the first step of the iteration

$$x_1 = \frac{7}{4}x_0(1-x_0)$$

hits the fixed point $x_1 = 3/7$. In other words, x_0 is a preimage of the fixed point. Solving the quadratic equation

$$\frac{3}{7} = \frac{7}{4}x_0(1-x_0)$$

we obtain the two solutions, $x_0 = 3/7$, which is the fixed point itself, and $x_0 = 4/7$. In terms of graphical iteration, this means that the parabola has the same height at both points. The two sub-valleys of the time profile are located precisely at these two points.

The Center Peak in the Time Profile

More generally, let us ask, when does the first iterate x_1 coincide with the fixed point $(a-1)/a$? The answer is obtained from the equation

$$\frac{a-1}{a} = ax_0(1-x_0)$$

or

$$x_0^2 - x_0 + \frac{a-1}{a^2} = 0$$

which is solved by

$$x_0 = \frac{a-1}{a} \text{ and } x_0 = \frac{1}{a}.$$

Based on these results let us rearrange the left and right intervals L and R and introduce a third interval C:

$$L = \left(0, \frac{a-1}{a}\right), \quad C = \left(\frac{a-1}{a}, \frac{1}{a}\right), \quad R = \left(\frac{1}{a}, 1\right).$$

The graphical iteration for initial values x_0 in the center interval C produces poly-lines that creep down to the intersection point of parabola and bisector, which represents the attractor. The number of iterations needed to get close to p_a will go up as x_0 is chosen closer to 1/2, and it will decrease as x_0 gets closer to the endpoints of the interval C. This explains the central mountain around $x = 1/2$ in the time profile. Finally, we note that initial points x_0 chosen from the interval R will be carried in one step of the iteration into the interval L and then follow the pattern which we have discussed already. This explains the extreme right slope in the time profile.

Beyond the Super Attractive Case

In the super attractive case ($a = 2$) the two end points $(a-1)/a$ and $1/a$ of the center interval are identical. Thus the interval C vanishes and the two sub-valleys form one big valley. Let us now see what happens when we increase the parameter a further. Figure 11.14 shows the result for $a = 2.75$. Here we see many sub-valleys (theoretically there are infinitely many) and intermediate mountain tops. Further experiments would show that for all parameters a between 2 and 3 there are similar results. Again let us see what the graphical iteration can teach us in this situation.

While for parameters a below 2 the fixed point $p_a = (a-1)/a$ is always to the left of 1/2 (i.e., on the left side of the vertex of the parabola in the graphical iteration), it is now to the right of 1/2 (i.e., beyond the apex of the parabola); and that has remarkable consequences. Also the preimage of the fixed point has reversed sides. Now $1/a$ is left of 1/2, and this implies that the bisector lies below the parabola at this point. This suggests that now there are preimages of $1/a$ (i.e., solutions to the equation $ax(1-x) = 1/a$). If x_0 is such a solution, then $x_1 = 1/a$, and $x_2 = (a-1)/a$. In other words, starting the iteration with x_0 would lead exactly to the fixed point in just two

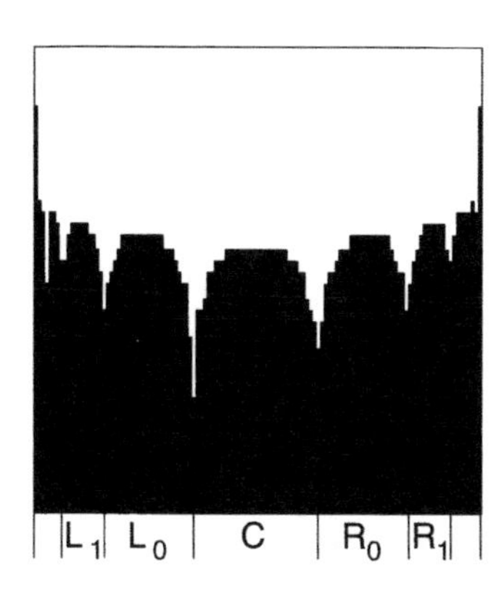

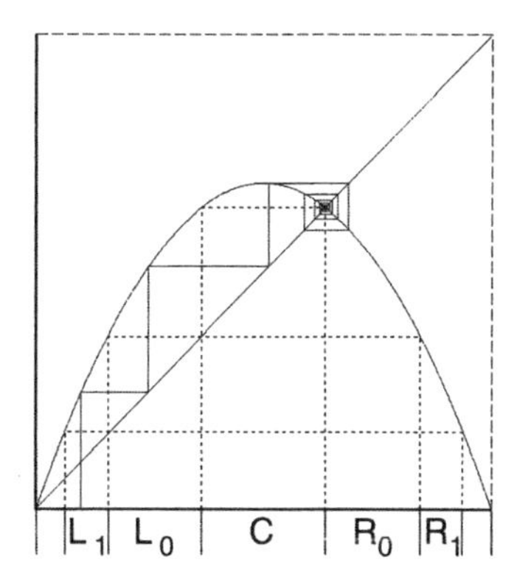

Time Profile $a = 2.75$

Time profile for $a = 2.75$ and graphical iteration of a typical orbit ($x_0 = 0.1$).

Figure 11.14

steps. Indeed, for $a > 2$ the quadratic equation $ax(1-x) = 1/a$ yields the solutions

$$\alpha_1 = \frac{1}{2} - \frac{1}{2a}\sqrt{a^2 - 4}$$

$$\beta_1 = \frac{1}{2} + \frac{1}{2a}\sqrt{a^2 - 4}$$

both symmetrically located around 1/2 for $a > 2$. Note that these solutions are defined only for parameters $a \geq 2$. This establishes three interesting intervals:

$$L_0 = \left(\alpha_1, \frac{1}{a}\right), \quad C = \left(\frac{1}{a}, \frac{a-1}{a}\right), \quad R_0 = \left(\frac{a-1}{a}, \beta_1\right)$$

which belong to the three central mountains of the time profile. Now we can look for the preimages of α_1: α_2 and β_2, then for the preimages of α_2, and so on. The valleys of the time profile are precisely at these values α_k and β_k. Given an initial point $x_0 = \alpha_k$ or $x_0 = \beta_k$, the first $k+1$ iterations are

$$\alpha_{k-1}, \alpha_{k-2}, \ldots, \alpha_1, \frac{1}{a}, \frac{a-1}{a}.$$

Thus, after $k+1$ iterations, the fixed point is reached.

Preimages of p_a and Iteration Patterns

For a between 2 and 3 the iteration behavior is as follows: if we are in C then the following step will be in R_0 and the next step back in C, and so on. This explains the spiral part of the poly-line in figure 11.14. If we start in L_0 then the next step will also be in C. We can extend this observation using the iterated preimages of p_a to obtain a complete decomposition of the unit interval. With α_0 and β_0 as above, we define sequences $\alpha_1, \alpha_2, \ldots$ and $\beta_1, \beta_2, \ldots$ implicitly by

$$f_a(\alpha_k) = f_a(\beta_k) = \alpha_{k-1}, \;\; \alpha_k < \beta_k, \;\; k = 1, 2, \ldots$$

These equations can also be solved explicitly,

$$\begin{aligned}\alpha_k &= \tfrac{1}{2} - \sqrt{\tfrac{1}{4} - \tfrac{\alpha_{k-1}}{a}} \\ \beta_k &= \tfrac{1}{2} + \sqrt{\tfrac{1}{4} - \tfrac{\alpha_{k-1}}{a}}\end{aligned}$$

for $k = 1, 2, \ldots$ There is an ordering

$$\cdots < \alpha_3 < \alpha_2 < \alpha_1 < \alpha_0 < \beta_0 < \beta_1 < \beta_2 < \beta_3 < \cdots.$$

These preimages are the locations of the small valleys in the time profile. These values define intervals $L_k = (\alpha_{k+1}, \alpha_k)$ and $R_k = (\beta_k, \beta_{k+1})$. The intervals L_k cover the left part of the unit interval from 0 to $1/a$, while the intervals R_k cover the right part from $(a-1)/a$ to 1.

Together with these intervals the iteration behavior is easily described:

$$\begin{aligned}&L_k \to L_{k-1} \to L_{k-2} \to \cdots \to L_0 \to C \\ &C \to R_0 \to C \\ &R_k \to L_{k-1}.\end{aligned}$$

This yields an interpretation of the infinite number of mountains in our time profile. Each one corresponds to one of the intervals L_k, R_k, or C. Let us contrast this behavior to the dynamics of the quadratic iterator for parameters below 2. For $1 < a < 2$ we have:

$$\begin{aligned}&L \to L \\ &C \to C \\ &R \to L.\end{aligned}$$

For the super attractive case ($a = 2$) we have only

$$\begin{aligned}&L \to L \\ &R \to L.\end{aligned}$$

The iteration behavior switches to much more complicated patterns when the parameter a crosses the super attractive case $a = 2$.

11.2 The Next Step from Order to Chaos: The Period-Doubling Scenario

Having studied the dynamics of the quadratic iterator f_a in detail for parameters between 1 and 3, we continue to increase a beyond 3. For such large parameters the fixed point $p_a = (a-1)/a$ is not stable anymore; it is a repeller. Is there a different attractor that takes over the role of p_a?.

Stable Oscillation

Let us see what happens, for example, if $a = 3.1$ (see figure 11.15). We obtain a time series which exhibits an entirely new behavior. There is oscillation as in the case of a between 2 and 3, but it does not finally settle down to one single point. Rather, it stabilizes in oscillating between two values, a low number $x_l(a)$, and a high value $x_h(a)$. Thus, in the final-state diagram we obtain just these two points at parameter $a = 3.1$.

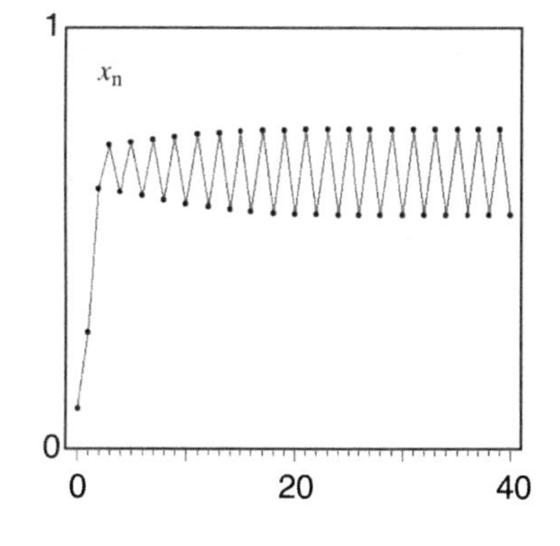

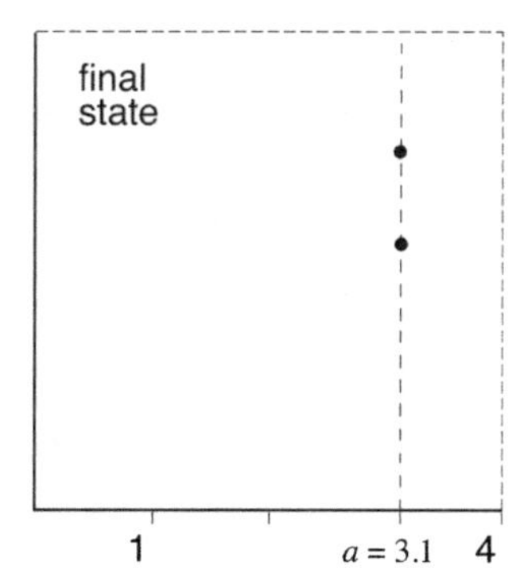

Time Series $a = 3.1$

Time series for $a = 3.1$ and initial value 0.1. The iteration leads to a final state which consists of two points $x_l(a)$ and $x_h(a)$.

Figure 11.15

Again we turn to graphical iteration to obtain some further insight. In figure 11.16 we consider initial values $x_0 = 0.075$ (left) and $x_0 = 0.65$ (right). Indeed, we observe a new phenomenon. In the left graph we first notice a familiar staircase. But then the poly-line turns into an inward spiral which slowly runs into a repeating loop. In the right plot the initial value x_0 is close to the unstable fixed point p_a. Its orbit spirals outward towards the same loop as seen on the left. In other words, while for $a < 3$ the fixed point p_a attracts all iterations, it turns into a repeller when a grows larger than 3. Close to p_a, iterations will be pushed away. The fixed point p_a loses its stability as a crosses the border $b_1 = 3$. This particular parameter value is called a *bifurcation point*.

Let us summarize our findings for the final-state diagram (see figure 11.17); $p_a = (a-1)/a$ is an attractor for all iterations starting in the interval (0,1) for parameters $1 < a < 3$. Formally, the attractor is a single point,

$$A(a) = \{p_a\} \text{ for } 1 < a < 3.$$

For $a > 3$, p_a still exists. Thus, an iteration started precisely at this point

Graphical Iteration $a = 3.1$

The periodic cycle $\{x_l, x_h\}$ is the attractor for the quadratic iterator at parameter $a = 3.1$. On the left the initial point for the iteration is $x_0 = 0.075$, while on the right $x_0 = 0.65$.

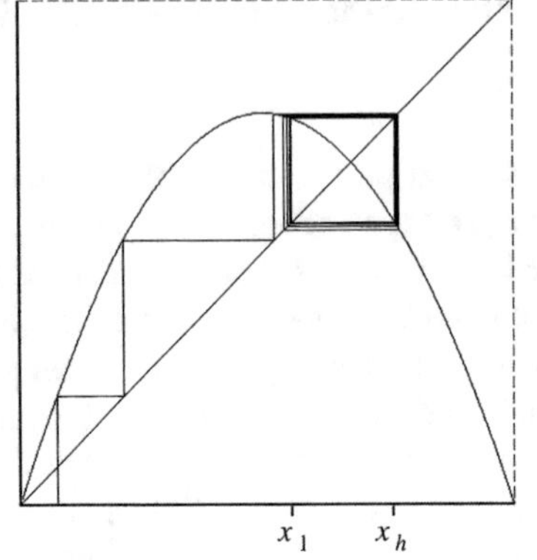

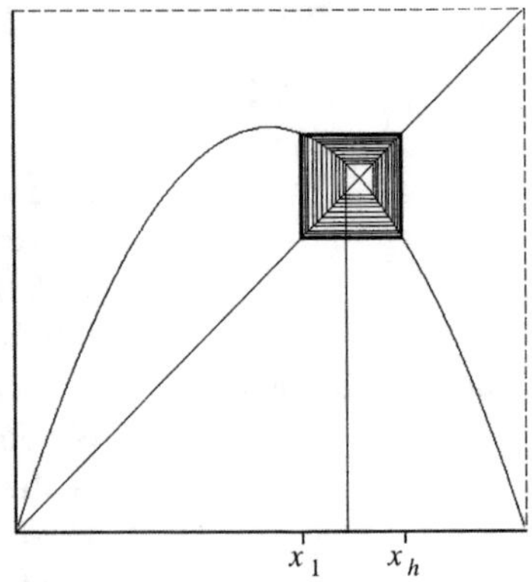

Figure 11.16

remains there forever,[4] $p_a = x_0 = x_1 = \cdots$. However, p_a is a repeller, and therefore is *not* part of the final state $A(a)$. The attractiveness has been taken over by the loop which oscillates between the two values $x_l(a)$ and $x_h(a)$. Thus, the final state is the attractor made of two points,

$$A(a) = \{x_l(a), x_h(a)\} \text{ for } a > 3.$$

We call the pair $\{x_l(a), x_h(a)\}$ a *2-cycle* or an *orbit of period two*. It is characterized by the fact that x_l is transformed into x_h and vice versa. This can be used for an explicit calculation of these numbers, see below. This periodic orbit exists for all parameters $3 < a \leq 4$. However, as the fixed point p_a loses stability at $b_1 = 3$, also the 2-cycle loses stability at a certain parameter value $b_2 > 3$, and this will be discussed below.

Explicit Calculation of Bifurcations

We have already calculated the fixed points of the quadratic iterator. Also for the 2-cycle we can carry out a direct calculation. Writing again $f_a(x) = ax(1-x)$ we need to find the solutions of the equation $f_a(f_a(x)) = x$. This is a fourth order equation:

$$-a^3x^4 + 2a^3x^3 - (a^2 + a^3)x^2 + (a^2 - 1)x = 0.$$

This looks complicated. But luckily we already know two solutions, namely, the fixed points of the iteration which solve $f_a(x) = x$ (i.e., 0 and $(a-1)/a$). The solution $x = 0$ allows us to simplify the equation to a third-order equation

$$-a^3x^3 + 2a^3x^2 - (a^2 + a^3)x + (a^2 - 1) = 0.$$

Knowing the second solution, $(a-1)/a$, we now divide this equation by $x - (a-1)/a$, which gives the second-order equation

$$-a^3x^2 + (a^2 + a^3)x - (a^2 + a) = 0$$

[4] The same holds for the other fixed point of the iterator, $p_0 = 0$, which is unstable for parameters $a > 1$.

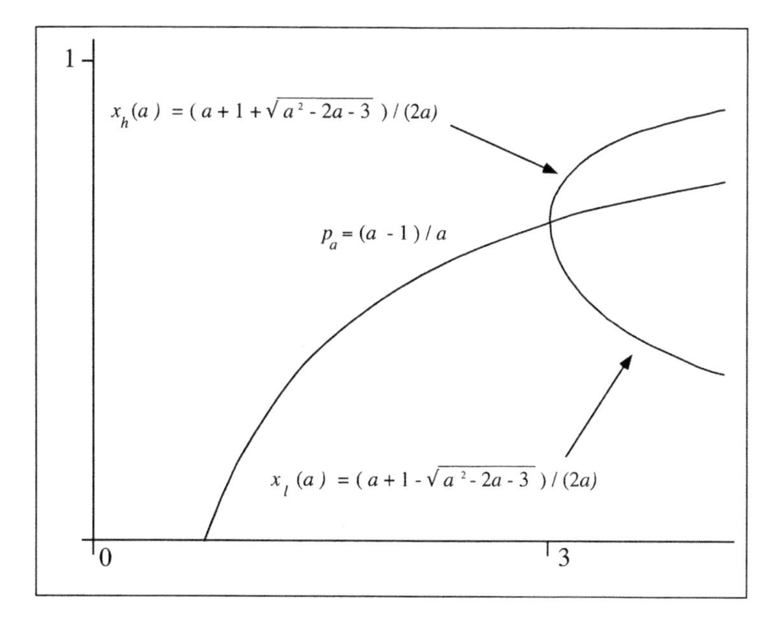

Figure 11.17 : In this bifurcation diagram the fixed points p_a and the 2-cycles $\{x_l(a), x_h(a)\}$ are shown.

or, dividing by $-a^3$,

$$x^2 - \frac{a+1}{a}x + \frac{a+1}{a^2} = 0.$$

The roots of this quadratic equation are

$$x_h(a) = \frac{a+1+\sqrt{a^2-2a-3}}{2a}$$

and

$$x_l(a) = \frac{a+1-\sqrt{a^2-2a-3}}{2a}.$$

Considering only parameters between 0 and 4, we note that these solutions are defined only for $a \geq 3$. Moreover, at $a = 3$ we get $x_l(a) = x_h(a) = (a-1)/a$, i.e., the two solutions bifurcate from the fixed point p_a. Figure 11.17 shows a bifurcation diagram of the explicitly calculated solutions.

Note that in the final-state diagram you do not see the solution $p_a = (a-1)/a$ for $a > 3$. This corresponds to the fact that although the fixed point continues to exist, it has become unstable (i.e., the iteration is pushed away from this point).

More Time Profiles

Figure 11.18 shows an experiment which is related to the time profiles we have computed so far. Here we take again sample starting points from the unit interval and run the iteration, counting the number of iterations until the orbit comes close to the 2-cycle. In practice, the iteration can be stopped as soon as x_k and x_{k+2} have become very close,

$$|x_{k+2} - x_k| < \varepsilon$$

where ε is some small number, for example $\varepsilon = 1/6000$. In the figure we choose three parameters,

$$a = 3.1, \quad a = 1 + \sqrt{5} \approx 3.236, \quad a = 3.4.$$

The results are much more complex than the time profiles for parameters between 1 and 3. In particular, there are two striking phenomena.

- The sub-valleys found for parameters just below $a = 3$ (see figure 11.14) have turned into spikes.
- The central areas between the two inner spikes at $1/a$ and $(a-1)/a$ look like compressed versions of the complete diagrams found in figures 11.12, 11.13, and 11.14, corresponding to the present cases $a = 3.1$, $a \approx 3.236$, and $a = 3.4$.

The Spikes at α_k and β_k

There must be a reason for this striking similarity. First, we recall that 0 is a repeller for the iteration if $a > 1$, and 1 is its preimage. For $a > 3$ we now find that $p_a = (a-1)/a$ is also a repeller, and $1/a$ is its preimage. This already explains to some degree that the 'spikes' at 0 and 1 in figures 11.12 to 11.14, can now be found at $1/a$ and $(a-1)/a$. The other 'spikes' in the new diagrams are at the points α_k and β_k, which we already have introduced as iterated preimages of the fixed point p_a. Spikes correspond to high iteration counts, which, however, seem to raise a contradiction. Starting the iteration exactly at one of the points $x_0 = \alpha_k$ or $x_0 = \beta_k$, we arrive exactly at the fixed point p_a after only $k+1$ iterations, $x_{k+1} = p_a$. Then the next two iterations are $x_{k+2} = x_{k+3} = p_a$, and, thus, the test

$$|x_{k+3} - x_{k+1}| < \varepsilon$$

is satisfied. This yields an iteration count

$$N_a(\alpha_k) = N_a(\beta_k) = k + 3,$$

a rather low number compared to the spikes observed in the numerical studies. But since the fixed point is unstable, this behavior is not observable. To be visible, infinitely precise computation would be required, but also the specification of the starting points with infinite precision. This is not possible, and therefore it is not visible. On the contrary, starting the iteration close to one of the points α_k or β_k produces a $(k+1)^{\text{st}}$ iterate close to the fixed point p_a. The closer it gets, the more further iterations it takes to move the orbit away from the repulsive fixed point and to come close to the attractive periodic cycle. This results in the spikes as seen in the figures.

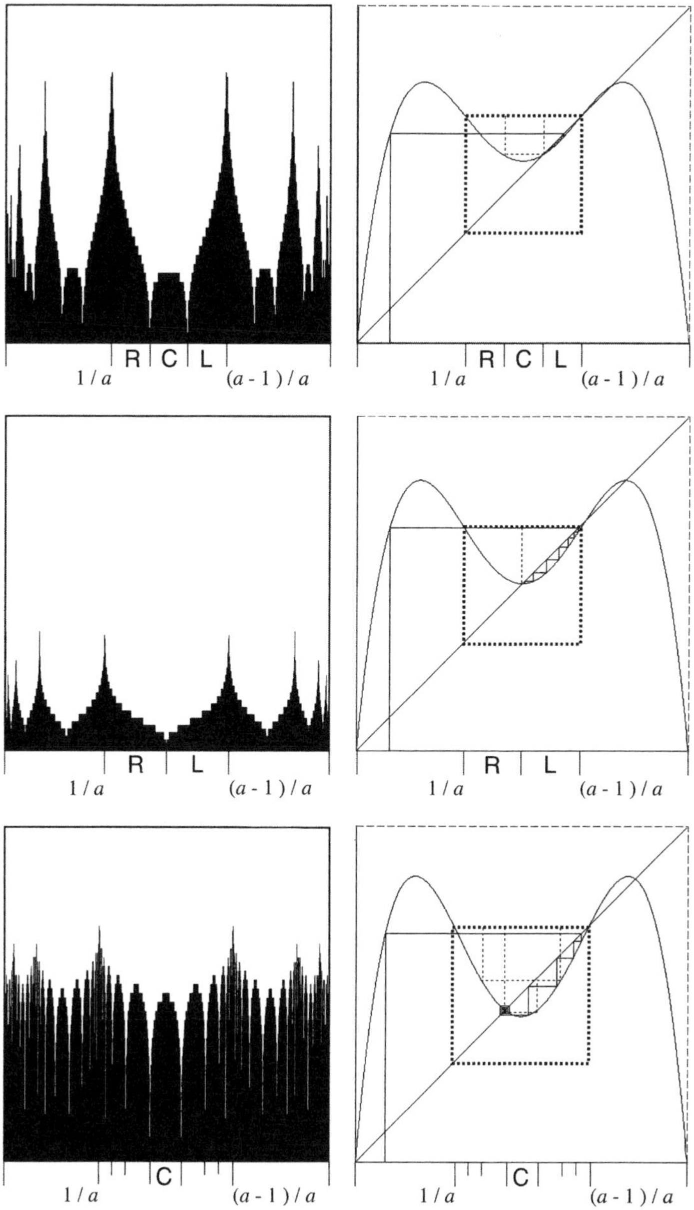

Time Profile $a > 3$

Time profiles for $a = 3.1$ (top), $a = 3.236$ (middle) and $a = 3.4$ (bottom). Note the similarities of the portion between $1/a$ and $(a-1)/a$ to the preceding time profiles 11.12, 11.13, and 11.14 which implies the reversal of the labels L and R.

R C L
1 / a (a - 1) / a
R C L
1 / a (a - 1) / a
R L
1 / a (a - 1) / a
R L
1 / a (a - 1) / a
C
1 / a (a - 1) / a
C
1 / a (a - 1) / a

Figure 11.18

The Iterate of f_a

But there is more to our observation of similarity, and that is revealed in the graphical iterations to the right of the time profiles. They show a typical orbit, not for the quadratic iterator $x_n = f_a(x_{n-1})$, but for its second-order composition $f_a(f_a(x))$ (i.e., we look at the iterator $x_n = f_a(f_a(x_{n-1}))$. For

f_a and f_a^2

Comparing $f_a(x)$ (left) and $f_a^2(x)$ (right) for $a = 1$ and $a = 2$.

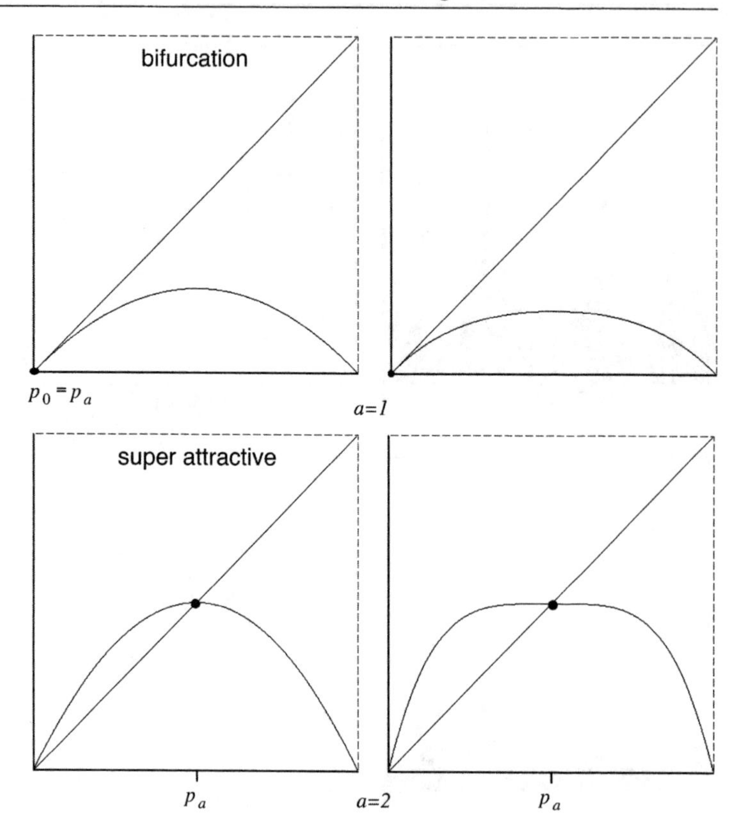

Figure 11.19

short we call this composition the *second iterate of* f_a and write this as

$$f_a^2(x) = f_a(f_a(x)).$$

You should not, however, confuse this with the square of the values of $f_a(x)$! The graph of $f_a^2(x)$ is given by the fourth-degree polynomial

$$\begin{aligned} f_a^2(x) &= a^2x(1-x)(1-ax(1-x)) \\ &= -a^3x^4 + 2a^3x^3 - (a^2+a^3)x^2 + a^2x \end{aligned}$$

shown in the figures. The second iterate of f_a has four fixed points: $\{0, p_a, x_l(a), x_h(a)\}$. These are the fixed points of f_a and the two elements of the 2-cycle. They can be seen in the figures as intersections of the graph with the bisector. Between $1/a$ and $(a-1)/a$ we have outlined a square to draw your attention to this section of the graph (see figure 11.18). Doesn't it look like the familiar parabola of $f_a(x)$ upside-down? In fact, we also observe that the poly-line representing the iteration behaves in this region similar to what we have seen for the iteration of $f_a(x)$.

Let us compare the graphs of $f_a(x)$ and $f_a^2(x)$ more systematically (see figures 11.19 and 11.20). We start at $a = 1$, which is the parameter where

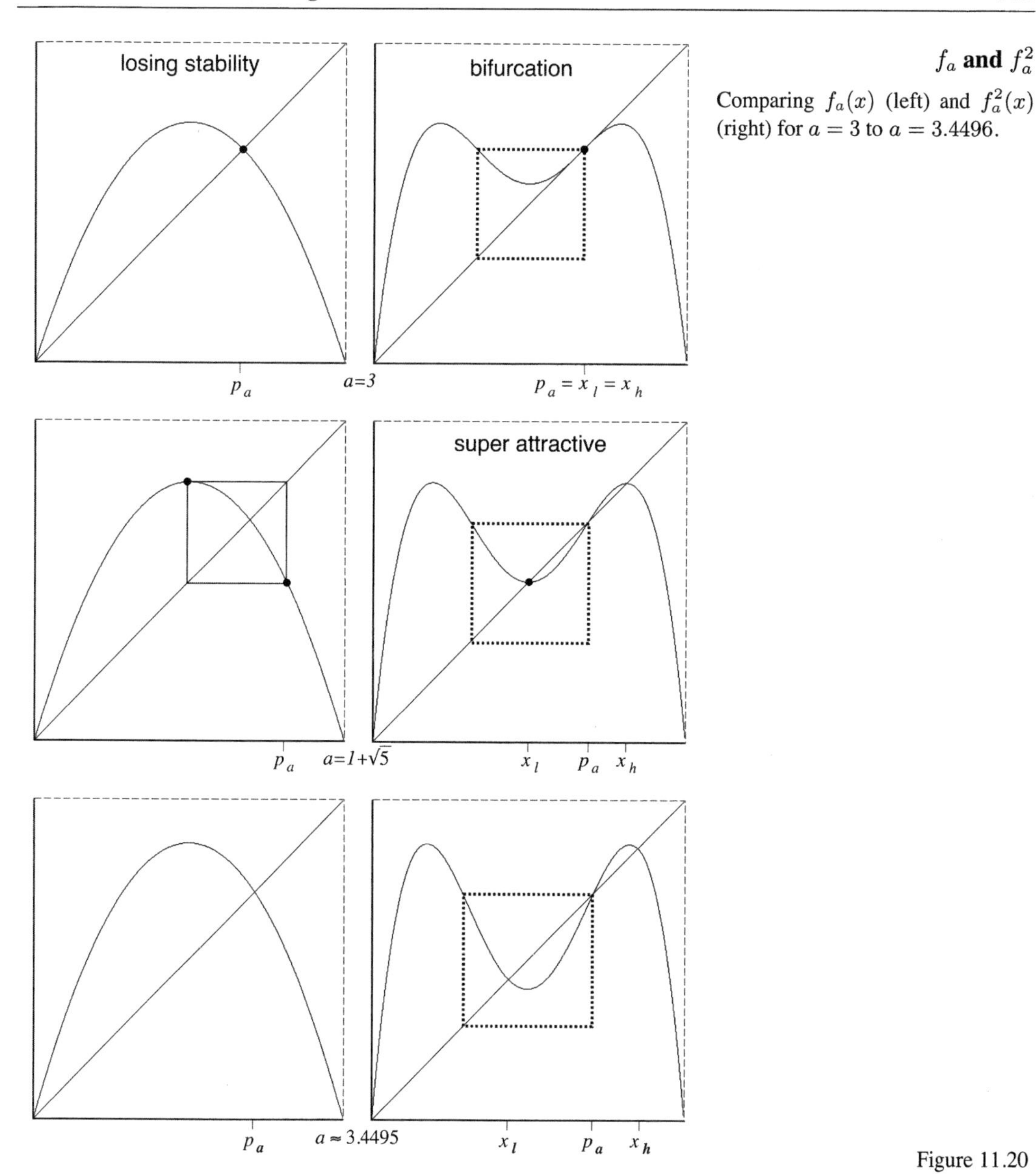

f_a and f_a^2

Comparing $f_a(x)$ (left) and $f_a^2(x)$ (right) for $a = 3$ to $a = 3.4496$.

Figure 11.20

the fixed point $p_0 = 0$ of $f_a(x)$ becomes unstable and a new fixed point $p_a = (a-1)/a$ begins to exist (for $a > 1$). Here the graph of $f_a^2(x)$ looks a bit lower and the fixed points are identical to those of $f_a(x)$. Then at $a = s_1 = 2$ we reach the super attractive case for $f_a(x)$. The new fixed point p_a and the critical point $x_{\max} = 1/2$ coincide. The graph of $f_a^2(x)$ has now reached the same height, but its top is almost flat.

At $a = b_1 = 3$ the fixed point p_a for $f_a(x)$ loses its stability.[5] Also for $f_a^2(x)$ the fixed point p_a loses its stability; but here two new, additional fixed points begin to exist (for $a > 3$): $x_l(a)$ and $x_h(a)$. Note that the portion of the graph which is enclosed by the dashed square looks like the graph of $f_a(x)$ for $a = 1$. The bifurcation at $b_1 = 3$ is called a *period-doubling bifurcation*; a fixed point becomes unstable and gives birth to a 2-cycle.

Then at $a = s_2 = 1+\sqrt{5}$ we obtain the super attractive case for $f_a^2(x)$: the fixed point $x_l(a)$ and the critical point $x_{\text{crit}} = 0.5$ are identical. For $f_a(x)$ this means that iterating x_{crit} for two steps brings us back to x_{crit}. If we increase the parameter further to $a = b_2 \approx 3.4495$, the fixed point $x_l(a)$ of $f_a^2(x)$ also becomes unstable. In other words, all the changes which we observe for $f_a(x)$ while varying the parameter a between 1 and 3 can also be found for $f_a^2(x)$ in the parameter range $a = b_1 = 3$ to $a = b_2$. This is recapitulated in the following table:

a	$f_a(x)$	$f_a^2(x)$
1	bifurcation of p_a	
2	super attrac., $p_a = x_{\text{crit}}$	
3	p_a becomes unstable	bifurcation of x_l and x_h
3.236	$(f_a(f_a(x_{\text{crit}})) = x_{\text{crit}})$	super attrac., $x_l = x_{\text{crit}}$
3.449		x_l becomes unstable

Now you can guess what happens if we increase a beyond b_2. Yes, we find fixed points of $f_a^2(f_a^2(x))$ which bifurcate off from $x_l(a)$ (and also from $x_h(a)$). This composition of f_a^2 (applied to a point x) is nothing else but the fourth iterate of f_a, i.e., $f_a(f_a(f_a(f_a(x))))$, and will be written for short f_a^4. The new stable fixed points of f_a^4 are equivalent to the birth of a stable cycle of period four for f_a. If we increase the parameter even further, stability is lost again, which marks the birth of a period 8 cycle for f_a and so on ...Again, the bifurcations are called period-doubling bifurcations. The periods of the attractive cycles are $1, 2, 4, 8, 16, 32, \ldots$

Period-Doubling Bifurcation Series

This process establishes two sequences of important parameters.

- the parameters $s_1, s_2, s_3, \ldots$ for which we obtain a super attractive case for $f_a, f_a^2, f_a^4, \ldots$ For these the critical point $x_{\text{crit}} = 0.5$ is a fixed point of $f_{s_1}, f_{s_2}^2, f_{s_3}^4$ etc.
- the sequence $b_1, b_2, b_3, \ldots$ of parameter values for which we have a period-doubling bifurcation.

We have seen $s_1 = 2, s_2 = 3.236\ldots, b_1 = 3$ and $b_2 = 3.449490\ldots$ and we already know where these sequences lead to: the Feigenbaum point s_∞ (recall our observation of self-similarity in the final-state diagram). Let us discuss the sequence of period-doubling bifurcations a bit further. It appears

[5] The derivative is $|f_a'(p_a)| = 1$.

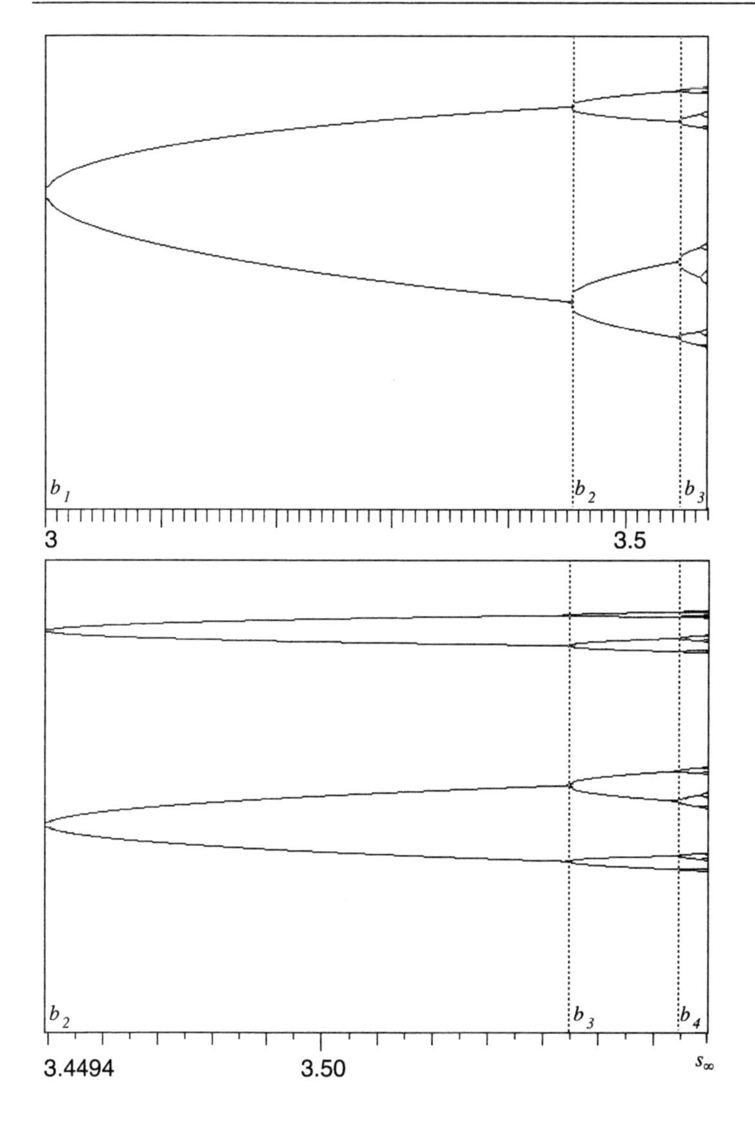

Period-Doubling Tree

Two close-ups of the period-doubling tree. Note the relative position of the points b_1, b_2, b_3 and b_4, which indicates that $(b_2 - b_1)/(b_3 - b_2) \approx (b_3 - b_2)/(b_4 - b_3)$.

Figure 11.21

that the distance d_k between two successive bifurcation points

$$d_k = b_{k+1} - b_k,\ k = 1, 2, 3, \ldots,$$

decreases rather rapidly. This is also visible in figure 11.21 where we have enlarged the period-doubling tree to show some more of its bifurcation points. A first guess would be that the decrease is geometric, i.e., that

$$\frac{d_k}{d_{k+1}} = \delta.$$

In that case the bifurcation values b_k would form a converging sequence

$$b_k = b_1 + d_1\left(1 + \frac{1}{\delta} + \cdots + \frac{1}{\delta^{k-2}}\right), \quad k = 2, 3, \ldots$$

with limit

$$b_\infty = b_1 + d_1\frac{\delta}{\delta - 1}.$$

Unfortunately things are not that easy. For example,

$$\begin{aligned}
d_1 &= b_2 - b_1 &= 3.449489\ldots - 3.0 &\approx 4.4949 \cdot 10^{-1}\\
d_2 &= b_3 - b_2 &= 3.544090\ldots - 3.449490\ldots &\approx 9.4611 \cdot 10^{-2}\\
d_3 &= b_4 - b_3 &= 3.564407\ldots - 3.544090\ldots &\approx 2.0316 \cdot 10^{-2}\\
d_4 &= b_5 - b_4 &= 3.568759\ldots - 3.564407\ldots &\approx 4.3521 \cdot 10^{-3}\\
d_5 &= b_6 - b_5 &= 3.569692\ldots - 3.568759\ldots &\approx 9.3219 \cdot 10^{-4}\\
d_6 &= b_7 - b_6 &= 3.569891\ldots - 3.569692\ldots &\approx 1.9964 \cdot 10^{-4}.
\end{aligned}$$

Based on these values we compute

$$\begin{aligned}
d_1/d_2 &= 4.7514\ldots\\
d_2/d_3 &= 4.6562\ldots\\
d_3/d_4 &= 4.6682\ldots\\
d_4/d_5 &= 4.6687\ldots
\end{aligned}$$

Thus, the decrease in d_k from bifurcation to bifurcation is not exactly geometric but only approximately geometric (i.e., the ratio $\delta_k = d_k/d_{k+1}$ converges with increasing k) and Feigenbaum computes[6]

$$\lim_{k\to\infty} \delta_k = \delta = 4.6692016091029\ldots$$

A Universal Constant

At first, it seems that this is just another number which somehow documents the behavior of our particular example, the quadratic iteration. It certainly has to be expected that the number δ depends on this specific model, just as the value of the Feigenbaum point $s_\infty = 3.5699456\ldots$ does. Indeed, if we just changed the scale of a, then naturally the value of s_∞ would turn out different. However, the nature of Feigenbaum's number $\delta = 4.669202\ldots$, discovered in October 1975, is quite different. It is universal, i.e., it is the same for a wide range of different iterators. Due to his achievement, this number is now called the *Feigenbaum constant*.

Computation of the Feigenbaum Constant

At first glance it may seem very difficult to compute the bifurcation parameter sequence $b_1, b_2, b_3, \ldots$ from which δ can be estimated. However, Feigenbaum did his first experiments merely using a pocket calculator, and it is not black magic at all. In this technical part we present an algorithm to compute δ from the sequence $s_1, s_2, s_3, \ldots$ of super

[6]See M. J. Feigenbaum, *Quantitative universality for a class of nonlinear transformations*, J. Stat. Phys. 19 (1978) 25–52.

attractive parameters, which are more accessible than the bifurcation parameters.[7] The computation is based on the formula

$$\delta = \lim_{n\to\infty} \frac{s_n - s_{n-1}}{s_{n-1} - s_{n-2}}$$

where $s_1, s_2, s_3, \ldots$ are the super attractive parameters corresponding to periodic cycles of length $1, 2, 4, 8, \ldots$ In other words, we may replace the bifurcation parameters in the usual approach described in the main text above by the super attractive parameters and arrive at the same constant, δ. We already know the first two parameters

$$s_1 = 2, \quad s_2 = 1 + \sqrt{5} \approx 3.236067978.$$

The following parameters $s_3, s_4, \ldots$ can be computed numerically using Newton's method for solving the appropriate nonlinear equations. We recall that the super attractive parameter s_n is characterized by a corresponding periodic orbit of minimal period 2^{n-1} with the property that the critical point $x_{\text{crit}} = \frac{1}{2}$ is a member of the cycle. Thus, s_n is a solution of the equation

$$f_a^{2^{n-1}}\left(\frac{1}{2}\right) = \frac{1}{2} \tag{11.3}$$

where the unknown variable is the parameter a. However, note that also $a = s_1, s_2, \ldots, s_{n-1}$ solves the same equation, because at these parameter values the point $x_0 = \frac{1}{2}$ is also periodic — however, with a minimal period smaller than 2^{n-1}. Thus, a numerical method for solving eqn. (11.3) for $a = s_n$ must be carefully tuned in order not to produce one of the irrelevant solutions.

By Newton's method we can approximate zeroes of differentiable functions, and in the case at hand the function can be taken as

$$g(a) = f_a^{2^{n-1}}\left(\frac{1}{2}\right) - \frac{1}{2}.$$

The method consists of iterating the transformation

$$\mathcal{N}(a) = a - \frac{g(a)}{g'(a)}$$

where $g'(a)$ denotes the derivative of the function g with respect to the variable a. In other words, for a given initial guess $a^{(0)}$ we compute the sequence $a^{(0)}, a^{(1)}, a^{(2)}, \ldots$ where

$$a^{(k+1)} = \mathcal{N}(a^{(k)}), \quad k = 0, 1, 2, \ldots$$

If this sequence converges to some number, then the limit is a solution to the equation $g(a) = 0$. Under suitable conditions, convergence can be guaranteed provided that the initial guess is sufficiently close

[7] We assume here that the reader is familiar with basic calculus (in particular the product rule) and Newton's method in one dimension.

to the solution. At this point we need to address the following three questions:

- Given some parameter a, how do we compute $g(a)$ and $g'(a)$?
- What is a good starting point $a^{(0)}$ near the (unknown) solution s_n?
- How do we decide to stop the iteration in Newton's method? In other words, under what conditions do we accept $a^{(k)}$ as a sufficient approximation of s_n?

Let us discuss these questions one by one. We begin by introducing the notation $x_0, x_1, x_2, \ldots$ for the critical orbit, i.e.,

$$x_0 = \frac{1}{2}, \quad x_{k+1} = f_a(x_k) = ax_k(1 - x_k) \tag{11.4}$$

for $k = 0, 1, 2, \ldots$ Note that these numbers depend on the parameter a. More precisely, we may think of them as functions of a, $x_k = x_k(a)$. Setting

$$N = 2^{n-1}$$

we can now write

$$g(a) = x_N - \frac{1}{2}$$

and, thus, $g(a)$ may be computed using the iteration in eqn. (11.4). To compute the derivative of g we need to differentiate $x_N = x_N(a)$. We use the prime notation to indicate derivatives with respect to the parameter a,

$$g'(a) = x'_N.$$

Again we have to use some iteration. First we note that x_0 is the constant $\frac{1}{2}$ and does not depend on a. Thus, $x'_0 = 0$. Then we consider $x_{k+1} = ax_k(1 - x_k)$, use the product rule applied to the three factors a, x_k, and $1 - x_k$, and obtain

$$\begin{aligned} x'_{k+1} &= (ax_k(1 - x_k))' \\ &= x_k(1 - x_k) + a(1 - x_k)x'_k - ax_k x'_k \\ &= x_k(1 - x_k) + a(1 - 2x_k)x'_k \end{aligned}$$

for $k = 0, 1, 2, \ldots$ In summary, we have to simultaneously carry out the iterations

$$\begin{aligned} x_{k+1} &= ax_k(1 - x_k) & x_0 &= \tfrac{1}{2} \\ x'_{k+1} &= x_k(1 - x_k) + a(1 - 2x_k)x'_k & x'_0 &= 0 \end{aligned}$$

for $k = 0, \ldots, N - 1$.

Good starting points for the method are easy to obtain. If the sequence of super attractive parameters has already been computed up to $s_1, \ldots, s_n$, then we can estimate the Feigenbaum constant by

$$\delta_n = \frac{s_{n-1} - s_{n-2}}{s_n - s_{n-1}}.$$

Using this result, we can produce an estimate $s_{n+1}^{(0)}$ for the next parameter s_{n+1} to be computed,

$$s_{n+1}^{(0)} = s_n + \frac{s_n - s_{n-1}}{\delta_n}.$$

This is very close to the true value of s_{n+1} and only a couple of Newton iterations suffice to bring the estimate to the highest possible precision. Initially, before any computations are done, we have only two numbers available, s_1 and s_2, which are not sufficient to compute the estimate δ_2 of δ. Here we simply set $\delta_2 = 4$.

The last point is related to error estimation in Newton's method. The usual approach is to make use of the relation

$$s_n - s_n^{(k)} \approx s_n^{(k+1)} - s_n^{(k)}$$

Thus, we may stop the iteration as soon as the estimate $(s_n^{(k+1)} - s_n^{(k)})/s_n^{(k)}$ for the relative error is of the same order as that of the machine unit.[8] Further iterations would not improve the quality of the result anymore. Thus, we set $s_n = s_n^{(k+1)}$.

The following table records the results using double-precision calculations. The column labeled # lists the number of Newton iterations that were necessary to compute the super attractive parameters.

n	s_n	#	δ_n
1	2.000000000000000000		
2	3.236067977499789696		4.0000000000
3	3.498561699327701520	6	4.7089430135
4	3.554643880189573995	1	4.6805191559
5	3.566667594798299166	1	4.6642974062
6	3.569243531637110338	4	4.6677055227
7	3.569795293749944621	4	4.6685641853
8	3.569913465422348515	3	4.6691571813
9	3.569938774233305491	3	4.6691910025
10	3.569944194608064931	3	4.6691994706
11	3.569945355486468581	3	4.6692011346
12	3.569945604111078447	3	4.6692015094
13	3.569945657358856505	3	4.6692015880
14	3.569945668762899979	3	4.6692016018
15	3.569945671205296863	2	4.6692016148

Convergence is very rapid; only a few Newton iterations suffice. The major computational burden lies in the number of iterations for obtaining x_N and x'_N. Already at $n = 15$ we obtain an approximation of δ correct in the leading 8 digits. The computer used in this experiment has a machine unit of $2^{-63} \approx 1.08 \cdot 10^{-19}$, i.e., 19 significant

[8] The machine unit is the smallest number 2^i such that the expression $1 + 2^i$ can still be distinguished from the number 1 using machine arithmetic.

decimal digits. At this precision it is not sensible to continue the table for larger indices n, because the change in consecutive values of s_n, i.e., $s_n - s_{n-1}$ will have less than half as many significant digits, which signals large errors in the computation of the approximations δ_n due to cancellation of digits.

Universality of δ

The Feigenbaum constant $\delta = 4.6692\ldots$ is universal, i.e., it is exactly the same for a whole class of iterations generated by functions similar to a quadratic function, such as $f(x) = \sin(\pi x)$ or $g_a(x) = ax^2 \sin(\pi x)$, to take an example which is not symmetric (figure 11.22 shows this function for $a = 2.3$). This class is better described by the following properties:

- f is a smooth function from $[0,1]$ into the real numbers.
- f has a maximum at x_m, which is quadratic, i.e., $f''(x_m) \neq 0$.
- f is monotone in $[0, x_m)$ and in $(x_m, 1]$.
- f has a negative Schwarzian derivative, i.e., $S_f(x) < 0$ for all x in $[0,1]$, where

$$S_f(x) = \frac{f'''(x)}{f'(x)} - \frac{3}{2}\left(\frac{f''(x)}{f'(x)}\right)^2 .$$

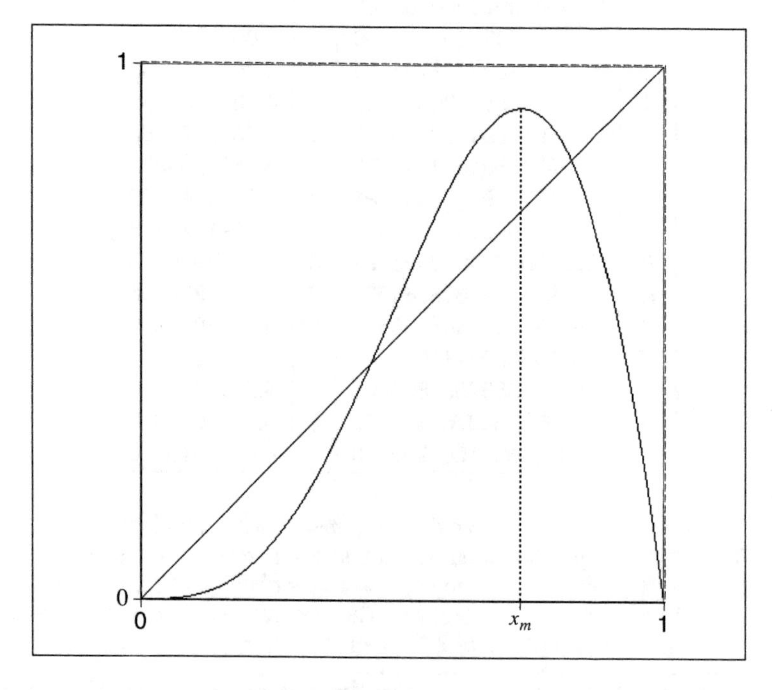

Figure 11.22 : Graph of $g_a(x) = ax^2 \sin(\pi x)$ for $a = 2.3$.

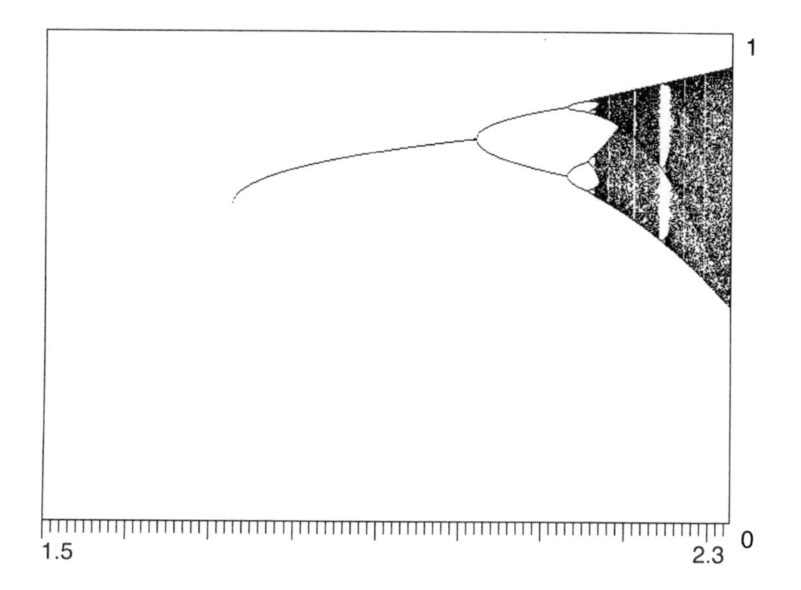

Figure 11.23 : The final-state diagram of $g_a(x) = ax^2 \sin(\pi x)$ for a between 1.5 and 2.3265. Again we observe the scenario familiar from the quadratic iterator (see figure 11.2). There is a break point at about $a = 1.7263$, where the period-doubling tree begins. For parameters less than that the final state is the attractive fixed point 0.

If one computes the bifurcation points of the period-doubling tree for a function from this class the difference d_k of its bifurcation parameters b_k will form a sequence which is asymptotically geometric. Again the ratios d_k/d_{k+1} converge to δ:

$$\lim_{k\to\infty} \frac{d_k}{d_{k+1}} = \delta = 4.6692\ldots$$

Figure 11.23 shows the final-state diagram for the iteration of $g_a(x)$ from figure 11.22.

For a while this universality appeared as a mathematical mystery, but it turned out that this number was much more important than that. It was conjectured that this number should also be verifiable in real physical experiments. This seemed to be really a little too far-fetched because there is no reason whatsoever that a real physical experiment should have anything in common with the simple-minded iteration process $x_{n+1} = ax_n(1 - x_n)$. However, the idea was right. In the early 1980's physicists carried out a whole variety of extremely sophisticated experiments in hydrodynamics, electronics, laser physics, and acoustics and found period-doubling bifurcations, with the surprising result that the associated numbers d_k/d_{k+1} did in fact numeri-

Universality of the Feigenbaum Constant

Results from experiments wherein period-doubling plays a role. The numbers in the third column are to be compared with the Feigenbaum constant $\delta = 4.669\ldots$ Table adapted from P. Cvitanović, *Universality in Chaos,* Adam Hilger, Bristol, 1984.

Table 11.24

Experimental Measurements of Period-Doublings		
Experiment	Number of Period Doublings	δ
Hydrodynamic:		
water	4	4.3 ± 0.8
helium	4	3.5 ± 0.15
mercury	4	4.4 ± 0.1
Electronic:		
diode	5	4.3 ± 0.1
transistor	4	4.7 ± 0.3
Josephson	4	4.4 ± 0.3
Laser:		
laser feedback	3	4.3 ± 0.3
Acoustic:		
helium	3	4.8 ± 0.6

cally show a remarkable degree of agreement with Feigenbaum's constant $\delta = 4.669\ldots$; see table 11.24.

11.3 The Feigenbaum Point: Entrance to Chaos

We started the discussion of this chapter with the self-similarity features in the final-state diagram of the quadratic iterator (see figure 11.3). This kind of self-similarity is already contained in the first part of the diagram, the period-doubling tree, ranging from $a = 1$ to $a = s_\infty$, the Feigenbaum point. However, the self-similarity in either case is not strict: although the branches of the tree look like small copies of the whole tree there are parts, like the stem of the tree, which clearly do not. Moreover, even the branches of the tree are not exact copies of the entire tree. Here we have to use the term 'self-similarity' in a more intuitive sense without being precise. By contrast, the self-similarity of the address tree of the Sierpinski gasket or the Cantor set was described very precisely in chapters 2, 3 and 5. In those cases the 'small copies' really are exact copies of the whole.

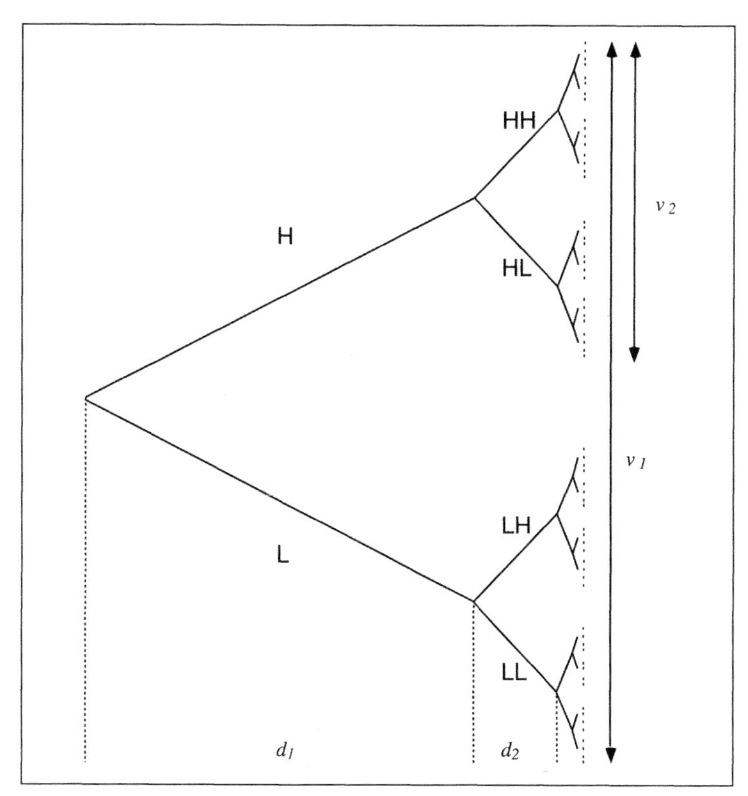

Period-Doubling Tree

Schematic representation of the period-doubling tree with scaling factors $d_1/d_2 = 4.669$. The vertical range of the complete tree is v_1 and the range of the upper main branch v_2. The ratio is $v_1/v_2 = 2.3$. Note that the leaves of this tree form a strictly self-similar Cantor set.

Figure 11.25

Self-Similarity of the Period-Doubling Tree

For the period-doubling tree everything is more complicated. First, we have noted that the sequence of differences d_k between the parameters of the bifurcation points is not precisely geometric. In other words, when we make close-ups as in figure 11.3, the scaling factor slightly changes from close-up to close-up approaching the factor $\delta = 4.669\ldots$ But this is only true for the scaling in the horizontal direction of the parameter a. With respect to the vertical direction we have to scale (in the limit) with approximately 2.3.

In figure 11.25 we have used these scaling factors 4.669... and 2.3 to obtain a schematic representation of the period-doubling tree which exhibits these limiting scaling properties in all stages. Note that the leaves of this tree form a strictly self-similar Cantor set.[9]

When comparing this tree with the original bifurcation tree, the non-linear distortion becomes apparent. Here branches of the same stage are exactly the same. In the original period-doubling tree, branches have different sizes. Nevertheless, we can identify corresponding branches. Also, the leaves of the original tree form a Cantor set. This happens right at the Feigenbaum point s_∞, where the final-state diagram reaches a new stage which is much more delicate than the situation for parameter values less than s_∞.

Dynamics at the Feigenbaum Point

For all parameters a between 3 and s_∞ we observe stable periodic orbits as final states. Now the natural question arises: what kind of dynamics do we have for $a = s_\infty$? This is a difficult problem. Considering a starting point x_0 from the final state, the Cantor set, the complete orbit of x_0 is also in the Cantor set. This reminds us of the chaos game for the Sierpinski gasket. In chapter 6 we used infinite symbolic addresses to identify points of the Sierpinski gasket and to describe the iteration process. It turns out that the same technique can be used to analyze the dynamics on the Cantor set at the Feigenbaum point.

Addresses for the Branches

We introduce addresses for the branches and the leaves of the period-doubling tree as we did previously for other binary trees. First, we label the lower main branch of the tree with L (for low) and the top branch with H (for high). When the two branches split into four we label the upper two parts with HH and HL and the lower two parts with LH and LL. This is the second stage of our addressing hierarchy. In figure 11.25 we have already indicated these labels for the schematic tree. The branches of the third stage would obtain the labels HHH, HHL, HLH, HLL, LHH, LHL, LLH, and LLL. In general we would obtain 2^k sub-branches labeled with k-letter addresses for stage k. The leaves would have infinite address strings. The first k letters of these addresses indicate the sub-branch of stage k to which a particular leaf belongs.

Dynamics in Terms of Addresses

Now we can start to discuss the dynamics of orbits on the Cantor set in terms of addresses. For all parameters a between $b_1 = 3$ and $b_2 = 3.4495$ we have a stable periodic oscillation of period two. This is the range of parameters were one-letter addresses are sufficient: we oscillate between the H-branch and the L-branch.

In the right half of figure 11.26 we show the super attractive case as an example of this oscillation. The diagram on the left shows the corresponding mapping of addresses: L is mapped to H (this relation is indicated by the upper left-hand grey box) and H is mapped to L (which is marked by the lower right-hand grey box).

[9] The fractal dimension D of of this special Cantor set has been estimated as $0.5376 < D < 0.5386$ by Peter Grassberger. See P. Grassberger and I. Procaccia, *Measuring the strangeness of strange attractors*, Physica 9D (1983) 189–208.

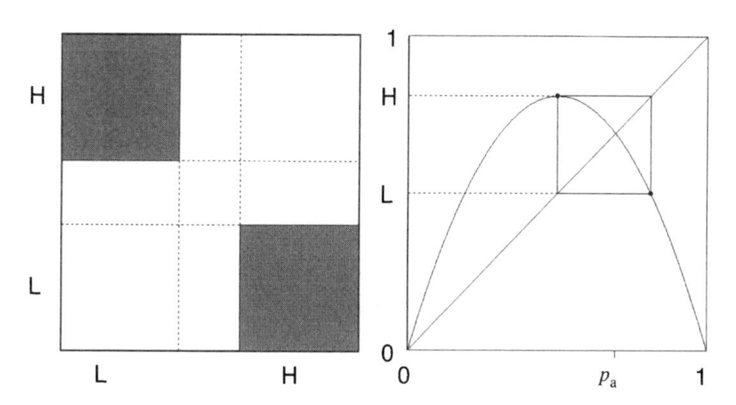

Period 2 Dynamics — Stage 1

Period 2 oscillation: L is mapped to H and H is mapped to L.

Figure 11.26

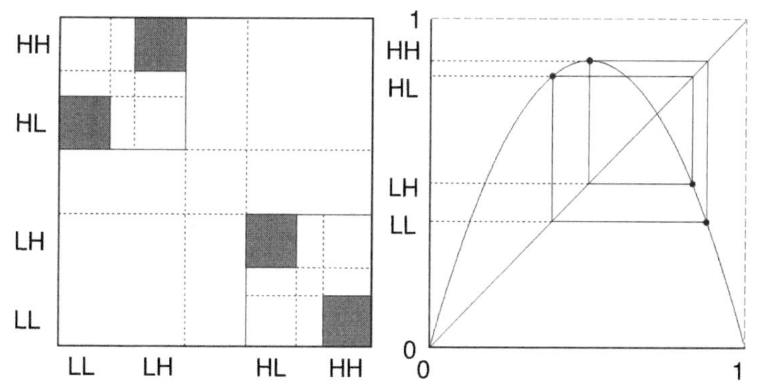

Period 4 Dynamics — Stage 2

Period 4 oscillation: $HH \to LL \to HL \to LH \to HH$.

Figure 11.27

Now we turn to the next stage ($b_2 < a < b_3$), which represents the oscillation between four different values. Here we need two-letter addresses. The right half of figure 11.27 shows the super attractive case. If we trace the poly-line, which represents the iteration, we can read off the mapping of addresses

$$HH \to LL \to HL \to LH \to HH.$$

This transformation is shown in the left diagram.

Now you might wonder whether there is a reason why we see precisely this sequence of addresses, and not, for example, $HH \to HL \to LH \to LL \to HH$ or $HH \to LH \to HL \to LL \to HH$. Indeed, we can rule out these other cases as follows.

First, we observe that the iteration always has to oscillate up and down between addresses which start with H and addresses which start with L. This carries over from the oscillation between H and L from which the 4-cycle bifurcated off. Next we note that the topmost address HH must be followed by the lowest, i.e., LL. These two observations already determine the complete

From Period 8 to the Feigenbaum Point

Diagrams for period 8 (upper left), 16 (upper right), and 32 (lower left). The lower right-hand diagram shows the transformation of infinite addresses, i.e., describes the symbolic dynamics at the Feigenbaum point.

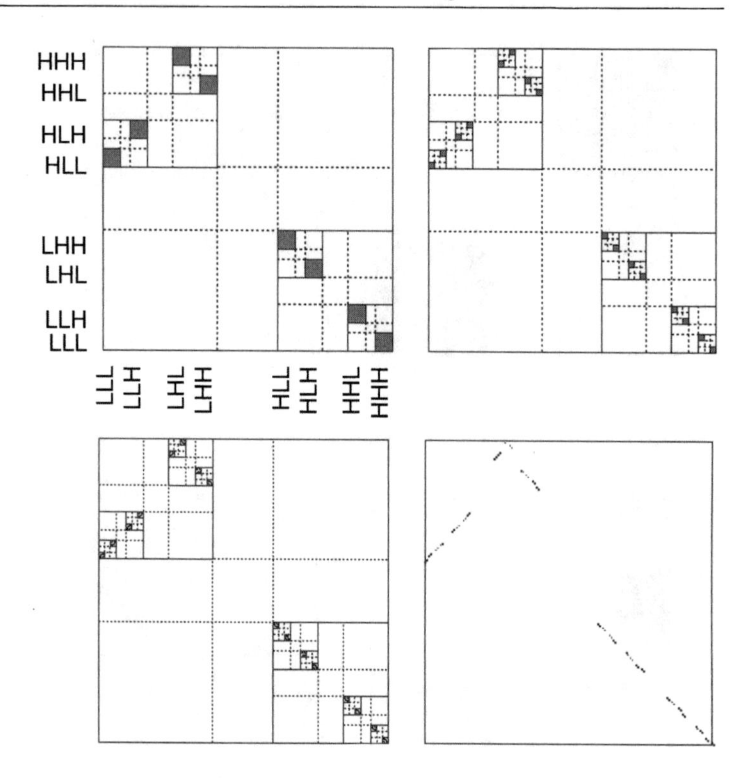

Figure 11.28

sequence. Note that these facts are clearly visible in the diagram. The grey boxes of the preceding diagram 11.26 now are refined, corresponding to the fact that the orbit oscillates between the main high to the main low branch as before. Furthermore, the box in the lower right-hand corner indicates that the highest address is mapped to the lowest.

The same kind of argument allows us to determine the address sequence which describes the orbit of period 8, which bifurcates from the 4-cycle. In figure 11.28 we show the related diagram (upper left) and the diagrams for the following two stages. The last diagram shows the transformation of infinite addresses, a transformation from points in a Cantor set to other points in the Cantor set. It describes the symbolic dynamics of our quadratic iterator at the Feigenbaum point.

Diagrams and Symbolic Dynamics

Let A_k be the set of all k-letter addresses (formed by H and L) and A_∞ the set of all infinite addresses. The dynamics of a periodic orbit with respect to point addresses is described by a transformation f_k: $A_k \to A_k$. For example, the 4-cycle can be described by

$$\begin{aligned} f_2(HH) &= LL, & f_2(LL) &= HL, \\ f_2(HL) &= LH, & f_2(LH) &= HH. \end{aligned}$$

This transformation is visualized in the stage-2 transformation diagram of figure 11.27. For each stage, we divide the axis of the transformation diagrams as in a typical Cantor set construction. Thus in the limit we obtain a diagram which visualizes the transformation $f_\infty : A_\infty \to A_\infty$ as a transformation of points of a Cantor set. Already when comparing the transformation diagrams of the first stages, it becomes apparent that there is a clear structure in these diagrams. First, let us discuss the refinement of the lower right grey box of the stage-1 diagram. In stage 3 it becomes apparent that here a diagonal of boxes begins to form. Thus, for the corresponding addresses starting with H we have the transformation rule:

$$f_\infty(HX_2X_3X_4\ldots) = LX_2^TX_3^TX_4^T\ldots \tag{11.5}$$

where X^T denotes the complement of X (i.e., X^T is H if X is L and X^T is L if X is H). Next, let us examine the refinement of the two upper boxes from the stage-2 diagram 11.27. The refinement of the leftmost grey box again leads to a diagonal of boxes (see figure 11.28). Again we can write down a transformation rule, now for all addresses which start with LL:

$$f_\infty(LLX_3X_4X_5\ldots) = HLX_3X_4X_5\ldots \tag{11.6}$$

The refinement of the topmost grey box is not as simple, but even more striking since here self-similarity is built in. In fact, the refinement of this box shown in stage 3 is just a scaled down copy of the stage-1 diagram, and in general, at stage k, this is a scaled down copy of the complete diagram for stage $k-2$. In the limit this leads to the self-similarity of the transformation diagram for f_∞: the graph of f_∞ for the

Step	Address	Rule	Step	Address	Rule
0	LLLLL	(11.6)	16	$LLLLH$	(11.6)
1	$HLLLL$	(11.5)	17	$HLLLH$	(11.5)
2	$LHHHH$	(11.7)	18	$LHHHL$	(11.7)
3	$HHLLL$	(11.5)	19	$HHLLH$	(11.5)
4	$LLHHH$	(11.6)	20	$LLHHL$	(11.6)
5	$HLHHH$	(11.5)	21	$HLHHL$	(11.5)
6	$LHLLL$	(11.7)	22	$LHLLH$	(11.7)
7	$HHHLL$	(11.5)	23	$HHHLH$	(11.5)
8	$LLLHH$	(11.6)	24	$LLLHL$	(11.6)
9	$HLLHH$	(11.5)	25	$HLLHL$	(11.5)
10	$LHHLL$	(11.7)	26	$LHHLH$	(11.7)
11	$HHLHH$	(11.5)	27	$HHLHL$	(11.5)
12	$LLHLL$	(11.6)	28	$LLHLH$	(11.6)
13	$HLHLL$	(11.5)	29	$HLHLH$	(11.5)
14	$LHLHH$	(11.7)	30	$LHLHL$	(11.7)
15	$HHHHL$	(11.5)	31	$HHHHH$	(11.5)
			32	LLLLL	(11.6)

Table 11.29 : Example transformation of an address.

addresses which start with LH is a scaled down copy of the complete graph. To compute the transformation of an address beginning with $LHX_3X_4X_5\ldots$ we first write down HH, then we drop the first two letters of the original address and apply f_∞ to the remaining letters, thus we compute $f_\infty(X_3X_4X_5\ldots)$. Finally we append the result of this evaluation to the initial letters HH. Thus,

$$f_\infty(LHX_3X_4X_5\ldots) = HHf_\infty(X_3X_4X_5\ldots). \tag{11.7}$$

Let us demonstrate the transformation of an address keeping track only of the first five digits. The recursive rule (11.7) is applied twice in steps 14 and 30. Moreover, in step 30 ($LHLHL$) the rule may possibly be applied even more times depending on the letters following the first five. Note that after 32 iterations the initial 5-letter address repeats. However, this is only true for the first five digits; the others have changed. In fact, it is true, that any number n of leading digits in the address of an orbit must go through a cycle of all possible 2^n combinations. Therefore, the orbit is not periodic. Moreover, the orbit gets arbitrarily close to any point in address space.

At the Feigenbaum point the final state of the iterator is given by an infinitely long, nonperiodic orbit in a Cantor set which gets arbitrarily close to every point of the Cantor set. We may say that this is a first sign of chaos; we are at the entrance to chaos.

Let us now turn to the self-similarity features related to the change of the dynamics as the parameter a increases. You will recall that for a in the interval $(1, 3)$ we have just one attractive fixed point at $p_a = (a-1)/a$, and all orbits belonging to initial values between 0 and 1 converge to that attractor. The way initial values are attracted changes at $a = s_1 = 2$ which marks the super attractive case. For parameters below s_1 initial values are attracted directly (monotonically) while for parameters above s_1 the orbit spirals around the fixed point. At $a = b_1 = 3.0$ the fixed point p_a becomes unstable, and an attractive 2-cycle is born. The old fixed point p_a continues to exist, but it now repels. Now the 2-cycle undergoes all the changes which we have seen for the fixed point. Especially to be noted is the fact that there is again the super attractive case at $a = s_2$. The 2-cycle finally becomes unstable at $a = b_2$. Then the story repeats for a 4-cycle, and so on.

Similarity Mechanisms in the Change of Dynamics

At each period-doubling bifurcation the dynamics of the iteration becomes dramatically more complex, though the mechanism is always the same. This is related to the 'similarity' of the graph of $f_a(x) = ax(1-x)$ (parabola) to sections of the graphs of the iterated transformations $f_a^2(x), f_a^4(x), f_a^8(x)$, and so on at higher parameters a. Figure 11.30 illustrates this similarity between the graphs of f_a (for $a = s_1 = 2$), $f_a^2(x)$ (for $a = s_2 \approx 3.236$) and $f_a^4(x)$ (for $a = s_3 \approx 3.498$).

A New Universal Function

We can make the similarity of the graphs even more apparent if we make a close-up of the squares outlined in figure 11.30. We enlarge the squares such that they match the unit square which encloses the whole graph. This

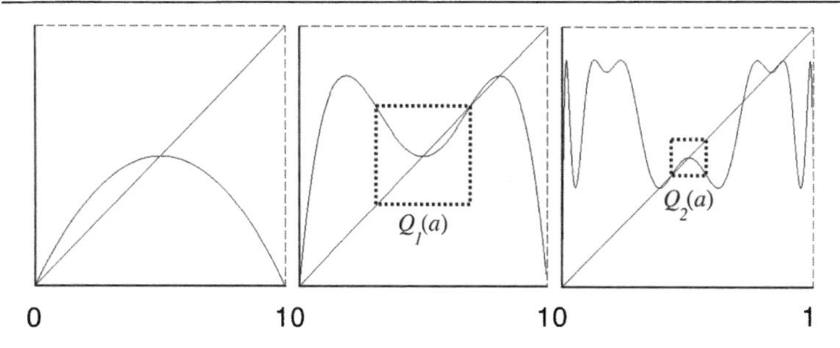

Similarity of the Parabola

$f_{s_1}(x)$ and parts of the graph of $f^2_{s_2}(x)$ and $f^4_{s_3}(x)$ at the super attractive parameters s_1, s_2 and s_3.

Figure 11.30

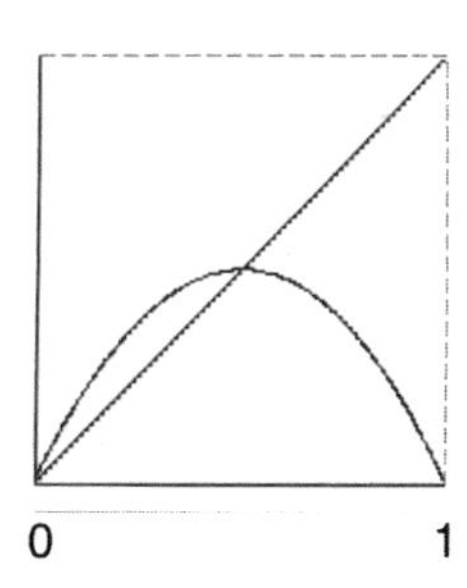

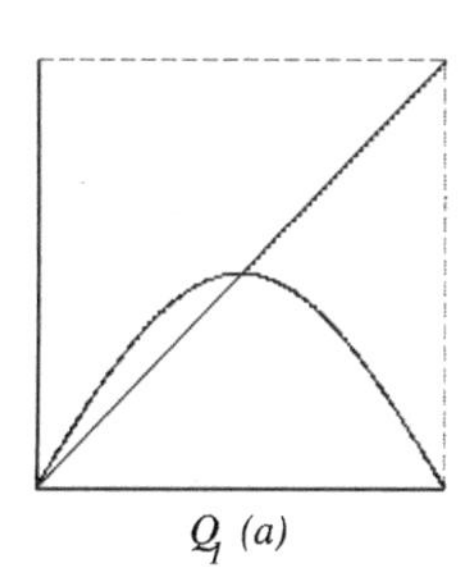

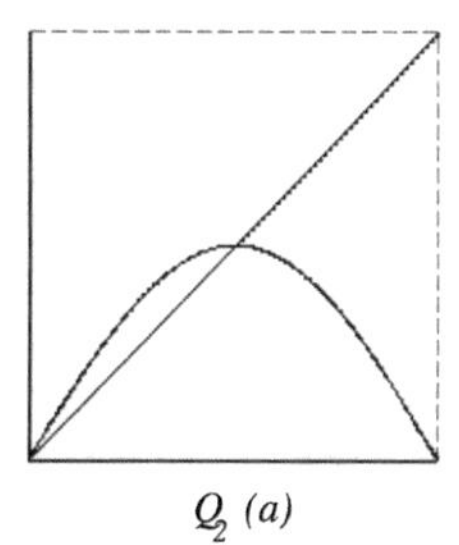

Enlargements

Similarity between the graph of f_{s_1} and 'close-ups' of $f^2_{s_2}$ (inverted) and $f^4_{s_3}$. (See the dotted squares in figure 11.30). Only the left graph, however, is a parabola, the other two represent fourth-order or eighth-order polynomials.

Figure 11.31

is demonstrated in figure 11.31, which shows, from left to right, the graph of $f_{s_1}(x)$ and magnifications of $f^2_{s_2}(x)$ and $f^4_{s_3}(x)$. Note that when magnifying $f^2_{s_2}(x)$ we also flipped the graph horizontally and vertically. If we did this for all values s_k and the corresponding compositions of f_{s_k}, we would obtain a sequence of close-ups settling down on the graph of a new function f_∞. When comparing the graphs shown in 11.31, it is already hard to see differences. The fact is that this new function is as universal as the constant $\delta = 4.669\ldots$

A Rescaling Operator

Let us describe the magnification process a bit more carefully and from a slightly different point of view: how do we get from f_{s_1} to the 'close-up-graph' of $f^2_{s_2}(x)$? This is shown in figure 11.32.

We can formalize this process using an operation on functions like f_a. This operator would turn the graph of f_a into the graph of $\Phi(f_a)$. In this respect it is like the Hutchinson operator (see chapter 5), which also is not a plain function of numbers but operated on images. But there is another similarity between these operators. When iterated, a particular Hutchinson operator H leads to a final image I which does not depend on the starting image. Rather, it is determined by the operator H alone, namely, as its fixed point $I = H(I)$. For Φ there also exists such an invariance property and this is:

$$f_\infty = \Phi(f_\infty).$$

Furthermore, we could start with graphs of other functions (like $g_a(x) = ax^2 \sin \pi x$), the iteration of Φ would lead to the universal function f_∞

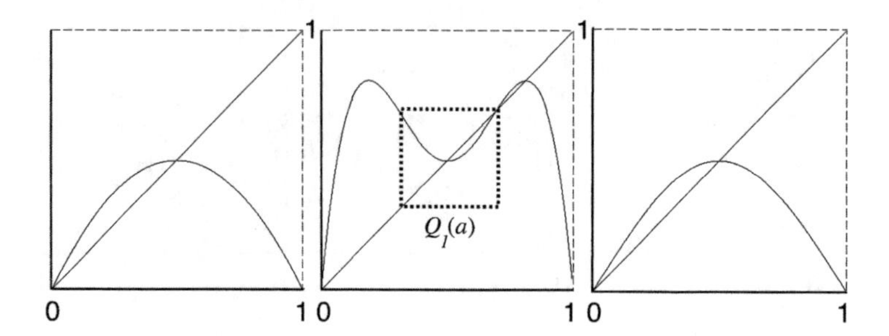

Figure 11.32 : We start with the graph of f_{s_1} (left). In the next step (center) we go to f_{s_2} and consider the composite of the function (i.e., we form $f_{s_2}(f_{s_2}(x))$). The square outlined is enlarged on the right using the magnification factor $s_2/(s_2 - 2)$ and additionally reflected both horizontally and vertically.

again. And δ is also related to the operator. It is a so called eigenvalue and controls the way the iteration approaches the fixed point.

In other words, the secret behind the universality of f_∞ and δ (and therefore also behind the self-similarity of the final-state diagram at the Feigenbaum point) is this operator Φ. It captures the essentials of this process. But we have to skip the mathematical details since the required techniques are definitely beyond the scope of this book.[10]

Again this means that if we had investigated for example the iteration $x_{n+1} = ax_n^2 \sin(\pi x_n)$ instead of the quadratic iteration, then all our observations made so far would have had exact analogies, except that the crucial parameters of a, b_k and s_k would have different values. But the constant δ and the function f_∞ would be the same. To show this kind of universality would require some very deep methods from current mathematics and would need the ingenuity of several of the best living mathematicians and physicists aided by extensive computer studies.

Let us return to our initial observation of self-similarity in the final-state diagram. We have seen that the scaling which is necessary to make the close-ups (as shown in figure 11.3) for the direction of the parameter a is essentially the universal constant $\delta = 4.669\ldots$ Now we can also point out what lies behind the scaling in the vertical direction. In fact, these are exactly the same scaling factors which we used to make the close-ups of $f_{s_2}^2$ or $f_{s_3}^4$ to be able to compare f_{s_1} and $f_{s_2}^2$, $f_{s_2}^2$ and $f_{s_3}^4$, etc.

[10]See P. Collet and J.-P. Eckmann, *Iterated Maps on the Interval as Dynamical Systems,* Birkhäuser, Boston, 1980, and M. Feigenbaum, *Universal behavior in nonlinear systems,* Physica 7D (1983) 16–39, also in: D. Campbell and H. Rose (eds.), *Order in Chaos* North-Holland, Amsterdam, 1983.

11.4 From Chaos to Order: A Mirror Image

Let us now turn to the second part of the final-state diagram, the parameter range between the Feigenbaum point s_∞ and the value $a = 4$ (see figure 11.33). We call this part the chaotic mirror image of the period-doubling tree. Indeed, there are features of the period-doubling, though in reversed order, but that is not all. Where chaos reigns, everything becomes infinitely more complicated. While for the first part of the final-state diagram we can predict for each parameter a exactly what the dynamics are, here we have great difficulty even distinguishing stable periodic from chaotic behavior.

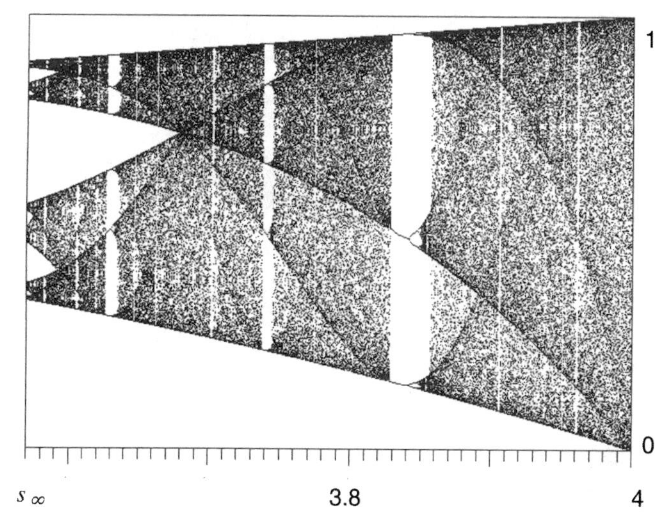

Final-State Diagram, Part Two

The second part of the final-state diagram of the quadratic iterator.

Figure 11.33

Band Splitting

We have already investigated the situation for the parameter value $a = 4$ in great detail in chapter 10. This is the parameter where the graph of $f_a(x) = ax(1-x)$ spans the unit square and we can observe chaos in the whole unit interval. In the final-state diagram this is represented by the random-looking distribution of dots which vertically span the range between 0 and 1. This kind of chaotic dynamics is not present for all parameters in the second part of the diagram. The chaos seems to be interrupted by windows of order where the final state again collapses to only a few points, corresponding to attractive periodic orbits. Furthermore, there seems to be an underlying structure of bands resulting from points not being uniformly distributed in each vertical line. Points seem to condense at certain lines which border bands that encapsulate the chaotic dynamics. For $a = 4$ there is only one band spanning the whole unit interval. As a decreases this band slowly narrows. Then at the parameter labeled with m_1, it splits into two parts; and at $a = m_2$ the two split into four parts (see figure 11.34).

Merging Points

Now we magnify the diagram between the parameters s_∞ and m_1 at the window shown in figure 11.34. There are more band splitting points. In fact, there is an infinite, decreasing sequence of parameter values $m_1, m_2, m_3, \ldots$

Band Splitting

Magnifying the diagram at the ' dow framed in black next to Feigenbaum point reveals fur band splitting parameters.

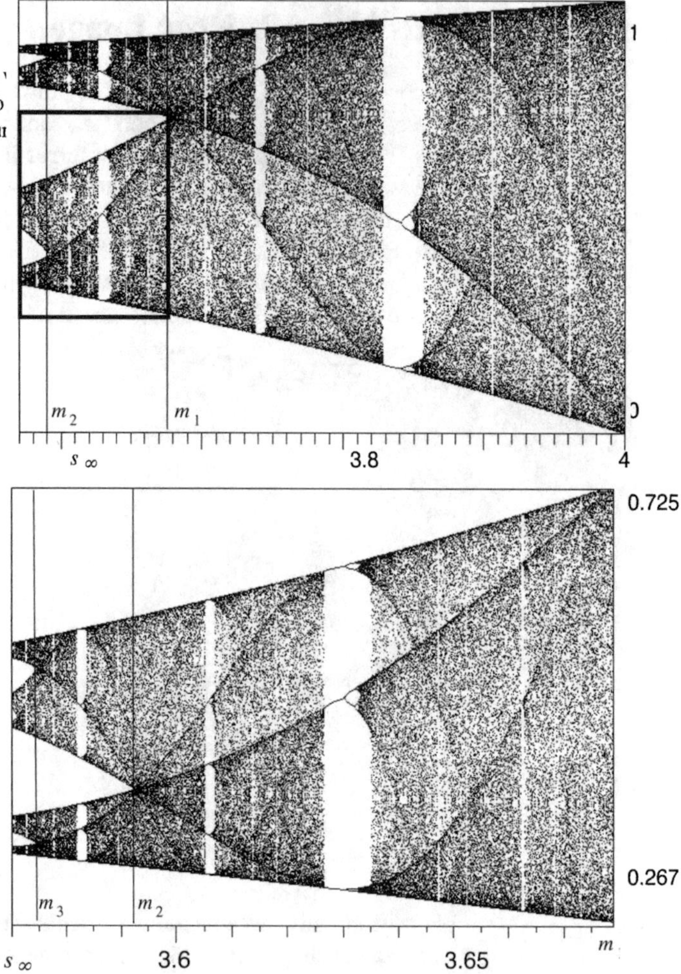

Figure 11.34

at which one observes the splitting into $2, 4, 8, \ldots$ (in general 2^k) bands. This can be interpreted as another consequence of the self-similarity of the final-state diagram at the Feigenbaum point. Thus, this sequence leads exactly to the limit $m_\infty = s_\infty$ (i.e., to the Feigenbaum point). Moreover, we can guess what the result of the following experiment by Großmann and Thomae should be.[11] They tried to find out whether the distances of the band merging points $d_k = m_{k+1} - m_k$ obey a particular growth law. They guessed that this would be similar to what had been found for the sequences s_k (the parameters of super attractiveness) and b_k (the parameters of the period-doubling bifurcations).

[11] See S. Großmann and S. Thomae, *Invariant distributions and stationary correlation functions of one-dimensional discrete processes* Zeitschrift für Naturforschung 32 (1977) 1353–1363.

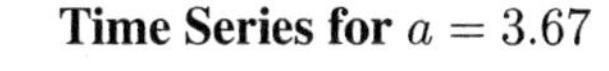

Time Series for $a = 3.67$

Time series for $a = 3.67$ (just below m_1). The orbit oscillates from step to step between the two marked bands (top). Within each band the dynamics look chaotic. In the lower two images f_a^2 is shown.

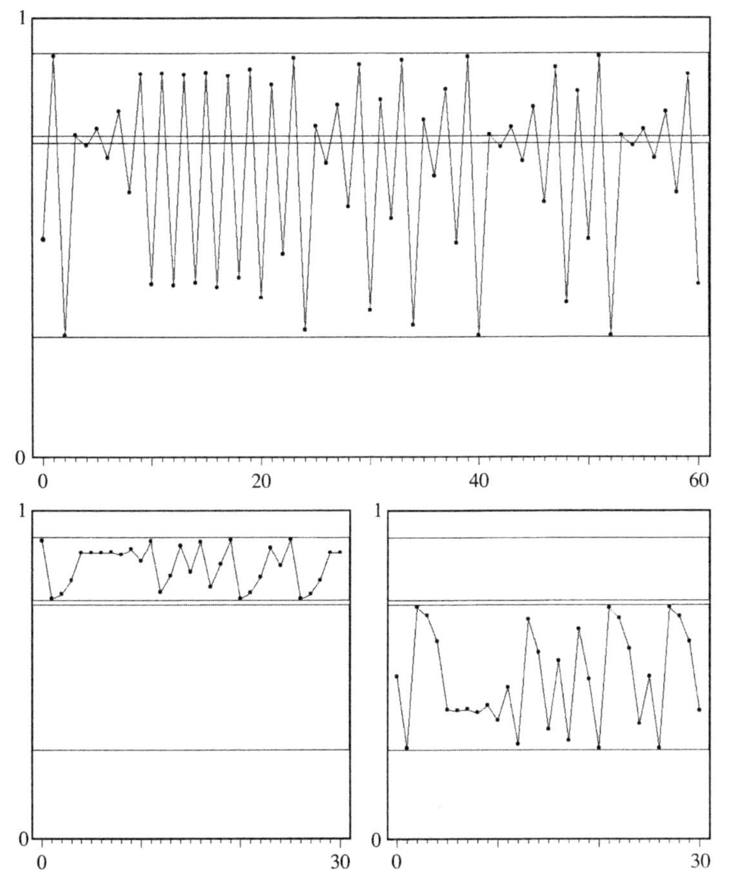

Figure 11.35

And indeed, they were able to confirm that the ratio d_k/d_{k+1} converges to the universal constant $\delta = 4.669\ldots$ as the number k increases.

Let us explore these bands a little bit further. What kind of change lies behind the splitting (or merging) of bands? Figure 11.35 (top) shows a typical time series of f_a for the parameter $a = 3.67$, which is slightly below $a = m_1$. It becomes immediately apparent what the two bands mean. Although the dynamics behave chaotically, it oscillates from step to step back and forth between two distinct bands. In other words, when we look at the dynamics of f_a^2 we see points only moving chaotically either in the upper or in the lower band. This is shown in the lower two images of figure 11.35. In summary, the first band splitting is also a kind of period-doubling bifurcation.

You will recall our histogram experiment from chapter 10. We visualized the mixing property of the chaos parabola by measuring how frequently the iteration of an orbit visits the different points of the unit interval (we had broken the unit interval into equally spaced sub-intervals, then we counted how often

Mixing Histograms

The histogram for $a = m_1 = 3.6785$ (bottom) looks like two scaled down copies of the histogram for $a = 4$ (top) fitted together.

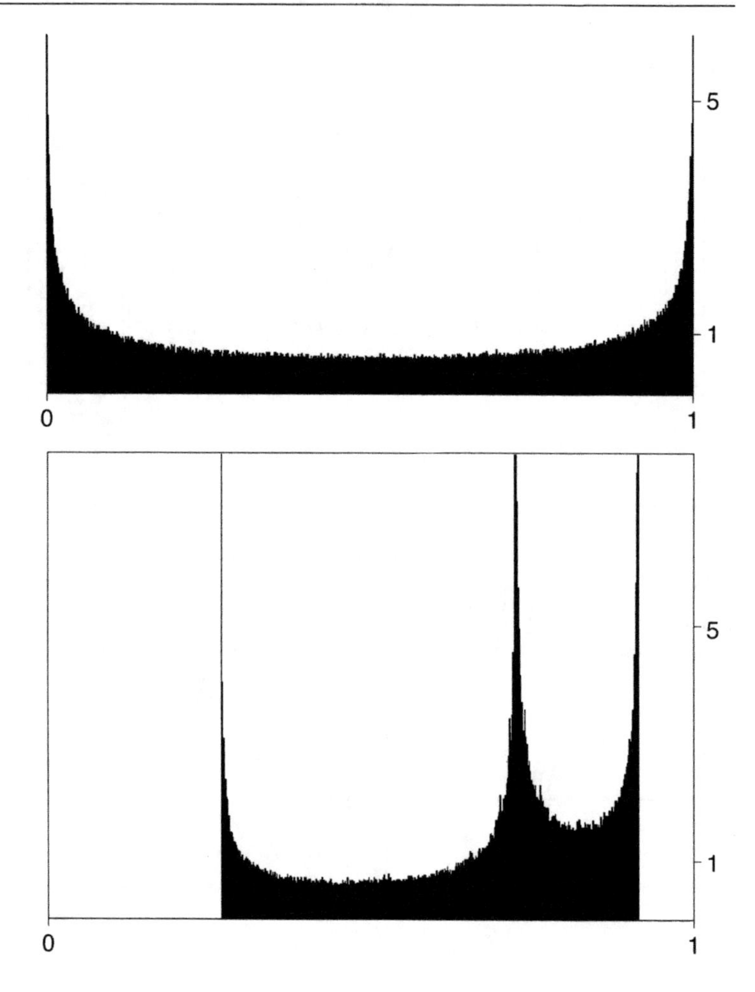

Figure 11.36

a given orbit had hit each of these intervals). Figure 11.36 shows the same experiment for $a = m_1$ and compares it with the histogram for $a = 4$. The range of the two bands at $a = m_1$ is clearly visible. Indeed, this looks a bit like two differently scaled copies of the diagram for $a = 4$ fitted together.

Comparing f_a^2

Let us again compare f_a and f_a^2 using graphical iteration. Figure 11.38 shows the result for $a = 4$ (top) and for $a = m_1$ (bottom). Let us call the parabola in the upper left graph ($4x(1 - x)$) a *generic parabola*. Generic parabolas are characterized by the fact that their graph precisely fits into a square which has one of its diagonals on the bisector of the x-y-coordinate system. Note that for $a = m_1$ we can also find a generic parabola in the graph of f_a^2 (lower right of figure 11.38). However, this is not quite correct, because f_a^2 is not really a parabola, but rather a fourth-degree polynomial having a graph that only looks parabolic in the outlined region enclosed by a dashed square. Once the iteration of $f_{m_1}^2(x)$ has led into this region it is

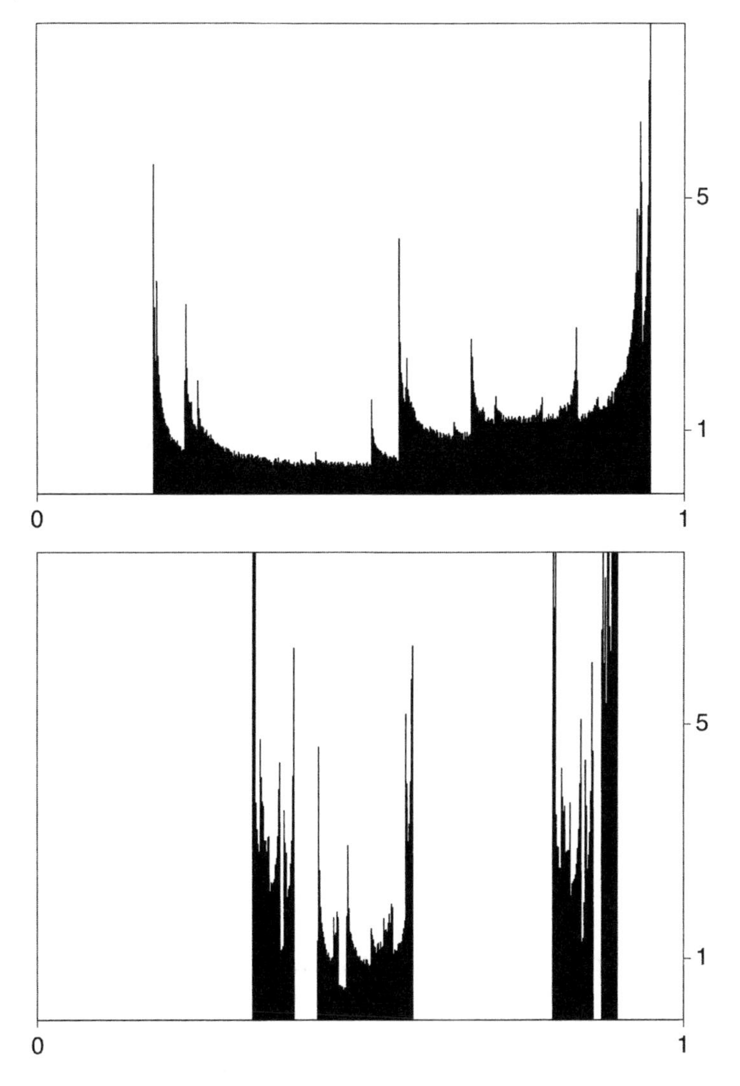

Another Histogram

Mixing histograms for a = 3.7 (above) which is a bit above m_1 and for $a = 3.585$ (below) which is smaller than m_2.

Figure 11.37

trapped, and we should expect to see chaotic behavior which spans the interval $(1/a, (a-1)/a)$. This corresponds to the lower band visible in the final-state diagram right at $a = m_1$. The upper band corresponds to the part of the graph of f_a^2 enclosed by the small dotted square. Also in this region the iteration is trapped and spans the interval $((a-1)/a, a/4)$.

Now you guess what the situation for all other parameters $a = m_k$ will be. In all these cases we find a generic parabola (i.e., in the graph of $f_{m_2}^4(x)$, in the graph of $f_{m_3}^8(x)$, etc.). Of course this explains only what we see at the special parameter values $a = m_k$. On the other hand it seems possible to trace these bands also in between. Somehow they shine through the whole second

f_a and f_a^2

Comparing f_a and f_a^2 at $a = 4$ and at $a = m_1$. f_a^2 forms two small versions of graphs similar to the parabola for $a = 4$ (enclosed by the dashed square).

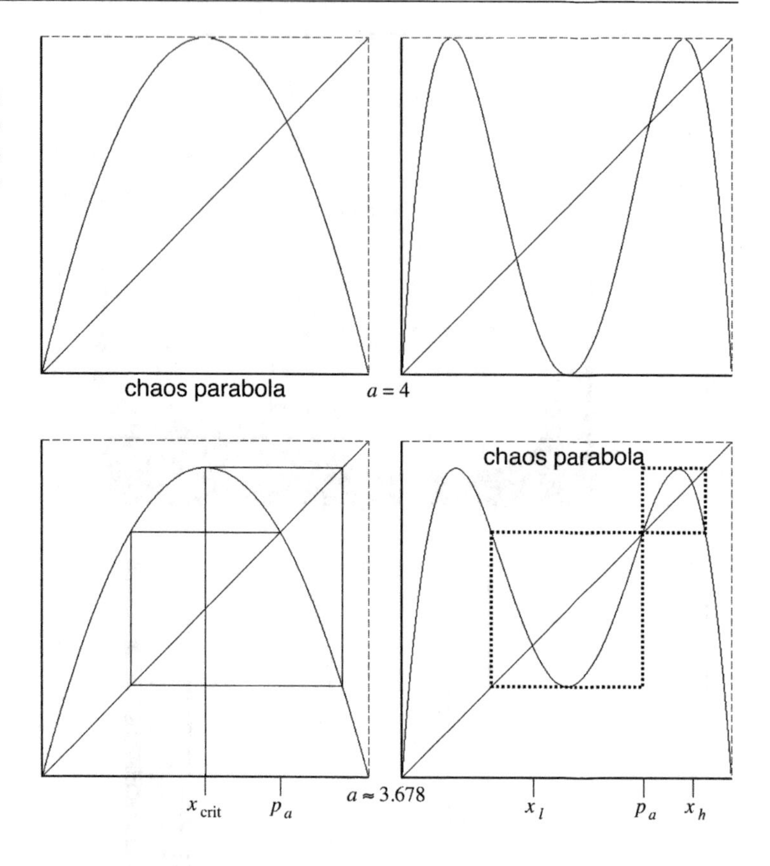

Figure 11.38

part of the final-state diagram; there is a mechanism behind this observation.

Critical Value Lines

In figure 11.39 we show the graphical iteration of a few initial values which we have chosen to be equally spaced near 0.5. For each initial value we have performed three iterations and drawn the corresponding outcome on the right side of the graph. First we note, that the iteration never leaves the outlined square (i.e., the points of the final-state diagram have to lie within the interval between the critical value $v_a = f_a(0.5)$ and $f_a(v_a)$). Furthermore, we observe that the values of the iteration condense a bit at these points. This happens because the parabola has its vertex at 0.5 which squeezes nearby orbits together.

In the histogram, we thus expect a spike at $v_a = f_a(0.5)$. Moreover, there should be another spike at the next iterate, $f_a(v_a) = f_a^2(0.5)$, and also at the following one, $f_a^2(v_a)$, and so on. For $a = 4$, however, $v_a = 1$ and all further iterates are 0. Thus, it is reasonable to expect only the two spikes at 0 and 1. For $a = m_1$, on the other hand, we have $f_a^2(v_a) = p_a = (a-1)/a$, the fixed point of f_a, and all further iterates are the same. Therefore, there should be three spikes, at v_a, $f_a(v_a)$, and $f_a^2(v_a) = p_a$. This, in fact, is just what figure 11.36 shows. In summary, this leads to the conjecture that what we see

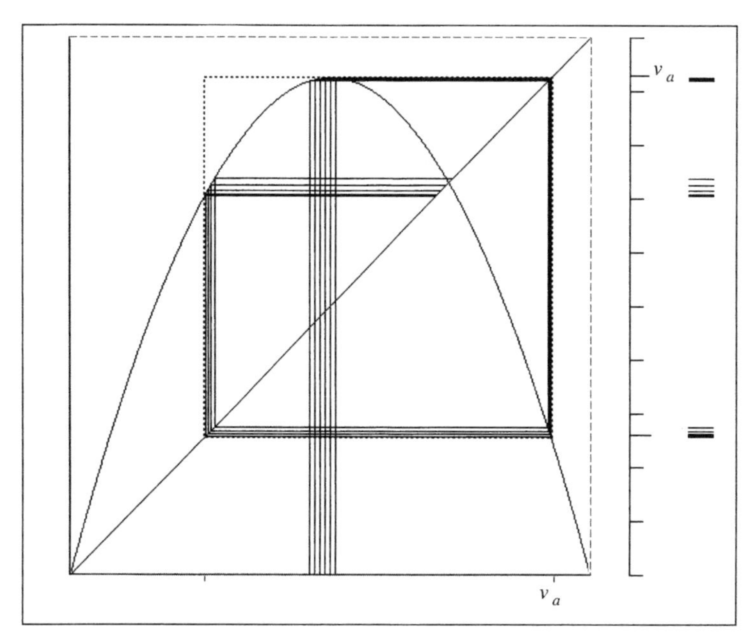

Initial Values Near 0.5

Graphical iteration of some equally spaced initial values near 0.5. The first three iterates condense at $v_a = f_a(0.5)$, $f_a(v_a)$ and $f_a^2(v_a)$

Figure 11.39

shining through as lines of condensation in figures 11.33 and 11.34 could be the trace of the iterates of the critical value v_a.

Figure 11.40 shows the experiment which confirms this conjecture. We compute the first eight iterates of 0.5 for the parameter range from s_∞ to 4. The upper plot shows the first four iterates (i.e., v_a to $f_a^3(v_a)$). These lines apparently correspond to the main bands (or stripes) which shine through the final-state diagram in figures 11.33 and 11.34. The lower plot shows all eight iterates exhibiting more of the relation to finer band structures.

Although these critical lines (i.e., the iteration $f_a^k(v_a)$ of the critical value v_a) explain our perception of a band structure in the final-state diagrams, this does not mean that the complete lines as shown are part of the final state. Already for v_a this is not true, as demonstrated in figure 11.41, which shows a close-up of the final-state diagram next to the line of critical values v_a. The final states are bound by this line, but we undoubtedly can see that from a certain parameter value[12] (about $a \approx 3.82843$) the final states consist of a stable attracting periodic cycle, of which only one point is shown in the blow-up.

Periodic Windows

In fact, this close-up shows a small part of one of the white windows which interrupt the chaotic region of the final-state diagram.[13] There are an infinite number of such windows, which all correspond to stable periodic cycles. This one between $a \approx 3.828$ and $a \approx 3.857$ is the most prominent one; it is the so

[12]More precisely, the parameter is $a = 1 + \sqrt{8} \approx 3.828427125$ (compare section 11.5).

[13]The name 'window' was used first in R. M. May's remarkable paper, *Simple mathematical models with very complicated dynamics,* Nature 261 (1976) 459–467.

Critical Lines

Iterates of the critical value v_a for the second part of the final-state diagram: (top) v_a, $f_a(v_a)$, $f_a^2(v_a)$ and $f_a^3(v_a)$, (bottom) v_a to $f_a^7(v_a)$.

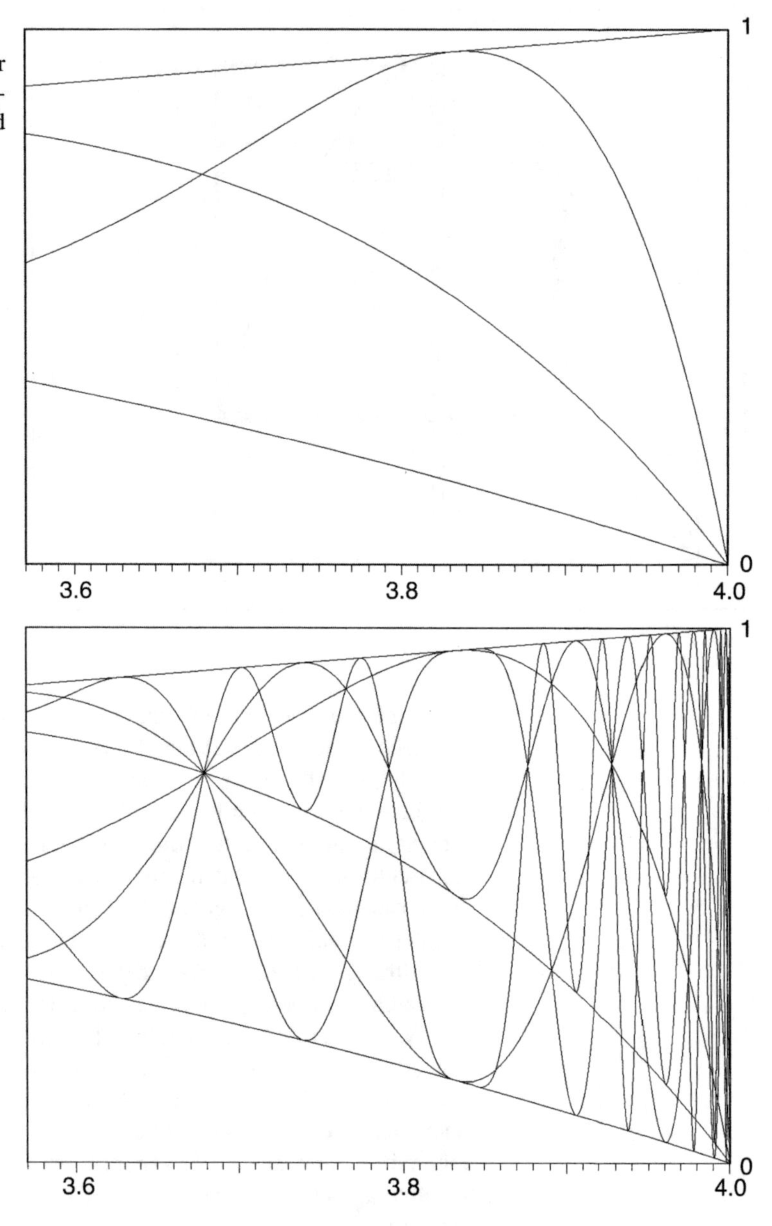

Figure 11.40

called period-three window. In figure 11.42 we have indicated not only this window but also the windows of period 5, period 7 and the window of period 6. But let us first examine the period-3 window a bit closer. The bottom part of the figure shows two successive close-ups of the part which is marked by the black frame.

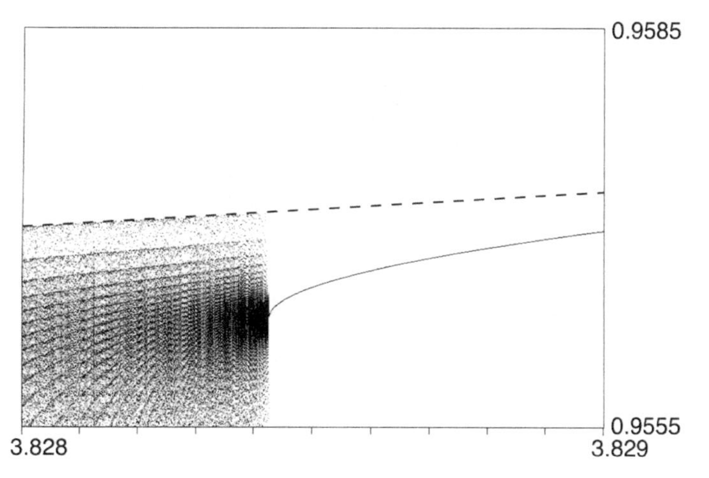

Start of the Period-3 Window

Close-up of the final-state diagram between $a = 3.828$ and $a = 3.829$ next to the line of critical values v_a which is shown in dashed patterns.

Figure 11.41

Self-Similarity Again

Again we discover self-similarity. We see smaller and smaller copies of the whole final-state diagram. And indeed we can find the complete scenario of period-doubling, chaos and band splitting again, however, on a much smaller scale. And again the mechanisms behind this are the same as before. There is only one important difference; instead of $f_a(x)$, here everything is based on $f_a^3(x)$. The period-doubling begins when the three fixed points of $f_a^3(x)$ lose their stability and six new fixed points of $f_a^6(x)$ are born (i.e., for f_a we have a 6-cycle). As a increases further, each of the stable fixed points of f_a^6 will undergo a period-doubling bifurcation (i.e., for f_a we will obtain attracting cycles of length $3 \cdot 2^2$), and so on. The relative length of the intervals for which these stable cycles exist will be governed once more by the universal number $\delta = 4.669\ldots$ At the end of this period-doubling scenario, near $a = 3.8415\ldots$, there will again be a transition to chaotic behavior very much like that at the Feigenbaum point s_∞.

Let us take a look at some graphs of f_a^3. In figure 11.43 (left) we have drawn the super attractive case. At the center again we observe a segment which looks like a small parabola. Indeed, the changes of this small part are responsible for the complete scenario of period-doubling which ends at $a \approx 3.857$ in fully developed chaos as it is shown in the center part of figure 11.42. The corresponding graph of f_a^3 is shown on the right-hand side of figure 11.43. And indeed, at the center a generic parabola is visible.

If we magnified any of the other periodic windows, we would indeed make exactly the same finding; but everything would be on an even smaller scale. In fact, between the period-3 window and the band merging point m_1 there are an infinite number of windows for all odd integers, i.e., for $3, 5, 7, 9, 11, \ldots$, which can be found in reversed order (i.e., 3 is right of 5, etc.). But as the period increases the size of these windows rapidly decreases and the period-9 window is already hard to find.

Windows of Periodicity

Two successive close-ups of the period-3 window. It starts at $a = 1 + \sqrt{8} \approx 3.8284$ and extends up to $a \approx 3.857$.

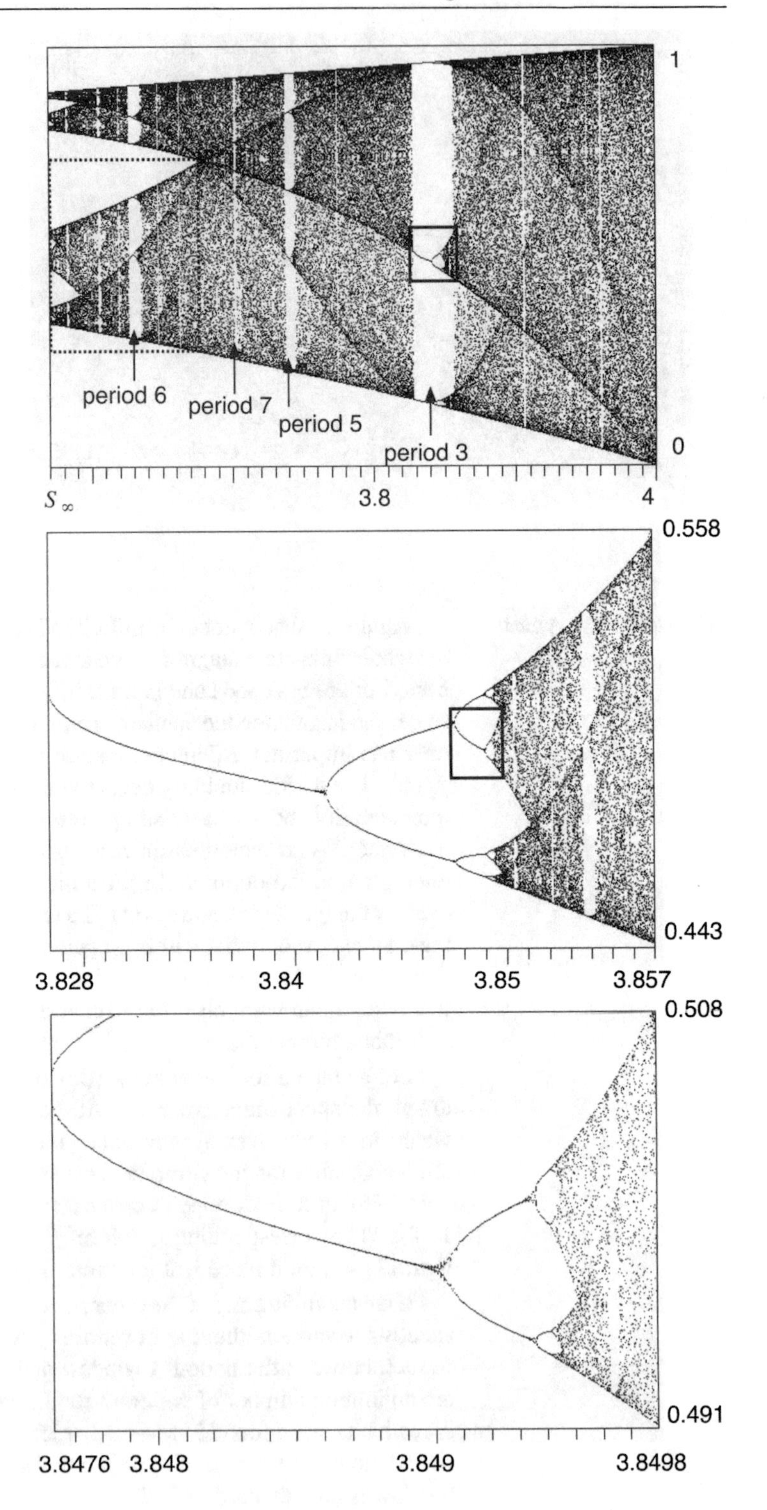

Figure 11.42

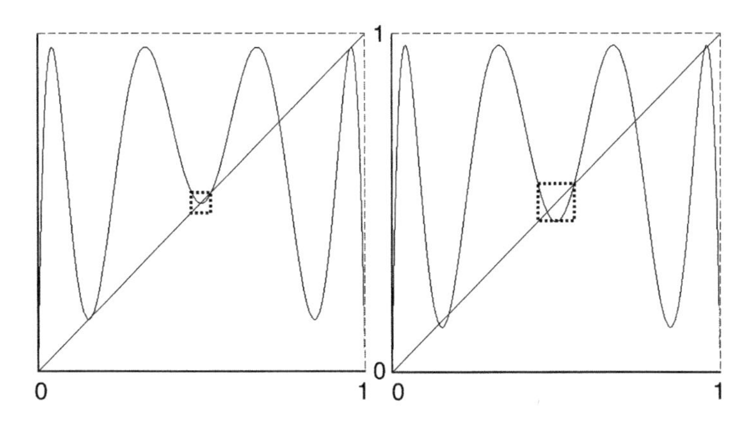

The Third Iterate f_a^3

The graphs of the third iterate f_a^3 of f_a. Left: super attractive case; right: fully developed chaos.

Figure 11.43

The Charkovsky Sequence

Are these all the periodic windows? Certainly not! You have heard of self-similarity — haven't you? Look at the left part of the diagram in figure 11.42 (top) which we have enclosed by a dashed rectangle. Here we find everything once again, but now with a doubled period. In other words, it starts with a period-6 window, then we find a period-10 window, etc. Expressed generally, in this step we find windows of period $2 \cdot k$, for all odd integers $k \geq 3$. In summary, self-similarity reveals a sequence of windows with period

$3, 5, 7, 9, 11, 13, \ldots$	(all odd integers)
$2 \cdot 3, 2 \cdot 5, 2 \cdot 7, 2 \cdot 11, \ldots$	(all $2 \cdot k$, k odd)
$4 \cdot 3, 4 \cdot 5, 4 \cdot 7, 4 \cdot 11, \ldots$	(all $2^2 \cdot k$, k odd)
$\ldots$	(all $2^n \cdot k$, k odd)
$\ldots 2^4, 2^3, 2^2, 2, 1$	(all powers of 2)

exactly in this order from right to left in the final-state diagram. Actually the last row of this sequence is special. It represents the period-doubling at the beginning of the diagram (we could say that this is the last periodic window). This strange sequence of numbers is named after the Russian mathematician Alexander N. Charkovsky and is the heart of some remarkable results on periodic points of feedback systems. But let us return to the question: are these all the periodic windows? You probably already suspect that this is not everything. Indeed, it is known that theoretically in any parameter interval we can find a periodic window (i.e., a stable periodic cycle).

Charkovsky Sequence and Periodic Points

In 1964 Charkovsky introduced his famous sequence

$$\begin{aligned}
&3 \rhd 5 \rhd 7 \rhd 9 \rhd \cdots \rhd \\
&2 \cdot 3 \rhd 2 \cdot 5 \rhd 2 \cdot 7 \rhd 2 \cdot 9 \rhd \cdots \rhd \\
&4 \cdot 3 \rhd 4 \cdot 5 \rhd 4 \cdot 7 \rhd 4 \cdot 9 \rhd \cdots \rhd \\
&\cdots \rhd 2^5 \rhd 2^4 \rhd 2^3 \rhd 2^2 \rhd 2 \rhd 1.
\end{aligned}$$

It is ordered by the symbol $\rhd$ (i.e., a comes before b is written as

$a \rhd b$). He was able to prove some very remarkable results such as the following.

Assume that f transforms an interval I onto itself and has a point of period k (i.e., there is x in I such that $f^k(x) = x$). Then f has points of period m for every m such that $k \rhd m$.

The consequences of this amazing result are manifold: if f has a point of period 3, then f has periodic points of any period. If f has a point of period $k \neq 2^n$, then f has infinitely many periodic points.

11.5 Intermittency and Crises: The Backdoors to Chaos

Periodic windows interrupt the chaotic region in an extreme way. At the beginning of such a window there is a sudden and dramatic change in the long-term behavior of the quadratic iterator. For example, let us look at the neighborhood of the parameter

$$a = w_3 = 1 + \sqrt{8} \approx 3.82843$$

at the start of the period-3 window. For parameters a slightly above w_3, a perfectly stable cycle of period 3 exists. On the other hand, as soon as we decrease the parameter a bit below the value w_3, we immediately stumble into chaos spanning the whole interval $(f_a(v_a), v_a)$.[14] This is quite a different route to chaos compared to the orderly path of period-doubling bifurcations.

The Vexatious Path to Chaos

But the situation is even more troublesome. The chaos for parameters below w_3 reveals itself only in the long term. The short-term behavior wrongly suggests stable periodic orbits. This is demonstrated in figure 11.44, which shows the time series of $x_0 = 0.5$ for $a = 3.82812 < w_3$. For more than 50 iterations the orbit looks perfectly stable and predictable, but then it rapidly falls into chaotic oscillation.

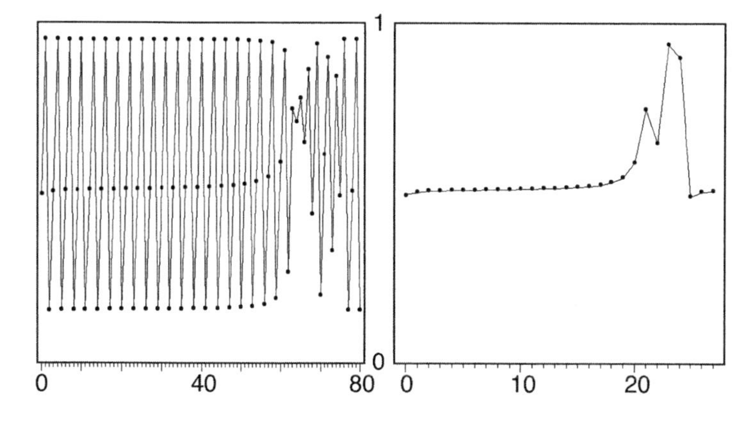

Intermittent Time Series

Time series of $x_0 = 0.5$ for $a = 3.82812$: (left) iteration of f_a, (right) iteration of f_a^3. In the right-hand graph the time axis is scaled up by 3 to allow a better comparison.

Figure 11.44

This experiment gives us a first impression of what is called *intermittency* — a new type of iterative behavior. What are the ingredients of intermittency? The first one is a so-called *tangent bifurcation of fixed points*. Let us explain by means of an example which is already familiar from section 11.2. It is the iterator

$$g_a(x) = ax^2 \sin \pi x$$

which we used to illustrate the universality of δ. Now please take another look at figure 11.23, the final-state diagram for g_a. Note that the period-doubling

[14] Recall that v_a is the critical value $v_a = f_a(0.5)$.

Graphical Iteration for $g_a(x)$

This series of graphs shows the tangent bifurcation for $g_a(x) = ax^2 \sin \pi x$ near $a^* \approx 1.7264$. For parameters below this value (top left and right) all orbits converge to the origin. Right at the bifurcation (lower left) the bisector becomes tangential to the graph of g_a. There is a new fixed point which attracts orbits from the right and repels them on the left. For parameters beyond a^* there are two new fixed points, one is stable and the other is unstable.

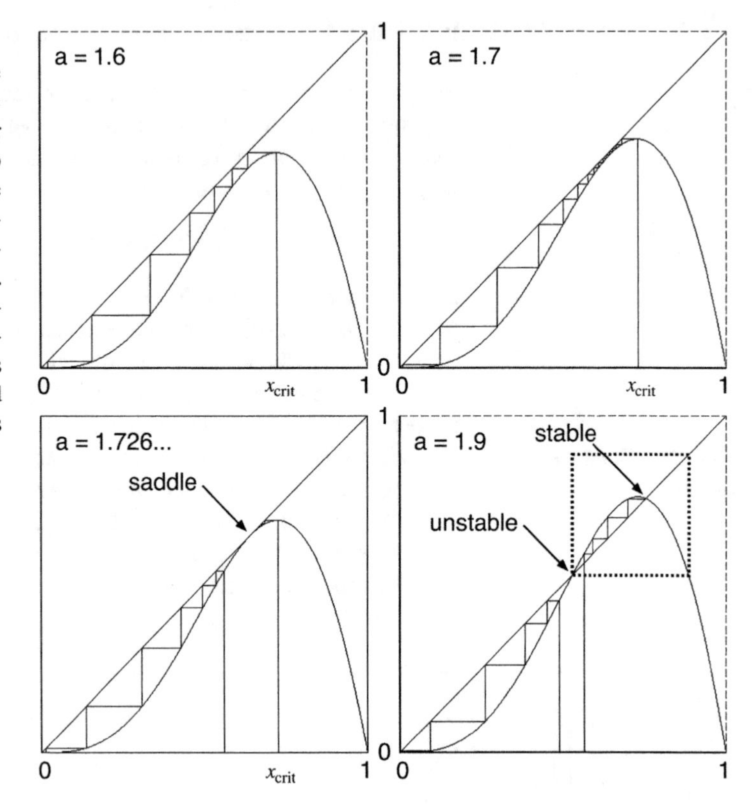

Figure 11.45

bifurcation tree seems to start out of nothing at about $a \approx 1.7264$. But what is going on for parameters less than 1.7264? Well, let us take a look at the corresponding graphical iteration.

First, figure 11.45 shows the situation for $a = 1.6$ (top left). In this case, the iteration of all initial values will eventually lead to the attractive fixed point 0. Now we increase the parameter a and observe the graph of g_a getting closer and closer to the bisector (top right). Finally, at

$$a = a^* \approx 1.7264289398722975$$

(bottom left) the bisector touches the graph tangentially at

$$x = x^* \approx 0.6457736765434055.$$

If we increase the parameter even further (bottom right), we see that the bisector intersects the graph near x^*. There are two new fixed points, a stable and an unstable one.

Tangent Bifurcation

The new stable fixed point corresponds to the stem of the period-doubling scenario which we have found in the final-state diagram of g_a. When the parameter a increases further, this stable fixed point will become unstable at

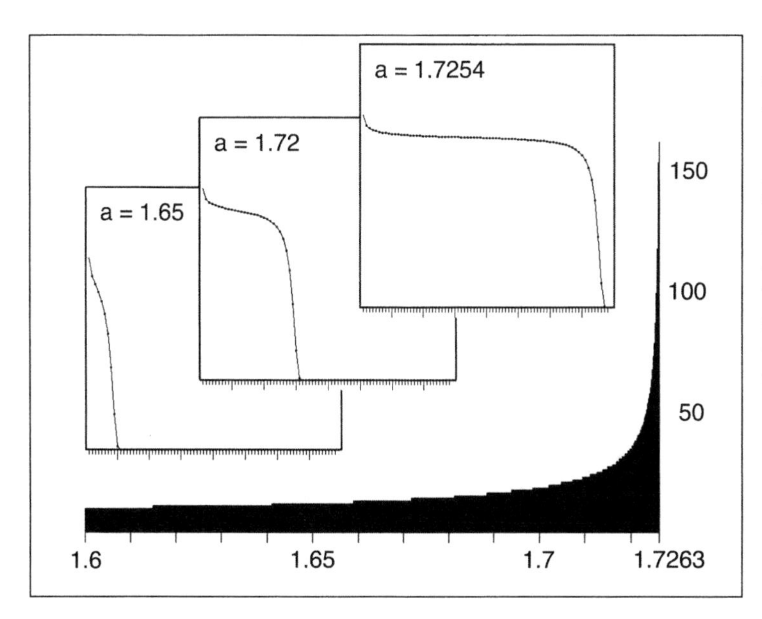

Evaluating Time Series

The number of iterates required to bring x_k into a ε-neighborhood of 0 ($\varepsilon = 10^{-8}$), when $x_0 = x_{\text{crit}}$, for $x_{n+1} = g_a(x_n) = ax_n^2 \sin \pi x_n$. The inserts show the corresponding time series for $a = 1.65$, $a = 1.71$ and $a = 1.7254$ illustrating how the iteration is trapped for many iterations near the point x^*, where the graph of g_a touches the bisector when $a^* \approx 1.7264$

Figure 11.46

a certain parameter, giving birth to a stable cycle of period 2. Later stable cycles of period 4 will appear, and so on in the familiar fashion. The square outlined in figure 11.45 (bottom right) encloses a part of the graph which again looks similar to a parabola. Indeed, this part is responsible for the familiar appearance of the final-state diagram. But let us return to the two fixed points. If we reduce the parameter to a^* these two fixed points join into a single point at x^*, called a *saddle point*. If we decrease a even further this fixed point also vanishes. This mechanism is called *tangent bifurcation*.[15]

Note that the point x^* is neither attracting nor repelling. Rather, it attracts values which lie above x^* and repels for points below x^*. In particular if we start the iteration with the initial value $x_0 = x_{\text{crit}}$ (i.e., the x-value belonging to the maximum of g_a), the orbit converges to the fixed point x^* and not to 0 (see figure 11.45, bottom left). But as soon as we decrease the parameter only a little bit, the fixed point disappears. At first only an extremely narrow channel opens (top right). If we again trace the iteration of x_{crit} we observe that the orbit is trapped for quite a while between the graph of g_a and the bisector; it is tacking like a sailboat in a narrow channel. But eventually the orbit leaves this channel and reaches the fixed point 0.

Counting Tacks

If we decrease the parameter a further, the width of the channel becomes larger; and the orbit may pass through the channel with fewer tacks, thus approaching 0 much faster. This behavior is measured in the experiment shown in figure 11.46. We vary the parameter a from 1.6 to a^* and run iterations always starting with $x_0 = x_{\text{crit}} \approx 0.7286$. We count the number of iterations $N(a)$ required for the orbit to reach a small neighborhood of the

[15] In some texts the term *saddle-node bifurcation* is used instead.

Counting Intermittent Iterations

For a sequence of parameters $a_k, k = 0, \ldots, 13$, approaching the saddle-node bifurcation at a^* from below we compute the number of iterations $N(a_k)$ of the orbit starting at the critical point $x_{\text{crit}} \approx 0.7286$ until it enters a small neighborhood of 0. The last column records the product $N(a_k)\sqrt{a^* - a_k}$ revealing the square root power law governing the exploding numbers of iterations.

Table 11.47

k	$a_k = a^* - 10^{-k}$	$N(a_k)$	$N(a_k)\sqrt{a^* - a_k}$
0	0.7264289398723	4	4.00000
1	1.6264289398723	10	3.16228
2	1.7164289398723	26	2.60000
3	1.7254289398723	77	2.43496
4	1.7263289398723	238	2.38000
5	1.7264189398723	749	2.36855
6	1.7264279398723	2 363	2.36300
7	1.7264288398723	7 467	2.36127
8	1.7264289298723	23 608	2.36080
9	1.7264289388723	74 649	2.36061
10	1.7264289397723	236 056	2.36056
11	1.7264289398623	746 469	2.36054
12	1.7264289398713	2 360 537	2.36054
13	1.7264289398722	7 464 659	2.36053

attractive fixed point 0. As a gets close to a^* the number of iterations increases rapidly. Just how rapidly the number of iterations explode is explored in the experiment which is reported in table 11.47. The last column lists the product $N(a)\sqrt{a^* - a}$ and reveals that it converges to approximately 2.36 as the parameter a approaches a^*. From this we conclude that there is a square root power law which describes the increase of the iteration count,

$$N(a) \approx \frac{2.36}{\sqrt{a^* - a}}. \tag{11.8}$$

In other words, as we reduce the distance of a from a^* by a factor of 1/100, the number of iterations goes up by the factor of 10.

What do we learn from this experiment? If an iterator is close to a tangent bifurcation, it is nearly impossible in a numerical study to distinguish between transient and long term behavior. The final state becomes apparent only after many iterations, perhaps many more than the time frame of the experiment and the computer allow.

Let us now return to our quadratic iterator. Figure 11.48 shows the graph of f_a^3 for

$$a = w_3 = 1 + \sqrt{8} \approx 3.828427125$$

and for the slightly smaller parameter value $a = 3.81$. Indeed, we observe a tangent bifurcation at $a = w_3$, with a saddle point at x^*. Note that this point lies slightly to the right of 1/2. Note also that the graph touches the bisector at two more points, namely, $f_a(x^*)$ and $f_a^2(x^*)$, which makes up the corresponding 3-cycle for f_a. In figure 11.43 we have already seen what happens if we increase the parameter a even further: the saddle splits into a stable and an unstable fixed point, and finally we could identify a small copy of the generic parabola. In other words, the tangent bifurcation at $a = w_3$

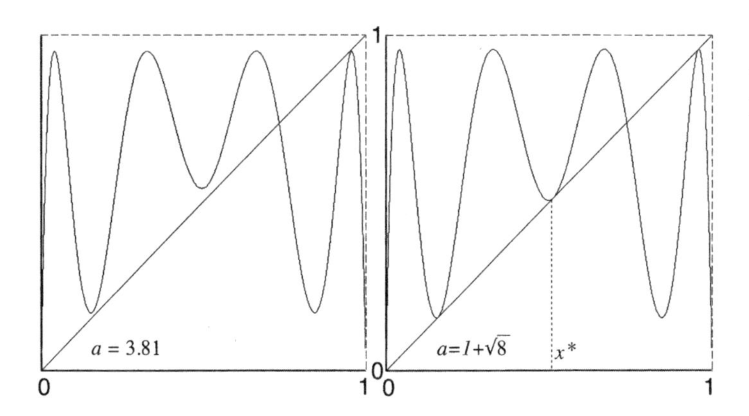

Tangent Bifurcation

Tangent bifurcation in f_a^3, at $a = 1 + \sqrt{8}$.

Figure 11.48

gives birth to the period-doubling scenario, chaos, and band-merging, which we observe in the period-3 window.

Intermittency

To investigate the behavior of the quadratic iterator for parameters slightly below the tangent bifurcation, we compute a long time series (see figure 11.49). The experiment shows recurrent long phases of almost resting behavior for f_a^3 corresponding to almost perfect cyclic behavior of period 3 for f_a. These phases of 'stability' — also called *laminar phases* — alternate with erratic and chaotic behavior. This dramatic interplay between bursts of chaos and almost periodic behavior is called *intermittency*.

The orderly parts of the time series correspond to phases of the iteration close to x^*, where the orbit is cruising against the wind in a narrow channel. But what happens after the orbit has escaped from these narrows? And what is the mechanism which always forces the iteration back to the orderly phase? This brings us to the second ingredient of intermittency: *homoclinic points*.

Homoclinic Points

Homoclinic points were discovered by the great French mathematician Henry Poincaré in his famous studies of the stability of the solar system at the end of last century. Poincaré already understood very well that homoclinic points generate chaos. However, it was not until the 1960's that Stephen Smale created the appropriate mathematical framework. Homoclinic points can occur in connection with saddle points. We have seen that saddles have unstable and stable parts. Vaguely speaking a homoclinic point is a point which belongs to both parts.

Let us return to our concrete example, the saddle at x^* of f_a^3 for $a = w_3$. In this case a homoclinic point is defined to be any point x_0 on the repelling side of the saddle, which, when iterated, is eventually transformed onto x^* or which gets back to the stable side of the saddle. Formally, either there is a number k such that $x_k = x^*$, or $x_k \to x^*$ as $k \to \infty$. Figure 11.50 shows an example for each case. It is a fact that there exist an infinite number of homoclinic points, and that they densely fill a small interval on the unstable side of the saddle point.

Intermittency in Time Series

Time series of $x_0 = 0.5$ for $a = 3.82812$: (top) some 300 iterations of f_a, (bottom) the corresponding 100 iterations of f_a^3 (note: time axis is scaled up by 3 to allow a better comparison).

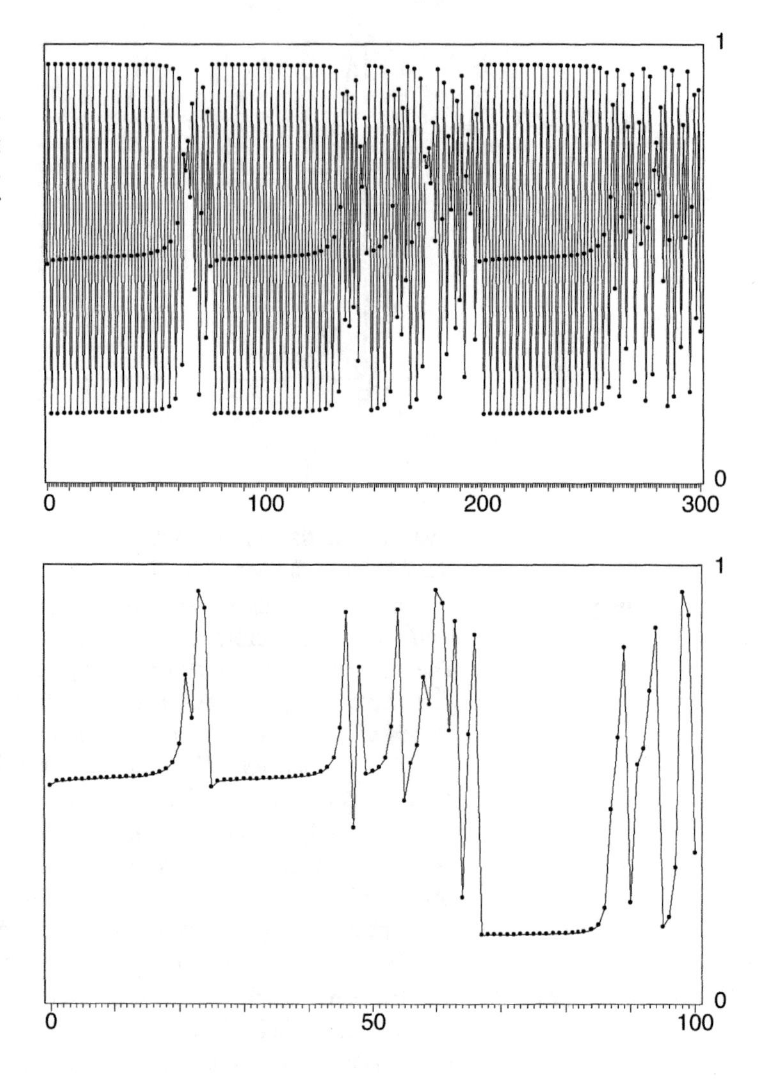

Figure 11.49

Now consider a homoclinic point as a starting value for the iteration at a slightly reduced parameter $a < w_3$. The saddle point is gone, but the orbit of the formerly homoclinic point does not change very much (at least not for an initial finite number of iterations). This orbit will lead from the formerly unstable side of x^* to the stable side. Now imagine the iteration of an initial value x_0 which has just traveled through the narrow channel. Unavoidably it will get close to a homoclinic point on the unstable side, which then guides it back to the stable side. In other words, we will see intermittent behavior. There is only one possibility for the iteration to escape from this behavior, namely, when the orbit hits one of the unstable fixed or periodic points (for example $p_a = (a-1)/a$). But the chance that this will happen is zero.

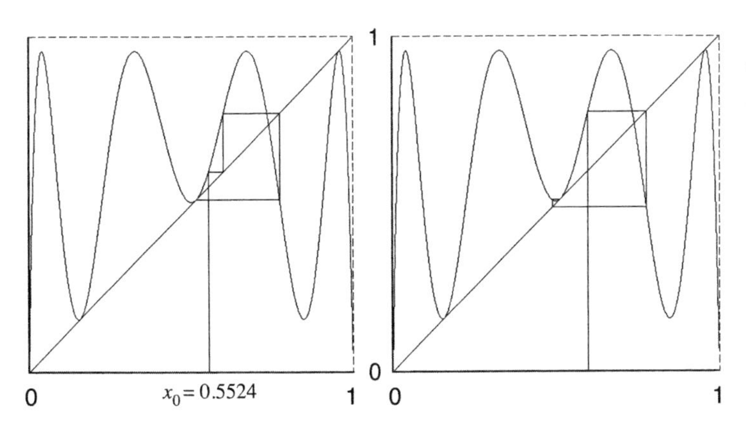

Homoclinic Points

Two homoclinic points for $a = 1 + \sqrt{8}$.

Figure 11.50

The Scaling Law of Intermittency

As the parameter a approaches w_3 from below the laminar phases become longer and longer. Eventually, the intermittent chaotic bursts disappear altogether and only the asymptotic periodic behavior remains. The average number of iterations that an orbit spends in the laminar phase thus tends to infinity. Moreover, it obeys a power law, which is of the same quality as the one which we found for the tangent bifurcation for g_a in eqn. (11.8); it is proportional to

$$\frac{1}{\sqrt{w_3 - a}}.$$

This relation can be derived analytically.[16]

Of course the vicinity of w_3 is only one example of the intermittency route to chaos. The same findings can be made at all other periodic windows.

Breakdown of Chaoticity via Crisis

The period-doubling scenario is the primary route to chaos. As a first alternative we have presented intermittency, a backdoor to chaos. There is one more important route which can be regarded as a close relative of intermittency: crisis. At the onset of a crisis — when a parameter of the system is varied appropriately — a chaotic region can be turned into a chaotic repeller. The typical phenomenon is that orbits, started in the previously chaotic region, behave chaotically only for a finite number of iterations. Eventually the chaoticity must break down and the long-term fate of the orbit is very definite and predictable. A crisis can be studied for the quadratic system $x \to ax(1-x)$ at the borderline parameter $a = 4$. All our previous work on the quadratic iterator was limited right at this point of crisis. But what happens for $a > 4$. A quick look at the corresponding graphical iteration (see figure 11.51) reveals that the unit square which housed the model of pure chaos at $a = 4$ breaks together. The parabola is no longer confined to the unit interval

[16]See Y. Pomeau and P. Manneville, *Intermittent transition to turbulence in dissipative dynamical systems,* Commun. Math. Phys. 74 (1980) 189–197. See also R. W. Leven, B.-P. Koch and B. Pompe, *Chaos in Dissipativen Systemen,* Vieweg, Braunschweig, 1989.

Escaping Orbits

For $a > 4$ in the quadratic iterator $x \to ax(1-x)$ most orbits escape from the unit interval through the gap at the vertex of the parabola.

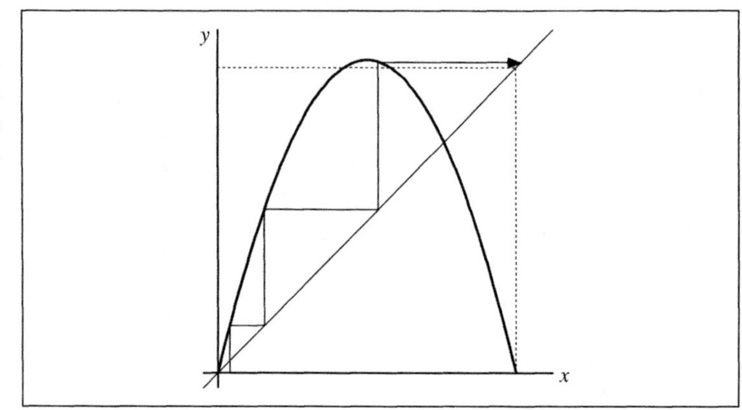

Figure 11.51

Example Orbit

Time series of an orbit started at $x_0 \approx 0.38$ for $a = 4.001$.

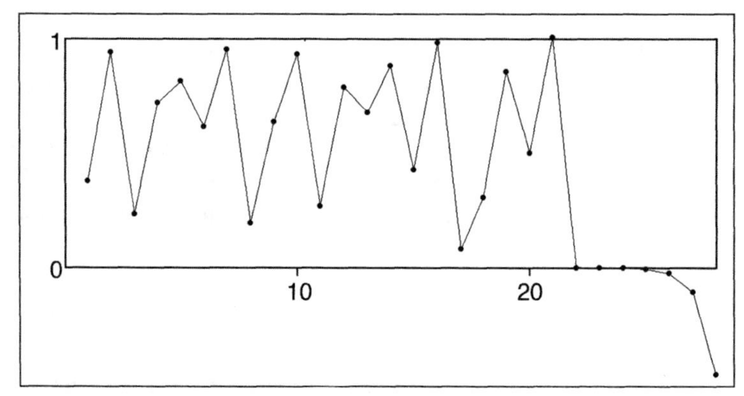

Figure 11.52

which causes points in a small interval around the critical point $x = 0.5$ to be transformed outside of the unit interval from where orbits rapidly escape to negative infinity. But also all preimages of that interval escape leaving almost no orbits which remain in the unit interval for all times.[17]

Chaotic Transients

Figure 11.52 displays a typical orbit for the parameter $a = 4.001$, slightly above 4. We observe that for a large number of iterations the orbit looks chaotic until eventually it escapes the unit interval and diverges. Orbits of this type are called *chaotic transients*. In other words, the chaos — which has ceased to exist for this parameter value — still casts a spooky shadow on the orbits. Only in the long term the phantom disappears. But how many iterations are necessary for this to happen? Or what is the lifetime of a chaotic transient? This number clearly depends on the choice of the initial point as well as on how close the parameter a is to the value 4. For example, the initial point $x_0 = 0.5$ escapes the unit interval in just one step, while the orbits of the fixed points 0 and 1 as well as their preimages, of course, cannot escape.

[17] In fact, only a Cantor set remains. We will pick up this scenario in more detail again in section 13.8.

Moreover, if $a - 4$ is large then large portions of the unit interval escape in only a few iterations since only a small portion of the parabola remains in the unit square. The first deficiency can be removed by considering an average escape time of orbits with initial points uniformly distributed over the entire unit interval. The dependence of this average escape time on the parameter a is established in the form of another interesting power law.

Lifetime of a Chaotic Transient

Let us make a numerical experiment to uncover this power law for parameters a beyond the crisis. We take 10,000 initial points, equally spaced in the unit interval, and compute their orbits until they have escaped the unit interval. Table 11.53 lists the average escape times, obtained for a decreasing sequence of parameters.

k	$a = 4 + 10^{-k}$	E_a	$E_a\sqrt{a-4}$
0	5.000000	2.307	2.307
1	4.100000	7.728	2.444
2	4.010000	26.996	2.700
3	4.001000	93.432	2.955
4	4.000100	307.607	3.076
5	4.000010	1017.663	3.218
6	4.000001	3118.888	3.119

Average Escape Times Near Crisis

Escape times E_a averaged over 10,000 orbits for a decreasing sequence of parameters $a = 4+10^{-k}$, $k = 0, \ldots, 6$.

Table 11.53

Plotted on a doubly logarithmic scale the data reveals the power law (see figure 11.54) represented by a straight line fit. The resulting slope is about 1/2. Thus, the power law is

$$\text{Average escape time } E_a \propto (a-4)^{-\frac{1}{2}},$$

and we say that $\frac{1}{2}$ is the critical exponent of the chaotic transient. The last column of table 11.53 reveals that the product $E_a\sqrt{a-4}$ tends to about 3.1 as the parameter a approaches a = 4. Thus, this is the factor of proportionality in the power law,

$$E_a \approx 3.1(a-4)^{-\frac{1}{2}}.$$

We can give an argument which supports these numerics.[18] It is based on the expectation that the relative number of orbits which escape the unit interval is approximately proportional to the length of the small interval around $x = \frac{1}{2}$, in which the parabola $ax(1-x)$ surpasses the value 1, i.e., to the fraction of the unit interval which escapes in one iteration. In conclusion, the average escape time should be inversely proportional to the length of the interval. To compute that interval we need to solve the equation $ax(1-x) = 1$ which yields

$$x_{1,2} = \frac{1}{2} \pm \sqrt{\frac{1}{4} - \frac{1}{a}}.$$

[18] Compare with the similar reasoning on page 489 for the lifetime of an orbit in the mixing experiment.

Escape Time Versus Parameter

Plot of the escape times E_a from table 11.53 versus $1/(a-4)$ using a logarithmic scale (base 10) on both axes. The slope of the line fit is about $1/2$.

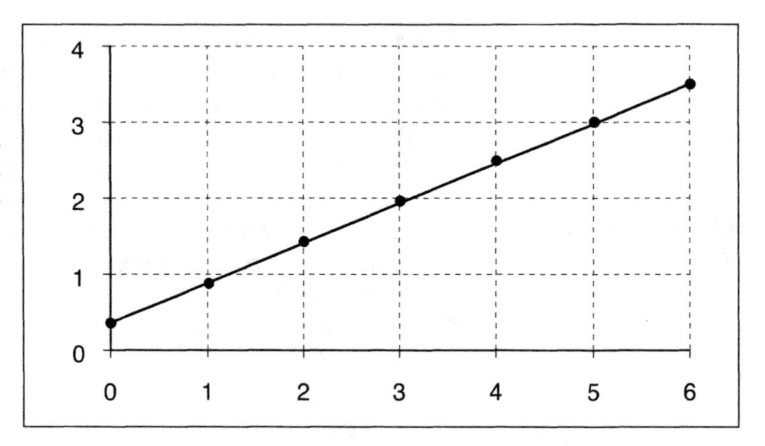

Figure 11.54

The sought interval thus is $[x_1, x_2]$. Letting $a = 4 + \varepsilon$ we find that its length is

$$x_2 - x_1 = 2\sqrt{\frac{1}{4} - \frac{1}{a}} = \frac{\sqrt{\varepsilon}}{2} \cdot \frac{1}{\sqrt{1 + \frac{\varepsilon}{4}}}$$

The second factor converges to 1 as ε tends to 0. Thus, the length of the interval is (asymptotically) proportional to $\sqrt{\varepsilon} = \varepsilon^{\frac{1}{2}}$. This yields the same critical exponent $\frac{1}{2}$ as our numerical test above.

Boundary Crisis

Let us present another view of the crisis as the parameter a passes through the critical value $a = 4$. Below the critical parameter, orbits are confined to the interval which is bounded by the critical value $v_a = a/4$ and its image $f_a(v_a)$. Orbits started between 0 and 1 outside of this interval rapidly iterate to the final states which are inside the interval. There is one exception, namely, the repelling fixed point at 0. It is right on the boundary between initial points whose orbits tend to the invariant set in the unit interval and those which lead to diverging sequences.[19] Precisely at $a = 4$ the final states collide with the repeller at 0 and the chaotic region suddenly disappears. This type of crisis is therefore also called a boundary crisis.[20]

To conclude let us remark that the phenomena of intermittency and crisis are presented here only for the simplest possible model, the quadratic iterator. Of course, they also occur in many other mathematical systems which are far from this simple case. Moreover, intermittency and crises have been observed in physical experiments, for example, in pipe flows, a compass forced by a magnetic field, electronic oscillators, lasers, thermal convection in liquid crystals, and more.[21]

[19] These are *negative* initial values.

[20] There are other types of crises, for example, the *interior crisis* in which a sudden widening of chaotic bands occurs. Such a crisis happens at the parameter which limits the period-3 window. For a review of crises see Grebogi, C., E. Ott, J. A. Yorke, *Crises, sudden changes in chaotic attractors, and transient chaos,* Physica 7D (1983) 181–200.

[21] For references and a review of the theory of chaotic transients see T. Tél, *Transient chaos,* to be published in: *Directions in Chaos III,* Hao B.-L. (ed.), World Scientific Publishing Company, Singapore.

Chapter 12

Strange Attractors: The Locus of Chaos

Never in the annals of science and engineering has there been a phenomenon so ubiquitous, a paradigm so universal, or a discipline so multidisciplinary as that of chaos. *Yet chaos represents only the tip of an awesome iceberg, for beneath it lies a much finer structure of immense complexity, a geometric labyrinth of endless convolutions, and a surreal landscape of enchanting beauty. The bedrock which anchors these local and global bifurcation terrains is the omnipresent* nonlinearity *that was once wantonly linearized by the engineers and applied scientists of yore, thereby forfeiting their only chance to grapple with reality.*

Leon O. Chua[1]

Having discussed the phenomena of chaos and the routes leading to it in 'simple' one-dimensional settings, we continue with the exposition of chaos in dynamical systems of two or more dimensions. This is the relevant case for models in the natural sciences since very rarely can processes be described by only one single state variable. One of the main players in this context is the notion of *strange attractors*.

To talk about strange attractors we have to consider a particular kind of dynamical systems: dissipative dynamical systems, i.e., systems with some sort of friction. The chief feature of dissipative systems is loss of energy. For example, a real pendulum swinging in air will have dissipation. Energy is lost continuously through the various kinds of friction which the pendulum experiences. In contrast, we speak of conservative dynamical systems when energy is maintained. This is the case in systems without friction. For example, the friction which heavenly bodies sustain is so little that we think of their motion as conservative; no energy is lost.

[1] In: International Journal of Bifurcation and Chaos, Vol. 1, No. 1 (1991) 1–2.

Guided by mathematical development physicists and mathematicians were led to believe that the long-term behavior of dissipative systems would always run into simple patterns of motion such as a rest point or a limit cycle. In contrast, strange attractors are those patterns which characterize the final state of dissipative systems that are highly complex and show all the signs of chaos. They very strongly defy the power of an intuitive understanding, and yet they now are proven to be all around us. It seems as if all of a sudden a whole new world of previously invisible beings is flying around us. Moreover, strange attractors are the point where chaos and fractals meet in an unavoidable and most natural fashion: as geometrical patterns, strange attractors are fractals; as dynamical objects, strange attractors are chaotic. There is now a whole new experimental and theoretical science dealing with strange attractors, their classification, the measurements of their quantitative properties, their reconstruction from physical data, and so on. But undoubtedly the mathematical understanding of strange attractors is just in its infancy and they will be one of the great challenges of future mathematical generations. It is by no means easy to understand even the notion of a strange attractor. In fact, strange attractors still have not received a final mathematical definition. Mathematics is sometimes described as the science which generates eternal notions and concepts for the scientific method: derivatives, continuity, powers, logarithms are examples. The notions of chaos, fractals and strange attractors are not yet mathematical notions in that sense, because their final definitions are not yet agreed upon.

Strange attractors have, however, become a *very* popular topic which has drawn interest not only from physics and mathematics but also from all other natural sciences and even the social sciences. The reason for the overwhelming popularity of chaos and strange attractors lies in the great expectations with which people come to the topic. Scientists hope to be able to crack the mysteries of our planet's climate, or human brain activity, as well as the secrets of turbulence through the metaphor of strange attractors. Fluid turbulence — one of the great unsolved problems in theoretical physics — occurs even in common daily routines, for example, when we open the water tap at the kitchen sink. First a smooth and regular flow of water appears. But as the water flow is increased the fluid starts to forcefully splash out without any regularity: turbulence. More important and with relevant technical applications is the turbulence occurring in the turbo-prop engines of an airplane or at the propeller of a ship or in large water pumps, where turbulence can actually eat away the metal impeller blades. There are also some chemical reactions which are periodic in time. In 1971, David Ruelle, one of the scientific notables in chaos theory, asked a specialist in these periodic reactions if he thought that one would find chemical reactions with chaotic time dependence. Ruelle recalls that he answered that if an experimentalist obtained a chaotic record in the study of a chemical reaction, he would throw away the record, saying that the experiment was unsuccessful.[2] This attitude, of course, was characteristic

[2]From D. Ruelle, *Strange Attractors*, Math. Intelligencer 2 (1980) 126–137.

Deterministic Nonperiodic Flow

EDWARD N. LORENZ

Massachusetts Institute of Technology

(Manuscript received 18 November 1962, in revised form 7 January 1963)

ABSTRACT

Finite systems of deterministic ordinary nonlinear differential equations may be designed to represent forced dissipative hydrodynamic flow. Solutions of these equations can be identified with trajectories in phase space. For those systems with bounded solutions, it is found that nonperiodic solutions are ordinarily unstable with respect to small modifications, so that slightly differing initial states can evolve into considerably different states. Systems with bounded solutions are shown to possess bounded numerical solutions.

A simple system representing cellular convection is solved numerically. All of the solutions are found to be unstable, and almost all of them are nonperiodic.

The feasibility of very-long-range weather prediction is examined in the light of these results.

Figure 12.1 : The abstract of Lorenz's paper which pioneered chaotic strange attractors.

not only of experimental chemistry, but also of all other natural sciences. But soon after the news of strange attractors had spread around the scientific laboratories of the world in the 1970's, things changed fundamentally. Researchers became aware of the subject and concentrated on the irregular patterns of processes which they previously had dismissed as misfits. Now we know several examples of chaotic behavior even in simple reaction systems. There have also been numerous strange attractors discovered in physics which are similar to the Lorenz attractor (discussed later in this chapter), and the concept of strange attractors is used in the sciences ranging from astronomy almost all the way to zoology.

A part of the drawing power in the concepts of chaos and strange attractors is probably due to the choice of the catchy names for these phenomena. The word 'chaos' was introduced in an article by Tien-Yien Li and James A. Yorke entitled *Period 3 implies chaos*,[3] while the term 'strange attractor' even goes back to 1971. Let us again quote from Ruelle's exposition in the *Mathematical Intelligencer*. "It seems that the phrase 'strange attractor' first appeared in print in a paper by Floris Takens (of Groningen) and myself.[4] I asked Floris Takens if he had created this remarkably successful expression. Here is his answer. 'Did you ever ask God whether he created this damned universe? ...I don't remember anything...I often create without remembering it ...' The creation of strange attractors thus seems to be surrounded by clouds and thunder. Anyway, the name is beautiful, and well suited to these astonishing objects, of which we understand so little."

[3]Li, T. Y. and Yorke, J. A., *Period 3 implies chaos,* American Mathematical Monthly 82 (1975) 985–992.

[4]D. Ruelle and F. Takens, *On the nature of turbulence,* Comm. Math. Phys. 20 (1971) 167–192 and 23 (1971) 343–344.

The first strange attractor ever recognized as such in the natural sciences is the Lorenz attractor, discovered in 1962. However, the work was published in the *Journal of the Atmospheric Sciences* which is not usually read by physicists and mathematicians. So the research on chaos was unnecessarily delayed by a decade or so until the real implications of Lorenz's achievement became clear. Although the Lorenz attractor is one of the 'oldest' known strange attractors, answers to some very basic questions about it are still outstanding. Recently, on the occasion of his sixtieth birthday, the great mathematician Stephen Smale, who has been one of the leading pioneers in dynamical systems and chaos theory for several decades, posed ten major open research problems. One of them asks for a proof that the geometric model of the Lorenz attractor proposed by John Guckenheimer and Philip Holmes[5] is true.

Given a dynamical system such as the Lorenz system, we can see the attractors on our computer graphics screens. This is fine. However, when physicists, for example, make measurements in some real-world experiment, they only obtain long and messy sequences of numbers, not equations. Then they must answer the question of what kind of dynamical system is behind the scene or perhaps even whether there is a strange attractor lurking behind their irregular and noisy data. One of the most fascinating achievements of chaos theory is that it has made available a tractable numerical method to attack this problem, the *reconstruction of strange attractors*. It even leads to algorithms which can compute numerical quantities such as dimensions and Ljapunov exponents that specify the degree of strangeness and 'chaoticity' of the attractor.

Strange attractors offer some new understanding of nonlinear effects. Moreover, they can be aesthetically pleasing. Thus, it is no wonder that the subject attracts researchers from all disciplines. It has been and continues to be a *hot topic*. The amount of literature that has been written on the topic of chaos surely surmounts any individual's reading capacity. Computers are needed to study dynamics on strange attractors, and computers are equally needed to maintain data bases of that vast literature. An (incomplete) list published by Hao Bai-Lin contains 117 books, conference proceedings, and collections of papers, and an unbelievable number of 2244 technical papers. It is therefore an outstanding achievement for Hao to have compiled the most influential papers, along with some of an introductory type, and the bibliography in one large volume of reprints entitled *Chaos II*.[6] It contains, for example, the original papers by Feigenbaum, Hénon, Lorenz, and May, to name just a few.

[5] J. Guckenheimer and P. Holmes, *Nonlinear Oscillations, Dynamical Systems, and Bifurcations of Vector Fields*, Springer-Verlag, New York, 1983. See section 5.7 therein.

[6] Hao, B. L., *Chaos II*, World Scientific, Singapore, 1990. For an even larger bibliography on chaos containing over 7000 references see Zhang Shu-yu, *Bibliography on Chaos – Directions in Chaos Vol. 5*, World Scientific, Singapore, 1991.

12.1 A Discrete Dynamical System in Two Dimensions: Hénon's Attractor

In chapter 10 we built our analysis of chaos for the iteration of the quadratic map starting from the paradigm of the kneading of dough (see section 10.4). The dynamics could be modeled by a stretch-and-fold operation. Of course real dough has some thickness which must be ignored in this approach. Here we learn about a particularly simple transformation which does not neglect this extra dimension. It was suggested by the French astronomer Michel Hénon[7] in 1976 as a simplified model for the dynamics of the Lorenz system which is the topic section 12.4 below. Because of its simplicity, it lends itself to computer studies and numerous investigations followed. Moreover, the gently swirling, boomerang-like shape of the attractor that arises through the dynamics is very appealing aesthetically. This object is now known as the *Hénon attractor*. In fact, it has become another icon of chaos theory next to the Mandelbrot set, the Feigenbaum diagram, and the Lorenz attractor.

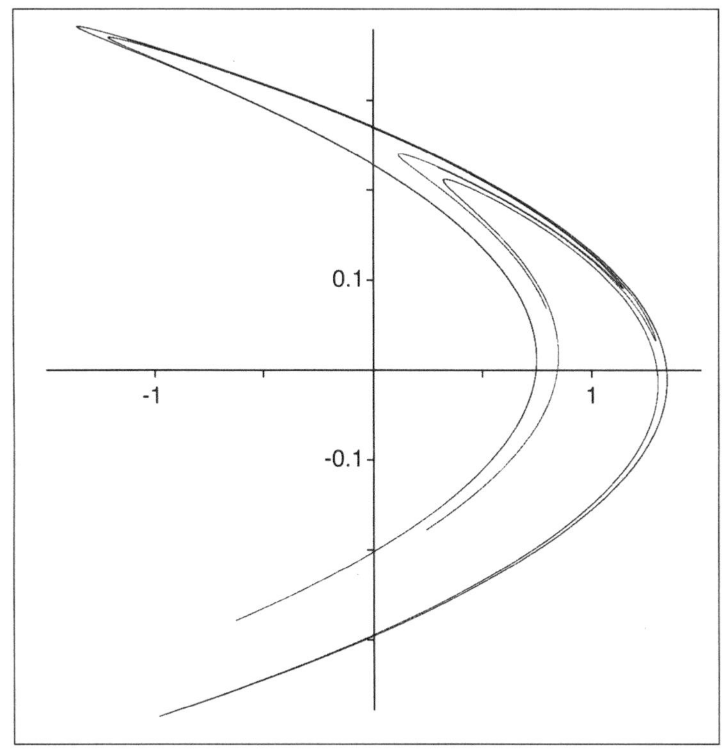

The Hénon Attractor

The figure shows 100,000 computed points of the orbit of the initial point (0,0) of Hénon's system (the first 100 points are omitted). The region shown is $-1.5 \leq x \leq 1.5$ and $-0.4 \leq y \leq 0.4$.

Figure 12.2

[7]See M. Hénon, *A two-dimensional mapping with a strange attractor,* Comm. Math. Phys. 50 (1976) 69–77.

Decomposition of Hénon's Transformation

The gridded square in the upper left is transformed in three steps: a nonlinear bending (upper right) in the y-direction, the contraction towards the y-axis (lower left) and a reflection at the diagonal (lower right). The region shown is $-2.2 \leq x \leq 2.2$ and $-2.2 \leq y \leq 2.2$.

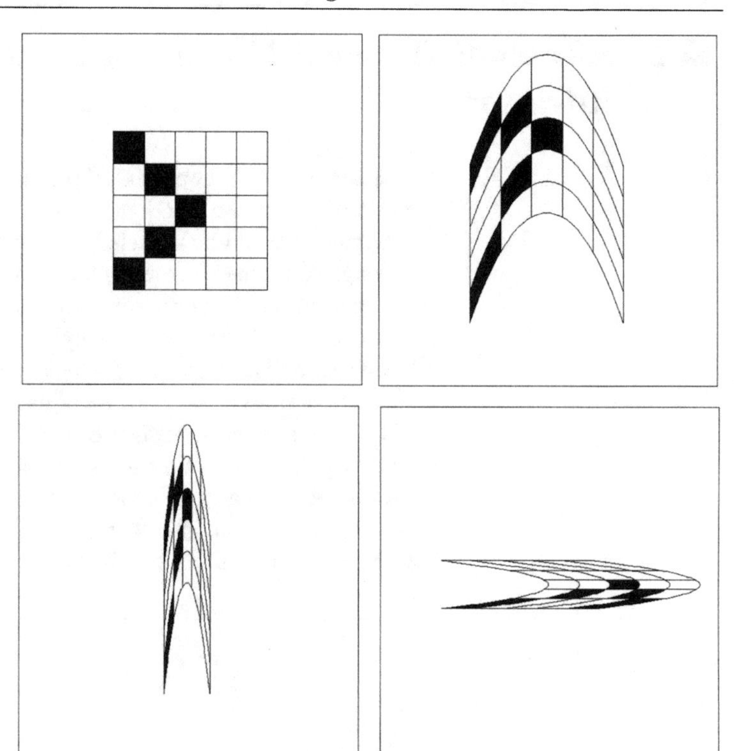

Figure 12.3

In a way which we will specify, the Hénon system leads from the one-dimensional dynamics of the quadratic transformation to higher-dimensional strange attractors. It is simple enough to allow an analysis similar to the analysis of chaos in the logistic transformation, yet it possesses features inherent in more complicated attractors such as the Lorenz attractor, about which we do not know nearly as much.

The Model

The stretch-and-fold action of the Hénon system happens in two dimensions, with coordinates denoted by x and y. The transformation, thus, is a transformation in the plane which operates just like *one* of the affine transformations from our paradigm, the Multiple Reduction Copying Machine (MRCM) from the first chapter. Explicitly, Hénon suggested a transformation

$$H(x, y) = (y + 1 - ax^2, bx) \tag{12.1}$$

where a and b are adjustable parameters. An orbit of the system consists of a starting point (x_0, y_0) and its iterated images, i.e.,

$$(x_{k+1}, y_{k+1}) = (y_k + 1 - ax_k^2, bx_k), \quad k = 0, 1, 2, \ldots$$

Similar to the logistic equation, these dynamics depend dramatically on the choice of the constants a and b besides that of the starting point. For some

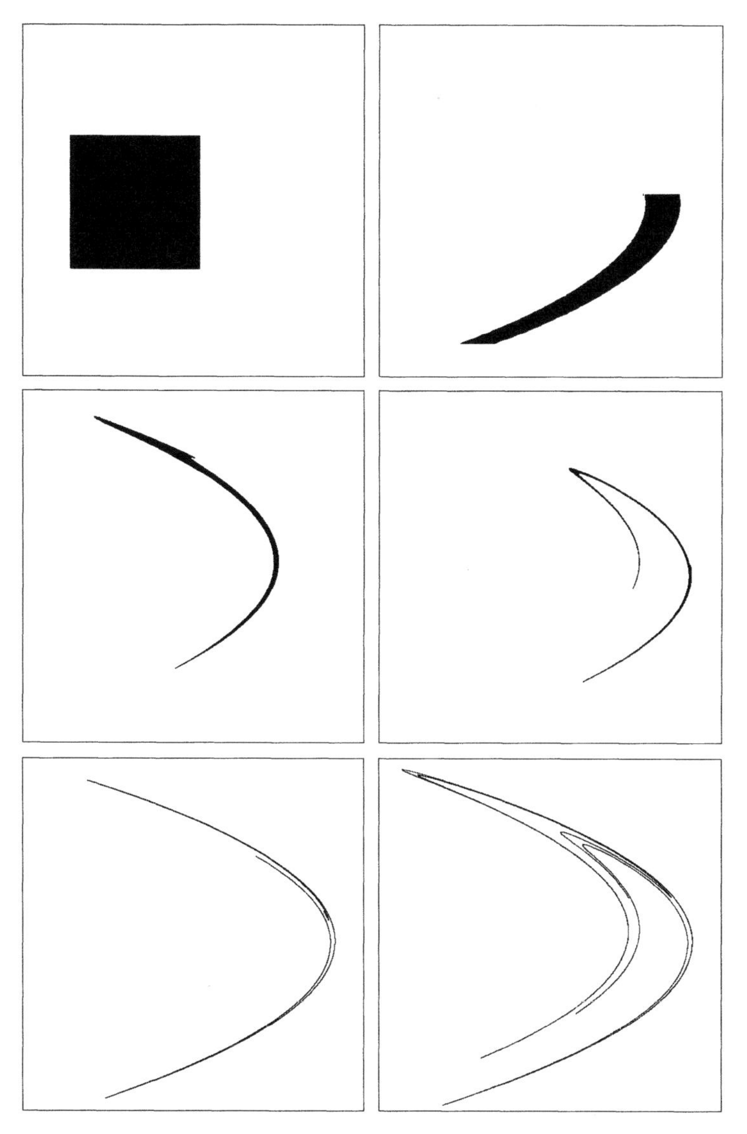

The 'MRCM' for Hénon

The Hénon transformation is applied to a square (digitized by an array of 100×100 regularly spaced points and shown in the upper left). The results of $0, 1, 2, 3, 5, 10$ iterations are shown from upper left row by row. Note that the initial square in the upper left portion of the figure is different from that in figure 12.3, which implies that the folding becomes visible only after several applications of the transformation. The region shown in each part is $-1.5 \leq x \leq 1.5$ and $-0.4 \leq y \leq 0.4$.

Figure 12.4

parameters almost all orbits tend to a unique periodic cycle, while chaos seems to reign for other choices. Hénon used the values

$$a = 1.4 \text{ and } b = 0.3.$$

Let us study the transformation to see the correspondence to the stretch-and-fold action. We can partition the application of the transformation H into three steps, visualized in figure 12.3.

Invariance of the Hénon Attractor

Slow motion action of the Hénon transformation applied to the attractor itself. The upper two rows of figures show how the attractor (top left) is stretched according to $H_1(x, y) = (x, y + 1 - ax^2)$. The bottom row continues with the folding part of the Hénon transformation ($H_2(x, y) = (bx, y)$). When the bottom right figure is reflected at the diagonal (or turned 90 degrees clockwise and flipped horizontally) the attractor from the top left is exactly reproduced. Figure 12.6 shows how the transformation acts on some example points. The region is $-1.5 \leq x \leq 1.5$, $-1.5 \leq y \leq 1.5$.

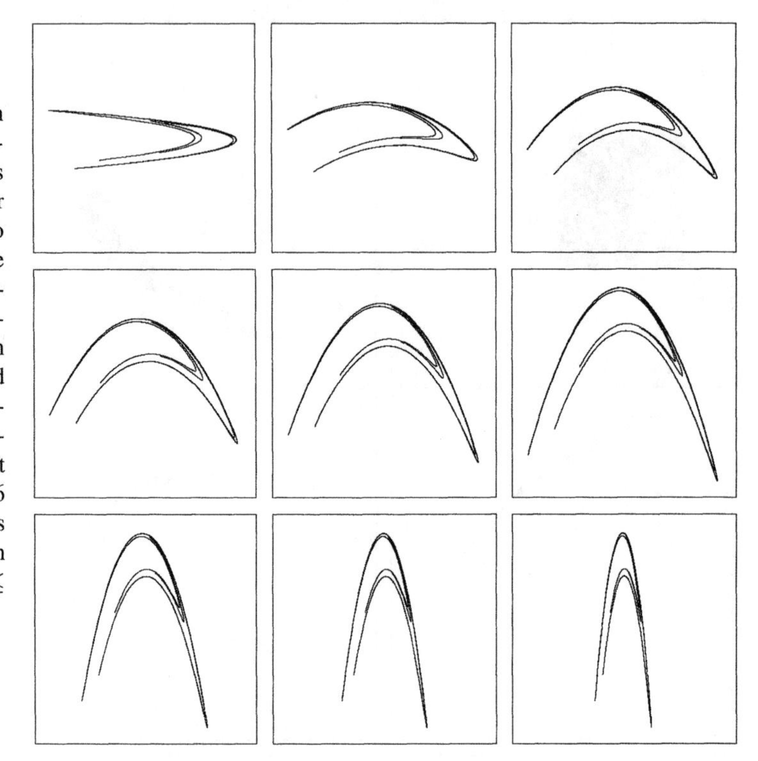

Figure 12.5

Three Phases of Stretch-and-Fold

1. **Bend up.** The first step consists of a nonlinear bending in the y-coordinate given by

   $$H_1(x, y) = (x, y + 1 - ax^2).$$

   For example, a horizontal line (y = constant) becomes a parabola with vertex at $(0, y + 1)$ and opening up at the bottom. In contrast, the remaining two steps are linear transformations.
2. **Contract in x.** Next a contraction in the x-direction is applied,

   $$H_2(x, y) = (bx, y).$$

   The contraction factor is given by the parameter b, which is 0.3 for Hénon's attractor.
3. **Reflect.** Finally a reflection at the diagonal,

   $$H_3(x, y) = (y, x)$$

   is in order.

The result of the concatenation is the same as applying the original transformation once, i.e.,

$$H(x, y) = H_3(H_2(H_1(x, y))).$$

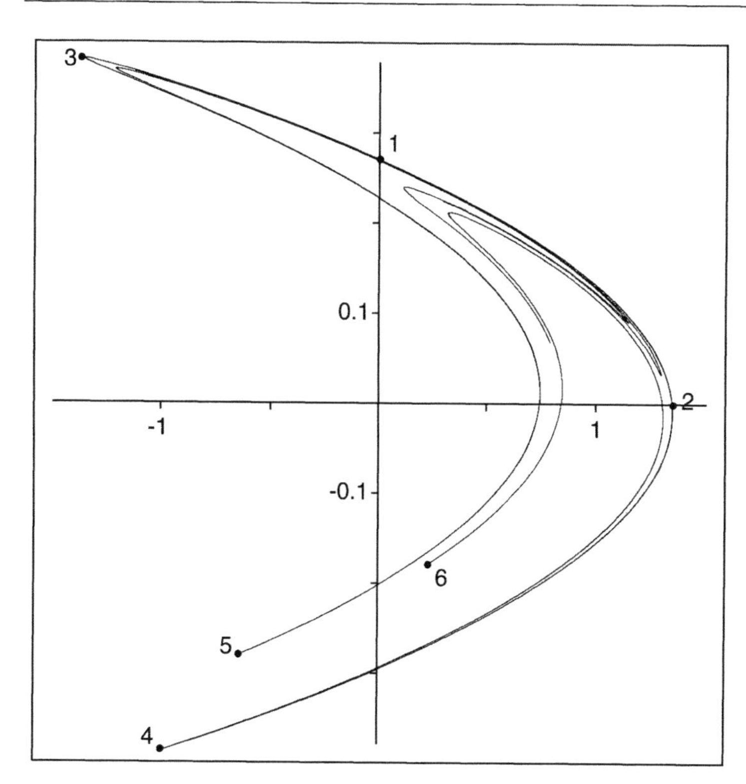

Invariance of the Hénon Attractor

Six points on the Hénon attractor are labeled by numbers 1 to 6. After the stretching, folding, and reflection of the Hénon transformation point 1 arrives at point 2, 2 arrives at 3, and so on. In other words, the initial points are chosen as consecutive points from an orbit. Try to follow these dynamics in figure 12.5. The region is $-1.5 \le x \le 1.5$, $-0.4 \le y \le 0.4$.

Figure 12.6

Speaking in terms of the kneading paradigm, we may say that the first step is the stretching of the dough. One person holds up the dough at the center while another person pulls down at the two ends of the dough. Step 2 folds the dough together, i.e., the second person moves both ends toward each other. Then at the end the dough is put back on the table and turned over.

The Attractor Appears

Let us see the effect of repeated application of these transformations. We take an arbitrary shape, for example a square, and apply the transformation H iteratively (see figure 12.4). Of course, the square is severely deformed and not even recognizable after only a couple of steps. After a few dozen more steps a curious shape emerges — the Hénon attractor. This attractor must be a subset of the plane which is invariant with respect to the kneading. The sequence in figure 12.5 demonstrates how the various parts of the attractor are transformed into each other.

Just as the logistic iterator may have periodic attractors which attract orbits of nearby starting points, the Hénon attractor pulls in nearby orbits. Moreover, these orbits typically fill up the attractor densely.

Not All Points Are Attracted

But how can we be sure that this phenomenon is not due to our particular choice of the initial point? After all, the transformation given by Hénon is quadratic and if x_0 is large, then x_1 will be much larger, and repeated applications of H drive the orbit beyond all bounds. For example, starting

The Trapping Region

The trapping region i the quadrilateral with vertices P_1 = $(-1.33, 0.42)$, $P_2 = (1.32, 0.133$ $(P_3 = 1.245, -0.14)$, and P_4 = $(-1.06, -0.5)$. Its image is als shown; it lies entirely inside the trap ping region. An orbit of an initi point within the trapping region car not escape the region.

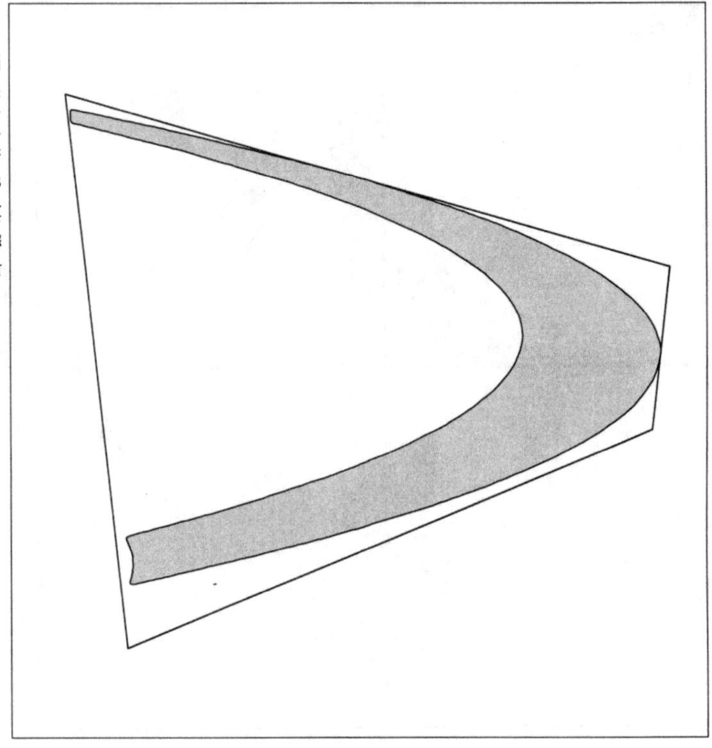

Figure 12.7

with $x_0 = 10$ and $y_0 = 0$ produces

$$\begin{aligned} x_0 &= 10.00 & y_0 &= 0.00 \\ x_1 &= -139.00 & y_1 &= 3.00 \\ x_2 &= -27\,045.40 & y_2 &= -41.70 \\ x_3 &= -1\,024\,035\,166.32 & y_3 &= -8113.62 \end{aligned}$$

and clearly the orbit escapes to 'infinity'. By the way, the initial value of x_0 does not even have to be very large for this effect to take place. The orbit of $(1.292, 0.0)$ also escapes, although the starting point is already quite close to the attractor.

The Trapping Region

Even though many orbits do escape to infinity we may still speak of an attractor because there is a so-called *trapping region* R from which no orbit can escape, thus orbits started within the region must converge to some limit set. The region is a quadrilateral carefully designed by Hénon and shown in figure 12.7. It can be verified using elementary algebra that the image of the region R obtained from one application of Hénon's transformation H does, in fact, lie entirely within the trapping region ($H(R) \subset R$). Thus, repeated application of H must always produce subsets of the region; no orbits can escape. Of course, the Hénon attractor lies in this trapping region. We may

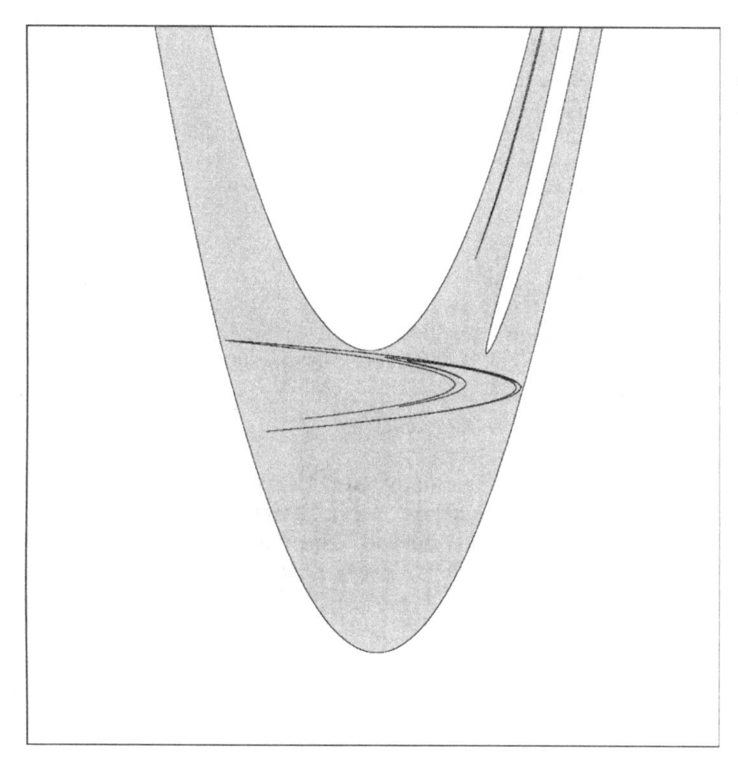

Basin of Attraction

The boundary of the basin of attraction for Hénon's attractor (shaded) which is shown in the center. The region shown is the square $-3 \leq x \leq 3$ and $3 \leq y \leq 3$.

Figure 12.8

now define it as

$$A = \bigcap_{k=0}^{\infty} H^k(R)$$

where H^k means the k-fold composition of H, as usual.

The Basin of Attraction

Now that we know that on the one hand there are escaping points and on the other there is a trapping region for the Hénon attractor, the question arises, which points in the plane have orbits that are eventually caught by the trapping region? The set of all such points is called the basin of attraction of A. Of course, the trapping region itself must be contained in the basin of attraction. Figure 12.8 provides a plot.[8] Basins of attraction will be in the center of interest in the two remaining chapters.

The Area Shrinks

Looking again at figure 12.7 we notice how the area of the quadrilateral shrinks when we apply the transformation. The same holds true for the square in figure 12.4. This observation can be mathematically verified (see the following technical section). The result is that any area considered shrinks by the factor of $b = 0.3$ when iterated once. Thus, taking a region of area 1, we obtain after two iterations an area of only 0.09. After k iterations the area

[8] Pictures of this sort were first published in S. D. Feit, *Characteristic exponents and strange attractors,* Comm. Math. Phys. 61 (1978) 249–260.

has reduced to 0.3^k. When we apply this to the trapping region, we arrive at the conclusion that the attractor A, which must reside in all the iterates of the region, can only cover a subset of the plane with an area equal to 0.

The Area Reduction

Consider a matrix

$$T = \begin{pmatrix} a & b \\ c & d \end{pmatrix}$$

and the parallelogram spanned by its two column vectors. It is a result from basic linear algebra that the corresponding area is

$$A = |\det T| = |ad - bc|,$$

the absolute value of the determinant of the matrix T. This is also the factor by which an area grows or shrinks when the linear transformation given by the matrix is applied. The Hénon transformation is not linear, but a similar result holds locally, based on the linearization of the transformation. A small area near a point $P = (x, y)$ is reduced by the factor given by the absolute value of the determinant of the derivative (the Jacobian matrix) of the transformation at that point. For the Hénon transformation this is

$$|\det DT(x,y)| = \left|\det \begin{pmatrix} -2ax & 1 \\ b & 0 \end{pmatrix}\right| = |b|.$$

Since $|b|$ is a constant which does not depend on the location of P the area changes uniformly by that factor.

The Strangeness of the Attractor

Let us now come to the chaoticity and the strangeness of the Hénon attractor. We will highlight two corresponding aspects: the sensitive dependence on initial conditions and the fractal structure. Thereafter we present an initial mathematical definition of the concept of a strange attractor.

Sensitive Dependence on Initial Conditions

To obtain a picture of the attractor, it is sufficient to compute just a single orbit of an initial point picked at random somewhere within the trapping region. Picking a different random initial point does not change the visual result in any way if we ignore the first hundred points or so which are needed to bring the orbit sufficiently close to the attractor. However, although two different orbits generate the same limit set, typically there is no correlation between them, even if the initial points are chosen very close to each other (see figures 12.9 and 12.10). Strictly speaking, this is not entirely true. For example, if the second initial point is a point from the first orbit (which may be as close to the first initial value as we want to), then both orbits of course are correlated as demonstrated in figure 12.11. What we mean is that almost all random points chosen arbitrarily close to the first initial value produce orbits which do not correlate with the first one.

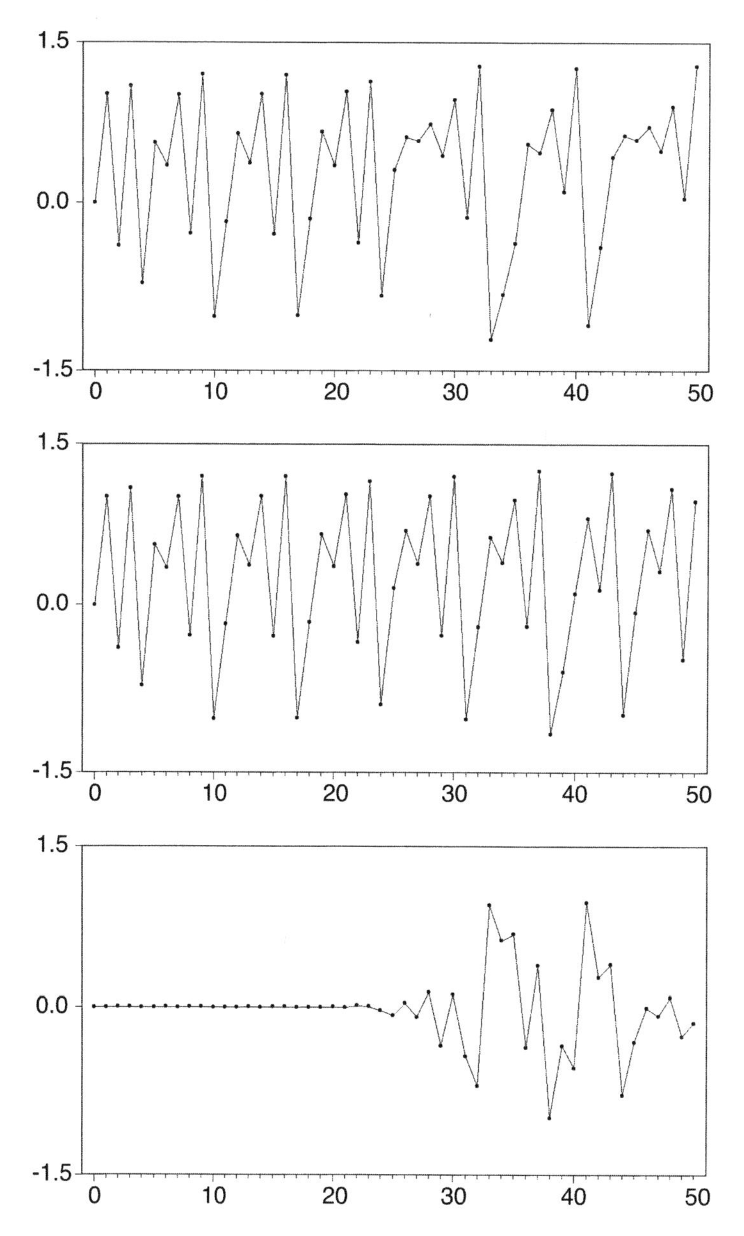

Sensitivity

To demonstrate the sensitivity to initial conditions, we compute two orbits $(x_0, y_0), (x_1, y_1), \ldots$ and (x'_0, y'_0), (x'_1, y'_1), $\ldots$ with initial points $(x_0, y_0) = (0, 0)$ and $(x'_0, y'_0) = (0.00001, 0)$. We plot three time series: the values x_k (top), x'_k (center) and the difference $|x_k - x'_k|$ (bottom). In the beginning the time series are undistinguishable, but after a number of iterations, the difference between them builds up rapidly and this 'error' becomes as large as the 'signal' itself — a consequence of sensitivity.

Figure 12.9

Zoom into the Fractal Structure

The first look at the Hénon attractor gives the appearance of a collection of a few curves which look like sections of parabolas. But this impression could not be further from the truth as the enlargements in figure 12.12 shows. The more we magnify a portion of the attractor the more 'curves' become visible. Thus, the Hénon attractor consists of an infinite number of parabola-

No Correlation Between Orbits

To show another effect of sensitivity to initial conditions, we compare again the two orbits (x_0, y_0), (x_1, y_1), ... and (x'_0, y'_0), (x'_1, y'_1), ... with initial points as in figure 12.9. We plot points (x_k, x'_k) for the first 50, 200, 1,000 and 100,000 iterations (from upper left to lower right). These points densely fill a square. Similar results would be obtained by plotting the second coordinates of the orbits against each other. The region shown is the square $-1.5 \leq x \leq 1.5$ and $1.5 \leq y \leq 1.5$.

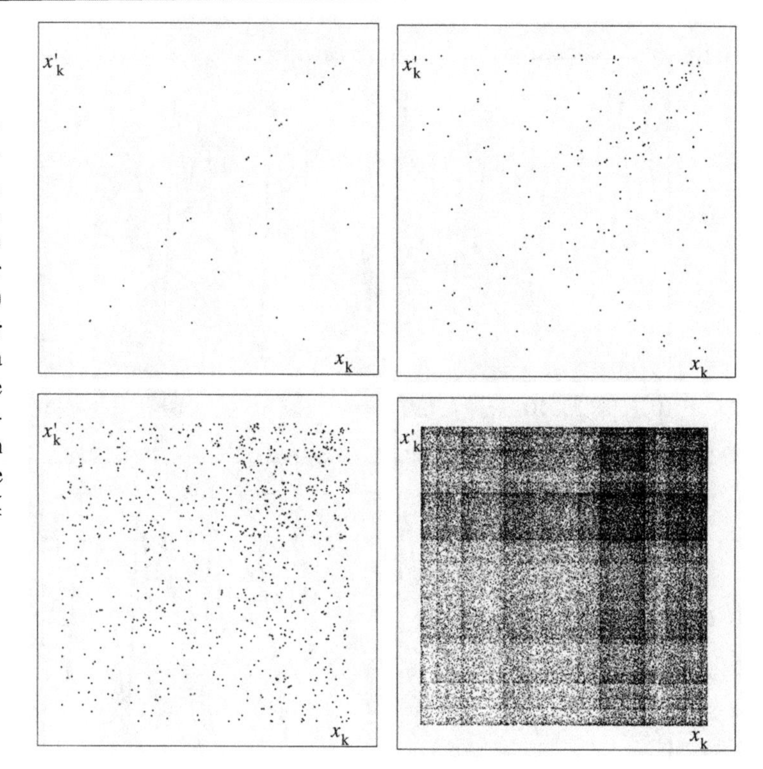

Figure 12.10

like layers. When we think about the kneading action of the transformation, this infinitely detailed puff pastry structure becomes, in fact, quite reasonable. Assume that there is at least one parabola-like curve in the attractor. The stretching and folding of this curve produces a curve composed of a parabola with two layers which must consequently also be part of the attractor.[9] The next iteration generates a structure with four layers, then we get eight, and so on. Thus, after considering all the images of the initial parabola we obtain an infinity of layers, all of which belong to the attractor. This is what we can see in the graphics when we zoom in on a point of the attractor. Due to the folding in the transformation, a cross-section of the attractor looks much like a Cantor set. We may say that sections of the Hénon attractor as in figure 12.12 are cross products of an interval with a Cantor set.[10]

Backward Orbits

In contrast to its one-dimensional cousin, the quadratic iterator, the Hénon transformation has an inverse. This expresses the fact that for any initial point there is not only a unique forward orbit but also a unique sequence of

[9] These two layers are separate, they must not intersect or overlay each other. The reason for this lies in the fact that no two points are transformed to the same image point. Hénon's transformation is one-to-one (see below).

[10] This means that we replace each point in the Cantor set by a small vertical line segment.

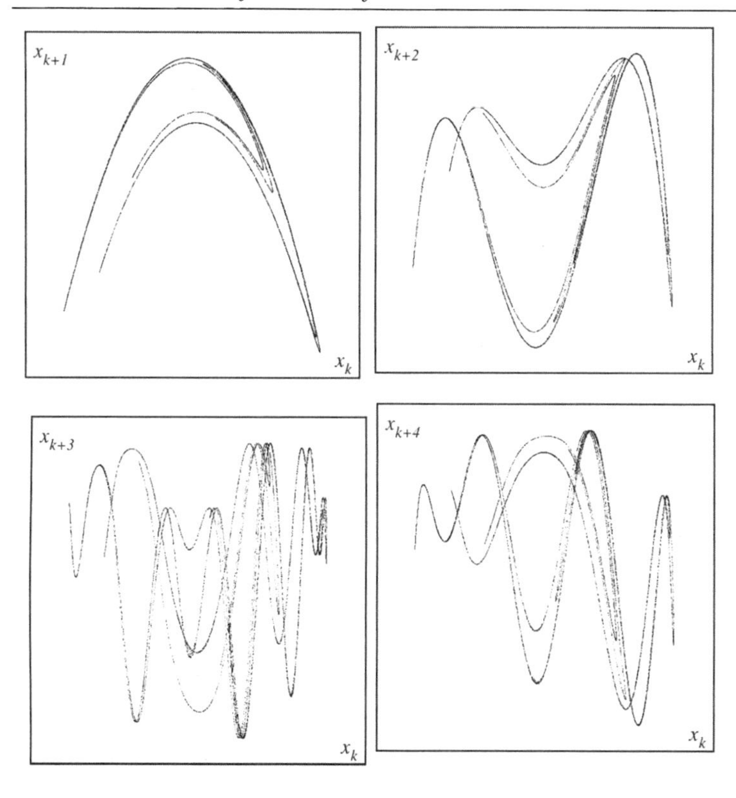

Correlation Is Possible

This plot demonstrates that the result from figure 12.10 depends on the choice of initial conditions. Here we chose the initial condition (x_0', y_0') of the second orbit to be equal to the mth iterate of the original orbit, i.e., (x_m, y_m). In effect, we are plotting points $(x_k, x_{k+m}), k = 0, 1, 2, \ldots$ The number m is 1 (upper left), 2 (upper right), 3 (lower left), and 4 (lower right). Note that we can choose the initial points of all orbits considered as close to each other as we wish by placing them in a neighborhood of a (repelling) fixed point. In all graphs there is a clear structure of the collection of points. In fact, they represent just modified versions of the Hénon attractor (compare the account of the reconstruction of strange attractors from time series in section 12.7).

Figure 12.11

predecessors. The forward iteration of a point is given by

$$\begin{aligned} x_{n+1} &= y_n + 1 - ax_n^2 \\ y_{n+1} &= bx_n \end{aligned}$$

which we now solve for x_n and y_n:

$$\begin{aligned} x_n &= \frac{y_{n+1}}{b} \\ y_n &= x_{n+1} + \frac{a}{b^2} y_{n+1}^2 - 1. \end{aligned}$$

Using these last two equations we can compute the *backward orbit* for any initial point (x_0, y_0), i.e., (x_{-k}, y_{-k}) for $k = 0, 1, 2, \ldots$ It is not a proven fact, but from numerical studies it seems apparent that all backward orbits must escape to infinity except those started in the attractor or in an unstable invariant set. Since these exceptional points cover a region of area 0, this case should almost never be observed numerically when iterating backwards. Even the smallest roundoff error in the course of the computation will throw such an orbit off the true one, outside of the attractor, and ultimately to infinity. More precisely, orbits diverge such that $y_k \to \infty$ as $k \to -\infty$.

Enlargements of the Hénon Attractor

Two successive enlargements of the Hénon attractor. The same 100,000 points as in Figure 12.2 are used. The regions shown are $0 \leq x \leq 1$ and $0 \leq y \leq 0.3$ on the left and $0.7 \leq x \leq 0.8$ and $0.15 \leq y \leq 0.18$ on the right.

Figure 12.12

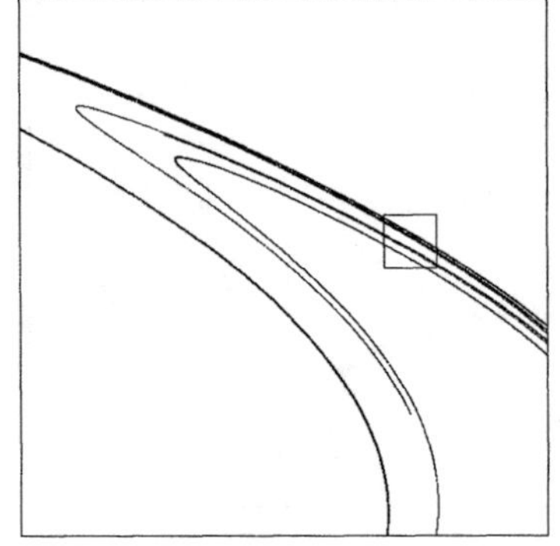

The Fractal Dimension

Again, the result of the magnifications in figure 12.12 shows a Cantor-like structure of parallel lines. The attractor definitely is a fractal. We can compute the box-counting dimension as usual, overlaying the attractor with grids of varying lattice sizes and counting cells which cover part of the attractor (see chapter 4). The result is a dimension of about 1.28, a value well above 1. We will return to the topic of dimension in section 12.6. It may be a surprise that this number is not the same as the dimension of the Cantor set, raised by 1, i.e.,

$$1 + \frac{\log 2}{\log 3} \approx 1.6309.$$

However, $\log 2 / \log 3$ is only the dimension of one particular Cantor set, namely, the standard one obtained by recursively deleting the middle thirds of intervals. We can change the construction by subdividing each interval into p equal parts of which we keep the first and last one while deleting all the others. This produces a Cantor set with dimension $\log 2 / \log p$. For example, for $p = 12$ we get that the dimension raised by 1 is

$$1 + \frac{\log 2}{\log 12} \approx 1.2789.$$

which is much closer to the numerically computed dimension of the Hénon attractor.

Characterization of Strange Attractors

Up to this point we have discussed a number of properties of the Hénon transformation. Let us summarize: there is a trapping region within which all initial points have orbits leading to the attractor. These orbits show sensitive dependence on initial conditions, and a single orbit seems to get close to all points of the attractor. Moreover, the attractor exhibits a fractal structure. These are the four chief properties of strange attractors which we note in the list below. Although strange attractors typically exist in spaces of more than two dimensions, we will restrict our presentation for simplicity to the two-dimensional case applicable to the Hénon attractor. Thus, let $T(x, y)$ be a given transformation in the plane with coordinates x and y. A bounded subset

A of the plane is a chaotic and strange attractor for the transformation T if there exists a set R with the following properties.[11]

Four Properties

1. **Attractor.** R is a neighborhood of A, i.e., for each point (x, y) in A there is a small disk centered at (x, y) which is contained in R. This implies in particular that A is in R. R is a trapping region, i.e., each orbit started in R remains in R for all iterations. Moreover, the orbit becomes close to A and stays as close to it as we desire. Thus, A is an *attractor*.
2. **Sensitivity.** Orbits started in R exhibit sensitive dependence on initial conditions. This makes A a *chaotic* attractor.
3. **Fractal.** The attractor has a fractal structure and is therefore called a *strange attractor*.
4. **Mixing.** A cannot be split into two different attractors.[12] There are initial points in R with orbits that get arbitrarily close to any point of the attractor A.

We need to point out that the above attempt at a definition is indeed only a first try. The discussion about what should be the most appropriate definition mathematically is still going on, and it seems that we will have to wait for the final clarification until some kind of breakthrough in understanding strange attractors has been achieved.[13]

In fact, the situation is even worse. Up to now no one knows whether the attractor in Hénon's transformation for $a = 1.4$ and $b = 0.3$ really is a strange attractor according to the above or a similar definition even though very extensive numerical checks have been performed which all indicate a positive answer. This underlines the incomplete state of affairs. For example, we could speculate that the experimental observations are due to an attractive periodic orbit with a *very* long period. If that is really the case, then we will never be able to compute that period because rounding errors in the computation will disturb the orbit too much.

Lozi's Piecewise Linear Model

However, the situation is perhaps not as bleak as it may now appear. One first step towards a solution of the problem has been carried out by Michal Misiurewicz.[14] He proved an earlier conjecture by René Lozi,[15] who hypothesized that a simplified version of Hénon's transformation in fact admits a strange attractor. The transformation is given by

$$\tilde{H}(x, y) = (1 + y - a|x|, bx), \tag{12.2}$$

and Lozi suggested the parameter values $a = 1.7$ and $b = 0.5$ (see figure 12.13). The only difference between Lozi's and Hénon's transformations is

[11] Except for part 3, which we added here, this definition has been given in D. Ruelle, *Strange attractors,* Math. Intelligencer 2 (1980) 126–137.

[12] Caution; this does not imply that the attractor must be a connected set.

[13] See the discussion on pages 255–259 in J. Guckenheimer and P. Holmes, *Nonlinear Oscillations, Dynamical Systems, and Bifurcations of Vector Fields,* Springer-Verlag, New York, 1983.

[14] M. Misiurewicz, *Strange attractors for the Lozi mappings,* in *Nonlinear Dynamics,* R. H. G. Helleman (ed.), Annals of the New York Academy of Sciences 357 (1980) 348–358.

[15] R. Lozi, *Un attracteur étrange (?) du type attracteur de Hénon,* J. Phys. (Paris) 39 (Coll. C5) (1978) 9–10.

The Lozi Strange Attractor

The attractor for Lozi's transformation eqn. (12.2) with $a = 1.7$ and $b = 0.5$. The region shown is $-1.5 \leq x \leq 1.5$ and $-0.75 \leq y \leq 0.75$. 100,000 points are plotted.

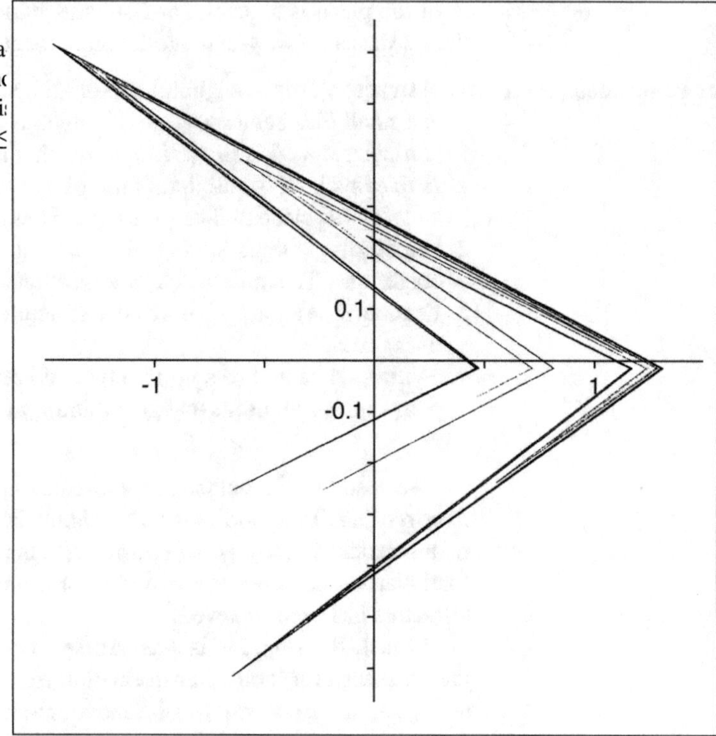

Figure 12.13

that the x^2 term is replaced by $|x|$. The fact that this modification makes the transformation linear for $x > 0$ and $x < 0$ allowed Misiurewicz to complete the analysis and the confirmation of a strange attractor.

The Dynamics

One of the purposes of the Hénon transformation is to provide a generalization of the quadratic iterator to two dimensions. In fact, for the special choice $b = 0$, the Hénon transformation reduces to

$$\begin{aligned} x_{n+1} &= 1 - ax_n^2 \\ y_{n+1} &= 0. \end{aligned}$$

The two coordinates are decoupled, and only the x-coordinate is relevant for the dynamics for $b = 0$. This iterator

$$x \to 1 - ax^2$$

is quadratic and equivalent to the logistic iterator

$$z \to rz(1 - z)$$

where we have chosen the symbol r for the parameter to avoid confusion with the parameter a from Hénon's transformation. Therefore, in the case $b = 0$, the Hénon transformation presents just another version of the Feigenbaum scenario (see chapter 11) of the quadratic iterator as the parameter a varies.

Equivalence of $1 - ax^2$ and $rz(1-z)$

In chapter 1 we identified the equivalence of the quadratic iteration $x \to x^2 + c$ with the logistic iterators $p \to p + rp(1+p)$ and $z \to rz(1-z)$ (see pages 63 and 68). Here we proceed along the same lines of argument to show that also the family $x \to 1 - ax^2$, which is nothing but Hénon's transformation eqn. (12.1) for the special choice $b = 0$, is equivalent to the other quadratic iterators. We show that

$$x_{n+1} = 1 - ax_n^2 \tag{12.3}$$

is identical to

$$z_{n+1} = rz_n(1 - z_n) \tag{12.4}$$

when using the setting

$$x_n = \frac{r}{a}(z_n - \frac{1}{2}) \quad \text{and} \quad a = \frac{r(r-2)}{4}. \tag{12.5}$$

To verify this statement, we compute x_{n+1} in terms of z_n using equations (12.3) and (12.5) and then compare the result with what we get from equations (12.4) and (12.5). This yields

$$\begin{aligned} x_{n+1} &= 1 - ax_n^2 = 1 - a\frac{r^2}{a^2}\left(z_n - \frac{1}{2}\right)^2 \\ &= -\frac{r^2}{a}z_n^2 + \frac{r^2}{a}z_n + 1 - \frac{r^2}{4a} \end{aligned}$$

and on the other hand

$$\begin{aligned} x_{n+1} &= \frac{r}{a}\left(z_{n+1} - \frac{1}{2}\right) = \frac{r}{a}\left(rz_n(1-z_n) - \frac{1}{2}\right) \\ &= -\frac{r^2}{a}z_n^2 + \frac{r^2}{a}z_n - \frac{r}{2a}. \end{aligned}$$

It remains to be shown that

$$1 - \frac{r^2}{4a} = -\frac{r}{2a}$$

which readily follows from $a = r(r-2)/4$.

Thus, the case $b = 0$ does not provide anything new. But what happens when $b \neq 0$? Several interesting questions come up. For example, what effect does this choice have on the dynamics? Is there still the Feigenbaum scenario present? If so, what about the universal Feigenbaum constant? How does the Hénon attractor fit into this picture? And are there perhaps other attractors possible besides those indicated by the Feigenbaum scenario? Since the Hénon transformation is so simple to implement on a computer, these questions can be followed up experimentally to some degree.

The Period-Doubling Cascade

Parameters a_k for the period-doubling cascade in Hénon's transformation (12.1) for $b = 0.3$. This sequence of parameters is approximately geometric. The last column specifies the estimates for the corresponding number δ. In the kth row this estimate is computed by $(a_k - a_{k-1})/(a_{k+1} - a_k)$.

k	Period	Parameter a_k	δ
0	1	−0.122 5	
1	2	0.367 5	
2	4	0.912 5	4.844
3	8	1.026	4.3269
4	16	1.051	4.696
5	32	1.056 536	4.636
6	64	1.057 730 83	4.7748
7	128	1.057 980 893 1	4.6696
8	256	1.058 034 452 15	4.6691
9	512	1.058 045 923 04	4.6691
10	1024	1.058 048 379 80	4.6694
11	2048	1.058 048 905 931	

Table 12.14

The Tangent Bifurcation

We begin by studying the fixed points of the Hénon transformation given by the solutions to the systems of equations

$$\begin{aligned} x &= 1 + y - ax^2 \\ y &= bx. \end{aligned}$$

There are two solutions (x_1, y_1) and (x_2, y_2), namely,

$$\begin{aligned} x_{1,2} &= \frac{b - 1 \pm \sqrt{(b-1)^2 + 4a}}{2a} \\ y_{1,2} &= bx_{1,2}. \end{aligned}$$

From now on we will consider only the choice $b = 0.3$, which is the parameter in the attractor suggested by Hénon. The discriminant $(b-1)^2 + 4a$ is negative, if

$$a < a_0 = -\frac{(b-1)^2}{4} = -0.1225.$$

No fixed points exist in this case. Only when a grows above $a_0 = -0.1225$, do both fixed points (x_1, y_1) and (x_2, y_2) exist and the first one of them is attracting. This is exactly the scenario of the tangent bifurcation (or saddle-node bifurcation) discussed in chapter 11, page 596.

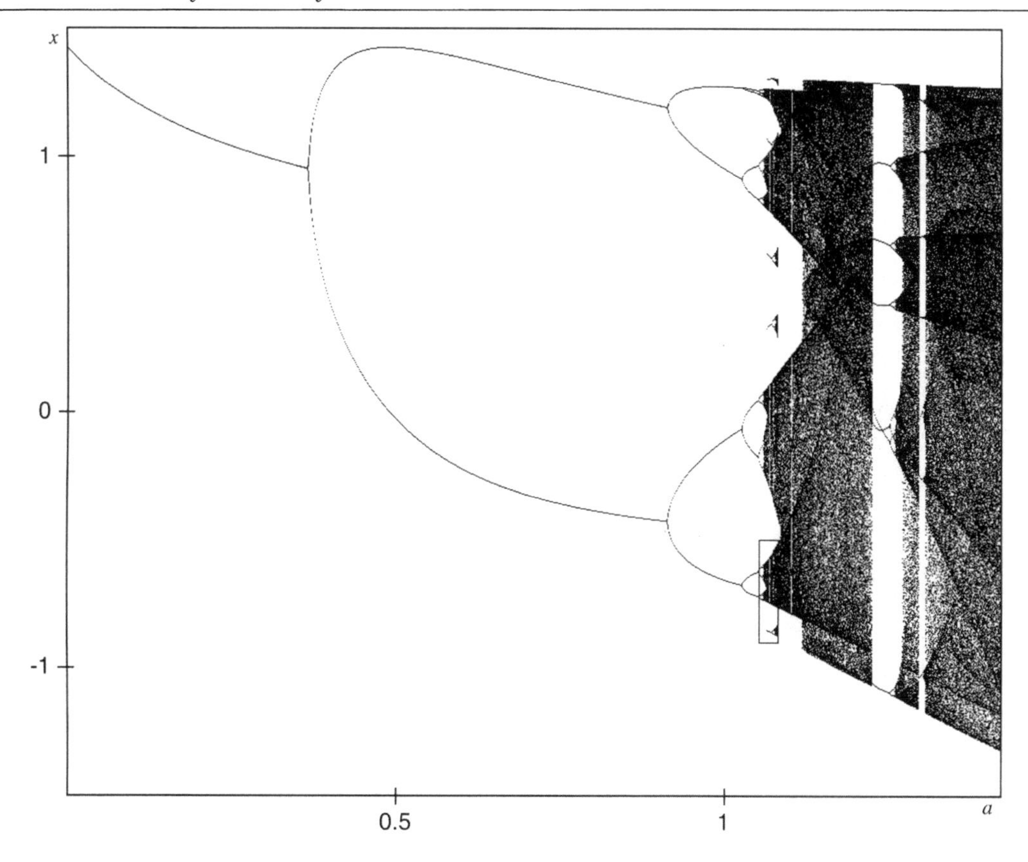

Figure 12.15 : We fix the parameter $b = 0.3$ in Hénon's transformation (12.1). The graph has as horizontal axis the range of parameters $0 < a < 1.5$. For each parameter value of a several initial points were taken and iterated. Note the spurious small specs at parameter value about 1.08. The bottom one is in the the boxed detail. This is enlarged in figure 12.16 revealing that the small spec corresponds to another complete Feigenbaum scenario.

The Feigenbaum Scenario

When the parameter a further increases, the attractive fixed point eventually becomes unstable and gives rise to an attractive cycle of period 2. This transition happens at the value $a_1 = 0.3675$.[16] As expected, the whole sequence of period-doubling bifurcations which is so familiar from the quadratic iterator also appears here in the Hénon transformation for $b = 0.3$ as a increases (see figure 12.15 and table 12.14).[17] The sequence of parameters for the period-doubling is almost geometric as for the quadratic iteration. And the table confirms that the 'universal' Feigenbaum constant 4.6692... is also encountered in this two-dimensional system.

[16]This value can be derived explicitly using the linearization (i.e., the derivative) of Hénon's transformation at the attractive fixed point (x_1, y_1). Its eigenvalues are given by $\lambda_{1,2} = -ax_1 \pm (a^2x_1^2 + b)^{1/2}$, and precisely at $a_1 = 3(b-1)^2/4 = 0.3675$ the second eigenvalue passes through $\lambda_2 = -1$, which signals a period-doubling bifurcation.

[17]The table is adapted from B. Derrida, A. Gervois, Y. Pomeau, *Universal metric properties of bifurcations of endomorphisms,* J. Phys. A: Math. Gen. 12, 3 (1979) 269–296.

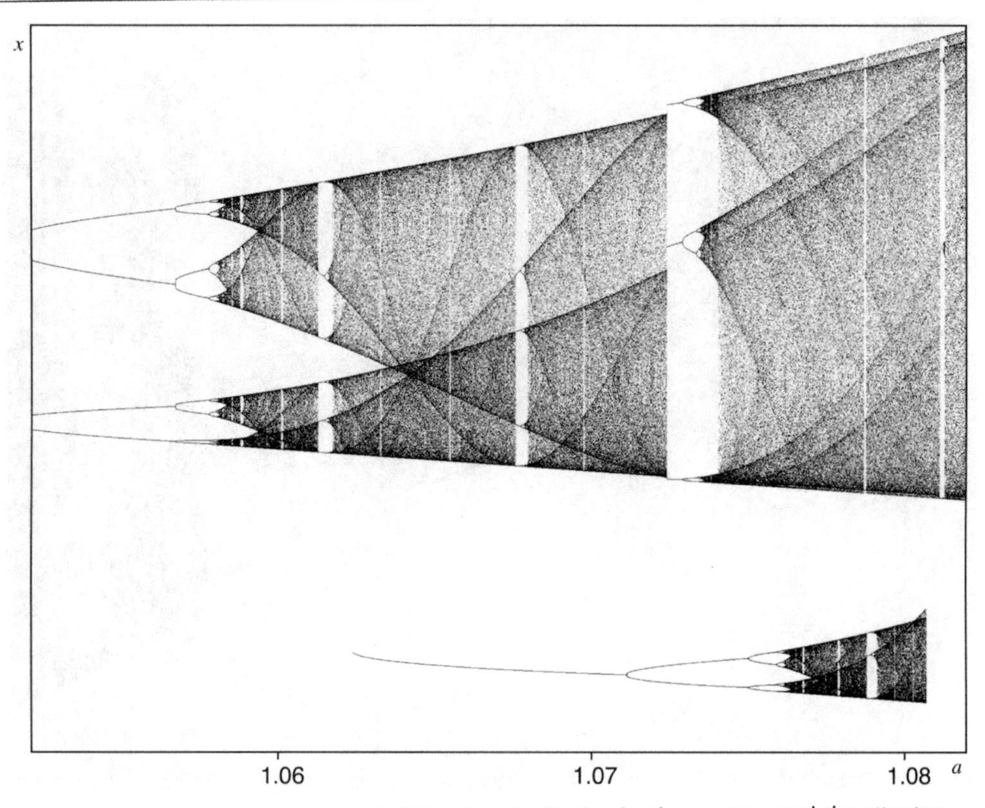

Figure 12.16 : The small boxed region in figure 12.15 is enlarged indicating that there are two coexisting attractors which are shown in the following figure.

Two Attractors

We are tempted to conjecture that everything is the same as in the one-dimensional case. But this is wrong. One important feature which can only arise here is included in the final-state diagram. This is the small version of the Feigenbaum tree in the lower part of the diagram in figure 12.16. It corresponds to a separate attractor. Thus, for the parameters about 1.08 there are two attractors with two corresponding basins of attraction (see figure 12.17). This cannot happen in the one-dimensional quadratic transformation.[18]

The Next Step

A spectacular success towards proving what the computer experiments seem to reveal were reported by Michael Benedicks and Lennart Carleson in 1991.[19] They were able to show that the chaotic dynamics which are present in the case $b = 0$ carry over to small values $b > 0$. For each sufficiently small $b > 0$ there are many parameters a such that the dynamics are beyond any doubt chaotic. In fact, when choosing parameter a at random, there is a

[18] A very extensive analysis of the scenario of appearing and disappearing strange attractors and periodic orbits has been carried out in C. Simó, *On the Hénon-Pomeau attractor,* Journal of Statistical Physics 21,4 (1979) 465–494. See also F. R. Marotto, *Chaotic behavior in the Hénon mapping*, Comm. Math. Phys. 68 (1979) 187–194.

[19] M. Benedicks and L. Carleson, *The dynamics of the Hénon map,* Annals of Mathematics 133,1 (1991) 73–169.

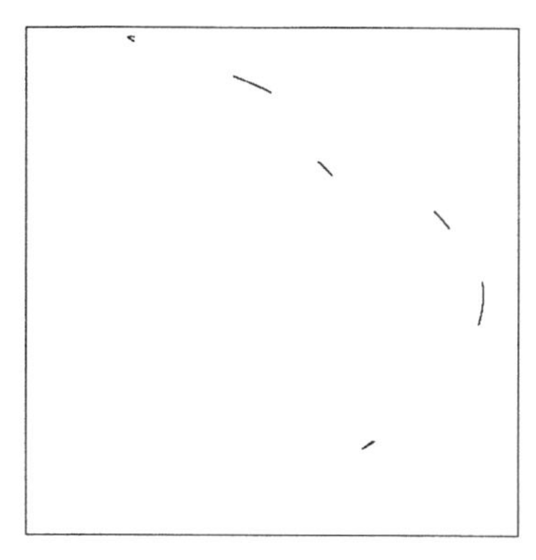
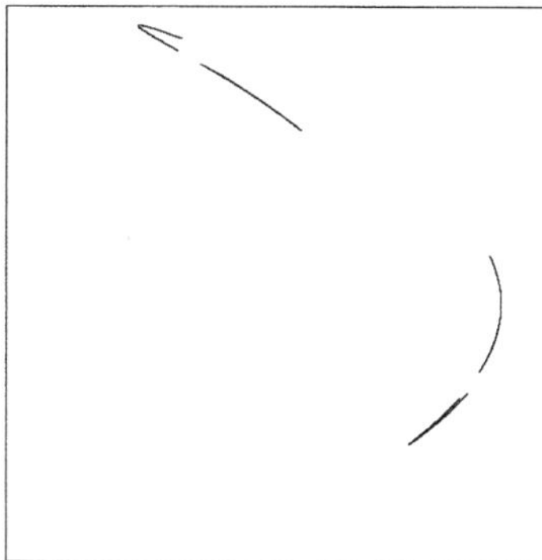

Coexistence of Two Attractors

For $a = 1.08$, $b = 0.3$ two attractors coexist. The left attractor belongs to the lower piece of figure 12.16 (initial point $(0.5, 0.0)$) while the right attractor corresponds to the main branch in the final-state diagram (initial point $(-0.1111, -0.105555)$).

Figure 12.17

positive probability that we get chaotic dynamics. The proof, however, does not reveal what these parameter values precisely are. And perhaps we will never know.

12.2 Continuous Dynamical Systems: Differential Equations

Differential equations have become the language in which modern science encodes the laws of nature. The victories of this approach are numerous. They reach from the laws governing the motion of the planets, to the laws of electromagnetism describing the orbit of an elementary particle in an accelerator, to the air flow carrying an airplane, to the models for the generation of blood cells in the bone marrow, to the mathematical model for numerical weather prediction. The mathematics of differential equations is not elementary. It is one of the great achievements made possible by calculus. Lorenz's discovery of a strange attractor was made in the numerical study of a set of differential equations which he had refined from mathematical models used for testing weather prediction. Although the topic of differential equations is some 300 years old and the results have filled libraries, nobody would have thought it possible that differential equations could behave as chaotically as Lorenz found in his experiments. Moreover, the computer was more than an aid in Lorenz's discovery; it was absolutely crucial, as crucial as for the discovery of Feigenbaum's universality. Up to today a rigorous mathematical understanding of Lorenz's discovery is still open. Lorenz is a meteorologist with a strong mathematical background. In fact one of his teachers at Harvard was George D. Birkhoff who was one of the historical fathers of modern chaos theory.

It is not our aim to introduce a theory of differential equations in this book. After all, we have tried to carefully avoid using methods of calculus wherever we thought it was possible. This is not because we do not appreciate calculus. How could we not? In fact, much of our own research work has been and is in differential equations. But we felt that we should try to explain the major thoughts of fractals and chaos without reference to that more advanced branch of mathematics as far as we could. Here we cannot avoid the subject any more, and we will try to provide a glimpse at what differential equations are about, though without history and any breadth.

Time Steps for the Logistic Equation

Since Lorenz's discovery was made in numerical simulations of differential equations, we think it is a good idea to approach differential equations from the numerical side. We will do that with an old friend: the logistic equation

$$p_{n+1} = p_n + rp_n(1 - p_n) \tag{12.6}$$

with initial value p_0. We rewrite this equation in the form

$$p_{n+1} - p_n = rp_n(1 - p_n). \tag{12.7}$$

Here the left-hand side of the equation is the growth of the population from one generation to the next, and p_n denotes the population size at time n, where time is measured in generations. Let us now use a slightly different way to indicate the same dependence substituting $p(n)$ for p_n. When time is measured in discrete steps like $n = 1, 2, 3, \ldots$ we prefer the old notation. We now interpret time as a continuous entity and choose the alternative notation. Let us assume we are given the population sizes $p(t)$ and $p(t + \Delta t)$, where

Δt denotes a small increment. Based on these numbers we can estimate the total population change per unit time, i.e., per generation, as

$$\frac{p(t + \Delta t) - p(t)}{\Delta t}. \tag{12.8}$$

When substituting t for n in this expression for the left side in eqn. (12.7) we obtain a modified logistic equation

$$\frac{p(t + \Delta t) - p(t)}{\Delta t} = rp(t)(1 - p(t)).$$

Note that in eqn. (12.7) $\Delta t = 1$.

Computing Time Series for Different Step Sizes

Now we can compute the population size $p(t + \Delta t)$ from that at time t by solving the above equation,

$$\begin{aligned} p(t + \Delta t) &= p(t) + \Delta t r p(t)(1 - p(t)) \\ p(0) &= p_0. \end{aligned} \tag{12.9}$$

Given an initial value p_0, a parameter r, and the time step size Δt, this formula generates the complete list of population sizes $p(k\Delta t), k = 0, 1, 2, \ldots$ Figure 12.18 provides graphs for $r = 3$. The plots show population sizes versus time for four different step sizes and three initial points each. As Δt becomes smaller and smaller, the resulting iteration will converge towards 1 in smaller and smaller steps. In fact, the time series which we will observe will look more and more like a smooth curve and there is a deep reason for that. As $\Delta t \to 0$ the iteration in eqn. (12.9) makes a transition into the world of differential equations and in turn can be seen as a numerical approximation of that differential equation. In fact, what we are looking at is the famous Euler's method, which is only one of a whole variety of numerical schemes.

Transition to the Differential Equation

Let us assume that the population is developing along a smooth curve $p(t)$. Then the estimate (12.8) for the population change per unit time can be interpreted graphically as the slope of the secant going through the points $(t, p(t))$ and $(t + \Delta t, p(t + \Delta t))$ (see figure 12.19). As we let $\Delta t \to 0$, the resulting secants appear to approximate the tangent to the curve $p(t)$ in the point $(t, p(t))$. The slope of that tangent would be obtained as we let $\Delta t \to 0$ in eqn. (12.8). In fact, we say that $p(t)$ has a tangent in $(t, p(t))$ with slope m provided the limit

$$\lim_{\Delta t \to 0} \frac{p(t + \Delta t) - p(t)}{\Delta t} = m$$

exists. This limit is denoted by $p'(t)$ and is called the *derivative* of $p(t)$ at t. Replacing the expression $p(t + \Delta t) - p(t)/\Delta t$ by the derivative $p'(t)$, we arrive at the limit equation, which is a *differential equation* for $p(t)$,

$$\begin{aligned} p'(t) &= rp(t)(1 - p(t)) \\ p(0) &= p_0 \end{aligned} \tag{12.10}$$

In other words, eqn. (12.10) means that we are looking for a function $p(t)$, such that $p(0) = p_0$ and such that for each $t > 0$ the graph $(t, p(t))$ of $p(t)$ has

Verhulst Dynamics

In each diagram, several initial conditions are taken and iterated. The step sizes are $\Delta t = 1.0, 0.5, 0.1, 0.001$ (from top to bottom).

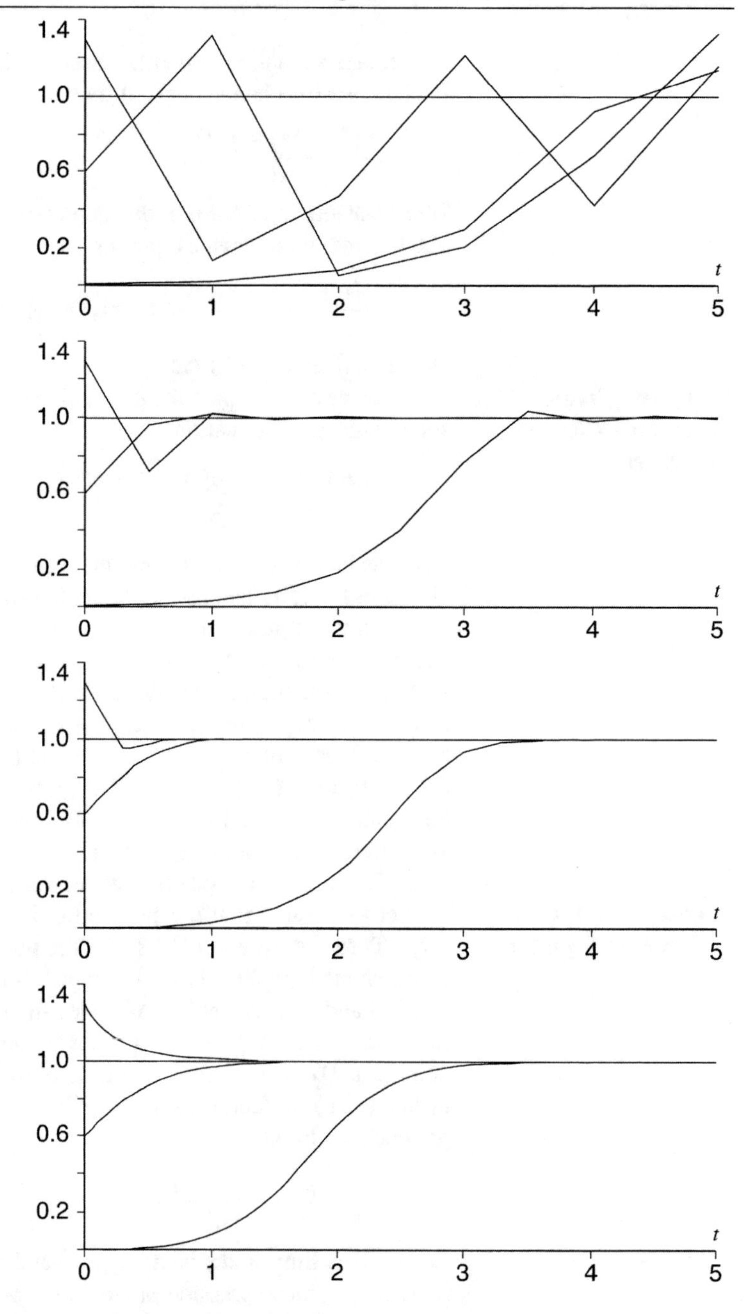

Figure 12.18

a tangent with slope $rp(t)(1 - p(t))$. Such a function $p(t)$ is called a solution of the differential equation (12.10) with initial value $p(0) = p_0$. A problem like eqn. (12.10) is called an *initial value problem*. This particular equation

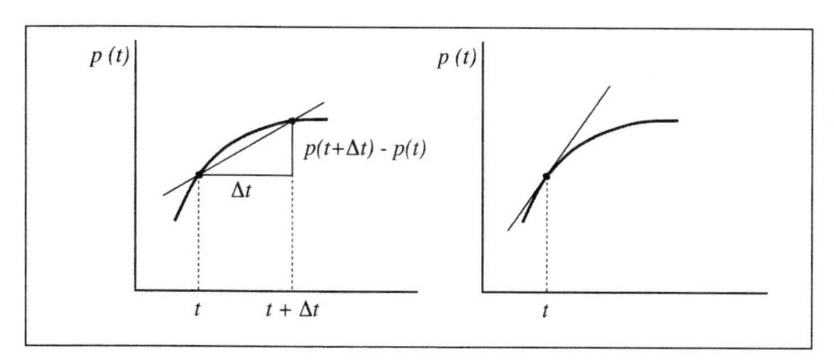

Secant and Tangent

Slope of secant and tangent as limit of secants as $\Delta t \to 0$.

Figure 12.19

can be solved analytically. The result is the same regardless of the parameter r; the solution for any given initial value $p_0 > 0$ monotonically[20] tends to the fully saturated population size $p = 1$. This is illustrated in the bottom part of figure 12.18, which shows close approximations to the true solutions.

Solving the Differential Equation

Since the right-hand side of eqn. (12.10) is nonlinear, this is an example of a nonlinear differential equation. Usually it is hard or impossible to provide an explicit solution for nonlinear differential equations, while there is a complete theory to solve linear differential equations. In our case we are lucky to be able to relate eqn. (12.10) to a linear differential equation by introducing the variable

$$x(t) = \frac{1}{p(t)}.$$

Then formal calculation using eqn. (12.10) gives

$$x'(t) = -\frac{p'(t)}{(p(t))^2} = -\frac{rp(t)(1-p(t))}{(p(t))^2} = -rx(t) + r.$$

When setting $x_0 = 1/p_0$ we obtain the initial value problem

$$\begin{aligned} x'(t) &= -rx(t) + r \\ x(0) &= x_0. \end{aligned}$$

This linear differential equation has the explicit solution

$$x(t) = 1 - (1 - x_0)e^{-rt}.$$

From this formula we read off that if $0 < x_0 < 1$, then $0 < x(t) < 1$ for all times $t > 0$ and $r > 0$. Moreover, the solution approaches the constant 1, $x(t) \to 1$ as $t \to \infty$. Likewise, if $x_0 > 1$, then $x(t) > 1$ for all $t > 0$ and again $x(t) \to 1$ as $t \to \infty$. This translates directly to the populations $p(t) = 1/x(t)$, the solutions of the differential equation (12.10). We have

$$p(t) = \frac{1}{x(t)} = \frac{1}{1 - (1 - \frac{1}{p_0})e^{-rt}} = \frac{p_0 e^{rt}}{p_0 e^{rt} - p_0 + 1}$$

[20] If $p_0 < 1$, then the solution $p(t)$ is a monotonically increasing function. For initial values $p_0 > 1$ it is monotonically decreasing. In both cases $\lim_{t\to\infty} p(t) = 1$.

Graphical Iteration

Iteration of $p(t + \Delta t) = p(t) + \Delta trp(t)(1 - p(t))$ for different values of Δtr which correspond to the values used in the upper three plots in figure 12.18.

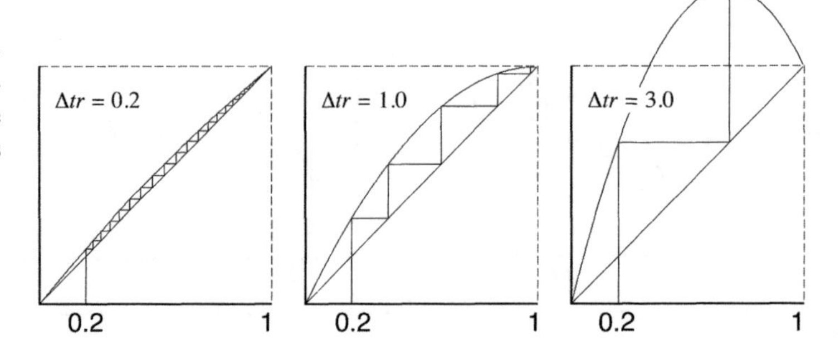

Figure 12.20

and in either case the population will go into saturation, $p(t) \to 1$. The various choices of $r > 0$ will only affect how fast saturation is reached.

True Solution Versus Approximation

This is dramatically different from the discrete model eqn. (12.9), where we see that the numerical solution substantially depends on the choice of the step size Δt. If the step size is large, as in the upper plot of figure 12.18, the solution is not monotonic and does not even approach the saturation $p = 1$. Let us explain why this is the case. Compare the numerical solution

$$p(t + \Delta t) = p(t) + \Delta trp(t)(1 - p(t))$$

with the original logistic equation

$$p_{n+1} = p_n + rp_n(1 - p_n)$$

Note that these formulas coincide for $\Delta t = 1$. Or interpreted differently, that means that the growth law for arbitrary time steps Δt reduces to the original Verhulst model by replacing the expression Δtr by the parameter r. We know quite well from chapters 10 and 11 that changing the parameter r may dramatically effect the behavior of the corresponding orbits. In other words, if we let $r = 3$, for example,[21] then we have chaotic orbits for $\Delta t = 1$, while we have orbits converging to 1 for $0 < \Delta t < 2/3$. Look at figure 12.20 to understand this change in behavior in terms of graphical iteration.

Stability Condition

In terms of the numerical approximation for the differential equation this means that for a given parameter r we have to restrict the step size Δt so that

$$\Delta tr < 2 \text{ or } \Delta t < \frac{2}{r} \tag{12.11}$$

in order to achieve at least the convergence to the saturation $p = 1$. A condition like this is called a *stability condition* for the numerical approximation. Conditions of this sort were first observed by Richard Courant, Hans Lewy,

[21] Recall that the logistic equation is equivalent to the quadratic iterator $x_{n+1} = ax_n(1 - x_n)$ using $x_n = rp_n/(r + 1)$ and $a = r + 1$ (see chapter 1, page 58).

and Kurt Otto Friedrichs and were the initial impulse in the development of the field of *numerical stability analysis*.

The relation of the differential equation (12.10) to its numerical approximation in eqn. (12.6) is very delicate. The stability condition eqn. (12.11) is a reflection of that. Another one lies in the fact that the continuous model eqn. (12.10), though formally nonlinear, is actually related to a linear problem, much unlike the discrete approximation eqn. (12.6) which is really nonlinear. This shows dramatically that passing to limits — or passing to discrete approximations — may change the nature of a problem significantly, a fact which has only entered the conscience of numerical analysts quite recently. This is another merit of chaos theory.

Numerical Methods

Differential equations such as $x'(t) = f(x(t))$ are one of the most important tools for modeling processes in the natural sciences, particularly in physics. Thus, there is a continuously growing body of mathematical research on the various different types of differential equations. However, for most equations considered, there is no known solution that we could write down in terms of a common formula. Only the numerical approximation with the help of the computer is possible; and, in fact, widely used. There are plenty of numerical methods available for this task. The first one, given by

$$x_{k+1} = x_k + \Delta t f(x_k),$$

which we have already introduced above. It is called Euler's method, and the step from one point x_k to the following, x_{k+1} is called an Euler step. The smaller the step size Δt the more accurate is the solution. Euler's method is of first order, meaning that the error of the numerical solution relative to the true solution of the differential equation is proportional to the step size Δt. In a second-order method, the error is proportional to the square Δt^2. Such methods are much superior to Euler's method because here the error decreases by a factor of $1/4$ each time the step size is halved. In other words, in order to achieve the same precision we may choose a much larger step size and thus may save a lot of computational effort in comparison to Euler's method. As an example for a second-order method we mention

$$x_{k+1} = x_k + \frac{\Delta t}{2}(f(x_k) + f(x_k + \Delta t f(x_k))),$$

which is called the trapezoidal or Heun's method. Both methods may also be used when the variable x is a vector of several components, which is the case considered in the following. Almost all textbooks on numerical methods present an analysis of such methods and some others. There exist many computer codes for higher-order methods which even adjust the step size to the local properties of the solution.[22]

[22] For most figures and computations presented in this book, we have used the code in *Numerical Recipes in C* by W. H. Press, B. P. Flannery, S. A. Teukolsky, W. T. Vetterling, Cambridge University Press, Cambridge, 1988.

No Chaos in Dimensions 1 and 2

One of the crucial differences between the discrete system eqn. (12.6) and the continuous counterpart eqn. (12.10) is the fact that it is plainly impossible for the dynamics of the differential equation to be chaotic. The reason is that no two trajectories (as shown in the lower part of figure 12.18 when $\Delta t \to 0$) can cross each other. This is important.

Life in the real world, however, is not so simple. In almost any physical system the state cannot be described by a single variable such as a population in the Verhulst model. Usually there are two, three and more (sometimes many more) variables necessary. For example, with a pendulum which is allowed to swing in a plane, we need to know angle and angular momentum (besides the configuration of the pendulum) in order to be able to forecast the motion. Or to compute the orbit of an asteroid, we need to know position and velocity, each one of which has three components for each of the spatial directions, giving a total of six variables. It turns out that in order to find chaos in such continuous dynamical systems we need to consider at least three-dimensional systems. As in the one-dimensional case, trajectories must not cross; and this implies that trajectories in a plane cannot act chaotically. They typically converge to a point, or escape to infinity, or perhaps spiral around closing in on some loop.[23] But in three dimensions chaos may reign; and it is the general consensus that in dynamical systems in nature, chaos is typical rather than the exception.

Approximation Method in 3 Dimensions

Let us therefore now briefly introduce a more general viewpoint, which is necessary to comprehend Lorenz's work. Consider the differential equation

$$\begin{aligned} x'(t) &= f(x(t)) \\ x(0) &= x_0 \end{aligned}$$

where $f(x)$ is some function. How would we think about this equation? Just like we did in the above special case. We would use the numerical Euler approximation

$$\begin{aligned} \frac{x(t+\Delta t) - x(t)}{\Delta t} &= f(x(t)) \\ x(0) &= x_0 \end{aligned}$$

for Δt small and pass to the limit $\Delta t \to 0$. With this in mind we can easily go one step further and look at systems of differential equations. Let us take real-valued functions $f(x, y, z)$, $g(x, y, z)$, and $h(x, y, z)$ and consider the system

$$\begin{aligned} x'(t) &= f(x(t), y(t), z(t)), \quad x(0) = x_0 \\ y'(t) &= g(x(t), y(t), z(t)), \quad y(0) = y_0 \\ z'(t) &= h(x(t), y(t), z(t)), \quad z(0) = z_0. \end{aligned}$$

As for one equation, we would just set up a system of numerical approxima-

[23]The basic tool for understanding planar dynamical systems is the Poincaré-Bendixson theory (see chapter 11 in M. W. Hirsch and S. Smale, *Differential Equations, Dynamical Systems, and Linear Algebra,* Academic Press, New York, 1974).

tions

$$\begin{aligned}
\frac{x(t+\Delta t)-x(t)}{\Delta t} &= f(x(t),y(t),z(t)), \quad x(0)=x_0\\
\frac{y(t+\Delta t)-y(t)}{\Delta t} &= g(x(t),y(t),z(t)), \quad y(0)=y_0\\
\frac{z(t+\Delta t)-z(t)}{\Delta t} &= h(x(t),y(t),z(t)), \quad z(0)=z_0,
\end{aligned}$$

and construct solutions by letting $\Delta t \to 0$.

12.3 The Rössler Attractor

The Rössler Model

In 1976 Otto E. Rössler found a particularly simple system, which is probably the most elementary geometric construction of chaos in continuous systems. Thus, before we start to discuss the Lorenz attractor in detail, let us follow Rössler's ideas.[24] His system of differential equations is

$$\begin{aligned} x' &= -(y+z) \\ y' &= x+ay \\ z' &= b+xz-cz \end{aligned} \tag{12.12}$$

where the three coefficients a, b, c are adjustable constants. In this section, we will fix the parameters a and b and change only c.

$$a = 0.2, \quad b = 0.2.$$

Otto E. Rössler

Figure 12.21

This system is the same type as that of the continuous Verhulst system in eqn. (12.10), with the difference that here we are given three variables x, y and z instead of the one for the population, p. Thus, we can interpret the system in eqn. (12.12) as a collection of laws of motion for a point at coordinates (x, y, z) in three-dimensional space. For any given initial coordinates (x_0, y_0, z_0) the system defines a unique trajectory which is parametrized by time t and satisfies the equations at all times. Denoting the coordinates of this trajectory by $(x(t), y(t), z(t))$ for time $t \geq 0$ this means

$$\begin{aligned} x'(t) &= -(y(t)+z(t)) & x(0) &= x_0 \\ y'(t) &= x(t)+ay(t) & y(0) &= y_0 \\ z'(t) &= b+x(t)z(t)-cz(t) & z(0) &= z_0. \end{aligned}$$

[24] Rössler, O. E., *An equation for continuous chaos,* Phys. Lett. 57A (1976) 397–398.

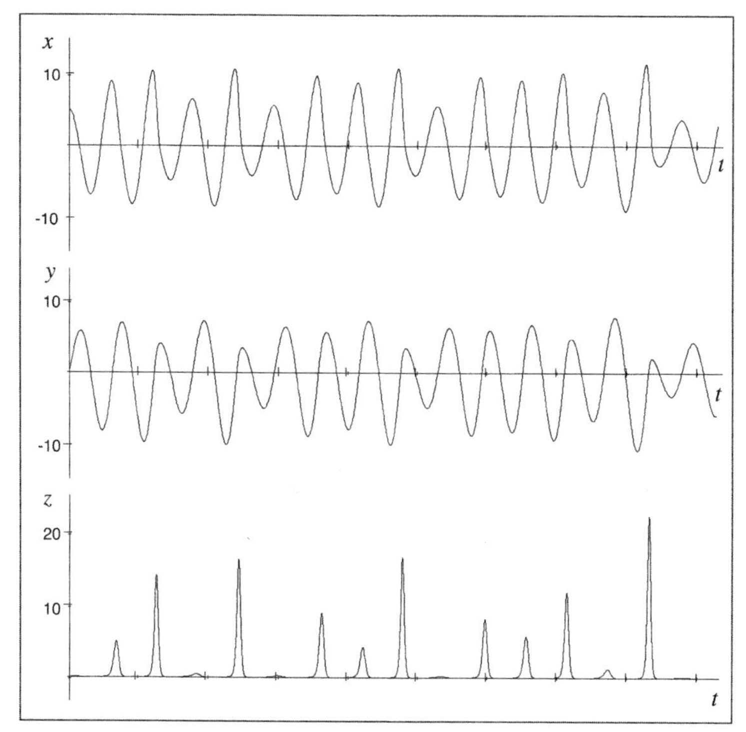

A Trajectory

The initial condition $(-1, 0, 0)$ for Rössler's system eqn. (12.12) produces these plots of $x(t)$, $y(t)$ and $z(t)$ versus time t. The parameter is $c = 5.7$.

Figure 12.22

Figure 12.22 shows a plot of the three components $x(t)$, $y(t)$, $z(t)$ versus time t. However, it is more instructive to plot this trajectory in three-dimensional space with coordinates (x, y, z) and consecutive points connected by short line segments; we obtain a first picture of the Rössler attractor (see figure 12.23). Orbits on the attractor spend most of their time near the xy-plane spiraling out from the origin. When an orbit has attained some critical distance from the origin it is first lifted away from the xy-plane. Then, after reaching some maximal z-value, it is reinserted into the spiraling piece of the attractor close to the plane. The larger the z-amplitude of this excursion has been, the closer to the origin the orbit will land, and the spiraling process followed by ejection and reinsertion repeats (see figure 12.25).

The Attractive Property

Let us first take a look at the phenomenology of the attractor by means of some simple numerical experiments before we try to understand how the equations provide the foundations for these effects. First of all we note that we are indeed dealing with an attractor. When we start the solution of the differential equation at some other initial point somewhere in the vicinity of the structure shown in figure 12.23, we get essentially the same result. Only the first part of the trajectory is noticeably different stemming from the transitional period necessary for the solution to get close to the attractor.

The Rössler Attractor

The trajectory from figure 12.22 is plotted in three-dimensional space revealing a first picture of the Rössler attractor. Two projections are given in figure 12.24.

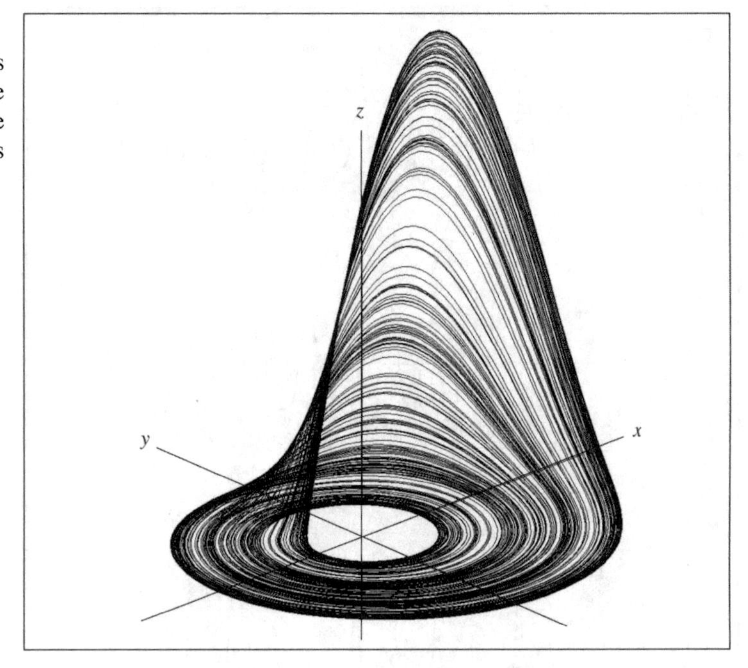

Figure 12.23

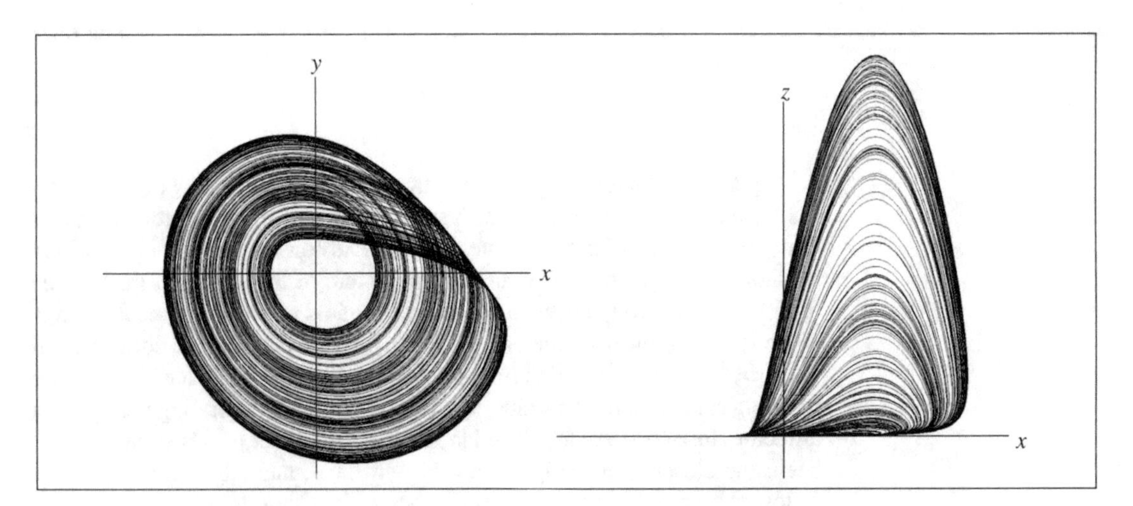

Figure 12.24 : The Rössler attractor, top view (left) and side view (right).

The Kneading Transformation

In order to understand the chaotic behavior of the dynamics in Rössler's attractor, we begin by showing how the nonlinear stretch-and-fold operation is hidden in the system. From the pictures, it appears that the attractor has the structure of a *folded band*. Starting from a section across the band near the negative x-axis we can observe two effects (see figure 12.26).

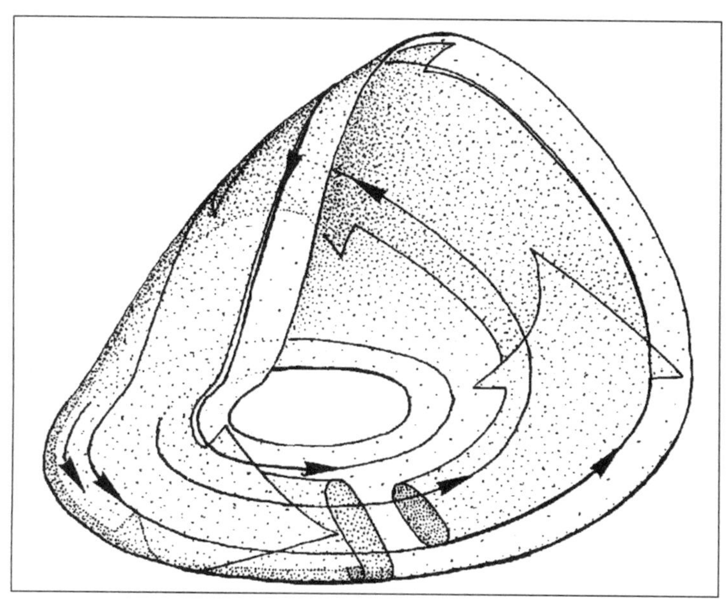

Dynamics on the Attractor

Schematic graph of the dynamics on the Rössler attractor. Reprinted with permission from R. H. Abraham, C. D. Shaw, *Dynamics, The Geometry of Behavior,* Part Two: *Chaotic Behavior,* 1983, Aerial Press, Santa Cruz, second edition Addison-Wesley, 1992.

Figure 12.25

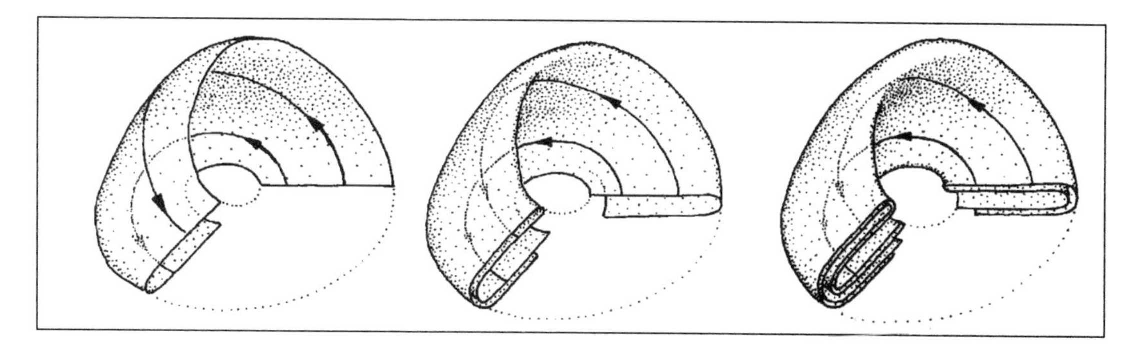

Figure 12.26 : Three iterations of the stretch-and-fold action are illustrated here for an initial line segment. Reprinted with permission from R. H. Abraham, C. D. Shaw, *Dynamics, The Geometry of Behavior,* Part Two: *Chaotic Behavior,* 1983, Aerial Press, Santa Cruz, second edition Addison-Wesley, 1992.

1. As the band winds around the center for about half of a turn its width increases. This corresponds to the stretching in the dough analogy.
2. Near the positive x-axis the band is more than twice as wide and the outer part begins to fold over and eventually covers the inner part and also part of the 'hole' at the center. This folding action is completed after about another half of a turn and the process then repeats.

The One-Dimensional Return Map

In other words, what we see here in three dimensions is essentially the same as the iteration of stretch-and-fold operations. Therefore, there must

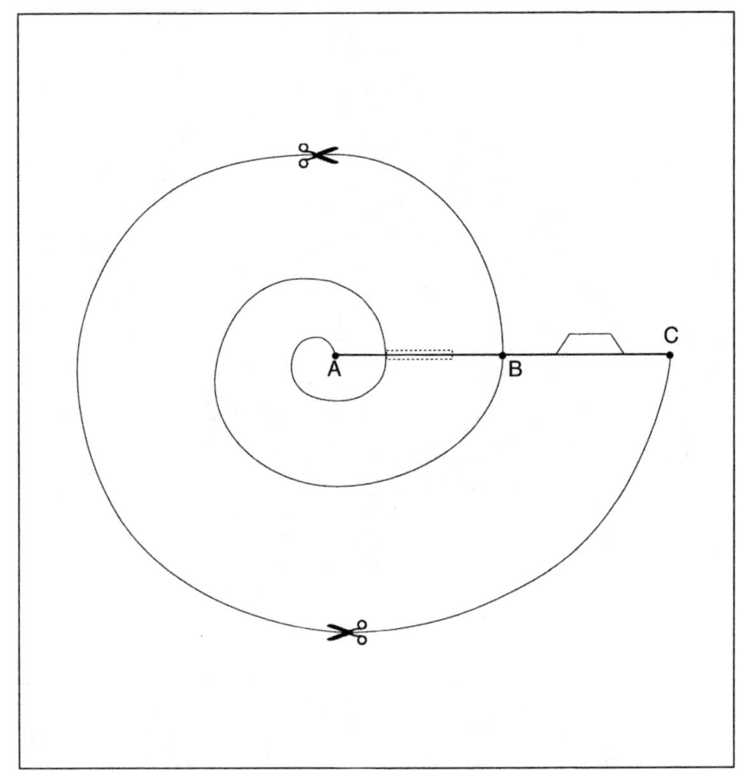

Paper Model of Rössler Attractor

To attain a three-dimensional paper model of the Rössler attractor follow these steps: 1. Make a photo copy of the figure enlarging it by a factor of two. 2. Using scissors cut the figure along the outline. 3. Cut a small slit at the position indicated between points A and B. 4. Fold at point B so that point C comes to lie on point A. 5. Insert the little flap between points B and C through the slit and fix it to the underside of the figure with some glue. Now we can follow around the trajectories of the system. The one already drawn is a periodic one which circles around the center 3 times before repeating. Hint: the lines are appearing only on one side of the paper, it may be more instructive to also draw them on the reverse side before starting the paper construction.

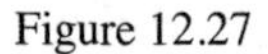
Figure 12.27

also be a direct relation to the quadratic iteration[25]

$$x \to rx(1-x)$$

because its dynamics are also nothing but stretch-and-fold operations. Let us uncover this relation and make it explicit. Recall that the relation between the quadratic transformation and the kneading is very elementary: the value x specifies a particle in the dough and $rx(1-x)$ is its position after one stretch-and-fold operation. Here we can proceed in the same manner. We define a cross-section of the Rössler band as a reference frame — for example, the part of the attractor which lies along the negative x-axis where the x-values of trajectories are at their minimum.[26] The band is quite flat, and therefore we can identify points in the intersection of the attractor and the half-plane simply by the absolute values of their x-coordinates. We regard such points as initial points and follow their trajectories around the band for one complete turn until they reenter the half plane. We arrive at a new position with a new x-coordinate. In order to record the stretch-and-fold operation we can mark in a

[25] Here we have called the parameter r in order to avoid confusion with the parameter a of the Rössler system.

[26] We consider the plane given by the equation $y + z = 0$. This implies that at the points where a trajectory pierces through the plane the value of $x'(t)$ is zero. Thus, the plane cuts the trajectories precisely at their minima and maxima of the x-value.

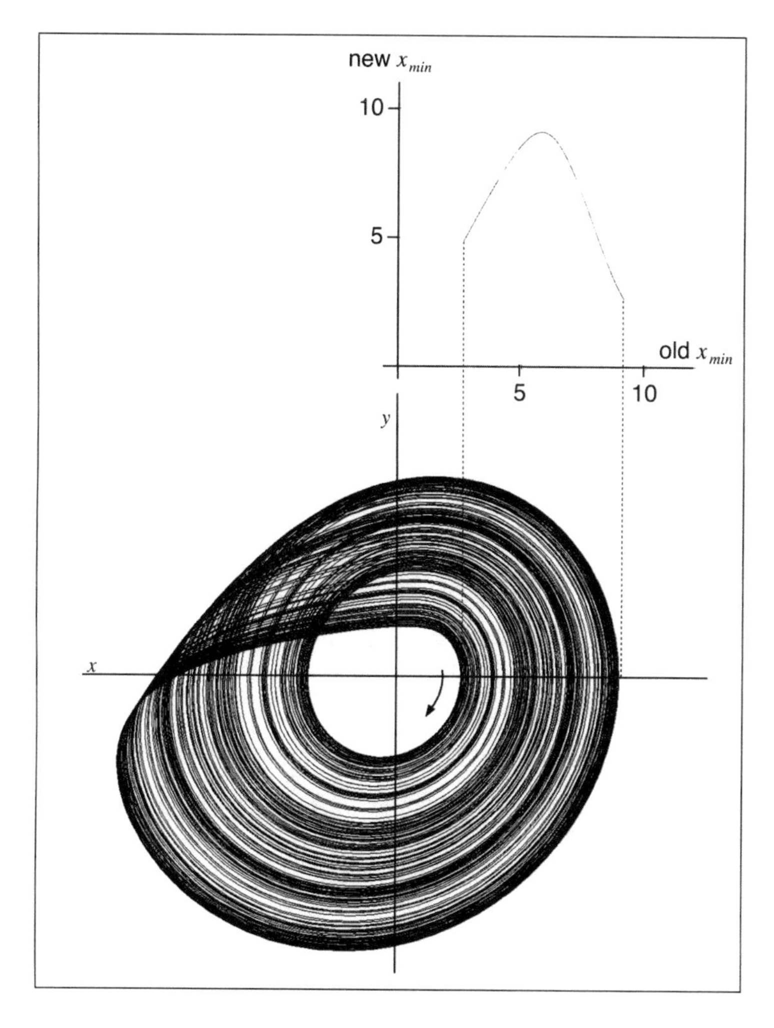

The Lorenz Map for Rössler's Attractor

A trajectory has local minima of x-values near the negative x-axis. After one turn around the attractor the next minimum is attained. The graph above the attractor, the Lorenz map of this system, displays this new x-value as a 'function' of the old one.

Figure 12.28

diagram the absolute value of this new x-value compared to the old one. When performing this procedure for a long trajectory on the Rössler band, we should obtain a good representation of the stretch-and-fold dynamics, which is in the same spirit as the parabola in the graphical iteration method (see figure 12.28). And moreover, the figure indeed reveals a shape which closely resembles a parabola! A function modeling the plot is called a *Lorenz map*. It provides a link between the dynamics of a continuous system and the discrete dynamics of transformations of an interval.[27] Once this link is known, it provides a shortcut for computing the dynamics of the underlying system. Instead of following a trajectory of the differential equation with possibly many steps and a large computational effort, we may simply evaluate the Lorenz map once (or

[27] The discovery of this quasi-one-dimensional character of the dynamics of a system of differential equations was made by Lorenz (see the following section 12.4).

A Periodic Solution

An attracting periodic solution of the Rössler system. $c = 8.0$.

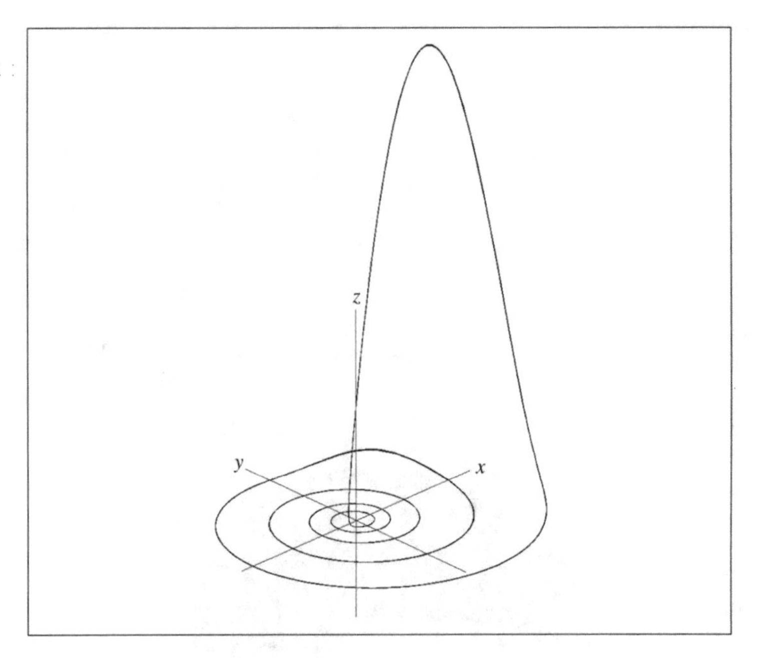

Figure 12.29

perform one step of graphical iteration) and arrive at the same result. What is more, it is usually much easier to analyze the properties of a Lorenz map than the dynamics for a differential equation. Thus, the results of chapters 10 and 11 regarding the chaos for transformations of intervals and the routes leading to it help to understand the dynamics of the more complicated continuous system.

We may conclude that the chaos present in the iteration of the quadratic iterator carries over to the Rössler band by means of the Lorenz map. Of course, when we change the parameters a, b or c in Rössler's system the parabola-like graph of the Lorenz map must change accordingly. In particular, we expect that the band can be replaced by periodicity, implying a periodic attractor in the Lorenz map corresponding to a periodic loop in phase space (x, y, z) for the differential equation (see figure 12.29).

The Feigenbaum Experiment

Let us continue these experiments with the next logical step: drawing a Feigenbaum-type diagram. Does the analogy between the quadratic iterator and the Rössler system go as far as to reproduce the period-doubling cascade or even the Feigenbaum constant 4.6692...? We do not claim to have performed this experiment with utmost numerical care, but the result shown in figure 12.30 seems to suggest that it is true. We retrieve the fundamentals of the spectrum of dynamics from the quadratic iterator.

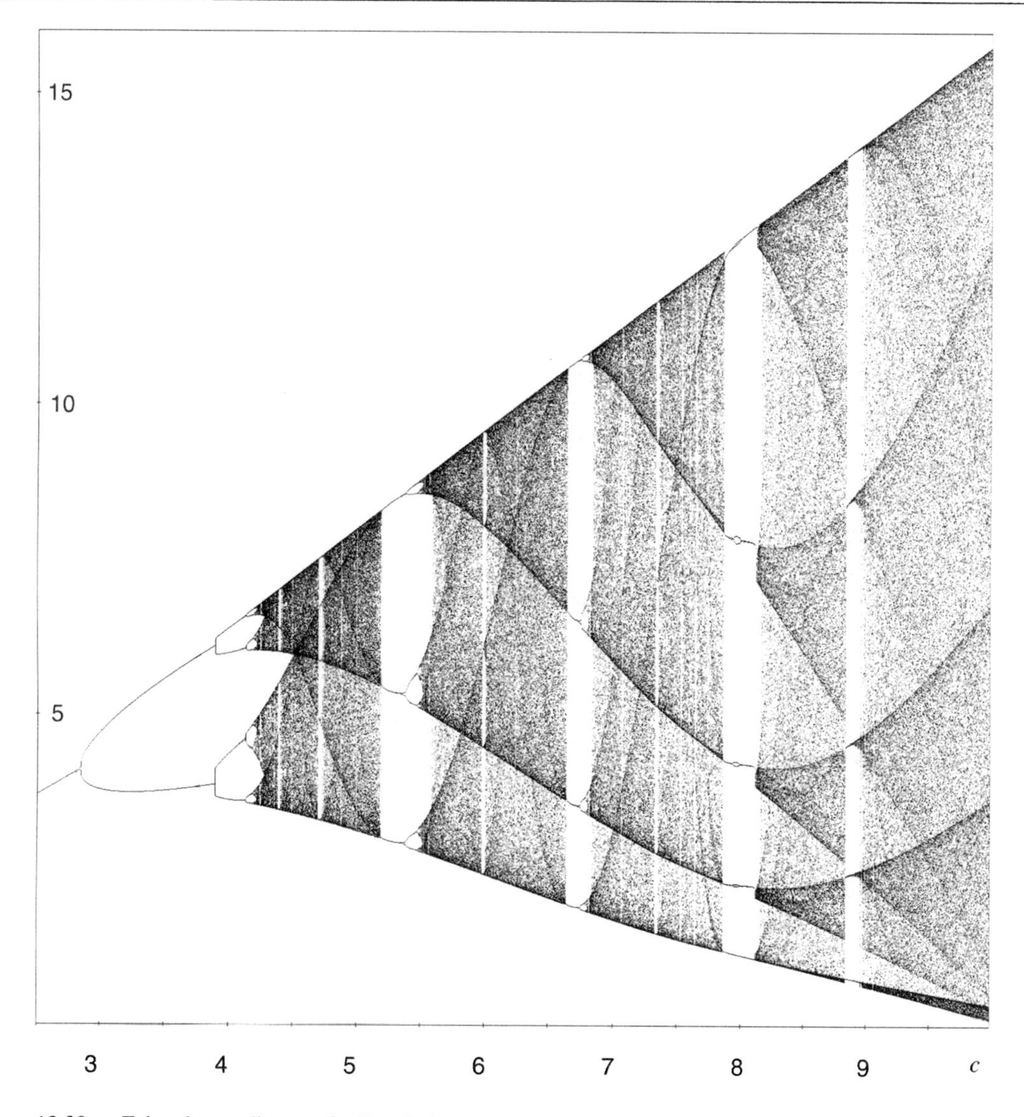

Figure 12.30 : Feigenbaum diagram for Rössler's system. The plot shows the parameter range $2.5 < c < 10.0$. Vertically the absolute values of the minimal x-values of the corresponding trajectories from the attractor are shown. This corresponds to a projection of a Lorenz map diagram such as given in figure 12.28 on the vertical axis. Initially, for small values of c, the attractor consists of a periodic orbit which has only one local minimum of x-values, i.e., it is a single loop. As the parameter c increases, this periodic orbit undergoes a period-doubling bifurcation. Check that for $c = 8$ there are 5 points in the diagram corresponding to the periodic solution shown in figure 12.29 which has 5 loops.

The Strangeness of the Attractor

This may seem like the end of the story of the chaotic attractor, but we have omitted one crucial limitation of its one-dimensional model, the Lorenz map; *it is false*. If it were exact, then this would imply that initial points in the cut through the band come in pairs which would land at precisely the same point when iterated once with the Lorenz map. (A parabola transforms two points into one.[28]) This would violate the fundamental property of the continuous systems discussed here, namely, that trajectories must not cross. If they share a single point then they must be identical for all times from that point on. And this applies in both directions of a trajectory forward and backward in time. So what is really happening? Our experiments are certainly not completely wrong. Similar studies which yield the same qualitative results have, however, been carried out by numerous researchers with more precision and effort. Thus, we are forced to conclude that since the folded part of the Rössler band cannot exactly merge with the other part, it can at least come close to it. Indeed, the two parts of the band can come *very* close to each other, as close as we want them to. Thus, what we see in the pictures after one turn around the attractor is not one band but two tightly packed layers of the band. After another rotation, both of these layers (which still must be separate) fold over and form four layers. Then we get eight, then sixteen and so on. Thus, in effect we should see an infinite number of layered bands stacked up somewhat like a pile of strangely intertwined extra thin pancakes which, moreover, have been subjected to a compactor. Expressed more academically, it is a *Cantor set of sheets*. And because these sheets are so close to each other, we cannot see them in the phase plots or in the figure 12.28, from which we assumed the Lorenz map. Only if we work with extraordinary precision, do we have a chance to see a glimpse of these layers. In other words, the fractal dimension of a chaotic attractor from the Rössler family must be $2 + \varepsilon$ where ε is a very small number.[29] It is this fractal character that qualifies Rössler's attractor as a *strange* one.

Poincaré Sections

The one-dimensional Lorenz map is a simplification of the real dynamics happening in the Rössler system. The dots which accumulate in the diagram of a section of the Rössler band in figure 12.28 do not exactly lie on a curve. They are very close, almost indistinguishably close, to a curve which looks like a parabola. To create a discrete transformation which reflects the true dynamics let us consider a rectangle containing the Rössler band in the section given by the z-axis and the negative part of the x-axis. This rectangle is transversal to the flow of the system which means that trajectories pierce right through it and do not approach it tangentially. A surface with this property is also called a Poincaré section.[30]

The dynamics of the system can be described with a transformation T defined on such a section. Given an initial point p on the section we follow the corresponding trajectory until the section is entered again. That point is taken as $T(p)$ and by the same procedure an image point $T(p)$ is defined for all initial points p of the section. The transformation T is called a Poincaré map (see figure 12.31). In place of a 'complete' trajectory, we may consider the corresponding iterates

[28] When using $rx(1-x)$ these points are $x_0 = 0.5 + d$ and $x_0 = 0.5 - d$, where d can be any number.

[29] The dimension can be estimated between 2.01 and 2.02.

[30] After Jules-Henri Poincaré (1854–1912) who laid the mathematical foundations of modern dynamical systems theory.

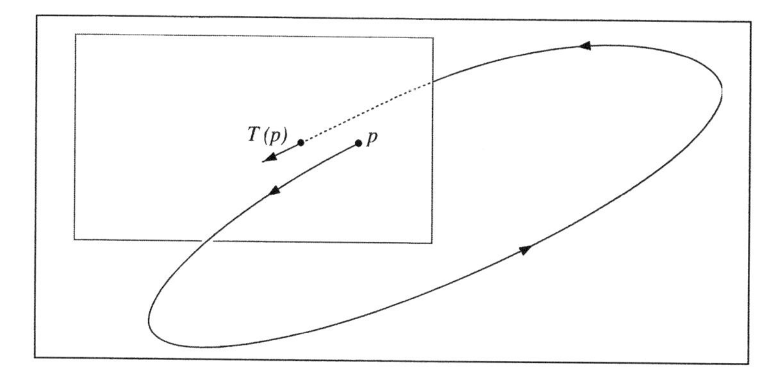

Figure 12.31 : The Poincaré map transforms the point p into $T(p)$ defined by the first reentry point of the trajectory starting at p in the shown cross-section.

of the Poincaré map. For example, a periodic trajectory matches up with a fixed point of the Poincaré map or one of its iterates.

How the Differential Equation Operates

To explain how the differential equations of the Rössler system

$$\begin{aligned} x' &= -(y+z) \\ y' &= x + ay \\ z' &= b + xz - cz \end{aligned}$$

can generate trajectories of the type discussed, we reproduce here the explanation of J. M. T. Thompson and H. B. Stewart from their book Nonlinear Dynamics and Chaos, Wiley, Chichester, 1986, pp. 235–236. This simple, autonomous system has only a single nonlinear term, the product of x and z in the third equation. Each term in these equations serves its function in generating the desired global structure of trajectories. Considering the first two equations, let us for the moment suppose that z is negligibly small. Then the subsystem

$$\begin{aligned} x' &= -y \\ y' &= x + ay \end{aligned}$$

can be transformed to the second-order linear oscillator

$$x'' - xx' + x = 0.$$

With positive a, this oscillator has negative damping, and the origin is an unstable focus for $0 < a < 2$. Thus in the full system of three first-order equations, trajectories near the (x, y) plane spiral outwards from the origin. This produces a spreading of adjacent trajectories, which is the first ingredient in the mixing action of chaos.

This spreading is achieved with only linear terms. But if the full system of three equations were linear, the spreading would merely continue as all trajectories diverge to an infinite distance from the origin. To confine the spreading action within a bounded attractor, the nonlinear term is required. The constant c in the third equation acts as a threshold for switching on the nonlinear folding action. Considering the third equation alone, whenever the value of x is less than the constant c, the coefficient of z is negative, and the z subsystem is stable, tending to restore z to a value near $-b/(x-c)$. However, if x should exceed c, then z will appear in the third equation multiplied by a positive factor, and the previously self-restoring z subsystem diverges. Choosing $b > 0$ ensures that this divergence will be towards positive z.

The effect of this is shown in figure 12.23. A trajectory spirals outwards while appearing to remain in a plane near to and parallel to the (x, y) plane. When x becomes large enough, the z subsystem switches on and the trajectory leaps upwards. Once z becomes large, the z term in the first equation comes into play, and x becomes large and negative, throwing the trajectory back towards smaller x. Eventually x decreases below c, the z variable becomes self-restoring, and the trajectory lands near the (x, y) plane again. Through the feedback of z to the x equation, trajectories are folded back and reinserted closer to the origin, where they begin an outward spiral once more.

12.4 The Lorenz Attractor

The Rössler system is an artificial system designed solely with the purpose of creating a model for a strange attractor which uses only the simplest chaos-generating mechanism, stretch-and-fold. Of course, Rössler knew about the Lorenz system, which had been published 13 years before. In fact, we may say the Rössler attractor is a model of the Lorenz model.

Edward N. Lorenz

Figure 12.32

The Lorenz System

The system of equations that Lorenz proposed does not look any more complicated than that of Rössler. Here it is:

$$\begin{aligned} x' &= -\sigma x + \sigma y \\ y' &= Rx - y - xz \\ z' &= -Bz + xy. \end{aligned} \tag{12.13}$$

The numbers σ, B, and R are the system's physical parameters, which Lorenz fixed at

$$\sigma = 10, \quad B = \frac{8}{3}, \quad R = 28.$$

Figure 12.33 shows the corresponding attractor, which is now called the Lorenz attractor. Clearly the geometry is more involved than for the Rössler band. There are two sheets in which trajectories spiral outwards. When the distance from the center of such a spiral becomes larger than some particular threshold, the solution is ejected from the spiral and attracted by the other spiral, where it again begins to spiral out and the game is repeated. The number of turns that a trajectory spends in one spiral and then in the other is not specified. It may wind around one spiral two times, then three times around the other, then

The Lorenz Attractor

Some trajectories from the Lorenz attractor.

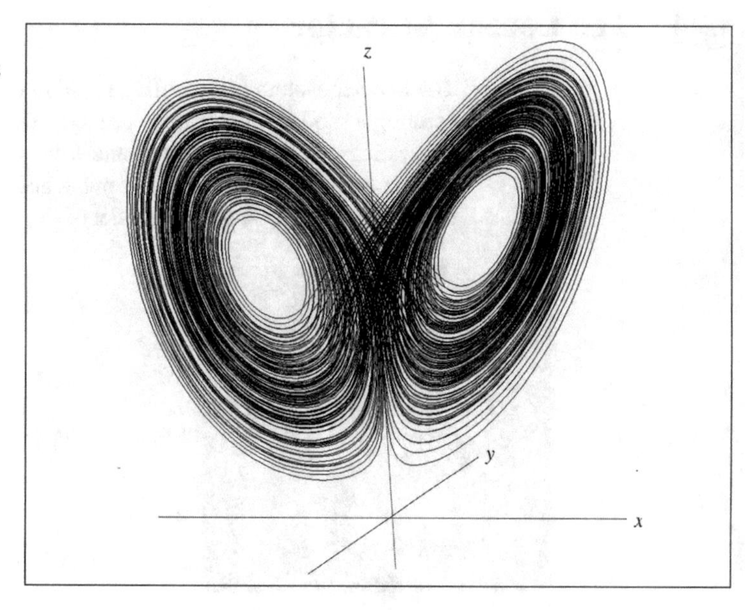

Figure 12.33

ten times around the first and so on. In fact, we believe that for any sequence of positive numbers which are not too large, for example 3, 11, 7, ... there exists a trajectory on the Lorenz attractor with precisely these numbers as turns around the spirals. Thus, there is a solution that turns 3 times around the right spiral, then 11 times around the left, then 7 times around the right again and so on.

Lorenz's Physical Model

What is the connection between these wildly spinning solutions and weather forecasting which is what Lorenz was interested in? Certainly, the trajectories should not be mistaken for the paths of air currents! If this were the case then the Lorenz attractor would act similar to a black hole in astrophysics, sucking in all the atmosphere — leaving nothing but emptiness around it and laying waste to the whole planet Earth. But we are not far from the truth. The Lorenz system is in fact a model of thermal convection which, however, includes not only a description of the motion of some viscous fluid or atmosphere but also the information about distribution of heat, the driving force of thermal convection.

When air is warmed near the Earth's surface it rises. This is an important factor in the atmospheric weather factory. Convection air currents may accumulate and give rise to convection cells of several types and, when forced more vigorously, may produce very turbulent motion in the atmosphere. Examples of convection cells are cylindrical rolls and structures, which are called *Bénard cells*, resembling a honeycomb from above. In these hexagonal cells the warmed portions of the fluid rise in the center, get colder near the top and sink back down to the surface around the boundary of the cell. The Lorenz system is related more to the cylindrical roll type of fluid motion in which one of the dimensions can be disregarded pretending that these rolls extend to infinity. The mathematical model of the fluid motion had originally been developed by Lord Rayleigh[31] in 1916. It assumes that all convection happens in a rectangular region whose bottom is heated such that the temperature difference between bottom and top remains constant. With certain parameter configurations in this model it turns out that the solutions to the model equations have a rather special form which was already known to Rayleigh. Lorenz took these special solutions, regarded their amplitudes as time-dependent, inserted them in the Rayleigh model, disregarded all terms that are not in this special form,[32] and arrived at a system of differential equations for the time-dependent amplitudes, eqn. (12.13).[33]

The Meaning of x, y and z

It is almost impossible to find out what the variables x, y, and z in the Lorenz system precisely stand for without consulting Lorenz's original paper. Let us give some technical details and visualizations of the convection process which was investigated by Lorenz. As already stated we deal with convection currents and temperature distribution in a rectangular region where the temperature difference ΔT between the bottom and the top is kept constant. The dynamics are assumed to be identical in all slices parallel to the rectangular region. The governing equations for the more general three-dimensional problem were worked out by Lord Rayleigh. The simplification to the two-dimensional case considered here is by B. Saltzman.[34]

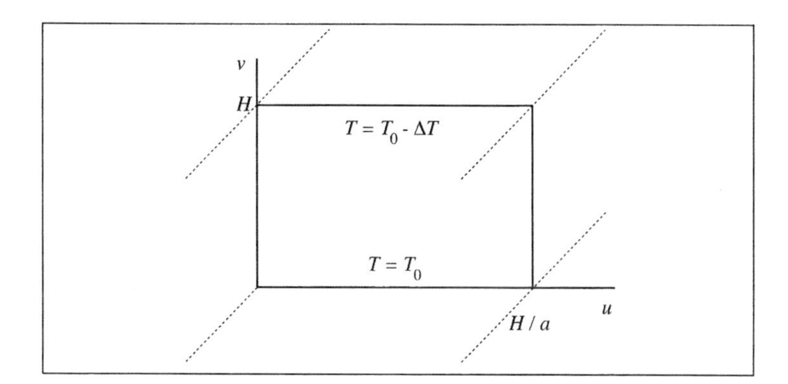

Figure 12.34 : The coordinate system in the cross-section of a bar where the Lorenz equations present a model for fluid flow and temperature.

[31] Lord Rayleigh, *On convective currents in a horizontal layer of fluid when the higher temperature is on the under side,* Phil. Mag. 32 (1916) 529–546.

[32] Except for one particular term related to the temperature distribution.

[33] Another interpretation is that the solutions of the Rayleigh equations can be written as Fourier series with time-dependent coefficients. Using these series in place of the original variables produces a system containing infinitely many equations. When keeping only the three most significant of these we again obtain the Lorenz system given in eqn. (12.13).

[34] B. Saltzman, *Finite amplitude free convection as an initial value problem,* J. Atmos. Sci. 19 (1962) 329–341.

We do not include these formulas but instead, present the type of approximation of a solution used by Lorenz. It involves a so-called stream function Ψ and a temperature function. The variables in these functions are the spatial coordinates u and v, and time t. In place of the actual temperature, the difference Θ with a temperature profile belonging to the state of no convection is used, i.e., where the temperature decreases linearly from some value T_0 at the bottom to $T_0 - \Delta T$ at the top (see figure 12.34).

Let us give these complicated-looking equations and then explain.

$$\begin{aligned} \frac{a}{(1+a^2)\kappa}\Psi(u,v,t) &= x(t)\sqrt{2}\sin\left(\frac{\pi a}{H}u\right)\sin\left(\frac{\pi}{H}v\right) \\ \frac{\pi R_a}{R_c \Delta T}\Theta(u,v,t) &= y(t)\sqrt{2}\cos\left(\frac{\pi a}{H}u\right)\sin\left(\frac{\pi}{H}v\right) \\ &\quad - z(t)\sin\left(\frac{2\pi}{H}v\right). \end{aligned} \tag{12.14}$$

The symbols used have the following interpretation.

$\Psi = \Psi(u,v,t)$	stream function
$\Theta = \Theta(u,v,t)$	local temperature difference
u	horizontal spatial coordinate
v	vertical spatial coordinate
$x(t), y(t), z(t)$	time-dependent coefficients (amplitudes)
t	time
H	depth of fluid layer (maximum of v)
a	parameter of geometry (fixed at $a = 1/\sqrt{2}$)
κ	thermal conductivity
R_a	Rayleigh number
R_c	critical value of R_a ($R_c = \pi^4(1+a^2)^3/a^2$)
ΔT	total temperature difference

The rectangular region has coordinates u ranging from 0 to $H/a = \sqrt{2}H$ and v ranging from 0 to H. The stream function Ψ is interpreted in the following sense. Ψ is a scalar field and the fluid motion at a time t_0 occurs along the isolines $\Psi(u,v,t_0) = \text{const}$. Thus, to obtain a picture of the fluid motion we can simply plot these isolines (see figure 12.35). More precisely, the velocity V of the fluid at a given point (u,v) in space and t in time is

$$V(u,v,t) = \begin{pmatrix} -\dfrac{d}{dv}\Psi(u,v,t) \\ \dfrac{d}{du}\Psi(u,v,t) \end{pmatrix}.$$

The corresponding temperature profile is given by $\Theta(u,v,t)$ and can be read directly from the formula (12.14).

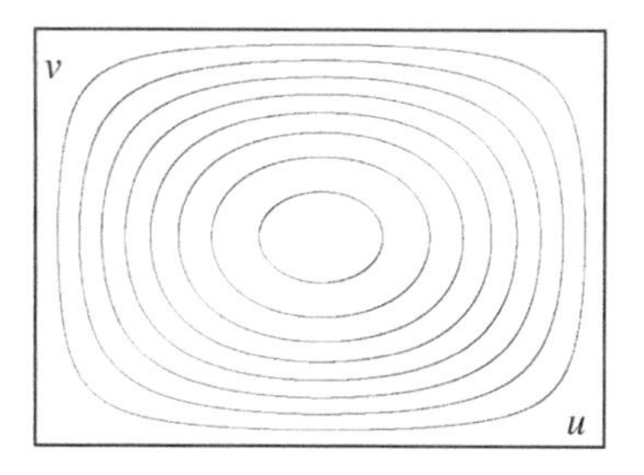

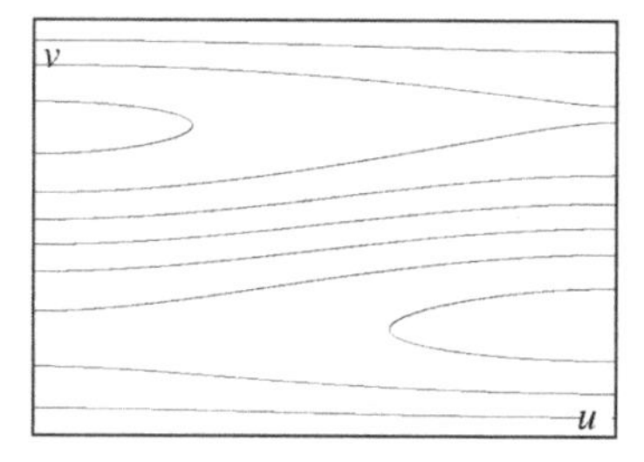

Figure 12.35 : Streamlines of convection currents (left) and corresponding temperature profile (right) at steady state.

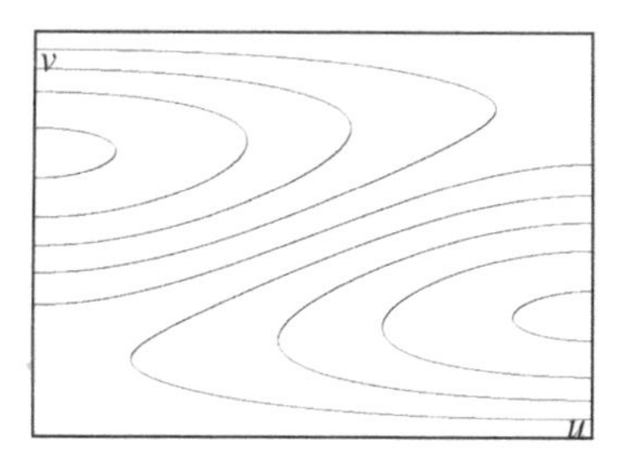

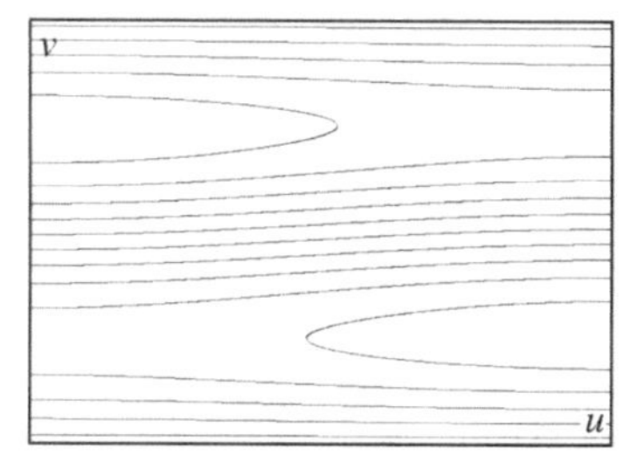

Figure 12.36 : The corresponding state variables (x, y, z) are $(2.403, 4.892, 4.673)$ (left) and $(16.610, 7.428, 45.428)$ (right).

Using this form of solution, Saltzman's equations reduce to the system of Lorenz which contains only the unknown time-dependent coefficients. Recall

$$\begin{aligned} x' &= -\sigma x + \sigma y \\ y' &= Rx - y - xz \\ z' &= -Bz + xy \end{aligned}$$

which includes three parameters: σ, the Prandtl number, $R = R_a/R_c$ and $B = 4/(1+a^2) = 8/3$.[35] Lorenz explains: 'In these equations x is proportional to the intensity of the convective motion, while y is proportional to the temperature difference between the ascending and the descending currents, similar signs of x and y denoting that warm fluid is rising and cold fluid is descending.'

Lorenz chose the parameters $\sigma = 10$ and $R = 28$. For this setting there are two steady states of the differential equation, i.e., values of x, y, z which remain constant (besides the origin $(0,0,0)$). These are

$$(6\sqrt{2}, 6\sqrt{2}, 27) \text{ and } (-6\sqrt{2}, -6\sqrt{2}, 27)$$

and correspond to the centers of the two 'holes' in the attractor shown in figure 12.33. Associated with these solutions are steady states of convection. We illustrate the first of these in figure 12.35. Other

[35]The derivatives x', y', z' are with respect to reparametrized time τ, namely, $\tau = \pi^2(1+a^2)\kappa t/H^2$.

of convection. We illustrate the first of these in figure 12.35. Other points in phase space correspond to other convection currents and temperature profiles. Two more examples for points on the Lorenz attractor are given in figure 12.36. Thus, when we follow a point along its trajectory on the Lorenz attractor, we must interpret its coordinates in this sense as given by the formulas in eqn. (12.14) and the figures. With these remarks we conclude the discussion of the modeling aspect of the Lorenz system.[36]

Model of the Dynamics with Lorenz Map

A schematic diagram of the stretch-split-and-merge process in the dynamics on the Lorenz attractor is shown at the top. The Lorenz map of the system shown below models the stretch-split-and-merge process as observed at the interval I. Each (graphical) iteration corresponds to a turn around one of the lobes of the attractor.

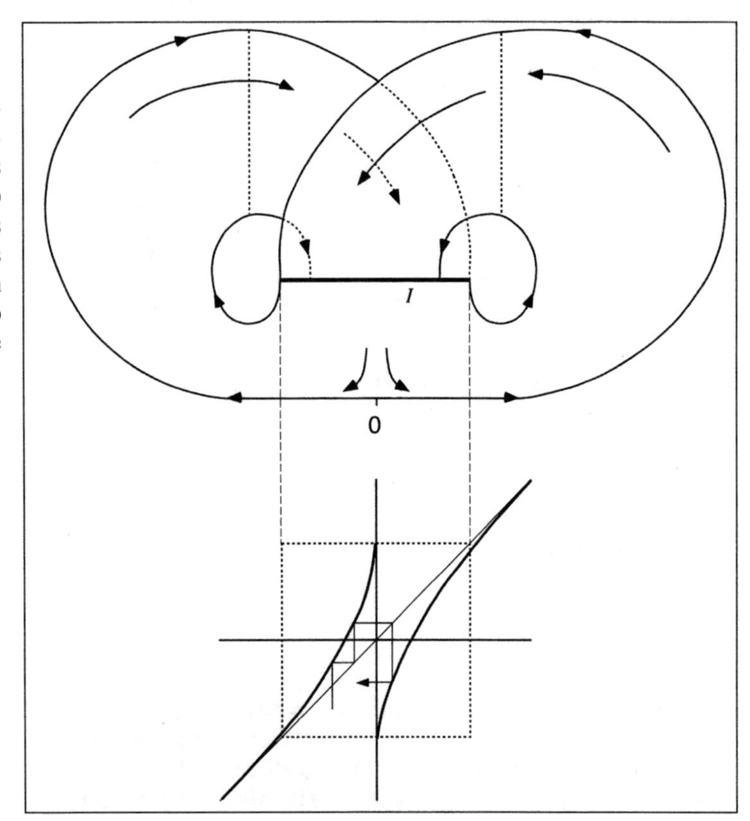

Figure 12.37

As outlined, there are several severe simplification steps before we get to the final set of equations, and we may rightfully say that the solutions of the system may not bear any significance for the real convection process. But it was not Lorenz's intention to be as precise as possible in the modeling. On the contrary, after having discovered a strange attractor (in a more complex system) he strived for the most elementary system that can be derived from the

[36]For more details and references see the original paper of E. N. Lorenz, *Deterministic nonperiodic flow,* J. Atmos. Sci. 20 (1963) 130–141. The book by J. M. T. Thompson and H. B. Stewart, *Nonlinear Dynamics and Chaos,* Wiley, Chichester, 1986, contains a broad introduction and much more material about the geometry underlying the attractor and its route to chaos.

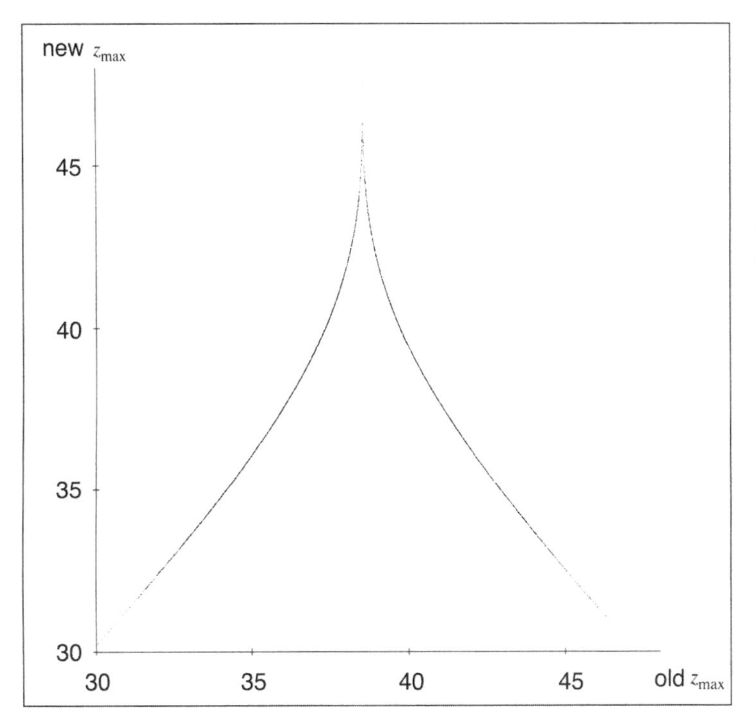

Lorenz Map

This Lorenz map for the Lorenz system models the dynamics of the attractor as observed at the dotted vertical lines pictured in figure 12.37. From this point of view it can be described by a stretch-and-fold process.

Figure 12.38

convection equations and that would still demonstrate the extreme sensitivity to initial conditions, which since has become the trademark of chaos.

A Model for the Lorenz Dynamics

The chaos-generating mechanism in the Lorenz system is a bit more involved than the one in the Rössler system. Rather than featuring a stretch-and-fold action, we have a stretch-split-and-merge operation as shown in the model in figure 12.37. Around the two spirals a stretching takes place. Both stretched bands split near the horizontal line in the center, one half of them returning to the left spiral, the other to the right. During the subsequent turn, the two bands on each part of the attractor merge, and the cycle is completed. Similar to the Lorenz map for the folded band, a Lorenz map for points on the central line segment can now be defined. The result is shown in the lower part of the figure.

The graphical iteration using this graph corresponds to the dynamics of points (x, y, z) on the attractor. The left half of it belongs to the left spiral of the attractor, the other to the right. There is a connection between this one-dimensional model and the shift transformation in chapter 10, $x \to \text{Frac}\,(2x) = 2x \bmod 1$, from which we deduced the essential properties of chaos: sensitive dependence on initial conditions, dense periodic points, and mixing.

A Periodic Solution

A stable, periodic solution to Lorenz's system at parameter $R = 100.5$, projected onto the xz-plane.

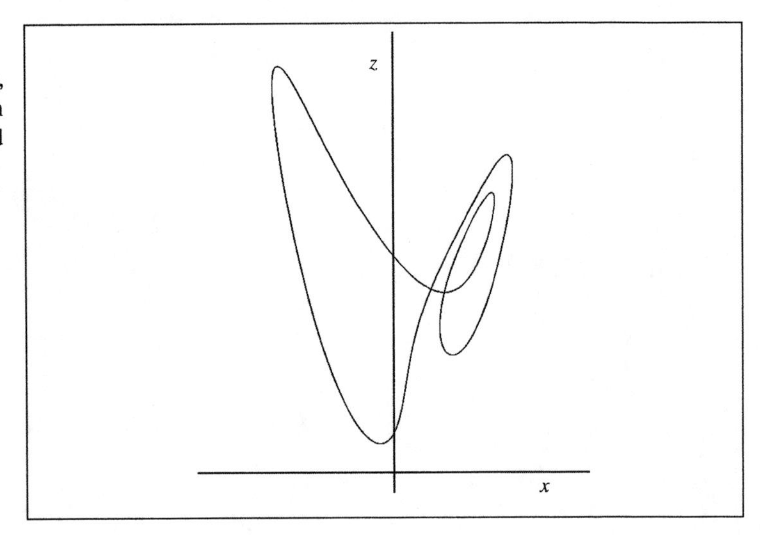

Figure 12.39

Another Lorenz Map

Originally Lorenz had proposed a different one-dimensional model. He observed that a trajectory "leaves one spiral only after exceeding some critical distance from its center. Moreover, the extent to which this distance is exceeded appears to determine the point at which the next spiral is entered; this in turn seems to determine the number of circuits to be executed before changing spirals again." He then concluded that the maximum z-value alone suffices to predict the maximum for the following circuit. To check this idea, he plotted many points with coordinates being two consecutive maximum z-values of a trajectory as in figure 12.38; thus was born the first Lorenz map.[37] The points appear to lie on a curve and graphical iteration can be used to predict the maximum z-value of the following spiral turns if the current one is given. The similarity of the graph to the tent function is quite apparent and Lorenz's paper continues with a short study of the chaotic dynamics associated with the tent function and concludes that there must be an infinity of trajectories corresponding to what he called nonperiodic deterministic flow, i.e., that one now calls chaos.

Cantor Set Structure and Fractal Dimension

Again, it is true that the Lorenz maps from figures 12.37 and 12.38 are only models for a truly two-dimensional Poincaré map. If such one-dimensional models were exact, this would imply a perfect merging of the two 'spiral surfaces' which contradicts the uniqueness of solutions of Lorenz's system of differential equations. Similar to the bands in the Rössler attractor, these surfaces come very, very close to each other, indistinguishable to the eye, but they cannot completely merge. Again, there is a Cantor set structure of these

[37]There is a surface which contains all points (x, y, z) with a maximum z-value of a corresponding trajectory. At such a maximum the derivative z' must necessarily vanish. Thus, from eqn. (12.13), $xy - Bz = 0$, or $z = xy/B$. This equation describes a surface, the graph of the function xy/B. A portion of this surface may be interpreted as a proper Poincaré section of the Lorenz system.

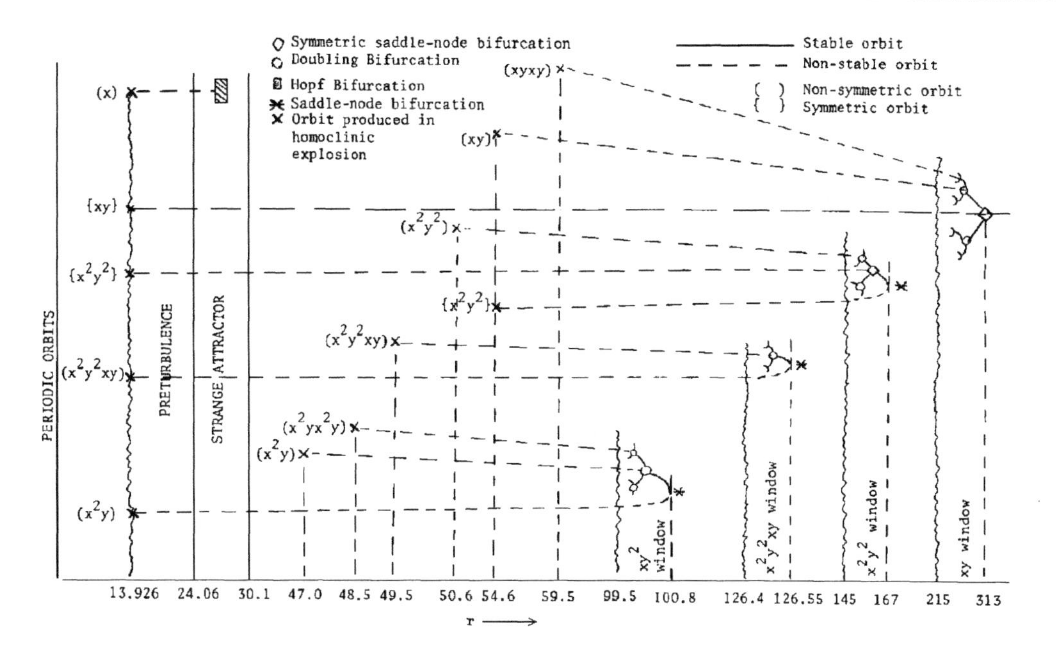

Figure 12.40 : Bifurcation diagram showing a few of the shorter periodic trajectories. Reproduced with permission of the publisher from C. Sparrow, *The Lorenz Equations: Bifurcations, Chaos, and Strange Attractors,* Springer-Verlag, New York, 1982, page 99.

surfaces leading to a fractal dimension only slightly above 2.[38]

Changing the Parameter R

Let us conclude this section by looking at phenomena brought to light by changing the parameter R in the Lorenz system. There are a large number of periodic solutions, some of which are stable and attracting, while others are unstable and repelling. For some parameters, chaos and stable equilibria coexist. The whole palette of features worth discussing cannot be included here; it would take up too much space. Here we can only highlight a couple of aspects; for further details we can only point to the literature.[39]

Period Doubling Window and Feigenbaum Number

We begin by looking at the range of parameters $99.524 < R < 100.795$. Here we can observe a period-doubling scenario similar to the one in the Rössler attractor. This cascade of bifurcations occurs as we decrease the parameter R. Figure 12.39 shows an attractive periodic orbit obtained for $R = 100.5$. It spirals around twice in the 'positive' half-space $x > 0$, then one time in the negative half-space $x < 0$ before it repeats. This solution would be called an x^2y-solution following the naming conventions in some of the literature.[40] When we lower the parameter below $R \approx 99.98$ this solution

[38] The dimension has been estimated as 2.073. See E. N. Lorenz, *The local structure of a chaotic attractor in four dimensions,* Physica 13D (1984) 90–104.

[39] For example, see the book by C. Sparrow, *The Lorenz Equations: Bifurcations, Chaos, and Strange Attractors,* Springer-Verlag, New York, 1982.

[40] In the code x^ky^l an x stands for a turn around the fixed point in the half-space $x > 0$, while the symbol y denotes a turn in the other half-space. Thus, in this example there are k turns of type x followed by l turns of type y. The notation can be extended

Intermittency

Intermittency in the Lorenz system. For the parameter $R = 166$ there is a periodic solution (top), while for $R = 166.2$ solutions appear similar, however, interrupted by sudden chaotic bursts (bottom).

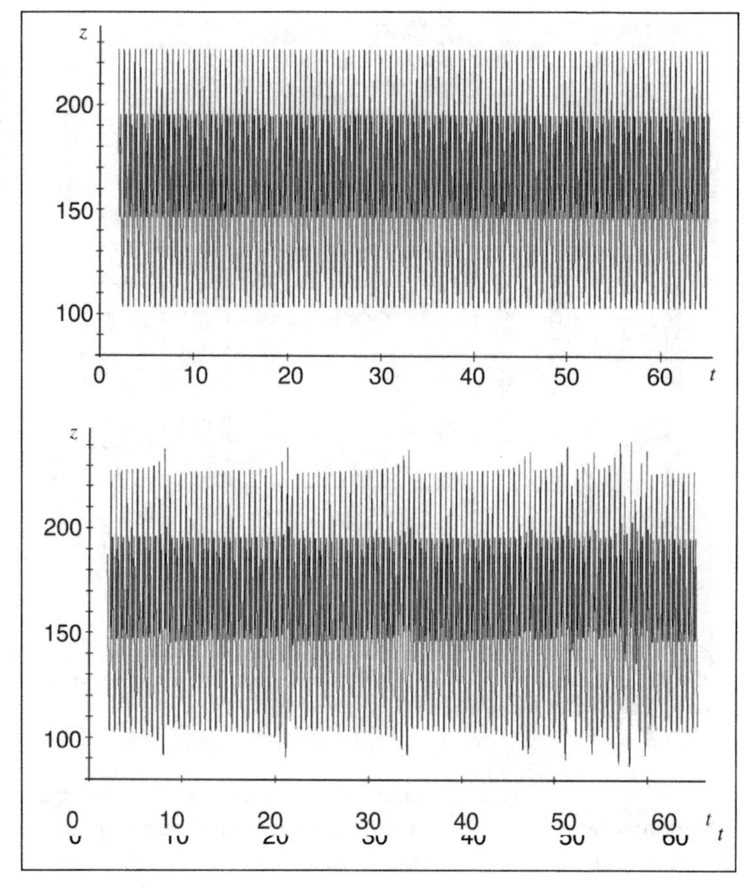

Figure 12.41

doubles up to a periodic solution of twice the period and type $x^2yx^2y = (x^2y)^2$. This happens in the same spirit as in the bifurcation of periodic solutions in Rössler's system. Further bifurcations of period-doubling can be observed when we continue to lower the parameter, as listed in the following table.

Parameter R	Type of Solution
100.795	x^2y
99.98	$(x^2y)^2$
99.629	$(x^2y)^4$
99.547	$(x^2y)^8$
99.529	$(x^2y)^{16}$
99.5255	$(x^2y)^{32}$

to longer symbol strings such as xy^2xy^3 and so on.

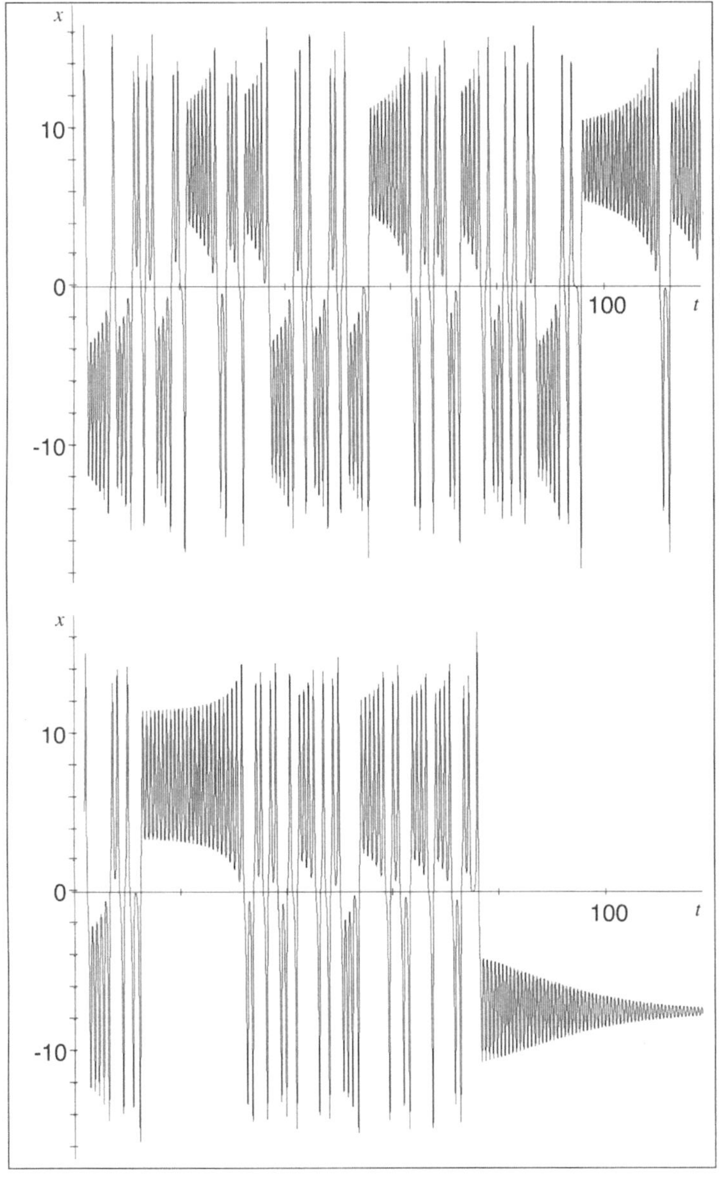

A Crisis of the Lorenz Attractor

For $R = 25$ (top) there is a strange attractor for the Lorenz system. When the parameter is lowered to $R = 22.4$ (bottom) the attractor undergoes a crises. Trajectories appear chaotic only in an initial transient phase. After that one of the attractive rest points is approached.

Figure 12.42

Based on more precise calculations the Feigenbaum number δ belonging to this sequence has been estimated.[41] The result is $\delta \approx 4.67$, a number which at this precision is indistinguishable from Feigenbaum's famous ratio 4.6692... There are other windows in the parameters R with period-doubling bifurcations. To give an impression of the complicated scenarios, we repro-

[41] V. Franceschini, *A Feigenbaum sequence of bifurcations in the Lorenz model*, Jour. Stat. Phys. 22 (1980) 397–406.

duce a chart (figure 12.40) compiled by Colin Sparrow.

Intermittency

Knowing that the dynamics of Lorenz's and Rössler's systems have a strong connection to the one-dimensional iteration of quadratic transformations, it is no longer a surprise to again find the back doors to chaos that we discussed in chapter 11: intermittency and crises. Recall that intermittency means that a solution spends most of its time near a periodic solution but is interrupted by sudden and erratic chaotic bursts. For an only slightly perturbed parameter value, these bursts disappear and only the periodic behavior remains. Exactly the same can be observed, for example, in the Lorenz system for the parameters $R = 166.2$ and $R = 166.0$ (see figure 12.41).

Crises

When a periodic solution 'disturbs' the chaos, an intermittent trajectory is produced. In this case, the chaos prevails. However, the chaotic attractor deteriorates, when a periodic trajectory steals the attractivity from the attractor. What remains of the chaos, is the long transient chaotic behavior of solutions, all of which approach the periodic one asymptotically. The chaotic attractor is said to be in a crisis. And again this can also be discovered in the Lorenz system (see figure 12.42). These phenomena are by no means restricted to the quadratic iterator and the Lorenz system; they have been identified in many other mathematical and physical dynamical systems.[42]

[42] See T. Tél, *Transient chaos*, in *Directions in Chaos III*, Hao B.-L. (ed.), World Scientific Publishing Company, Singapore.

12.5 Quantitative Characterization of Strange Chaotic Attractors: Ljapunov Exponents

We have come to live in a world where almost everything can and must be measured or estimated in terms of dollars or billions of dollars, seconds or centuries, inches or light-years, milliliters or gallons, ... It is only logical and in fact necessary to also define quantitative properties of attractors in dynamical systems. Physicists, especially, have developed that approach to the subject matter to a rather specialized form of an art difficult for the uninitiated to keep up with.

A Gallon of Attraction?

What are the questions we can ask about attractors which lead to some hopefully meaningful quantitative description? Dealing with chaotic attractors, immediately two aspects come to mind. Firstly, an *attractor* pulls in neighboring points, and secondly, in a truly *chaotic* attractor, orbits of nearby points must diverge from each other due to the sensitive dependence on initial condition. This leads to the measurement of some averaged attraction and repulsion by the so-called Ljapunov numbers and exponents, which we have already encountered in chapter 10. Attractors in two or more dimensions always have several Ljapunov exponents. We will see how a negative Ljapunov exponent is necessary for any attractor, while an additional, positive exponent qualifies the attractor as a chaotic one.

An Ounce of Strangeness?

The next question to be answered is, how *strange* is a strange attractor? The fractal character of the attractor should somehow come in at this point. Therefore, the quantity of the corresponding fractal dimension is crucial when attempting to quantify strangeness. These dimensions are the topic of the following section.

The First Ljapunov Exponent for the Hénon Attractor

To begin with, we will aim at quantifying the sensitivity on initial conditions for a discrete transformation such as the Hénon transformation from the first section of this chapter. We will use the same technique that we have already applied for the quadratic transformation $4x(1-x)$ in section 10.1.[43] Let us briefly recall the main idea. By comparing an orbit belonging to some initial condition with an orbit for an initial condition which carries an error E_0, we can record how the error amplifies during the course of the iteration to $E_1, E_2, \ldots$ The error amplification factor $|E_n/E_0|$ is written as a 'telescope product'

$$\left|\frac{E_n}{E_0}\right| = \left|\frac{E_n}{E_{n-1}}\right| \left|\frac{E_{n-1}}{E_{n-2}}\right| \cdots \left|\frac{E_1}{E_0}\right| .$$

The Ljapunov exponent characterizes the average logarithmic growth of the relative error per iteration. To arrive at a well-defined exponent we must let the size of the initial error go to zero,

$$\lambda = \lim_{n\to\infty} \lim_{E_0\to 0} \frac{1}{n} \sum_{k=1}^{n} \log\left|\frac{E_k}{E_{k-1}}\right|$$

[43] Perhaps it would be a good idea at this point to reread the last pages of that section.

In practice we can renormalize the size of the error in each iteration to some convenient number ε. This should lead to a Ljapunov exponent λ, which characterizes for a given orbit how fast nearby orbits get closer or move away. In other words, a small error in an initial point will be scaled by the factor of e^{λ} (on the average) in each iteration. Thus, a positive exponent means that nearby orbits move away, while a negative exponent means that nearby orbits are attracted, which we would not expect for a chaotic attractor.

However, for the Hénon transformation the situation is more complex than in the one-dimensional case given by the quadratic iteration. In one dimension an error may be either positive or negative, and we can expect in either case that the error amplification factors are the same. But here — for the Hénon system — a point is given by two numbers (x, y) and an error of ε still leaves the choices

$$(\tilde{x}, \tilde{y}) = (x + \varepsilon \cos \phi, y + \varepsilon \sin \phi) \tag{12.15}$$

where ϕ is an arbitrary angle. A priori it is not clear that errors in different directions cause the same effect. Thus, let us numerically test different choices of ϕ using the following procedure. For a set of given parameters ε, ϕ, and number N of iterations, we perform the following steps:

The Algorithm

1. **Initialization.** Choose the initial point $(0, 0)$ and iterate the Hénon transformation 100 times. The point arrived at, (x, y), is precisely on the Hénon attractor (at least as far as the precision of the machine allows for). Initialize an accumulator to the value zero (for storage of the sum of the logarithms of the error amplification factors).
2. **Initial error.** Compute the perturbed point $(\tilde{x}, \tilde{y})$ according to eqn. (12.15).
3. **Transformation.** Iterate both points one time, i.e., obtain the points $H(x, y)$ and $H(\tilde{x}, \tilde{y})$ and compute the distance d between these two points.
4. **Error amplification.** The error of ε has increased (or decreased) by the factor of d/ε. We accumulate the logarithm of this factor.
5. **Renormalization.** Move the point $H(\tilde{x}, \tilde{y})$ on the line to $H(x, y)$ so that the distance of both points becomes equal to ε (see figure 12.43). Replace the old (exact) point (x, y) by its successor $H(x, y)$ and replace $(\tilde{x}, \tilde{y})$ by the above constructed point which will serve as the displaced point for the next iteration.
6. **Loop.** Go back to step 3 until N iterations have been performed.
7. **Result**. The average of the logarithms of the amplification factor is the exponent. Thus, divide the contents of the accumulator by N.

Directions of Errors Must Be Maintained

It it important to stress that in the renormalization step 5, we need to *maintain the direction of the error* while reducing the actual error to a size of ε. Otherwise the error amplification factor $|E_{k+1}|/|E_k|$ would not come out correctly. Errors in different directions produce different amplification factors. Let us demonstrate this with just one example. For simplicity, take the initial point (0,0) and consider perturbed initial points $(\varepsilon, 0)$ and $(0, \varepsilon)$

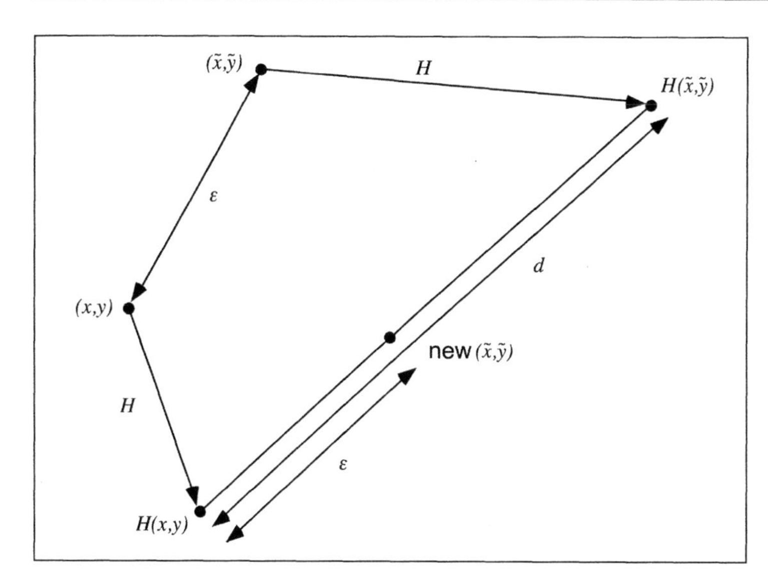

Renormalization

The renormalization step in the calculation of a single Ljapunov exponent.

Figure 12.43

Iterations	$\phi = 0$	$\phi = \pi/2$	$\phi = \pi$	$\phi = 3\pi/2$
10	0.31253	0.23086	0.31261	0.23078
100	0.40709	0.39891	0.40708	0.39892
1 000	0.42403	0.42321	0.42403	0.42321
10 000	0.41867	0.41858	0.41866	0.41859
100 000	0.42015	0.42013	0.42014	0.42014

The calculation of a Ljapunov exponent using different initial errors of magnitude $\varepsilon = 0.00001$ and increasing numbers of iterations.

Table 12.44

which both carry an error of size ε. The transformed points are

$$\begin{aligned} H(0,0) &= (1,0) \\ H(\varepsilon,0) &= (1-a\varepsilon^2, b\varepsilon) \\ H(0,\varepsilon) &= (1+\varepsilon,0). \end{aligned}$$

The size of the error in the iterate of $(\varepsilon, 0)$ is the distance between $H(0,0)$ and $H(\varepsilon,0)$ which is approximately $b\varepsilon$. On the other hand the size of the error in $H(0,\varepsilon)$ is equal to ε. Thus, these error sizes are clearly different and would produce different amplification factors. Therefore, when we replace an actual error by a smaller one, we are not allowed to change the direction of the error.

Ljapunov Number Computation

Table 12.44 lists the results when using the four angles 0, $\pi/2$, π, and $3\pi/2$. Thus, initial points are displaced in step 2 in the east, north, west and south directions by an amount of ε.

The result is surprising but clear. As the number of iterations increases the computed exponents converge to the same number no matter what initial error direction, i.e., angle ϕ, is used. Moreover, a slightly modified experiment indicates that it does not matter the least at which point on the attractor we start

The Unstable Directions

Two points of an orbit on the Hénon attractor are shown along with line segments indicating the directions in which deviations are amplified most. These directional vectors arise naturally during the computation of the Ljapunov exponent (compare figure 12.43).

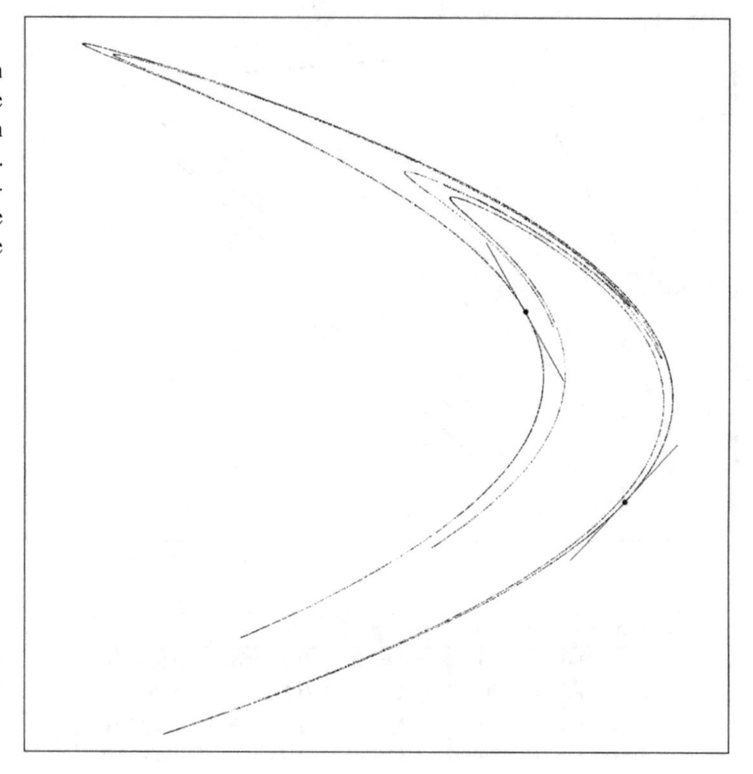

Figure 12.45

the computation of the exponent; the convergence is to the same number.[44] This Ljapunov exponent is $\lambda \approx 0.42$ which we can interpret in the sense that small deviations are amplified by the factor of $e^{\lambda} \approx 1.52$ on the average, where the average is taken over a long orbit which 'covers' the whole Hénon attractor. We have remarked above that deviations are amplified differently in different directions. What this Ljapunov exponent measures is the maximal average amplification. In figure 12.45 we visualize these directions as line segments at two points of an orbit on the attractor. Apparently, we have that these most unstable directions are 'tangential' to the attractor.

The Exact Method for Calculating the Exponent

The calculation of the Ljapunov exponent according to the algorithm presented above is unsatisfactory in one respect. Namely, the choice of the size ε of the error to which we renormalize after each iteration may effect the result. What we are really interested in is the amplification factor for infinitesimally small errors, which we obtain when we let $\varepsilon \to 0$. To overcome this ambiguity, we need the concept of the derivative

[44]For example, we can choose a different initial point in place of $(0,0)$ in step 1, or we allow more than 100 preparatory iterations. Of course, if we start exactly at a fixed point or a periodic point of the attractor, and if the machine is not thrown off this periodic orbit by round-off errors, then we expect to arrive at a different exponent. But the likelihood of this happening is practically zero.

of the Hénon transformation. In matrix notation it is

$$DH(x,y) = \begin{pmatrix} -2ax & 1 \\ b & 0 \end{pmatrix}.$$

Using the derivative, we can compute how an infinitesimally small error in a point (x, y) of the attractor is transformed by one iteration. If the error is in the direction given by the vector (dx, dy), then we multiply this vector by the derivative matrix

$$\begin{pmatrix} -2ax & 1 \\ b & 0 \end{pmatrix} \begin{pmatrix} dx \\ dy \end{pmatrix} = \begin{pmatrix} -2ax \cdot dx + dy \\ b \cdot dx \end{pmatrix}. \quad (12.16)$$

The amplification factor is determined by the quotient of the lengths of the two vectors (dx, dy) and $(-2ax \cdot dx + dy, b \cdot dx)$. Using this approach, the algorithm for the computation is modified as follows.

Iterations	$\varepsilon = 0.1$	$\varepsilon = 0.01$	$\varepsilon = 0.001$	$\varepsilon \to 0$
10	0.41660	0.26115	0.30848	0.31257
100	0.45323	0.40951	0.40766	0.40709
1,000	0.42586	0.42356	0.42403	0.42403
10,000	0.41669	0.41740	0.41885	0.41866
100,000	0.41936	0.42030	0.42021	0.42015
1,000,000	0.41906	0.41918	0.41922	0.41924

Table 12.46 : The calculation of a Ljapunov exponent using differently sized errors is compared with the method using the derivative of the transformation (last column). Using errors $\varepsilon > 0$ the exponent is slightly underestimated.

1. **Initialization.** Iterate the initial point $(0, 0)$ 100 times to arrive at (x, y). Initialize an accumulator to zero.
2. **Initial error.** For an arbitrary angle ϕ consider $(\cos\phi, \sin\phi)$, the direction of the error.
3. **Transformation.** Compute the transformed error according to eqn. (12.16) and iterate (x, y), i.e., obtain the point $H(x, y)$ and the transformed error $(-2ax \cdot \cos\phi + \sin\phi, b \cdot \cos\phi)$.
4. **Error amplification.** The error has increased (or decreased) by the factor $d = \sqrt{(-2ax \cdot \cos\phi + \sin\phi)^2 + (b \cdot \cos\phi)^2}$. Accumulate the logarithm of this factor.
5. **Renormalization.** Replace the old point (x, y) by its successor $H(x, y)$ and replace the error direction $(\cos\phi, \sin\phi)$ by the new directional vector $(-2ax \cdot \cos\phi + \sin\phi, b \cdot \cos\phi)/d$.
6. **Loop.** Go back to step 3 until N iterations have been performed.
7. **Result.** Divide the contents of the accumulator by N.

Transformation of a Disk

A small disk centered at a point of the Hénon attractor and its image under the transformation.

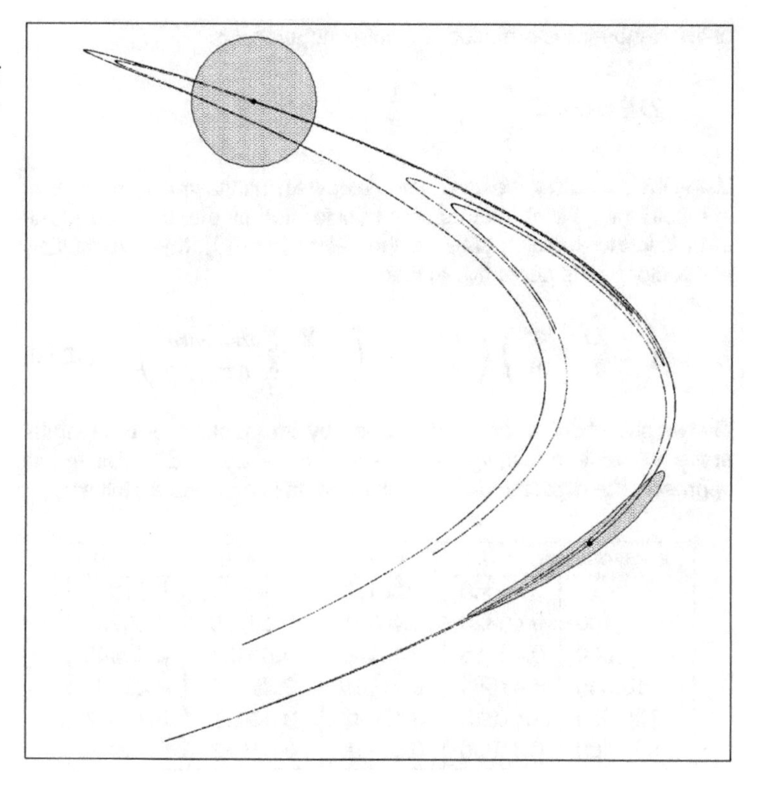

Figure 12.47

The Second Ljapunov Exponent

Thus, the chaoticity of the attractor can be measured the same way as for the quadratic transformation and expressed as a positive Ljapunov exponent. Now we can come to the second question regarding the speed at which points are pulled in by the attractor.

This can be expressed in the form of a second Ljapunov exponent. Imagine a small disk of diameter d centered at a point on the attractor and consider its transformed image (see figure 12.47), which looks similar to a slightly deformed ellipse. The center (x, y) of the disk is transformed to the next point $H(x, y)$ on the attractor, the center of the ellipse. The diameter of the disk which is in the direction of the strongest error amplification is transformed to the long axis of the ellipse. The short axis gives a measurement of how much points near (x, y) are attracted in one iteration. In fact, we have all the information to compute the average attraction already in our hands! There are two pieces of information:

1. On the average, the long axis of the ellipse results from a stretching of a diameter of the disk and the magnification factor is the Ljapunov number $e^\lambda \approx 1.52$.
2. The area of the ellipse must be exactly 0.3 times the area of the disk (see page 615).

Since the area of an ellipse is proportional to the product of both axis lengths, the length of the small axis of the ellipse must be equal to about $0.3/1.52 \approx 0.197$. Now the result can be stated. On the average, small disks centered at points on the Hénon attractor are elongated in one direction by the factor of $e^{0.42} \approx 1.52$ and compressed in another direction by the factor of 0.197. This motivates us to relabel the Ljapunov exponent 0.42 as λ_1 and to introduce a second Ljapunov exponent $\lambda_2 < \lambda_1$ such that the corresponding Ljapunov number is $e^{\lambda_2} = 0.197$. In other words, we arrange the new exponent so that the product $e^{\lambda_1+\lambda_2}$ equals the factor 0.3 by which the area changes per iteration. These definitions are summarized in the following table.

Ljapunov Exponent	Ljapunov Number	Interpretation
$\lambda_1 \approx 0.42$	$e^{\lambda_1} \approx 1.52$	error amplification
$\lambda_2 \approx -1.62$	$e^{\lambda_2} \approx 0.197$	contraction factor
$\lambda_1 + \lambda_2 \approx -1.20$	$e^{\lambda_1+\lambda_2} = 0.3$	area reduction factor

There are two Ljapunov exponents for attractors of systems in two dimensions. If a transformation is defined in three dimensions then there are three Ljapunov exponents $\lambda_1, \lambda_2, \lambda_3$. Intuitively, we can define them by requiring that

1. e^{λ_1} is the maximal average factor by which an error is amplified.
2. $e^{\lambda_1+\lambda_2}$ is the maximal average factor by which an area changes.
3. $e^{\lambda_1+\lambda_2+\lambda_3}$ is the maximal average factor by which a volume changes.

The exponents are ordered,

$$\lambda_1 \geq \lambda_2 \geq \lambda_3.$$

Positive exponents mean expansion along a certain direction, while negative exponents characterize contractions. Algorithms for their computation can be based on the above definitions.[45]

The First Ljapunov Exponent for Continuous Systems

Instead of going into the technical details of an implementation of numerically computing several exponents let us continue to discuss Ljapunov exponents for attractors for differential equations in three dimensions such as the Lorenz and the Rössler systems. We proceed along the same lines as in the case of discrete systems. Given a trajectory on the attractor we can perturb its initial condition and observe how the trajectory starts to diverge from the reference trajectory. After some time interval, say τ, has elapsed we can stop,

[45] See for example:
1. Benettin, G. L. and L. Galgani, A. Giorgilli, J.-M. Strelcyn, *Lyapunov characteristic exponents for smooth dynamical systems and for Hamiltonian systems; a method for computing all of them. Part 1: Theory, Part 2: Numerical application,* Meccanica 15, 9 (1980) 21.
2. Eckmann, J.-P. and D. Ruelle, *Ergodic theory of chaos and strange attractors,* Reviews of Modern Physics 57, 3 (1985) 617–656.
3. Leven, R. W., B.-P. Koch and B. Pompe, *Chaos in Dissipativen Systemen,* Vieweg, Braunschweig, 1989.
4. Parker, T. S. and L. O. Chua, *Practical Numerical Algorithms for Chaotic Systems,* Springer-Verlag, New York, 1989. This last reference also contains pseudo code for an implementation.

Renormalization

Renormalization of errors along a trajectory.

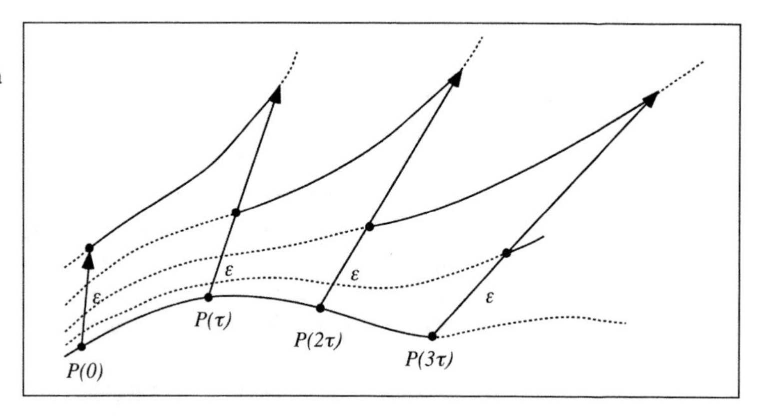

Figure 12.48

Ljapunov Exponent for Lorenz System

$\Delta t = 1/N$	$\varepsilon = 0.1$	$\varepsilon = 0.01$	$\varepsilon = 0.001$	$\varepsilon = 0.0001$
1/50	1.478	1.463	1.458	1.458
1/100	1.048	1.048	1.044	1.043
1/200	0.970	0.956	0.954	0.953
1/400	0.939	0.940	0.936	0.935

Table 12.49

evaluate the accumulated difference, renormalize that error and continue (see figure 12.48).

This amounts to considering the transformation that moves an initial point along its trajectory a distance corresponding to the time interval τ. This operator can be treated just like an ordinary discrete transformation. Its iteration produces points on the attractor which eventually should fill up the whole attractor. Moreover, it has three Ljapunov exponents, the largest of which we can compute using the same algorithm as above (with the only difference being that we are dealing here with three in place of two dimensions). Table 12.49 lists the results of such a computation for $\tau = 1$. There are two parameters of the numerical method, the size ε to which an error is renormalized in each step (see column headings), and the Euler step size Δt in the integration of the differential equation (see first column). Choosing $\Delta t = 1/N$ means that in order to follow a trajectory for a time interval of $\tau = 1$ a total of N Euler steps are necessary. Depending on these choices, we get differing results computed from a total of 10,000 time units for the first Ljapunov exponent. In another test run, we experimented with increasing lengths of the trajectory on which the calculation is based.[46] In particular we see that the numerical results depend most strongly on the choice of the Euler step length. At this point we can

[46]Using $\varepsilon = 0.00001$ and $N = 400$ Euler steps per time unit the results are $\lambda_1 = 0.93061$ after 100 units, 0.93456 after 200 units, 0.93470 after 10,000 units, and 0.93572 after 100,000 units of time.

only say that the largest Ljapunov exponent seems to be about $\lambda_1 \approx 0.9$. The numerical shortcomings can be removed from the method almost completely by the use of more sophisticated techniques.

Removing the Discrepancy

The numerical problems can be solved with two techniques. The first one implements the limiting case $\varepsilon = 0$. As in the case of Ljapunov exponents for discrete transformations, we have to resort to a derivative. If Φ_τ denotes the transformation that shifts an initial point τ time units down its trajectory (this is called the flow), we need its derivative to compute by how much an infinitesimal error in the initial condition is amplified. Just as Φ_τ is not given explicitly but only numerically as the outcome of some integration procedure, so is the development of infinitesimal errors. In other words, we must solve another initial value problem to compute the error amplification factors. This extra system is called the variational equation. It involves derivatives of the terms occurring in the original equation. We cannot explain the derivation of this system here;[47] instead we will exemplify the variational equation for the Lorenz system to give you a flavor of the procedure. Recall the Lorenz system

$$\begin{aligned} x' &= -\sigma x + \sigma y \\ y' &= Rx - y - xz \\ z' &= -Bz + xy. \end{aligned}$$

Let $x = x(t), y = y(t), z = z(t)$ denote a particular solution, a trajectory of the attractor, for example. The development of an error (dx, dy, dz) along this trajectory is governed by the variational equation which we represent here in matrix notation

$$\begin{pmatrix} dx' \\ dy' \\ dz' \end{pmatrix} = \begin{pmatrix} -\sigma & \sigma & 0 \\ R - z & -1 & -x \\ y & x & -B \end{pmatrix} \begin{pmatrix} dx \\ dy \\ dz \end{pmatrix}$$

This is a linear system of differential equations where the coefficients in the matrix depend on time (the symbols x, y, z must be replaced by the coordinates of the trajectory to evaluate the variational equation correctly). In a computer implementation both the trajectory and the error development can be computed simultaneously using an initial value solver. This brings us to the second topic. In place of the primitive Euler method for the initial value problem, we should use a better initial value solver. For example, the adaptive step size Runge-Kutta method is an option. Such an improved method increases the precision significantly without requiring more computer time.

[47] The theory of differential equations is needed here. See for example M. W. Hirsch, S. Smale, *Differential Equations, Dynamical Systems, and Linear Algebra,* Academic Press, New York, 1974.

The Second Ljapunov Exponent for Continuous Systems

We conclude the part about the Ljapunov exponents with some arguments for the second and third exponent of the Lorenz system. As we have seen in the numerical experiments leading to the table 12.49 an arbitrary deviation of an initial value on the attractor magnifies exponentially. On the average, an error is amplified by the factor of $e^{0.9} \approx 2.5$ during one time unit. However, this is only the 'generic' result which we expect to see in simulations with randomly chosen initial errors. But there are also special directions of errors so that perturbed trajectories in the ideal mathematical system (without any systematic error from a numerical procedure) do not actually diverge from the reference trajectory. The first such direction is the one given by the velocity vector of the trajectory. To clarify the point, consider a trajectory $(x(t), y(t), z(t))$ on the attractor and the 'perturbed' trajectory $(\tilde{x}(t), \tilde{y}(t), \tilde{z}(t))$ with initial condition

$$\tilde{x}(0) = x(\delta),\ \tilde{y}(0) = y(\delta),\ \tilde{z}(0) = z(\delta)$$

with $\delta > 0$. In other words, we choose the error in the initial condition precisely such that the new trajectory starts on the reference trajectory, only at a little time δ away. Now do the two trajectories diverge from each other? The answer is a clear 'no'. Both trajectories must be identical except for a small time shift, i.e.,

$$\begin{aligned}\tilde{x}(t) &= x(t+\delta)\\ \tilde{y}(t) &= y(t+\delta)\\ \tilde{z}(t) &= z(t+\delta).\end{aligned}$$

Thus, the two points $(x(t), y(t), z(t))$ and $(\tilde{x}(t), \tilde{y}(t), \tilde{z}(t))$ move around the spiraling attractor in great harmony, sometimes getting closer to each other only to eventually spread out again. However, *on the average* the trajectories can neither spread away from each other nor can they get closer together. Thus, a Ljapunov exponent calculation with such data would result in an average amplification factor equal to exactly 1, the logarithm of which is 0. This is the second Ljapunov exponent, $\lambda_2 = 0$, of the Lorenz system. This says that there is a direction in which errors are preserved. This direction is the one along the tangent of the trajectory.

Another interpretation of the second Ljapunov exponent is in terms of how an area is transformed along the flow of the system. We consider a small quadrilateral with one side along the tangent of the trajectory and the other side in the direction corresponding to the maximal Ljapunov exponent. Moving along the flow of the system, one side elongates by a factor of $e^{\lambda_1} > 1$, while the other remains constant (on the average). Thus, $\lambda_1 + \lambda_2$ can be considered as the exponent by which an area changes on the average.[48] By the way, these arguments are always applicable when the attractor arises from a dynamical system given by a differential equation. Thus, for example, also for the Rössler attractor, there is one Ljapunov exponent equal to zero.

[48]We have already mentioned this above. In fact, when following a randomly chosen element of area along a trajectory (with a properly adjusted renormalization technique), it converges to the one described here.

The Third Ljapunov Exponent for Continuous Systems

All that remains to be measured now is the rate by which nearby points are attracted by the Lorenz attractor. Similar to the method used for the Hénon attractor, we can discuss this rate using the factor r by which volumes are contracted. Extending the concept from above, we reconsider the small quadrilateral and place another vector perpendicular to the other two, thus creating a small volume element. In this third direction, distances must shrink by the factor of e^{λ_3} where the product

$$e^{\lambda_1} e^{\lambda_2} e^{\lambda_3} = r$$

yields the volume contraction factor. From this formula, it is easy to extract the third Ljapunov exponent λ_3.

In the case of the Lorenz attractor, we do not even have to numerically compute the volume contraction factor, because there is a basic result from vector calculus[49] which when applied to our system states that this rate is constant and independent of the particular choice of the volume. Furthermore, this rate is

$$r = e^{-(\sigma+B+1)} = e^{-(10+8/3+1)}.$$

In other words, the sum of the three Ljapunov exponents must be equal to the exponent in the above formula,

$$\lambda_1 + \lambda_2 + \lambda_3 = -\left(13 + \frac{2}{3}\right).$$

Thus, with $\lambda_1 \approx 0.9$, $\lambda_2 = 0$ we obtain the result

$$\lambda_3 \approx -12.8.$$

In other words, a trajectory in the vicinity of the Lorenz attractor approaches this strange attractor incredibly fast; the distance to the attractor reduces by a factor of $e^{-12.8} \approx 2.8 \cdot 10^{-6}$ per unit of time!

Invariance of the Exponents

One last remark should be made regarding different coordinate systems. For example, we may stretch the Lorenz attractor by a factor of 2 in the x-direction. Certainly, we would not say that this creates an entirely new object. Therefore, we expect that the dynamical features represented by the set of Ljapunov exponents also do not change. Indeed this important invariance property is guaranteed, even when the change of coordinates is not simply linear.

Summary

Let us summarize the main aspects of Ljapunov exponents relevant to strange attractors

1. It is recognized that an attractor is chaotic if it has one positive Ljapunov exponent.
2. In discrete systems, as well as in continuous systems, there are as many Ljapunov exponents as there are dimensions of the underlying space.

[49] The rate of change of a volume element in a vector field is given by the divergence. If $x' = F_1(x, y, z)$, $y' = F_2(x, y, z)$, $z' = F_3(x, y, z)$ is a system in three dimensions, then the divergence is div $F = \partial F_1/\partial x + \partial F_2/\partial y + \partial F_3/\partial z$. See V. I. Arnold, *Ordinary Differential Equations*, MIT Press, Cambridge, 1973.

3. For a chaotic attractor belonging to a system of three differential equations, the second Ljapunov exponent is equal to zero.
4. The sum of all Ljapunov exponents characterizes how fast area (in two-dimensional systems) or volume (in three-dimensional systems) shrinks. The sum is negative.
5. We have presented straightforward numerical methods for the computation of the first, largest Ljapunov exponent.
6. The Ljapunov exponents are independent of the choice of coordinates.

In some cases such as those presented here all Ljapunov exponents can be deduced from the knowledge of the largest one, which is easily estimated from numerical experiments. There exist more complicated algorithms for their computation when such information is not available (see the cited literature).[50]

[50] A recent study and comparison of various algorithm is given in K. Geist, U. Parlitz, W. Lauterborn, *Comparison of Different Methods for Computing Lyapunov Exponents*, Progress of Theoretical Physics 83,5 (1990) 875–893.

12.6 Quantitative Characterization of Strange Chaotic Attractors: Dimensions

While the main purpose of the Ljapunov exponents is to characterize the dynamical properties of orbits and trajectories on attractors the fractal dimension focuses on the geometry of the attractor. However, it turns out that the usual box-counting dimension is extremely difficult to compute for strange attractors. Thus, researchers have devised other forms of fractal dimension more suitable in this context. Two new dimensions, called information and correlation dimensions, have joined the box-counting dimension. They are designed to reflect not only the fractal geometry of the underlying objects, but also the dynamics which take place in them.

Once there were several definitions of dimension in competition, it became a hot topic of research to bring order to these numbers. Relations between the different dimensions and the Ljapunov exponents were stated. Some of them were obvious, some needed a proof, while another one has remained as the famous Kaplan-Yorke conjecture, which has allowed only a partial proof up to this date. Moreover, the various dimensions are linked to each other by a whole continuous spectrum of dimensions.

Let us start the discussion by recalling the box-counting dimension of chapter 4 (in our first volume) and applying it to the Hénon attractor. The figures presented in section 12.1 indicate that the attractor has the shape of a Cantor set of lines, from which we conjecture that its fractal dimension should lie somewhere between 1 and 2.

Box-Counting the Hénon Attractor

To compute this dimension we first find a rectangular region that contains the whole attractor.[51] This region is subdivided into square tiles of linear size s. Then we count those tiles which contain a part of the Hénon attractor and call that number $N(s)$. When repeating the same procedure using smaller sizes s we expect to find that the count $N(s)$ scales like a power of s,

$$N(s) \propto s^{-D_f},$$

where D_f is the box-counting dimension.[52] The attractor is usually computed using a great many iterations of Hénon's transformation. After each such iteration (except for the first few dozen or so which are needed to get sufficiently close to the attractor) we check if the current point is in a tile that we have not yet visited, in which case we increase our count by 1. After we have visited all tiles that cover the attractor we stop the iteration, repeat the whole procedure for a different size s and finally compute D_f from the power law as the slope of the graph of $\log N(s)$ versus $\log 1/s$. This sounds quite straightforward and innocent, doesn't it? So let's do it!

Well, there is a problem. Assume we decide to carry out the iteration for a fixed number of times, say 1,000,000. When counting boxes for decreasing

[51] An example is the rectangle given by $-1.5 \leq x \leq 1.5$ and $-0.4 \leq y \leq 0.4$.

[52] In this section we use the symbol D_f for the box-counting dimension, which is the version that most commonly used for the fractal dimension.

Log/Log Plot

The plot shows the box count $\log_2 N(s)$ versus $\log_2 1/s$ for the Hénon attractor. Note how the curve flattens off for small values of s, i.e., just where the dimension (slope of the curve) could be calculated relatively precisely. Thus, data such as this cannot be used in the straightforward manner for dimension calculations. To demonstrate this effect more dramatically we have based the figure on data obtained from an orbit of only 1000 points.

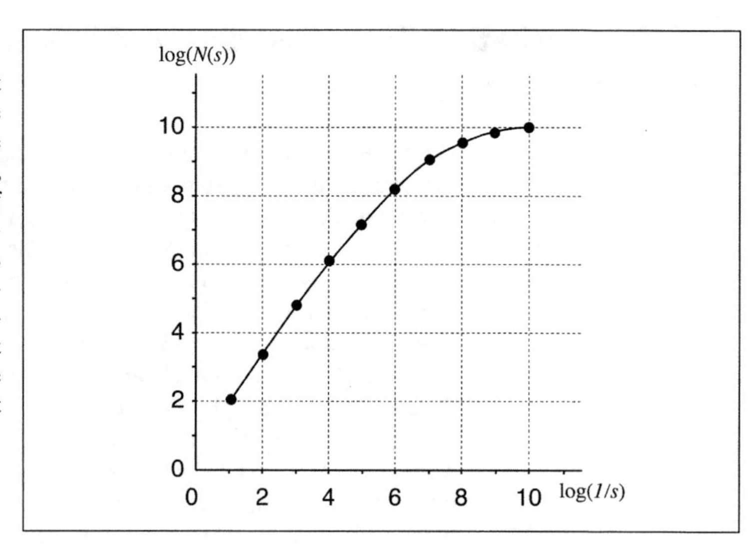

Figure 12.50

box sizes s, we initially get an increasing count. However, there is a definite small size s^* such that any box count $N(s)$ with $s < s^*$ will yield the same number, namely, 1,000,000. This is so because eventually each iterate is in a box by itself.[53] We have carried out the computation for only 1,000 points and have summarized the result in figure 12.50. The diagram shows points that seem to be on a curve rather than a straight line. How can we fit a line to that data? We have three options, all of which, however, are unsatisfactory:

1. If we ignore the flat part of the curve stemming from the high-resolution grids, we will get a systematic error because we are using only the very coarse grids.
2. Fitting the line only to the data for the high-resolution grids yields a slope of about zero. Thus, the measured fractal dimension is also $D_f = 0$. However, this is the correct box-counting dimension only for the collection of 1,000,000 points and not the dimension for the infinite number of points that form the Hénon attractor.
3. We can leave the decision to some automatic program which treats all data points equally. However, the result will be some interpolation between the two wrong values from the above methods.

Thus, in any case we get a result, some number for a dimension, but we do not know whether we are really measuring the dimension of the attractor or just seeing an artifact of the limitations of the method.

The Problem with Box-Counting Dimensions

Thus, there is a serious problem, which even caused researchers to give up computing box-counting dimensions for many attractors they were studying. So where exactly is the impediment in this can of worms? It seems that we have to adapt the number of iterations to the resolution s. The smaller the size

[53] To see this, just set s^* to the minimum distance between any two points of the computed orbit.

s the more iterations we should perform to measure $N(s)$, the total number of boxes. But how many iterations are enough? Well, let us assume that after, say, 14,365,811 iterations we have found a box not previously visited. But is it the last one we will find if we continue to iterate? There is no way to tell. But still we must make a decision. For example, we can propose to perform another 100,000 iterations; and if we do not find any more new tiles, we are content and stop the iteration. But probing only the next iteration beyond the 100,000 may prove that we are wrong in this assumption. The lesson to learn is that we are not able to count $N(s)$ directly; all we can expect is a count $N(s,n)$ which depends on the number n of iterations performed.[54] Table 12.51 examines this dependence. $N(s,n)$ is the count of tiles of linear size s that contain one or more iterates from an orbit on the Hénon attractor computed for a length n.[55]

Counting Boxes and Iterations

Iterations n	$N(s,n)$ $s=1/8$	$N(s,n)$ $s=1/16$	$N(s,n)$ $s=1/32$	$N(s,n)$ $s=1/64$
100	27	41	64	77
1,000	28	68	151	310
10,000	28	69	168	380
100,000	28	69	169	391
1,000,000	28	69	169	395

Table 12.51

Now what can we say about the results in the table? The numbers confirm our reservations about the method. At resolution $s = 1/8$ we get 27 boxes after 100 iterations, only one more box in the next 900 iterations, and no more in almost another million iterations. Thus, we would assume that $N(1/8) = 28$. The picture becomes worse when we improve the resolution. At $s = 1/64$ (which is not yet a small enough value for a serious study) we find 391 boxes in the first 100,000 iterations and another four boxes in the next 900,000. How many more iterations should we perform? In the case of the Hénon attractor iterations are very cheap, the numbers cited in the research literature are typically in the order of tens of millions.[56] But for many other systems one million iterations may be well above any reasonable computational time frame. And for the Hénon transformation, even 100 million iterations are probably insufficient when we require results for even smaller resolutions. Thus, the problem outlined above really exists in practice, and a solution should be sought.

Analysis of the Problem

New phenomena raise new questions, and this also is true in the case

[54]To a lesser degree the box count also depends on the choice of the initial point of the orbit used and the positioning of the lattice which defines the boxes.

[55]To be more precise, the numbers in this table have been computed using rectangular boxes. The value $s = 1/16$, for example, means that we have subdivided the region shown in figure 12.2 into 16×16 rectangles, which would be represented as squares in the figure.

[56]In the case presented here even 10 million iterations produce the same result as the first million listed in the table.

Box-Count Versus Number of Iteration

Figure adapted from P. Grassberger, *On the fractal dimension of the Hénon attractor,* Physics Letters 97A (1983) 224–226. Changes $\Delta N(s,n)/\Delta n$ averaged over five runs and for five different box sizes. Clearly a line fit is appropriate for large n verifying the power law conjecture in eqn. (12.17).

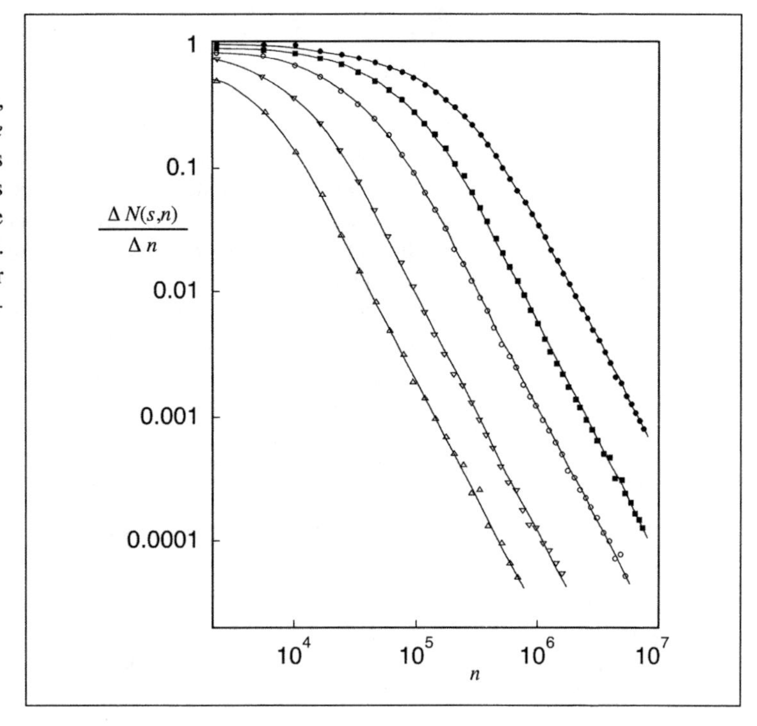

Figure 12.52

considered here. Specifically we may now ask what the dependence of $N(s,n)$ on the number n of iterations is. If we know this dependence up to some precision, then perhaps we can extrapolate from our counts $N(s,n)$ to arrive at an estimate for

$$N(s) = \lim_{n\to\infty} N(s,n)$$

which is needed for the dimension calculation. These issues were addressed in 1983 in a paper by Peter Grassberger.[57] His tabulated data suggested a behavior

$$N(s,n) \approx N(s) - \text{const} \cdot s^{-\alpha} n^{-\beta} \tag{12.17}$$

for large numbers n of iterations.[58] In order to test this conjecture, consider the growth rates of $N(s,n)$ as n increases. If eqn. (12.17) is true, then these rates should scale according to

$$\frac{\Delta N(s,n)}{\Delta n} \approx \text{const} \cdot s^{-\alpha} n^{-\beta-1}$$

[57] P. Grassberger, *On the fractal dimension of the Hénon attractor,* Physics Letters 97A (1983) 224–226.

[58] A heuristic derivation of such scaling behavior has been provided by W. E. Caswell and J. A. Yorke in the paper *Invisible errors in dimension calculations: geometric and systematic effects,* in: *Dimensions and Entropies in Chaotic Systems,* G. Mayer-Kress (ed.), Springer-Verlag, Berlin, 1989 (second edition), pp. 123–136.

where the constant is different from that in eqn. (12.17). Given a table of values of $N(s,n)$ for different box sizes s and iteration counts n, we can extract the growth rates using a not too small increment Δn. This data can be plotted in log/log diagrams of $\Delta N(s,n)/\Delta n$ versus s and n. If these plots reveal straight lines, then the conjecture is supported, and the exponents α and β can be obtained from the corresponding slopes (see figure 12.52). Grassberger performed such an analysis with five runs of 7.5 million iterations each and using five different box sizes s. His findings are positive, and his estimate for the exponents is

$$\alpha = 2.42 \pm 0.15 \text{ and } \beta = 0.89 \pm 0.03.$$

How to Get the Dimension from α and β

Having calculated α and β we can now compute an estimate for $N(s)$ based on two measurements, for example, $N(s_0, n_0)$ and $N(s_0, 2n_0)$, as follows. Let us denote the constant in eqn. (12.17) by γ_1. Then we assume

$$\begin{aligned} N(s_0, n_0) &= N(s_0) - \gamma_1 s_0^{-\alpha} n_0^{-\beta} \\ N(s_0, 2n_0) &= N(s_0) - \gamma_1 s_0^{-\alpha} 2^{-\beta} n_0^{-\beta}. \end{aligned}$$

We estimate the constant γ_1 by solving these equations for γ_1,

$$\gamma_1 = \frac{N(s_0, 2n_0) - N(s_0, n_0)}{(1 - 2^{-\beta}) s_0^{-\alpha} n_0^{-\beta}}$$

and obtain the result (with variable s and n)

$$N(s) = N(s,n) + \gamma_1 s^{-\alpha} n^{-\beta}$$

where the parameters α, β, and γ_1 on the right-hand side are known. For example, Grassberger computed $N(s) = 238,513 \pm 200$ for $s = 0.0169/60$. The calculation of the box-counting dimension can then proceed along the same line. With another constant of proportionality γ_2 we assume

$$\begin{aligned} N(s) &= \gamma_2 s^{-D_f} \\ N(2s) &= \gamma_2 2^{-D_f} s^{-D_f}. \end{aligned}$$

Thus,

$$\frac{N(s)}{N(2s)} = 2^{D_f}$$

and it follows

$$D_f = \frac{\log N(s) - \log N(2s)}{\log 2}.$$

Several such evaluations can be made and compared to estimate what the dimension D_f should be when s tends to 0.

The Result

These estimates for α and β along with the measurements for $N(s,n)$ are sufficient to first extrapolate $N(s)$ and then to calculate the box-counting dimension for these extrapolated values. Grassberger reported the number

$$D_f = 1.28 \pm 0.01$$

for the dimension of the Hénon attractor.[59]

Why We Need an Alternative to Box-Counting

It is clear that the procedure outlined above is very elaborate even in the relatively simple two-dimensional case considered here. More typical attractors arise from more complex systems where even the mere computation of trajectories requires some numerical approximation techniques. Moreover, in other cases the dimension may be much larger than 1.28, perhaps on the order of 4, 5 or even more, which implies that a very much larger effort is necessary. All in all, the box-counting dimension is nice to have, but there are too many problems with its computation for the numerical results to be dependable.

There is another reason why a different dimension might be preferable. When looking at an orbit which densely fills up the attractor we notice that it visits some areas of the attractor much more frequently than others. For example, at an intermediate box size s Grassberger counted about 40,000 visited boxes in a run of 7,500,000 points. That gives rise to an average of about 190 points per box. But clearly figure 12.52 (center curve) shows that during the last million iterations or so, we can expect to hit one empty box in about every 10,000 iterations. At the end of the computation these boxes have a much lower than average count. The box-counting dimension ignores such differences; if a box is visited it is counted only once no matter how many million more times the orbit passes through that box. In other words, the box-counting dimension does not reflect the distribution of points from an orbit on the attractor (compare figure 12.53).

The Natural Measure

To overcome this shortcoming of fractal dimension, boxes should be weighted according to how many times an orbit visits them. Let us go one step further, introducing a more general concept called measure.[60] Consider an open subset B of a space X in which an attractor lies, for example, a subset B of the plane or of the Euclidean three-dimensional space. Orbits that are typically observed in computer studies seem to eventually fill up the attractor densely. If almost all exact orbits which start in or near the attractor fill up the attractor densely, then the system is called *ergodic*. We can count the number of times an orbit $x_0, x_1, x_2, \ldots \in X$ enters the subset B, and it is natural to assume that the percentage of all points which are in B stabilizes as we perform more and more iterations. This percentage is called the *natural*

[59] Previous computations using the straightforward box-counting procedure resulted in the underestimate $D_f = 1.26$. See D. A. Russell, J. D. Hanson, E. Ott, *Dimension of strange attractors,* Phys. Rev. Lett. 45 (1980) 1175–1178.

[60] For a mathematical treatise on ergodic theory including measures (and Ljapunov exponents) see the book R. Mañé, *Ergodic Theory and Differentiable Dynamics,* Springer-Verlag, Heidelberg, 1987.

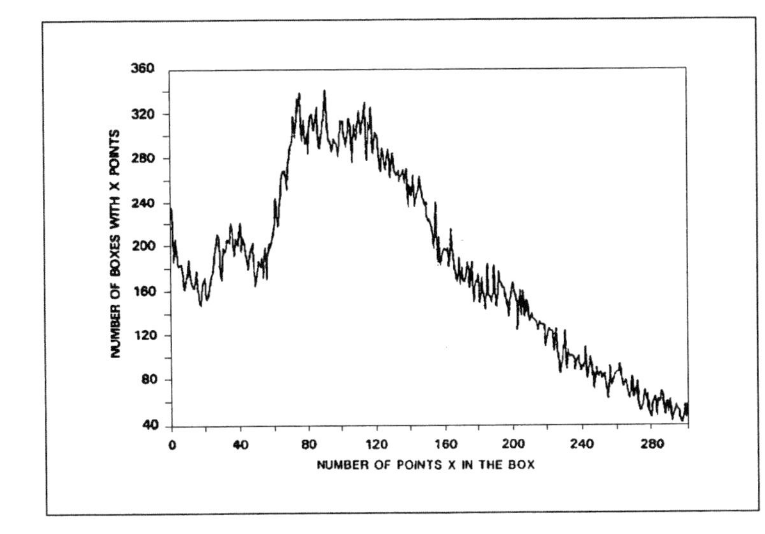

Density Distribution

An orbit on the Hénon attractor consisting of 44 million points is computed, and frequencies for boxes of linear size 0.000125 are counted. Such a density distribution may serve as the basis for dimension calculations. The figure is adapted from W. E. Caswell and J. A. Yorke, *Invisible errors in dimension calculations: geometric and systematic effects,* in: *Dimensions and Entropies in Chaotic Systems,* G. Mayer-Kress (ed.), Springer-Verlag, Berlin, 1986 and 1989, p. 123–136.

Figure 12.53

measure $\mu(B)$ for the system. Formally,

$$\mu(B) = \lim_{n\to\infty} \frac{1}{n+1} \sum_{k=0}^{n} \mathbf{1}_B(x_k) \tag{12.18}$$

where $\mathbf{1}_B(x)$ is a function which is 1 or 0 when x is in B or not. In other words,

$$\mathbf{1}_B(x) = \begin{cases} 1 & \text{if } x \in B \\ 0 & \text{otherwise} \end{cases}$$

and $\sum_{k=0}^{n} \mathbf{1}_B(x_k)$ is the number of points from the (finite) orbit $x_0, \ldots, x_n$ which fall in the set B.

The natural measure can be understood as a means of quantifying the *mass* of a portion of the attractor. Consider an orbit on the Hénon attractor and imagine that we drop a grain of sand at every point of the orbit. In some places the sand will accumulate faster than in other places. The weight of the sand collected in the subset B is approximated by $\mu(B)$ times the total weight of the distributed sand. While following the orbit and distributing more and more grains of sand this approximation becomes more and more exact.

More on the Natural Measure

The natural measure is an invariant measure. For the system

$$x_{k+1} = f(x_k)$$

this means that if $f^{-1}(B)$ denotes the set of points which after one iteration land in B (the preimage of B) then its measure must be the same as that of B itself,

$$\mu(f^{-1}(B)) = \mu(B).$$

This is clear when we think of the natural measure being generated from the statistics of a long orbit: for each point of the orbit which falls in $f^{-1}(B)$ there is also a point in B, namely, its successor. Note that only when the transformation f is one-to-one, i.e., for every point y there is one and only one point x such that $f(x) = y$, we have that the above condition is equivalent to

$$\mu(B) = \mu(f(B)).$$

Given a continuous transformation f there is at least one invariant measure. In general, however, there are many invariant measures. For example, if f has a fixed point $x^* = f(x^*)$ and $f(x) \neq x^*$ for all $x \neq x^*$, then the Dirac measure δ, defined by $\delta(B) = 1$ if x^* is in B and $\delta(B) = 0$ otherwise, is invariant. However, if this fixed point is a repeller, which is the typical case when x^* is in a chaotic attractor, then the Dirac measure is of no practical importance, because almost no orbits generate this measure in the sense of eqn. (12.18). A dynamical system given by a transformation f and an associated invariant measure μ is called ergodic if almost all orbits generate μ.[61] In this case μ is called the natural measure.

The natural measure can also be defined for continuous dynamical systems. In that case the measure $\mu(B)$ is the relative time that a trajectory $x(t)$ spends in the region B. Formally,

$$\mu(B) = \lim_{t\to\infty} \frac{1}{t} \int_0^t \mathbf{1}_B(x(s))ds.$$

The natural measure captures the long-term statistics of the iteration process, and we would like to see it entered in a new kind of fractal dimension. The logical approach would be to replace the simple box-count by a counting procedure in which each box is weighted according to its natural measure. Thus, places which the orbit passes through very frequently have a stronger impact on the calculation than boxes which the orbit rarely visits. In terms of a formula, we replace $\log N(s)$ by

$$I(s) = \sum_{k=1}^{N(s)} \mu(B_k) \log_2 \frac{1}{\mu(B_k)}. \tag{12.19}$$

[61] More precisely, let f be a transformation defined on a subset A of the Euclidean n-dimensional space and μ an invariant measure on A. Further let g be an arbitrary function on A taking real values (for example, the function $\mathbf{1}_B$ from above). Then the system given by f, A, and μ is ergodic if for almost all orbits (i.e., for all initial points x_0 in some set $A_0 \subset A$ with $\mu(A_0) = 1$) the average value of g is the same as the value of g averaged with respect to the measure μ. In terms of a formula, this is

$$\lim_{m\to\infty} \frac{1}{m+1} \sum_{k=0}^{m} g(x_k) = \int_A g(x)d\mu(x).$$

Here the sum ranges over all $N(s)$ boxes B_k of linear size s that cover the attractor.[62] This quantity $I(s)$ specifies the amount of information necessary to specify a point of the attractor to within an accuracy of s; or in other words, it is the information obtained in making a measurement that is uncertain by an amount s.

How Much Is 4.3 Bits of Information?

Before presenting the results of a computation let us give an interpretation of this information using a little game. One point in the attractor is chosen at random on the basis of the natural measure. For example, we may run the iteration of the corresponding dynamical system on a computer. At some point we decide to push a button which halts the program and outputs the current point of the orbit. This point is in a certain box B_k. Assume there is an 'investigator' outside who attempts to determine this box B_k. This person knows all the boxes and their corresponding probabilities. He may ask questions about the outcome of the experiment, which we are allowed to answer only with 'yes' or 'no'. For example, he may ask, 'is the box in question among the boxes B_1 to B_{100}?' After some finite number of questions the investigator knows the box B_k in which the random point lies, and the game can be repeated. There is an optimal strategy for the investigator to formulate the questions, and the information $I(s)$ is a measure of how successful this strategy is. Namely, the information is the *average number of questions* asked until the right box is discovered. Here is a simple example. There are eight boxes, numbered B_0 to B_7 which contain an equal amount of the measure μ, i.e., $\mu(B_k) = 1/8$ for $k = 0, \ldots, 7$. What is the optimal strategy to discover the box containing a randomly chosen point? The investigator would intuitively try to pose the questions such that after each answer the number of possibilities is reduced by a factor of one-half. Thus, we ask if the number of the box is $0, 1, 2$, or 3. Let us assume that answer is 'yes'. Then the next question would ask whether the number is 0 or 1. If the answer to the second question is 'no', we finish up with the question, is it the box number 2? Three questions have sufficed to find the correct box. Moreover, it does not matter the least in which box the unknown point lies; in any case three questions suffice to identify the box. There is no better strategy. Thus, the information content is $I = 3$.

To complicate the matter let us remove one of the boxes and adjust the probabilities to $\mu(B_k) = 1/7$ for all boxes. Using the same strategy as before, we find again that three questions will always suffice. Moreover, in one case we are finished with only two questions.[63] Thus, on the average some number between two and three questions will be used (about 2.857). We can further reduce the average number of questions when considering several unknown boxes at once (see below). The optimal result is $\log_2 7 \approx 2.807$, which is precisely what formula (12.19) prescribes.

[62] The logarithm is taken with base 2. Often the definition is made with natural logarithms. This merely introduces a factor of proportionality. The advantage of the base-2 logarithm is that the result $I(s)$ is given in units of bits (see below).

[63] The first question determines in which of the sets $\{0, 1, 2, 3\}$ and $\{4, 5, 6\}$ the unknown number is. The second question differentiates between the sets $\{0, 1\}$, $\{2, 3\}$, $\{4, 5\}$, and $\{6\}$. In the last case (6), we are already done. If the number is less than 6, we need one more question. The average number of questions needed thus is $3 \cdot 6/7 + 2 \cdot 1/7 = 20/7 \approx 2.857 < 3$.

The Information Theory of Shannon

Why do we call the quantity $I(s)$ in eqn. (12.19) a measurement of information? And what are the physical units of such information? The theoretical foundation of information had been given by Claude Shannon[64] in 1948. Let us give some heuristic arguments starting again with the eight boxes from above. We can rewrite the list of indices of the boxes, the numbers 0,..., 7, in binary notation using three bits per number as follows:

$$000, 001, 010, 011, 100, 101, 110, 111.$$

With this notation the investigator's questions are: Is the first bit in the code of the unknown box set, i.e., equal to 1? The second? The third? After these three questions the box is known. In other words, we need three binary digits to code the eight different outcomes. Therefore, the information is $I = \log_2 8 = 3$, and it is measured in units of bits. In other words, the information I specifies the average word length when we encode the possible events in binary numbers. In the above case the result is rather obvious because the number of possible events is a power of 2 and, moreover, all events are equally likely.

Let us take the first step to generalize to other numbers of events with equal probabilities. For example, consider a setup with three outcomes, A, B, and C, each one with probability 1/3. Using not more than two questions, we can find out which of the three events happened. The first one differentiates between $\{A, B\}$ and $\{C\}$, and the second determines whether A or B was chosen in the first case. We summarize the procedure in the following table where the two questions are 'Is the set C?' and 'Is the set B?' 0 stands for the answer 'no', and 1 for 'yes'.

Event	Prob.	Questions		Number of
		1	2	Questions
A	1/3	0	0	2
B	1/3	0	1	2
C	1/3	1		1

To compute the average number of questions asked we sum up the products of the probabilities and number of questions from each line,

$$2 \cdot \frac{1}{3} + 2 \cdot \frac{1}{3} + 1 \cdot \frac{1}{3} = \frac{5}{3} \approx 1.667.$$

The formula in eqn. (12.19) tells us, however, that an optimal strategy will require only

$$I = \sum_{k=1}^{3} \frac{1}{3} \log_2 3 = \log_2 3 \approx 1.585$$

[64]Claude Elwood Shannon, an engineer and mathematician, became one of the pioneers of computer science introducing a mathematical theory of the capacity of communication channels. He worked at the Bell Laboratories and since 1956 as a professor at the Massachusetts Institute of Technology.

questions on the average. How can we improve the strategy? Since we are dealing with averages, we may assume that the experiment is carried out many times and we may group pairs of events and ask questions to determine both events. In our case these pairs are AA, AB, AC, BA, ..., CB, CC. Then we can use the strategy outlined in the following table.

Event	Prob.	Questions				Number of
		1	2	3	4	Questions
AA	1/9	0	0	0		3
AB	1/9	0	0	1		3
AC	1/9	0	1	0		3
BA	1/9	0	1	1		3
BB	1/9	1	0	0		3
BC	1/9	1	0	1		3
CA	1/9	1	1	0	0	4
CB	1/9	1	1	0	1	4
CC	1/9	1	1	1		3

There are nine different cases. In seven of them three questions suffice, while for the remaining two we need four questions. Thus, the average is

$$7 \cdot 3 \cdot \frac{1}{9} + 2 \cdot 4 \cdot \frac{1}{9} = \frac{29}{9}.$$

Since we always determine two events simultaneously, we divide by two to get an average of

$$\frac{29}{18} \approx 1.611$$

questions per event. This is an improvement over 1.667 from the simple strategy but still not optimal. We may now continue to consider triple events in place of pairs, then quadruple and so on, improving our average. The way to pose a question is to always group all possible events into two groups such that the total probability in each one of them is as close to 0.5 as possible. Shannon's results indicate that at the end we get $\log_2 3$ as the optimal average.

Moreover, it does not matter whether or not the events are equally likely. Let us reconsider the setup of $N = 8$ boxes and ask how much information there is in a message which says that the event was a '2 or 3'. This information should carry the same number of bits as outcomes '0 or 1', '4 or 5', and '6 or 7'. There are four such cases, and therefore the number of bits in this information is $\log_2 4 = 2$. Note that $p = 2/N = 1/4$ is the probability of the event, and another way to express our result is to say that the information contains

$$\log_2 \frac{1}{p} \text{ bits.}$$

In fact, this can be applied to more general situations. For example, consider the information that the outcome of a measurement is one of M specified events out of a total of N equally likely events. The probability of such a result is $p = M/N$ and the information content is again $\log_2 1/p$ bits. To illustrate, take $M = 3$ and $N = 24$. Then the message has $\log_2 24/3 = \log_2 8 = 3$ bits. Or, setting $N = 22$, we would obtain $\log_2 22/3 \approx 2.874$ bits.

We can now make the step towards a collection of mutually exclusive events or measurements which occur with different probabilities. Let us use the symbols $B_1, \ldots, B_N$ for the events and $p_1, \ldots, p_N$ for the probabilities, where we assume that they sum up to 1. According to our reasoning above, we may say that the information contained in the measurement B_k is $\log_2 1/p_k$ bits. Now we can do the last step and ask what the average information content would be in this case. This average is not simply the arithmetic mean, $1/N \sum \log_2 1/p_k$, because the events are not equally likely. Thus, the expectation of the information we get is the weighted sum

$$I = \sum_{k=1}^{N} p_k \log_2 \frac{1}{p_k}$$

and its units of measurements are bits.[65]

Let us take, for example, the independent events A with probability 1/8 and B with 7/8. The information, according to Shannon, is

$$\frac{1}{8} \log_2 8 + \frac{7}{8} \log_2 \frac{8}{7} \approx 0.544.$$

Grouping three consecutive events the following strategy can be applied.

Event	Prob.	Questions					Number of Questions
		1	2	3	4	5	
AAA	1/512	0	0	0	0	0	5
AAB	7/512	0	0	0	0	1	5
ABA	7/512	0	0	0	1	0	5
BAA	7/512	0	0	0	1	1	5
ABB	49/512	0	0	1			3
BAB	49/512	0	1	0			3
BBA	49/512	0	1	1			3
BBB	343/512	1					1

The result is an average of one-third of

$$\frac{5}{512} + 3 \cdot \frac{5 \cdot 7}{512} + 3 \cdot \frac{3 \cdot 49}{512} + \frac{343}{512} = \frac{894}{512}$$

which is $298/512 \approx 0.582$ and quite close to the theoretical optimal value.

[65] When working with the natural logarithm the information is given in units of $1/\log 2 \approx 1.44$ bits.

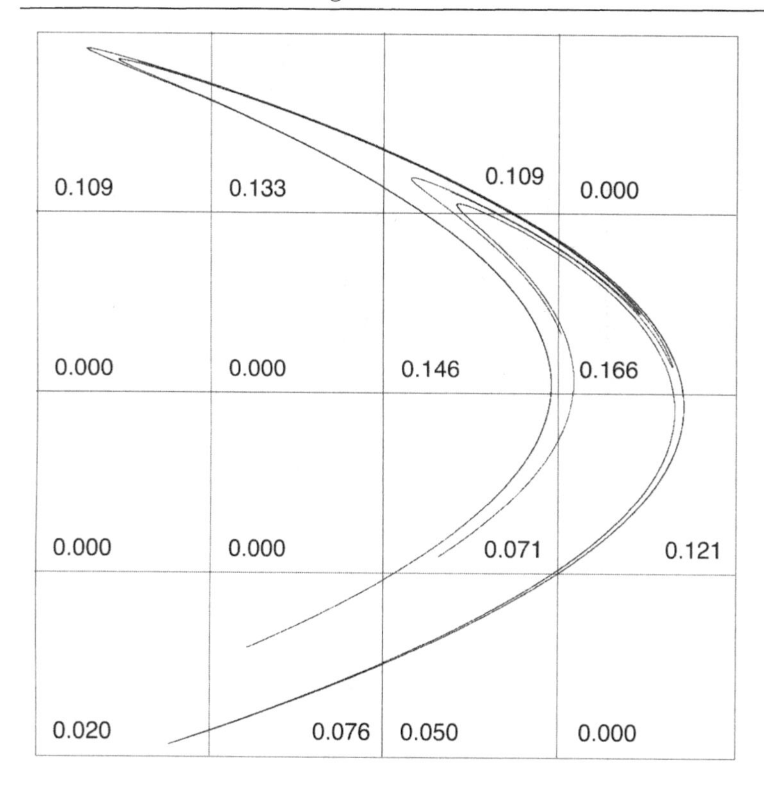

Figure 12.54 : Partitioning of the Hénon attractor into 16 boxes. The numbers indicate the natural measure of the boxes.

Let us finally give an example of the formula as applied to the natural measure of a strange attractor. Figure 12.54 shows the subdivision of the Hénon attractor into 16 boxes. We have computed a long orbit and collected its statistics resulting in the probabilities attached to the boxes as shown. In this configuration the information that a point is in the upper left box has a value of $\log_2 1/0.109 \approx 3.198$ bits. The average amount of information per measurement is

$$\begin{aligned}
&0.109 \log_2 1/0.109 + 0.133 \log_2 1/0.133 + \\
&0.109 \log_2 1/0.109 + 0.146 \log_2 1/0.146 + \\
&0.166 \log_2 1/0.166 + 0.071 \log_2 1/0.071 + \\
&0.121 \log_2 1/0.121 + 0.020 \log_2 1/0.020 + \\
&0.076 \log_2 1/0.076 + 0.050 \log_2 1/0.050 = 3.168 \text{ bits.}
\end{aligned}$$

This result is the same as the number $I(s_2)$ in table 12.55.

Clearly, as the box size s decreases the information $I(s)$ must increase. Table 12.55 present a first rough numerical draft for the evaluation of this information for the case of the Hénon attractor. In figure 12.56 we plot the

Using 10,000,000 iterations and grids of up to 1024 × 1024 boxes covering the region $-1.5 \leq x \leq 1.5$ and $-0.4 \leq y \leq 0.4$, we compute the number of nonempty boxes $N(s_k)$ and the information $I(s_k)$ using eqn. (12.19). The last column contains estimates for the information dimension. The entry in row k of the last column (D_I) is computed by $I(s_k) - I(s_{k-1})$.

Table 12.55

k	s_k	$N(s_k)$	$I(s_k)$	D_I
1	1/2	4	1.840	
2	1/4	10	3.168	1.328
3	1/8	28	4.481	1.313
4	1/16	69	5.624	1.143
5	1/32	169	6.877	1.253
6	1/64	395	8.145	1.268
7	1/128	917	9.371	1.226
8	1/256	2181	10.580	1.209
9	1/512	5053	11.797	1.217
10	1/1024	12064	13.039	1.242

Information Versus Scale

The information $I(s_k)$ is plotted versus the index k in table 12.55. The slope of the line is about 1.2.

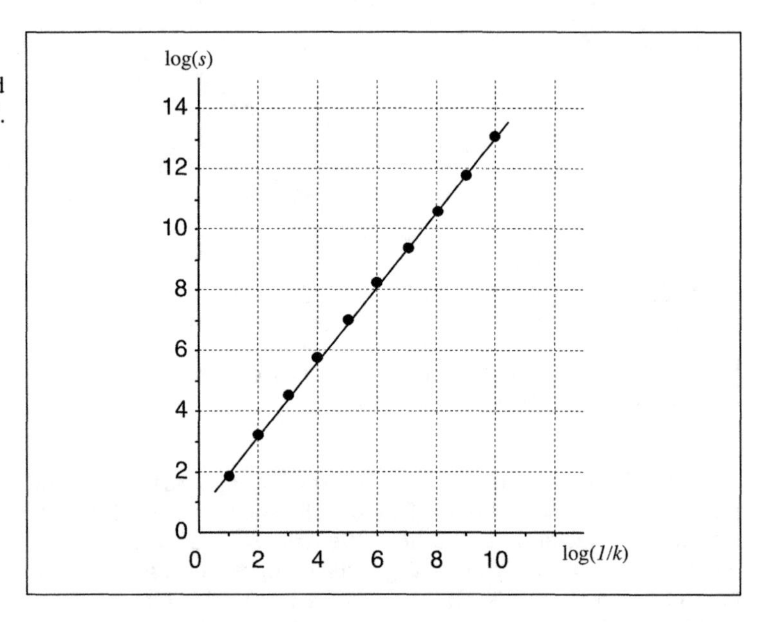

Figure 12.56

information $I(s_k)$ versus the logarithmic inverse scale $\log_2(1/s_k) = k$.[66] These data reveal another power law, namely, that $I(s)$ increases logarithmically with $1/s$ as $s \to 0$. In other words,

$$I(s) \approx I_0 + D_I \log_2 \frac{1}{s}$$

where I_0 is a constant and D_I the slope of the line in the plot of $I(s)$ versus $\log_2 1/s$, the number D_I characterizes this information growth. It is the additional amount of information obtained when doubling the resolution of

[66] To be precise, note that the boxes used are not squares, but rectangles chosen so that 2^k by 2^k of them cover the indicated region with the attractor. Thus, $1/s_k$ is not equal to the horizontal or vertical size of a box but a certain multiple of that. This is done only for convenience of notation and does not have an effect on the dimension calculation.

the subdivision of the attractor. D_I is called the *information dimension*. In our case it ranges from 1.21 to 1.24. Just like the fractal (box-counting) dimension it is not an integer, which provides another reason to call the Hénon attractor a fractal and a strange attractor.[67] In fact, the information dimension D_I is a lower bound for the fractal (box-counting) dimension D_f,

$$D_I \le D_f.$$

This is confirmed by our crude estimate, $D_I = 1.23 \pm 0.02$, and Grassberger's result, $D_f = 1.28 \pm 0.01$.

Information and Box-Counting Dimension

The relation $D_I \le D_f$ for the information and box-counting dimension can be shown directly. This follows from basic inequalities for the arithmetic and geometric means. Let $a_1, \ldots, a_N$ be positive numbers. Then their geometric mean is less than or equal to the arithmetic mean,

$$\sqrt[N]{\prod_{k=1}^{N} a_k} \le \frac{1}{N} \sum_{k=1}^{N} a_k.$$

We need a generalization of this fact. If p_k denote positive numbers (probabilities) with $p_1 + \cdots + p_N = 1$, then

$$\prod_{k=1}^{N} a_k^{p_k} \le \sum_{k=1}^{N} p_k a_k.$$

Furthermore, equality holds if and only if all numbers a_k are the same. In our application we set $a_k = 1/p_k$ and obtain

$$\prod_{k=1}^{N(s)} \left(\frac{1}{p_k}\right)^{p_k} \le \sum_{k=1}^{N(s)} p_k \frac{1}{p_k} = N(s).$$

Taking the logarithm on both sides we arrive at

$$I(s) = \sum_{k=1}^{N(s)} p_k \log_2 \frac{1}{p_k} \le \log_2 N(s).$$

Using this result and the definition of the box-counting and the infor-

---

[67] The box dimension of a nonfractal such as an interval or a square is an integer. The corresponding statement holds true for the information dimension. Consider for example the unit interval with the uniform measure μ. Let us choose $s_k = 2^{-k}$ and compute the information $I(s_k)$. For any interval (box) B of length s_k we have $\mu(B) = 2^{-k}$. Thus,

$$I(s_k) = \sum_{l=1}^{2^k} \mu(B_l) \log_2 \frac{1}{\mu(B_l)} = \sum_{l=1}^{2^k} 2^{-k} \log_2 2^k = k = \log_2 \frac{1}{s_k}$$

and therefore the information dimension D_I is equal to 1.

mation dimension in the form of

$$D_I = \lim_{s \to 0} \frac{I(s)}{\log_2 1/s} \text{ and } D_f = \lim_{s \to 0} \frac{\log_2 N(s)}{\log_2 1/s}$$

we are done with the argument for $D_I \leq D_f$.

The Mass Dimension

There is another interesting version of dimension which has a strong relation to the information dimension. It is the pointwise or mass dimension. Consider a point (x, y) in the attractor and disks $B_r(x, y)$ of radius r centered at (x, y). The probability or mass contained in this disk is $\mu(B_r(x, y))$. The pointwise dimension at the point (x, y) is the exponent α in the power law which specifies how fast this mass decreases as the radius r decreases to 0,

$$\mu(B_r(x, y)) \propto r^\alpha.$$

In other words,

$$\alpha = \lim_{r \to 0} \frac{\log \mu(B_r(x, y))}{\log r}.$$

This exponent α is also called the Hölder exponent at the point (x, y). If this scaling law holds for all points on the attractor with the same α, then the attractor is a *homogeneous fractal*. However, it actually happens very often that the Hölder is not the same for all points, and in this case the attractor is called an *inhomogeneous fractal* or a *multifractal* (see the appendix on multifractals).[68]

A Spectrum of Dimensions

The box-counting and information dimensions can be embedded in a collection of dimensions called Rényi dimensions, which have attracted much attention from physicists in the last 10 years.[69] The idea is to modify the formula for the information $I(s)$ in eqn. (12.19) obtaining the Rényi information of the qth-order

$$I_q(s) = \frac{1}{1-q} \log_2 \sum_{k=1}^{N(s)} p_k^q \tag{12.20}$$

for $q \geq 0$ where we have abbreviated $p_k = \mu(B_k)$. Only boxes B_k with positive measure ($p_k > 0$) need to be considered in this sum. The Rényi dimension D_q for the parameter q is defined as

$$D_q = \lim_{s \to 0} \frac{I_q(s)}{\log_2 1/s}. \tag{12.21}$$

Let us start the discussion with two special cases, $q = 0$ and $q = 1$. It follows directly from the above definitions that $I_0(s) = \log_2 N(s)$,

[68] Compare D. Ruelle, *Chaotic Evolution and Strange Attractors*, Cambridge University Press, Cambridge, 1989.

[69] This part of our exposition is written along the lines presented in R. W. Leven, B.-P. Koch and B. Pompe, *Chaos in Dissipativen Systemen*, Vieweg, Braunschweig, 1989.

the logarithm of the number of nonempty boxes of linear size s. in other words, the Rényi dimension D_0 is nothing but the usual fractal (box-counting) dimension,

$$D_0 = D_f.$$

Let us now consider $q = 1$. Here eqn. (12.20) is not directly applicable because the denominator $1 - q$ vanishes. This problem can be solved by using the limit of $I_q(s)$ as $q \to 1$. According to the rule of l'Hospital[70] we have

$$\lim_{q\to 1} I_q(s) = -\lim_{q\to 1} \frac{\sum p_k^q \log p_k}{\sum p_k^q \log 2} = \sum_{k=1}^{N(s)} p_k \log_2 \frac{1}{p_k}.$$

This is just the information $I(s)$ discussed in detail above! Thus, it follows

$$D_1 = D_I.$$

We have already shown that $D_0 \geq D_1$. The fact is that the Rényi dimension is a decreasing function, i.e.,

$$D_p \geq D_q \text{ if } p < q.$$

If D_q decreases strictly for increasing parameter $q \geq 0$, then the measure on the attractor is called inhomogeneous or multi-fractal (compare appendix A.2). This may be the case even though the pointwise dimension D_p may be the same almost everywhere.

When computing D_q for integers $q \geq 2$, there is an alternative to the direct approach using equations (12.20) and (12.21). The term p_k^q is the probability that q points drawn randomly from the attractor according to its natural measure fall in the same box B_k. Furthermore, the sum $\sum p_k^q$ can be interpreted as the probability that the q random points mentioned above together fall into any one of the boxes B_k of size s. How can we create such samples? A representation of the natural measure of the attractor can be obtained from almost all orbits on it, and we may simply draw q samples of points from such an orbit. The probability that q points fall in the same box is proportional to the relative number $C_q(s)$ of q-tuples where all pairwise distances are bounded by the size s. But let us be more precise. We consider a finite portion of an orbit $x_0, x_1, \ldots, x_m$ (where each x_k is a point in a Euclidean space) and define the number $T_q(m, s)$ of q-tuples $(x_{i_1}, x_{i_2}, \ldots, x_{i_q})$ of different points from the orbit satisfying

$$d(x_{i_k}, x_{i_l}) < s \text{ for } k, l = 1, \ldots, q$$

where $d(x, y)$ denotes the Euclidean distance between points x and y. Then we let

$$C_q(s) = \lim_{m\to\infty} \frac{T_q(m, s)}{m^q}.$$

[70] The rule of l'Hospital is useful when evaluating indeterminate forms. If $\lim_{x\to a} f(x) = \lim_{x\to a} g(x) = 0$, and $f'(x)/g'(x)$ approaches a limit as x tends to a, then $f(x)/g(x)$ approaches the same limit.

Now we have

$$\sum_{k=1}^{N(s)} p_k^q \propto C_q(s)$$

and it follows that

$$D_q = \frac{1}{1-q} \lim_{s\to 0} \frac{\log_2 C_q(s)}{\log_2 1/s}. \tag{12.22}$$

In the special case $q = 2$ we deal with pairs of points and $C_2(s)$ is a kind of correlation in pairs of points. Therefore, D_2 is called the correlation dimension.

Based on numerical experiments made around 1978 James L. Kaplan and James A. Yorke came to the conclusion that in most cases it is possible to predict the dimension of a strange attractor from the knowledge of the Ljapunov exponents of the corresponding transformation.[71] Although their formula for the dimension has not been rigorously proven (except for some special cases) it opens up the door to the experimental study of dimensions for many dynamical systems. This so called *Kaplan-Yorke conjecture* has been tested and discussed in many research papers.[72] It is such an important topic because in many dynamical systems the various dimensions of the attractors are hard to compute, while the Ljapunov exponents are relatively accessible.

The Ljapunov Dimension

Now what is the Kaplan-Yorke conjecture? Let us consider a dynamical system

$$x_{k+1} = f(x_k)$$

where f is a transformation in n-dimensional Euclidean space. Let us assume that f has n Ljapunov exponents, ordered by magnitude,

$$\lambda_1 \geq \lambda_2 \geq \cdots \geq \lambda_n.$$

We define a function

$$k \to \gamma(k) = \lambda_1 + \lambda_2 + \cdots + \lambda_k$$

for integers $k = 1, \ldots, n$. Moreover we set $\gamma(0) = 0$ and extend γ to all real numbers $0 \leq d \leq n$ such that its graph is linear between two adjacent integers (see figure 12.57). Because of the ordering of the exponents the graph of $\gamma(d)$ is concave, and there is exactly one point $d^* > 0$ at which $\gamma(d^*) = 0$. This value is called the *Ljapunov dimension*

$$D_L = \max\{d > 0 \mid \gamma(d) \geq 0\}.$$

---

[71] J. L. Kaplan and J. A. Yorke, *Chaotic behavior of multidimensional difference equations*, in: *Functional Differential Equations and Approximation of Fixed Points*, H.-O. Peitgen and H. O. Walther (eds.), Springer-Verlag, Heidelberg, 1979.

[72] See for example D. A. Russell, J. D. Hanson, E. Ott, *Dimension of strange attractors*, Phys. Rev. Lett. 45 (1980) 1175–1178.

Expressed more explicitly, the Ljapunov dimension is

$$D_L = m + \frac{1}{|\lambda_{m+1}|} \sum_{k=1}^{m} \lambda_k \tag{12.23}$$

where m is the maximum integer with $\gamma(m) = \lambda_1 + \cdots + \lambda_m \geq 0$. (In the case $\lambda_1 < 0$ we simply set $D_L = 0$.)

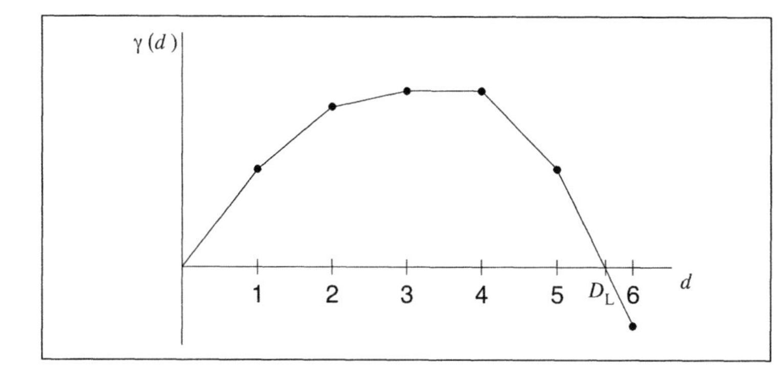

Defining the Ljapunov Dimension

Determination of the Ljapunov dimension D_L. The function $\gamma(d)$ is an interpolation of the sum of Ljapunov exponents.

Figure 12.57

The Kaplan-Yorke Conjecture

The conjecture proposed by Kaplan and Yorke claims that generally the Ljapunov dimension is equal to the information dimension D_I.[73] Instead of going into the heuristic motivation for the conjecture given in the literature cited, we concentrate on a special case, namely, continuous dynamical systems in three dimensions, such as the Lorenz system. This is not a loss, because this case already contains the main ingredients of the flavor of the original derivation. We may assume without doing any harm that the three Ljapunov exponents satisfy

$$\lambda_1 > 0, \ \ \lambda_2 = 0, \ \ \lambda_3 < 0$$

and

$$\lambda_1 + \lambda_2 + \lambda_3 < 0.$$

This is the typical situation in, for example, the Lorenz or the Rössler systems. We consider a small cube with side length s and how this cube evolves under the flow of the system (see figure 12.58). For simplicity we assume that such cubes behave according to the Ljapunov exponents as follows. One side of the cube is expanded to $se^{\lambda_1 t}$ where λ_1 is the largest positive exponent in the system and $t \geq 0$ denotes time. The next side does not change because we assume that the second exponent is $\lambda_2 = 0$. The remaining side contracts to length $se^{\lambda_3 t}$. We now fill such a distorted cube by smaller cubes. To provide a

[73]In their first paper in 1979 they believed that $D_L = D_f$, the fractal (box-counting) dimension. However, later they dropped this idea and conjectured that $D_L = D_I$. Such historical information and a heuristic argument (too technical to be included here) for the conjecture can be found in P. Frederickson, J. L. Kaplan, S. D. Yorke, J. A. Yorke, *The Liapunov dimension of strange attractors,* Journal of Differential Equations 49 (1983) 185–207.

The Flow of a Box

A cube is positioned on a trajectory of a strange attractor such as the Lorenz attractor. After time t it is a deformed box with approximate average side lengths as indicated.

Figure 12.58

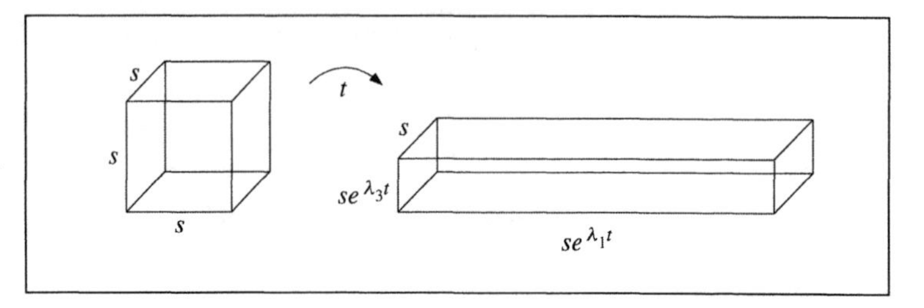

nice fit we choose the side length of the small cubes to be equal to the shortest side of the deformed cube, $s(t) = se^{\lambda_3 t}$. How many of these small cubes do we need? The volume of the original cube is reduced by a factor of

$$e^{(\lambda_1+\lambda_2+\lambda_3)t} = e^{(\lambda_1+\lambda_3)t},$$

thus the volume is

$$e^{(\lambda_1+\lambda_3)t}s^3$$

The small cubes have volume

$$\left(e^{\lambda_3 t}\right)^3 s^3 = e^{3\lambda_3 t}s^3.$$

To get an estimate the number $N(s(t))$ of small cubes necessary to cover the transformed original cube after time t, we have to divide these two numbers and obtain

$$N(s(t)) = e^{(\lambda_1 - 2\lambda_3)t}.$$

As we let the time t grow, two things happen:

- The image of the original cube of size s covers and approximates the whole attractor. The longer we wait the better the approximation becomes.
- The size of the small cubes used to fill the image of the first one rapidly decreases to zero, $s(t) = se^{\lambda_3 t} \to 0$ as $t \to \infty$ because $\lambda_3 < 0$.

Thus, we can estimate the box-counting dimension by comparing $N(s(t))$ with $1/s(t)$ for $t \to \infty$ on a double logarithmic scale. In other words, we expect

$$D_f = \lim_{t\to\infty} \frac{\log N(s(t))}{\log 1/s(t)} = \lim_{t\to\infty} \frac{(\lambda_1 - 2\lambda_3)t}{-\lambda_3 t - \log s} = 2 + \frac{\lambda_1}{|\lambda_3|}.$$

The right-hand side is exactly the Ljapunov dimension as given in eqn. (12.23). This supports the original conjecture that $D_f = D_L$. However, it seems that the cubes used in this approach tend to cover the regions of space with large natural measure better than those with extremely low measure. Thus, the conjecture was changed to $D_I = D_L$.

An Exact Confirmation of the Kaplan-Yorke Conjecture

For a further support and insight we discuss an example of a two-dimensional transformation for which the Ljapunov dimension is exactly equal to the box-counting dimension and the information dimension.[74] A system is created with

$$\lambda_1 > 0 \text{ and } \lambda_1 + \lambda_2 < 0$$

so that the Ljapunov dimension becomes

$$D_L = 1 + \frac{\lambda_1}{|\lambda_2|}.$$

For the notation of the transformation we use the operator Frac which denotes the fractional part of its argument[75] and its counterpart Int which stands for the corresponding integer part. Thus,

$$\mathrm{Int}(x) \text{ integer, } 0 \leq \mathrm{Frac}\,(x) < 1, \text{ and } \mathrm{Int}(x) + \mathrm{Frac}\,(x) = x.$$

With these definitions the transformation is

$$\begin{aligned} x_{k+1} &= \mu_2 x_k + \frac{\mathrm{Int}(\mu_1 y_k)}{\mu_1} \\ y_{k+1} &= \mathrm{Frac}\,(\mu_1 y_k). \end{aligned}$$

We choose an integer $\mu_1 > 1$ and a number μ_2 with

$$0 < \mu_2 < \frac{1}{\mu_1}.$$

More instructive than the formulas is the figure 12.59, which illustrates the operation of this transformation in two steps for the example $\mu_1 = 3$ and $\mu_2 = 1/6$. The unit square is transformed into itself yielding μ_1 vertical strips, each one of width μ_2. Further applications produce more and thinner strips always within those from the previous stages; after k transformations we have μ_1^k such strips of width μ_2^k. The attractor thus consists of an infinite number of vertical lines. A horizontal cut through the attractor is obviously a Cantor set. To compute the box-counting dimension, cover the attractor by squares of linear size $s = \mu_2^k$. A number of $1/\mu_2^k$ squares stacked up on top of each other will cover a complete vertical strip of the kth stage approximation of the attractor. Thus, the total number of squares required to cover the whole attractor is

$$\frac{\mu_1^k}{\mu_2^k}.$$

Letting $k \to \infty$ we obtain for the dimension

$$\begin{aligned} D_f &= \lim_{k\to\infty} \frac{\log \mu_1^k/\mu_2^k}{\log \mu_2^{-k}} = \lim_{k\to\infty} \frac{k\log\mu_1 - k\log\mu_2}{-k\log\mu_2} \\ &= 1 + \frac{\log\mu_1}{|\log\mu_2|}. \end{aligned}$$

[74] This example has been furnished in D. A. Russell, J. D. Hanson, E. Ott, *Dimension of strange attractors,* Phys. Rev. Lett. 45 (1980) 1175-1178.

[75] This was introduced on page 504 in chapter 10.

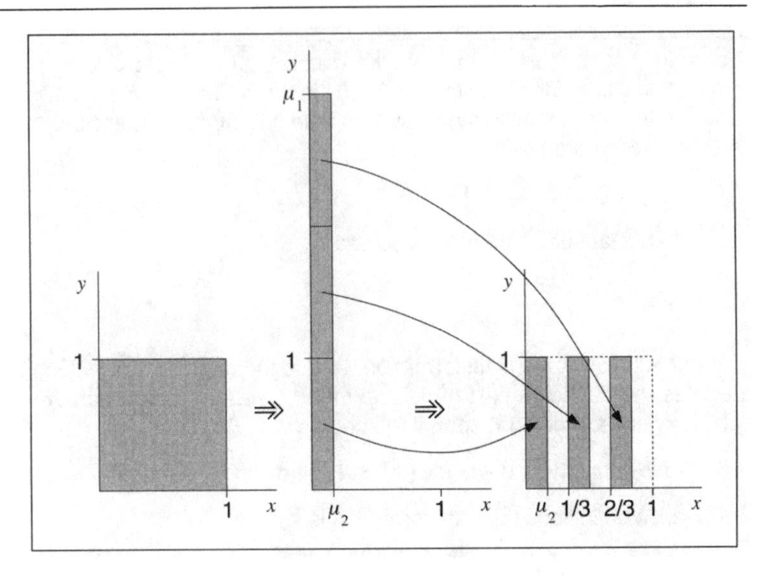

Figure 12.59 : An example for a transformation with explicitly calculable Ljapunov exponents. In the intermediate stage pictured in the middle, the unit square undergoes an affine transformation. The width is reduced by the factor $\mu_2 = 1/6$ and the height is enlarged by a factor $\mu_1 = 3$. In the final step the long strip is cut in μ_1 pieces which are placed back into the unit square as shown on the right.

To determine the Ljapunov exponents of the transformation, we note that deviations in the x-direction are damped by the factor $\mu_2 < 1$, while errors in the y-direction are amplified by the factor $\mu_1 > 1$. In other words, the derivative matrix is diagonal with entries μ_2 and μ_1. Thus, the Ljapunov exponents are $\lambda_1 = \log \mu_1$ and $\lambda_2 = \log \mu_2$. We note that our result for the box-counting dimension coincides with the Ljapunov dimension. In this case it is also equal to the information dimension because the natural invariant measure on the Cantor set of lines is 'uniform'.

The Problem of High Dimensionality

The main uses of Ljapunov dimension arise in systems where the attractor has a very high dimension, say 10 and above. In these cases it is nearly hopeless to estimate the box-counting or information dimension in the usual way. However, Ljapunov exponents may still be calculated in sufficient numbers and precision.[76] Let us discuss the basic problem with high dimensions. Consider the box-counting procedure in its crudest form. We compute an orbit of, say a total of M points, using a given discrete transformation and look at the numbers $N(s)$ and $N(s/2)$ of boxes of size s and $s/2$ needed to cover the computed orbit. From these two numbers the dimension D can be estimated using the usual procedure. In order to achieve a result in which we can trust

[76] See D. Farmer, *Chaotic attractors of an infinite-dimensional system*, Physica 4D (1982) 366–393.

to some extent, we should require that $N(s)$ is not too small. For example, $N(s)$ should be of the order 2^{10} (about 1000). Then we must expect that

$$N(s/2) \approx 2^D N(s) \approx 2^{D+10}.$$

where D is the fractal dimension. To obtain that many boxes we obviously need at least the same number of iterations in the orbit. Thus,

$$M \geq 2^{D+10}.$$

Now let us assume that the attractor in question has a very high dimension, say $D = 20$. This implies that the length of the orbit must be at least 2^{30} which is about one billion, a number that is almost impossible to reach in such complex systems using today's technology. And even if we could compute such long orbits it would be very questionable to conclude a dimension of $D = 20$ from a single scaling ratio of 1,000 to 1,000,000,000 boxes. Thus, we indeed find that high dimensions are a problem, and only the Ljapunov dimension seems to be a candidate for a feasible means of measurement.

How Precise Is a Dimension?

Another problem with dimension calculation, even for small dimensions, is the accuracy of the results. In many publications we can read statements like 'the dimension of such and such an attractor is $D = 1.234 \pm 0.005$'. We may rightfully ask what this statement really means. The typical procedure when computing dimensions is to fit a line to measured data in a double logarithmic scale. Estimating the slope of this line (and thus the dimension) is normally done by solving a so-called least-squares problem. The underlying assumption is that the measured data points are distributed about a line with errors due to fluctuations which are normally distributed. The error bounds such as ± 0.005 then indicate some standard deviation from the estimated slope. However, none of the assumptions of this model are verified for the application at hand. Only in its pure mathematical idealization do we have that the data comes from a straight line with imperfections only due to inaccurate measurement techniques. In practice there are upper and lower cutoffs at large scales and small scales s. Even between these limits different slopes may be appropriate in different ranges. This is especially true when evaluating measurements coming from a concrete physical experiment. Moreover, there is no theory yet for analyzing the statistical fluctuations of these measurements about the fitted line. All in all, it is true that dimension measurements should be interpreted with a healthy amount of scepticism.[77]

[77]Regarding this topic see for example W. E. Caswell and J. A. Yorke, *Invisible errors in dimension calculations: geometric and systematic effects*, in: *Dimensions and Entropies in Chaotic Systems*, G. Mayer-Kress (ed.), Springer-Verlag, Berlin, 1986 and 1989, p. 123–136, and K. Judd and A. I. Mees, *Estimating dimensions with confidence*, International Journal of Bifurcation and Chaos 1,2 (1991) 467–470.

12.7 The Reconstruction of Strange Attractors

Understanding natural processes does not start with a set of equations for a dynamical system. On the contrary such models are usually obtained at the end of a long course of action consisting of the identification of the phenomena to be studied, conducting series of often difficult and elaborate experiments, running trial and error computer simulations, and finally making a mathematical analysis. Somewhere in this process the question about order or chaos arises. How can we determine from measured data whether there is some underlying deterministic governing equation for the phenomena observed, or whether the data is merely noise without any structure? In other words, we want to know from a given sequence of numbers whether they come from an attractor, and if so, we also need to quantify this attractor in terms of Ljapunov exponents and dimension so that we may speak of a chaotic and strange attractor.

Let us imagine a somewhat simpler situation; we have a black box in which some continuous dynamical system is running. We may only probe the system at discrete time intervals and obtain the value of one of the state variables of the system. For example, choose one of the variables, calling it $z(t)$, and a time interval τ. Our examination of the black box would yield the sequence of numbers

$$z_0 = z(0), z_1 = z(\tau), z_2 = z(2\tau), z_3 = z(3\tau), \dots \tag{12.24}$$

Given such data, can we reconstruct some meaningful picture of some underlying attractor? Can we say something about its dimension and the Ljapunov exponents? At first thought this seems a rather hopeless undertaking.

Distinguishing Random from Deterministic Behavior

The prospects, however, are not as bleak as they seem. Here is an example. Figure 12.60 shows three time series all of which look more or less random. However, if one of them really is deterministic, then the numbers must follow some rule. In other words, z_k may be determined from the past of the sequence, and we may hope to be able to put it in the form of

$$z_k = \Psi(z_{k-1}, z_{k-2}, z_{k-3}, \dots)$$

where Ψ denotes some (so far) unknown transformation. Let us be even more optimistic and assume that z_k strongly depends on its predecessor, z_{k-1}, and only mildly (or not at all) on all earlier predecessors. To check this assumption we produce plots of z_k versus z_{k-1} as shown in figure 12.61. The result is quite clear. There is no evident structure in the first set of data. The points obtained from the second series clearly lie on a section of what appears to be a parabola. This tells us that this set of data can be generated by means of graphical iteration of a function whose graph is the parabola pictured in the figure. In fact, we used the formula $4x(1-x)$ of the generic parabola to produce the data. Thus, this simple procedure already enables us to completely unravel the random-looking data and to uncover its deterministic quadratic generation process. But will such a cheap trick work in real applications where other variables are hidden in the 'black box' or the dependence of the presence on

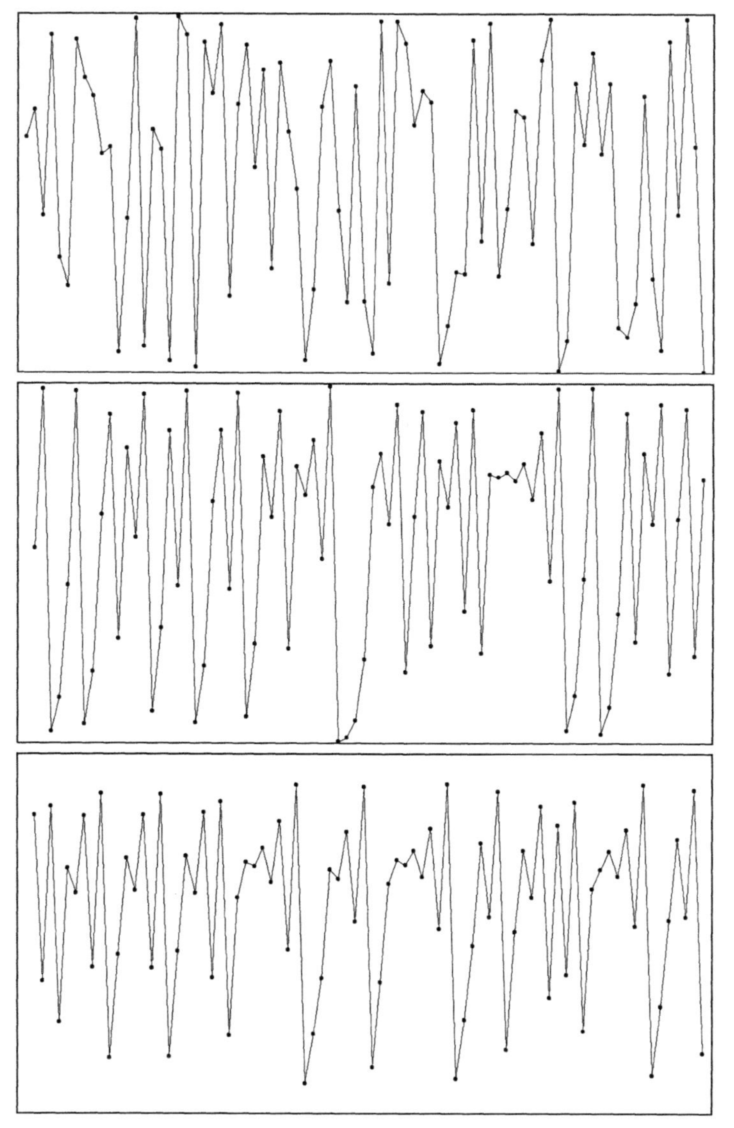

Random or Not Random Data?

Three series obtained from a 'black box'. Which one is random, which is deterministic? Or are they all random?

Figure 12.60

the past is more complicated? The third data set presents such a case. The corresponding plot of z_k versus z_{k-1} shows a collection of points which are not distributed throughout the entire square in contrast to the points of the first random data set. They form a clear structure which, however, cannot be obtained as the graph of a function. But this structure can be interpreted as an attractor of some underlying system which is a crucial insight opening the door to further numerical investigations.[78]

[78]The structure seen in the bottom plot of figure 12.61 seems to be related to the Hénon attractor. In fact, the underlying

Reconstruction Attempts

The plots show the points z_k versus z_{k-1} from the three time series in figure 12.60. This test shows that the first series is apparently random, while the second comes from a simple one-dimensional deterministic system. In the third, there is some structure indicating a strange attractor.

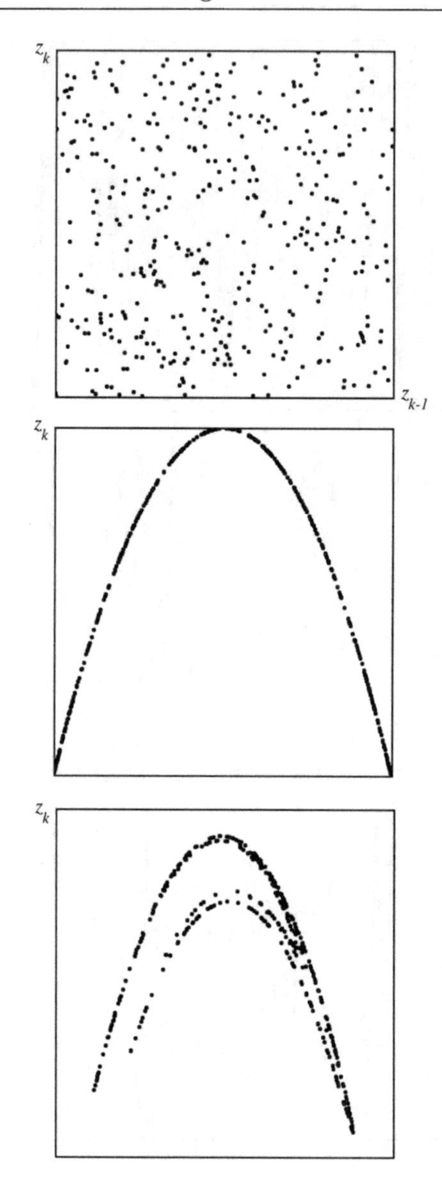

Figure 12.61

The Phase Space Reconstruction

Thus, this method of analysis of time series may lead to useful results. The fact is that a straightforward extension of this very simple procedure allows us to retrieve the geometric structure of *any* underlying attractor.[79] Let us hide

sequence $z_0, z_1, z_2, \ldots$ has been generated using the Hénon system $(x_{k+1}, y_{k+1}) = H(x_k, y_k)$ (see section 12.1) and setting $z_k = x_k, k = 0, 1, 2, \ldots$

[79]This procedure was suggested by David Ruelle, see N. H. Packard, J. P. Crutchfield, J. D. Farmer, R. S. Shaw, *Geometry from a time series,* Phys. Rev. Lett. 45 (1980) 712–716.

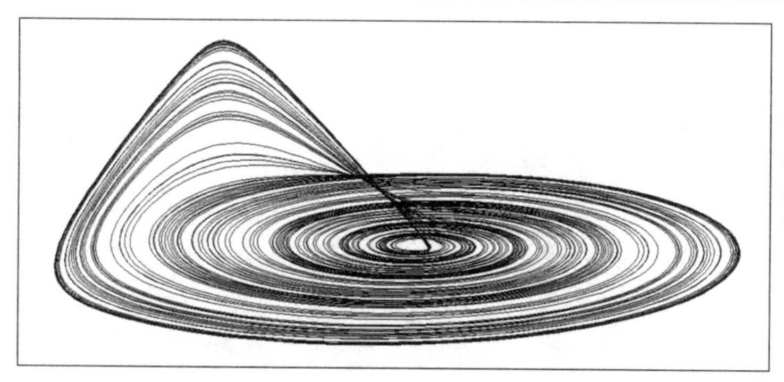

Reconstruction of the Rössler Attractor

Using a time delay of $T = 0.5$ we obtain a picture of phase space with a reconstruction of the Rössler attractor.

Figure 12.62

the Rössler system in the 'black box', run the system, and again assume that the sequence in eqn. (12.24) is extracted from the machine. We now choose a time delay T (a multiple of τ) and look at the following sequence of vectors

$$\begin{array}{l} (z(0), z(T), z(2T)) \\ (z(\tau), z(\tau+T), z(\tau+2T)) \\ (z(2\tau), z(2\tau+T), z(2\tau+2T)) \\ \vdots \\ (z(k\tau), z(k\tau+T), z(k\tau+2T)). \end{array}$$

Plotting these points in three-dimensional space with connecting line segments, we obtain figure 12.62. Clearly, the essential features of the Rössler attractor are apparent.

The Reconstruction Principle

This is not an accident! Strange attractors theoretically can always be faithfully reconstructed using the above procedure. However, working in three dimensions, we cannot expect the procedure to perform when the dimension of the attractor surpasses 3. In such a case a dense subset of the three-dimensional space would be filled. We may, however, simply work in higher-dimensional spaces using vectors

$$u(t) = (z(t), z(t+T), \ldots, z(t+2NT)) \tag{12.25}$$

with $2N + 1$ components. If N is chosen large enough the attractor will 'fit' in the chosen space. Following some theorems derived by Ricardo Mañé and Floris Takens, this can be guaranteed if the dimension of the attractor is not larger than N.[80] Here the choice of the time lag T is almost arbitrary. However, in practice there are limitations. If T is quite small, then the vectors to be plotted will have components which are almost identical, resulting in a reconstructed attractor which will be very close to the 'diagonal' of the space. On the other hand if T is very large, then there is only very little correlation

[80] Mañé, R., *On the dimension of the compact invariant set of certain nonlinear maps,* in: *Dynamical Systems and Turbulence, Warwick 1980,* Lecture Notes in Mathematics 898, Springer-Verlag (1981) 230–242. Takens, F., *Detecting strange attractors in turbulence,* in: *Dynamical Systems and Turbulence, Warwick 1980,* Lecture Notes in Mathematics 898, Springer-Verlag (1981) 366–381.

Different Time Lags

The time lag in the reconstruction process should not be too small or too large. Here $T = 10$ (top) and $T = 0.1$ (bottom) for the Rössler attractor.

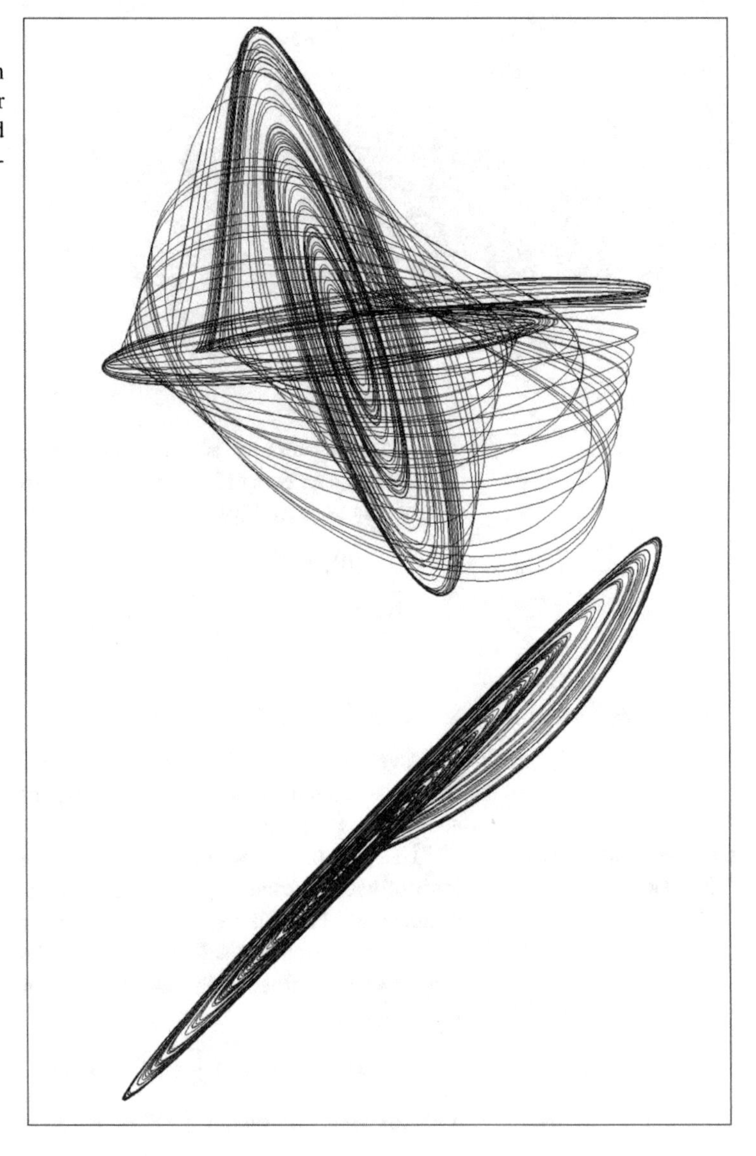

Figure 12.63

between the components of the vectors, and trajectories on the attractor appear to wander all around phase space such that the structure is hard to detect. In figure 12.63 we have chosen two different values of the time lag T to illustrate these points.

The reconstruction of strange attractors can be interpreted as a change of coordinates. Often the attractor is defined in some infinite-dimensional space (e.g., a space of functions).[81] In this case the reconstruction amounts to a

[81]Doyne Farmer presents an in-depth study of reconstructions of attractors in an infinite-dimensional space with calculations

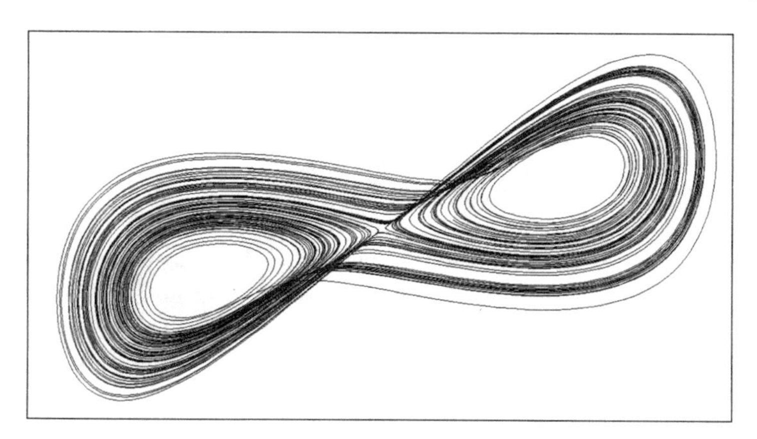

Reconstruction of the Lorenz Attractor

A reconstruction of the Lorenz attractor based on time series of the x-coordinate. We have used $T = 0.05$.

Figure 12.64

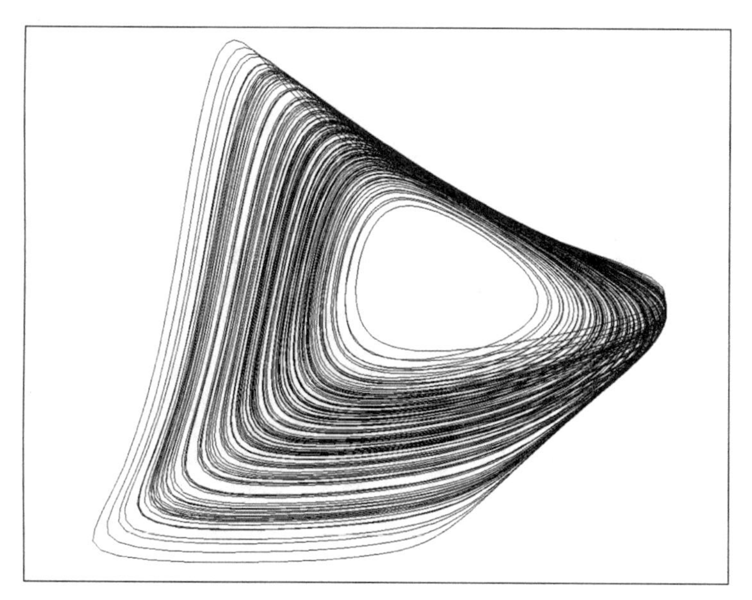

Reconstruction of the Lorenz Attractor

Another reconstruction of the Lorenz attractor based on time series of the z-coordinate. We have used $T = 0.2$. This image reveals that the folded band structure of the Rössler attractor is hidden in the Lorenz attractor.

Figure 12.65

projection of the original to a finite-dimensional Euclidean space. Choosing the dimension $2N + 1$ of the embedding space large enough guarantees that the projection is injective. This means that each point in the projected attractor corresponds to one and only one point in the original attractor. In other words, we see a truthful representation and not some image where parts of the attractor are collapsed onto each other. Thus, the reconstructed attractor is not identical to the original but a more or less distorted copy (see figures 12.64 and 12.65 for

of dimensions and Ljapunov exponents. See D. Farmer, *Chaotic attractors of an infinite-dimensional system,* Physica 4D (1982) 366–393. This study is continued in P. Grassberger and I. Procaccia, *Measuring the strangeness of strange attractors,* Physica 9D (1983) 189–208.

reconstructions of the Lorenz attractor). Changing coordinates moderately[82] does not effect the dimension or the Ljapunov exponents. Thus, we should be able to extract that information from a time series of a single variable.

In the following we sketch the computation of the Ljapunov exponents from time series. The dimension calculation based on the reconstruction does not yield any additional difficulties. Thus, the methods presented in section 12.6 may be used.

Ljapunov Exponents from Time Series

This problem was addressed and solved in the mid 1980's. The first algorithm was proposed by A. Wolf et al.[83] and is visualized in figure 12.66. It is a direct generalization of the method for explicitly given dynamical systems as described in our section 12.5. The method can be extended to compute the other Ljapunov exponents. This involves following more than one trajectory besides the reference orbit. An alternative and advanced method, proposed by Jean-Pierre Eckmann and David Ruelle uses least-squares approximations of derivative matrices.[84] It offers some advantages and is considered current state of the art.

Computing the Largest Exponent

For a given sequence from eqn. (12.24) we first construct an embedding according to eqn. (12.25) obtaining a finite sequence

$$u(0), u(\tau), u(2\tau), \ldots$$

of vectors with $2N+1$ components. This is the basic data upon which the method builds; we call it the reference orbit. If we had an explicit governing equation that would generate this orbit, then we could start with $u(0)$, prescribe a small error, add it to $u(0)$, obtaining $v(0)$. Furthermore, using the governing equation we could compute $v(\tau)$ and compare that with $u(\tau)$, compute the error amplification factor, and proceed in the same manner with the next time step. The main problem lies in the fact that we do not have a formula to compute $v(\tau)$ but only a finite sequence of points from the reference orbit. However, from this sequence we may choose a point $u(k_0\tau)$ which approximates the desired initial point $v(0)$, i.e.,

$$|u(k_0\tau) - u(0)| < \delta$$

for some a priori chosen tolerance $\delta > 0$. Let us rename this point

$$v_0(0) = u(k_0\tau).$$

For this point we know the successors, namely,

$$v_0(j\tau) = u((k_0 + j)\tau), \quad j = 1, 2, 3, \ldots$$

[82]Formally speaking, this means that the coordinate transformation and its inverse must both be Lipschitz continuous (see chapter 4).

[83]A. Wolf, J. B. Swift, H. L. Swinney, J. A. Vastano, *Determining Lyapunov exponents from a time series,* Physica 16D (1985) 285–317.

[84]See J.-P. Eckmann, S. O. Kamphorst, D. Ruelle, S. Ciliberto, *Liapunov exponents from time series,* Phys. Rev. 34A (1986) 4971–4979. The method was independently proposed by M. Sano and Y. Sawada, *Measurement of the Lyapunov spectrum from a chaotic time series,* Phys. Rev. Lett. 55 (1985) 1082.

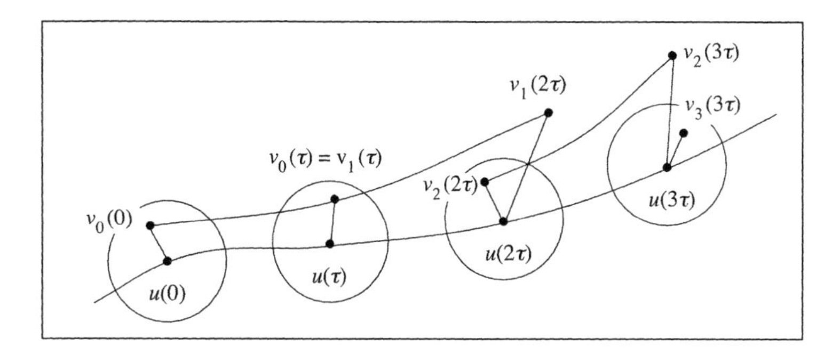

Figure 12.66 : For a discrete time series the renormalization procedure is modified.

Now we have two orbits to compare. The logarithmic error amplification factor for the first time interval is

$$l_0 = \frac{1}{\tau} \log \frac{||v_0(\tau) - v_0(0)||}{||u(\tau) - u(0)||}$$

where the norm $|| \cdot ||$ indicates the Euclidean length of a vector. Then we repeat the procedure for the next point $u(\tau)$ of the reference orbit. At that point we need to find another point $v_1(\tau)$ from the orbit which represents an error with a direction close to the one obtained from $v_0(\tau)$ relative to $u(\tau)$. In the event that our previous orbit is still close to the reference orbit, we may simply continue with it, thus setting $v_1(\tau) = v_0(\tau)$. This yields an error amplification factor l_1. Continuing we obtain factors $l_2, l_3, \ldots, l_{m-1}$ until the data is exhausted. Next we average the logarithmic error amplification factor

$$\lambda = \frac{1}{m} \sum_{j=0}^{m-1} l_j$$

over the whole reference orbit. This gives an approximation of the largest Ljapunov exponent.

There are a couple of important technical remarks to make about this method. First of all, since the data usually stems from physical measurements there is some amount of noise present in the numbers obtained. Therefore, we must not choose our perturbed points $v_k(k\tau)$ too close to the reference point $u(k\tau)$, because then the noise would dominate the stretching effect on the chaotic attractor. On the other hand we should not allow the error to become too large in order to avoid nonlinear effects. Thus, in practice we prescribe some minimal error ε and a maximal error δ and require

$$\varepsilon < |u(k\tau) - v_k(k\tau)| < \delta$$

as shown in figure 12.66. When choosing $v_k(k\tau)$, the direction of the error should match that of the previous point, $v_{k-1}(k\tau)$. This is accomplished by minimizing the angle θ in figure 12.67. However, it

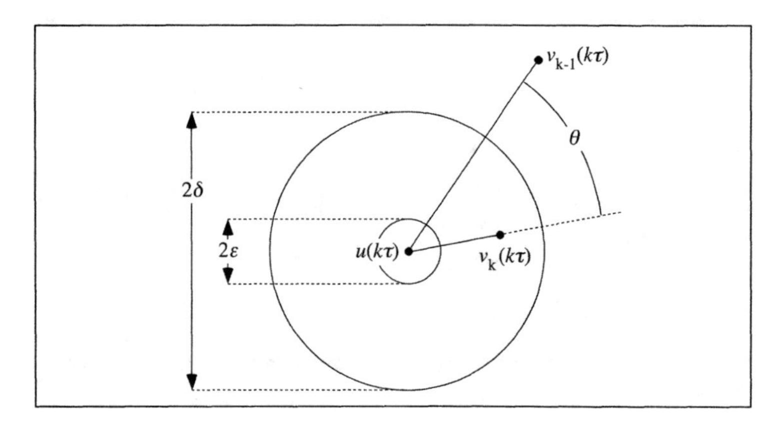

Figure 12.67 : Selecting a point from the reference orbit in a cone.

may be the case that there are no points at all in the reference orbit which lie in the annulus. Then one has to start anew with just any point close to $u(k\tau)$. If this occurs only seldom, then it should not have a dramatic negative effect on the result. Moreover, the result depends on the choice of the time step τ. When we replace τ by a multiple of itself, we get a different Ljapunov exponent approximation. All in all the method has many parameters that need to be tuned properly.[85]

Michael Faraday's Pioneering Experiment

We conclude this section with an impressive application of the time delay reconstruction method to a physical system termed *acoustic turbulence*. This example is very interesting from several perspectives. On the one hand, it provides a convincing reconstruction along with a successful calculation of dimensions and exponents.

On the other, there is a link to the first historical observation of the period-doubling route to chaos. This is an experiment by Michael Faraday[86] with a periodically driven nonlinear system and dates back to the year 1831.[87] Faraday constructed a large, 18-foot-long vibrating plate which held a shallow layer of water. He observed how 'heaps' of liquid were oscillating in a sloshing motion. In particular he reported that "each heap recurs or is re-formed in two complete vibrations of the sustaining surface." In other words, he observed the first step in a period-doubling cascade. Later, in 1883, Lord Rayleigh confirmed these results, and today the sophisticated methods of chaos theory and modern computer-controlled experiments are performed to study these transitions from order to chaos in similar systems.[88]

[85] The time delay τ may be chosen optimally in the reconstruction process. See A. M. Fraser and H. L. Swinney, *Independent coordinates for strange attractors from mutual information,* Phys. Rev. A 33 (1986) 1034–1040.

[86] Michael Faraday (1791–1867), British physicist, pioneered electromagnetism and invented the dynamo.

[87] M. Faraday, *On a peculiar class of acoustical figures, and on certain forms assumed by groups of particles upon vibrating elastic surfaces,* Phil. Trans. Roy. Soc. London 121 (1831) 299–340.

[88] For references see the historical notes in W. Lauterborn and J. Holzfuss, *Acoustic chaos,* International Journal of Bifurcation and Chaos 1,1 (1991) 13–26.

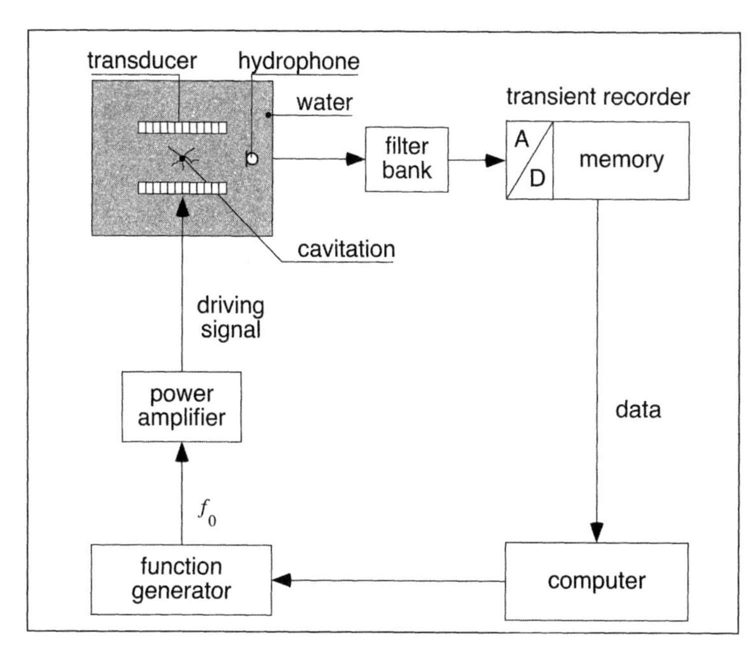

Setup for Acoustic Chaos

Experimental arrangement for acoustic measurement of driven liquids.

Figure 12.68

Acoustic Chaos in Ultrasonic Cavitation

For display in this section we have chosen one of these experiments reported by Werner Lauterborn and Joachim Holzfuss. Its purpose is to study acoustic chaos in ultrasonic cavitation, which appears as a hissing noise when a liquid is bombarded with sound of high intensity. This noise stems from the rupture of the liquid structure which results in an organized cloud of small bubbles undergoing complicated dynamics. The driving signal is of a constant frequency and the intensity increases from zero to a high value within a fraction of a second. A spectral analysis of the emitted noise shows strong subharmonic spectral lines at 1/2, 1/4, and also 1/3 of the driving frequency before the broadband hissing noise occurs at high-intensity values of the driving sound near the end of the experiment.[89] These spectral lines are a sign of a period-doubling cascade and the Feigenbaum scenario.

The experimental setup is shown in figure 12.68.[90] The water in the container must be specially prepared with certain additions. The high-intensity sound which drives the experiment is produced in a cylindrical transducer of about 3 inches in length and diameter. The voltage applied to the transducer is computer controlled. Finally, the sound produced is picked up by the hydrophone, appropriately 'cleaned' by the filter bank and stored in the computer memory for later processing. The duration of such an experiment typically is about the order of a quarter of a second.

For the phase space reconstruction of the pressure data $p(t)$, it is suffi-

[89] The first observations of this sort go back to R. Esche in 1952.

[90] Our exposition is based on the above mentioned paper by W. Lauterborn and J. Holzfuss. The figures 12.68, 12.69, and 12.70 are reproduced with their kind permission.

Reconstruction

Phase space reconstruction of periodic trajectories and a chaotic attractor.

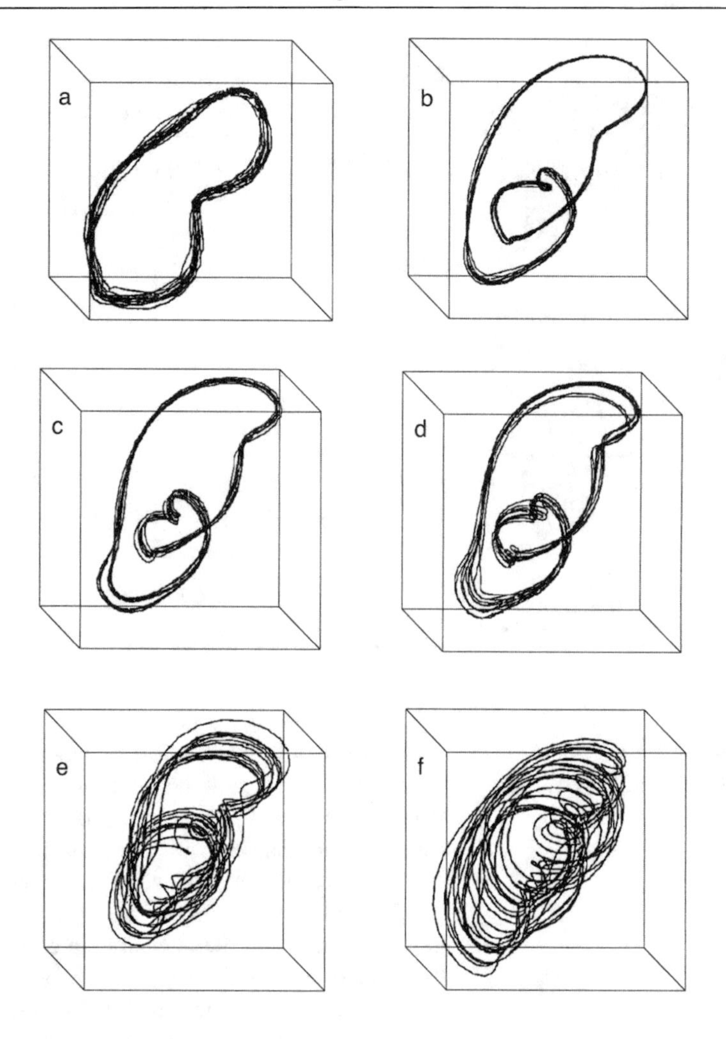

Figure 12.69

cient to consider only a three-dimensional setting, i.e., vectors $(p(t), p(t + T), p(t + 2T))$, where the time delay T is about one-tenth of the period of the driving signal. Figure 12.69 shows six such reconstructions for increasing driving intensities. Following the interpretation of Lauterborn and Holzfuss, the periodic trajectory in part (a) splits into two bands in (b) and four bands in (c). This structure breaks up when the voltage is further increased in parts (d) to (f), where a strange attractor appears. Thus, the hissing sound in the experiment clearly is not some stochastic noise but linked to pronounced effects in nonlinear dynamics. This has been further supported by the computation of dimensions (see figure 12.70) and Ljapunov exponents for the reconstructions. For instance, the first three exponents for the strange attractor from figure 12.69(f) are $\lambda_1 = 1.9$, $\lambda_2 = 0$, and $\lambda_3 = -3.0$. These values yield a Ljapunov dimension of $D_L \approx 2.7$ in agreement with figure 12.70.

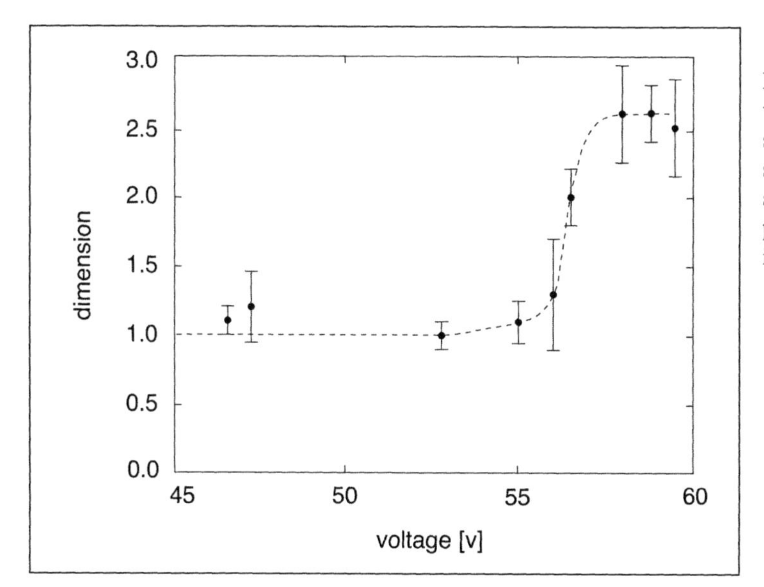

Dimension

Fractal dimension versus driving voltage. At high voltages the dimension is between 2 and 3. The dimension has been calculated as an average of the pointwise dimension (see page 686) evaluated at several points in the attractor.

Figure 12.70

Acoustic chaos is only one of the many examples which convincingly demonstrate that the methods and notions of chaos theory can be successfully applied to problems which only a couple of decades ago would have been discarded as intractable.[91] Chaos theory has opened our eyes and shown us where to look to find those rich structures which indicate once hidden laws of nature.

[91]To give one example, reconstructions of chaotic attractors in hydrodynamics have been reported by Tom Mullin, in: *Chaos in physical systems,* in: *Fractals and Chaos,* A. J. Crilly, R. A. Earnshaw, H. Jones (eds.), Springer-Verlag, New York, 1991. Several examples for chaos in mechanical systems is described in the introductory text F. C. Moon, *Chaotic Vibrations,* John Wiley & Sons, New York, 1987.

12.8 Fractal Basin Boundaries

Sensitive dependence on initial conditions is one of the central properties in chaos. In this last section of this chapter we present a different kind of sensitivity, namely, the so-called *final-state sensitivity*. This phenomenon may occur whenever there are several coexisting attractors. We have seen such a case already for the Hénon transformation in certain parameter ranges.[92] These may be strange attractors or perhaps simply attractive fixed points of a transformation. With such a transformation on hand, the orbit (or trajectory) for a given initial point will typically converge to either one of the attractors. Therefore, there must be a *boundary* of the corresponding basins of attractions. Such boundaries often are fractals. Physically, and also numerically, an initial point can only be specified up to some precision s. If all orbits started within the distance of s from the initial point converge to the *same* attractor, then there is no problem regarding the prediction of the final state. However, if some of these orbits converge to one attractor and the rest of them to the other attractor, we have a problem. No longer can we safely predict the final state belonging to the given initial point. Clearly, the severity of this problem becomes worse when the fractal dimension of the basin boundaries gets larger. Thus, we have that fractal basin boundaries with a large fractal dimension present an increased obstruction to the predictability in nonlinear systems with several attractors. In other words, in this situation the fractal dimension obtains an immediate dynamical interpretation.

How Large Is the Unsafe Region?

Given an uncertainty s, we call an initial point unsafe, if there is another initial point at a distance less than s, which converges to a different attractor. We may now ask how large the area (or the volume) $A(s)$ formed by unsafe initial points is. In the simple situation pictured in figure 12.71, this area is given by a strip of width $2s$ around the (smooth) basin boundary and is thus proportional to s. This means that when we improve our precision of initial points by a factor of 2, the region of unsafe initial points reduces by the factor 1/2.

However, if the basin boundary is fractal, then the scaling law should be more complicated. To see this, recall the box-counting dimension (see chapter 4). The size of the region of unsafe initial points is about the same as the area (or volume) $A(s)$ of the $N(s)$ boxes of width s needed to cover the basin boundary. If D_f denotes the fractal dimension and d the Euclidean dimension of the embedding space, we get that the number $N(s)$ is proportional to $1/s^{D_f}$ and

$$A(s) \approx N(s) \cdot s^d \propto s^{d-D_f} = s^{\alpha}.$$

The exponent

$$\alpha = d - D_f$$

is called the uncertainty exponent, and because $D_f \geq d-1$, we get that $0 < \alpha \leq 1$.

[92] See figure 12.15.

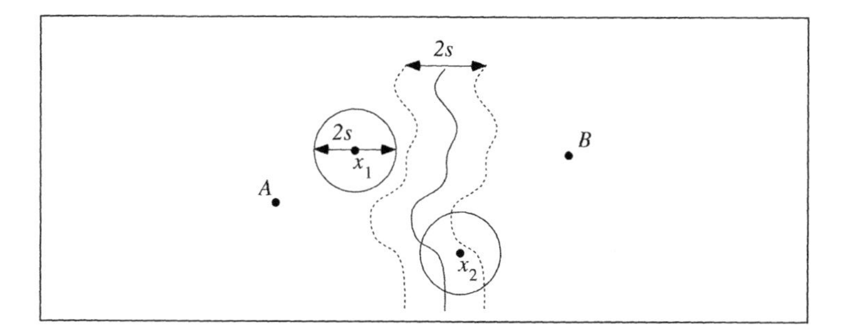

Figure 12.71 : Safe and unsafe initial points. Two attractors (A and B) with basins bounded by a smooth curve. The initial point x_1, which has an uncertainty of amount s, is safe, while the point x_2 is unsafe. The region of unsafe initial points is the strip of width $2s$ around the basin boundary.

Let us present another interpretation of this exponent. Assume that we intend to reduce the size of the region of unsafe initial points by a factor of 1/10 by improving the precision with which we measure and approximate the initial points. How many more significant digits do we need? If r is the necessary precision then $r^\alpha = s^\alpha/10$. Taking logarithms (with base 10) we get

$$\log_{10} r = \log_{10} s - \frac{1}{\alpha}.$$

Thus, we need $1/\alpha$ additional digits of precision to achieve our goal. If the uncertainty exponent α is small then this increase in precision can become very expensive.

The Pendulum Experiment

In the remainder of this section we present a study of such fractal basin boundaries arising in a physical experiment. Imagine a metal ball tied to a string which is attached to the ceiling. This pendulum may swing in all directions, and we adjust the length of the string so that the ball is close to the ground, where we place three strong magnets near the resting position of the pendulum (see figure 12.72). Holding the ball close to each of the three magnets, we can feel the attracting forces; when we let go of the ball close to one of the magnets, it will stay there. When we release the ball somewhere else, we can observe how the pendulum swings back and forth; and when the course of the ball comes close to one of the magnets, the direction of movement may drastically change due to the attracting force. In general, the pendulum's swings seem erratic and quite unpredictable. But eventually, the ball comes to rest due to the friction involved (air resistance and internal friction in the string, the attachment, and so on). You can play a little game to forecast which of the magnets the ball will finally be attracted to. For some initial positions it is easy to make a reliable prediction; for others it is nearly impossible. In fact, the experiment is one exhibiting the final-state sensitivity described above.

The Pendulum Setup

The metal ball of the pendulum swings over three magnets.

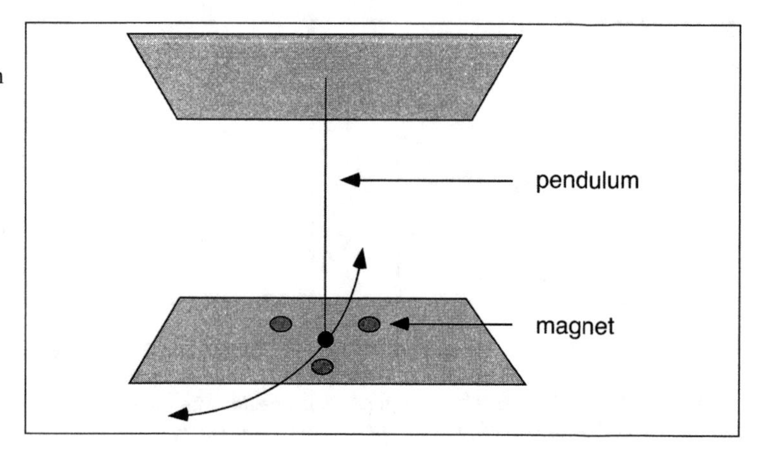

Figure 12.72

A Computer Model Is Necessary

In spite of this inherent difficulty, we may try to make many experiments with the pendulum in order to draw a map depicting the final rest point of the ball for all the chosen initial positions. Of course, it would be a very long and tedious physical experiment to do this for an array of, say, 100×100 initial conditions. Therefore, we prefer to employ a computer simulation for the experiment. This involves setting up a mathematical model for the pendulum motion and the implementation of routines for the solution of the underlying equations. Our central assumptions are the following:

- The pendulum length is long compared to the spacing of the magnets. Thus, we may assume for simplicity that the ball moves about on a plane rather than on a sphere with a large radius.
- The magnets are point attractors positioned a short distance below the pendulum plane at the vertices of an equilateral triangle.
- The force applied to the ball by a particular magnet is proportional to the inverse of its squared distance from the magnet.

System Parameters

The conclusion of the premises is given by a differential equation of second order in two variables. There are several parameters in the model. For example, the amount of friction in the pendulum is a parameter that can easily be changed in a computer program even while it is running, whereas it would be very hard, if not impossible, to change the friction in a physical experiment (we might have to consider building a pendulum suspended in a liquid). Another parameter is the strength of the force pulling the pendulum to its mid-point position. The figures 12.73 and 12.74 list the outcomes of some trajectories of solutions for different parameter settings.

Basins of Attraction

We now come to the central experiment with the pendulum, plotting the basins of attraction. Here we can produce three diagrams (one for each of the three magnets) collecting all the initial points which lead the pendulum to one of the three resting positions. These sets of points are called the basins of attraction. The three magnets are in competition over all points in the plane,

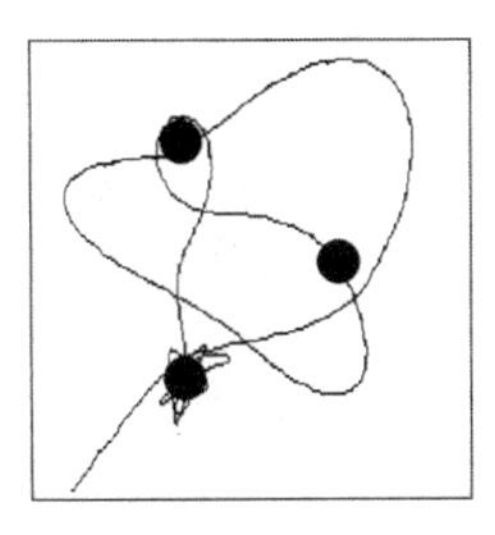
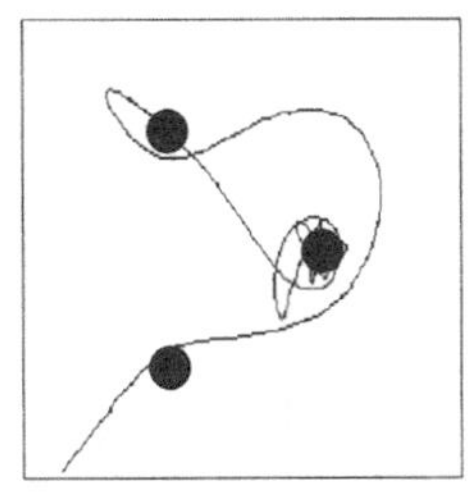
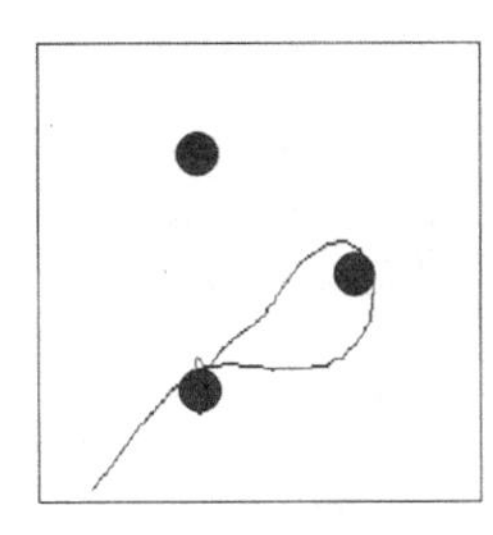

Trajectories and Friction

Top view of the paths of the pendulum. Same initial point in the lower left, but different friction parameters (low friction on the left and high friction on the right). Note that the final rest point of the pendulum is not the same in the three experiments.

Figure 12.73

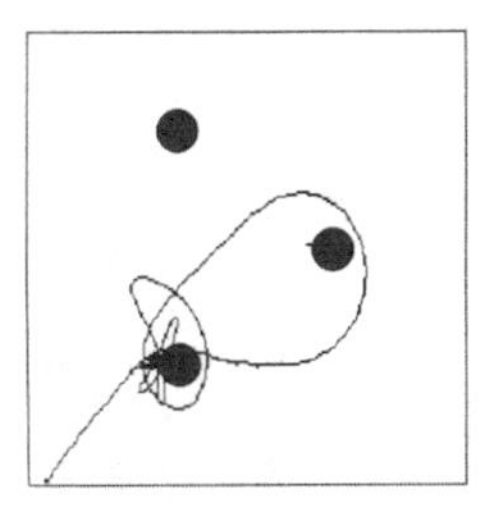
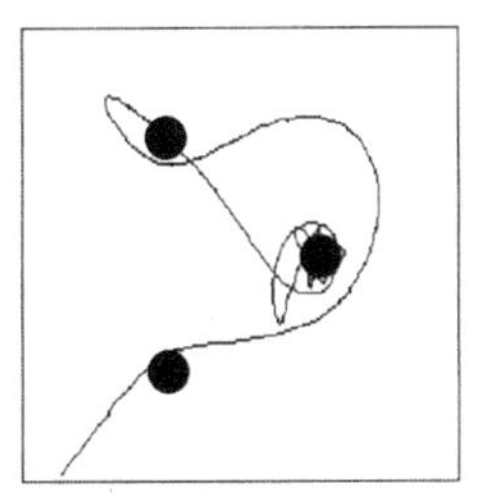
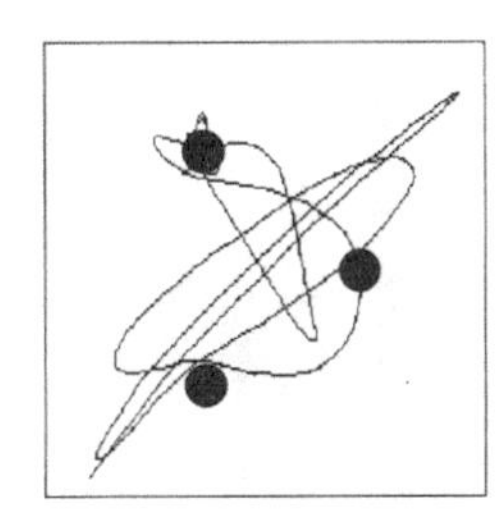

Trajectories and Gravitational Force

Top view of the paths of the pendulum. Same initial point and different force parameters with low force pulling the pendulum to its midpoint position (left), and strong force (right).

Figure 12.74

and it is not at all clear how the plane is divided up between them. Figure 12.75 shows the outcome of the experiment. Note, that the union of the three shaded regions essentially fills up the whole square.[93] The boundaries of the basins are shown in the fourth plot of the figure. Although it looks as if the boundaries are made up of just a few line segments, it is demonstrated in a close-up that they really have a very complicated structure (see figure 12.80) which has similarity to a Cantor set. In other words, wherever two basins seem to meet, we discover upon closer examination that the third basin is there in between them, and so ad infinitum.

We can also observe that the fractal structure of the basin boundaries becomes more apparent when we reduce the friction parameter in the computer simulation (see figure 12.77).

[93]Theoretically it is also possible that the pendulum comes to rest at another fourth position. For example, if we choose very weak magnets and place them at a certain distance from the natural resting position of the pendulum, then the ball may come to a full stop close to this resting position as if the magnets did not exist. In the experiments discussed here, the magnets are assumed to be so strong that this fourth resting position is not stable. Any arbitrarily small deviation will be magnified, and the ball will be driven to one of the other three resting positions above the magnets. Therefore, this unstable fourth resting position is not numerically observable. It is located exactly in the centers of the plots in figure 12.75.

Basins of Attraction

Basins of attraction for the pendulum over three magnets. For each of the three magnets, one of the above figures shows the basin shaded in black. The fourth picture displays the borders between the three basins. This border is not a simple line; but within itself it has a Cantor-like structure, as the enlargement in figures 12.79 and 12.80 show (see also the color plates 27 and 28).

Figure 12.75

The Equation of Motion for the Pendulum Over Three Magnets

The pendulum point mass moves about in the xy-plane, and the magnets are positioned below the xy-plane, say at a distance of d from the plane. Thus, assume the pendulum at position $(x, y, 0)$ and a magnet at $(x_1, y_1, -d)$. According to the second law of Coulomb, we assume that the force that the magnet applies to the pendulum is inversely proportional to the square of the distance between the two

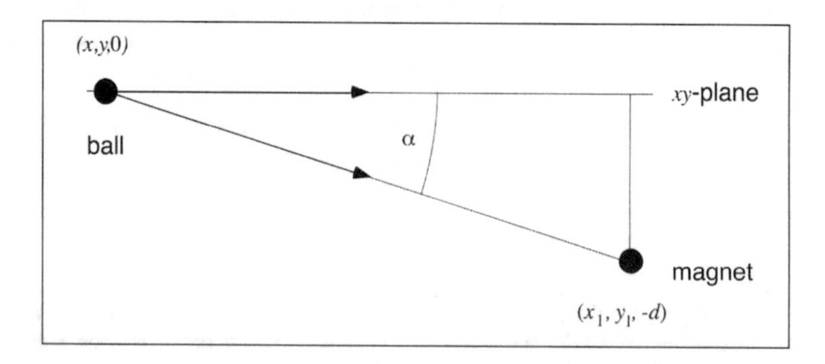

Figure 12.76 : The force of a magnet applied to the pendulum.

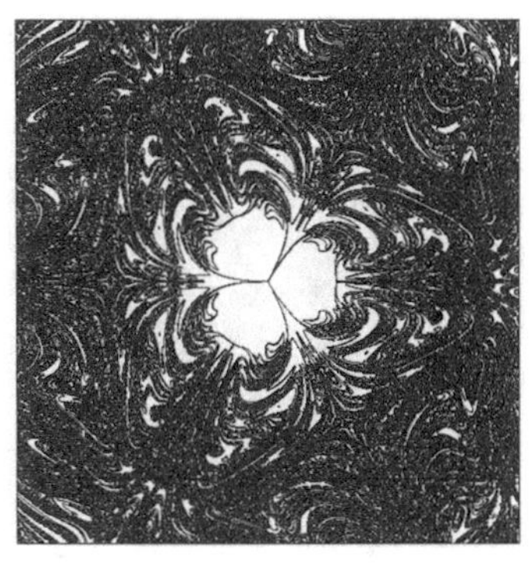

Basin Boundaries with Low Friction

A sequence depicting the fractal basin boundaries for increasing friction parameters (0.2, 0.3, 0.4, 0.5, from upper left to lower right). Increasing the friction results in smaller areas of uncertain initial points. Systems with high friction are more predictable than those with little friction

Figure 12.77

given points, i.e., proportional to

$$\frac{1}{(x_1 - x)^2 + (y_1 - y)^2 + d^2}.$$

However, the pendulum motion is restricted to the xy-plane, and therefore we should multiply the force by the cosine of the angle α, which is indicated in the figure 12.76.

After some transformations using the elementary expressions for the cosine, we arrive at a force in the xy-plane proportional to

$$\frac{1}{\left(\sqrt{(x_1 - x)^2 + (y_1 - y)^2 + d^2}\right)^3} \begin{pmatrix} x_1 - x \\ y_1 - y \end{pmatrix}.$$

This is a vector in the xy-plane. There are two other forces that must be considered, namely, the gravitational force that pulls the pendulum ball back to the center of the xy-plane, and the friction force. The gravitational force can be modeled simply as a force proportional to

$$-\begin{pmatrix} x \\ y \end{pmatrix}$$

Basin Boundaries with Low Spring Constant

The fractal basin boundaries change with the choice of the spring constant C. Here we have used the values $0.3, 0.2, 0.1, 0.0$ (from upper left to lower right.) The pendulum introduces complexity in the system. The more pronounced the self-restoring force of the pendulum the larger the region of unsafe initial points becomes.

Figure 12.78

whereas the friction force acts in opposition to the direction of movement and is proportional to the speed. Thus, the force vector is taken to be proportional to

$$-\begin{pmatrix} x' \\ y' \end{pmatrix}.$$

The above forces can be summed up in a differential equation using Newton's law, which relates the total force to the acceleration of the mass. With a setup of three magnets at positions (x_1, y_1), (x_2, y_2), (x_3, y_3), the force from the magnets becomes a sum over three terms. After moving all terms onto the left side of the equation we get

$$\begin{aligned} x'' + Rx' - \sum_{i=1}^{3} \frac{x_i - x}{\left(\sqrt{(x_i - x)^2 + (y_i - y)^2 + d^2}\right)^3} + Cx &= 0 \\ y'' + Ry' - \sum_{i=1}^{3} \frac{y_i - y}{\left(\sqrt{(x_i - x)^2 + (y_i - y)^2 + d^2}\right)^3} + Cy &= 0 \end{aligned}$$

which is a system of two ordinary differential equations of second order. The solution of the corresponding initial value problem requires specification of the position (x, y) and velocity (x', y'). In our experiment

Figure 12.79 : Same as the lower right picture in figure 12.75, however at the fourfold resolution of 2048 × 2048 pixels.

the position varies over a square region around the magnet positions and the initial velocity is always chosen to be zero.

The numerical procedure used to solve such an initial value problem can take many different forms. Any standard textbook on numerical analysis will have a number of methods to choose from, starting from the simple Euler scheme, which is not recommended here because of a serious lack of stability, up to more complex algo-

Blowup of Basin Boundaries

This enlargement of a portion of figure 12.75 reveals the Cantor-set like structure of the boundaries of the basins of attraction in the pendulum experiment.

Figure 12.80

rithms such as adaptive Runge-Kutta methods.[94]

[94] A good source of reference and algorithms, including codes in three programming languages, is given in *Numerical Recipes*, W. H. Press, B. P. Flannery, S. A. Teukolsky, W. T. Vetterling, Cambridge University Press, Cambridge, 1986.

Chapter 13

Julia Sets: Fractal Basin Boundaries

I must say that in 1980, whenever I told my friends that I was just starting with J. H. Hubbard a study of polynomials of degree 2 in one complex variable (and more specifically those of the form $z \to z^2 + c$), they would all stare at me and ask: Do you expect to find anything new? It is, however, this simple family of polynomials which is responsible for producing these objects which are so complicated — not chaotic, but on the contrary, rigorously organized according to sophisticated combinatorial laws.[1]

Adrien Douady

The goal of this chapter is to demonstrate how genuine mathematical research experiments open a door to a seemingly inexhaustible new reservoir of fantastic shapes and images. Their aesthetic appeal stems from structures which are beyond imagination and yet, at the same time, look strangely familiar. The ideas we present here are part of a world wide interest in so called *complex* dynamical systems. They deal with chaos and order, both in competition and coexistence. They show the transition from one condition to the other and how magnificently complex the transitional region generally is. One of the things many dynamical systems have in common is the competition of several centers for the domination of the plane. A single boundary between territories is seldom the result of this contest. Usually, an unending filigree entanglement and unceasing bargaining for even the smallest areas results. We studied the quadratic iterator in chapters 1, 10 and 11 and learned that it is the most prominent and important paradigm for chaos in deterministic dynamical systems. Now we will see that it is also a source of fantastic fractals. In fact the most exciting discovery in recent experimental mathematics, i.e., the Mandelbrot set, is an offspring of these studies. Now, about 10 years after Adrien Douady and John Hamal Hubbard started their research on the Mandelbrot

[1] Adrien Douady, *Julia sets and the Mandelbrot set*, in: The Beauty of Fractals, H.-O. Peitgen, P. H. Richter, Springer-Verlag, Heidelberg, 1986.

set, many beautiful truths have been gained about this 'most complex object mathematics has ever seen'. Almost all of this progress stems from their work.

This chapter begins with an informal discussion of basin boundaries (section 13.1) and a short introduction to complex numbers (section 13.2) and methods for complex quadratic equations (section 13.3). Thus, readers who are already familiar with these notions and tools may want to only briefly scan through these sections and then start reading section 13.4 where Julia sets finally come in.

13.1 Julia Sets as Basin Boundaries

Pixel Game Rules

In section 12.8 we have presented the pendulum over three magnets as a physical example of competition between centers of attraction. The corresponding dynamical system is given by means of certain physical laws, which manifest themselves in the form of a differential equation. We now consider a sort of game where the dynamical laws are much simpler. They are given as a table of rules. Imagine a large square board where the fields are assigned labels as in the game of chess. For example A1 denotes the lower left square, and C8 is the eighth square of the third row. In each square we write the coordinates of a follow-up square as shown in figure 13.1. The game simply consists in following the instructions: we place a peg on an initial square at pleasure, read the coordinates from that square and move the peg to the indicated destination. Then the procedure is repeated.

Pixel Game Scenarios

Having understood the simple rules of the game, we now may ask what the possible dynamical patterns in this game are. The answer is quite easy. Assuming that the player is never required to leave the domain given by the board, i.e., all destinations are coordinates on the board, the journey of the peg has the following alternatives:

- The path comes to a halt, i.e., at some point the peg arrives on a square whose destination is itself. Such a square may be called a fixed point.
- The above is not true. In that case the sequence of visited squares must be periodic from some point on. This is true because there are only finitely many squares on the board; and thus, after a finite number of moves, the peg will arrive on a square that was already visited before. From then on the sequence becomes periodic.

Let us carry out this procedure for all squares on the board. The result is shown in figure 13.2. In summary, the figure shows:

- There are three fixed squares, namely, B3, F10 and K3.
- All trajectories, i.e., all sequences starting anywhere on the board, terminate at one of the three above fixed squares. There are no cycles.
- For each fixed square (for example, B3), there is a *basin of attraction* which consists of all initial squares which eventually lead to the fixed square. The basins of attraction can have several components.

We may add one additional piece of information on the entries of the board, namely, the number of moves necessary to advance the peg from an initial square to the final fixed square. This is also shown in figure 13.2; see the numerical entries in the squares. It seems to be suggested here that from squares near the boundary of a basin it takes more moves.

Julia Sets and the Pixel Game

This simple game already shows the basic procedure used in the computation of Julia sets. The squares correspond to the finite, though possibly very large, number of points in the plane that can be represented in the computer. The rules that prescribe the transition from one square to the next are given in terms of a formula, and not as a table. In fact, as the reader probably has

Pixel Game Transitions

Each square contains the coordinates of a next square. Pick an initial square at pleasure and follow the sequence of squares. Does this journey ever end; and if so, where does it end?

L	K 2	K 3	K 3	K 4	K 4	I 5	I 6	I 7	I 8	I 8	I 9
K	K 3	K 3		K 3	I 3	H 4	H 5	G 7	H 8	H 9	H 9
I	I 3	I 3	K 4	K 3	I 2	G 3	F 5	F 7	F 8	G 9	G 10
H	H 3	I 4	K 5	L 3	K 1	D 2	B 5	D 7	F 9	F 9	G 10
G	G 4	H 5	K 7	K 6	L 2	A 2	C 7	C 10	E 10	F 10	F 10
F	F 4	F 6	F 9	F 10	F 11	F 11	F 11	F 11	F 11		F 10
E	E 4	D 5	B 7	B 6	A 2	L 2	I 7	I 10	G 10	F 10	F 10
D	D 3	C 4	B 5	A 3	B 1	H 2	K 5	H 7	F 9	F 9	E 10
C	C 3	C 3	B 4	B 3	C 2	E 3	F 5	F 7	F 8	E 9	E 10
B	B 3	B 3		B 3	C 3	D 4	D 5	E 7	D 8	D 9	D 9
A	B 2	B 3	B 3	B 4	B 4	C 5	C 6	C 7	C 8	C 8	C 9
	1	2	3	4	5	6	7	8	9	10	11

Figure 13.1

already guessed, the transition rules of our pixel game have been derived from a formula corresponding to a Julia set with three basins of attraction.

Complex Polynomial Transition Functions

The mathematical analysis of the dynamical properties analogous to our simple pixel game is carried out in dynamical systems theory. There are many different classes of dynamical systems. In section 12.8, for example, we looked at a mathematical model of a pendulum — in mathematical terms, a so called ordinary differential equation. Population models in biology have led to another example, namely, discrete dynamical systems based on the iteration of a real function (a function with one real variable and real values, e.g., $z \to z + rz(1-z)$, such as those discussed in chapter 1). When the transition function from one state of the system to the next is given as a complex polynomial or complex rational function, then the corresponding background is provided by the theory of Julia sets.

Gaston Julia and Pierre Fatou

The name Julia set stems from the French mathematician Gaston Julia (1893–1978), who developed much of the theory while he was recovering from his war wounds in an army hospital. During World War I he served as an officer. On 25 January 1915 the German headquarters decided to celebrate the Kaiser's birthday by organizing an attack on the French front with the intention of taking 1000 prisoners. The attack took place and was extremely violent, but the goal was not achieved. Many on both sides were killed or

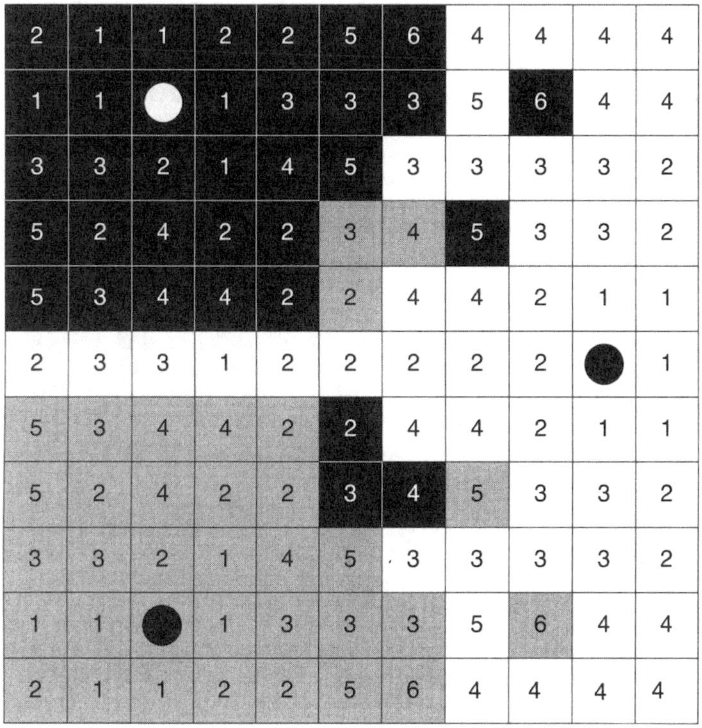

2	1	1	2	2	5	6	4	4	4	4
1	1		1	3	3	3	5	6	4	4
3	3	2	1	4	5	3	3	3	3	2
5	2	4	2	2	3	4	5	3	3	2
5	3	4	4	2	2	4	4	2	1	1
2	3	3	1	2	2	2	2	2		1
5	3	4	4	2	2	4	4	2	1	1
5	2	4	2	2	3	4	5	3	3	2
3	3	2	1	4	5	3	3	3	3	2
1	1		1	3	3	3	5	6	4	4
2	1	1	2	2	5	6	4	4	4	4

Results of the Pixel Game

This figure summarizes the results for the Pixel Game of the previous figure. There are three fixed squares (marked by a large dot) with three corresponding basins of attraction (black, white and grey) around them. The numbers in the squares indicate the number of moves necessary to advance to the final fixed square.

Figure 13.2

wounded. Julia was one of those who were badly wounded. He lost his nose during the aforementioned attack, and thereafter wore a black leather strap across his face. Simultaneously Julia's competitor Pierre Fatou (1878–1929) created another huge volume of results in the area of complex iteration.[2] These early results were almost forgotten for many years and became popular again only in the 1980s through Mandelbrot's work. The immense progress that Julia and Fatou were able to make must be valued all the more because in those days there were no computers to aid in the understanding of the complicated matter; instead they had to rely completely on their imagination. Much of Julia's work was motivated by a one page article by the famous British mathematician Sir Arthur Cayley, who published a problem in 1879, the 'Newton-Fourier Imaginary Problem' (see figure 13.3).[3] When computing the roots of $z^3 - 1 = 0$ using Newton's method one is led to a competition situation which is very similar to our pendulum experiment.

Cayley's Problem

Cayley's problem is related to the pendulum over three magnets from the last chapter in a very nice way. We considered the question of which

[2]In addition to his work in astronomy at the Paris observatory Pierre J. L. Fatou was very productive in mathematics, delivering numerous results, in particular in complex analysis.

[3]Arthur Cayley, *The Newton-Fourier imaginary problem,* American Journal of Mathematics 2, 1879.

Cayley's Paper

The original paper entitled *The Newton-Fourier Imaginary Problem* of Sir Arthur Cayley, 1879.

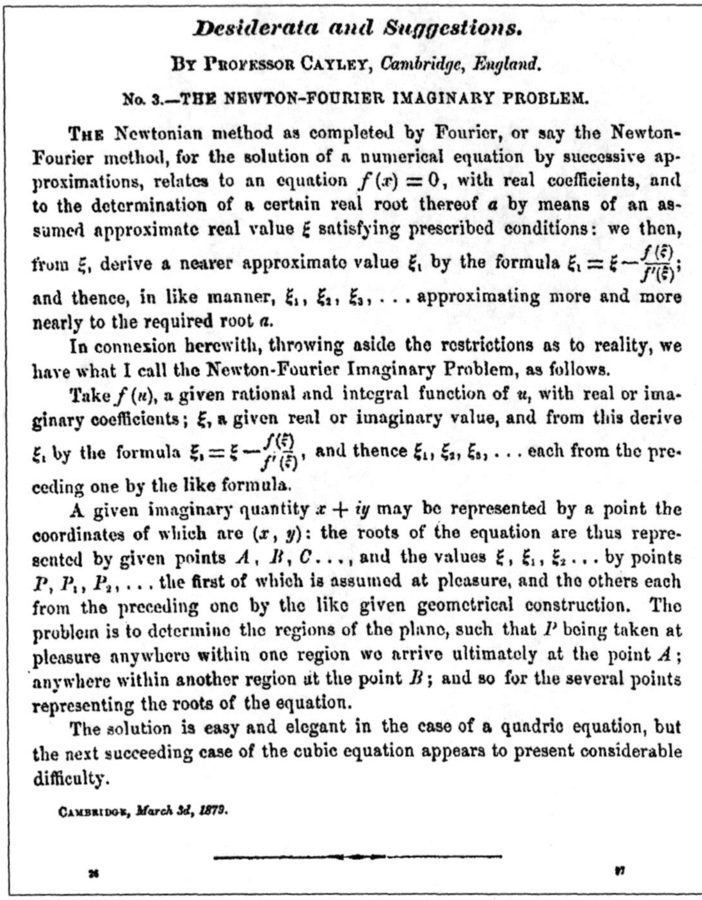

Desiderata and Suggestions.

By Professor Cayley, *Cambridge, England.*

No. 3.—THE NEWTON-FOURIER IMAGINARY PROBLEM.

The Newtonian method as completed by Fourier, or say the Newton-Fourier method, for the solution of a numerical equation by successive approximations, relates to an equation $f(x) = 0$, with real coefficients, and to the determination of a certain real root thereof a by means of an assumed approximate real value ξ satisfying prescribed conditions: we then, from ξ, derive a nearer approximate value ξ_1 by the formula $\xi_1 = \xi - \frac{f(\xi)}{f'(\xi)}$; and thence, in like manner, $\xi_1, \xi_2, \xi_3, \ldots$ approximating more and more nearly to the required root a.

In connexion herewith, throwing aside the restrictions as to reality, we have what I call the Newton-Fourier Imaginary Problem, as follows.

Take $f(u)$, a given rational and integral function of u, with real or imaginary coefficients; ξ, a given real or imaginary value, and from this derive ξ_1 by the formula $\xi_1 = \xi - \frac{f(\xi)}{f'(\xi)}$, and thence $\xi_1, \xi_2, \xi_3, \ldots$ each from the preceding one by the like formula.

A given imaginary quantity $x + iy$ may be represented by a point the coordinates of which are (x, y): the roots of the equation are thus represented by given points $A, B, C \ldots$, and the values $\xi, \xi_1, \xi_2 \ldots$ by points $P, P_1, P_2, \ldots$ the first of which is assumed at pleasure, and the others each from the preceding one by the like given geometrical construction. The problem is to determine the regions of the plane, such that P being taken at pleasure anywhere within one region we arrive ultimately at the point A; anywhere within another region at the point B; and so for the several points representing the roots of the equation.

The solution is easy and elegant in the case of a quadric equation, but the next succeeding case of the cubic equation appears to present considerable difficulty.

Cambridge, *March 3d, 1879.*

26 97

Figure 13.3

magnet the pendulum would come to rest over. Cayley considered Newton's method and asked, which solution of the equation $z^3 - 1 = 0$ in the complex plane[4] would the method converge towards if one starts with an arbitrary initial guess. The equation $z^3 - 1 = 0$ has three solutions, $\{1, e^{2\pi i/3}, e^{4\pi i/3}\}$. Newton's method for this equation is given by the feedback iteration

$$z_{n+1} = z_n - \frac{z_n^3 - 1}{3z_n^2}. \tag{13.1}$$

Its basic properties are designed so that the three solutions of the cubic equation behave like magnets for the iteration. Julia understood that the three corresponding basins of attraction have a common boundary. Any boundary point is like a three-corner point between three countries, i.e., arbitrarily close to any boundary point we can find points in each

[4]Complex numbers will be introduced in the next section.

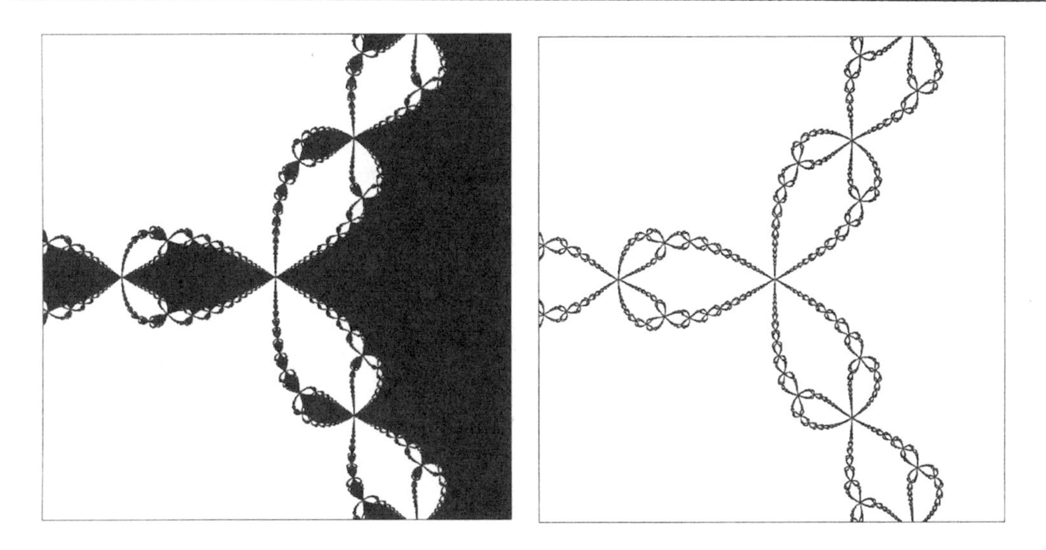

Figure 13.4 : The figure shows the basins of attraction for Newton's method applied to the equation $z^3 - 1 = 0$. Left, the basin for the solution $z = 1$ is shown in black, while at right only the basin boundaries are pictured. It is an example of a Julia set.

of the three basins of attraction.[5]

Actually, our pixel game of figures 13.1 and 13.2 is based on a variant of Newton's method. The pixels correspond to squares in the complex plane, their centers being taken as points z_n, and the coordinates of the target square being derived from the complex number z_{n+1} according to equation 13.1.

Before being able to understand and compute Julia sets you must feel comfortable with complex numbers. On the following few pages we provide the reader with a short introduction that contains the most important facts. Readers already basically familiar with complex numbers may skip to the following section dealing with complex square roots.

[5]A recent collection of papers discussing Newton's method as dynamical systems is in *Newton's Method and Dynamical Systems*, H.-O. Peitgen (ed.), Kluver Academic Publishers, Dordrecht, 1989. See also H.-O. Peitgen P. H. Richter, *The Beauty of Fractals*, Springer-Verlag, Heidelberg, 1986, chapters 6 and 7.

13.2 Complex Numbers — A Short Introduction

Complex numbers are an extension of real numbers. The historical motivation for their invention stems from the desire to be able to solve algebraic equations that normally, i.e., by using traditional real numbers, have no solution. For example, $x^2 + 1 = 0$ has no real solution and, thus, a symbolic solution was created and called the *imaginary unit* i with the postulated property

$$i^2 = -1.$$

A *complex number* z has two components which are called the *real* and the *imaginary* part. We write

$$z = x + yi,$$

where RE $(z) = x$ denotes the real, IM $(z) = y$ the imaginary part, and i the imaginary unit.

The arithmetic of complex numbers is a straightforward extension of the arithmetic of real numbers. We add two numbers $z = x + yi$ and $w = u + vi$ by

$$z + w = (x + yi) + (u + vi) = (x + u) + (y + v)i,$$

i.e., by adding the real and imaginary parts separately. We multiply the two numbers by

$$\begin{aligned} z \cdot w &= (x + yi)\,(u + vi) = xu + xvi + yui + yvi^2 \\ &= (xu - yv) + (xv + yu)i. \end{aligned}$$

Note that the term yvi^2 contributes to the real part of the product, since $i^2 = -1$.

Embedding of the Real Numbers

Let us look at a special case, namely, choosing z and w with imaginary component being zero, i.e.,

$$z = x + 0i = x \text{ and } w = u + 0i = u.$$

For these numbers the addition and multiplication rules say that

$$\begin{aligned} z + w &= (x + u) + (0 + 0)i = x + u, \\ z \cdot w &= (xu - 0 \cdot 0) + (x \cdot 0 + u \cdot 0)i = xu. \end{aligned}$$

This indicates that the real numbers can be regarded also as complex numbers, namely, as those having imaginary components being zero. All the arithmetic laws for real numbers also apply for the real numbers interpreted as complex numbers. However, there is an important distinction between real and complex numbers. There are positive real numbers, and this idea induces an ordering of all real numbers. In other words, given two distinct real numbers, x and y, we have either $x < y$ or $y < x$. Such an ordering does not exist in the complex numbers.

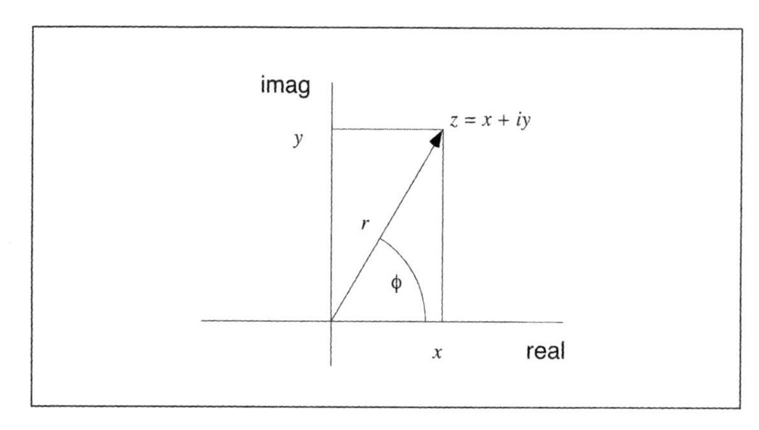

Complex Number

The complex number z corresponds to a point in the plane with coordinates x and y. The length of the vector from the origin to the point is $r = |z| = \sqrt{x^2 + y^2}$ and is called the absolute value or modulus of z. The counterclockwise angle ϕ that the vector makes with the real positive axis is called the argument $\arg z$.

Figure 13.5

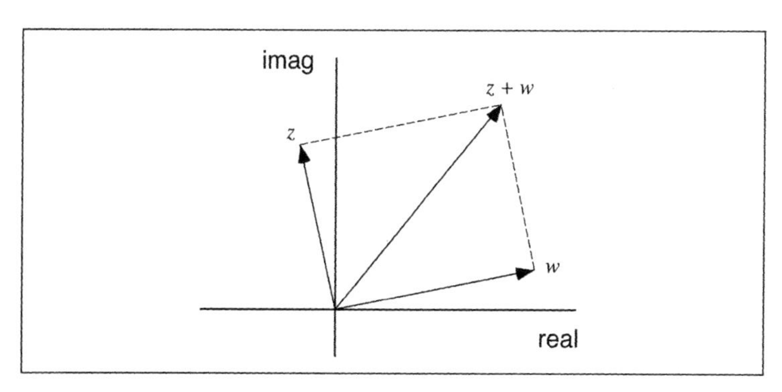

Complex Addition

The addition of two complex numbers is identical to the addition of two point vectors from the origin to the numbers.

Figure 13.6

Complex Numbers as Points in the Plane

While real numbers can be geometrically interpreted as points on a line, it is common to identify complex numbers as points in a plane. The coordinates of a complex point are its real and imaginary parts (see figure 13.5). Thus, we may write a complex number $z = x + yi$ as a vector

$$z = \begin{pmatrix} x \\ y \end{pmatrix} = \begin{pmatrix} \text{real part} \\ \text{imaginary part} \end{pmatrix}$$

and x and y are also called the *Cartesian coordinates* of z.

Polar Coordinates

The length of the point vector corresponding to a complex number z is called its *absolute value* or *modulus* $r = |z|$. For $z = x + yi$ it is given by

$$r = |z| = \sqrt{x^2 + y^2}.$$

The counterclockwise angle ϕ that the vector makes with the real positive axis is called the argument

$$\phi = \arg z.$$

Thus, there are two ways to specify a complex number z: either by the Cartesian coordinates $x = \mathrm{RE}\,(z)$ and $y = \mathrm{IM}\,(z)$, or using its so called *polar coordinates*, i.e., its modulus $r = |z|$ and argument $\phi = \arg z$. Using the sine and cosine trigonometric functions we may write

$$z = r(\cos\phi + i\sin\phi)r.$$

Thus, the conversion from polar to Cartesian coordinates is given by

$$\begin{aligned} x &= r\cos\phi \\ y &= r\sin\phi. \end{aligned}$$

The conversion of Cartesian to polar coordinates, however, is more complicated.

Converting to Polar Coordinates

A point in the xy-plane can be specified as usual by its x- and y-coordinates. An alternative way is given by the polar coordinates which characterize a point (x, y) (respectively a complex number $z = x+yi$) by the absolute value $r = |z|$ and the angle ϕ between the vector from the origin to the point (x, y) and the x-axis. This angle is also called the argument of z. Thus,

$$z = r(\cos\phi + i\sin\phi).$$

Care must be taken since the argument is not uniquely defined: if the point z is zero, all angles can be used, if z is not zero, then the argument is defined only up to integer multiples of 2π. Thus, ϕ and $\phi + 2\pi$ describe the same. As a formula, the polar coordinate transformation point is given by

$$\begin{aligned} &r = \sqrt{x^2+y^2} \\ &\sin\phi = \frac{y}{r} \text{ and } \cos\phi = \frac{x}{r}, \end{aligned}$$

The second formula is not explicit, it defines the argument ϕ in terms of $\sin\phi$ and $\cos\phi$. In practice we would use the arctan function applied to y/x to obtain an angle between $-\pi/2$ and $\pi/2$ (assuming $x \neq 0$). Then it is left to decide whether the argument is this angle or the angle plus π. Explicitly, the algorithm is:

```
   if (x > 0) then
      phi = arctan(y/x)
   else if (x < 0) then
      phi = pi + arctan(y/x)
   else if (y > 0) then
      phi = pi/2
   else if (y < 0) then
      phi = -pi/2
   else
      print "Error: x = y = 0, argument unde-
fined."
   end if
```

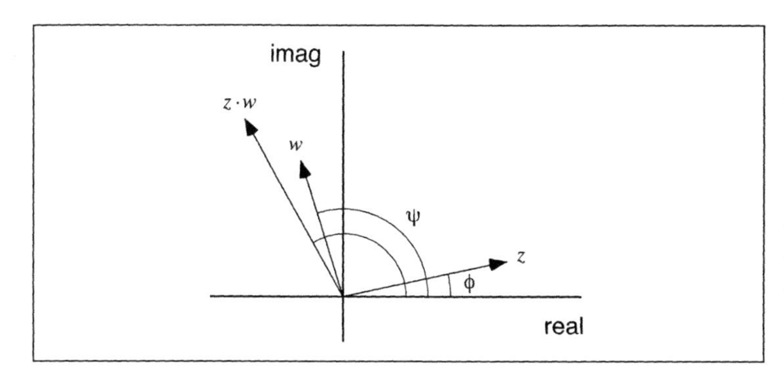

Complex Multiplication

Multiplication of two complex numbers. The product is given by a point whose argument is the sum of the arguments of the factors and whose absolute value, i.e., its distance to the origin of the plane, is given by the product of the absolute values of the factors.

Figure 13.7

Geometry of Multiplication

Using polar coordinates we can present an elegant geometrical interpretation of multiplication of complex numbers. Consider two complex numbers

$$\begin{aligned} z &= r(\cos\phi + i\sin\phi), \\ w &= s(\cos\psi + i\sin\psi). \end{aligned}$$

and their product zw. Then the polar coordinates of the product are easy to obtain. The modulus of the product is equal to the product of the moduli r and s, and its argument is simply the sum of the arguments of the factors, $\phi + \psi$. Thus,

$$z \cdot w = rs(\cos(\phi + \psi) + i\sin(\phi + \psi)). \tag{13.2}$$

In other words, multiplying two complex numbers means adding the corresponding angles and multiplying the lengths of the associated vectors (see figure 13.7).

Derivation

It remains to be shown that the geometric interpretation agrees with the initial definition of the product of two numbers. We let

$$\begin{aligned} z &= x + yi = r(\cos\phi + i\sin\phi) \\ w &= u + vi = s(\cos\psi + i\sin\psi) \end{aligned}$$

and make use of the double angle identities

$$\begin{aligned} \cos(\phi + \psi) &= \cos\phi\cos\psi - \sin\phi\sin\psi \\ \sin(\phi + \psi) &= \sin\phi\cos\psi + \sin\psi\cos\phi. \end{aligned}$$

We show eqn. (13.2) by computing

$$\begin{aligned} zw &= (xu - yv) + (xv + uy)i \\ &= r\cos\phi \cdot s\cos\psi - r\sin\phi \cdot s\sin\psi \\ &\quad + (r\cos\phi \cdot s\sin\psi + s\cos\psi \cdot r\sin\phi)i \\ &= rs(\cos\phi\cos\psi - \sin\phi\sin\psi) \\ &\quad + rs(\sin\phi\cos\psi + \sin\psi\cos\phi)i \\ &= rs(\cos(\phi + \psi) + \sin(\phi + \psi))i. \end{aligned}$$

Complex Conjugate

For a complex number $z = x + yi$ the complex conjugate is $\bar{z} = x - yi$, and their product is $z \cdot \bar{z} = x^2 + y^2 = |z|^2$, the squared modulus of z.

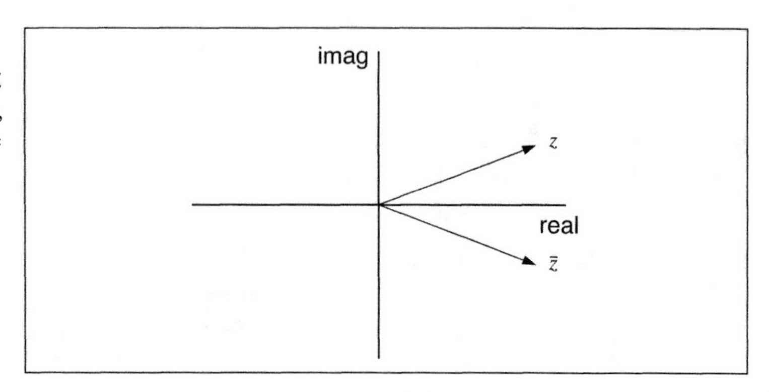

Figure 13.8

Euler Notation

Closely related to the polar representation is the Euler notation of complex numbers. In this notation a complex number with modulus 1 and argument ϕ can be written as

$$e^{i\phi} = \cos\phi + i\sin\phi.$$

The polar representation of a number with modulus r and argument ϕ is conveniently expressed as $re^{i\phi}$. With this definition the laws for the exponential function with real exponents carry over to complex exponents. For example,

$$\begin{aligned} e^{a+bi} &= e^a(\cos b + i\sin b) \\ re^{i\phi} \cdot se^{i\psi} &= rs\, e^{(\phi+\psi)i}. \end{aligned}$$

The remarkable fact about complex numbers is that we can compute with them in almost the same fashion as with ordinary real numbers. For example, the common laws such as the commutative and the distributative laws apply.

Subtraction and Division

Let us finally state the arithmetic rules for subtraction and division of complex numbers. Subtraction is straightforward just like addition

$$z - w = (x - u) + (y - v)i$$

while division is somewhat more complicated. Division, of course, is the inverse of multiplication. Thus, geometrically, dividing by a complex number with modulus $r > 0$ and argument ϕ should correspond to a scaling by the inverse factor $1/r$ and a *clockwise* rotation by the angle ϕ. Thus, in polar coordinates,

$$\begin{aligned} \frac{z}{w} &= \frac{r(\cos\phi + i\sin\phi)}{s(\cos\psi + i\sin\psi)} \\ &= \frac{r}{s}\left(\cos(\phi - \psi) + i\sin(\phi - \psi)\right). \end{aligned}$$

To derive the formula for division in terms of Cartesian coordinates, it is helpful to introduce the so-called *complex conjugate* number $\bar{z}$. If $z = x + yi$, then $\bar{z} = x - yi$. Thus, in the complex plane $\bar{z}$ is the point z mirrored at the real axis (see figure 13.8). To be specific, we have that $z = \bar{z}$ if and only if z is a real number. Also,

$$z \cdot \bar{z} = x^2 + y^2 = |z|^2.$$

Now, we compute for $w \neq 0$

$$\frac{z}{w} = \frac{z\bar{w}}{w\bar{w}} = \frac{z\bar{w}}{|w|^2} = \frac{xu + yv}{u^2 + v^2} + \frac{-xv + uy}{u^2 + v^2}i.$$

Thus, a division is computationally more expensive than a multiplication.

On the Cost of a Division

In the complex division scheme presented above, a total of 8 multiplication and division steps are used. It is possible to reduce this number to 6 as follows. First compute

$$q = \frac{v}{u}$$

and then multiply nominator and denominator by $1 - qi$, i.e., we have

$$\frac{z}{w} = \frac{x + yq}{u + vq} + \frac{y - xq}{u + vq}i.$$

This procedure is numerically stable for $|u| \geq |v|$. In the other case, for $|u| < |v|$, multiplication of nominator and denominator by $u/v - i$ will do the job. It has been proven that we cannot further reduce the number of multiplications and divisions in a complex division operation. Whether this improvement is of practical value or not depends very much on the computer hardware used and on the precision desired in the calculation. Note that we have not counted additions, which on some computers are about as costly as a multiplication.

The Point at Infinity

Mathematicians like to think about the complex plane as a sphere, called the *Riemannian sphere*. This is justified if we carry out an identification of a punctured sphere with a plane, much like when we create a map of our planet (see figure 13.9). In this identification every point in the plane corresponds to a unique point on the sphere, and every point on the sphere, except the north pole, corresponds to a point in the plane. Observe that points in the plane which are further and further away from the origin correspond to points on the sphere located closer and closer to the north pole. In this way mathematicians regard the north pole on the sphere as the *point at infinity*.

The Mathematics of Stereographic Projection

There is a transformation that defines the complex number $u + vi$ corresponding to a point (x, y, z) on the sphere as shown. To be specific, we choose the radius of the sphere equal to 1 and fix the south pole of the sphere at the origin of the complex plane. The z-axis points upwards through the poles of the sphere. Then the transformation is

$$P(x, y, z) = \frac{2}{2 - z}(x + yi).$$

Stereographic Projection

Stereographic projection is a method to create a planar map of a sphere. Here we use a sphere of radius 1 and the complex plane which touches the south pole of the sphere at the origin. For each point (x, y, z) on the sphere there is a corresponding projected point $u + vi$ in the plane. To construct this projection consider the ray from the north pole through (x, y, z). The intersection of the ray with the complex plane defines the stereographic projection $u + vi$ of (x, y, z).

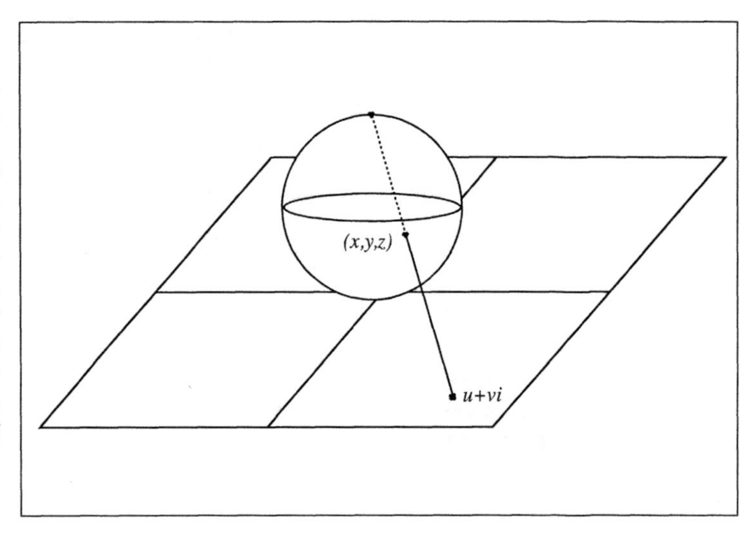

Figure 13.9

Hence,

$$u = \frac{2}{2-z}x \text{ and } v = \frac{2}{2-z}y.$$

The inverse of this transformation is required for the computation of the point (x, y, z) on the sphere when $u + vi$ is given. The formula is

$$P^{-1}(u+iv) = \frac{1}{u^2+v^2+4}(4v, 4u, 2u^2+2v^2).$$

Let us take as an example the complex number $u + vi = \sqrt{2}(1-i)$. The last formula for the inverse specifies the corresponding point on the sphere

$$\begin{aligned} x &= \sqrt{2}/2 \\ y &= -\sqrt{2}/2 \\ z &= 1. \end{aligned}$$

It is located on the equator. The transformation P applied to this point produces the original complex number $\sqrt{2}(1-i)$, of course.

The z-component of the inverse transformation P^{-1} is

$$z = \frac{2(u^2+v^2)}{u^2+v^2+4}.$$

When $u + vi$ tends to infinity, i.e., $u^2 + v^2$ grows without bounds, then the z-component tends to 2,

$$\lim_{u^2+v^2\to\infty} \frac{2(u^2+v^2)}{u^2+v^2+4} = 2.$$

Thus, the point on the sphere slides up to the north pole, which therefore is called the point at infinity.

13.3 Complex Square Roots and Quadratic Equations

Complex Square Roots With the above explanation of complex multiplication, it is also geometrically clear how to compute a square root of a complex number $z = r(\cos\phi + i\sin\phi)$. We must divide the argument by 2 and take the square root of the absolute value:

$$\sqrt{z} = \sqrt{r}\left(\cos\frac{\phi}{2} + i\sin\frac{\phi}{2}\right).$$

Except for $z = 0$ there are always two square roots, $\sqrt{z}$ as in the formula, and $-\sqrt{z}$,

$$\begin{aligned} -\sqrt{z} &= -\sqrt{r}\left(\cos\frac{\phi}{2} + i\sin\frac{\phi}{2}\right) \\ &= \sqrt{r}\left(\cos\left(\frac{\phi}{2} + \pi\right) + i\sin\left(\frac{\phi}{2} + \pi\right)\right). \end{aligned}$$

However, in most applications relevant to Julia sets we will work with the Cartesian coordinates (x, y) rather than the polar coordinates (r, ϕ). It is possible to first make a change of coordinates to obtain the polar coordinates, then carry out the square root as described and finally make another change of coordinates to get back the usual Cartesian coordinates. This is rather cumbersome. Therefore, let us derive a more efficient algorithm for complex square root computation. Let

$$w = u + vi$$

be the complex number for which the square root

$$z = \sqrt{w} = x + yi$$

is desired. We have that

$$z^2 = (x^2 - y^2) + (2xy)i$$

and to satisfy $z^2 = w$ the two equations

$$\begin{aligned} u &= x^2 - y^2 \\ v &= 2xy \end{aligned}$$

must hold. We now solve this system of equations for x and y. The second equation solved for y yields

$$y = \frac{v}{2x}$$

and, substituting this into the first one we obtain

$$u = x^2 - \frac{v^2}{4x^2}.$$

Sorting terms in this equation then gives rise to a biquadratic equation for x,

$$x^4 - ux^2 - \frac{v^2}{4} = 0.$$

We now solve this for x^2 using the standard formula for quadratic equations. There is a special case, namely, when $v = 0$, which we postpone until later, so let us assume for now that v is not zero. Then we have

$$x^2 = \frac{u + \sqrt{u^2 + v^2}}{2}.$$

Notice that the right hand side is strictly positive since the square root of $u^2 + v^2$ is always greater than $|u|$. Here we use only the plus sign before the square root because a minus sign would give us a negative right side, which does not lead to a real solution x. Taking the square root again and inserting the result into the equation for y yields the final result

$$\begin{aligned} x &= \pm\frac{\sqrt{2}}{2}\sqrt{u + \sqrt{u^2 + v^2}}, \\ y &= \frac{v}{2x}. \end{aligned}$$

Note that this denotes two symmetric solutions, a pair (x, y) for each sign in the first expression.

The special case $v = 0$ is very simple, of course. It corresponds to the root of a real (not necessarily positive) number u. In this case, when $u \geq 0$ we have

$$\begin{aligned} x &= \sqrt{u}, \\ y &= 0, \end{aligned}$$

while

$$\begin{aligned} x &= 0, \\ y &= \sqrt{-u}, \end{aligned}$$

when $u < 0$.

Errors in the Computation

The derived formulas, although being mathematically correct, may exhibit a severe limitation once implemented on a computer. Suppose that v is very small in magnitude as compared to a negative value of u, say, for example $u = -10^6$, and $v = 10^{-6}$. Then $u^2 + v^2 = 10^{12} + 10^{-12}$ and although u^2 and v^2 are computed exactly on the machine, their sum cannot be represented correctly as a 32 bit single-precision or even as a 64 bit double-precision floating point number. This is so because $10^{12} + 10^{-12}$ would require 24 significant decimal digits in the machine representation of numbers. Thus, the machine rounds off to obtain the result $u^2 + v^2 = 10^{12}$. As a conclusion

$$x = \frac{\sqrt{2}}{2}\sqrt{-10^6 + \sqrt{10^{12} + 10^{-12}}}$$

evaluates to zero. This is the wrong answer, since

$$\sqrt{10^{12} + 10^{-12}} \approx 10^6 + 0.5 \cdot 10^{-18}$$

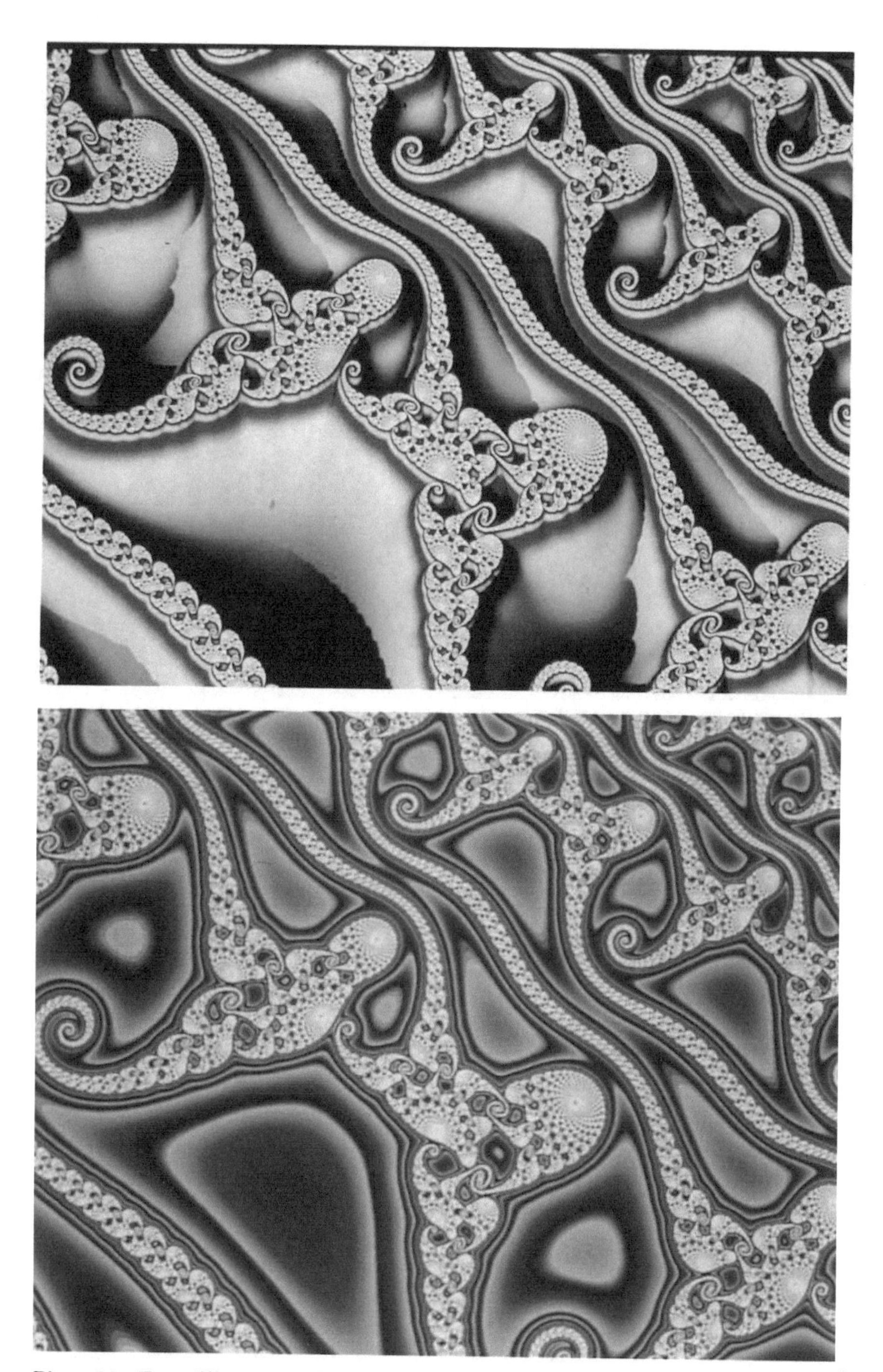

Plate 29: Two different renderings of a detail of the Mandelbrot set. The 3D-rendering shows the height corresponding to the distance to the set (top). The 2D-rendering uses colors to represent the distance (bottom).

Plate 30: 3D-rendering of the distance to the Mandelbrot set (height corresponds to distance). The insert shows an enlargement near the boundary of the Mandelbrot set.

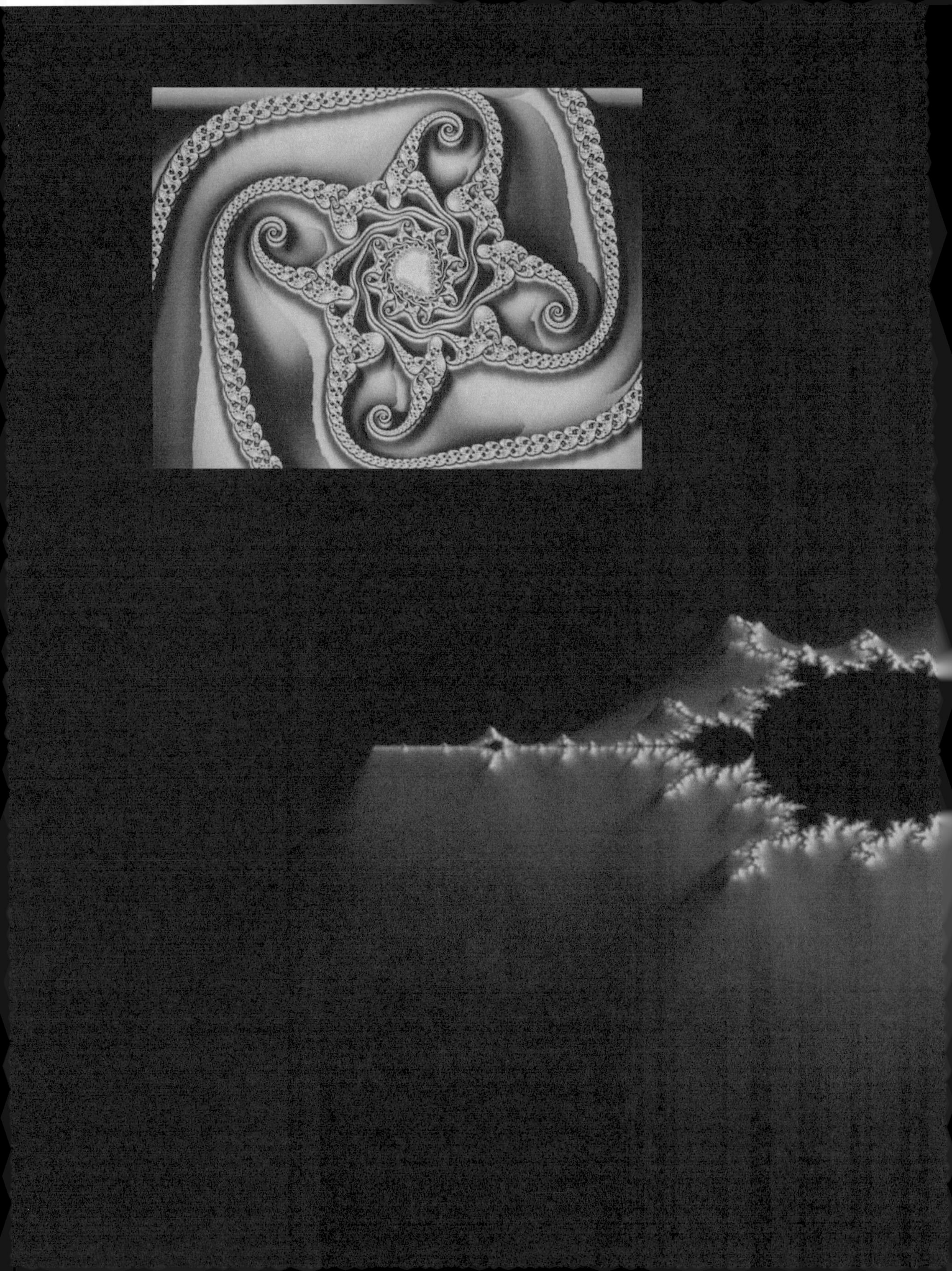

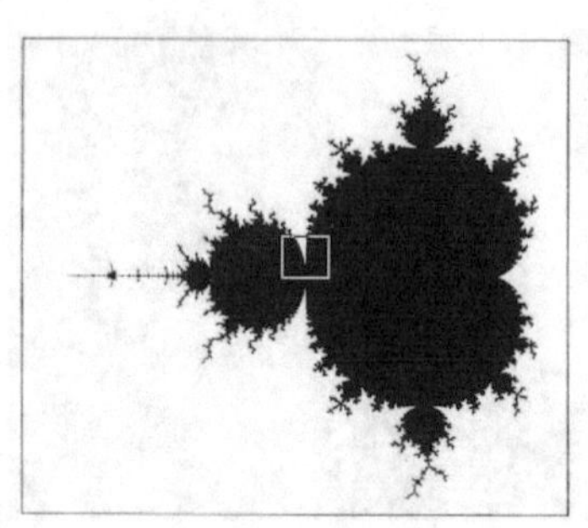

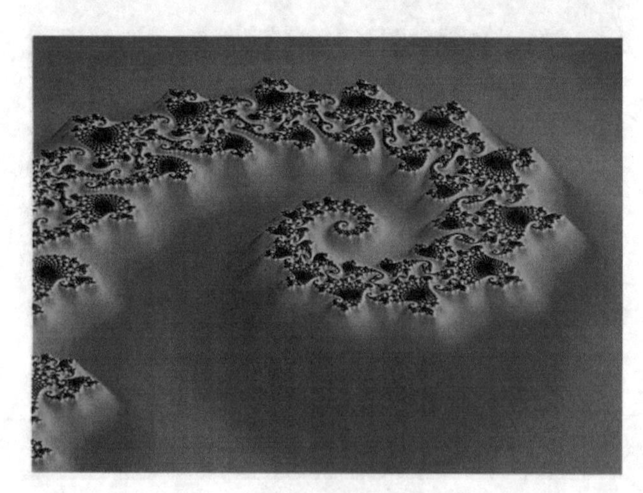

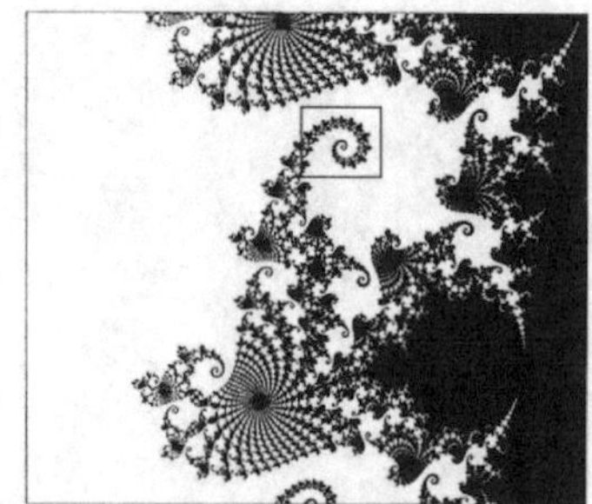

Plate 31: 3D-zoom sequence of details of the Mandelbrot set. Height corresponds to distance. The black-and-white images indicate the location of the closeups.

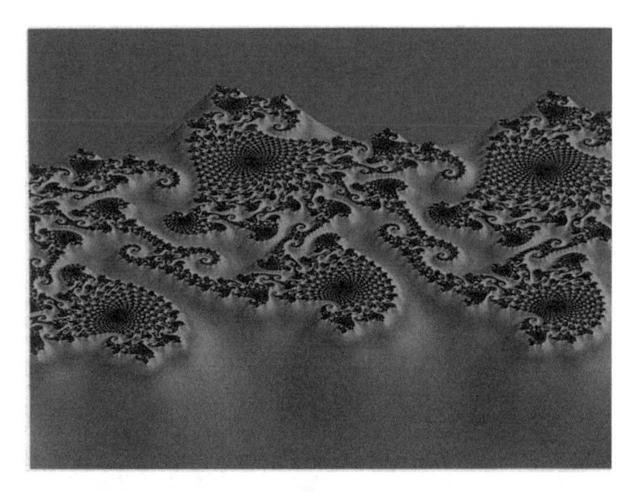

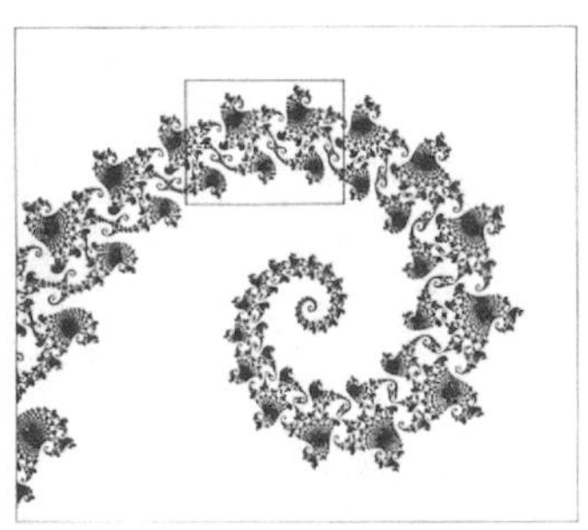

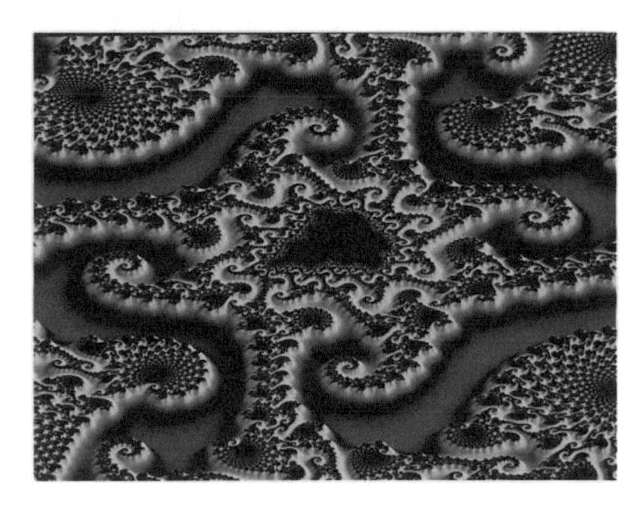

Plate 32: Continuation of the 3D-zoom sequence of plate 31. The size of the area shown in the last closeup is approximately 0.000015.

Plate 33: A 50,000 particle DLA cluster initialized at a point on a square lattice. © B.B. Mandelbrot, C.J.G. Evertsz, C. Kolb.

Plate 34: DLA cluster starting from a line with fuzzy regions of low growth probability. © B.B. Mandelbrot, C.J.G. Evertsz, C. Kolb.

Plate 35: Combination of two DLA clusters at different potentials with field lines. © C.J.G. Evertsz, B.B. Mandelbrot, F. Normant.

Plate 36: The boundary lines between the grey and black bands are equipotential lines. © B.B. Mandelbrot, C.J.G. Evertsz, C. Kolb.

Plate 37: The configuration of plate 35 with rendering showing equipotentials.

Plate 38: Exactly self-similar Koch tree with equipotential lines and field lines. © C.J.G. Evertsz, B.B. Mandelbrot, F. Normant.

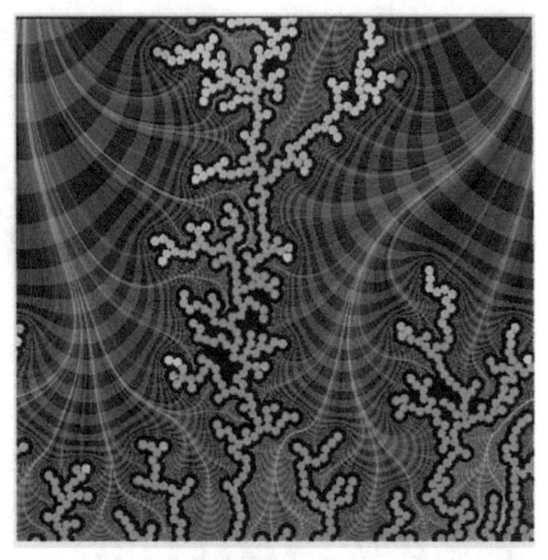

Plate 39: Off-off-lattice DLA cluster with particle color indicating time of addition. © L. Woog, C.J.G. Evertsz, B.B. Mandelbrot.

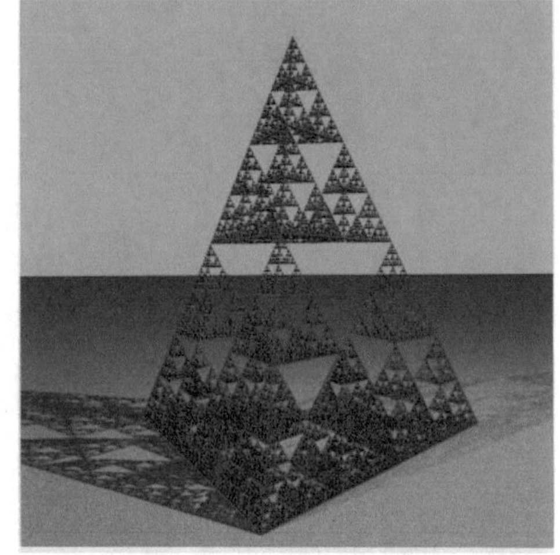

Plate 40: Ray Tracing images of the construction of a 3D-Sierpinski gasket at stage 5 (left) and stage 7 (right). © R. Lichtenberger.

and, thus,

$$x \approx \sqrt{\frac{10^{-18}}{4}} = 0.5 \cdot 10^{-9}.$$

The relative error in the computed value of x is therefore 100 percent. This may not sound so bad, since the absolute error is indeed very small. But the worst is yet to come. The evaluation of $y = v/(2x)$ yields a division by zero — an error which usually terminates the program rather unexpectedly. The problem encountered here is a consequence of what has been termed by numerical analysts *loss of significance* or *cancellation of significant digits*.

Fortunately, in our case there is a remedy for the problem. The solution to the square root problem $u = x^2 - y^2, v = 2xy$ can also be written as

$$\begin{aligned} y &= \pm\frac{\sqrt{2}}{2}\sqrt{-u + \sqrt{u^2 + v^2}}, \\ x &= \frac{v}{2y}. \end{aligned}$$

This is derived in much the same way as our first solution, and we skip these details. Let us again assume that $|v|$ is small in comparison to $|u|$. The second solution also exhibits loss of significance, namely, when $u > 0$, while in the previous solution cancellation occurs for $u < 0$. Thus, we may effectively eliminate loss of significance by choosing one or the other method depending on the sign of u. You can check that for our specific example $u = -10^6$, $v = 10^{-6}$ we indeed get $y = 10^3$ and $x = 0.5 \cdot 10^{-9}$.

The Complete Algorithm

Summarizing, a complete algorithm which takes into account the special cases $u = 0$ and $v = 0$ is given below. Here only one root is computed, namely, the one with a positive real component (if the real component is not zero). The other root is identical except for the signs of x and y. This algorithm is very fast. It essentially involves just substituting two real square roots for one complex square root while avoiding loss of significant digits.

Algorithm for the Complex Square Root of $u + vi$

This algorithm computes the complex square root with nonnegative real part.

```
if (u > 0) then
   x = sqrt((u + sqrt(u*u + v*v))/2)
   y = v / (2*x)
else if (u < 0) then
   y = sign(v) * sqrt((-u + sqrt(u*u + v*v))/2)
   x = v / (2*y)
else
   x = sqrt(abs(v)/2)
   if (x > 0) then
      y = v / (2*x)
   else
      y = 0
   end if
end if
```

The functions `sqrt`, `sign`, `abs` are the real positive square root, the sign, and the absolute value of real numbers. The sign of v is $+1$ if $v \geq 0$ and -1 otherwise. The factor `sign(v)` ensures that the root computed has a nonnegative real component.

Solving Complex Quadratic Equations

With this tool for calculating the complex square root we can solve any quadratic equation in complex numbers using the same formula as for real ones. If

$$az^2 + bz + c = 0,$$

where all numbers are complex, must be solved, then we can write the solutions as

$$z_{1,2} = \frac{-b \pm \sqrt{b^2 - 4ac}}{2a}.$$

Similar to the case of a complex square root, cancellation may occur when $|4ac|$ is small compared to $|b^2|$.

As demonstrated, polynomial equations of degree two have two complex roots. In general, any polynomial of degree $n \geq 1$ has exactly n complex zeros, where the zeros must be counted with proper multiplicities. For example, if $b^2 = 4ac$ in the quadratic equation, the root $z_1 = z_2$ is counted as a double root.

Let us present an example that will come up again in a section further below. Up to three decimal places, solve the quadratic equation

$$z^2 - z + c = 0$$

where $c = -1/2 + (1/2)\,i$. The recipe formula prescribes the computation (set $a = 1$ and $b = -1$)

$$z_{1,2} = \frac{1 \pm \sqrt{1 - 4c}}{2} = \frac{1 \pm \sqrt{3 - 2i}}{2}.$$

Now the algorithm for the computation of the square root of $3 - 2i$ can be applied (here we have $u = 3$ and $v = -2$). Thus,

$$x = \sqrt{\frac{u + \sqrt{u^2 + v^2}}{2}} = \sqrt{\frac{3 + \sqrt{13}}{2}} \approx 1.817$$

$$y = \frac{v}{2x} \approx \frac{-2}{2 \cdot 1.817} \approx -0.550$$

and thus

$$\sqrt{3 - 2i} \approx \pm(1.817 - 0.550i).$$

Substituting this result in the expression for z_1 and z_2, we obtain

$$z_{1,2} \approx \frac{1 \pm (1.817 - 0.550i)}{2},$$

i.e., $z_1 \approx 1.408 - 0.275i$ and $z_2 \approx -0.408 + 0.275i$.

13.4 Prisoners versus Escapees

We are now well prepared to come to the real stuff of this chapter: Julia sets. The simplest example of a nonlinear iteration procedure in the complex numbers is given by the transformation

$$z \to z^2.$$

Geometrically the squaring of a complex number means that the corresponding length of z is squared in the ordinary sense and that the corresponding angle $\arg z$ of z is doubled (mod 2π). The following table lists three examples. We take an initial point inside the unit circle, i.e., a complex number with absolute value less than 1, another initial point on the unit circle, and finally one outside the unit circle.[6] Squaring z gives z^2, squaring it again gives z^4, and so on as listed.

	length	angle	length	angle	length	angle
z	0.8	10°	1.0	10°	1.5	50°
z^2	0.64	20°	1.0	20°	2.25	100°
z^4	0.4096	40°	1.0	40°	5.06	200°
z^8	0.1678	80°	1.0	80°	25.63	40°
z^{16}	0.0281	160°	1.0	160°	656.90	80°
z^{32}	0.0008	320°	1.0	320°	431439.89	160°

Dynamics of $z \to z^2$

The iteration of three initial points using the simple squaring operation $z \to z^2$.

Table 13.10

Table 13.10 shows the dynamical behavior of the generated points. In all cases we observe that the sequences of points circle around the origin. However the initial point inside the unit circle leads to a sequence which converges to the origin, the point exactly on the unit circle leads to a sequence which remains there forever, and the point outside leads to a sequence which escapes to infinity, i.e., the absolute value of the iterates become larger and larger (see also figure 13.11). In fact, if the initial point has absolute value of 10, say, then the absolute values of the iterates are $100 = 10^2$ for the first, $10,000 = 10^4$ for the second, $100,000,000 = 10^8$ for the third, $10,000,000,000,000,000 = 10^{16}$ for the fourth, $100,000,000,000,000,000,000,000,000,000,000 = 10^{32}$ for the fifth, and so on. In other words, the absolute values of the iterations literally explode within a few steps.

Prisoner and Escape Set

This leads us to an important dynamic dichotomy: the complex plane of initial values is subdivided into two subsets. The first one collects points for which the iteration escapes. It is called the *escape set* E. The iteration for all other initial values remains in a bounded region forever, and we collect these points in the so-called *prisoner set* P. Note that P is the disk around zero with radius 1 and that E is the outside of that disk.[7] The boundary between E and P is the unit circle. In this context it is called the *Julia set* of the iteration.

[6] The unit circle is the set of points in the complex plane having distance 1 to the origin, i.e., $\{x + yi \mid x^2 + y^2 = 1\}$.

[7] Compare the section about the Cantor set as a prisoner set in chapter 2.

Iterating $z \to z^2$

The three initial points from table 13.10 are iterated.

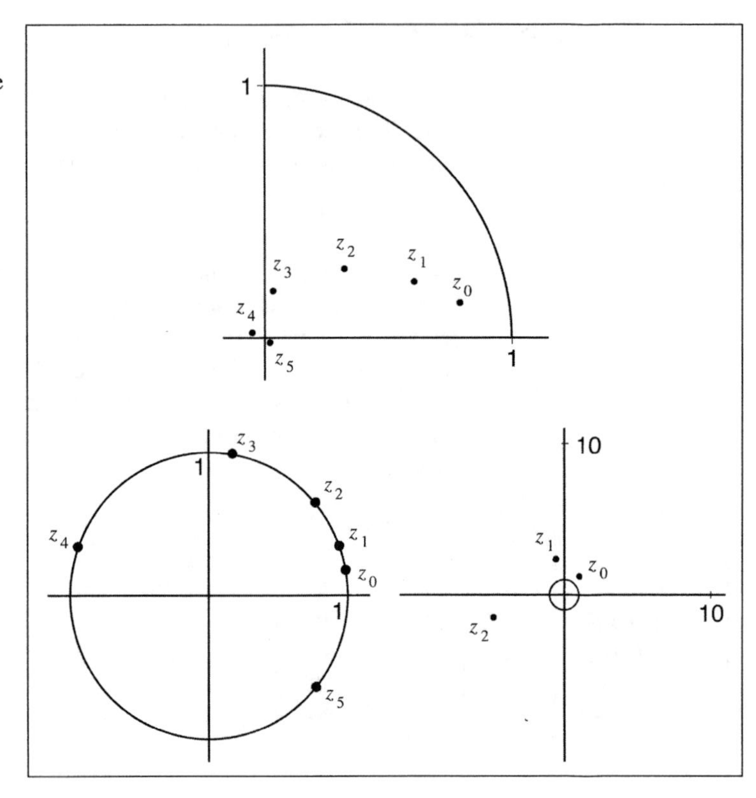

Figure 13.11

For initial values in the Julia set, i.e., $|z| = 1$, the iteration generates only points which again lie on the unit circle. In other words, the Julia set is *invariant* under iteration.

Two Basins of Attraction

Note that the interior of P can be interpreted as a basin of attraction, the attractor being the point 0. If we just restrict attention to real initial points z_0, then the behavior can be visualized by graphical iteration (see figure 13.12).

Fixed Points and Basins of Attraction

Note that the iteration has two fixed points; 0 and 1. However, 0 is attracting and 1 is repelling, i.e., if we start near 1 as, for example with $z_0 = 1 + \varepsilon$, then z_1 is about twice as far from 1 than z_0, $z_1 = (1+\varepsilon)^2 \approx 1+2\varepsilon$. If we consider the iteration of z^2 on the sphere as explained in the previous section, the north pole is an attractor for the dynamics of $z \to z^2$. Likewise the escape set E can be interpreted as a basin of attraction, the basin of the point at infinity. Thus the dynamic dichotomy is that we have orbits $z_0, z_1, z_2, \ldots$ which escape to infinity, and orbits which do not.

We have two attractors, the origin 0 and the point at infinity, whose basins of attraction are the open unit disk and everything outside the unit disk respectively, while the boundary points, which are all located on the unit circle, wander around on the circle forever. It is the boundary, which is shared between the basins of attraction, that is called the Julia set. Here in this simple

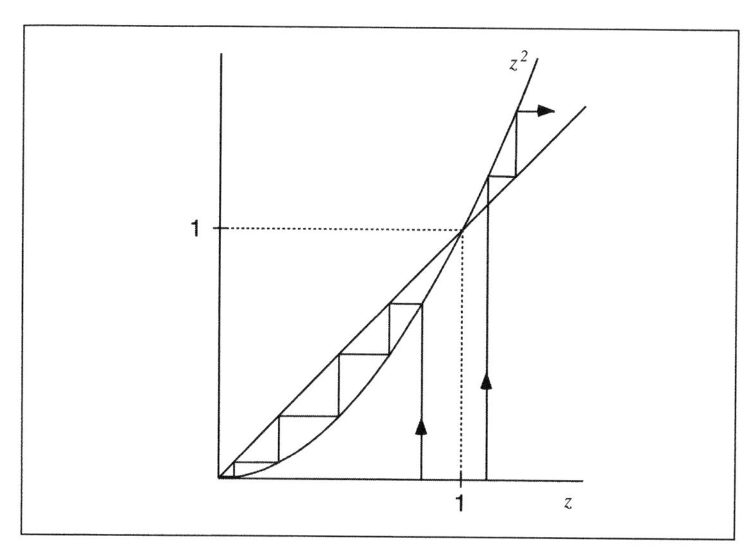

Graphical Iteration for $z \to z^2$

Initial points z_0 with $|z_0| < 1$ rapidly converge to the origin, while points with $|z_0| > 1$ escape to infinity.

Figure 13.12

example, the Julia set is a circle and, thus, a geometrical object from classical Euclidean geometry. However, it is very special because it is not a fractal, and most Julia sets are. Besides that property, however, it shares the important typical properties of many Julia sets: it is the boundary of basins of attraction; and the dynamics on it are chaotic.

Julia Sets for the Quadratic Family $f_c(z) = z^2 + c$

The special case $z \to z^2$ is the entrance door into an amazing zoo of beautiful fractal Julia sets. One department of that zoo of Julia sets is built on the iteration of $f_c(z) = z^2 + c$, where c is some complex parameter. The Julia set of $z \to z^2$ is right in the center of this class; set $c = 0$. As the unit circle is the boundary of the escape set for $f_0(z) = z^2$, the other Julia sets are the boundaries of the escape set of $z \to z^2 + c$. This we can view as their definition. To see these other Julia sets we need a special viewing device. Let us therefore learn how to build this apparatus. In reality it is only a visualization of the escape set E for an arbitrary choice of c, because whatever remains will then be the prisoner set P, and the common boundary between E and P is the Julia set.

The Definition

Let us first make some more precise definitions. The escape set for the parameter c is

$$E_c = \{z_0 \ : \ |z_n| \to \infty \text{ as } n \to \infty\}\,.$$

In this definition the orbit $z_0, z_1, z_2, \ldots$ of the initial point z_0 of course is given by

$$z_{n+1} = z_n^2 + c, \ n = 0, 1, 2, \ldots$$

The prisoner set for parameter c is

$$P_c = \{z_0 \mid z_0 \notin E_c\}\,;$$

Three Escaping Points

The iteration of three initial points for $z \to z^2 + c$, $c = -0.5 + 0.5i$. All three orbits escape to infinity.

	Orbit 1		Orbit 2		Orbit 3	
	x	y	x	y	x	y
z_0	1.00	0.00	0.50	0.25	0.00	0.88
z_1	0.50	0.50	-0.31	0.75	-1.27	0.50
z_2	-0.50	1.00	-0.96	0.03	0.87	-0.77
z_3	-1.25	-0.50	0.43	0.44	-0.34	-0.85
z_4	0.81	1.75	-0.51	0.88	-1.12	1.07
z_5	-2.90	3.34	-1.01	-0.39	-0.41	-1.90
z_6	-3.26	-18.91	0.37	1.30	-3.93	2.04
z_7	-347.46	123.68	-2.04	1.46	10.79	-15.52
z_8			1.53	-5.46	-124.77	-334.49
z_9			-28.01	-16.27		

Table 13.13

Three Prisoner Points

The iteration of three initial points for $z \to z^2 + c$, $c = -0.5 + 0.5i$. All three orbits do not escape to infinity. Rather they seem to converge to a point $z \approx 0.41 + 0.28i$.

	Orbit 1		Orbit 2		Orbit 3	
	x	y	x	y	x	y
z_0	0.000	0.000	0.500	-0.250	-0.250	0.500
z_1	-0.500	0.500	-0.313	0.250	-0.688	0.250
z_2	-0.500	0.000	-0.465	0.344	-0.090	0.156
z_3	-0.250	0.500	-0.402	0.180	-0.516	0.472
z_4	-0.688	0.250	-0.371	0.355	-0.456	0.013
z_5	-0.090	0.156	-0.488	0.237	-0.292	0.488
z_{100}	-0.473	0.291	-0.393	0.290	-0.438	0.217
z_{200}	-0.394	0.279	-0.411	0.271	-0.409	0.290
z_{300}	-0.411	0.273	-0.409	0.276	-0.407	0.272
z_{400}	-0.408	0.276	-0.409	0.275	-0.409	0.276
z_{500}	-0.409	0.275	-0.409	0.275	-0.409	0.275

Table 13.14

it is the complement of E_c. The Julia set for parameter c is the boundary of the escape set E_c. A point is a boundary point if arbitrarily close to it we can find initial points with escaping orbits as well as points from the prisoner set.

A First Fractal Julia Set

For a first example let us choose $c = -0.5 + 0.5i$ and a few initial points (see tables 13.13 and 13.14). We can observe two basic behaviors. In the first table, the iterated points escape to infinity. In the second table, the iterated points do not escape but eventually settle down to a certain point, namely, $z \approx -0.408 + 0.275i$. This indicates that we again have two basins of attraction, but zero is no longer one of the attracting points. Further below, in section 13.7, we will compute this fixed point of the iteration as a solution of an equation.

Figure 13.15 shows the result of an extended experiment which tests all points in the complex plane. The prisoner set is pictured in black, while everything white belongs to the escape set. Consequently, the boundary of the black region is the Julia set. This bordering curve is obviously a typical

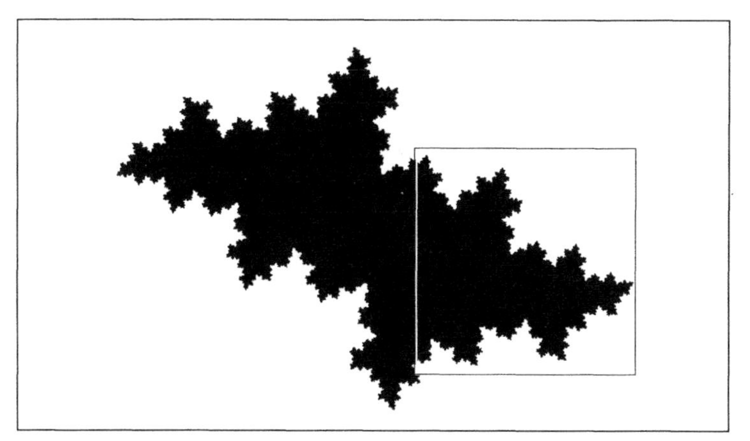

Prisoner Set for $c = -0.5 + 0.5i$

The prisoner set for $z \to z^2 + c$, $c = -0.5 + 0.5i$ is shown in black. Points outside escape to infinity. The framed region is enlarged in figure 13.16.

Figure 13.15

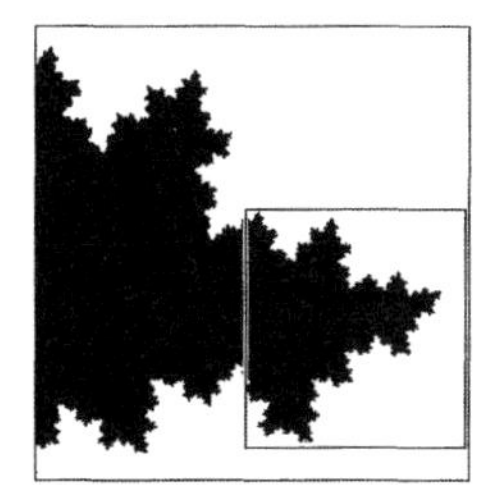

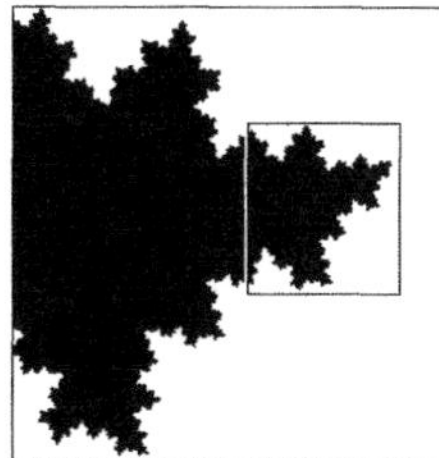

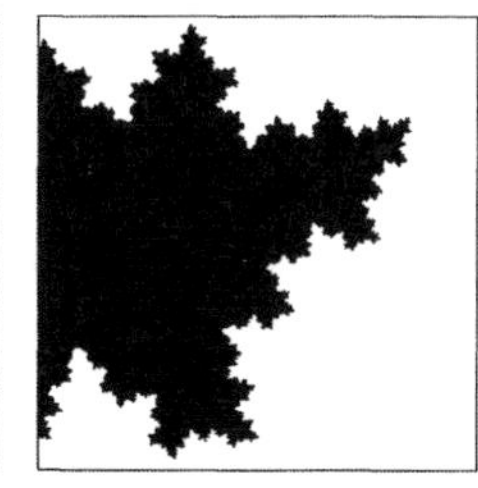

Blowups of Prisoner Set for $c = -0.5 + 0.5i$

The prisoner set for $z \to z^2 + c$, $c = -0.5+0.5i$ from figure 13.15 is successively enlarged near a boundary point. Each picture (from left to right) is a computation of the small framed region in the previous one.

Figure 13.16

fractal. No matter how much we magnify a region near the Julia set, there is always detail which looks similar (see the next figure 13.16).

The Point of No Return

The key to the computation of the escape set E_c is the observation that points z_k from an orbit will escape to infinity with certainty once their absolute value is large enough. This seems clear because the square of a large number is much larger and adding the constant c will be rather insignificant. Then in the next step of the iteration of $z \to z^2 + c$ this effect is even more pronounced. As a result we can see that in the iteration for large z, the constant c can be neglected, and we are left with $z \to z^2$, which we understand very well. But how large must an iterate be so that we can decide that the orbit will definitely escape to infinity? Fortunately, there is an optimal answer to this question. It is given by a computable number $r(c)$ which depends on the parameter c. We can show that one may choose the number $r(c)$ as the maximum of the absolute value $|c|$ and 2:

$$r(c) = \max(|c|, 2).$$

Thus, if $|z_k|$ exceeds $r(c)$ in absolute value, we can be confident that the iteration escapes to infinity. The algorithm which classifies an initial point as a member of the escape set or the prisoner set may therefore proceed as follows.

If the absolute value exceeds $r(c)$ at some iteration, the algorithm terminates and returns the result that the initial point is in the escape set. Clearly, the iterates of a point that escapes to infinity must exceed $r(c)$ in magnitude at some iteration. Otherwise, the points from the orbit would never be able to move far away, and the initial point would have to be a prisoner. Therefore, the criterion really catches all escaping points. However, in practice it may take a very long time until the orbit escapes a disk of radius $r(c)$. Thus, we prescribe a maximal number of iterations; and if the iterated point does not exceed $r(c)$ in absolute value during these iterations, then we must assume — up to the precision of the algorithm — that the initial point does not belong to the basin of infinity, but to the prisoner set.

The Threshold Radius

Here we demonstrate the following proposition:

Let z be a complex number not less than $|c|$ and greater than 2 in absolute value. Then z is an escaping point for the iteration $z \to z^2+c$.

Let a parameter c be given and set $r(c) = \max(|c|, 2)$. For the starting point z of the iteration we assume that

$$|z| \geq |c| \text{ and } |z| > 2$$

holds. Then there exists a (possibly) small but positive number $\varepsilon > 0$ with $|z| = 2 + \varepsilon$. The triangle inequality for complex numbers implies

$$|z^2| = |z^2 + c - c| \leq |z^2 + c| + |c|.$$

Solving this inequality for $|z^2 + c|$ and continuing, we derive

$$\begin{aligned} |z^2 + c| &\geq |z^2| - |c| \\ &= |z|^2 - |c| \\ &\geq |z|^2 - |z| \\ &= (|z| - 1)|z| \\ &= (1 + \varepsilon)|z|. \end{aligned}$$

Thus, if we iterate once, the absolute value will increase by at least a factor $1 + \varepsilon$. The kth iterate of z will thus be at least $(1 + \varepsilon)^k$ times as large as z in magnitude. Therefore, the absolute values clearly tend to infinity and z is in the basin of attraction of infinity.

From this proposition it immediately follows that the orbit must escape to infinity if any one point in an orbit for $z \to z^2 + c$ is larger than $r(c) = \max(|c|, 2)$ in magnitude; the initial point of that orbit is in the escape set.

Encirclement of the Prisoner Set

Let us demonstrate how an algorithm makes use of this important fact in order to encircle the prison closer and closer. To begin with we choose a parameter c and compute the threshold radius $r(c) = \max(|c|, 2)$. We also define a region of interest in the complex plane and place it on a grid of pixels for viewing on a computer graphics screen. The pixels correspond to initial points for the orbits which we will test next. First we check if there are some

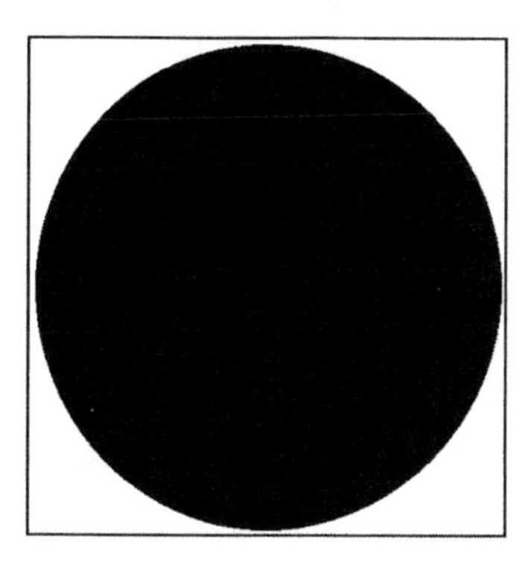

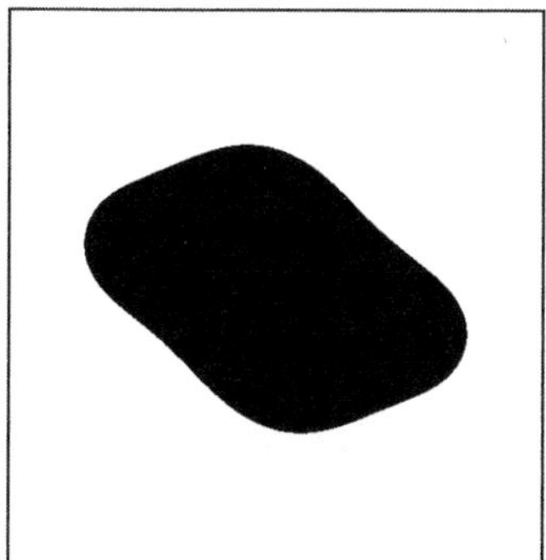

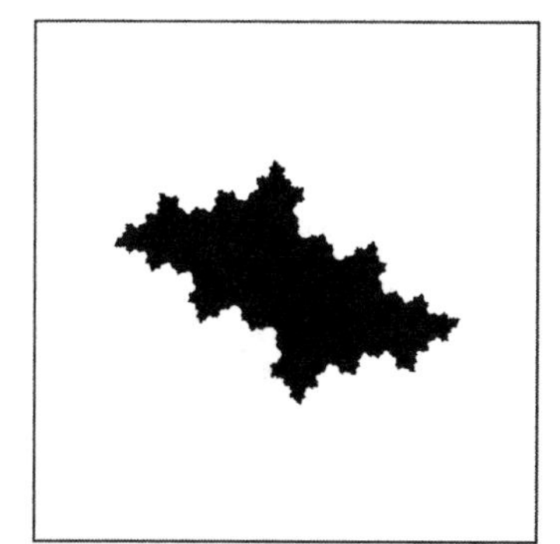

Encirclement for $c = -0.5 + 0.5i$

Encirclement for $c = -0.5 + 0.5i$. The initial approximation $Q_c^{(0)}$ is shown in the upper left (a disk of radius 2). The other approximations are $Q_c^{(-k)}$ for $k = 1$ (upper right), $k = 2$ (lower left), and $k = 10$ (lower right).

Figure 13.17

initial points z_0 which are greater in magnitude than the threshold radius $r(c)$. All these can be discarded; they have already been identified as points of the escape set E_c. Thus, our initial approximation of the prisoner set P_c is simply the disk of radius $r(c)$ centered at 0. For this approximation, we introduce the notation $Q_c^{(0)}$,

$$Q_c^{(0)} = \{z_0 \ : \ |z_0| \le r(c)\}\,.$$

We now allow one iteration for each of the remaining points or pixels, obtaining a complex number z_1 for each pixel. Again we can apply the criterion for escaping orbits. If the absolute value $|z_1|$ is larger than $r(c)$, then the corresponding point is identified as an escapee and can be removed from further consideration. In effect we obtain an improved approximation containing the prisoner set. It consists of all points which after one iteration of f_c are still in $Q_c^{(0)}$. We say that they form the preimage of $Q_c^{(0)}$ and use notation $Q_c^{(-1)}$.

$$Q_c^{(-1)} = \{z_0 \ : \ |z_1| \le r(c)\}\,.$$

Next we repeat the procedure for all remaining points, which yields the next approximation, and so on. In each iteration we remove points and thus move in on the prisoner set more and more closely. Formally, we write

$$Q_c^{(-k)} = \{z_0 \mid |z_k| \le r(c)\}\,, k = 0, 1, 2, \ldots$$

Encirclement in Alternating Colors

In the upper figure an overlay of encirclements for $c = -0.5 + 0.5i$ is drawn in alternating colors. In black: $Q_c^{(k)}$ with $k = 0, -2, -4, -20$; in white: $Q_c^{(k)}$ with $k = -1, -3, -5$. The lower figure shows a detail of the encirclement. In black: $Q_c^{(k)}$ with $k = 0, -2, -4, -6, -20$; in white: $Q_c^{(k)}$ with $k = -1, -3, -5, -7$.

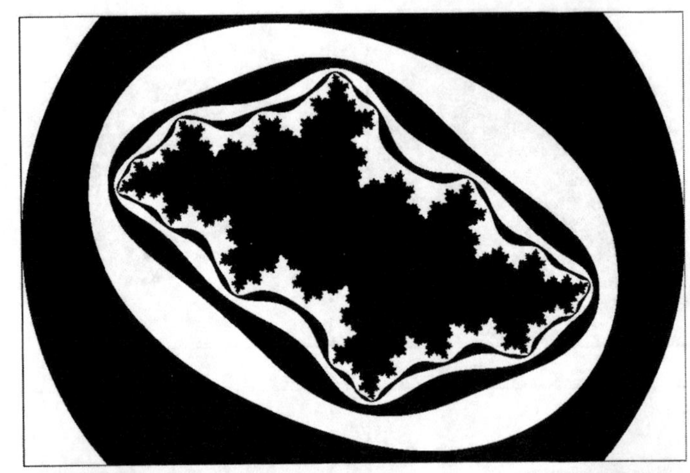

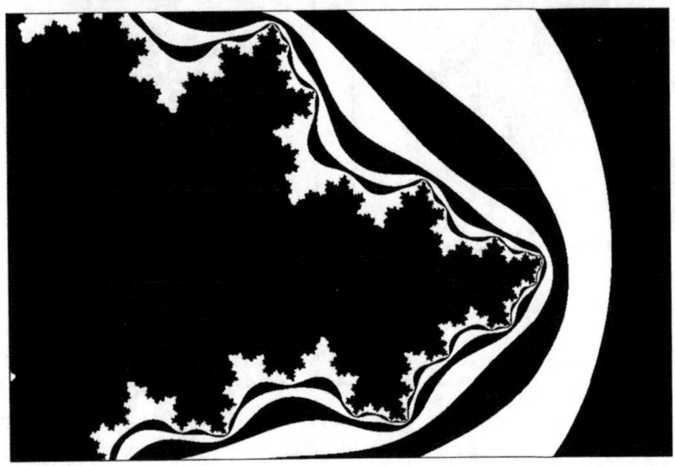

Figure 13.18

and

$$\lim_{k \to \infty} Q_c^{(-k)} = P_c.$$

The Case $c = 0$

For the case $c = 0$ we can write down explicitly what these approximating sets are. We have $r(c) = 2$ and the initial approximation $Q_c^{(0)}$ is the disk with radius 2. The following approximations are also disks. For example, $Q_c^{(-1)}$ is the collection of all points z that satisfy $|z^2| \leq 2$. This is the set of points z with $|z| \leq \sqrt{2}$. Thus, $Q_c^{(-1)}$ is a disk of radius $\sqrt{2}$. The radius of the next approximation, $Q_c^{(-2)}$, is $\sqrt{\sqrt{2}} = 2^{1/4}$. A general formula is

$$Q_c^{(-k)} = \left\{ z_0 \ : \ |z_0| \leq 2^{1/2^k} \right\}.$$

This sequence of radii rapidly approaches 1, the radius of the prisoner set P_c. The first few radii are: $2, 1.414, 1.189, 1.090, 1.044, 1.022, 1.011, 1.005, \ldots$

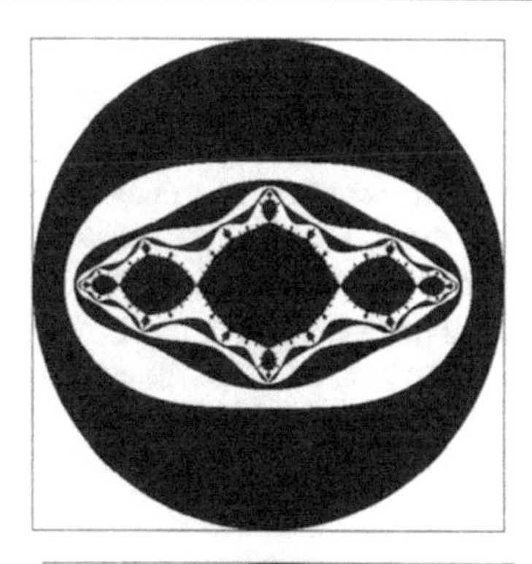

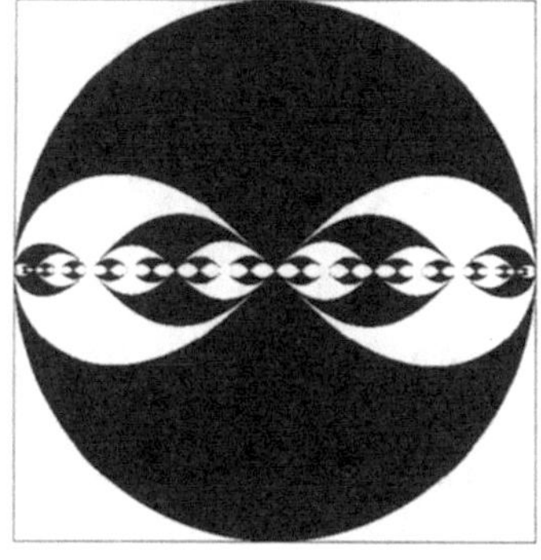
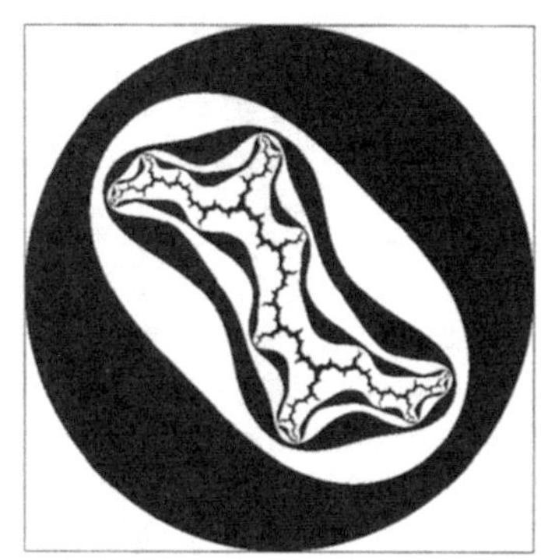

Encirclements

Encirclements for $c = -1, 0.4, -2$, and $c = i$ (from upper left to lower right).

Figure 13.19

The Case $c = -0.5 + 0.5i$

For the case $c = -0.5 + 0.5i$ from figures 13.15 and 13.16, we cannot give explicit formulas for the approximations of the prisoner set, but we show a sequence of figures illustrating how the prisoner set is encircled more and more closely (see figure 13.17).

Stacking Encirclements

It is possible to stack encirclements on top of each other, allowing a better comparison and insight into how the prisoner set is approximated. For this purpose it is necessary to shade the encirclements differently, for example, using alternating black and white sets. In our next figure 13.18 we show encirclements drawn on top of each other in this way. $Q_c^{(0)}, Q_c^{(-2)}, Q_c^{(-4)}$, and $Q_c^{(-20)}$ are black while $Q_c^{(-1)}, Q_c^{(-3)}$, and $Q_c^{(-5)}$ are white.

In figure 13.19 this is done for some further choices of the parameter c. Indeed, we can observe a variety of rather different results. For $c = -1$ we have a connected (one piece) Julia set which is the common boundary of two basins of attraction (the basin of the infinite attractor and the basin of a finite attractor, which in this case is the period-2 cycle: $-1, 0, -1, 0, \ldots$). In the other cases there is no finite attractor, the prisoner set has no interior points and is equal to the Julia set. For $c = 0.4$ the prisoner set has dissolved into a dust of points, while for $c = -2$ and $c = i$ we observe boundary cases: the Julia set is a single connected set. We are about to discover an important dichotomy: prisoner sets are either connected or a dust of points. This is explained further in section 13.8.

Implementation Details for the Pixel Game

Pixel game algorithms to compute Julia sets (and the Mandelbrot set) have become very popular through an article in the mathematics column in the Scientific American.[8] Here we want to visualize the prisoner set approximation $Q_c^{(k)}$ by coloring an array of pixels like the one shown in figure 13.2. A pixel with coordinates x and y is colored black if $z = x + yi$ is an initial point belonging to $Q_c^{(k)}$. At the beginning we present an overview of the algorithm using complex notation for the variables c and z. In a computer implementation the complex addition and multiplication rules must be properly used.

```
R = max(|c|,2)
i = 0
while (i < k)
   if (|z| > R) then
      return (z belongs to the escape set)
   end if
   z = z*z + c
   i = i + 1
end while
return (z belongs to Q_c^(k))
```

The algorithm requires the iteration of `z*z + c`. Here some care must be taken. Let x and y be the real and imaginary part of z, and cr and ci those of c. People have often erroneously written

```
x = x*x - y*y + cr
y = 2*x*y + ci
```

The mistake is that when the new value of y is computed, the old value of x has already been lost (overwritten by its new value); and thus `2*x*y` is not what one expects. The correct version, of course, uses a temporary variable to hold the new value of x as in

```
temp = x*x - y*y + cr
y = 2*x*y + ci
x = temp
```

One hint for an efficient implementation is in order here. It is far more efficient to check whether $|z|^2 > R^2$ than to check if $|z| > R$, because $|z|^2 = x^2 + y^2$ and the absolute value of z requires a costly square root for the computation. Moreover, the numbers x^2 and y^2 can be reused for the next iteration, resulting in an additional speed-up. Here are the details:

```
R2 = max(|c|,2) * max(|c|,2)
i = 0
x2 = x*x
y2 = y*y
while (i < k)
   if (x2 + y2 > R2) then
      return (z belongs to the escape set)
   end if
```

[8] A. K. Dewdney, *Computer Recreations: A computer microscope zooms in for a look at the most complex object in mathematics*, Scientific American (August 1985) 16–25.

```
      y = 2*x*y + ci
      x = x2 - y2 + cr
      i = i + 1
      x2 = x*x
      y2 = y*y
   end while
   return (z belongs to Q_c^(k))
```

13.5 Equipotentials and Field Lines for Julia Sets

Capturing the prisoner set P_c by the encirclements $Q_c^{(-k)}$ has turned out to be a very fruitful idea. We will now see how we can refine it to an even more powerful tool. This leads us to the work of Douady and Hubbard. More than 60 years had passed since the outstanding work of Julia and Fatou at the beginning of the century before Douady and Hubbard developed new methods to continue the unveiling of the secrets of Julia sets. They found a beautiful way to do this using an analogy from electrostatics.

The Electrostatic Field

Think of the prisoner set as a piece of metal charged with electrons. This charge produces an *electrostatic field* in the surrounding space, resulting in an attracting force on any small test charge of the opposite polarity. The field is given by vectors indicating the direction and the strength of the force per unit test charge (see figure 13.20). The lines which follow the vectors from any given point to the charged prisoner set are those that infinitesimally small test particles would travel when exposed to the field. These lines are called *field lines*.

Field of a Wire

The electrostatic field of a wire shown in a cross-section which contains the charged wire.

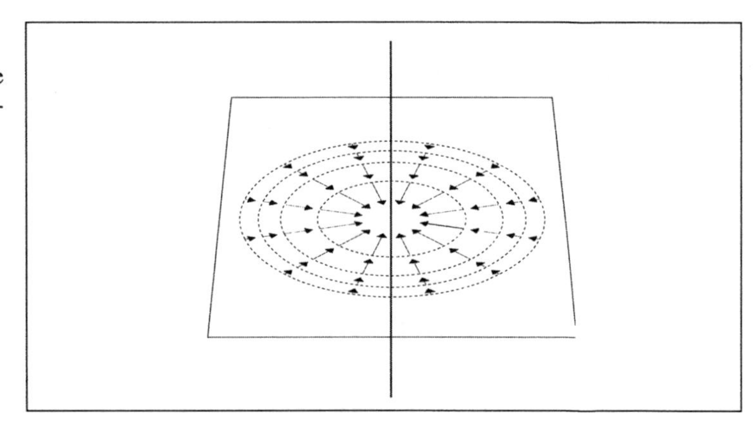

Figure 13.20

Now the theory of electrostatics applies to objects and charges in three-dimensional space, while prisoner and Julia sets are embedded in the two-dimensional space of complex numbers. However, a flat prisoner set placed in three-dimensional space has an electrostatic field which varies in all three directions of space. To get back to a theory which is completely two-dimensional in nature, we extend the prisoner set in the third dimension; we consider an infinitely long cylinder-like set whose cross-section is the prisoner set. On it we imagine an infinite amount of charge uniformly distributed along its entire length. The electrostatic field of this extended prisoner set is identical in all cross-sectional slices, any one of which completely describes the whole field.

The Field of an Infinite Cylinder

In principle we can compute the electrostatic field from Coulomb's law, which states that the force between two charged particles is proportional to the product of the two charges and inversely proportional to the square of their distance. Thus, the field of a single point charge is spherical: all field vectors point to the center, and the field strength is the same for all points on a sphere centered at the point charge (inversely proportional to the square of the radius). From this basic principle it follows that the electrostatic field of an infinite straight is cylindrical and the strength is inversely proportional to the distance to the wire (and not the square of the distance). Moreover, the electrostatic field is the same when we consider a wire which has some thickness, for example with a unit disk as the cross-section, and when the distance is measured to the center of the wire.

The Potential Function ...

For a given force field such as an electrostatic field it is possible and useful to discuss the *potential energy* of an object in terms of the external work necessary to move it from place to place against the forces of the field. The idea is familiar from Newtonian mechanics and the gravitational force field. In this case the potential is proportional to height; the work necessary to lift an object from height h_0 to height h_1 is proportional to the height *difference* $h_1 - h_0$. In an electrostatic field there is also a potential function which allows us to compute the work required to move a test charge from one point to another as the difference of the potentials at the two points. This implies, of course, that this work is independent of the particular path chosen to move the test charge. Moreover, we can retrieve the energy by allowing the particle to return to its initial position. We say that the electrostatic field is *conservative*.

...and its Equipotential Surfaces

Of special importance are the equipotential surfaces. These are defined as surfaces on which the potential is constant.[9] For example, the equipotentials of a point charge are spheres, those of the infinite wire are cylinders. Equipotential surfaces are perpendicular everywhere to the direction of the electrostatic field. In many cases, a system of equipotentials and field lines can be regarded as a very special system of polar coordinates. By the definition of equipotentials it is clear that there is no external work required to move a test charge to any other place as long as the destination point is on the same equipotential surface as the initial point. But more than this, the equipotential surfaces give an idea of the intensity of force: the intensity of the field is inversely proportional to the distance between equipotential surfaces when they are drawn for equally spaced values of the potential. Crowded equipotentials mean relatively high force, and sparse equipotentials, relatively low force.

The computation of the electrostatic potential requires advanced techniques from calculus. The work to move a particle in a field is proportional to the distance traveled and the active field force. However, because the field force is not a constant, this calculation must be done in practice as a summation over many steps, each one representing the work performed to move a particle a small distance in which the field is approximately constant. The background for this computation is the theory of line integrals; and we have chosen not to

[9] In the following we often call these surfaces just 'equipotentials'.

present these details here.[10] In general the complexity of the computation is too great to allow for explicit formulas for the electrostatic potential. But there are some special cases, namely, the potential of a point charge is proportional to the inverse of the distance, and the potential of the infinitely long wire is proportional to the logarithm of the distance.[11]

Potential of the Unit Disk

Let us now return to prisoner sets and how their potential functions can aid in the understanding of the dynamics of the quadratic iteration. This ground-breaking connection between dynamics and potentials was made in Douady and Hubbard's theory. It applies to cases where the prisoner and Julia sets are *connected*. This implies that typically field lines converge to the Julia set. We begin by discussing the important example of the unit disk, the prisoner set of f_c, $c = 0$, $z \to z^2$, corresponding to an infinite wire with a unit disk cross-section. The electrostatic field and the potential are not affected by the third (vertical) dimension. Thus, in the following we will ignore this third dimension of the physical setting and concentrate on the two-dimensional complex plane. Of course, the field lines in this plane cannot leave the plane. Moreover, the equipotential surfaces intersect the complex plane in curves, which we call equipotential curves or simply equipotentials. As already explained above the potential function is logarithmic, we may write

$$p(z) = \log |z|,$$

for the potential, where the base of the logarithm does not matter.[12] The equipotential curves are concentric circles (given by $|z| = r$, where r is the radius with $r > 1$) and the field lines are straight lines (given by $\arg z = \phi$, where ϕ is an angle between 0 and 2π). In other words, we choose angle zero for the radial line attached to the disk at $z = 1$ and going counterclockwise we identify each field line by an angle between 0 and 2π. In figure 13.21 we also show some equipotential lines. These are circles with $\log |z| = 2^k \log 2$. Stepping from circle to circle outward the potential $p(z)$ becomes twice as large. Note that if we use logarithms to the base 2, the equation for the circles becomes $\log_2 |z| = 2^k$. In the following we will therefore use base-2 logarithms.

[10] Any college level physics text book on electricity and magnetism should have a section on the electrostatic potential and some examples for its computation.

[11] The reader familiar with calculus will see a connection with the corresponding force fields. If r denotes the distance to the point charge or the center of the infinite wire, then the electric fields are proportional to $1/r^2$ and $1/r$ respectively. Integrating these functions gives factors $1/r$ and $\log r$.

[12] In any case the formula ignores the factor of proportionality which depends on the density of charge considered for the infinite wire, among other things. Changing the base of the logarithm is the same as introducing a different factor which affects neither the equipotentials nor the field lines.

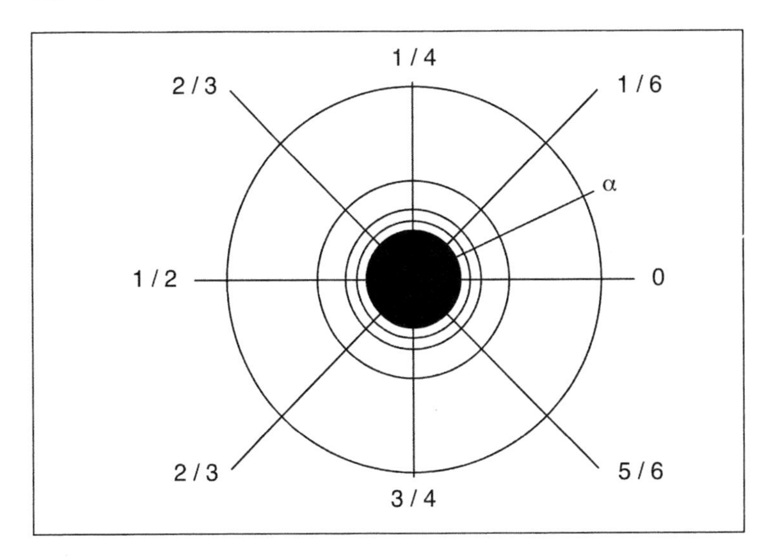

Potential of Unit Disk

Field lines and equipotential lines for the unit disk. The angles of the field lines are given in multiples of 2π (i.e., $\phi = 2\pi\alpha, 0 \leq \alpha \leq 1$).

Figure 13.21

Connected versus Disconnected

A set is called connected provided it cannot be decomposed into two disjoint, non-empty subsets (which are both open and closed in the topology of the set). There are several other mathematical notions of connectedness. For example, a set is called pathwise connected provided any two of its points can be connected by a continuous path which is entirely within the set. The property of being connected is not the same as pathwise connected. For example, take as a set A the graph of the function $f(x) = \sin 1/x$ together with the line segment $\{(0, y) \mid -1 \leq y \leq +1\}$, see figure 13.22.

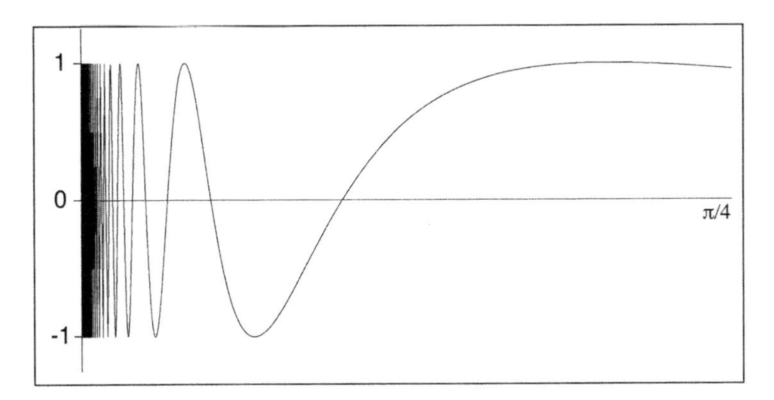

Figure 13.22 : The graph of the function $\sin 1/x$ together with a piece of the y-axis form a set that is connected, but not pathwise connected.

A is connected but not pathwise connected, because a point on the graph and a point on the line segment cannot be connected by a continuous path which is entirely in A! On the other hand, a set which is not connected can be decomposed into disjoint parts. In particular, a set is called totally disconnected provided its connected components (i.e., the maximal connected subsets) are single points. Any finite set of points is totally disconnected. Infinite point sets can also be totally disconnected, for example, the set $\{1/n \mid n = 1, 2, 3 \ldots\}$ or the set of rational numbers within all real numbers. These are examples where the sets are still countable. The Cantor set is an example of a set which is both totally disconnected and uncountable (see chapter 2).

Iteration of Field Lines

The field lines of the unit disk are closely related to the dynamics of f_c, $c = 0$, $z \to z^2$. From our introduction to complex numbers, we know that squaring a number z means doubling its argument ϕ. In other words, the iteration of all initial points from a field line with argument ϕ produces results which are all on the field lines with twice that argument. Thus, field lines are transformed into field lines. This transformation is given by

$$\phi \to 2\phi \bmod 2\pi = \begin{cases} 2\phi \text{ if } \phi < \pi \\ 2\phi - 2\pi \text{ otherwise} \end{cases}$$

When defining $\phi = 2\pi\alpha$ with $0 \le \alpha \le 1$, this is nothing else but our familiar *shift transformation*

$$\alpha \to \text{Frac}\,(2\alpha)$$

which we studied extensively in chapter 10. There it turned out that it was fruitful to use the binary expansion of numbers to reveal properties of the iteration. Let us apply this to our situation. For example, we have the binary expansion $1/3 = 0.\overline{0101}$. Applying the shift transformation means that we have to shift digits:

$$\begin{aligned} \text{step 1: } 0.1\overline{01} &= 2/3 \\ \text{step 2: } 0.\overline{01} &= 1/3. \end{aligned}$$

After two iterations an initial point is back on the same field line. This corresponds to a 2-cycle in the transformation of field lines

$$\alpha_1 = \frac{1}{3} \to \alpha_2 = \frac{2}{3} \to \alpha_1 = \frac{1}{3} \to \cdots,$$

and this periodicity is apparent in the binary expansion of the angle. From the expansion $1/6 = 0.0\overline{01}$ we read off that $\alpha = 1/6$ iterates in one step to the cycle just discussed. Indeed, all kinds of iterative behavior of the shift transformation can also be found in the transformation of field lines: periodic angles (like 1/3), pre-periodic ones (like 1/6), but also sensitive dependence on initial conditions, and so on. These dynamics also apply for points on the unit circle (the Julia set), where the field lines terminate.

Periodic and Pre-Periodic Angles

Rational numbers $\alpha = p/q$ in the unit interval $(0 < \alpha < 1)$ can be written as infinite periodic decimal or binary fractions. In other words, we can write α as a decimal of the form

$$\alpha = 0.d_1 d_2 \ldots d_l \overline{d_{l+1} d_{l+2} \ldots d_{l+m}},$$

where the d_i are decimal or binary digits and the overlining denotes periodic repetition. Here we consider only binary digits.

What happens if we iterate $\alpha \to 2\alpha \bmod 1$? This can be read off directly from the binary expansion. In the case $l = 0$ we obtain an orbit of period m:

$$\begin{aligned} \alpha_0 &= 0.\overline{d_1 d_2 \ldots d_m} \\ \alpha_1 &= 0.\overline{d_2 \ldots d_m d_1} \\ &\vdots \\ \alpha_{m-1} &= 0.\overline{d_m d_1 \ldots d_{m-1}} \\ \alpha_m &= 0.\overline{d_1 d_2 \ldots d_m} = \alpha_0. \end{aligned}$$

If $l > 0$, then α is pre-periodic; l iterations will lead to the periodic cycle. We can also write α in the form

$$\alpha = \frac{k}{2^l(2^m - 1)}$$

where k is an integer less than $2^l(2^m - 1)$. Let us briefly verify this notation. First, we again consider the case $l = 0$, thus $\alpha = 0.\overline{d_1 d_2 \ldots d_m}$. We let

$$k = d_1 d_2 d_3 \ldots d_m = \sum_{i=1}^{m} d_i 2^{m-i}$$

and obtain $2^m \alpha = k + \alpha$. Solving this equation for α yields $\alpha = k/(2^m - 1)$. The case $l > 1$ can be verified in just the same way (let k be the integer $k = d_1 \ldots d_{l+m} = \alpha 2^l (2^m - 1)$).

From the Potential of P_0 to that of P_c

The electrostatic potential has turned out to be the crucial tool for the mathematical analysis of prisoner sets. This is the fundamental work of Adrien Douady and John H. Hubbard. If the prisoner set P_c is connected, then the escape set E_c carries a system of field lines and equipotentials (see figure 13.24 for $c = -1$). Recall that the potential of the prisoner set P_0, the unit disk, can be interpreted as an ordinary polar coordinate system equipped with a dynamics given by $z \to z^2$. The beauty of Douady and Hubbard's work[13] lies in the fact that the potential of any connected prisoner set P_c can be interpreted as a particular polar coordinate system for the escape set E_c equipped with dynamics given by $z \to z^2 + c$. Underlying this work is a famous result of the German mathematician Bernhard Riemann, the Riemann mapping theorem,

[13] A. Douady, J. H. Hubbard, *Étude dynamique des pôlynomes complexes,* Publications Mathematiques d'Orsay 84-02, Université de Paris-Sud, 1984.

Riemann Mapping Theorem

A one-to-one correspondence between the potential of the unit disk and the potential of any connected prisoner set.

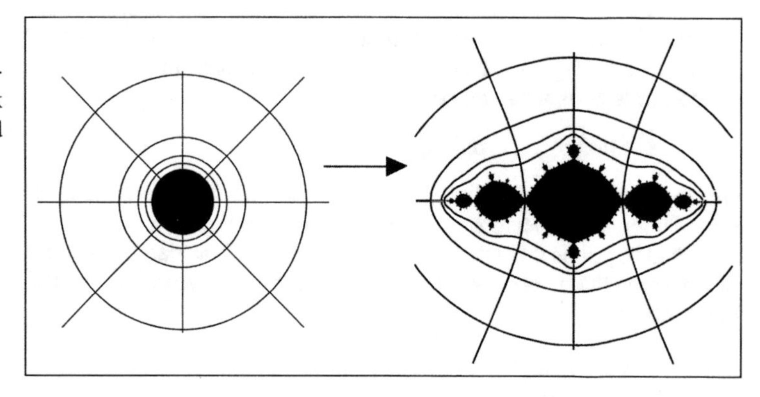

Figure 13.23

which allows to relate the potential of the unit disk to that of any connected prisoner set: equipotentials and field lines of P_0 and P_c can be brought into a one-to-one correspondence. Moreover, the dynamics of field lines of P_c can be played back to the dynamics of the field lines of the unit disk P_0, which is governed by the binary arithmetic for the transformation $\alpha \to \text{Frac}\,(2\alpha)$.

Field Lines for $z \to z^2 + c$

Let us begin to work out the procedure for labeling the field lines for the prisoner set P_c corresponding to $z \to z^2 + c$. There are many possible choices but there is only one which relates the dynamics of $z \to z^2$ in E_0 to the dynamics of $z \to z^2 + c$ in E_c. The goal is a labeling of field lines by angles $\alpha \in [0, 1)$ such that a field line with angle α is transformed to another with angle $\text{Frac}\,(2\alpha)$ under $z \to z^2 + c$. The crucial field line is the one corresponding to the line $\alpha = 0$ of the unit disk that lands at the point $z = 1$ on the boundary of the disk. The property of that point in this context is that it is a fixed point of $z \to z^2$. Thus, the entire field line landing at $z = 1$ is transformed onto itself under the dynamics of $z \to z^2$. Moreover, note that $z = 1$ is a repelling fixed point. Now we turn to $z \to z^2 + c$ using the example $c = -1$ (see figure 13.24). There are two fixed points,

$$z_1 = \frac{1 - \sqrt{5}}{2}, \text{ and } z_2 = \frac{1 + \sqrt{5}}{2}.$$

They are located at the left far end of P_{-1} and at the pinching point marked in figure 13.24. The derivatives of $f_c(z) = z^2 + c$ at z_1 and z_2 are $|1 \mp \sqrt{5}| > 1$. Thus, both fixed points are repelling and consequently points of the Julia set, which is the boundary of P_{-1}. Therefore, we expect that each will identify a field line. It would be natural to define angles for the field lines of P_{-1} so that angle $\alpha = 0$ corresponds to either z_1 or z_2. But which one? Well, if we want to be consistent with the dynamics of $z \to z^2 - 1$ we have to be careful. There is one field line landing at z_2. That is the set $\{z = x + 0i \mid x \geq z_2\}$. We verify that if z is from that line then $z^2 - 1$ is from that line as well.[14] In other words, we label that field line with $\alpha = 0$, because it remains invariant

[14] Let $x \geq (1 + \sqrt{5})/2$. Then $x^2 - 1 \geq (6 + 2\sqrt{5})/4 - 1 = (1 + \sqrt{5})/2$.

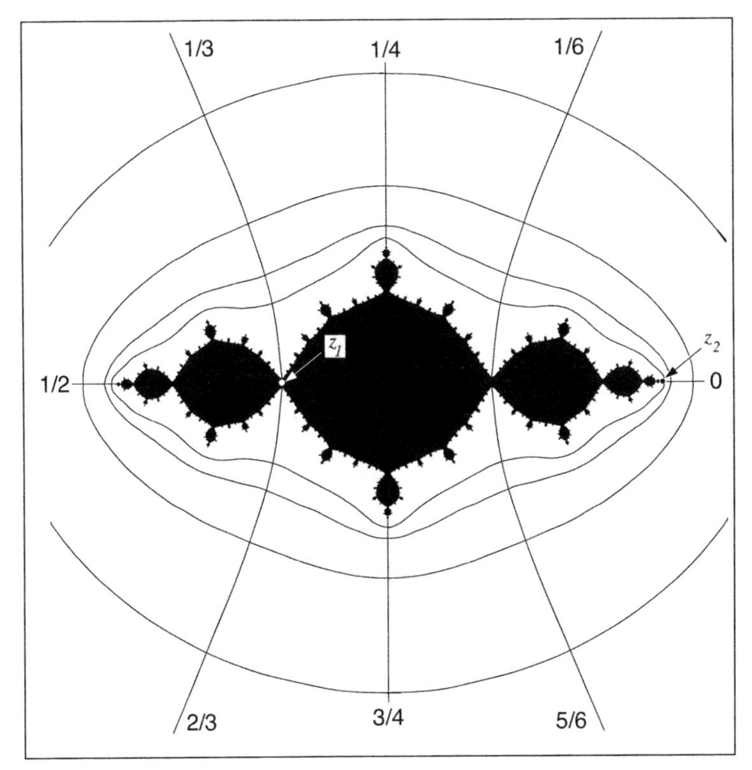

Potential and Field Lines

Equipotentials and field lines for $c = -1$. The angles of the field lines are given in multiples of 2π.

Figure 13.24

under $z \to z^2 - 1$, exactly as the corresponding field line of $z = 1$ under $z \to z^2$. But now there seems to be a problem. What happens with the field line(s) landing at z_1? First of all, if there was only one line landing at z_1, then it would also remain invariant under $z \to z^2 - 1$, following the dynamics associated with field lines. But that would trouble our choice of labeling the field line for z_2, because there is only one angle α such that $\alpha = \text{Frac}(\alpha)$, namely, $\alpha = 0$ which we have already used for the field line which lands at z_2. However, there is a way out of the dilemma allowing two field lines landing at z_1, which is just what we see in figure 13.24. This means that P_{-1} must be pinched at z_1. Now $z \to z^2 - 1$ sends field lines to field lines by simply doubling the angle. Thus, the two field lines at z_1 must be transformed to each other. There are only two angles α_1 and α_2 having the property that $\alpha_1 = \text{Frac}(2\alpha_2)$, and $\alpha_2 = \text{Frac}(2\alpha_1)$, namely, $\alpha_1 = 1/3$ and $\alpha_2 = 2/3$. Thus, we label the two lines landing at the fixed point z_1 by $1/3$ and $2/3$. Continuing in this way one can understand all the pinching points which give rise to an interpretation of the prisoner set P_{-1} itself. The same procedure applies to other parameter values c as long as the associated prisoner set is connected.

Let us summarize. The comprehension of connected Julia sets requires a study of the potential and an appropriate analysis of the dynamics of field

Encirclement $Q_c^{(k)}$

Overlay of encirclements for $c = -0.5+0.5i$ drawn in alternating colors: $Q_c^{(k)}$, for $k = 2, 1, 0, -1, -2$ and -3. Observe that only $Q_c^{(0)}$ is a circle.

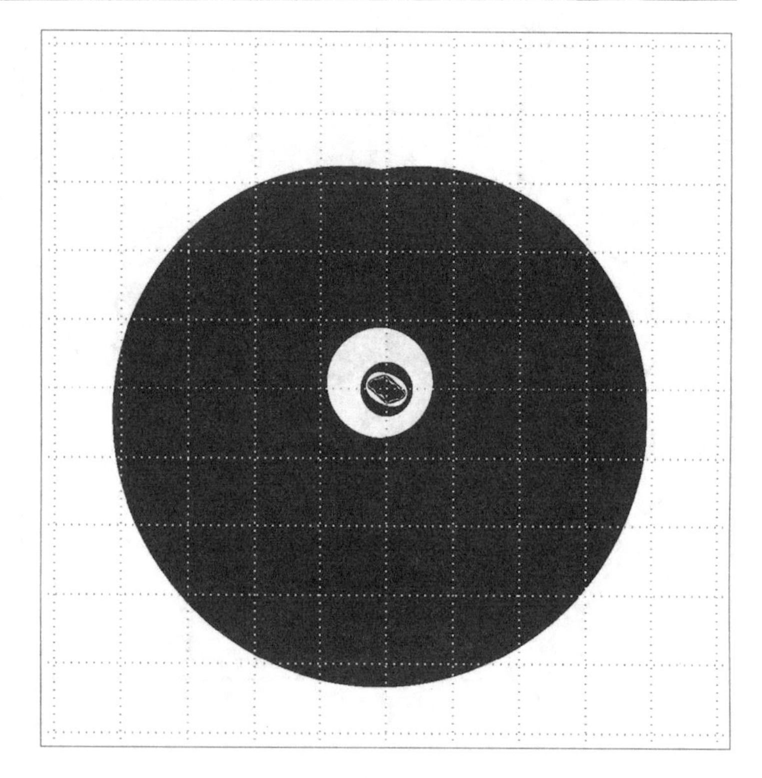

Figure 13.25

lines under $z \to z^2 + c$. That dynamics is just given by an angle-doubling. To work out the details it is necessary to understand where on the Julia sets field lines land. A priori it is not even clear that field lines land at all. It turns out that for a special class of Julia sets[15] all field lines land. For all other cases it is only known that field lines corresponding to *rational* angles α land on the boundary of the prisoner set.

Towards the Potential of Prisoner Sets

The computation of a potential function for connected prisoner sets is strongly related to encirclements of the prisoner sets. Recall that the encirclements of Julia sets from the previous section are approximations of the prisoner sets given by

$$Q_c^{(-k)} = \{z_0 \mid |z_k| \leq r(c)\}, \text{ with } z_{l+1} = z_l^2 + c,$$

which also can be written as

$$Q_c^{(-k)} = \left\{z_0 \mid z_k \in Q_c^{(0)}\right\}.$$

In other words, the $Q_c^{(-k)}$ are the iterated *preimages* of $Q_c^{(0)} = \{z \mid |z| \leq r(c)\}$, which can be called a reference or *target set*. Now we can also look at the

[15] The class of locally connected Julia sets, see section 14.2.

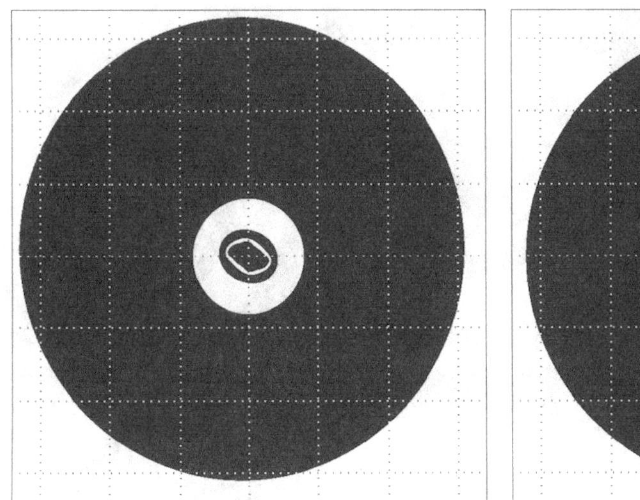
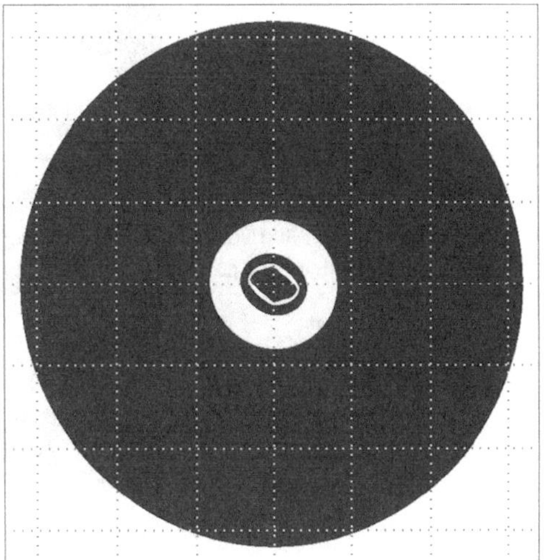

Figure 13.26 : Approximation of $P_c^{(k)}$ by $Q_c^{(k-1)}(D^{(1)})$ (left) and $Q_c^{(k-2)}(D^{(2)})$ (right).

images of $Q_c^{(0)}$, denoted by

$$Q_c^{(k)} = \left\{ z_k \mid z_0 \in Q_c^{(0)} \right\}.$$

The same construction can be carried out using an arbitrary target set T, setting

$$Q_c^{(k)}(T) = \{z_k \mid z_0 \in T\}$$

for all integers k. When k is negative, z_k denotes a preimage of z_0, i.e., k iterations started at z_k produce z_0. If we look at the special case $c = 0$ and the target set $T = Q_c^{(0)}$, then the images and preimages $Q_c^{(k)}(T)$ are just the disks which are bound by the equipotential lines shown in figure 13.21, denoted by

$$D^{(k)} = \left\{ z \mid \log_2 |z| \le 2^k \right\}.$$

Choosing the Right Target Set

If, however, we choose $c \neq 0$, then things become a bit more complicated and interesting. We can observe that in this case only the target set $T = Q_c^{(0)}$ is a disk (see figure 13.25). Indeed most published images of Julia sets show the sets $Q_c^{(k)}(D^{(0)})$ in black and white (or with an appropriate color coding). In other words, they use the disk with radius $r = 2$ as the target set. Such images may look nice, but they have a major deficiency. We cannot interpret the boundaries of $Q_c^{(k)}$ as approximations of equipotential lines when $c \neq 0$. This is clear because we cannot expect that the circle with radius 2 is an equipotential curve for all prisoner sets. There is only a single case where this holds, namely, for $c = 0$. For all other cases the equipotentials are not circles. Only when we consider equipotentials distant from the prisoner set, it acts approximately like the unit disk and the equipotentials are very close to circles. The defect is visually apparent. For example, observe that in figure

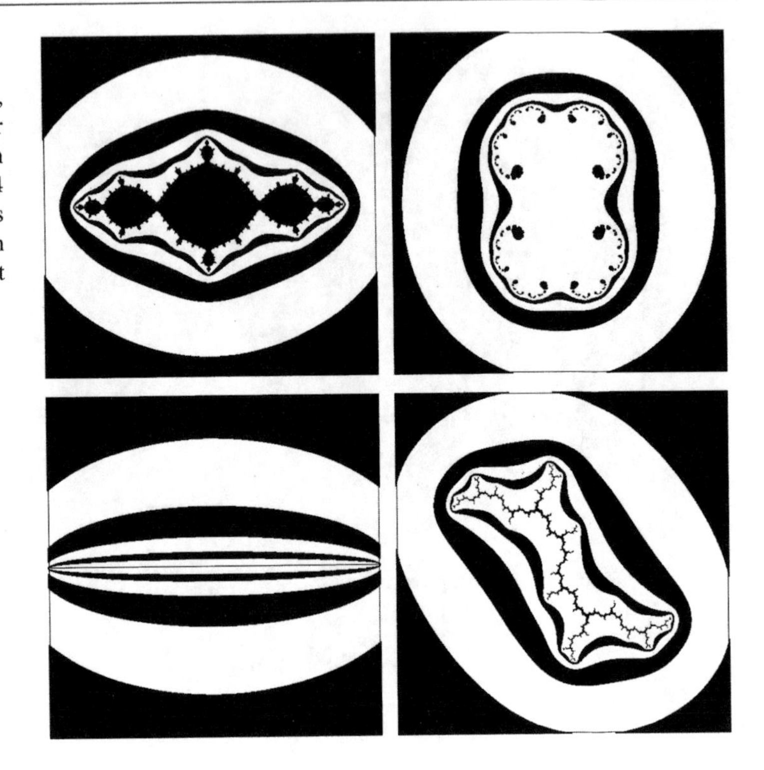

Encirclements

Encirclements by $P_c^{(k)}$ for $c = -1$, 0.4, -2, i (from upper left to lower right). Compare these images with figure 13.19. Note that for $c = 0.4$ the Julia set is not connected. Thus the image has to be interpreted with care. In this case field lines are not even defined.

Figure 13.27

13.18 the boundaries of the $Q_c^{(k)}$ come closer in some places and diverge in other places in a rather artificial way which is not related to the approximated prisoner set. Equipotential lines should not exhibit such a behavior. The problem arises from the choice of the target set T. Let us again look at the case $c = 0$. For this parameter the target set $T = Q_c^{(0)} = D^{(0)}$ is not very special. We could also take $T = Q_c^{(1)} = D^{(1)}$ or $T = Q_c^{(2)} = D^{(2)}$ and obtain essentially the same result. In fact one has the equality

$$Q_c^{(k-l)}(D^{(l)}) = Q_c^{(k)}(D^{(0)})$$

for the special case $c = 0$. In other words, we can say that the boundary of $Q_c^{(k-l)}(D^{(l)})$ is a potential curve independent of the value of l.

For $c \neq 0$ this is certainly not true. But what we might consider to be next best in fact holds true. The limit, denoted by

$$P_c^{(k)} = \lim_{l \to \infty} Q_c^{(k-l)}(D^{(l)})$$

does converge to a set with an equipotential as its boundary.[16] Figure 13.26 shows the approximations of a prisoner set by $Q_c^{(k-l)}(D^{(l)})$ for $l = 1$ and

[16] A. Douady, J. H. Hubbard, *Étude dynamique des pôlynomes complexes,* Publications Mathematiques d'Orsay 84-02, Université de Paris-Sud, 1984.

$l = 2$. In fact, even for $l = 3$ it is hard to see any differences as compared to $l = 2$. Note that the radius of the disk $D^{(k)}$ is 2^{2^k}. Thus the radius of $D^{(3)}$ is already 256. In practice we can use $Q_c^{(k-3)}(D^{(3)})$ as a rather good approximation of $P_c^{(k)}$. More explicitly, we can write

$$P_c^{(k)} = \left\{ z_0 \;\middle|\; \lim_{l\to\infty} \frac{\log_2 |z_l|}{2^l} \le 2^k \right\}. \tag{13.3}$$

In fact, the boundary of $P_c^{(k)}$ is an equipotential curve, and the potential function is given by

$$p_c(z_0) = \lim_{l\to\infty} \frac{\log_2 |z_l|}{2^l}.$$

Using this function has another advantage; it allows us to drop the special considerations with respect to the threshold $r(c)$ (which is 2 if $|c|$ is small, but is $|c|$ if $|c| > 2$). It is no longer required in the definition of the sets $P_c^{(k)}$ which also provide an encirclement of the prisoner set P_c, with the additional effect of simultaneously approximating the equipotentials.

Decomposition into Level Sets

The potential function p_c and the sets $P_c^{(k)}$ induce a natural decomposition of the escape set E_c into *level sets*

$$L_c^k = \left\{ z \mid 2^{k-1} < p_c(z) \le 2^k \right\}.$$

For any integer k this is what remains of $P_c^{(k)}$, when we take away $P_c^{(k-1)}$. In other words, $z \to z^2 + c$ transforms L_c^k into L_c^{k+1}. Note that these level sets have indices k that range throughout all integers. The level sets with large positive indices are very large rings, while those close to the prisoner set have negative indices k with large absolute value.

The level sets capture one important aspect of the dynamics, namely, the magnitude of the iterates. The other aspect, corresponding to field lines, can be visualized using a suitable decomposition of the level sets, presented in the following section.

13.6 Binary Decomposition, Field Lines and Dynamics

In the last section we subdivided the escape set into an infinite collection of level sets. Now we can turn to the binary decomposition of these level sets L_c^k, which provide a means of identifying field lines and dynamics.[17]

Binary Decomposition of Level Sets — $c = 0$

Again, let us first look at the case $c = 0$ and then generalize. We start by decomposing a level set L_c^k into two sets

$$\begin{aligned} L_c^k(0) &= \{z \in L_0^k \mid \mathrm{Im} z \geq 0\} \\ L_c^k(1) &= \{z \in L_0^k \mid \mathrm{Im} z \leq 0\} \end{aligned}$$

which are the upper and lower halves of L_c^k.[18] We call these subsets the *stage-1 cells* of the level set L_c^k, with associated labels 0 and 1. We can do this for all level sets, i.e., for all integers k. Now let us find out how we can obtain the next stage of binary decomposition. We refine $L_c^k(0)$ and $L_c^k(1)$ into four cells with two-digit labels written as

$$L_c^k(00), L_c^k(01), L_c^k(10), L_c^k(11).$$

These four sets are defined as the preimages of the stage-1 cells of the next level set L_c^{k+1}. More precisely, we let $L_c^k(00)$ and $L_c^k(10)$ be the two preimages of $L_c^{k+1}(0)$, and we let $L_c^k(01)$ and $L_c^k(11)$ be the two preimages of $L_c^{k+1}(1)$. With $f_c(z) = z^2 + c$ as usual and $c = 0$ we have

$$\begin{aligned} f_c(L_c^k(00)) &= f_c(L_c^k(10)) = L_c^{k+1}(0), \\ f_c(L_c^k(01)) &= f_c(L_c^k(11)) = L_c^{k+1}(1). \end{aligned}$$

Furthermore, we order these cells counterclockwise starting at angle $\alpha = 0$ so that the labels appear in the order $00, 01, 10$ and 11. Thus, we have indeed defined a refinement of the stage-1 cells into stage-2 cells with

$$\begin{aligned} L_c^k(00) \cup L_c^k(01) &= L_c^k(0), \\ L_c^k(10) \cup L_c^k(11) &= L_c^k(1). \end{aligned}$$

There are 2^n stage-n cells in a level set. They have n-digit labels $b_n \ldots b_1$. When these stage-n cells are known, the stage-$(n+1)$ cells can be defined. A stage-n cell $L_c^{k+1}(b_n \ldots b_1)$ of the next level has two preimages which are the stage-$(n+1)$ cells

$$L_c^k(0b_n \ldots b_1) \text{ and } L_c^k(1b_n \ldots b_1).$$

Of these two cells, the first one, $L_c^k(0b_n \ldots b_1)$ comes first in counterclockwise orientation starting at $\alpha = 0$. The other one $L_c^k(1b_n \ldots b_1)$ is just $L_c^k(0b_n \ldots b_1)$ reflected at the origin. Binary decomposition is clearly a recursive process. Thus, if we start with a stage-1 decomposition of say L_c^1, this defines the stage-2 cells of L_c^0, which in turn define the stage-3 cells of L_c^{-1}, etc. This is shown in figure 13.28.

[17] The idea of binary decomposition was introduced in H.-O. Peitgen, F. v. Haeseler, D. Saupe, *Cayley's problem and Julia sets*, Mathematical Intelligencer 6.2 (1984) 11–20 and H.-O. Peitgen and P. Richter, *The Beauty of Fractals*, Springer-Verlag, 1986.

[18] These sets are not disjoint; they intersect along the real line. But that is not relevant for our discussion.

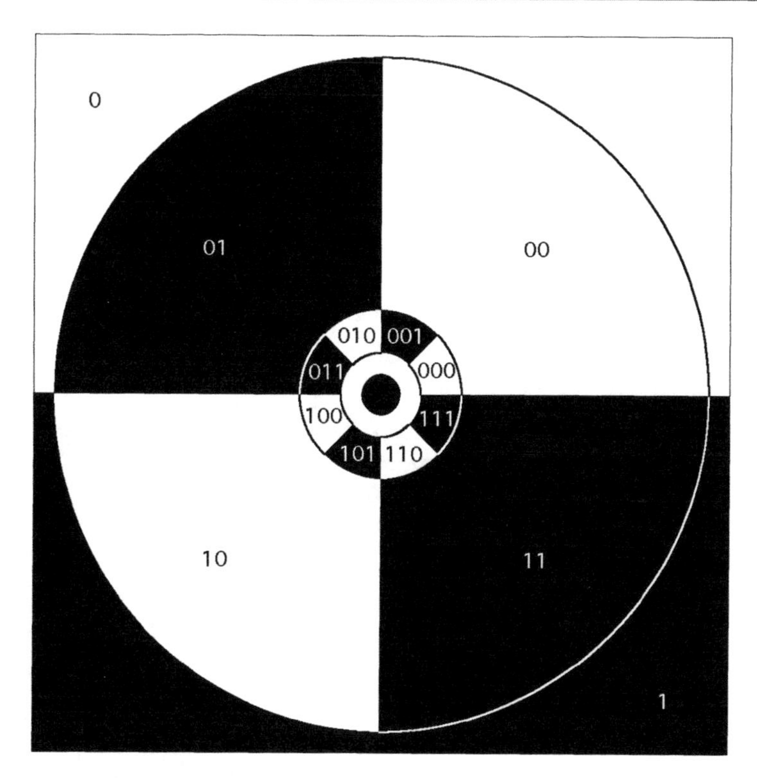

Binary Decomposition for $c = 0$

The small black disk in the center is the unit disk, the large circle has radius 16. Pictured are the two stage-1 cells of level set L_c^3, the four stage-2 cells of L_c^2, and the eight stage-3 cells of L_c^1. Cells are shaded according to their labels; even labels make white cells, odd labels give black cells.

Figure 13.28

Labels Show Dynamics

Note that we have chosen the labeling of cells in such a way that the transformations of cells can be read directly from the labels: The cell of stage n with label $b_n \dots b_1$ from the kth level set is transformed by $z \longrightarrow z^2$ into the cell of stage $n-1$ with label $b_{n-1} \dots b_1$ in the next level set. In terms of a formula we have

$$f_c(L_c^k(b_n bn - 1 \dots b_1)) = L_c^{k+1}(b_{n-1} \dots b_1).$$

The effect of applying the quadratic transformation to a cell is represented by a shift of the labels to the left (and dropping the first digit).

Approximation of Field Lines

Binary decomposition allows us to approximate arbitrary field lines of the potential. As we proceed to higher and higher stage subdivisions of level sets, the labeling of these cells converges to the binary expansion of the angles of the field lines passing through the cells. For example let us identify the field line ϕ_α with binary expansion $\alpha = 0.01011\dots$. This field line must pass through the cells labelled 0, then 01, then 010, then 0101, then 01011, and so on. In figure 13.29 you can try to read off the angles for $\alpha = 0.\overline{01} = 1/3$ and for $\alpha = 0.\overline{10} = 2/3$.

Binary Decomposition of Level Sets — $c \neq 0$

Let us now allow the parameter c to vary. If $|c|$ is small then the prisoner set is close to a disk, and the field lines and equipotentials should be about the same as for the unit disk at parameter $c = 0$. In fact, essentially everything

Binary Decomposition and Field Lines

Binary decomposition allows us to identify the angle of field lines. Shown are the two angles 1/3 and 2/3 with binary expansions 0.0101...and 0.1010..., which can also be read off from the labels of the cells that the field line passes through.

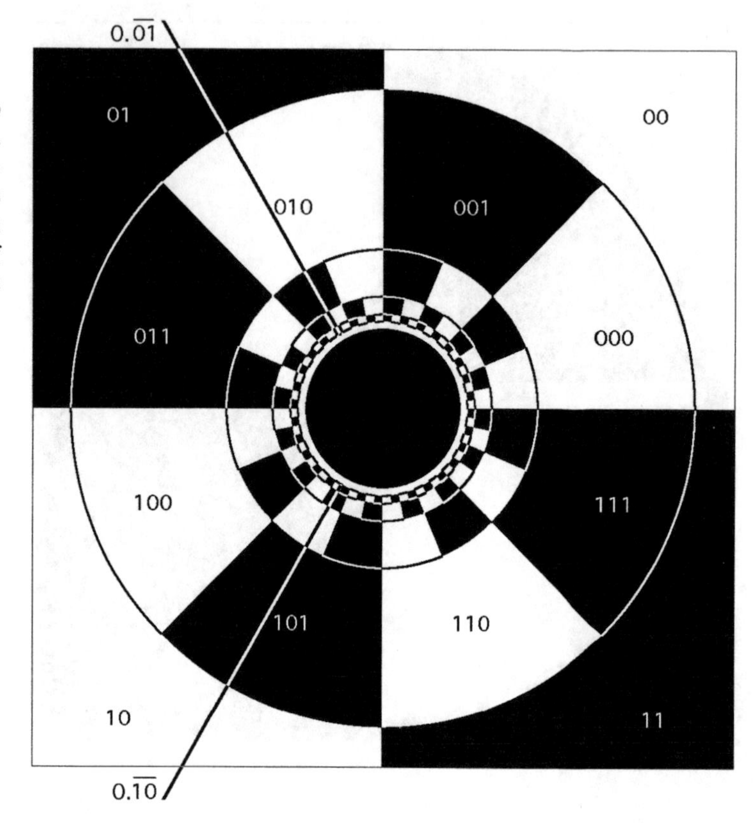

Figure 13.29

follows the same scheme as for $c = 0$. First, we divide L_c^k into $L_c^k(0)$ and $L_c^k(1)$ just as before. But when computing the stage-2 cells we encounter a complication: when continuing with the construction exactly as for $c = 0$ it is neither true that

$$L_c^k(00) \cup L_c^k(01) = L_c^k(0)$$

nor can we find a cell which is aligned precisely at the x-axis providing a natural start for the labeling of cells. The problem is due to the fact that the field lines are generally not straight lines, and we want to subdivide level sets along field lines. Only in the limit, far beyond the prisoner set, do field lines become approximately straight. Therefore, the solution to the problem is to consider starting our decomposition for a level set L_c^m with m sufficiently large. We take the intersection of L_c^m and the positive part of the real-axis as line segment AB, which is equal to, or at least very, very close to a portion of a field line. We may associate this field line with the angle $\alpha = 0$ because far away from the origin the constant c does not matter much in the iteration of f_c, and thus the situation is the same as in the potential of the unit disk for which the zero angle field line is on the positive real axis.

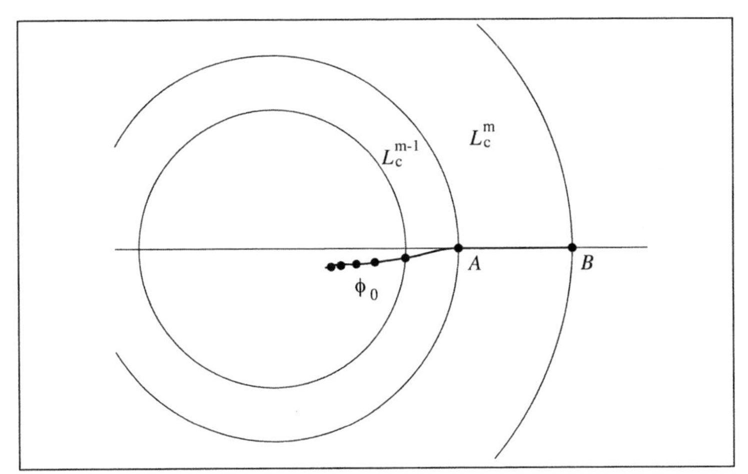

Angle 0

ϕ_0 is constructed as a chain of curve segments which are fitted from the (iterated) preimages of the line segment AB.

Figure 13.30

By construction we have that

$$f_c(A) \approx B$$

where the approximation can be made as precise as desired by considering an initial level set L_c^m, with m sufficiently large. The line segment AB has two preimages, which are curves that lie in the level set L_c^{m-1}. One of the preimages of B is A, and thus, one of the two preimage curves links up with the segment AB. Next we can consider the preimages of that curve; one of them links up with the curve (see figure 13.30). Continuing this process of fitting curve segments together we form a chain of curves which approaches the prisoner set P_c. This is the field line ϕ_0 with angle $\alpha = 0$. Having produced this crucial field line we can start to subdivide the level sets and label the cells starting with the cell which is attached to the field line with angle 0.

But there is yet another complication. Generally (for $c \neq 0$) it is not true that a cell has two disjoint preimages, cells of the next level closer to the prisoner set. But this is required for the construction of the binary decomposition of level sets. It turns out that this requirement is true for all parameter values c corresponding to connected Julia sets. We discuss disconnected Julia sets at the end of this chapter.

Decoding the Structure of Julia Sets

We are now prepared to apply the above method to decompose the levels in the electrostatic potential of non-trivial (i.e., for $c \neq 0$) connected prisoner sets. Our first choice is $c = -1$. Figure 13.31 shows how the binary decomposition of escape set E_c works in this case. First of all, one of the complications mentioned above is not yet present in this example. The field line with angle 0 is a part of the positive real axis, as is the case for the unit disk and $c = 0$. This is true because $c = -1$ is a real parameter, and preimages of real numbers $a > 0$ under $f_c(z) = z^2 - 1$ are real.

The cells in the level sets are coded by the same rules as for the special case $c = 0$, and in figure 13.31 we have assigned binary labels going coun-

Binary Decomposition for $c = -1$

The two field lines with angles $1/3 = 0.0101\ldots$ and $2/3 = 0.1010\ldots$ are shown.

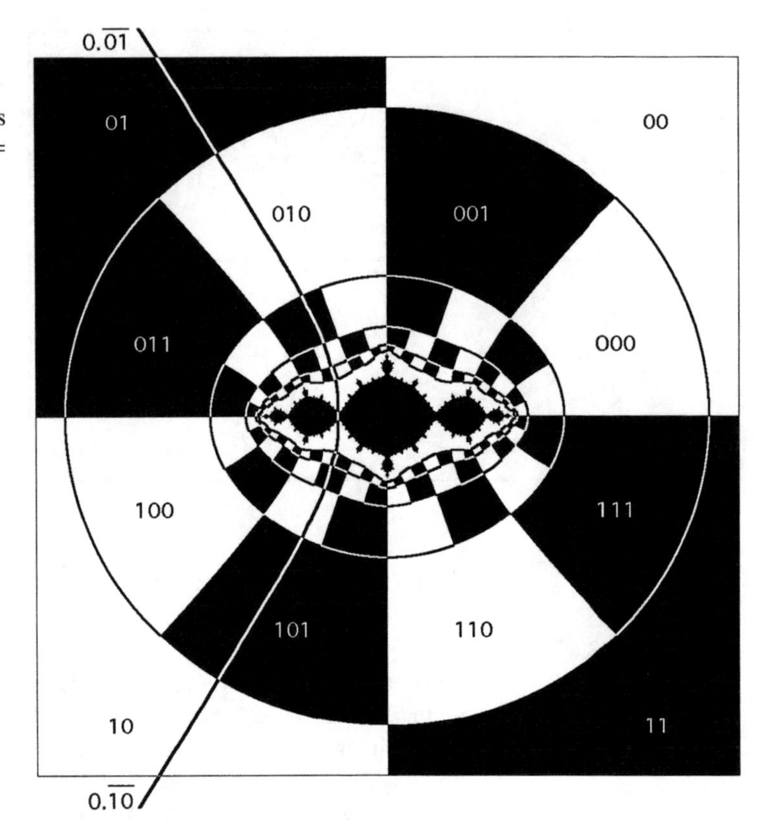

Figure 13.31

terclockwise starting at the crucial field line with angle $\alpha = 0$. Based on the construction we can approximate field lines, read off the binary expansion of their angles and determine how a field line is transformed under $z \rightarrow z^2 + c$. For example, we may read off the binary expansion

$$\alpha = 0.d_1 d_2 d_3 \ldots d_n \ldots$$

for the field line ϕ_α from the binary decomposition. Now the question is, to which field line ϕ_β will it be transformed? Well, obviously we have chosen the labeling of the cells of our binary decomposition appropriate to the dynamics of $z \rightarrow z^2 + c$. Thus, the line ϕ_β will be exactly the field line for the angle

$$\beta = 0.d_2 d_3 \ldots d_n \ldots .$$

In other words, we have just confirmed one of the crucial observations of Douady and Hubbard: from the point of view of field line dynamics in E_c the dynamics of $z \rightarrow z^2 + c, c \neq 0$, acts like angle doubling, just as for $c = 0$.

Equivalence to $z \rightarrow z^2$

Using the Riemann mapping theorem Douady and Hubbard showed in a mathematically rigorous way that as long as the prisoner set P_c is connected, the dynamics of $z \rightarrow z^2 + c$ in the escape set E_c is equivalent to the dynamics of $z \to z^2$ outside the unit disk. The binary decomposition of level sets provides a visual approach to this work and allows us to follow Douady and Hubbard for a while on their way of understanding Julia sets (and even the Mandelbrot set).

Pinching Model for $c = -1$

Let us continue our discussion of pinching disks by field lines started in section 13.5. First we look at the Julia set for $c = -1$. We already discussed the field line ϕ_0 which ends in the repelling fixed point $z_2 = (1 + \sqrt{5})/2$. Now we consider the field line with angle $\alpha = 1/2$. According to field line dynamics, this field line must be transformed into the field line with angle 0. Therefore, it must land at a preimage of z_2. The two preimages of z_2 are z_2 itself (it is a fixed point) and $-z_2$. Thus, the field line $\phi_{1/2}$ lands at $-z_2$.

We also discussed the field lines $\phi_{1/3}$ for $\alpha = 1/3$ and $\phi_{2/3}$ for $\alpha = 2/3$ ending in the second fixed point $z_1 = (1 - \sqrt{5})/2$, which is also repelling. Thus, the prisoner set P_c had to be pinched at z_1. The two field lines $\phi_{1/3}$ and $\phi_{2/3}$ are transformed into each other by $z \to z^2 + c$.

We can continue to understand the pinching process by just looking for field lines which are transformed into $\phi_{1/3}$ and $\phi_{2/3}$, or by looking for the iterated preimages of z_1. Exploiting field line dynamics we find more places to pinch the unit disk, as listed in table 13.32.

α	angle	binary	β	angle	binary
α_0	0	0.0	α_0	0	0.0
α_1	$1/2$	0.1	α_0	0	0.0
α_2	$1/3$	$0.\overline{01}$	α_3	$2/3$	$0.\overline{10}$
α_3	$2/3$	$0.\overline{10}$	α_2	$1/3$	$0.\overline{01}$
α_4	$1/6$	$0.0\overline{01}$	α_2	$1/3$	$0.\overline{01}$
α_5	$5/6$	$0.1\overline{10}$	α_3	$2/3$	$0.\overline{10}$
α_6	$5/12$	$0.01\overline{10}$	α_5	$5/6$	$0.1\overline{10}$
α_7	$7/12$	$0.10\overline{01}$	α_4	$1/6$	$0.0\overline{01}$
α_8	$5/24$	$0.001\overline{10}$	α_6	$5/12$	$0.\overline{01}$
α_9	$7/24$	$0.010\overline{01}$	α_7	$7/12$	$0.10\overline{01}$

Table 13.32 : Field lines with angle α are transformed to field lines with angle β using $z \to z^2 + c$ with $c = -1$. Corresponding field lines with pinching are shown in figure 13.33.

They explain the pinching in figure 13.33. If we start with a disk and pinch first following field lines $\phi_{1/3}$ and $\phi_{2/3}$, then following field lines $\phi_{1/6}$ and $\phi_{5/6}$, and so on, we obtain a model of P_c with $c = -1$.

Pinching Model for $c = i$

Let us finally turn to figure 13.34 to carry out the procedure for a different example, $c = i$. Note that in this case the field line with angle 0 is not a straight line. From the figure we may guess the binary expansion for the field line landing in the point $z_0 = i$, and obtain $\alpha = 0.0\overline{01} = 1/6$. Now note that in terms of angles the transformation

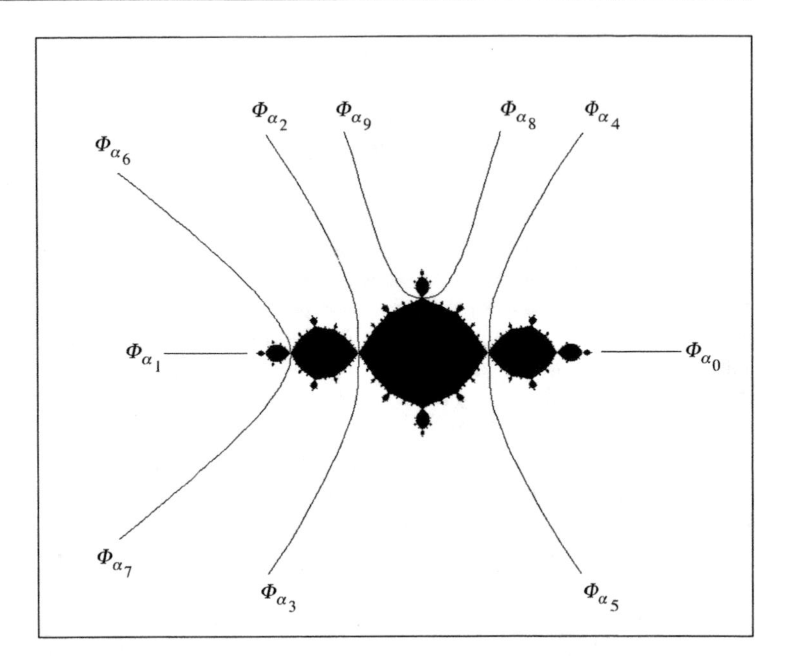

Figure 13.33 : Pinching of a disk by field lines which meet in a repelling fixed point and their iterated preimages.

of this field line under $z^2 + i$ gives

$$1/6 \to 1/3 \to 2/3 \to 1/3.$$

If we follow the dynamics of $z_0 = i$ under $z \to z^2 + i$, we obtain

$$i \to -1 + i \to -i \to -1 + i$$

which is exactly the same dynamics — in both cases a pre-periodic point of period two! This was expected, of course, and confirms our guess of the angle 1/6.

Moreover, just as in the case of $c = -1$, there are two repelling fixed points

$$z_{1,2} = \frac{1 \pm \sqrt{1 - 4i}}{2}$$

One of these fixed points corresponds to the fixed point $z = 1$ for $z \to z^2$ and therefore has angle $\alpha = 0$. The other one is a point at which three field lines meet and consequently these three field lines $\phi_{1/7}$, $\phi_{2/7}$, $\phi_{4/7}$ must be permuted among themselves, thus we compute that $0.\overline{001} = 1/7$, $0.\overline{010} = 2/7$, and $0.\overline{100} = 4/7$.

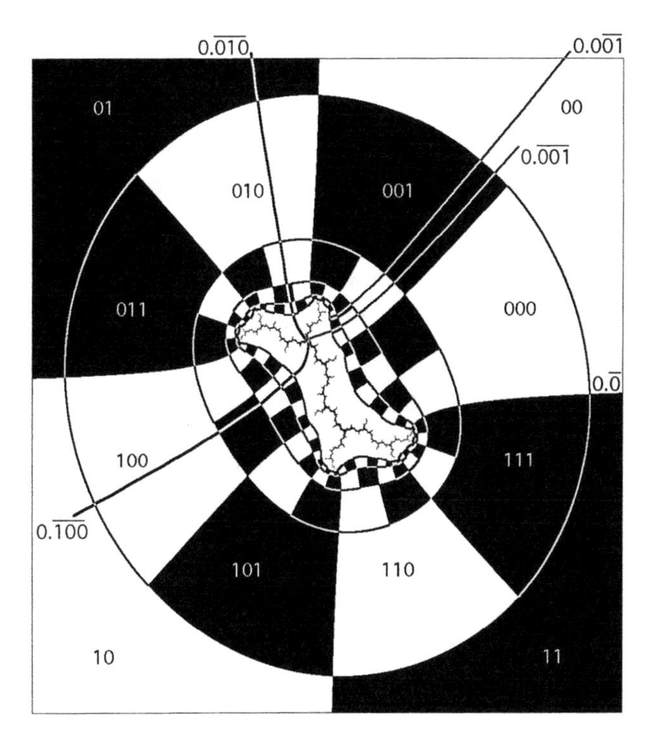

Figure 13.34 : Binary decomposition for $c = i$ and field lines with angle $1/6$ (landing in point $z_0 = i$) and $1/7$, $2/7$, and $4/7$.

13.7 Chaos Game and Self-Similarity for Julia Sets

The method for the computation of the prisoner sets and the Julia sets as their boundaries using the pixel game is quite slow. Even on modern PCs and workstations, the resulting image will not appear in a matter of seconds. To obtain a faster, but usually sufficiently detailed, picture of the Julia set we can use the chaos game as introduced in chapter 6. For this purpose we need a set of transformations that when applied iteratively to a point will generate an approximation of the Julia set. In contrast to chapter 6, however, these transformations cannot be affine transformations because one look at our figures reveals that Julia sets do not possess the affine self-similarity properties characteristic of iterated function systems.

Julia Sets as Attractors

The idea of the chaos game for Julia sets is not as far fetched as we might think. When points from the escape set close to the Julia set are iterated using the rule $z \longrightarrow z^2 + c$, they move away from the Julia set and approach infinity. Thus, we may call the Julia set a repeller with regard to the transformation $z \rightarrow z^2 + c$. Now, we can think about the inverted transformations, i.e., the transformations that take a point w to the point z, where $w = z^2 + c$, thus effectively iterating backwards. With respect to such inverse transformations the character of the Julia set must be different, it cannot be a repeller anymore but now plays the role of a new kind of attractor. The situation is completely analogous to the Multiple Reduction Copy Machine (MRCM), where the attractor in the chaos game is also a more or less complicated geometric point set in the plane. Now the remaining question is: what are the correct inverse transformations, that must be applied for this kind of chaos game?

Two Nonlinear Transformations

The problem to be solved is the following: given a complex number w, what are its preimages z such that $w = z^2 + c$? This is an exercise in equation solving for

$$z^2 - w + c = 0.$$

We need not go back to the general formula for quadratic equations from section 13.3 since the above equation does not contain a linear z-part. We simply rewrite it as $z^2 = w - c$ and apply the complex square root to obtain

$$z_{1,2} = \pm\sqrt{w - c}.$$

Taking the plus-minus sign into account we have two solutions to the equation. This means there are generally two preimages z for each point w (except for $w = c$). Thus, there are two transformations in the MRCM:[19]

$$w \ \longrightarrow \ +\sqrt{w - c} \ \text{ and } \ w \ \longrightarrow \ -\sqrt{w - c}$$

The chaos game for these two transformations then works as usual. First, we pick an arbitrary initial point w. Then one of the two preimages according to the above formulas is selected at random and w is replaced by this preimage and shown on the monitor. The process is repeated until enough points have

[19] Heuristically, the situation is quite clear. Taking square roots, for example, $\sqrt{100}$, means to reduce, at least for large numbers.

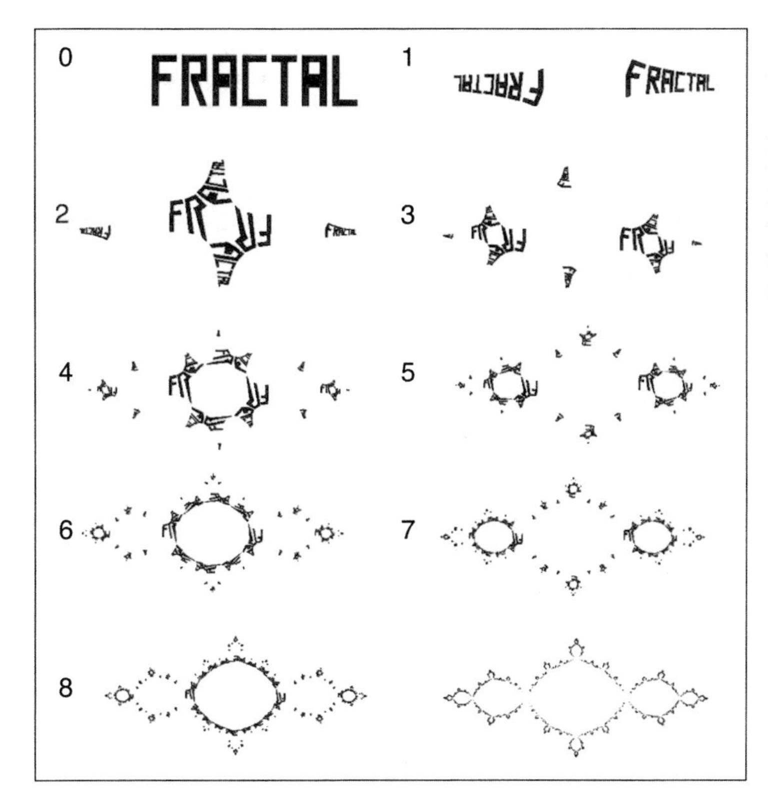

MRCM for Julia Set

The MRCM with the two nonlinear lenses $\pm\sqrt{w-c}$ with $c = -1$ is applied to an initial image consisting of the sequence of letters 'FRACTAL'. In each step two deformed copies of the input image are composed which rapidly converge to the corresponding Julia set.

Figure 13.35

been collected. The first few computed points should not be displayed as they come from the transitional period necessary for the random initial point to be attracted close enough to the Julia set. We can skip this phase if we start directly on the Julia set. Fortunately there is one point of the Julia set which can be easily determined: a repelling fixed point.[20]

A fixed point can be computed directly from the equation $z = z^2 + c$, without iterating, so

$$z^2 - z + c = 0$$

must be solved. Let us do this for $c = -0.5+0.5i$. This is exactly the exercise from section 13.3. The result was two solutions

$$\begin{aligned} z_1 &= 1.408 - 0.275i, \\ z_2 &= -0.408 + 0.275i. \end{aligned}$$

The second solution z_2 is an attracting fixed point and certainly not a point of the Julia set. However, the other fixed point z_1 is a repeller (i.e., points nearby are pushed away by the iteration); and z_1 is in the Julia set.

[20] It is already known from the work of Julia and Fatou that any repelling periodic point belongs to the Julia set. In fact, the repelling periodic points are dense in the Julia set.

A Derivative Criterion for Repelling and Attracting Fixed Points

The distinction between attractive and repelling fixed points can be made mathematical so that one does not have to test numerically for the attractive or repelling property. However, this test involves a tool from calculus because the essence of the method lies in the derivative of the transformation $z \to z^2 + c$ at the fixed point z. This derivative is computed the same way as for real functions; it is $2z$. Given this number the fixed point can be classified: the fixed point is attractive if the absolute value of the derivative at that fixed point is less than 1; it is repelling if the derivative is greater than 1 in absolute value. The remaining case, in which the absolute value of the derivative is exactly 1 is undecided; such fixed points are called indifferent and have been the source of rather deep mathematical research. The criterion is also applicable for all other underlying transformations besides the quadratic $z \to z^2 + c$. The proof of this criterion is not difficult, and we omit it.

In our example of the quadratic transformation with $c = -0.5+0.5i$ and its two fixed points $z_1 = 1.408-0.275i$ and $z_2 = -0.408+0.275i$, we compute the absolute values of the derivatives up to three decimals of precision

$$|2z_1| = |2.816 - 0.550i| = \sqrt{2.816^2 + 0.550^2} = 2.869$$

and

$$|2z_2| = |-0.816 + 0.550i| = \sqrt{0.816^2 + 0.550^2} = 0.984.$$

The result is clear; the first fixed point z_1 is repelling, and the other one is attracting.

Depending on the choice of the parameter c, the performance of the chaos game algorithm may be more or less satisfactory. In some cases there are regions in the Julia set that are hard to get to. Then the Julia set may look as if it is composed of several parts, while it really is connected. There are modifications possible to improve the method, called the *Modified Inverse Iteration Method*.[21] The next figure illustrates the progress of the algorithm as more and more points are computed.

The Invariance of Julia Sets

Julia sets can be seen as attractors in the chaos game, and this reveals an important fact about the self-similarity of Julia sets. Recall that for the attractor of an MRCM the whole attractor is covered by small copies of itself. These small copies are nothing but the images of the whole attractor under the transformations in the iterated function system.[22] The same property should be true here. This says specifically that as we apply one of the two transformations of our iterated function system for the Julia set

$$w \to +\sqrt{w-c}$$
$$w \to -\sqrt{w-c}$$

[21] See page 178 in *The Science of Fractal Images*, H.-O. Peitgen, D. Saupe (eds.), Springer-Verlag, New York, 1988.

[22] Recall from chapter 5 that if $w_1, w_2, \ldots, w_n$ are the contractions and A is the attractor, then $A = \bigcup_{k=1}^{n} w_k(A)$.

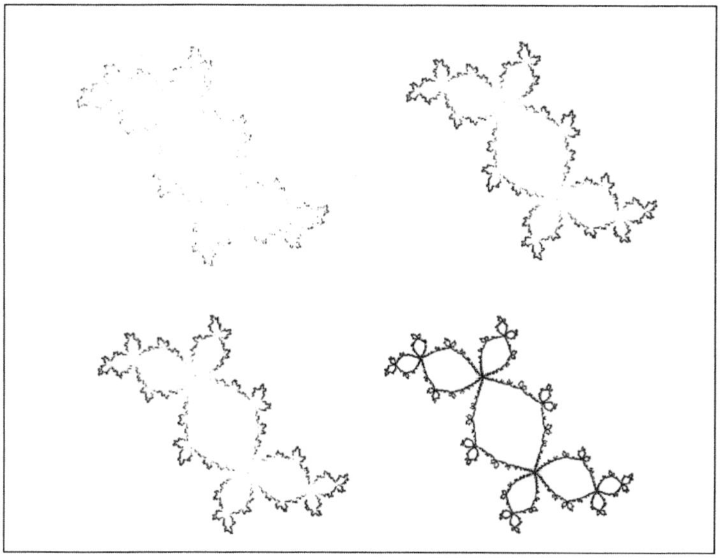

Progress of the Chaos Game

Computation of the Julia Set for $c = -0.12 + 0.74i$ (termed the *rabbit* by Douady and Hubbard). Although the performance of the Chaos Game is in this case not too satisfactory, a first overview of the Julia set appears rather rapidly. The top left image shows 1,000 points, the top right one shows 10,000. In the bottom left one even 100,000 points of the Chaos Game are plotted. For the bottom right image the Modified Inverse Iteration Method was used. It requires only 4,750 points.

Figure 13.36

to any point w of the Julia set, then we obtain yet another point of the Julia set. Therefore, the Julia set is called an *invariant set* with respect to the inverse transformations of $z \to z^2 + c$.

Moreover, if again z is a point from the Julia set, we may ask the question: What kind of point is the image of z, i.e., $z^2 + c$? It cannot be in the basin of attraction of infinity because then the initial point z would also have to be an escaping point, but z was chosen in the Julia set. On the other hand, $z^2 + c$ cannot be in the interior of the prisoner set; it must be on the boundary. The reason for this lies in the continuity of the quadratic transformation: arbitrarily close to z there are escaping initial points, and continuity of $z \to z^2+c$ implies that this neighborhood relation must also hold for the set of transformed points. Summarizing, this observation is expressed in the statement that the Julia set is invariant not only with respect to the inverse transformations $w \to +\sqrt{w-c}$ and $w \to -\sqrt{w-c}$, but also with respect to the transformation $z \to z^2 + c$ itself.

The Self-Similarity of Julia Sets

Thus, the Julia set remains invariant under forward iteration (using $z \to z^2 + c$) as well as under backwards iteration (using either of the two inverse transformations). This property is called *complete invariance* and describes one of the key properties of Julia sets. Therefore, the global structure of the Julia sets must also appear in the images and preimages of the Julia set, which explains the apparent self-similarity. This self-similarity is not to be confused with the strict or affine self-similarity that is discussed in earlier chapters of this book. The similarity here is based on a *nonlinear* transformation, and thus, the smaller copies of the Julia set contained in itself are not exact copies but strongly distorted ones, which are even folded back on themselves.

Self-Similarity of a Julia Set

The self-similarity of the Julia sets. These two pictures show how a very small section of the Julia set, denoted by R_{-7}, is transformed several times. In each transformation the covered portion of the Julia set indicated by the bold black parts labeled R_{-6} to R_{-1} increases. After six iterations the result R_{-1} is already one half of the Julia set; one more application of $z \to z^2 + c$ yields the whole set R_0.

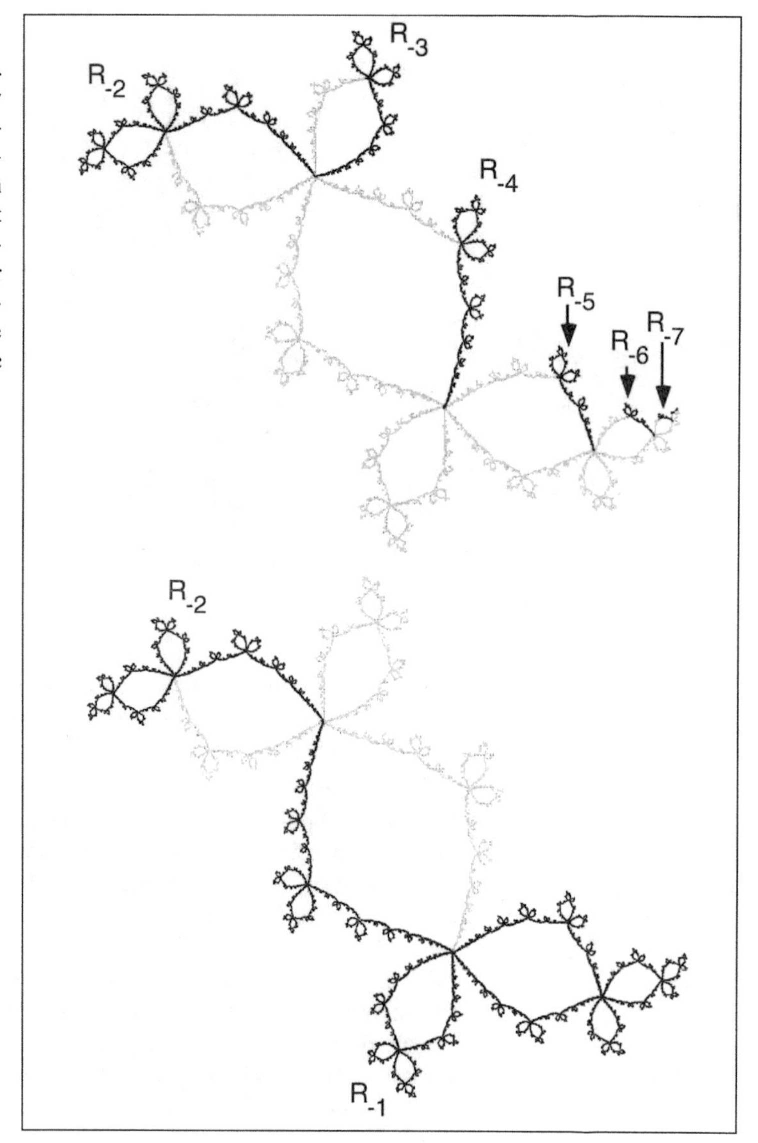

Figure 13.37

Nevertheless, the following amazing property has been shown to be true (see figure 13.37). Take any small section of the Julia set, e.g., intersect a small disk with the Julia set and assume that this intersection is not empty. Then we apply the iteration $z \to z^2 + c$ to every point in this set. We obtain a new, typically larger, subset of the Julia set. Iterating this procedure a *finite* number of times will result in the complete Julia set! This says that the immensely complicated global structure of the Julia set is already contained in any arbitrarily small section of it.

13.8 The Critical Point and Julia Sets as Cantor Sets

We have visualized Julia sets using encirclements $Q_c^{(k)}$ and $P_c^{(k)}$ leading to the powerful tools of potential functions and field lines. These, however, apply only to connected Julia and prisoner sets. Let us now develop an understanding of the reasons for this limitation.

For this purpose we return for a moment to the discussion of the iterative behavior of $x \to ax(1-x)$ for real numbers a and x. You will recall that the dynamics of this iteration is equivalent to that of $z \to z^2 + c$, again for real numbers z and c. If $c \leq 1/4$ we can translate one into the other by

$$x_n = \frac{1}{2} - \frac{z_n}{a} \text{ and } a = 1 + \sqrt{1-4c}. \tag{13.4}$$

We will use this equivalence later to interpret the discussion that follows. What are the prisoner sets for $x \to ax(1-x)$, and how do they change with the parameter a. Let us adapt our encirclement experiments to the case of interest. We start with some large interval $Q_a^{(0)}$ guaranteed to enclose the prisoner set P_a. Then we look at the iterated preimages $Q_a^{(-k)}$, which in the limit tend to the prisoner set P_a. This would not be difficult by explicit calculation, but we prefer to carry out the evaluation graphically to make the issue visually apparent. For this task we introduce *backward graphical iteration.* We are given a value, say y, and we want to find x so that $ax(1-x) = y$. This amounts to solving a quadratic equation which we do graphically as follows.

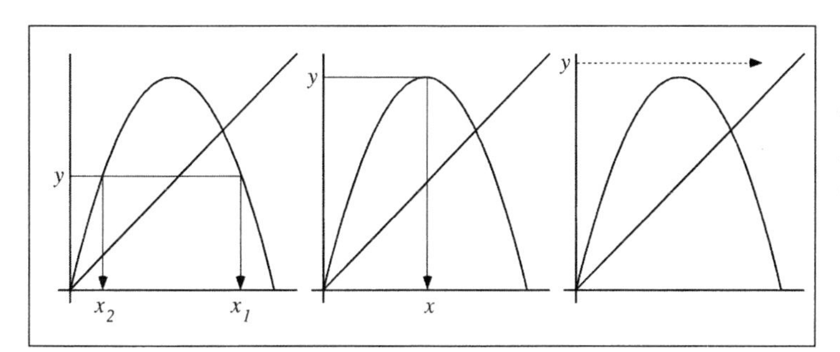

Backward Iteration — One Stage

Three cases for the backward iteration: two solutions, one solution, and no solution.

Figure 13.38

Backward Graphical Iteration

We pick y on the y-axis and draw the horizontal line through this point. It may intersect the graph of $g(x) = ax(1-x)$ in two points, one point, or no point at all. Then we draw vertical lines from these intersection points to the x-axis. There we can read off the solutions to the equation. These are the preimages of y. The procedure really is nothing but carrying out the usual graphical iteration backwards. If we want to repeat this procedure, we draw the vertical lines only up to the diagonal and then draw horizontal lines again to meet the graph (see figure 13.39). In this way we generate backward orbits, which can be described by a tree: given y, we may find two, one, or no preimages. For each of those we may again find two, one, or no preimages, and

Backward Iteration — Two Stages

Backward iteration for two stages. There are two preimages of y and there are four points which, when iterated twice, land in y.

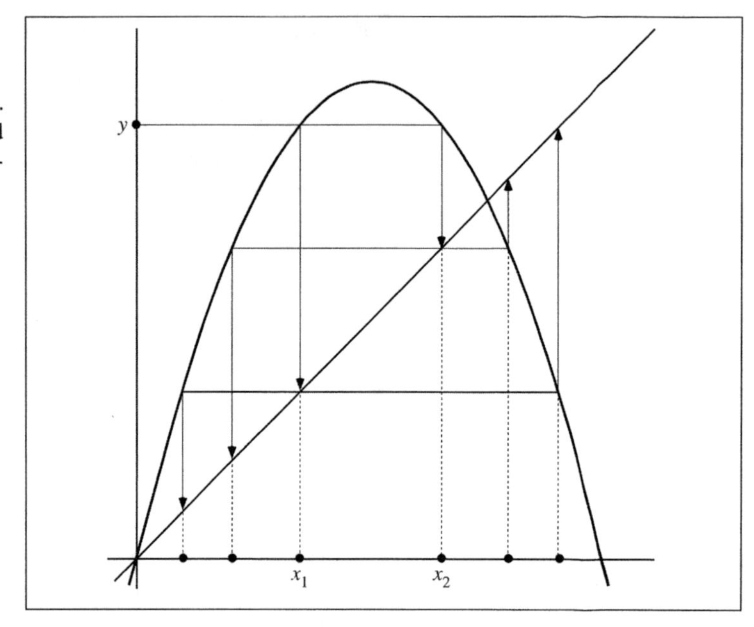

Figure 13.39

so on. When there is no preimage, then the tree is pruned at the corresponding branch.[23]

Now we use this method to find $Q_a^{(1)}$ when $Q_a^{(0)}$ is given as an initial large interval.[24] Figure 13.40 shows the result for $a = 3.5$. We observe that the $Q_a^{(-k)}$ are a nested sequence of intervals which shrink towards the unit interval; and in fact, if we restrict attention to real numbers only, then we may conclude that the prisoner set is $P_a = [0, 1]$.

Indeed, for this example ($a = 3.5$) we could have obtained the result much more cheaply. Observe first, that any $x_0 < 0$ leads to an orbit escaping to infinity. Then also orbits starting at $x_0 > 1$ escape because the first iterate is $x_1 < 0$, and thus the following iterates grow negatively without bound as in the first case. Moreover, an orbit for $0 \leq x_0 \leq 1$ cannot escape the unit interval because $0 \leq x_1 = ax_0(1 - x_0) \leq 1$.

However, if we take $a = 4.5$, for example, we cannot argue that way, although it is true that orbits starting outside the interval $[0, 1]$ do escape to infinity as before. Let us look at the backward graphical iteration of a large interval. Its preimage consists of *two* disjoint parts (refer to figure 13.41). Thus, we are led to conjecture that the prisoner set is not connected.

To investigate further let us iterate backwards a few times (see figure 13.42). We position several graphs of $g(x) = ax(1 - x)$ for $a = 4.5$ in such a way

[23] Using complex numbers in place of real ones, branches are never pruned since we always have two preimages (except for $y = c$ where only one preimage $x = 0$ exists). Thus, when computing the backward iteration in the (complex) chaos game it is always possible to choose randomly between two options.

[24] In fact, one can take any interval which encloses the unit interval $[0, 1]$.

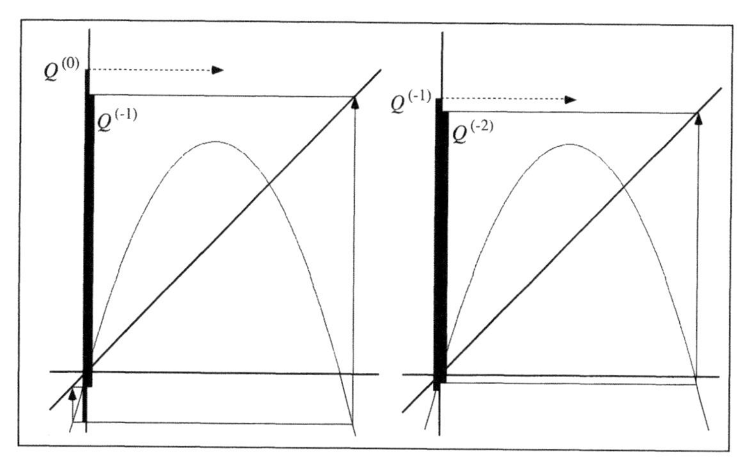

Backward Iteration of an Interval

Encirclement by backward iteration. Here an entire interval indicated by the thick line on the y-axis labeled $Q^{(0)}$ is iterated backwards once (left) and again (right). ($a = 3.5$)

Figure 13.40

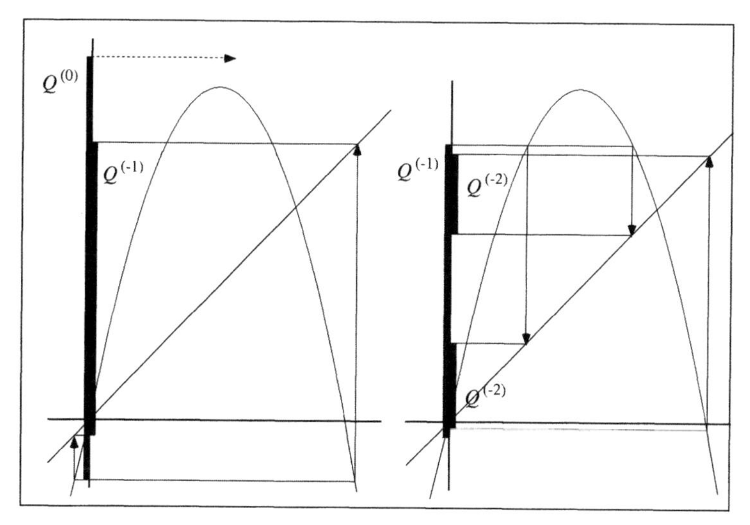

Encirclement by Backward Iteration

Two cycles of backward iteration of the interval $Q^{(0)}$ yield two intervals. ($a = 4.5$).

Figure 13.41

that the result of the first step feeds into the second step and so on. We make a very important observation: the resulting encirclements $Q_a^{(-i)}$ correspond to a Cantor set construction. In other words, P_a will be a Cantor set. Why do we call P_a a 'Cantor set'? Usually one refers to the Cantor set as an interval from which the (open) middle thirds are recursively removed. Thus, in the usual Cantor set all pieces in a given construction stage have the same length; and the resulting limit object is strictly self-similar. In our construction of encirclements we obtain something very similar, but the pieces of a given step have different sizes and the limiting object is not self-similar in the strict sense. It is a Cantor set — but slightly distorted.

Backward Iteration Generating Cantor Set

Encirclement of Cantor set through backward iteration of the unit interval. ($a = 4.5$)

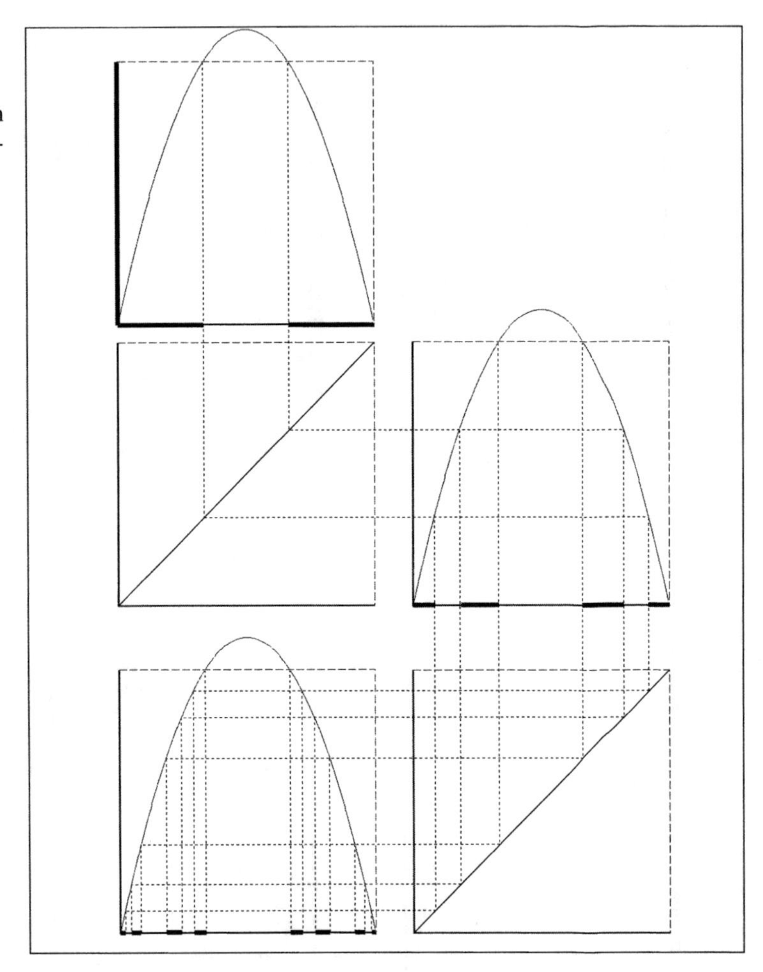

Figure 13.42

The Fate of the Critical Point Is Decisive

Let us summarize. We have just seen that for $a = 3.5$ the prisoner set on the real line is an interval, while for $a = 4.5$ the prisoner set is a Cantor set. In other words, as we increase a from 3.5 to 4.5 something comparable to a phase transition occurs. Exactly at which parameter a is the transition? To answer this question we observe that the disintegration of the preimage of the unit interval into two parts is caused by the fact that the vertex of the parabola of the graph of $g(x) = ax(1-x)$ is *above* the unit square $[0, 1] \times [0, 1]$. The coordinates of the vertex are $(x, y) = (1/2, a/4)$. The x-coordinate is usually called the *critical point*, while the y-coordinate is referred to as the *critical value* of the function $g(x) = ax(1-x)$. Both are characterized by the fact that the critical value has only one preimage, namely, the critical point (compare figure 13.38). The fate of the critical point, i.e., the long-term behavior of the

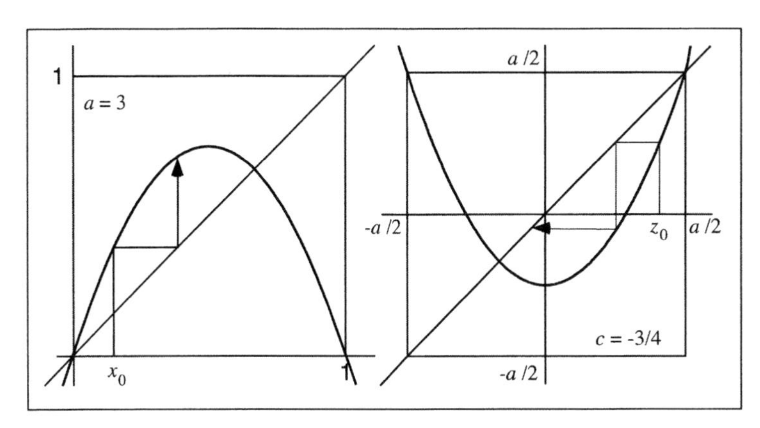

Essential Squares

Graphical iteration of $x \to ax(1-x)$ (left) and $z \to z^2 + c$ (right).

Figure 13.43

orbit of $x_0 = 1/2$, called the *critical orbit*

$$\frac{1}{2} \to \frac{a}{4} \to \frac{4a^2 - a^3}{16} \to \cdots,$$

determines whether P_a is an interval (one connected piece) or a Cantor set. When $a \leq 4$ then the critical value does not exceed 1, $a/4 \leq 1$, and thus the critical orbit remains in the unit interval for all iterations; P_a is an interval. However, if $a > 4$ then $a/4 > 1$, the critical orbit goes to $-\infty$, and P_a is a Cantor set.

Translation to $z^2 + c$

So far we have verified only the real case and the iteration $x_{n+1} = ax_n(1-x_n)$. Let us now begin to translate this into the iteration $z_{n+1} = z_n^2 + c$ for real z_n. Is there a square which corresponds to the unit square for the x_n-iteration? Indeed, this is easy to locate, using either the explicit transformation rules in eqn. (13.4) or by an analysis of the graphical iteration (see figure 13.43). The square here is determined by the far right intersection of the graph of $f(z) = z^2 + c$ with the diagonal, which is at the positive solution of $z^2 + c = z$, i.e.,

$$z = \frac{1 + \sqrt{1 - 4c}}{2}$$

as long as $c < 1/4$. This number is equal to $a/2$ according to eqn. (13.4). Thus, the square extends from $-a/2$ to $a/2$; and we call it the *essential square*.

What is the critical orbit here? Again we can use the transformation rules in eqn. (13.4) and compute the corresponding initial point: $z_0 = a(1/2 - x_0)$ for $x_0 = 1/2$, which is $z_0 = 0$. Alternatively, we can observe that the prisoner set P_c is the interval $[-a/2, a/2]$ as long as the minimum of the graph of $f(z) = z^2 + c$ does not exceed the essential square in figure 13.43. In other words, the critical orbit is now given by the orbit generated by the critical point $z_0 = 0$,

$$0 \to c \to c^2 + c \to \cdots$$

At $c = 1/4$

When c is large, all orbits escape to infinity (right). Precisely at $c = 1/4$ the parabola touches the bisector and some orbits do not escape (left).

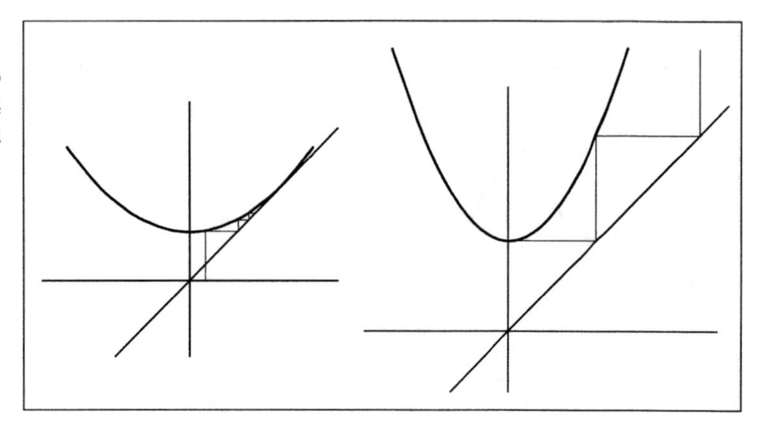

Figure 13.44

And again we have to check whether the critical orbit goes to infinity or not. Also note that if the critical value c leaves the essential square we will have a Cantor set for P_c as before.

This observation allows us to make a conjecture about all real c-values corresponding to connected prisoner sets. First, we observe that the minimum of the graph of $f(z)$ touches the base of the essential square when $c = -a/2$. Using the transformation rules, this equation is the same as

$$c = -\frac{1 + \sqrt{1 - 4c}}{2} \tag{13.5}$$

and also

$$\sqrt{1 - 4c} = -2c - 1.$$

Squaring the last equation yields

$$1 - 4c = 4c^2 + 4c + 1$$

and

$$4c^2 + 8c = 0$$

or, equivalently,

$$4c(c + 2) = 0.$$

Clearly, there are two solutions to this last equation, $c = 0$ and $c = -2$. But only the latter one solves the original eqn. (13.5).[25] Thus, the vertex of the parabola exceeds the essential square when $c < -2$. On the other hand, if c is very large then the parabola is completely above the bisector as in figure 13.44 (left), and then the critical orbit definitely goes to infinity. The transition between this case and the case where the critical orbit will remain

[25] The other solution $c = 0$ is an artefact of the squaring of the equation.

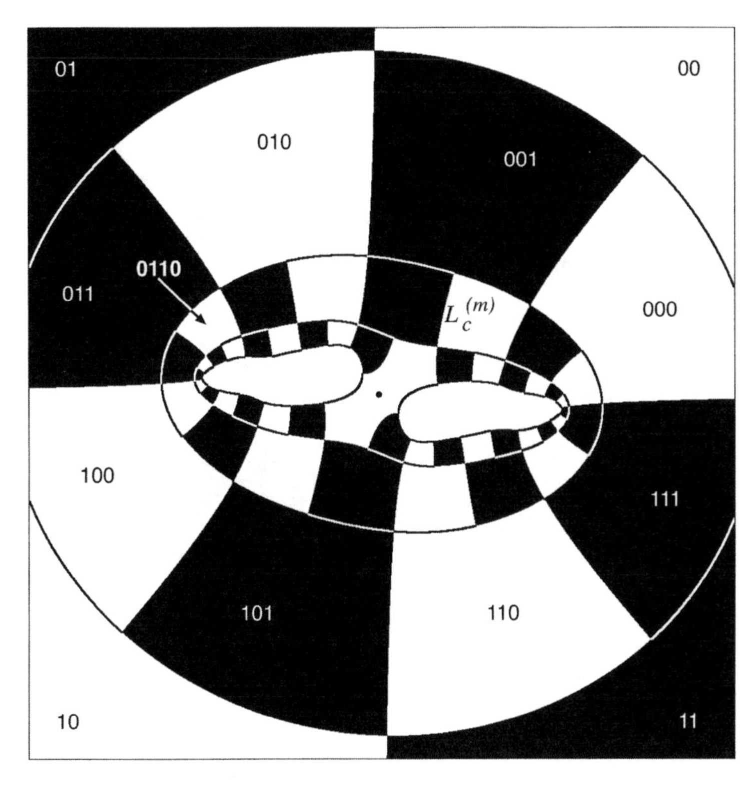

Figure-Eight Level Set

If not all preimages of the cells of $L_c^{(m)}$ are disjoint the level set $L_c^{(m-1)}$ forms a figure-eight shape.

Figure 13.45

in the prisoner set occurs at a certain parameter c. Precisely at this parameter c the parabola just touches the bisector (see the right hand graph in figure 13.44). This configuration occurs when the two points of intersection of the graph of $f(z)$ with the diagonal coincide, i.e., when

$$\frac{1+\sqrt{1-4c}}{2} = \frac{1-\sqrt{1-4c}}{2}.$$

This is the case when $4c = 1$, thus $c = 1/4$. In summary, we expect that the prisoner set P_c is connected provided c is in the interval $[-2, 1/4]$.

Complex Parameters c

Let us now take the last step and allow the parameter c to be complex and see how the critical orbit determines which Julia sets are Cantor sets. When we looked at the binary decomposition of level sets we made the assumption that each cell of, say, level set L_c^m would have two disjoint preimages. This provided the decomposition of L_c^{m-1}, which then again formed a ring of cells encircling the prisoner set. But what happens when the preimages of a cell are not disjoint? This is shown in figure 13.45. The cell labeled 0110 has only one cell as a preimage. Because of the symmetry with respect to 0, this preimage includes the critical point 0. Thus, the original cell with label 0110 must contain the critical value c. If L_c^{m-1} is glued together at one cell, this level set forms a kind of figure-eight shape. This figure-eight encircles two

A Cantor Set in the Complex Plane

For $c = -2.2$ the prisoner set is a Cantor set.

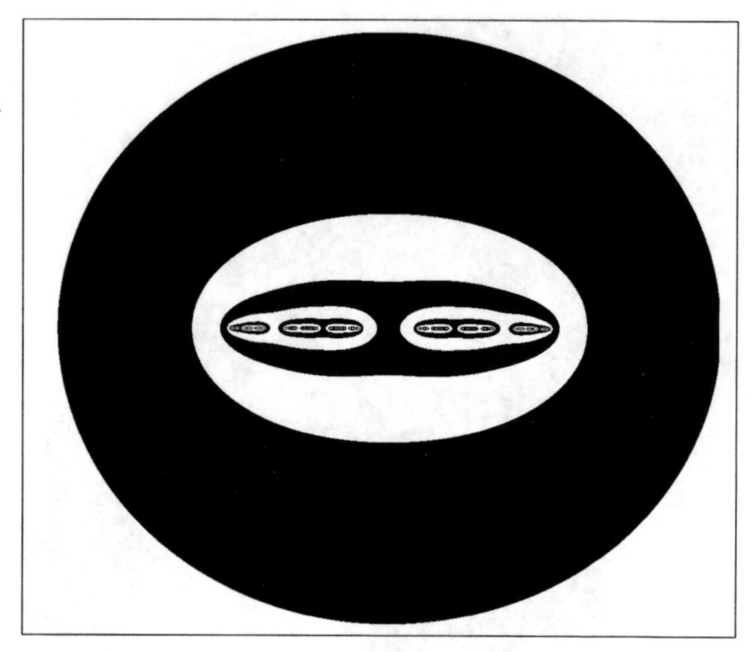

Figure 13.46

disjoint parts of $P_c^{(m-2)}$. Thus the prisoner set P_c decomposes into at least two parts.

Cantor Set Construction in the Complex Plane

Now we look at the level set $L_c^{(m-2)}$. It obviously also decomposes into two disjoint parts, and we find that each of them again encloses two parts of $P_c^{(m-3)}$, making four parts all together, each of which encloses a part of the prisoner set, and so on. We again observe, as in the case of real numbers, a typical Cantor set construction. In other words, in this construction we end up with a prisoner set which is a Cantor set in the complex plane (see figure 13.46).

Now what happens to the critical point 0 in this case? We have found that the critical point is contained in $L_c^{(m-1)}$. Therefore, its first iterate is in $L_c^{(m)}$, its second in $L_c^{(m+1)}$, and so on. In other words, its absolute value iterates towards infinity. We see that in the complex case, the fate of the critical point is decisive. In fact, these observations have been crucial for the understanding of what is now called the *structural dichotomy of Julia sets* in the complex plane and were explored in the early work of Julia and Fatou. Julia sets for $z \to z^2 + c$ are either connected or Cantor sets. This dichotomy will be a major theme in the last chapter.

Encircling Prisoner Sets and the Critical Point

Let us briefly demonstrate how the fate of the critical point becomes decisive for the structure of Julia sets. We have seen some initial arguments for the following facts:

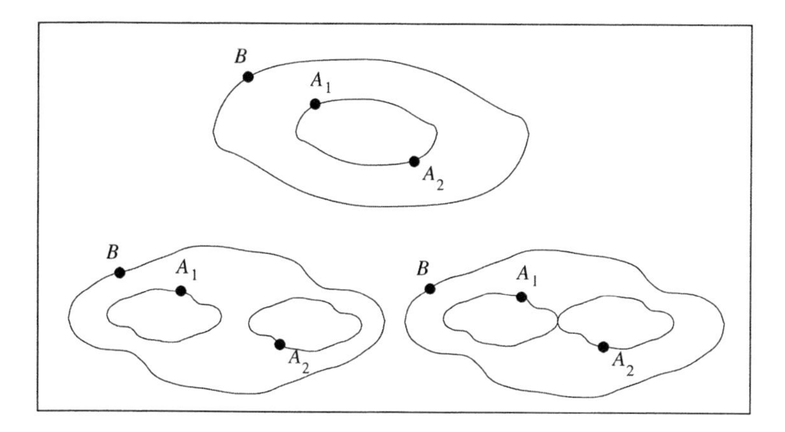

Figure 13.47 : The three possible preimages of a disk-like set.

- The Julia set is a Cantor set if, and only if, the iteration of the critical point 0 leads to infinity (in absolute value).
- The Julia set is one piece (connected) if, and only if, the iteration of the critical point 0 is bounded.

The prisoner set P_c is a subset of $P_c^{(m)}$ whose boundary curve, for sufficiently large m, is a circle (or at least as close to a circle as we would like). The encirclement $P_c^{(m)}$ is a connected set, and it is symmetric with respect to 0. Now, what is the shape of $P_c^{(k)}$ for $k < m$? Well, for k close to m it is certainly still very similar to a disk, but is this true for all values of k?

The crucial observation is that if $P_c^{(k)}$ is disk-like (i.e., a deformed disk), then there are exactly three possible cases for the next approximation $P_c^{(k-1)}$ of the prisoner set: $P_c^{(k-1)}$ is either

- disk-like (and contains the critical point 0),
- made up of two disjoint disk-like subsets (and does not contain the critical point),
- made up of two disk-like subsets which touch exactly at one point (the critical point 0), i.e., its boundary forms a figure-eight.

Assume we were to move a point B around the boundary of $P_c^{(k)}$. With respect to $z \rightarrow z^2 + c$, the point B has two symmetric preimages, namely, the two square roots $A_1(B)$ and $A_2(B) = -A_1(B)$ of $B - c$. Imagine the point B starting at B_0 makes a full turn, then the points $A_1(B)$ and $A_2(B)$ completely trace out the boundary of $P_c^{(k-1)}$, which must be symmetric with respect to the origin 0. Let us take a closer look at the trace of, say, $A_1(B)$. There are two cases. Either the root returns to the initial position $A_1(B_0)$ (case 1) or it ends at the symmetric point $A_2(B_0)$ (case 2).

Case 1. In the first case the root has circumscribed a disk-like set D_1. Since the other root is symmetric it encircles a set D_2, symmetric

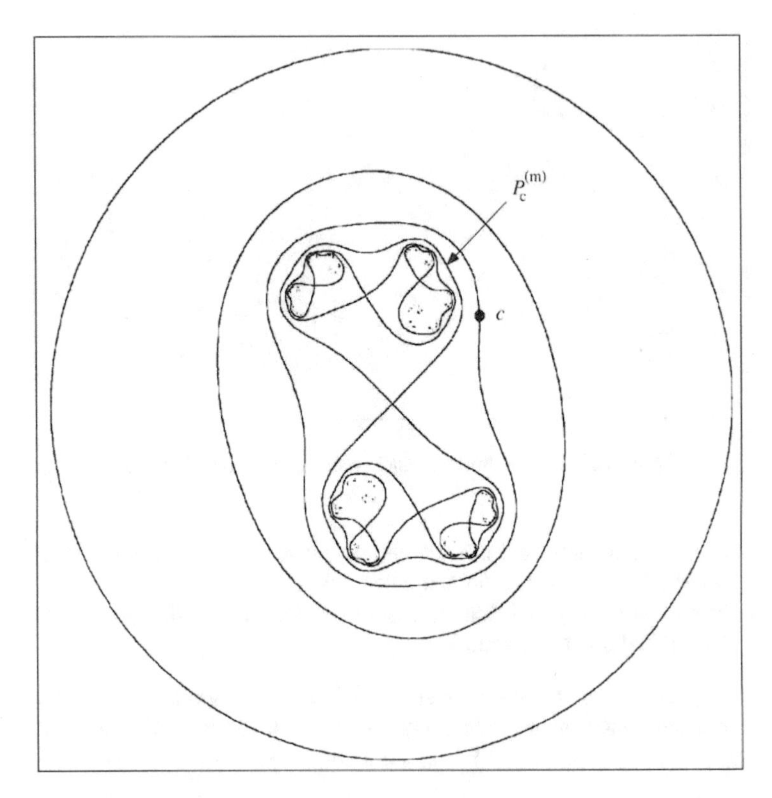

Figure 13.48 : This figure demonstrates the Cantor set property of a Julia set. The Julia set is located in the lobes of corresponding figure-eight shapes.

to D_1. If D_1 and D_2 are disjoint, then $P_c^{(k-1)}$ is made up of two disk-like sets; and the critical point 0 is not in $P_c^{(k-1)}$. If D_1 and D_2 are not disjoint, they can meet at just one point, namely, the critical point 0 (which then is on the boundary of $P_c^{(k-1)}$).

Case 2. One of the roots traces a path from $A_1(B_0)$ to $A_2(B_0)$, while the other root traces the symmetric path from $A_2(B_0)$ to $A_1(B_0)$. These paths cannot intersect, and, thus, together they form the boundary of another disk-like and symmetric set, $P_c^{(k-1)}$. In this case the critical point 0 is in the interior of $P_c^{(k-1)}$.

Figure 13.47 summarizes these three possible configurations.

Now let us assume that all $P_c^{(k)}$ are connected, disk-like sets. Then the critical point is included in all these sets. Since the prisoner set P_c is approximated by the sets $P_c^{(k)}$, we can conclude in this case that P_c is also connected and that the critical point is in the prisoner set (otherwise it cannot not be in all sets $P_c^{(k)}$). On the other hand, if the orbit of the critical point does not leave the prisoner set, then the critical point 0 is an element of all sets $P_c^{(k)}$; thus, they all must be disk-like.

Let us finally look at the alternative. Assume that while $P_c^{(k)}$ for $k > m$ is disk-like, $P_c^{(m)}$ is not, i.e., it is made up of two disjoint disk-like sets or a figure-eight shape. In this case, the critical point is either not in $P_c^{(m)}$ or it is on the boundary of $P_c^{(m)}$. In both cases, it is in the escape set. On the other hand, if the iteration of the critical point leads to infinity (in absolute value), then it cannot be an element of all $P_c^{(k)}$, which implies that not all these sets are disk-like. Now if $P_c^{(m)}$ is made up of two disjoint disk-like sets, $P_c^{(m-1)}$ must be made of four disjoint sets (one pair in the D_1 component and one pair in D_2). Then $P_c^{(m-2)}$ is made up of 8 components, and so on. This is a typical Cantor set construction. Figure 13.48 shows the construction for the case where $P_c^{(m)}$ is a figure-eight.

13.9 Quaternion Julia Sets

Complex numbers are a two-dimensional extension of real numbers. It is possible to extend the space of complex numbers further. However, the attempt by the Irish physicist and mathematician William R. Hamilton to create a space of numbers with three components failed. Instead he had to resort directly to a space of numbers with four components. This space, invented in the year 1843, is called the space of quaternions H. A quaternion $x \in H$ can be represented by the symbol

$$x = x_0 + x_1 i + x_2 j + x_3 k$$

where j and k denote two additional imaginary units and x_0 to x_3 denote the four components of the quaternion. Thus, x_0 is the real part of x and $x_0 + x_1 i$ is the complex part of the quaternion x. Almost all the rules for real and complex number hold also in the space of quaternions. The only exception is the multiplication, which is not commutative, i.e., if x and y are two quaternions, then generally $xy \neq yx$.

Rules for Quaternions

Let x and y be two quaternions

$$\begin{aligned} x &= x_0 + x_1 i + x_2 j + x_3 k \\ y &= y_0 + y_1 i + y_2 j + y_3 k. \end{aligned}$$

Then the sum is

$$x + y = (x_0 + y_0) + (x_1 + y_1)i + (x_2 + y_2)j + (x_3 + y_3)k.$$

The product xy is computed by formally multiplying x and y by use of the distributive law and the conventions

$$\begin{aligned} &i^2 = j^2 = k^2 = -1 \\ &ij = -ji = k \\ &jk = -kj = i \\ &ki = -ik = j. \end{aligned}$$

The division of quaternion numbers is also defined but not relevant to this section.

With these definitions we can interpret the iteration $x \to x^2 + c$ for quaternions. Moreover, we can extend the notion of prisoner and escape sets from the complex to the quaternion space. A quaternion Julia set consequently is defined as the boundary of a quaternion prisoner set. However, the visualization of quaternion Julia sets is significantly more demanding compared to the complex case, because we have to deal with fractal objects in four dimensions. Rendering techniques have been pioneered by V. Alan Norton in 1982, who produced a collection of amazing images.[26] In the color section of this book we have included two images of this type.[27]

Let us explain some more details which are needed to understand what these images are displaying. First we note that a simplification can be achieved by choosing the quaternion parameter c in the iteration of $x \to x^2 + c$ as a complex number, i.e., $c = x_0 + x_1 i$. If the initial point $x \in H$ is complex (i.e., with third and fourth component being 0) then the orbit remains in the complex plane and we have the same outcome as if we had worked in the complex space from the beginning. In other words, the quaternion Julia set, restricted to the subset which corresponds to the complex plane, is identical to the traditional Julia set. Moreover, the set

$$H_0 = \{x \in H \mid x = x_0 + x_1 i + x_2 j\},$$

which contains all quaternions with the last component $x_3 = 0$, is invariant under the iteration. In other words, if $x \in H_0$ then $x^2 + c \in H_0$ also. Thus, in a first attempt, we may ignore the fourth component of quaternions and work only with the first three. The result is a Julia set in three dimensions. An example is shown in color plate 16 for the case $c = 0.2809 + 0.53i$. The object is cut open at the complex plane revealing a familiar looking rabbit-type Julia set (compare figure 13.36). In the third dimension the Julia set seems to have a very complicated structure, which, however, also reveals some elements of regularity.[28] This picture shows only a three-dimensional section of the entire Julia set which is a subset of the four-dimensional quaternion. To get a feeling for the whole structure of the Julia set we may display a different section (see color plate15). Note that again the rabbit-type complex Julia set is contained as a cross-section of the object displayed.

[26] See B. B. Mandelbrot, *The Fractal Geometry of Nature,* W. H. Freeman and Co., New York, 1982, and V. A. Norton, *Generation and display of geometric fractals in 3-D,* Computer Graphics 16, 3 (1982) 61–67. A more recent reference with an advanced rendering technique is J. C. Hart, D. J. Sandin, L. H. Kauffman, *Ray tracing deterministic 3-D fractals,* Computer Graphics 23, 3 (1989) 91–100.

[27] The pictures stem from the computer science diploma thesis (1991) of our student Ralph Lichtenberger, which is based on the ray tracing technique introduced by Hart, Sandin, and Kauffman mentioned in the above footnote.

[28] The Julia set contains circles. For a discussion of this effect see V. A. Norton, *Julia sets in the quaternions,* Computers and Graphics 13, 2 (1989) 267–278.

Chapter 14

The Mandelbrot Set: Ordering the Julia Sets

In the Mandelbrot set, nature (or is it mathematics?) provides us with a powerful visual counterpart of the musical idea of 'theme and variation': the same shapes are repeated everywhere, yet each repetition is somewhat different. It would have been impossible to discover this property of iteration if we had been reduced to hand calculation, and I think that no one would have been sufficiently bright or ingenious to 'invent' this rich and complicated theme and variations. It leaves us no way to become bored, because new things appear all the time, and no way to become lost, because familiar things come back time and time again. Because this constant novelty, this set is not truly fractal by most definitions; we may call it a borderline fractal, a limit fractal that contains many fractals. Compared to actual fractals, its structures are more numerous, its harmonies are richer, and its unexpectedness is more unexpected.

Benoit Mandelbrot[1]

The Mandelbrot set is certainly the most popular fractal, probably the most popular object of contemporary mathematics of all. Some people claim that it is not only the most beautiful but also the most complex object which has been seen, i.e., made visible. Since Mandelbrot made his extraordinary experiment in 1979, it has been duplicated by tens of thousands of amateur scientists around the world.[2] They all like to delve into the unlimited variety

[1]Edited from an interview in the video film: *Fractals, An Animated Discussion,* by H.-O. Peitgen, H. Jürgens, D. Saupe, C. Zahlten, W. H. Freeman and Company, New York, 1990.

[2]See B. B. Mandelbrot, *Fractal aspects of the iteration of* $z \rightarrow \lambda z(1 - z)$ *for complex* λ *and* z, Annals New York Academy of Sciences 357 (1980) 249–259. For an historical account of the discovery read Mandelbrot's contributions in *Fractals for the Classroom, Part One,* H.-O. Peitgen, H. Jürgens, D. Saupe, Springer-Verlag, New York, 1991, p. 1–16, and *Fractals and the Rebirth of Iteration Theory* in *The Beauty of Fractals*, H.-O. Peitgen, P. H. Richter, Springer-Verlag, Heidelberg, 1986.

Benoit B. Mandelbrot

Figure 14.1

of pictures which can develop on a computer screen. Sometimes many hours are required for their generation; but this is the price you have to pay for the adventure of finding something new and fantastic where nobody has looked before.

Is this wealth just a generous gift from mathematics to those who like to marvel at beautiful pictures, or does this apparent beauty and complexity have a deeper meaning? Do the apparent pictorial features of the Mandelbrot set have an equal counterpart in its mathematical beauty? In other words, does the Mandelbrot set present a glimpse of what mathematicians sometimes call the aesthetics of mathematics? The answer is a vigorous 'yes' indeed.

14.1 From the Structural Dichotomy to the Binary Decomposition

We learned in chapter 13 that for each parameter value of c in the iteration of $z \to z^2 + c$ there is a unique prisoner set P_c and a corresponding escape set E_c. Moreover, we also learned how to draw images of P_c and E_c. The equipotentials and field lines reveal the natural structure of E_c, leading to a deeper understanding of the dynamics of $z \to z^2 + c$ in the escape set and its boundary, the Julia set J_c. But so far we have only seen a glimpse of the infinite variety of Julia sets that can be explored.

Speaking in terms of a metaphor, we are confronted with an infinite picture book, each page of which holds the image of one particular Julia set J_c and where the page numbers are the complex parameters c belonging to the Julia sets. How many chapters does this infinite book have? And is there a principle of order? The answer to the latter query is 'yes' and follows from results already contained in the mathematical masterpieces of Julia and Fatou. The key is the structural dichotomy, which states that for any choice of c the associated Julia set J_c and the prisoner set P_c are both

- either one piece (mathematically: connected)
- or a dust (mathematically: totally disconnected).

We already discussed this dichotomy in the previous chapter. In fact, in the case of totally disconnected prisoner sets the study of their encirclements explained that the corresponding Julia sets can be understood as generalized Cantor sets. In other words, the infinite book of Julia sets can be organized in two chapters: the first for all connected Julia sets and the other for those that are totally disconnected. Around 1979 Mandelbrot had the idea of picturing this dichotomy within the set of all parameters c varying in the complex plane **C** . This led directly to the *Mandelbrot set*

$$M = \{c \in \mathbf{C} \mid J_c \text{ is connected}\}.$$

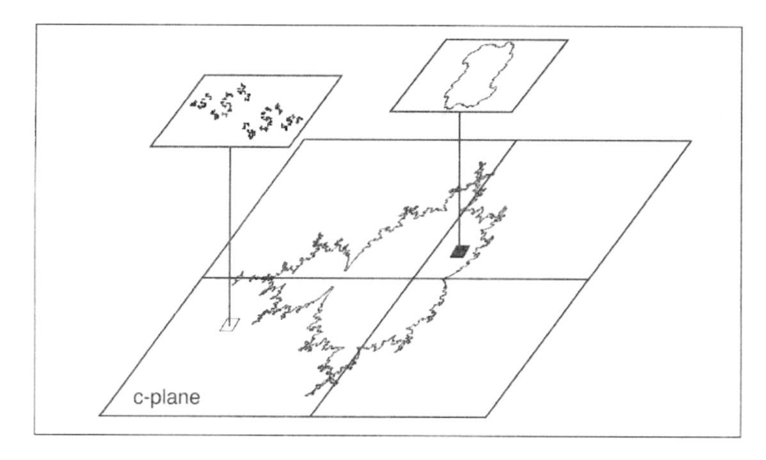

The Mandelbrot Set — Dichotomy of Julia Sets

Any point in the c-plane, interpreted as a parameter c for the iteration of $z \to z^2 + c$, corresponds to a Julia set. The point is colored black if the corresponding Julia set is connected, and white if the set is disconnected. This is the essence of Mandelbrot's experiment from 1979.

Figure 14.2

The Mandelbrot Set — Old and New Rendering

The insert shows an original printout from Mandelbrot's experiment. We have produced the large Mandelbrot set using a modern laser printer and a more accurate mathematical algorithm.

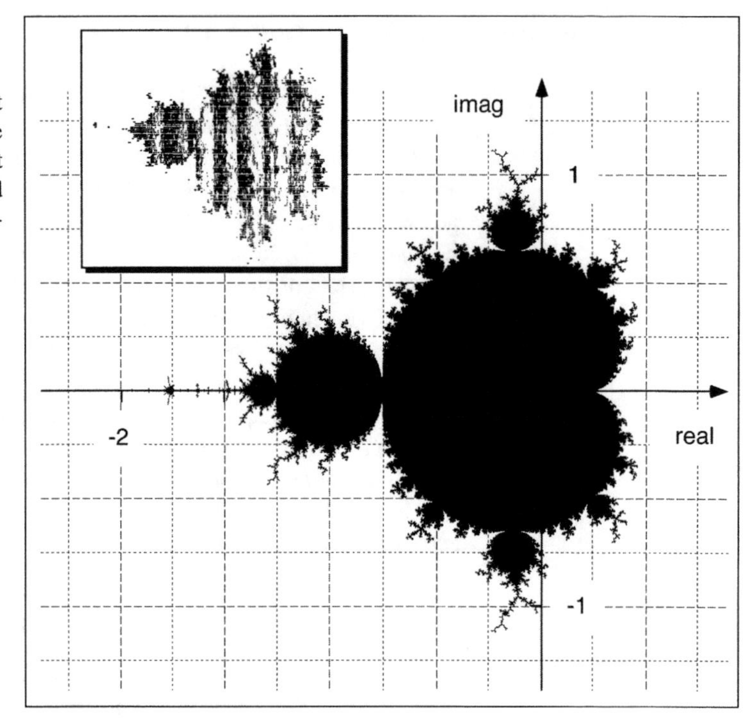

Figure 14.3

He colored each point (pixel of a computer screen) in the plane of c-values black or white depending on whether the associated Julia set turned out to be one piece or a dust (see figure 14.2). The result was a black and white image (see insert in figure 14.3) which, given the graphics technology of that time, did not look very impressive or promising. But Mandelbrot realized that he had discovered one of the gems of mathematics and pushed further. Thus, he initiated a revitalization of a mathematical field which had been dormant for nearly 60 years.

Characterization by the Orbit of the Critical Point

But how did he actually let the computer make the decision whether a parameter c belongs to the Mandelbrot set or not? He was one of the few people at that time who knew the works of Julia and Fatou very well. In particular, he was aware of the fact that there is a tight interrelation between the dichotomy of Julia sets and the fate of the critical point. Let us recall this fact from section 13.8:

Fact. *The prisoner set P_c is connected if and only if the the critical orbit $0 \to c \to c^2 + c \to \cdots$ is bounded.*

This fact provides an alternative definition for the Mandelbrot set. In other words, Mandelbrot used

$$M = \{c \in \mathbf{C} \mid c \to c^2 + c \to \cdots \text{ remains bounded}\} \qquad (14.1)$$

as the definition for M in his 1979 experiments.

Figure 14.4 : A connected and a disconnected Julia set.

Note that this definition is very similar to that of the prisoner set P_c (see page 735), written as

$$P_c = \{z_0 \in \mathbf{C} \mid z_0 \to z_0^2 + c \to \cdots \text{ remains bounded}\} .$$

However, while the Julia set is part of the plane of initial values z_0 whose orbits reside in the same complex plane, the Mandelbrot set is in the plane of parameter values c, and it is not appropriate to plot any orbits from the iteration of $z \to z^2 + c$ in this plane.

The First Clues About M

From our studies we already have a few clues about the Mandelbrot set. First of all, everything outside of a disk of radius 2 is not part of the Mandelbrot set, because, if $|c| > 2$, then the critical point escapes to infinity,[3] and the Julia set is a Cantor set. Further, in chapter 13 we saw that for real parameters c with $-2 \le c \le 0.25$, the iteration of the critical point is bounded and the Julia set is connected. Thus, the interval $[-2, 0.25]$ on the real axis belongs to M, which is contained in a disk of radius 2 centered at the origin.[4]

Encirclement of M

Given a parameter c, how can we computationally decide whether the orbit of c is bounded or not, i.e., whether $c \in M$? Theoretically, this might require knowledge of the complete critical orbit, i.e., an infinite number of iterations. The problem is the same as computing whether an initial point z_0 is in the prisoner set or not. Therefore, we again look at encirclements M_k of M,

[3] See the technical section on page 738.

[4] In fact, the point $c = -2$ is the only point of the Mandelbrot set that has an absolute value equal to 2.

which we may define analogous to the encirclements of prisoner sets, namely, as

$$M_k = \left\{ c \in \mathbf{C} \;\middle|\; \lim_{l \to \infty} \frac{\log_2 |z_l|}{2^l} \leq 2^k,\; z_0 = c \right\} \tag{14.2}$$

where

$$z_{l+1} = z_l^2 + c, \;\; l = 0, 1, 2, \ldots$$

Encirclements of M

In this technical section we derive the formula (14.2). Since M is contained in a disk of radius 2 around the origin we may use a disk of radius greater than or equal to 2 as a target set T. We set

$$R^{(-k)}(T) = \{c \in \mathbf{C} \mid z_k \in T,\; z_0 = c\}\,,\; k = 0, 1, 2, \ldots$$

where $z_0, z_1, z_2, \ldots$ denotes the critical orbit, i.e.,

$$z_{k+1} = z_k^2 + c, \;\; z_0 = c, \;\; k = 0, 1, 2, \ldots$$

Let us first point out in this paragraph that $z_k \in T$, $k \geq 1$, implies that all the previous iterates $z_0, z_1, \ldots, z_{k-1}$ are also in T. Let $r \geq 2$ be the radius of the disk T and consider the case $|c| \geq r$ with initial point $z_0 = c$. Let z_m be any point of the orbit of z_0 with $|z_m| \geq |c|$. For the next point z_{m+1} of the orbit it follows that

$$\begin{aligned} |z_{m+1}| &= |z_m^2 + c| \geq |z_m^2| - |c| \\ &\geq |z_m|^2 - |z_m| = (|z_m| - 1)|z_m| \\ &\geq (r-1)|z_m| \geq |z_m|. \end{aligned}$$

Since already $|z_0|$ fulfills the assumption for $|z_m|$ we conclude

$$r \leq |c| = |z_0| \leq |z_1| \leq |z_2| \leq \cdots$$

In other words, the sequence $|z_0|, |z_1|, |z_2|, \ldots$ is monotonically increasing. Thus, if z_m is in the disk T, then also z_0 to z_{m-1} must be in T. Let us now consider the remaining case $|c| < r$ and assume that $|z_m| > r$, i.e., $z_m \notin T$. Then

$$\begin{aligned} |z_{m+1}| &= |z_m^2 + c| \geq |z_m^2| - |c| \\ &> r^2 - r = (r-1)r \\ &= (1+\varepsilon)r > r. \end{aligned}$$

By induction it follows that also all the following points in the orbit are not in T, i.e.,

$$z_m, z_{m+1}, z_{m+2}, z_{m+3}, \ldots \notin T.$$

Thus, if $z_k \in T$, then also all iterates z_0 to z_{k-1} must be in T, which was to be shown.

In other words, the encirclement $R^{(-k)}(T)$ is given by precisely those parameter values c for which all of the first k iterations of $z_0 = c$ hit the target set T. Recall from the characterization in equation

(14.1) that c belongs to the Mandelbrot set if the entire critical orbit is in the target set. In other words, the encirclement $R^{(-k)}(T)$ is an approximation of M which improves as the number k increases. Let us elaborate this important point.

We already pointed out above that M is contained in a disk of radius 2. This implies that M is also contained in the target set T (which must contain the disk). Expressed in a formula,

$$M \subset T = R^{(0)}(T).$$

Now $R^{(-1)}(T)$ is the set of all parameters c for which the critical value $z_0 = c$ is in T and the first iterate $z_1 = z_0^2 + c = c^2 + c$ is also in T. Thus, $R^{(-1)}(T)$ is a subset of $R^{(0)}(T)$. Moreover, M is contained in $R^{(-1)}(T)$:

$$M \subset R^{(-1)}(T) \subset R^{(0)}(T).$$

Now $R^{(-2)}(T)$ is the set of all parameters c for which $z_0 = c$ is in T, $z_1 = z_0^2 + c$ is in T, and the second iterate $z_2 = z_1^2 + c$ is also in T. Thus, $R^{(-2)}(T)$ is a subset of $R^{(-1)}(T)$ and contains M:

$$M \subset R^{(-2)}(T) \subset R^{(-1)}(T) \subset R^{(0)}(T).$$

This reasoning can go on producing better and better encirclements $R^{(-k)}(T)$ for $k = 3, 4, 5$ and so on. In the limit we obtain the Mandelbrot set itself,

$$\bigcap_{k=0}^{\infty} R^{(-k)}(T) = M. \tag{14.3}$$

There is an important difference between the encirclements of prisoner sets and those of the Mandelbrot set, which must be kept in mind when interpreting pictures of the Mandelbrot set. This is the fact that the encirclements $R^{(-k)}(T)$ of M are not iterated preimages of T with respect to some fixed transformation (i.e., $R^{(-k)}(T)$ is not an image of $R^{(-k-1)}(T)$).

The encirclements $R^{(-k)}(T)$ of the Mandelbrot set depend on the choice of the target set T. Besides disks of radius greater than or equal to 2 we could as well have chosen ellipses, squares, or any other shape as long as the property in eqn. (14.3) is guaranteed. In order to remove this ambiguity, we proceed along the same line as carried out for the prisoner sets. We use the disks

$$D^l = \left\{ z \mid \log_2 |z| \leq 2^l \right\}, \ l = 0, 1, 2, \ldots$$

as target sets T and define the kth encirclement of M as

$$M_k = \lim_{l \to \infty} R^{(k-l)}(D^l)$$

for any integer k. Let us put this rather abstract formula into a more accessible form. The parameter c is in $R^{(k-l)}(D^l)$, if

$$\log_2 |z_{l-k}| \leq 2^l$$

or, after dividing by 2^{l-k},

$$\frac{\log_2 |z_{l-k}|}{2^{l-k}} \leq 2^k.$$

In this case taking the limit $l \to \infty$ is the same as letting $l - k \to \infty$. Thus,

$$M_k = \left\{ c \in \mathbf{C} \;\middle|\; \lim_{l \to \infty} \frac{\log_2 |z_l|}{2^l} \leq 2^k,\ z_0 = c \right\},$$

where

$$z_{l+1} = z_l^2 + c, \quad z_0 = c.$$

This is now in the form suitable for computer implementation. Moreover, we clearly see the analogy to the encirclements $P_c^{(k)}$ of the prisoner set P_c (compare eqn. (13.3)).

For positive indices k, the encirclements M_k are almost identical to large disks. This is due to the fact that $z \to z^2 + c$ acts like $z \to z^2$ when $|z|$ is large. For example, M_3 is approximately a disk with radius $2^8 = 256$. In other words, we can accept a parameter c as an element of M_k if $k + 3$ iterations, started at $z_0 = c$, do not escape the disk of radius 256. The result is given in figure 14.5, which shows the boundaries of the encirclements M_0 through M_{-10}.

A Simple Algorithm for the Mandelbrot Set

For a given parameter c the fate of the critical point $z = 0$ must be determined by the algorithm. If $|c| > 2$, we already know that the orbit must escape, thus an algorithm can terminate immediately and return the result that the Julia set is disconnected. So let us assume that $|c| \leq 2$. In the most simple algorithm we would iterate $z \to z^2 + c$, starting with $z_0 = c$, and check the points on the orbit. If a point from the orbit is outside a disk of radius

$$R = \max(2, |c|) = 2,$$

we are sure that the orbit must escape to infinity,[5] and again the algorithm may terminate with the same result (i.e., c is not in M). Of course, some maximum number m of iterations must be prescribed to avoid infinite loops. But computing the parameters for which the iteration does not leave $T = \{c \in \mathbf{C} \mid |c| \leq 2\}$ within m steps is nothing other than computing $R^{(-m)}(T)$, which is an approximation of M. Instead of using 2 as the threshold radius of the target set, we propose using 256, which ensures that the resulting boundaries of encirclements can be interpreted as equipotentials of the Mandelbrot set.

[5] See the derivation on page 737.

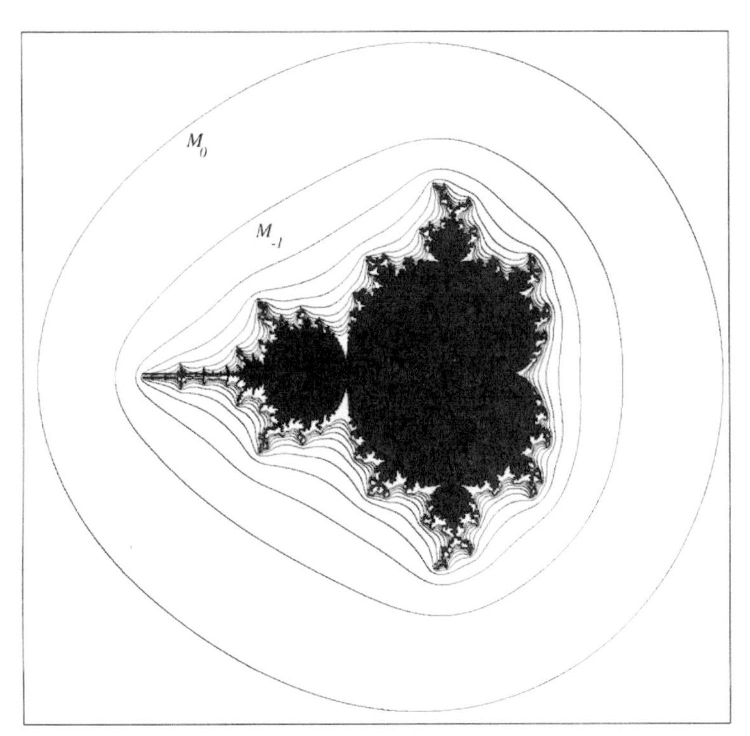

Encirclement of the Mandelbrot Set

The Mandelbrot set M and its approximation by encirclements M_0 through M_{-10}.

Figure 14.5

```
   k = 0
z = c
while (k < m)
    if (|z| > 256) then
       return (disconnected, c in M_{4-k})
    end if
    z = z*z + c
    k = k + 1
end while
return (result = c in M_{4-m})
```

The Mandelbrot Set is Connected

The concept of a potential has been the key tool in the mathematical analysis of the Mandelbrot set. A necessary prerequisite for that is the important fact that the Mandelbrot set is connected. This has been known since 1982 through a paper of Douady and Hubbard,[6] in which they showed that the encirclement of the Mandelbrot set always generates domains which are bounded by circle-like curves. If the encirclement is properly manufactured it can be shown that the bounding curves are in fact equipotentials of the Mandelbrot

[6] A. Douady, J. H. Hubbard, *Iteration des pôlynomes quadratiques complexes,* CRAS Paris 294 (1982) 123–126.

Equipotentials and Field Lines of Mandelbrot Set

The system of equipotentials and field lines provides a polar coordinate system for the complement of the Mandelbrot set.

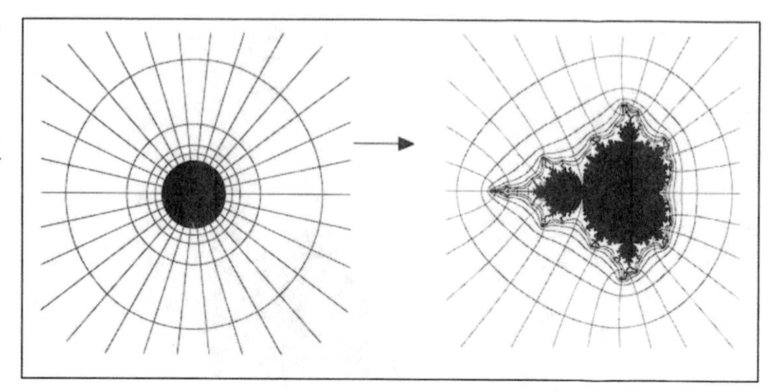

Figure 14.6

set. This is the case for the encirclements

$$M_k = \left\{ c \in \mathbf{C} \;\middle|\; \lim_{l \to \infty} \frac{\log_2 |z_l|}{2^l} \le 2^k,\ z_0 = c \right\}.$$

The boundaries of the sets M_k are equipotentials of the explicit potential function

$$p_M(c) = \lim_{l \to \infty} \frac{\log_2 |z_l|}{2^l},$$

where

$$z_l = z_{l-1}^2 + c, \quad z_0 = c.$$

From Field Lines to the Pinching Model

There is, however, a major difference to the encirclements of the prisoner sets. When computing those encirclements, we keep the parameter c fixed and work with only one fixed transformation $z \to z^2 + c$, while in the case of the Mandelbrot set, we keep the initial point $z_0 = 0$ fixed, but change the value of c from pixel to pixel. This means that we cannot associate a dynamics with the encirclements.

In a way the system of equipotentials and field lines can be viewed as a particular polar coordinate system for the complement of the Mandelbrot set (see figure 14.6). More precisely, there is a one-to-one correspondence between the equipotentials and field lines of the unit disk and those of the Mandelbrot set. Each field line is then given by an angle α, where $0 \le \alpha < 1$. In fact, this correspondence can be made in such a way that the binary expansion carries essentially the information about where the field line will land, as long as α is a rational number. For example, the point where the major period two bud is attached to the cardioid is a pinching point where two field lines land on the boundary of the Mandelbrot set and the angles are $\alpha = 1/3$ and $\alpha = 2/3$.

Details of this field line model yield an almost complete understanding of the Mandelbrot set as a pinching model.[7] Incidentally, it is known that each field line with a rational angle α lands on the Mandelbrot set, while for irrational angles this is still unknown in general. This is related to the fundamental problem of whether the Mandelbrot set is not only connected but also locally connected, which is one of a whole variety of unsolved research problems. A conjecture, also related to this problem, was recently proven positively by M. Shishikura, a young Japanese mathematician.[8] The conjecture was that the boundary of the Mandelbrot set has fractal dimension 2, which would somehow characterize the incredible complexity of the magnifications which we have seen.

Potential Function and Level Sets of the Mandelbrot Set

Let us now discuss the level sets for the Mandelbrot set

$$L^k = \left\{c \in \mathbf{C} \mid 2^{k-1} < p_M(c) \leq 2^k\right\}.$$

Such a level set L^k simply is the difference between the encirclements M_k and M_{k-1}.[9] There is a major distinction between these level sets and those of prisoner sets. When computing those level sets, we keep the parameter c fixed and work with only one fixed transformation $z \to z^2 + c$, while in the case of the Mandelbrot set, we keep the initial point $z_0 = 0$ fixed, but change the value of c from pixel to pixel. This means that we cannot make the interpretation that the kth level set L^k is transformed to the next one, L^{k+1}, as was the case for prisoner sets. Nevertheless, we can define a binary decomposition of each level into cells, even though there is no dynamics defined in them.

Binary Decomposition of M

Let us generate a decomposition of a level set L^k into 2^n cells which we label as $L^k(b_1 b_2 \ldots b_n)$, where the digits b_i are 0 or 1. The decomposition can be defined by two criteria:

- For all $c \in L^k$ we compute the nth iterate z_n of $z \to z^2 + c$ starting at $z_0 = c$ and check its imaginary component $\text{Im}(z_n)$. If $\text{Im}(z_n) \geq 0$, then the label of the corresponding cell has the form $b_1 \ldots b_{n-1}0$; and if the imaginary component is negative, then the form of the label is $b_1 \ldots b_{n-1}1$.
- We order the 2^n cells lining them up in counter-clockwise direction starting at that cell which is aligned at the real axis (at angle $\phi_0 = 0$) and assign corresponding labels $0\ldots00, 0\ldots01, 0\ldots10, 0\ldots11$, and so on.

Observe that for real, positive parameters $c > 0$ the iteration of $z_0 = c$ produces only real and positive numbers. Therefore, we always have one cell which is aligned at the real axis as assumed in the second criterion. When decomposing the level sets for a prisoner set in the previous chapter, such an argument could not have been made. The computation of the special field line

[7] See A. Douady, J. H. Hubbard, *Étude dynamique des pôlynomes complexes,* Publications Mathematiques d'Orsay 84-02, Université de Paris-Sud, 1984. See also H.-O. Peitgen and P. H. Richter, *The Beauty of Fractals*, Springer-Verlag, Heidelberg, 1986.

[8] M. Shishikura, *The Hausdorff dimension of the boundary of the Mandelbrot set and Julia sets,* SUNY Stony Brook, Institute for Mathematical Sciences, Preprint #1991/7.

[9] Note that the simple algorithm for the Mandelbrot set as discussed above not only allows one to determine the encirclements M_k but also the level sets L_k. Actually the algorithm returns the message that the Julia set disconnected, and $c \in L_{4-k}$. Furthermore, it does not require the explicit evaluation of $p_M(c)$.

Figure 14.7 : All level sets are decomposed into 64 cells. Field lines become apparent. In the right image an alternating coloring scheme is applied making also the level sets visible.

ϕ_0 was necessary. In the case of the Mandelbrot set, we get the field line ϕ_0 for free![10] While we have a loss in same ways — no dynamic interpretation of binary cells — we have a gain in another.

Figure 14.7 displays the decomposition of the level sets L_2 to L_{-1} into 64 cells which gives a good idea of the field lines in the larger neighborhood of the Mandelbrot set M. However, if we want a closer look at the boundary of M, we need more and more field lines. A nice solution to this visualization problem is obtained if we decompose, say, level set L_k into 2 cells, L_{k-1} into 4 cells, L_{k-2} into 8 cells, and in general level set L_{k-l} into 2^{l+1} cells (see figure 14.8). In this way we automatically obtain more and more detail when looking closer at the boundary of M (see figure 14.9). As for the prisoner sets an appropriate pinching model has been worked out also for the Mandelbrot set. Here, however, since there is no dynamical interpretation for the cells a full understanding would require a deeper analysis of the work of Douady and Hubbard.[11]

Pinching for the Mandelbrot Set

In section 14.2 we will see that the buds that are attached to the main body of the Mandelbrot set correspond to prisoner sets that are the domains of attraction of stable periodic points. In those points where the buds are are pinching points, where two field lines meet. Their angles are always given by $\alpha = n/(2^m - 1)$. For example, the major bud

[10]This comes from the fact that the coefficients of the polynomials which must be iterated for the computation of ϕ_0 are real. The same applies in the case of Julia sets for real parameters c; their field lines for angle 0 are also on the positive real axis.

[11]See A. Douady, Systèmes dynamiques holomorphes, Expose no. 599 Séminaire Bourbaki 1982/83, Astérisque 105/106 (1983) 39–63.

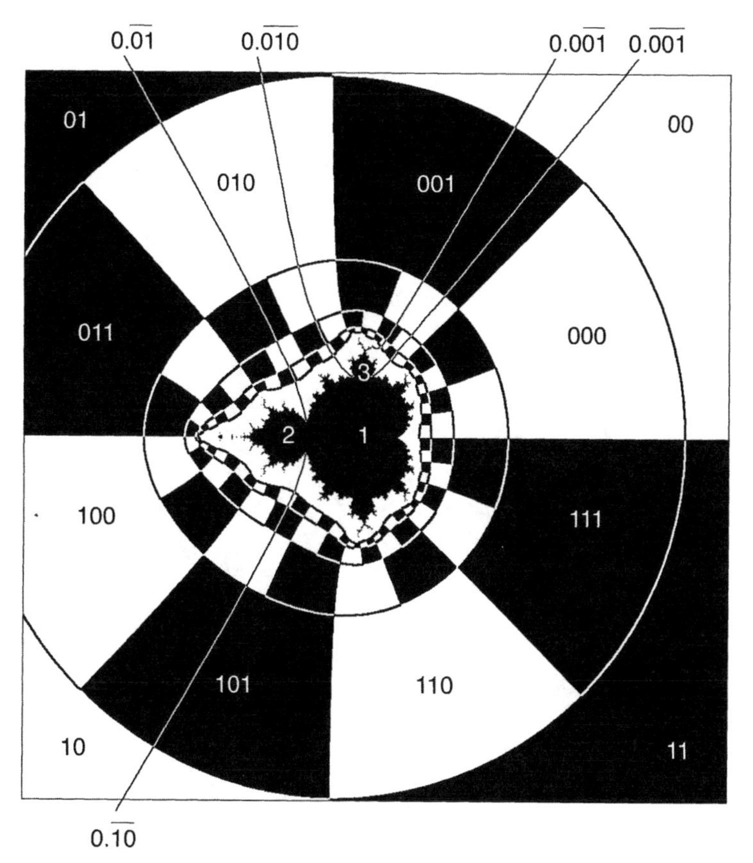

Binary Decomposition for Mandelbrot Set

The binary decomposition is shown for a few level sets with the corresponding labels of the cells. Some field lines are shown with their angle α given in binary form.

Figure 14.8

which describes the stable period 2 cycles is pinched by the field lines with angle $\alpha = 1/3 = 0.\overline{01}$ and $\alpha = 2/3 = 0.\overline{10}$. As Douady's and Hubbard's work shows, there is a deeper reason for the coincidence that the period 2 bud has $m = 2$. In fact this goes on as a firm rule: the period three buds have $m = 3$, i.e., for example, the upper period 3 bud has angles $\alpha = 1/7 = 0.\overline{001}$ and $\alpha = 2/7 = 0.\overline{010}$. In general a period m bud is characterized by two angles with $\alpha = n/(2^m - 1)$. Understanding these field lines is a first step to a pinching model of the Mandelbrot set. However, this approach would not give a complete understanding of the Mandelbrot set because it is not known whether all field lines land on the Mandelbrot set. If this were the case, then the boundary of the Mandelbrot set would be locally connected, and that is still an open conjecture. But the work of Douady and Hubbard has shown that all field lines with rational angles do, in fact, land in the Mandelbrot set, and that gives a very extended description. We have indicated the role of rational angles with odd denominator. Let us look at angles with even denominators, like $\alpha = 1/6 = 0.0\overline{01}$, for example.

Figure 14.9 : More details of the binary decomposition for the Mandelbrot set.

With respect to multiplication by 2 in mod 1 arithmetic we obtain

$$0.0\overline{01} \to 0.\overline{01} \to 0.\overline{10} \to 0.\overline{01},$$

i.e., a preperiodic cycle of length 2. Exactly the same behavior is obtained when we study the critical orbit $z_k = z_{k-1}^2 + c$, for $z_0 = i$, and $k = 1, 2, 3, \ldots$ and $c = i$. Indeed

$$i \to -1 + i \to -i \to -1 + i.$$

The general rules is this: if the critical orbit $z_k = z_{k-1}^2 + c$, for $z_0 = c$, and $k = 1, 2, 3, \ldots$ becomes periodic after l iterations and the period is equal to m or divides m, then the angle of the field line attached at the point c is $\alpha = n/[2^l(2^m - 1)]$.

14.2 The Mandelbrot Set — A Road Map for Julia Sets

The complexity of the Mandelbrot set is in an altogether different class compared to that of Julia sets. On the one hand, the Mandelbrot set has a solid interior without any structure, and on the other hand it is bordered by a very complex boundary with an infinity of different shapes. For a first impression of this variety, we provide a selection of images around the boundary (figure 14.10) and a zoom sequence (figure 14.11).

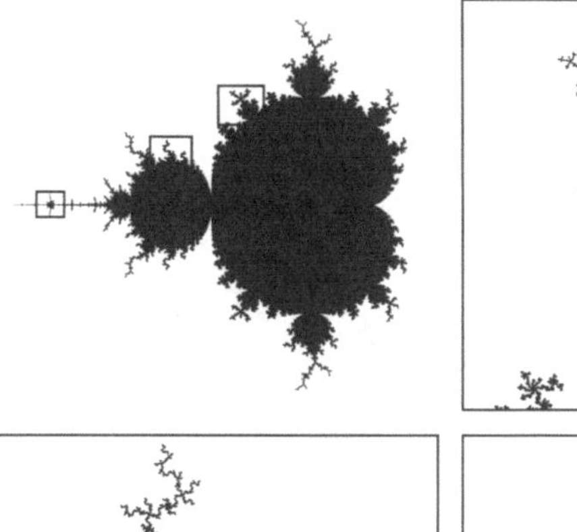

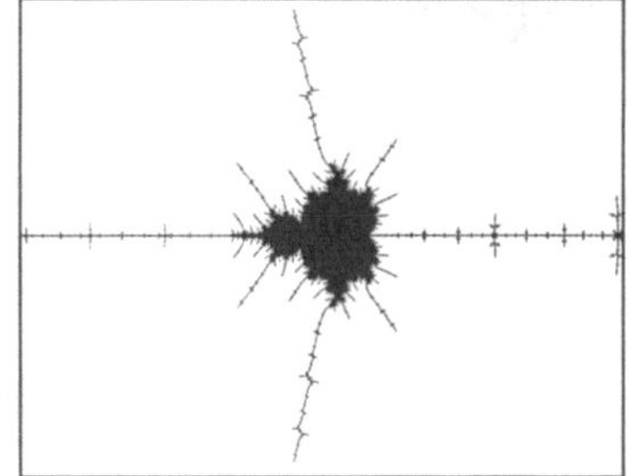

Journey Around the Mandelbrot Set

Journey around the Mandelbrot set with locations of the individual images being marked on the initial one.

Figure 14.10

The Buds in the Mandelbrot Set

The first striking feature of the Mandelbrot set are its small buds which are lined up along the big, heart-shaped, central region. These buds have a meaning for the associated Julia sets. Let us first take a look at the main body of the set in the center. This heart-like set intersects the real axis in the interval from -0.75 to 0.25. We recall that the Julia set for $c = 0$ is a circle with an attractive fixed point at the origin. This fixed point is *super attractive*; the critical point is equal to the fixed point (see chapter 11). It is a fact that the parameters c on the line between -0.75 and 0.25 are precisely those real parameters for which one of the fixed points of $z \rightarrow z^2 + c$ is an attractor.[12] Therefore, it is no surprise that the big heart-shaped region is the set of all (complex) parameters c for which one of the two fixed points of $z \rightarrow z^2 + c$ is attractive. Figure 14.12 shows two examples of Julia sets of this type (see also figure 13.15). Observe the position of the fixed points.

[12]This c-interval $(-0.75, 0.25)$ corresponds to the a-interval $(1, 3)$ for $x \rightarrow ax(1 - x)$ (see chapter 1).

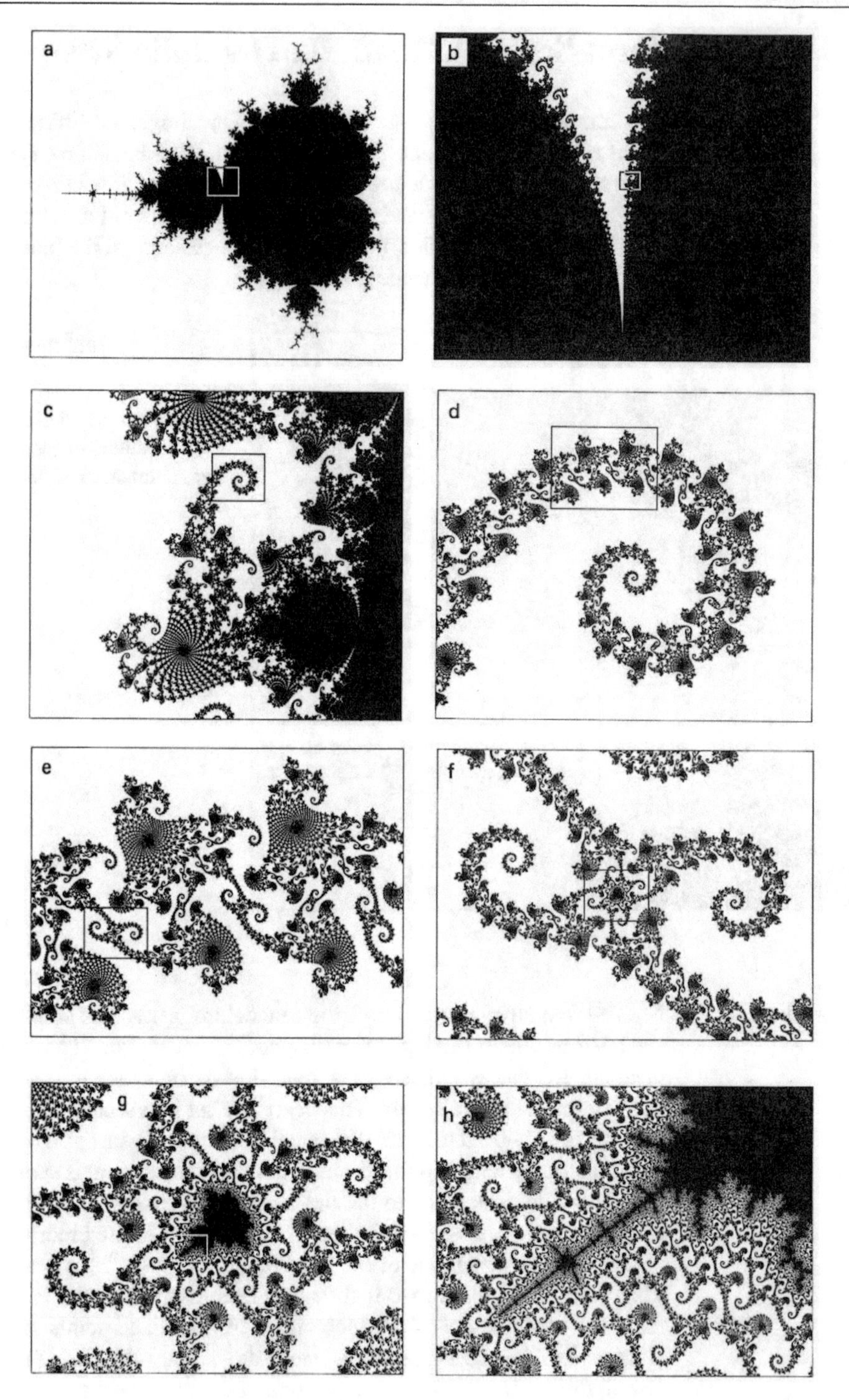

Figure 14.11 : Zoom into the Mandelbrot set.

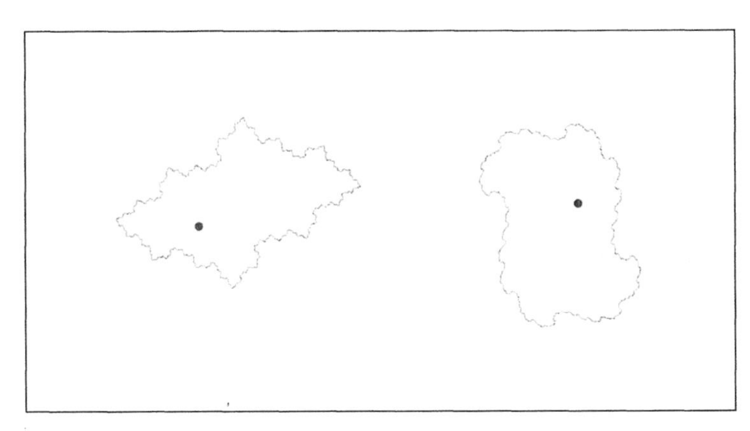

Basin of Attractive Fixed Points

Julia sets that bound the basin of attraction of an attractive fixed point marked by the dot. Note that the other fixed point is in the Julia set and is repelling. The parameters are $c = -0.55 - 0.3i$ (left) and $c = 0.28 + 0.2i$ (right).

Figure 14.12

The Real Parameters $-0.75 < c < 0.25$

The fixed points of the quadratic iteration $z \to z^2 + c$ are given by the formula

$$z_{1,2} = \frac{1 \pm \sqrt{1-4c}}{2}.$$

The derivative of $z^2 + c$ at $z_{1,2}$ is just twice that number. As c varies from -0.75 to 0.25, we have that $1 - 4c$ varies from 4 to 0, and that the root $\sqrt{1-4c}$ goes from 2 to 0. Thus, $1 - \sqrt{1-4c}$ is between -1 and 1, and smaller than 1 in absolute value which identifies an attractive fixed point. Thus, the fixed point $(1 - \sqrt{1-4c})/2$ is an attractor for c between -0.75 and 0.25.

The Boundary of the Heart-Like Central Region

We can determine an explicit formula for the outline of the 'heart' of the Mandelbrot set using the derivative criterion as follows. On the outline we have that the derivative of $z \to z^2 + c$ at one of the fixed points is equal to 1 in absolute value (the interior of the heart consists of parameters for which one of the fixed points has a derivative less than 1 in absolute value).

Assume that z is a fixed point of $z \to z^2 + c$, i.e., z solves

$$z^2 - z + c = 0.$$

The derivative at z is given by $2z$ which we write in polar coordinates as

$$2z = re^{i\phi}$$

with $r \geq 0$ and $0 \leq \phi < 2\pi$. We combine the two equations and arrive at

$$\left(\frac{re^{i\phi}}{2}\right)^2 - \frac{re^{i\phi}}{2} + c = 0.$$

This last equation is now solved for c:

$$c = \frac{1}{2}re^{i\phi} - \frac{1}{4}r^2e^{2i\phi}.$$

Given an arbitrary number $re^{i\phi}$ this result specifies a parameter c such that the derivative of $z \to z^2 + c$ at one of the corresponding fixed points matches the given number. For example, in order to obtain a representation of the interior of the heart-like center of the Mandelbrot set we consider $r < 1$, while $r = 1$ yields its boundary curve. Writing $c = x + yi$, we can split the real and the imaginary components of the above equation (for $r = 1$):

$$\begin{aligned} x &= \frac{\cos\phi}{2} - \frac{\cos 2\phi}{4} \\ y &= \frac{\sin\phi}{2} - \frac{\sin 2\phi}{4}. \end{aligned} \tag{14.4}$$

These final equations readily produce a complex point for any given argument ϕ. Such a representation of a curve is called a parametrization (ϕ being the parameter of the curve in this case). Here are examples of points on the curve for 5 values of ϕ:

ϕ	x	y
0	0.25	0.0
$2\pi/5$	0.35676	0.32858
$2\pi/4$	0.25	0.5
$2\pi/3$	-0.12500	0.64952
π	-0.75	0.0

It turns out that at the parameter values

$$\phi = \frac{2\pi}{k}, \quad k = 2, 3, 4, 5, 6, \ldots$$

one of the main buds of the Mandelbrot set is attached to the heart-shaped center. Moreover, the period of the attractive cycles that belong to these buds is given by the number k in $\phi = 2\pi/k$.

Let us make a final remark. The fixed points of $z \to z^2 + c$ are

$$z_{1,2} = \frac{1 \pm \sqrt{1-4c}}{2}.$$

and the derivatives at these fixed points are twice that, namely, $1 \pm \sqrt{1-4c}$. From this representation it follows that if one derivative is inside the unit circle, then the other must be outside. Thus, if one fixed point is attracting, then the other must be repelling. Also, if one fixed point is indifferent, then the other must be repelling (except in the case where the fixed points are identical, i.e., for the case $c = 1/4$).

π and the Mandelbrot Set

Let us briefly interrupt our tour of the Mandelbrot set to present a splendid observation that relates the Mandelbrot set to the constant $\pi = 3.141592\ldots$ It is a result of a small computer experiment, carried out by Dave Boll in 1991 and communicated by him using an electronic bulletin board called USENET which is read worldwide at most universities and schools. The following paragraphs are taken directly from his message.[13]

Boll's Observation

"I was writing a 'quick-n-dirty' program to verify that the 'neck' of the Mandelbrot set at $c = -0.75+0i$ is actually of zero thickness.[14] Accordingly, I was testing the number of iterations that points of the form $-0.75 + \varepsilon i$ (ε being a small number) went through before escaping. Here's a quick list for special values of ε:

ε	Iterations
0.1	33
0.01	315
0.001	3143
0.0001	31417
0.00001	314160
0.000001	3141593
0.0000001	31415928

Does the number of iterations strike you as a suspicious number? How about the product of the number of iterations with ε? It's π, to within $\pm\varepsilon$. My initial reaction was 'What the HELL is π doing here?'.

Adopting the motto 'When in doubt, keep going', I tried the same experiment at the 'butt' of the Mandelbrot set,[15] located at $c = 0.25 + 0i$. I was now trying points of the form $c = 0.25 + \varepsilon + 0i$, with ε again a small number. Here are some more results for various values of ε:

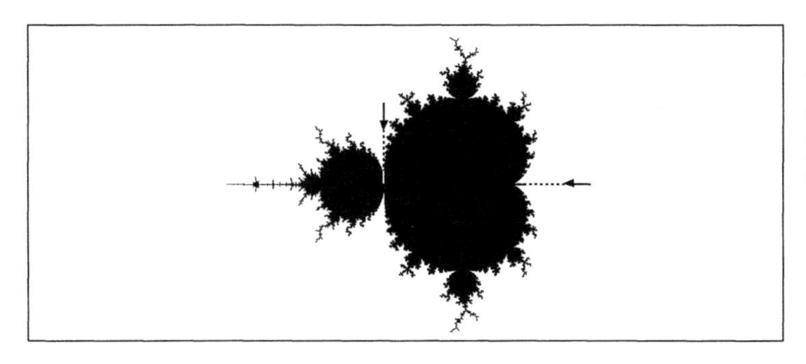

Two Paths of Approach

The tables in Boll's experiment record the number of iterations for parameters that approach the Mandelbrot set along the dotted lines.

Figure 14.13

[13] Slightly edited by the authors to adapt to the notation used in this book.

[14] This refers to the fact that the heart-shaped part of the Mandelbrot set and the major disk-shaped bud to the left touch in precisely one point, $c = -0.75 + 0i$.

[15] This refers to the cusp in the heart-shaped part of the Mandelbrot set.

ε	Iterations
0.1	8
0.01	30
0.001	97
0.0001	312
0.00001	991
0.000001	3140
0.0000001	9933
0.00000001	31414
0.000000001	99344
0.0000000001	314157
0.00000000001	993457
0.000000000001	3141625

Again, we get the same type of relationship, this time it is

$$\lim_{\varepsilon \to 0} N(\varepsilon) \cdot \sqrt{\varepsilon} = \pi$$

where $N(\varepsilon)$ is the number of iterations. [...] Has anyone seen this? What's going on?"

So far Boll's experiment. Initially, he did not get any responses, except for a couple saying in effect 'that's pretty strange'. A year later he re-posted his findings on the bulletin board,[16] and this time, there were some reactions with attempts for an explanation. In fact, it turned out that the second occurrence of π near $c = 1/4$ had already been noticed even long before 1991 in the context of intermittency. This can be illustrated by graphical iteration since the computations in the experiment near $c = 1/4$ involve only real numbers (see figure 14.14). Clearly there is a connection to the tangent bifurcation we considered in the study of intermittency in section 11.5. For that case we already demonstrated that $N(\varepsilon)\sqrt{\varepsilon}$ tends to a constant for the iteration of $g_a(x) = ax^2 \sin(\pi x)$ as a approaches the bifurcation point a^* ($\varepsilon = a^* - a$). However, the constant is about 2.36 while it is a surprise that for the system belonging to the Mandelbrot set we get π. In the following section we provide a heuristic explanation for this astonishing fact.[17]

Why is it π?

We consider the iteration

$$x_{k+1} = x_k^2 + \frac{1}{4} + \varepsilon, \quad x_0 = 0 \tag{14.5}$$

[16] D. Boll, *Pi and the Mandelbrot set (again)*, USENET article <1992Feb26.222630.36612@yuma.acns.colostate. edu>.

[17] It stems from J. Guckenheimer, P. Holmes, *Nonlinear Oscillations, Dynamical Systems, and Bifurcations of Vector Fields*, Springer-Verlag, New York, 1983 (see page 344) and is in the spirit of the paper by Y. Pomeau and P. Manneville, *Intermittent transition to turbulence in dissipative dynamical systems*, Commun. Math. Phys. 74 (1980) 189–197, which initiated the discussion of intermittency.

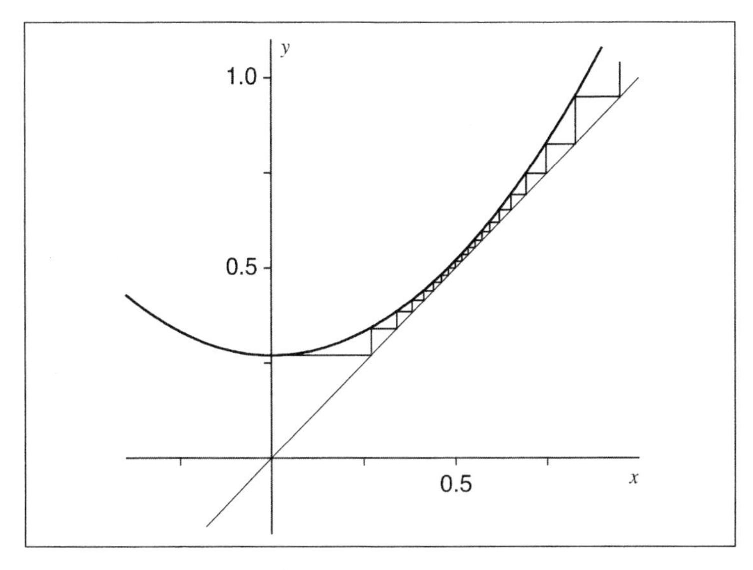

Tangent Bifurcation at $c = 1/4$

In the second experiment reported by Boll the number $N(\varepsilon)$ of iterations of $x \to x^2 + 1/4 + \varepsilon$ starting at $x = 0$ is measured in relation to ε obtaining $N(\varepsilon)\sqrt{\varepsilon} \to \pi$ as $\varepsilon \to 0$.

Figure 14.14

and interpret this as a sequence of Euler steps for a corresponding differential equation

$$x'(t) = f_\varepsilon(x(t)).$$

Thus, we claim that the iteration can be written in the form

$$x_{k+1} = x_k + hf_\varepsilon(x_k).$$

for some function f_ε with parameter ε and step size h.[18] In fact, using the step size $h = 1$, we can solve this equation for the function f_ε, obtaining

$$f_\varepsilon(x_k) = x_{k+1} - x_k = x_k^2 + \frac{1}{4} + \varepsilon - x_k.$$

We arrive at the differential equation

$$x'(t) = x^2(t) - x(t) + \frac{1}{4} + \varepsilon. \tag{14.6}$$

Starting a solution at the initial condition $x(0) = 0$, we now compute the time t that the solution of the differential equation spends until it arrives at $x(t) = 1$. Translated into the context of the discrete iteration in eqn. (14.5), the time t is the number of Euler steps until $x_k \geq 1$, which corresponds to the number of iterations in the computer experiment. As we let $\varepsilon \to 0$ we will see that $\sqrt{\varepsilon}t \to \pi$. We divide eqn. (14.6) by the right hand side and integrate from time 0 to time t.

$$\int_0^t \frac{x'(s)}{x(s)^2 - x(s) + \frac{1}{4} + \varepsilon} ds = \int_0^t ds.$$

[18] Then x_k approximates the solution $x(kh)$ for $k = 0, 1, 2, \ldots$

Using the substitution rule we replace $x'(s)ds$ by dx and t by $x(t)$ on the left.

$$\int_0^{x(t)} \frac{dx}{x^2 - x + \frac{1}{4} + \varepsilon} = t. \tag{14.7}$$

An antiderivative is given by

$$\frac{1}{\sqrt{\varepsilon}} \arctan\left(\frac{x}{\sqrt{\varepsilon}} - \frac{1}{2\sqrt{\varepsilon}}\right).$$

Using this to evaluate the definite integral in eqn. (14.7) and multiplying by $\sqrt{\varepsilon}$ we obtain

$$\arctan\left(\frac{x(t)}{\sqrt{\varepsilon}} - \frac{1}{2\sqrt{\varepsilon}}\right) - \arctan\left(-\frac{1}{2\sqrt{\varepsilon}}\right) = \sqrt{\varepsilon}t.$$

We are interested in the time t when $x(t) = 1$. And we get

$$2\arctan\left(\frac{1}{2\sqrt{\varepsilon}}\right) = \sqrt{\varepsilon}t.$$

Letting $\varepsilon \to 0$, the left hand side tends to π, and we see the result

$$\lim_{\varepsilon \to 0} \sqrt{\varepsilon}t = \pi.$$

Although this is not a rigorous proof for the observed phenomenon, it provides a supporting argument for it. To be rigorous, it must still be shown that passing from the differential equation to the corresponding discrete iteration using the Euler method in fact does not destroy the asymptotics that we have derived above. Concerning the other experiment for the point $c = -3/4$ a similar but more complicated approach can be taken.[19]

The Period-Two Disk

We now return to the discussion of the Mandelbrot set itself. At the left end of the heart-shaped region (at $c = -0.75$, where the number π can be seen in the iteration counts) there is a bud. It is a perfect disk of radius 0.25 centered at $c = -1$. Figure 14.15 shows two Julia sets for parameters from this disk. For such parameters neither one of the two fixed points of $z \to z^2 + c$ can be attractive because c is outside of the heart-shaped center of M. Now what are the dynamics of the iteration within the prisoner set in this case? Let us check this with an experiment. For $c = -1$ we pick two initial points close to the fixed points

$$z_{1,2} = \frac{1 \pm \sqrt{5}}{2}$$

which are

$$\begin{aligned} z_1 &= +1.61803398\ldots \\ z_2 &= -0.61803398\ldots \end{aligned}$$

[19] See G. Edgar, *Pi and the Mandelbrot set,* USENET article <1992Mar27.135743.28423@zaphod.mps.ohio-state. edu>.

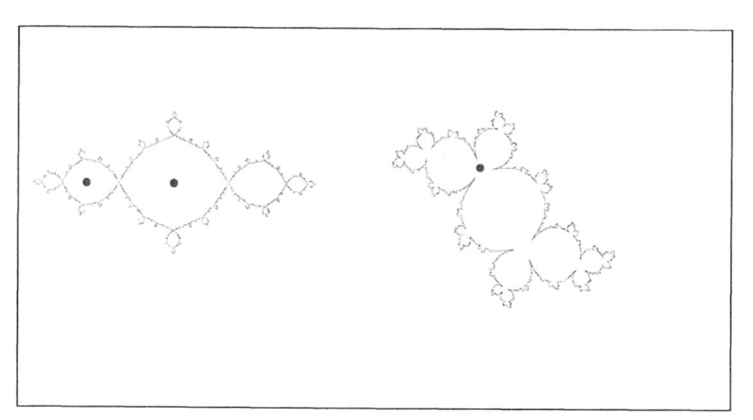

Two Julia Sets from the Big Bud of M

Julia sets that bound the basin of attraction of an attractive cycle (marked by large dots) of period 2. Left $c = -1$ (the super attractive case), right $c = -0.83 + 0.16i$.

Figure 14.15

	Orbit 1		Orbit 2	
	x	y	x	y
z_0	1.62	0.00	−0.62	0.00
z_1	1.6244	0.00	−0.6156	0.00
z_2	1.63868	0.00	−0.62104	0.00
z_3	1.68523	0.00	−0.61431	0.00
z_4	1.84009	0.00	−0.62262	0.00
z_5	2.38593	0.00	−0.61235	0.00
z_6	4.69268	0.00	−0.62503	0.00
z_7	21.69268	0.00	−0.60933	0.00
z_8	440.89443	0.00	−0.62871	0.00
z_9	194386.8964	0.00	−0.60472	0.00
z_{10}		0.00	−0.63431	0.00
z_{11}		0.00	−0.59765	0.00
z_{12}		0.00	−0.64282	0.00
z_{13}		0.00	−0.58679	0.00
z_{14}		0.00	−0.65568	0.00
z_{15}		0.00	−0.57008	0.00
z_{16}		0.00	−0.54437	0.00

Dynamics Near the Fixed Points for $c = -1$

The iteration of two initial points for $z \to z^2 + c$, $c = -1$ started close to the fixed points $z_1 = 1.61803398\ldots$ and $z_2 = -0.618033988\ldots$ Both orbits diverge.

Table 14.16

First, the iteration confirms that *neither* one of the fixed points is attractive.[20] Table 14.16 lists the first 16 iterations of two initial points (the fixed points, however, are *rounded* to two decimal places).

While the iteration of the first orbit leads to infinity, the iteration of the second orbit reveals the essential dynamics in the prisoner set. Table 14.17 lists another 18 iterations of this orbit. It is dominated by the attractive orbit

[20] Again the derivative criterion for attracting and repelling fixed points can be applied to verify our findings. The absolute values of the derivatives are $|2z_1|$ and $|2z_2|$. In fact, both of these numbers are greater than 1. Thus, both fixed point are repelling.

Dynamics Near the Fixed Points (Continued)

Continuation of the computation of the second orbit of the last table 14.16. It converges to the periodic orbit which oscillates between 0 and −1.

Table 14.17

	Orbit 2 (continued)			Orbit 2 (continued)	
	x	y		x	y
z_{17}	−0.70367	0.00	z_{26}	−0.11015	0.00
z_{18}	−0.50485	0.00	z_{27}	−0.98787	0.00
z_{19}	−0.74512	0.00	z_{28}	−0.02412	0.00
z_{20}	−0.44479	0.00	z_{29}	−0.99942	0.00
z_{21}	−0.80216	0.00	z_{30}	−0.00116	0.00
z_{22}	−0.35654	0.00	z_{31}	−1.00000	0.00
z_{23}	−0.87288	0.00	z_{32}	0.00000	0.00
z_{24}	−0.23808	0.00	z_{33}	−1.00000	0.00
z_{25}	−0.94332	0.00	z_{34}	0.00000	0.00

of period 2:

$$0 \to -1 \to 0 \to \cdots$$

All initial values of the interior of the prisoner set are attracted by this orbit, and the Julia set is the boundary of this basin of attraction.

In fact, for all parameters c from the interior of the 'period-two disk' of the Mandelbrot set we obtain an attractive orbit of period two, and the Julia set is the boundary of its basin of attraction. Note that in this disk of parameters, the value $c = -1$ is special. Here the critical point coincides with one of the periodic points. Therefore, this point is called super attractive.

The Derivative Criterion for the Periodic Cycle

The derivative criterion for repelling and attracting fixed points (see page 766) can be generalized to periodic cycles. Let us take for example $c = -1$. Here the points

$$z_0 = 0 \text{ and } z_1 = -1$$

are periodic points for $z \to z^2 - 1$. Thus, they are fixed points of the transformation iterated twice, i.e., of

$$z \to (z^2 - 1)^2 - 1 = z^4 - 2z^2.$$

Now the same criterion applies as in the case of fixed points for the quadratic transformation. If the absolute value of the derivative of this transformation at the fixed point exceeds 1, then the fixed point is a repeller. If the absolute value is less than 1, then the fixed point is an attractor. In our case the derivative is $4z^3 - 4z$. The evaluation of the derivative at the two fixed points gives

$$|4z_0^3 - 4z_0| = 0 < 1$$

and

$$|4z_1^3 - 4z_1| = |-4 + 4| = 0 < 1.$$

Thus, the fixed points are attractive for $z \longrightarrow z^4 - 2z^2$. This translates to the original situation as claimed. Namely, that the periodic sequence of points $0, -1, 0, -1, \ldots$ is also attractive with respect to the original transformation $z \longrightarrow z^2 - 1$.

The derivative criterion allows us also to compute the period-2 disk, i.e., the locus of all parameters c such that an orbit of period 2 is attractive. First we derive a quadratic equation for the periodic points with (minimal) period 2. By definition a point of period 2 must satisfy

$$(z^2 + c)^2 + c - z = 0.$$

We note that also the two fixed points, given by $z^2 + c - z = 0$, solve this equation. Therefore, the polynomial $z^2 + c - z$ must be a factor of the polynomial $(z^2 + c)^2 + c - z$ of degree 4. In fact,

$$(z^2 + c)^2 + c - z = (z^2 + z + 1 + c) \cdot (z^2 + c - z).$$

From this it follows that the solutions z_1 and z_2 of

$$z^2 + z + 1 + c = 0 \tag{14.8}$$

are precisely the points of period 2. In other words, we have

$$\begin{aligned} z_1^2 + c &= z_2 \\ z_2^2 + c &= z_1 \end{aligned}$$

which, of course, can also be checked directly, which is, however, more cumbersome. We now compute the derivative of the twice iterated transformation $z \longrightarrow (z^2 + c)^2 + c$ at the periodic point z_1 (using the chain rule),

$$2(z_1^2 + c)2z_1 = 4z_1 z_2.$$

(The derivative at z_2 is identical.) It follows from Vieta's law applied to the solutions z_1, z_2 of eqn. (14.8) that

$$z_1 z_2 = 1 + c.$$

Thus, the above computed derivative is equal to $4(1 + c)$, a quantity that is less than 1 in magnitude provided that

$$|1 + c| < \frac{1}{4}.$$

This inequality describes a disk of radius $1/4$ centered at $c = -1$, and gives us the result that precisely for parameters c in this disk the periodic points z_1 and z_2 are attractive.

Two Julia Sets from the Next Buds of M

Julia sets that bound the basin of attraction of an attractive cycle of period 3 and 4. Left $c = -0.13 + 0.76i$, right $c = 0.28 + 0.53i$.

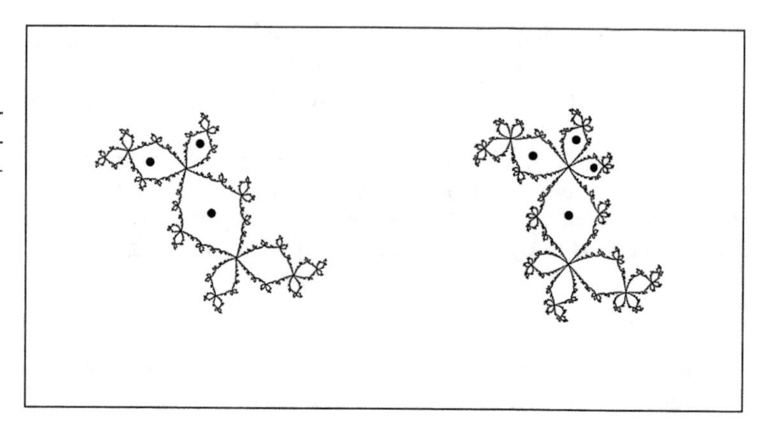

Figure 14.18

The next big buds attached at the edge of the heart-shaped center of M correspond to period-three behavior; then there are buds which house parameters belonging to attractive cycles of period 4, and so on. Figure 14.18 shows Julia sets which bound basins of attraction for period-3 and period-4 cycles. Figure 14.19 presents an overview of the periodic behavior associated with the buds, or *atoms* as Mandelbrot calls them. Clearly they are strictly organized. Each bud carries on its boundary another complete set of smaller buds with corresponding sequences of periodic attractive cycles. Note that there is an amazing rule for the periods corresponding to the buds. Two given buds of periods p and q at the cardioid determine the period of the largest bud in between them as $p + q$. Similar rules are true for the buds on buds.

The Touching Points of Buds

Now let us take a look at the points of the cardioid again. The fixed point which has been attractive for parameters within the heart-shaped region loses this property right on the cardioid. Here the corresponding fixed point is said to be indifferent. In other words, the iteration $z \to z^2 + c$ is at the fixed point neither attractive nor repelling; it is related to a rotation (see eqn. (14.4)). Depending on whether that rotation is given by a rational or irrational number, the associated prisoner set is dramatically different. Moreover, when the rotation is irrational, several striking cases can be distinguished. In the following we will not make an attempt to discuss the classification to the extent it is known. Rather we will pick some particular cases to demonstrate the intricacies which evolve in understanding the prisoner sets. For the points where the buds are attached this rotation is given by an angle $\phi = 2\pi\alpha$, which is a rational multiple of 2π (i.e., α is a rational number). For example, we have $\alpha = 1/2$ at the point where the 'period-two disk' is attached and we have $\alpha = 1/3$ where the 'period-three bud' above the cardioid is attached. Figure 14.20 shows the Julia sets for these two examples.

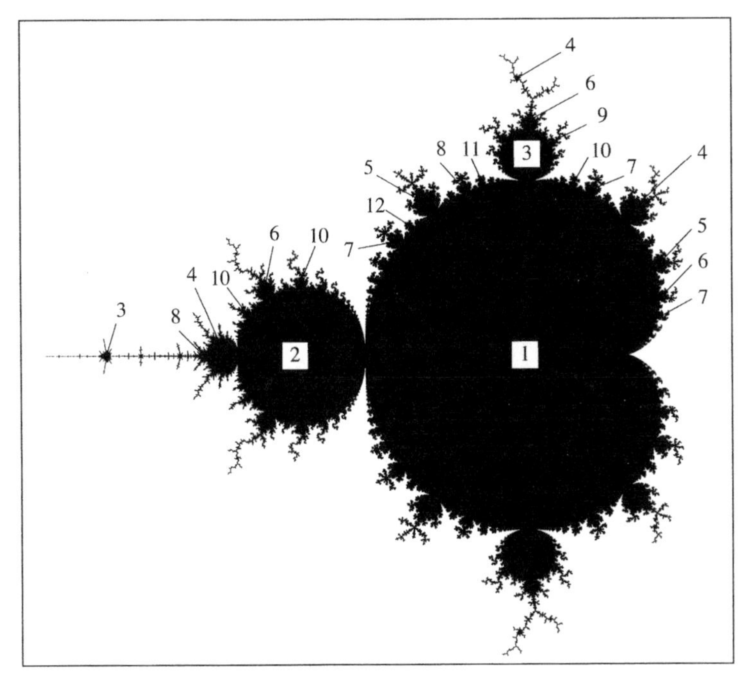

The Mandelbrot Set and Its Atoms

The buds of the Mandelbrot set correspond to Julia sets that bound basins of attraction of periodic orbits. The numbers in the figure indicate the periods of these orbits.

Figure 14.19

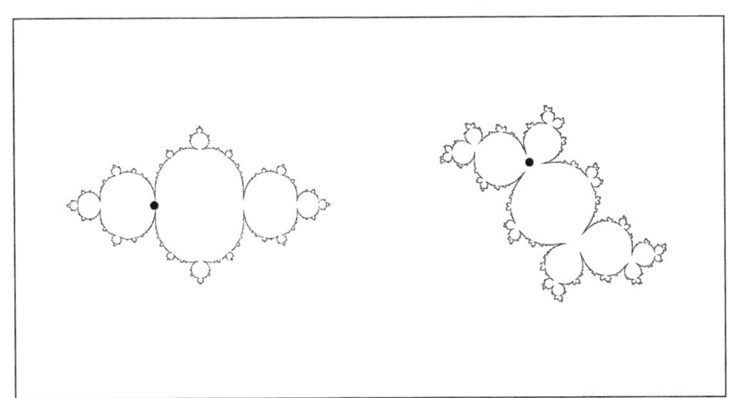

Parabolic Fixed Points

Julia sets corresponding to parabolic fixed points. $c = 0.75 + i$ (left) and $c = -0.125 + 0.64925i$ (right), the point where the 'period-three bud' is attached to the heart-shaped center of M.

Figure 14.20

The Derivative Criterion for Indifferent Fixed Points

We have characterized a fixed point z of $z \to z^2 + c$ by its derivative $\lambda = 2z$. It is

- attractive if $|\lambda| < 1$ and repelling if $|\lambda| > 1$.

Now we turn to the intermediate case. We call the fixed point indifferent if the derivative $|\lambda|$ is equal to 1 in absolute value. In this case the derivative is determined by the argument $\phi = \arg \lambda$,

$$\lambda = \cos \phi + i \sin \phi$$

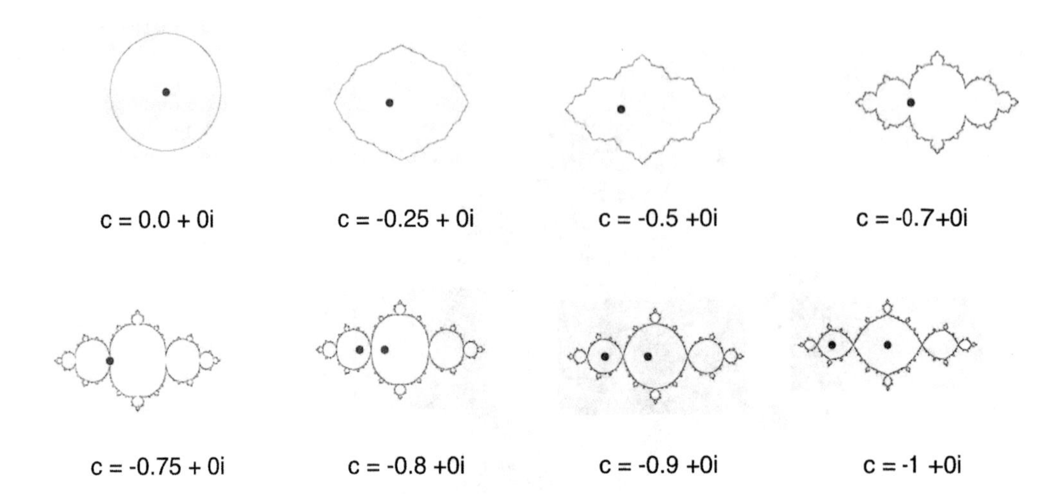

Figure 14.21 : Starting from the Julia set for $c = 0$ (the circle) we decrease the parameter to $c = -1$. The Julia set develops a pinching point for $c = -0.75$ and is the boundary of a period-2 attractor for the remaining plots.

with $0 \leq \phi < 2\pi$. Setting as usual $\phi = 2\pi\alpha$, we call an indifferent fixed point

- rationally indifferent (or parabolic) if α is a rational number and
- irrationally indifferent if α is an irrational number.

The characterization of periodic orbits can be extended in the same way. In this case we check the derivative of the corresponding iterated map.

But what happens if the fixed point is indifferent and the rotation angle α (of the derivative) is irrational? This is really a complicated case and leads to Julia sets which are only slightly understood. For a moment let us stay with those cases which are more accessible. We know that any irrational number can be approximated by sequences of rational numbers. Some sequences approach a given irrational number faster than others. Some irrational numbers admit approximating sequences which converge very rapidly, others only admit sequences which converge rather slowly.[21] It turns out that these differences matter substantially for the character of a Julia set corresponding to an angle α. If α is an irrational number such that any approximating sequence of rational numbers converges very slowly (in some precise sense), then the prisoner set is a so called *Siegel disk*.[22] It turns out that among all irrational numbers there is one which stands out as the one which is the hardest to approximate by

[21] The technical section on page 812 explains in what sense the approximation of irrationals by rationals can be poor.

[22] This case is named after the German mathematician Carl Ludwig Siegel who established characteristics of that case in his paper *Iteration of analytic functions*, Ann. of Math. 43 (1942) 607–616. More precisely, the Siegel disk is only the disk-like component of the prisoner set which contains the fixed point. The other components are preimages of that disk.

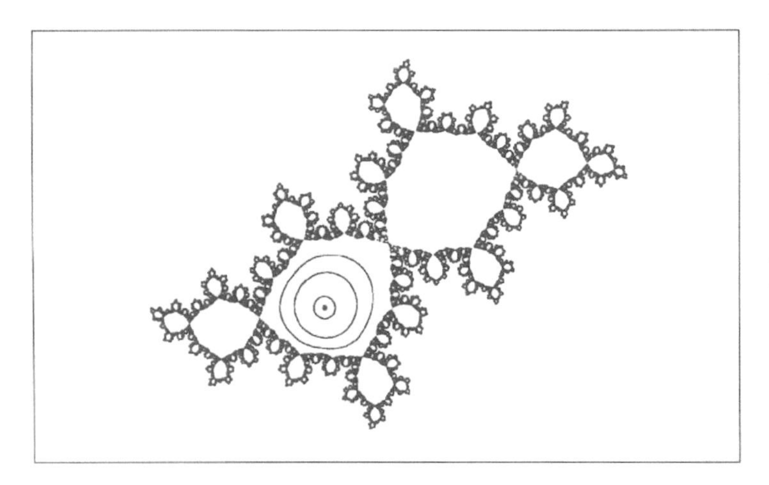

A Siegel Disk

The Julia sets that belongs to $c \approx -0.3905407802 - 0.5867879073i$ is a Siegel disk. With α being the golden mean, use $\phi = 2\pi\alpha \approx 3.883222077$ in eqn. (14.4) on page 800 to obtain the real and imaginary components of c. The dynamics near the fixed point is characterized by invariant curves on which the iteration acts like a rotation by the angle α.

Figure 14.22

rational numbers. This number is the golden mean $\alpha = (\sqrt{5} - 1)/2$. Figure 14.22 shows the corresponding, most prominent Siegel disk. Note that in this case the indifferent fixed point lies in the interior of the prisoner set, while it is known that for the rationally indifferent case it is on the boundary (i.e., part of the Julia set). Thus, if α_n, $n = 1, 2, \ldots$ is a sequence of rational numbers which approaches α, the golden mean, the associated Julia sets are fundamentally different from the one corresponding to the limit α.

What we have looked at so far are the Julia sets that correspond to the attractive and the super attractive case, the rationally indifferent case and the Siegel disk case.[23] All these cases are characterized by the fact that the prisoner set has non-empty interior. Besides those prisoner sets which are totally disconnected (with parameter values c that are not in M) and those which have a non-empty interior, there are also boundary cases: prisoner sets with no interior point, but which, however, are still connected. These typically have many branches at all levels of details and are called *dendrites*. The most popular, certainly, is the dendrite for $c = i$. But also the prisoner set for $c = -2$, which is simply the real interval $[-2, 2]$ (which has no branches), is in this class. See figures 14.23 and 14.28 for more examples. While these Julia sets are quite easy to compute and draw, this class of dendritic Julia sets also contains an infinity of monsters for which it probably is impossible to provide a valid computer graphic representation.

[23]For more general than quadratic iterations, given by rational functions $f(x) = p(x)/q(x)$, with p and q being polynomials, there is even a fifth different type of prisoner set with non-empty interior. The other four cases are: attractive case, super-attractive case, rationally indifferent case, and Siegel disk case. This is a result which partially goes back to Julia and Fatou and was completed by Dennis Sullivan (see *Quasiconformal homeomorphisms and dynamics I*, Annals of Math. 122 (1985) 401–418). This fifth type is a so-called *Herman ring*, named after M. R. Herman who constructed the first example (see M. Herman, *Exemples de fractions rationelles ayant une orbite dense sur la sph/'ere de Riemann*, Bull. Soc. Math. France 112 (1984) 93–142). See also A. F. Beardon, *Iteration of Rational Functions*, Springer-Verlag, New York, 1991.

Dendrites

Julia sets for $c = -2$ (left), a line segment, and $c = i$ (right), a more typical dendrite.

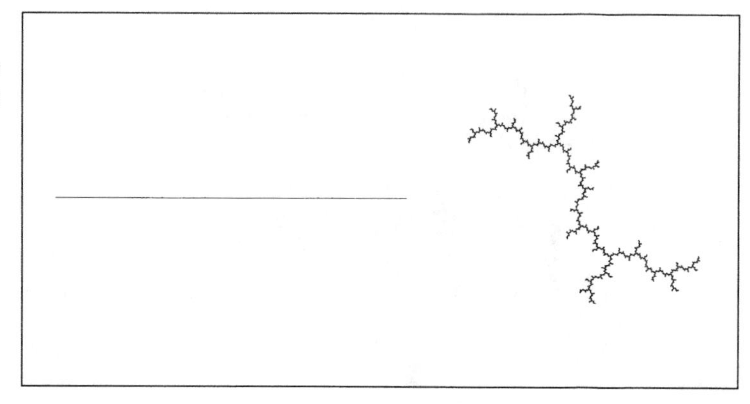

Figure 14.23

Irrational and Liouville Numbers

A number α is said to be badly approximable by rational numbers provided α satisfies the following number theoretical condition. There are $\varepsilon > 0$ and $\mu > 0$ such that

$$\left|\alpha - \frac{p}{q}\right| > \frac{\varepsilon}{q^\mu} \tag{14.9}$$

holds for all integers p and positive integers q. The famous result of Joseph Liouville dating from 1844 states that any (irrational) algebraic number[24] α of degree $n \geq 2$ is badly approximable where the constant μ in eqn. (14.9) is $\mu = n$. For example, the golden mean $\alpha = (\sqrt{5} - 1)/2$ is algebraic of degree 2 (it solves the equation $x^2 + x - 1 = 0$) and fulfills eqn. (14.9).

Liouville's estimate allows the explicit construction of transcendental numbers, i.e., irrational numbers which are not algebraic. Most real numbers are transcendental, but only relatively few are explicitly known. Indeed, algebraic numbers are countable, as Cantor argued by observing that algebraic equations with integer coefficients are countable.

For the explicit construction of a transcendental number, take for example, the number

$$\alpha = a_1 10^{-1!} + a_2 10^{-2!} + \cdots + a_k 10^{-k!} + \cdots, \tag{14.10}$$

where the coefficients a_i are arbitrary digits between 1 and 9 and $k!$ is 'k factorial' ($k! = k \cdot (k-1) \cdots 2 \cdot 1$). Thus, we may write the decimal expansion of α as

$$\alpha = 0.a_1 a_2 000 a_3 00000000000000000 a_4 000 \ldots$$

The blocks of zeroes in the decimal expansion of α grow rapidly. Now we introduce α_k to be the rational number obtained by only considering

[24] A number α is called *algebraic* of degree n provided it solves an equation of the form $b_0 + b_1 x + \cdots + b_n x^n = 0$ with integer coefficients b_i and where n is minimal.

terms up to (and including) the kth term in the definition of α in eqn. (14.10). Then we have that

$$|\alpha - \alpha_k| < \frac{10}{10^{(k+1)!}}.$$

Now let us assume that α is algebraic of some degree n. Setting

$$\alpha_k = \frac{p}{q} = \frac{p}{10^{k!}}$$

we would have from Liouville's result that there is $\varepsilon > 0$ and

$$|\alpha - \alpha_k| > \frac{\varepsilon}{10^{nk!}}.$$

Combining the two estimates for $|\alpha - \alpha_k|$ yields

$$\frac{\varepsilon}{10^{nk!}} < |\alpha - \alpha_k| < \frac{10}{10^{(k+1)!}}$$

or

$$\frac{(10^{k!})^{k+1}}{(10^{k!})^n} = (10^{k!})^{k+1-n} < \frac{10}{\varepsilon}.$$

However, the last inequality cannot be right; as k grows the left hand expression grows beyond all bounds. This means that α cannot be algebraic of degree n; it must be transcendental. This number is an example of a so called Liouville number. It does not satisfy the above number theoretical condition in eqn. (14.9), which characterizes poor approximation by rational numbers.

Liouville Monster

It is known that Julia sets which belong to fixed points around which $z \to z^2 + c$ is not linearizable are not locally connected.[25] An example is given by $z \to z^2 + c$ such that there is a fixed point which is irrationally indifferent and α is a Liouville number, i.e., it can be approximated well by rational numbers (see the technical section above). To get some flavor of the unimaginable complexity of such a Julia set, let us discuss this case from the viewpoint of field lines of the potential p_c as introduced in the previous chapter. The Julia set is not locally connected; and according to a result of Constantin Carathéodory, this implies that not all field lines land in a point of the Julia set.[26] However, it is known that this is not the rule; on the contrary, almost all field lines land in points of the Julia set. Douady and Hubbard have

[25] The transformation $f_c(z) = z^2 + c$ is called *linearizable* around a fixed point z_0 provided in a neighborhood of that fixed point the mapping is (analytically) equivalent to λz, where λ is the derivative of f_c at the point z_0. Examples are fixed points for which there is a Siegel disk, or attracting, or repelling fixed points, see A. F. Beardon, *Iteration of Rational Functions*, Springer-Verlag, New York, 1991, page 133. A set X is called *locally connected* provided every point p in X and every neighborhood U of p contains an open set V with $p \in V$ which is connected. For the theorem about the absent local connectedness see D. Sullivan, *Conformal dynamical systems*, Lecture Notes in Mathematics 1007 (1983) 725–752.

[26] See C. Carathéodory, *Untersuchungen über die konformen Abbildungen von festen und veränderlichen Gebieten*, Math. Ann. 72 (1912).

succeeded in showing that in particular all field lines corresponding to angles which are rational multiples of 2π must land in a point of the Julia set. It is beyond imagination to picture the complexity which such a strange condition must enforce: all rational field lines land but at least some irrational field lines do not! Note, that it is also not known whether Julia sets associated to Siegel disks (i.e., their boundaries) are locally connected in general. In other words, we begin to see how important the idea of the potential of prisoner sets is for understanding the structure of Julia sets. Moreover, concepts such as local connectedness are not just a phantasy of too theoretically minded mathematicians. They are decisive in understanding very physical properties of real objects if we accept Julia sets as real objects!

Do Field Lines Land?

It is not easy to imagine a structure of a Julia set so that it has field lines that do not land on the Julia set. In figure 14.24 we present a model (it is certainly not a Julia set) where the nature of this problem becomes apparent and understandable.

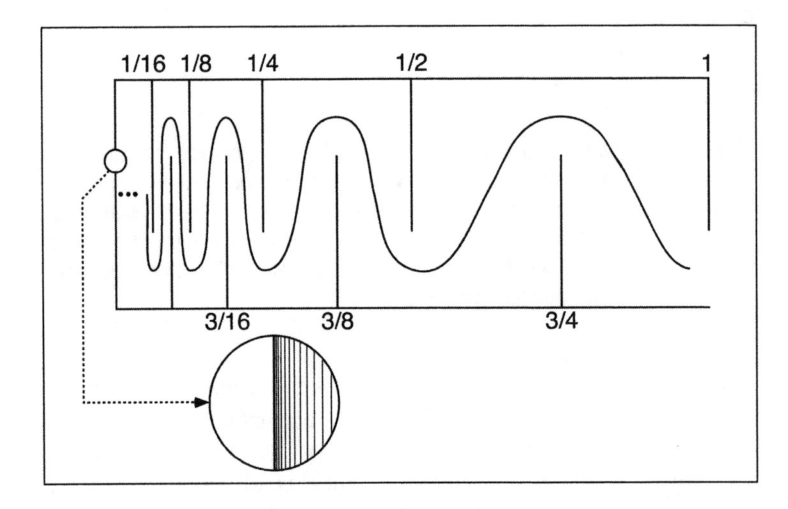

Figure 14.24 : For this model of a double-comb, there is a field line which oscillates infinitely often and which does not converge to a point at the left end of the comb. The double comb is not locally connected: in any small disk centered around the midpoint of the far left tooth, there is an infinite set of separate line segments piling up towards the left end.

We show a double comb-like structure which is essentially given by two combs with infinitely many intertwined teeth positioned as shown in the figure. The combs are connected by an additional, longer tooth at the left end. A field line which is trying to get to this far left tooth must wiggle more and more to get around the infinitely many teeth and therefore will never arrive at the goal. But there must be a field line which oscillates infinitely often and which separates lines that land on the upper comb from those that land on the lower one. Pathologies like this were studied around 1912 by Carathéodory, who found that a compact, connected subset K of the plane has a potential with field lines that land everywhere on the boundary of K provided that the boundary is locally connected.[27] This explains the problems in figure 14.24. Although the infinite double comb is connected, it is not locally connected.

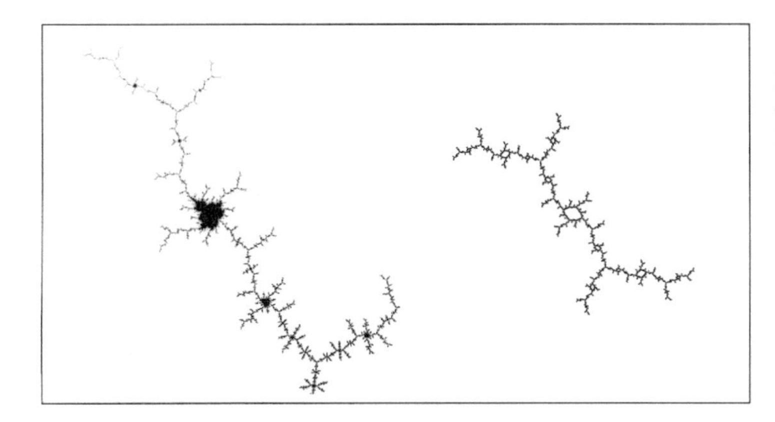

Miniature M-Set I

Enlargement of a secondary Mandelbrot set in the upper region of the Mandelbrot set. The Julia set is for $c = 0.159789 + 1.03332i$.

Figure 14.25

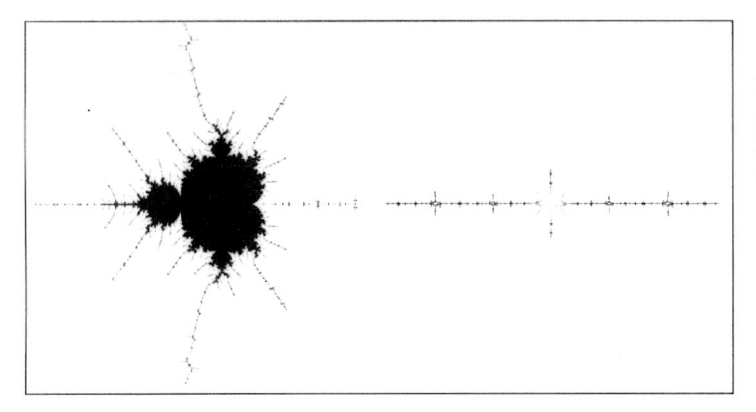

Miniature M-Set II

Enlargement of a secondary Mandelbrot set in the left region of the Mandelbrot set (the tip). The Julia set is for $c = -1.77578$.

Figure 14.26

Secondary Mandelbrot Sets

Let us now briefly look at those prisoner sets which appear to be hybrids of Julia sets which are dendrites and those which bound prisoner sets with interior points. We find such cases for parameter values c from so called *secondary Mandelbrot sets*. These are small copies of M which, however, are subsets of M. In fact, one can find such small copies of M in the neighborhood of any point of the boundary of the Mandelbrot set. In other words, this is

[27] Another condition that must be satisfied is that the exterior of K is simply connected, i.e., must not contain any 'holes'. See C. Carathéodory, *Über die Begrenzung einfach zusammenhängender Gebiete,* Math. Ann. 73 (1913).

Final-State Diagram and Mandelbrot Set

The final-state diagram of $z \to z^2 + c$ in comparison to the Mandelbrot set.

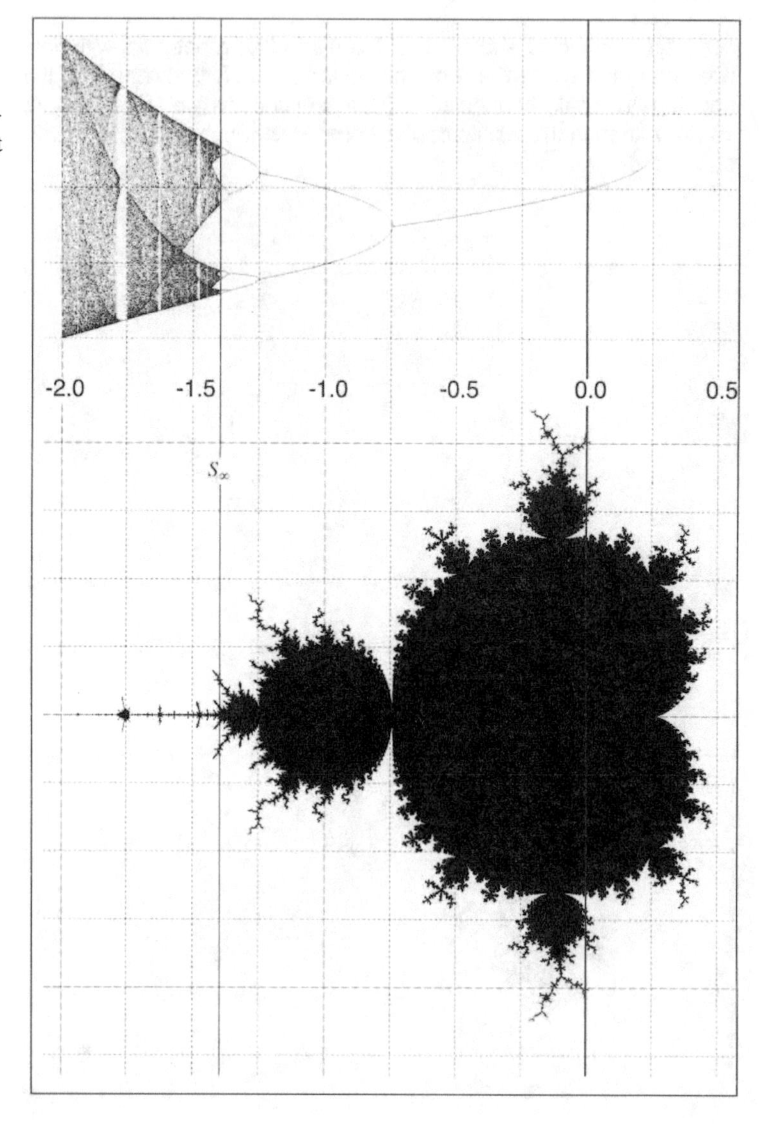

Figure 14.27

a self-similarity feature of M.[28] Figure 14.25 shows an enlargement of the upper region of M and a corresponding Julia set. The interior of the prisoner set is attracted by a period-3 cycle. The same is true for our other example of a secondary Mandelbrot set, which is located on the real axis and shown in figure 14.26.

This brings us back to the discussion of $z \to z^2 + c$ for real numbers z and c. In chapter 11 we used the final-state diagram as an ordering principle for the

[28] The exact relation of a copy with M is discussed in A. Douady, J. H. Hubbard, *On the dynamics of polynomial-like mappings*, Ann. Sci. Ecole Norm. Sup. 18 (1985) 287–344.

long term behavior of the quadratic iterator. Figure 14.27 shows the final-state diagram of $z \rightarrow z^2+c$ in comparison to the Mandelbrot set. At real parameters from the heart-shaped region of M we see just one branch in the diagram, which identifies the attractive fixed point for the varying parameters c. Reading from right to left this branch splits, representing the two points forming period-2 cycles. This corresponds to the real parameters of the 'period-2 bud' in M. We observe that the complete period-doubling tree corresponds to a stack of smaller and smaller buds aligned at the real axis starting at the heart-shaped region of M. Now look at the periodic windows in the final-state diagram. They correspond to the secondary Mandelbrot sets at the tip of M.

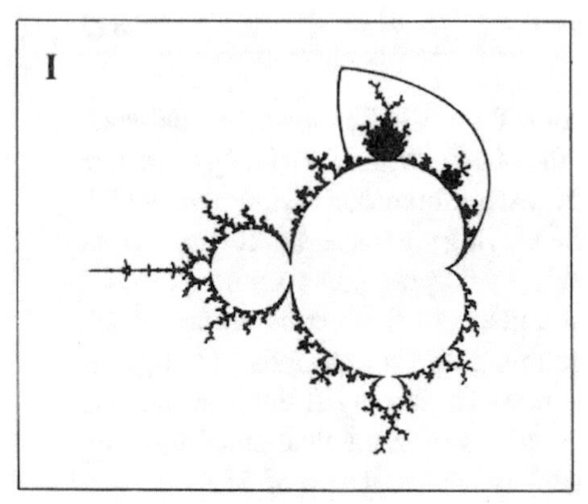
I

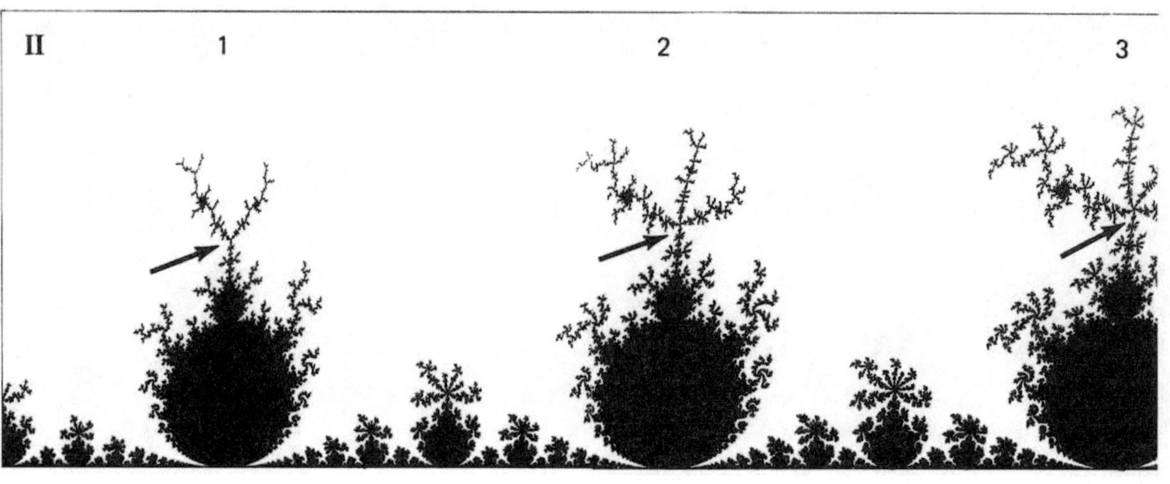
II
1
2
3

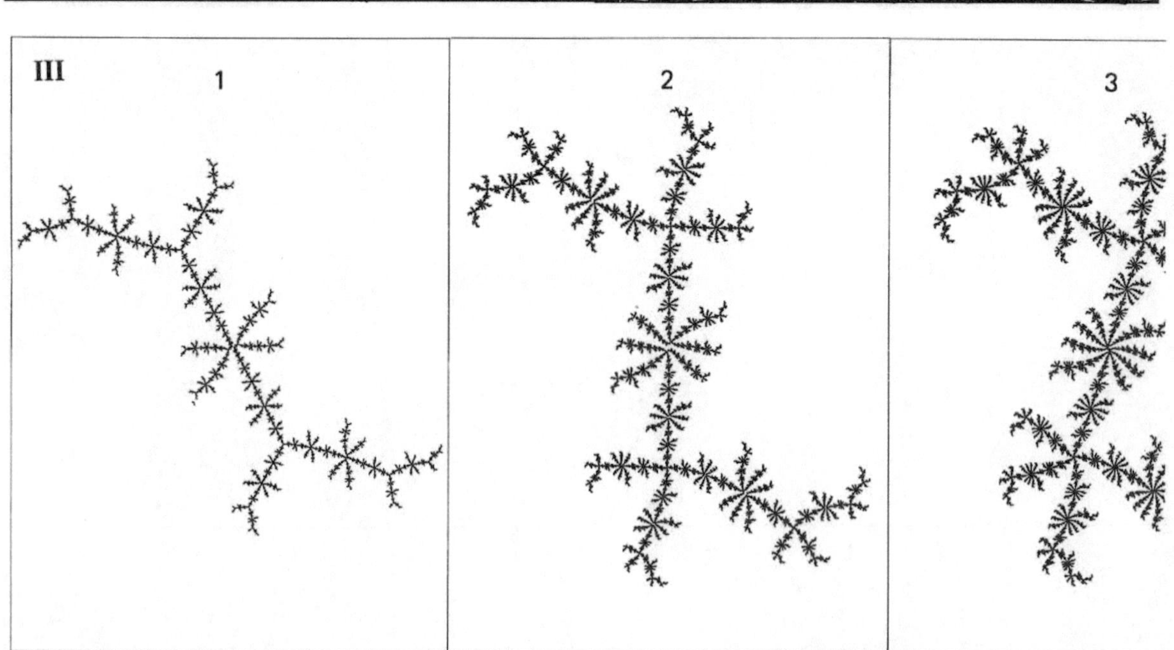
III
1
2
3

Correspondence between Mandelbrot set and Julia sets. The section of the Mandelbrot set indicated in part I (top) is enlarged in part II (center). The magnification factor is not uniform; it increases from left to right in order to make the disk-like parts of the Mandelbrot set appear about the same size. The Julia sets computed for the parameter values indicated by the arrows are shown below.

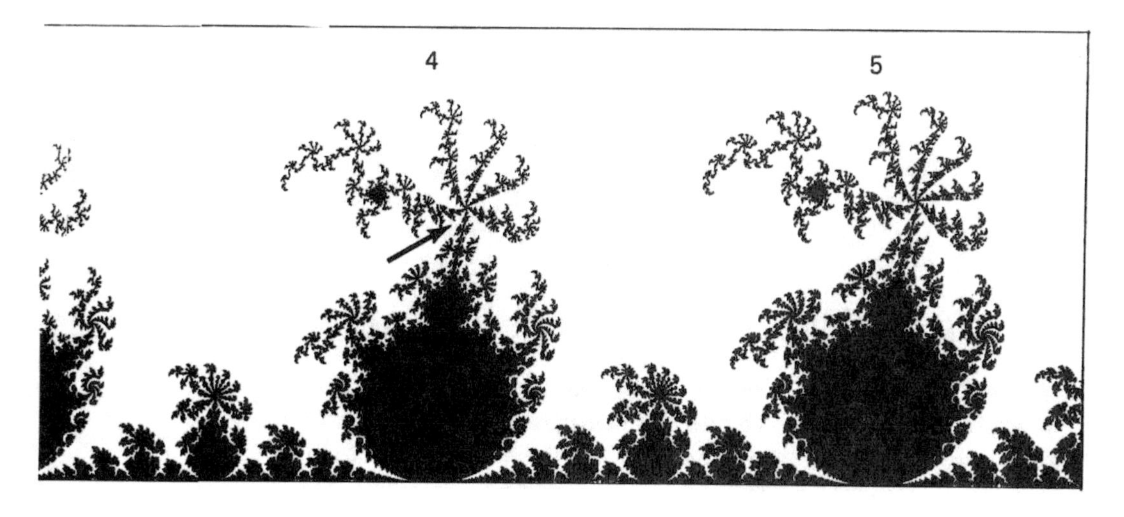

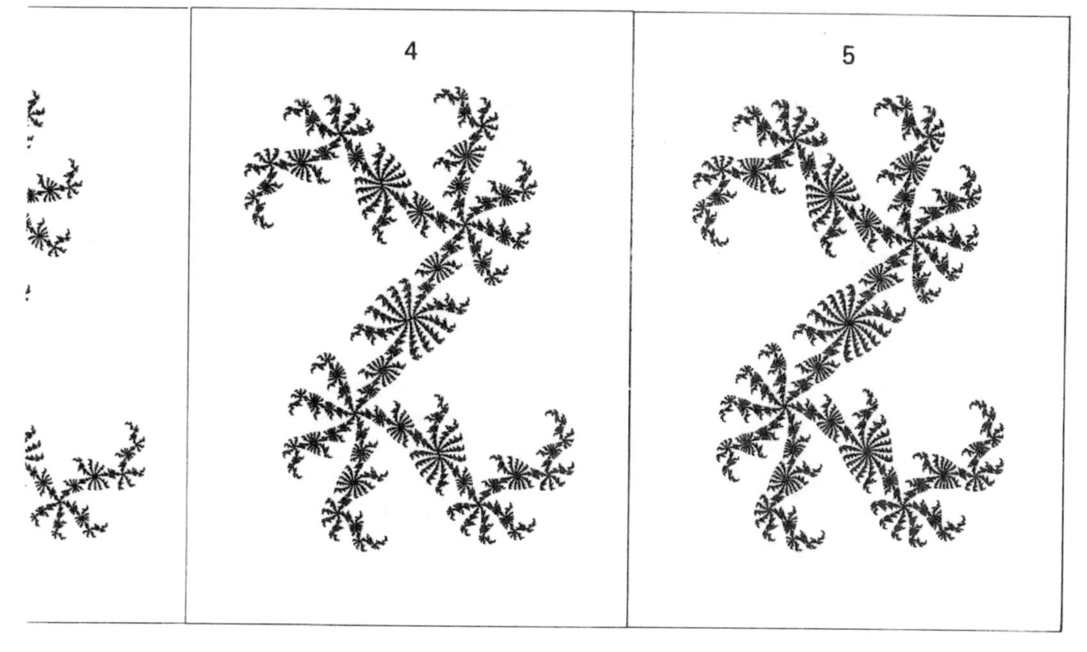

Figure 14.28

Zoom into the Mandelbrot Set

In these 9 images a zoom into the boundary of the Mandelbrot set is shown. The final magnification is 300,000,000 fold.

Figure 14.29

14.3 The Mandelbrot Set as a Table of Content

We saw in the last section that the Mandelbrot set is a visualization of the structural dichotomy of Julia sets for the quadratic iterator. It can also be used as a road map to all possible kinds of Julia sets. This road map is fantastically detailed, and this fascinating property of the Mandelbrot set — Mandelbrot already observed and documented it as early as 1980 — is the topic of this section.

Comparing the Boundary of M and Julia Sets

Let us look at the sequence of plots in figure 14.28. The upper portion of this figure shows a region along the boundary of the heart-shaped central region of the Mandelbrot set which is continuously enlarged (from left to right the magnification factor increases from 1 to about 6) and stretched out, so that the respective segment of the cardioid becomes a line segment. We see disk-like components out of which grow dendritic structures with 2, 3, ..., 6 major branches (from left to right). Now we choose as particular c-values the major branch points of these dendrites (see arrows). For each of these 5 c-values we compute the associated Julia set, displayed in the lower part of the figure.

We observe a striking resemblance in the structures and the combinatorics of the dendritic structures in the Mandelbrot set and the Julia sets. The geometry of these Julia sets is also visible in the Mandelbrot set when magnified appropriately at the corresponding parameter values. In this sense it is a visual *table of content* of the book of Julia sets (see also figures 14.29 and 14.30).

Figure 14.31 shows another more systematic example, which, however, still defies a rigorous mathematical understanding. The upper part shows the Mandelbrot set together with an enlargement around the c-value given by the

Zoom into a Julia Set

A parameter c is chosen from the center of the last image in the previous figure 14.29. We compute successive enlargements centered about this parameter c as for the Mandelbrot set in figure 14.29. Note how similar these Julia set sections are to the Mandelbrot set closeups. In the final images the objects are practically indistinguishable except for the scale and a rotation.

Figure 14.30

cross hair cursor. Incidentally, the magnification factor is about 10^6, and the result is identical with part (f) in figure 14.11. The bottom part of figure 14.31 shows the Julia set for the c-value which is determined by the cross hair cursor and a blow up of a tiny portion of that Julia set around the point of the cursor in the Julia set. The magnification factor is about 10^5 here, and the resulting image is rotated by approximately 55° clockwise. The resulting similarity is striking and cannot be an accident. But if there is a systematic relation between the enlargements seen in the Mandelbrot set and enlargements of corresponding Julia sets, it cannot be easy to grasp in this case.

Note that the center of the upper magnification in figure 14.31 is a tiny copy of the Mandelbrot set, as figure 14.32 reveals, while the center of the lower left enlargement in figure 14.31 is definitely not a Mandelbrot set, as figure 14.33 shows. Thus, the similarity between the Mandelbrot set and the particular Julia set seen in figure 14.31 is a transitional one, i.e., it depends on the magnification factor.

For a large variety of Julia sets, however, the similarity is now understood quite clearly and with all mathematical rigor. This is a result of the beautiful work of the Chinese mathematician Tan Lei.[29] To discuss her results we need a few new tools. In particular, we have to extend our concept of self-similarity.

[29]Tan, Lei, *Similarity between Mandelbrot set and Julia sets,* Commun. Math. Phys. 134 (1990) 587–617.

Figure 14.31 : Similarity of an enlargement of M and the closeup of the Julia set J_c at $c = -0.745429+0.113008i$.

Enlargement of Mandelbrot Set

An enlargement centered at $c = -0.7454285 + 0.1130089i$. The width of the figure is 0.000006.

Figure 14.32

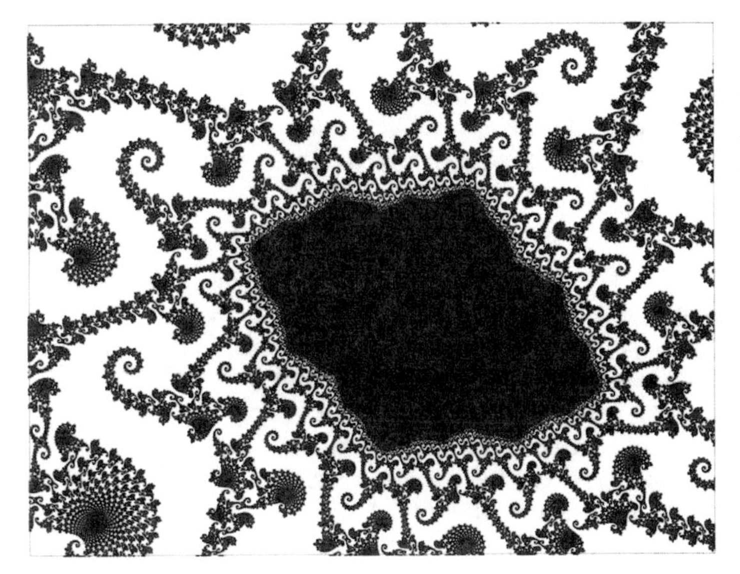

Enlargement of Julia Set

An enlargement for the Julia set for $c = -0.7454285 + 0.1130089i$. The figure is centered at c and has width 0.000045.

Figure 14.33

Magnification at a Particular Point

In order to discuss the similarity between Julia sets and the Mandelbrot set we have to magnify in a very particular fashion: we scale up around a given point, rotate by a certain angle, and repeat this procedure infinitely often. Recall that an image I (i.e., a subset of the plane) is *self-similar* if it is composed of parts which are small copies of the whole. Thus,

$$I = w_1(I) \cup w_2(I) \ldots \cup w_n(I)$$

where the transformations w_i are similarity transformations. Turned the other

Self-Similarity of Stars

The rays extend beyond all bounds. The stars are self-similar at the center.

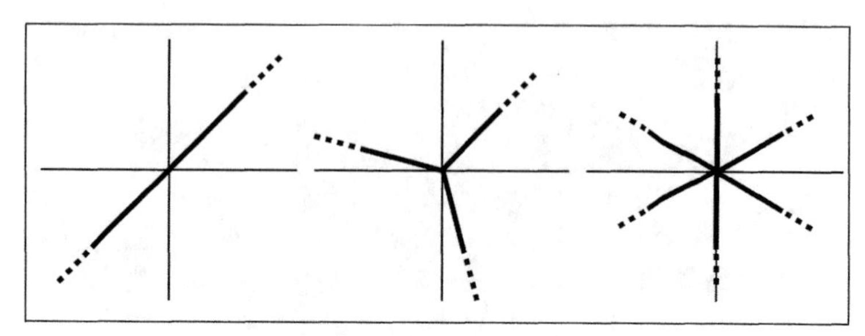

Figure 14.34

way around, a suitable magnification (i.e., scaling and rotation) of the parts produces the complete set I. We center the magnification at one point, for example at $z_0 = 0$. This can be described as a scaling operation using a factor $\rho > 1$. Then ρI denotes all points of I scaled up by ρ (i.e., $\rho I = \{\rho z \mid z \in I\}$). Now assume that we have an object I which is invariant with respect to scaling by ρ,

$$I = \rho I.$$

An object which satisfies this condition obviously must extend to infinity. The infinite stars shown in figure 14.34 are examples. But let us focus our attention on the neighborhood of the point z_0. We choose a disk $D_r(z_0)$ centered at z_0 with radius r as our area of interest. If there is a number $\varepsilon > 0$ such that

$$D_r(0) \cap I \;=\; D_r(0) \cap \rho I$$

holds for all radii $r < \varepsilon$, we say that the set I is *self-similar in 0*. The infinite and finite stars (see figure 14.34) are examples.[30] More generally, we may even allow that the scaling factor ρ is a *complex* number. In chapter 13 we showed that the multiplication by a complex number can be understood as scaling by $|\rho|$ and rotating by the angle $\phi = \arg \rho$.[31] In other words, the multiplication of an object (in the complex plane) by the complex number ρ with $|\rho| > 1$ can be interpreted as magnification in our sense. Look for example again at figure 14.34. The multiplication could involve a rotation by a multiple of 120 degrees (right) or 60 degrees (left). But let us go for more interesting examples.

Can you imagine an object which is self-similar in 0 under scaling with the golden mean $g = (\sqrt{5}-1)/2$ and rotation by 90 degrees (i.e., $\rho = ge^{i\pi/2}$) and which is not a star like the objects in figure 14.34? You have already seen such an object! It is the golden spiral familiar from chapter 4. Figure 14.35 shows a small collection of logarithmic spirals. You will recall that logarithmic spirals are best described in polar coordinates (r, ϕ). The equation for the logarithmic

[30] A cover of a book that shows a picture of a hand holding that very book provides another example (see figure 3.11).
[31] Recall that $\rho = |\rho|(\cos\phi + i\sin\phi) = |\rho|e^{i\phi}$.

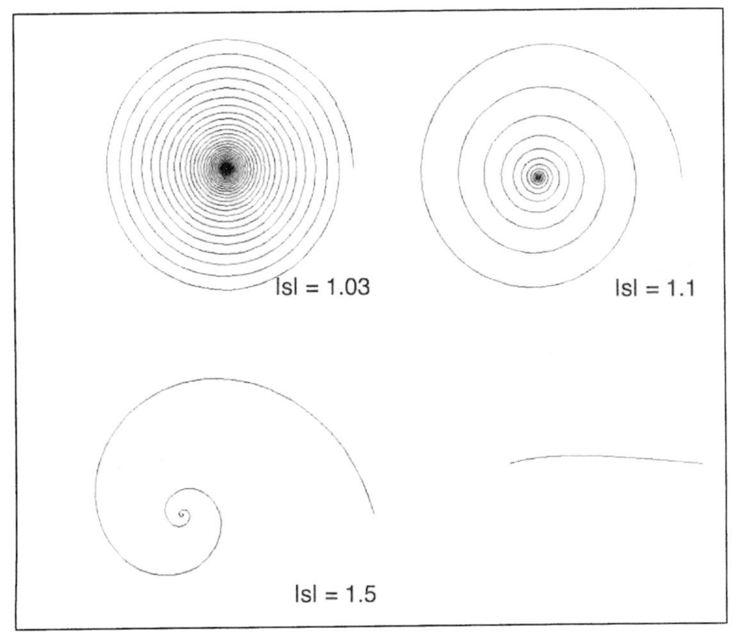

Different Scaling Values

The logarithmic spiral with $a = -0.019$, -0.061, -0.258 and -8.795 (from upper left to lower right) is invariant under rotating by 90 degrees (clockwise) and scaling by $|\rho| = 1.03, 1.1, 1.5$, and $1,000,000$.

Figure 14.35

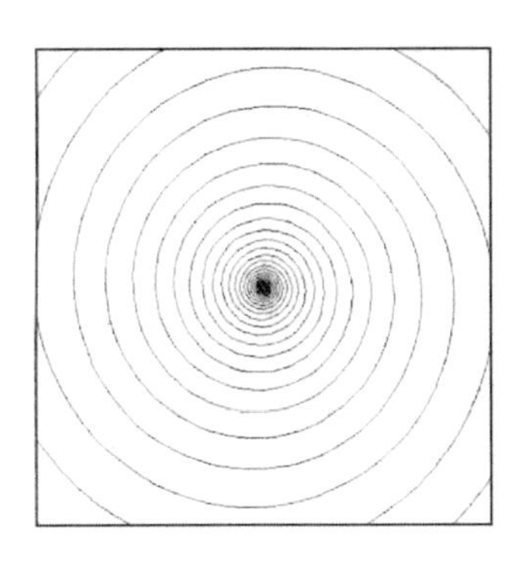

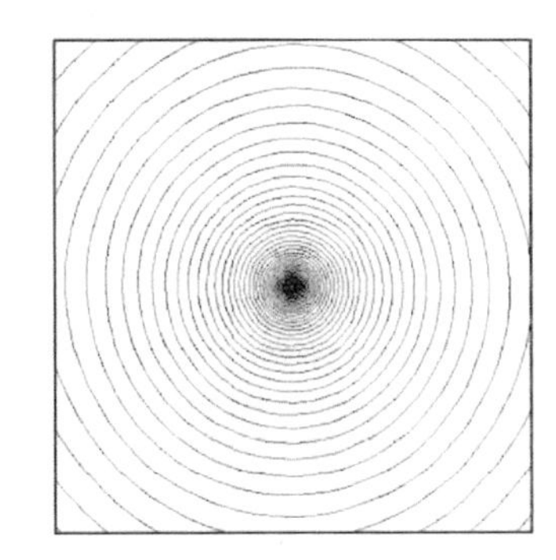

Interwoven Spirals

One of the two spirals shown is a simple spiral, while the other is a double spiral, i.e., it is a composition of two interwoven logarithmic spirals. Can you decide at a first glance which is which? If not, just follow the spirals for one turn.

Figure 14.36

spiral is

$$\log r = a\phi,$$

where r denotes the radius of the spiral at angle ϕ. From this notation we see that if we rotate by, for example, 90 degrees, we have to scale by $e^{a\pi/2}$ to obtain the same spiral. In other words, it is self-similar in 0 under the multiplication by $\rho = e^{a\pi/2}e^{i\pi/2}$.[32] Observe how the appearance of the spirals changes from almost circular to almost straight. In the latter case it is hard to see any rotation at all. The stretching is so large that it overrides the rotational effect in the visualization.

[32]This result can be extended; the spiral is self-similar under multiplication with $e^{a\phi+i\phi}$ for all angles ϕ.

Spiral or not Spiral?

The artwork is by Nicholas Wade. Reproduced with the kind permission of the artist. From: Nicholas Wade, The Art and Science of Visual Illusions, Routledge & Kegan Paul, London, 1982.

Figure 14.37

Logarithmic spirals are good models to describe some of the patterns we are going to discuss in the Mandelbrot set. Even compositions of spirals such as the double spiral from figure 14.36 and more complicated patterns can be found. But before we turn to those beautiful images, take a look at figure 14.37. This is indeed a fantastic spiral, don't you agree? Do you see the spiral? Which spiral? Is there a spiral at all? You might already begin to see that the world of spirals alone can be so confusing that an untrained eye can easily be fooled. Thus it is nice to have a bit of solid mathematics to hold onto when studying the wealth of structures in the Mandelbrot set.

We now return to the development of the description of the self-similarity features of the Mandelbrot and Julia sets. So far we have discussed the term self-similarity of a set I in a point z_0 for $z_0 = 0$. In this case the similarity transformation is simply described by the multiplication by a complex number ρ. If $z_0 \neq 0$, the transformation consisting of a magnification (with possible rotation) around z_0 is technically more complicated. We first have to translate the object by $-z_0$ in order to move the center z_0 of the magnification to the origin. For this purpose let us introduce the notation

$$I - z_0 = \{z - z_0 \mid z \in I\}$$

Then we proceed as before; we magnify and rotate using multiplication by a

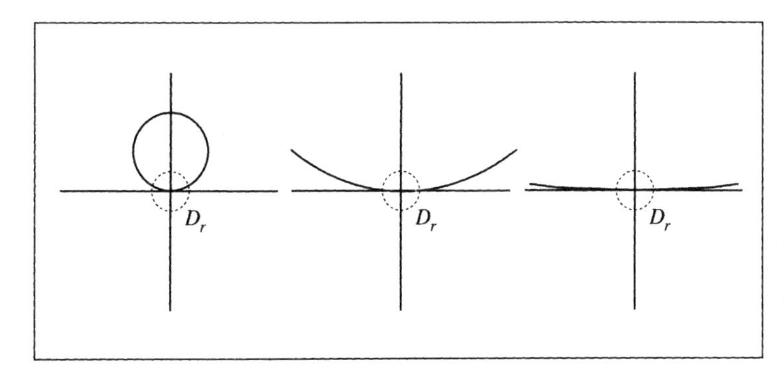

Self-Similarity of the Circle Segment

The circle is asymptotically self-similar at any of its points. The given circle of radius 1 around the center $z = i$ (left) is scaled by a factor of $\rho = 5$ (center) and again (right). In the limit, a straight line is approached.

Figure 14.38

complex number ρ. In other words, for the self-similarity property at a point z_0 we require that

$$D_r(0) \cap (I - z_0) = D_r(0) \cap \rho(I - z_0)$$

holds for all sufficiently small radii $r > 0$.[33]

Asymptotic Self-Similarity at a Point

Let us return to the Mandelbrot set M and the Julia sets J_c. Unfortunately, we cannot postulate the self-similarity of M or a Julia set J_c at any point. The local structures of these objects are somewhat too complex for plain self-similarity, but not by much. The self-similarity at a point does not become apparent after a single magnification and rotation. The magnified and rotated copy is not identical to the original, not even in the small neighborhood of the center of magnification. However, when we repeat this rescaling procedure over and over, the resulting objects *converge* to a set which is self-similar at a point. Therefore, we call this property of the Mandelbrot set and of Julia sets *asymptotic self-similarity at a point*. Formally we can describe this as follows: we call I asymptotically self-similar at the point z_0 if there are

- a complex scaling factor ρ, called *multiplier*, with $|\rho| > 1$,
- a small radius $r > 0$,
- and a limit object L (a subset of the complex plane) which is self-similar at the origin

such that the relation

$$\lim_{n \to \infty} D_r(0) \cap \rho^n(I - z_0) = L \cap D_r(0)$$

holds.

Let us consider a simple example: a circle. Figure 14.38 shows the circle S, given by the set of points $\{z \mid |z - i| = 1\}$, and its scaled copies, using the scaling factor $\rho = 5$. Observe that at the intersection of $\rho^n S$ and the disk $D_r(0)$ about the origin is an arc from the (rescaled) circle and straightens more and more from stage to stage. It converges to the line segment $L = D_r \cap \{(x + yi \mid y = 0\}$ as n increases.

[33] An alternative would be to use the similarity transformation $\theta(z) = \rho z + (1 - \rho)z_0$. This is a similarity transformation which has z_0 as fixed point. Then we require that $D_r(z_0) \cap I = D_r(z_0) \cap \theta(I)$.

Misiurewicz Points

The similarity between Julia sets and the Mandelbrot set is rigorously understood in the neighborhood of certain points along the boundary of the Mandelbrot set, the so called Misiurewicz points. Let us explain what these points are.

We consider the iteration $z_{n+1} = z_n^2 + c$ and call an initial point z_0 pre-periodic provided there exist $m \geq 1$ and $p \geq 1$ such that $z_m = z_{m+p}$. In other words, after m iterations the orbit becomes periodic with period p. It is important that *m must not be 0*, i.e., the initial point must not be part of the periodic orbit itself! In the case $m = 0$ the initial point is a point of the periodic cycle and would be periodic but not pre-periodic. Now we come to the definition of Misiurewicz points. A parameter value c is called a Misiurewicz point, if the critical point $z_0 = 0$ is pre-periodic when iterating $z_{n+1} = z_n^2 + c$.

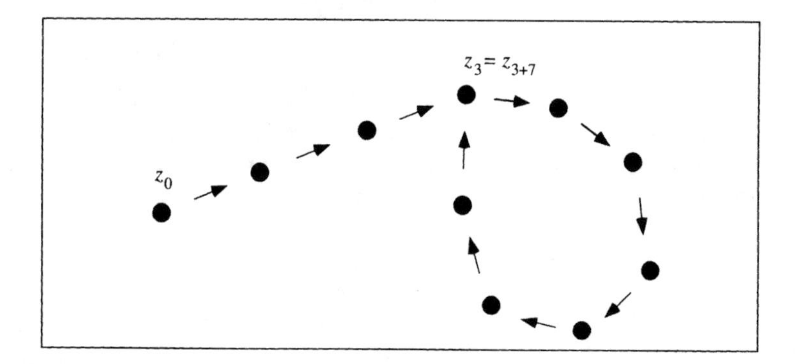

Figure 14.39 : z_0, z_1 and z_2 are pre-periodic points with period 7.

If c is a Misiurewicz point, then the critical orbit does not escape. Thus, all Misiurewicz points are in the Mandelbrot set. But there are many more facts about Misiurewicz points which have been proven.[34] For example,

- If c is a Misiurewicz point, then the corresponding periodic cycle $z_m, z_{m+1}, \ldots, z_{m+p-1}$ is repelling.
- If c is a Misiurewicz point, then the associated prisoner set is equal to the Julia set: $P_c = J_c$ (i.e., P_c has no interior),
- Misiurewicz points are dense at the boundary of the Mandelbrot set M. This means that if we take any point on the boundary of M and an arbitrarily small disk around that point, then there exists a Misiurewicz point in that disk.

Let us give two examples of Misiurewicz points. The first such point is $c = -2$, the point at the leftmost tip of the Mandelbrot set. The orbit

[34] See A. Douady, J. H. Hubbard, *Etudes dynamiques des polynômes complexes I, II*, Publ. Math. d'Orsay 84-02 (1984), 85-02 (1985).

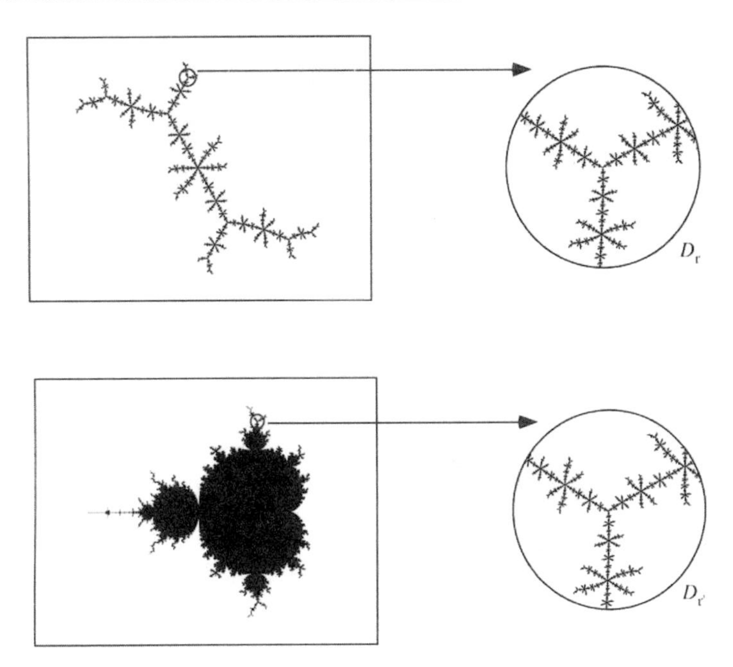

Similarity at Misiurewicz Point

The point $\gamma \approx -0.1011 + 0.9563i$ is a Misiurewicz point. This figure shows the Julia set for $c = \gamma$ (top) and the Mandelbrot set (bottom) with enlargements of small disks centered at the point $z = \gamma$ (upper right) and $c = \gamma$ (lower right). The proper rescaling operation, according to Tan Lei's theorem, is carried out only a couple of times at those points and the results can be compared on the right. Indeed, both structures are almost identical. Further rescaling operations would make the pictures even indistinguishable.

Figure 14.40

of the critical point is

$$\begin{aligned} z_0 &= 0 \\ z_1 &= c = -2 \\ z_2 &= (-2)^2 - 2 = 2 \\ z_3 &= 2^2 - 2 = 2 \end{aligned}$$

After only two iterations we hit a repelling fixed point. In the terminology of the above definition for a pre-periodic point, we have $m = 2$ and $p = 1$.

The second example is the parameter $c = i = \sqrt{-1}$. The orbit of the critical point is

$$\begin{aligned} z_0 &= 0 \\ z_1 &= c = i \\ z_2 &= i^2 + i = -1 + i \\ z_3 &= (-1+i)^2 + i = -i \\ z_4 &= (-i)^2 + i = -1 + i \end{aligned}$$

In this case we hit a cycle of period 2 again after two iterations. In the terminology of the definition for a pre-periodic point, we have $m = 2$ and $p = 2$.

Misiurewicz Point i

The Mandelbrot set near the Misiurewicz point $c = i$.

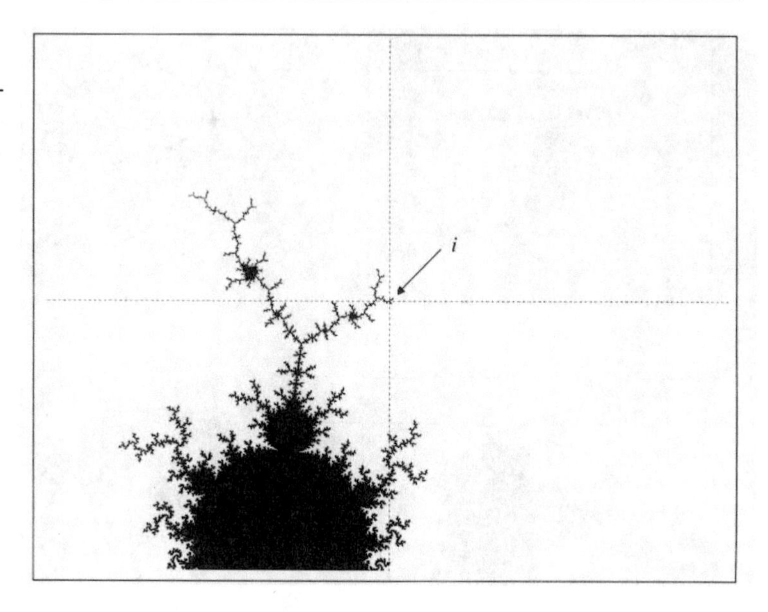

Figure 14.41

Now we can state Tan Lei's result:[35] Loosely speaking, if the parameter $c \in M$ is a particular type of point, called a Misiurewicz point[36] then the following is true.

- The Julia set J_c and the Mandelbrot set are both asymptotically self-similar in the point $z = c$ using the same multiplier ρ.
- The associated limit objects L_J and L_M are essentially the same; they differ only by some scaling and a rotation ($L_M = \lambda L_J$, where λ is a suitable complex number).

The second fact is illustrated in figure 14.40 for the Misiurewicz point $c \approx -0.1011 + 0.9563i$.

Since Misiurewicz points are dense at the boundary of the Mandelbrot set,[37] Tan Lei's theorem is quite powerful. In a way, near all these points in the Mandelbrot set we can see the shape of the corresponding Julia sets. Or we could say the boundary of M is a visual *table of contents* for infinitely many Julia sets.

[35] Tan Lei, *Similarity between the Mandelbrot set and Julia sets,* Report Nr 211, Institut für Dynamische Systeme, Universität Bremen, June 1989, and, Commun. Math. Phys. 134 (1990) 587–617.

[36] A parameter c is a Misiurewicz point, if the critical point $z_0 = 0$ of $z \to z^2 + c$ is pre-periodic — but not periodic itself, i.e., the iteration starting at z_0 leads to a periodic cycle, which does not contain z_0 itself (see the technical section above). Michal Misiurewicz is a Polish mathematician who became well known in the study of dynamics of one-dimensional maps like $x \to ax(1-x)$. Also he showed that the Lozi model of the Hénon system bears a strange attractor (see section 12.1).

[37] This means that in the neighborhood of any point of the boundary, there exists a Misiurewicz point. See A. Douady, J. H. Hubbard, *Étude dynamique des pôlynomes complexes,* Publications Mathematiques d'Orsay 84-02, Université de Paris-Sud, 1984.

Tan Lei's Theorem

The details of Tan Lei's theorem cannot be presented without using derivatives from calculus. Let c be a Misiurewicz point. Then by definition the critical point 0 is pre-periodic. Let us consider the orbit of the critical value $z_0 = c$ (i.e., we drop the initial point of the orbit of the critical point). Denote by $l \geq 0$ the minimal number of iterations such that z_l is periodic, and $z_{l+p} = z_l$ for some minimal period $p \geq 1$. Let $f_c(z) = z^2 + c$ and define the multiplier ρ of c by

$$\rho = (f_c^p)'(z_l)$$

where f_c^p denotes the p-fold composition of f_c. Note that z_l is a fixed point of f_c^p corresponding to a cycle of period p for the iteration of f_c. Recall that the derivative criterion states that the fixed point (and the cycle) is attractive if the derivative ρ is less than 1 in absolute value and repelling if it is greater than 1. Since the periodic orbit started at z_l is a repelling one (because c is a Misiurewicz point) we must have that $|\rho| > 1$.

To compute the multiplier of c we use the chain rule from calculus. Applied once, it yields

$$(f_c^p)'(z_l) = (f_c^{p-1})'(f(z_l)) \cdot f_c'(z_l).$$

Repeatedly applying the chain rule to $(f_c^{p-1})'(f(z_l))$ a total of $p-1$ times, we obtain

$$\rho = f_c'(z_{l+p-1}) \cdot f_c'(z_{l+p-2}) \cdots f_c'(z_l).$$

In other words, we have to compute the periodic orbit that is associated with a Misiurewicz point and then multiply the derivatives of $f_c(z) = z^2 + c$ at the points of the cycle. Since $f_c'(z) = 2z$, this yields a simple formula, namely,

$$\begin{aligned} \rho &= 2z_{l+p-1} \cdot 2z_{l+p-2} \cdots 2z_l \\ &= 2^p \cdot z_{l+p-1}\, z_{l+p-2} \cdots z_l. \end{aligned}$$

Let us compute the multipliers ρ for the examples $c = -2$ and $c = i$. In the first case, we have $l = p = 1$ and $z_l = 2$. Hence $\rho = 2^1 \cdot z_l = 4$. For $c = i$, we obtain $l = 1$, $p = 2$, $z_l = -1 + i$ and $z_{l+1} = -i$. Hence

$$\rho = 2^2 z_l z_{l+1} = 4 \cdot (-1 + i) \cdot (-i) = 4 + 4i.$$

Thus, $|\rho| = 4\sqrt{2} \approx 5.66$ and $\arg \rho = \pi/4$. The result is that for $c = i$ the multiplier ρ corresponds to a scaling by about 5.66 and a rotation by an angle of $\phi = 45$ degrees.

Tan Lei's Theorem can now be stated as follows. Let c^* be a Misiurewicz point with multiplier ρ.

- There exist closed subsets L_J and L_M of the complex plane such that

$$\rho L_J = L_J \text{ and } \rho L_M = L_M.$$

- The Julia set J_{c^*} is asymptotically self-similar at the point c^*, and there exists $r > 0$ such that

$$\lim_{n\to\infty} \rho^n(J_{c^*} - c^*) \cap D_r = L_J \cap D_r.$$

- Furthermore, the Mandelbrot set is asymptotically self-similar in the point c^*, and there exists $r' > 0$ such that $r > r'$ and

$$\lim_{n\to\infty} \rho^n(M - c^*) \cap D_{r'} = L_M \cap D_{r'}.$$

- Finally, there is a complex number $\lambda \neq 0$ such that

$$L_M \cap D_{r'} = \lambda L_J \cap D_{r'}$$

Thus, the limit objects are the same except for some scaling and a rotation.

Moreover, there is an explicit formula for λ:

$$\lambda = \frac{\frac{d}{dz} f^l_{c^*}(z)\big|_{z=c^*}}{\frac{d}{dc} f^l_c(c)\big|_{c=c^*} - \frac{d}{dc}\alpha(c)\big|_{c=c^*}} \tag{14.11}$$

where $\alpha(c)$ is the 'first point of the periodic orbit' of (minimal) period p. More precisely, $\alpha(c)$ is the analytic function which satisfies

$$\alpha(c) = f^p_c(\alpha(c)) \text{ with } \alpha(c^*) = f^l_{c^*}(c^*).$$

Implicit differentiation of this (left) formula can be used to compute the derivative of $\alpha(c)$ required in eqn. (14.11).

Is the Dendrite at $c = i$ a Spiral?

Let us close this section with two amazing observations which can be explained by means of Tan Lei's result. First, let us reconsider the Misiurewicz point at $c = i$ (see figure 14.41). In the technical section above we calculated the multiplier ρ of c. The result was $\rho = 4 + 4i$, which means that the similarity transformation that yields the self-similarity of the Mandelbrot set and the Julia set at $c = i$ is given by scaling factor of $\sqrt{32} \approx 5.66$ and a rotation by 45 degrees.

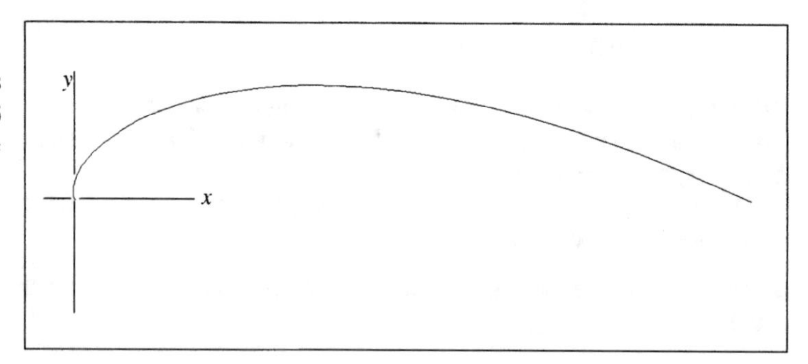

Fast Spiral

The figure shows a spiral which is invariant under scaling by $r \approx 5.66$ and simultaneous rotation by $\phi = 45$ degrees.

Figure 14.42

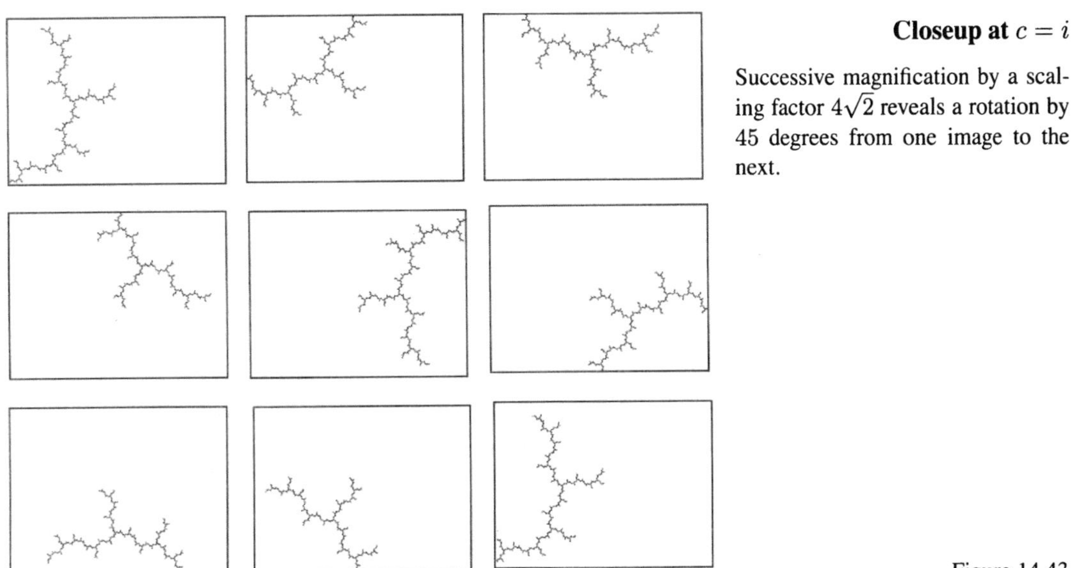

Closeup at $c = i$

Successive magnification by a scaling factor $4\sqrt{2}$ reveals a rotation by 45 degrees from one image to the next.

Figure 14.43

Inspecting the dendrite near $c = i$ in figure 14.41 we seem to encounter a contradiction. On the one hand there appears to be an ordinary dendrite with branches. On the other Tan Lei's result tells us that we really should see a spiral because the limit model L is invariant under multiplication with ρ with $\arg \rho \neq 0$. But what does that mean graphically? Let us look in figure 14.42 at a logarithmic spiral which is self-similar at 0 under multiplication with the same complex factor ρ. Now we understand: it is very hard to see that this is really a spiral, the rotational effect is dominated by the stretching too much. Looking back at the number ρ, this should really be no surprise, a 45 degree turn stretches the spiral by a factor of $4\sqrt{2} \approx 5.66$, and thus, a full turn means stretching by the same factor raised to the eighth power, $(4\sqrt{2})^8 = 2^{20}$, i.e., about a million times!

Now let us try to reveal the same spiraling structure that must be present at the tip of the dendrite at the parameter value $c = i$. In figure 14.43 we take successive magnifications centered at $c = i$, each one enlarged by a factor of $4\sqrt{2}$. However, we do not rotate! According to Tan Lei's result we should see that the resulting dendrite appears as if it were rotated by 45 degrees from step to step. Thus, after eight magnifications we should see a repetition; and indeed we do!

Finally we consider another amazing consequence from Tan Lei's result. As we have seen, close-ups of the Mandelbrot set near its boundary typically result in images which apparently contain small copies of the Mandelbrot set. In fact, it is known that there are infinitely many small copies of the Mandelbrot

Pollution With Small M-Sets

This
sequence of enlargements zooms in on a small copy of the Mandelbrot set. Its diameter is about 10^{-8}. At the center of the spiral in the upper left, there is a Misiurewicz point.

Figure 14.44

set populating the boundary region.[38] Figure 14.44 confirms that for a typical c-value.

The existence of small copies of the Mandelbrot set everywhere along its boundary seems to contradict Tan Lei's theorem, which postulates that near Misiurewicz points the Mandelbrot set and corresponding Julia sets look similar. Moreover, the similarity should become increasingly stronger when we magnify at a Misiurewicz point. On the other hand Julia sets cannot contain small copies of the Mandelbrot set; these would contradict the self-similarity of the Julia sets. Thus, continuous magnification of the Mandelbrot set at a Misiurewicz point will reveal copies of M without end, while magnification of the corresponding Julia set cannot produce these copies (see figure 14.45). Yet, according to Tan Lei's theorem, both structures should become indistinguishable as the magnification factor increases. How can we resolve this contradiction? The answer is surprising and is related to the meaning of asymptotic self-similarity at some point z_0.

Figure 14.46 shows a rather artificial but instructive example which exhibits the same 'contradiction'. We consider a straight line segment of length 1 placed with one end at the origin. At the other end we attach a square with side length $d_0 = 1$. Then we bisect the line segment. The left half interval is bisected again, and we repeat the procedure again and again. In other words, we construct the bisection points at

$$x_1 = \frac{1}{2},\ x_2 = \frac{1}{4},\ \ldots,\ x_n = \frac{1}{2^n},\ \ldots$$

[38] See A. Douady, J. H. Hubbard, *Etudes dynamiques des polynômes complexes I, II,* Publ. Math. d'Orsay 84-02 (1984), 85-02 (1985).

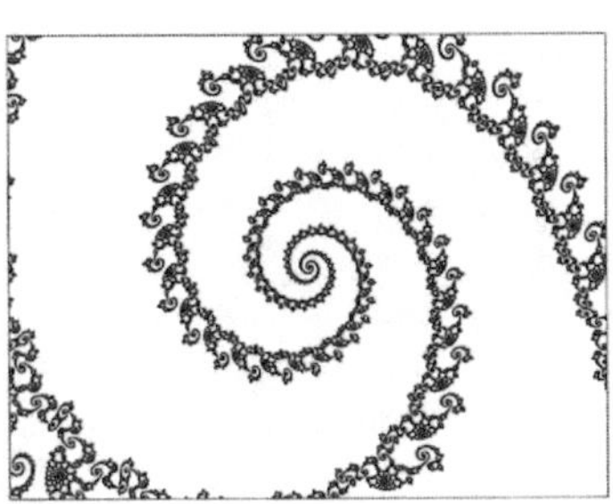

Two Spirals

An enlargement of the Julia set J_c with $c = -0.77568377 + 0.13646737i$, a Misiurewicz point, is shown on the left. The image is centered at c and has width 0.00036. Pictured on the right is an enlargement of the Mandelbrot set at the same Misiurewicz point (the width is 0.00048). The double spirals are almost identical, although the right one must contain infinitely many small copies of the Mandelbrot set, while the left one must not have any such copies in it (see figure 14.44.

Figure 14.45

At these bisection points we place squares T_n of side lengths

$$d_1 = \frac{1}{4},\ d_2 = \frac{1}{16},\ \ldots,\ d_n = \frac{1}{2^{2n}},\ \ldots$$

This is the set B. Now we successively magnify around 0, each time by a factor of $\rho = 2$, and watch what we see in the disk of radius 1. After each magnification the side length of the squares is doubled, but on the other hand they seem to move outward. Now look at a bisection point, say, $x = 1/2^n$. After the first magnification, we have the square T_{n+1} at this point, scaled by the factor $\rho = 2$. This square ρT_{n+1} has side length

$$2d_{n+1} = \frac{2}{2^{2(n+1)}} = \frac{1}{2^{2n+1}}.$$

After the second magnification we have the square $\rho^2 T_{n+2}$ at the point $x = 2^{-n}$, which has side length

$$2^2 d_{n+2} = \frac{2^2}{2^{2(n+2)}} = \frac{1}{2^{2n+2}}.$$

In general, after k such iterations we see the square $\rho^k T_{n+k}$ at the point $x = 2^{-n}$, which has side length

$$2^k d_{n+k} = \frac{2^k}{2^{2(n+k)}} = \frac{1}{2^{2n+k}}.$$

Thus in the limit for $k \to \infty$ these side lengths diminish and, moreover, this is true at all the bisection points. In other words, we have shown that our set B is asymptotically self-similar at 0 and the limit model L is just the straight line segment of length 1, which is quite a surprise when comparing B with L visually.

Asymptotic Self-Similarity

This object is asymptotically self-similar at the origin 0, and the limit object is a straight line. The lower part is obtained from the upper part by a scaling by a factor of 2.

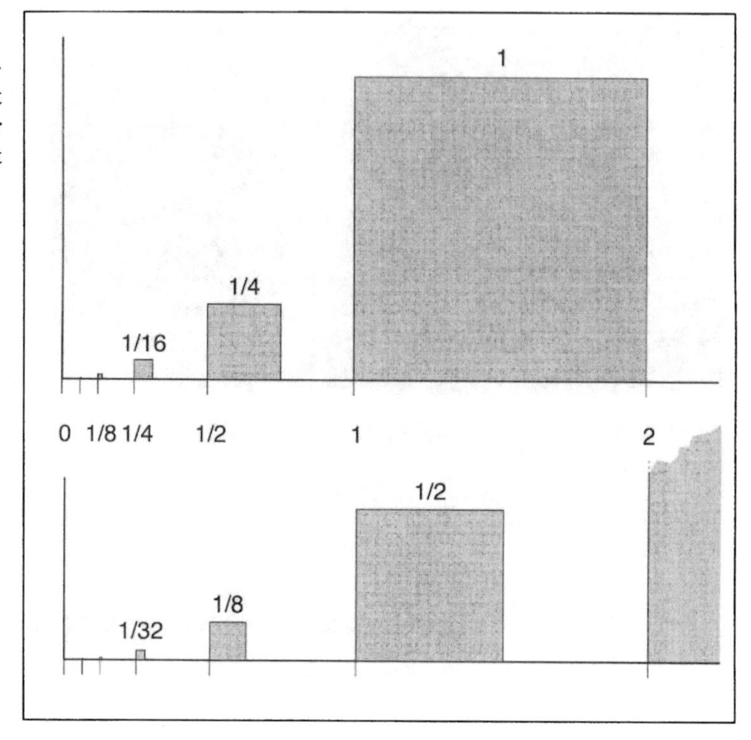

Figure 14.46

The reason for this effect lies obviously in the fact that the sizes of the squares in B decrease much more rapidly than the magnification can compensate for in successive magnifications. In other words, if the sizes of the squares were rather to go like 1/2, 1/4, 1/8, ..., for example, then each magnification would yield the same object in the disk of radius 1, i.e., in this case B would be even self-similar in 0.

Thus, we have learned by means of this construction that if we take a limit model which is self-similar at a point z_0 and pollute it with 'garbage' near z_0 making sure that the size of the 'garbage' decreases sufficiently rapidly close to z_0, then we obtain a structure which is still asymptotically self-similar in z_0! Since the 'garbage' can be arbitrary as long as it decreases in size sufficiently, that means that one limit model L stands for a whole class of such structures. We may replace some squares in our set B by triangles, others by hexagons, or whatever shape we like. As long as their sizes are chosen appropriately, the set is still asymptotically self-similar.

This explains why the upper left double spiral in figure 14.44 does not show any copies of the Mandelbrot set, and we will never see any such copies if we continue to magnify around the Misiurewicz point $c^* \approx -0.7756838 + 0.1364674i$ (the center of the double spiral). Yet there are infinitely many copies of the Mandelbrot set in the image. But we cannot see them because their sizes decrease so rapidly towards c^*. Or, to put it in another way, the two images in figure 14.45 are the same from any practical point of view, i.e., as an assembly of pixels. But for the underlying mathematical objects which are visualized here, there is a world of a difference.

It is interesting to turn the argument around: If a set B is asymptotically self-similar at a point with a limit object L, then whatever the pollution in B is, it must decrease in size sufficiently rapidly near that point. In other words, Tan Lei's result gives us an estimate for how small copies of the Mandelbrot set must be near the center c^* of the double spiral. The sizes of the small Mandelbrot set copies are beautifully balanced; they are just at the threshold between becoming submerged into invisibility and explosion upon successive magnification. These small copies can be seen when magnifying properly chosen regions. However, they cannot be seen when iteratively magnifying a region centered at the point c^*.

Bibliography

1. Books

[1] Abraham, R. H., Shaw, C. D., *Dynamics, The Geometry of Behavior,* Part One:*Periodic Behavior* (1982), Part Two: *Chaotic Behavior* (1983), Part Three: *Global Behavior* (1984), Aerial Press, Santa Cruz. Second edition Addison-Wesley, 1992.

[2] Aharony, A. and Feder, J. (eds.), *Fractals in Physics*, Physica D 38 (1989); also published by North Holland (1989).

[3] Allgower, E., Georg, K., *Numerical Continuation Methods — An Introduction,* Springer-Verlag, New York, 1990.

[4] Arnold, V. I., *Ordinary Differential Equations,* MIT Press, Cambridge, 1973.

[5] Avnir, D. (ed.), *The Fractal Approach to Heterogeneous Chemistry: Surfaces, Colloids, Polymers,* Wiley, Chichester, 1989.

[6] Banchoff, T. F., *Beyond the Third Dimension,* Scientific American Library, 1990.

[7] Barnsley, M., *Fractals Everywhere,* Academic Press, San Diego, 1988.

[8] Beardon, A. F., *Iteration of Rational Functions,* Springer-Verlag, New York, 1991.

[9] Becker K.-H., Dörfler, M., *Computergraphische Experimente mit Pascal*, Vieweg, Braunschweig, 1986.

[10] Beckmann, P., *A History of Pi,* Second Edition, The Golem Press, Boulder, 1971.

[11] Bélair, J., Dubuc, S., (eds.), *Fractal Geometry and Analysis,* Kluwer Academic Publishers, Dordrecht, Holland, 1991.

[12] Bondarenko, B., *Generalized Pascal Triangles and Pyramids, Their Fractals, Graphs and Applications,* Tashkent, Fan, 1990, in Russian.

[13] Borwein, J. M., Borwein, P. B., *Pi and the AGM — A Study in Analytic Number Theory,* Wiley, New York, 1987.

[14] Briggs, J., Peat, F. D., *Turbulent Mirror,* Harper & Row, New York, 1989.

[15] Bunde, A., Havlin, S. (eds.), *Fractals and Disordered Systems,* Springer-Verlag, Heidelberg, 1991.

[16] Campbell, D., Rose, H. (eds.), *Order in Chaos,* North-Holland, Amsterdam, 1983.

[17] Chaitin, G. J., *Algorithmic Information Theory,* Cambridge University Press, 1987.

[18] Cherbit, G. (ed.), *Fractals, Non-integral Dimensions and Applications,* John Wiley & Sons, Chichester, 1991.

[19] Collet, P., Eckmann, J.-P., *Iterated Maps on the Interval as Dynamical Systems,* Birkhäuser, Boston, 1980.

[20] Crilly, A. J., Earnshaw, R. A., Jones, H. (eds.), *Fractals and Chaos,* Springer-Verlag, New York, 1991.

[21] Cvitanović, P. (ed.), *Universality in Chaos,* Second Edition, Adam Hilger, New York, 1989.

[22] Devaney, R. L., *An Introduction to Chaotic Dynamical Systems, Second Edition,* Addison-Wesley, Redwood City, CA, 1989.

[23] Devaney, R. L., *Chaos, Fractals, and Dynamics,* Addison-Wesley, Menlo Park, 1990.

[24] Durham, T., *Computing Horizons,* Addison-Wesley, Wokingham, 1988.

[25] Dynkin, E. B., Uspenski, W., *Mathematische Unterhaltungen II,* VEB Deutscher Verlag der Wissenschaften, Berlin, 1968.

[26] Edgar, G., *Measures, Topology and Fractal Geometry,* Springer-Verlag, New York, 1990.

[27] Engelking, R., *Dimension Theory,* North Holland, 1978.

[28] Escher, M. C., *The World of M. C. Escher,* H. N. Abrams, New York, 1971.

[29] Falconer, K., *The Geometry of Fractal Sets,* Cambridge University Press, Cambridge, 1985.

[30] Falconer, K.,*Fractal Geometry, Mathematical Foundations and Applications,* Wiley, New York, 1990.

[31] Family, F., Landau, D. P. (eds.), *Aggregation and Gelation,* North-Holland, Amsterdam, 1984.

[32] Family, F., Vicsek, T. (eds.), *Dynamics of Fractal Surfaces,* World Scientific, Singapore, 1991.

[33] Feder, J., *Fractals,* Plenum Press, New York 1988.

[34] Fleischmann, M., Tildesley, D. J., Ball, R. C., *Fractals in the Natural Sciences,* Princeton University Press, Princeton, 1989.

[35] Garfunkel, S., (Project Director), Steen, L. A. (Coordinating Editor) *For All Practical Purposes, Second Edition,* W. H. Freeman and Co., New York, 1988.

[36] GEO Wissen — Chaos und Kreativität, Gruner + Jahr, Hamburg, 1990.

[37] Gleick, J., *Chaos, Making a New Science,* Viking, New York, 1987.

[38] Golub, G. H., Loan, C. F. van, *Matrix Computations,* Second Edition, Johns Hopkins, Baltimore, 1989.

[39] Guckenheimer, J., Holmes, P., *Nonlinear Oscillations, Dynamical Systems, and Bifurcations of Vector Fields,* Springer-Verlag, New York, 1983.

[40] Guyon, E., Stanley, H. E., (eds.), *Fractal Forms,* Elsevier/North-Holland and Palais de la Découverte, 1991.

[41] Haken, H., *Advanced Synergetics,* Springer-Verlag, Heidelberg, 1983.

[42] Haldane, J. B. S., *On Being the Right Size,* 1928.

[43] Hall, R., *Illumination and Color in Computer Generated Imagery,* Springer-Verlag, New York, 1988.

[44] Hao, B. L., *Chaos II,* World Scientific, Singapore, 1990.

[45] Hausdorff, F., *Grundzüge der Mengenlehre,* Verlag von Veit & Comp., 1914.

[46] Hausdorff, F., *Dimension und äußeres Maß,* Math. Ann. 79 (1918) 157–179.

[47] Hirsch, M. W., Smale, S., *Differential Equations, Dynamical Systems, and Linear Algebra,* Academic Press, New York, 1974.

[48] Hommes, C. H., *Chaotic Dynamics in Economic Models,* Wolters-Noordhoff, Groningen, 1991.

[49] Jackson, E. A., *Perspectives of Nonlinear Dynamics,* Volume 1 and 2, Cambridge University Press, Cambridge, 1991.

[50] Knuth, D. E., *The Art of Computer Programming, Volume 2, Seminumerical Algorithms,* Addison-Wesley, Reading, Massachusetts.

[51] Kuratowski, C., *Topologie II,* PWN, Warsaw, 1961.

[52] Lauwerier, H., *Fractals,* Aramith Uitgevers, Amsterdam, 1987.

[53] Lehmer, D. H., Proc. 2nd Symposium on Large Scale Digital Calculating Machinery, Harvard University Press, Cambridge, 1951.

[54] Leven, R. W., Koch, B.-P., Pompe, B., *Chaos in Dissipativen Systemen,* Vieweg, Braunschweig, 1989.

[55] Lindenmayer, A., Rozenberg, G., (eds.), *Automata, Languages, Development,* North-Holland, Amsterdam, 1975.

[56] Mandelbrot, B. B., *Fractals: Form, Chance, and Dimension,* W. H. Freeman and Co., San Francisco, 1977.

[57] Mandelbrot, B. B., *The Fractal Geometry of Nature,* W. H. Freeman and Co., New York, 1982.

[58] Mañé, R., *Ergodic Theory and Differentiable Dynamics,* Springer-Verlag, Heidelberg, 1987.

[59] McGuire, M., *An Eye for Fractals,* Addison-Wesley, Redwood City, 1991.

[60] Menger, K., *Dimensionstheorie,* Leipzig, 1928.

[61] Mey, J. de, *Bomen van Pythagoras,* Aramith Uitgevers, Amsterdam, 1985.

[62] Moon, F. C., *Chaotic Vibrations,* John Wiley & Sons, New York, 1987.

[63] Parchomenko, A. S., *Was ist eine Kurve,* VEB Verlag, 1957.

[64] Parker, T. S., Chua, L. O., *Practical Numerical Algorithms for Chaotic Systems,* Springer-Verlag, New York, 1989.

[65] Peitgen, H.-O., Richter, P. H., *The Beauty of Fractals,* Springer-Verlag, Heidelberg, 1986.

[66] Peitgen, H.-O., Saupe, D., (eds.), *The Science of Fractal Images,* Springer-Verlag, 1988.

[67] Peitgen, H.-O. (ed.), *Newton's Method and Dynamical Systems,* Kluver Academic Publishers, Dordrecht, 1989.

[68] Peitgen, H.-O., Jürgens, H., *Fraktale: Gezähmtes Chaos,* Carl Friedrich von Siemens Stiftung, München, 1990.

[69] Peitgen, H.-O., Jürgens, H., Saupe, D., *Fractals for the Classroom, Part One,* Springer-Verlag, New York, 1991.

[70] Peitgen, H.-O., Jürgens, H., Saupe, D., Maletsky, E., Perciante, T., Yunker, L., *Fractals for the Classroom, Strategic Activities, Volume One,* and *Volume Two,* Springer-Verlag, New York, 1991 and 1992.

[71] Peters, E., *Chaos and Order in the Capital Market,* John Wiley & Sons, New York, 1991.

[72] Press, W. H., Flannery, B. P., Teukolsky, S. A., Vetterling, W. T., *Numerical Recipes,* Cambridge University Press, Cambridge, 1986.

[73] Preston, K. Jr., Duff, M. J. B., *Modern Cellular Automata,* Plenum Press, New York, 1984.

[74] Prigogine, I., Stenger, I., *Order out of Chaos,* Bantam Books, New York, 1984.

[75] Prusinkiewicz, P., Lindenmayer, A., *The Algorithmic Beauty of Plants,* Springer-Verlag, New York, 1990.

[76] Rasband, S. N., *Chaotic Dynamics of Nonlinear Systems,* John Wiley & Sons, New York, 1990.

[77] Richardson, L. F., *Weather Prediction by Numerical Process,* Dover, New York, 1965.

[78] Ruelle, D., *Chaotic Evolution and Strange Attractors,* Cambridge University Press, Cambridge, 1989.

[79] Sagan, C., *Contact,* Pocket Books, Simon & Schuster, New York, 1985.

[80] Schröder, M., *Fractals, Chaos, Power Laws,* W. H. Freeman and Co., New York, 1991.

[81] Schuster, H. G., *Deterministic Chaos,* VCH Publishers, Weinheim, New York, 1988.

[82] Sparrow, C., *The Lorenz Equations: Bifurcations, Chaos, and Strange Attractors,* Springer-Verlag, New York, 1982.

[83] Stauffer, D., *Introduction to Percolation Theory,* Taylor & Francis, London, 1985.

[84] Stauffer, D., Stanley, H. E., *From Newton to Mandelbrot,* Springer-Verlag, New York,1989.

[85] Stewart, I., *Does God Play Dice,* Penguin Books, 1989.

[86] Stewart, I., *Game, Set, and Math,* Basil Blackwell, Oxford, 1989.

[87] Thompson, D'Arcy, *On Growth an Form,* New Edition, Cambridge University Press, 1942.

[88] Thompson, J. M. T., Stewart, H. B., *Nonlinear Dynamics and Chaos,* Wiley, Chichester, 1986.

[89] Toffoli, T., Margolus, N., *Cellular Automata Machines, A New Environment For Modelling,* MIT Press, Cambridge, Mass., 1987.

[90] Vicsek, T., *Fractal Growth Phenomena,* World Scientific, London, 1989.

[91] Wade, N., *The Art and Science of Visual Illusions,* Routledge & Kegan Paul, London,1982.

[92] Wall, C. R., *Selected Topics in Elementary Number Theory,* University of South Caroline Press, Columbia, 1974.

[93] Wegner, T., Peterson, M., *Fractal Creations,* Waite Group Press, Mill Valley, 1991.

[94] Weizenbaum, J., *Computer Power and Human Reason,* Penguin, 1984.

[95] West, B., *Fractal Physiology and Chaos in Medicine,* World Scientific, Singapore, 1990.

[96] Wolfram, S., Farmer, J. D., Toffoli, T., (eds.) *Cellular Automata: Proceedings of an Interdisciplinary Workshop,* in: Physica 10D, 1 and 2 (1984).

[97] Wolfram, S. (ed.), *Theory and Application of Cellular Automata,* World Scientific, Singapore, 1986.

[98] Zhang Shu-yu, *Bibliography on Chaos,* World Scientific, Singapore, 1991.

2. General Articles

[99] Barnsley, M. F., *Fractal Modelling of Real World Images,* in: The Science of Fractal Images, H.-O. Peitgen, D. Saupe (eds.), Springer-Verlag, New York, 1988.

[100] Cipra, B., A., *Computer-drawn pictures stalk the wild trajectory,* Science 241 (1988) 1162–1163.

[101] Davis, C., Knuth, D. E., *Number Representations and Dragon Curves,* Journal of Recreational Mathematics 3 (1970) 66–81 and 133–149.

[102] Dewdney, A. K., *Computer Recreations: A computer microscope zooms in for a look at the most complex object in mathematics,* Scientific American (August 1985) 16–25.

[103] Dewdney, A. K., *Computer Recreations: Beauty and profundity: the Mandelbrot set and a flock of its cousins called Julia sets,* Scientific American (November 1987) 140–144.

[104] Douady, A., *Julia sets and the Mandelbrot set,* in: The Beauty of Fractals, H.-O. Peitgen, P. H. Richter, Springer-Verlag, 1986.

[105] Dyson, F., *Characterizing Irregularity,* Science 200 (1978) 677–678.

[106] Gilbert, W. J., *Fractal geometry derived from complex bases,* Math. Intelligencer 4 (1982) 78–86.

[107] Hofstadter, D. R., *Strange attractors : Mathematical patterns delicately poised between order and chaos,* Scientific American 245 (May 1982) 16–29.

[108] Mandelbrot, B. B., *How long is the coast of Britain? Statistical self-similarity and fractional dimension,* Science 155 (1967) 636–638.

[109] Peitgen, H.-O., Richter, P. H., *Die unendliche Reise,* Geo 6 (Juni 1984) 100–124.

[110] Peitgen, H.-O., Haeseler, F. v., Saupe, D., *Cayley's problem and Julia sets,* Mathematical Intelligencer 6.2 (1984) 11–20.

[111] Peitgen, H.-O., Jürgens, H., Saupe, D., *The language of fractals,* Scientific American (August 1990) 40–47.

[112] Peitgen, H.-O., Jürgens, H., *Fraktale: Computerexperimente (ent)zaubern komplexe Strukturen,* in: *Ordnung und Chaos in der unbelebten und belebten Natur,* Verhandlungen der Gesellschaft Deutscher Naturforscher und Ärzte, 115. Versammlung, Wissenschaftliche Verlagsgesellschaft, Stuttgart, 1989.

[113] Peitgen, H.-O., Jürgens, H., Saupe, D., Zahlten, C., *Fractals — An Animated Discussion,* Video film, W. H. Freeman and Co., 1990. Also appeared in German as *Fraktale in Filmen und Gesprächen,* Spektrum Videothek, Heidelberg, 1990. Also appeared in Italian as *I Frattali,* Spektrum Videothek edizione italiana, 1991.

[114] Ruelle, D., *Strange Attractors,* Math. Intelligencer 2 (1980) 126–137.

[115] Stewart, I., *Order within the chaos game?* Dynamics Newsletter 3, no. 2, 3, May 1989, 4–9.

[116] Sved, M. *Divisibility — With Visibility,* Mathematical Intelligencer 10, 2 (1988) 56–64.

[117] Voss, R., *Fractals in Nature,* in: *The Science of Fractal Images,* H.-O. Peitgen , D. Saupe (eds.), Springer-Verlag, New York, 1988.

[118] Wolfram, S., *Geometry of binomial coefficients,* Amer. Math. Month. 91 (1984) 566–571.

3. Research Articles

[119] Abraham, R., *Simulation of cascades by video feedback,* in: "Structural Stability, the Theory of Catastrophes, and Applications in the Sciences", P. Hilton (ed.), Lecture Notes in Mathematics vol. 525, 1976, 10–14, Springer-Verlag, Berlin.

[120] Aharony, A., *Fractal growth,* in: *Fractals and Disordered Systems,* A. Bunde, S. Havlin (eds.), Springer-Verlag, Heidelberg, 1991.

[121] Bak, P., *The devil's staircase,* Phys. Today 39 (1986) 38–45.

[122] Bandt, C., *Self-similar sets I. Topological Markov chains and mixed self-similar sets,* Math. Nachr. 142 (1989) 107–123.

[123] Bandt, C., *Self-similar sets III. Construction with sofic systems,* Monatsh. Math. 108 (1989) 89–102.

[124] Banks, J., Brooks, J., Cairns, G., Davis, G., Stacey, P., *On Devaney's definition of chaos,* American Math. Monthly 99,4 (1992) 332–334.

[125] Barnsley, M. F., Demko, S., *Iterated function systems and the global construction of fractals,* The Proceedings of the Royal Society of London A399 (1985) 243–275

[126] Barnsley, M. F., Ervin, V., Hardin, D., Lancaster, J., *Solution of an inverse problem for fractals and other sets,* Proceedings of the National Academy of Sciences 83 (1986) 1975–1977.

[127] Barnsley, M. F., Elton, J. H., Hardin, D. P., *Recurrent iterated function systems,* Constructive Approximation 5 (1989) 3–31.

[128] Bedford, T., *Dynamics and dimension for fractal recurrent sets,* J. London Math. Soc. 33 (1986) 89–100.

[129] Benedicks, M., Carleson, L., *The dynamics of the Hénon map,* Annals of Mathematics 133,1 (1991) 73–169.

[130] Benettin, G. L., Galgani,L., Giorgilli, A., Strelcyn, J.-M., *Lyapunov characteristic exponents for smooth dynamical systems and for Hamiltonian systems; a method for computing all of them. Part 1: Theory, Part 2: Numerical application,* Meccanica 15, 9 (1980) 21.

[131] Berger, M., *Encoding images through transition probabilities,* Math. Comp. Modelling 11 (1988) 575–577.

[132] Berger, M., *Images generated by orbits of 2D-Markoc chains,* Chance 2 (1989) 18–28.

[133] Berry, M. V., *Regular and irregular motion,* in: Jorna S. (ed.), Topics in Nonlinear Dynamics, Amer. Inst. of Phys. Conf. Proceed. 46 (1978) 16–120.

[134] Blanchard, P., *Complex analytic dynamics on the Riemann sphere,* Bull. Amer. Math. Soc. 11 (1984) 85–141.

[135] Borwein, J. M., Borwein, P. B., Bailey, D. H., *Ramanujan, modular equations, and approximations to π, or how to compute one billion digits of π,* American Mathematical Monthly 96 (1989) 201–219.

[136] Brent, R. P., *Fast multiple-precision evaluation of elementary functions,* Journal Assoc. Comput. Mach. 23 (1976) 242–251.

[137] Brolin, H., *Invariant sets under iteration of rational functions,* Arkiv f. Mat. 6 (1965) 103–144.

[138] Cantor, G., *Über unendliche, lineare Punktmannigfaltigkeiten V,* Mathematische Annalen 21 (1883) 545–591.

[139] Carpenter, L., *Computer rendering of fractal curves and surfaces,* Computer Graphics (1980) 109ff.

[140] Caswell, W. E., Yorke, J. A., *Invisible errors in dimension calculations: geometric and systematic effects,* in: *Dimensions and Entropies in Chaotic Systems,* G. Mayer-Kress (ed.), Springer-Verlag, Berlin, 1986 and 1989, p. 123–136.

[141] Cayley, A., *The Newton-Fourier Imaginary Problem,* American Journal of Mathematics 2 (1879) p. 97.

[142] Charkovsky, A. N., *Coexistence of cycles of continuous maps on the line,* Ukr. Mat. J. 16 (1964) 61–71 (in Russian).

[143] Corless, R. M., *Continued fractions and chaos,* The American Math. Monthly 99, 3 (1992) 203–215.

[144] Corless, R. M., Frank, G. W., Monroe, J. G., *Chaos and continued fractions,* Physica D46 (1990) 241–253.

[145] Coven, E., Kan, I., Yorke, J. A., *Pseudo-orbit shadowing in the family of tent maps,* Trans. Amer. Math. Soc. 308,1 (1988) 227–241.

[146] Cremer, H., *Über die Iteration rationaler Funktionen,* Jahresberichte der Deutschen Mathematiker Vereinigung 33 (1925) 185–210.

[147] Crutchfield, J., *Space-time dynamics in video feedback,* Physica 10D (1984) 229–245.

[148] Dekking, F. M., *Recurrent Sets,* Advances in Mathematics 44, 1 (1982) 78–104.

[149] Derrida, B., Gervois, A., Pomeau, Y., *Universal metric properties of bifurcations of endomorphisms,* J. Phys. A: Math. Gen. 12, 3 (1979) 269–296.

[150] Devaney, R., Nitecki, Z., *Shift Automorphism in the Hénon Mapping,* Comm. Math. Phys. 67 (1979) 137–146.

[151] Douady, A., Hubbard, J. H., *Iteration des pôlynomes quadratiques complexes,* CRAS Paris 294 (1982) 123–126.

[152] Douady, A., Hubbard, J. H., *Étude dynamique des pôlynomes complexes,* Publications Mathematiques d'Orsay 84-02, Université de Paris-Sud, 1984.

[153] Douady, A., Hubbard, J. H., *On the dynamics of polynomial-like mappings,* Ann. Sci. Ecole Norm. Sup. 18 (1985) 287–344.

[154] Dress, A. W. M., Gerhardt, M., Jaeger, N. I., Plath, P. J, Schuster, H., *Some proposals concerning the mathematical modelling of oscillating heterogeneous catalytic reactions on metal surfaces,* in: L. Rensing, N. I. Jaeger (eds.), Temporal Order, Springer-Verlag, Berlin, 1984.

[155] Dubuc, S., Elqortobi, A., *Approximations of fractal sets,* Journal of Computational and Applied Mathematics 29 (1990) 79–89.

[156] Eckmann, J.-P., Ruelle, D., *Ergodic theory of chaos and strange attractors,* Reviews of Modern Physics 57, 3 (1985) 617–656.

[157] Eckmann, J.-P., Kamphorst, S. O., Ruelle, D., Ciliberto, S., *Liapunov exponents from time series,* Phys. Rev. 34A (1986) 4971–4979.

[158] Elton, J., *An ergodic theorem for iterated maps,* Journal of Ergodic Theory and Dynamical Systems 7 (1987) 481–488.

[159] Faraday, M., *On a peculiar class of acoustical figures, and on certain forms assumed by groups of particles upon vibrating elastic surfaces,* Phil. Trans. Roy. Soc. London 121 (1831) 299–340.

[160] Farmer, D., *Chaotic attractors of an infinite-dimensional system,* Physica 4D (1982) 366–393.

[161] Farmer, J. D., Ott, E., Yorke, J. A., *The dimension of chaotic attractors,* Physica 7D (1983) 153–180.

[162] Fatou, P., *Sur les équations fonctionelles,* Bull. Soc. Math. Fr. 47 (1919) 161–271, 48 (1920) 33–94, 208–314.

[163] Feigenbaum, M. J., *Universality in complex discrete dynamical systems,* in: Los Alamos Theoretical Division Annual Report (1977) 98–102.

[164] Feigenbaum, M. J., *Quantitative universality for a class of nonlinear transformations,* J. Stat. Phys. 19 (1978) 25–52.

[165] Feigenbaum, M. J., *Universal behavior in nonlinear systems,* Physica 7D (1983) 16–39. Also in: Campbell, D., Rose, H. (eds.), *Order in Chaos,* North-Holland, Amsterdam, 1983.

[166] Feigenbaum, M. J., *Some characterizations of strange sets,* J. Stat. Phys. 46 (1987) 919–924.

[167] Feit, S. D., *Characteristic exponents and strange attractors,* Comm. Math. Phys. 61 (1978) 249–260.

[168] Fine, N. J., *Binomial coefficients modulo a prime number,* Amer. Math. Monthly 54 (1947) 589.

[169] Fisher, Y., Boss, R. D., Jacobs, E. W., *Fractal Image Compression: Image and Text Compression,* J. Storer (ed.), Kluwer Academic Publishers, Norwell, MA, 1992.

[170] Fournier, A., Fussell, D., Carpenter, L., *Computer rendering of stochastic models,* Comm. of the ACM 25 (1982) 371–384.

[171] Franceschini, V., *A Feigenbaum sequence of bifurcations in the Lorenz model,* Jour. Stat. Phys. 22 (1980) 397–406.

[172] Fraser, A. M., Swinney, H. L., *Independent coordinates for strange attractors from mutual information,* Phys. Rev. A 33 (1986) 1034–1040.

[173] Frederickson, P., Kaplan, J. L., Yorke, S. D., Yorke, J. A., *The Liapunov dimension of strange attractors,* Journal of Differential Equations 49 (1983) 185–207.

[174] Geist, K., Parlitz, U., Lauterborn, W., *Comparison of Different Methods for Computing Lyapunov Exponents,* Progress of Theoretical Physics 83,5 (1990) 875–893.

[175] Goodman, G. S., *A probabilist looks at the chaos game,* in: *Fractals in the Fundamental and Applied Sciences,* H.-O. Peitgen, J. M. Henriques, L. F. Peneda (eds.), North-Holland, Amsterdam, 1991.

[176] Grassberger, P., *On the fractal dimension of the Hénon attractor,* Physics Letters 97A (1983) 224–226.

[177] Grassberger, P., Procaccia, I., *Measuring the strangeness of strange attractors,* Physica 9D (1983) 189–208.

[178] Grassberger, P., Procaccia, I., *Characterization of Strange Attractors,* Phys. Rev. Lett. 50 (1983) 346.

[179] Grebogi, C., Ott, E., Yorke, J. A., *Crises, sudden changes in chaotic attractors, and transient chaos,* Physica 7D (1983) 181–200.

[180] Grebogi, C., Ott, E., Yorke, J. A., *Attractors of an N-torus: quasiperiodicity versus chaos,* Physica 15D (1985) 354.

[181] Grebogi, C., Ott, E., Yorke, J. A., *Critical exponents of chaotic transients in nonlinear dynamical systems,* Physical Review Letters 37, 11 (1986) 1284–1287.

[182] Grebogi, C., Ott, E., Yorke, J. A., *Chaos, strange attractors, and fractal basin boundaries in nonlinear dynamics,* Science 238 (1987) 632–638.

[183] Großmann, S., Thomae, S., *Invariant distributions and stationary correlation functions of one-dimensional discrete processes,* Z. Naturforsch. 32 (1977) 1353–1363.

[184] Haeseler, F. v., Peitgen, H.-O., Skordev, G., *Pascal's triangle, dynamical systems and attractors,* Ergod. Th. & Dynam. Sys. 12 (1992) 479–486.

[185] Haeseler, F. v., Peitgen, H.-O., Skordev, G., *Linear cellular automata, substitutions, hierarchical iterated function systems and attractors,* in: *Fractal Geometry and Computer Graphics,* J. L. Encarnacao, H.-O. Peitgen, G. Sakas, G. Englert (eds.), Springer-Verlag, Heidelberg, 1992.

[186] Hammel, S. M., Yorke, J. A., Grebogi, C., *Numerical orbits of chaotic processes represent true orbits,* Bull. Amer. Math. Soc. 19,2 (1988) 465–469.

[187] Hart, J. C., DeFanti, T., *Efficient anti-aliased rendering of 3D-linear fractals,* Computer Graphics 25, 4 (1991) 289–296.

[188] Hart, J. C., Sandin, D. J., Kauffman, L. H., *Ray tracing deterministic 3-D fractals,* Computer Graphics 23, 3 (1989) 91–100.

[189] Hénon, M., *A two-dimensional mapping with a strange attractor,* Comm. Math. Phys. 50 (1976) 69–77.

[190] Hentschel, H. G. E., Procaccia, I., *The infinite number of generalized dimensions of fractals and strange attractors,* Physica 8D (1983) 435–444.

[191] Hepting, D., Prusinkiewicz, P., Saupe, D., *Rendering methods for iterated function systems,* in: *Fractals in the Fundamental and Applied Sciences,* H.-O. Peitgen, J. M. Henriques, L. F. Peneda (eds.), North-Holland, Amsterdam, 1991.

[192] Hilbert, D., *Über die stetige Abbildung einer Linie auf ein Flächenstück,* Mathematische Annalen 38 (1891) 459–460.

[193] Holte, J., *A recurrence-relation approach to fractal dimension in Pascal's triangle,* International Congress of Mathematicians, Kyoto, Aug. 1990.

[194] Hutchinson, J., *Fractals and self-similarity,* Indiana University Journal of Mathematics 30 (1981) 713–747.

[195] Jacquin, A. E., *Image coding based on a fractal theory of iterated contractive image transformations,* to IEEE Transactions on Image Processing (1992) 18–30.

[196] Judd, K., Mees, A. I. *Estimating dimensions with confidence,* International Journal of Bifurcation and Chaos 1,2 (1991) 467–470.

[197] Julia, G., *Mémoire sur l'iteration des fonctions rationnelles,* Journal de Math. Pure et Appl. 8 (1918) 47–245.

[198] Jürgens, H., *3D-rendering of fractal landscapes,* in: *Fractal Geometry and Computer Graphics,* J. L. Encarnacao, H.-O. Peitgen, G. Sakas, G. Englert (eds.), Springer-Verlag, Heidelberg, 1992.

[199] Kaplan, J. L., Yorke, J. A., *Chaotic behavior of multidimensional difference equations,* in: *Functional Differential Equations and Approximation of Fixed Points,* H.-O. Peitgen, H. O. Walther (eds.), Springer-Verlag, Heidelberg, 1979.

[200] Kawaguchi, Y., *A morphological study of the form of nature,* Computer Graphics 16,3 (1982).

[201] Koch, H. von, *Sur une courbe continue sans tangente, obtenue par une construction géometrique élémentaire,* Arkiv för Matematik 1 (1904) 681–704.

[202] Koch, H. von, *Une méthode géométrique élémentaire pour l'étude de certaines questions de la théorie des courbes planes,* Acta Mathematica 30 (1906) 145-174.

[203] Kummer, E. E., *Über Ergänzungssätze zu den allgemeinen Reziprozitätsgesetzen,* Journal für die reine und angewandte Mathematik 44 (1852) 93–146.

[204] Lauterborn, W., *Acoustic turbulence,* in: *Frontiers in Physical Acoustics,* D. Sette (ed.), North-Holland, Amsterdam, 1986, pp. 123–144.

[205] Lauterborn, W., Holzfuss, J., *Acoustic chaos,* International Journal of Bifurcation and Chaos 1, 1 (1991) 13–26.

[206] Li, T.-Y., Yorke, J. A., *Period three implies chaos,* American Mathematical Monthly 82 (1975) 985–992.

[207] Lindenmayer, A., *Mathematical models for cellular interaction in development, Parts I and II,* Journal of Theoretical Biology 18 (1968) 280–315.

[208] Lorenz, E. N., *Deterministic non-periodic flow,* J. Atmos. Sci. 20 (1963) 130–141.

[209] Lorenz, E. N., *The local structure of a chaotic attractor in four dimensions,* Physica 13D (1984) 90–104.

[210] Lovejoy, S., Mandelbrot, B. B., *Fractal properties of rain, and a fractal model,* Tellus 37A (1985) 209–232.

[211] Lozi, R., *Un attracteur étrange (?) du type attracteur de Hénon,* J. Phys. (Paris) 39 (Coll. C5) (1978) 9–10.

[212] Mandelbrot, B. B., Ness, J. W. van, *Fractional Brownian motion, fractional noises and applications,* SIAM Review 10,4 (1968) 422–437.

[213] Mandelbrot, B. B., *Fractal aspects of the iteration of $z \mapsto \lambda z(1-z)$ for complex λ and z,* Annals NY Acad. Sciences 357 (1980) 249–259.

[214] Mandelbrot, B. B., *Comment on computer rendering of fractal stochastic models,* Comm. of the ACM 25,8 (1982) 581–583.

[215] Mandelbrot, B. B., *Self-affine fractals and fractal dimension,* Physica Scripta 32 (1985) 257–260.

[216] Mandelbrot, B. B., *On the dynamics of iterated maps V: conjecture that the boundary of the M-set has fractal dimension equal to 2,* in: Chaos, Fractals and Dynamics, Fischer and Smith (eds.), Marcel Dekker, 1985.

[217] Mandelbrot, B. B., *An introduction to multifractal distribution functions,* in: *Fluctuations and Pattern Formation,* H. E. Stanley and N. Ostrowsky (eds.), Kluwer Academic, Dordrecht, 1988.

[218] Mañé, R., *On the dimension of the compact invariant set of certain nonlinear maps,* in: *Dynamical Systems and Turbulence, Warwick 1980,* Lecture Notes in Mathematics 898, Springer-Verlag (1981) 230–242.

[219] Marotto, F. R., *Chaotic behavior in the Hénon mapping,* Comm. Math. Phys. 68 (1979) 187–194.

[220] Matsushita, M., *Experimental Observation of Aggregations,* in: *The Fractal Approach to Heterogeneous Chemistry: Surfaces, Colloids, Polymers,* D. Avnir (ed.), Wiley, Chichester 1989.

[221] Mauldin, R. D., Williams, S. C., *Hausdorff dimension in graph directed constructions,* Trans. Amer. Math. Soc. 309 (1988) 811–829.

[222] May, R. M., *Simple mathematical models with very complicated dynamics,* Nature 261 (1976) 459–467.

[223] Menger, K., *Allgemeine Räume und charakteristische Räume, Zweite Mitteilung: Über umfassenste n-dimensionale Mengen,* Proc. Acad. Amsterdam 29 (1926) 1125–1128.

[224] Misiurewicz, M., *Strange Attractors for the Lozi Mappings,* in Nonlinear Dynamics, R. H. G. Helleman (ed.), Annals of the New York Academy of Sciences 357 (1980) 348–358.

[225] Mitchison, G. J., Wilcox, M., *Rule governing cell division in Anabaena,* Nature 239 (1972) 110–111.

[226] Mullin, T., *Chaos in physical systems,* in: *Fractals and Chaos,* Crilly, A. J., Earnshaw, R. A., Jones, H. (eds.), Springer-Verlag, New York, 1991.

[227] Musgrave, K., Kolb, C., Mace, R., *The synthesis and the rendering of eroded fractal terrain,* Computer Graphics 24 (1988).

[228] Norton, V. A., *Generation and display of geometric fractals in 3-D,* Computer Graphics 16, 3 (1982) 61–67.

[229] Norton, V. A., *Julia sets in the quaternions,* Computers and Graphics 13, 2 (1989) 267–278.

[230] Olsen, L. F., Degn, H., *Chaos in biological systems,* Quarterly Review of Biophysics 18 (1985) 165–225.

[231] Packard, N. H., Crutchfield, J. P., Farmer, J. D., Shaw, R. S., *Geometry from a time series,* Phys. Rev. Lett. 45 (1980) 712–716.

[232] Peano, G., *Sur une courbe, qui remplit toute une aire plane,* Mathematische Annalen 36 (1890) 157–160.

[233] Peitgen, H. O., Prüfer, M., *The Leray-Schauder continuation method is a constructive element in the numerical study of nonlinear eigenvalue and bifurcation problems,* in: *Functional Differential Equations and Approximation of Fixed Points,* H.-O. Peitgen, H.-O. Walther (eds.), Springer Lecture Notes, Berlin, 1979.

[234] Pietronero, L., Evertsz, C., Siebesma, A. P., *Fractal and multifractal structures in kinetic critical phenomena,* in: *Stochastic Processes in Physics and Engineering,* S. Albeverio, P. Blanchard, M. Hazewinkel, L. Streit (eds.), D. Reidel Publishing Company (1988) 253–278. (1988) 405–409.

[235] Pomeau, Y., Manneville, P., *Intermittent transition to turbulence in dissipative dynamical systems,* Commun. Math. Phys. 74 (1980) 189–197.

[236] Prusinkiewicz, P., *Graphical applications of L-systems,* Proc. of Graphics Interface 1986 – Vision Interface (1986) 247–253.

[237] Prusinkiewicz, P., Hanan, J., *Applications of L-systems to computer imagery,* in: "Graph Grammars and their Application to Computer Science; Third International Workshop", H. Ehrig, M. Nagl, A. Rosenfeld and G. Rozenberg (eds.), (Springer-Verlag, New York, 1988).

[238] Prusinkiewicz, P., Lindenmayer, A., Hanan, J., *Developmental models of herbaceous plants for computer imagery purposes,* Computer Graphics 22, 4 (1988) 141–150.

[239] Prusinkiewicz, P., Hammel, M., *Automata, languages, and iterated function systems,* in: *Fractals Modeling in 3-D Computer Graphics and Imaging,* ACM SIGGRAPH '91 Course Notes C14 (J. C. Hart, K. Musgrave, eds.), 1991.

[240] Rayleigh, Lord, *On convective currents in a horizontal layer of fluid when the higher temperature is on the under side,* Phil. Mag. 32 (1916) 529–546.

[241] Reuter, L. Hodges, *Rendering and magnification of fractals using iterated function systems,* Ph. D. thesis, School of Mathematics, Georgia Institute of Technology (1987).

[242] Richardson, R. L., *The problem of contiguity: an appendix of statistics of deadly quarrels,* General Systems Yearbook 6 (1961) 139–187.

[243] Rössler, O. E., *An equation for continuous chaos,* Phys. Lett. 57A (1976) 397–398.

[244] Ruelle, F., Takens, F., *On the nature of turbulence,* Comm. Math. Phys. 20 (1971) 167–192, 23 (1971) 343–344.

[245] Russell, D. A., Hanson, J. D., Ott, E., *Dimension of strange attractors,* Phys. Rev. Lett. 45 (1980) 1175–1178.

[246] Salamin, E., *Computation of π Using Arithmetic-Geometric Mean,* Mathematics of Computation 30, 135 (1976) 565–570.

[247] Saltzman, B., *Finite amplitude free convection as an initial value problem — I,* J. Atmos. Sci. 19 (1962) 329–341.

[248] Sano, M., Sawada, Y., *Measurement of the Lyapunov spectrum from a chaotic time series,* Phys. Rev. Lett. 55 (1985) 1082.

[249] Saupe, D., *Efficient computation of Julia sets and their fractal dimension,* Physica D28 (1987) 358–370.

[250] Saupe, D., *Discrete versus continuous Newtons method : A case study,* Acta Appl. Math. 13 (1988) 59–80.

[251] Saupe, D., *Point evalutions of multi-variable random fractals,* in: *Visualisierung in Mathematik und Naturwissenschaften - Bremer Computergraphiktage 1988,* H. Jürgens, D. Saupe (eds.), Springer-Verlag, Heidelberg, 1989.

[252] Sernetz, M., Gelléri, B., Hofman, F., *The Organism as a Bioreactor, Interpretation of the Reduction Law of Metabolism in terms of Heterogeneous Catalysis and Fractal Structure,* Journal Theoretical Biology 117 (1985) 209–230.

[253] Siegel, C. L., *Iteration of analytic functions,* Ann. of Math. 43 (1942) 607–616.

[254] Sierpinski, W., *Sur une courbe cantorienne dont tout point est un point de ramification,* C. R. Acad. Paris 160 (1915) 302.

[255] Sierpinski, W., *Sur une courbe cantorienne qui contient une image biunivoquet et continue detoute courbe donnée,* C. R. Acad. Paris 162 (1916) 629–632.

[256] Simó, C., *On the Hénon-Pomeau attractor,* Journal of Statistical Physics 21,4 (1979) 465–494.

[257] Shanks, D., Wrench, J. W. Jr., *Calculation of π to 100,000 Decimals,* Mathematics of Computation 16, 77 (1962) 76–99.

[258] Shaw, R., *Strange attractors, chaotic behavior, and information flow,* Z. Naturforsch. 36a (1981) 80–112.

[259] Shishikura, M., *The Hausdorff dimension of the boundary of the Mandelbrot set and Julia sets,* SUNY Stony Brook, Institute for Mathematical Sciences, Preprint #1991/7.

[260] Shonkwiller, R., *An image algorithm for computing the Hausdorff distance efficiently in linear time,* Info. Proc. Lett. 30 (1989) 87–89.

[261] Smith, A. R., *Plants, fractals, and formal languages,* Computer Graphics 18, 3 (1984) 1–10.

[262] Stanley, H. E., Meakin, P., *Multifractal phenomena in physics and chemistry,* Nature 335 (1988) 405–409.

[263] Stefan, P., *A theorem of Šarkovski on the existence of periodic orbits of continuous endomorphisms of the real line,* Comm. Math. Phys. 54 (1977) 237–248.

[264] Stevens, R. J., Lehar, A. F., Preston, F. H., *Manipulation and presentation of multidimensional image data using the Peano scan,* IEEE Transactions on Pattern Analysis and Machine Intelligence 5 (1983) 520–526.

[265] Sullivan, D., *Quasiconformal homeomorphisms and dynamics I,* Ann. Math. 122 (1985) 401–418.

[266] Sved, M., Pitman, J., *Divisibility of binomial coefficients by prime powers, a geometrical approach,* Ars Combinatoria 26A (1988) 197–222.

[267] Takens, F., *Detecting strange attractors in turbulence,* in: *Dynamical Systems and Turbulence, Warwick 1980,* Lecture Notes in Mathematics 898, Springer-Verlag (1981) 366–381.

[268] Tan Lei, *Similarity between the Mandelbrot set and Julia sets,* Report Nr 211, Institut für Dynamische Systeme, Universität Bremen, June 1989, and, Commun. Math. Phys. 134 (1990) 587–617.

[269] Tél, T., *Transient chaos,* to be published in: *Directions in Chaos III,* Hao B.-L. (ed.), World Scientific Publishing Company, Singapore.

[270] Velho, L., de Miranda Gomes, J., *Digital halftoning with space-filling curves,* Computer Graphics 25,4 (1991) 81–90.

[271] Voss, R. F., *Random fractal forgeries,* in : Fundamental Algorithms for Computer Graphics, R. A. Earnshaw (ed.), (Springer-Verlag, Berlin, 1985) 805–835.

[272] Voss, R. F., Tomkiewicz, M., *Computer Simulation of Dendritic Electrodeposition,* Journal Electrochemical Society 132, 2 (1985) 371–375.

[273] Vrscay, E. R., *Iterated function systems: Theory, applications and the inverse problem,* in: Proceedings of the NATO Advanced Study Institute on Fractal Geometry, July 1989. Kluwer Academic Publishers, 1991.

[274] Wall, C. R., *Terminating decimals in the Cantor ternary set,* Fibonacci Quart. 28, 2 (1990) 98–101.

[275] Williams, R. F., *Compositions of contractions,* Bol. Soc. Brasil. Mat. 2 (1971) 55–59.

[276] Willson, S., *Cellular automata can generate fractals,* Discrete Appl. Math. 8 (1984) 91–99.

[277] Witten, I. H., Neal, M., *Using Peano curves for bilevel display of continuous tone images,* IEEE Computer Graphics and Applications, May 1982, 47–52.

[278] Wolf, A. Swift, J. B., Swinney, H. L., Vastano, J. A., *Determining Lyapunov exponents from a time series,* Physica 16D (1985) 285–317.

[279] Yorke, J. A., Yorke, E. D., *Metastable chaos: the transition to sustained chaotic behavior in the Lorenz model,* J. Stat. Phys. 21 (1979) 263–277.

[280] Young, L.-S., *Dimension, entropy, and Lyapunov exponents,* Ergod. Th. & Dynam. Sys. 2 (1982) 109.

[281] Zahlten, C., *Piecewise linear approximation of isovalued surfaces,* in: *Advances in Scientific Visualization, Eurographics Seminar Series,* F. H. Post, A. J. S. Hin (eds.), Springer-Verlag, Berlin, 1992.

Index

Printed in the United States
By Bookmasters